SciDAC 2009

Journal of Physics: Conference Series Volume 180

San Diego, California, USA
14-18 June 2009

ISBN: 978-1-61738-304-5

Printed from e-media with permission by:

Curran Associates, Inc.
57 Morehouse Lane
Red Hook, NY 12571
www.proceedings.com

Some format issues inherent in the e-media version may also appear in this print version.

Printed by Curran Associates, Inc. (2010)

Institute of Physics
Dirac House, Temple Back
Bristol BS1 6BE UK

Phone: 44 1 17 929 7481
Fax: 44 1 17 920 0979

techtracking@iop.org

TABLE OF CONTENTS

COMBUSTION

COMPUTER SCIENCE/HIGH PERFORMANCE COMPUTING

DATA AND DISTRIBUTED COMPUTING

FUSION ENERGY

GEOSCIENCES

LATTICE QUANTUM CHROMODYNAMICS

MATERIALS

NUCLEAR PHYSICS

VISUALIZATION

By almost any measure, the SciDAC community has come a long way since DOE launched the SciDAC program back in 2001. At the time, we were grappling with how to efficiently run applications on terascale systems (the November 2001 TOP500 list was led by DOE's ASCI White IBM system at Lawrence Livermore achieving 7.2 teraflop/s). And the results stemming from the first round of SciDAC projects were summed up in two-page reports. The scientific results were presented at annual meetings, which were by invitation only and typically were attended by about 75 researchers.

Fast forward to 2009 and we now have SciDAC Review, a quarterly magazine showcasing the scientific computing contributions of SciDAC projects and related programs, all focused on presenting a comprehensive look at Scientific Discovery through Advanced Computing. That is also the motivation behind the annual SciDAC conference that in 2009 was held from June 14–18 in San Diego.

The annual conference, which can also be described as a celebration of all things SciDAC, grew out those meetings organized in the early days of the program. In 2005, the meeting was held in San Francisco and attendance was opened up to all members of the SciDAC community. The schedule was also expanded to include a keynote address, plenary speakers and other features found in a conference format.

This year marks the fifth such SciDAC conference, which now comprises four days of computational science presentations, multiple poster sessions and, since last year, an evening event showcasing simulations and modeling runs resulting from SciDAC projects. The fifth annual SciDAC conference was remarkable on several levels. The primary purpose, of course, is to showcase the research accomplishments resulting from SciDAC programs in particular and computational science in general. It is these accomplishments, represented in 38 papers and 52 posters, that comprise this set of conference proceedings. These proceedings can stand alone as evidence of the success of DOE's innovative SciDAC efforts.

But from the outset, a critical driver for the program was to foster increased collaboration among researchers across disciplines and organizations. In particular, SciDAC wanted to engage scientists at universities in the projects, both to expand the community and to develop the next generation of computational scientists. At the meeting in San Diego, the fruits of this emphasis were clearly visible, from the special poster session highlighting the work of the DOE Computational Science Graduate Fellows, to the informal discussions in hotel hallways, to focused side meetings apart from the main presentations. A highlight of the meeting was the keynote address by Dr Ray Orbach, until recently the DOE Under Secretary for Science and head of the Office of Science. It was during his tenure that the first round of projects matured and the second set of SciDAC projects were launched. And complementing these research projects was Dr Orbach's vision for INCITE, DOE's Innovative and Novel Computational Impact on Theory and Experiment program, inaugurated in 2003. This program allocated significant HPC resources to scientists tackling high-impact problems, including some of those addressed by SciDAC teams. Together, SciDAC and INCITE are dramatically accelerating the field of computational science.

As has been noted before, the SciDAC conference celebrates progress in advancing science through large-scale modeling and simulation. Over 400 people registered to attend this year's talks, poster sessions and tutorials, all spanning the disciplines supported by DOE. While the principal focus was on SciDAC accomplishments, this year's conference also included invited presentations and posters from colleagues whose research is supported by other agencies. At the 2009 meeting we also formalized a developing synergy with the Department of Defense's HPC Users Group Meeting, which has occasionally met in parallel with the SciDAC meeting. But in San Diego, we took the additional steps of organizing a joint poster session and a joint plenary session, further advancing opportunities for broader networking. Throughout the four-day program, attendees at both meetings had the option of sitting in on sessions at either conference. We also included several of the NSF Petascale applications in the program, and have also extended invitations to our computational colleagues in other federal agencies, including the National Science Foundation, NASA, and the National Oceanographic and Atmospheric Administration, as well as international collaborators to join us in San Diego.

In 2009 we also reprised one of the more popular sessions from Seattle in 2008, the Electronic Visualization and Poster Night, during which 29 scientific visualizations were presented on high-resolution large-format displays. The best entries were awarded one of the coveted 'OASCR Awards.' The conference also featured a session about breakthroughs in computational science, based on the 'Breakthrough Report' that was published in 2008, led by Tony Mezzacappa (ORNL). Tony was also the chair of the SciDAC 2005 conference.

For the third consecutive year, the conference was followed by a day of tutorials organized by the SciDAC Outreach Center and aimed primarily at students interested in scientific computing. This year, nearly 100 participants attended the tutorials, hosted by the San Diego Supercomputer Center and General Atomics.

This outreach to the broader community is really what SciDAC is all about – Scientific Discovery through Advanced Computing. Such discoveries are not confined by organizational lines, but rather are often the result of researchers reaching out and collaborating with others, using their combined expertise to push our boundaries of knowledge. I am happy to see that this vision is shared by so many researchers in computational science, who all decided to join SciDAC 2009.

While credit for the excellent presentations and posters goes to the teams of researchers, the success of this year's conference is due to the strong efforts and support from members of the 2009 SciDAC Program Committee and Organizing

Committee, and I would like to extend my heartfelt thanks to them for helping to make the 2009 meeting the largest and most successful to date.

Program Committee members were: David Bader, LLNL; Pete Beckman, ANL; John Bell, LBNL; John Boisseau, University of Texas; Paul Bonoli, MIT; Hank Childs, LBNL; Bill Collins, LBNL; Jim Davenport, BNL; David Dean, ORNL; Thom Dunning, NCSA; Peg Folta, LLNL; Glenn Hammond, PNNL; Maciej Haranczyk, LBNL; Robert Harrison, ORNL; Paul Hovland, ANL; Paul Kent, ORNL; Aram Kevorkian, SPAWAR; David Keyes, Columbia University; Kwok Ko, SLAC; Felice Lightstone, LLNL; Bob Lucas, ISI/USC; Paul Mackenzie, Fermilab; Tony Mezzacappa, ORNL; John Negele, MIT; Jeff Nichols, ORNL; Mike Norman, UCSD; Joe Oefelein, SNL; Jeanie Osburn, NRL; Peter Ostroumov, ANL; Valerio Pascucci, University of Utah; Ruth Pordes, Fermilab; Rob Ross, ANL; Nagiza Samatova, ORNL; Martin Savage, University of Washington; Tim Scheibe, PNNL; Ed Seidel, NSF; Arie Shoshani, LBNL; Rick Stevens, ANL; Bob Sugar, UCSB; Bill Tang, PPPL; Bob Wilhelmson, NCSA; Kathy Yelick, NERSC/LBNL; Dave Zachmann, Vista Computational Technology LLC.

Organizing Committee members were: Communications: Jon Bashor, LBNL. Contracts/Logistics: Mary Spada and Cheryl Zidel, ANL. Posters: David Bailey, LBNL. Proceedings: John Hules, LBNL. Proceedings Database Developer: Beth Cerny Patino, ANL. Program Committee Liaison/Conference Web Site: Yeen Mankin, LBNL. Tutorials: David Skinner, NERSC/LBNL. Visualization Night: Hank Childs, LBNL; Valerio Pascucci, Chems Touati, Nathan Galli, and Erik Jorgensen, University of Utah.

Again, my thanks to all.

Horst Simon
San Diego, California
June 18, 2009

Welcome to San Diego and the 2009 SciDAC conference. Over the next four days, I would like to present an assessment of the SciDAC program. We will look at where we've been, how we got to where we are and where we are going in the future.

Our vision is to be first in computational science, to be best in class in modeling and simulation. When Ray Orbach asked me what I would do, in my job interview for the SciDAC Director position, I said we would achieve that vision. And with our collective dedicated efforts, we have managed to achieve this vision.

In the last year, we have now the most powerful supercomputer for open science, Jaguar, the Cray XT system at the Oak Ridge Leadership Computing Facility (OLCF). We also have NERSC, probably the best-in-the-world program for productivity in science that the Office of Science so depends on. And the Argonne Leadership Computing Facility offers architectural diversity with its IBM Blue Gene/P system as a counterbalance to Oak Ridge.

There is also ESnet, which is often understated—the 40 gigabit per second dual backbone ring that connects all the labs and many DOE sites. In the President's Recovery Act funding, there is exciting news that ESnet is going to build out to a 100 gigabit per second network using new optical technologies. This is very exciting news for simulations and large-scale scientific facilities.

But as one noted SciDAC luminary said, it's not all about the computers—it's also about the science—and we are also achieving our vision in this area. Together with having the fastest supercomputer for science, at the SC08 conference, SciDAC researchers won two ACM Gordon Bell Prizes for the outstanding performance of their applications. The DCA++ code, which solves some very interesting problems in materials, achieved a sustained performance of 1.3 petaflops, an astounding result and a mark I suspect will last for some time. The LS3DF application for studying nanomaterials also required the development of a new and novel algorithm to produce results up to 400 times faster than a similar application, and was recognized with a prize for algorithm innovation—a remarkable achievement.

Day one of our conference will include examples of petascale science enabled at the OLCF. Although Jaguar has not been officially commissioned, it has gone through its acceptance tests, and during its shakedown phase there have been pioneer applications used for the acceptance tests, and they are running at scale. These include applications in the areas of astrophysics, biology, chemistry, combustion, fusion, geosciences, materials science, nuclear energy and nuclear physics.

We also have a whole compendium of science we do at our facilities; these have been documented and reviewed at our last SciDAC conference. Many of these were highlighted in our Breakthroughs Report. One session at this week's conference will feature a cross-section of these breakthroughs. In the area of scalable electromagnetic simulations, the Auxiliary-space Maxwell Solver (AMS) uses specialized finite element discretizations and multigrid-based techniques, which decompose the original problem into easier-to-solve subproblems. Congratulations to the mathematicians on this. Another application on the list of breakthroughs was the authentication of PETSc, which provides scalable solvers used in many DOE applications and has solved problems with over 3 billion unknowns and scaled to over 16,000 processors on DOE leadership-class computers. This is becoming a very versatile and useful toolkit to achieve performance at scale.

With the announcement of SIAM's first class of Fellows, we are remarkably well represented. Of the group of 191, more than 40 of these Fellows are in the 'DOE space.' We are so delighted that SIAM has recognized them for their many achievements.

In the coming months, we will illustrate our leadership in applied math and computer science by looking at our contributions in the areas of programming models, development and performance tools, math libraries, system software, collaboration, and visualization and data analytics. This is a large and diverse list of libraries. We have asked for two panels, one chaired by David Keyes and composed of many of the nation's leading mathematicians, to produce a report on the most significant accomplishments in applied mathematics over the last eight years, taking us back to the start of the SciDAC program. In addition, we have a similar panel in computer science to be chaired by Kathy Yelick. They are going to identify the computer science accomplishments of the past eight years. These accomplishments are difficult to get a handle on, and I'm looking forward to this report.

We will also have a follow-on to our report on breakthroughs in computational science and this will also go back eight years, looking at the many accomplishments under the SciDAC and INCITE programs. This will be chaired by Tony Mezzacappa.

So, where are we going in the SciDAC program? It might help to take a look at computational science and how it got started. I go back to Ken Wilson, who made the model and has written on computational science and computational science education. His model was thus: The computational scientist plays the role of the experimentalist, and the math and CS researchers play the role of theorists, and the computers themselves are the experimental apparatus. And that in simulation science, we are carrying out numerical experiments as to the nature of physical and biological sciences.

Peter Lax, in the same time frame, developed a report on large-scale computing in science and engineering. Peter remarked, 'Perhaps the most important applications of scientific computing come not in the solution of old problems, but in the discovery of new phenomena through numerical experimentation.'

And in the early years, I think the person who provided the most guidance, the most innovation and the most vision for where the future might lie was Ed Oliver. Ed Oliver died last year. Ed did a number of things in science. He had this personality where he knew exactly what to do, but he preferred to stay out of the limelight so that others could enjoy the fruits of his vision. We in the SciDAC program and ASCR Facilities are still enjoying the benefits of his vision. We will miss him.

Twenty years after Ken Wilson, Ray Orbach laid out the fundamental premise for SciDAC in an interview that appeared in SciDAC Review:

'SciDAC is unique in the world. There isn't any other program like it anywhere else, and it has the remarkable ability to do science by bringing together physical scientists, mathematicians, applied mathematicians, and computer scientists who recognize that computation is not something you do at the end, but rather it needs to be built into the solution of the very problem that one is addressing. '

As you look at the Lax report from 1982, it talks about how 'Future significant improvements may have to come from architectures embodying parallel processing elements—perhaps several thousands of processors.' And it continues, 'esearch in languages, algorithms and numerical analysis will be crucial in learning to exploit these new architectures fully.'

In the early '90s, Sterling, Messina and Smith developed a workshop report on petascale computing and concluded, 'A petaflops computer system will be feasible in two decades, or less, and rely in part on the continual advancement of the semiconductor industry both in speed enhancement and cost reduction through improved fabrication processes.' So they were not wrong, and today we are embarking on a forward look that is at a different scale, the exascale, going to 1018 flops.

In 2007, Stevens, Simon and Zacharia chaired a series of town hall meetings looking at exascale computing, and in their report wrote,

'Exascale computer systems are expected to be technologically feasible within the next 15 years, or perhaps sooner. These systems will push the envelope in a number of important technologies: processor architecture, scale of multicore integration, power management and packaging.'

The concept of computing on the Jaguar computer involves hundreds of thousands of cores, as do the IBM systems that are currently out there. So the scale of computing with systems with billions of processors is staggering to me, and I don't know how the software and math folks feel about it.

We have now embarked on a road toward extreme scale computing. We have created a series of town hall meetings and we are now in the process of holding workshops that address what I call within the DOE speak 'the mission need,' or what is the scientific justification for computing at that scale. We are going to have a total of 13 workshops. The workshops on climate, high energy physics, nuclear physics, fusion, and nuclear energy have been held. The report from the workshop on climate is actually out and available, and the other reports are being completed. The upcoming workshops are on biology, materials, and chemistry; and workshops that engage science for nuclear security are a partnership between NNSA and ASCR. There are additional workshops on applied math, computer science, and architecture that are needed for computing at the exascale. These extreme scale workshops will provide the foundation in our office, the Office of Science, the NNSA and DOE, and we will engage the National Science Foundation and the Department of Defense as partners.

We envision a 10-year program for an exascale initiative. It will be an integrated R&D program initially—you can think about five years for research and development—that would be in hardware, operating systems, file systems, networking and so on, as well as software for applications. Application software and the operating system and the hardware all need to be bundled in this period so that at the end the system will execute the science applications at scale. We also believe that this process will have to have considerable investment from the manufacturers and vendors to be successful.

We have formed laboratory, university and industry working groups to start this process and formed a panel to look at where SciDAC needs to go to compute at the extreme scale, and we have formed an executive committee within the Office of Science and the NNSA to focus on these activities. We will have outreach to DoD in the next few months. We are anticipating a solicitation within the next two years in which we will compete this bundled R&D process.

We don't know how we will incorporate SciDAC into extreme scale computing, but we do know there will be many challenges. And as we have shown over the years, we have the expertise and determination to surmount these challenges.

Journal of Physics: Conference Series **180** (2009) 011003 doi:10.1088/1742-6596/180/1/011003

The scaling issue: scientific opportunities

Raymond L. Orbach

3001 Veazey Terrace NW, #525, Washington DC 20008

E-mail: rorbach@earthlink.net

Abstract. A brief history of the Leadership Computing Facility (LCF) initiative is presented, along with the importance of SciDAC to the initiative. The initiative led to the initiation of the Innovative and Novel Computational Impact on Theory and Experiment program (INCITE), open to all researchers in the U.S. and abroad, and based solely on scientific merit through peer review, awarding sizeable allocations (typically millions of processor-hours per project). The development of the nation's LCFs has enabled available INCITE processor-hours to double roughly every eight months since its inception in 2004. The "top ten" LCF accomplishments in 2009 illustrate the breadth of the scientific program, while the 75 million processor hours allocated to American business since 2006 highlight INCITE contributions to U.S. competitiveness. The extrapolation of INCITE processor hours into the future brings new possibilities for many "classic" scaling problems. Complex systems and atomic displacements to cracks are but two examples. However, even with increasing computational speeds, the development of theory, numerical representations, algorithms, and efficient implementation are required for substantial success, exhibiting the crucial role that SciDAC will play.

1. Introduction and background

The stunning achievement of the Earth Simulator's official opening on March 11, 2002 changed the approach to high end computation forever. Not only did its peak speed of 40 TeraFlops eclipse any other supercomputer, but its efficiency on real scientific problems (e.g., the Atmospheric General Circulation Model) was in the double digit range (65% in this case). The fastest U.S. computers, based on commodity chips, not only had substantially lower peak speeds, but also exhibited efficiencies only in the single digit range. As a consequence, the Earth Simulator *sustained speeds* were nearly two orders of magnitude faster than for any U.S. computer.

I was rather shaken by the gulf in capability between the Earth Simulator and anything deployed in the United States. As luck would have it, I was invited to give the Commencement Address at the RAND Graduate School on June 22, 2002 [1]. The title of my talk, inspired by the Earth Simulator, was "Computation for the 21st Century: A New Paradigm." I wrote:

> We are now in an era where computational simulation can inform our approach to science, and I believe the social sciences and humanities. We are now able to contemplate exploration of worlds never before accessible to mankind … we can simulate systems to discover physical laws for which there are no known predictive equations.… This approach to understanding complex systems is to be thought of in the same vein as experiment and analytic theory. In science of the 21st century, simulation and high-end computation are equal partners with theory and experiment … this critical third leg of the triad.

I also wrote:

> Bluntly put, we are out of business in some critical areas of computational science.... What does it mean to lose scientific leadership, to be #2?... The great scientific discoveries will take place in Yokohama, Japan, and not here.... I believe our country cannot afford to be second best, to be "good enough." Our economy, our intellectual environment, literally our national security depends upon our scientific primacy.... The United States needs to face up to our dilemma, a construct of our own doing.... As a product of the *Sputnik* generation, I can personally attest to the vigor and vitality of the U.S. response. It is now incumbent upon us to repeat this dedication in this new era of *computnik*, to regain our scientific leadership and primacy.

The ensuing seven years (almost to the date of this conference) can only be described as a sea change. I believe it is fair to say that a revolution in thinking and performance has taken place. The policy of the Office of Science in 2002, in my view, was inward looking. Though we spoke about access, in reality the only way in which a researcher could gain time on our computers was either directly through a DOE contract or grant, or in combination with a researcher supported by DOE.

On top of this insularity, there were so many DOE researchers applying for time on our one supercomputer, the National Energy Research Scientific Computing Center (NERSC) at Lawrence Berkeley National Laboratory, that the number of cycles available for any specific researcher was limited. From its web site, "... what distinguishes NERSC is its success in creating an environment that makes these [scientific computing facilities] effective for scientific research." That was true then, and continues to be true. But nevertheless, it was a *capacity* machine, lacking the *capability* that was exhibited by the Earth Simulator.

All was not bleak, however. A year before the Earth Simulator, the Scientific Discovery through Advanced Computing (SciDAC) program began as a five-year program "to develop the scientific computing software and hardware infrastructure needed to use terascale computers to advance DOE research programs in basic energy sciences, biological and environmental research, fusion energy sciences, and high-energy and nuclear physics." Serendipitously, I had included in my RAND Graduate School address:

> We need to begin with the science, and scientists.... Bring to the table the computer scientists, the applied mathematicians, those who are good at algorithms, together with the chip makers and computer architects.... We shall be able to investigate systems of great complexity, and understand predictive laws for their behavior. We will ... learn the rules of collective behavior on a scale previously unknown.

Putting all this together, in 2002:

- The Earth Simulator outpaced U.S. computers in sustained speed by almost two orders of magnitude;
- In practice, only DOE supported researchers had access to DOE computers;
- Users received only small portion of the cycles required to address their problem of choice;
- SciDAC had begun a year earlier on a five-year trial basis.

The Office of Science addressed these four issues directly. Though grid computation was popular at the time, we recognized that a large fast mainframe was the only route to rapid sustained speeds. Time has shown that this was the correct choice. The Department of Energy put in place a U.S. competition for a National Leadership Computing Facility. A peer reviewed process led to investment in the first Leadership Computing Facility at Oak Ridge National Laboratory in 2004. A second Leadership Computing Facility was established at Argonne National Laboratory in 2006. These facilities are designated as *capability* facilities, while the facility at NERSC, along with that at Pacific Northwest National Laboratory, are designated as a *capacity* facilities.

All Office of Science computational facilities were opened to researchers worldwide since 2003. Allocations are based solely on the quality of the proposal. Industry is encouraged to make use of the DOE leadership class computational facilities.

In order to advance scientific discovery, the number of users at the capability facilities is limited. Allocations are typically in the millions of processor hours. The Leadership Computing Facilities at

Oak Ridge and Argonne National Laboratories will provide over one billion processor hours this year to the INCTE program.

Finally, SciDAC was recognized as an essential element for efficient use of leadership class computing facilities, and its first five-year existence has been extended indefinitely.

What has happened is truly astonishing [2]. "Long term partnerships have been forged, exciting science is being done, and high-performance simulation on leadership-class computers is keeping the United States at the forefront of science and technology." A panel of computational scientists, applied mathematicians, and computer scientists gathered together in February 2008 at the invitation of the Director of the Office of Advanced Scientific Computer Research, Dr. Michael Strayer, to identify recent breakthroughs in computational science and enabling technologies supported by INCITE, SciDAC and/or the base program. They identified the 10 most significant science accomplishments over the past 18 months, singular achievements that tell an incredible story of computational discovery [3], exhibited in figure 1.

Rank	Title
1	Modeling the Molecular Basis of Parkinson's Disease (Tsigelny)
2	Discovery of the Standing Accretion Shock Instability and Pulsar Birth Mechanism in a Core-Collapse Supernova Evolution and Explosion (Blondin)
3	Prediction and Design of Macromolecular Structures and Functions (Baker)
4	Understanding How Lifted Flame Stabilized in a Hot Coflow (Yoo)
5	New Insights from LCF-enable advanced kinetic simulations of global turbulence in fusion systems (Tang)
6	High Transition Temperature Superconductivity: A High-Temperature Superconducting State and a Pairing Mechanism in 2-D Hubbard Model (Scalapino)
7	PETSc: Providing the Solvers for DOE High-Performance Simulations (Smith)
8	Via Lactea II, A Billion Particle Simulation of the Dark Matter Halo of the Milky Way (Madau)
9	Probing the properties of water through advanced computing (Galli)
10	First Provably Scalable Maxwell Solver Enables Scalable Electromagnetic Simulations (Kovel)

Figure 1. Top 10 Computational Science Accomplishments. INCITE: 1, 3, 4, 8. SciDAC: 2, 5, 6, 7, 9, 10.

U.S. industry has successfully competed for INCITE resources. Over 75 million processor hours have been allocated to American business since 2006. Some of their accomplishments are exhibited in figure 2.

The combination of new architectures and SciDAC programs has allowed INCITE to rapidly increase its allocations, from 4.9 million processor hours in 2004 to 1.3 billion processor hours in 2010. Figure 3 exhibits the actual annual INCITE allocations from 2004 through 2010, and then extrapolates the allocations through 2015.

- **General motors (25 million processor hours in 2008 and 2009): thermoelectric materials capable of turning waste heat into electricity**
- **Proctor & Gamble (11.1 million processor hours in 2007-2009): bubble formation, dynamics, and stability for dissolving of soap and forming of suds**
- **General Atomics (5.4 million processor hours from 2006-2009): workings of turbulence in a fusion reactor**
- **Pratt & Whitney (>4.5 million processor hours from 2006-2009): computational fluid dynamics to better understand gas-turbine engines, improve fuel efficiency, control emissions, reduce noise, and boost operability**
- **DreamWorks Animation (1.8 million processor hours in 2006 and 2007): pushing the limits of real-time ray tracing to improve computer graphics imagery**
- **Corning Inc. (3.2 million processor hours in 2007-2009): simulation of dense suspensions, calculations of glass transition behavior for silica and silicate glasses**
- **Boeing (1.8 million processor hours from 2006-2009): next-generation aero-elastic wing simulation capabilities using computational fluid dynamics**
- **Gene Network Sciences (22,200,000 processor hours in 2009): simulating cardiac rhythm disturbances to identify underlying electrical mechanisms for arrhythmias**
- **Fluent Inc. (>300,000 processor hours in 2007 and 2008): large-scale engineering simulation early in the process of automotive design**

Figure 2. Selection of American business use of INCITE, and accomplishments, since 2006.

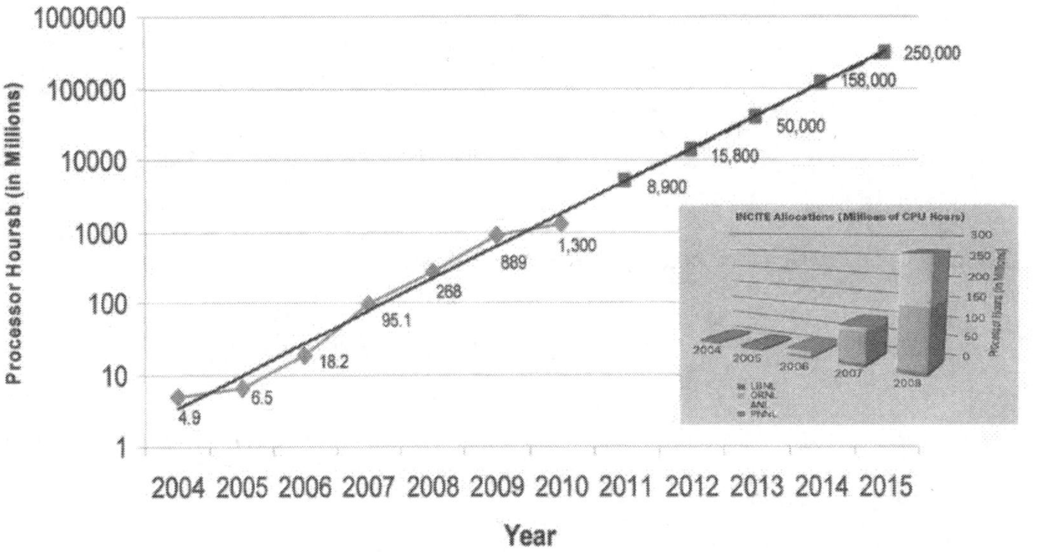

Figure 3. Number of projected INCITE processor hours allocated each year from 2004–2015: ♦ actual, ■ extrapolated. INCITE in 2010 uses only Leadership Class Facilities at ORNL and ANL.

The line through the points in figure 3 can be written as,

$$Log_{10}(processor - hours) = 0.47 years,$$

corresponding to a doubling of processor hour allocations every 7.7 months. This doubling rate is actually three times faster than Moore's law. Hopefully, it will have the same impact [4]: "the driving force of technological and social change in the late 20th and early 21st centuries."

2. Scientific opportunities

With the current peak speeds (~1.3 PetaFlops at ORNL) and INCITE allocations, and the prospect of ExaFlop peak speeds within the next decade [5], it is important to think of the scientific opportunities that can be addressed. There have been three workshops that addressed this issue [6], and I would like to focus one of the most exciting prospects – scaling. There are two aspects in particular to which I would like to draw attention: scaling in complex systems, and scaling from atomic displacement to cracks. While important advances have been made in both fields, they have been hampered by computational constraints that can be rectified. As I shall discuss, this will involve both increases in computational speeds and in SciDAC advances.

The fundamental problem with advance in scaling computations is computational cost. If the cost is proportional to $O(N^3)$, then a computer that is $1000X$ faster can only run a calculation of $10X$ larger. As we shall see, the two reasonably typical problems cited in this paper require *at least* 8 orders of magnitude improvement over current scaling ranges. As noted by R.J. Harrison [7], some combination of theory, numerical representation, algorithm development, and efficient implementation are all required to extend current scaling time limits to laboratory measurement times (equivalently, current scaling length scale limits to laboratory measurement length scales). An example of progress along these lines can be found in the paper titled "Extending the time scale in atomistic simulation of materials" by Voter et al. [8].

2.1 Scaling in complex systems

"Spin Glasses" [9] are deceptively simple. The stereotypical example is a dilute magnetic alloy, say 6% of Mn dissolved randomly in a Cu matrix: Cu:Mn 6 at.%. The characteristic individual exchange spin flip time is of the order of 10^{-12} sec, but the characteristic times for reaching equilibrium as measured in the laboratory are greater than the age of the universe. This is thought to be caused by a highly degenerate set of ground states well separated in phase space so that "communication" between states is strongly hindered.

The localized magnetic moments (here Mn spins) are randomly spatially distributed and are dilute, and there is no net moment in the absence of a magnetic field. This randomness causes the usual two spin correlation function to vanish when ensemble averaged over allowable configurations. Instead, a four spin magnetic correlation function is introduced [10]

$$G(x,t) = V^{-1}\overline{\sum_i < \sigma_{i+x}\tau_{i+x}\sigma_i\tau_i >_t}$$

where the brackets $< >$ indicate a thermal average, V is the volume of the sample, x is the distance from a spin at site-label i to a spin at site-label $i+x$, t is the evolution time for the system, σ_i is the Ising magnetization at the i th site for realization σ, τ_i is the Ising magnetization at the i th site for realization τ, and the overbar represents the ensemble average over quenched states.

The calculations are performed using Monte Carlo methods, so the time t equates to the number of Monte Carlo cycles. The characteristic time for Cu:Mn 6 at. % is 2.44×10^{-13}, corresponding to a single Monte Carlo cycle. The maximum number of Monte Carlo cycles in [10] was ~10^6, equivalent to a maximum physical time of $2.44\times10^{-7}\,\text{sec}$, while laboratory time scales [11] are typically between $10\,\text{sec}$ and $10^5\,\text{sec}$, or some 10^8 to 10^{12} times longer! The Monte Carlo results are well represented by the functional form,

$$G(x,t) = \frac{A(T)}{x^{\alpha}} \exp\left\{ -\left[\frac{x}{\xi(T,t)} \right]^{\delta} \right\},$$

with $\alpha = 0.5$ and $\delta = 1.48$, and $A(T)$ a temperature dependent constant. The correlation length $\xi(T,t)$ fits to power law dynamics,

$$\xi(T,t) = B(T)\,t^{\lambda(T)},$$

with $\lambda(T) = 0.16T$. Experimentally [11], at $B(T) = 0.653$ at $T = 0.72T_g = 28K$, where T_g is the "glass temperature" where the transition to the spin-glass state takes place ($T_g = 31.5K$ for Cu:Mn 6 at.%), and $\lambda(T) = 0.17T$, in remarkable agreement with the Monte Carlo results, especially because the time scales between theory and experiment differ by at least 8 orders of magnitude.

A competing theory [12], the so-called "droplet model", uses activated dynamics with

$$\xi(T,t) = C\left(t\ell nt\right)^{1/0.21},$$

that also fits experiment, but with a very small coefficient $C = 10^{-5}$. Figure 4 exhibits the prediction of both theories, along with the experimental results ($N_s \approx \xi^3$).

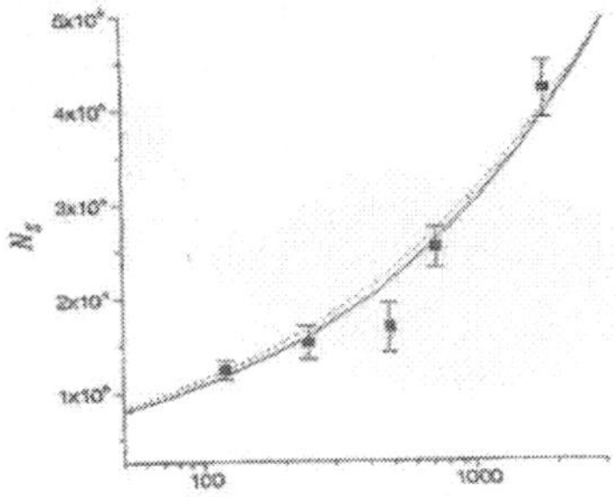

Figure 4. A plot of the number of spins within the correlation length ξ as a function of time. The solid curve drawn through the points is the prediction for power law dynamics, while the dashed curve is the prediction for activated dynamics.

2.2. Scaling from atomic displacements to cracks

Radiation damage is inherently multiscale with interacting phenomena ranging from picoseconds to decades, and nanometers to meters. The difficulty is that first principle simulations are limited to hundreds of atoms for tens of picoseconds. Figure 5 exhibits the requisite length and time scales for simulations to overlap experiment. If one drops the accuracy to the level of empirical interatomic potentials, millions of atoms for microseconds can be simulated [13]. However, this still falls far short of the length scale and especially the time scale of experiments.

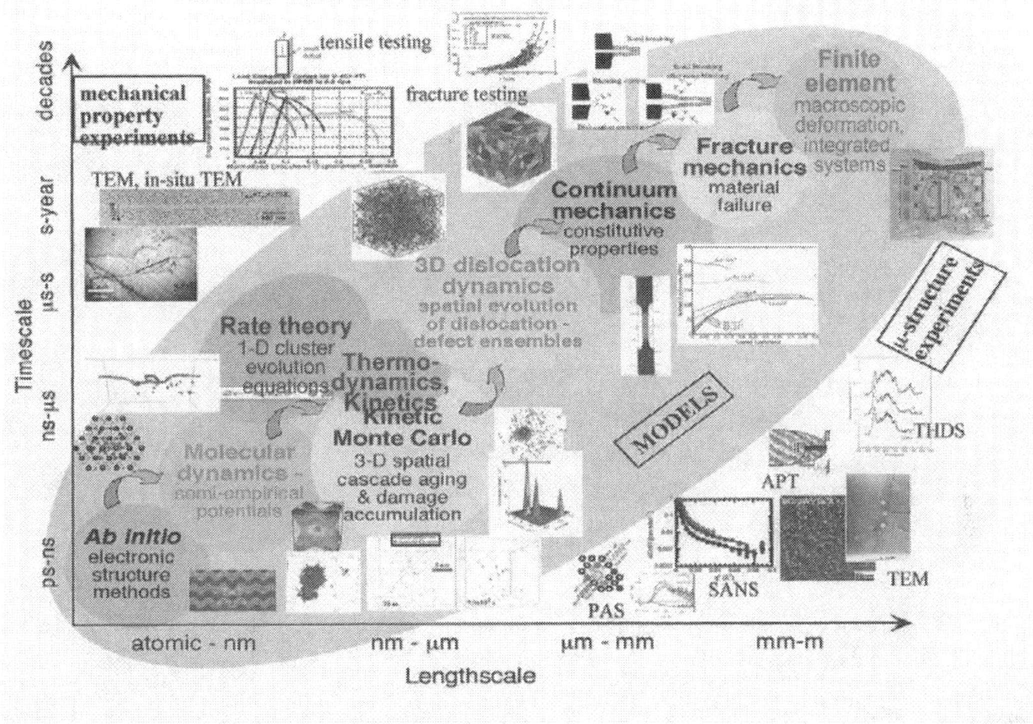

Figure 5. Radiation damage: time and length scales for mechanical property experiments, and for theoretical methods (from [13]).

Perhaps the simplest simulation with relevance to experiment is that of crack propagation in silicon. Instead of trying to simulate the entire system with a common level of accuracy, the atoms are divided into different regions where accurate descriptions are feasible. Of course, the difficulty occurs at the boundaries between these regions. Figure 6 exhibits the geometrical decomposition of a propagating crack tip in a silicon slab into three distinct dynamic regions [14]: a continuum region where finite-element description is accurate (FE), a region describable by atomistic molecular-dynamics (MD), a quantum tight-binding region (TB), and the two "handshaking" regions between: FE-MD and MD-TB. This approach enables a quantum mechanical treatment of the crack tip, where its use is critical, and a mean-field treatment well away from the crack tip where a quantum mechanical treatment is far too costly in computer time, and not necessary.

New methods are being developed that offer promise to increase the number of atoms in an atomistic description of diffusion dynamics [15]. The eventual aim is to describe accurately more complex materials so that simulation can be used to describe materials properties in real systems. For example, understanding irradiation effects on reactor pressure vessels is critical for nuclear reactor design and licensing [16]. A common alloy is Fe-C(0.05-0.2%)-Mn(0.7-1.6%)-Ni(0.2-1.4%)-Si(0.2-0.6%)-Cr(0.05-0.5%)-Cu(0.05-0.4%)-P(0.005-0.025%). Neutron collisions result of hundreds of displacements per atom over the life of the pressure vessel. Reactions take place on the scale of the nuclear to the size and service lifetimes of structural components, spanning factors in excess of 10^{14} (length) and 10^{22} (time). The problem is really one of [16]

> … "multi-physics" involving numerous basic nuclear, atomic and solid-state physics processes linked to complex nano and microstructural evolutions in multi-constituent, multi-phase engineering materials through non-equilibrium thermodynamics and accelerated kinetics, leading to structure-property and property-property relations described by micro and macro mechanics models.

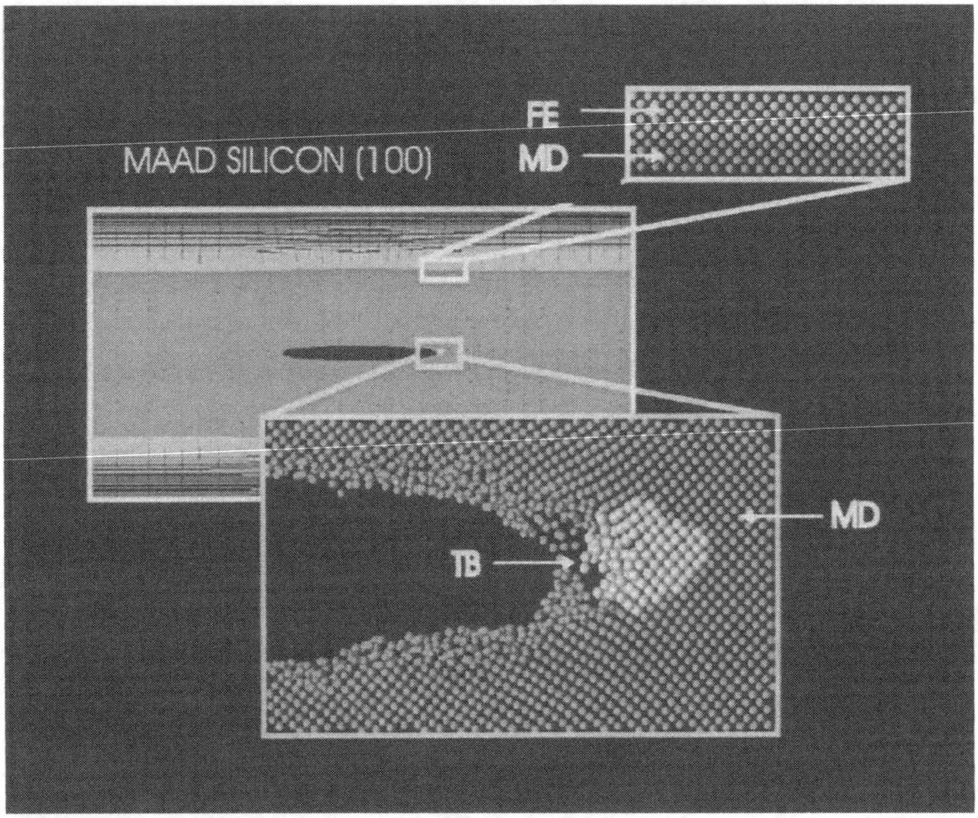

Figure 6. The geometrical decomposition of a periodic thin (10.9Å) Si slab with a thin crack exposing (100) faces placed under uniaxial tension imposed by initializing a velocity gradient along the pulling direction and constraining the outermost surfaces to move at constant speed [14]. The slab is divided into five different dynamic regions for the simulation: the continuum finite-element region (FE); the atomistic molecular-dynamics region (MD); the quantum tight-binding region (TB); the FE-MD "handshaking" region; and the MD-TB "handshaking region [14]. The image is the simulated silicon slab, with expanded views of the FE-MD (orange notes-blue atoms) interface and the TB (yellow atoms) region surrounded by MD (blue) atoms. Note that the TB region surrounds the crack tip with broken-bond MD atoms trailing behind this region. The acronym, MAAD, implies "macroatomistic *ab-initio* dynamics."

3. Summary

The history and future of INCITE, and the concomitant importance of SciDAC, promises to address a most recalcitrant problem in science: scaling. The dynamics of complex systems and crack formation and propagation are but two examples where both must work together, even with the prospect of remarkable advances in computational speeds. Theory, numerical representation, algorithm development, and efficient implementation are necessary ingredients for scientific discovery in the age of progression from PetaFlop to ExaFlop computational speeds.

References

[1] Orbach R 2002 Computation for the 21st century: a new paradigm *RAND Graduate School Commencement Address* (available from rorbach@earthlink.net)

[2] *SciDAC-2:Advancing SCIENCE through Large-Scale MODELING and SIMULATION* (SciDAC Review, **Special Issue 11** 2009) p 8

[3] *SciDAC-2:Advancing SCIENCE through Large-Scale MODELING and SIMULATION* (SciDAC Review, **Special Issue 11** 2009) p 33

[4] *Moore's law – Wikipedia, the free encyclopedia* p 1

[5] *SciDAC-2:Advancing SCIENCE through Large-Scale MODELING and SIMULATION* (SciDAC Review, **Special Issue 11** 2009) p 5

[6] *Exascale for Energy, Ecological Sustainability and Global Security (E^3SGS) Town Hall Meeting: 2007 April 17-18Lawrence Berkeley National Laboratory; 2007 May 17-18 Oak Ridge National Laboratory; 2007 May 31-June 1 Argonne National Laboratory* (available from www.sc.doe.gov/ascr/WorkshopsConferences)

[7] Harrison R J 2009 (American Chemical Society presentation March 26)

[8] Voter A F , Montalenti F and Germann T C 2002 *Annu. Rev. Mater. Res.* **46** 321

[9] Binder K and Young A P 1986 *Rev. Mod. Phys.* **58** 801; Mydosh J A 1993 *Spin glasses, an Experimental Introduction* (London:Taylor & Francis); Mezard M, Parisi G and Virasoro M A 1987 *Spin Glass Theory and Beyond* (Singapore: World Scientific)

[10] Marinari E, Parisi G, Ruiz-Lorenzo J and Ritort F 1996 *Phys. Rev. Lett.* **76** 843; Kisker J, Santen L, Schreckenberg M and Rieger H 1996 *Phys. Rev.* B **53** 6418

[11] Joh Y G, Orbach R, Wood G G, Hammann J and Vincent E 1999 *Phys. Rev. lett.* **82** 438

[12] Wirth B D, Odette G R, Maroudas D and Lucas G E 1997 *J. Nucl. Mater.* **244** 185; private communication

[13] Bernstein N, Kermode J R and Csányi G 2009 *Rep. Prog. Phys.* **72** 026501

[14] Broughton J Q, Abraham F F, Bernstein N and Kaxiras E 1999 *Phys. Rev.* B **60** 2391

[15] Csányi G, Albaret T, Payne M C and De Vita A 2004 *Phys. Rev. Lett.* **93** 175503

[16] Odette G R and Wirth B D 2005 *Handbook of Materials Modeling* (Berlin: Springer) pp 999-1037

Journal of Physics: Conference Series **180** (2009) 012001 doi:10.1088/1742-6596/180/1/012001

Large scale shape optimization for accelerator cavities

Volkan Akcelik, Lie-Quan Lee, Zenghai Li, Cho Ng, Liling Xiao and Kwok Ko

SLAC National Accelerator Laboratory, Menlo Park, CA 94025, USA

E-mail: volkan@slac.stanford.edu

Abstract. We present a shape optimization method for designing accelerator cavities with large scale computations. The objective is to find the best accelerator cavity shape with the desired spectral response, such as with the specified frequencies of resonant modes, field profiles, and external Q values. The forward problem is the large scale Maxwell equation in the frequency domain. The design parameters are the CAD parameters defining the cavity shape. We develop scalable algorithms with a discrete adjoint approach and use the quasi-Newton method to solve the nonlinear optimization problem. Two realistic accelerator cavity design examples are presented.

1. Introduction

Computer science research and algorithm development are crucial for the advancement of accelerator R&D. Simulation plays an important role in the design of accelerators because of its impact on performance improvement and cost reduction. However, traditional trial and error approaches are expensive and time consuming even with the help of computing. We use PDE constrained optimization techniques to provide a more efficient and effective method for the shape design of accelerating cavities. The objective is to determine the best design parameters giving the cavity shape with the desired spectral response. The constraint is the large scale Maxwell eigenvalue problem. In the following, we first discuss the discrete adjoint methods used in the shape optimization algorithm. Then, we present two realistic examples: optimization of the choke-mode cavity and the crab cavity.

2. Maxwell eigenvalue shape optimization

We consider the Maxwell eigenvalue shape optimization problem with the following form

$$\underset{\mathbf{e}_i, k_i, \mathbf{d}}{\text{minimize}} \quad \mathcal{J}(\mathbf{e}_i, k_i, \mathbf{d})$$

$$\text{subject to} \quad \mathbf{K}\mathbf{e}_i + jk_i\mathbf{W}\mathbf{e}_i - k_i^2\mathbf{M}\mathbf{e}_i = \mathbf{0} \tag{1}$$

$$\mathbf{e}_i^H\mathbf{M}\mathbf{e}_i = 1 \tag{2}$$

$$\Re(\mathbf{e}_i)^T\mathbf{M}\Im(\mathbf{e}_i) = 0 \tag{3}$$

$$\mathbf{l} \leq \mathbf{d} \leq \mathbf{u} \tag{4}$$

where $\mathbf{e}_i \in \mathbb{C}^{n_e}$ and k_i are discrete eigenvector and eigenvalues for the ith mode, n_e is the number of degrees of freedom, $\mathbf{d}$ is the design variable vector, j is the complex number,

$\mathbf{K}, \mathbf{M}, \mathbf{W} \in \mathbb{R}^{n_e \times n_e}$ are the stiffness, mass and waveguide matrices, $\mathbf{e}_i^H$ is the complex conjugate transpose of $\mathbf{e}_i$, $\mathbf{l}$ and $\mathbf{u}$ are the lower and upper bounds for the design variable $\mathbf{d}$ [1]. Equations 2 and 3 are the normalization conditions, $\Re(\mathbf{e}_i)$ and $\Im(\mathbf{e}_i)$ are real and complex parts of eigenvector $\mathbf{e}_i$. The objective function $\mathcal{J}$ may include more than one eigenmode.

To derive the optimality conditions we introduce the following Lagrange multipliers: adjoint vectors $\mathbf{t}_i \in \mathbb{C}^{n_e}$, and adjoint constants $\xi_i \in \mathbb{R}$ and $\eta_i \in \mathbb{R}$. Taking inner products of the constraints with the adjoint variables we obtain the real valued Lagrangian functional:

$$\begin{aligned}\mathcal{L}(\mathbf{e}, k, \mathbf{t}, \xi, \eta, \mathbf{d}) \;=\;& \mathcal{J} + \sum_{i=1}^{n_a} \left[\frac{1}{2} \mathbf{t}_i^H (\mathbf{K}\mathbf{e}_i + jk_i \mathbf{W}\mathbf{e}_i - k_i^2 \mathbf{M}\mathbf{e}_i) + \frac{1}{2} \text{c.c.} \right. \\ & \left. + \frac{1}{2} \xi_i \left(\mathbf{e}_i^H \mathbf{M} \mathbf{e}_i - 1 \right) + \eta_i \Re(\mathbf{e}_i)^T \mathbf{M} \Im(\mathbf{e}_i) \right] \qquad (5)\end{aligned}$$

where c.c. is the complex conjugate of the previous term.

Optimality Conditions. The first order optimality conditions require stationary Lagrangian with respect to its arguments [2, 3, 4]. These variations result in the state, adjoint and inversion equations. The state equation is the forward Maxwell eigenvalue problem (Equations 1-3). The adjoint equations are obtained by taking variations of Lagrange multipliers with respect to the state variables $\mathbf{e}$ and k for each mode:

$$\begin{aligned}\mathbf{K}\mathbf{t} + (jk)^* \mathbf{W}\mathbf{t} - \left(k^2\right)^* \mathbf{M}\mathbf{t} + \xi_i \mathbf{M}\mathbf{e} + j\eta \mathbf{M}\mathbf{e}^* &= \frac{\partial \mathcal{J}}{\partial \mathbf{e}} \\ j\mathbf{t}^T \mathbf{W}\mathbf{e}^* + 2k^* \mathbf{t}^T \mathbf{M}\mathbf{e}^* &= \frac{\partial \mathcal{J}}{\partial k}\end{aligned}$$

Adjoint equations are linear system of equations. Each mode involved in the objective function yields one adjoint equation. To obtain the design equation, we take variation with respect to the pth component of the design variable and obtain the expression for the pth component of the design equation:

$$\begin{aligned}\frac{\partial \mathcal{L}}{\partial d_p} =\;& \frac{\partial \mathcal{J}}{\partial d_p} + \sum_{i=1}^{n_a} \left[\frac{1}{2} \mathbf{t}_i^H \left(\frac{\partial \mathbf{M}}{\partial d_p} \mathbf{e}_i + jk_i \frac{\partial \mathbf{W}}{\partial d_p} \mathbf{e}_i - k_i^2 \frac{\partial \mathbf{K}}{\partial d_p} \mathbf{e}_i \right) + \frac{1}{2} \text{c.c.} \right. \\ & \left. + \frac{1}{2} \xi \mathbf{e}_i^H \frac{\partial \mathbf{M}}{\partial d_p} \mathbf{e}_i + \eta_i \Re(\mathbf{e}_i)^T \frac{\partial \mathbf{M}}{\partial d_p} \Im(\mathbf{e}_i) \right]\end{aligned}$$

Given a shape estimate $\mathbf{d}$, to compute the reduced gradient we solve the state and adjoint equations, and evaluate the inversion equation. We use a quasi-Newton method to find the optimum shape and to solve the nonlinear optimization problem [5].

3. Shape optimization for choke-mode cavity

The first shape optimization application is the design of choke-mode cavity. This structure is an example of damped accelerator cavities proposed for high gradient accelerator research. In this design the accelerating mode is trapped inside the cavity by means of a choke structure while unwanted higher-order-modes (HOM) are coupled out from the cavity through radial waveguides and absorbed in matched loads. There are several objectives that the optimum shape of the choke-mode cavity has to achieve:

- Meet the accelerating mode frequency to 11.424 GHz.

- Satisfy field flatness for the accelerating mode.
- Maximize external Q value of the accelerating mode.
- Minimize external Q value of the higher-order-modes.
- Constrain the $\frac{R}{Q}$ (shunt impedance) of the accelerating mode.

In the design of the choke-mode cavity, the accelerating model and 18 higher-order-modes are used for optimization. Design parameters have simple bound constraints and the objective function is formulated to satisfy the required objectives. The CAD model of the choke-mode cavity with the definition of design variables are shown in Figure 1. The cavity consists of 9 cells, and the 7 center cells are required to be identical. The optimization problem has 21 design parameters. We use a penalty method to enforce the eigenvalue constraint for the accelerating mode. The nonlinear problem is solved using a quasi-Newton method with active set strategy.

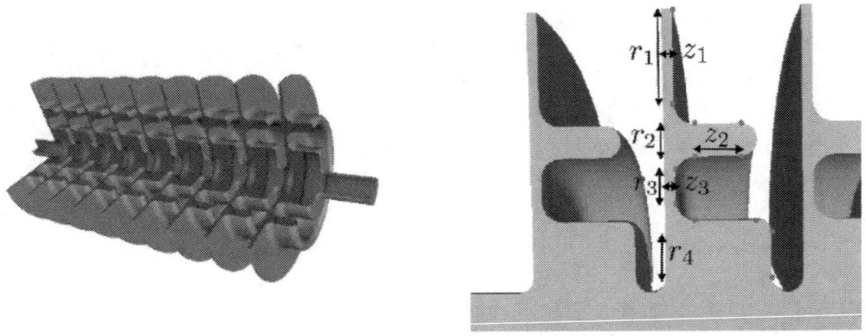

Figure 1. CAD model of the choke-mode cavity (left), and definition of design parameters for the choke-mode cavity (right).

The resulted optimum design parameters are tabulated in Table 1. In addition, the HOM frequency spectrum for the initial and optimum cavities, and Q values are presented in Table 2. The original and optimum cavities are shown in Figure 2. In conclusion, shape optimization decreases Q values of higher-order-modes in the choke-mode structure considerably while the accelerating frequency is fixed at 11.424 GHz with the desired field flatness is satisfied and the $\frac{R}{Q}$ value is constrained.

Table 1. Optimum design parameters for the choke-mode cavity in μm.

Cell 1						
Δ r1	Δ r2	Δ r3	Δ r4	Δ z1	Δ z2	Δ z3
592	-1040	-308	0.2	1696	590	445
Cell 2-8						
Δ r1	Δ r2	Δ r3	Δ r4	Δ z1	Δ z2	Δ z3
712	-1712	1410	-157	1794	634	4
Cell 9						
Δ r1	Δ r2	Δ r3	Δ r4	Δ z1	Δ z2	Δ z3
0.7	-1107	207	-192	1800	1081	1797

Table 2. Frequency and Q values for the original and optimized cavity.

	Original			Optimized		
Mode number	Frequency	Q-value	$\frac{R}{Q}$	Frequency	Q-value	$\frac{R}{Q}$
1	14892740171	305.55	330	14978085631	42.16	534
2	14937523596	253.10	115	15035058182	36.12	949
3	15005849652	190.64	3553	15144908731	28.89	3320
4	15088362142	135.78	221	15172877655	17.79	894
5	15172741072	95.86	17808	15343691411	24.40	2833
6	15246074352	70.55	27920	14519190264	7.57	18793
7	15299862686	56.13	12930	14528647324	7.55	111636
8	15332802296	48.62	2658	14543776889	7.49	102575
9	15355243313	46.55	8024	14562027197	7.39	24278
10	15852388614	26.38	141	14577899185	7.23	17278
10	15867396841	27.08	793	14586453979	7.07	5082
12	15894595624	28.53	2069	14588721890	6.95	1819
13	15936710950	31.30	5016	15579178277	24.14	39917
14	15996493807	36.61	40031	15833022245	26.50	39520
15	16074011649	47.16	52354	16059703082	35.05	5406
16	16160965257	69.73	20842	16247173522	42.95	3217
17	16239011987	124.65	1201	16378170780	80.48	2358
18	16290249709	264.02	1865	16239569727	9.44	7315

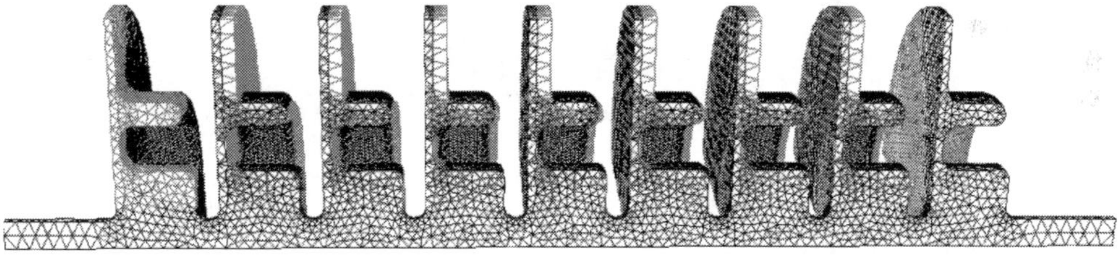

Figure 2. FEM meshes for original (black) and optimized (red) choke-mode cavities.

4. Shape optimization of crab cavity

The second shape optimization example is the design of a crab cavity which will be a potential upgrade of the Large Hadron Collider (LHC) at CERN. The main design goal is to minimize the peak value of the magnetic field on the cavity surface, and constrain cavity frequency at 800 MHz. This is a challenging problem, since as the cavity deforms the location of the peak value can change in a non-smooth fashion so that the derivative of the objective function with respect to design parameters is not continuous. As a result, we use L_p norm in the objective function. We start with small p and increase this value and follow a continuation strategy. As p approaches infinity, L_p norm approaches to infinity norm. Numerical care must be taken to prevent overflow when p is too large.

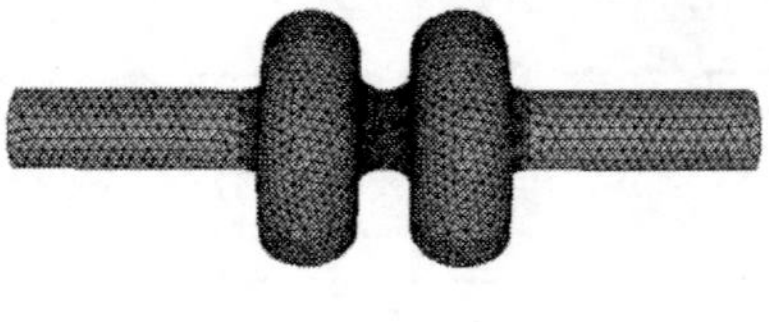

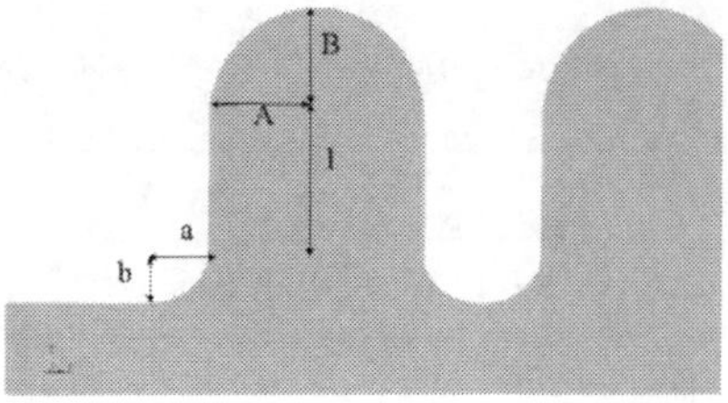

Figure 3. The FEM mesh for squash cavity (left), and definition of the design parameters (right).

The FEM mesh and definition of design parameters are shown in Figure 3. The sum of CAD parameters a and A fixed, and there are 4 independent design parameters. We consider two cases, with different bounds for design variables. In the first case the bound is set to $0.01m$, and in the second to $0.03m$. Table 3 shows stages of continuation for Case 1. Optimization start with $p = 64$, and terminates when $p = 16384$. Tables 4 shows maximum magnetic field values for initial and optimum cavities for Case 1 and 2. Finally optimum design parameters are listed in Table 5. The optimization procedure succeeds in decreasing the maximum value of magnetic field considerably while constraining the operating frequency.

Table 3. Initial and optimized magnetic $|\mathbf{B}|_p$ and electric $|\mathbf{E}|_p$ values with increasing p values and corresponding infinity norms.

p	$\lvert\mathbf{B}\rvert_p$	$\lvert\mathbf{E}\rvert_p$	$\lvert\mathbf{B}\rvert_\infty$	$\lvert\mathbf{E}\rvert_\infty$
64	$9.975e-8$	28.753	$1.131e-7$	34.919
	$9.404e-8$	29.417	$1.100e-7$	35.904
256	$1.045e-7$	34.068	$1.100e-7$	35.904
	$1.012e-7$	33.233	$1.063e-7$	35.009
1024	$1.049e-7$	34.556	$1.063e-7$	35.009
	$1.041e-7$	34.331	$1.054e-7$	34.779
4096	$1.050e-7$	34.667	$1.054e-7$	34.779
	$1.049e-7$	34.656	$1.053e-7$	34.768
16384	$1.052e-7$	34.740	$1.053e-7$	34.768
	$1.051e-7$	34.741	$1.052e-7$	34.769

Table 4. Initial and optimum field values for Case 1 and Case 2.

	$\lvert\mathbf{B}\rvert_\infty$	$\lvert\mathbf{E}\rvert_\infty$
Initial	$1.131e-7$	34.919
Case 1	$1.052e-7$	34.769
Case 2	$1.055e-7$	34.858

Table 5. Optimum design parameters in terms of change in CAD parameters.

	ΔA	Δa	ΔB	Δb	Δl
Case 1	-0.00307	0.0307	-0.01	0.004105	0.00390
Case 2	-0.00305	0.0305	-0.03	0.004084	0.01991

References

[1] Lee LQ, and *et al.* 2005 *J. Phys.: Conf Series* **16** 205–09.
[2] Akcelik V, Biros G, Ghatas O, Hill J, Keyes D, and B. van Bloemen Waanders 2006, Parallel algorithms for PDE-constrained optimization, In: M. Heroux, P. Raghaven and H. Simon, Editors, *Parallel Processing in Scientific Computing* SIAM.
[3] Akcelik V, Ko K, Lee LQ, Li Z and Ng CK 2008, Shape determination for deformed electromagnetic cavities *J. Comput Phys* **227-3**, 1722–38.
[4] Akcelik V, and *et al.* 2005 *J. Phys.: Conf. Series* **16** 435–45
[5] Nocedal J and Wright N 2000 *Numerical Optimization* (Springer).

Advanced computations of multi-physics, multi-scale effects in beam dynamics

J F Amundson, A Macridin, P Spentzouris and E G Stern
Fermi National Accelerator Laboratory, Batavia, IL 60510, USA
E-mail: amundson@fnal.gov

Abstract. Current state-of-the-art beam dynamics simulations include multiple physical effects and multiple physical length and/or time scales. We present recent developments in Synergia2, an accelerator modeling framework designed for multi-physics, multi-scale simulations. We summarize recent several recent results in multi-physics beam dynamics, including simulations of three Fermilab accelerators: the Tevatron, the Main Injector and the Debuncher.

1. Introduction

Early accelerator simulations focused on single-particle dynamics. To a first approximation, the forces on the particles in an accelerator beam are dominated by the external fields due to magnets, RF cavities, etc., so the single-particle dynamics are the leading physical effects. Detailed simulations of accelerators must include collective effects such as the space-charge repulsion of the beam particles, the effects of wake fields in the beam pipe walls and beam-beam interactions in colliders. These simulations require the sort of massively parallel computers that have only become available in recent times.

We give an overview of the accelerator framework Synergia2[1], which was designed to take advantage of the capabilities of modern computational resources and enable simulations of multiple physical effects. We also summarize some recent results utilizing Synergia2 and BeamBeam3d[2], a tool specialized for beam-beam simulations.

2. Overview of Synergia2

Synergia2 is a framework for combining modules for the simulation of individual physical effects into a unified system capable of multi-physics simulations. Figure 1 shows an overview of Synergia2, including current components (non-linear dynamics, space charge, resistive wall) as well as components that are currently under development (electron cloud) and planned to be integrated (beam-beam). It is an ultimate goal of the Synergia project to include multiple implementations of each component, so that optimal choices may be tailored for each individual simulation.

Synergia2 is also designed for computations on massively parallel machines. Figure 2 shows scaling results for a fixed-size space charge calculation of a single bunch utilizing a $64 \times 64 \times 1024$ grid and 200M particles, with significant reduction from the expected power-law behavior only seen above 1024 cores. Work to extend Synergia to simulate $n \sim \mathcal{O}(100)$ loosely-coupled bunches

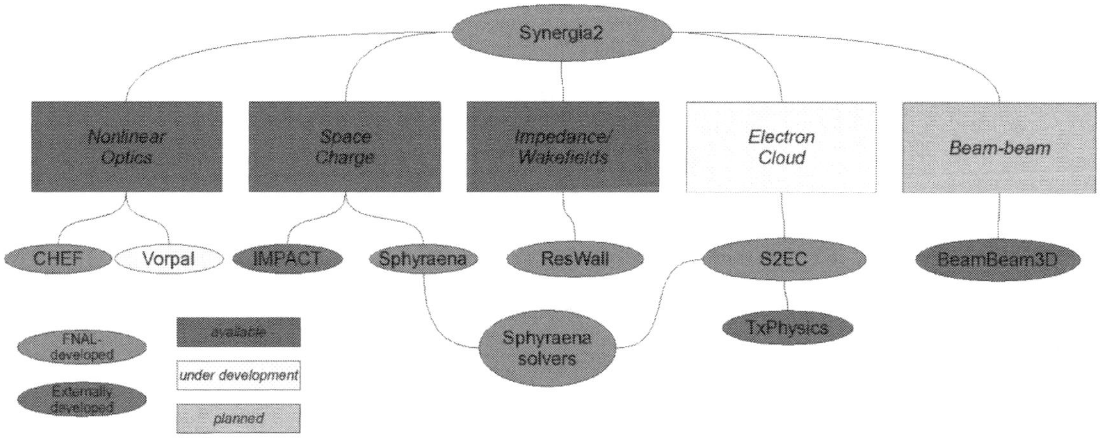

Figure 1. Overview of Synergia2.

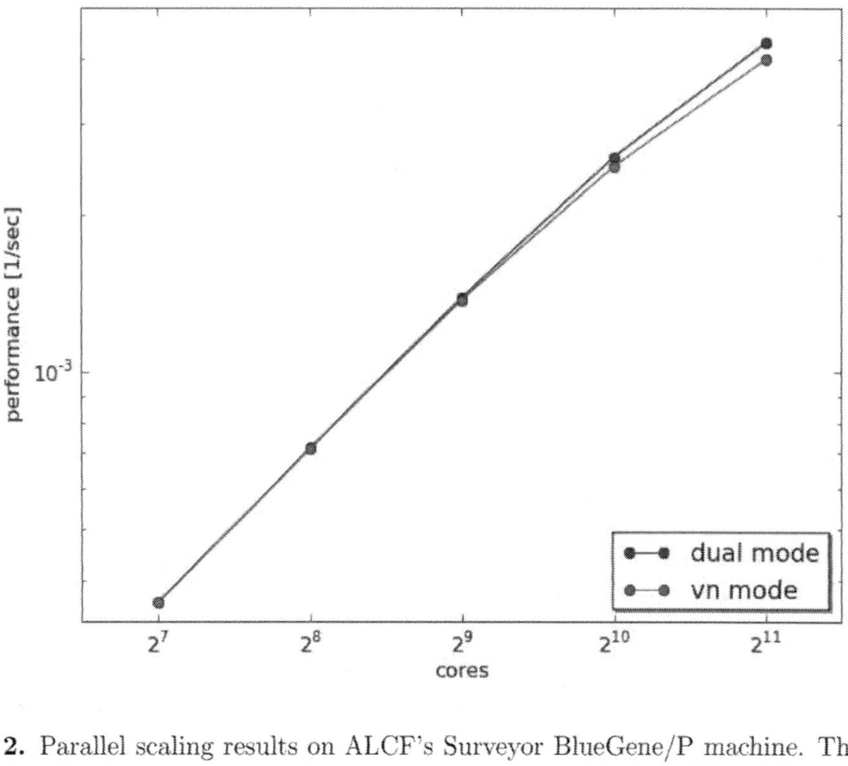

Figure 2. Parallel scaling results on ALCF's Surveyor BlueGene/P machine. The two curves were produced using two computational cores per processor (dual mode) and four computational cores per processor (vn mode), respectively.

is underway. We expect the resulting simulations to efficiently utilize n times as many cores as the single-bunch case.

3. Resistive wall module

The newest addition to Synergia2 is the resistive wall module. It simulates the effect of the current induced by the beam in the surrounding pipe. The current implementation contains a simple dipole approximation for resistive-wall wakefields in the limit of walls of infinite thickness ([3]):

$$\frac{\Delta \vec{p}_\perp}{p} = \frac{2}{\pi b^3}\sqrt{\frac{c}{\sigma}}\frac{N_i\langle \vec{r}_i\rangle}{\beta\gamma}\frac{L}{\sqrt{z_i}} = W_0 L\langle \vec{r}\rangle. \tag{1}$$

Extensions to higher-order multipoles and thin walls are straightforward and underway. The module was originally developed as an extension to BeamBeam3d, where it was extensively benchmarked against analytical models of resistive wall effects[4]. The module as included in Synergia2 is capable of modeling resistive wall effects within a single bunch as well as bunch-to-bunch effects.

4. Poisson solver development

Poisson solvers are used in multiple contexts in beam simulations. They are required for the calculation of space charge effects, electron-cloud effects, and beam-beam interactions. The most recent extension of the Poisson solver suite in Synergia2 is a a solver for elliptical boundary conditions. Elliptical boundary conditions are required for detailed simulations of accelerators with elliptical pipes; the Fermilab Main Injector is one such machine. Figure 3 shows the output

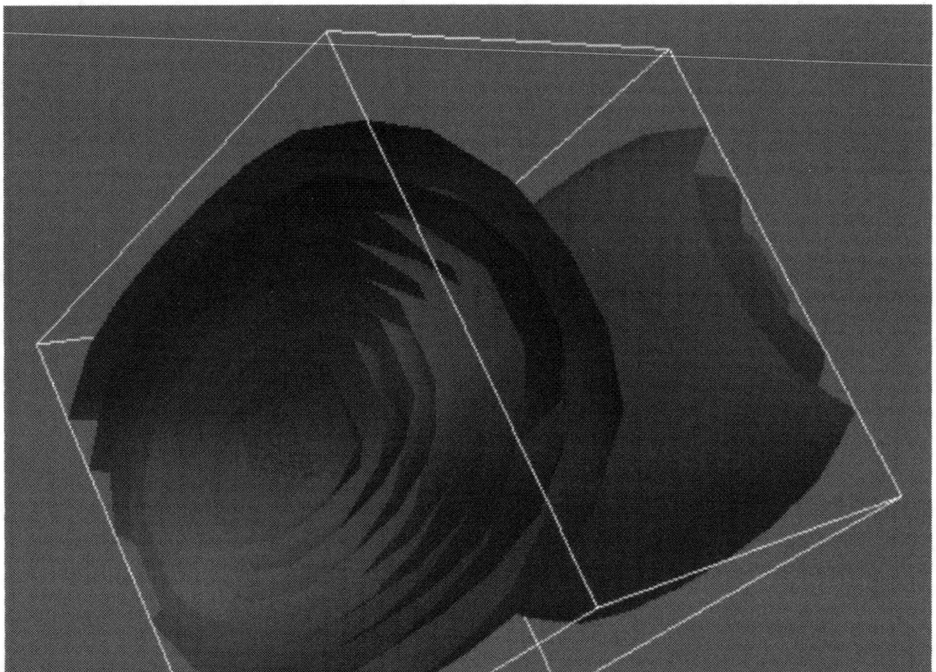

Figure 3. Test field solution using the Sphyraena elliptical solver.

of the Synergia2 elliptical solver for a test problem.

5. Beam-beam and resistive wall effects in the Fermilab Tevatron

Our first recent beam dynamics application is simulation of the Fermilab Tevatron. A detailed understanding of the behavior of the Tevatron requires the simulation of the interaction of the beams with each other (beam-beam) as well as the interaction between the beams and the currents they induce in the beam pipe (resistive wall.) These simulations are described in detail in Ref. [4].

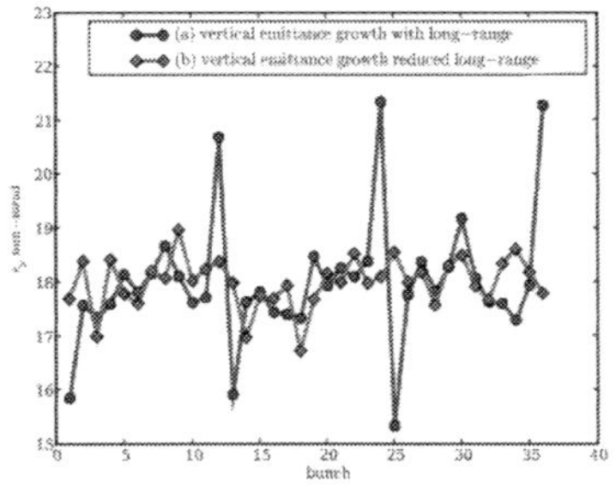

Figure 4. Simulations of emittance growth vs. bunch number due to beam-beam and resistive wall interactions in the Tevatron.

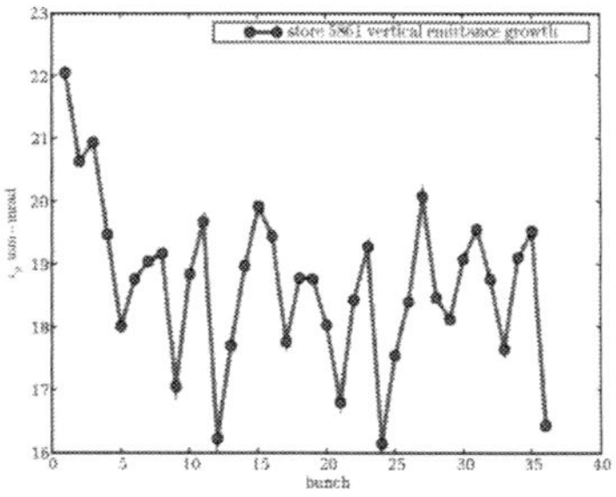

Figure 5. Experimental measurements of emittance growth in the Tevatron showing similar bunch-to-bunch variation as the simulations with the long-range beam-beam effects included.

One of the main successes of Ref. [4] is the correlation between the simulation results shown in Figure 4 and the experimental results shown in Figure 5.

6. Space charge and resistive wall effects in the Fermilab Main Injector

The first application of the new resistive wall module in Synergia2 has been for simulations of the the Fermilab Main Injector[5]. The resistive wall instability in the Main Injector was discovered many years ago; an active damping system is required to keep the machine stable. Future Synergia2 simulations will include a model of the full damping system. The injection energy of the Main Injector is 8 GeV. This low energy, coupled with the high intensity, makes space charge an important effect in addition to the resistive wall effect. Synergia2 is able to effectively combine simulations of these effects.

7. Space charge and non-linear optics in the Fermilab Debuncher

A current topic of simulation for Synergia2 is the beam for the proposed Mu2e experiment at Fermilab. The Mu2e project will require running the Fermilab Debuncher at intensities 10^5 times higher than it has previously handled. At the same time, a highly non-linear resonant extraction system is planned to send the beam to the experiment. Synergia2 is uniquely qualified to simulate the ramped non-linear optics of the extraction system in the presence of a significant space charge effect.

The simulations start with a completely linear lattice. The linear lattice is both used as a benchmark to compare against non-linear effects and as a starting point for a smooth transition to the nonlinear lattice. Figure 7 shows the horizontal phase space distribution for the linear case, in contrast to the nonlinear case in Figure 8.

The most important result of the simulation is the identification of resonant behavior in the nonlinear lattice. Figures 9 and 10 show the transition from smoothly distributed fractional

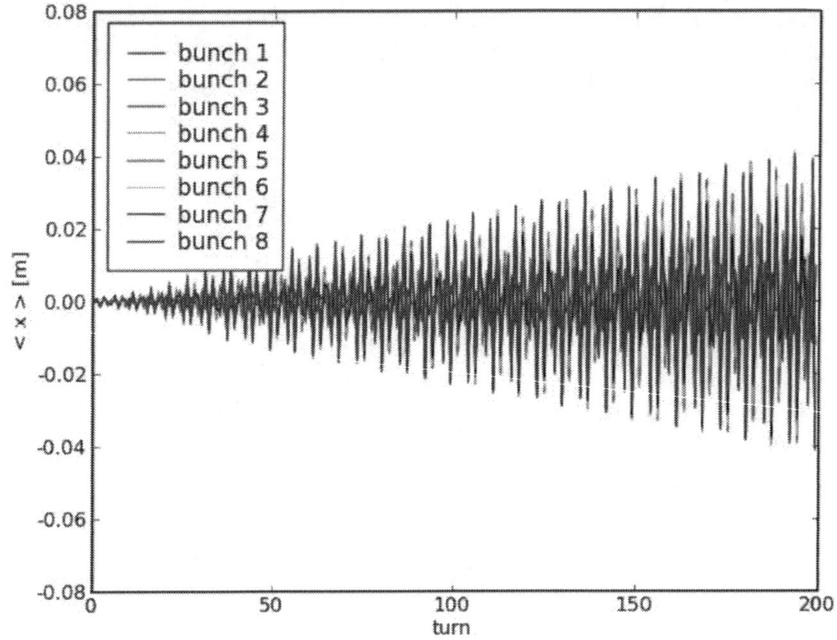

Figure 6. Resistive wall instability and space charge modeled with eight coupled bunches in Fermilab's Main Injector.

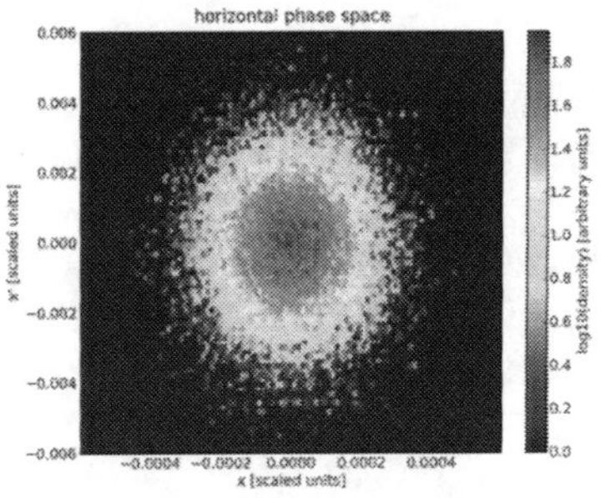

Figure 7. Horizontal phase space distribution for the linear Debuncher lattice.

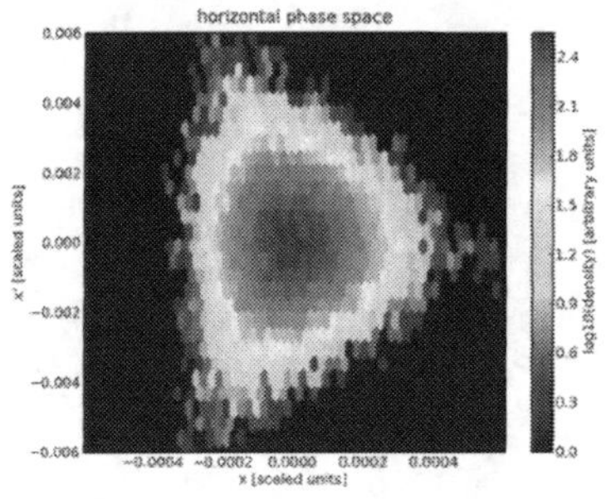

Figure 8. Horizontal phase space distribution for the nonlinear Debuncher lattice.

tunes in the linear case to particles bunched along resonant areas in the nonlinear case. The effect on individual particle trajectories can be seen in the phase space portrait of a single particle shown in Figure 11. The particle exhibits a quasi-periodic behavior consistent with a 13/21 resonance.

Acknowledgments
This work was supported by the United States Department of Energy under contract DE-AC02-07CH11359 and the ComPASS project funded through the Scientific Discovery through

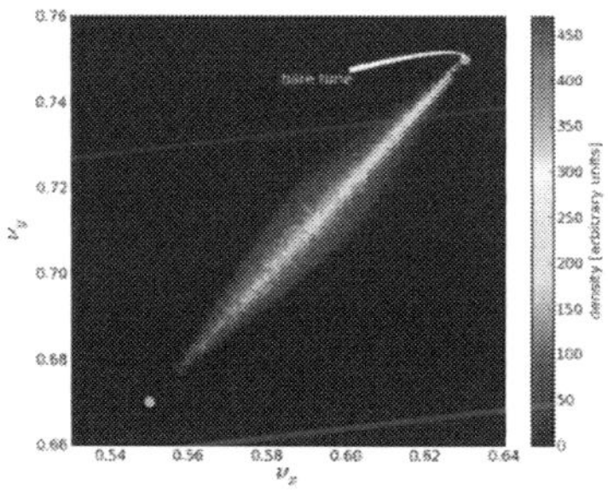

Figure 9. Transverse fractional tune distribution for the linear Debuncher lattice.

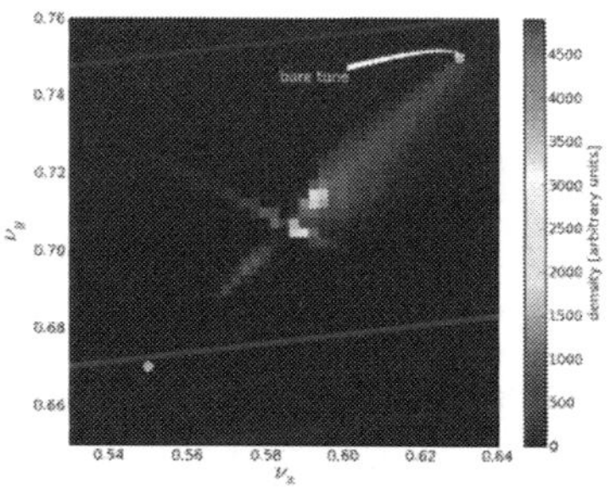

Figure 10. Transverse fractional tune distribution for the nonlinear Debuncher lattice.

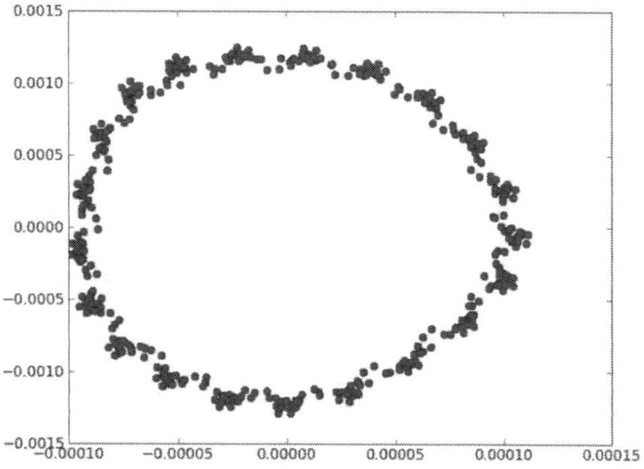

Figure 11. Individual particle phase space trajectory trapped near an $n/21$ resonance, which is compatible with an expected 13/21 resonance.

Advanced Computing program in the DOE Office of High Energy Physics. This research also used resources of the Argonne Leadership Computing Facility at Argonne National Laboratory, which is supported by the Office of Science of the U.S. Department of Energy under contract DE-AC02-06CH11357.

8. References

[1] J. Amundson, P. Spentzouris, J. Qiang, and R. Ryne *J. Comp. Phys.* **211** 229–248 (2006).

[2] J. Qiang, M. A. Furman, R. D. Ryne, *J. Comp. Phys.* **198**, 278–294 (2004). J. Qiang, M. A. Furman, R. D. Ryne, *Phys. Rev. ST Accel. Beams*, 104402 (2002).

[3] A. Chao, *Physics of Collective Beam Instabilities in High Energy Accelerators.*, John Wiley and Sons, Inc., (1993)

[4] E. G. Stern, A. A. Valishev, J. F. Amundson, and P. Spentzouris, "Fully 3D Multiple Beam Dynamics Processes Simulations for the Tevatron," FERMILAB-PUB-08-393-APC-CD.

[5] J. F Amundson, P. Spentzouris, and E. G. Stern, "Simulation Studies of the Interplay Between Space-Charge and Impedance Effects of the Fermilab Main Injector," *Presented at 42nd ICFA Advanced Beam Dynamics Workshop on High-Intensity, High-Brightness Hadron Beams, Nashville, TN, August 2008.*

Validation of Broadly Filtered Diagonalization Method for Extracting Frequencies and Modes from High-Performance Computations

T. M. Austin[1], J. R. Cary[1,2], G. R. Werner[2] and L. Bellantoni[3]

[1]Tech-X Corporation, 5621 Arapahoe Ave., Boulder, CO 80303, USA
[2]Center for Integrated Plasma Studies, University of Colorado, Boulder CO 80309, USA
[3]Fermi National Accelerator Laboratory, Batavia, IL, 60510, USA

E-mail: `austin@txcorp.com`

Abstract. Recent developments have shown that one can get around the difficulties of finding the eigenvalues and eigenmodes of the large systems studied with high performance computation by using broadly filtered diagonalization [G. R. Werner and J. R. Cary, J. Comp. Phys. 227, 5200 (2008)]. This method can be used in conjunction with **any** time-domain computation, in particular those that scale very well up to 10000s of processors and beyond. Here we present results that show that this method accurately obtains both modes and frequencies of electromagnetic cavities, even when frequencies are nearly degenerate. The application was to a well-characterized Kaon separator cavity, the A15. The computations are shown to have a precision to a few parts in 10^5. Because the computed frequency differed from the measured frequency by more than this amount, a careful validation study to determine all sources of difference was undertaken. Ultimately, more precise measurements of the cavity showed that the computations were correct, with remaining differences accounted for by uncertainties in cavity dimensions and atmospheric and thermal conditions. Thus, not only was the method validated, but it was shown to have the ability to predict differences in cavity dimensions from fabrication specifications.

1. Introduction

Electromagnetic cavities are essential to many high-energy physics experiment. They can be tuned to deliver a desired performance by adjusting physical parameters like shape, curvature, and size. Yet, it is a non-trivial task to determine the dependency of such physical parameters on cavity performance. Also, usage of expensive metals such as niobium means design errors can be particularly costly. Coupled with the knowledge that fabrication cannot deliver exact specifications, it is critical to perform computational studies to assist in cavity design and avoid design errors.

Finite element simulations are presently used to compute frequencies and eigenmodes of electromagnetic cavities within the US Department of Energy's science program [1]. Recently, a novel approach that uses the Finite-Difference Time-Domain (FDTD) method applied to Maxwell's equations for computation of frequencies and eigenmodes of electromagnetic cavities was reported [2]. The standard approach to using the FDTD method applied to Maxwell's equations for frequency extraction is to use narrowly-filtered states combined with post-simulation FFTs. The limitations of this approach are that it cannot account for degenerate frequencies or reconstruct spatial mode patterns. The new broadly filtered *Filter Diagonalization Method* (FDM) presented by Werner and Cary in [2] overcomes these limitations.

We present validation results for the broadly filtered FDM applied to a 3.9 GHz Kaon separator fabricated at Fermilab in 1999. Previously, experimental frequency measurements obtained from bead pull experiments were reported in [3] in addition to computational frequency measurements obtained using the Finite Integration Technique. We show here that calculations using the VORPAL computational framework [4] along with the broadly filtered FDM method predicted differences in cavity dimensions from fabrication specifications.

2. Broadly-Filtered FDM

The second-order form of Maxwell's equations,

$$-\frac{\partial^2}{\partial t^2}\mathbf{E}(x,y,z,t) = \nabla \times \nabla \times \mathbf{E}(x,y,z,t) = L\mathbf{E}(x,y,z,t), \tag{1}$$

permits the eigenvalue problem for $\nabla \times \nabla \times$ to be viewed as an eigenvalue problem for the second-order time evolution operator. The standard approach to finding the eigenspace of the $\nabla \times \nabla \times$ operator is to work directly with a discretized form of $\nabla \times \nabla \times$ whereby a large scale eigenvalue problem, $\mathbf{A}\vec{x} = \lambda\vec{x}$, generated from, e. g., the finite element method is solved [1].

The broadly filtered FDM is based on the concept that L can be transformed into block-diagonal form with one large block filtered out and the remaining small block diagonalized by the singular value decomposition (SVD). The eminently scalable FDTD method is employed along with targeted excitations that excite eigenmodes inside of a range of interest. The Dey-Mittra boundary algorithm [5] is employed to model the curved domains that arise in electromagnetic cavities.

To excite the modes inside the frequency range $[\omega_L, \omega_U]$, we define a time-varying current as $\mathbf{J}(x,y,z,t) = g(t)\hat{\mathbf{J}}(x,y,z)$ given that

$$g(t) = \begin{cases} 2\left[\frac{\sin(\omega_L(t-T/2))}{t-T/2} - \frac{\sin(\omega_U(t-T/2))}{t-T/2}\right]\exp^{-\sigma_\omega^2(t-T/2)^2/2} & 0 \le t \le T, \\ 0 & \text{otherwise} \end{cases} \tag{2}$$

and $\hat{\mathbf{J}}(x,y,z)$ is a spatial field that promotes extraction of the eigenmodes. This current is applied to Maxwell's equations as

$$\begin{aligned} \frac{\partial \mathbf{B}(x,y,z,t)}{\partial t} &= -\nabla \times \mathbf{E}(x,y,z,t) \\ \frac{\partial \mathbf{E}(x,y,z,t)}{\partial t} &= -\nabla \times \mathbf{B}(x,y,z,t) - \mathbf{J}(x,y,z,t). \end{aligned} \tag{3}$$

The parameter σ_ω in Eq. (2) is calculated via the distance of the frequency nearest to the frequency range $[\omega_L, \omega_U]$. Also, if a K-degeneracy exists, then K different spatial patterns, $\hat{\mathbf{J}}$, are used yielding K simulations to extract the K-degeneracy.

The targeted excitation acts as the filtering component of the broadly filtered FDM. Small scale linear algebra is then employed for a small scale eigenvalue problem that only requires minutes to solve on a single CPU core. This is in contrast to the large scale eigenvalue problem discussed above that demands 10000s of cores and significant run time. In the method discussed here, the bulk of the work goes into running the FDTD simulations out to time T, during which time the fields are sampled to construct the small scale eigenvalue problem.

We now briefly describe the generation of the small scale eigenvalue problem. The reader can refer to [2] for more details. Once the simulation is ran to time T, the field of interest (either the electric field or the magnetic field) is sampled during the next few cycles at P grid points and at L times. Referring to these sampled discrete fields as $\mathbf{s}_\ell$, we construct the relationship,

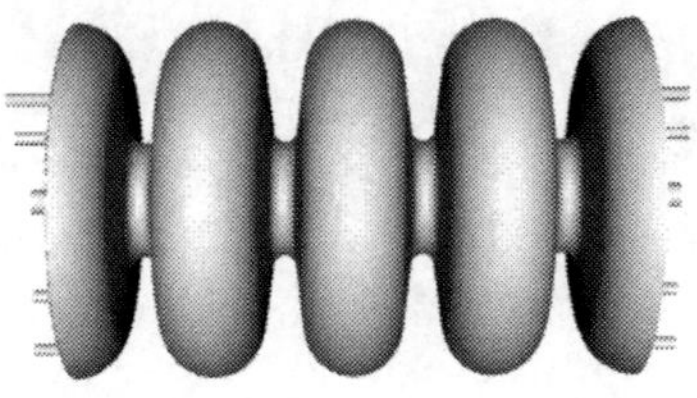

Figure 1. A side view of the 3.9 GHz A15 crab cavity that was fabricated in 1999 for the development of a separated K^+ beam.

$\mathbf{r}_\ell = L\mathbf{s}_\ell$ where L denotes taking the second-order time derivative of the $\mathbf{s}_\ell$. We then recognize that any eigenvector $\mathbf{v}_m$ can be expressed as

$$\mathbf{v}_m = \sum_{\ell=1}^{L} a_{m,\ell}\mathbf{s}_\ell, \tag{4}$$

and that the application of L yields

$$\lambda_m \mathbf{v}_m = \sum_{\ell=1}^{L} a_{m,\ell}\mathbf{r}_\ell. \tag{5}$$

We then have the generalized eigenvalue problem

$$R\mathbf{a}_m = \lambda_m S\mathbf{a}_m, \tag{6}$$

where the number of columns of R and S correspond to L and the number of rows corresponds to P. Instead of solving Eq. (6) we solve the generalized eigenvalue problem

$$S^\dagger R\mathbf{a}_m = \lambda_m S^\dagger S\mathbf{a}_m, \tag{7}$$

using the SVD for $S^\dagger S = VD^2V^\dagger$, where D is diagonal and V is orthogonal. The eigenvalues are related to frequencies by $\omega_m = \sqrt{\lambda_m}/2\pi$ and the eigenvectors of Eq. 7 represent coefficients used in reconstructing the corresponding eigenmodes of Maxwell's equations.

3. A15 Cavity

3.1. Background

To validate the broadly filtered FDM, we examined a well-characterized Kaon separator cavity, the A15 cavity. This cavity, which is pictured in Fig. 1, was designed in 1999 as a Kaon separator [6]. Its shape is determined by physical parameters like the equatorial radius, iris radius, iris curvature, equatorial curvature, and cell half length. These parameter values are 47.19 mm, 15.00 mm, 3.31 mm, 13.6 mm, and 19.2 mm yielding a 3.9 GHz crab cavity. In [3], Bellantoni reported on experimental frequency measurements using bead-pull experiments and also preliminary calculations using a Finite Integration Technique. The TM_{010} deflecting modes had the experimentally-determined frequency values given in Table 1.

3.2. Initial Calculations

With a lower frequency bound of $f_L = 3902$ MHz and an upper frequency bound of $f_U = 4110$ MHz, the correct form of Eq. (2) was determined given that the separation between the frequency range, $[f_L, f_U]$, and the nearest mode was 210 MHz. The nearest mode, $\hat{f}$, at 4320 MHz

Table 1. Frequencies (in MHz) of the five TM_{010} modes of the A15 cavity. Measurements were reported in 2000 by Bellantoni in the unpublished note [3].

f_0	f_1	f_2	f_3	f_4
3902.810	3910.404	3939.336	4001.342	4106.164

Table 2. Calculated frequencies (in MHz) using Richardson extrapolation of the five TM_{010} modes of the A15 cavity using the original specifications for the A15 cavity reported in [3].

f_0	f_1	f_2	f_3	f_4
3900.512	3908.552	3937.325	4000.107	4103.616

was reduced by h = 1e-7 by running the excitation for at least 200 cycles at 4000 MHz. The simulation was ran for another π cycles during which time the fields were sampled. At these parameter values, the value of σ_ω in Eq. (2) was $2\pi|\hat{f} - f_U|/5.68$ and the corresponding excitation time was set to be $T = 11.4/\sigma_\omega$.

In Table 2, we present results obtained using the broadly filtered FDM and Richardson extrapolation from computational results at longitudinal resolutions from 0.533 mm to 0.267 mm, with transverse resolution arbitrarily 25% greater. A comparison of Tables 1 and 2 show that results are only accurate to 1 part in 10^4 which exceeds the computational uncertainty. Possible errors include error in the computational software, missing physics effects, and/or error in the geometrical model. Since the method had been previously verified for a rectangular cavity in [2], errors in the computational software were unlikely. Missing physics effects were found to be insufficient to account for the error so we focused on errors in the geometrical model.

3.3. Validation

In [7], the authors reported on the sensitivity of frequency values to various cavity parameters like equatorial radius, iris radius, and cell half length. The sensitivity of the frequency with respect to the equatorial radius was found to be -80.6 MHz/mm while the sensitivity to iris radius and cell half length was -25.6 MHz/mm and 17.5 MHz/mm, respectively. Knowing that a typical machining error could be 1 mil or 0.0254 mm, it seemed likely that a machining error could produce the discrepancies between computational results and experimental measurements.

Preliminary simulations were performed for an equatorial radius reduced by $\sim$ 1 mil. Results at the fundamental frequency were found to be 3902.514 MHz as compared to the original computed value of 3900.512 MHz, implying accuracy to 1 part in 10^5, which is within computational uncertainty. The results for all five frequencies are found in Table 3. As a result of these findings, a careful re-measurement of the A15 was made at Fermilab. New measurements showed that the equatorial radius was 47.165 ± 0.007 mm in contrast to the original 47.19 mm. Thus, the method predicted differences in cavity dimensions from fabrication specifications.

3.4. Mode Reconstruction

The method can also be used for reconstruction of the modes. In Fig. 2, we show a slice of the B_z component of the magnetic field for the five TM_{010} modes. The reconstruction of the modes only requires a few minutes on a single core. Fields are described in the Hierarchical Data Format (HDF5) following the VizSchema [8] conventions. This standard permits dumping of fields at arbitrary times for viewing with tools like VisIt.

Table 3. Calculated frequencies (in MHz) using Richardson extrapolation of the five TM_{010} modes of the A15 cavity. The altered specifications for the A15 cavity were used in the simulations whereby the equatorial radius was $\sim$ 1 mil smaller.

f_0	f_1	f_2	f_3	f_4
3902.514	3910.509	3939.155	4001.757	4105.940

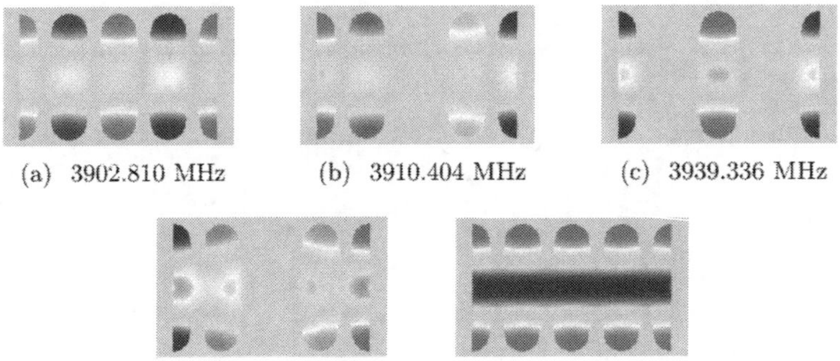

(a) 3902.810 MHz (b) 3910.404 MHz (c) 3939.336 MHz

(d) 4001.342 MHz (e) 4106.164 MHz

Figure 2. Slice of the B_z component of the magnetic field for the five TM_{010} modes of the A15 cavity. Simulations were ran at a resolution of $1.07 \times 1.3 \times 1.3$ mm^3 resulting in approximately 1.05 million grid points.

4. Summary

Validation of the broadly filtered FDM introduced in [2] was performed for the 3.9 GHz Kaon separator cavity. Original calculations revealed that the specifications were not delivered and that adjustments to the equatorial radius were needed to deliver accurate frequency calculations due to machining error. The broadly filtered FDM was also used for reconstruction of the spatial mode patterns.

5. Acknowledgements

This work was supported by the U.S. Department of Energy grants DE-FG02-04ER41317, DE-FC02-07ER41499, and DE-AC02-07CH11359.

References

[1] Ko K, Folwell N, Ge L, Guetz A, Lee L, Li Z, Ng C, Prudencio E, Schussman G, Uplenchwar R *et al.* 2006 *Physica C: Superconductivity and Its Applications* **441** 258–262

[2] Werner G and Cary J 2008 *Journal of Computational Physics* **227** 5200–5214

[3] Bellantoni L Characterization of the aluminum models of the 3.9 GHz separators for a K^+ beam at Fermilab unpublished note

[4] Nieter C and Cary J 2004 *Journal of Computational Physics* **196** 448–473

[5] Dey S and Mittra R 1997 *IEEE Microwave and Guided Wave Letters* **7** 273–275

[6] Bellantoni L, Edwards H, McAshan M and Wanzenberg R 2001 Design and measurements of a deflecting mode cavity for an RF separator *Conference: Particle Accelerator Conference, Chicago, IL (US), 06/18/2001–06/22/2001*

[7] Burt G, Bellantoni L and Dexter A 2007 *EUROTeV-Report-2007-003*

[8] VizSchema `https://ice.txcorp.com/trac/vizschema` last accessed April 16, 2009.

State of the art in electromagnetic modeling for the Compact Linear Collider

Arno Candel[1], Andreas Kabel, Lie-Quan Lee, Zenghai Li, Cho Ng, Greg Schussman and Kwok Ko

SLAC National Accelerator Laboratory, Menlo Park, CA 94025, USA

E-mail: [1] candel@slac.stanford.edu

Abstract. SLAC's Advanced Computations Department (ACD) has developed the parallel 3D electromagnetic time-domain code T3P for simulations of wakefields and transients in complex accelerator structures. T3P is based on state-of-the-art Finite Element methods on unstructured grids and features unconditional stability, quadratic surface approximation and up to 6th-order vector basis functions for unprecedented simulation accuracy. Optimized for large-scale parallel processing on leadership supercomputing facilities, T3P allows simulations of realistic 3D structures with fast turn-around times, aiding the design of the next generation of accelerator facilities. Applications include simulations of the proposed two-beam accelerator structures for the Compact Linear Collider (CLIC) – wakefield damping in the Power Extraction and Transfer Structure (PETS) and power transfer to the main beam accelerating structures are investigated.

1. Introduction

Particle accelerators have been essential tools of scientific discovery for decades. Future linear e+e- colliders operating at the next energy frontier above 1 TeV are expected to complement the Large Hadron Collider (LHC) and lead to further insights which reach far beyond the Standard Model in particle physics. Such accelerators are among the world's largest, most complex facilities and their system requirements continually push the envelope of accelerator science and technology.

The construction and operation costs of next-generation linear colliders have become almost prohibitive, and simple scaling of previous accelerator designs to larger energies might not be optimal. Innovative designs can significantly improve the price/performance ratio. For example, the Compact Linear Collider (CLIC) design improves the power transfer efficiency of accelerating energy to the particle beam.

The CLIC is a design study for a future e+e- collider based on normal-conducting X-band (12 GHz) technology. The current CLIC design consists of two accelerators, one providing the other with RF power. A low energy, high-current drive beam passes through the passive Power Extraction and Transfer Structures (PETS) and supplies intense electromagnetic fields to the accelerating structures for the main beam. The conventional technology, based on klystrons, is only used to power the drive beam. Figure 1 shows the layout of one basic unit of the two-beam accelerator which is replicated many thousand times for each of the two 21 kilometer-long CLIC linacs. For this novel two-beam accelerator scheme to work, numerous technical problems must first be resolved (e.g., RF breakdown and dark current issues at high power and high

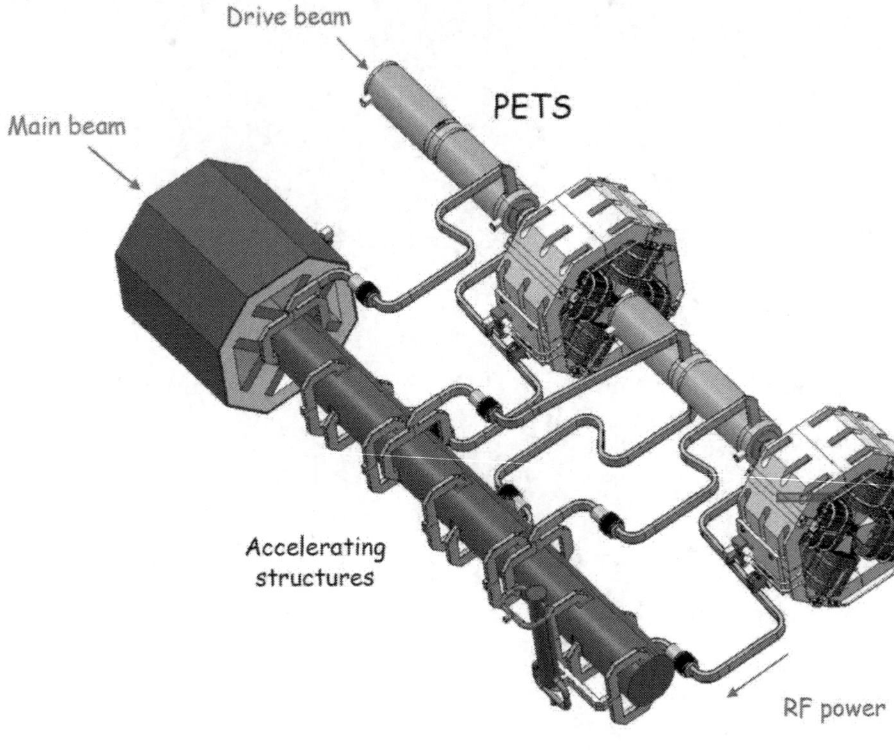

Figure 1. Compact Linear Collider two-beam accelerator: RF power to accelerate the main beam in the accelerating structures is provided by a high current drive beam passing through the power extraction and transfer structures (PETS). Each PETS is connected to two accelerating structures through waveguides. Schematic courtesy of CERN.

gradients, wakefield damping and coupling between the PETS and the accelerator structures, etc.). Thorough understanding of these challenges is essential for the accelerator R&D to develop a more cost-effective and reliable design. A strong collaborative effort between multiple research labs has been put forward to address these issues through numerical and experimental means.

This paper presents our contribution as part of the CLIC R&D effort. We have devoted our focus to the analysis of wakefield effects in the PETS and accelerator structures through advanced computing. In the PETS structure, high drive beam current requirements can lead to strong wakefields. If these wakefields are not dissipated, they can cause drive beam instabilities. Further, parasitic fields propagating between the PETS and accelerating structures may cause harmful effects to the main colliding beams and need to be mitigated in the design. We use the time-domain code T3P from the parallel Finite Element code suite ACE3P to study these transient effects.

Section 2 outlines the numerical methods used in T3P, and sections 3 and 4 present numerical validation results of wakefield damping and power transfer in the CLIC two-beam accelerator structures.

2. Parallel electromagnetic time-domain code T3P

The simulation code T3P solves the full set of Maxwell's equations in the time domain and includes retardation and boundary effects from first principles. T3P uses unstructured meshes of curved (second-order) tetrahedral elements for modeling of small geometric features and scattering effects with high accuracy. In combination with higher-order field representation (up to 6th order), unprecedented simulation accuracy is obtained through efficient use of computational resources.

In our approach, Ampère's and Faraday's laws are combined and integrated over time to yield the inhomogeneous vector wave equation for the time integral of the electric field $\mathbf{E}$:

$$\left(\varepsilon\frac{\partial^2}{\partial t^2} + \sigma_{\text{eff}}\frac{\partial}{\partial t} + \nabla\times\mu^{-1}\nabla\times\right)\int^{\text{t}}\mathbf{E}\,\mathrm{d}\tau = -\mathbf{J}, \tag{1}$$

with permittivity $\varepsilon{=}\varepsilon_0\varepsilon_r$ and permeability $\mu{=}\mu_0\mu_r$. In the current implementation, a constant value of the effective conductivity $\sigma_{\text{eff}}{=}\tan\delta\cdot 2\pi\text{f}\cdot\varepsilon$ is assumed by fixing a frequency f, and the losses are specified by the loss tangent $\tan\delta$. As is common for wakefield computations of rigid beams, the electric current source density $\mathbf{J}$ is given by a one-dimensional Gaussian particle distribution, moving at the speed of light along the beam line.

The computational domain is discretized into curved tetrahedral elements and $\int^t\mathbf{E}\,\mathrm{d}\tau$ in Eq. (1) is represented as an expansion in hierarchical Whitney vector basis functions $\mathbf{N}_i(\mathbf{x})$

$$\int^t\mathbf{E}(\mathbf{x},\tau)\,\mathrm{d}\tau = \sum_{i=1}^{N_p} e_i(t)\cdot\mathbf{N}_i(\mathbf{x}), \tag{2}$$

up to order p within each element. For illustration, the numbers of basis functions for first, second and sixth order are $N_1 = 6$, $N_2 = 20$ and $N_6 = 216$, respectively. Higher-order elements (both curved and with higher-order basis functions) not only significantly improve field accuracy and dispersive properties [1], but also generally lead to higher-order accurate particle-field coupling equivalent to, but much less laborious than, complicated higher-order interpolation schemes commonly found in finite-difference methods.

The Galerkin method is used for spatial discretization. The resulting set of global expansion coefficients e_i represents the field degrees of freedom (DOFs) of the system. The Newmark-Beta scheme is employed for temporal discretization, where $t = n\cdot\Delta t$:

$$\begin{aligned}\left(\mathbf{M}+\frac{\Delta t}{2}(\mathbf{R}+\mathbf{Q})+\beta(\Delta t)^2\mathbf{K}\right)\mathbf{e}^{n+1} &= (\Delta t)^2\left(\beta\mathbf{f}^{n+1}+(1-2\beta)\,\mathbf{f}^n+\beta\mathbf{f}^{n-1}\right)\\ &+\left(2\mathbf{M}-(1-2\beta)(\Delta t)^2\mathbf{K}\right)\mathbf{e}^n-\left(\mathbf{M}-\frac{\Delta t}{2}(\mathbf{R}+\mathbf{Q})+\beta(\Delta t)^2\mathbf{K}\right)\mathbf{e}^{n-1}\end{aligned} \tag{3}$$

with matrices
$$M_{i,j} = \int\varepsilon\mathbf{N}_i\cdot\mathbf{N}_j\,\mathrm{d}\Omega, \qquad Q_{i,j} = \int\frac{1}{c\mu}(\mathbf{n}\times\mathbf{N}_i)\cdot(\mathbf{n}\times\mathbf{N}_j)\,\mathrm{dS},$$
$$R_{i,j} = \int\sigma_{\text{eff}}\mathbf{N}_i\cdot\mathbf{N}_j\,\mathrm{d}\Omega, \qquad K_{i,j} = \int\frac{1}{\mu}(\nabla\times\mathbf{N}_i)\cdot(\nabla\times\mathbf{N}_j)\,\mathrm{d}\Omega,$$
and rhs vector
$$f_i = \int\mathbf{N}_i\cdot\mathbf{J}\,\mathrm{d}\Omega.$$

Solution of the implicit system (3) with a linear solver leads to an unconditionally stable scheme for $\beta\geq 0.25$ [2], where the largest allowable time step does not depend on the smallest mesh size in the overall computational domain. However, it is reasonable to use corresponding scales for the discretization of time and space. The simulation time step Δt determines the temporal field resolution, and both the local mesh size h and the local order of the basis functions p determine the spatial resolution of the fields. Depending on the local speed of light and the resulting wave numbers for the frequency range of interest, static or adaptive *hp*-refinement of both the mesh size and the basis order can significantly improve the modeling accuracy.

The electric field **E** and the magnetic flux density **B** are easily obtained from the solution vector $\mathbf{e}(t)$ at a given (discrete) time t:

$$\mathbf{E}(\mathbf{x},t) = \sum_i \partial_t e_i(t) \cdot \mathbf{N}_i(\mathbf{x}) \qquad \text{and} \qquad \mathbf{B}(\mathbf{x},t) = -\sum_i e_i(t) \cdot (\nabla \times \mathbf{N}_i)(\mathbf{x}). \tag{4}$$

T3P is optimized for large-scale parallel operation on leadership supercomputing facilities, and the performance is mostly dominated by the scalability of the linear solver used for the field updates. Due to the large memory requirements of direct solvers, large-scale problems are typically solved with iterative methods (e.g., the conjugate gradient (CG) algorithm in combination with suitable preconditioners). Figure 2 shows the strong scalability of T3P on a Cray XT4 system.

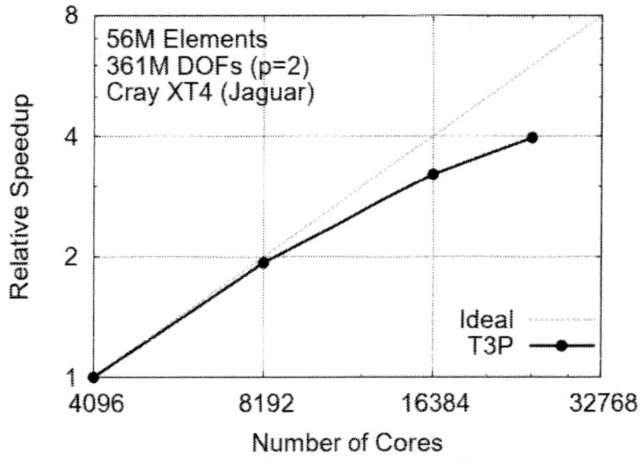

Figure 2. Strong scalability of T3P on the Jaguar Cray XT4 system for a problem with 360M DOFs, normalized to a reference run on 4096 cores with about 18 seconds of walltime per time step. The scalability is dominated by the linear solver for the field updates (here: CG with incomplete Cholesky preconditioner).

More detailed information about the employed methods, validation, and applications of T3P have been published earlier [3, 4, 5, 6].

3. PETS wakefield damping calculations

The PETS is a periodically loaded traveling wave structure with an active length of 21.3 cm, a period of 6.253 mm and an aperture of 23 mm. It is assembled from 8 identical sectors separated by radial slots with embedded lossy dielectric loads to dampen the transverse wakefields.

Figure 3 shows a detailed view of a high-quality mesh model of the PETS with 34 regular cells, 2 matching cells, an outer tank and the output coupler. A model of one symmetric quadrant of the PETS requires roughly 9 million tetrahedral elements, with an average mesh resolution of below 1 mm in the vacuum region and 0.3 mm for the dielectric loads (ε_r=24, μ_r=1) which require a locally refined mesh for proper resolution of the smaller resulting wavelengths. Using this PETS mesh model, wakefield calculations have been performed with T3P. It is sufficient for purposes of this study to only model one symmetric quarter of the structure. To excite transverse dipole wakefields, a rigid Gaussian beam is driven along the z-direction with an x-offset of 2.5 mm, in combination with electric boundary conditions on the yz-symmetry plane, and magnetic (symmetric) boundary conditions on the xz-symmetry plane. The beam profile has a length of σ_z=2 mm and is truncated at $\pm 5\sigma_z$. For simplicity, the same basis order p is used for all mesh elements and the effective conductivity is calculated with f=12 GHz, see Eq. (1).

Figure 4 shows one half of the structure with a snapshot of the dipole wakefields excited by the transit of the beam. The wake potential is a measure of how much a trailing particle is deflected by the wakefields caused by a leading bunch when passing through the structure. It is

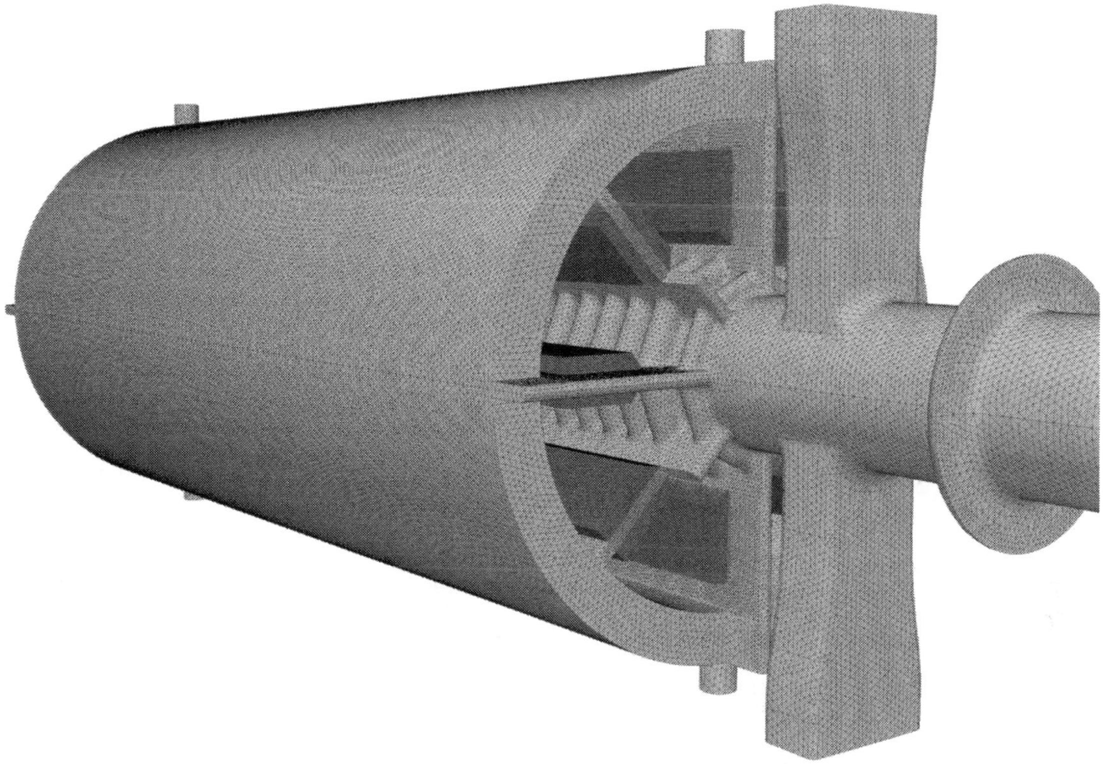

Figure 3. External view of the unstructured conformal tetrahedral mesh model of the PETS. For T3P calculations, only one symmetric quadrant consisting of approx. 9 million tetrahedral quadratic elements is used. The drive beam travels from left to right. The dielectric loads are highlighted in red and the corresponding mesh region is locally refined for improved resolution of the smaller resulting wavelengths.

defined as the work performed by the wakefields on the trailing particle of unit charge during transit through the structure: $\mathbf{w} = \int_{-\infty}^{\infty} (\mathbf{E} + \mathbf{v} \times \mathbf{B})\, dz$. It depends on the cavity geometry, the distance from the leading bunch, and the charge and offset of the leading bunch. The transverse wake potential is typically normalized by the leading bunch charge and offset.

Figure 5 shows the convergence of the transverse wake potential as a function of the order p of the Finite Element basis functions used in T3P calculations, and fast convergence is observed. Figure 6 shows the convergence behavior of the transverse impedance, the Fourier transform of the wake potential of a point charge. Figure 7 shows the effect of the lossy dielectric loads by comparing the transverse wake potential with and without losses (tanδ=0.32 vs. tanδ=0, c.f. Eq. (1)). The dielectric properties and permeability of the absorber material are kept constant (i.e., ε_r=24, μ_r=1). A drop of about one order of magnitude in the wakefield amplitudes is observed after about one active length if dissipation losses are included in the simulation.

For both the wake potential and the impedance, results from p=1 calculations are the least accurate and show differences to calculations with p=2 and p=3, which are almost identical. In a previous study [6] with a slighly simplified geometry, T3P results were compared to results obtained with the commercial parallel finite-difference code GdfidL, and good general agreement between the codes was found. As expected for a conventional finite-difference code, based on a structured mesh and cut-cell boundary approximation, GdfidL results show similar deviations to the converged results as the T3P results obtained with p=1.

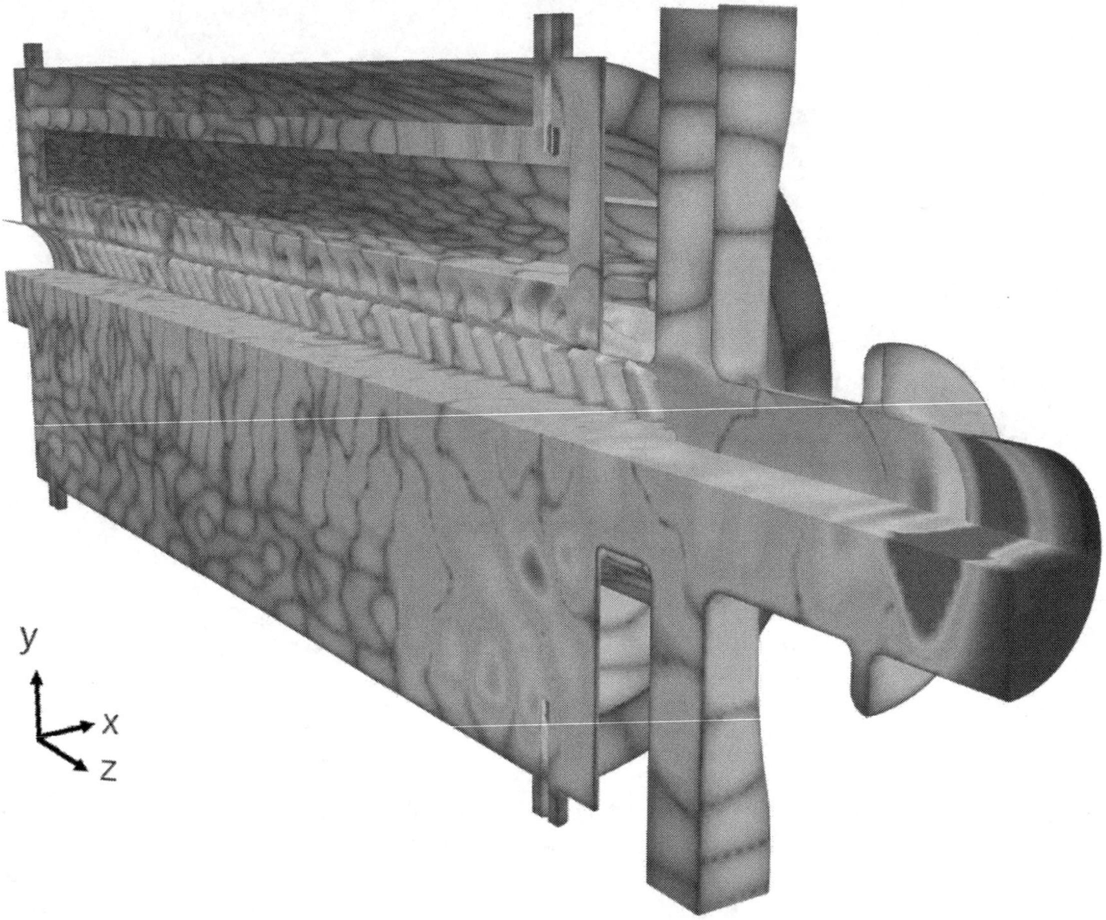

Figure 4. Snapshot of excited dipole wakefields in the PETS, as calculated with T3P. The electric field magnitude is shown in a cut view. On the bottom half, the fields on the symmetry planes are shown as well, and the beam with an offset of x=2.5 mm is visible near the end of the PETS. Strong damping in the lossy dielectric loads (ε_r=24, tanδ=0.32) is directly observed, as well as some fields in the output coupler and choke.

Table 1 shows the runtime parameters for the PETS wakefield damping simulations with T3P for p=1...3, performed on the Jaguar Cray XT4 machine at NCCS. The time scale for wakefield

Table 1. Runtime parameters for the PETS wakefield damping studies with T3P.

Basis Order	Degrees of Freedom	CPUs	Walltime/step
p=1	10M	1152	0.30 secs
p=2	54M	1152	1.63 secs
p=3	159M	4096	10.6 secs

damping in the PETS is on the order of several nanoseconds. Consequently, meaningful results for p=1 can be obtained within a few hundred CPU hours, while p=3 requires in the order of a

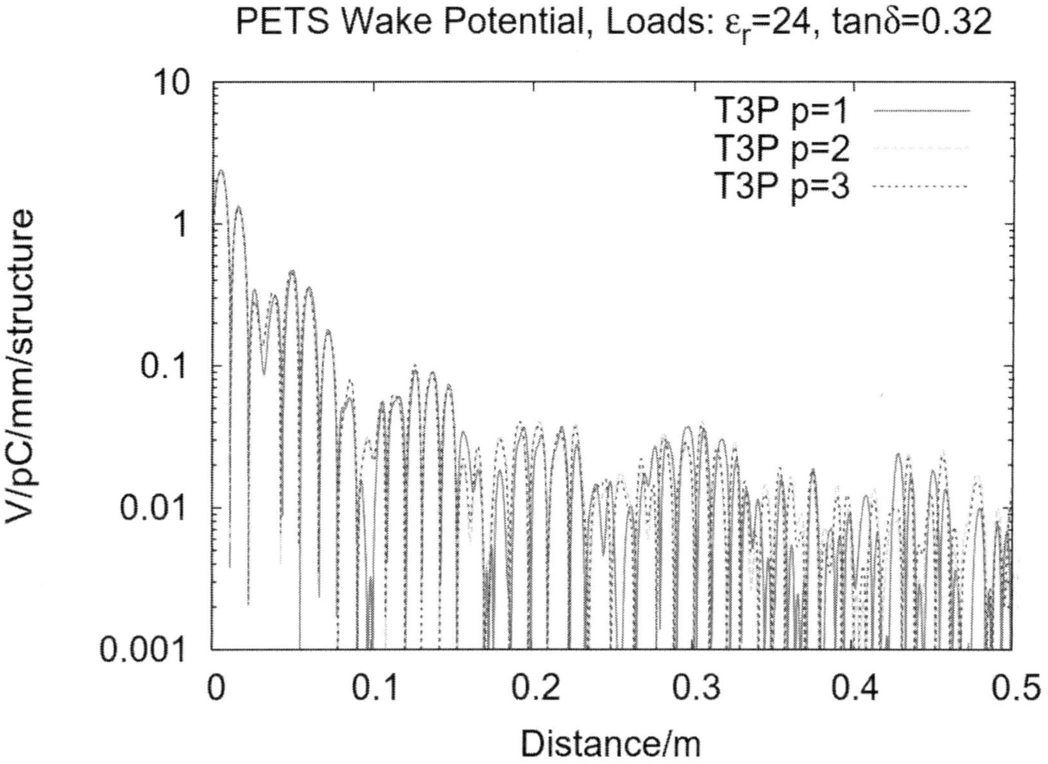

Figure 5. Convergence of transverse wake potential in the PETS as a function of the order p of the Finite Element basis functions used in T3P calculations. A Gaussian dipole current of σ_z=2 mm is used to excite the wakefields.

hundred thousand CPU hours. Results with p=2 require on the order of ten thousand CPU hours and represent a reasonable compromise in terms of accuracy vs. computational requirements.

For wakefield damping calculations with p=2, a simulation time step of 0.5 picoseconds is sufficient to reach convergence in the results. This time step corresponds to a wavelength of 150 microns and frequencies in the terahertz range. It is small enough to model both the frequency content of the beam (σ_z=2 mm), and the smallest features in the geometry with high accuracy.

4. Power transfer from PETS to the accelerating structures

The previous section has shown that T3P is able to reach numerical convergence in the wakefield damping results for the PETS. It is now conceivable for the first time to perform numerical simulations of a coupled structure of both the PETS and the accelerating structure in great detail.

In recent years evidence has been found that the maximum sustainable gradient in an accelerating structure depends on the RF power flow through the structure. The high-gradient CLIC accelerating structures are consequently designed to have a very low group velocity, in the order of only 2% of the speed of light, which results in long filling times. To obtain useful simulation results, control of numerical dispersion errors and long-term stability are essential.

Figure 8 shows the simulation model used for T3P calculations of the power transfer from the PETS to the accelerating structure. A total of 18M tetrahedral elements are used to discretize the coupled structures. In the present study, fundamental mode power transfer from the PETS

Journal of Physics: Conference Series **180** (2009) 012004 doi:10.1088/1742-6596/180/1/012004

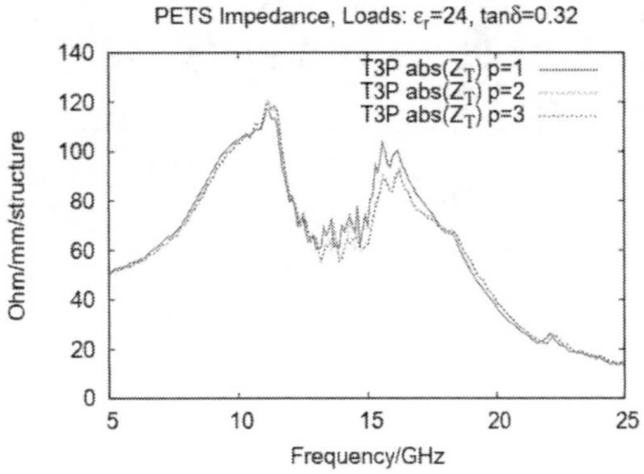

Figure 6. Convergence of transverse impedance in the PETS as a function of the order p of the Finite Element basis functions used in T3P calculations. The impedance is the Fourier transform of the wake potential of a point charge, and is calculated from the transverse wake potential shown in figure 5.

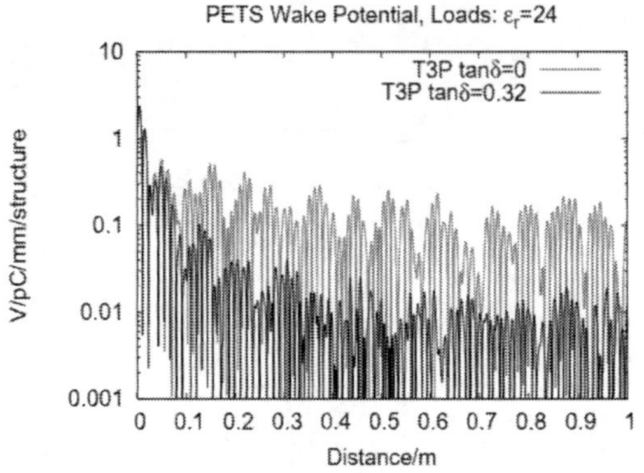

Figure 7. Impact of the lossy absorbers on the damping of the transverse wakefields in the PETS. The transverse wake potential for the lossless case (tanδ=0) is compared to the lossy case (tanδ=0.32), as calculated with T3P with second-order basis functions (p=2).

to the accelerating structure is investigated numerically with T3P. The drive beam in the PETS has no transverse offset and excites monopole modes which couple through the output coupler and the connecting waveguides to the accelerating structure. Since the waveguides are specified to have near perfect transmission of the fundamental mode, their detailed layout is not important for the purposes of this study. A quarter model of the PETS is connected to a half model of the accelerating structure, and the use of symmetric (magnetic) boundary conditions leads to one PETS being connected to two accelerating structures – in accordance with the CLIC design.

Figure 9 shows a time sequence of the power flow from the PETS to the accelerating structure. The low group velocity ($\approx$ 0.02c) of the accelerating structure leads to an effective pulse compression which produces high gradients. Dispersion effects in the accelerating structure might have led to a broadening of the RF pulse due to the high-frequency content of the single bunch profile, while the current CLIC design uses multiple bunches for the PETS drive beam. Multi-bunch simulations will be performed to verify the two-beam accelerator power transfer scheme.

5. Summary

SLAC's parallel 3D electromagnetic time-domain code T3P employs state-of-the-art parallel Finite Element methods on curved conformal unstructured meshes with higher-order field representation. T3P allows large-scale time-domain simulations of complex, realistic 3D structures with unprecedented accuracy, aiding the design and operation of the next generation of accelerator facilities.

T3P was applied to calculate transverse wakefield damping effects in the CLIC power and extraction structure (PETS), and convergence in the results was found. Furthermore, a two-structure time-domain simulation of the power transfer from the PETS to the main CLIC accelerating structures was performed. Future applications will include transverse wakefield coupling effects from the PETS to the accelerating structures. Further code development is necessary to accommodate accurate numerical termination of the waveguides that are essential for wakefield damping.

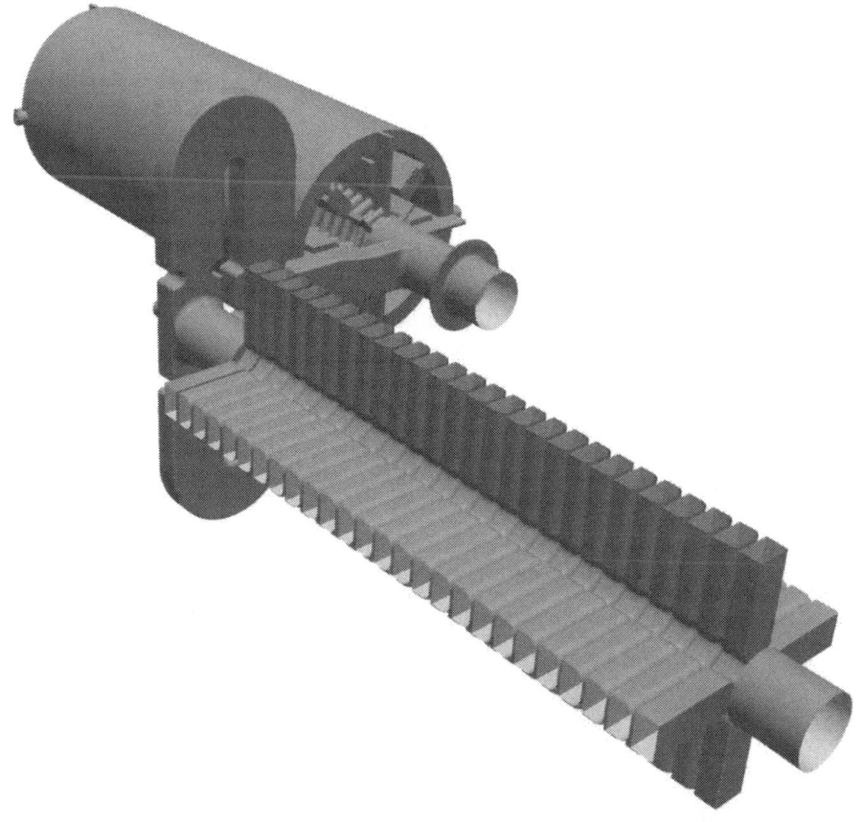

Figure 8. Overview of the coupled model of the PETS (left) and the accelerating structure (right), used for T3P calculations of RF power transfer. An on-axis drive beam in the PETS excites monopole modes in the PETS which are transferred through coupling waveguides to the accelerating structure, a low group-velocity design based on four connected waveguides and slightly detuned cells for operating at high gradients. Each PETS is connected to two accelerating structures, but only one accelerating structure is shown here.

Acknowledgments
This work was supported by the US DOE ASCR, BES, and HEP Divisions under contract No. DE-AC002-76SF00515. This research used resources of the National Energy Research Scientific Computing Center, and of the National Center for Computational Sciences at Oak Ridge National Laboratory, which are supported by the Office of Science of the U. S. Department of Energy under Contract No. DE-AC02-05CH11231 and No. DE-AC05-00OR22725. – We also acknowledge the contributions from our SciDAC collaborators in numerous areas of computational science and A. Grudiev, I. Syratchev and W. Wuensch from the CLIC RF structure development program at CERN.

References

[1] Ainsworth M 2004 Dispersive properties of high-order Nedelec/edge element approximation of the time-harmonic Maxwell equations, *Philos. trans.-Royal Soc., Math. phys. eng. sci.*, **362**, no. 1816
[2] Gedney S, Navsariwala U 1995 An unconditionally stable finite element time-domain solution of the vector wave equation *Microwave and Guided Wave Letters, IEEE*, **5**, no. 10

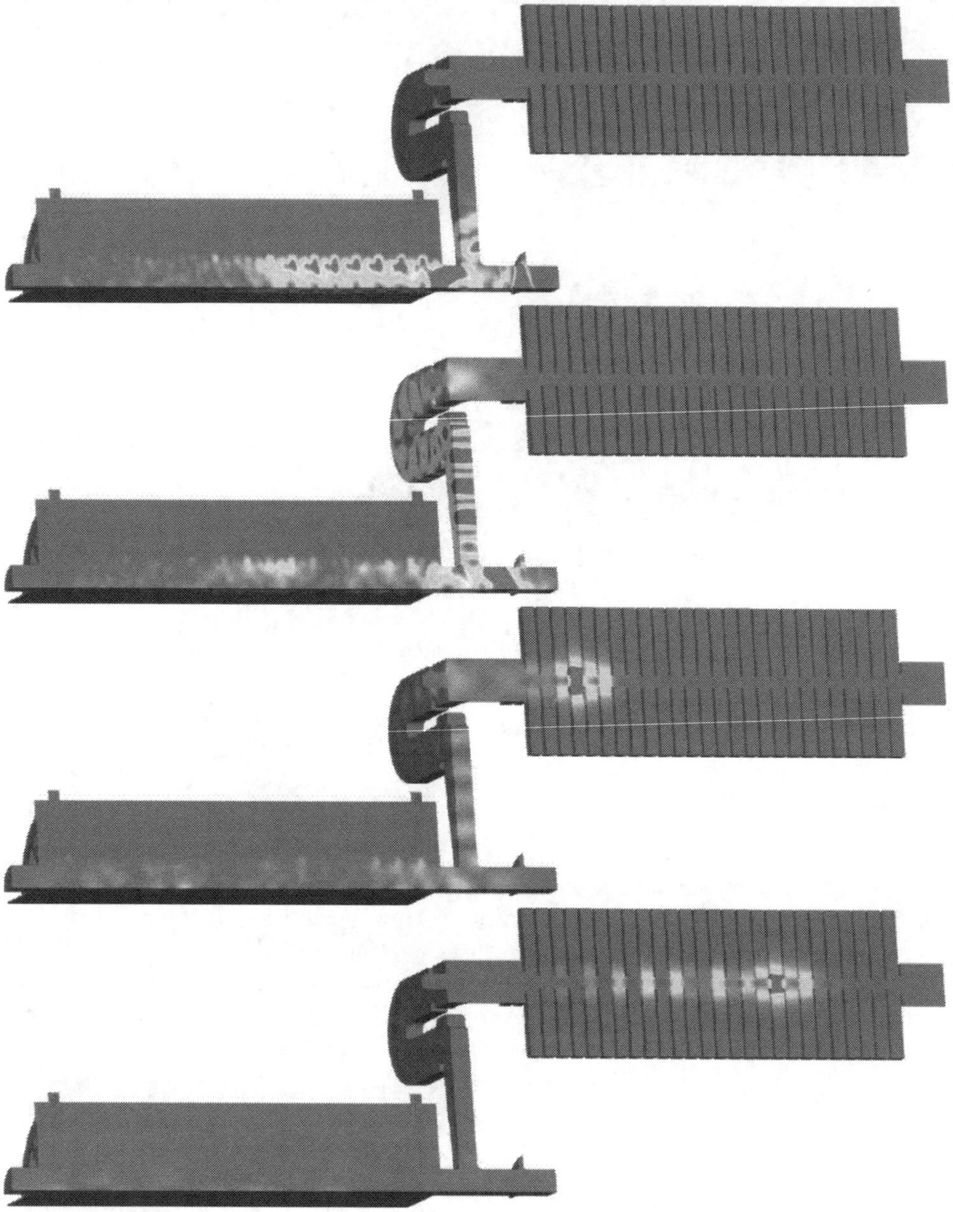

Figure 9. Power transfer from the PETS to the main beam accelerating structure, as simulated with T3P (p=1) for a single drive bunch. The low group velocity ($\approx$ 0.02c) of the accelerating structure leads to an effective pulse compression which produces high gradients.

[3] Lee L *et al* 2007 Enabling technologies for petascale electromagnetic accelerator simulation *J. Phys.: Conf. Ser.* **78** *012040*
[4] Lee L *et al* 2008 Computational science research in support of petascale electromagnetic modeling *J. Phys.: Conf. Ser.* **125** *012077*
[5] Ng C *et al* 2008 Design and optimization of large accelerator systems through high-fidelity electromagnetic simulations, *J. Phys.: Conf. Ser.* **125** *012003*
[6] Candel A *et al* 2008 Wakefield Computations for the CLIC PETS using the Parallel Finite Element Time-Domain Code T3P, *Proc. X-band RF Structure & Beam Dynamics Workshop*, Cockcroft Institute, UK

Journal of Physics: Conference Series **180** (2009) 012005 doi:10.1088/1742-6596/180/1/012005

Recent results and future challenges for large scale particle-in-cell simulations of plasma-based accelerator concepts

C Huang[1], W An[1], V K Decyk[1], W Lu[1], W B Mori[1], F S Tsung[1], M Tzoufras[2], S Morshed[3], T Antonsen[3], B Feng[4], T Katsouleas[5], R A Fonseca[6], S F Martins[6], J Vieira[6], L O Silva[6], E Esarey[7], C G R Geddes[7], W P Leemans[7], E Cormier-Michel[7], J-L Vay[7], D L Bruhwiler[8], B Cowan[8], J R Cary[8,9] and K Paul[8]

[1] University of California, Los Angeles, CA 90095, USA
[2] University of Oxford, Clarendon Laboratory, Parks Road, Oxford OX1 3PU, UK
[3] University of Maryland, College Park, MD 20742, USA
[4] University of Southern California, Los Angeles, CA 90089, USA
[5] Duke University, Durham, NC 27708, USA
[6] Instituto Superior Tecnico, Lisbon 1049-001, Portugal
[7] Lawrence Berkeley National Lab, Berkeley, CA 94720, USA
[8] Tech-X Corporation, Boulder, CO 80303, USA
[9] University of Colorado, Boulder, CO 80309, USA

Email: huangck@ee.ucla.edu

Abstract. The concept and designs of plasma-based advanced accelerators for high energy physics and photon science are modelled in the SciDAC COMPASS project with a suite of Particle-In-Cell codes and simulation techniques including the full electromagnetic model, the envelope model, the boosted frame approach and the quasi-static model. In this paper, we report the progress of the development of these models and techniques and present recent results achieved with large-scale parallel PIC simulations. The simulation needs for modelling the plasma-based advanced accelerator at the energy frontier is discussed and a path towards this goal is outlined.

1. Introduction

Particle accelerators have been the major tool for experimental high energy physics and photon science. The largest particle accelerator ever built to date, the Large Hadron Collider (LHC) at CERN, is based on the superconducting RF accelerator technology. It is 27 kilometers in length and cost 6 billion dollars. Similarly the world's first hard X-ray free electron laser, the Linac Coherent Light Source (LCLS), also relies on a kilometer-long conventional RF accelerator to accelerate an electron beam to 14.35 GeV for x-ray FEL production. A possible way to reduce the size and cost of next generation accelerators by an order of magnitude or more is to use relativistically moving plasma waves as the accelerating structure [1, 2]. Relativistic plasma waves can be excited by the space charge forces of a charged particle beam in the Plasma Wakefield Acceleration (PWFA) scheme or by the radiation pressure of a high intensity laser in the Laser Wakefield Acceleration (LWFA) scheme.

Recent experiments on plasma-based advanced accelerators have achieved 30~50 GeV/m acceleration gradient and produced GeV class electron beams with percents of energy spread from the

LWFA scheme over cm distances. In recent PWFA experiments similar gradients were demonstrated over meter distances and 80 GeV electrons were produced. Simulations show that properly phased electrons can be accelerated while keeping the energy spread near or below a percent using the PWFA scheme. However, the energy frontier for high energy physics is in the TeV range, while 1-20 GeV electron beams with small angular divergence are desirable for applications in photon science. The key challenge to use plasma-based accelerators for these applications is to extend the acceleration length over meters of plasma with stringent requirements on the beam parameters and quality. It is also a challenge to accurately simulate these next generation experiments.

The interactions among the plasma particles and with the driver in the self-consistent electromagnetic fields they produce are fully modeled with Particle-In-Cell (PIC) codes [3, 4]. These models provide the most fundamental, microscopic level description of the interactions through Maxwell's equations and Newton's laws; therefore, they are also very computation intensive. PIC models are differentiated by the kinds of fields retained in the model, the manner in which the fields are solved, and the charge density and current density are deposited.

Within the SciDAC COMPASS (Community Petascale Project for Accelerator Science and Simulation) project, a suit of parallel PIC codes (VORPAL, OSIRIS, WARP, QuickPIC) are actively maintained and enhanced to model particle beam generation, acceleration for the energy frontier and radiation sources for biological and material applications using the plasma-based accelerator concepts. VORPAL [5], OSIRIS [6], Warp [7], are state-of-the-art, fully explicit, multi-dimensional, fully parallelized, fully relativistic, PIC codes. These PIC codes are capable of reproducing the detailed physics at the expense of computation power. They have also been successfully benchmarked against each other and used to validate other reduced algorithms [8]. Recently the speed of these PIC codes has been greatly enhanced through a boosted frame algorithm by utilizing the fact that the range of space/time scale is not a Lorentz invariant [9]. This speed improvement enables 3D full scale simulations of near term LWFA experiments in great detail.

For the LWFA problem, the ponderomotive guiding center approximation can be used to remove the need of resolving the finest time scale of the laser oscillation. In the VORPAL code, this is implemented in the laser envelope model, which applies the approximation without loss of kinetic effects such as particle trapping [10]. Pulse depletion is also modeled accurately for long propagation distances, although the corresponding spectral broadening eventually breaks the assumptions of the algorithm. The model is therefore suitable to simulate both long propagation distances and injection mechanisms, with orders of magnitude speedup relative to the standard PIC algorithm, depending on application. The QuickPIC code is a highly efficient, fully parallelized, fully relativistic, three-dimensional PIC code for simulating LWFA/PWFA schemes [11] that utilizes the quasi-static approximation together with the ponderomotive guiding center approximation. Under the quasi-static approximation, the time scale of the evolution of the driver is separated out from the plasma evolution and a fully three-dimensional electromagnetic field solve and particle push is reduced to a sequence of transverse two-dimensional field solves and particle pushes. Overall, this algorithm speeds up the computational time by 2 to 4 orders of magnitude without losing accuracy for problems of interest.

Using this collection of PIC codes with peta-scale computers, we can carry out high resolution high fidelity full-scale simulations to determine if there are physics obstacles to a plasma based accelerator design before large capital expenditures are spent building key components. Currently, a 1-5 GeV single stage LWFA can be modeled in 3D using full PIC codes. Reduced models and techniques are being developed to allow simulation of multi-GeV stages, including the boosted frame algorithm, density scaled simulation, and the envelope or quasi-static models, and with careful benchmarking these are making it possible to simulate 10~50 GeV single stage LWFA/PWFA in 3D. In this paper, we describe recent large-scale kinetic simulations of plasma-based accelerators by the SciDAC COMPASS project. Section 2 describes recent progress and results from simulations to model future PW class LWFA experiments towards a Laser-Plasma Linear Collider (LPLC). Section 3 describes modeling of a TeV PWFA Linear Collider (PWFA-LC) concept using the quasi-static model. Section

4 provides an overview of the needs for modeling LWFA/PWFA for the energy frontier. Finally, the path towards peta-scale and beyond simulations of LWFA/PWFA is outlined.

2. Recent progress and results of simulation of LWFA

2.1. Modelling next generation LWFA experiments and LPLC designs

The simulation of the long distances associated with the next generation of lasers systems can be achieved with the use of reduced codes and methods including QuickPIC, VORPAL envelope, and also with the boosted frame approach. These techniques, combined with benchmarkings against full PIC simulations for short distances and scaled simulations, allow one to address the range of physics required for LPLC modules and other applications.

Three dimensional simulation with QuickPIC in the laboratory frame and OSIRIS in the boosted frame [12] were performed for different possible configurations for next generation experiments with a ~300J laser using weakly-nonlinear/nonlinear regimes and self-injected/externally-injected electron beams. QuickPIC results for self-injection and external-injection/external-guiding regimes are shown in figure 1. Self-injection (modeled with test particles) is obtained with a moderate intensity pulse (a_0=5.8) propagating in a 2.7×10^{17} cm^{-3} density plasma. The external-injection case uses a_0=2.0, and a plasma density of 2.2×10^{16} cm^{-3}. Results agree with theoretical scalings [13] for the accelerating gradients and the final electron beam energies. These scalings show how to scale results from one density to another for matched laser spot sizes. We also note the extremely stable propagation and wake excitation in the external injection scenario was seen for the meter scale distance.

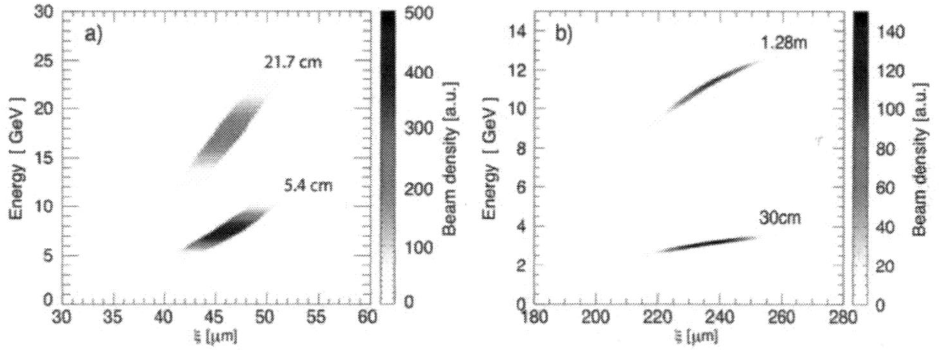

Figure 1. QuickPIC results for (a) self-injection regime with theoretical scaling predicting 13 GeV after 22 cm propagation, and (b) an ongoing simulation of the external-injection/ external-guiding configuration, with theoretical scaling predicting 53 GeV after 5.28 m.

The same scenarios were simulated in OSIRIS in the boosted frame, since a common full PIC simulation cannot be accomplished with the computational power currently available: this would take more that 1 year on hundreds of processors. The boosted frame scheme leverages on the Lorentz transformation of the laser pulse and plasma column to reduce the numerical requirements, in particular the number of time steps involved. This enables fully kinetic three-dimensional simulation of LWFA with considerably smaller computational cost [14]. The scheme can be applied to the design of current experiments in the mm scale, allowing fast parameter scans, but can also be used to simulate new scenarios to be available with the next generation of laser systems. Fully kinetic simulations in the Lorentz boosted frame were performed for the configurations described above, and confirmed the possibility to achieve energies in excess of 10 GeV with self-injected electrons, and nearly 50 GeV for externally-injected bunches with a 300J class laser.

Designs for an LPLC [15] have indicated that accelerator stages in the 10 GeV range may be optimal for such a machine, and next generation lasers including the proposed BELLA project at

LBNL will access this regime. Stages were simulated in the less strongly driven, quasi-linear regime, where dynamics are similar for electrons and positrons, using a combination of scaled, envelope, and Lorentz boosted simulations to access the required physics. VORPAL simulations demonstrated that scaling the laser spot size and pulse length with the plasma wake wavelength allows a shorter simulation at high density using explicit PIC to deduce the properties of 10 GeV stages, and characterized the wake structure and evolution [16, 17, 18]. A wide range of laser and plasma parameters were simulated, establishing conditions for a stage that efficiently transfers laser energy into the particle bunch while maintaining high beam quality, including efficient depletion of the laser energy into the accelerating field of the wake and high transfer of laser energy deposited in the wake to the particles. These simulations showed that low energy spread 10 GeV bunches of both electrons and positrons can be obtained using a petawatt laser. In this regime, the transverse fields of the wake are also shaped by the shape of the laser spot. Simulations are now using this feature, and further tailoring of the laser and plasma, to control beam propagation and further increase achievable performance [19]. An advantage of scaled simulation using a full PIC code is that because the laser period is resolved, laser evolution up to depletion can be modeled to evaluate efficiency.

To accurately model laser and electron bunch oscillation, VORPAL envelope simulations have been used to model 0.6m of a 10 GeV stage at the design density of $10^{17}/cm^{-3}$ [10]. Figure 2 compares the transverse profiles of the laser pulse using the envelope model and a scaled simulation; one sees that the envelope model correctly captures the self-focusing oscillations. These simulations also established designs for other applications, including Thomson gamma sources [20].

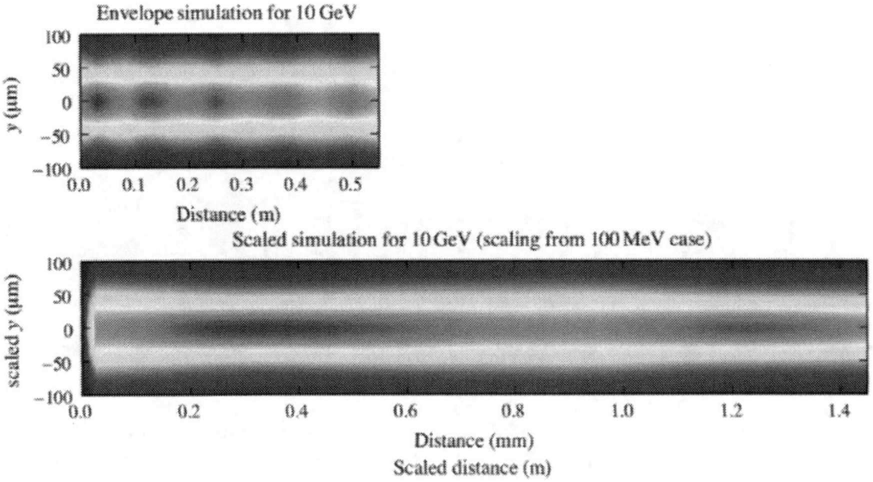

Figure 2. Evolutions of the transverse profiles of the laser pulse for a 10 GeV LWFA stage with the envelope model (top) and the scaled simulation (bottom) using VORPAL.

Simulations of 10 GeV stages in the boosted frame are also being conducted in Warp and VORPAL. Warp simulations of scaled 10 GeV stages at plasma density $n_e=10^{19}cm^{-3}$ were performed in 2-1/2D (see [21]) and 3-D (see figure 3) using Lorentz boosted reference frames with relativistic factor γ between 1 (laboratory frame) and 10. The high density simulations result in short run time for effective benchmarking between the algorithms. Two figures of merit are considered (both compared in the laboratory frame): (a) the peak energy of the accelerated electron beam; (b) the average energy history of the electron beams. Agreement within a few percent is observed on the beam peak energy between calculations in all frames, with a speedup of 100 measured between the calculations in the laboratory frame (total time ~35,000 sec. ~10 hours) and in the frame at γ=10 (total time ~350 sec. ~6 min.). The average beam energy history reveals agreement at a few percents level for the accelerating phase, followed by a growing discrepancy during the decelerating phase, when using a low dispersion electromagnetic solver [14]. The agreement in the deceleration phase is much improved if the standard Yee solver is used, at a cost in CPU time, but nonetheless achieving a maximum measured speedup of 10. The gain in efficiency scales roughly as 1/n where n is the plasma density. Studies are hence in progress to use boosted frame simulations to directly simulate 10 GeV stages at plasma densities of $10^{17}cm^{-3}$, which are not presently computationally accessible using conventional explicit simulations.

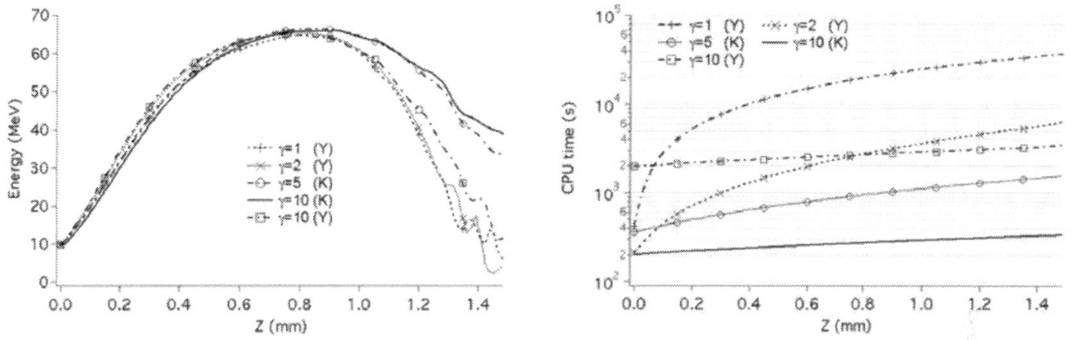

Figure 3. (Left) average beam energy versus longitudinal position in the laboratory frame from 3-D Warp calculations in a frame moving at γ=1, 2, 5 and 10; (right) CPU time recorded as the beam crosses successive stations in the laboratory frame (all runs performed using 256 cores on LBNL Linux cluster). The symbols (Y) and (K) indicate respectively whether the standard Yee scheme or a low-dispersion solver was used.

For collider and many light source applications, advances are required in all of the methods to accurately model the very tight beam specifications required. Momentum errors were reduced using high order force weighting and smoothing, improving simulation of experimental momentum spread [22], and further development may be required for collider applications with extremely low emittance.

We collaborated with researchers in the VACET project and at Tech-X to produce VisIt parallel 3D renderings allowing visualization of large simulation outputs, and to improve analytics by beginning to automate detection, characterization, and fast parallel tracking of particle bunches in the simulations [18, 23, 24].

2.2. Modelling radiation emission from LWFA

The self-injected electrons in a LWFA perform betatron oscillations in the ion channel associated with the blowout region created by the laser [25, 26], and will therefore radiate. These oscillations occur as the particle is gaining energy from the wakefield, and can be observed in the trajectories depicted in figure 4. The radiation is in the X-ray to gamma ray range and could be used as a light source.

By post-processing the particles trajectories, estimates for the radiation can be obtained. A detailed discussion of the radiation calculations using particle tracking in OSIRIS and the visualization package VisXD [27] will be presented in a future publication [28]. We note that, as conditions close to the energy frontier are reached, the energy losses associated with the betatron motion of the electrons in the ion channel should be considered. For these scenarios where radiation losses are important, numerical codes should include the radiation damping physics [29], as QuickPIC and OSIRIS already do. Particle tracking results in VORPAL have been used to analyze injection trajectories, formation of high quality bunches, and betatron oscillation [18, 23].

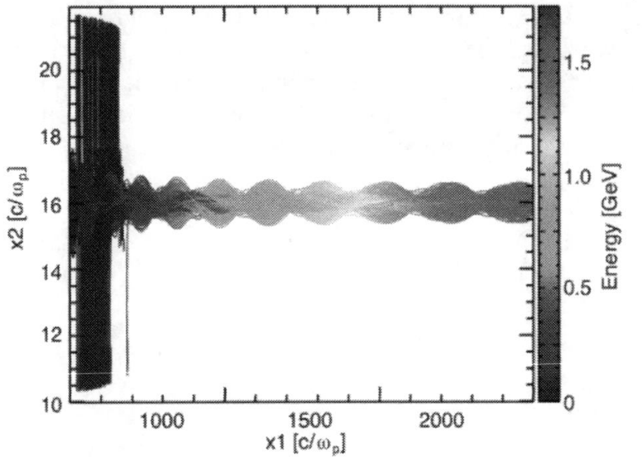

Figure 4. Projection of the trajectories (color represents energy) of the self-injected particles for the ~1.5GeV LWFA configuration. Particles (moving from left of right) are mostly injected off-axis.

3. Recent progress and results of simulation of PWFA-LC

For the next TeV scale linear collider, high energy (500 GeV) electron/positron beams with high charge and high quality are required. One possible design based on PWFA is the PWFA-LC concept [30]. In this concept, 19 beam trains each with 250 (125×2) bunches and energy of 25 GeV are used. These beam trains are fed into 19 meter-long plasma cells and drive the plasma wakefields in each cell, respectively. The main beams that are used for the collider also have initial energies of 25 GeV. Each main beam immediately follows a drive beam in the plasma cell and extract energy from the wakefield. It is designed that the main beams will gain 25 GeV in each cell while achieving small energy spread (< 1%) and emittance preservation. By staging the 19 plasma cells, the main beams will be accelerated to 500 GeV required for collision. Preliminary and non-optimized simulations show that >50% energy transfer efficiency is obtained while keeping the energy spread < 1%.

The next linear collider also requires transverse beam size on the order of 10 nm to achieve high luminosity, demanding high transverse resolution in the simulations. Furthermore, 3D effects such as the hosing instability [31] and asymmetric main beam spot sizes require a 3D model. The quasi-static model is a practical choice for such simulations due to the elimination of the Courant condition. Recently the scalability of QuickPIC has been improved with a pipeline algorithm and a new 2D domain decomposition strategy for particle beams, significantly increasing the simulation throughput and enable higher transverse resolution and therefore more realistic spot sizes in the simulation.

3.1. Scaling quasi-static algorithm to 10,000+ processors

The quasi-static code QuickPIC is built on the UPIC Framework [32] with spectral space solver. A strong scaling test has shown that the UPIC Framework can scale to more than 8,000 processors for a 3D electromagnetic problem of 512×256×512 grids and 16 particles/cell on the latest homogeneous computer architecture with 2D spatial domain decomposition. The original implementation of QuickPIC with 1D spatial domain decomposition has similar but worse scaling property in comparison to the UPIC Framework. In the QuickPIC computation cycle, the plasma response to the driver is evaluated by sweeping a slice of plasma along the longitudinal direction and calculating the trajectories of the plasma particles as the slice moves from the front of the driver to the back [11, 33]. The latency in the Fourier space field solver for each update of the transverse 2D slice eventually limits the number of processors QuickPIC will scale to. For example, for $10^3\times10^3$ grids and 4 particles/cell in the 2D slice, the field solver can scale to 512 homogeneous processors.

In order to scale to 10,000+ processors, an algorithm that can allocate more computation tasks to the problem is needed. For a fully-explicit PIC code, this can be done through dividing the simulation domain to finer grains or in more spatial dimensions (at most 3 spatial dimensions). While for the quasi-static algorithm, a software pipelining technique can be used [34]. The pipelining algorithm adds another level of parallelism to QuickPIC by dividing the job into many pipeline stages along the direction of propagation. Each pipeline stage consists of a group of processors that works on its own domain. However, unlike conventional domain decomposition for which each processor does the calculation on its own input and communicates with its neighbors through the boundary, the initial state of the whole 2D plasma slice for a particular pipeline stage is transferred from an updated 2D slice from the previous stage (the exception is the first pipeline stage which always starts with a fresh uniform plasma slice). Therefore, pipeline stages are separated not only spatially but also in time.

The pipeline algorithm enables the efficient use of more than 1,000 processors, dramatically decreasing the turn-around time of a PWFA simulation. It has been successfully implemented into both the basic and full quasi-static versions of QuickPIC for the particle beam driver and verified to produce the same result as the non-pipelining version. The HDF I/O routine has also been modified to allow data merging of each processor group on the fly without post-processing. Performance measurement shows that the speedup of the 2D plasma solver in the pipelining mode is nearly ideal. This is because the data transfer between two successive pipeline stages is relatively inexpensive compared to the time spent on the solver itself and the transfer also overlaps with the computation. For the whole computation cycle including the 3D beam update, the overall efficiency of each processor group in the pipelining mode reaches 85% (with 2048 processors in 64 pipeline stages) relative to the non-pipelining mode. Thus using 64 pipeline stages leads to a 54 times reduction in turn-around time for a long simulation (assuming filling time of the pipeline is negligible).

The pipeline algorithm works without particular requirements on the domain decomposition, therefore one can still choose an appropriate domain decomposition strategy within each pipeline stage. In the first implementation of the pipelined QuickPIC code, 1D domain decomposition the longitudinal direction was used, same as the pipeline decomposition direction. This setup eventually limits the maximum number of processors to the number of grid points available in the longitudinal direction which is typically on the order of 1000. Recently, we further improved the pipeline algorithm with a 2D domain decomposition in the pipeline stages. Preliminary scaling tests show that it will scale to at least 16,384 processors for a PWFA simulation with 2048×2048×256 grid points (figure 5).

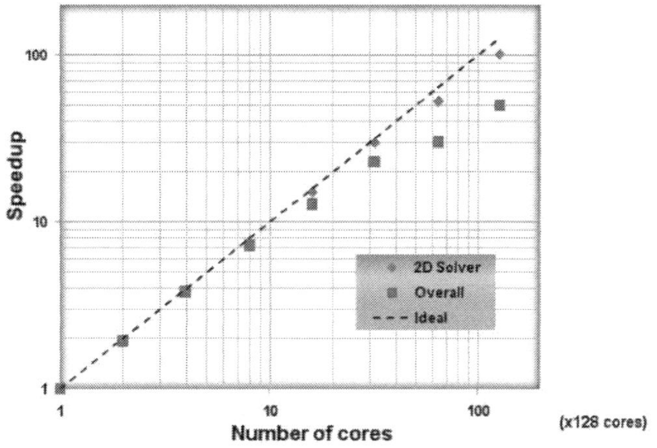

Figure 5. Strong scaling of QuickPIC on the NERSC Franklin platform with 2048×2048×256 grids and 4 particles/cell. 128 cores per pipeline stage are used and 1 to 128 stages are tested. The smallest domain size is 2048×16×2. The blue diamonds and the red squares are measured speedup for the calculation of a plasma slice in the 2D solver and for one 3D step, respectively.

3.2. High resolution simulation of the PWFA-LC concepts

A PWFA stage that can provide a large acceleration gradient and high total efficiency (30% ~ 90% from drive beam to main beam) and accelerate sufficient charge (~10^{10} electrons) with high beam quality and low energy spread (0.1% ~1%) is essential to the PWFA-LC concept. An ideal design for the first stage (25GeV) and the last stage (475GeV) of the PWFA-LC was investigated using QuickPIC [35]. In this design, the beam loading theory [36] based on the theoretical framework [37] of nonlinear wake excitation was used to derive the optimal current profiles of the drive and main beams. In the simulation, the main beam with 1.73×10^{10} electrons gains 25 GeV in 0.6 meter with an acceleration gradient of 42.7 GeV/m. Less than 1% energy spread of the main beam was achieved and the efficiency of energy transfer from the drive beam to the main beam is 51%.

Using the QuickPIC code with the latest improvements in simulation capability described in section 3.1, a more detailed study of a PWFA-LC stage with main beam parameters close to the requirements of a future TeV collider can be carried out. For such a matched beam with tight spot sizes, the peak beam density would be 10^3~10^4 times higher than the plasma density, possibly causing the plasma ions to move even on the sub-picosecond time scale of the beam duration [38]. This issue also needs to be considered for LWFA collider design not using a weakly nonlinear wake. The ions are attracted to the center of the beam, therefore, to include the ion dynamics in the simulation also demands a transverse resolution of ~10 nm. As a first step, a preliminary high resolution simulation of a nominal PWFA-LC stage has been conducted. The drive and main beams have 2.9×10^{10} electrons and 1×10^{10} electrons respectively and they are both modeled with ~4,200,000 macro-particles. Their initial energies are 25 GeV. The emittance of the main beam is 0.093 mm·mrad and the matched spot sizes are 100 nm for $n_e=1\times10^{17}$ cm^{-3}. The drive beam spot size can be larger as it is not used in the final collision. An emittance of 10 mm·mrad is chosen which is typical for current state-of-the-art electron beams from a linac. The matched spot size is 1 μm. The resolution of the simulation is 49nm×49nm×304nm and it is carried out with 8192×8192×1024 grid points using 8192 processors on Franklin. There are 4 particles per cell for the plasma electron and ion respectively. Their separation is comparable to the real atom separation of ~20 nm at 1×10^{17} cm^{-3}. We note that this is the first time in PIC simulation of PWFA that one can simulate nearly all the particles in a real plasma. Figure 6 shows the results from the simulation.

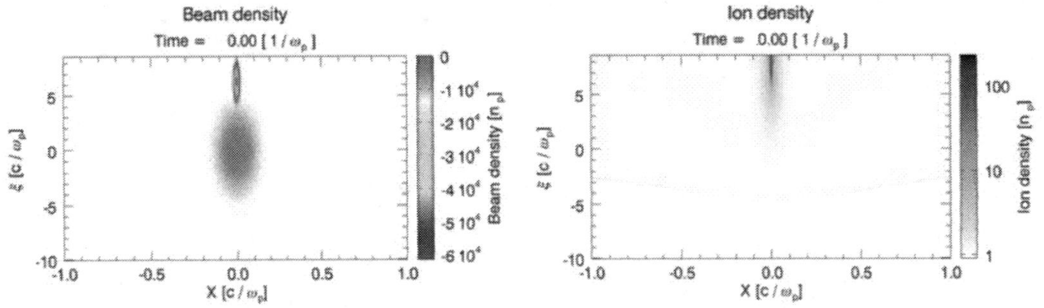

Figure 6. High resolution QuickPIC simulation of the PWFA-LC concept with mobile ions. The spot sizes of the beams are matched to the plasma density. Left: the drive and the main beam (moving down); right: the density of the ion background showing ion concentration near the axis.

4. Future challenges of PIC simulations and the path towards the energy frontier

The recent progress on the development of the boosted frame technique and reduced models (envelope model for LWFA only and quasi-static model for LWFA and PWFA) and the usage of massive parallel computing resources have enabled simulation studies of the near term LWFA/PWFA experiments and conceptual designs of the next generation facilities. However, the needs to faithfully simulate the laser/beam driver evolution and the main beam dynamics in a future plasma-based collider are still challenging. The computation cost of the former simulation can be reduced by using the boosted frame technique or envelope/quasi-static model. Here we estimate the computation requirements for the main beam. For example, the smallest emittance of the main beam in a TeV collider is 0.04 mm·mrad. The matched spot size of the main beam at 500 GeV in a plasma of 1×10^{17} cm^{-3} will be 30 nm, which is three orders of magnitudes smaller than the longitudinal spot size or the plasma wavelength. Furthermore, at the final focus where a plasma lens may be used to focus the beam to the interaction point, the smallest beam size in the transverse dimension is 6 nm requiring nm resolution in the simulation. However, the transverse box size needs to be around 20 collisonless skin depths, which is ~300 microns. This is 10^5 times larger than the required resolution. In the longitudinal direction, the plasma wavelength needs to be well resolved using O(1000) grids. Therefore a realistic 3D high-resolution simulation of the accelerated beam would need $10^5\times10^5\times1000 = 1\times10^{13}$ grids and 4×10^{13} particles (assuming 4 particles/cell), demanding extremely large amount of memory and processors. The time step and the number of time steps for meter-long propagation distance are ~ 0.002 fs and 1×10^9 for a full PIC simulation, or ~ 16 ps and 208 for a quasi-static simulation. Such simulation is difficult to run even using the quasi-static model and the largest computers in the world. Alternatively, as described above, in the linear regime, the focusing fields can be shaped, and simulations are investigating how this can allow use of larger diameters beams [19].

To increase the capability of large-scale PIC simulation for the energy frontier, a comprehensive approach needs to be taken. This will include algorithm development to enhance the single core performance of the PIC algorithm and to optimize PIC codes to scale efficiently on many-core system and to adapt them to heterogeneous architecture. On the other hand, mesh refinement, further development of the boost frame algorithm and the envelope/quasi-static models might also play important roles to reduce the computational demands substantially.

The suit of PIC codes in the SciDAC COMPASS project are constantly being improved to run efficiently on latest homogenous platforms. For example, OSIRIS performs well on a single processor while maintaining high parallel efficiency with load-balancing, i.e., >85% for strong scaling and >95% for weak scaling on greater than 32,000 processors of the Argonne BlueGene Intrepid Computer (figure 7). Additional work on performance also focuses on better utilization of the vector unit of modern CPUs. Preliminary work has shown a 2-2.5x speedup on the Nehalem architecture with vectorized field interpolation, particle push and current deposit routines.

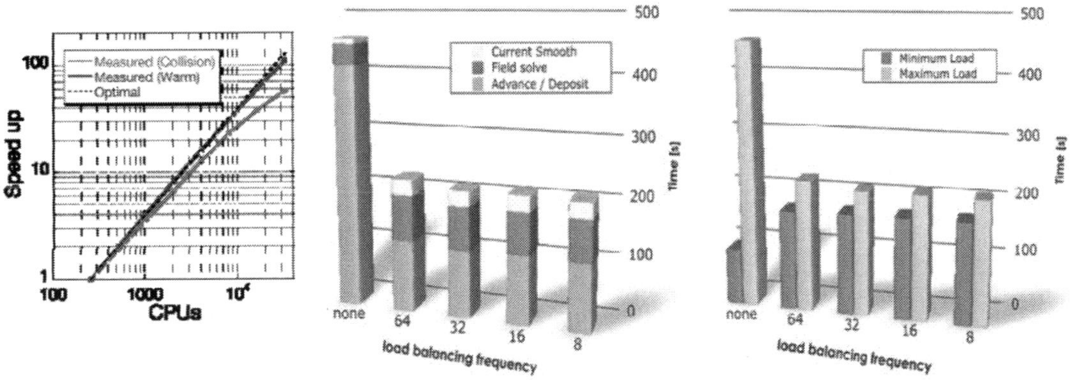

Figure 7. Strong scaling for OSIRIS on up to 32,768 processors on the Intrepid BlueGene machine (left) and timing results (middle and right) vs. load balance frequency when the particle push and field solve are load balanced together.

However, near-term petascale platforms will likely make a paradigm shift to the heterogeneous architecture with millions threads and processing units. One has to ensure that the key PIC algorithms will continue to scale well on such a platform. Currently, efforts are being made at UCLA to rewrite critical pieces of the PIC algorithm to take advantage of new architectures such as GPUs. As a first step, the kernel of the UPIC framework has been ported to the GPU and optimization of the key routines is under way. On the other hand, the latest version of OSIRIS already supports arbitrary precision floating points (which allows the data to be moved to and from GPU's).

In addition, the Adaptive Mesh Refinement (AMR) technique can be used to significantly reduce the requirements for computer memory and the number of operations for large scale simulations. Its application to the fully self-consistent modeling of beams and plasmas is especially challenging, due to properties of the Vlasov-Maxwell system of equations, but successful PIC simulations using AMR have been reported [39]. The application of AMR to the modeling of LWFA is being explored with Warp. For the simulation of a 10GeV LWFA stage, the wake wavelength is O(100μm) while the electron bunch and laser wavelength are typically submicron in size. As a result, the resolution required for different parts of the problem may vary by more than two orders of magnitude in each direction, corresponding to up to 6 orders of magnitude of possible (theoretical) savings by use of mesh refinement. While algorithm limitations will most probably limit the actual speedup to lower values, we anticipate that speedups of one order of magnitude or more are achievable. We note that the savings offered by mesh refinement will apply in addition to the savings provided by the various techniques already employed (envelope, quasi-static, fluid, boosted frame).

5. Conclusions

The plasma-based advanced accelerator is a promising technology to reduce the size and cost of a linear collider or a X-ray light source by providing orders of magnitude higher accelerating gradient than the conventional RF technology. Large scale PIC simulations have provided insight into the proof-of-principle LWFA/PWFA experiments and elucidated the relevant physics. Recently, developments of various PIC models and techniques, such as the implementation of the envelope model and boosted frame approach, the enhancement of the quasi-static model with a pipelining algorithm and the performance and scaling optimization of the full PIC model, pave the way of using large-scale PIC simulations to model the next generation LWFA/PWFA experiments with PW class lasers or state-of-the-art electron beam drivers. Conceptual designs of 10 GeV LWFA module or 25 GeV PWFA module are now being developed and refined with PIC simulations as a guide. In the future, adaptation of PIC codes to the latest vector units in a modern processor and innovative many-core/heterogeneous computing architecture and algorithm development such as the AMR technique will be pursued to meet the simulation requirement for the energy frontier of 1 TeV or more.

6. Acknowledgements

This work is supported by Department of Energy, Office of Science, through the SciDAC program, and by NNSA through NA-22. Simulations are conducted at the NERSC platforms, the UCLA Dawson cluster, the LBNL Linux cluster, the LLNL ATLAS facility and the Argonne ALCF facility.

References

[1] T. Tajima, J. M. Dawson, Phys. Rev. Lett., 43, 267 (1979).
[2] P. Chen et al, Phys. Rev. Lett., 55, 1537 (1985).
[3] C. K. Birdsall et al., Plasma Physics via Computer Simulation, McGraw-Hill, New York, 1985.
[4] J. M. Dawson, Rev. Mod. Phys., 55, 403 (1983).
[5] Nieter et al., Journal of Computational Physics, 2004.
[6] R. A. Fonseca et al., Lec. Notes in Comp. Sci., 2329, 342 (2002); R. G. Hemker, Ph.D. Thesis, UCLA (2000).
[7] D. P. Grote, A. Friedman, J.-L. Vay, I. Haber, AIP Conf. Proc. 749, 55 (2005).
[8] K. Paul et al., AIP Conf. Proc. 1086, 315 (2009).

[9] J.-L. Vay, Phys. Rev. Lett. 98, 130405 (2007).
[10] B. Cowan et al., AIP Conf. Proc. 1086, 208 (2009).
[11] C. Huang et al., J. Computational Phys. 217, 658 (2006).
[12] S. F. Martins et al., in preparation.
[13] W. Lu et al, Phys. Rev. STAB, 10, 061301 (2007).
[14] J.-L. Vay et al., These proceedings.
[15] C.B. Schroeder et al., AIP Conf. Proc. 1086, 309 (2009).
[16] E. Cormier-Michel et al., AIP Conf. Proc. 1086, 297 (2009).
[17] C.G.R. Geddes et al., Proc. Particle Accel. Conf., Vancouver, Canada (2009) WE6RFP075.
[18] C.G.R. Geddes et al, SciDAC review, accepted.
[19] E. Cormier-Michel et al., in preparation.
[20] C.G.R. Geddes et al., Proc. Conf. on Applications of Accelerators in Research & Industry 2008.
[21] J.-L. Vay et al., Proc. Particle Accel. Conf., Vancouver, Canada (2009) TU1PBI04.
[22] C. G. R. Geddes et al., AIP Conf. Proc. 1086, 12 (2009).
[23] O. Ruebel et al., Proc. Supercomputing Conference 2008.
[24] D.M. Ushizima et al., International Conference on Machine Learning & Applications, 2008.
[25] D. H. Whittum et al., Phys. Rev. Lett. 64, 2511 (1990).
[26] S. Kneip et al., Phys. Rev. Lett. 100, 105006 (2008).
[27] R A Fonseca et al., Plasma Physics and Controlled Fusion 50, 124034 (2008).
[28] J. L. Martins et al., in preparation.
[29] S. Kiselev et al., Phys. Rev.Lett., 93, 135004 (2004).
[30] A. Seryi et al., Proc. Particle Accel. Conf., Vancouver, Canada (2009), WE6PFP081.
[31] C. Huang et al., Phys. Rev. Lett. 99, 255001 (2007).
[32] V. K. Decyk, Computer Phys. Comm. 177, 95 (2007).
[33] C. Huang et al., J. Phys.: Conf. Ser. 46 190-199, 2006.
[34] B. Feng et al., Journal of Computational Physics, In Press, DOI: 10.1016/j.jcp.2009.04.019.
[35] C. Huang et al., Proc. Particle Accel. Conf., Vancouver, Canada (2009), WE6RFP097.
[36] M. Tzoufras et al., Phys. Rev. Lett. 101, 145002 (2008).
[37] W. Lu et al., Phys. Rev. Lett. 96, 165002 (2006).
[38] J. B. Rosenzweig et al., Phys. Rev. Lett. 95, 195002 (2005).
[39] J.-L. Vay et al., Phys. Plasmas 11 (2004).

Simulating relativistic beam and plasma systems using an optimal boosted frame.

J.-L. Vay[1], D. L. Bruhwiler[2], C. G. R. Geddes[1], W. M. Fawley[1], S. F. Martins[3], J. R. Cary[2,3], E. Cormier-Michel[1], B. Cowan[2], R. A. Fonseca[4], M. A. Furman[1], W. Lu[5], W. B. Mori[5], L. O. Silva[4]

[1]LBNL, Berkeley, CA, USA
[2]Tech-X Corp., Boulder, CO, USA
[3]Department of Physics, University of Colorado, Boulder, CO, USA
[4]Instituto de Plasmas e Fusão Nuclear, Instituto Superior Técnico, Lisbon, Portugal
[5]UCLA, Los Angeles, CA, USA

E-mail: jlvay@lbl.gov

Abstract. It was shown recently that it may be computationally advantageous to perform computer simulations in a Lorentz boosted frame for a certain class of systems. However, even if the computer model relies on a covariant set of equations, it was pointed out that algorithmic difficulties related to discretization errors may have to be overcome in order to take full advantage of the potential speedup. In this paper, we summarize the findings, the difficulties and their solutions, and review the applications of the technique that have been performed to date.

1. Introduction

In [1], we have shown that the ratio of longest to shortest space and time scales of a system of two or more components crossing at relativistic velocities is not invariant under a Lorentz transformation. This implies the existence of an "optimum" frame of reference minimizing a measure of the ratio of space and time scales. Since the number of computer operations (e.g., time steps), for simulations based on first-principle formulations, is proportional to the ratio of the longest to shortest time scale of interest, it follows that such simulations will eventually have different computer runtimes, yet equivalent accuracy, depending solely upon the choice of frame of reference. The scaling of theoretical speedup was derived for a generic case of two crossing identical rigid particle beams, and for three particle acceleration related problems: particle beams interacting with electron clouds [2], free electron lasers [3], and laser-plasma accelerators [4]. For all the cases considered, it was found that the ratio of space and time scales varied as γ^2 for a range of γ, the relativistic factor of the frame of reference relative to the optimum frame. For systems involving phenomena (e.g., particle beams, plasma waves, laser light in plasmas) propagating at large γ, theoretical speedup of simulations being performed in an optimum boosted frame could reach several orders of magnitude, as compared to the same simulation being performed in the laboratory frame, for example. As an example, a speed-up of 1,000 was demonstrated on a Particle-In-Cell (PIC) simulation of transverse instability of an ultra-relativistic (γ=500) proton beam interacting with a background of electrons. In the remainder of this paper, we summarize the difficulties and limitations of the method, the solutions that were developed, and review its applications performed to date.

2. Difficulties and limitations

Even if the fundamental electrodynamics and particles equations are written in a covariant form, the numerical algorithms that are derived from them may not retain this property. As an example, we considered in [5] an isolated beam propagating in the laboratory frame at relativistic velocity. When applying the effect of the beam field on itself using the Newton-Lorentz equation of motion, the contribution from the radial electric field is largely canceled by the contribution from the azimuthal magnetic field. However, we showed that the so-called 'Boris particle pusher' [6] (which is widely used in PIC codes), does make an approximation in the calculation of the Lorentz force which leads to an inexact cancellation of the electric component by the magnetic component. The magnitude of the error grows with the beam relativistic factor and in practice, it is unacceptably large for simulations of ultra-relativistic charged beams, where the cancellation needs to be nearly complete. The issue was resolved by changing the form of the Lorentz force term in the Boris pusher, and solving analytically the resulting implicit system of equations (see [5] for details).

For electromagnetic calculations, a limitation may come from numerical dispersion in the electromagnetic solver. The standard Yee solver [7], also widely used in PIC codes, has a relatively high numerical dispersion that is anisotropic and depends on the cell aspect ratio and the ratio of the time step to the cell size in each direction. Short of using a different field solver, this limits the speedup that can be gained by performing the calculation in a boosted frame because the number of cells needs to be kept above a certain minimum level in the longitudinal direction to ensure that the numerical dispersion falls below an acceptable level in that direction. Solvers that produce no numerical dispersion along major axes have been tested in the Scidac codes Osiris [8], Vorpal [9] and Warp [10] with some success, but have yet to be fully validated.

An additional practical complication of numerical simulation in a boosted frame is that inputs and outputs are often specified (or desired) in the laboratory frame. For example, in LWFA simulations, laser and plasma parameters have to be transformed from the laboratory to the new relativistic boosted frame, so that the electromagnetic waves will be Doppler-shifted, and the background plasma, with higher density, is now drifting. Different strategies have been adopted in the three Scidac codes applied to LWFA for laser injection in a Lorentz boosted frame: in Osiris, the laser beam is initialized in the computational box, enlarging the box in the transverse direction and in front of the plasma to accommodate the higher focusing angle and longer wavelength; in Vorpal, the laser is injected from all faces of the computational box; in Warp, the laser is injected at a plane that is fixed in the laboratory frame and drifting in the boosted frame.

The initial phase-space distribution of a particle beam is generally known in the laboratory. For calculations in boosted frames of large γ, deriving the initial beam conditions at a given time can be easy if the initial conditions are simple (e.g., LWFA), or more difficult and/or computationally costly if injecting the beam in a particle accelerator for example, where its longitudinal extent in the boosted frame can cover several lattice periods. In order to circumvent this difficulty, a procedure was implemented in Warp which injects the beam through a transverse plane that is fixed in the laboratory, but drifting in the boosted frame, similarly to the laser injection method. Due to long range space charge forces, it is still necessary to provide a reasonable estimate of the beam distribution near the injection plane; this is accomplished by the use of "frozen" drifting macroparticles.

After the relativistic PIC algorithm evolves the system in the boosted frame, the results must be transformed back to the laboratory frame. To construct a single time shot in the laboratory, a range of boosted time shots is required. This analysis can be done either as a post-processing script or, more efficiently, at runtime. Nevertheless, we emphasize that the transformation of several relevant output quantities for LWFA is straightforward, by using relativistic invariants (e.g., total injected charge), or by simple Lorentz transformation (e.g., maximum particle energy, final laser energy). Finally, specific diagnostics might already contain all space and time information necessary for complete Lorentz transformation (e.g., particle tracking). As a set of additional diagnostics, we have found it convenient in Warp to record quantities at a number of regularly spaced "stations", immobile in the laboratory frame, at a succession of discrete times, for both detailed time histories and laboratory time-averages.

Details of the input and output procedures can be found in [11] for Warp, in [12] for Osiris and in [13] for Vorpal.

3. Examples of application

3.1. Electron cloud driven instabilities

Several existing and planned future particle accelerators have limitations due to the electron cloud instability that may negatively impact the beam quality and in some cases even lead to severe beam loss. A calculation of electron cloud driven instability [2] for an ultra-relativistic beam was performed with the Warp code framework in (a) standard PIC mode using the new particle pusher in a Lorentz boosted frame; (b) in quasistatic mode [14] using linear maps to push beam particles into the accelerator lattice. The two runs were in good agreement and completed using similar computer resources and runtimes. The speedup factor of the PIC boosted frame calculation compared to a PIC calculation in the laboratory frame was estimated at 500. For many calculations of electron cloud instability, the boosted frame approach may not resolve any additional physics not included in the quasistatic approach. We note, however, that the quasistatic method requires significant special coding for taking into account eventual longitudinal motion of electrons [14], as well as a special parallelization scheme [15] for parallelization along the axis of beam propagation, that are not standard to PIC codes. By contrast, the boosted frame method includes naturally the longitudinal dynamics and requires more modest modifications to an existing standard PIC code or framework (including none for full parallelization).

3.2. Laser wakefield acceleration

Laser driven plasma waves offer orders of magnitude increases in accelerating gradient over standard accelerating structures (which are limited by electrical breakdown), thus holding the promise of much shorter particle accelerators. Yet, computer modeling of the wake formation and beam acceleration requires fully kinetic methods and large computational resources due to the wide range of space and time scales involved [16]. For example, modeling 10 GeV stages for the LOASIS (LBNL) BELLA proposal [17] in one-dimension demanded as many as 5,000 processor hours on a NERSC supercomputer [13]. As discussed in [1], the range of scales can be greatly reduced if one adopts the common assumption that the backward-emitted radiation can be neglected, enabling, for the first time, the full-PIC simulation of the next generation of laser systems [15].

Warp simulations at plasma density $n_e = 10^{19}$ cm^{-3} were performed in 2-1/2D [11] and 3-D [15] using reference frames moving anywhere between $\gamma_f = 1$ (laboratory frame) and 10. These simulations are scaled replicas of 10 GeV stages that would operate at actual densities of 10^{17} cm^{-3} [18, 19] and allow short run times to permit effective benchmarking between the algorithms. Two figures of merit were considered (both evaluated in the laboratory frame): (a) the peak energy of the accelerated electron beam; (b) the average energy history of the electron beams. Agreement within a few percent was observed on the beam peak energy between calculations in all frames, with a speedup of 100 measured between the calculation in the frame at $\gamma = 10$ and the calculation in the laboratory frame. The average beam energy history reveals agreement at the few percent level for the accelerating phase, followed by a growing discrepancy during the decelerating phase. In order to obtain the speedup of 100, a low dispersion electromagnetic solver was used for $\gamma \geq 5$, while the standard Yee solver was used otherwise, for the reasons stated in the preceding section. Percentage level agreement was obtained during both accelerating and decelerating phases by using the Yee solver for all values of γ, at a cost, but nonetheless achieving a maximum measured speedup of 10. The reasons for the discrepancies at high values of γ between the two electromagnetic solver runs are currently under investigation.

The parallel VORPAL framework has been used to simulate proof-of-principle acceleration of test-particles to 10 GeV energies in one and two spatial dimensions, for plasma channel parameters in which pump depletion and dephasing lengths are equal [20]. The 1D simulations were done both in the laboratory frame and in a Lorentz boosted reference frame [3], showing qualitative agreement in the peak energy of the accelerated particles. The boosted-frame simulations were 1,500x faster, reduced errors in the laser group velocity, and eliminated problems with artificial dark current. The 2D simulations were not feasible in the laboratory frame, but the boosted-frame results [13] showed agreement with the predictions of scaling arguments [18], with an estimated 2,000x speedup.

A set of 3-D benchmarks was performed between OSIRIS boost and OSIRIS and QuickPIC[14] in the laboratory frame, for weakly-nonlinear/nonlinear regimes and self-injected/externally-injected electron beams, with very good qualitative and quantitative agreement. Particular emphasis was given to a 1.5 GeV self-injection stage of LWFA, successfully reproducing the results obtained with laboratory frame simulations presented in [21]. The boosted frame scheme was tested for a broad set of numerical parameters and algorithms, namely different boost speeds with varying resolutions, higher order field solvers [22], particle interpolation schemes [23], and modified Boris pusher [5], showing overall result convergence and consistency. For the particular case of a 1.5 GeV self-injection stage, results recover those in the laboratory frame presented in [21].

3.3. Free electron lasers

In a short wavelength free-electron laser (FEL), a high energy electron beam interacts with a static magnetic undulator. In the optimal boost frame with Lorentz factor γ, the down-shifted FEL radiation and up-shifted undulator have identical wavelengths and the number of required time-steps (presuming the Courant condition applies) decreases by a factor of $2\gamma^2$ for fully electromagnetic simulations. Examples of boosted-frame simulations are compared [25] to results obtained with the eikonal (i.e, SVEA) and wiggler-period averaged code Ginger [24]. It was concluded that if the necessary FEL physics can be studied with an eikonal code, it will run much faster than a full EM code in whatever frame. However, if there are important physical phenomena that cannot be resolved properly by an eikonal code, a boosted-frame electromagnetic code is a very attractive alternative to a brute force full EM calculation in the laboratory frame.

3.4. Coherent synchrotron radiation

Another application for which the Lorentz-boosted frame method might be useful is that of modeling coherent synchrotron radiation (CSR) [26] emitted by high current, high brightness relativistic electron beams. Because full scale electromagnetic simulation of CSR in the laboratory frame is difficult due to the wide range of scales (chicane lengths of order meters, radiation wavelengths of orders microns), in order to make the calculation tractable most CSR simulation codes apply simplifications such as ignoring transverse variation of CSR across the electron beam. We have begun preliminary work of simulating CSR emission with the boosted frame method in Warp, examining the behavior of a high current, short electron beam transiting a simple dipole magnet. Our early results show that upon exit from the undulator the electron beam shows the characteristic energy loss variation with longitudinal position that one expects from previous theoretical analyses of CSR. Further studies are currently underway.

4. Conclusion

The non-invariance of the range of scales of a physical system implies that the computational cost of a certain class of computer simulations depends strongly on the choice of the simulation frame of reference. Some algorithmic difficulties may arise due to the loss of covariance upon discretization of the Maxwell-Vlasov system of equations, and the need to transform input/output data between the laboratory frame and the Lorentz boosted frame. Most such difficulties have been overcome and no "show-stopper" has been identified at this time. First principle simulations in boosted frames have been performed successfully by the Scidac codes Warp, Osiris and Vorpal in application to electron cloud driven instabilities, laser wakefield acceleration and free electron lasers, with speedups ranging between a few and several orders of magnitude. Further testings and studies of advanced numerical techniques are underway to fully validate the method and push the limits toward higher speed-ups. Our recent progress shows that first principle modeling in a Lorentz boosted frame is a viable alternative or complement to using reduced descriptions like the quasistatic [14] or eikonal [24] approximations, or scaled parameters [19], when there is a need to study physics that is not accessible to the other descriptions, or for validation of their simplifications.

5. Acknowledgments

We are thankful to E. Esarey, B. A. Shadwick and C. B. Schroeder for insightful discussions, and to D. P. Grote for support with the Warp code. This work was supported by the U.S. DOE under contracts DE-AC02-05CH11231, DE-AC52-07NA27344, DE-FG03-92ER40727, DE-FC02-07ER41500, DE-FG02-03ER54721, the U.S.-LHC Accelerator Research Program (LARP), the U.S. DOE SciDAC program, the Fundação Calouste Gulbenkian, FCT under grants SFRH/BD/35749/2007 and PTDC/FIS/66823/2006 (Portugal), and by the European Community FP6 program (project EuroLeap, contract #028514). This research used resources of NERSC, supported by the U.S. DOE under Contract No. DE-AC02-05CH11231.

References

[1] J.-L. Vay, *Phys. Rev. Lett.* **98** (2007) 130405.
[2] *Proc. International Workshop on Electron-Cloud Effects*, Daegu, S. Korea (2007).
[3] N. Kroll, P. Morton, M. Rosenbluth, *IEEE J. Quantum Electron* **QE-17** (1981) 1436.
[4] T. Tajima and J. M. Dawson, *Phys. Rev. Lett.* **43** (1979) 267.
[5] J.-L. Vay, *Phys. Plas.*, bf 15 (2008) 056701.
[6] J. P. Boris, *Proc. Fourth Conf. Num. Sim. Plasmas*, Naval Res. Lab., Wash., D. C., 3-67, 2-3 November 1970.
[7] K. S. Yee, *IEEE Trans. Ant. Prop.* **14** (1966) 302-307
[8] Fonseca, R.A. *et al.*, *Proc. Int. Conf. on Computational Science*, Amsterdam, Netherlands (2002) 342-351.
[9] C. Nieter and J.R. Cary, *J. Comput. Phys.* **196** (2004) 448.
[10] D. P. Grote, A. Friedman, J.-L. Vay, I. Haber, *AIP Conf. Proc.* bf 749 (2005) 55.
[11] J.-L. Vay *et al.*, *Proc. Particle Accelerator Conference*, Vancouver, Canada (2009) TU1PBI04.
[12] R. A. Fonseca *et al.*, *Plasma Phys. and Control Fusion* **50** (2008) 124034.
[13] D. L Bruhwiler *et al.*, *Proc. 13th Advanced Accelerator Concepts Workshop*, Santa Cruz, CA (2008) 29.
[14] C. Huang *et al.*, *J. of Comput. Phys.* **217** (2006) 658-679.
[15] C. Huang, These proceedings
[16] C. G. R. Geddes *et al.*, *J. Phys. Conf. Series V* **125** (2008) 12002/1-11.
[17] http://loasis.lbl.gov
[18] E. Cormier-Michel, *et al. Proc. 13th Advanced Accelerator Concepts Workshop*, Santa Cruz, CA (2008) 297.
[19] C. G. R. Geddes *et al.*, *Proc. Particle Accelerator Conference*, Vancouver, Canada (2009) WE6RFP075.
[20] E. Esarey, *et al.*, *Proc. 11th Advanced Accelerator Concepts*, Stony Brook, NY (2004) 578.
[21] W. Lu *et al.*, *Phys. Rev. ST Accel. Beams* **10** (2007) 061301.
[22] A. D. Greenwood *et al.*, *J. Comp. Phys.* **201** (2004) 665-684.
[23] T. Esirkepov, *Comput. Phys. Comm.* **135** (2001) 144-53.
[24] W. M. Fawley, LBNL Tech. Rpt. LBNL-49625-Rev. 1 (2004); also SLAC Rpt. LCLS-TN-04-3.
[25] W. M. Fawley, J.-L. Vay, *Proc. Particle Accelerator Conference*, Vancouver, Canada (2009) WE5RFP029.
[26] M. Venturini, *et al.*, *Phys. Rev. ST Accel. Beams* **8** (2005) 014202

Computation of electron cloud diagnostics and mitigation in the main injector

S A Veitzer[1], P LeBrun[2], J R Cary[1], P Spentzouris[2], P H Stoltz[1] and J F Amundson[2]

[1] Tech-X Corporation, 5621 Arapahoe Ave., Suite A, Boulder, CO, 80303
[2] Fermi National Accelerator Laboratory, P.O. Box 500, Batavia, IL 60510-5011

E-mail: veitzer@txcorp.com

Abstract. High-performance computations on Blue Gene/P at Argonne's Leadership Computing Facility have been used to determine phase shifts induced in injected RF diagnostics as a function of electron cloud density in the Main Injector. Inversion of the relationship between electron cloud parameters and induced phase shifts allows us to predict electron cloud density and evolution over many bunch periods. Long time-scale simulations using Blue Gene have allowed us to measure cloud evolution patterns under the influence of beam propagation with realistic physical parameterizations, such as elliptical beam pipe geometry, self-consistent electromagnetic fields, space charge, secondary electron emission, and the application of arbitrary external magnetic fields. Simultaneously, we are able to simulate the use of injected microwave diagnostic signals to measure electron cloud density, and the effectiveness of various mitigation techniques such as surface coating and the application of confining magnetic fields. These simulations provide a baseline for both RF electron cloud diagnostic design and accelerator fabrication in order to measure electron clouds and mitigate the adverse effects of such clouds on beam propagation.

1. Introduction

Future experiments to study high energy physics require the construction of new, high-intensity particle accelerators that are pushing the frontiers of accelerator design. Electron cloud effects are considered to be one of the most important factors that will limit machine performance for high-intensity accelerators, and electron cloud mitigation methods can have a large influence on the cost and design of such accelerators. Recently, researchers have developed a new diagnostic technique that uses microwaves to measure electron cloud densities [1, 2, 3], and there are a number of experimental programs both in the United States (Fermilab, SLAC, LBL, Cornell) and abroad (CERN) that are using injected microwaves to measure both electron cloud densities and the effects of cloud mitigation techniques, such as coatings to reduce production of secondary electrons. At Fermilab, a planned upgrade of the Main Injector to 2 MW of beam power represents an important direction for the lab for producing new science. The performance-degrading aspects of electron clouds are a major concern for the Main Injector upgrade.

We have recently been developing the simulation tools needed to model microwave diagnostics of electron clouds and to simulate electron cloud buildup in the Main Injector. The simulations contain detailed multi-physics processes, including RF microwave propagation through a plasma in an elliptical beam pipe, modeling of electron cloud evolution, beam currents, static external fields, RF fields, space charge, and secondary electron emission. To achieve this, we use the plasma simulation package VORPAL [4]. The flexibility and parallel capabilities of VORPAL form a firm foundation for producing accurate and detailed results for this complex modeling challenge. We are currently performing detailed numerical simulations at the Blue Gene/P computational cluster located at the Argonne Leadership Computing Facility, as well as other high-performance clusters. We have verified the physical simulation components of VORPAL relevant to modeling electron clouds, and have validated integrated electron cloud simulations against experiments measuring cyclotron resonances and solenoidal fields [5]. We show preliminary results of these ongoing numerical studies as it relates to the proposed upgrade of the Main Injector at FermiLab. The work reported here is the result of a partnership between Tech-X Corporation and FermiLab through the COMPASS SciDAC project, in addition to funding provided by the Department of Energy through the Small Business Innovation Research (SBIR) program.

The buildup of electron clouds in high-intensity accelerators can negatively impact both beam quality and overall machine performance. Electrons can gain energy due to the potential of the beam, and they can interact with the beam pipe walls and produce more electrons through secondary emission. If enough electrons build up, they can cause an increase in beam emittance and possibly excite instabilities in the beam that will limit the current and luminosity. Zimmermann presents a comprehensive review of instabilities that are driven by electron clouds in reference [6].

Recently a new diagnostic has been developed whereby microwaves are injected into accelerator beam pipe sections during normal operation. It is well-known that a plasma will induce a phase shift, and to first order, the magnitude of the phase shift is linearly dependent on the plasma electron density. The amplitude of the We track the amplitude of the RF electric field a point at the end of the vacuum chamber in the simulation; first without any electrons, and subsequently with an electron cloud. We simplify the problem by assuming that there is no change in the wave amplitude, A, or wave frequency, ω, induced by the electron cloud. The phase difference is then derived by subtracting two waveforms,

$$A\sin(\omega t+\phi_1)-A\sin(\omega t+\phi_2) = 2A\sin[(\phi_1-\phi_2)/2]\cos[\omega t+(\phi_1+\phi_2)/2] \quad (1)$$

$$\approx A(\phi_1-\phi_2)\cos[\omega t(\phi_1+\phi_2)/2] \quad (2)$$

where ω is the angular frequency of the wave, ϕ_1 and ϕ_2 are the phase shifts of the wave for the two cases. The approximation is valid for small phase shifts. Squaring and taking a time-average over many wave periods gives the phase shift directly,

$$\frac{1}{2}(\phi_1-\phi_2)^2 \approx \left\langle (\sin(\omega t+\phi_1)-\sin(\omega t+\phi_2))^2 \right\rangle \quad (3)$$

This method works well for simulations, where electrons can be turned on and off. Experimentally, the phase shift must be deduced from other measurements, such as measuring sidebands produced by phase modulation due to gaps in the proton bunch trains.

In the context of simulating electron cloud effects in the Main Injector, we use an elliptical beam pipe with major radius 5.8801 cm, and minor radius 2.3876 cm, and a length of 1.0 m. The electron cloud is initially uniformly distributed throughout the volume of the beam pipe over a length of 0.5 m with no thermal energy. The simulations reported here have a total time of $1\mu s$, with a time step of about 2.7 ps. The frequency of the RF diagnostic wave is approximately 1.672 GHz, which is about 10% above the cutoff frequency of the beam pipe. We have also considered circular cross section beam pipes, additional electron cloud and magnetic field configurations, and the effects of higher order RF modes, with similar results.

2. Results

2.1. Beam current effects

An increase in the beam current can have significant effects on the long-term electron cloud evolution and spatial distribution of the electrons. Electrons are accelerated to the center of the beam pipe when the beam passes through the cloud, and they drift to the pipe walls inbetween bunches. A higher beam current has the potential to create more secondary electrons. However, depending on the bunch spacing, the beam potential may keep the electrons from drifting all the way to the walls, thereby reducing the overall cloud density. In our particular simulations shown here, a 5-fold increase in the beam current corresponds to a significant decrease in the overall density of electrons after many bunch crossings. This is partially due to the elliptical geometry of the simulations because in the long direction, the electrons will not have enough time between bunches to drift to the walls and create more electrons through secondary emission. Recent simulations in field-free regions of the Main Injector and circular cross sections at similar bucket frequencies have shown that 1×10^{11} protons/bunch creates a maximal cloud buildup [7]. Figure (1) shows a sample electron cloud density at approximately 0.485 μs after the initiation of the simulation. The top plot shows the case where there are 1×10^{10} protons per bunch, and for comparison, the bottom plot shows the case where there are 5×10^{10} protons per bunch. In the figures, a beam bunch can be seen entering the simulation region from the left moving towards the cloud. More than 25 bunches have passed through the cloud by this point in the simulation.

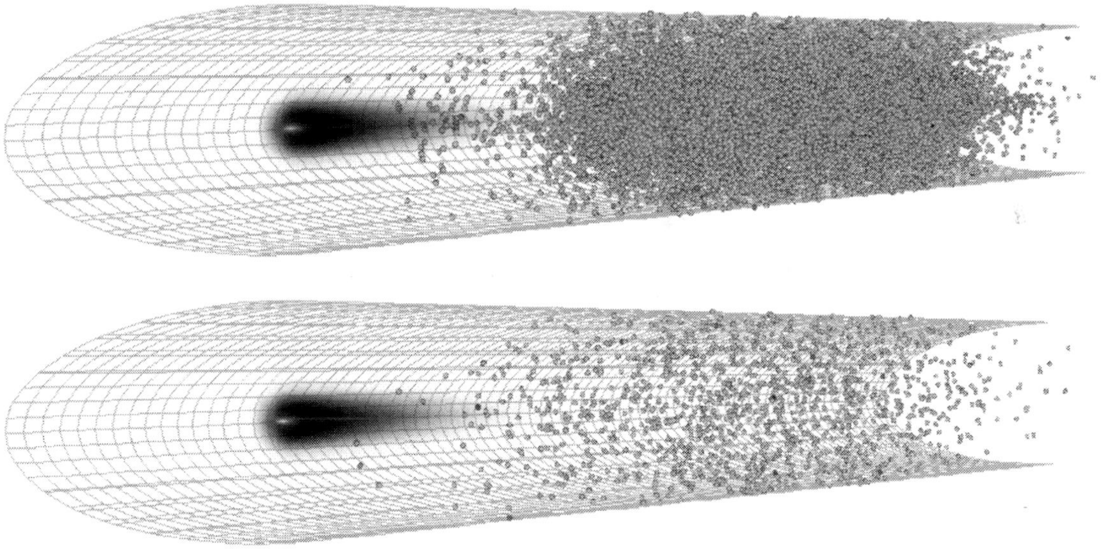

Figure 1. Sample electron cloud distribution at 0.484 μs in a Main Injector simulation with 1×10^{10} protons per bunch (top) and 5×10^{10} protons per bunch (bottom). The elliptical beam pipe is clipped to reveal a beam bunch entering the simulation domain to the left of the electron cloud. Electrons are colored by their along-beam velocity.

2.2. Phase shift measurements

We have measured phase shifts for a variety of different simulation parameters. In the current context, we compare phase shifts for TE_{11} modes injected into an elliptical beam pipe to simulate electron cloud evolution in the Main Injector, as described in the section above. In the first case, with 1×10^{10} protons per bunch, the average electron density over the entire simulation is about

4.75×10^{11} electrons/m^3, while in the second case, the average density is only 1.575×10^{11} electrons/m^3. The computed phase shift is 3.128×10^{-3} rad/m in the first case and 5.823×10^{-5} in the second case. It is observed that the overall phase shift is not an accurate measure of the average density in these simulations. We believe that this is due to non-uniformity in the spatial distribution of the electron clouds. The phase shift is induced in regions where the transverse electric field of the RF signal is large. In this case, for a TE_{11} mode, the transverse electric field is large near the center of the beam pipe. This produces a relatively larger phase shift in the first case because there is a relatively larger concentration of electrons near the center of the beampipe.

Figure (2) shows computed phase shifts as a function of time where we have used a moving window with width $0.012\mu s$, corresponding to 20 RF periods. While the overall phase shift is an indicator of the average cloud denisty in the region where the transverse electric field is large, time-resolved measures of the phase shift such as can be seen in Figure (2) are much better for understanding the buildup and evolution of electron clouds. The large deviations in phase shift shown in the left-hand plot in Figure (2) are due to the formation of vertical bands of electrons in the cavity, which have been seen in a number of simulations, especially in the context of applied external magnetic fields. When electrons in these bands are near the center of the beam pipe, the measured phase shift is quite large, and correspondingly, the electrons have a much higher potential for disrupting the beam because they are in the path of the beam at these times. Our simulations indicate the disruption is likely to be more pronounced at the tail than the head of the beam, at least in the regime where space charge from the cloud itself is not so large as to keep a significant density of electrons in the center of the beam pipe between bunch crossings.

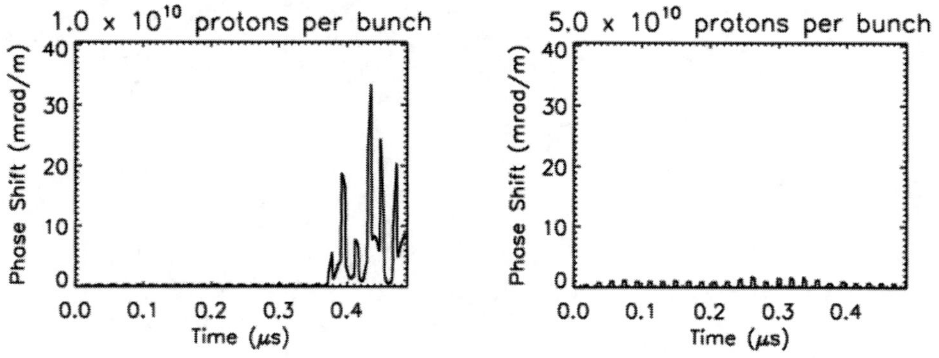

Figure 2. Computed phase shifts for a moving window of $0.012\mu s$ with a beam containing 1.0×10^{10} protons per bunch (left), and 5.0×10^{10} protons per bunch (right). The dramatic increase in phase shift starting at about $0.35\mu s$ in the left-hand plot is due to the emergence of non-uniform structure in the clouds. The overall number of electrons decreases only about 14% in the first case, but by nearly 30% in the second case.

2.3. Magnetic field effects

External magnetic fields can also have a significant effect on the buildup and evolution of electron clouds in accelerators. One potential technique for mitigating electron cloud buildup is to apply a solenoidal field which confines secondary electrons near the beam pipe walls. This serves two purposes; (1) electrons are confined in a region where they will not disrupt the beam, and (2) electrons do not gain enough energy from the beam potential to effectively increase the cloud density through multipacting and secondary electron emission. We have performed simulations with the Main Injector geometry including the effects of externally applied fields. Figure (3) shows results for two different magnetic field configurations at simulation times of 17 ns (left), $0.5\mu s$ (middle), and $1.0\mu s$ (right). The magnetic fields are uniform in space. The first bunch crossing occurs around 17 ns. Our simulations indicate that while modest solenoidal fields are effective in confining electrons near the walls, and thereby mitigating the effects of electron clouds, even a small transverse dipole field negates this effect, as can be seen in the figures. We conclude that using solenoidal fields to mitigate electron cloud buildup might be an effective technique in field-free regions. However, care needs to be exercised in dipole and quadrupole regions.

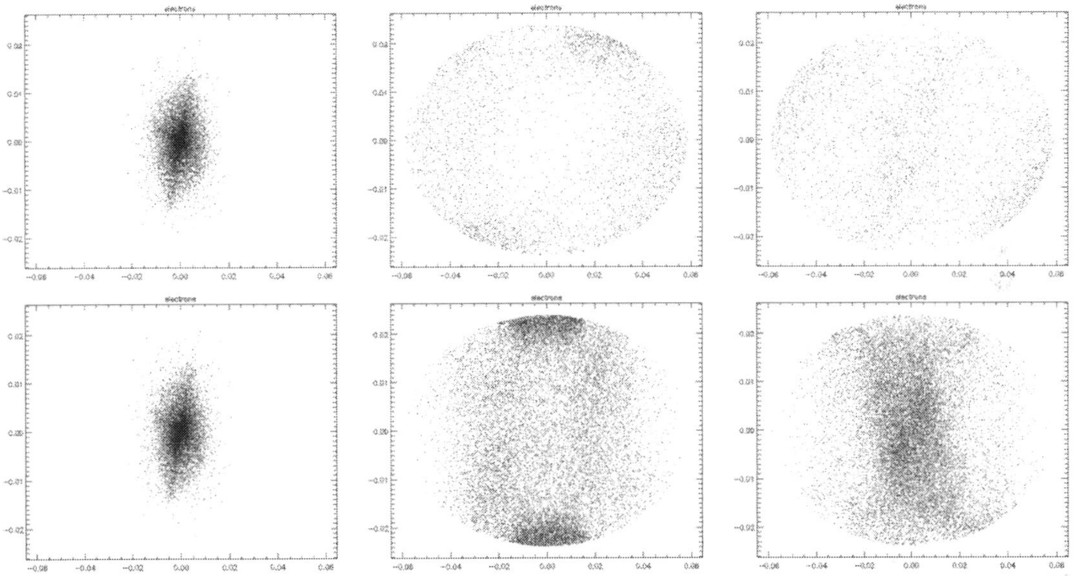

Figure 3. Transverse profile of electron cloud at three different simulation times. The top row is for the case where there is an applied 10 Gauss solenoidal external magnetic field, while the bottom row there is a 10 Gauss solenoidal field and an additional 10 Gauss dipole (transverse) field applied.

Acknowledgments
Much of this work was performed under the auspices of the Department of Energy under Small Business Innovation Research Contract DE-FG02-08ER85042. The authors would like to acknowledge the hard work of the VORPAL development team: T. Austin, G. I. Bell, D. L. Bruhwiler, R. S. Busby, J. Carlsson, M. Carey, B. M. Cowan, D. A. Dimitrov, A. Hakim, J. Loverich, S. Mahalingam, P. Messmer, P. J. Mullowney, C. Nieter, K. Paul, C. Roark, S. W. Sides, N. D. Sizemore, D. N. Smithe, D. J. Wade-Stein, G. R. Werner, M. Wrobel, N. Xiang, C. D. Zhou.

References

[1] Kroyer T 2005 *CERN-THESIS-2005-061*, Vienna Tech. U.
[2] Kroyer T, Caspers F, Mahner E and Wien T 2005 *Proc. of PAC05*
[3] Byrd J, De Santis S, Sonnad K and M. Pivi 2007 *Proc. of PAC07*
[4] Nieter C, Cary J R 2004 *J. Comput. Phys.*, **196**, pp. 448-73
[5] Veitzer S A, Sonnad K, Stoltz P H and Byrd A 2009 *Proc. of PAC09*
[6] Zimmermann F 2004 *Phys. Rev. ST Accel. Beams* **7**, 124801, DOI: 10.1103/PhysRevSTAB.7.124801
[7] Furman M A, Kourbanis I and Zwaska R M 2009 *Proc. of PAC09*

Journal of Physics: Conference Series **180** (2009) 012008 doi:10.1088/1742-6596/180/1/012008

Advances in parallel partitioning, load balancing and matrix ordering for scientific computing

Erik G Boman[1], Umit V Catalyurek[2], Cédric Chevalier[1], Karen D Devine[1], Ilya Safro[3], Michael M Wolf[4]

[1] Scalable Algorithms Dept., Sandia National Laboratories
[2] Biomedical Informatics and Electrical & Computer Engineering Depts., Ohio State Univ.
[3] Mathematics and Computer Science Division, Argonne National Laboratory
[4] Computer Science Dept., Univ. of Illinois at Urbana-Champaign

Abstract. We summarize recent advances in partitioning, load balancing, and matrix ordering for scientific computing performed by members of the CSCAPES SciDAC institute.

1. Introduction

Most large scientific computations are now carried out on parallel computers. As many of these architectures have distributed memory, it is important to distribute the data across the processors in a manner that achieves high parallel efficiency. Computing an effective distribution of the data is known as the load-balancing (or partitioning) problem.

A related problem is finding a good ordering of the unknowns of sparse matrices in simulations that use sparse direct solvers. Indeed, in Cholesky or LU factorizations, the factors are usually less sparse than the input matrix, and the fill depends on the order of the unknowns.

For both load balancing and sparse matrix ordering, we can use combinatorial approachs based on graph or hypergraph partitioning. Zoltan [10] is a toolkit that provides efficient and robust parallel data partitioning, as well as related functionalities like sparse matrix ordering.

A graph $G = (V, E)$ is a set V of vertices and a set E of edges made up of pairs of vertices ($E \subset V \times V$). A hypergraph is an extension of this model that allows each hyperedge to have two *or more* vertices. Each hyperedge is a subset of V.

As the general graph- or hypergraph-partitioning problem is NP-complete, several heuristics have been developed to provide good quality within a short execution time. The most popular scheme is the multilevel framework that coarsens the (hyper)graph to a smaller (hyper)graph with similar topological properties which is easily partitioned; it then projects a partition on the small (hyper)graph back to the original one, refining it to improve partition quality. The phase that reduces the (hyper)graph size is called coarsening; it is followed by the initial partitioning and the uncoarsening phases.

In this paper, we describe an improved coarsening method in multilevel graph partitioning. We discuss application of hypergraph and graph partitioning to distributing large sparse matrices to optimize matrix-vector multiplication. And we discuss progress in parallel sparse matrix ordering for direct solvers and present results from a SciDAC application.

2. Coarsening with a weighted aggregation scheme

Many common multilevel schemes for simple graphs are based on strict coarsening (Strict Aggregation SAG). It is carried out by matching groups (usually pairs) of vertices together and representing each group with a single vertex in the coarsened space (e.g., matching [13, 15], first choice [9]). Another class of multilevel schemes is based on the Algebraic Multigrid (AMG) and represents a *weighted aggregation* process (WAG) [14, 16].

By contrast to SAG, in WAG each node can be divided into fractions, and different fractions belong to different aggregates. As AMG solvers have shown, weighted, instead of strict aggregation is important in order to express the likelihood of nodes to belong together; these likelihoods will then accumulate at the coarser levels, reinforcing each other where appropriate. This enables more freedom at the coarser levels and avoid making hardened local decisions, such as edge matching, before accumulating the relevant global information, while a strict aggregation may lead to conflicts between local and global pictures.

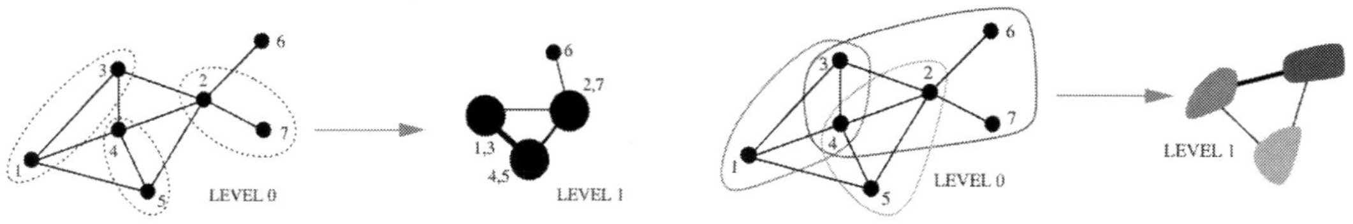

Figure 1. The left figure represents a SAG coarsening; each fine vertex belongs to one coarse vertex. The figure on the right is a WAG coarsening which allows fine vertices to be split across several coarse vertices.

We have shown [8] that the coarsening schemes in AMG can be adapted to graph partitioning; they produce higher quality partitions with less variation (i.e., greater robustness). A summary of these observations may be found in Figure 2. We compare WAG against the most popular technique in the graph partitioning community, Heavy Edge Matching (HEM) [15], which performs a strict pairing of vertices. We use two kinds of refinement during uncoarsening. The simplest is the *gradient* method, a local optimization without any hill-climbing possibility; the other is the more powerful Fidducia-Mattheyses (FM) [11] refinement, which allows hill-climbing.

We believe this approach is also more scalable (since it is similar to the approaches in multigrid codes like ML and Hypre), making it suitable for large problems. We have made steps toward extending the method from graphs to hypergraphs [3] for use in Zoltan's hypergraph partitioners.

3. Sparse matrix partitioning

A partitioning problem of particular importance is *sparse matrix partitioning*, where the goal is to reduce the communication in sparse matrix-vector multiplication. Since sparse matrix-vector multiplication is a kernel in many numerical algorithms, any improvement in parallel performance can impact a wide range of applications. We have performed an extensive empirical comparison of established and several new methods [5], demonstrating that 2D methods can reduce both the communication volume and the number of messages compared to traditional 1D row (or column) methods. Aggregated results for a set of 1413 test matrices are shown in Figure 3.

A drawback of 2D methods has been that most are quite expensive to compute. We have developed a new algorithm, nested dissection partitioning [1, 2] (see Figure 4), which is as effective as previous 2D methods to reduce communication but essentially as fast to compute as 1D methods.

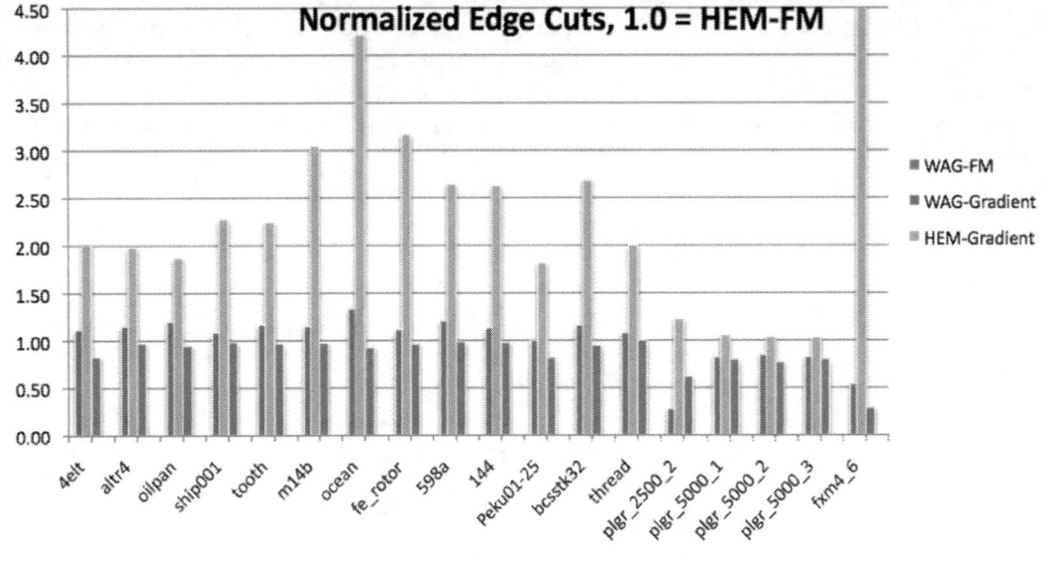

Figure 2. Edge cut (cost) for different coarsening schemes, scaled so that HEM with FM refinement is 1.0. We allow a fine vertex to be split over up to four coarse vertices.

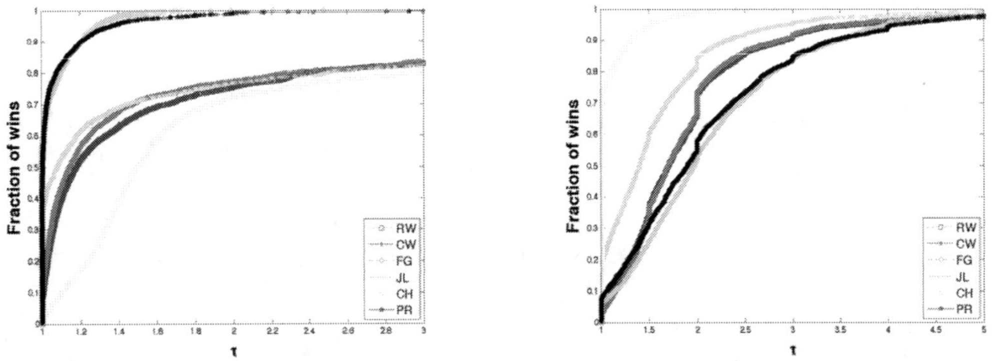

Figure 3. Performance profiles for comparison of common 1D (RW, CW) and some 2D (FG, JL, CH, PR) partitioning methods. We compared communication volume (left) and number of messages (right). Higher is better.

4. Sparse matrix ordering for direct solvers

Solving large, sparse linear systems of the type $Ax = b$ is a kernel that takes much compute time in scientific computing. Although iterative solvers are often preferred, sparse direct solvers are used for highly ill-conditioned problems or within preconditioners. We have investigated matrix reordering schemes to reduce fill in sparse matrix factorization and, thus, allow larger systems to be solved. We also provide ordering software (nested dissection) in Zoltan. Preliminary results indicate that ordering methods provided by PT-Scotch [7] can be significantly better than previous approaches. In an experiment using a large sparse matrix from a plasma code at Princeton Plasma Physics Lab (95 million nonzeros), we showed that the number of floating point operations in the factorization phase was reduced by up to 35% with Zoltan and PT-Scotch as compared to ParMetis (Figure 5). The memory footprint was also smaller with PT-Scotch. We are currently working with the SuperLU authors so that parallel Zoltan ordering may be

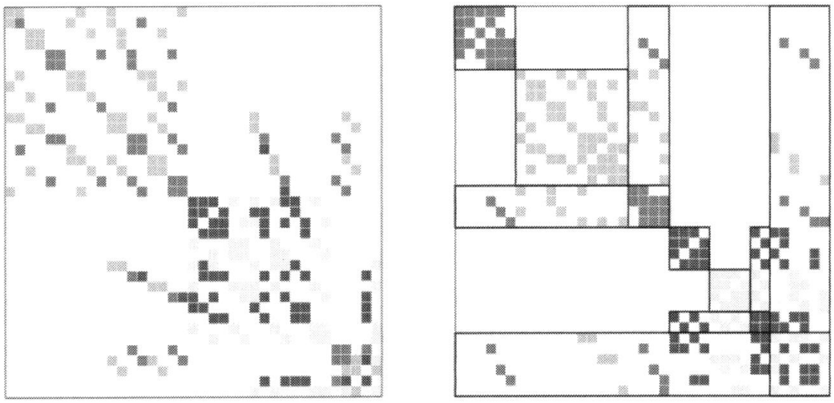

Figure 4. Nested dissection partitioning of a sparse matrix into four parts. Each nonzero is colored to show to which part (processor) it belongs. Original matrix (left) and permuted to reveal the nested structure (right).

supported in future versions of SuperLU_dist, which would make this new capability more easily accessible to a wide range of applications.

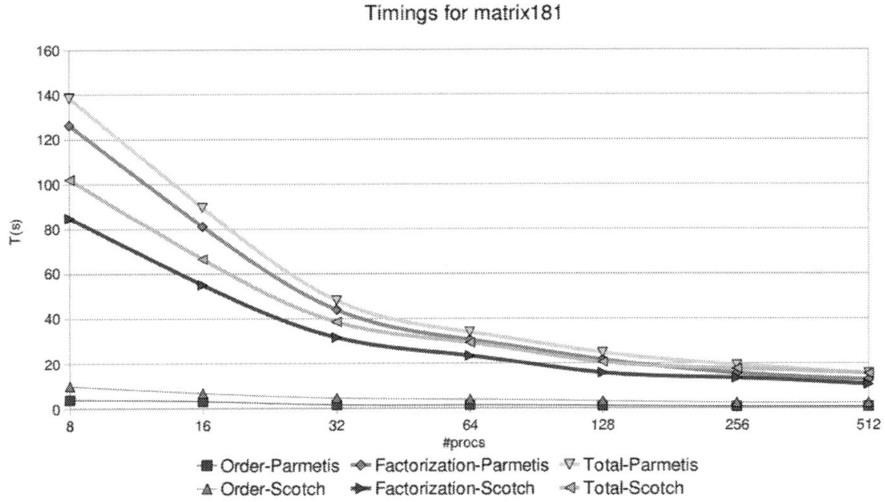

Figure 5. SuperLU_Dist 2.3 runtimes to factorize *matrix181*, for a Parmetis or a Scotch ordering of the unknowns. Typical use on this matrix is up to 64 processors.

We are also developing novel sparse matrix ordering approaches based on hypergraph partitioning [6, 4, 12]. For symmetric matrices, we have continued work on a novel hypergraph partitioning-based nested dissection approach that leverages structural decompositions of the form AA^T or ADA^T for achieving better orderings [6, 4]. Direct solutions of the systems in the form of $ADA^Tx = b$ requires triangular factorization of ADA^T, where A is a sparse and possibly rectangular matrix, and D is a diagonal matrix. We have proposed a new ordering method that applies hypergraph partitioning directly to A [4], as opposed to the standard approach of nested dissection (by vertex separator) on ADA^T. We also developed a simple, yet effective, structural factorization method that allows us to apply our ordering method to square symmetric matrices [4]. Furthermore, our hypergraph-based approach is better suited for the multilevel framework by accurately representing the separator at each level of the hierarchy.

For unsymmetric matrices, the reordering problem is even more challenging. Ordering is generally done as a preprocessing step, but in the unsymmetric case pivoting is needed at run-time to maintain numerical stability. Therefore, a column permutation is used to heuristically reduce fill, though row interchanges from pivoting may cause fill later. There are currently two approaches: Symmetrization (order $A + A^T$), or use a local ordering like COLAMD directly on A. Neither approach is satisfactory for large problems in parallel, because the first creates extra work and loss of concurrency, while the second is inherently serial. We have therefore developed a new algorithm, HUND [12] (Hypergraph Unsymmetric Nested Dissection), that is based on hypergraph partitioning and matching, both of which CSCAPES researchers are already working on. Major advantages of HUND over previous methods are that (i) it produces orderings of consistently high quality, (ii) it uses the structure of the original matrix (not symmetrized), and (iii) it is possible to parallelize. Parallel implementation of HUND in Zoltan is in progress.

Acknowledgments

Sandia is a multiprogram laboratory operated by Sandia Corporation, a Lockheed Martin Company, for the U. S. Department of Energy's National Nuclear Security Administration under contract DE-AC04-94AL85000. The CSCAPES Institute is supported by the U.S. Department of Energy's Office of Science through the SciDAC program.

References

[1] E. G. Boman. A nested dissection approach to sparse matrix partitioning. *Proc. of Applied Math. and Mechanics*, 7(1):1010803 – 1010804, 2007. ICIAM07, Zurich, Switzerland.

[2] E. G. Boman and M. M. Wolf. A nested dissection approach to sparse matrix partitioning for parallel computations. Technical Report 2008-5482J, Sandia National Laboratories, NM, 2008.

[3] A. Buluc and E. Boman. Towards scalable parallel hypergraph partitioning. In *Proc. of Computer Science Research Institute (CSRI) 2008*, Albuquerque, NM, 2008. Sandia National Labs.

[4] U. V. Catalyurek, C. Aykanat, and E. Kayaaslan. Hypergraph partitioning-based fill-reducing ordering. Technical Report OSUBMI-TR-2009-n02 and BU-CE-0904, The Ohio State University, Dept. of Biomedical Informatics and Bilkent University, Computer Engineering Dept., 2009. Submitted for publication.

[5] U. V. Catalyurek, C. Aykanat, and B. Uçar. On two-dimensional sparse matrix partitioning: Models, methods, and a recipe. Technical Report OSUBMI_TR_2008_n04, The Ohio State University, Department of Biomedical Informatics, 2008. Submitted for publication.

[6] U. V. Çatalyürek. *Hypergraph Models for Sparse Matrix Partitioning and Reordering.* PhD thesis, Bilkent University, Computer Engineering and Information Science, Nov 1999.

[7] C. Chevalier and F. Pellegrini. PT-SCOTCH: A tool for efficient parallel graph ordering. *Parallel Computing*, 34(6–8):318–331, 2008.

[8] C. Chevalier and I. Safro. Comparison of coarsening schemes for multilevel graph partitioning. In *Proc. of 3rd Workshop on Learning and Intelligent Optimization (LION-3)*, Trento, Italy, 2009.

[9] J. Cong and J. R. Shinnerl, editors. *Multilevel Optimization and VLSICAD.* Kluwer, 2003.

[10] K. Devine, E. Boman, R. Heaphy, B. Hendrickson, and C. Vaughan. Zoltan data management services for parallel dynamic applications. *Computing in Science and Engineering*, 4(2):90–97, 2002.

[11] C. M. Fiduccia and R. M. Mattheyses. A linear-time heuristic for improving network partitions. In *Proceedings of the 19th Design Automation Conference*, pages 175–181. IEEE, 1982.

[12] L. Grigori, E. Boman, S. Donfack, and T. Davis. Hypergraph unsymmetric nested dissection ordering for sparse LU factorization. Technical Report 2008-1290J, Sandia National Labs, 2008. Submitted to SIAM J. Sci. Comp.

[13] B. Hendrickson and R. W. Leland. A multi-level algorithm for partitioning graphs. In *Supercomputing*, 1995.

[14] Y. F. Hu and J. A. Scott. A multilevel algorithm for wavefront reduction. *SIAM J. Scientific Computing*, 23:2000–031, 2001.

[15] G. Karypis and V. Kumar. A fast and high quality multilevel scheme for partitioning irregular graphs. Technical Report 95-035, University of Minnesota, June 1995.

[16] I. Safro, D. Ron, and A. Brandt. Multilevel algorithms for linear ordering problems. *Journal of Experimental Algorithmics*, 13:1.4–1.20, 2008.

Journal of Physics: Conference Series **180** (2009) 012009 doi:10.1088/1742-6596/180/1/012009

ALPS: A framework for parallel adaptive PDE solution

Carsten Burstedde[⋆] **Martin Burtscher**[⋆] **Omar Ghattas**[⋆,†]
Georg Stadler[⋆] **Tiankai Tu**[⋆] **Lucas C. Wilcox**[⋆]

[⋆]Institute for Computational Engineering and Sciences (ICES), The University of Texas at Austin, 1 University Station C0200, Austin TX 78712, USA
[†]Jackson School of Geosciences and Department of Mechanical Engineering, The University of Texas at Austin, 1 University Station C0200, Austin TX 78712, USA

E-mail: {carsten,burtscher,omar,georgst,ttu,lucasw}@ices.utexas.edu

Abstract. Adaptive mesh refinement and coarsening (AMR) is essential for the numerical solution of partial differential equations (PDEs) that exhibit behavior over a wide range of length and time scales. Because of the complex dynamic data structures and communication patterns and frequent data exchange and redistribution, scaling dynamic AMR to tens of thousands of processors has long been considered a challenge. We are developing *ALPS*, a library for dynamic mesh adaptation of PDEs that is designed to scale to hundreds of thousands of compute cores. Our approach uses parallel forest-of-octree-based hexahedral finite element meshes and dynamic load balancing based on space-filling curves. *ALPS* supports arbitrary-order accurate continuous and discontinuous finite element/spectral element discretizations on general geometries. We present scalability and performance results for two applications from geophysics: seismic wave propagation and mantle convection.

1. Introduction

The advent of the age of petascale computing brings unprecedented opportunities for breakthroughs in scientific understanding and engineering innovation. However, the raw performance made available by petascale systems is by itself not sufficient to solve many challenging modeling and simulation problems. For example, the complexity of solving evolutionary partial differential equations often scales as $n^{\frac{4}{3}}$, where n is the number of unknowns.[1] Thus, the three-orders-of-magnitude improvement in peak speed of supercomputers over the past dozen years has meant just a factor of 5.6 improvement in spatio-temporal resolution—not even three successive refinements of mesh size. For many problems of scientific and engineering interest, there is a desire to increase the resolution of current simulations by several orders of magnitude. Thus, although supercomputing performance has outpaced Moore's Law over the past several decades due to increased concurrency [1], the curse of dimensionality imposes much slower scientific returns.

The work requirements of scientific simulations typically scale as n^α. The power α can be reduced through the use of optimal solvers such as multigrid for PDEs and fast multipole for integral equations and N-body problems. Once α has been reduced as much as possible, further

[1] Optimal solvers require $\mathcal{O}(n)$ work per time step, and time accurate integration often implies $\mathcal{O}(n^{\frac{1}{3}})$ time steps.

reductions in work can be achieved only by reducing n itself. This can be accomplished in two ways: through the use of adaptive mesh refinement/coarsening (AMR) techniques, and the use of higher order accurate discretizations (in space and time). AMR places mesh points only where needed to resolve features of the solution, whereas high-order approximations reduce the number of mesh points necessary to achieve a given accuracy.

Fortunately, many problems have local multiscale character, i.e., resolution is needed only in localized (possibly dynamically evolving) regions, such as near fronts, discontinuities, material interfaces, reentrant corners, boundary and interior layers, and so on. In this case, AMR methods can deliver orders-of-magnitude reductions in the number of mesh points. However, AMR methods can also impose significant overhead, in particular on highly parallel computing systems, due to their need for frequent re-adaptation and load-balancing of the mesh over the course of the simulation. Because of the complex data structures and communication patterns and frequent data exchange and redistribution, scaling dynamic AMR to tens of thousands of processors has long been considered a challenge. Space constraints preclude a proper review of existing methods and software for parallel AMR, but see the discussion and references in [2, 3].

We have developed the *ALPS* (**A**daptive **L**arge-scale **P**arallel **S**imulations) framework for parallel adaptive solution of PDEs [3]. *ALPS* includes the *octor* [4] and *p4est* libraries for parallel dynamic mesh adaptivity on single-octree-based and forest-of-octree-based geometries, respectively, and the *mangll* library for arbitrary-order hexahedral continuous and discontinuous finite/spectral element discretizations on general multi-octree geometries. *ALPS* has been shown to scale well weakly and strongly to over 60,000 processor cores [3]. In this paper, we describe several applications of the *ALPS* framework to two continuum mechanics problems—the propagation of elastic waves (with application to global seismology) and the buoyancy-driven creeping (Stokes) flow of a non-Newtonian incompressible fluid (with application to global mantle convection)—and provide sample performance results on Ranger, the 579 teraflops, 62,976-core Sun system at the Texas Advanced Computing Center (TACC).

2. Elastic wave propagation: Global seismology

The elastic wave equation models the propagation of longitudinal and shear waves in a linearly elastic medium. For isotropic media, the material stiffness tensor involves only the Lamé parameters λ and μ. In mixed velocity–stress form, the elastic wave equation reduces to

$$\rho \frac{\partial \boldsymbol{v}}{\partial t} = \nabla \cdot \boldsymbol{S} + \boldsymbol{f}, \tag{1a}$$

$$\frac{\partial \boldsymbol{S}}{\partial t} = \mu \left(\nabla \boldsymbol{v} + \nabla \boldsymbol{v}^{\top} \right) + \lambda (\nabla \cdot \boldsymbol{v}) \boldsymbol{I}, \tag{1b}$$

where $\boldsymbol{v}$ and $\boldsymbol{S}$ are the unknown velocity vector and Cauchy stress tensor of the medium, and ρ, t, f, and $\boldsymbol{I}$ are the mass density, time, body force, and identity tensor, respectively.

We have built an elastic wave propagation code based on the *mangll* and *p4est* libraries from *ALPS*. Velocity and stress fields are discretized using the discontinuous Galerkin (dG) method with a Godunov flux [5] in space and a five-stage fourth-order explicit Runge-Kutta (RK) method in time. The elements are the same as those used in the spectral element method [6], having nodes at the tensor product Legendre-Gauss-Lobatto (LGL) points. The dG method duplicates degrees of freedom between adjacent element faces. On nonconforming interfaces between elements (arising due to 2:1 subdivision between adjacent hexahedra), the numerical flux is computed by introducing a face integration mesh that incorporates the contributions from each smaller face individually using the two-dimensional tensor LGL quadrature based on the nodes of the smaller face. The LGL quadrature reduces the block diagonal dG mass matrix to a diagonal matrix, permitting faster application of its inverse.

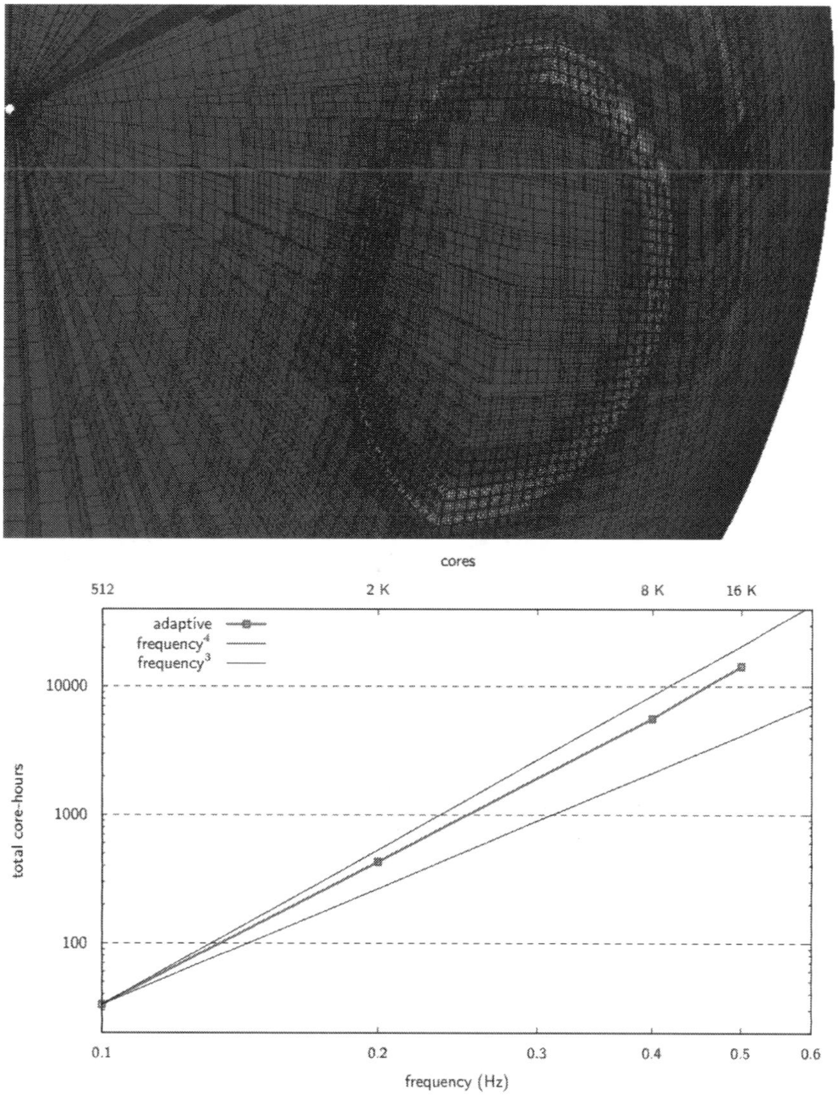

Figure 1. *Top:* cut of wavefront-adapted spherical mesh for a double-couple source. *Bottom:* scaling of run time with the number of cores for dynamic-adaptive wave propagation solver.

Figure 1 presents results from the solution of an elastic wave propagation problem in a spherical geometry. The material parameters approximate those of the Earth and the source is a simple double-couple. The top image depicts a snapshot of a mesh that has been adapted to propagating wavefronts. The significant reduction of mesh points can be seen in the image. The bottom plot presents scaling results to 16K cores, showing the growth in total processor-hours as a function of maximum source frequency. A perfectly scalable static mesh explicit wave propagation solver would have a total solution time that scales as the fourth power of frequency. The scaling is shown as the top black line (with 512 cores as the base case). Dynamic adaptivity has the potential to reduce the complexity to be closer to the third power of frequency (shown as the bottom black line), since adapting to moving wavefronts effectively eliminates one dimension from the mesh. The cubic scaling cannot be achieved for problems on finite domains or with interfaces, due to reflections and generation of new waves (the number of wavefronts

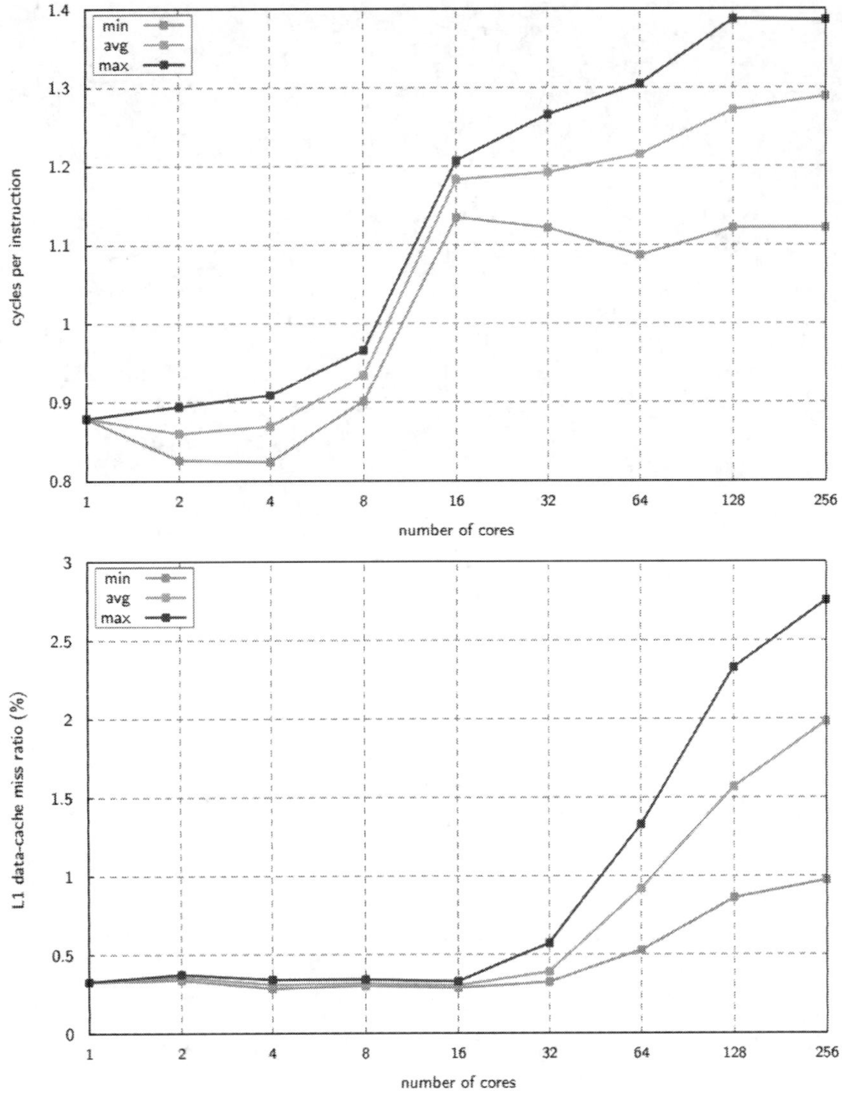

Figure 2. CPU performance for wave propagation code with 10K total elements of polynomial degree 6. *Top:* cycles per instruction. *Bottom:* L1 data-cache miss ratio in percent.

that must be tracked does not remain constant). The actual observed scaling is plotted as the red line, showing that, even in a layered Earth model, the "optimal" fourth power scaling can be improved using adaptivity.

We also studied the efficiency of the implementation using performance counter measurements from the *PAPI* library [7]; the results are summarized in Figure 2. Up to eight cores, more than one instruction is executed per cycle on average, indicating that there is significant instruction-level parallelism. As the compute nodes become saturated when using all 16 cores per node, we observe a jump in the cycles per instruction (CPI). Yet, even above 16 cores, the CPI is low because we carefully tuned important sections of the source code so that the compiler can vectorize them [8]. Beyond 16 cores, the CPI grows due to an increase in cache misses, which is caused by an increase in data that need to be exchanged between compute nodes, which results in less regular memory access. Nevertheless, the L1 data-cache miss ratio is surprisingly low for

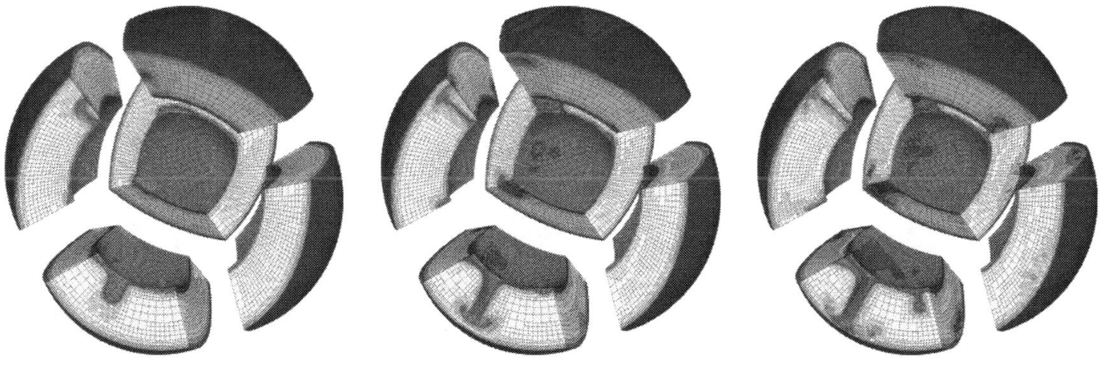

Figure 3. Snapshots of three time steps from a mantle convection simulation using *Rhea*, showing dynamically-adapted refinement near rising thermal plumes and thermal boundary layers.

code that processes about 1.75, 1.12 and 0.96 million degrees of freedom per second per core on 1, 32 and 256 cores, respectively. This low ratio is due to data being accessed in mostly regular patterns, which allows data to be prefetched effectively.

3. Non-Newtonian Stokes flow: Mantle convection

Our second example is flow of a viscous incompressible creeping non-Newtonian fluid, which is governed by equations for mass, momentum, and energy balance,

$$\nabla \cdot \boldsymbol{v} = 0, \tag{2a}$$

$$-\nabla \cdot \left[\eta \left(\nabla \boldsymbol{v} + \nabla \boldsymbol{v}^{\top}\right) - \boldsymbol{I}p\right] = \rho \boldsymbol{g}, \tag{2b}$$

$$\rho c_p \left(\frac{\partial T}{\partial t} + \boldsymbol{v} \cdot \nabla T\right) - \nabla \cdot (k \nabla T) = \rho H, \tag{2c}$$

where $\boldsymbol{v}$, p, and T are the unknown velocity, pressure, and temperature; and $\eta = \eta(T, \boldsymbol{v})$, ρ, $\boldsymbol{g}$, c_p, k and H are the viscosity, density, gravitational acceleration, specific heat, thermal conductivity, and internal heat generation rate. We are interested in convection in Earth's mantle, where the flow is driven by buoyancy. Thus we invoke the Boussinesq approximation, which replaces the density in the gravitational force term in (2b) by $\rho = \rho_0 [1 - \alpha(T - T_0)]$, where ρ_0 and T_0 denote reference temperature and density and α is the coefficient of thermal expansion. Various constitutive laws are used for the mantle, but in general the viscosity depends nonlinearly on both temperature and second invariant of the deviatoric strain rate tensor.

We have built a parallel AMR mantle convection code, *Rhea*, that solves (2a)–(2b) with appropriate boundary conditions and temperature- and strain-rate-dependent viscosity [3] using *ALPS* components. The images in Figure 3 illustrate dynamic mesh adaptivity in resolving thermal upwellings and boundary layers. The first version of *Rhea* discretizes the velocity, pressure, temperature fields with mapped trilinear hexahedral finite elements. Mantle flows are strongly advection-dominated, and therefore we employ the streamline-upwind Petrov-Galerkin (SUPG) scheme to stabilize the energy equation (2c). The equal-order discretization of the Stokes equations (2a), (2b) is stabilized with pressure projection [9]. Explicit integration of the energy equation decouples the temperature update from the nonlinear Stokes solve; the latter is carried out at each time step using the updated temperature via a lagged-viscosity (Picard) iteration. Each Picard iteration requires a variable-viscosity Stokes solve, which is discussed below.

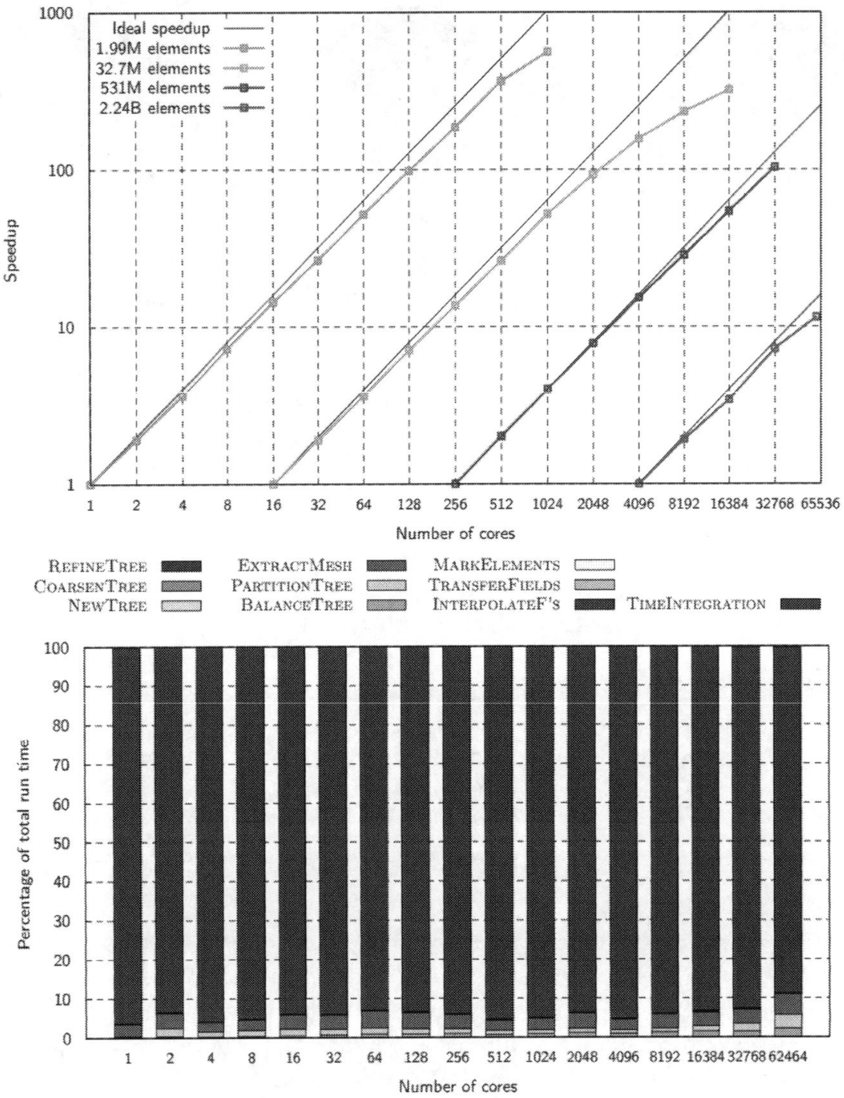

Figure 4. *Top:* fixed-size (strong) scalability for adaptive solution of energy equation for four problem sizes on cubic domain. Mesh is adapted every 32 time steps. *Bottom:* weak scalability for adaptive solution of energy equation. Total run time is broken down into numerical PDE solution (blue) and AMR functions (all other colors). Problem size increases isogranularly at 131,000 elements per core; largest problem has approximately 7.9 billion elements.

First, we study the scalability of AMR for solution of the energy equation (2c). This low-order-discretized, explicitly-solved, scalar, linear equation is a severe test of AMR, since there is very little numerical work over which to amortize AMR. As can be seen from the top plot in Figure 4, the fixed-size scaling speedups are nearly optimal over a wide range of core counts. For instance, solving the problem with 531 million elements (blue line) exhibits a speedup of 101 on 32,768 cores over the same problem running on 256 cores (128 would be optimal). Moreover, scaling weakly from 1 to 62,464 cores (bottom of Figure 4), we see that all mesh adaptation functions—including coarsening, refinement, interpolation, rebalancing and repartitioning of the mesh—together impose little overhead on the PDE solver. Only for 62K cores does the total

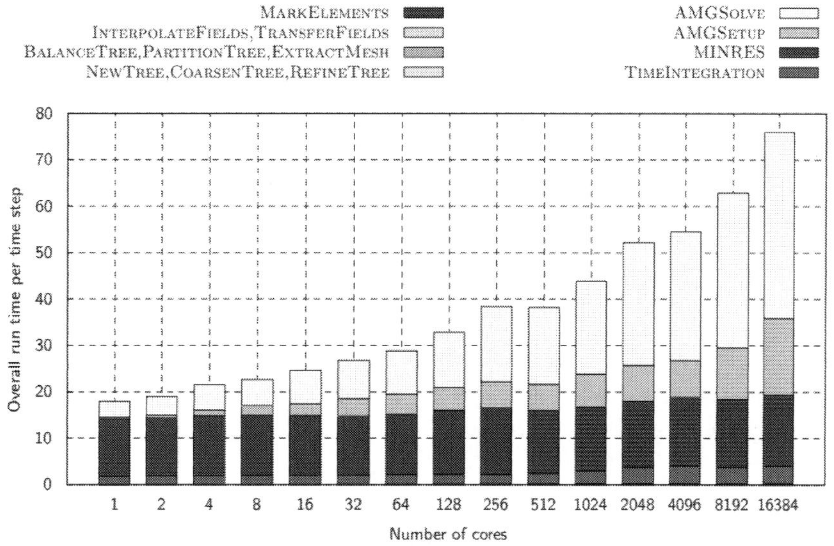

Figure 5. Breakdown of mantle convection run time per time step into AMG setup (gray), AMG V-cycle (yellow), MINRES iteration (blue), and time integration of the energy equation (red) using *BoomerAMG* preconditioning. AMR operations (all other colors) have negligible cost. Mesh is adapted every 16 time steps, and problem size increases isogranularly at 50K elements per core (largest problem has 815 million elements).

cost of AMR exceed 10% of the end-to-end run time, and even then just barely.

Next, we study the algorithmic and parallel scalability of the variable-viscosity Stokes solver, which is invoked at each Picard iteration within each time step. We use MINRES to solve the symmetric indefinite Stokes system; the preconditioner is a block-diagonal matrix, which invokes one V-cycle of an algebraic multigrid (AMG) method for the viscous (1,1) block, and an inverse-viscosity-scaled mass matrix for the pressure Schur-complement (2,2) block (see [10] for details). For Cartesian geometry, the viscous vector operator can be approximated by a scalar, variable-coefficient Poisson operator associated with each of the three velocity components (provided viscosity gradients are not too large). The Poisson operators are approximately inverted by one V-cycle of *BoomerAMG* from the *hypre* library [11]. The bar chart in Figure 5 shows the breakdown of different components of an adaptive variable-viscosity Stokes solve using *BoomerAMG* as we scale weakly from 1 to 16K cores. Spherical geometry, on the other hand, induces additional coupling among the velocity components that cannot be effectively approximated by the decomposition into scalar Poisson solves. Thus we employ one V-cycle of the adaptive smoothed aggregation AMG solver *ML* from *Trilinos* [12] on the entire viscous vector block. Table 1 presents weak scalings of a single variable-viscosity Stokes solve on the spherical domain using *ML*. We draw several conclusions from Figure 5 and Table 1. First, for the full AMR mantle convection simulations with *Rhea* (i.e. including energy and nonlinear Stokes), the cost of parallel AMR is in the noise. Second, using a V-cycle of either *BoomerAMG* or *ML* as a preconditioner for the viscous block, in conjunction with a scaled mass matrix approximation of the Schur complement, results in excellent algorithmic scalability, as evidenced by the near-insensitivity of MINRES iterations to a 16,000-fold increase in problem size and number of cores. Third, while both AMG implementations exhibit excellent algorithmic scalability, there is an opportunity for improvement of the parallel scalability of the AMG setup and V-cycle components at $O(10^4)$ cores, based on our tests on a cluster system and viscous operators with several-orders-of-magnitude variation in coefficients. Finally, the examples demonstrate that

#cores	#dofs	MINRES #iterations	AMG setup (s)	AMG V-cycle (s)	MINRES matvec (s)	ζ_A	ζ_I	ζ_V	ζ
8	1.89M	77	7.12	41.3	130.8	1.00	1.00	1.00	1.00
64	12.8M	76	8.47	38.9	114.9	1.01	1.12	1.05	1.10
512	99.7M	78	10.2	48.1	129.9	0.99	1.02	0.87	0.95
4096	672M	109	36.5	189.8	208.7	0.70	0.89	0.31	0.41
16384	2.43B	109	66.0	211.3	216.5	0.70	0.86	0.28	0.36

Table 1. Weak scaling of Stokes solve on spherical shell domain using *ML*, with 10^3 viscosity contrast, on a 3 times locally refined mesh. The numbers of cores, degrees of freedom, and MINRES iterations are tabulated, along with the time for AMG setup, AMG V-cycle, and MINRES matvecs. Also shown are algorithmic efficiency ζ_A (based on MINRES iterations), implementation efficiency ζ_I (based on MINRES timing, excluding V-cycle), V-cycle efficiency ζ_V (based on *ML* timing), and overall efficiency ζ (based on overall wall clock).

our framework for parallel forest-of-octrees AMR with high-order-accurate discretization and general geometries scales very well to at least $O(10^4)$ processor cores.

Acknowledgement

This work was partially supported by NSF (OCI-0749334, CCF-0427985, CNS-0540372, CNS-0619838, DMS-0724746) and DOE (06ER25782, 08ER25860). Computing resources at TACC were provided under TeraGrid award MCA04N026. We acknowledge many helpful discussions with Rob Falgout, Ulrike Yang, Rahul Sampath, George Biros, and Tim Warburton. We thank TACC for their outstanding support, in particular Bill Barth, Jay Boisseau, Tommy Minyard, Romy Schneider, and Karl Schulz.

References

[1] Colella P, Dunning T H, Gropp W D and Keyes D E 2004 A Science-Based Case for Large-Scale Simulation, Volume 2 Office of Science, U.S. Department of Energy URL `http://www.pnl.gov/scales/`
[2] Diachin L F, Hornung R, Plassmann P and Wissink A 2006 *Parallel Adaptive Mesh Refinement* (SIAM) chap 8 (*Software, Environments, and Tools* no 20)
[3] Burstedde C, Ghattas O, Gurnis M, Tan E, Tu T, Stadler G, Wilcox L C and Zhong S 2008 Scalable adaptive mantle convection simulation on petascale supercomputers *Proceedings of SC08* (IEEE/ACM)
[4] Tu T, O'Hallaron D R and Ghattas O 2005 Scalable parallel octree meshing for terascale applications *Proceedings of SC2005*
[5] Hesthaven J S and Warburton T 2008 *Nodal Discontinuous Galerkin Methods: Algorithms, Analysis, and Applications* (*Texts in Applied Mathematics* vol 54) (Springer)
[6] Deville M, Fischer P and Mund E 2002 *High-Order Methods for Incompressible Fluid Flow* (*Cambridge Monographs on Applied and Computational Mathematics* vol 9) (Cambridge University Press)
[7] Performance applications programming interface (PAPI) URL `http://icl.cs.utk.edu/papi/`
[8] Diamond J, Kim D, Burtscher M, Keckler S, Pingali K and Browne J 2009 Multicore optimization for Ranger *Proceedings of Teragrid '09*
[9] Dohrmann C and Bochev P 2004 *International Journal for Numerical Methods in Fluids* **46** 183–201
[10] Burstedde C, Ghattas O, Stadler G, Tu T and Wilcox L C 2009 *Computer Methods in Applied Mechanics and Engineering* **198** 1691–1700
[11] Center for Applied Scientific Computing, Lawrence Livermore National Laboratory 2007 hypre. *High Performance Preconditioners, User Manual* `https://computation.llnl.gov/casc/linear_solvers/`
[12] Gee M, Siefert C, Hu J, Tuminaro R and Sala M 2006 ML 5.0 smoothed aggregation user's guide Tech. Rep. SAND2006-2649 Sandia National Laboratories

High-order finite-volume adaptive methods on locally rectangular grids

P Colella[1], M Dorr[2], J Hittinger[2], D F Martin[1], P McCorquodale[1]

[1] Applied Numerical Algorithms Group, Lawrence Berkeley National Laboratory, 1 Cyclotron Road Mail Stop 50A1148, Berkeley, CA 94720-8142

[2] Center for Applied Scientific Computing, Lawrence Livermore National Laboratory, Box 808, L-561, Livermore, CA 94551-0808

E-mail: `PWMcCorquodale@lbl.gov`

Abstract. We are developing a new class of finite-volume methods on locally-refined and mapped grids, which are at least fourth-order accurate in regions where the solution is smooth. This paper discusses the implementation of such methods for time-dependent problems on both Cartesian and mapped grids with adaptive mesh refinement. We show 2D results with the Berger–Colella shock-ramp problem in Cartesian coordinates, and fourth-order accuracy of the solution of a Gaussian pulse problem in a polytropic gas in mapped coordinates.

1. High-order finite-volume methods on Cartesian grids

In the finite-volume approach, the spatial domain in $\mathbb{R}^D$ is discretized as a union of control volumes that covers the domain. With Cartesian grids, a control volume $V_{\boldsymbol{i}}$ takes the form

$$V_{\boldsymbol{i}} = [\boldsymbol{i}h, (\boldsymbol{i} + \boldsymbol{u})h], \boldsymbol{i} \in \mathbb{Z}^D, \boldsymbol{u} = (1, 1, \ldots, 1),$$

where h is the grid spacing.

A finite-volume discretization of a partial differential equation is based on averaging that equation over control volumes, applying the divergence theorem to replace volume integrals by integrals over the boundary of the control volume, and approximating the boundary integrals by quadratures. For example, for time-dependent problems of the form

$$\frac{\partial U}{\partial t} + \nabla \cdot \vec{F} = 0 \tag{1}$$

the discretized solution in space is the average of U over a control volume:

$$\langle U \rangle_{\boldsymbol{i}}(t) = \frac{1}{h^D} \int_{V_{\boldsymbol{i}}} U(\boldsymbol{x}, t) d\boldsymbol{x}. \tag{2}$$

In that case, we can compute the evolution of the spatially discretized system by a method-of-lines approach:

$$\frac{d\langle U \rangle_{\boldsymbol{i}}}{dt} = -\frac{1}{h^D} \int_{V_{\boldsymbol{i}}} \nabla \cdot \vec{F} d\boldsymbol{x} = -\frac{1}{h} \sum_{d=1}^{D} (\langle F^d \rangle_{\boldsymbol{i}+\frac{1}{2}\boldsymbol{e}^d} - \langle F^d \rangle_{\boldsymbol{i}-\frac{1}{2}\boldsymbol{e}^d}) \tag{3}$$

$$\langle F^d \rangle_{\boldsymbol{i}\pm\frac{1}{2}\boldsymbol{e}^d} = \frac{1}{h^{D-1}} \int_{A_d^{\pm}} F^d dA, \tag{4}$$

where $A_d^{\pm}$ are the high and low faces bounding V_i with normals pointing in the direction of unit vector e^d. In the right-hand side of (4), the integrals over faces are approximated using some quadrature rule. Conserved quantities in the original PDE then satisfy an analogous conservation law in the discretized system.

Most finite-volume methods use the midpoint rule to approximate the flux integrals in (4) leading to a second-order accurate method. We are developing higher-order methods (fourth order or better) derived from replacing the integrand on the right-hand side of (4) with a Taylor expansion about the center of the face. This paper will be concerned with fourth-order methods, in which

$$\langle F^d \rangle = F^d(\boldsymbol{x}_0) + \frac{h^2}{24} \sum_{d' \neq d} \frac{\partial^2 F^d}{\partial x_{d'}^2} + O(h^4). \tag{5}$$

In computing fluxes on faces, we use face-averaged variables, which can be obtained to fourth-order accuracy with the standard deconvolution algorithm [6]:

$$\langle w \rangle_{i+\frac{1}{2}e^d} = \frac{7}{12}(\langle w \rangle_i + \langle w \rangle_{i+e^d}) - \frac{1}{12}(\langle w \rangle_{i-e^d} + \langle w \rangle_{i+2e^d}) \tag{6}$$

For hyperbolic problems, it is also necessary to introduce limiters to suppress oscillations in the presence of discontinuities and underresolved gradients. To do this, we apply an extension of the PPM limiter [6, 5] to values obtained from the deconvolution formula (6). In this approach, we view the PPM limiter as producing limited, and possibly doubly-valued face averages of the variable w. In the latter case, we solve a Riemann problem at the face to compute the flux. The use of the limiter [5] leads to an algorithm that preserves fourth-order accuracy in the presence of smooth extrema. If, in addition, there is a constraint that w remain non-negative, we can use a version of the FCT limiter in [7] to impose that constraint, with the low-order method given by the CTU scheme in [3]. Since we do not use the FCT limiter to impose any bounds other than positivity, the method remains fourth-order accurate for smooth data.

For the case of the compressible Euler equations, we also include an artificial viscosity, replacing the flux $\vec{F}$ by $\vec{F} - \nu \nabla U$, where

$$\nu = K_1 h^2 \min\{h^2 \frac{|\nabla \cdot \vec{u}|^2}{K_2 c^2}, 1\} |\nabla \cdot \vec{u}| \tag{7}$$

in which $\vec{u}$ is the velocity, c is the speed of sound, and K_1 and K_2 are constants. Then ν is cubic in $h|\nabla \cdot \vec{u}|$ when $h|\nabla \cdot \vec{u}|$ is small, and linear in $h|\nabla \cdot \vec{u}|$ when $h|\nabla \cdot \vec{u}|$ is large, relative to the speed of sound. In the latter case, (7) reduces to the artificial viscosity in the second-order method of [6]. For smooth flows, (7) leads to a fourth-order accurate method.

The equation (3) is of the form

$$\frac{dU}{dt} = f(t, U) \tag{8}$$

and we solve this with a standard fourth-order Runge–Kutta method to integrate from $U(t^n)$ to $U(t^n + \Delta t)$, using:

$$f_0 = f(t^n, U(t^n)); \quad f_1 = f(t^n + \frac{\Delta t}{2}, U(t^n) + f_0 \frac{\Delta t}{2}); \tag{9}$$

$$f_2 = f(t^n + \frac{\Delta t}{2}, U(t^n) + f_1 \frac{\Delta t}{2}); \quad f_3 = f(t^n + \Delta t, U(t^n) + f_2 \Delta t); \tag{10}$$

$$U(t^n + \Delta t) = U(t^n) + \frac{\Delta t}{6}(f_0 + 2f_1 + 2f_2 + f_3) + O((\Delta t)^5). \tag{11}$$

2. Multilevel algorithm

In a multilevel algorithm, for each pair of successive levels, we shall need to interpolate from data on coarser-level patches at the beginning and end of each coarse timestep, to ghost cells of finer-level patches at intermediate times. After advancing the coarse solution one timestep, from time t^n to time $t^n + \Delta t_c$, we implement this interpolation in the following stages:

(i) Copy data: Copy $U(t^n)$ and $U(t^n + \Delta t_c)$ from the original coarser-level patches to a new set of patches made from coarsening the finer-level patches by the refinement ratio n_{ref} and then expanding each patch by three layers of ghost cells.

(ii) Interpolate in time: Using the intermediate results f_0, f_1, f_2, f_3 from (9)–(10) in the Runge–Kutta method as applied on the coarser level, interpolate from data at times t^n and $t^n + \Delta t_c$ on the expanded coarsened fine patches to times $t^n + k\Delta t_c/n_{ref}$ for all $k = 1, 2, ..., n_{ref} - 1$, on the same patches. Do this using a Taylor series about the midpoint in time:

$$\begin{aligned} U(t^n + \alpha\Delta t) =& U(t^n) + \frac{\Delta t}{24}(5f_0 + 4f_1 + 4f_2 - f_3) + (\frac{f_1 + f_2}{2})((\alpha - \frac{1}{2})\Delta t) \\ &+ (\frac{f_3 - f_0}{2\Delta t})((\alpha - \frac{1}{2})\Delta t)^2 + (\frac{f_0 - f_1 - f_2 + f_3}{6(\Delta t/2)^2})((\alpha - \frac{1}{2})\Delta t)^3 + O((\Delta t)^4) \end{aligned} \quad (12)$$

for $0 \leq \alpha \leq 1$.

(iii) Interpolate in space: At all of the intermediate times, interpolate from data on the expanded coarsened finer-level patches to ghost cells of the original finer-level patches. For coarse cell $\boldsymbol{i}$, solve a constrained linear least-squares problem for $a_{\boldsymbol{i},\boldsymbol{p}}$ in the overdetermined system

$$\sum_{p_d \geq 0, \|\boldsymbol{p}\|_1 \leq 3} a_{\boldsymbol{i},\boldsymbol{p}} \langle (\boldsymbol{x} - \boldsymbol{x_i})^{\boldsymbol{p}} \rangle_{\boldsymbol{j}}^c = \langle u \rangle_{\boldsymbol{j}}^c \text{ for all } \boldsymbol{j} \in N(\boldsymbol{i}) - \{\boldsymbol{i}\} \quad (13)$$

where $N(\boldsymbol{i})$ is a set of neighboring cells to $\boldsymbol{i}$, and the conservation constraint is

$$\sum_{p_d \geq 0, \|\boldsymbol{p}\|_1 \leq 3} a_{\boldsymbol{i},\boldsymbol{p}} \langle (\boldsymbol{x} - \boldsymbol{x_i})^{\boldsymbol{p}} \rangle_{\boldsymbol{i}}^c = \langle u \rangle_{\boldsymbol{i}}^c \quad (14)$$

where $\boldsymbol{x_i}$ is the center of cell $\boldsymbol{i}$. Then for each fine cell $\boldsymbol{k}$ within coarse cell $\boldsymbol{i}$:

$$\langle u \rangle_{\boldsymbol{k}}^f = \sum_{p_d \geq 0, \|\boldsymbol{p}\|_1 \leq 3} a_{\boldsymbol{i},\boldsymbol{p}} \langle (\boldsymbol{x} - \boldsymbol{x_i})^{\boldsymbol{p}} \rangle_{\boldsymbol{k}}^f. \quad (15)$$

Aside from the details outlined in this section, the algorithm is the same as in [2]. In particular, conservation at refinement boundaries is exactly the same.

3. Finite-volume methods on mapped grids

We can extend this approach to the case of mapped grids. We assume that we have a smooth mapping $\boldsymbol{X}$, independent of time, from some abstract coordinate space into physical space:

$$\boldsymbol{X} = \boldsymbol{X}(\boldsymbol{\xi}) \ , \ \boldsymbol{X} : [0, 1]^D \to \mathbb{R}^D.$$

Given this mapping, the divergence of a vector field in physical space can be written in terms of derivatives in the mapping space, that is

$$\nabla_{\boldsymbol{x}} \cdot \vec{F} = \frac{1}{J} \nabla_{\boldsymbol{\xi}} \cdot (\boldsymbol{N}^T \vec{F}), \quad (16)$$

$$J = det(\nabla_{\boldsymbol{\xi}} \boldsymbol{X}) \ , \ \boldsymbol{N}_{p,q}^T = det((\nabla_{\boldsymbol{\xi}} \boldsymbol{X})(p|\boldsymbol{e}^q)), \quad (17)$$

where $\boldsymbol{A}(p|\boldsymbol{v})$ denotes the matrix obtained by replacing the pth row of the matrix $\boldsymbol{A}$ by the vector $\boldsymbol{v}$. The relationship (16) is a consequence of the chain rule, equality of mixed partials, and Cramer's rule. We can also rewrite (1) in terms of the independent variable $\boldsymbol{\xi}$:

$$\frac{\partial(JU)}{\partial t} + \nabla_{\boldsymbol{\xi}} \cdot (\boldsymbol{N}^T \vec{F}) = 0. \tag{18}$$

The semi-discrete form of (18) corresponding to (3)–(4) is then

$$\frac{d}{dt}\langle JU\rangle_{\boldsymbol{i}} = -\frac{1}{h^D}\int\limits_{V_{\boldsymbol{i}}} \nabla \cdot (\boldsymbol{N}^T \vec{F}) d\boldsymbol{\xi} = -\frac{1}{h}\sum_{d=1}^{D}(\langle(\boldsymbol{N}^T\vec{F})_d\rangle_{\boldsymbol{i}+\frac{1}{2}\boldsymbol{e}^d} - \langle(\boldsymbol{N}^T\vec{F})_d\rangle_{\boldsymbol{i}-\frac{1}{2}\boldsymbol{e}^d}) \tag{19}$$

$$\langle(\boldsymbol{N}^T\vec{F})_d\rangle_{\boldsymbol{i}+\frac{1}{2}\boldsymbol{e}^d} = \frac{1}{h^{D-1}}\int\limits_{A_d}(\boldsymbol{N}^T\vec{F})_d dA. \tag{20}$$

For a single-level problem, we solve using a generalization of the advection method in [4] to nonlinear hyperbolic systems. With mapped grids and multiple levels of refinement, we can use a straightforward extension of the approach described for the Cartesian case in section 2.

4. Results

Figure 1 shows a color plot and contour lines for density with the shock-ramp problem of Berger and Colella [2], implemented in Cartesian coordinates with two levels of refinement, by a factor of 4, with an effective resolution of 1024 × 256. We use artificial viscosity coefficients of $K_1 = K_2 = 0.3$ in (7).

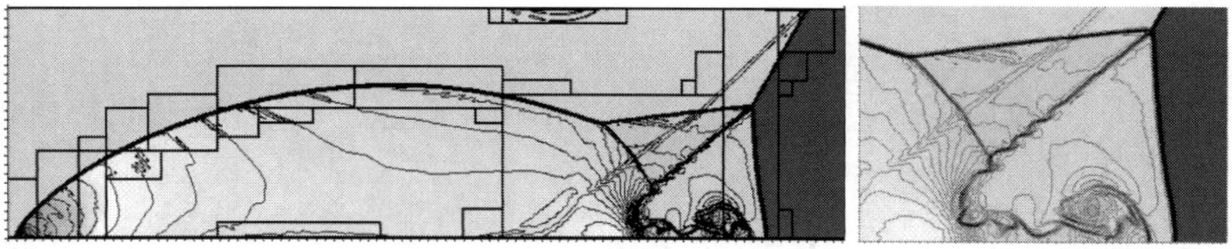

Figure 1. Density and contour plot for ramp problem on two levels in Cartesian coordinates. Outlines of grid boxes are shown with thick black lines. The right-hand plot is a close-up of the left-hand plot showing the double Mach reflection region. This calculation demonstrates the ability of the limiters and the artificial viscosity to produce a robust treatment of problems with strong shocks.

As an example using mapped coordinates, Figure 2 shows a density plot of a Gaussian acoustic pulse in a 2D polytropic gas in a periodic domain, implemented with a mapped coordinate system and two levels of refinement, by a factor of 4. The coordinate mapping is

$$\boldsymbol{X}(\xi,\eta) = (\xi + \sin(2\pi\xi)\sin(2\pi\eta)/10, \eta + \sin(2\pi\xi)\sin(2\pi\eta)/10). \tag{21}$$

There is no artificial viscosity. Table 1 shows fourth-order convergence of the component $\langle J\rho\rangle$, where ρ is density.

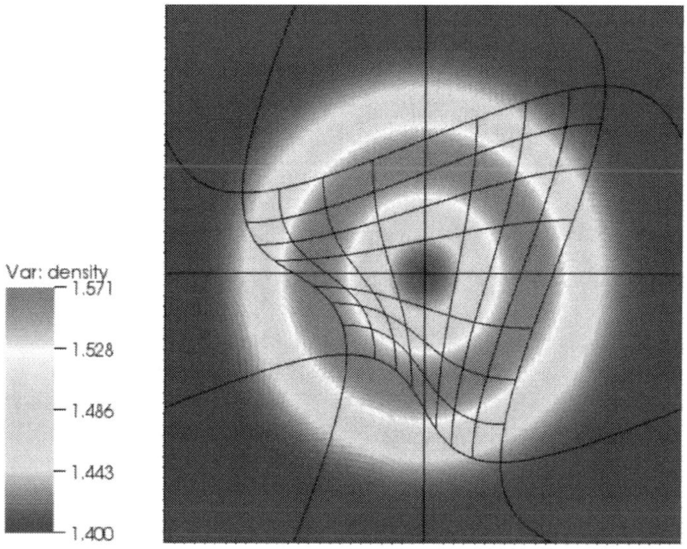

Figure 2. Plot of density with a Gaussian acoustic pulse in a polytropic gas in a periodic domain, using two levels of refinement with the mapped coordinates of (21), after 240 coarse-level time steps each of length 0.0006. At the initial time, the density profile is a circularly symmetric Gaussian with its peak at the center of the domain. Each grid box consists of 32×32 cells; the outlines of these boxes are shown in black.

Table 1. Convergence of $\langle J\rho \rangle$ in two-level Gaussian acoustic pulse example with mapped coordinates of (21). The simulation was run with different mesh spacings, and the difference between simulations with coarse-level mesh spacing (in abstract space) h and $h/2$ was taken at time 0.144. This table shows the norms of those differences in $\langle J\rho \rangle$, and the rates of convergence.

norm	1/32 to 1/64	rate	1/64 to 1/128	rate	1/128 to 1/256	rate	1/256 to 1/512
L_1	1.3072e-04	3.82	9.2338e-06	3.93	6.0686e-07	3.97	3.8791e-08
L_2	2.4617e-04	3.73	1.8504e-05	3.91	1.2332e-06	3.97	7.8870e-08
L_∞	1.5054e-03	3.44	1.3899e-04	3.84	9.6921e-06	3.92	6.3868e-07

References

[1] Barad M and Colella P 2006 *J. Comput. Phys.* **208** 1
[2] Berger M J and Colella P 1989 *J. Comput. Phys.* **82** 64
[3] Colella P 1990 *J. Comput. Phys.* **87** 171
[4] Colella P, Dorr M, Hittinger J and Martin D F 2009 submitted to *J. Comput. Phys.*
[5] Colella P and Sekora M 2008 *J. Comput. Phys.* **227** 7069
[6] Colella P and Woodward P R 1984 *J. Comput. Phys.* **54** 174
[7] Zalesak S T 1979 *J. Comput. Phys.* **35** 335

Interoperable mesh components for large-scale, distributed-memory simulations

K Devine[1], L Diachin[2], J Kraftcheck[4], K E Jansen[3], V Leung[1], X Luo[3], M Miller[2], C Ollivier-Gooch[5], A Ovcharenko[3], O Sahni[3], M S Shephard[3], T Tautges[6], T Xie[3] and M Zhou[3]

[1] Sandia National Labs, Albuquerque, NM, [2] Lawrence Livermore National Lab, Livermore, CA, [3] Rensselaer Polytechnic Institute, Troy, NY, [4] University of Wisconsin, Madison, WI, [5] University of British Columbia, Vancouver, BC, [6] Argonne National Laboratory, Argonne, IL

E-mail: kddevin@sandia.gov, diachin2@llnl.gov, kraftche@cae.wisc.edu, jansen@rpi.edu, vjleung@sandia.gov, xluo@scorec.rpi.edu, miller86@llnl.gov, cfog@mech.ubc.ca, shurik@scorec.rpi.edu, osahni@scorec.rpi.edu, shephard@scorec.rpi.edu, tautges@mcs.anl.gov, txie@scorec.rpi.edu, zhou@scorec.rpi.edu

Abstract. SciDAC applications have a demonstrated need for advanced software tools to manage the complexities associated with sophisticated geometry, mesh, and field manipulation tasks, particularly as computer architectures move toward the petascale. In this paper, we describe a software component — an abstract data model and programming interface — designed to provide support for parallel unstructured mesh operations. We describe key issues that must be addressed to successfully provide high-performance, distributed-memory unstructured mesh services and highlight some recent research accomplishments in developing new load balancing and MPI-based communication libraries appropriate for leadership class computing. Finally, we give examples of the use of parallel adaptive mesh modification in two SciDAC applications.

1. Introduction

Many simulations for solving partial differential equations (PDEs) require the ability to model complex geometries defined by CAD or other engineering design tools. Such geometries in turn require the generation of meshes that have high fidelity to both the computational domain description as well as to the features of interest in the numerical solution, e.g., regions of high gradients. This process is further complicated by the need to perform simulations on petascale computers where additional issues such as load balance, communication costs, and optimal data decompositions come into play. Because these operations are common to many simulations, re-usable software for these tasks can be shared across many application codes which could reduce the time, effort, and expertise required to develop and maintain simulation software.

Reuse of scientific software has traditionally relied on the use of *frameworks* or *library-based* implementations and many useful tools have been developed over the years for both unstructured (e.g., [1, 2, 3, 4, 5]) and structured meshes (e.g., [6, 7, 8]). In addition, libraries that provide mesh quality improvement [9, 10], front tracking [11], mesh refinement [12], and parallel partitioning and load balancing [13] have been developed as well. However, the use of libraries or frameworks can significantly hamper experimentation with different software instances that provide similar functionality. In particular, libraries of similar purpose often package functionality in very different ways. Consequently, data structures shared between the application and library and even the control flow between the application and library may need to be totally re-designed. This is especially true for meshing and geometry libraries where applications often directly access the underlying data structures, which can be quite different from implementation to implementation. Thus, using different libraries interchangeably or interoperably for this functionality has proven difficult at best and has hindered the wide spread use of advanced meshing and geometry tools developed by the research community.

Components represent a higher level of abstraction than libraries. Essentially, a component defines both a *specification* for an application programming interface (API) and an abstract *data model* defining the semantics of the data that is passed through the interface. There are several key advantages to using a component-based approach in that the focus is on interfaces rather than on data structures or file formats. This allows any application using a component to use another implementation of the same component API, because all implementations have substantially equivalent functionality.

In this paper, we describe a parallel unstructured mesh component developed by the Interoperable Technologies for Advanced Petascale Simulations (ITAPS) project (Section 2). As we have worked to scale the software that uses our mesh component implementations to 100,000 processors and beyond, several interesting research projects to improve load balance and reduce communication costs have arisen. We briefly describe our work in these areas and show both weak and strong scaling results for implicit finite element/volume solvers and adaptive mesh refinement services (Section 3). We conclude by showcasing the use of the adaptive mesh refinement service in accelerator and fusion SciDAC applications (Section 4).

2. The ITAPS parallel mesh component

The ITAPS parallel mesh component builds on previous work that resulted in the definition of a serial abstract data model and interfaces for serial mesh data. In this section, we briefly describe the key concepts from that earlier work that are germaine to the current discussion, and describe the parallel data model (Section 2.1) and language- and data-structure-independent interface to support query and modification of meshes on distributed memory computers (Section 2.2). We then highlight several software implementations of the interface and services that use the serial and parallel interfaces to provide key functionality to application codes (Section 2.3).

2.1. The abstract data model

At a high level, the ITAPS data model divides the data required by a simulation into three *core data types*: the geometric data, the mesh data, and the field data. These core data types are associated with each other through *data relation managers*. The data relation managers control the relationships among two or more of the core data types, resolve cross references between entities in different groups, and can provide additional functionality that depends on multiple core data types. Key building blocks within these data models are the concepts of *entities*, *entity sets*, and *tags*.

- *Entities* are used to represent atomic pieces of information such as vertices in a mesh or edges in a geometric model. Entity adjacency relationships define how the entities connect to each other and both first-order and second-order adjacencies are supported.
- *Entity sets* are arbitrary collections of entities that may be an ordered list or unordered. The two primary supported relationships among entity sets are *contained in* and *parent/child* to allow for subsetting and hierarchical applications. In addition, entity sets also have "set operation" capabilities such as set subtraction, intersection, or union.

- *Tags* are used as containers to attach user-defined data to ITAPS entities and entity sets. Tags can be multi-valued which implies that a given tag handle can be associated with many different entities. We support specialized tag types for improved performance as well as the more general opaque case that allows any type of data to be attached.

To support many of the services that applications desire, such as adaptive mesh refinement, the data model includes the concept of modification to allow changes to geometry, topology, or set structure. In the case of the mesh, capabilities include changing vertex coordinates and adding or deleting entities. Modification often requires interactions between the mesh, geometry and field data models and is one of the primary uses for the data relations manager.

The parallel ITAPS data model extends the concepts described above to handle the requirements of distributed memory applications. In such applications, the unstructured mesh is typically divided or *partitioned* over the independent processor memories of the computer. To be useful to applications, we must maintain information about mesh entities and their adjacencies that is "shared". In addition, applications require the ability to move mesh entities and their associated information between different processors to update the mesh partition while maintaining this shared information. Moreover, applications expect that the mesh data model defined previously will continue to work as expected within a process or for global address spaces and shared memory paradigms. These requirements are addressed through the following additional core concepts.

- A mesh *partition* is a decomposition of the mesh entities (e.g., vertices, edges, faces, and regions) into subsets called *parts*. The partition is responsible for mapping the entities to parts and for mapping the parts to processes. We note that each process may have one or more parts and that each part is wholly contained within a process. Parts are identified globally by unique part IDs and, within a process, by opaque part handles. A partition has a communicator associated with it. Thus "global" operations are performed with respect to data in all parts in the partition's communicator and "local" operations are performed with respect to either a part's or process's data.
- Mesh entities are *owned* by exactly one part in the partition where ownership imbues the right to modify. It is important to note that ownership is not necessarily static during the course of a computation and can be changed due to a repartitioning of the mesh or due to local micro-migration operations. In addition, some entities will have read-only copies on other parts, for example, along part boundaries and for ghosting operations. No globally unique entity IDs are required or supplied by the data model although they can be constructed by the user as a pair *[part ID, entity handle]*.
- Mesh entities can be further classified as an *internal entity* (an owned entity not on an interpart boundary), *a part-boundary entity* (an entity on an interpart boundary which are shared between parts), or a *ghost entity* (a non-owned, non-part-boundary entity); see Figure 1. *Copies* are defined to be all ghost entities plus all non-owned part-boundary entities. The data model defines rules for the amount of information about copies that an implementation must manage. For example, an entity's owner must store information about all copies of the entity, and a copy must store information about the entity's owner. Remote parts and entities are computed after mesh modification so that queries for remote data do not require communication.

2.2. The ITAPS iMeshP interface

Once the abstract data model is defined, the next step to creating interoperable technologies is to define common interfaces that support its functionality. A key aspect of the ITAPS approach is that we do not enforce any particular data structure or implementation with our interfaces,

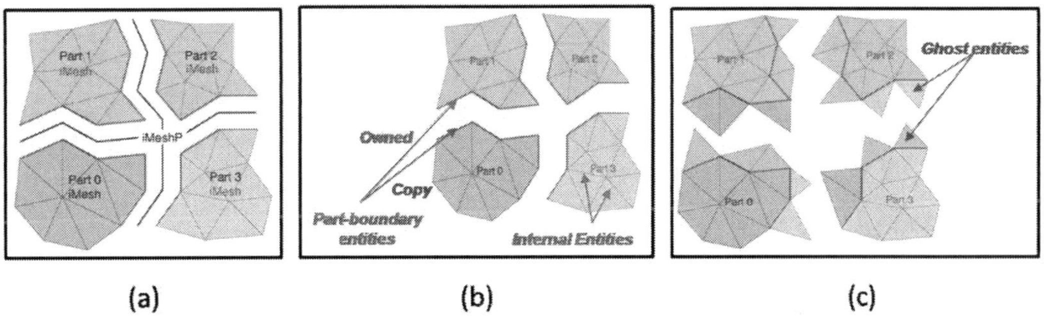

Figure 1. A simple example showing the iMeshP datamodel. Image (a) shows the relationship bewteen iMesh and iMeshP, (b) shows entity classification, and (c) shows ghost entities.

requiring only that certain questions about the geometry, mesh, or field data can be answered through calls to the interface. All data passed through the interface is in the form of opaque handles to objects defined in the data model. One of the most challenging aspects of this effort remains balancing performance of the interface with the flexibility needed to support a wide variety of data types. Further challenges arise when considering the support of many different scientific programming languages which we address using a two-pronged approach. First, we provide a C-language binding for our interfaces that is compatible with most needs in scientific computing. Additional flexibility, albeit at a somewhat higher cost, is supported through the use of the SIDL/Babel technology [14] provided by the Common Component Architecture Forum.

In previous work, a full specification for the serial mesh interface, called iMesh, was completed and implemented by several institutions. The extension to iMeshP, the parallel mesh interface, required the definition of a number of additional functions; for example, functions to easily create and modify partitions, create ghost entities, retrieve ghost and owner entity tag data, and determine an entity's ownership status. To simplify the iMeshP interface, we allow part handles to be substituted for entity-set handles in all serial iMesh functions. Thus, operations such as adding entities to parts and querying the number of entities in a part can be achieved using the same interface as adding entities to and querying entity sets. Additional iMeshP functions provide information about part boundaries and neighboring parts. Furthermore, the iMeshP interface supports parallel operations needed for efficient computation, load balancing and mesh modification. By necessity, these operations involve parallel communication and both synchronous and asynchronous parallel operations are supported. This design enables such things as updates of tag data in ghost entities during computation, large- or small-scale entity migration for dynamic load balancing or edge swapping, updates of vertex coordinates in non-owned vertices for mesh smoothing, and coordination in the creation of new entities along part boundaries for mesh refinement. For more information see [15].

To illustrate iMesh and iMeshP interface usage, we provide a simple example of using the C-binding version in Figure 2. Line 14 shows the creation of a new mesh instance which creates the local opaque handle `mesh` that is used in later calls to refer to this instance of the interface on this process. Likewise, line 15 shows the creation of the `root_set` which contains all the mesh data on the processor once it is loaded. Line 20 shows the call to create the partition handle on all processors and associate the MPI communicator with it. Line 21 shows the use of the `iMeshP_load` function to populate the mesh and partition interface on all processors using a string name identifier. In this example, the data is loaded from file `125hex.vtk`, but `iMeshP_load` can also be used for on-the-fly mesh creation. Line 24 shows the global query to retrieve the number of parts in the partition. Line 25 shows the call to get the global number of

```
#include ''iMesh.h''
#include ''iMeshP.h''
#include <mpi.h>

int main( int argc, char *argv[] )
{
    // create and populate the Mesh instance
  iMesh_Instance mesh;
  iBase_EntitySetHandle root_set;
  iMeshP_PartitionHandle partition;
  int num_parts, num_vtx, ierr;

  iMesh_newMesh("", &mesh, &ierr, 0);
  iMesh_getRootSet(mesh, &root_set, &ierr);

  MPI_Init(&argc, &argv);

   // create the partition and load the mesh
  iMeshP_createPartitionAll(mesh, MPI_COMM_WORLD, &partition, &ierr)
  iMeshP_load(mesh, partition, root_set, "125hex.vtk", "", &ierr, 10, 0);

   // get the number of parts and number of vertices in the partition
  iMeshP_getNumParts(mesh, partition, &num_parts, &ierr);
  iMeshP_getNumOfTypeAll(mesh, partition, root_set, iBase_VERTEX, &num_vtx, &ierr);
}
```

Figure 2. Example use of the C-binding of the iMeshP interface.

vertices in the partition; this call may require global communication if it is not stored locally.

2.3. ITAPS software using iMesh and iMeshP

The ITAPS consortium has produced four implementations of the iMesh interface based on pre-existing mesh databases. Each of the four has its own particular strengths and so are useful in different application settings. In addition, two of the four iMesh implementations also have at least partial iMeshP implementations available. These implementations are listed in Table 1. In addition, the ITAPS team is developing a number of *component services* that use the iMesh and iMeshP interfaces to support simulations involving complex domains, adaptive techniques, and high-order methods. Specific tools include mesh quality improvement through smoothing with Mesquite [9] and swapping [10], high-order mesh curve correction tools [16], adaptive mesh refinement through MeshAdapt [12], front tracking through FronTier [11], dynamic load balancing services through Zoltan [13], and visualization plug-ins into VisIt [17]. These component services can be used directly by applications and can also be integrated to form higher-level integrated services such as shape-optimization and AMR-Front tracking technologies. Each of these services has been demonstrated to work with multiple implementations of the iMesh and iMeshP component APIs. We generally see little reduction in the overall efficiency of an application using the iMesh or iMeshP interfaces compared to using native data structures when following the ITAPS "best practices" implementation guidelines. For example, we tested the time to partition a MOAB mesh using Zoltan through the MOAB native interfaces and compared this with the time required to partition the same mesh through the iMesh interface. We found the overhead ranged from under 1% for small problem sizes to about 2% for larger problems sizes. Similarly, when building a stiffness matrix associated with a simple finite element solver, the overhead costs associated with the use of iMesh ranged from 2% to 11% depending on the access pattern chosen.

Table 1. iMesh and iMeshP Implementations

Implementation	**Emphasis**	**Parallel Capability**	**Applicaitons**
FMDB: Flexible Mesh Database [1, 2] (iMesh/iMeshP)	Adaptively changing meshes (entity addition or removal)	Scalable to 32K procs; 2B elements	fusion, accelerators CFD, solid mechanics multiphase flow
MOAB: Mesh Oriented dAtaBase [3] (iMesh/iMeshP)	Low memory usage first; then CPU time	Up to 64 procs	nuclear reactors, accelerators, rad. transport, inertial confinement fusion
GRUMMP: Generation and Refinement of Mixed-Element Meshes in Parallel [4] (iMesh)	Fast adjacency retrieval for mesh generation and improvement, adaptation	In development	CFD, biological systems, structural mechanics
NWGrid: Northwest Grid Generation Code [5] (iMesh)	Simplicial meshes; parallel generations of unstructured, hybrid, meshes	Parallelism based on Global Arrays; Scalable to at least 10K procs	CFD, subsurface transport, biological systems

3. Scalability of ITAPS software

In this section, we hightlight recent results on the scalability of ITAPS tools that use the iMeshP interface. In particular, in this paper we focus on the combintation of the FMDB mesh database, Zoltan dynamic load balancing, and MeshAdapt services used in implicit finite element and finite volume applications. We are interested in both strong and weak scaling to minimize run time and maximize resolution, respectively. This has driven the need to research new load balancing technologies and communication tools; particularly as we scale to O(100,000) processors and beyond. We describe these research efforts in some detail and highlight the resulting scalability of our tools and solvers that use our tools on a number of different leadership class computers.

The ability to scale implicit finite element and finite volume computations such as those used in the accelerator and fusion SciDAC application efforts, requires ensuring both the system formulation and solution are effectively load balanced. Graph-based partitioners are well known to produce a partitioning of the mesh into parts that are well balanced in terms of the specified partition object type while also minimizing inter-part communications. However, traditional graph-partitioners consider a single objective optimization subject to a single balance constraint. In the case of mesh-based analysis, the defined graph nodes are often mesh regions. This selection does an excellent job of balancing the number of regions (elements) and therefore the workload for the construction of the part-level finite element system. However, in the case of C^0 interpolating basis functions, for example, the workload balance for the iterative solution (e.g. matrix vector product and vector norms) of the resulting system is proportional to the number of mesh vertices per part. Since mesh vertices are not the objects in the original partitioning, the balance may not be optimal, particularly when the numbers of mesh entities per part is relatively low (e.g., a few thousand). We are developing two multiple compute-object based partition improvement algorithms to reduce the vertex imbalance thereby improving the overall balance of two mesh entities as required by a scalable implicit solve. They are referred to as local iterative inter-part boundary modification (LIIPBMod) and heavy part split (HPS).

- *The LIIPBMod Algorithm.* LIIPBMod locally migrates small numbers of mesh regions from parts that are relatively heavily loaded with respect to mesh vertices to neighboring parts which are relatively lightly loaded with respect to mesh entities. On the heavily loaded part, the mesh vertices on the part boundary are traversed and the ones bounding a small number of elements are identified. If the neighboring part is lightly loaded, the whole "cavity" (all the adjacent elements of the picked vertex) is migrated to the neighboring part. By this minor inter-part boundary adjustment, the vertex imbalance is improved while only modestly perturbing the good element balance. This procedure may need to be repeated for several iterations to achieve desired vertex balance.

- *The HPS Algorithm,* Our studies of the mesh partitions given by a graph-based partitioner show that the percentage of heavily loaded parts (more than 10% imbalance) is usually less than 1%. The idea of HPS is that, if the desired number of parts is numP, first distribute the mesh to 99%*numP parts by a graph-based partitioner and leave the other 1%*numP parts empty. Then, select the 1%*numP parts with the highest vertex load and split them into two parts (i.e. migrate roughly half of the mesh entities from them to one of the empty parts). The splitting makes the heavily loaded parts become lightly loaded. Since the peak of the imbalance determines the scalability, HPS lowers the peak and hence improves the performance.

Figure 3 shows the results of applying the two algorithms to a 16.7M element anisotropically adapted mesh used in the simulation of an abdominal aorta aneurism (left image). The two graphs (center for LIIPBmod and right graph for HPS) indicate the number of vertices per part divided by the average per part before (red dots) and after (blue dots) application of the algorithms. Note in both cases the spikes (red dots in the upper parts of the graphs) that reduce scalability are dramatically lowered after the algorithms have been applied.

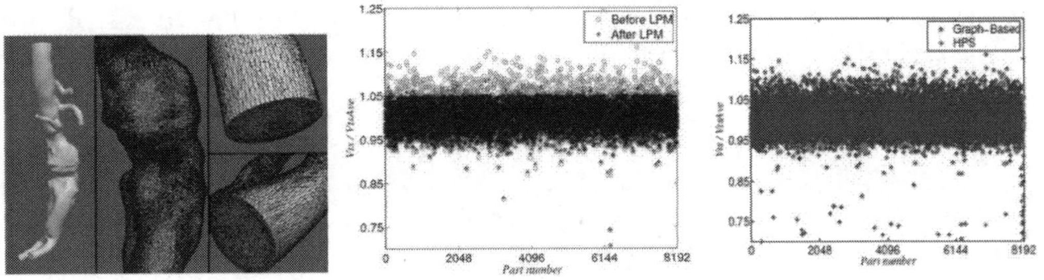

Figure 3. Vertex imbalance before and after LIIPBMod and HPS.

Using the results of the multi-compute object load balancer, we showed that unstructured mesh solvers based on implicit methods can scale to very large numbers of processors [18]. Table 2 illustrates the parallel efficiency of PHASTA, a parallel, unstructured and implicit flow solver developed at RPI on up to O(100,000) cores of IBM BG and Cray XT systems [19]. PHASTA uses FMDB and the MeshAdapt service for refinement on up to 32K processors and iMeshP and iZoltan to partition the resulting meshes into 128K parts for distribution to almost the full machine for the analysis step. Near-perfect (linear) strong scaling of the analysis step over multiple doublings of cores can be clearly seen on various systems with a slight decrease in parallel efficiency on the largest core counts due to the fact that computational load per core becomes insignificant (for a fixed-size problem). More information on PHASTA's scaling studies can be found in [20].

We have also examined the scaling of the MeshAdapt service in both the weak and strong sense. Mesh adaptation is characterized by small, but variable, work per operation which implies that achieving "perfect scaling" can be very costly. In particular, our research has found that

Table 2. Strong scaling results of PHASTA up to O(100,000) cores on IBM BlueGene (BG) and Cray XT systems (1 implies perfect scaling with 100% parallel efficiency)

105 Million Elements			
Core Count	BGL-CCNI RPI	BGP-ALCF ANL	XT4-NERSC LBNL
512	1.00	1.00	1.00
1,024	1.01	1.02	1.03
2,048	1.00	0.99	1.16
4,096	0.99	0.99	1.00
8,192	1.02	0.95	0.77
16,384	1.03	0.95	
32,768	0.93	0.88	

1 Billion Elements	
Core Count	BGP-ALCF ANL
16,384	1.00
32,678	0.99
65,536	0.97
131,072	0.89

we can run adaptive mesh refinement algorithms on the large numbers of parts typically used in a simulation analysis, and the overall time will still be a small percentage of the overall total solution time. For example, for the weak scaling results shown in Table 3 for uniform adaptive refinement, the time required by the adaptivity algorithms for the largest case on 32768 processors is only 0.04% of the total time.

That said, the scaling efficiency is clearly decreasing as the number of processors increases, and to improve the scaling characteristics of the MeshAdapt and other services, we are developing a new general-purpose, MPI-based communication package called the Inter-Processor Communication Manager (IPComMan). This package aims to reduce data exchange costs by exploiting communications in a local neighborhood for each processor. The neighborhood is the subset of processors exchanging messages with each other during a specific communication round, which in many applications is bounded by a constant, typically under 40, independent of the total number of processors. Strong scaling of uniform adaptive mesh refinement using the IPComMan Message Passing library for a mesh starting with 4.3M elements and ending with 2.2B elements is shown in the right of Table 3. Significant improvements in scaling efficiency compared to the weak scaling case are evident due to decreased communication costs.

Table 3. Uniform adaptive mesh refinement: weak scaling using the AutoPack Message Passing library and strong scaling using the IPComMan library

Weak Scaling				
Number of Parts	Initial Mesh	Adapted Mesh	Time (s)	Scaling Factor
2048	17M	128M	5.0	1.0
4096	34M	274M	4.8	1.05
8192	65M	520M	5.1	0.97
16384	520M	1.1B	6.1	0.82
32768	274M	2.2B	7.4	0.68

Strong Scaling		
Number of Parts	Time (s)	Scaling Factor
2048	21.5	1.0
4096	11.2	0.96
8192	5.67	0.95
16384	2.73	0.99

4. Use of ITAPS parallel adaptive software in SciDAC applications

ITAPS technologies have impacted DOE applications in a number of ways including direct use of the services and software described in this paper (adaptive algorithms, mesh quality improvement, partitioning, front tracking), technology advancement through the demonstration and insertion of key new technology areas (shape optimization and petascale mesh generation), and by looking ahead and anticipating the needs of application teams through the development of new services (mesh to mesh transfer for coupling multiphysics applications). In this section we highlight a few key examples of the use of ITAPS software in SciDAC applications. While we focus on our interactions with accelerator and fusion applications to insert adaptivity into their codes in this paper, more extensive interactions funded under the auspices of the SciDAC and other DOE programs include work with multiple accelerator and fusion teams along with subsurface flow and nuclear energy application teams.

Accelerator Design. The ITAPS team is working extensively with the "Community Petascale Project for Accelerator Science and Simulation (COMPASS)" SciDAC project, and in particular researchers at the SLAC National Accelerator Laboratory, to provide high-order mesh generation and adaptive control methods to improve the processes for the design and optimization of accelerator cavities. For example, in calculating the short-range wakefield inside an accelerator structure, only the small region in the vicinity of the moving particle beam is required to have a highly refined mesh in the simulation. The moving curved mesh adaptation procedure, which refines a small region of interest near the beam (see Figure 4), greatly reduces the computational effort required for a given level of accuracy. In particular, using such techniques has resulted in a tenfold reduction in the computational cost of these simulations [21, 22].

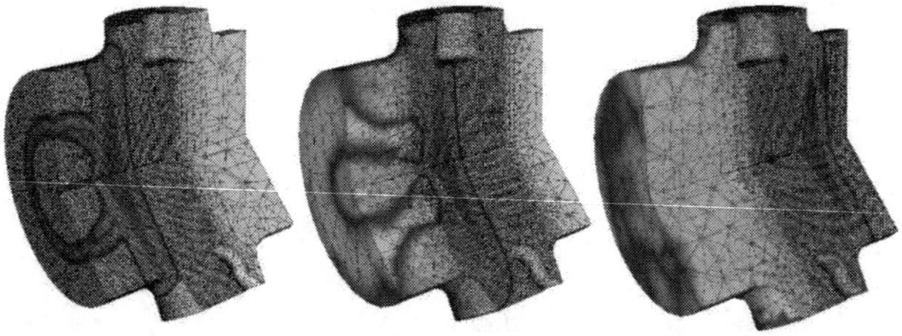

Figure 4. The electric fields on tree-refined curved meshes around the moving beam

Fusion. The SciDAC-funded "Center for Extended MHD modeling (CEMM)" has extensively used a 3D MHD code to simulate global instabilities in magnetic fusion devices. The ITAPS team is working with CEMM fusion scientists to extend the MHD high-order finite element software (M3D-C1) to interface with unstructured mesh adaptation technologies. This enables them to gain the efficiencies of using adaptive meshes (see Figure 5 for isotropic and anisotropic adapted meshes [23, 24]) and allows them to model general curved reactor domains (see the rightmost image in Figure 5).

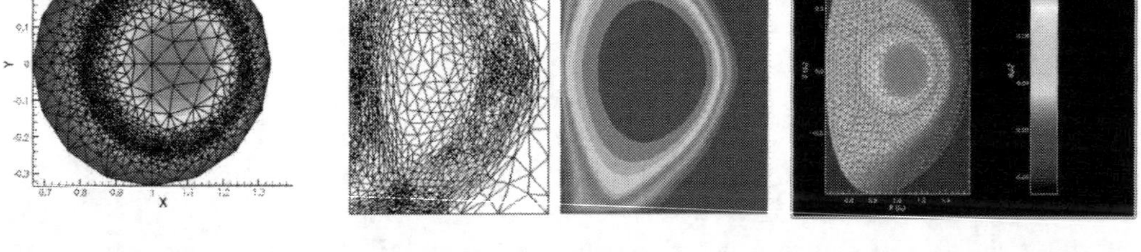

Figure 5. Isotropic and anisotropic mesh adaptation to increase computational efficiency in fusion applications. On the far right we show the vorticity contours in a curved reactor domain.

More information about ITAPS

More information on the ITAPS project including detailed descriptions of the ITAPS services, interfaces, software, and interactions with SciDAC application teams can be found at `http://www.itaps-scidac.org`.

Acknowledgments

This work was performed under the auspices of the U.S. Department of Energy by Lawrence Livermore National Laboratory under Contract DE-AC52-07NA27344; the Canadian Natural Sciences and Engineering Research Council under Special Research Opportunities Grant SRO-299160; by Rensselaer Polytechnic Institute under DOE grant number DE-FC02-01ER25460 and the NSF PetaApps project OCI-0749152, and by Sandia National Laboratory; a multiprogram laboratory operated by Sandia Corporation, a Lockheed Martin Company, for the U. S. Department of Energy's National Nuclear Security Administration under contract DE-AC04-94AL85000. Computing Resources used in this paper include the Rensselaer Computational Center for Nanotechnology Innovations (BG/L) and DOE INCITE (Intrepid - BG/P).

References

[1] Remacle J F and Shephard M 2003 *International Journal for Numerical Methods in Engineering* **58** 349–374
[2] Seol E and Shephard M 2006 *Engineering with Computers* **22** 197–213
[3] Tautges T J, Meyers R E, Merkley K, Stimpson C and Ernst C 2004 *Sandia report SAND 2004-1592* (Sandia National Laboratories)
[4] Ollivier-Gooch C F 1998–2005 GRUMMP — Generation and Refinement of Unstructured, Mixed-element Meshes in Parallel http://tetra.mech.ubc.ca/GRUMMP
[5] Trease H 2006 The NWGrid mesh generation system Pacific Northwest National Laboratory - http://www.emsl.pnl.gov/nwgrid
[6] Collela P 2009 Chombo home page Tech. Rep. https://seesar.lbl.gov/anag/chombo Lawrence Berkeley National Laboratory
[7] Garaizar X, Hornung R and Kohn S 1999 Structured adaptive mesh refinement applications infracture Tech. Rep. http://www.llnl.gov/casc/SAMRAI Lawrence Livermore National Laboratory
[8] Bill Henshaw 2009 Overture home page Tech. Rep. http://www.llnl.gov/CASC/Overture.html Lawrence Livermore National Laboratory
[9] 2009 Mesquite download page http://www.cs.sandia.gov/ web9200/9200_download.html
[10] Freitag L A and Ollivier-Gooch C F 1997 *International Journal for Numerical Methods in Engineering* **40** 3979–4002
[11] Bo W, Fix B, Glimm J, Li X, , Liu X, Samulyak R and Wu L 2007 *Proceedings in Applied Mathematics and Mechanics*
[12] Li X, Shephard M and Beall M 2005 *to appear Comp. Meth. Appl. Mech. Engng.*
[13] Devine K, Boman E, Heaphy R, Hendrickson B and Vaughan C 2002 *Computing in Science and Engineering* **4** 90–97
[14] Dahlgren T, Epperly T, Kumfert G and Leek J 2005 *Babel User's Guide* CASC, Lawrence Livermore National Laboratory Livermore, California version 0.10.10
[15] iMeshP Interface Documentation 2009 http://www.itaps-scidac.org/software/imeshp_html/index.html
[16] Luo X, Shephard M, O'Bara RMand Rocco R and Beall M 2004 *Engineering with Computers* **20** 273–285
[17] Childs H, Brugger E S, Bonnell K S, Meredith J S, Miller M, Whitlock B J and Max N 2005 *Proceedings of IEEE Visualization 2005* pp 190–198
[18] Shephard M, Jansen K, Sahni O and Diachin L 2007 *Journal of Physics: Conference Series* vol 78-012053
[19] Whiting C H and Jansen K E 2001 *International Journal of Numerical Methods in Fluids* **35** 93–116
[20] Sahni O, Carothers C, Shephard M and Jansen K 2009 *Scientific Programming*
[21] Luo X J, Shephard M, Lee L, Ge L and Ng C 2008 *17th International Meshing Roundtable Conference* (Pittsburgh, PA)
[22] Lee L, Akcelik V, Ng C and Ko K 2008 *Journal of Physics: Conference Series* **125**
[23] Jardin S C, Ferraro N, Breslau J, Bauer A and Shephard M 2007 *European Physical Society meeting on Plasma Physics*
[24] Jardin S, Ferraro N, Luo X, Chen J, Breslau J, Jansen K and Shephard M 2008 *Journal of Physics:Conference Series* vol 125

Parallel multiphysics algorithms and software for computational nuclear engineering

D Gaston, G Hansen, S Kadioglu, D A Knoll, C Newman, H Park, C Permann and W Taitano

Multiphysics Methods Group
Idaho National Laboratory, Idaho Falls, ID 83415-3840

E-mail: `dana.knoll@inl.gov`

Abstract. There is a growing trend in nuclear reactor simulation to consider multiphysics problems. This can be seen in reactor analysis where analysts are interested in coupled flow, heat transfer and neutronics, and in fuel performance simulation where analysts are interested in thermomechanics with contact coupled to species transport and chemistry. These more ambitious simulations usually motivate some level of parallel computing. Many of the coupling efforts to date utilize simple code coupling or first-order operator splitting, often referred to as loose coupling. While these approaches can produce answers, they usually leave questions of accuracy and stability unanswered. Additionally, the different physics often reside on separate grids which are coupled via simple interpolation, again leaving open questions of stability and accuracy. Utilizing state of the art mathematics and software development techniques we are deploying next generation tools for nuclear engineering applications. The Jacobian-free Newton-Krylov (JFNK) method combined with physics-based preconditioning provide the underlying mathematical structure for our tools. JFNK is understood to be a modern multiphysics algorithm, but we are also utilizing its unique properties as a scale bridging algorithm. To facilitate rapid development of multiphysics applications we have developed the Multiphysics Object-Oriented Simulation Environment (`MOOSE`). Examples from two `MOOSE`-based applications: `PRONGHORN`, our multiphysics gas cooled reactor simulation tool and `BISON`, our multiphysics, multiscale fuel performance simulation tool will be presented.

1. Introduction

The use of multiphysics simulation is on the increase in computational nuclear engineering. The increase in this activity is typically attributed to increasing computer power, but in truth, advanced numerical methods are playing an equal role. The term multiphysics simulation is used to describe simulations which include some level of disparate physical phenomena. Typical examples of multiphysics problems in nuclear engineering include coupling fluid flow, heat transfer and neutron kinetics for reactor dynamics, coupling fluid flow and structural dynamics to consider fluid-structure interactions for fuel rod fretting, and coupling nonlinear thermomechanics with contact, fission product behavior, and species transport to study fuel performance. In addition to multiphysics coupling, most of these problems also have multiscale issues to resolve.

In this endeavor the computational nuclear engineer is faced with choices of time integration and spatial discetization of the coupled system. Frequently, first-order operator splitting, or even explicit coupling of different codes, is used to perform the time integration. While this approach can produce results, it can also produce significant time integration error [1] and stability issues [2].

In this paper we discuss a modern multiphysics algorithm, the Jacobian-Free Newton-Krylov method (JFNK) combined with physics based preconditioning [3]. This approach has many of the desirable features of operator splitting, but can significantly reduce the time integration error and stability issues. We describe the JFNK methodology and its use in both a multiphysics and multiscale setting. Next we will describe and evolving software framework (`MOOSE`[4]) which utilizes this algorithmic framework and enables rapid development of multiphysics engineering analysis tools. Finally, we describe two evolving multiphysics engineering analysis tools; one for high temperature gas cooled reactors (`PRONGHORN`[5]) and one for fuel performance simulation (`BISON`[6]).

2. Multiphysics methods

As a nuclear engineering model problem used for the purposes of algorithm discussion, consider thermal conduction coupled to a one group diffusion model of neutron kinetics, ignoring delayed neutrons,

$$\rho C_p \frac{\partial T}{\partial t} - \nabla \cdot (K(T)\nabla T) = w\Sigma_f(T)\phi, \tag{1}$$

$$\frac{1}{v}\frac{\partial \phi}{\partial t} - \nabla \cdot (D(T)\nabla \phi) + \Sigma_a(T)\phi = \nu\Sigma_f(T)\phi. \tag{2}$$

In this highly simplified homogeneous picture of a reactor, nuclear fission generates heat, and this heat is removed from the system by thermal conduction. Here ρ is material density, C_p is material heat capacity, T is material temperature, K is the material thermal conductivity, w is the energy released per fission event, Σ_f is the macroscopic fission cross-section, Σ_a is the macroscopic absorption cross-section ϕ is the neutron scalar flux, v is the neutron velocity, D is the neutron diffusion coefficient, and ν is the number of neutrons produced per fission event. In this model D, K, and Σ_f all all functions of T. This simple model problem is a nonlinear PDE system with multiple time and space scales. We can use this problem to help highlight the difference in tightly coupled multiphysics solution algorithms and loosely coupled multiphysics solution algorithms. Without loss of generality we ignore spatial discretization issues.

It is our opinion that modern multiphysics simulations should be second-order accurate in time, or higher. However, for simplicity of presentation we will discuss some issues within the context of a first-order time integration approach. A fully implicit first-order in time discretization of this problem is

$$\rho C_p \frac{T^{n+1} - T^n}{\Delta t} - \nabla \cdot (K(T^{n+1})\nabla T^{n+1}) = w\Sigma_f(T^{n+1})\phi^{n+1}, \tag{3}$$

$$\frac{1}{v}\frac{\phi^{n+1} - \phi^n}{\Delta t} - \nabla \cdot (D(T^{n+1})\nabla \phi^{n+1}) + \Sigma_a(T^{n+1})\phi^{n+1} = \nu\Sigma_f(T^{n+1})\phi^{n+1}. \tag{4}$$

This specific time discretization will require nonlinear iteration within a time step.

The actual process of first-order splitting, or coupling two separate codes together, does not solve the above time discretization requiring nonlinear iteration. Typically D, K, Σ_f and Σ_a are evaluated from the previous time step. When coupling existing codes to execute a time step, sources are evaluated with an explicit dependence of variables from the other code, to simplify data exchange between codes. Such a time discretization would look like

$$\rho C_p \frac{T^{n+1} - T^n}{\Delta t} - \nabla \cdot (K(T^n)\nabla T^{n+1}) = w\Sigma_f(T^n)\phi^n, \tag{5}$$

$$\frac{1}{v}\frac{\phi^{n+1}-\phi^{n}}{\Delta t}-\nabla\cdot(D(T^{n})\nabla\phi^{n+1})+\Sigma_{a}(T^{n})\phi^{n+1}=\nu\Sigma_{f}(T^{n})\phi^{n+1}. \tag{6}$$

An analysis can be performed of this time discretization to further understand the time integration error [1]. Frequently this error can be significant, especially if many time steps are considered. Additionally, this type of splitting and linearization can produce new time step constraints for nonlinear stability. Iteration on sources and temperature dependent coefficients within a time step can help minimize these errors and stability issues.

2.1. Jacobian-free Newton-Krylov methods

Newton's method requires the solution of the linear system

$$\mathbf{J}^{k}\delta\mathbf{u}^{k}=-\mathbf{F}(\mathbf{u}^{k}), \tag{7}$$

followed by the update

$$\mathbf{u}^{k+1}=\mathbf{u}^{k}+\delta\mathbf{u}^{k}, \tag{8}$$

where $\mathbf{J}$ is the Jacobian matrix, $\mathbf{F}(\mathbf{u})$ is the nonlinear system of equations, $\mathbf{u}$ is the state vector, and k is the nonlinear iteration index . In vector notation, the $(i,j)^{th}$ element of the Jacobian matrix is,

$$J_{i,j}=\frac{\partial F_{i}(\mathbf{u})}{\partial u_{j}}. \tag{9}$$

Forming each element of $\mathbf{J}$ requires taking analytic or numerical derivatives of $F_i(\mathbf{u})$ with respect to $\mathbf{u}$ at each grid point. This can be difficult, time consuming, and error-prone.

In JFNK, a Krylov method [7] is used to solve Eq. (7). An initial linear residual, $\mathbf{r}_0$, given an initial guess, $\delta\mathbf{u}_0$, is evaluated as

$$\mathbf{r}_{0}=-\mathbf{F}(\mathbf{u})-\mathbf{J}\delta\mathbf{u}_{0}. \tag{10}$$

Note that the nonlinear iteration index, k, has been dropped. This is because the Krylov iteration is performed at a fixed k. The approximate solution at the l- th Krylov iteration, $\delta\mathbf{u}_l$, is constructed from a linear combination of the Krylov vectors (search directions) $\{\mathbf{r}_0,\mathbf{J}\mathbf{r}_0,(\mathbf{J})^2\mathbf{r}_0,...,(\mathbf{J})^{l-1}\mathbf{r}_0\}$, which were constructed during the previous $l-1$ Krylov iterations. This linear combination of Krylov vectors can be written as,

$$\delta\mathbf{u}_{l}=\delta\mathbf{u}_{0}+\sum_{j=0}^{l-1}\alpha_{j}(\mathbf{J})^{j}\mathbf{r}_{0}, \tag{11}$$

where evaluating the scalars α_j is part of the Krylov iteration. Upon examining Eq. (11) we see that a Krylov method requires the action of the Jacobian only in the form of matrix-vector products, which can be approximated by [8];

$$\mathbf{J}\mathbf{v}\approx\left[\mathbf{F}(\mathbf{u}+\epsilon\mathbf{v})-\mathbf{F}(\mathbf{u})\right]/\,\epsilon, \tag{12}$$

where $\mathbf{v}$ is a Krylov vector (i.e. one of $\{\mathbf{r}_0,\mathbf{J}\mathbf{r}_0,(\mathbf{J})^2\mathbf{r}_0,...,(\mathbf{J})^{l-1}\mathbf{r}_0\}$), and ϵ is a small perturbation. This matrix-free approach, besides its obvious memory advantage, has many unique capabilities. Namely, Newton-like nonlinear convergence *without forming* the true Jacobian. A fair amount of literature exists on JFNK motivated by ODEs and boundary value problems, while less exists for time-dependent PDEs. Standard "PDE motivated" references are [8, 9, 3]. It is also possible to form $\mathbf{J}\mathbf{v}$ analytically without forming $\mathbf{J}$. This approach has a number of positive features to consider, but we are not employing it in our work yet.

The JFNK method also has its advantages and disadvantages. The main advantage is that for a tight convergence tolerance there is no splitting or linearization error [1]. The algorithm provides a clean way to include other nonlinear phenomena. Another advantage is that one can implement a variety of time discretizations such as higher-order backward difference formula (BDF) methods or implicit Runge-Kutta methods.

The main disadvantage of the JFNK method is that one is solving a nonlinear problem, with iteration, to advance a time step. Furthermore, one is solving the full dimensional system implicitly. Thus a large matrix system must be approximately inverted on each Newton iteration. In order for this approach to be tractable one must produce an effective preconditioner.

2.2. Preconditioning of JFNK

The purpose of preconditioning a Krylov method is to efficiently cluster eigenvalues of the iteration matrix, which in turn will reduce the required number of Krylov iterations. Traditionally, standard iterative methods such as Jacobi, SSOR, or ILU are applied to the system matrix when constructing a preconditioner [7]. We refer to this as algebraic preconditioning methods. When applying physics-based preconditioning to the JFNK method the system matrix, $\mathbf{J}$ is never formed.

When employing right preconditioning to solve the Newton step this is typically expressed as,

$$(\mathbf{J}\mathbf{M}^{-1})(\mathbf{M}\delta\mathbf{u}) = -\mathbf{F}(\mathbf{u}). \tag{13}$$

$\mathbf{M}^{-1}$ is a linear operator which symbolically represents the preconditioning process, and $\mathbf{M}$, the inverse of the preconditioning process, is functionally not required. In the idea of physics-based preconditioning (to be discussed), $\mathbf{M}^{-1}$ represents a linearized time step. The right preconditioned version of the JFNK matrix-vector multiply approximation is:

$$\mathbf{J}\mathbf{M}^{-1}\mathbf{v} \approx [\mathbf{F}(\mathbf{u} + \epsilon\mathbf{M}^{-1}\mathbf{v}) - \mathbf{F}(\mathbf{u})] \;/\; \epsilon. \tag{14}$$

This process is actually done in two steps;

(i) Perform $\mathbf{y} = \mathbf{M}^{-1}\mathbf{v}$

(ii) Perform $\mathbf{J}\mathbf{y} \approx [\mathbf{F}(\mathbf{u} + \epsilon\mathbf{y}) - \mathbf{F}(\mathbf{u})] \;/\; \epsilon$.

2.3. Physics-based preconditioning

There exist numerous legacy algorithms to solve nonlinear systems, and these algorithms are typically based on linearization and time splitting. This linearization and time splitting has usually been developed with significant insight into the time scales or physical behavior of the problem. As a benefit of this insight, a reduced implicit system, or a sequence of segregated implicit systems may be solved in place of the fully coupled system. In the end, these legacy algorithms may make excellent preconditioners for the Jacobian-free Newton-Krylov method.

This algorithmic concept is referred to as physics-based preconditioning. Here, the Jacobian matrix is not directly manipulated algebraically to find an approximate inverse; rather, approximations are made to the original differential system that result in a modified, more tractable, approximate Jacobian system for preconditioning.

Consider using the first-order, explicit coupling, operator split method for our model problem as a preconditioner. The Newton step, $\mathbf{J}\delta\mathbf{u} = -\mathbf{F}(\mathbf{u})$, of Eqs. (3) and (4) can be expressed as:

$$\begin{bmatrix} J_{\phi,T} & J_{\phi,\phi} \\ J_{T,T} & J_{T,\phi} \end{bmatrix} \begin{bmatrix} \delta\phi \\ \delta T \end{bmatrix} = - \begin{bmatrix} F_{\phi} \\ F_{T} \end{bmatrix},$$

with

$$\delta \mathbf{u} = [\delta\phi, \delta T]^T, \ \ \mathbf{F}(\mathbf{u}) = [F_\phi, F_T]^T.$$

Typically in problems such as this, simplification is made to the operators $J_{T,T}$ and $J_{\phi,\phi}$ within the preconditioner. As an example, one can linearize the temperature dependent coefficients, $K(T), D(T)$, as $K(T^n), D(T^n)$, simplifying the evaluation of $J_{T,T}$ and $J_{\phi,\phi}$. After enough simplification it may be more appropriate to call the preconditioning operators $M_{T,T}$ and $M_{\phi,\phi}$. A few v-cycles of a multigrid method are used to smooth the two separate elliptic systems in the preconditioner. This process is equivalent to using code coupling (ie the time discretization of Eqs. (5) and (6) as the preconditioner). Thus only simple matrices are formed for preconditioning purposes. One could add some source coupling while retaining a block lower triangular system such that:

$$\mathbf{M} = \begin{bmatrix} M_{\phi,\phi} & 0 \\ M_{T,\phi} & M_{T,T} \end{bmatrix}.$$

Then a preconditioning process $\delta\mathbf{u} \approx \mathbf{M}^{-1}(-\mathbf{F}(\mathbf{u}))$ is done in two steps

(i) Approximate inversion of a parabolic system of order N (number of grid points) to solve for $\delta\phi$.

$$M_{\phi,\phi}\delta\phi = -F_\phi \tag{15}$$

(ii) Approximate inversion of a parabolic system of order N (number of grid points) to solve for δT.

$$M_{T,T}\delta T = -F_T - M_{T,\phi}\delta\phi \tag{16}$$

Other forms of physics-based preconditioning are discussed in [3].

2.4. Advanced uses of JFNK

In addition to the standard multiphysics solver application of JFNK we have started using the method to improve the convergence rates of other computational nuclear engineering problems [10]. We have successfully used the JFNK method to accelerate the convergence of the classic power iteration used to evaluate the eigenvalue in criticality searches. Here we use the standard power iteration as a preconditioner. The constraint equation for the fundamental mode is added as an additional nonlinear function, and the eigenvalue is added as an unknown to the system.

We have also started using JFNK as a multilevel solver. In reactor physics, a low-order diffusion-like solver on a coarse mesh is often used to accelerate the convergence of a higher-order transport method on a fine mesh. Both models can produce a version of the scalar flux. These two models are typically coupled via a fixed point iteration. We have employed JFNK where the lower order diffusion-like problem is used as the nonlinear function on the coarse mesh. Then, inside this function evaluation we execute a self-consistent transport sweep on a fine mesh. In this context the JFNK method has become a nonlinearly consistent scale-bridging algorithm. We are also using this algorithmic concept within `BISON`to do multiscale thermal transport in nuclear fuel.

3. Software and results

3.1. `MOOSE`

Motivated by the existence of strong multiphysics coupling in many computational nuclear engineering processes, we are developing the *Multiphysics Object Oriented Simulation Environment*, or `MOOSE` [4]. Algorithmically `MOOSE` is based on solving all systems in a fully coupled manner using JFNK and physics-based preconditioning. `MOOSE` is designed to support the development of *engineering analysis applications*. Unlike a "research code," the architecture and development methodology of `MOOSE` must focus on more than sheer performance. `MOOSE` was developed with the following in mind:

- Massively parallel performance and scalability is needed for large problems, but economics mandate that `MOOSE` applications also perform well on desktop computers and modestly parallel machines.
- `MOOSE` must be developed using modern software engineering principles, as its value as a framework is directly related to the cost of maintaining and extending applications based on it.
- `MOOSE` is a component of an engineering analysis system that includes domain geometry construction, mesh generation, mesh quality and adaptation, and visualization and analysis of the final results. As such, the usability of `MOOSE` applications depends on the efficacy of the entire simulation system; including the robustness of system-wide uncertainty quantification and the ability to assess the propagation of uncertainty throughout.

The design of `MOOSE` capitalizes on the fact that the (unpreconditioned) JFNK implementation requires only residual evaluations of the discrete system. Strategic use of this results in a modular, pluggable architecture that greatly simplifies adding and coupling new physics together. Further, nonlinear material properties and boundary conditions are handled consistently.

`MOOSE` is based on a layered approach to providing the core services in support of multiphysics simulation. Underneath `MOOSE` is the `libMesh` finite-element framework developed by the CFDLab at the University of Texas in Austin [11]. `libMesh` provides a set of utilities for massively parallel finite-element based computations, including mesh I/O, a finite-element library, and interfaces to solver packages. Utilizing the interfaces in `libMesh` provides `MOOSE`, and subsequently the applications built using `MOOSE`, a considerable amount of flexibility including the ability to swap out solver libraries such as PETSc [12] and Trilinos [13], and to utilize diverse large scale parallel computing resources.

3.2. BISON

`BISON` [6], the fuel performance code developed using `MOOSE`, is designed for fully coupled steady and transient analysis and to be efficient on both desktop computers and in massively parallel environments. This discussion summarizes the current status of the `BISON` fuel performance analysis code and demonstrates results on selected three dimensional capabilities that support fuel performance calculations.

While the equation sets in BISON could be applied to a variety of fuel types, here we will focus on the application to traditional light water reactor fuel. The problem consists of three fully coupled partial differential equations for heat conduction, oxygen nonstoichiometry and linear elastic solid mechanics. Let $\Omega \in \mathbb{R}^d$, $d = 2, 3$ define the computational domain (the fuel pellet).

The heat conduction model assumes fission reactions generate heat at a uniformly distributed constant rate, Q,

$$\rho C_p \frac{\partial T}{\partial t} + \nabla \cdot \mathbf{q} - Q = 0, \tag{17}$$

where T, ρ and C_p are temperature, density and specific heat of UO_2, respectively. The heat flux $\mathbf{q}$ within the UO_2 domain Ω may be written as

$$\mathbf{q} = -k\nabla T, \tag{18}$$

where k denotes thermal conductivity, A hyperstoichiometric model for oxygen is included as described in [6]. Let $\mathbf{J}$ denote the oxygen flux in the hyperstoichiometric regime with,

$$\mathbf{J} = -D\left(\nabla x - \frac{xQ^*}{FRT^2}\nabla T\right), \tag{19}$$

where D is diffusivity of UO_2, F is the thermodynamic factor of oxygen, Q^* is the heat of transport of oxygen, and R is the universal gas constant. The nonstoichiometric model for oxygen diffusion, x, is given by

$$\frac{\partial x}{\partial t} + \nabla\cdot\mathbf{J} = 0. \tag{20}$$

A linear elastic model for the displacement $\mathbf{u}$, using the closure relationships discussed in [6] is the current mechanics model, in this study as

$$\frac{\partial^2\mathbf{u}}{\partial t^2} - \mathrm{A}^{\mathrm{T}}\mathrm{D}\mathrm{A}\mathbf{u} + f = 0 \tag{21}$$

with

$$\mathrm{A} = \begin{bmatrix} \partial_x & 0 & 0 \\ 0 & \partial_y & 0 \\ 0 & 0 & \partial_z \\ \partial_y & \partial_x & 0 \\ 0 & \partial_z & \partial_y \\ \partial_z & 0 & \partial_x \end{bmatrix}, \quad \mathrm{D} = c_1\begin{bmatrix} 1 & c_2 & c_2 & 0 & 0 & 0 \\ c_2 & 1 & c_2 & 0 & 0 & 0 \\ c_2 & c_2 & 1 & 0 & 0 & 0 \\ 0 & 0 & 0 & c_3 & 0 & 0 \\ 0 & 0 & 0 & 0 & c_3 & 0 \\ 0 & 0 & 0 & 0 & 0 & c_3 \end{bmatrix},$$

and

$$c_1 = \frac{E(1-\nu)}{(1+\nu)(1-\nu)}, \quad c_2 = \frac{\nu}{(1-\nu)}, \quad c_3 = \frac{(1-2\nu)}{2(1-\nu)}.$$

The coefficients E and ν are Young's modulus and Poisson's ratio. The forcing term f weakly enforces thermal expansion.

Figure 1 displays two items. The first is a convergence plot for GMRES iteration on a specific Newton iteration for a time dependent problem of thermal expansion of a fuel pellet. One can see that the simplest of physics-based preconditioning options can provide significant acceleration of GMRES in `BISON`. The second item is a graphic of a pre-cracked fuel pellet which has undergone thermal expansion.

Figure 2 displays parallel scaling plot from BISON on a multiscale problem [14]. Here, thermal conduction is being solved on the engineering scale and a mesoscale simulation is being performed inside the nonlinear function evaluation to self-consistently evaluate a lower-length scale informed thermal conductivity.

3.3. PRONGHORN

Very-high Temperature Gas-cooled Reactors (VHTR) are intrinsically different from conventional Light Water Reactors (LWR). For example, a large temperature difference between inlet and outlet creates a significant density variation in the coolant. In this case, an incompressible assumption with Boussinesq approximations for buoyancy may not give accurate heat transfer. The strong temperature gradients introduce strong nonlinear couplings between heat-conduction and neutronics calculations. The traditional approach to modeling these complex multiphysics phenomena via operator-splitting or code coupling may introduce significant numerical and/or modeling errors.

`PRONGHORN` [5] is being developed as a tightly coupled multiphysics 3-D code which enables simulations of complex transient phenomena that are encountered in VHTRs for both

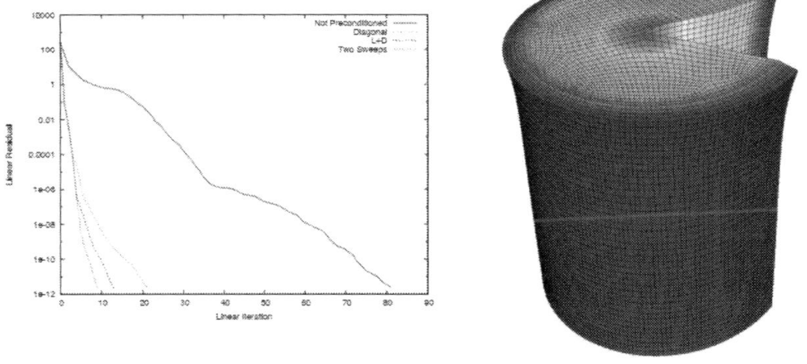

Figure 1. Results of physics-based preconditioning in `BISON`and a cracked fuel pellet undergoing thermal expansion.

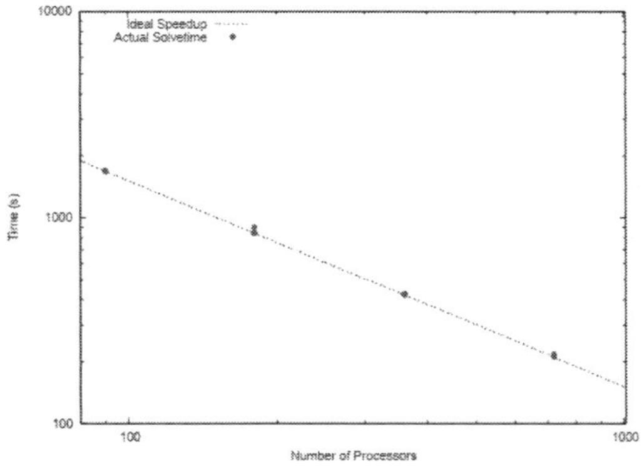

Figure 2. Parallel scaling study with `BISON`.

operational and accident scenarios. VHTRs are expected to have long transients ($\sim$ days). To model this, eliminating time step restrictions due to explicit discretization and operator-split errors by employing fully implicit discretizations is the only viable route. `PRONGHORN` is a two and three dimensional (with two dimensional cylindrical coordinate) pebble bed reactor simulation application. Current capabilities include heat conduction and transport using methods developed for porous media applications. The reactor fluid medium model is described by

$$\frac{\partial \epsilon \rho_f}{\partial t} + \nabla \cdot \epsilon \rho_f \vec{u} = 0, \tag{22}$$

$$\frac{\partial \epsilon \rho_f \vec{u}}{\partial t} + \epsilon \nabla P - \epsilon \rho_f \vec{g} + W \rho_f \vec{u} = 0, \tag{23}$$

$$\frac{\partial}{\partial t}\left[\epsilon \rho_f c_{pf} T_f\right] + \nabla \cdot (\epsilon \rho_f c_{pf} \vec{u} T_f) - \nabla \cdot \epsilon \kappa_f \nabla T_f + \alpha (T_f - T_s) = 0, \tag{24}$$

where ϵ is porosity, ρ is density, $\rho_f \vec{u}$ is momentum, P is pressure, W is fluid resistivity coefficient, c_p is heat capacity, T is temperature, κ is thermal conductivity, α is the heat transfer coefficient and the subscripts f and s denote fluid and solid properties. The reactor solid medium is governed by

$$\frac{\partial}{\partial t}\left[(1-\epsilon)\rho_s c_{ps} T_s\right] - \nabla \cdot \kappa_{s\,\mathrm{eff}} \nabla T_s + \alpha(T_s - T_f) - Q = 0, \tag{25}$$

where $\kappa_{s\,\mathrm{eff}}$ is the effective thermal conductivity of the solid and Q is the heat source. The reactor neutronics is given by

$$\frac{1}{v_g}\frac{\partial \phi_g}{\partial t} - \nabla \cdot D_g \nabla \phi_g + \Sigma_{Rg}\phi_g - (1-\beta)\chi_g \sum_{g'=1} \nu\Sigma_{fg'}\phi_{g'} - \sum_{g'=1, g'\neq g} \Sigma_s^{g'\to g}\phi_{g'} - \sum_k \lambda_k C_k = 0, \tag{26}$$

$$\frac{dC_k}{dt} + \lambda_k C_k - \beta_k \sum_{g'} \nu\Sigma_{fg'}\phi_{g'} = 0, \tag{27}$$

where v is neutron speed, ϕ is neutron flux, D is diffusion coefficient, Σ_R is removal cross–section, β is delayed neutron fraction, χ is fission spectrum, ν is average number of neutrons emitted per fission reaction, Σ_f is the fission cross–sections, Σ_s is the scattering cross–sections, λ is the decay constant, C is precursor concentrations. The subscripts g and k represent energy group indices for prompt and delayed neutrons, respectively.

Equations (22)–(27) represent the continuity, momentum, fluid and solid energy, and neutron diffusion equations, respectively. The coefficients (e.g. thermal conductivities, heat transfer coefficients, and neutron cross sections) are typically strong functions of the independent variables, which introduces significant nonlinearity. A closure relation is obtained by the equation of state,

$$P = \rho_f R T_f. \tag{28}$$

Figure 3 displays two items. The first is a materials geometry for a 2-D benchmark PBMR 400 test case [5]. The second is a steady state multiphysics results which shows the variation of the eigenvalue, k_{eff} with prescribed steady state power. As the desired steady-state power level is varied, the reactor temperature profile will vary. Since the neutronics cross-section are a function of temperature, the eigenvalue will change. The plot was made using a standard accelerated power iteration but in a multiphysics setting.

Figure 4 displays two items from a multiphysics transient. A power profile and a zoom of this profile using a variety of time integration approaches. The power transient is activated by removing a control rod. We can see that the second-order in time tightly coupled solver produces the most accurate answer.

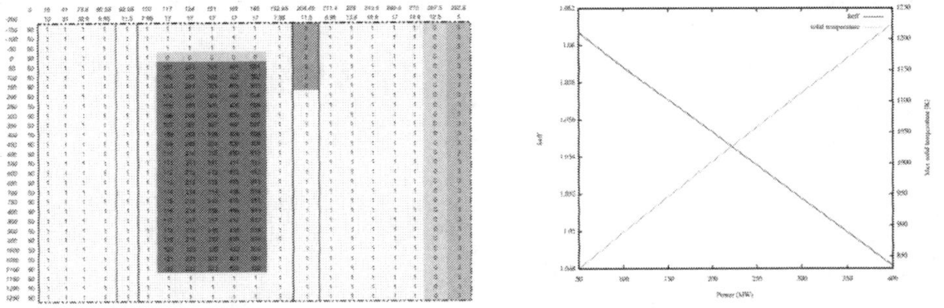

Figure 3. Steady-State Multiphysics PRONGHORN Result

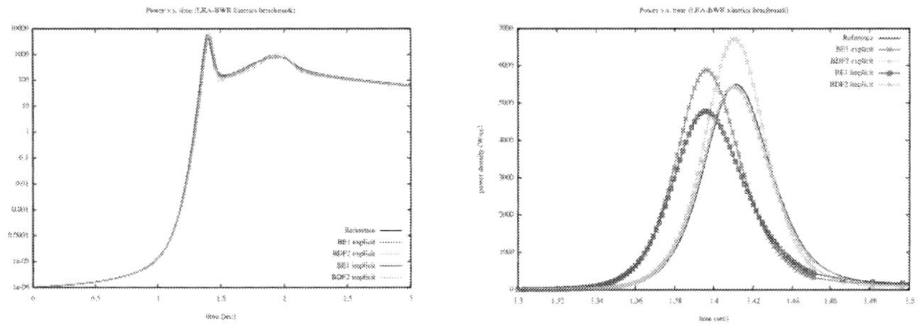

Figure 4. Transient Multiphysics PRONGHORN Result

4. Conclusions

We have discussed the algorithmic issues related to accurate multiphysics time integration on problems in computational nuclear engineering. We have described the modern algorithmic approach of JFNK and physics-based preconditioning. We have presented the results of our software framework development activities that have occurred over the last year and we have presented results from two recently developed engineering application codes.

5. References

[1] D. A. Knoll, L. Chacón, L. G. Margolin, and V. A. Mousseau. On balanced approximations for the time integration of multiple time scale systems. *J. Comput. Phys.*, 185:583–611, 2003.

[2] D.L. Ropp and J. N. Shadid. Stability of operator splitting methods for systems with indefinite operators: reaction-diffusion systems. *J. Comput. Phys.*, 203:449–466, 2005.

[3] D. A. Knoll and D. E. Keyes. Jacobian-free Newton-Krylov methods: A survey of approaches and applications. *J. Comput. Phys.*, 193:357–397, 2003.

[4] D. Gaston, C. Newman, G. Hansen, and D. Lebrun-Grandié. MOOSE: A parallel computational framework for coupled systems of nonlinear equations. *Nuclear Engineering and Design*, 2009. in press.

[5] H. Park, D. Gaston, S. Kadiouglu, D. Knoll, D. Lebrun-Grandie, R. Martineau, and W. Taitano. Tightly coupled multiphysics simulation for pebble bed reactors. In *American Nuclear Society 2009 International Conference on Advances in Mathematics, Computational Methods, and Reactor Physics*, Saratoga Springs, NY, May 3–7 2009.

[6] C. Newman, G. Hansen, and D. Gaston. Three dimensional coupled simulation of thermomechanics, heat, and oxygen diffusion in UO_2 nuclear fuel rods. *Journal of Nuclear Materials*, 2009. In press, http://dx.doi.org/10.1016/j.jnucmat.2009.03.035.

[7] Y. Saad. *Iterative Methods for Sparse Linear Systems.* PWS Publishing Company, Boston, 1996.

[8] P. N. Brown and Y. Saad. Hybrid Krylov methods for nonlinear systems of equations. *SIAM J. Sci. Stat. Comput.*, 11:450–481, 1990.

[9] Tony F. Chan and Kenneth R. Jackson. Nonlinearly preconditioned Krylov subspace methods for discrete Newton algorithms. *SIAM J. Sci. Stat. Comput.*, 5:533–542, 1984.

[10] D.A. Knoll, H. Park, and K. Smith. Application of the jacobian-free newton-krylov method in computational reactor physics. In *American Nuclear Society 2009 International Conference on Advances in Mathematics, Computational Methods, and Reactor Physics*, Saratoga Springs, NY, May 3–7 2009.

[11] Benjamin S. Kirk, John W. Peterson, Roy H. Stogner, and Graham F. Carey. libMesh: a C++ library for parallel adaptive mesh refinement/coarsening simulations. *Eng Comput-Germany.*

[12] Satish Balay, Kris Buschelman, Victor Eijkhout, William D. Gropp, Dinesh Kaushik, Matthew G. Knepley, Lois Curfman McInnes, Barry F. Smith, and Hong Zhang. PETSc users manual. Technical Report ANL-95/11 - Revision 2.1.5, Argonne National Laboratory, 2004.

[13] M. Heroux et al. Trilinos: an object-oriented software framework for the solution of large-scale, complex multi-physics engineering and scientific problems. http://trilinos.sandia.gov, 2008.

[14] M.R. Tonks, G. Hansen, D. Gaston, C. Permann, P. Millett, and D. Wolf. Fully coupled engineering and mesoscale simulations of thermal conductivity in uo_2 fuel using an implicit multiscale approach. *Journal of Physics: Conference Series*, 2009, this conference.

Journal of Physics: Conference Series **180** (2009) 012013 doi:10.1088/1742-6596/180/1/012013

Computational Science: Emerging Opportunities and Challenges

Bruce Hendrickson

Senior Manager for Computer Science & Mathematics, Sandia National Laboratories[1]
Affiliated Faculty in Computer Science, University of New Mexico

E-mail: bah@sandia.gov

Abstract. In the past two decades, computational methods have emerged as an essential component of the scientific and engineering enterprise. A diverse assortment of scientific applications has been simulated and explored via advanced computational techniques. Computer vendors have built enormous parallel machines to support these activities, and the research community has developed new algorithms and codes, and agreed on standards to facilitate ever more ambitious computations. However, this track record of success will be increasingly hard to sustain in coming years. Power limitations constrain processor clock speeds, so further performance improvements will need to come from ever more parallelism. This higher degree of parallelism will require new thinking about algorithms, programming models, and architectural resilience. Simultaneously, cutting edge science increasingly requires more complex simulations with unstructured and adaptive grids, and multi-scale and multi-physics phenomena. These new codes will push existing parallelization strategies to their limits and beyond. Emerging data-rich scientific applications are also in need of high performance computing, but their complex spatial and temporal data access patterns do not perform well on existing machines. These interacting forces will reshape high performance computing in the coming years.

1. Introduction

Computational modeling and simulation are widely acknowledged as an essential component of many areas of science and engineering. This acceptance is built upon an intellectual foundation emerging from decades of research in applied mathematics and computer science. Computational modeling is essential in settings where experiments are impossible (e.g. astrophysics and climate) or highly dangerous (e.g. weapons safety or nuclear waste storage). More broadly, computing provides the ability to reduce design and experimental costs and to provide depth of insight that is not attainable in other ways. Gradually, the old paradigm of "design, test, analyze" is being replaced by a new paradigm of "analyze, design, test".

Although computation has been used for decades, the recent explosion of interest in computational simulation is closely tied to advances in computer hardware and software. The power of parallel computers has enabled simulations that have sufficient fidelity and sophistication to provide deep insight into scientific and engineering questions. The parallel computing landscape has been quite stable for the past 15 years, with distributed memory

[1] Sandia is a multipurpose laboratory operated by Sandia Corporation, a Lockheed-Martin Company, for the United States Department of Energy under contract DE-AC04-94AL85000.

clusters and MPI providing a consistent model for software and algorithm development. This stability has been essential to the growth and success of computational science.

However, this paper will argue that the stability of the recent past is coming to an end, and that the computational science world will become more complicated. This looming increase in complexity is driven by several external forces. Most familiarly, the nodes of our future parallel machines will have many processing units, and this will force changes in our algorithms and programming models. But as discussed in Section 3, multicore nodes are really just a symptom of a larger and more profound trend. Due to the relentless progress of Moore's law, space on a chip is now abundant, which will allow for innovative experimentation with new architectural ideas. The likely flurry of new architectures will provide both new opportunities and new challenges to the computational science community.

Simultaneous with this architectural upheaval, the demands of scientific computations are changing rapidly as discussed in Section 2. Traditional applications are becoming more complex with unstructured and adaptive data structures, and multiple scales and types of phenomena to model. These evolutionary changes reflect the growing sophistication of the community, but they will tax existing parallel computing methodologies. Less appreciated is the growing set of non-traditional applications, particularly in the social sciences. These emerging applications have very different computational characteristics than traditional applications, and will likely require profound revisions to many of our cherished assumptions about high performance computing.

This paper will review this changing landscape and discuss some of the likely consequences for computational science and high performance computing.

2. Trends in Computational Science Applications

2.1. Evolving Applications

Many areas of science and engineering have been dramatically altered over the past two decades by the growing use of computational methods. Computer models allow for the study of phenomena that are impossible or dangerous to study experimentally, and enable detailed and controlled investigation of complex phenomena. The impact of computational science is too vast to detail here, but has been well documented by others (e.g. [1]).

The growth of computational science has been built on several inter-related developments. First among these is the increasing performance and price-performance of high end computers. Second is the adoption of standard programming models and languages that have facilitated large investments in software. Third is the invention of a range of sophisticated parallel algorithms to solve key problems in computational simulation.

All of these trends are visible in the evolution of the winners of the annual Gordon Bell Prize for outstanding achievement in high-performance computing applications. But these prize winners also tell a different, less encouraging tale. Gordon Bell Prize winners invariably have kernels with very high spatial and temporal locality, which can make excellent use of traditional memory hierarchies. Although there are important exceptions, many prize winners have had either dense matrix operations or particle simulations at their heart, and these kernels are well known to have excellent locality properties. But only a subset of science can be performed with these or similarly high-locality computational kernels. Many important computations with different kernels perform much less well.

State–of–the–art computational science and engineering codes increasingly rely upon unstructured or adaptive meshes. The former require more frequent and more complex memory access patterns, while the latter generate difficult dynamic load balancing challenges. Key scientific questions increasingly span multiple physical scales and involve the interactions between multiple physical phenomena. The computational methods required to answer such questions are still being developed, but they are much more complex than many of the techniques that have been parallelized successfully in the past. Although heroic calculations have been done on tens of thousands of processors, much of the most important science involves complex phenomena that cannot currently be scaled to large processor counts.

There are many challenges associated with improving the performance and scalability of complex simulations. Most of these applications are limited by the time required to access memory, not by the performance of the processor. Unfortunately, multicore processors will likely make this problem worse by adding computational power but not increasing the off-chip memory bandwidth commensurately. New algorithms are needed to improve memory performance, but this may not be sufficient. As discussed below, new processor designs and parallel programming models may also need to be considered.

Almost all successful MPI programs are written in a *bulk synchronous* style in which the computation proceeds in phases. Within each phase, the processors first participate in a communication operation. For the remainder of the phase, each processor can perform computation without any communication. Virtually all of the highly scalable scientific kernels can be formulated this way including matrix-vector multiplication, explicit time integration and particle simulation. Important kernels that don't fit this model, like sparse factorizations, typically don't scale beyond a few hundred processors.

For bulk synchronous computations the challenge is to balance the computational load so that each processor stays busy for the duration of the computational part of each phase. Sophisticated techniques and tools have been developed to solve this problem, often involving graph or hypergraph partitioning. But these work best when the computational requirements remain steady for many phases. Adaptive grids and multi-physics methods require expensive repartitioning to retain balance. Operations that cannot be fit into a bulk synchronous framework suffer even more. For example, sparse direct methods have frequent partial synchronizations, and processors are often idle while waiting for others to catch up. An overall balance of the workload is not sufficient. The work must be balanced at every synchronization point. New algorithms may improve the situation, but more radical changes to processors and programming models may also be needed.

2.2. Emerging Applications

The phrase "computational science" is usually associated with the modeling and simulation of physical phenomena. Most physics and engineering simulations involve two or three spatial and one temporal dimensions. Much of the phenomena is short range, and longer range forces can be handled via hierarchical methods. These properties greatly constrain the amount and type of interaction that occurs between different parts of the computation. All of these factors contribute to the ability to perform successful parallel simulations.

But science is much broader than just the modeling of physical phenomena, and our definition of "computational science" should reflect this. Webster's defines science as "a branch of knowledge or study dealing with a body of facts or truths systematically arranged and showing the operation of general laws." By this definition, science is much broader than the disciplines that have received the bulk of computational attention. In particular, many of the social sciences are only beginning to discover the power of computation. However, there are early indications that computing will reshape these fields as much as it has transformed the physical sciences.

As one salient example, consider the study of social communities. Historically, this involved labor-intensive interviews and a good bit of luck. One iconical instance is Zachary's 1977 study of the social dynamics in a karate club [2]. Through surveys, interviews and direct observation, Zachary painstakingly constructed the network of friendships among the 34 club members. When, serendipitously, the club split in two, the friendship network provided a novel window into social dynamics.

These days, a sociologist can instead tap into rich and voluminous data from internet-enabled social networks. Text messages, twitter feeds, blogs and chat rooms leave electronic artifacts that can be mined for sociological insight. Compared to survey or observational data, internet data comes from more spontaneous activities, and has rich semantic and temporal structure. It can also be huge and of variable quality. In a recent study, Leskovec and Horvitz analyzed a month of instant-messaging traffic which involved 240 million people and 30 billion interactions [3]. The questions that a sociologist can ask and the methodologies required to answer them are profoundly different than they were in Zachary's time only 30 years ago. The data can only be studied through the use of advanced computational methods. But these techniques are very much in their infancy

This emerging style of computational science places very different demands on machines and programming models than traditional modeling and simulation. Social networks have much lower spatial locality than the interactions in a finite element model or a particle simulation, and they do not naturally live in a low dimensional space. Complex networks cannot be partitioned nicely across the processors of a parallel cluster. Many network analysis algorithms involve complex traversals of the network structure with extremely low spatial and temporal locality. These computations cannot make good use of caches and other features of a memory hierarchy, and so a processor spends the vast bulk of its time waiting for data. Furthermore, key algorithms cannot easily be expressed in a bulk synchronous way. Rather, parallelism is fine-grained and highly dynamic. Thus social network analysis violates many of the properties that have enabled the effective parallelization of physical simulations.

The demands of these types of emerging applications will require the high performance computing community to revisit all aspects of our existing infrastructure from processor design, to parallel architectures, programming models and languages, and algorithms.

3. Trends in Computer Architectures

3.1. Commodity Clusters and MPI

The late eighties and early nineties witnessed considerable experimentation in parallel architectures. Vendors like Thinking Machines, Kendall Square, nCUBE and others produced machines that reflected different visions for harnessing the power of parallelism. This era of exploration came to an end with the rise of the "killer micros", and the dominance of distributed memory clusters made principally from commodity components. These machines have compelling cost advantages, and they have dominated the marketplace for the past 15 years, providing a valuable degree of stability in parallel programming models.

Also in the mid-nineties, MPI replaced a host of proprietary programming models and became the ubiquitous method for writing parallel applications. Although it has its detractors, MPI is well suited for expressing coarse grained, bulk synchronous parallelism. By explicitly managing message traffic, the programmer is able to maximize locality and thereby gain high performance.

However, MPI and commodity clusters are less well suited for problems with fine grained or dynamic parallelism, or applications with complex, non-local memory access patterns. A number of alternative programming techniques have been proposed including partitioned global address spaces [4], active messages [5], and Cilk [6]. Each of these has some advantages for fine-grained computations with complex access patterns, and some of their ideas can be found in recent parallel language efforts like Chapel [7], X10 [8], and Fortress [9]. But none of these approaches has become mainstream. MPI's portability minimizes risk to developers, and it allows many applications to get high performance on existing machines.

Thus, MPI and clusters each reinforce the other. With existing applications written in MPI, there is minimal incentive for vendors to support advanced capabilities that would require more expensive hardware. Meanwhile, application developers are loathe to explore programming approaches that reduce portability and probably don't improve performance on existing machines.

As discussed above, there are applications that are poorly served by MPI on clusters, but these applications do not currently have much impact on the high performance computing marketplace.

3.2. Multicore and Beyond

With feature size shrinking inexorably, space on an integrated circuit is no longer scarce. For the first time we are in an era of plentiful transistors, and the ramifications of this change are profound and not yet well understood. One immediate consequence is the opportunity to fit multiple processing cores on a chip. These multi-core (and soon to be many-core) components are the commodity units out of which most future parallel machines will be built. An important driver behind multicore is the need to limit clock speeds due to power and heat considerations. Much of the historical improvement in processor performance came from faster clock rates, but this can no longer be sustained since a faster clock consumes more power and generates more heat. Instead, future improvements in processor performance will increasingly have to come from the exploitation of parallelism.

This need for on-chip parallelism has many positive aspects for the high performance computing community. Unlike many other software communities, we have considerable experience with parallelism, and we have a large investment in parallel software. Furthermore, the need for parallelism for the masses will lead to new tools and techniques for expressing and exploiting parallelism. We can, perhaps, contribute to these, and we can certainly benefit from them in our work on multi-core parallelism.

But the move to many-core nodes will also provide many challenges for our community. Parallelism within a node will be different from parallelism between nodes. Communication between cores on a chip will likely have higher bandwidth and lower latency than off-chip communication. On-chip parallelism might be best expressed with a shared memory programming model like OpenMP, while inter-chip parallelism will still be expressed with MPI. Thus, to obtain peak performance, developers may need to explicitly program for two (or more) levels of parallelism. The use of multi-core nodes is a very active area of research and it is not yet clear what programming models are best [10]. Meanwhile, MPI is being enhanced to be more suited to multi-core nodes [11]. At the very least, it is safe to say that the multi-core phenomena has energized a community that has enjoyed 15 years of stability in its programming models.

As argued above, multi-core nodes are a natural response to the growth in available space on a chip. But frankly, this is not a particularly innovative response to such a major change, and we are likely to see more interesting alternatives. Architects may choose to provide specialized hardware to compute common functions, particularly if this will improve performance on important benchmarks. Another possibility is heterogeneous nodes in which different processing units are specialized for different tasks. An important example of this kind of architecture is the Cell processor [12] which was designed for gaming and graphics applications but which has a niche in high performance computing and provides the horsepower for today's fastest computer [13]. General purpose graphics processors are also showing excellent potential for delivering high performance as nodes of distributed memory machines. The challenge of programming heterogeneous nodes is a big impediment, but the opportunity for greatly enhanced performance has appeal. But these approaches merely provide faster compute nodes, and do not address the challenges of inter-node communication and load balancing.

As discussed above, a growing fraction of scientific applications are limited by memory performance, not processing power. The traditional method for addressing this problem is the use of a memory hierarchy in which fast (but expensive and therefore small) memory holds data that is likely to be needed soon. Commodity processors have multiple layers of cache to reduce latency and fancy techniques for performing out-of-order computations to keep a processor busy whenever possible. This is largely invisible to the programmer, but locality enhancing algorithms are central to achieving high performance. In essence, distributed memory machines merely add an additional layer of memory hierarchy and MPI is a user-managed mechanism for prefetching. But as discussed in Section 2, the expanding use of unstructured and adaptive data structures reduces exploitable locality. And emerging applications like graph analysis often have very complex global memory access patterns with little opportunity to make use of traditional memory hierarchies [14]. Radically different architectures to address the problem of the memory wall are needed to provide high performance for applications that run poorly on current machines.

As an interesting example of the potential of a radically different architecture, consider the Cray XMT [15], a follow-on to the Tera MTA [16]. The MTA/XMT has a non-commodity processor which has no support for a memory hierarchy. Instead, it has hardware support for 128 very light-weight threads. The processor executes an instruction from one thread, and in the next clock cycle it executes an instruction from a different thread. If the first instruction involved reading or writing memory, the memory operation gets started while the processor executes instructions from other threads. By the time the processor gets back to this thread, the memory operation will have been completed. In this way, the processor can be kept busy doing useful work even for very complex data access patterns. The MTA/XMT is able to *tolerate* latency via the use of massive multithreading.

The potential for the MTA/XMT to address the challenges of evolving applications was hinted at in a study by Oliker and Biswas [17]. They implemented an adaptive mesh code on three different machines – a traditional parallel cluster, a cache–coherent shared memory machine, and an early MTA. They observed that the MTA code was significantly shorter and simpler than code for the other machines, and that the MTA exhibited much better scalability on small numbers of processors. Performance significantly decreased on the other machines when going from one to two processors due to the introduction of expensive techniques for dynamic load balancing. Due to this cost, parallelism only improved performance for large computations on many processors. The MTA did not suffer from this problem and exhibited excellent scalability even for small numbers of processors. However, only a small MTA was available at the time, so the authors could not assess the scalability of the MTA approach for large numbers of processors.

Emerging applications which are dominated by memory access costs are excellent candidates for the MTA/XMT or other innovative approaches to improve memory performance. Lumsdaine, et al. studied the performance of the MTA-2 and a cluster on several graph algorithms [14]. Their most complex example was Δ-stepping, an efficient algorithm for computing single-source shortest paths for graphs with non-negative edge weights [18]. Lumsdaine, et al. made two observations. First, the performance of the MTA-2 implementation was insensitive to problem instance, but the distributed memory code was unable to solve hard problems. Specifically, when the graph had vertices with very many neighbors, the use of ghost nodes by the distributed memory code led to memory requirements that exceeded the memory available on a node. Second, on easy problem instances (Erdös–Renyi random graphs), both implementations scaled well, but the distributed memory code required about 10 times as many processors to achieve the same performance as the MTA-2 code. This was despite the fact that the MTA-2 clock was 220 MHz, while the cluster used 2.0 GHz Opterons. Thus, processor utilization on the MTA-2 was 100 times greater than on the cluster.

The MTA/XMT also provides a useful case study about alternative programming models. The machine does not run MPI, and instead uses a proprietary, pragma-based, shared-memory language that has similarities to OpenMP. A sophisticated compiler and runtime environment identify and schedule a large number of virtual threads on the physical hardware. Parallelism can be finer-grained and more dynamic. The programmer does not have to explicitly manage interprocessor communication, manual handle load balancing, or tether the degree of parallelism to the number of processing units. Thus, the programming model affords a very different way of expressing and exploiting parallelism than MPI.

Although the MTA/XMT approach is attractive for very unstructured applications, it has important limitations. Since the processor is low-volume, Cray cannot afford to optimize its design and layout, so the clock speed on the XMT is a slow 400 MHz, nearly an order of magnitude slower than commodity processors. This slow clock combined with a reliance on multiple threads means that the processor runs serial code extremely slowly, which can be a bottleneck for some applications. Also, the processor has very limited ability to exploit any locality that may be present in a computation. So the utility of the MTA/XMT as a general purpose supercomputer is uncertain. Rather, it complements the existing landscape by supporting applications for which mainstream architectures perform poorly. For the purposes of this paper, it provides an interesting example of the potential impact of radical architectural innovation.

Future heterogeneous architectures could perhaps exploit locality when it is available to reduce latency, and use massive multithreading to tolerate latency for code segments with parallelism but poor locality. At least as likely, significantly new ideas are waiting to be discovered and implemented by architects working under quite different design constraints than those that have driven past designs.

4. Conclusions

After 15 years of comparative stasis, the high performance computing world is on the cusp of major changes. Computer architects are facing both new freedoms and new constraints, which will demand and enable innovative new ideas. These will largely be driven by the mass market, but since scientific computing relies on commodity components, our world will be greatly impacted. Programming models will need to evolve to respond to new architectures. At a minimum, MPI will be expanded to better support many-core nodes. But significantly different architectures may require more radical changes to parallel tools and models.

In the midst of this architectural change, existing applications are becoming increasingly memory-limited, and struggling with major load balancing challenges. MPI may be able to help with improved mechanisms for light-weight and asynchronous communication, but the challenges are difficult. Even more daunting, newly emerging applications like social network analysis are profoundly different from those that have defined scientific computing in the past. To successfully address these data-centric applications, the high performance computing community will have to rethink many of our cherished presumptions about node architectures, programming models and overall system design.

In aggregate, these forces will reshape high performance computing in coming years. Perhaps the current hegemony of clusters will yield to a more diverse world with different machines targeting different classes of applications. Or maybe a new model will emerge which supports a broader range of needs, perhaps involving heterogeneous architectures.

However the future unfolds, by responding to these external forces, scientific computing will be greatly enriched. The challenge of new problems and the infusion of ideas from entirely different communities will bring fresh intellectual vigor to scientific computing and impel it forwards towards even greater successes.

References

[1] Keyes D, Colella P, Dunning T and Gropp W 2003 A science-based case for large-scale simulation, Volume 1 Office of Science, U.S. Department of Energy
[2] Zachary R W 1977 *J. Anthropological Research* **33** 452–473
[3] Leskovec J and Horvitz E 2008 Planetary-scale views on a large instant-messaging network *Proc. World Wide Web Conference* (Beijing, China)
[4] El-Ghazawi T, Carlson W, Sterling T and Yelick K 2005 *UPC: Distributed Shared-Memory Programming* (Wiley-Interscience)

[5] von Eicken T, Culler D, Goldstein S and Schauser K 1992 Active messages: A mechanism for integrated communication and computation *Proc. 19th International Symp. Computer Arch.* (Queensland, Australia)
[6] Blumofe R, Joerg C, Kuszmaul B, Leiserson C, Randall K and Zhou Y 1995 *ACM SIGPLAN Notices* **30** 207–216
[7] Chamberlain B, Callahan D and Zuma H P 2007 *International J. High Performance Computing Applications* **21** 291–312
[8] Ebcioglu K, Saraswat V and Sarkar V 2005 X10: An experimental language for high productivity programming of scalable systems *Proc. P-HEC Workshop*
[9] Allen E, Chase D, Hallett J, Luchangco V, Maessen J W, Ryu S, Steele G L and Topin-Hochstadt S 2007 The Fortress language specification Tech. rep. Sun Microsystems available at http://research.sun.com/projects/plrg/Publications/index.html
[10] Heroux M 2008 *J. Physics: Conference Series* **125** 1082–1104
[11] Mpi forum http://www.mpi-forum.org/
[12] Williams S, Shalf J, Oliker L, Kamil S, Husbands P and Yelick K 2006 The potential of the cell processor for scientific computing *Proc. 3rd Conf. Computing Frontiers* (ACM) pp 9–20
[13] Crawford C, Henning P, Kistler M and Wright C 2008 Accelerating computing with the cell broadband engine processor *Proc. 5th Conf. Computing Frontiers* (ACM) pp 3–12
[14] Lumsdaine A, Gregor D, Hendrickson B and Berry J 2007 *Parallel Processing Letters* **17** 5–20
[15] 2009 Cray XMT: massively multithreaded platform http://www.cray.com/products/xmt
[16] Anderson W, Briggs P, Hellberg C, Hess D, Khokhlov A, Lanzagorta M and Rosenberg R 2003 Early experiences with scientific programs on the Cray MTA-2 *Proc. SC'03*
[17] Oliker L and Biswas R 2000 *IEEE Trans. Parallel and Distrib. Sys.* **11** 931–940
[18] Meyer U and Sanders P 1998 Delta-stepping: A parallel single source shortest path algorithm. *Proc. 6th European Symp. on Algorithms* (London, UK: Springer–Verlag) pp 393–404

Applications and algorithms for mixed integer nonlinear programming

Sven Leyffer[1], Jeff Linderoth[2], James Luedtke[2], Andrew Miller[3], Todd Munson[1]

[1] Mathematics and Computer Science Division, Argonne National Laboratory, Argonne, IL 60439, USA

[2] Department of Industrial and Systems Engineering, University of Wisconsin-Madison, Madison, WI 53706, USA

[3] IMB, Université Bordeaux 1; RealOpt, INRIA Bordeaux Sud-Ouest, 33405 Talence, France

E-mail: `linderoth@wisc.edu`

Abstract. The mathematical modeling of systems often requires the use of both nonlinear and discrete components. Discrete decision variables model dichotomies, discontinuities, and general logical relationships. Nonlinear functions are required to accurately represent physical properties such as pressure, stress, temperature, and equilibrium. Problems involving both discrete variables and nonlinear constraint functions are known as *mixed-integer nonlinear programs* (MINLPs) and are among the most challenging computational optimization problems faced by researchers and practitioners. In this paper, we describe relevant scientific applications that are naturally modeled as MINLPs, we provide an overview of available algorithms and software, and we describe ongoing methodological advances for solving MINLPs. These algorithmic advances are making increasingly larger instances of this important family of problems tractable.

1. Introduction

A mixed-integer nonlinear program (MINLP) is the numerical optimization problem of finding

$$z_{\text{MINLP}} = \underset{x\in X, y\in Y\cap\mathbb{Z}^p}{\text{minimize}} \; f(x,y) \;\; \text{subject to } g(x,y) \leq 0, \tag{MINLP}$$

where $f : \mathbb{R}^{n+p} \rightarrow \mathbb{R}$ and $g : \mathbb{R}^{n+p} \rightarrow \mathbb{R}^m$ are continuously differentiable functions; x and y are continuous and discrete variables, respectively; and X and Y are compact polyhedral subsets of $\mathbb{R}^n$ and $\mathbb{R}^p$, respectively. If the objective function $f(x,y)$ and the constraint function $g(x,y)$ are convex functions, then the problem is known as a *convex MINLP*, otherwise the problem is a *nonconvex MINLP*.

Given their generality and flexibility, MINLPs have been proposed for many diverse and important scientific applications, including the efficient management of electricity transmission [1], contingency analysis and blackout prevention of electric power systems [2], the design of water distribution networks [3], operational reloading of nuclear reactors [4], and minimization of the environmental impact of utility plants [5].

Unfortunately, current algorithms and available software are often unable to solve practically-sized instances of these important models. Our research is aimed at correcting the mismatch between natural optimization models and available robust optimization solvers. We hope to enable application scientists to confidently employ MINLP tools to address their challenging scientific problems.

2. Applications

Design of thermal insulation layer for the large hadron collider: A thermal insulation system uses a series of heat intercepts and surrounding insulators to minimize the power required to maintain the heat intercepts at certain temperatures. Problems of this type arise in the design of superconducting magnetic energy storage systems and have been used in the Large Hadron Collider project. Figure 1-(a) shows the conceptual design.

The designer chooses the maximum number of intercepts, the thickness and area of each intercept, and the material of intercept from a discrete set of materials. The choice of material affects the thermal conductivity and total mass of the insulation system. Nonlinear functions in the model are required to accurately model system characteristics such as heat flow between the intercepts, thermal expansion, and stress constraints. Integer variables are used to model the discrete choice of the type of material to use in each layer.

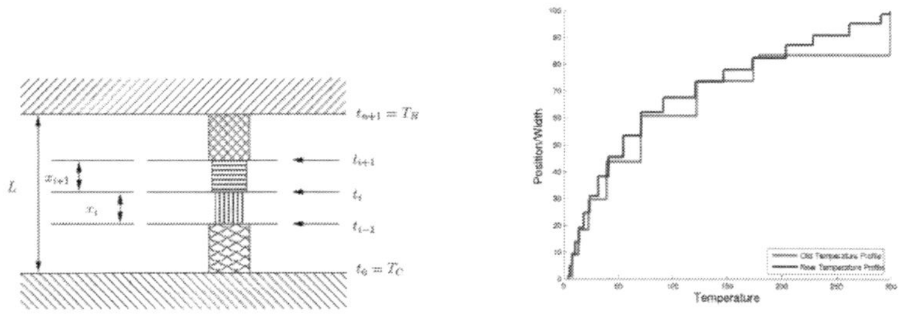

(a) Conceptual Design of Insulation Layer (b) Optimal Temperature Profile

Figure 1. Design of Thermal Insulation

In [6], by using MINLP, we have identified solutions that were as much as 4% better than those identified by any previously available method. The improvement is largely due to the ability to handle a larger number of intercepts than previously possible. The optimal temperature profile of the new design is shown as the blue curve in Figure 1-(b), compared to the red curve for the previous suboptimal design. The area between the two profiles corresponds to the savings in cooling power.

Nuclear core-reload: The fuel elements in a nuclear reactor core are divided into groups of different ages. At the end of each cycle the oldest group is removed, the remaining groups are reshuffled, and a fresh group is brought in. Finding a pattern for how to move the elements to maximize fuel efficiency, subject to power safety constraints, leads to a mixed-integer nonlinear problem. The model includes dependent variables that describe physical properties such as neutron flux, burn-up, and yield. The neutron transport equations are converted to a set of algebraic equations using Green's functional theory, giving rise to a stationary description of the neutron flux in the core. The fuel burn-up is approximated by discretizing the differential equation. The resulting MINLP contains binary variables that model the reloading operation and algebraic equations arising from the discretization of the PDE. Figure 2-(a) shows an initial design that ignores the integrality of the reloading operation, and Figure 2-(b) shows the optimal reloading pattern, where different colors represent the different age groups. It is important to note that simple rounding of the solution in Figure 2-(a) does not produce a valid reload-pattern.

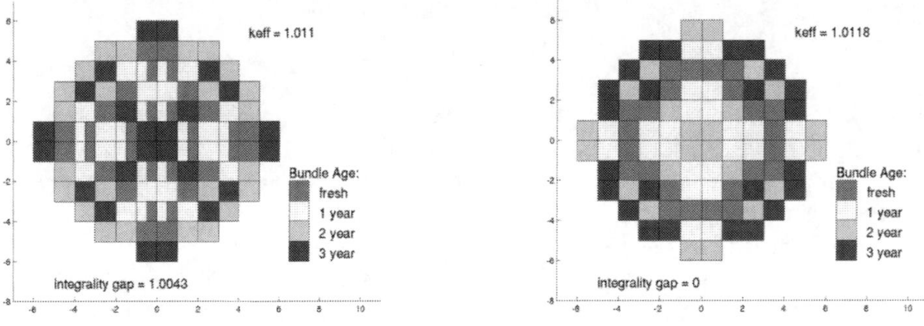

(a) Solution to NLP Relaxation (b) Solution to MINLP

Figure 2. Core-reload Problem Solutions

3. Algorithms for MINLP

The basic operations common to algorithms for solving (MINLP) are *relax* and *search*. Algorithms differ in the manner these operations are performed. For convex MINLPs, the most natural relaxation is to disregard the integrality constraints $y \in \mathbb{Z}^p$. The resulting nonlinear program (NLP), when solved to global optimality, provides a lower bound on the optimal solution value z_{MINLP}. If the solution $(\hat{x}, \hat{y})$ to the relaxation has $y \in \mathbb{Z}^p$, then $(\hat{x}, \hat{y})$ must be the optimal solution to (MINLP), otherwise there is a $j \in \{1, \dots, p\}$ with $\hat{y}_j \notin \mathbb{Z}$. In *branch-and-bound*, the search continues by creating two new subproblems, one with the additional constraint $y_j \le \lfloor \hat{y}_j \rfloor$, and one with the additional constraint $y_j \ge \lceil \hat{y}_j \rceil$. Each of these two subproblems is then (recursively) solved by the same procedure, resulting in a search-tree of subproblems that are evaluated. Software that implements a branch-and-bound algorithm for convex MINLPs is listed in Table 1. The software packages differ in the way the search-tree is created and in the NLP algorithm used to solve relaxations.

Unfortunately, if $f(x, y)$ or $g(x, y)$ are not convex functions, then standard algorithms for solving NLP are able to guarantee convergence only to a local minimum. Thus, the solution of the relaxation *does not* provide a lower bound on z_{MINLP}. In this case of non-convex MINLP, an additional relaxation step is required. Typically, the relaxation is based on a decomposition of the nonlinear functions into components, and for each component, a "convex envelope" relaxation is built [7]. Branch and bound is again applied to the relaxation, with the additional complication that continuous variables x may require branching to improve the convex relaxation of the original nonconvex functions. Solvers of this class (listed in Table 1) differ in the manner in which relaxations are created. The impact of creating strong relaxations for nonconvex functions is briefly demonstrated in Section 4.2.

The final class of algorithms developed for solving convex MINLP are those based on linearizations. If $f(x, y)$ and $g(x, y)$ are convex, then for any point $(\hat{x}, \hat{y})$ the inequalities

$$f(\hat{x}, \hat{y}) + \nabla f(\hat{x}, \hat{y})^T \begin{bmatrix} x - \hat{x} \\ y - \hat{y} \end{bmatrix} \le f(x, y) \quad \text{and} \quad g(\hat{x}, \hat{y}) + \nabla g(\hat{x}, \hat{y})^T \begin{bmatrix} x - \hat{x} \\ y - \hat{y} \end{bmatrix} \le 0, \quad (1)$$

underestimate the objective function and form a linear outer approximation of the feasible region, respectively. In linearization-based methods, a linear approximation to (MINLP) is dynamically created by applying the inequalities (1) at many different points $(\hat{x}, \hat{y})$. The approximation is solved by a branch-and-bound procedure to ensure convergence. There is significant flexibility in where linearizations are taken and how they are combined with the branch-and-bound search. Software packages with implementations of linearization-based algorithms are listed in Table 1. Most of the codes listed in Table 1 are available for use free of charge via the NEOS server: `http://www-neos.mcs.anl.gov/`.

Table 1. Software for solving MINLP

Algorithm	**Software**
Branch and Bound, Convex	Bonmin, KNITRO, MINLP-BB, SBB
Branch and Bound, Nonconvex	BARON, Couenne, LINDO-Global
Linearization-Based, Convex	Alpha-ECP, Bonmin, Dicopt, FilMINT

4. Strong reformulations

Recognizing and exploiting instance-specific structure in (MINLP) to create strong reformulations can have a huge impact on an algorithm's ability to solve an instance. We conclude with two brief examples demonstrating this point.

4.1. Strong reformulations for convex MINLPs

Integer variables in MINLPs are often used as "indicator variables." For example, consider

$$S = \{(x, y, z) \in \mathbb{R}^2 \times \{0, 1\} \, : \, y \geq x^2, \quad uz \geq x \geq lz, \; x \geq 0\}, \text{ where } u, l \in \mathbb{R}. \qquad (2)$$

Setting the indicator variable $z = 0$ forces $x = 0$, but when $z = 1$, the inequalities $x \in [l, u], y \geq x^2$ are enforced. We have shown in [8] that the convex hull of S is given by the *perspective reformulation*: $\text{conv}(S) = \left\{(x, y, z) \in \mathbb{R}^3 \, : \, yz \geq x^2, \; uz \geq x \geq lz, \; 1 \geq z \geq 0, \; x, y \geq 0\right\}$. Reformulating simple sets S that appear in (MINLP) in the form conv(S) will significantly strengthen the relaxation and improve algorithm performance.

As a concrete example, consider a Separable Quadratic Uncapacitated Facility Location Problem (SQUFL) [9]. The MINLP formulation of SQUFL has constraints of the form

$$x_{ij}^2 - y_{ij} \leq 0 \quad \text{and} \quad y_{ij} \leq z_i \quad \forall i, j. \qquad (3)$$

The indicator variable z_i should take the value 1 if facility i is opened. If $z_i = 0$, then the constraints (3) force $y_{ij} = x_{ij} = 0$, meaning that no demand for customer j can be served from facility i. Since inequalities (3) are of the form (2), the perspective reformulation may be applied, replacing $x_{ij}^2 - y_{ij} \leq 0$ with $x_{ij}^2 z_i - y_{ij} \leq 0 \; \forall i, j$.

In [10], an instance with 30 facilities and 100 customers was solved by applying Bonmin to the original reformulation. The solution required 16,697 CPU seconds and 45,901 nodes in the search-tree. By applying the perspective reformulation, the same instance was solved in *23 CPU seconds*, enumerating 44 nodes of the search tree [8]. The speedup factor is more than 700.

4.2. Strong relaxations of multilinear forms

A particular nonconvex structure we are studying is the case of *multilinear* functions of the form $\phi(x) = \sum_{i=1}^{m} a_i \prod_{j \in S_i} x_j$ where $a_i \in \mathbb{R}$ and $S_i \subseteq N := \{1, \ldots, n\}$ and we assume $l \leq x \leq u$ for some $l, u \in \mathbb{R}^n$. Many applied problems have terms of this form, either explicitly (e.g. the pooling problem [11]) or after reformulation. The "textbook" approach for obtaining relaxations of such problems is to reformulate the problem (by adding new variables) into a problem in which the only nonconvexities remaining are bilinear constraints of the form $x_k = x_i x_j$, and then to relax each of these constraints using a set of simple linear inequalities known as the the McCormick inequalities [12]. We consider an alternative approach in which we use the convex hull of the set $\{z \in \mathbb{R}, x \in \mathbb{R}^n \mid z = \phi(x), \; l \leq x \leq u\}$, which can be obtained by writing the vector x as a convex combination of the vertices of the hypercube containing x [13].

To demonstrate the impact of this reformulation technique, we conducted a preliminary test on the following example problem:

$$\underset{l \leq x \leq u}{\text{minimize}} \; 4(x_1 + x_2 + x_3) + 3(x_4 + x_5) + 3.5(x_6 + x_7) + 2.5x_8$$

$$\text{subject to } x_1 x_2 x_3 x_4 - x_1 x_2 - x_4 x_5 + x_5 + x_6 \geq 230, \; x_3 x_4 x_5 x_6 - x_1 x_4 - x_6 x_7 + x_2 + x_8 = -2,$$

where $l = [-1, -2.5, -0.5, -0.5, 1, 1, 1, 1]$ and $u = [4, 4, 6, 6, 6, 5, 3]$. For this example, the McCormick relaxation based on a bilinear reformulation yields a lower bound of 2.33 on the value of the best possible solution, compared to a significantly better lower bound of 36.33 obtained by solving the convex hull relaxation. We then passed the solution of each of these relaxations to the NLP solver IPOPT [14] in an attempt to find a good feasible solution. IPOPT failed to find a feasible solution when given the McCormick relaxation solution as an initial point, but found a feasible solution of value 72.1 when starting from the convex hull relaxation solution. Solving this problem with the global optimization solver BARON [7] confirmed that this solution is optimal. This example suggests that the convex hull approach may be useful both for yielding stronger lower bounds and for finding better feasible solutions.

Acknowledgments
This work was supported in part by National Science Foundation Grant CCF-0830153 and by the Office of Advanced Scientific Computing Research, Office of Science, U.S Department of Energy, under Contract DE-AC02-06CH11357 and Grants DE-FG02-08ER25861 and DE-FG02-09ER25869.

References

[1] Momoh J, Koessler R, Bond M, Stott B, Sun D, Papalexopoulos A and Ristanovic P 1997 *IEEE Transaction on Power Systems* **12**(1) 444–455

[2] Donde V, Lopez V, Lesieutre B, Pinar A, Yang C and Meza J 2005 Identification of severe multiple contingencies in electric power networks *Proceedings 37th North American Power Symposium* lBNL-57994

[3] Bragalli C, D'Ambrosio C, Lee J, Lodi A and Toth P 2006 An MINLP solution method for a water network problem *Algorithms - ESA 2006 (14th Annual European Symposium. Zurich, Switzerland, September 2006, Proceedings)* (Springer) pp 696–707

[4] Quist A J, van Gemeert R, Hoogenboom J E, Ílles T, Roos C and Terlaky T 1998 *Annals of Nuclear Energy* **26** 423–448

[5] Eliceche A M, Corvalán S M and Martínez P 2007 *Computers and Chemical Engineering* **31** 648–656

[6] Abhishek K, Leyffer S and Linderoth J T 2007 Modeling without categorical variables: A mixed-integer nonlinear program for the optimization of thermal insulation systems Preprint ANL/MCS-P1434-0607 Argonne National Laboratory Argonne, IL

[7] Sahinidis N V 1996 *Journal of Global Optimization* **8** 201–205

[8] Günlük O and Linderoth J 2008 Perspective relaxation of mixed integer nonlinear programs with indicator variables *IPCO 2008: The Thirteenth Conference on Integer Programming and Combinatorial Optimization* vol 5035 ed Lodi A, Panconesi A and Rinaldi G pp 1–16

[9] Günlük O, Lee J and Weismantel R 2007 MINLP strengthening for separable convex quadratic transportation-cost UFL Tech. Rep. RC24213 (W0703-042) IBM Research Division

[10] Lee J 2007 *IBM Journal of Research & Development* **51** 489–497

[11] Floudas C A and Aggarwal A 1990 *ORSA Journal on Computing* **2** 225–235

[12] McCormick G P 1976 *Mathematical Programming* **10** 147–175

[13] Sherali H 1997 *Acta Mathematica Vietnamica* **22** 245–270

[14] Wächter A and Biegler L T 2006 *Mathematical Programming* **106** 25–27

Factorization-based sparse solvers and preconditioners

X S Li[1], M Shao[2], I Yamazaki[1] and E G Ng[1]

[1]One Cyclotron Road, Lawrence Berkeley National Laboratory, MS 50F-1650, Berkeley, CA 94720.
[2]School of Mathematical Sciences, Fudan University, Shanghai 200433, China.

E-mail: `xsli@lbl.gov`

Abstract. Efficient solution of large-scale, ill-conditioned and highly-indefinite algebraic equations often relies on high quality preconditioners together with iterative solvers. Because of their robustness, the factorization-based algorithms could play a significant role when they are combined with iterative methods, particularly in the development of scalable solvers. We present our recent work in using the direct solver SuperLU code base to develop a new supernode-based ILU preconditioner and a domain-decomposition hybrid solver. Our ILU preconditioner is a modification of the classic ILUTP approach, incorporating a number of techniques to improve robustness and performance, which include new dropping strategies that accommodate the use of supernodal structure in the factored matrix. Our hybrid solver is based on the Schur complement method. We use parallel graph partitioning to obtain hierarchical interface/domain decomposition, and multiple parallel direct solvers to solve the subdomain problems simultaneously, and parallel preconditioned iterative solvers to solve the interface problem. We will demonstrate the effectiveness of our new techniques by applying them to two SciDAC applications, modeling next-generation particle accelerators and fusion devices.

1. A supernodal approach to incomplete LU factorization

1.1. Introduction

A critical component of the iterative solution techniques is the construction of effective preconditioners. Physics-based preconditioners are quite effective for structured problems, such as those arising from discretized partial differential equations. On the other hand, a class of methods based on incomplete LU decomposition are still regarded as the generally applicable "black-box" preconditioners for unstructured systems arising from a wide range of applications areas. A variety of ILU techniques have been studied extensively in the past, including distinct strategies of dropping elements, such as the level-of-fill structure-based approach (i.e., ILU(k)) [1], numerical threshold-based approach [2], and more recently, numerical inverse-based multilevel approach [3].

The value-based threshold method is often more reliable than the level-based method, but it is harder to implement efficiently. One of the most sophisticated value-based methods is ILUTP proposed by Saad [2, 1], which combines a dual dropping strategy with numerical pivoting ("T" stands for threshold, and "P" stands for pivoting). The dual dropping rule in ILU(τ, p) first removes the elements that are smaller than τ from the current factored row or column. It then keeps only the largest p elements to control the memory requirement.

Our method can be considered to be a variant of the ILUTP approach, and we modified our high-performance direct solver SuperLU [4] to perform incomplete factorization. One key component in SuperLU is supernode, which gives significant performance advantages over a non-supernodal (e.g., column-wise) algorithm on modern cache-based architectures, and on GPU-type accelerators. Although the average size of the supernodes in an incomplete factor is expected to be smaller than in a complete factor because of dropping, we attempt to retain supernodes as much as possible. We have adapted the dropping rules to incorporate the supernodal structures as they emerge during factorization. Therefore, our new algorithm has the combined benefits of retaining numerical robustness of ILUTP as well as achieving fast construction and application of the ILU preconditioner.

Our main contributions can be summarized as follows. We adapted the classic dropping strategies of ILUTP in order to incorporate supernode structures and to accommodate dynamic supernodes due to partial pivoting. For the secondary dropping strategy, we proposed an area-based fill control method, which is more flexible and numerically robust than the traditional column-based scheme. Furthermore, we incorporated several heuristics for adaptively modifying various threshold parameters as the factorization proceeds, which improves the robustness of the algorithm.

1.2. Sketch of the supernodal ILU algorithm

Our base algorithm framework is the left-looking, partial pivoting, supernodal sparse LU factorization algorithm implemented in SuperLU [4]. A key concept in SuperLU is to exploit dense blocks appearing in the L and U factors. In particular, we define a supernode in L to be a range $(r : t)$ of columns with the triangular block on the diagonal being full, and the identical nonzero structure elsewhere among the columns. Using the same supernode partition to the rows of U, the nonzero structure of each column in U consists of a number of dense segments. Thus, the compressed data structure for L consists of a collection of supernodes as dense submatrices, and that for U consists of a collection of dense subvectors.

The factorization algorithm is left-looking, with a supernode-panel update kernel. A panel is a set of consecutive columns, and the size of panel is an algorithmic blocking parameter used to enhance data reuse in the memory hierarchy; it enables use of Level 3 BLAS. At each step of panel factorization, we obtain a panel in the U factor and a panel in the L factor.

Our incomplete factorization algorithm retains most of the algorithmic ingredients from SuperLU, with added dropping rules that are applied to the L and U factors on-the-fly. The description of the high level algorithm is given in Algorithm 1. The steps marked as **bold** correspond to the new steps introduced to perform ILU. Since partial pivoting with row interchange is used, the resulting factorization is performed on the matrix $P_rP_0D_rAD_cP_c^T$, where D_r and D_c are diagonal scaling matrices, P_0 is the row permutation matrix returned from MC64 [5] that is meant to permute large magnitude entries on the main diagonal (this step is optional), P_c is the column permutation matrix for sparsity preservation, and P_r is the row permutation matrix from partial pivoting. The matrices D_r, D_c, P_0 and P_c are obtained before factorization, and P_r is obtained during factorization. In the following sections, we describe our adaptation of the dropping rules to the situation when supernodes are present.

1.3. Value-based dropping criteria

Our primary dropping criteria are threshold-based and akin to the ILUTP variants [2, 6]. That is, while performing Gaussian elimination with partial pivoting, we set to zero the entries in L and U with modulus smaller than a prescribed threshold τ, where $\tau \in [0, 1]$.

Since our compressed storage is column oriented for both L and U, the dropping rule is also column oriented. The upper triangular matrix U is stored in a normal compressed column format, we can easily remove the small elements while storing the newly computed column into the compressed storage, using the first criterion given in Figure 1.

The lower triangular matrix L is stored as a collection of supernodes. Our goal is to retain the supernodal structure to the largest extent as in the complete factorization. In a naive implementation of ILU, we may apply the traditional dropping to each individual column. But after dropping, the nonzero structures among the columns in the original supernode will be different, then we will need to regroup the columns into smaller supernodes, resulting in a performance penalty. Instead, we adopt an alternative approach that retains the original supernode partition as much as possible. That is, we either keep or drop an entire row in a supernode when it is formed at the current step. This is similar to what was proposed by Gupta and George in the context of incomplete Cholesky factorization [7]. Our dropping criterion is the second rule shown in Figure 1. Since we use partial pivoting, the magnitude of the elements in L is bounded above by one, and so the absolute quantity is the same as the relative quantity. The use of

∞-norm for row i of a supernode implies that when row i is dropped, the magnitude of every element in this row is smaller than τ. Therefore, in a traditional column-wise algorithm, these elements should be dropped as well. We did an experiment to compare the supernodal ILU and the column-wise ILU (setting maximum supernode size to be one). For 54 matrices, GMRES with supernodal ILU converged for 47 cases, and the column-wise ILU succeeded with only 42 matrices. This shows that our supernodal version is numerically superior.

Algorithm 1. *Left-looking, supernode-panel ILU algorithm*

(i) Preprocessing

1.1) **(optional)** Use MC64 to find a row permutation P_0 and row and column scaling factors D_r and D_c such that $P_0 D_r A D_c$ is an I-matrix;

1.2) If step 1.1) is not performed, do a simple LAPACK-style row/column equilibration to obtain $D_r A D_c$;

1.3) Compute a column permutation P_c to preserve sparsity of the LU factorization of $P_0 D_r A D_c P_c^T$;

(ii) Factorization of $P_0 D_r A D_c P_c^T$

```
FOR each panel of columns DO
2.1) Symbolic factorization: determine which supernodes to the left will update the current panel and
     a topological order of updates;
2.2) Panel factorization:
       FOR each updating supernode DO
           Apply triangular solve to obtain the U part;
           Apply matrix-matrix multiplication to obtain the L part;
       END FOR
2.3) Inner factorization:
       FOR each column j in the panel DO
           Update the current column j;
           Apply the dropping rule to the U part;
           Find pivot in this column;
           (optional) Modify the diagonal entry to handle zero-pivot breakdown;
           Determine supernode boundary;
           IF column j starts a new supernode THEN
             Apply the dropping rule to the newly formed supernode L(:, r : j − 1);
           END IF
       END FOR
END FOR
```

Value-based dropping criteria for ILU(τ)

1) Dropping elements in U: If $|u_{ij}| < \tau \|A(:, j)\|_\infty$, we set u_{ij} to zero.

2) Dropping elements in L: In a supernode $L(:, r : t)$, if $\|L(i, r : t)\|_\infty < \tau$, we set the entire i-th row to zero.

Figure 1. The value-based dropping criteria.

1.4. Secondary dropping to control fill-in adaptively

ILU(τ) works well if there is sufficient memory, but it may still have too much fill. A secondary dropping can be used to alleviate the problem. In Saad's ILU(τ,p) approach [2], p is the largest number of nonzeros (not the level-of-fill) allowed in each row of F (in a row-wise algorithm). Gupta and George suggested using $p(j) = \gamma \cdot \mathrm{nnz}(A(:, j))$ for the j-th column instead of a constant, where γ is an upper bound of the fill ratio defined by a user [7]. They also proposed a method of computing a secondary dropping tolerance by an interpolation formula rather than sorting the largest p entries. But Gupta's heuristic depends largely on the distribution of the nonzero modulos in F.

We now present a new strategy for choosing p. Given a user-desired upper bound of the overall fill ratio γ, we define an upper bound function $f(j)$ for each column j, $f : [1, n] \rightarrow [1, \gamma]$, which satisfies $f(n) \leq \gamma$. Then at the j-th column, if the current fill ratio

$$\frac{\text{nnz}(F(:, 1 : j))}{\text{nnz}(A(:, 1 : j))} \tag{1}$$

exceeds $f(j)$, we choose a maximum possible value p such that when we keep the largest p elements, the current fill ratio is bounded by $f(j)$. This criterion can be adapted to our supernodal algorithm as follows. For a supernode with k columns, p may be computed as

$$p = \max\left\{\frac{f(j) \cdot \text{nnz}(A(:, 1 : j)) - \text{nnz}(F(:, 1 : j - k))}{k},\ k\right\}. \tag{2}$$

In other words, if we keep the largest p rows of this supernode, the current fill ratio is guaranteed not to exceed $f(j)$. The second k term in $\max\{\ldots\}$ is to ensure that we do not drop any row in the diagonal block of the supernode.

This is also an ILU(τ,p) approach with adaptive p, similar to Gupta's scheme. However, our fill ratio definition (1) is *area-based* instead of column-based, because we count all the fill-ins from column 1 to column j. That is, we only monitor the overall memory growth instead of that of each individual column. This is more flexible than the column-based method in that it allows larger amount of fill for certain columns so long as the cumulative fill ratio in the previous columns is small. At the end of factorization, the total fill ratio is still bounded by γ because of the condition $f(n) \leq \gamma$.

Since L and U are stored in different data structures, we may choose two functions, $f_L(j)$ for L and $f_U(j)$ for U, so long as $f_L(n) + f_U(n) \leq \gamma$. A simple way is to assign $f_L(n)$ and $f_U(n)$ to be the areas of $L(:, j)$ and $U(:, j)$ relative to $F(:, 1 : j)$, as follows:

$$f_U(j) = \frac{j}{2n}\gamma, \quad f_L(j) = \left(1 - \frac{j}{2n}\right)\gamma. \tag{3}$$

Then we split the fill quota proportionally with $f_U(j) : f_L(j)$ ratio.

In conjunction with the dynamic, area-based strategy for choosing p, we devised an adaptive scheme for choosing τ as well. Specifically, let $\tau(1) = \tau_0$ be the user-input threshold, at column j, if the fill ratio given by Equation (1) is larger than $f(j)$, we set $\tau(j + 1) = \min\{1, 2\,\tau(j)\}$, otherwise, we set $\tau(j + 1) = \max\{\tau_0, \tau(j)/2\}$. That is, we maintain $\tau(j) \in [\tau_0, 1]$.

1.5. Numerical experiments

We tested our algorithms on an Opteron cluster running a Linux operating system at NERSC.[5] Each node contains dual Opteron 2.2 GHz processors, with 5 GBytes usable memory. We use only one processor of a node. The processor's theoretical peak floating-point performance is 4.4 Gflops/sec. We use PathScale `cc` compiler with the following optimization flags: `-O3 -OPT:IEEE_arithmetic=1 -OPT:IEEE_NaN_inf=ON`, which conforms to the IEEE-754 standard. We have chosen 54 test matrices: 5 are from the M3D-C^1 code for extended MHD modeling in fusion energy study [8], which is being developed in the CEMM SciDAC project; 49 are from Matrix Market [9] and the University of Florida Sparse Matrix collection [10]. These are all unsymmetric matrices of medium to large size. The iterative solver is restarted GMRES with our ILU as a right preconditioner (i.e. solving $PAM^{-1}y = Pb$). The stopping criterion is $\|r_k = b - Ax_k\|_2 \leq \delta\,\|b\|_2$, here we use $\delta = 10^{-8}$ which is in the order of the square

[5] `http://www.nersc.gov/nusers/systems/jacquard`

root of IEEE double precision machine epsilon. We set the dimension of the Krylov subspace to be 50 and maximum iteration count to be 1000.

We now present the results of the tests comparing various parameter settings. The ILU configurations include:

- ILU(τ), $\tau = 10^{-4}$;
- ILU(τ, p), $\tau = 10^{-4}$ or 10^{-8}, $p = \gamma \cdot \mathrm{nnz}(A)/n$;
- column-based adaptive p, $\tau = 10^{-4}$ or 10^{-8};
- area-based adaptive p, $\tau = 10^{-4}$ or 10^{-8};
- area-based adaptive $\tau(j)$, $\tau_0 = 10^{-4}$, no secondary dropping

Figure 2 shows the performance profiles of the fill ratio and the time ratio for the 54 test matrices. The former shows the fraction of the problems that a solver could solve within the fill ratio x, and the latter shows the fraction of the problems that a solver could solve within a multiple of x of the best time among all the solvers. We can see that a small τ such as 10^{-8} is generally not good, that is, it is not efficient to use the secondary dropping rule only. The value-based dropping criterion in Figure 1 should play a significant role.

A key conclusion is that our new area-based scheme is much more robust than the column-based scheme; it is also better than ILU(τ) when the fill ratio does not exceed the user-desired γ. ILU(τ) becomes better only when the fill ratio is unbounded (i.e., allow it to exceed γ). This is consistent with the intuition that an ILU preconditioner tends to be more robust with more fill-ins.

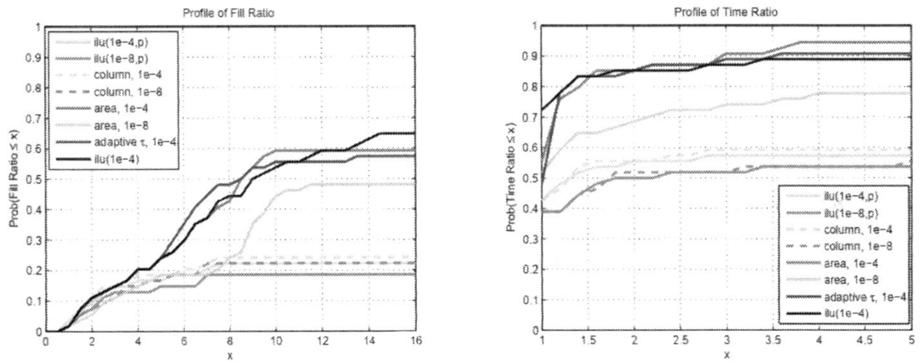

(a) Profile of fill ratio (b) Profile of time ratio

Figure 2. Performance profiles after incorporating the secondary dropping rules; $\gamma = 10$.

Figure 2(b) shows the runtime comparison of the solvers. In this plot, the matrices with fill ratio larger than 10 are considered as failure. Thus, the comparison is made under the same memory constraint, and none of the solvers are allowed to consume much more memory than the others. The top three solvers are much better than the others. Our area-based adaptive-p or adaptive-τ schemes have similar performance, with the adaptive-p scheme having a slight edge over the other one.

Taking into account both memory and time, we can see that the secondary dropping helps achieve a good trade-off, with controlled fill-in and the solver not being much slower. Either our "red" scheme or "blue" scheme can be used as a default setting in the code.

Table 1 compares the new ILU-preconditioned GMRES solver to SuperLU for a sequence of matrices from the fusion M3D-C^1 code. The ILU parameters are: $\tau = 10^{-4}$, and $\gamma = 10$. The advantage of GMRES+ILU over SuperLU is remarkable, both in time and in memory: SuperLU fails for the two larest problems; for **matrix61**, the largest that both solvers succeed, ILU only incurs one-tenth the amount of fill of SuperLU, and GMRES+ILU is about 10x faster than SuperLU. For all problems, GMRES can succeed with very low fill ratio in ILU—at or below 3.0.

Problems	Order	Nonzeros (millions)	ILU		GMRES		SuperLU	
			time	fill ratio	time	iters	time	fill ratio
matrix31	17,298	2.7 m	8.2	2.7	0.6	9	33.3	13.1
matrix41	30,258	4.7 m	18.6	2.9	1.4	11	111.1	17.5
matrix61	66,978	10.6 m	54.3	3.0	7.3	20	612.5	26.3
matrix121	263,538	42.5 m	145.2	1.7	47.8	45	fail	–
matrix181	589,698	95.2 m	415.0	1.7	716.0	289	fail	–

Table 1. Results of the test matrices from the fusion simulation code M3D-C^1.

2. Hybrid linear solver

2.1. Introduction

Although significant progress has been made in the development of high performance direct solvers [11, 12, 13], the size of the systems that can be directly factored is limited due to the large memory requirement. Preconditioned iterative solvers [14, 1, 15] can reduce the memory requirement, but they often suffer from slow convergence due to the ill-conditioning and highly-indefinite nature of the systems. To address these challenges, a number of parallel hybrid solvers have been developed based on a domain decomposition idea called the Schur complement method [16, 17]. In this method, the unknowns in interior domains are first eliminated directly using techniques from the direct solvers, and the remaining Schur complement system is solved approximately using a preconditioned iterative solver. This method has the potential of balancing the robustness of the direct solver with the efficiency of the iterative solver since the unknowns in each interior domain can be eliminated efficiently, and in parallel, while the sparsity can be enforced for solving the Schur complement system, where most of fill occurs. Furthermore, for a symmetric positive definite system, it can be shown that the Schur complement has a smaller condition number than the original coefficient matrix [18]. Consequently, the preconditioned iterative solver often requires fewer iterations to solve the Schur complement system. Unfortunately, for a general linear system, this method can still suffer from slow convergence as the size of the Schur complement increases, especially with the existing parallel hybrid solvers which are designed primarily to achieve good scalability of time to compute the preconditioners.

To overcome these drawbacks, as part of our SciDAC project (TOPS) [19], we have been developing a new parallel implementation of the Schur complement method which provides the robustness and flexibility to solve large highly-indefinite linear systems on a large number of processors. In this section, we outline our implementation and present our preliminary results for solving linear systems of this type, which arise from two SciDAC applications: the next-generation particle accelerators modeling (ComPASS) [20] and the fusion devices simulation (CEMM) [21].

2.2. Schur complement method

The Schur complement method is a non-overlapping domain decomposition method, which is also referred to as iterative substructuring. Specifically, the original linear system is first reordered into a 2×2 block system of the following form,

$$\begin{pmatrix} A_{11} & A_{12} \\ A_{21} & A_{22} \end{pmatrix} \begin{pmatrix} x_1 \\ x_2 \end{pmatrix} = \begin{pmatrix} b_1 \\ b_2 \end{pmatrix}, \tag{4}$$

where A_{11} is a block-diagonal matrix, each of whose diagonal blocks represents an *interior domain*, A_{22} represents *separators*, and A_{12} and A_{21} are the *interfaces* between A_{11} and A_{22}. After one step of the block Gaussian elimination, the 2×2 block system (4) becomes

$$\begin{pmatrix} A_{11} & A_{12} \\ 0 & S \end{pmatrix} \begin{pmatrix} x_1 \\ x_2 \end{pmatrix} = \begin{pmatrix} b_1 \\ \widehat{b}_2 \end{pmatrix}, \tag{5}$$

where S is the Schur complement defined as

$$S = A_{22} - A_{21}A_{11}^{-1}A_{12}, \tag{6}$$

and $\widehat{b_2} = b_2 - A_{21}A_{11}^{-1}b_2$. Hence, the solution of the linear system (4) can be computed by

(i) first solving the Schur complement system

$$Sx_2 = \widehat{b_2}, \tag{7}$$

(ii) then solving the interior system

$$A_{11}x_1 = b_1 - A_{12}x_2. \tag{8}$$

Since most of the fill occurs in the Schur complement S, our hybrid solver follows the approach, in which the interior system is solved directly, while the Schur complement system is solved approximately.

2.3. Parallel implementation

Our high performance implementation takes full advantage of the state-of-the-art software. For example, we form the 2×2 block system (4) based on the hierarchical interface decomposition (HID), which is constructed by Hybrid Iterative Parallel Solver (HIPS) [22]. We then invoke either a serial direct solver SuperLU [12] or parallel direct solver SuperLU_DIST [13] to compute the LU factorization of each interior domain in parallel. To preserve sparsity of the LU factors, each interior domain is permuted with the nested dissection ordering obtained either by METIS [23] or by PT-SCOTCH [24], which is a serial or parallel algorithm, respectively. Next, the Schur complement system (7) is solved using a Krylov subspace method from PETSc [25] combined with an ILU preconditioner. Finally, the solution of the interior system (8) can be computed efficiently, and in parallel, since the LU factorizations of each interior domains have already been computed.

Since computing the preconditioner for solving the Schur complement system is often the computational and memory bottleneck, we give a brief description of how the preconditioner is computed. In our current implementation, the preconditioners are the exact LU factors of a sparsified Schur complement. Specifically, let us denote the ℓ-th interior domain and the corresponding interfaces by $A_{11}^{(\ell)}$, $A_{12}^{(\ell)}$, and $A_{21}^{(\ell)}$, respectively, such that the coefficient matrix in the 2×2 block system (4) can be written as

$$\left(\begin{array}{c|c} A_{11} & A_{12} \\ \hline A_{21} & A_{22} \end{array}\right) = \left(\begin{array}{cccc|c} A_{11}^{(1)} & & & & A_{12}^{(1)} \\ & A_{11}^{(2)} & & & A_{12}^{(2)} \\ & & \ddots & & \vdots \\ & & & A_{11}^{(k)} & A_{12}^{(k)} \\ \hline A_{21}^{(1)} & A_{21}^{(2)} & \dots & A_{21}^{(k)} & A_{22} \end{array}\right). \tag{9}$$

If each interior domain $A_{11}^{(\ell)}$ is factored by a single processor, then the ℓ-th processor stores the nonzeros of $A_{11}^{(\ell)}$ and $A_{21}^{(\ell)}$ in a row-wise order, and the nonzeros of $A_{12}^{(\ell)}$ in a column-wise order. If multiple processors are used for each interior domain, then the rows of $A_{11}^{(\ell)}$ and $A_{12}^{(\ell)}$, and the columns of $A_{21}^{(\ell)}$ are evenly distributed among the processors. Furthermore, the rows of A_{22} are evenly distributed among the processors that solve the Schur complement system (7). To simplify the discussion, we assume a single processor is used to factor each interior domain. Hence, the ℓ-th processor computes the LU factorization of $A_{11}^{(\ell)}$, i.e., $A_{11}^{(\ell)} = L^{(\ell)}U^{(\ell)}$.

The following pseudocode shows how the ℓ-th processor computes the corresponding rows of the approximate Schur complement $\widetilde{S}$ when the processor owns the k_1-th to k_2-th rows of A_{22}.

1. $\widetilde{F}^{(\ell)} \approx (L^{(\ell)})^{-1}A_{12}^{(\ell)}$
2. $\widetilde{E}^{(\ell)} \approx (U^{(\ell)})^{-T}(A_{21}^{(\ell)})^T$

3. $T^{(\ell)} = (E^{(\ell)})^T F^{(\ell)}$
4. $S(k_1 : k_2, :) = A_{22}(k_1 : k_2, :)$
5. **for** $i = 1 \ldots k$
6. $\quad S(k_1 : k_2, :) = S(k_1 : k_2, :) - T^{(i)}(k_1 : k_2, :)$
7. **end for**
8. $\widetilde{S}(k_1 : k_2, :) \approx S(k_1 : k_2, :)$

On lines 1 and 2 of the above pseudocode, the lower-triangular systems are solved with regard to the sparsity of the interfaces $A_{12}^{(\ell)}$ and $A_{21}^{(\ell)}$. Then, the nonzeros with magnitude less than a user-specified threshold σ_1 are discarded to enforce sparsity of $\widetilde{E}^{(\ell)}$ and $\widetilde{F}^{(\ell)}$. On line 3, each processor computes the contribution $T^{(\ell)}$, and on lines 4 to 7, the contributions from other processors are gathered to compute the k_1-th to k_2-th rows of the Schur complement S. Finally, on line 8, after S is scaled and permuted to improve the numerics, nonzeros with the magnitude less than a user-specified threshold σ_2 are discarded to compute the approximate Schur complement $\widetilde{S}$.

After $\widetilde{S}$ is computed, SuperLU_DIST is invoked to compute the LU factor of $\widetilde{S}$, which is used as the preconditioner. We note that the matrix-vector product with the Schur complement S within the Krylov method is computed by applying the sequence of the sparse matrix operations (6) on the vector, and hence, the exact Schur complement S does not have to be stored explicitly for this phase of the solver.

2.4. Numerical experiments

We present preliminary results of our hybrid solver to solve large highly-indefinite linear systems of equations. For our numerical experiments, we used two test matrices from different SciDAC applications, the **tdr190k** matrix of dimension 1,100,242 from the next-generation particle accelerators modeling (ComPASS) [20], and the **matrix211** matrix of dimension 801,378 from the fusion M3D-C^1 code (CEMM) [21]. All the experiments were conducted on the NERSC Cray XT4 machine.

To demonstrate the effectiveness of our hybrid solver, we compare its performance with that of a direct solver SuperLU_DIST [13], and that of another hybrid solver HIPS [16]. The primary difference between our hybrid solver and HIPS is the way the preconditioner is computed for solving the Schur complement system (7). HIPS computes the preconditioner based on the ILU factorization of S, but sparsity of the preconditioner is enforced based on both numerical values and locations of nonzeros. Specifically, the fill is allowed only between separators adjacent to the same domain. Because of this, even though the numerical drop tolerance in our numerical experiments was set to be zero, HIPS still enforces the sparsity of the preconditioner based on the locations of nonzeros. On the other hand, our hybrid solver currently computes the exact LU factorization of $\widetilde{S}$.

In Table 2, we compare the total numbers of nonzeros in the LU and ILU factors. The sparsity of the matrices $\widetilde{E}$ and $\widetilde{F}$ is enforced using the drop tolerances $\sigma_1 = 10^{-6}$ and 10^{-7} for the matrices **tdr190k** and **matrix211**, respectively. An additional drop tolerance $\sigma_2 = 10^{-5}$ was used in our hybrid solver to enforce the sparsity of $\widetilde{S}$. In these numerical experiments, a single processor is used to factor each interior domain. We first note that the total number of nonzeros with SuperLU_DIST increases as the number of processors increases. This is because the quality of the nested dissection ordering degrades with the increase in the number of processors.[6] For the **tdr190k** and **matrix211** matrices, HIPS failed to converge within 1,000 unrestarted GMRES iterations using 32 and 128 processors, respectively. On the other hand, our hybrid solver (denoted by Hybrid in the table) is flexible enough to achieve convergence within 30 iterations for all the numerical experiments presented here.

Figure 3 shows the total solution times as a function of the number of processors in a strong scaling study. We see that SuperLU_DIST does not scale beyond 128 processors for these test matrices. On the other hand, our hybrid solver could still reduce the solution time using 512 processors. Furthermore, even when HIPS converged, our total solution time was less than theirs since our hybrid solver takes full advantage of the state-of-the art software. These numerical results demonstrate the potential of our

[6] PT-SCOTCH and ParMETIS are used for the **tdr190k** and **matrix211** matrices, respectively.

	tdr190k					matrix211			
# procs	2	8	32	128	512	8	32	128	512
Hybrid	750	673	545	621	735	491	461	504	593
HIPS	987	718	719	--	--	875	680	440	--
SuperLU_DIST	768	787	937	1192	1466	1590	1751	1957	2010

Table 2. Total number of nonzeros in the LU and ILU factors.

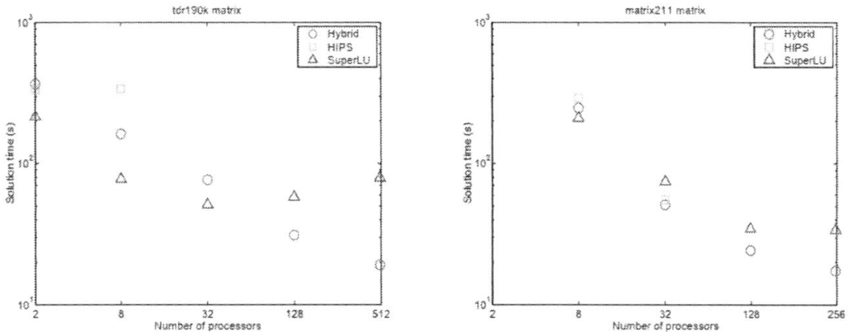

Figure 3. Scalability of time to solve the linear systems on a distributed memory computer.

hybrid solver to be more flexible and robust for solving large highly-indefinite systems on a large number of processors.

We note that HIPS only exploits single-level parallelism, that is, each subdomain solution is performed by only one processor. Then the following dilemma would be faced: when the problem size is large and we need to use large processor count, we will need to generate large number of subdomains, leading to an increase of the Schur complement size as well as the iteration count (sometimes divergence). Our remedy for this problem is to exploit two-level parallelism. That is, we keep the number of interior subdomains fixed and relatively small, and so is the size of the Schur complement, while allowing multiple processors to solve each subdomain problem. The number of processor groups is the same as the number of subdomains. We also use multiple processors to solve the Schur complement system. This ensures that the convergence rate is independent of the number of processors used.

Figure 4 shows the performance of our hybrid solver using multiple processors to factor each interior subdomain. Here, we use larger drop tolerance, $\sigma_1 = 10^{-5}$ and $\sigma_2 = 10^{-4}$, in order to reduce the memory requirement. In the "two-level" case, we fixed the number of subdomains to be 8. The "single-level" case also refers to our code, but keeps the number of subdomains the same as the number of processors. Our two-level parallel scheme also uses less memory than the single-level scheme; it needs less than one-third of the memory needed by SuperLU_DIST. These results show the great potential of this "two-level parallelization" to reduce the memory requirement while achieving good parallel scalability.

3. Conclusions

We are developing a high performance incomplete LU factorization preconditioner and a hybrid direct and iterative solver based on the Schur complement method. The experiments showed that our supernode-based ILUTP is superior to the classic ILUTP. For some fusion problems, it can reduce memory by nine fold, while solving the problem ten times faster. Our hybrid solver is more flexible in exploiting multiple levels of parallelism, and numerically more reliable than the state-of-the-art hybrid solver HIPS. For the highly indefinite accelerator and fusion SciDAC problems, our hybrid solver scales beyond 512 processors, much faster than the parallel direct solver. The memory reduction can be two to four fold.

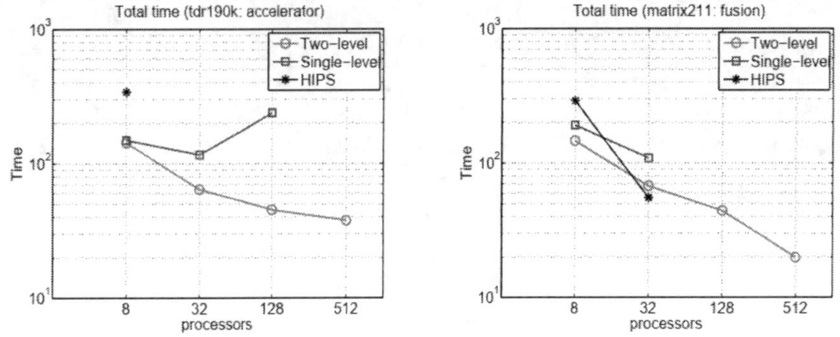

Figure 4. Comparison of single-level and two-level parallel hybrid solver.

Acknowledgments

This research was supported in part by the Director, Office of Science, Office of Advanced Scientific Computing Research, of the U.S. Department of Energy under Contract No. DE-AC02-05CH11231. We used the resources at the National Energy Research Scientific Computing Center.

References

[1] Saad Y 2004 *Iterative methods for sparse linear systems* (Philadelphia: SIAM)
[2] Saad Y 1994 *Numerical Linear Algebra with Applications* **1** 387–402
[3] Bollhöfer M and Saad Y 2006 *SIAM J. Scientific Computing* **27** 1627–1650
[4] Demmel J W, Eisenstat S C, Gilbert J R, Li X S and Liu J W H 1999 *SIAM J. Matrix Analysis and Applications* **20** 720–755
[5] Duff I S and Koster J 1999 *SIAM J. Matrix Analysis and Applications* **20** 889–901
[6] Saad Y 1996 *Iterative methods for sparse linear systems* (Boston, MA: PWS Publishing Company)
[7] Gupta A and George T 2008 Adaptive techniques for improving the performance of incomplete factorization preconditioning Tech. Rep. RC 24598(W0807-036) IBM Research Yorktown Heights, NY
[8] Jardin S C, Breslau J and Ferraro N 2007 *Journal of Computational Physics* **226** 2146–2174
[9] Matrix Market `http://math.nist.gov/MatrixMarket/`
[10] Davis T A University of Florida Sparse Matrix Collection `http://www.cise.ufl.edu/research/sparse/matrices`
[11] Amestoy P, Duff I, Koster J and L'Excellent J Y 2001 *SIAM Journal on Matrix Analysis and Applications* **23** 15–41
[12] Demmel J, Eisenstat S, Gilbert J, Li X and Liu J 1999 *SIAM J. Matrix Analysis and Applications* **20** 720–755
[13] Li X and Demmel J 2003 *ACM Trans. Mathematical Software* **29** 110–140
[14] Axelsson O 1994 *Iterative solution methods* (New York: Cambridge University Press)
[15] van der Vorst H 2003 *Iterative Krylov methods for large linear systems* (New York: Cambridge University Press)
[16] Gaidamour J and Henon P 2008 HIPS: a parallel hybrid direct/iterative solver based on a schur complement *Proc. PMAA*
[17] Giraud L, Haidar A and Watson L T 2008 *Parallel Computing* **34** 363–379
[18] Smith B, Bjorstad P and Gropp W 1996 *Domain Decomposition* (New York: Cambridge University Press)
[19] Towards Optimal Petascale Simulation (TOPS) http://www.scalablesolvers.org/
[20] Community Petascale Project for Accelerator Science and Simulation (ComPASS) https://compass.fnal.gov
[21] Center for Extended MHD Modeling (CEMM) URL: http://w3.pppl.gov/cemm/
[22] Henon P and Saad Y 2006 *SIAM J. Sci. Comput* **28** 2266–2293
[23] Karypis Lab, Digital Technology Center, Department of Computer Science and Engineering, University of Minesota METIS - Serial Graph Partitioning and Fill-reducing Matrix Ordering http://glaros.dtc.umn.edu/gkhome/metis/metis
[24] Laboratoire Bordelais de Recherche en Informatique (LaBRI) SCOTCH - Software package and libraries for graph, mesh and hypergraph partitioning, static mapping, and parallel and sequential sparse matrix block ordering http://www.labri.fr/perso/pelegrin/scotch/
[25] Mathematics and Computer Science Division, Argonne National Laboratory The portable, extensible, toolkit for scientific computation (PETSc) www.mcs.anl.gov/petsc

Large-scale electromagnetic modelings based on high-order methods: nanoscience applications

Misun Min[†], Paul Fischer[†], Jason Montgomery[‡], Stephen Gray[‡]

[†]Mathematics and Computer Science Division, Argonne National Laboratory, IL 60439, USA
[‡]Chemistry Division, Argonne National Laboratory, IL 60439, USA

E-mail: mmin@mcs.anl.gov

Abstract. This paper presents large-scale computations and theoretical or computational aspects of the spectral element methods for solving Maxwell's equations that have potential applications in nanoscience for surface-enhanced Raman scattering (SERS) and solar cell devices. We study the surface-enhanced electromagnetic fields near the surface of metallic nanoparticles using spectral element discontinuous Galerkin method. We solve Maxwell's equations in time-domain and provide accuracy and efficiency of our method compared to the conventional finite difference method. We demonstrate light transmission properties for nanoslab and nanoslits, and time-averaged electric fields over the cross sections of nanoholes in a hexagonal array.

1. Introduction

This paper presents numerical algorithms for solving strongly enhanced scattering fields refracted from metallic device at nanoscale. An important contribution to the enhanced fields is due to surface plasmon (SP) excitations near the metal surface. SP excitations change light transmission or reflection properties when the metallic nanostructures such as nanohole or nanowell arrays exposed to different chemicals. The surface sensitive technology will be able to detect unique signatures of chemical and biological structures at molecular level.

The numerical challenge is to efficiently resolve the enhancement interactions with as few as grid points possible and to accurately represent the sharp jumps in material properties that arise at metal interfaces. Lower-order methods such as first- or second-order approximation, including the popular finite-difference time-domain (FDTD) method, have limitations in representing the materials accurately for complex geometries because of staircase discretization errors. Optimizing response of such nanomaterial devices involves many variables in the structure with size Very small grid spacing and very large computational resource are required to obtain accurate fields. Our studies shows that FDTD method faild to provide reasonable profiles near the metal surface even with very fine resolution.

Here we use the spectral element discontinuous Galerkin (SEDG) method with meshes conforming the complex geometry of nanostructured materials with sharp edges and nanosholes with multilayer composite of metals and dielectrics. The enhanced local field are dependent on the polarization of the incident light, nanoparticle size, the thickness of layers, etc. We will demonstrate numerical studies of the different parameters and compared with FDTD results.

2. Formulation

We consider the Maxwell's equations with Drude model in SI unit. With an auxiliary differential equation (ADE) for the Drude model defined as in [3], [4], our governing equations are written as follows:

$$\epsilon\frac{\partial H}{\partial t} = -\nabla \times E, \tag{1}$$
$$\mu\frac{\partial E}{\partial t} = \nabla \times H - J, \tag{2}$$
$$\frac{\partial J}{\partial t} = \alpha J + \beta E, \tag{3}$$

where the field vectors are defined as $E = (E_x, E_y, E_z)$, $H = (H_x, H_y, H_z)$, and $J = (J_x, J_y, J_z)$ for electric density, magnetic density, and current density, respectively. The phenomenological parameters in dielectric media are

$$\epsilon = \epsilon_0\epsilon_r, \quad \mu = \mu_0\mu_r, \quad \alpha = 0, \quad \text{and} \quad \beta = 0 \tag{4}$$

and in the metallic region are

$$\epsilon = \epsilon_0\epsilon_r, \quad \mu = \mu_0\mu_r, \quad \alpha = -\Gamma_p, \quad \text{and} \quad \beta = \epsilon_0\omega_p^2. \tag{5}$$

The parameters ϵ_0, μ_0, ϵ_r, μ_r, Γ_p, and ω_p indicate permittivity of free space, permeability of free space, relative permittivity, relative permeability, Drude damping coefficient, and plasmon frequency. The relative permeability is $\mu_r = 1$ for most of the cases and the parameters ϵ_r, μ_r, Γ_p, and ω_p are defined as follows:

$$\epsilon_r = 1.0 \text{ (air)}, \ \epsilon_r = 1.33^2 \text{ (water)}, \ \text{and} \ \epsilon_r = 1.5^2 \text{ (glass)}, \tag{6}$$
$$\epsilon_r = 1.0, \ \Gamma_p = 0.2367 \text{ eV}, \text{ and } \omega_p = 7.32318 \text{ eV (gold)}. \tag{7}$$

Our simulations are based on the spectral element discontinuous Galerkin (SEDG) code, NekCEM, which has been desigend for computational electromagnetics on large-scale parallel platforms. Our SEDG method uses tensor-product of N-th order Legendre-Lagrange interpolation polynomials based on the Gauss-Legendre-Lobatto grids, resulting in $n \approx EN^3$ degrees of freedom for each field component [1].

3. Numerical convergence

We present validation of NekCEM with light transmission calculations for 2D nanoslab and 2D nanoslits provided with comparison to the FDTD results. Figures 1 shows nanoslab with 24 nm thickness. Figure 2 shows nanoslits with 24 nm thicness and 100 nm distance between slits. The nanoslab and nanoslits are placed between air and glass. For the configuration of Figure 1, we can obtain analytic solutions for the transmission coefficient for various range of wavelength. Figures 3 and 4 show the transmission coefficient curves for the wavelengths between 300 and 1000 on different mesh sizes with NekCEM and FDTD. We observe that NekCEM results of $N = 5$ are equivalent to the results of FDTD with 0.5 nm case. From the Figures 1-2, one can see the spectral element mesh. For N=5, it is approximately equal to 4 or 5 nm mesh size in FDTD. But the level of accuracy is close to the case of 0.5 nm of FDTD. Tables 1 and 2 show the convergence of NekCEM and FDTD for the wavelength ω=501.11017 nm. The coarse grid with $N = 3$ or $N = 5$ already reached to 6 digit accuracy for NekCEM, whereas FDTD results are only 3 digit accuracy even with 5 times finer grids with 0.5 nm grid spacings.

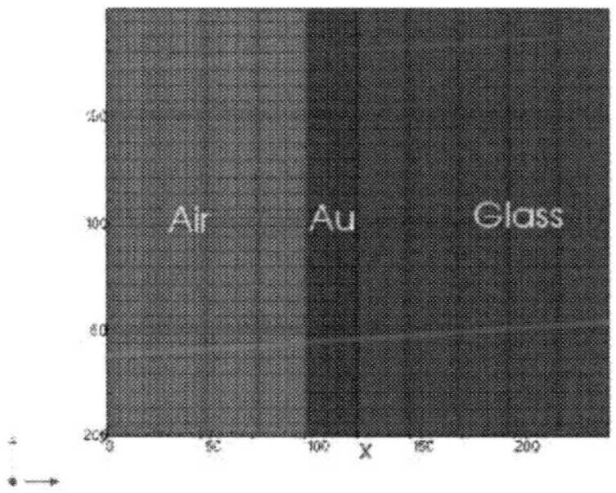

Figure 1. 2D nanoslab with size of 24 nm.

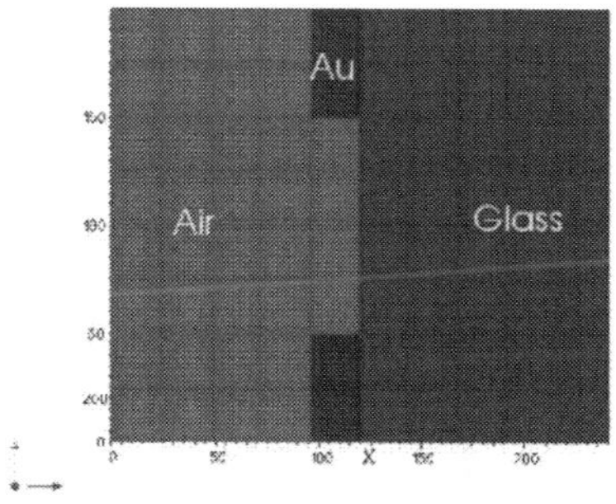

Figure 2. 2D nanoslits with size of 24 nm.

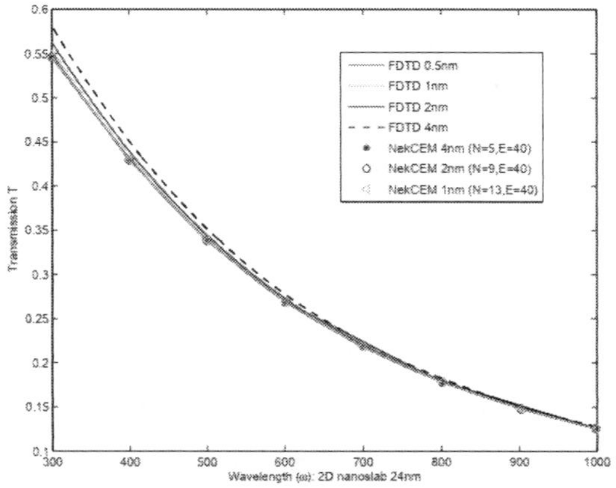

Figure 3. 2D nanoslab: Transmission T for wavelengths ω=300-1000.

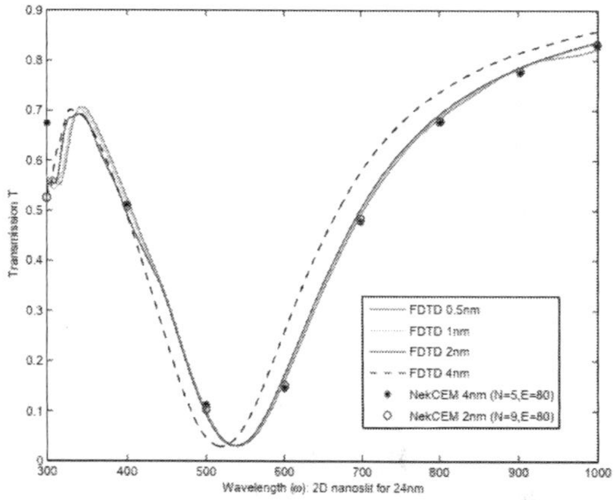

Figure 4. 2D nanoslits: Transmission T for wavelengths ω=300-1000.

Table 1. Convergence comparison for 2D nanoslab

	T= 0.3375175655	(analytic solution)	for ω= 501.11017 nm	
	NekCEM		FDTD	
N= 3	T= 0.33743298	err=8.45e-05		
N= 5	T= 0.33751212	err=5.43e-06	T= 0.35095084 (dx=4.0 nm)	err=1.34e-02
N= 9	T= 0.33751590	err=1.66e-06	T= 0.34368324 (dx=2.0 nm)	err=6.16e-03
N=13	T= 0.33751589	err=1.67e-06	T= 0.34046561 (dx=1.0 nm)	err=2.94e-03
N=15	T= 0.33751589	err=1.67e-06	T= 0.33895832 (dx=0.5 nm)	err=1.44e-03

4. Nanohole applications and parallel performance

We use meshes that conform to the materialwith high-order polynomial approximations [1] which enable us to handle complex geometry of nanostructured materials with sharp edges with

Table 2. Comparison for 2D nanoslits

	T = N/A (analytic solution)	for ω= 501.11017 nm
	NekCEM	FDTD
N= 3	T= 0.1090288060495094	
N= 5	T= 0.1081468232317703	T= 4.43053916E-02 (dx=4.0 nm)
N= 9	T= 9.9669399705373168E-002	T= 8.96057561E-02 (dx=2.0 nm)
N=11	T= 9.8431225244869594E-002	T= 9.35624540E-02 (dx=1.0 nm)
N=13	T= 9.7112908185410532E-002	T= 9.56578478E-02 (dx=0.5 nm)

multilayer composite of metals. The enhancement factor and local field distribution are studied depending on the polarization of the incident excitation light, nanoparticle size, the thickness of layers, etc. We have simulated a unit-cell of 3D nanohole system on hex-lattice. The radii of nanoholes are 50 nm and their center-to-center distances are 200 nm. The thickness of nanogold is 25 nm with the radii of nanoholes of 50 nm and their center-to-center distances of 200 nm. The thin nanosilver with holes are placed on the glass substrate and filled with water on the top. Figure 5 shows time-averaged E^2 profile at the bottom of silver with y-polarized incident field. Our SEDG simulations show that the SP excitations around the metallic nanohole surface are well-captured with physically reasonable profiles as shown in the Figures 5-6 and have good agreement with the results from FDTD method on finer grids.

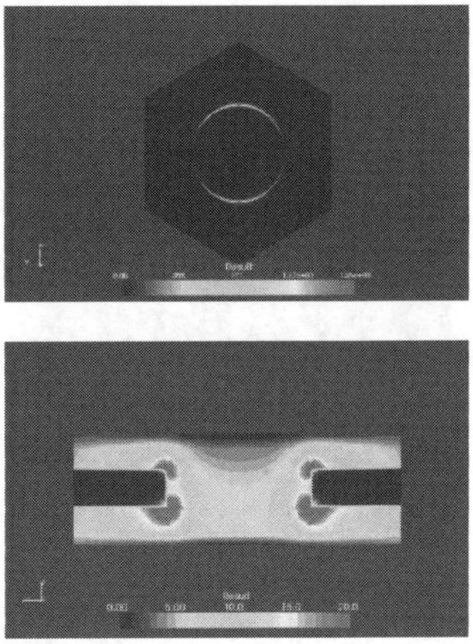

Figure 5. 3D nanohole with radius of 50 nm.

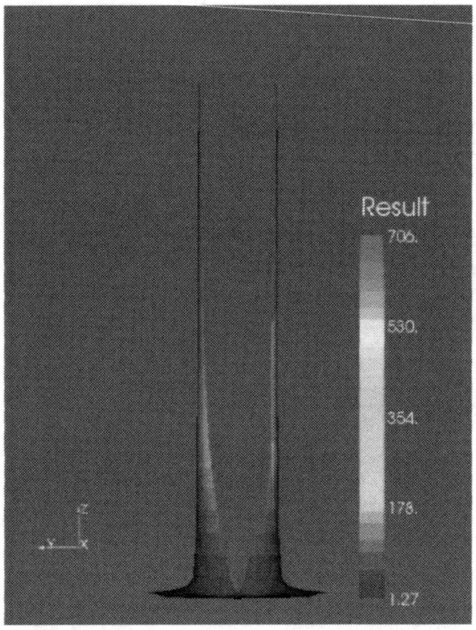

Figure 6. 3D nanohole with radius of 50 nm.

Table 3 shows its parallel performance and scalability for NekCEM. NekCEM achieves 73 percent efficiency at P = 4096 for n/P = 2734, which is a relatively small number of points per processor. Scaling studies of the latest version of Nek5000 reported in the 2008 SciDAC article showing $>$ 50 percent efficiency for Navier-Stokes computations on P = 65,536 for

$n/P = 7,300$ [2]. For similar n and P, we expect NekCEM to acheive > 75 percent efficiency given the improved communication charateristics of the DG formulations which has no edge or corner-exchanges) and absence of elliptic solves in NekCEM.

Table 3. Parallel Performance of NekCEM for a Strong-Scaling Test

nProc	Time (sec)	Ideal	Speedup (Ideal)	Efficiency
64	3.305E+3	3.305E+3	64.0 (64)	1.000
128	1.702E+3	1.652E+3	124.2 (128)	0.970
256	9.520E+2	8.262E+2	222.1 (256)	0.867
512	4.814E+2	4.132E+2	439.5 (512)	0.858
1024	2.831E+2	2.065E+2	747.1 (1024)	0.729
2048	1.400E+2	1.032E+2	1510.9 (2048)	0.737
4096	6.870E+1	5.164E+1	3078.9 (4096)	0.751

5. Conclusion

We studied the surface-enhanced electromagnetic fields near the surface of metallic nanoparticles using spectral element discontinuous Galerkin method. We solve Maxwell's equations in time-domain and provide accuracy and efficiency of our method compared to the conventional finite difference method. We demonstrate time-averaged electric fields over the cross sections of nanoholes in a hexagonal array. We demonstrate parallel performance on the ALCF BG/P showing 75 percent efficiency.

Acknowledgments

This work was supported by the Office of Advanced Scientific Computing Research, Office of Science, U.S. Department of Energy, under Contract DE-AC02-06CH11357.

References

[1] M.O. Deville, P.F. Fischer, E.H. Mund, High-order methods for incompressible fluid flow, *Cambridge University Press, Cambridge, UK*, (2002).

[2] Paul Fischer, James Lottes, David Pointer, Andrew Siegel, Petascale Algorithms for Reactor Hydrodynamics, *Proceedings of SciDAC*, 2008. *Cambridge University Press, Cambridge, UK*, (2002).

[3] S.K. Gray, T. Kupka, Propagation of light in metallic nanowire arrays: Finite-difference time domain studies of silver cylinders, *Physical Review B*, **68**, pp. 045415/1–045415/11, (2003).

[4] J.M. Oliva, S.K. Gray, Theoretical study of dielectrically coated metallic nanowires, *Chemical Physics Letters*, **379**, pp. 325–331, (2003).

[5] M. S. Min, T. W. Lee, P. F. Fischer, S. K. Gray, Fourier spectral simulations and Gegenbauer reconstructions for electromagnetic waves in the presence of a metal nanoparticle, *Journal of Computational Physics*, **213 (2)**: pp. 730-747, (2006).

[6] A. Taflove, S.C. Hagness, Computational electrodynamics, the finite difference time domain method, *Artech House, Norwood, MA*, (2000).

[7] J.S. Hesthaven, T. Warburton, Nodal discontinuous Galerkin methods, Springer, New York, (2008).

Scalable parallel solution coupling for multiphysics reactor simulation

Timothy J. Tautges and Alvaro Caceres

Argonne National Laboratory, Argonne, IL 60439, USA

{tautges, acaceres}@mcs.anl.gov

Abstract. Reactor simulation depends on the coupled solution of various physics types, including neutronics, thermal/hydraulics, and structural mechanics. This paper describes the formulation and implementation of a parallel solution coupling capability being developed for reactor simulation. The coupling process consists of mesh and coupler initialization, point location, field interpolation, and field normalization. We report here our test of this capability on an example problem, namely, a reflector assembly from an advanced burner test reactor. Performance of this coupler in parallel is reasonable for the chosen problem size and range of processor counts. The runtime is dominated by startup costs, which amortize over the entire coupled simulation. Future efforts will include adding more sophisticated interpolation and normalization methods, to accommodate different numerical solvers used in various physics modules and to obtain better conservation properties for certain field types.

1. Introduction

Simulation applications are evolving toward more complexity in terms of both the physics modeled and the length scales at which they are modeled. These multiphysics and multiscale codes share a common need to transfer solutions from one discretization to another. In many cases, a multiphysics or multiscale code is assembled from several, often stand-alone codes. For these codes, solution transfer must consider the overall process, including not only the mathematical details of mapping from one mesh to another, but also the efficient communication and mapping between meshes whose distribution on the parallel machine may not have been chosen to optimize solution transfer. In this paper, we discuss a modular approach to solution transfer that accommodates a variety of mesh types, mapping functions, and multi-physics solution frameworks.

The Interoperable Technologies for Advanced Petascale Simulations (ITAPS) project is developing common interfaces to mesh (iMesh), geometry (iGeom), and other data, and services based on them [1]. The interfaces are designed to be data-structure-neutral and can be called directly from C, C++, and Fortran applications. iMesh defines a data model consisting of entities (the finite element plus polygons and polyhedra), entity sets (arbitrary groupings of entities and other sets), the interface instance or "root set," and tags (data annotated on members of any of the other data types). Services are available for mesh smoothing [2] and adaptive mesh refinement [3], to name a few. The ITAPS interfaces are being used as the basis for the SHARP reactor simulation project [4], which performs the coupled solution of thermal/hydraulics and neutronics for fast reactor cores. Solution transfer plays a major role in this effort.

The solution mapping process consists of transferring a field from a source mesh to a target mesh, where the source and target meshes are each used by a specific physics, are represented in iMesh, and are distributed over a parallel machine. In this work we focus on solution transfer for overlapping 3D domains. Motivated by a component-based approach to multiphysics code development, we make the following assumptions when designing solution transfer capability:

1. Each physics component is free to choose the mesh type and parallel distribution best for that physics, without regard for its suitability for solution transfer.

2. Each physics type and mesh are distributed across a set of processors, defined by an MPI communicator for each mesh

3. If multiple physics types are assigned to a given processor, the mesh for all physics on that processor is stored in a single iMesh instance, and that instance communicates with all other processors containing pieces of any of those meshes.

Considerable work has been done on the topic of solution transfer (e.g., see [5-6]). Many papers focus on the mathematical properties of mapping solutions, with various constraints imposed which are specific to the physics being solved. Also, various mapping functions are used, depending on the shape functions used to compute the field. Despite these variations, the various approaches to parallel solution transfer have many similarities. For example, most approaches require locating one mesh in another, communicating fields between processors, and normalizing results after interpolation (though the method actually used in that normalization may vary). Hence, we approach the problem of solution transfer from a process point of view, where distinct steps are implemented in a modular fashion. This allows substitution of one or more specific steps, e.g. mapping function, while reusing other parts of the implementation.

2. Parallel solution transfer algorithm

The parallel solution transfer process comprises several functions.

Initialization: We assume the mesh has been read onto the machine and is initialized (with parallel interfaces) according to the assumptions above. Parallel details of the mesh (interface and partitioning sets, communicator, etc.) are encapsulated in the iMesh instance. In order to facilitate point location on a source mesh, a kdtree decomposition is computed for the local part of the source mesh on each processor. In addition, the bounding box corner points for the root node of this tree on each processor is sent to all processors holding target mesh. These box coordinates are used to narrow the search for source elements containing each target point, which reduces parallel communication during this process.

Point location: Point location is the process of choosing points at which data will be coupled (usually derived from a target mesh), and locating the processor(s), element(s), and parametric coordinates in those element(s) that contain each target point. The source element(s) containing a given point can be on the local or remote processor. The algorithm used in this phase is shown in figure 1. In this case, target points are derived from the target mesh but need not be restricted to vertices in that mesh. This algorithm reduces data communication by storing point location data on the source mesh processor and passing an index into that list to the target mesh processor requesting the interpolation point. This strategy incurs a small extra storage cost (one tuple of three integers per target point) but reduces communication during the interpolation phase. All processors holding source and target mesh participate in the scatter/gather communication. In theory this communication could be optimized in cases where the target and source processor sets were disjoint (no processor held both source and target mesh). However, it is unclear whether that approach would save much communication time, or whether it would be worth the added complexity in the application.

Choose target points t_i
$\forall$ t_i:
$\forall$ root box B_j containing t_i:
If *j != p* add tuple *Tu1(j, i, xyz)*, where *xyz* are the coordinates of t_i
Else add tuple *Tu_tmp(i)*
Scatter/gather *Tu1(j, i, xyz)*, using *j* as destination processor; *j* gets replaced on destination processor with source processor
$\forall$ *tu1* $\in$ *Tu1*
Traverse the source mesh kdtree to find the leaf node(s) whose boxes contain t_i
$\forall$ element *e* in each leaf node, find the natural coordinates *pqr* in *e*
If *pqr* $\in$ *[0, 1]*, add tuple *Tu2(e, pqr)[k]*, and tuple *Tu3(j, i, k)[l]*
Else add tuple *Tu3(j, i, -1)[l]*
Scatter/gather *Tu3(j, i, k)*, using *j* as destination processor; *j* gets replaced on destination processor with source processor
$\forall$ *tu3* $\in$ *Tu3(j, i, k)*, if *k* != -1, add tuple *Tu4(j, i, k, f)[l]*, where *f* is a double-precision parameter used in the interpolation phase
Repeat steps 4a-c for all tuples in *Tu_tmp*; in step 4c, put contained points in *Tu5(i,k)* instead of *Tu3*

Figure 1. Algorithm for point location phase of solution coupling.

Interpolation: During the execution of a physics component, field values are computed and stored on various entities in the source mesh on each processor. This process may be repeated for iterations working toward some steady state or as part of a time advancement calculation. The field values on this source mesh must be interpolated onto the target mesh. The algorithm used in the interpolation stage is described in figure 2. As described in the introduction, our approach is modular, allowing the application to choose the type of shape function used for the interpolation. In our current implementation, the shape functions are implemented in the iMesh/MOAB library (discussed later). The only requirement in this implementation is that the location of the target point within the element be represented using three double precision numbers whose bounds are between -1 and 1.

Choose field to interpolate, *f*
Scatter/gather *Tu4(j, i, k, f)* using *j* as the destination processor; *j* is replaced with the source processor *i*
$\forall$ *Tu4(j, i, k, f)[l]*, interpolate *f(e, pqr)*, *e, pqr* $\in$ *Tu3(e, pqr)[k]* $\rightarrow$ *f* $\in$ *Tu4(j, i, k, f)[l]*
Scatter/gather *Tu4(j, i, k, f)* using *j* as the destination processor; *j* is replaced with the source processor *i*
$\forall$ *Tu4(j, i, k, f)[l]*, *f* $\rightarrow$ *f[i]*
$\forall$ *Tu5(i, k)[l]*, interpolate *f(e, pqr)*, *e, pqr* $\in$ *Tu3(e, pqr)[k]* $\rightarrow$ *f[i]*

Figure 2. Algorithm for interpolation phase of solution coupling.

Normalization: After interpolation, a field is assigned values at target points. If repeated interpolations are done from one mesh to another and back again, the values of the field can "drift" because of numerical roundoff error and other factors, depending on the shape function evaluations. In some cases this drift can be disregarded, but in other cases it can add or remove energy in the physical system. To avoid this situation, the coupler also provides for normalizing some integrated quantity over the target mesh to match the same integrated quantity on the source mesh. Currently, we implement only the global integrated quantity, according to the integration function associated with the shape function used in the interpolation (discussed later).

3. Coupling results

Figure 3 shows the ABTR reflector assembly model, with different colors used to indicate different geometric volumes. The source mesh consists of 12k hexahedral elements, while the target mesh contains 130k tetrahedral elements. For these examples, an analytic function is defined at hexahedral mesh vertices, and this field is mapped to the target mesh vertices based on trilinear basis functions.

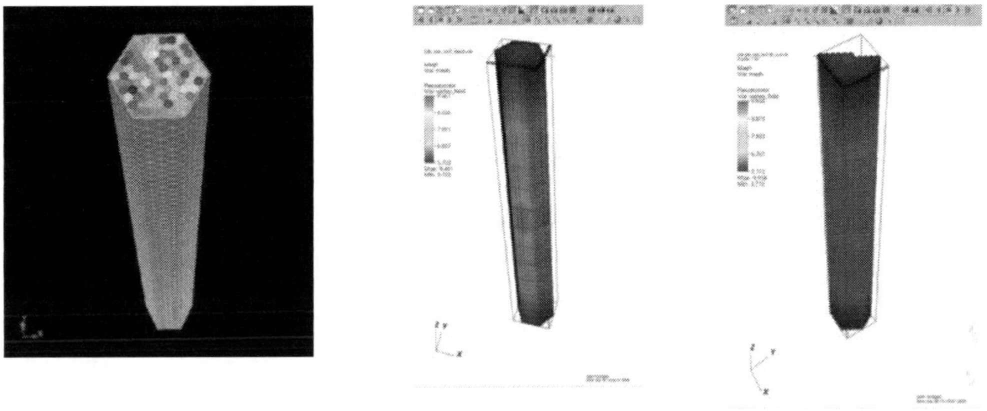

Figure 3. Geometry and target mesh for the ABTR reflector (left); source field defined on hexahedral mesh (middle); transferred field on target tetrahedral mesh (right). Note that this shows a "pseudocolor" plot, which interpolates the vertex values to show a smoothly varying field.

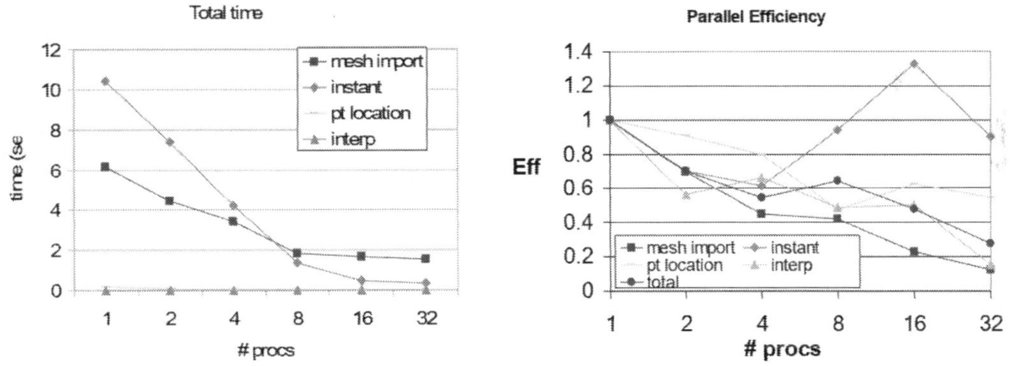

Figure 4. Execution time for coupling (left), parallel efficiency (right).

Run times for the various coupling steps are shown in Figure 4; normalization time was not measured. Note that the "Initialization" step described in Section 2 includes both "mesh import" and "instantiation" items in Figure 4. The data shown is for coupling the same meshes on increasing numbers of processors. In absolute terms, the initialization of the problem dominates the other phases in this problem, with the interpolation phase accounting for a maximum of 0.06%, for the 32-processor run. For problems where the mesh is not deforming, the interpolation stage is the only one repeated for each coupling step, therefore these times will be acceptable. Parallel efficiency shows mixed results, with poor scaling for mesh import and interpolation and moderate to good scaling for instantiation and point location. Although interpolation time appears to scale quite poorly, absolute times for this portion of the example were below 0.001 seconds, which is close to the clock tick size. Timing results should be generated on larger problems to better assess scaling of this step.

Several other conclusions can be drawn from this data. First, the mesh import time is not being reduced as quickly as other times with increasing processor counts. The capability to read only those portions of mesh assigned to a given processor is already under development.

Second, the point location algorithm described in figure 1 includes a sequential search over the bounding box extents for each processor for each point being interpolated. This step will grow in cost as the number of processors increases. We intend to develop a tree-based search of those boxes as well, bringing scaling closer to O(log(p)) than to O(p) for this part of point location. Since the tree structure is already implemented, the development of this capability will be straightforward.

Third, we note that normalization of the solution over all processors is currently not done because a vertex-based field is being mapped in these examples. That step should not require a great deal of execution time, however, since most of the element-based calculations are local to each processor, followed by a reduction of results over all processors

4. Summary and future work

This paper describes the formulation and implementation of a parallel solution coupling capability. The coupling process consists of mesh and coupler initialization, point location, interpolation, and normalization. Coupling results show good qualitative agreement with the field being coupled. Scaling varies for the process steps; the interpolation step scales poorly but takes a negligible amount of absolute time compared to other steps. In the case of static mesh, this step is the only one repeated for each iteration or time step. Larger examples must be run to investigate whether this step scales well.

The implementation of solution transfer in MOAB will be extended in various ways. More interpolation methods will be added as needed, based on the physics components being coupled. Normalization over specific subsets in the mesh is needed, to focus on specific subregions in the problem. This extension will be straightforward, based on material sets defined in the source and target meshes. We will also study tuning the interpolation and normalization methods to achieve various conservation properties. This is a research topic that will be of great interest to the scientific community as coupled multiphysics simulations become more common.

Acknowledgments

This work was supported by the U.S. Department of Energy's Scientific Discovery through Advanced Computing program under Contract DE-AC02-06CH11357.

References

[1] K.K. Chand, L.F. Diachin, X. Li, C. Ollivier-Gooch, E.S. Seol, M.S. Shephard, T. Tautges, and H. Trease, "Toward interoperable mesh, geometry and field components for PDE simulation development," *Engineering with Computers*, vol. 24, 2008, pp. 165-182.

[2] Patrick M. Knupp, "Applications of mesh smoothing: copy, morph, and sweep on unstructured quadrilateral meshes," *International Journal for Numerical Methods in Engineering*, vol. 45, 1999, pp. 37-45.

[3] M.S. Shephard, K.E. Jansen, O. Sahni, and L.A. Diachin, "Parallel adaptive simulations on unstructured meshes," *Journal of Physics: Conference Series*, vol. 78, 08, 2007, p. 012053.

[4] A. Siegel, T. Tautges, A. Caceres, D. Kaushik, P. Fischer, G. Palmiotti, M. Smith, and J. Ragusa, "Software design of SHARP," *Joint International Topical Meeting on Mathematics and Computations and Supercomputing in Nuclear Applications, M and C + SNA 2007*, La Grange Park, IL 60526, United States: American Nuclear Society, 2007, p. 13.

[5] J. Grandy, "Conservative Remapping and Region Overlays by Intersecting Arbitrary Polyhedra," *Journal of Computational Physics*, vol. 148, 1999, pp. 433-466.

[6] L. Margolin, "Second-order sign-preserving conservative interpolation (remapping) on general grids," *Journal of Computational Physics*, vol. 184, 2003, pp. 266-298.

2D and 3D core-collapse supernovae simulation results obtained with the CHIMERA code

S. W. Bruenn[1], A. Mezzacappa[2], W. R. Hix[2], J. M. Blondin[3], P. Marronetti[1], O. E. B. Messer[4], C. J. Dirk[1] and S. Yoshida[5]

[1] Physics Department, Florida Atlantic University, 777 W. Glades Road, Boca Raton, FL 33431-0991

[2] Physics Division, Oak Ridge National Laboratory, Oak Ridge, TN 37831–6354

[3] Department of Physics, North Carolina State University, Raleigh, NC 27695-8202

[4] Center for Computational Sciences, Oak Ridge National Laboratory, Oak Ridge, TN 37831–6354

[5] Max-Planck-Institut fur Gravitationsphysik, Albert Einstein Institut, Golm, Germany

E-mail: bruenn@fau.edu

Abstract. Much progress in realistic modeling of core-collapse supernovae has occurred recently through the availability of multi-teraflop machines and the increasing sophistication of supernova codes. These improvements are enabling simulations with enough realism that the explosion mechanism, long a mystery, may soon be delineated. We briefly describe the CHIMERA code, a supernova code we have developed to simulate core-collapse supernovae in 1, 2, and 3 spatial dimensions. We then describe the results of an ongoing suite of 2D simulations initiated from a 12, 15, 20, and 25 $M_\odot$ progenitor. These have all exhibited explosions and are currently in the expanding phase with the shock at between 5,000 and 20,000 km. We also briefly describe an ongoing simulation in 3 spatial dimensions initiated from the 15 $M_\odot$ progenitor.

1. Introduction

The neutrino transport mechanism is still the leading explosion mechanism candidate for core-collapse supernova arising from stars with initial mass of $\sim 8 - 25\ M_\odot$. Originally proposed by (1), it asserts that the fraction of neutrinos absorbed by the outer layers that are released by the inner core during and following its collapse deposit sufficient energy to power the explosion. For many years this mechanism was investigated by computer modeling with the assumption of spherical symmetry, and by the turn of this century these simulations had achieved considerable realism with codes employing Boltzmann neutrino transport, state-of-the-art neutrino interactions, and general relativity (2; 3; 4). These simulations found that the shock wave, launched by the bounce of the core, stalled at a radius between $\sim$ 100 - 200 km rather than giving rise to an explosion.

While the spherically symmetric simulations failed to explode, other investigations taking place during this period of time pointed to the importance of multi-dimensional effects. In particular, a variety of fluid instabilities are present or quickly develop in the immediate post-bounce core (5). The most important of these for the neutrino heating mechanism is neutrino-driven convection—convection driven by the neutrino heating above the neutrinosphere. Because neutrinos heat the bottom of the heating layer most intensely a negative entropy gradient builds up which renders the layer convectively unstable. For convection to grow, however, the fluid must remain in the heating layer for a critical length of time. In particular, the ratio of the advective timescale, $\tau_{\rm adv}$, to some averaged timescale of convective growth, $\tau_{\rm cv}$, must be $\gtrsim 3$ (6). If convection gets established in the heating layer, hot gas from the neutrino-heating region will be transported directly to the shock, while downflows simultaneously will carry cold, accreted matter to the layer of strongest neutrino heating where a part of this gas, being cold, readily absorbs more energy from the neutrinos. The loss of energy accompanying the advection of matter through the gain radius is thereby reduced and more energy stays in the heating layer.

Attaining the magic value 3 of $\tau_{\rm adv}/\tau_{\rm cv}$ is aided by the instability of the stalled shock to low-mode aspherical oscillations, which are referred to as the standing accretion shock instability or 'SASI.' Discovered by (7), the development of the SASI leads to an enlargement of the heating layer in one region and its diminution in another. This results in an enhanced efficiency of neutrino heating and more favorable conditions for convection in the enlarged region, and conditions more favorable for the establishment of down-flows or return-flows in the constricted region. In addition to promoting an explosion, neutrino-driven convection and the SASI can lead to the large asymmetries observed for SN 1987A and other supernovae, and might account for the large observed velocities of neutron stars (8; 9).

2. The CHIMERA Code

The CHIMERA code is designed to simulate core-collapse supernovae in 1, 2, and 3 spatial dimensions and with realistic spectral neutrino transport from the onset of collapse to the order of 1 sec post bounce with the present generation of supercomputers. It conserves total energy (gravitational, internal, kinetic, and neutrino) to within ± 0.5 B, given a conservative gravitational potential. The code currently has three main components: a hydro component, a neutrino transport component, and a nuclear reaction network component. In addition there is a Poisson solver for the gravitational potential and a sophisticated equation of state. A preliminary version of the code was briefly described in (10; 11; 12) and we briefly highlight the main features here.

The hydrodynamics is performed via a Lagrangian remap implementation of the Piecewise Parabolic Method (PPM) (13). A moving radial grid option wherein the radial grid follows the average radial motion of the fluid makes it possible for the core infall phase to be followed with good resolution. Following bounce, an adaptive mesh redistribution algorithm keeps the radial grid between the core center and the shock structured so as to maintain approximately constant $\Delta\rho/\rho$. For 256 and 512 radial zones, this ensures that there are at least 15 and 30 radial zones per decade in density, respectively. The equation of state (EOS) is standard and is described in (10). The algorithm described in (14), modified for greater robustness, was employed to stabilize shocks oriented along grid lines. Gravity is computed by an approximate general relativistic potential for spherical gravity and Newtonian 2D/3D spectral Poisson solver for the higher moments..

Neutrino transport is implemented by a "ray-by-ray-plus" approximation (cf. 15) whereby the lateral effects of neutrinos such as lateral pressure gradients (in optically thick conditions), neutrino advection, and velocity corrections are taken into account, but transport is performed only in the radial direction. Transport is computed by means of multigroup flux-limited diffusion with a sophisticated flux limiter that has been tuned to reproduce results of a general relativistic Boltzmann transport solver to within a few percent (16). All $O(v/c)$ observer corrections have been included. The transport solver is fully implicit and solves for four neutrino flavors simultaneously (i.e., $\nu_{\rm e}$'s, $\bar{\nu}_{\rm e}$'s, ν_μ's and ν_τ's (collectively $\nu_{\rm x}$'s), and $\bar{\nu}_\mu$'s and $\bar{\nu}_\tau$'s (collectively $\bar{\nu}_{\rm x}$'s)), allowing for neutrino-neutrino scattering and pair-exchange, and different ν and $\bar{\nu}$ opacities. State-of-the-art neutrino interactions are included with full energy dependences.

The nuclear composition in the non-NSE regions of these models is evolved by the thermonuclear reaction network of (17). This is a fully implicit general purpose reaction network, however in these models only reactions linking the 14 alpha nuclei from ^{4}He to ^{60}Zn are used. Data for these reactions is drawn from the REACLIB compilations (18).

3. Simulation Results

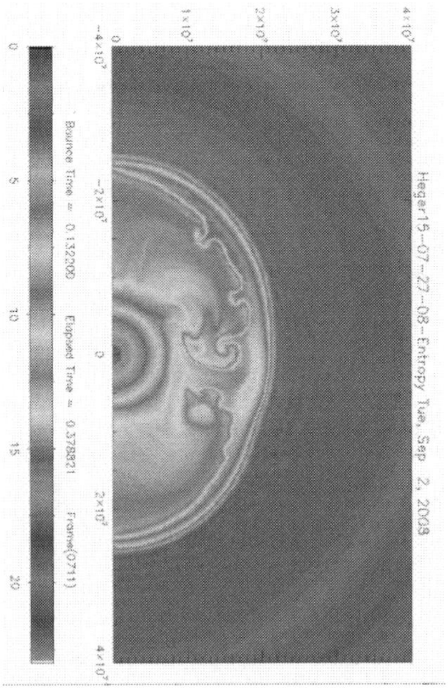

Figure 1. Snapshot of the entropy distribution of the 15 $M_{\odot}$ 2D simulation at 132 ms post-bounce time.

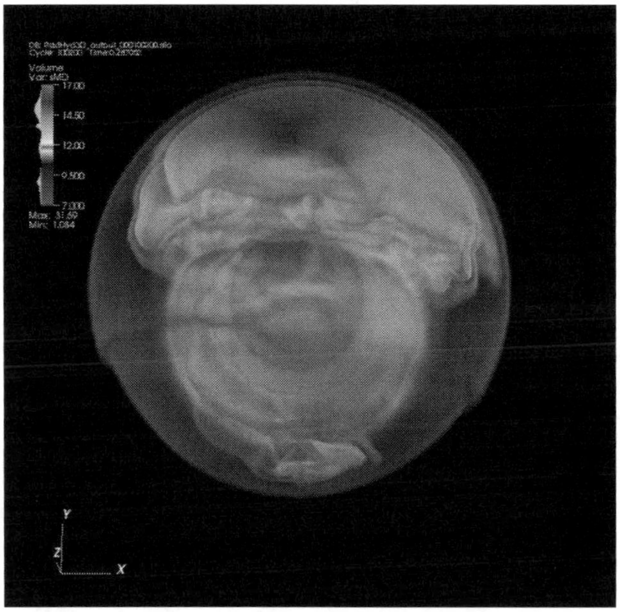

Figure 2. Snapshot of the entropy distribution of the 15 $M_{\odot}$ 3D simulation at 132 ms post-bounce time.

Simulations were performed for 12, 15, 20, and 25 $M_{\odot}$ progenitors (19). A resolution of 256x256 radial and angular zones was used. The simulations were repeated for the 15 $M_{\odot}$ progenitor with resolutions of 512x128 and 512x256. With the many improvements in the CHIMERA code and additions of more realistic physics (e.g., GR corrections, state-of-the-art neutrino microphysics) since our first results were reported in (10) we find that all our models explode as before, but now improved neutrino microphysics rather than nuclear reactions play a key role. For a period of time after bounce (65 ms for the 12 and 15 $M_{\odot}$ models, 105 ms and 130 ms respectively for the 20 and 25 $M_{\odot}$ models) the 2D and corresponding spherically symmetric simulations track each other very closely. During this period the heating layer becomes convectively unstable but its growth is suppressed in the 2D simulations because of the too rapid inflow of material. At the end of this period $\tau_{\rm adv}/\tau_{\rm cv}$ exceeds 3 in some regions and convective mushrooms begin to appear in the heating layer and at almost the same time the shock begins to exhibit SASI dipole and quadrupole deformations along the polar axis. This expands the heating layer in the polar regions and the enhanced convection in these regions quickly develops into a large-scale overturn (Fig. 1), with material rising and expanding in the polar regions and flowing towards the equatorial region. By 225 ms post-bounce for the 15 $M_{\odot}$, as an example, large lobes of high-entropy neutrino heated gas are pushing the shock outwards in the polar regions while an equatorial accretion funnel has developed which is channeling lower-entropy newly shocked material down to the gain layer. The narrow end of the accretion funnel tends to oscillate slowly from one hemisphere to the other, seeming thereby to alternately pump up one hemisphere and then the other with energy by the enhanced neutrino emission at its base.

Eventually a runaway condition is met and the shock begins to accelerate rapidly outwards. Measured by the first signs of positive radial post-shock velocities (a different criterion than that used by the Garching group) explosions commence at roughly 300 ms from bounce for the four models. The 12, 15, and 25 $M_\odot$ models exhibit a pronounced quadrupolar prolate configuration of the shock as the explosion commences, while the 20 $M_\odot$ the configuration of the shock is more dipolar. The ratio of the major to minor axis of the shock ranges from 1.5 (25 $M_\odot$ model) to 2.0 (15 $M_\odot$ model). The explosion energies, computed conservatively as the sum of the kinetic, internal thermal, and kinetic energies of outward moving material is shown in Figure 3. These explosion energies are clearly asymptoting to observed supernovae energies.

A medium resolution simulation in 3 spatial dimensions with 304 radial zones, 76 angular, and 152 azimuthal zones initiated from the 15 $M_\odot$ progenitor is also in progress. Figure (2) shows a the core configuration at the same post-bounce time as the corresponding 2D simulation shown in Figure (1). The 3D simulation is clearly exhibiting large scale convection patterns emerging from the poles at this time. It's further evolution should validate or throw into question conclusions drawn from 2D simulations.

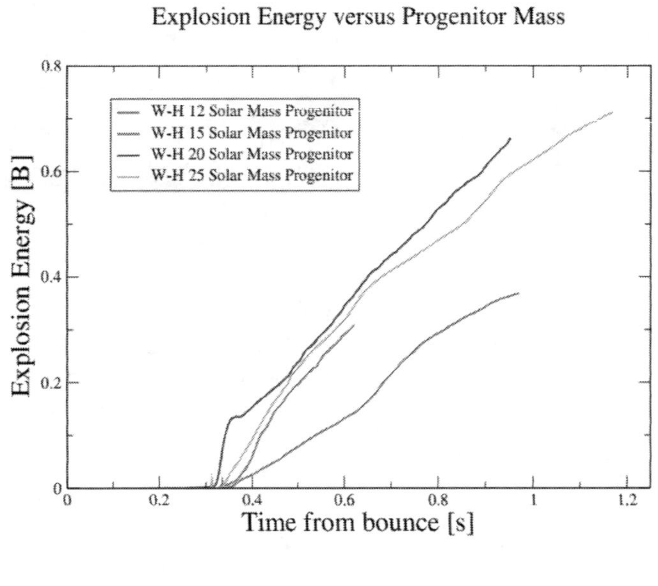

Figure 3. Explosion energies as a function of post-bounce time.

Acknowledgments

The authors would like to acknowledge partial funded by the U.S. Department of Energy under Contract No. DEAC05-00OR22725 at Oak Ridge National Laboratory, the resources of the National Center for Computational Sciences at Oak Ridge National Laboratory, and a TACC Ranger (TG-MCA08X010) computational award. S. W. B., P. M. and O. B. E. M. acknowledge partial support from an NSF-OCI-0749204 award, S. W. B., P. M., O. B. E. M., W. R. H., and A. M. acknowledge partial support from NASA award (07-ATFP07-0011). A.M., W.R.H., and O.E.B.M. are supported at the Oak Ridge National Laboratory, managed by UT-Battelle, LLC, for the U.S. Department of Energy under contract DE-AC05-00OR22725.

References

[1] Colgate S A and White R H 1966 *Astrophysical Journal* **143** 626–681

[2] Mezzacappa A, Liebendörfer M, Messer O E, Hix W R, Thielemann F K and Bruenn S W 2001 *Physical Review Letters* **86** 1935–1938

[3] Liebendörfer M, Mezzacappa A, Thielemann F K, Messer O E, Hix W R and Bruenn S W 2001 *Physical Review D* **63** 103004

[4] Liebendörfer M, Rampp M, Janka H T and Mezzacappa A 2005 *The Astrophysical Journal* **620** 840–860

[5] Bethe H A 1990 *Reviews of Modern Physics* **62** 801–866

[6] Foglzzo T, Scheck L and Janka H T 2006 *The Astrophysical Journal* **652** 1436–1450

[7] Blondin J M, Mezzacappa A and DeMarino C 2003 *Astrophysical Journal* **584** 971–980

[8] Kifonidis K, Plewa T, Janka H T and Müller E 2006 *Astronomy and Astrophysics* **453** 661–678

[9] Scheck L, Plewa T, Janka H T, Kifonidis K and Müller E 2007 *Astronomy and Astrophysics* **457** 963–986

[10] Bruenn S W, Dirk C J, Mezzacappa A, Hayes J C, Blondin J M, Hix W R and Messer O E B 2006 *Journal of Physics* Conference Series (IOP) pp 393–402

[11] Mezzacappa A, Bruenn S W, Blondin J M, Hix W R and Messer O B E 2006 *The Multicolored Landscape of Compaot Objects and their Explosive Origins* (*AIP Conference Proceedings* vol 924) (American Institute of Physics) pp 234–242

[12] Bruenn S W, Mezzacappa A, Hix W R, Blondin J M, Marronetti P, Messer O E B, Dirk C J and Yoshida S 2009 *Proceedings of the Cefalu', Sicily International Astronomy Meeting "Probing Stellar Populations out to the Distant Universe", September 7-19, 2008, in press*

[13] Colella P and Woodward P R 1984 *Journal of Computational Physics* **54** 174–201

[14] Sutherland R S, Bisset D K and Bicknell G V 2003 *Astrophysical Journal Supplement* **147** 187–195

[15] Buras R, Rampp M, Janka H T and Kifonidis K 2006 *Astronomy and Astrophysics* **447** 1049–1092

[16] Liebendörfer M, Messer O E B, Mezzacappa A, Bruenn S W, Cardall C Y and Thielemann F K 2004 *Astrophysical Journal Supplement* **150** 263–316

[17] Hix W R and Thielemann F K 1999 *Journal of Computational and Applied Mathematics* **109** 321–351

[18] Rauscher T and Thielemann F K 2000 *Atomic Data and Nuclear Data Tables* **75** 1–351

[19] Woosley S E and Heger A 2007 *Physics Reports* **442** 269–283

Journal of Physics: Conference Series **180** (2009) 012019 doi:10.1088/1742-6596/180/1/012019

Hybrid petacomputing meets cosmology: the Roadrunner Universe project

Salman Habib, Adrian Pope, Zarija Lukić, David Daniel, Patricia Fasel, Nehal Desai, Katrin Heitmann, Chung-Hsing Hsu, Lee Ankeny, Graham Mark, Suman Bhattacharya and James Ahrens

Los Alamos National Laboratory, Los Alamos, New Mexico, NM 87545

E-mail: habib@lanl.gov

Abstract. The target of the Roadrunner Universe project at Los Alamos National Laboratory is a set of very large cosmological N-body simulation runs on the hybrid supercomputer Roadrunner, the world's first petaflop platform. Roadrunner's architecture presents opportunities and difficulties characteristic of next-generation supercomputing. We describe a new code designed to optimize performance and scalability by explicitly matching the underlying algorithms to the machine architecture, and by using the physics of the problem as an essential aid in this process. While applications will differ in specific exploits, we believe that such a design process will become increasingly important in the future. The Roadrunner Universe project code, MC^3 (Mesh-based Cosmology Code on the Cell), uses grid and direct particle methods to balance the capabilities of Roadrunner's conventional (Opteron) and accelerator (Cell BE) layers. Mirrored particle caches and spectral techniques are used to overcome communication bandwidth limitations and possible difficulties with complicated particle-grid interaction templates.

1. Introduction

1.1. Motivation for petascale cosmological simulations

Cosmology has made tremendous progress over the last two decades, driven by the development of precision tools for studying the cosmic microwave background (CMB) and the large scale structure of the Universe, as well as its expansion history. These precision probes have been associated with large-volume cosmological surveys: all-sky CMB satellite missions, very large sky area and volume galaxy surveys, as well as observations of the sky in the time domain. Results from modern surveys have cemented a cross-validated cosmological "Standard Model" presenting a comprehensive picture of the evolution history of the Universe and its fundamental make-up: 23% in dark matter which only interacts gravitationally (and a large fraction of which is in localized clumps called halos), 72% in a smooth "dark energy" component described by a cosmological constant, adiabatic Gaussian random initial density fluctuations, and flat spatial geometry [1]. Although this is a great triumph, it has exposed some of the biggest puzzles in physical science: What is dark matter made of? What drives the accelerated expansion of the Universe? Does general relativity need to be modified? What is the origin of primordial fluctuations? Is the Universe exactly spatially flat?

In order to help answer these questions, the observational state of the art continues its rapid progress: surveys now coming on line and those planned for the next decade represent an improvement in capability by roughly two orders of magnitude. This translates into a determination of cosmological parameters at the 1% level, however, such an advance will only be possible if the underlying theory can be controlled to the sub-percent level. This is a severely demanding task [2].

With one or two exceptions, all cosmological probes are based on (statistical) studies of fluctuations, whether temperature in the case of the CMB, or of the mass distribution in the case of weak gravitational lensing. The fundamental task for theory is to produce accurate predictions for the CMB, for the mass density and velocity field (along with associated statistical measures such as the fluctuation power spectrum), and for tracers of mass and velocity, such as galaxies and galaxy clusters. Cosmological information on large spatial scales is (relatively) easier to interpret because on these scales the physics is essentially linear, however, observable quantities can have a large variance due to finite sampling limitations. At smaller, more nonlinear scales, the statistical limitations can be essentially removed, but modeling becomes significantly more complicated; keeping the associated systematic errors in check is a difficult challenge. There is currently no alternative to precision simulation as the theoretical tool of choice for dealing with the nonlinear regime of structure formation.

Numerical simulations are also essential in providing the underlying skeleton of cosmological structure from which object catalogs (for galaxies, clusters, etc.) can be built. The resulting mock catalogs have many uses: to design and test observational strategies, to understand possible systematic errors therein, and to confront theoretical predictions with observations.

The large-scale distribution of matter – on scales greater than several Mpc (megaparsec; 1 parsec=3.26 light-years) – is determined primarily by gravity. On smaller scales, complex baryonic physics requires modeling of (magneto) hydrodynamic, thermal, chemical, and radiative processes. Because dark matter outweighs baryons by a factor of five, gravity-only N-body simulations provide the bedrock on which all other techniques rest. Therefore, the first set of simulation desiderata are determined by what is required of N-body simulations. These are: (i) ability to resolve dark matter halos and subhalos hosting galaxies, (ii) sufficient volume to control sample variance, and (iii) enough information to model galaxy bias. For next-generation surveys, this translates into simulations with multi-Gpc box-sizes and particle counts in the 10^{11-12} range, all with ~kpc force resolution. These requirements are $1-2$ orders of magnitude beyond the current state of the art.

Aside from sheer size, the time to completion for a single run is a significant factor. Current "hero" simulations take months to complete. Such long times are quite unacceptable if simulations are to play an important role in the inverse analysis of observational data – as they must for next-generation surveys. Brute force Markov Chain Monte Carlo as applied to parameter estimation requires $O(10^{4-5})$ simulation runs. Although sophisticated statistical techniques can be used to lower this requirement by $2-3$ orders of magnitude [3], one still desires an individual run to take a few days at most, i.e., an order of magnitude faster than present practice. (There are other reasons for requiring large numbers of simulations: exploration of different physical mechanisms, testing at scale, computations of covariance matrices., etc.) Therefore, in terms of overall throughput, simulations need to be improved by $2-3$ orders of magnitude. This is the petascale computing challenge for precision cosmology.

1.2. Equations and computational techniques

Structure formation in the Universe is driven by the gravititional instability. In the current theoretical picture, initial density perturbations collapse and merge in a hierarchical fashion to form dark matter halos within a global "cosmic web" structure. Baryonic matter collects in halos, eventually forming stars and galaxies. The collisionless evolution of matter subject only to gravity is given by the Vlasov-Poisson equation in an expanding Universe (with scale factor

$a(t)$ and in comoving coordinates, i.e., with the metric $ds^2 = dt^2 - a(t)^2 d\mathbf{x} \cdot d\mathbf{x}$):

$$\frac{\partial f}{\partial t} + \dot{\mathbf{x}} \cdot \frac{\partial f}{\partial \mathbf{x}} - \nabla\phi \cdot \frac{\partial f}{\partial \mathbf{p}} = 0, \qquad \mathbf{p} = a^2 \dot{\mathbf{x}} \tag{1}$$

$$\nabla^2 \phi = 4\pi G a^2 (\rho(\mathbf{x}, t) - \rho_b(t)), \qquad \rho(\mathbf{x}, t) = a^{-3} m \int d^3\mathbf{p} f(\mathbf{x}, \dot{\mathbf{x}}, t). \tag{2}$$

The phase-space distribution $f(\mathbf{x}, \dot{\mathbf{x}}, t)$ satisfies a flow (Vlasov) equation (1) with a self-consistent force determined by the Poisson equation (2) sourced by the matter inhomogeneity (ρ_b is the background density). To include hydrodynamics, the baryons are split off as a separate species subject to pressure-gradient forces as well as to gravity. We do not consider this extension here.

The problem with the Vlasov-Poisson equation is that it is six-dimensional and forms small structures even with smooth initial conditions, rendering it impossible to solve directly. The way out of this problem is to use N-body techniques. Here tracer particles represent the distribution function and their evolution is generated by self-consistent evaluation of the inter-particle forces and time-stepping of the particle equations of motion. The scale factor itself can be taken as the time variable, and in this case the nontrivial equation of motion is $d\mathbf{p}/da = \nabla\phi/(aH)$, where the Hubble parameter, $H = \dot{a}/a$. In the case of a spatially flat Friedmann-Robertson-Walker cosmology with a cosmological constant, $H = H_0 a^{-3/2}\sqrt{\Omega_0 + \Omega_\Lambda a^3}$ so that H is completely determined by its present value (H_0) and by the corresponding values of the matter and dark energy densities (as a fraction of the critical density), Ω_0 and Ω_Λ.

Inter-particle forces can be computed directly (using approximate methods to avoid the naive N_p^2 scaling) or by using grid-based Poisson solvers, the Particle-Mesh (PM) method [4]. Here the particles are deposited on a regular three-dimensional grid, followed by a Poisson solve implemented via an FFT, the resulting grid force then being interpolated back onto the particles. This method reduces the direct $O(N_p^2)$ computation to $O(N_g \log N_g) + O(N_p)$ where N_g is the number of grid points. Unfortunately, the force resolution is now controlled by the grid size, and leads to a runaway memory requirement. Gpc-scale cosmology runs require a force resolution of order 10 kpc, a spatial dynamic range requirement of $> 10^5$. This imposes in turn a memory requirement of 10-100 PBytes. One solution to this problem is the use of superposition: the PM algorithm contributes the medium/long range force and a direct or tree-based solver handles the near-field regime. (Other approaches include AMR methods and global tree-based algorithms.) MC^3 splits the inter-particle force problem into two parts, a medium resolution solver based on FFTs augmented by a direct particle-particle short-range solver. The biggest FFT provides up to four orders of magnitude of dynamic range, the remaining factor of $10 - 100$ coming from the short-range force evaluations.

2. The Roadrunner architecture

Since the Roadrunner architecture is unusual, we present a short review of it here (Fig. 1). At the top level, Roadrunner is a set of "Connected Units" (CUs), each CU having 180 compute nodes and 12 I/O nodes. Each compute node has two dual-core Opterons clocked at 1.8 GHz and four IBM PowerXcell 8i processors (Cell BEs) coupled to the Opterons via 8×PCIE. Each Cell BE has a peak performance of 100 GFlops (double precision) and twice that in single precision. A Cell BE consists of eight Synergistic Processor Units (SPUs) and one Power Processing Unit (PPU), an in-line CPU with 512KB of L2 cache. The SPUs are connected via the Element Interconnect Bus with a bandwidth of $\sim$ 200 GB/s per Cell BE. The SPUs do not access system memory directly and there is no L1 cache; each SPU has 256K of SRAM-based locally addressable memory for both code and data (not coherent with main memory). Control of memory access is via a DMA-based interface and data transfers can be initiated either by the SPU or the PPU. The central idea is to obtain performance through the use of SIMD parallelism, low memory latency and high memory bandwidth between SPUs.

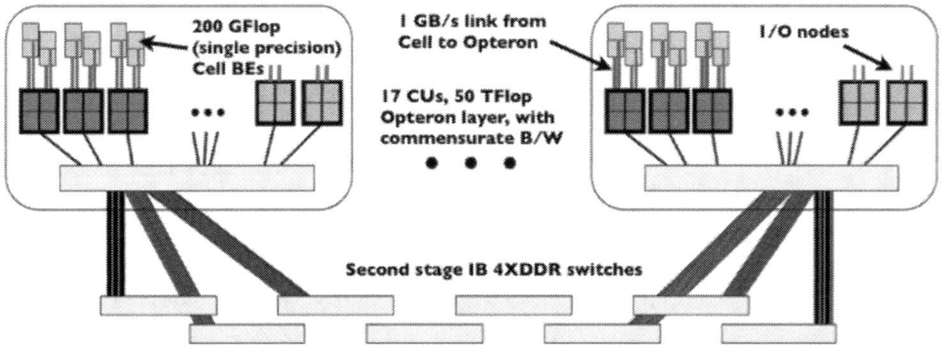

Figure 1. Roadrunner architecture schematic, see text for details.

Local RAM is 16 GB each for the Opterons and for the Cell BEs. The Cell BEs are paired in blades, each blade having 8 GB of NUMA shared memory. There is no local disk storage. The inter-node communication is a fat-tree 4×DDR Infiniband network. For our application, each node is best thought of as two "doublets" each, with one dual-core Opteron and one Cell blade. The reason for this is that the Cell blade pairs share NUMA memory with a ~5 GB/s bandwidth. Going across blades is much slower – by more than an order of magnitude. The communication bottleneck between the Cell BEs and the Opterons is due to the PCIE bandwidth being ~ 1 GB/s. Thus, to fully utilize the power of the Cell BEs, they must be given very compute-intensive tasks with only limited (two-way) communication. Roughly speaking, Roadrunner is a factor of 50-100 out of balance in terms of matching bandwidth to raw flops. Consequently, the code design for MC^3 is based on two guiding principles: (i) Simplify and reduce communication as much as possible, compensating with increased computation and memory overhead, and (ii) Match the code and algorithm design to the general machine architecture.

3. MC^3: Cell-accelerated P^{3}M

The Roadrunner Universe project code, MC^3, is a hybrid design: At the top level of code organization, the medium/long range force is handled by an FFT-based method that operates at the Opteron layer. At this layer, only grid information is stored and manipulated (except when carrying out analysis steps). Particles live at the Cell layer. There is a rough memory balance between the grid and particle data (in most cases, $N_p = N_g$) matching well to the memory organization on the machine. The particle-grid deposition and grid-particle interpolation steps are performed on the Cells with only grid information passing between the two layers. This compensates for the limited bandwidth available between the Cell BEs and the Opterons. The local force calculations stay at the Cell level. Because implementing complicated data structures at the Cell level is difficult, and conditionals can severely degrade performance, our choice for the local force solve is a direct particle-particle interaction, i.e., a P^{3}M code [4] with hardware acceleration.

At the second level, our approach has two key aspects. (i) Since particle communication across the Cell layer is slow relative to the computational performance of the Cell processors, we have essentially eliminated it as a factor by using *particle overloading*, a mirrored particle cache (details below). (ii) Conventional P^{3}M codes use a mixture of spatial filtering/differencing and spectral techniques; complicated communication templates can be problematic, particularly at the Cell level. To combat this, MC^3 uses *digital filtering and differencing* in the spectral domain, and only the simplest spatial deposition technique, Cloud-In-Cell (CIC), is used (Cell layer only). The combination of particle overloading and spectral techniques can eliminate essentially all communication (or at least reduce it to a negligible overhead, if one wants to be more memory efficient), except that in the parallel FFT. The enhanced filter optimization possible with spectral techniques allows for a simple P^3M implementation, with good error control and performance.

In general, P^3M codes perform best when the particle distribution is not very clustered (Molecular Dynamics-like), so that the short-range force calculations do not become overwhelmingly time-consuming. However, extreme clustering is precisely the signature of cosmological structure formation so we must investigate whether Cell-acceleration is up to the task. Given the typical size of the simulation cube (~Gpc) and the biggest FFTs that can be performed (10^4 cubed), we are led to the conclusion that a typical particle-particle interaction domain will be of size 1 Mpc or less. Given the total number of particles that we can run (about 512^3/doublet), and knowing the mass profile of halos from previous simulations, we find that the worst case scenario is roughly 10^5 particles in the interaction region, which translates into roughly 3×10^{11} floating point operations. Since these computations can be carried out in local coordinate single precision in each doublet node on the Cell BEs, the timing estimate works out to a few seconds per step. The typical number of timesteps is $O(10^4)$, which amounts to roughly 10 hours of run-time. Since code scalability is guaranteed by the particle overloading, these are very encouraging numbers: what currently takes months on other machines, can – potentially – be done on Roadrunner on the timescale of a day.

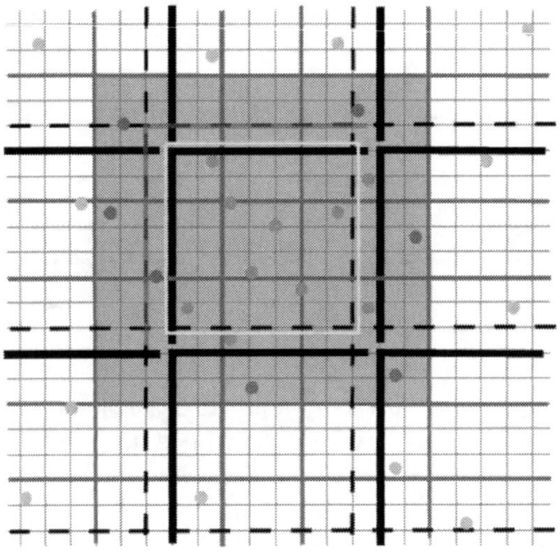

Figure 2. Particle overloading in 2-D: A node holds particles within its spatial domain ("active"/green within yellow boundary) and copies of particles within a given distance from its boundary ("passive"/red in outer blue square). More than one particle copy can be held by neighboring processors – overloading is a mirrored particle cache, not a conventional guard zone.

3.1. Particle overloading

An important aspect of large-volume cosmological simulations is that the density distribution is very highly clustered with an overall topology descriptively referred to as the "cosmic web". The clustering is such that the maximum distance moved by a particle is roughly 30 Mpc, much smaller than the overall scale of the simulation box (~Gpc). With a 3-D domain decomposition, each doublet node will hold a roughly 100-500 Mpc box on a side, depending on the simulation run size. The idea behind particle overloading is for a given node to also own particles belonging to a zone of size roughly 30 Mpc (or some similar range) extending out from the nominal spatial boundary for that node (so-called "passive" particles). Note that copies of these particles – essentially a replicated particle cache – will also be held by other processors, in only one of which will they be "active" in the sense of contributing to the density field, hence the term overloading (Figure 2).

The point of this cache is that for a large number of time-steps – potentially for the entire simulation if the cache is big enough – *no explicit particle communication* across nodes is required (from knowledge of the particle's local position, each processor knows whether its copy is active or passive, the particle state being switched as domain boundaries are crossed). The cached particle 'skin' also allows particle deposition and force interpolation to be done using information entirely local to a node, thus grid communication is also reduced.

The size of the overloading layer need not be large enough to avoid communication entirely. A more practical idea is to use a smaller size, say 5 Mpc, and refresh the particle cache as needed. This involves only 26-way near-neighbor communication and we have verified that the penalty is a small fraction of the time spent in a single global (long-range) time step.

This reduction in communication has two associated costs: (i) loss of memory efficiency because of the overloading, (ii) numerical error for the short-range force that will slowly leak in from the edge of the outer (passive particle) nodal boundary (the medium/long-range force remains exact). In our applications, the memory inefficiency can be tolerated to the point where the memory in passive and active particles is roughly equal, but this is not expected to be a limitation in the majority of the cases of interest. In any case, this problem can be completely side-stepped using cache refreshes as mentioned above. The second problem can be mitigated by increasing the boundary thickness. However, in this case also, a better method is to use the cache refresh, "recycling" the passive particles after some finite number of steps: each processor gets rid of all of its passive particles and then refreshes the passive zone with (high accuracy) active particles exchanged from nearest-neighbor nodes, incurring only a small and intermittent communication overhead.

3.2. Spectral filtering and differentiation

Conventional P^3M codes employ a combination of spatial and spectral techniques [4]. The spatial operations are the particle deposition, the force interpolation, and the finite-differences for computing field derivatives. The spectral operations include the influence (or pseudo-Green) function FFT solve, digital filtering, and spectral differentiation techniques. In general, spatial techniques are less flexible and more tedious to implement. Also, templates for higher-order spatial operators can be complicated and lead to messy and dense communication patterns, increasing the communication load across the Opteron and Cell BE layers, and being very unsuited to particle operations on the Cell BE.

Accurate P^3M codes are usually run with Triangle-Shaped-Cloud (TSC), a higher-order spatial deposition/filtering scheme and with higher-order spatial differencing templates. In grid units, the CIC deposition kernel filters at the level of two grid cells with a large amount of associated "anisotropy noise"; TSC filters at about three grid cells with much reduced noise. MC^3 uses only CIC deposition/interpolation with a Gauss-sinc spectral filter to mimic TSC-like behavior. In addition, a fourth-order Super-Lanczos spectral differentiator [5] is used, along with a sixth-order influence function. This approach has two key benefits: (i) no spatial differentiation is needed reducing work at the Cell level involving data motion and a potential source of communication is eliminated, and (ii) the filtering is flexible and tunable, allowing an excellent force match at only three grid cells, and reducing the anisotropy noise of the basic CIC scheme by better than an order of magnitude. The "Poisson-solve" in the RRU code is the composition of all the kernels above in one single Fourier transform. Note that each component of the field gradient requires an independent FFT.

Figure 3 shows the excellent force-matching due to the spectral techniques; the multiple realizations in a test code use particle pairs at fixed distances but with random orientations of the separation vector in order to sample the anisotropy imposed by the grid-based calculation of the force. To obtain the short-range force, one subtracts the filtered grid force from the exact

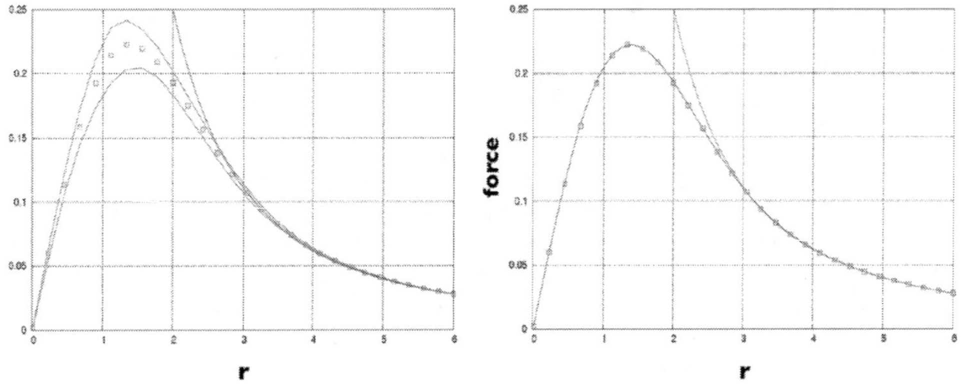

Figure 3. Force between two particles using digital filtering and spectral (Super-Lanczos) differentiation with CIC deposition matching the exact $1/r^2$ behavior at a separation of three grid cells, along with low anisotropy noise – the bracketing red lines showing the 1-σ deviation (left panel). The worst case noise for the total force is only at the ~percent level. The right panel shows the accuracy of the fitting function used to determine the short range force.

Newtonian force. In our case, this is best done by fitting an expression which has the correct asymptotic behaviors at small and large separation distances. The detailed form of the fit is not particularly illuminating, but its high accuracy can be seen in Fig. 3.

3.3. Time-stepping

The time-stepping in MC^3 is based on the widely employed symplectic scheme. The basic idea behind symplectic integration is not to finite-difference the equations of motion, but to view evolution as a symplectic map on phase space, written in the exponentiated form familiar from Lie methods:

$$\zeta(t) = \exp(-t : H :)\zeta(0) \tag{3}$$

where ζ is a phase-space vector $(\mathbf{x}, \mathbf{v})$ for a particle, H is the effective single-particle Hamiltonian, and the operator, $: H := [H, \;]_P$, denotes the action of taking the Poisson bracket with the Hamiltonian. Suppose that the Hamiltonian can be writen as the sum of two parts; then by using the Campbell-Baker-Hausdorff (CBH) theorem:

$$\exp(X)\exp(Y) = \exp(X + Y + \frac{1}{2}\{X, Y\} + \frac{1}{12}(\{X, \{X, Y\}\} + \{\{Y, X\}, Y\} + \cdots)) \tag{4}$$

(where the commutator, $\{X, Y\} = XY - YX$), we can build an integrator for the time evolution due to that Hamiltonian. Repeated application of the CBH formula can be used to show that

$$\exp(-t(: H_1 : + : H_2 :)) = \exp(-t : H_1 : /2)\exp(-t : H_2 :)\exp(-t : H_1 : /2) + O(t^3), \tag{5}$$

a second order symplectic integrator. In the basic PM application, the Hamiltonian H_1 is the free particle (kinetic) piece while H_2 is the one-particle effective potential; corresponding respectively to the 'stream' and 'kick' maps $M_1 = \exp(-t : H_1 :)$ and $M_2 = \exp(-t : H_2 :)$. In the stream map, the particle position is drifted using its known velocity, which remains unchanged; in the kick map, the velocity is updated using the force evaluation, while the position remains unchanged. This way of writing one symmetric 'split-operator' step is termed SKS (stream-kick-stream). A KSK step is another way to have a second-order symplectic integrator.

In the presence of both short and long-range forces, we split the Hamiltonian into two parts, $H = H_{sr} + H_{lr}$ where H_{sr} contains the kinetic energy term and the direct particle-particle interaction potential (with an associated map M_{sr}), whereas H_{lr} is just the long range potential, corresponding to the map M_{lr}. Since the long range force varies relatively slowly, we construct a single time-step map by subcycling M_{sr}:

$$M_{full}(t) = M_{lr}(t/2)(M_{sr}(t/n_c))^{n_c} M_{lr}(t/2), \tag{6}$$

the total map M_{sr} being a usual second-order symplectic integrator. This corresponds to a KSK step, where the S is not an exact stream step as in the PM case, but has enough M_{sr} steps composed together to obtain the required accuracy [6].

For typical problems the number, n_c, of short time steps for each long time step will range between $10 - 100$. In gravity-only mode the total time in the code is spent roughly equally between the long and short time steps. Once hydrodynamics, cooling, and feedback terms are added, the short-range part of the code dominates and the effective efficiency increases (although so does the workload).

4. Notes on implementation

MC^3 is written in C/C++/MPI and is highly optimized for performance and efficient memory use. For enhanced scalability, the parallelization is based on a 3-D domain decomposition of the periodic simulation cube. The FFT implementation is based on a 2-D domain decomposition with repeated application of optimized serial 1-D FFTs (we use FFTW [7]). This methodology is known to scale to thousands of processors. The initializer for MC^3 has been implemented and is based on spectral methods for obtaining realizations of a Gaussian random field. A sufficiently large FFT ($N_p = N_g$) can be performed directly at the Opteron level even for the largest problems of interest, with $N_g = 10^4$.

Although Roadrunner is one machine, it has three programming levels. At the top level are the Opterons connected via Infiniband – general purpose CPUs with MPI as the communication library. The Cell level itself consists of programming the PPU and the SPUs. Communication across the top level and the Cells is handled by the Data Communication and Synchronization (DaCS) library which provides mechanisms for managing, synchronizing, and sharing data. The PPU is a stripped down PowerPC without an instruction pool or branch prediction but clocked at 3.2 GHz. The PPU functions as a bridge from the Opterons to the eight SPU (in-order) vector co-processors and can also carry out independent computations. Altivec registers are avalable to the PPU, and we have found that compiler optimizations can have major impacts on performance resulting in speedups by a factor of 3. With a lightweight programming style, the PPU can yield good performance. Thus, it is worthwhile to look for a role for the PPU beyond that simply of work partitioning, launching tasks for SPUs, and retrieving results from them.

Particle-grid interactions can be complicated to handle on the SPUs because of the way arrays are stored (column/row order, Fortran/C) since spatial locality of particle information does not map to memory locality for the grid. Data reordering by the PPU leads to too high an overhead, and we did not try data movement at the SPU level due to coherence issues. Because of its access to all available memory, the PPU is our choice to carry out the particle deposition and force interpolation steps (CIC and inverse CIC). Compiler optimization results in an Opteron level of performance for this task.

In MC^3, data transfers consist of a DaCS launch and SPU DMA – we do not use the PPU (Cf. Section 2). To maximize performance in the absence of a cache, we use asynchronous memory access to overlap data movement and computation ('cache by hand'). We found the availability of high-performance libraries for the SPU a necessary aid in obtaining good performance (many useful intrinsics exist, e.g., vector instructions, memory control). MC^3 employs the SPUs for the "hot" part of the computation – the direct particle-particle force calculation and the associated particle position and velocity updates. The number of pairwise force computations in the P^3M algorithm is controlled by the size of the matching zone where the short-range force computation is handed over to the long-range, grid-based, PM force calculation. This local range control is affected in MC^3 by a chaining mesh with a (radial) bucket size of three grid cells. Our results show that the short-range computation can be run on the Cells more than fifty times faster than on the Opterons.

Because some of the techniques used in MC^3 are new, a data-parallel PM/P^3M reference code has been implemented that, aside from the overloading, is a close algorithmic copy. This reference code is based on our earlier MC^2 (Mesh-based Cosmology Code) suite which has been tested and used extensively [8]. The reference code is fully operational and is being used for validation of MC^3 in a phased manner. An example is shown in Fig. 4, where the density fluctuation power spectrum is compared (as a ratio) for a medium-resolution test run, primarily as a check of the particle overloading implementation.

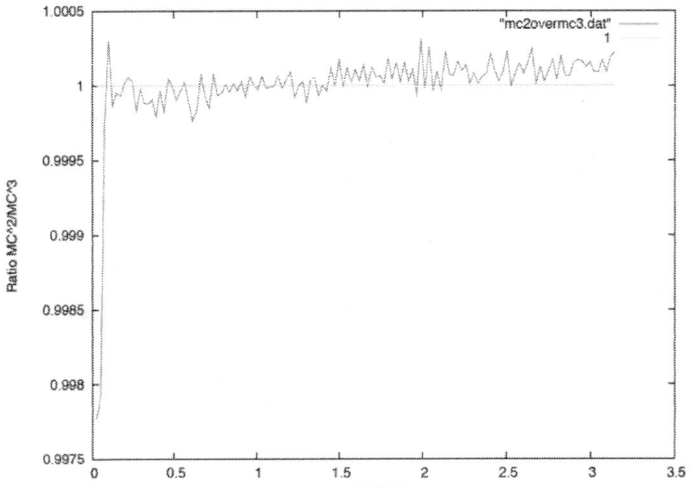

Figure 4. Ratio of the matter power spectrum for the reference code relative to MC^3 as a function of wave number for a medium resolution run (particle overloading test). The relative error is only 2×10^{-4}. The greater error at the lowest k modes is a finite sampling artifact.

A key aspect of the MC^3 code suite is a capability for very fast "on the fly" data analysis. Because the raw data from each run can easily exceed hundreds of TB, it is essential that as much analysis as possible be already performed before the code output is dumped to the file system. The data reduction required is roughly a factor of two orders of magnitude. This data reduction is achieved primarily by a very fast halo finder that can identify clustered regions and distill the particle information contained therein into a much-reduced data vector for individual halos. The halo finder runs on the Opteron layer: The serial part of this halo finder is based on a KD tree algorithm and the parallelization is based on the particle overloading already implemented for the time-stepping. The halo finder's overloading zone is taken to be sufficiently large compared to the size of the largest halos, ~ 5 Mpc being typically sufficient. Because only near-neighbor communication is involved (26 nodes for a given single node), the halo finder is arbitrarily scalable: scaling tests indicate that a trillion particle analysis on Roadrunner should take no more than several minutes.

5. Summary and discussion

To summarize, it is often stated that programming the Roadrunner hybrid system proceeds in either a top-down fashion (put code on the Opterons and transfer the computationally intensive parts to the Cell) or via a bottom-up approach (view the machine primarily as an MPI cluster of SPUs). MC^3 represents a third alternative where the machine is considered as an entire system and the code is designed around it, not based on the particular ideologies mentioned above, but using elements of both. If this is done flexibly enough, successful ideas can be reused – perhaps in different forms – on other architectures. In our case, the particle overloading infrastructure can be used to make MC^3 run on an architecture like that of the Cray XT5 with a different serial solution for the local force solver, such as a tree-based method, instead of the direct particle-particle force evaluation.

Validation against the reference code and convergence testing of MC^3 is currently in progress, as the Roadrunner computer comes closer to achieving full operational status. Timing tests have so far yielded results very close to the initial estimates. The science program for the Roadrunner Universe runs is described in the project Blue Book to be released shortly.

Future development of MC^3 will be in several different directions. The two most obvious capability extensions are the addition of hydrodynamics (both SPH and grid-based) solvers for the Cell BE and the ability to handle small-volume simulations with very high mass and force resolution. In addition, we aim to add sub-grid modules (e.g., star formation, supernova feedback) and post-processing capabilities such as mock catalogs for galaxies and the modeling of galaxy clusters based on semi-analytic prescriptions [9]. More radical extensions include modifications of general relativity on large scales (as a possible explanation of cosmic acceleration) and self-consistent inclusion of dark energy fluctuations. Different methods for introducing non-Gaussian initial conditions will also be implemented.

6. Acknowledgements

We are indebted to Cornell Wright, Sriram Swaminarayan, and Scott Pakin for their generous help with Roadrunner programming issues. SH wishes to acknowledge many discussions on particle codes with Andreas Adelmann, Viktor Decyk, Robert Ryne, and Martin White, and to extend a special note of appreciation to Andy White for his encouragement. AP was supported by a LANL Director's Fellowship. This work was supported by the LDRD program at Los Alamos National Laboratory.

[1] Komatsu E et al. [WMAP Team] 2009 *Astrophys. J. Supp.* **180**, 330
[2] See, e.g., the report of the tri-agency Dark Energy Task Force, Albrecht A et al. 2006 *arXiv:astro-ph/0609591*
[3] Habib S, Heitmann K, Higdon D, Nakhleh C and Williams B 2007 *Phys. Rev.* D **76**, 083503
[4] Hockney R W and Eastwood J W 1988 *Computer Simulation Using Particles* (New York: Adam Hilger)
[5] Hamming R W 1998 *Digital Filters* (Mineola: Dover)
[6] For a similar application in beam physics, see Qiang J, Ryne R D, Habib S and Decyk V 2000 *J. Comp. Phys.* **163**, 434
[7] Frigo M and Johnson S G 2005 *Procedings of the IEEE* **93** 216
[8] Heitmann K et al. 2008 *Comp. Sci. Disc.* **1**, 015003
[9] See, e.g., Bode P, Ostriker J P and Vikhlinin A 2009 *arXiv:0905.3748*

Simulating Supersonic Turbulence in Magnetized Molecular Clouds

Alexei G Kritsuk[1], Sergey D Ustyugov[2,1], Michael L Norman[1] and Paolo Padoan[1]

[1] University of California - San Diego, 9500 Gilman Drive, Mail Code 0424, La Jolla, CA 92093-0424, USA

[2] Keldysh Institute of Applied Mathematics, Russian Academy of Sciences, Miusskaya Sq. 4, Moscow 125047, RUSSIA

E-mail: akritsuk@ucsd.edu

Abstract. We present results of large-scale three-dimensional weakly magnetized supersonic turbulence simulations with an isothermal equation of state at grid resolutions up to 1024^3 cells with the Piecewise Parabolic Method on a Local Stencil. The turbulence is driven by a large-scale isotropic solenoidal force in a periodic computational domain and fully develops in a few flow crossing times. We then evolve the flow for a number of flow crossing times and analyze various statistical properties of the saturated turbulent state. We show that the energy transfer rate in the inertial range of scales is surprisingly close to a constant, indicating that Kolmogorov's phenomenology for incompressible turbulence can be extended to magnetized supersonic flows. We also discuss numerical dissipation effects and convergence of different turbulence diagnostics as grid resolution refines from 256^3 to 1024^3 cells.

1. Introduction

Supersonic turbulence plays an important role in shaping hierarchical internal substructure of molecular clouds (MCs) [1, 2]. This turbulence is the key ingredient in statistical theories of star formation as it controls initial conditions for gravitationally collapsing objects [3, 4, 5]. The nature of molecular cloud turbulence is still poorly understood and no simple conceptual theory exists akin to that of Kolmogorov's phenomenology for turbulence in fluids. Two major complications are *strong compressibility* of interstellar gas mediated by radiative energy losses and presence of dynamically important *magnetic fields*. While these two factors bring into play additional variables and equations rendering the modeling hardly tractable analytically and complicating numerical analysis, the nature of underlying nonlinearity contained in the advection terms remains the same. From a mathematician's perspective, the general problem is essentially the same as in the Navier-Stokes turbulence, one of the remaining six unsolved Millennium Prize Problems for which the Clay Institute of Mathematics is offering a bonus of $1 million.[1] Thus hopes to find universality in interstellar turbulence are not entirely unfounded, see e.g. [6], although some researchers believe that breakdown of universality in incompressible isotropic magnetohydrodynamic (MHD) turbulence is possible, e.g., due to nonlocality of interactions [7].

[1] http://www.claymath.org/millennium/Navier-Stokes_Equations/

The supersonic regime typical of MC turbulence is extremely hard to achieve in the laboratory and the information available from astronomical observations is rather limited [6]. This makes numerical simulations, perhaps, the only tool available to explore the statistics of supersonic turbulence in detail. High dynamic range simulations can shed light on the energy transfer between scales and on the key spatial correlations of the relevant fields in these flows only if they provide sufficient scale separation to resolve the dynamics of prevailing nonlinear interactions in the inertial range. In addition, low numerical dissipation and high-order numerical methods for three-dimensional compressible MHD are needed to achieve an accurate description of the inertial range of supersonic turbulence.

Stable, accurate, divergence-free simulation of magnetized supersonic turbulence is a severe test of numerical MHD schemes and has been surprisingly difficult to achieve due to the range of flow conditions present. Over the last two years we have developed Piecewise Parabolic Method on a Local Stencil (PPML), a new higher order-accurate, low dissipation numerical method which requires no additional dissipation or local "fixes" for stable execution [8, 9, 10]. PPML is a local stencil variant of the popular PPM algorithm [11] for solving the equations of compressible ideal magnetohydrodynamics. The principal difference between PPML and PPM is that cell interface states are "evolved" rather that reconstructed at every timestep, resulting in a compact stencil. Interface states are evolved using Riemann invariants containing all transverse derivative information. The scheme is fully multidimensional as the conservation laws are updated in an unsplit fashion, including the terms corresponding to the tangential directions in the amplitude equations. This helps to avoid numerous well-known pathologies, such as "carbuncles", etc. [12]. Divergence-free evolution of the magnetic field is maintained using the higher order-accurate constrained transport technique [13]. The low dissipation and wide spectral bandwidth of this method make it an ideal choice for direct simulations of compressible turbulence.

In Ref. [14] we reported results from a set of three simulations of driven isothermal supersonic turbulence at a sonic Mach number of 10 on 512^3 meshes demonstrating the performance of our PPML solver on models with different degrees of magnetization from super-Alfvénic through trans-Alfvénic regimes. Turbulence statistics derived from these three-dimensional models show that the density, velocity, and magnetic energy spectra vary strongly with the strength of turbulent magnetic field fluctuations. At the same time, the three cases suggest that a linear scaling in the 4/3-law of incompressible MHD turbulence [15, 16] is preserved even in the strongly compressible regime at $M_s = 10$, if the proper density weighting is applied to the Elsässer fields following the 1/3-rule introduced in Refs. [14, 17, 21], see Section 4.2 for more detail.

In this paper we present results from PPML simulations of statistically isotropic MHD turbulence at a sonic Mach number $M_s = 10$ and Alfvénic Mach number $M_A = 3$ on grids from 256^3 to 1024^3 cells. We use this set of simulations to check the (self-)convergence rates for various statistical measures with improved grid resolution. We also confirm the universal linear scaling for the mixed third-order structure functions of modified Elsässer fields discussed earlier in [14].

2. Numerical Model and Parameters

We use PPML to solve numerically the MHD equations for an ideal isothermal gas in a cubic domain of size $L = 1$ with periodic boundary conditions

$$\frac{\partial \rho}{\partial t} + \nabla \cdot (\rho \mathbf{u}) = 0, \tag{1}$$

$$\frac{\partial \rho \mathbf{u}}{\partial t} + \nabla \cdot \left[\rho \mathbf{u}\mathbf{u} - \mathbf{B}\mathbf{B} + \left(p + \frac{\mathbf{B}^2}{2} \right) \mathbf{I} \right] = \mathbf{F}, \tag{2}$$

$$\frac{\partial \mathbf{B}}{\partial t} + \nabla \cdot (\mathbf{u}\mathbf{B} - \mathbf{B}\mathbf{u}) = 0. \tag{3}$$

Here ρ and $\mathbf{u}$ are the gas density and velocity, $\mathbf{B}$ is the magnetic field, the gas pressure $p \equiv c_s^2\rho$ and the isothermal sound speed $c_s \equiv 1$, $\mathbf{I}$ is the unit tensor. PPML conserves mass, momentum and magnetic flux, and keeps $\nabla \cdot \mathbf{B} = 0$ to the machine precision. A fixed large-scale ($k \leq 2$) isotropic nonhelical solenoidal force $\mathbf{F} \equiv \rho\mathbf{a} - \langle\rho\mathbf{a}\rangle$, where $\mathbf{a}$ is the acceleration and $\langle\ldots\rangle$ indicates averaging over the entire computational domain, is used to stir the gas keeping M_s close to 10 during the simulation. The models are initiated with a uniform density $\rho_0 \equiv 1$, large-scale velocity field $\mathbf{u_0} = \tau\mathbf{a}$, where τ is a constant such that $\langle\mathbf{u}_0^2\rangle^{1/2} = 8$, and a uniform magnetic field $\mathbf{B_0}$ aligned with the x-coordinate direction. The magnetic field strength is parameterized by the ratio of thermal-to-magnetic pressure $\beta_0 \equiv p_0/\mathbf{B_0}^2$. In Ref. [14] we carried out three pilot simulations with $\beta_0 = 0.2$, 2, and 20 at 512^3. Here we present results for three simulations with $\beta_0 = 2$ and grid resolutions 256^3, 512^3, and 1024^3 cells. The largest of these three experiments was carried out using 2048 cores of *Ranger* supercomputer at TACC.

Our numerical experiments covered the transition to turbulence and evolution for up to eight dynamical times, $t_d \equiv L/2M_s$. The magnetic field was not forced and receives energy passively through interaction with the velocity field. This random forcing does not generate a mean field, but still leads to amplification of the small-scale magnetic energy through compression in shocks in combination with flux freezing and, possibly to some extent, via a process known as small-scale dynamo, e.g. [18] and references therein. It is implicitly assumed that the effective magnetic Prandtl number $P_M \approx 1$ in our numerical models. This is far from the realistic values for the interstellar conditions, where $P_M \gg 1$. The limitations imposed by the available computational resources allow us to reach effective Reynolds numbers on the order of 10^4 in our largest MHD simulations, while the realistic values for molecular clouds can be as large as $\sim 10^8$.

3. The Saturated Turbulent State and Convergence

After the initial stir-up phase, our simulations reach a statistical steady state. A snapshot of the projected (column) density in such a saturated state is shown in Fig. 1. The morphology of structures visible in this projected density distribution is surprisingly similar to that from a non-magnetized simulation at $M_s = 6$ presented in Fig. 1. of Ref. [14], even though the power spectrum index is about -1.7 here (see Section 4.1 and Fig. **3d**) and -2.0 there (see Fig. **2a** in Ref. [14]) and the morphology of the density distribution in thin slices (not shown) is clearly different. This illustrates the loss of information in the projection procedure that makes it difficult to measure physical parameters of supersonic turbulent flows from the observed projected density distributions alone.

The properties of this saturated state representing fully developed macroscopically isotropic compressible MHD turbulence are the main focus of this paper. While some level of physical understanding of small-scale rms field amplification and its saturation exists, the structure of the saturated state is poorly known, even in the incompressible case [7].

We compared the saturation levels for several integral characteristics of turbulence for models with three different resolutions: 256^3, 512^3, and 1024^3. Since in the $\beta_0 = 2$ case the kinetic energy $E_K \equiv \int \frac{1}{2}\rho u^2 dV$ dominates over the magnetic energy $E_M \equiv \int \frac{1}{2}B^2 dV$ by a factor of about 3, it is practical to consider E_K and E_M separately instead of concentrating on the total energy $E_T = E_K + E_M$. Figure **2a** shows that the kinetic energy is already converged on 256^3 grids, in agreement with [19]. In contrast, the saturation level of magnetic energy at $\beta_0 = 2$ clearly depends on the grid resolution and there are good chances that convergence can be reached already at resolution of 2048^3 cells. The mean values of E_M for $t \in [3, 7]$ are 10.0, 12.3, and 13.5 with the highest corresponding to our finest grid resolution of 1024^3. The lack of convergence in E_M stems from numerical diffusion suppressing small-scale compressions and also the small-scale dynamo action, if any, in low-resolution runs. The mean value of E_K for the same averaging interval is 39.8, thus the magnetic energy contributes about one quarter of the total energy, $E_M/E_T \approx 0.25$, in the 1024^3 model.

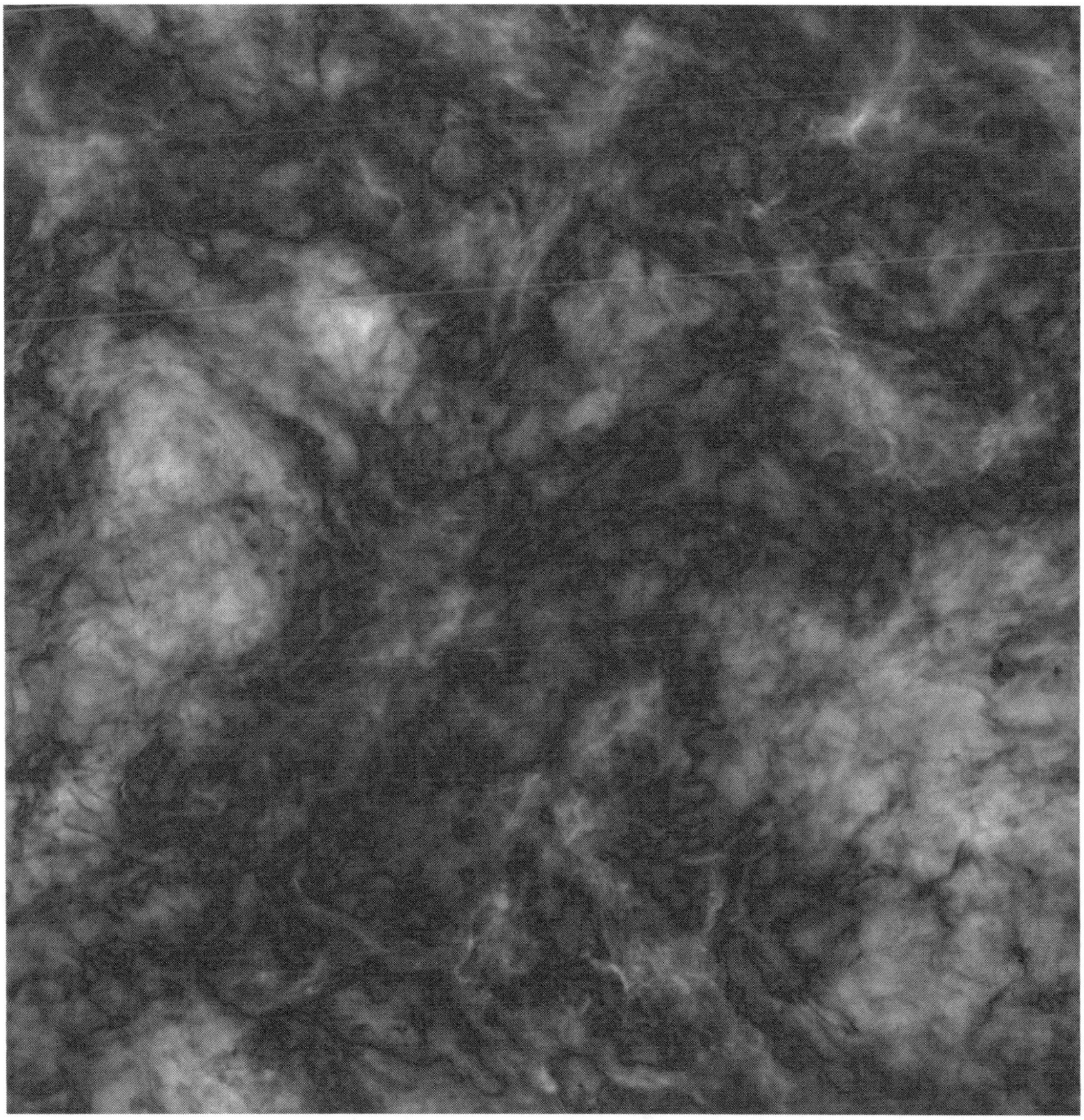

Figure 1. Projected (column) density image for a snapshot at $t = 6.35t_d$ from a 1024^3 PPML simulation of MHD turbulence at $M_s = 10$ and $M_A = 3$ carried out on *Ranger* (TACC) using 2048 cores. White-blue-yellow colors correspond to low-intermediate-high projected density values. The dynamic range of the image is about 100.

While the root mean square (rms) sonic Mach number M_s saturates at a level of about 9.2 independent of the grid resolution, the Alfvén Mach number $M_A \approx 4.0$, 3.1, and 2.8 for grids with 256^3, 512^3, and 1024^3 cells, respectively (see Fig. 2**b**). With the same initial uniform field $\mathbf{B}_0$, we get turbulent states with slightly different M_A values with a tendency towards less super-Alfvénic flows at higher grid resolutions.

Since we do not include the explicit dissipative terms in the equations we integrate numerically, the dissipation in our models is purely numerical and the higher resolution models

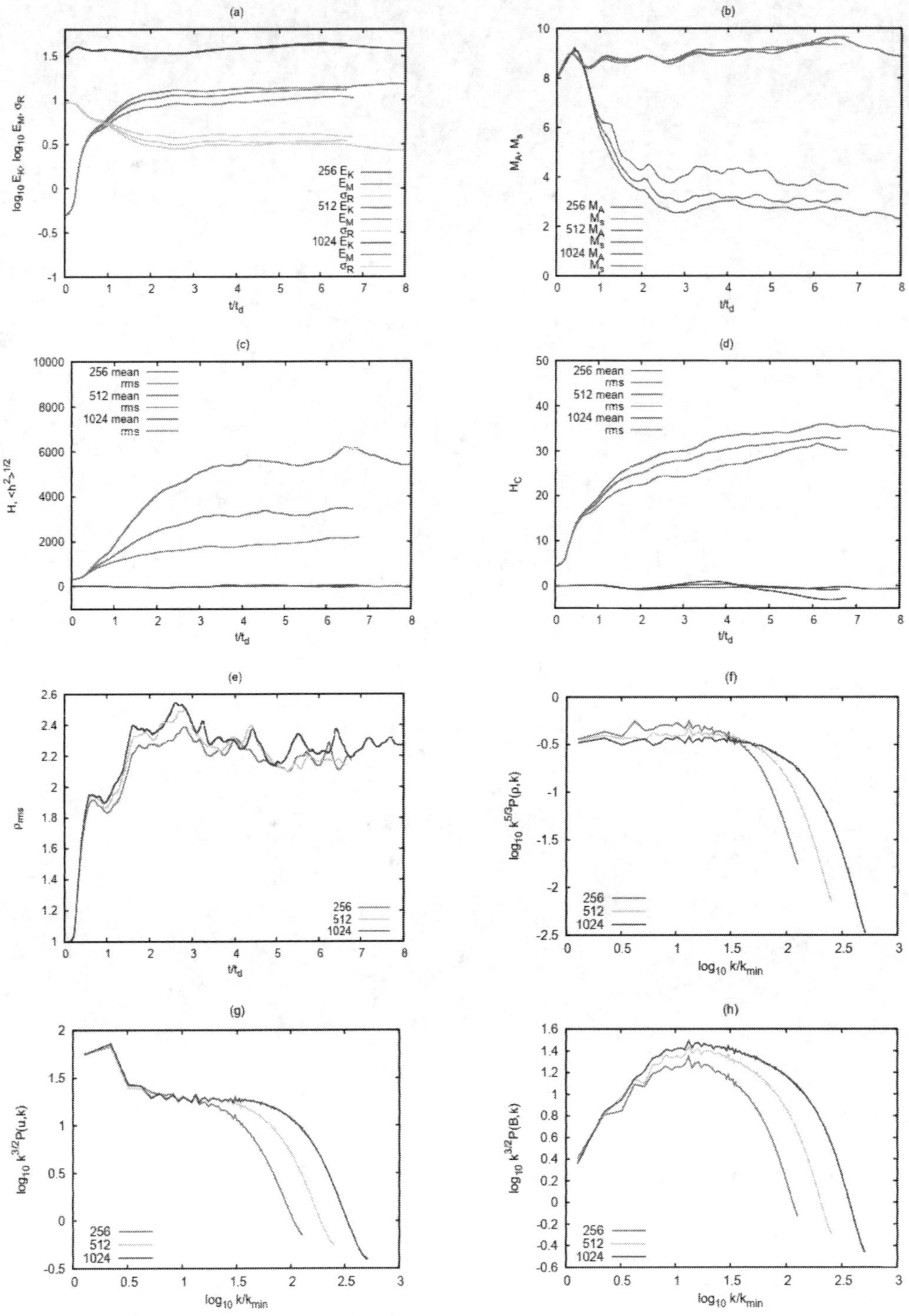

Figure 2. Statistics of MHD turbulence at $M_s = 10$ and $\beta_0 = 2$, from PPML simulations at grid resolution of 256^3, 512^3, and 1024^3 cells: (**a**) mean kinetic, magnetic, and residual energy ($\sigma_R \equiv (E_K - E_M)/E_T$) vs. time; (**b**) rms sonic and Alfvénic Mach numbers; (**c**) mean and rms helicity; (**d**) mean and rms cross-helicity; (**e**) rms gas density; (**f**) compensated density power spectra at $t = 3.5t_d$; (**g**) compensated velocity power spectra at $t = 3.5t_d$; (**h**) compensated magnetic energy power spectra at $t = 3.5t_d$.

correspond to higher effective Reynolds numbers.[2] This can be traced in the non-converging levels of rms helicity gradually growing with higher grid resolution, see Fig. 2**c**. Since our driving force is non-helical to a very good approximation, the mean helicity stays close to zero over the course of a simulation. Figure 2**d** shows a rather slow convergence in the levels of the rms cross-helicity and finite deviations of the mean cross-helicity $H_C \equiv \int \mathbf{u}\cdot\mathbf{B}dV$ from zero in low resolution runs. PPML does not exactly conserve cross-helicity, even though it is one of the ideal invariants in MHD. The normalized mean cross-helicity is, however, contained within $\pm 0.8\%$ for $\beta_0 = 2$ at 1024^3. Even though both the kinetic and the magnetic energy as well as the rms Alfvén Mach number reach saturation after about 3 dynamical times of evolution in our models with $\beta_0 = 2$, the rms cross-helicity continues to grow slowly for up to 4 or even 5 dynamical times meaning that stirring the flow for $3t_d$ might not be sufficient to reach a statistical steady state.

Another global diagnostic to look at is the rms density. In models with higher resolution the flow is better resolved and higher level of density fluctuations is observed. The rms denstity is, however, not very sensitive to the grid resolution and changes from 2.2 to 2.3 as the resolution jumps from 256^3 to 1024^3, see Fig. 2**e**.

Finally, to illustrate the convergence of spectral properties of turbulent fluctuations, in Fig. 2**f**, **g**, and **h** we show compensated power spectra of the density, the velocity, and the magnetic field for a set of snapshots from simulations with different resolutions taken early in the saturated state at $t = 3.5t_d$. The velocity spectra can be used to access the spectral bandwidth of PPML. As one can see from Fig. 2**g**, the velocity spectrum is resolved up to $\log_{10} k/k_{min} \approx 1.1$ at 256^3 and up to $\log_{10} k/k_{min} \approx 1.4$ at 512^3, consistent with the grid refinement by a factor of two.

4. Exploring Scaling Laws in Supersonic MHD Turbulence

Here we present time-averaged statistics for our highest resolution $\beta_0 = 2$ model at 1024^3 grid cells and compare these with our previous results for non-magnetized (HD) simulations at $M_s = 6$ and grid resolutions of 1024^3 and 2048^3 cells [17, 21, 14]. Our MHD data are based on 100 full data snapshots equally spaced in time for $t \in [3, 8]t_d$, i.e. we do the averaging over five dynamical times which makes it, perhaps, the best available sample for compressible MHD turbulence at this resolution. Our projected density spectra are based on a sample of 750 projections, including all three projection directions, covering the same time span $t \in [3, 8]t_d$. We first begin with the scaling properties of the basic fields, such as the density, velocity, and magnetic field and then discuss ideal invariants, such as the total energy and the energy transfer rate that may or may not display universal behavior.

4.1. Basic Fields

The probability density function (pdf) of the gas density is one of the most important inputs for the theories of star formation in molecular clouds as it determines the amount of the dense material available for gravitational collapse to form stars [3, 22, 5]. The ability to predict the pdf is important in other extreme astrophysical environments, such as, e.g., accretion onto "seed" black holes in the first galaxies [23]. It is also known as one of the most robust statistical characteristics for compressible isothermal turbulence as its lognormal shape [24, 25, 26, 27] gets quickly established in numerical simulations, e.g. [17]. Here we demonstrate the lognormal form of the pdf in our $\beta_0 = 2$ model which gives a perfect fit for a range of $\sim 10^8$ in probability at high densities (Fig. 3**a**). Compared to the unmagnetized cases at $M_s = 6$ studied in [17, 14], the lognormal distribution in the MHD case with $M_s = 10$ is, as expected at higher Mach numbers,

[2] We follow here the approach developed in Ref. [20] for nonmagnetized simulations with PPM and assume that the nature of dissipation (purely numerical in our case) does not affect the dynamics in the inertial range of scales. Resolving the explicit dissipative terms in Eqs. (2) and (3) would require a much higher grid resolution to obtain an equivalent scale separation.

somewhat wider. If we parametrize the standard deviation σ as a function of the rms sonic Mach number, $\sigma^2 \equiv \ln(1+b^2M_s^2)$, we get $b = 0.22$ and 0.27 as the best fit values for $\log_{10}\rho \in [-1, 2.8]$ in the MHD and HD cases, respectively. The mean of the distrubution is also a function of the standard deviation, $\langle \ln\rho \rangle = -\frac{1}{2}\sigma^2$. For more detail on the dependence of the pdf on the value of β_0 see [14].

The density power spectrum (Fig. 3**b**) scales roughly as $k^{-2/3}$, shallower than the slope of -1 in the HD case at a smaller $M_s = 6$ [17]. The power spectrum of the logarithm of density (Fig. 3**c**) scales somewhat steeper, $k^{-4/3}$, but still shallower than $-5/3$ found at $M_s = 6$ [14]. The power spectrum of the logarithm of projected gas density roughly follows the scaling $k^{-5/3}$, which is somewhat steeper than the scaling of the density power spectrum (Fig. 3**d**), but still shallower than -2 measured for the HD case at $M_s = 6$ [14]. The sensitivity of the density spectra to the value of β_0 is discussed in detail in [14].

The velocity power spectrum as well as the results of the Helmholtz decomposition of the velocity $\mathbf{u}$ into the solenoidal ($\nabla \cdot \mathbf{u}_s \equiv 0$) and dilatational ($\nabla \times \mathbf{u}_c \equiv 0$) parts, $\mathbf{u} = \mathbf{u}_s + \mathbf{u}_c$, are shown in Fig. 3**e**. The velocity power index is very close to $-3/2$ at k/k_{min} around 30, the value proposed for an isotropic incompressible case with turbulent energy equipartition [28, 29], and the spectrum is dominated by the solenoidal component which follows the same scaling. Note that in the HD simulation of Ref. [17] this range of wavenumbers is buried under the bottleneck bump which is known to be much weaker, if exists at all, in the MHD models. At lower wavenumbers (but still within the inertial range), the velocity spectrum tends to get steeper, as should be expected for a strongly supersonic regime. The spectrum for the dilatational component of velocity is generally slightly steeper, as was previously reported for non-magnetized turbulence [30, 17]. Figure 3**f** compares the fraction of dilatational component in the velocity power, $\chi_c(k) \equiv P(\mathbf{u}_c, k)/P(\mathbf{u}, k)$, from our PPML MHD model at $\beta_0 = 2$ with those in our previous PPM $M_s = 6$ HD runs at resolutions of 1024^3 and 2048^3 cells [17, 21, 14]. In both HD runs $\chi_c(k) \approx 1/3$ in the inertial range, even though we used different driving forces and different parameterizations for PPM diffusion in these simulations. The fact that $\chi_c(k)$ tends to be close to $1/3$ in isotropic flows at high Mach numbers can be explained by purely geometric considerations. At the same time, in our MHD model $\chi_c(k) \approx 1/4$ or even somewhat lower. A lower mean level of saturation for $\chi_c(k)$ in magnetized flows at high Mach numbers can be explained by the additional magnetic pressure term in the momentum conservation equation. A similar value for $\chi_c(k)$ was obtained in Ref. [19] for a model with stronger field ($\beta_0 = 0.02$) at the same grid resolution, cf. [31]. The k^2-compensated power spectra of dilatational velocities in Fig. 14 of Ref. [19], however, demonstrate an unusual rise of power near the Nyquist frequency that we do not observe in our simulations with PPML. Neither do we see such rise in our HD models with PPM (Fig. 3**e**). The origin of this sharp rise in compressions at high wavenumbers in simulations with the *Athena* code [32] is unclear.

4.2. Ideal Invariants

Since the total energy E_T is a conserved quantity, it is instructive to follow its distribution as a function of scale, $E_T(k) \equiv \frac{1}{2}\left[P(\rho^{1/2}\mathbf{u}, k) + P(\mathbf{B}, k)\right]$, see Fig. 3**g**. It appears that to a good approximation the total energy follows a simple scaling law $E_T(k) \sim k^{4/3}$. It is hard to believe, though, that this scaling exponent is universal. We have shown that in non-magnetized turbulence the power index for $P(\rho^{1/2}\mathbf{u}, k)$ does depend on the rms Mach number of the flow [17]. In MHD models, the scaling of $E_T(k)$ would also depend on the saturation level of the magnetic energy. Nevertheless, it is interesting to see how a combination of the magnetic and kinetic energy, each of which do not display a clear extended scaling range (Fig. 3**g**) add up to result in an extended flat stretch of the total energy spectrum. Is this coincidental? Or is this a result of self-organization in supersonic super-Alfvénic MHD turbulence?

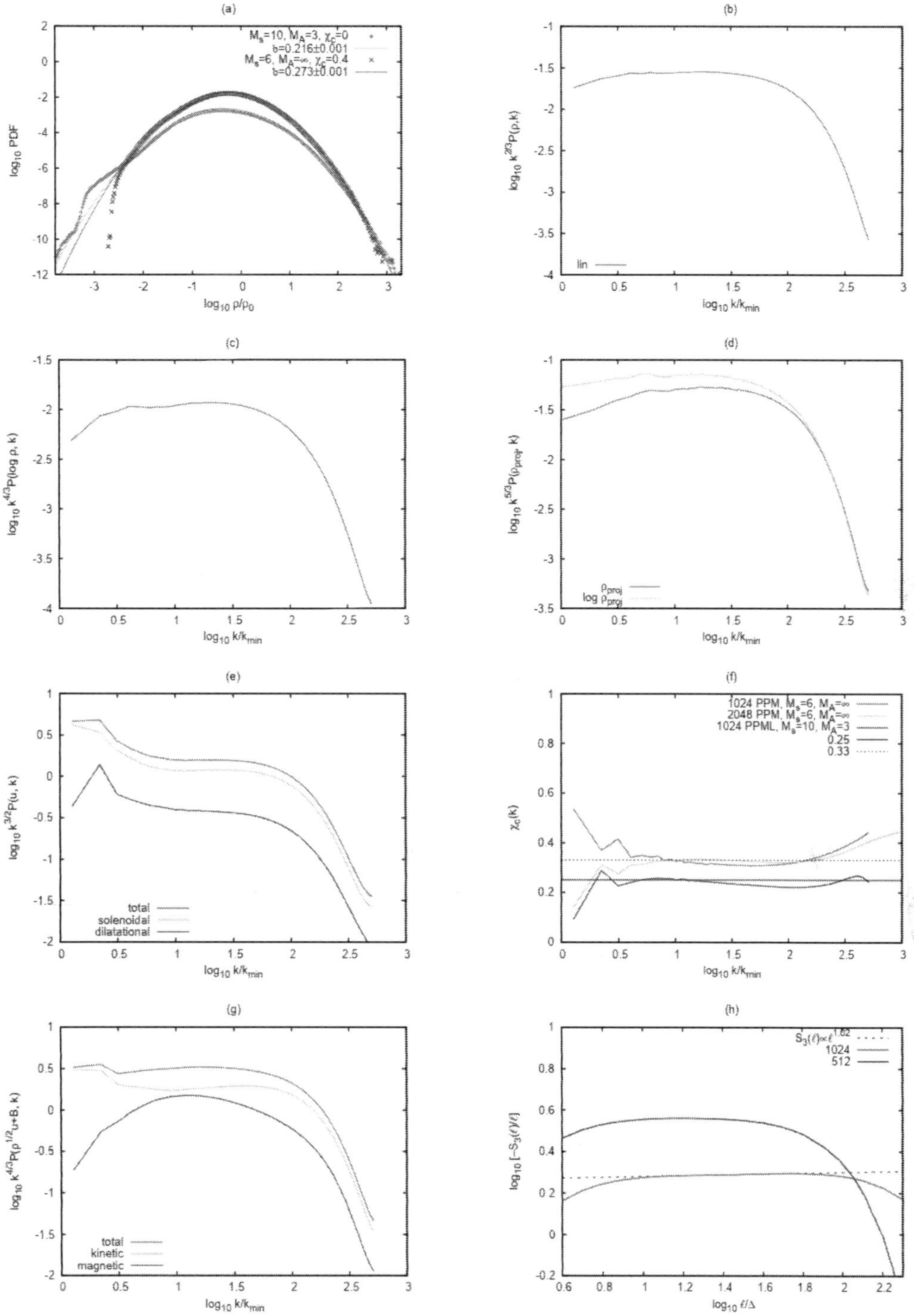

Figure 3. Time-average statistics of MHD turbulence at $M_s = 10$ and $\beta_0 = 2$, from PPML simulation at grid resolution 1024^3 cells: (**a**) the gas density pdf and lognormal fits for $\log_{10}\rho \in [-1, 2.8]$; (**b**) compensated power spectrum of the gas density; (**c**) as **b**, but for the logarithm density; (**d**) as **b**, but for the projected density and for the logarithm of projected density; (**e**) as **b**, but for the velocity and its solenoidal and dilatational parts; (**f**) ratio of dilatational to total velocity power χ_C as a function of wavenumber k; (**g**) compensated power spectra of total, kinetic, and magnetic energy. (**h**) compensated third-order structure functions $S_3(\ell)$ for generalized Elsässer fields $\mathbf{Z}^{\pm}$ from 512^3 and 1024^3 simulations.

A better candidate for universal scaling is the energy transfer rate. Indeed, our HD simulations have shown that $P(\rho^{1/3}\mathbf{u}, k)$ follows the Kolmogorov $-5/3$ scaling at $M_s = 6$ [17]. There is a set of exact scaling laws for homogeneous and isotropic incompressible MHD turbulence analogous to the 4/5-law of Kolmogorov for ordinary turbulence in neutral fluids [33, 15, 16]. The MHD laws can be expressed in terms of Elsässer fields, $\mathbf{z}^{\pm} \equiv \mathbf{u} \pm \mathbf{B}/\sqrt{4\pi\rho}$ [34], as $S^{\pm}_{\parallel,3} \equiv \left\langle \delta z^{\mp}_{\parallel}(\ell)[\delta z^{\pm}_i(\ell)]^2 \right\rangle = -\frac{4}{d}\epsilon^{\pm}\ell$, where $\delta \mathbf{z}_{\parallel}(\ell) \equiv [\mathbf{z}(\mathbf{x}+\hat{\mathbf{e}}\ell) - \mathbf{z}(\mathbf{x})] \cdot \hat{\mathbf{e}}$, d is the space dimension, $\hat{\mathbf{e}}$ is a unit vector with arbitrary direction, $\hat{\mathbf{e}}\ell$ is the displacement vector, the brackets denote an ensemble average, and summation over repeated indices is implied. Equivalently, the scaling laws can be rewritten in terms of the basic fields ($\mathbf{u}$, $\mathbf{B}$), but in MHD there are no separate *exact* laws for the velocity or the magnetic field alone. It is important not to neglect the correlations between the $\mathbf{u}$ and $\mathbf{B}$ fields or $\mathbf{z}^+$ and $\mathbf{z}^-$ fields constrained by the invariance properties of the equations in turbulent cascade models. In supersonic turbulence, it is equally important to properly account for the density–velocity and density–magnetic field correlations. In incompressible MHD, the total energy E_T and the cross-helicity H_C play the role of ideal invariants and, thus, the (total) energy transfer rate $\epsilon^T = (\epsilon^+ + \epsilon^-)/2$ and the cross-helicity transfer rate $\epsilon^C = (\epsilon^+ - \epsilon^-)/2$. For vanishing magnetic field $\mathbf{B}$, one recovers the Kolmogorov 4/5-law, $\langle[\delta \mathbf{u}_{\parallel}(\ell)]^3\rangle = -4/5\epsilon\ell$, where ϵ is the mean rate of kinetic energy transfer [16].

Numerical simulations generally confirm these incompressible scalings, although the Reynolds numbers are perhaps still too small to reproduce the asymptotic linear behavior with a desired precision and the results are sensitive to statistical errors [35, 30, 36]. Often, the absolute value of the longitudinal difference is used and still a linear scaling is recovered numerically, while there is no rigorous result for the normalization constant in this case. The third-order transverse velocity structure functions also show a linear scaling in simulations of nonmagnetized turbulence [30, 17] and the difference of scaling exponents for longitudinal and transverse structure functions can serve as a robust measure of statistical uncertainty of the computed exponents [37]. In MHD simulations, the third-order structure functions of the Elsässer fields $\left\langle |\delta z^{\mp}_{\parallel}(\ell)|^3 \right\rangle$ were shown to have an approximately linear scaling too [35, 38], but again there is no reason for this result to universally hold in all situations since the correlations between the $\mathbf{z}^{\pm}$ fields play an important role in nonlinear transfer processes.

Can the 4/3-law for incompressible MHD turbulence be extended to supersonic regimes in molecular clouds? As we discussed in Ref. [14], a proper density weighting of the velocity, $\rho^{1/3}\mathbf{u}$, preserves the approximately linear scaling of the third-order structure functions at high Mach numbers in the nonmagnetized case (for a more involved approach to density weighting see Ref. [39]). It is straightforward to redefine the Elsässer fields for compressible flows using this "1/3-rule", $\mathbf{Z}^{\pm} \equiv \rho^{1/3}(\mathbf{u} \pm \mathbf{B}/\sqrt{4\pi\rho})$, so that they match the original $\mathbf{z}^{\pm}$ fields in the incompressible limit and reduce to $\rho^{1/3}\mathbf{u}$ in the limit of vanishing $\mathbf{B}$. The new $\mathbf{Z}^{\pm}$ fields can have universal scaling properties in homogeneous isotropic turbulent flows with a broad range of sonic and Alfvénic Mach numbers.

We use data from our 1024^3 MHD simulation to support the conjecture proposed in Ref. [14] and compute the third-order structure functions $S^{\pm}_{\parallel,3}$ and $S^{\pm}_{\perp,3}$ defined above using the $\mathbf{Z}^{\pm}$ fields. Since we are interested in the energy transfer through the inertial interval, we compute the sum of $S^{-}_{\parallel,3}$ and $S^{+}_{\parallel,3}$ which determines the energy transfer rate ϵ^T in the incompressible limit. To further reduce statistical errors, we combine the transverse and longitudinal structure functions, but the absolute value operator was not applied to the field differences. Figure 3**h** shows the compensated scaling for $S_3(\ell) \equiv (S^{-}_{\parallel,3} + S^{+}_{\parallel,3} + S^{-}_{\perp,3} + S^{+}_{\perp,3})/4$ averaged over 100 flow snapshots and the best linear fit $S_3(\ell) \sim \ell^{1.020\pm0.001}$ for $\log \ell/\Delta \in [1.1, 1.8]$. For comparison, we overplot data from our 512^3 simulation discussed in Ref. [14]. Clearly, at 1024^3, we get a more extended scaling range and confirm the linear scaling found in [14]. Howefer, further work is needed to verify this result with high resolution simulations at different levels of magnetization.

5. Conclusions

We verified the applicability of the 1/3-rule of hydrodynamic supersonic turbulence to super-Alfvénic regimes of MHD turbulence using data from a new simulation with the Piecewise Parabolic Method on a Local Stencil on a grid of 1024^3 cells. Our results suggest that the energy transfer rate is approximately constant accross the inertial interval and that the 4/3-law of incompressible MHD can be extended to the supersonic turbulent flows.

Acknowledgments

This research was supported in part by the National Science Foundation through grants AST-0607675 and AST-0808184, as well as through TeraGrid resources provided by NICS, PSC, SDSC, and TACC (MCA07S014 and MCA98N020).

References

[1] Elmegreen B G and Scalo J 2004 *ARA&A* **42** 211–73 (*Preprint* arXiv:astro-ph/0404451)
[2] Scalo J and Elmegreen B G 2004 *ARA&A* **42** 275–316 (*Preprint* arXiv:astro-ph/0404452)
[3] Padoan P and Nordlund Å 2002 *ApJ* **576** 870–79 (*Preprint* arXiv:astro-ph/0011465)
[4] McKee C F and Ostriker E C 2007 *ARA&A* **45** 565–687
[5] Padoan P, Nordlund Å, Kritsuk A G, Norman M L and Li P S 2007 *ApJ* **661** 972–81 (*Preprint* arXiv:astro-ph/0701795)
[6] Heyer M H and Brunt C M 2004 *ApJL* **615** L45–8 (*Preprint* arXiv:astro-ph/0409420)
[7] Yousef T A, Rincon F and Schekochihin A A 2007 *Journal of Fluid Mechanics* **575** 111
[8] Popov M V and Ustyugov S D 2007 *Computational Mathematics and Mathematical Physics* **47** 1970–89
[9] Popov M V and Ustyugov S D 2008 *Computational Mathematics and Mathematical Physics* **48** 477–99
[10] Ustyugov S D, Popov M V, Kritsuk A G and Norman M L 2009 *Preprint* arXiv:0905.2960
[11] Colella P and Woodward P R 1984 *J. Comp. Phys.* **54** 174–201
[12] Quirk J J 1994 *International Journal for Numerical Methods in Fluids* **18** 555–74
[13] Gardiner T A and Stone J M 2005 *J. Comp. Phys.* **205** 509–39 (*Preprint* arXiv:astro-ph/0501557)
[14] Kritsuk A G, Ustyugov S D, Norman M L and Padoan P 2009 *ASP Conference Series* **406** 15–22
[15] Politano H and Pouquet A 1998 *Phys. Rev. E* **57** 21
[16] Politano H and Pouquet A 1998 *Geophysical Research Letters* **25** 273–6
[17] Kritsuk A G, Norman M L, Padoan P and Wagner R 2007 *ApJ* **665** 416–31 (*Preprint* arXiv:0704.3851)
[18] Schekochihin A A and Cowley S C 2007 *Turbulence and Magnetic Fields in Astrophysical Plasmas* (Magnetohydrodynamics: Historical Evolution and Trends) p 85
[19] Lemaster M N and Stone J M 2009 *ApJ* **691** 1092–108 (*Preprint* arXiv:0809.4005)
[20] Sytine I V, Porter D H, Woodward P R, Hodson S W and Winkler K H 2000 *J. Comp. Phys.* **158** 225–38
[21] Kritsuk A G, Padoan P, Wagner R and Norman M L 2007 *AIP Conf. Proc.* **932** 393–9
[22] Krumholz M R and McKee C F 2005 *ApJ* **630** 250–268 (*Preprint* arXiv:astro-ph/0505177)
[23] Milosavljević M, Bromm V, Couch S M and Oh S P 2009 *ApJ* **698** 766–780 (*Preprint* arXiv:0809.2404)
[24] Vazquez-Semadeni E 1994 *ApJ* **423** 681
[25] Padoan P, Nordlund A and Jones B J T 1997 *MNRAS* **288** 145–52
[26] Nordlund Å K and Padoan P 1999 The Density PDFs of Supersonic Random Flows *Interstellar Turbulence* ed Franco J and Carraminana A p 218
[27] Biskamp D 2003 *Magnetohydrodynamic Turbulence* (Cambridge University Press)
[28] Iroshnikov P S 1963 *Astronomicheskii Zhurnal* **40** 742
[29] Kraichnan R H 1965 *Physics of Fluids* **8** 1385–87
[30] Porter D H, Pouquet A and Woodward P R 2002 *Phys. Rev. E* **66** 026301
[31] Boldyrev S, Nordlund Å and Padoan P 2002 *Phys. Rev. Lett.* **89** 031102 (*Preprint* arXiv:astro-ph/0203452)
[32] Stone J M, Gardiner T A, Teuben P, Hawley J F and Simon J B 2008 *ApJS* **178** 137–77 (*Preprint* arXiv:0804.0402)
[33] Chandrasekhar S 1951 *Proc. Royal Society London, Ser. A* **204** 435–49
[34] Elsässer W M 1950 *Physical Review* **79** 183
[35] Biskamp D and Müller W C 2000 *Physics of Plasmas* **7** 4889–900
[36] Boldyrev S, Mason J and Cattaneo F 2006 *Preprint* arXiv:astro-ph/0605233
[37] Kritsuk A G and Norman M L 2004 *ApJL* **601** L55–8 (*Preprint* arXiv:astro-ph/0310697)
[38] Momeni M and Mahdizadeh N 2008 *Physica Scripta Volume T* **131** 014004
[39] Pan L, Padoan P and Kritsuk A G 2009 *Phys. Rev. Lett.* **102** 034501–4 (*Preprint* arXiv:0808.1330)

Journal of Physics: Conference Series **180** (2009) 012021 doi:10.1088/1742-6596/180/1/012021

Baryon acoustic oscillations in the Lyman alpha forest

Michael L Norman, Pascal Paschos and Robert Harkness

Laboratory for Computational Astrophysics, SERF 0424, University of California San Diego, 9500 Gilman Drive, La Jolla, CA 92093

mlnorman@ucsd.edu

Abstract. We use hydrodynamic cosmological simulations in a (600 Mpc)^3 volume to study the observability of baryon acoustic oscillations (BAO) in the intergalactic medium as probed by Lyman alpha forest (LAF) absorption. The large scale separation between the wavelength of the BAO mode (~150 Mpc) and the size of LAF absorbers (~100 kpc) makes this a numerically challenging problem. We report on several 2048^3 simulations of the LAF using the ENZO code. We adopt WMAP5 concordance cosmological parameters and power spectrum including BAO perturbations. 5000 synthetic HI absorption line spectra are generated randomly piercing the box face. We calculate the cross-correlation function between widely separated pairs. We detect the BAO signal at z=3 where theory predicts to moderate statistical significance.

1. Introduction

Baryon Acoustic Oscillations (BAO) is a promising standard ruler technique in observational cosmology to measure the expansion history of the universe and thereby a probe of cosmic dark energy [1]. BAO refers to acoustic waves in the cosmic photon-baryon plasma prior to decoupling at z=1080. They have a characteristic wavelength which is the sound speed in the plasma $\sim c/3^{1/2}$ times the age of the universe at decoupling. For the concordance cosmological model this is about 150 comoving Mpc [2].

BAO modulates the power spectrum of density fluctuations that seed structure formation. Modes of this wavelength have been detected in the cosmic microwave background (CMB) anisotropies and galaxy large scale structure (LSS)[1]. In principle, the BAO should be imprinted on all forms of cosmic structure including the distribution of intergalactic gas as probed by the Lyman alpha forest (LAF).

The LAF is a thicket of narrow absorption lines seen in the spectra of high redshift quasars [3]. It is caused by resonant Ly α absorption in intergalactic hydrogen clouds spread out in wavelength by the cosmological redshift. A high resolution quasar absorption line spectrum thus probes the detailed spatial distribution of intergalactic gas along a sightline many hundreds of Mpc in length. The best way to detect BAO in the LAF is by cross-correlating absorption spectra in widely separated quasar pairs. To date, the BAO has not been detected in the LAF for lack of sufficient data (it is a statistical detection; see below), although surveys are being designed to remedy this [4]. [5] have used analytic arguments to estimate how much data might be needed to reach the precision required in dark energy studies. In this work we use hydrodynamical cosmological simulations of unprecedented size to explore the eventual detectability of BAO in the LAF.

2. Simulations

We have used the hydrodynamic cosmological code ENZO [6] to simulate the structure of the IGM on scales exceeding the BAO length scale. The large scale separation between the wavelength of the fundamental BAO mode (~150 Mpc) and the size of LAF absorbers (~100 kpc) makes this a numerically challenging problem. We have done two simulations at 2048^3 resolution in cubic computational domains 307 and 614 Mpc on a side. The cosmological parameters were taken to be the 5 –year WMAP concordance parameters: Ω_b=0.043, Ω_{dm}=0.207, Ω_Λ=0.75, h=0.72, σ_8=0.8, n=0.95. The first three are the fraction of the closure density in baryons, dark matter, and vacuum energy, respectively; h is the Hubble parameter in units of 100 km/s/Mpc; and σ_8 and n are the amplitude and primordial slope of the matter fluctuation power spectrum, respectively. Gaussian random field initial conditions were generated at the starting redhsift z=60 in the usual way using an input power spectrum obtained from CAMB08 [7] including the BAO features.

ENZO evolves both the dark matter and primordial hydrogen and helium gas in a cubic, triply periodic volume subject to self-gravity in an expanding universe. A photoionizing UV background radiation field is assumed which controls the ionization state of the gas. The equations of multispecies gas dynamics are evolved on a uniform Cartesian grid of size 2048^3 cells using the PPM method. This yields a co-moving grid resolution of 150(300) kpc for the smaller(larger) box. Dark matter is represented as 2048^3 collisionless particles, and is evolved using the particle-mesh technique. The Poisson for the self-consistent gravitational field is solved using FFTs. ENZO is parallelized for execution on distributed memory clusters with multicore nodes using a combination of MPI and OpenMP. See [6] for more details on the ENZO code.

3. Results

A volumetric rendering of the logarithm of the gas density at z=3 from the 300 Mpc simulation is shown in figure 1a. One recognizes the characteristic filament/void "cosmic web" structure imprinted by the clustering of cold dark matter [8]..Lines of sight intersecting the baryonic filaments create the LAF absorption seen in the spectra of distant quasars. Modulation of the cosmic web by BAO is too subtle to be detected by eye, so statistical techniques are employed.

To this end synthetic HI absorption spectra are generated using the method described in [9] for N=5000 lines of sight randomly piercing a face of the cube and running parallel to a coordinate axis. A sample spectrum is shown in figure 1b. Here the black line is the spectrum at the native resolution of the simulation (corresponding to a spectral resolution of 13000), and the red line is the spectrum convolved with the resolution of the SDSS spectrometer (spectral resolution 1800). The cross-correlation function among the N(N-1)/2 pairs is calculated and binned versus comoving separation on the sky (equivalent to angular separation). The result is shown in figure 2a. Here we have used the highest resolution spectra and ignore noise. The 1σ standard error to the mean of the cross correlation in 1 Mpc bins including the propagation of error due to the variance in the product of two sightlines is shown as error bars. The BAO signal is clearly evident at the predicted separation of 155 comoving Mpc. We estimate the statistical significance of this result to be 1.86σ by comparing the amplitude of the peak, after we subtract the upper level of error at the zero correlation level, to the size of the error bar at the peak. The bump at comoving separation of ~75 Mpc appears to be unrelated to the BAO phenomenon and is statistically insignificant within our margin of error. Note that the statistical significance depends on our sample size of Ly α skewers through the volume. The variance of the sightline pair product will not change using a larger sample, but the error to the mean per correlation bin (x) will decrease as $Np(x)^{-0.5}$, where Np(x) is the number of pairs per comoving separation.

In Fig 2b we explore the physical origin of the BAO signal. We plot the cross-correlation function of dark matter density, baryon density, and baryon temperature along same lines of sight used for figure 2a. We see a narrow peak in all three quantities at the BAO distance scale (155 Mpc). Temperature shows the strongest signal, followed by the dark matter and baryon densities at comparable amplitudes. The shoulder to the right of the BAO peak we interpret as due to a local maximum in the fluctuating signal at 170 Mpc; other peaks can be seen at separations of 75, 125, and 200 Mpc.

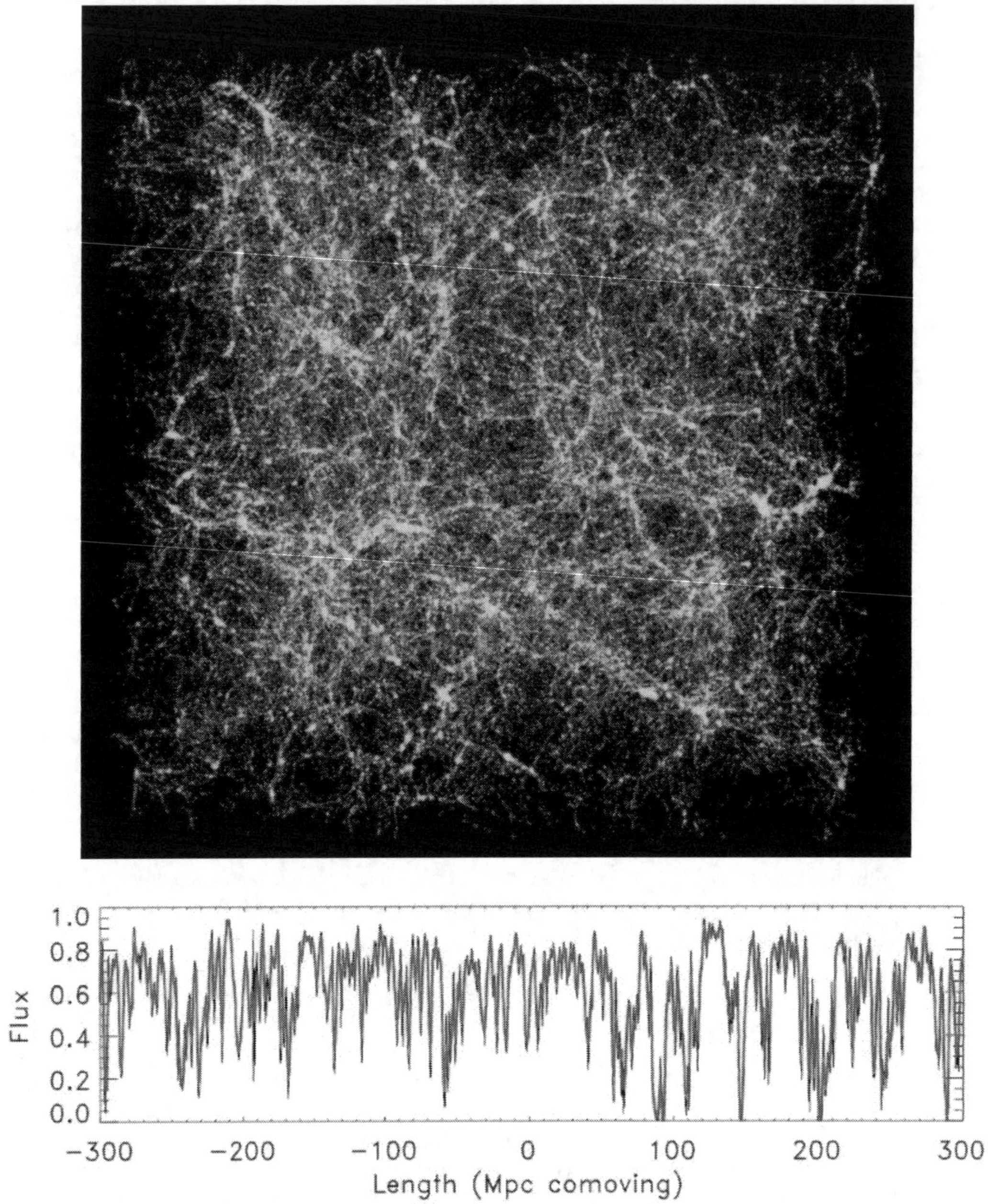

Figure 1. (a) 2048^3 hydrodynamic simulation of the Lyman alpha forest in a 307 Mpc cube of the universe. Rendering of log(baryon density), courtesy J. Insley, ANL. (b) synthetic neutral hydrogen absorption spectrum for one LOS through a 614 Mpc cube. Black and red lines are at spectral resolutions of 13,000 and 1800, respectively.

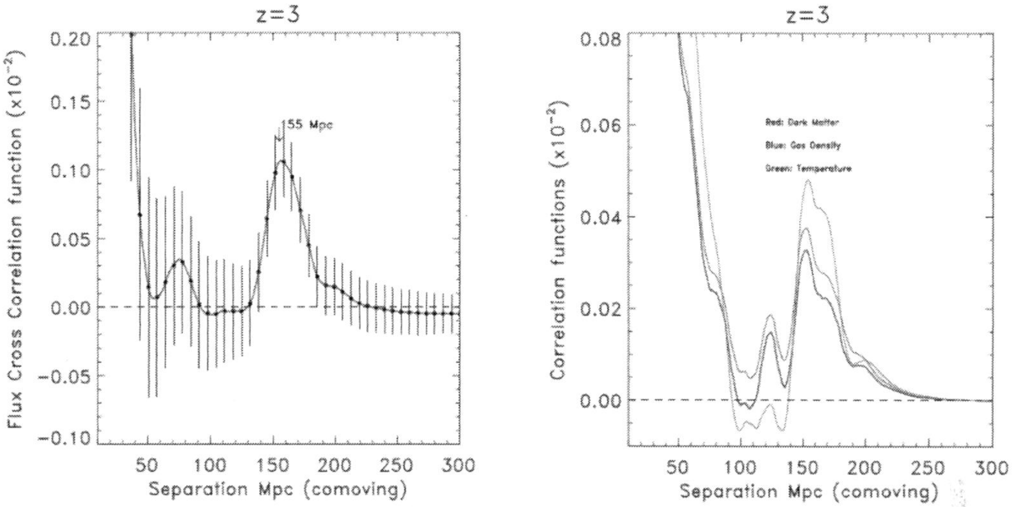

Figure 2. (a) cross correlation of transmitted flux at z=3 in quasar pairs versus separation on the sky, in comoving Mpc. (b) cross correlation of dark matter density, baryon density, and baryon temperature along same lines of sight used for figure 2a. The BAO signal is evident at ~150 Mpc.

4. Discussion and future work

We have simulated the observability of baryon acoustic oscillations (BAO) in the Lyman alpha forest using hydrodynamic cosmological simulations in large boxes (300 and 600 Mpc on a side) and high resolution (2048^3 particles and cells). With a wavelength of ~150 Mpc (comoving), the BAO is barely(easily) detectable in the 300(600) Mpc volumes. At a redshift of z=3, the BAO signal stands out in the cross correlation function of the transmitted flux in widely separated simulated quasar absorption line spectra. The signal is also present in skewers of dark matter density, and baryon density and temperature.

Restricting our attention to results from the 600 Mpc volume (figure 2), the statistical significance of the detection is about 1.86σ. However we have used ideal spectra in obtaining this result. In the future it will be important to study the effects of spectral resolution and noise on the BAO signal. We are currently repeating the calculation of the flux cross correlation function lowering the spectral resolution to that of the upcoming BOSS survey [4].

It is also important to study the effect of redshift evolution. The BAO signal should increase with decreasing redshift as the BAO mode grows in amplitude and the LAF absorption becomes less saturated. However, below z=2 the intergalactic LAF absorption begins to disappear while the circum-galactic absorption becomes more prominent [10]. The redshift at which the BAO signal is maximal is not known at the present time, but should be easily estimated with numerical simulations.

Finally, we also need to quantify the effect of numerical resolution on the BAO signal. At 2048^3 cells, our spatial resolution is only 300 kpc in the 600 Mpc volume. This is relatively poor resolution when it comes to resolving the internal structure of baryonic features which give rise to the LAF [11]. We are currently running a 4096^3 simulation in a 600 Mpc volume which should significantly improve the measurement and our understanding of the numerical systematics.

Acknowledgements

The authors acknowledge helpful discussions with Martin White. We wish to thank Joe Insley at Argonne National Lab for assistance with the rendering in figure 1. MLN and PP acknowledge partial support by NSF grants AST-0708960 and AST-0808184. Simulations were carried out on NERSC Seaborg under the INCITE program and NICS Kraken under LRAC allocation TG-MCA98N020.

References

[1] Eisenstein, D. and Bennett, C. Physics Today, April 2008, pp. 44-50; Eisenstein, D. et al. 2007, Astrophysical J., Vol. 633, p. 560
[2] Bashinsky,S.; Bertschinger, E. 2001. Physical Review Letters, vol. 87, Issue 8
[3] Rauch, M. 1998. Annual Review of Astronomy and Astrophysics, Volume 36, pp. 267-316.
[4] Schlegel, D. J.; Blanton, M.; Eisenstein, D.; Gillespie, B.; Gunn, J.; Harding, P.; McDonald, P.; Nichol, R.; Padmanabhan, N.; Percival, W.; and 8 coauthors, 2007. Bulletin of the American Astronomical Society, Vol. 39, p. 966
[5] McDonald, P. & Eisenstein, D. 2007. Physical Review D, Vol. 76, Issue 6
[6] Norman, M.L.; Bryan, G.L.; Harkness, R.; Bordner, J.; Reynolds, D.; O'Shea, B.; Wagner, R. 2007. in Petascale Computing: Algorithms and Applications, Ed. D. Bader, CRC Press LLC
[7] A. Lewis, A. Challinor, and A. Lasenby 2000. The Astrophysical Journal Vol. 538, p. 473, http://camb.info, astroph-ph/9911177
[8] Bond, J. R., Kofman, L., Pogosyan, D. 1996. Nature, Vol. 380, pp. 603
[9] Zhang, Y., Anninos, P., Norman, M. L., and Meiksin, A. 1997. The Astrophysical Journal Vol. 485, p.496
[10] Davé, R.; Hernquist, L.; Katz, N.; Weinberg, D. H. 1999. The Astrophysical Journal, Volume 511, Issue 2, pp. 521-545.
[11] Jena, T., Norman, M.L., Tytler, D.T., Kirkman, D., Suzuki, N., Chapman, A., Melis, C., Paschos, P., O'Shea, B, So, G. and 5 coauthors, 2005. Monthly Notices of the Royal Astronomical Society, Volume 361, Issue 1, pp. 70-96.

Journal of Physics: Conference Series **180** (2009) 012022 doi:10.1088/1742-6596/180/1/012022

Computational models of stellar collapse and core-collapse supernovae

Christian D Ott[1,2,3], Erik Schnetter[3,4], Adam Burrows[5], Eli Livne[6], Evan O'Connor[1], and Frank Löffler[3]

[1] TAPIR, Mailcode 350-17, California Institute of Technology, Pasadena, CA, USA
[2] Niels Bohr International Academy, Niels Bohr Institute, Copenhagen, Denmark
[3] Center for Computation & Technology, Louisiana State University, Baton Rouge, LA
[4] Department of Physics & Astronomy, Louisiana State University, Baton Rouge, LA, USA
[5] Department of Astrophysical Sciences, Princeton University, Princeton, NJ, USA
[6] Racah Institute of Physics, Hebrew University, Jerusalem, Israel

E-mail: `cott@tapir.caltech.edu`

Abstract. Core-collapse supernovae are among Nature's most energetic events. They mark the end of massive star evolution and pollute the interstellar medium with the life-enabling ashes of thermonuclear burning. Despite their importance for the evolution of galaxies and life in the universe, the details of the core-collapse supernova explosion mechanism remain in the dark and pose a daunting computational challenge. We outline the multi-dimensional, multi-scale, and multi-physics nature of the core-collapse supernova problem and discuss computational strategies and requirements for its solution. Specifically, we highlight the axisymmetric (2D) radiation-MHD code `VULCAN/2D` and present results obtained from the first full-2D angle-dependent neutrino radiation-hydrodynamics simulations of the post-core-bounce supernova evolution. We then go on to discuss the new code `Zelmani` which is based on the open-source HPC `Cactus` framework and provides a scalable AMR approach for 3D fully general-relativistic modeling of stellar collapse, core-collapse supernovae and black hole formation on current and future massively-parallel HPC systems. We show `Zelmani`'s scaling properties to more than 16,000 compute cores and discuss first 3D general-relativistic core-collapse results.

1. Introduction

Core-collapse supernova (CCSN) explosions are powered by the release of gravitational energy in the collapse of a massive star's core to a protoneutron star (PNS). While this general CCSN picture may be clear, its details have evaded understanding despite many decades of concerted theoretical and numerical effort.

When thermonuclear core burning ends, the core of a massive star (i.e., $M \gtrsim 8 - 10$ solar masses $[M_\odot]$) is comprised of iron-group nuclei (or O/Ne nuclei in the lowest-mass massive stars) and supported against gravity's pull primarily by the degeneracy pressure of relativistic electrons. Shell burning adds mass to this core and eventually pushes it over its maximum supportable mass. Gravitational collapse results and is accelerated by the capture of electrons on free and bound protons and by the photodissociation of heavy nuclei into alphas and nucleons. Core collapse continues until the central, subsonically-collapsing region (the *inner core* of $\sim 0.5\,M_\odot$) reaches nuclear density. There, the strong nuclear force kicks in, stiffening the nuclear equation of state (EOS) and halting collapse, resulting in the rebound of the inner core into the still collapsing outer core. A hydrodynamic shock is formed in this *core bounce* and initially propagates outward in mass and radius while losing energy to the dissociation of heavy infalling nuclei and neutrinos that stream away from the postshock region. The shock stalls within $\sim 10-20$ ms after bounce and at a radius of $\sim 100-200$ km. For a successful CCSN explosion, it must be re-energized, otherwise continued accretion will push the PNS over its mass limit. This results in a second phase of gravitational collapse, leading to the formation of a black hole, turning the stellar collapse event into a collapsar[1] [2].

Finding and understanding the mechanism of shock revival has been the key problem of core-collapse supernova theory in the past $\sim 45+$ years. Current theory and modeling suggests (see, e.g., [3, 4] and references therein) that there are (at least) three ways to blow up massive stars: (1) the neutrino mechanism, relying primarily on an imbalance between charged-current neutrino heating and cooling in the immediate postshock regions in combination with convection and the standing-accretion-shock instability (SASI) (e.g., [5–8]), (2) the magnetorotational mechanism, based on bipolar jets created by strong magnetic stresses in rapidly rotating cores (e.g., [9–11]) and (3) the acoustic mechanism recently proposed by Burrows et al. [12–14], which rests on the excitation of non-radial pulsations in the PNS by accretion and turbulence and their damping by strong sound waves that steepen into shocks, depositing energy very efficiently in the postshock region. The acoustic mechanism has not yet been confirmed by other groups and perturbative work suggests that the pulsation amplitudes may be limited by a parametric instability that transfers energy in faster damping daughter modes [15], but this mechanism remains a compelling possiblity.

In the three potential explosion mechanisms, the breaking of spherical symmetry is either fundamentally necessary or a key facilitating factor, making *multi-D modeling* a necessity. Furthermore, a CCSN simulation must not only spatially resolve the steep gradients at the PNS surface and the small to large scale turbulent eddies of overturn (typical $\Delta x \sim \mathcal{O}(100\,\mathrm{m})$ or less for magnetoturbulence [11, 16]), but the simulation domain must also extend to many thousand kilometers to encompass sufficient material to track a long-term postbounce accretion phase without boundary effects affecting the evolution and to allow a potential explosion to fully develop. Hence, *multi-scale modeling* is required and may be implemented via fixed and/or adaptive mesh refinement (FMR/AMR) or particle methods (smoothed particle hydrodynamics [SPH], e.g. [17]). Furthermore, for a complete model of stellar collapse and the CCSN postbounce phase, a broad spectrum of tightly-coupled physics must be included – ideally accurately, but at least approximately. This includes, but is not necessarily limited to, (magneto)hydrodynamics (MHD), general relativity (GR), nuclear physics (nuclear EOS and nuclear reactions), neutrino radiation-transport, and neutrino-matter interaction microphysics.

The multi-D, multi-scale, and multi-physics nature of the CCSN problem makes it complex and difficult to model and solve computationally. At the same time, however, owing to its complexity, physically accurate computational modeling, in combination with future detailed observations of CCSN neutrinos and gravitational waves[2], is our only chance of solving it.

In this contribution to the Scientific Discovery through Advanced Computing (SciDAC) Conference 2009, we discuss central aspects of our broad computational approach to stellar collapse and the CCSN problem and highlight recently obtained results. In section 2,

[1] By *collapsar* we mean a stellar collapse event that does not lead to a supernova explosion. If the progenitor star has the needed angular momentum distribution, a Collapsar-type Gamma-Ray Burst may occur in a collapsar, see e.g., [1].

[2] Gravitational waves are lowest order quadrupole waves. Hence, they are an intrinsically multi-D phenomenon and do not exist in spherical symmetry. See [18] for a thorough introduction to gravitational wave theory.

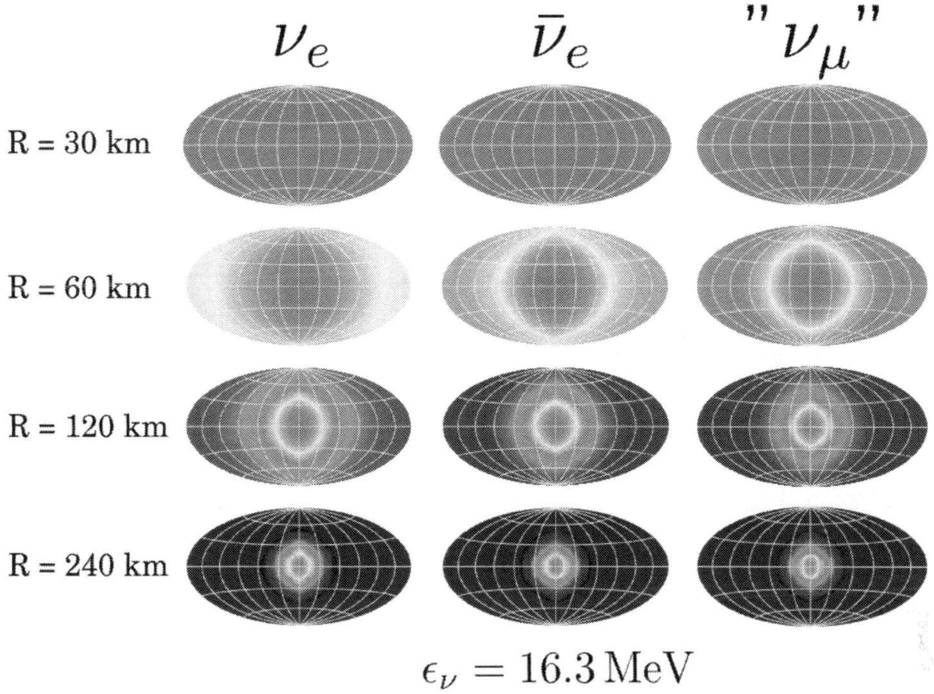

Figure 1: Hammer-type (smoothed) map projections of the normalized specific intensity $I_\nu(\vartheta, \varphi)/J_\nu$ (where J_ν is the mean intensity) in model s20.nr at 160 ms after bounce. The color map is logarithmic and each individual projection is set up to range from $\max(I_\nu/J_\nu)$ (red) to $10^{-4}\max(I_\nu/J_\nu)$ (black). Shown is the specific intensity of ν_e, $\bar{\nu}_e$, and "ν_μ" neutrinos at $\epsilon_\nu = 16.3\,\text{MeV}$ on the equator and at radii of 30, 60, 120, and 240 km. The Hammer projection is set up in such a way that ϑ varies in the vertical from 0° (top) to 180° (bottom) and φ varies horizontally from −180° (left) to +180° (right). Grid lines are drawn in ϑ- and φ-intervals of 30°. Note (a) that the neutrino radiation fields are isotropic at $R = 30\,\text{km}$, (b) the increasing forward-peaking of I_ν with increasing radius (and decreasing optical depth), and (c) that at any given radius I_ν of "ν_μ" is more forward-peaked than that of the $\bar{\nu}_e$ component, which, in turn, is always more forward-peaked than the ν_e component. This fact is a consequence of a transport mean-free path that varies with species (and energy; not shown here) and is smallest for the electron neutrinos.

we introduce the CCSN code `VULCAN/2D` and present results from the first long-term full-2D momentum-space angle-dependent radiation-hydrodynamics simulations of the postbounce phase in CCSNe. We go on to describe in section 3 our full-GR 3D stellar collapse simulation package `Zelmani` which is based on the `Cactus` computational framework [19] and designed for massively-parallel execution. `Zelmani` has already been applied to simulations of rapidly-rotating 3D core collapse for which we present results. In Section 4, we wrap up and present a forward-looking summary.

2. Angle-dependent neutrino radiation-hydrodynamic `VULCAN/2D` simulations

`VULCAN/2D` is a general Newtonian axisymmetric (2D) radiation-magnetohydrodynamics code described in [13, 20–22] and extended and applied to the stellar collapse and CCSN problems in a large number of studies (e.g., [12, 13, 21–27]). `VULCAN/2D` implements the arbitrary Lagrangian-Eulerian (ALE) technique with second-order TVD remap. The scheme is directionally unsplit and allows for arbitrary grids. Here we use `VULCAN/2D` in hydrodynamic mode which implements a finite-difference representation of the Newtonian Euler equations with artificial viscosity.

`VULCAN/2D` allows for the use of general EOS tables and for the present study we employ the finite-temperature nuclear Shen EOS [28] which is based on a relativistic mean-field model for nuclear interactions and transitions to an ideal gas of nuclei, nucleons, photons, and electrons at low densities.

`VULCAN/2D` implements neutrino transport in two different multi-group (and multi-species) ways. The module implementing time-implicit multi-group flux-limited diffusion (MGFLD) was described in [13] and evolves the angle-averaged mean radiation intensity J_ν, the zeroth angular moment of the specific intensity I_ν, and uses the flux limiter of [29]. The angle-dependent transport module implements the method of discrete ordinates (S_n) [22, 30] in time-implicit fashion at low optical depths and matches to MGFLD at optical depth $\tau \gtrsim 2$ for accelerated conversion at high optical depths where the radiation field is isotropic [21]. Both transport solvers employ the neutrino microphysics outlined in [31] (including nucleon-nucleon bremsstrahlung, e.g., [32]), use typically 16 energy groups logarithmically spaced from 2.5 to 220 MeV, and consider ν_e and $\bar{\nu}_e$ individually, while lumping together ν_μ, $\bar{\nu}_\mu$, ν_τ, and $\bar{\nu}_\tau$ into "ν_μ." In both MGFLD and S_n we neglect neutrino energy-bin coupling (relevant in inelastic scattering) and assume the slow-motion approximation to radiation transport appropriate in the postbounce phase, neglecting velocity-dependent terms of $O(v/c)$. As a consequence, neutrino advection, Doppler shifts and aberration effects are not considered. This greatly limits the computational complexity of the problem, but its impact on the transport solution depends on the particular choice of reference frame and was examined in [33]. Around core bounce and neutrino breakout, during the non-linear phase of the SASI hundreds of milliseconds after bounce, and in the case of very rapid rotation, including $O(v/c)$ terms is advisable. Full $O(v/c)$ Boltzmann transport with energy redistribution will be addressed in the future.

In the S_n solver, we discretize the angular radiation distribution evenly in $\cos\vartheta$ from -1 to 1 and make the number of φ-bins (running from 0 to π, because of axial symmetry) a function of $\cos\vartheta$ to tile the hemisphere more or less uniformly in solid angle. In our time-dependent S_n runs, we employ 8 $\cos\vartheta$ bins, resulting in a total of 40 angular zones. Steady-state radiation fields are computed either with 8 $\cos\vartheta$ bins, 12 $\cos\vartheta$ bins (92 total angular zones) or 16 $\cos\vartheta$ bins (162 total angular zones) at each spatial grid point.

The computational costs for a `VULCAN/2D` simulation can be estimated based on the single-zone update cost for one neutrino group/species, the number of required updates, and the number of zones, neutrino energy groups and species. For a typical MGFLD simulation, $N_{\text{zones}} = 40000$, $N_{\text{energy groups}} = 16$, $N_{\text{species}} = 3$ and the single-zone update cost including hydro, gravity, and one group of MGFLD is ~ 125 flop. Thus, a single timestep requires ~ 240 Gflop and a typical simulation lasts for ~ 1 million timesteps, leaving us with a total flop count of ~ 250 Pflop. In a S_n simulation, the same single-zone update flop count applies, but the number of zones is scaled by a factor equaling the number of momentum-space angular zones modified by an empirical correction factor of ~ 0.1 obtained through timing measurements. Thus, a S_8 simulation is about 4 times more expensive than a MGFLD run and requires ~ 1000 Pflop to complete.

In order to complete a simulation in reasonable time, we parallelize `VULCAN/2D` transparently and efficiently via MPI in neutrino groups (energy/species). Hence, for 16×3 groups we obtain a speed-up of almost 48 compared with a single-core calculation, since only scalars are communicated at the end of each timestep. Node-local OpenMP parallelism is an additional way for increasing performance and is currently under consideration. Domain decomposition, however, is not a viable option, since the communication overhead due primarily to the relatively small number of zones in a 2D simulation would quickly dominate over any performance gains.

Assuming a sustained performance of 1 Gflop/core, a MGFLD calculation is completed in ~ 60 days, while the S_n calculations presented here have required $90 - 120$ days on 48 cores, but were run for only a limited amount of physical postbounce time.

2.1. *Simulation setup*

We consider here two models based on the 20-$M_{\odot}$ supernova progenitor of [34]. Model s20.nr is mapped onto the `VULCAN/2D` grid without rotation while we impose a precollapse central angular velocity of π rad s^{-1} in model s20.π and decrease it slowly with distance from the rotation axis according to the simple rotation law specified in [23]. This results in an essentially rigidly-rotating PNS core with a period of $\sim$ 2 ms and strong rotational deformation of the entire PNS and the postshock region. We choose a `VULCAN/2D` grid setup with a pseudo-Cartesian central region that smoothly transitions to a spherical grid at 20 km. This not only removes the coordinate singularity at the origin of spherical grids, but also liberates the PNS core and allows for larger hydrodynamic timesteps [12, 13]. In the angular direction we employ 120 zones and there are 30 logarithmically-spaced radial zones interior to the grid transition and 200 (also logarithmically-spaced) radial zones from 20 km to 4000 km.

Both s20.nr and s20.π are run with MGFLD to 160 ms after core bounce. At this point, we solve for stationary-state angle-dependent S_n radiation fields, then evolve the simulations with S_8 (see [21] for resolution tests) over postbounce intervals of 340 ms and 390 ms for model s20.nr and s20.π, respectively. For comparison we also continue the MGFLD variants.

2.2. *Results*

The CCSN problem is particularly challenging in its radiation transport aspects, because the energy- and species-dependent neutrino radiation fields transition from being completely isotropic inside the PNS (where the diffusion limit applies) to being completely forward-peaked at $\tau \ll 1$ (the free-streaming limit). The transition between diffusion and free-streaming is handled in an approximate way via the flux limiter in MGFLD (e.g., [29]), but only angle-dependent transport can self-consistently and accurately capture the gradual change of the radiation field whose degree of forward-peakedness in the postshock heating region has an influence on the neutrino heating efficiency [21, 35]. In Fig. 1, we present map projections of the angular distribution of the specific neutrino radiation intensity I_{ν} as seen by an equatorial observer located at various radii. Shown are the I_{ν} of ν_e, $\bar{\nu}_e$, and "ν_{μ}" neutrinos at a neutrino energy of $\epsilon_{\nu} = 16.3$ MeV. Inside the PNS, at $R = 30$ km, all radiation fields are isotropic and become increasingly forward peaked with radius, corresponding to decreasing optical depth. At any given radius, "ν_{μ}"s are more forward-peaked than $\bar{\nu}_e$s which, in turn, are always more forward-peaked than ν_es. This hierarchy is characteristic for the postbounce phase of CCSNe and is a result of the species-dependent transport mean-free path systematics in CCSN matter (e.g., [36]).

In the left panel of Fig. 2 we plot the specific net gain, defined as neutrino heating Q^+_{ν} minus neutrino cooling Q^-_{ν} and contrast the S_n results for the nonrotating model s20.nr (left half) with the rapidly rotating model s20.π (right half) at 160 ms after bounce. In addition, we superpose velocity vectors to visualize the flow of matter. The qualitative and quantitative differences between the two models are large. In model s20.nr, there is strong neutrino heating in the immediate postshock region which drives strong convection. One also notes a slight deformation of the shock away from spherical symmetry which is indicative of the onset of the SASI. In model s20.π, on the other hand, the neutrino heating occurs very asymmetrically and primarily in polar regions where the neutrino flux is highest (see also the discussion in [21, 37, 38]). The shock is more extended along the poles, but there is a pronounced equatorial bulge with very little net heating. Convective overturn is confined to polar regions due to (1) a large positive specific angular momentum gradient at low latitudes, and (2) the absence of strong neutrino heating in these regions [37, 39, 40]. In addition, no signs of the SASI are apparent and the subsequent evolution of model s20.π suggests that rapid rotation delays and modifies the SASI in axisymmetry. The situation may be different in 3D (e.g., [41, 42]).

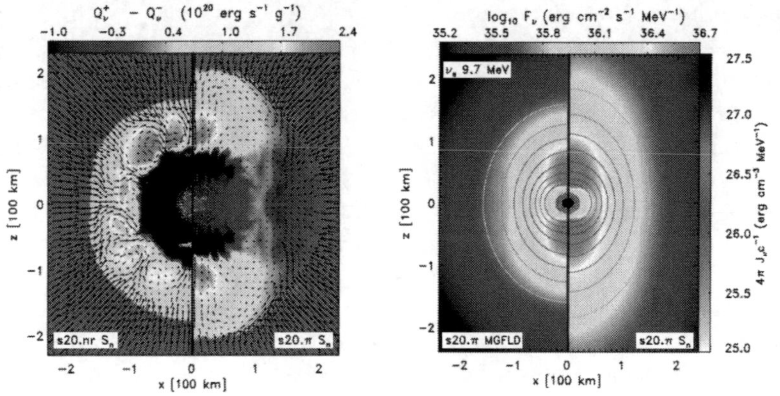

Figure 2: **Left:** Comparison of the net gain (neutrino heating minus cooling) in the nonrotating model s20.nr and in the rapidly-rotating model s20.π as obtained with 2D angle-dependent transport at 160 ms after bounce. Due to its rapid rotation, model s20.π has a larger shock radius, enhanced heating near the axis, but due to a positive specific angular momentum gradient in the postshock region at low latitudes, also much less convective overturn. **Right:** Comparison of the ν_e fluxes at $\epsilon_\nu = 9.7$ MeV in model s20.π at 160 ms after bounce and as obtained with MGFLD (left section) and S_n (right section). The radiation fields are oblate in the PNS core and deform to a prolate shape further out. Note that S_n predicts a prolateness of the radiation field to much greater radii than MGFLD. The latter leads to nearly spherically symmetric radiation fields at radii greater than $\gtrsim$ 150–200 km. This result is largely independent of neutrino species and ϵ_ν.

The right panel of Fig. 2 visualizes the spectral flux density of ν_e at $\epsilon_\nu = 9.7$ MeV and at 160 ms after bounce in the rapidly spinning model s20.π and contrasts the MGFLD result (left half) with that obtained with S_n (right half). In both variants, the radiation field is oblate inside the PNS core, but quickly transitions to a prolate shape further out. The density gradient in the polar regions of the core is much steeper, allowing for much smaller neutrino sphere radii ($R_\nu = R(\tau \approx 2/3)$) and resulting in a dramatic enhancement of the polar neutrino flux [21, 37, 38]. The MGFLD result captures the overall systematics, but due to the diffusive nature of the 2D MGFLD approach, the radiation field asymmetry is smoothed out at low τ and becomes nearly spherical at radii $\gtrsim$ 150 km. The angle-dependent variant, on the other hand, captures the true radiation-field asymmetry even at large radii. In the case of model s20.π, this results in polar neutrino heating that is locally larger by up to a factor of two than in the MGFLD calculation. The radiation field asymmetry in the nonrotating model s20.nr is much smaller and the radiation fields predicted by MGFLD and S_n are more similar, while the S_n calculation still yields locally $\sim$ 10% greater heating rates.

Figure 3 visualizes the dynamical effect of angle-dependent neutrino transport on the postshock evolution of CCSNe by comparing the time evolution of the angle-averaged shock radii of the MGFLD and S_n variants of models s20.nr and s20.π. In the latter, switching to S_n leads to a shock expansion primarily in the polar regions where the increase in neutrino heating is greatest. However, increased neutrino cooling from the now also extended cooling region counteracts the shock expansion and lets it settle at a new, somewhat higher average radius. The larger radius and the increased local heating lead to an earlier onset of the rotationally-modified SASI in the S_n variant. This is reflected in the earlier and more pronounced oscillations of the average shock radius. However, towards the end of the postbounce time covered by the simulations, the MGFLD variant's average shock radius catches up and there is no large overall difference between the dynamics seen in the S_n and MGFLD calculations of model s20.π.

The dynamical evolutions in the S_n and MGFLD variants of the nonrotating model s20.nr are

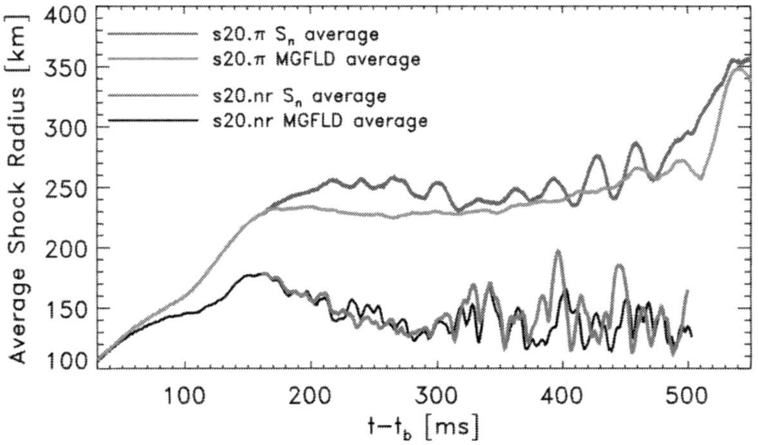

Figure 3: Average shock radius as a function of time after core bounce in MGFLD and S_n variants of models s20.nr and s20.π. The large oscillations seen in the curves of the nonrotating model s20.nr are due to strong SASI oscillations and are somewhat more pronounced in the S_n variant. The average shock radius of the rapidly rotating s20.π settles at larger values than in the nonrotating case, but the growth of the SASI is slowed down by rapid rotation and only the S_n model with its enhanced polar heating exhibits long-period SASI oscillations at intermediate postbounce times. See text and [21] for discussion.

even more similar and the SASI shock excursions in the two variants remain practically in phase for almost 200 ms. The S_n calculation exhibits larger SASI excursions at later times, but, as in the case of model s20.π, there is no qualitative change between the MGFLD and S_n postbounce evolutions despite stronger neutrino heating (by up to 30% of the total heating rate at late times) in the latter.

2.3. Discussion

Neutrinos carry away $\sim$ 99% of the gravitational energy of a neutron star formed in stellar collapse. Their transport and their interactions with matter are a central aspect of any CCSN model. Here, we have presented and summarized results from our recent `VULCAN/2D` simulations [21] that for the first time addressed for a multi-D simulation the long-term dependence of the postbounce dynamics of CCSNe on the neutrino transport technique employed. Comparing MGFLD and angle-dependent S_n transport, we find that the former has difficulties in capturing physical radiation field asymmetries and preserving them at low optical depth. The S_n approach self-consistently evolves the radiation field from the diffusion limit and isotropy to the free-streaming limit and forward-peakedness. S_n generally leads to locally stronger neutrino heating, but the feedback in the supernova engine is sufficiently strong to lead to an adjustment of the system to a new equilibrium configuration without a true qualitative change when compared with MGFLD. Thus, it appears unlikely that differences in neutrino transport and/or interactions that result in by 10% − 30% increased heating (as seen in our simulations) can have a strong impact on the CCSN dynamics. Larger effects seem necessary to turn the CCSN dud into an explosion.

The work of the Garching group [5, 43] and of the ORNL/FAU group [7] suggest that the inclusion of a general-relativistic monopole term in the otherwise Newtonian potential in combination with a soft[3] EOS can increase the neutrino heating efficiency due to hardened

[3] These groups use the $K = 180\,\mathrm{MeV}$ variant of the Lattimer-Swesty EOS which is too soft to support NSs with gravitational masses above $\sim 1.7\,M_\odot$.

neutrino spectra and may lead to explosion. Furthermore, [6] have shown in a simplified model with parametrized neutrino heating and cooling that explosions are more easily obtained in 2D than in 1D. This result may very well extend to 3D and supports and strengthens the CCSN community's motivation to move towards 3D simulations.

3. A new, fully general-relativistic approach in 3D

Today's technically most advanced and physically most complete stellar collapse and long-term postbounce CCSN simulations are carried out in axisymmetry (2D), use angle-dependent neutrino transport or MGFLD either in full 2D [11, 12, 21, 44] or in a ray-by-ray approximation [5, 7, 43, 45] and implement Newtonian (magneto)hydrodynamics and gravity either in purely Newtonian fashion [11, 12, 21, 44] or with a GR monopole term in an otherwise Newtonian gravitational potential [5, 7, 43, 45]. The current state-of-the-art for axisymmetric GR core-collapse calculations, which traditionally have focussed on estimates of the GW signal from rotating core collapse and bounce, is set by [46] who performed conformally-flat[4] calculations of the collapse and early postbounce phase with a finite-temperature (T) microphysical EOS and a simple deleptonization scheme [51] for the collapse phase in lieu of neutrino transport.

Current published 3D simulations do not yet rival their 2D counterparts in accuracy and physical completeness. Iwakami et al. [42, 52] investigated the SASI in 3D with steady-state initial conditions, spherical Newtonian gravity, a finite-T microphysical EOS, parametrized neutrino heating/cooling, a cut-out core, and a fixed high accretion rate at the outer boundary. The Basel group [53], focussing on the GW signal of stellar collapse, performed 3D calculations of the collapse phase with a finite-T microphysical EOS and the deleptonization scheme of [51] during collapse, but neglected neutrino transport and heating/cooling in the postbounce phase.

More physically accurate simulations with better neutrino physics and transport in the postbounce phase are required to address the explosion mechanism. Several groups are in the process of implementing codes with the necessary features (e.g., [7, 45, 54]).

In the following, we present our approach to 3D CCSN modeling which builds upon the tremendous recent progress in numerical relativity (see, e.g., [55]) and implements GR hydrodynamics and full GR curvature evolution in a variant of the Arnowitt-Deser-Misner (ADM) [56] $3+1$ formalism. This allows us to not only more accurately follow the CCSN hydrodynamics, but provides for the capability to form black holes dynamically and in 3D in failing core-collapse supernovae – something that is impossible in Newtonian or pseudo-GR formulations. Our approach makes heavy use of the `Cactus` computational toolkit [19] and is optimized for execution on supercomputers, implementing AMR, domain decomposition, parallel I/O, and hybrid MPI/OpenMP parallelism for improved scaling on massively-parallel systems.

In Section 3.1, we introduce our GR curvature and hydrodynamics formulation and discuss the computational infrastructure and the various physics components of our approach, and highlight parallel scaling results. In Section 3.2, we go on to discuss results obtained with our approach by [48, 49, 57] in the first set of GR simulations of rotating core collapse in 3D.

3.1. Computational approach, application codes and parallel performance

We employ the open source software framework `Cactus` [19, 58] designed for computational scientific and engineering problems. It has a modular structure and enables scalable parallel computation across different architectures, as well as collaborative code development between different research groups.

[4] The conformal flatness condition (CFC) is an approximation to GR in which the radiative degrees of freedom have been suppressed [47]. CFC is exact in spherical symmetry and is accurate to $\lesssim 5\%$ in the core-collapse scenario [48–50].

`Cactus` consists of a central part, called the *flesh*, that provides core routines, and of components, called *thorns*. The flesh is independent of all thorns and provides the main program, which parses input parameters and activates the appropriate thorns, passing control to thorns as required. By itself, the flesh does very little science; to do any computational task the user must compile in thorns and activate them at run time. Parallelism, communication, load balancing, memory management, and I/O are handled by a special component, the *driver*, which is not part of the flesh and which can be transparently exchanged. The flesh (and the driver) have complete knowledge about the state of the application, allowing inspection and introspection through generic APIs.

`Cactus` runs on all current mainstream architectures. Applications, developed on standard workstations or laptops, can be seamlessly run on clusters or supercomputers. `Cactus` provides easy access to many cutting-edge software technologies being developed in the academic research community, including the Globus Metacomputing Toolkit [59, 60], HDF5 [61] parallel file I/O, the PETSc scientific library [62], AMR, multi-block methods [63], web interfaces [64], and advanced visualization tools (e.g. VisIt [65]).

The `Einstein Toolkit` [66] is a set of `Cactus` thorns providing infrastructure and basic functionality for GR applications codes using the variables of the ADM formalism [56]. Within a simulation, the ADM variables are employed for coupling GR curvature evolution with matter and radiation variables and also serve in run-time analysis of the simulation results, such as e.g. evaluating constraints, locating apparent horizons [67, 68] or event horizons [69], or calculating gravitational wave signals.

3.1.1. GR curvature evolution. The Einstein equations are necessary for the correct description of gravity in the strong-field regime. They are a set of ten coupled, non-linear wave-type partial differential equations. We solve these equations using the *BSSN* formulation (e.g., [70] and references therein). This formulation, similar to ADM, breaks up the four-dimensional spacetime into $3 + 1$ dimensions, three spatial and one time dimensions. This leads to 25 hyperbolic time evolution equations coupled to 9 elliptic constraint equations.

Of the 25 evolution equations, 8 are not specified by the Einstein equations, but instead have to be chosen as gauge conditions to determine the time evolution of the curvilinear coordinates of spacetime. We choose the so-called $1 + \log$ slicing condition and the Γ-driver shift condition [57, 71], which are standard gauge conditions used BSSN. They ensure stable, long-term time evolutions. The 9 constraint equations have to be satisfied initially, requiring solving an elliptic system for setting up initial data, and remain then satisfied under time evolution up to within the discretization error. We monitor the constraints during time evolution and do not re-solve them. The resulting equations for the BSSN system and the gauge conditions can be time-evolved with standard discretization methods.

We are using the `Kranc` package [72, 73] for automatic code generation for the BSSN formulation, gauge conditions, and constraint equations. `Kranc` is a `Mathematica` package which generates `Cactus` thorns from equations. Starting from equations in `Mathematica` format which specify a system of PDEs in abstract index notation, `Kranc` discretizes the equations and generates a complete `Cactus` thorn that evaluates these equations. `Kranc`-generated thorns use all relevant `Cactus` APIs for initial data setup, analysis, time integration, and AMR.

Automatic code generation greatly reduces the time and effort necessary to implement the BSSN equations, since these contain about 5,000 individual terms. `Kranc` also allows us to experiment with modifications to the formulation, e.g. to increase accuracy near singularities, and with modifications to low-level implementation details (loop blocking, vectorization) to achieve higher efficiencies on modern computer architectures.

We have implemented the BSSN equations in an open-source `Cactus` thorn arrangement called

`McLachlan` [74, 75] in the XiRel [76, 77] project. `McLachlan` is a full-featured BSSN solver able to simulate not only stellar collapse, but also relativistic binary systems of black holes or neutron stars [75].

3.1.2. GR hydrodynamics, microphysics and neutrinos On the GR hydrodynamics (GRHD) side, we implement the Valencia formalism of GRHD [78, 79] which is based on a dimensionally-split flux-conservative high-resolution shock-capturing finite-volume approach. These methods are generally of lower order than those used for the spacetime curvature evolution, i.e., second-order in space and second-order in time. The main reason for this is not only that the discretization of the GR hydrodynamics problem becomes much more complicated with increasing order, but also that all higher-order methods must drop back to first order near discontinuities in the flow in order to prevent spurious oscillations in the solution. GR magnetohydrodynamics (GRMHD) is an extension to GRHD and can be straightforwardly implemented within the same general formalism [80, 81].

The `Zelmani` CCSN package proper is a collection of `Cactus` thorns implementing GRHD, a finite-temperature nuclear EOS, neutrino leakage, and heating [48, 49, 57, 82]. We generally use the name `Zelmani` synonymously to refer to the entire set of codes involved in our 3D GR CCSNe simulations, including `McLachlan`, `Carpet`, `CactusEinstein`, and `Cactus`. The GRHD module is based on a modified version of the open-source `Whisky` code [83, 84] that allows for general, finite-temperature EOS and neutrino-matter interactions. In our simulations, we employ PPM reconstruction of variables at cell interfaces and the approximate HLLE solver [85] for the relativistic Riemann problem.

Fully relativistic neutrino radiation transport is a formidable problem. Multi-D GR formulations exist (e.g., [86, 87]), but are still awaiting their first implementations. While we are actively exploring various ways to implement GR transport, we presently resort to the deleptonization scheme of [51] for collapse and employ neutrino leakage (e.g., [88]) and parametrized heating in the postbounce phase [6, 82]. Neutrino pressure contributions are included via the approximation discussed in [51].

3.1.3. Adaptive mesh refinement and parallel scaling. `Carpet` [63, 89, 90] is our AMR driver for the `Cactus` framework. `Carpet` acts as a driver layer for `Cactus`, providing adaptive mesh refinement, multi-patch capability, and efficient parallelization and I/O. We make both `Cactus` and `Carpet` publicly available as open source, and both are also used by a number of other numerical relativity and computational astrophysics groups.

`Carpet` provides spatial discretization based on highly efficient block-structured, decomposed, logically Cartesian grids with hybrid MPI/OpenMP [91, 92] parallelism. `Carpet` offers both AMR and multi-block capabilities, covering the domain with sets of distorted, logically rectangular blocks of grids. Time integration is performed via the recursive Berger–Oliger AMR scheme [93], including subcycling in time. As demonstrated in the left panel of Fig. 4, `Carpet` presently scales well to more than 16,000 cores with AMR on Leadership HPC systems.

`Carpet` employs HDF5 [61] for parallel, binary I/O, which is also used for checkpointing and restarting. The right panel of Fig. 4 depicts results of I/O benchmarks, comparing the achieved I/O bandwidth to the theoretical peak bandwidth on selected HPC systems. `Carpet` achieves a significant fraction of this already on about 1,000 cores.

Although the speed and performance of high-end computers have increased dramatically over the last decade, the ease of programming such parallel computers has not progressed. To address these issues in computational modeling in general and in numerical relativity and CCSN simulations in particular, we are developing the *Alpaca tools* [94, 95]. In contrast to existing debuggers and profilers, these tools work at the much higher level of the physical equations and

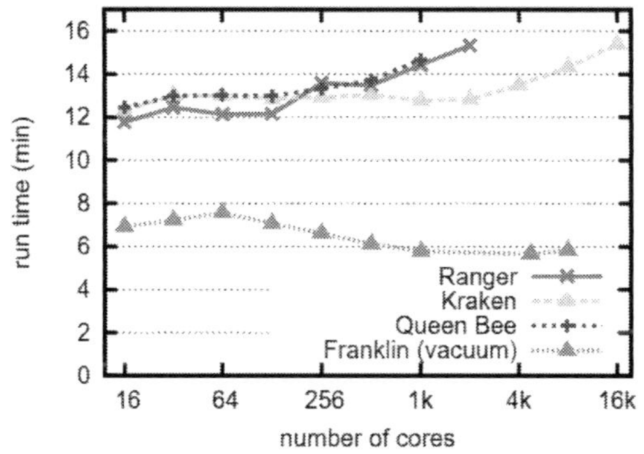

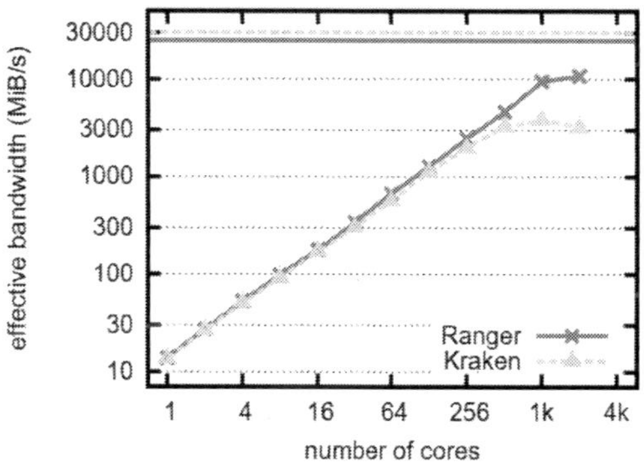

Figure 4: **Left:** Weak scaling benchmark for spacetime curvature and GRHD evolution of a single neutron star (Ranger, Kraken, Queen Bee) or of a single black hole (Franklin) with nine levels of mesh refinement. Ideal scaling would be an horizontal line in both cases. `Cactus/Carpet` scales well up to more than 16,000 cores. Ranger and Kraken are the currently largest machines on the NSF TeraGrid, Queen Bee is the largest machine on LONI and Franklin is a large DoE/NERSC system. **Right:** Cumulative I/O bandwidth for writing checkpoint files in parallel. This measurement includes application overhead. The theoretical peak bandwidth is shown as an horizontal line. `Cactus/Carpet` is able to achieve a significant fraction of this.

their discretizations and not at the level of individual lines of code or variables. The Alpaca tools are not external to the application, but are built-in, so that they have direct high-level access to information about the running application, and can interact with the user on a correspondingly high level.

3.2. Computational costs, simulations and first results

A typical simulation setup to track collapse and postbounce CCSN evolution with GR curvature evolution and GRHD uses a 9-level AMR grid hierarchy with $\sim 400^3$ computational zones each, providing high resolution near the center ($\delta x \lesssim 300\,\mathrm{m}$), while encompassing the inner $\sim 5000\,\mathrm{km}$ of the dying star. There are about 400 3D grid functions required for curvature and GRHD which translates to a memory footprint of $\sim 2\,\mathrm{TB}$ (including inter-process buffers assuming 1024 processes). A single point update requires $\sim 50\,\mathrm{kflop}$ and ~ 1 million fine grid updates are required, resulting in a total cost of $\sim 1500\,\mathrm{Pflop}$ for a single simulation. On 1024 compute cores and assuming a sustained performance of 1 Gflop/s per core, such a simulation requires ~ 17 days to complete. If radiation transport, even in approximate ray-by-ray fashion, is included, the memory footprint and the total number of required flops must be scaled by a factor of $10 - 100$.

As a first application of our `Cactus`-based 3D GR `Zelmani` CCSN code, we are considering the collapse and very early postbounce phase of rapidly rotating iron cores with an emphasis on the study of rotational multi-D dynamics leading to the emission of gravitational waves (GWs). The latter may, in combination with neutrinos, play an important future role as diagnostic tools for the CCSN mechanism [3, 96] and in the case of rotating collapse can provide information on the nuclear EOS, as well as on the rotation rate of the inner core at bounce [46].

We have carried out the first parameter study of rotating core collapse in 3D GR, investigating the dependence of dynamics and GW signal on progenitor stellar structure and precollapse rotational setup, specified by the degree of differential rotation and the initial central angular velocity. Since these simulations were run with an earlier version of `Zelmani`, postbounce deleptonization was neglected. The results of this study were extensively discussed in [48, 49, 57]

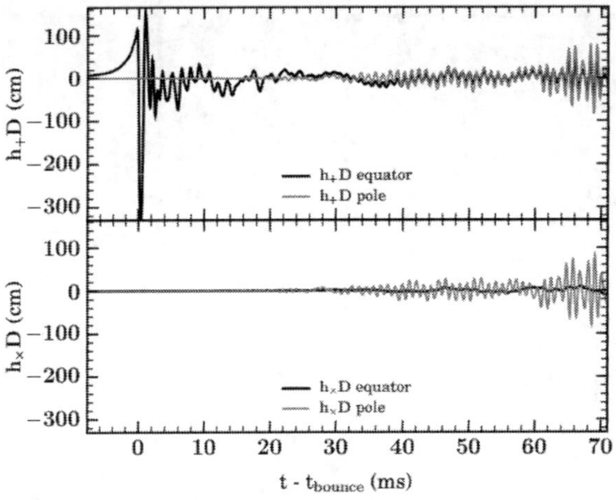

Figure 5: **Left:** Gravitational wave (GW) signals of the + polarization (top, $h_+ D$, where D is the source distance) and of the $\times$ polarization (bottom, $h_\times D$), emitted by model E20 as a function of postbounce time and as seen by polar (red lines) and equatorial (black lines) observers. Collapse and bounce are axisymmetric (only h_+ and emission only away from the symmetry axis) while at postbounce times nonaxisymmetric dynamics and GW emission develop. **Right:** Density colormap showing the inner 150x150 km of the equatorial plane ($z = 0$) at 71 ms after bounce in the 3D GR model E20A. Velocity vectors are superposed and mark spiral density waves that develop due to a nonaxisymmetric instability in the PNS and propagate through the postshock region.

and we highlight a number of the findings in the following.

In the left panel of Fig. 5 we plot the two GW polarizations h_+ and $h_\times$, both scaled by distance D to the source and in units of centimeters, as extracted from the multi-D dynamics in our model E20A that is based on the rotating 20-$M_\odot$ progenitor star of [97]. Its precollapse central angular velocity is $\sim 3.1\,\mathrm{rad\,s^{-1}}$ and its precollapse rotation rate $\beta = T/|W|$ is ~ 0.004. The corresponding values early after bounce are $\sim 4\,\mathrm{rad\,ms^{-1}}$ and ~ 0.09. The simulation is run in 3D from the onset of collapse to $\sim 70\,\mathrm{ms}$ after bounce.

We find that model E20A, like all other considered models, stays axisymmetric through collapse, core bounce, and the early postbounce phase. At bounce, the large deceleration of the inner core's infall leads to a large negative spike in the waveform clearly visible in h_+D as seen by an equatorial observer and shown in the left panel of Fig. 5. Due to the axisymmetry of the rotating collapse dynamics, GWs are emitted only in one polarization (linear polarization; here, due to the choice of source orientation, in the + polarization) and only away from the symmetry axis of the system. After bounce and on a timescale of $\gtrsim 20-40\,\mathrm{ms}$ nonaxisymmetric dynamics develops in the PNS and is likely due to a corotation-type instability in which an azimuthal mode picks up power from the axisymmetric background rotation at the point where its mode pattern speed is in corotation with the fluid (see, e.g., [98]). The quadrupole components ($\ell = 2, m = 2$) of the nonaxisymmetric dynamics are reflected in the late-time GW signal as a quasi-periodic, elliptically polarized signal, strongly correlated in h_+ and $h_\times$ and emitted at twice the $m = 2$ pattern speed. This nonaxisymmetric GW emission is smaller in amplitude than the burst associated with core bounce, but, due to its longer duration, the total energy emission is greater and the emission's narrowband nature favors detection by GW observatories such as LIGO [96, 99].

Corotation instabilities are well known from studies of astrophysical disks and belong to a class of dynamical shear instabilities that draw from the shear energy stored in differential rotation and transport angular momentum outward in spiral waves. Differential rotation is abundant in the postshock region [40, 57] and the spiral waves are apparent in the right panel of Fig. 5 that depicts the density distribution with superposed fluid velocity vectors in the equatorial plane of model E20A at $\sim 70\,\mathrm{ms}$ after bounce.

4. Summary and outlook

The challenging complexity and non-linearity of the CCSN problem and current technical and computational limitations call for a broad computational program with multiple modeling approaches. In this contribution to SciDAC 2009, we introduced two CCSN simulation programs and discussed their recent results. The first, `VULCAN/2D`, a full CCSN code with an implementation of the angle-dependent and MGFLD radiation-MHD equations, is capable of studying all presently discussed CCSN mechanisms, but is limited to axisymmetry and Newtonian gravity and dynamics. Being based on proven legacy technology and despite not employing the massively-parallel domain-decomposition paradigm, `VULCAN/2D` runs very efficiently on a modest number of compute cores and yields a high science output per flop (e.g., [11–13, 21–23, 25, 26, 37]). This includes the first long-term true multi-D angle-dependent neutrino transport CCSN calculations highlighted in this article. The result of this study is an example of *Mazurek's law*[5]. Applied to the present situation, it states that in the tightly-coupled CCSN phenomenon, even a rather significant ($\gtrsim 10\% - 30\%$) change of the postbounce conditions, in this case, of the neutrino heating rate, is absorbed by the strong feedback between radiation, hydrodynamics, EOS, and gravity and no qualitative change results.

The frontier of CCSN modeling is clearly 3D and the hope is that the additional degree of freedom and the more accurate representation of (turbulent) convection and SASI help produce successful and powerful explosions whose asymptotic energies match observations. We approach the computationally and technically challenging step to 3D with the code `Zelmani` that is based on the open-source `Cactus` framework and uses scalable AMR via the `Carpet` driver module that implements state-of-the-art HPC paradigms, including hybrid MPI/OpenMP parallelism for modern massively-parallel multi-core architectures. `Zelmani` is set apart from other 3D codes (e.g., [7, 45, 53]) by its full GR nature, evolving Einstein's equations from one spatial 3-hypersurface to the next and treating the dynamics fully relativistically instead of re-solving a Newtonian or pseudo-relativistic Poisson equation at every timestep coupled to Newtonian hydrodynamics. Dynamical spacetime evolution not only enables us to study the impact of GR on the CCSN evolution, but also allows us to investigate the dynamical formation of black holes in failing core-collapse supernovae. Such *collapsars* are considered as likely candidates for the central engines of gamma-ray bursts (GRBs; e.g., [1]), but their dynamical formation has not been modeled and remains to be understood.

While still under development towards a full GR radiation-MHD CCSN code, `Zelmani` has already been applied to 3D GR hydrodynamic studies of rotating stellar collapse, leading to an improved understanding of the gravitational wave signature of CCSNe [48, 49, 57]. In its completed form, `Zelmani` will allow us to take full advantage of petaflop supercomputers to comprehensively address the CCSN problem and the CCSN-GRB relationship.

Acknowledgments

It is a pleasure to acknowledge help from and stimulating conversations with J. Murphy, L. Dessart, E. Abdikamalov, G. Allen, D. Arnett, W. Benger, S. Bruenn, P. Diener, H. Dimmelmeier, I. Hawke, I. Hinder, K. Kotake, C. Meakin, B. Messer, A. Mezzacappa,

[5] Mazurek's law originated in the context of stellar collapse at Stony Brook University in the 1980's when Ted Mazurek was there. It is now used to generally refer to the strong feedback in a complicated astrophysical situation which dampens the effect of a change in any single parameter [100, 101].

J. Nordhaus, L. Rezzolla, B. Schutz, E. Seidel, S. Su, J. Tohline, and S. Woosley. The Cactus-based part of this work would not have been possible without the invaluable help of our late colleague Thomas Radke whom we miss dearly both personally and professionally. CDO is supported by a Sherman Fairchild Prize Fellowship at Caltech and by an Otto Hahn Prize of the Max Planck Society. AB is partially supported by the Scientific Discovery through Advanced Computing (SciDAC) program of the US Department of Energy under grant number DE-FC02-06ER41452. EO is supported by a Natural Sciences and Engineering Research Council of Canada postgraduate scholarship. The development of Cactus, the Einstein Toolkit, and the performance tools is supported by the NSF grants *XiRel* (no. 0701566), *Alpaca* (no. 0721915), and *Blue Waters* (no. 0725070). Simulations and benchmarks were performed on Queen Bee at LONI under allocation loni_numrel03, and on Kraken at NICS and Ranger at TACC under the NSF TeraGrid allocation TG-MCA02N014. Additional computations were performed at the National Energy Research Scientific Computing Center (NERSC), which is supported by the Office of Science of the US Department of Energy under contract DE-AC03-76SF00098.

References

[1] Woosley S E and Bloom J S 2006 *Ann. Rev. Astron. Astrophys.* **44** 507
[2] Heger A, Fryer C L, Woosley S E, Langer N and Hartmann D H 2003 *Astrophys. J.* **591** 288
[3] Ott C D 2009 *Submitted to Class. Quant. Grav. arXiv:0905.2797 [astro-ph]*
[4] Janka H T, Langanke K, Marek A, Martínez-Pinedo G and Müller B 2007 *Phys. Rep.* **442** 38
[5] Marek A and Janka H T 2009 *Astrophys. J.* **694** 664
[6] Murphy J W and Burrows A 2008 *Astrophys. J.* **688** 1159
[7] Bruenn S 2009 private communication
[8] Scheck L, Janka H T, Foglizzo T and Kifonidis K 2008 *Astron. Astrophys.* **477** 931
[9] LeBlanc J M and Wilson J R 1970 *Astrophys. J.* **161** 541
[10] Bisnovatyi-Kogan G S, Popov I P and Samokhin A A 1976 *Astrophys. Space Sci.* **41** 287
[11] Burrows A, Dessart L, Livne E, Ott C D and Murphy J 2007 *Astrophys. J.* **664** 416
[12] Burrows A, Livne E, Dessart L, Ott C D and Murphy J 2006 *Astrophys. J.* **640** 878
[13] Burrows A, Livne E, Dessart L, Ott C D and Murphy J 2007 *Astrophys. J.* **655** 416
[14] Ott C D, Burrows A, Dessart L and Livne E 2006 *Phys. Rev. Lett.* **96** 201102
[15] Weinberg N N and Quataert E 2008 *Mon. Not. Roy. Astron. Soc.* **387** L64
[16] Cerdá-Durán P, Font J A and Dimmelmeier H 2007 *Astron. Astrophys.* **474** 169
[17] Rosswog S 2009 *Preprint, ArXiv:0903.5075 [astro-ph]*
[18] Thorne K S 1987 *300 Years of Gravitation* ed Hawking S W and W I (Cambridge, UK: Cambridge University Press)
[19] Cactus Computational Toolkit home page URL http://www.cactuscode.org/
[20] Livne E 1993 *Astrophys. J.* **412** 634
[21] Ott C D, Burrows A, Dessart L and Livne E 2008 *Astrophys. J.* **685** 1069
[22] Livne E, Burrows A, Walder R, Lichtenstadt I and Thompson T A 2004 *Astrophys. J.* **609** 277
[23] Ott C D, Burrows A, Livne E and Walder R 2004 *Astrophys. J.* **600** 834
[24] Livne E, Dessart L, Burrows A and Meakin C A 2007 *Astrophys. J. Supp. Ser.* **170** 187
[25] Dessart L, Burrows A, Livne E and Ott C D 2006 *Astrophys. J.* **645** 534
[26] Dessart L, Burrows A, Ott C D, Livne E, Yoon S Y and Langer N 2006 *Astrophys. J.* **644** 1063
[27] Dessart L, Burrows A, Livne E and Ott C D 2008 *Astrophys. J. Lett.* **673** L43
[28] Shen H, Toki H, Oyamatsu K and Sumiyoshi K 1998 *Nucl. Phys. A* **637** 435
[29] Bruenn S W 1985 *Astrophys. J. Supp. Ser.* **58** 771
[30] Castor J I 2004 *Radiation Hydrodynamics* (Radiation Hydrodynamics, by John I. Castor, pp. 368. ISBN 0521833094. Cambridge, UK: Cambridge University Press, November 2004.)
[31] Burrows A, Reddy S and Thompson T A 2006 *Nuclear Physics A* **777** 356
[32] Thompson T A, Burrows A and Horvath J E 2000 *Phys. Rev. C.* **62** 035802
[33] Hubeny I and Burrows A 2007 *Astrophys. J.* **659** 1458
[34] Woosley S E, Heger A and Weaver T A 2002 *Rev. Mod. Phys.* **74** 1015
[35] Messer O E B, Mezzacappa A, Bruenn S W and Guidry M W 1998 *Astrophys. J.* **507** 353

[36] Thompson T A, Burrows A and Pinto P A 2003 *Astrophys. J.* **592** 434

[37] Walder R, Burrows A, Ott C D, Livne E, Lichtenstadt I and Jarrah M 2005 *Astrophys. J.* **626** 317

[38] Kotake K, Yamada S and Sato K 2003 *Astrophys. J.* **595** 304

[39] Fryer C L and Heger A 2000 *Astrophys. J.* **541** 1033

[40] Ott C D, Burrows A, Thompson T A, Livne E and Walder R 2006 *Astrophys. J. Suppl. Ser.* **164** 130

[41] Yamasaki T and Foglizzo T 2008 *Astrophys. J.* **679** 607

[42] Iwakami W, Kotake K, Ohnishi N, Yamada S and Sawada K 2008 *submitted to ApJ, arXiv:0811.0651 [astro-ph]*

[43] Buras R, Janka H T, Rampp M and Kifonidis K 2006 *Astron. Astrophys.* **457** 281

[44] Swesty F D and Myra E S 2009 *Astrophys. J. Supp. Ser.* **181** 1

[45] Bruenn S W, Dirk C J, Mezzacappa A, Hayes J C, Blondin J M, Hix W R and Messer O E B 2006 *J. Phys. Conf. Ser.* **46** 393

[46] Dimmelmeier H, Ott C D, Marek A and Janka H T 2008 *Phys. Rev. D.* **78** 064056

[47] Isenberg J A 2008 *International Journal of Modern Physics D* **17** 265

[48] Ott C D, Dimmelmeier H, Marek A, Janka H T, Hawke I, Zink B and Schnetter E 2007 *Phys. Rev. Lett.* **98** 261101

[49] Ott C D, Dimmelmeier H, Marek A, Janka H T, Zink B, Hawke I and Schnetter E 2007 *Class. Quant. Grav.* **24** 139

[50] Cerdá-Durán P, Faye G, Dimmelmeier H, Font J A, Ibáñez J M, Müller E and Schäfer G 2005 *Astron. Astrophys.* **439** 1033

[51] Liebendörfer M 2005 *Astrophys. J.* **633** 1042

[52] Iwakami W, Kotake K, Ohnishi N, Yamada S and Sawada K 2008 *Astrophys. J.* **678** 1207

[53] Scheidegger S, Fischer T, Whitehouse S C and Liebendörfer M 2008 *Astron. Astrophys.* **490** 231

[54] Liebendörfer M, Whitehouse S C and Fischer T 2009 *Astrophys. J.* **698** 1174

[55] Pretorius F 2007 *Lecture Notes to appear in the book: "Relativistic Objects in Compact Binaries: From Birth to Coalescence," arxiv:0710.1338 [gr-qc]*

[56] Arnowitt R, Deser S and Misner C W 1962 *Gravitation: An introduction to current research* ed Witten L (New York: John Wiley) pp 227–265 (*Preprint* gr-qc/0405109)

[57] Ott C D 2006 *Stellar Iron Core Collapse in 3+1 General Relativity and The Gravitational Wave Signature of Core-Collapse Supernovae* Ph.D. thesis Universität Potsdam Potsdam, Germany URL http://nbn-resolving.de/urn/resolver.pl?urn=urn:nbn:de:kobv:517-opus-12986

[58] Goodale T, Allen G, Lanfermann G, Massó J, Radke T, Seidel E and Shalf J 2003 *Vector and Parallel Processing – VECPAR'2002, 5th International Conference, Lecture Notes in Computer Science* (Berlin: Springer)

[59] Allen G, Dramlitsch T, Foster I, Goodale T, Karonis N, Ripeanu M and Toonen B 2000 *Proceedings of 4th Globus Retreat, July 30 to August 1 2000, Pittsburg* URL http://citeseer.nj.nec.com/article/allen00cactusg.html

[60] Globus Project Home Page URL http://www.globus.org

[61] HDF 5: Hierarchical Data Format Version 5 URL http://hdf.ncsa.uiuc.edu/HDF5/

[62] Balay S, Buschelman K, Gropp W D, Kaushik D, Knepley M, McInnes L C, Smith B F and Zhang H 2001 PETSc home page `http://www.mcs.anl.gov/petsc` URL http://www.mcs.anl.gov/petsc

[63] Schnetter E, Diener P, Dorband N and Tiglio M 2006 *Class. Quantum Grav.* **23** S553–S578

[64] Cactus web server (live demonstration) URL http://cactus.cct.lsu.edu:5555/

[65] VisIt Visualization Tool URL https://wci.llnl.gov/codes/visit/

[66] Schnetter E 2008 Multi-physics coupling of Einstein and hydrodynamics evolution: A case study of the Einstein Toolkit cBHPC 2008 (Component-Based High Performance Computing)

[67] Thornburg J 1996 *Phys. Rev. D* **54** 4899–4918

[68] Thornburg J 2004 *Class. Quantum Grav.* **21** 743–766

[69] Diener P 2003 *Class. Quantum Grav.* **20** 4901–4917

[70] Alcubierre M, Brügmann B, Dramlitsch T, Font J A, Papadopoulos P, Seidel E, Stergioulas N and Takahashi R 2000 *Phys. Rev. D* **62** 044034

[71] Alcubierre M, Brügmann B, Diener P, Koppitz M, Pollney D, Seidel E and Takahashi R 2003 *Phys. Rev. D* **67** 084023

[72] Husa S, Hinder I and Lechner C 2006 *Comput. Phys. Comm.* **174** 983–1004

[73] Kranc: Automated Code Generation URL http://numrel.aei.mpg.de/Research/Kranc/

[74] McLachlan, a Public BSSN Code URL http://www.cct.lsu.edu/~eschnett/McLachlan/index.html

[75] Brown D, Diener P, Sarbach O, Schnetter E and Tiglio M 2009 *Phys. Rev. D* **79** 044023

[76] XiRel: Next Generation Infrastructure for Numerical Relativity URL http://www.cct.lsu.edu/xirel/

[77] Tao J, Allen G, Hinder I, Schnetter E and Zlochower Y 2008 XiRel: Standard benchmarks for numerical relativity codes using Cactus and Carpet Tech. Rep. CCT-TR-2008-5 Louisiana State University URL http://www.cct.lsu.edu/CCT-TR/CCT-TR-2008-5

[78] Banyuls F, Font J A, Ibáñez J M, Martí J M and Miralles J A 1997 *Astrophys. J.* **476** 221

[79] Font J A, Goodale T, Iyer S, Miller M, Rezzolla L, Seidel E, Stergioulas N, Suen W M and Tobias M 2002 *Phys. Rev. D* **65** 084024

[80] Cerdá-Durán P, Font J A, Antón L and Müller E 2008 *Astron. Astrophys.* **492** 937

[81] Giacomazzo B and Rezzolla L 2007 *Class. Quantum Grav.* **24** S235–S258

[82] O'Connor E and Ott C D 2009 *in preparation*

[83] Baiotti L, Hawke I, Montero P J, Löffler F, Rezzolla L, Stergioulas N, Font J A and Seidel E 2005 *Phys. Rev. D* **71** 024035

[84] Whisky, EU Network GR Hydrodynamics Code URL http://www.whiskycode.org/

[85] Einfeldt B 1988 *SIAM J. Numer. Anal.* **25** 294–318

[86] Cardall C and Mezzacappa A 2003 *Phys. Rev. D.* **68** 023006

[87] Zink B 2008 *arXiv:0810.5349 [astro-ph]*

[88] Ruffert M, Janka H T and Schaefer G 1996 *Astron. Astrophys.* **311** 532–566

[89] Schnetter E, Hawley S H and Hawke I 2004 *Class. Quantum Grav.* **21** 1465–1488

[90] Mesh refinement with Carpet URL http://www.carpetcode.org/

[91] MPI: Message Passing Interface Forum URL http://www.mpi-forum.org/

[92] OpenMP: Simple, Portable, Scalable SMP Programming URL http://www.openmp.org/

[93] Berger M J and Oliger J 1984 *J. Comput. Phys.* **53** 484–512

[94] Alpaca: Tools for Application-Level Profiling and Correctness Analysis URL http://www.cct.lsu.edu/~eschnett/Alpaca/

[95] Schnetter E, Allen G, Goodale T and Tyagi M 2008 Alpaca: Cactus tools for application level performance and correctness analysis Tech. Rep. CCT-TR-2008-2 Louisiana State University URL http://www.cct.lsu.edu/CCT-TR/CCT-TR-2008-2

[96] Ott C D 2009 *Class. Quant. Grav.* **26** 063001

[97] Heger A, Langer N and Woosley S E 2000 *Astrophys. J.* **528** 368

[98] Watts A L, Andersson N and Jones D I 2005 *Astrophys. J. Lett.* **618** L37

[99] LIGO URL http://ligo.caltech.edu

[100] Lattimer J 2009 private communication

[101] Burrows A 2001 private communication

Type Ia Supernovae: Advances in Large Scale Simulation

S. E. Woosley[1], A. S. Almgren[2], A. J. Aspden[2], J. B. Bell[2], D. Kasen[1], A. R. Kerstein[3], H. Ma[1], A. Nonaka[2], and M. Zingale[4]
[1]Department of Astronomy and Astrophysics, University of California, Santa Cruz, CA 95064
[2]Center for Computational Science and Engineering, Lawrence Berkeley National Laboratory, Berkeley, CA, 94720
[3]Combustion Research Facility, Sandia National Laboratories, Livermore, CA
[4]Department of Physics and Astronomy, Stony Brook University, Stony Brook, NY, 11794-3800

E-mail: woosley@ucolick.org

Abstract. There are two principal scientific objectives in the study of Type Ia supernovae - first, a better understanding of these complex explosions from as near first principles as possible, and second, enabling the more accurate utilization of their emission to measure distances in cosmology. Both tasks lend themselves to large scale numerical simulation, yet take us beyond the current frontiers in astrophysics, combustion science, and radiation transport. Their study requires novel approaches and the creation of new, highly scalable codes.

1. Introduction

Type Ia supernovae (SN Ia) are among the brightest explosions in the modern universe and also the most regular, i.e., their brightness does not vary greatly from event to event. Those variations that do occur are correlated with other observables, such as their duration and decline rate, and this makes them very useful as "calibratable standard candles" in cosmology. The accepted paradigm that SN Ia come from the thermonuclear explosion of white dwarfs accreting mass in binary systems has been with us for almost fifty years [7], and for almost as long, scientists have tried to build realistic computer models [4]. It is a hard problem. The details of how the runaway ignites are important, yet the sparks that seed the first flames are embedded in a turbulent convective flow, and may not ignite first at the center of the star. Once ignited, the burning proceeds initially as a deflagration that is subject to several multi-dimensional instabilities which enhance the turbulence on its surface. At some density, observations strongly suggest that the flame makes a transition to detonation. A transition of this sort has no terrestrial analogue and understanding it takes us into territory unfamiliar to combustion scientists (high Karlovitz number and high Damköhler number). Finally the radiation transport is challenging. The supernova produces a glowing hot ball of radioactive ashes. The gamma-rays emitted in the decay are reprocessed into optical emission by the plasma. The debris are anisotropic and heterogeneous. It is this radiation that is ultimately observed and therefore most constraining on the models.

Our Computational Astrophysics Consortium studies all these questions and to do so is

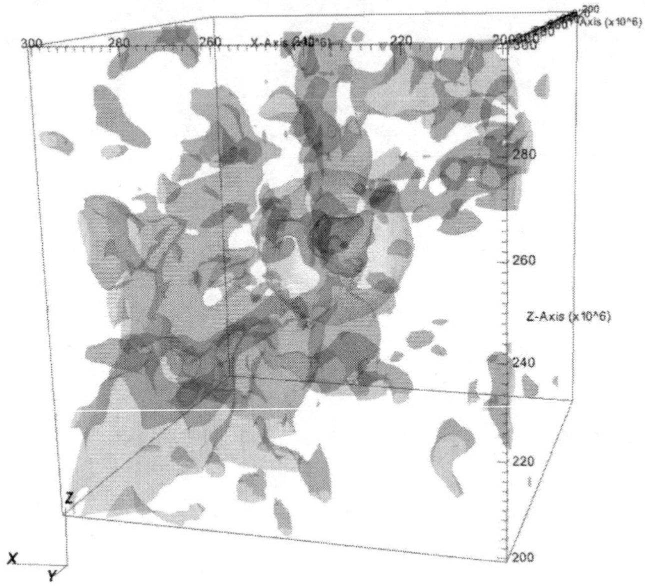

Figure 1. Convective motions and energy generation in a Chandrasekhar mass white dwarf at the time the first flame is born [17]. Radial velocity surfaces are colored red ($+4 \times 10^6$ cm s^{-1}) and blue (-4×10^6 cm s^{-1}). Energy generation rate surfaces are colored yellow (8×10^{12} erg g^{-1} s^{-1}) orange (8×10^{13} erg g^{-1} s^{-1}), and green (8×10^{14} erg g^{-1} s^{-1}). The domain shown (1000 km)^3 encompasses most of the convective region. Ignition occurs about 20 km from the center of mass of the star, but this is just the first ignition and other hot spots may still develop.

developing an ensemble of new scalable, multi-dimensional codes. In the following sections we briefly summarize our progress.

2. Ignition and the MAESTRO Code

The final explosive runaway that makes the supernova is preceded by about a century during which convection carries away an ever increasing energy generation. Finally, at some point, the time scale for further temperature increase becomes too short for convection to be efficient. It is the turbulent dynamics of this convective period that determine the initial number and location of the fusion flames that burn through the white dwarf. We have now performed the first 3D, full-star calculations of this convective phase using the new low Mach number hydrodynamics code, MAESTRO [1, 2, 3, 16, 17]. The low Mach number approximation is derived from the compressible hydrodynamics equations by decomposing the pressure into thermodynamic and dynamic components and recasting the equation of state as an elliptic constraint on the velocity field. This analytically removes acoustic waves, allowing the code to take time steps based on the advective speed rather than the sound speed.

We modeled a Chandrasekhar-mass white dwarf for its last 2 hours, watching the peak temperature rise from 6×10^8 K to over 8×10^8 K, when the first flame ignited near the center of the star. Figure 1 shows surfaces of radial velocity and nuclear energy generation rate at the point where the peak temperature crosses 8×10^8 K on its rapid rise to igniting the first flame. The reactions are strongly peaked. The radial velocity field shows a strong asymmetry—the dipole nature of this flow was first discussed in [11] and studied in [13]. Full details of this calculation are presented in [17]. The computational domain in the case shown here was 384^3, and this one calculation required approximately 600,000 CPU-hours on the Cray XT4 machine, Jaguar, at ORNL. In the near future, we plan to run at twice the resolution in order to increase the numerical Reynolds number of the turbulent convection.

3. Flame Propagation in the CASTRO and SNe Codes

Once born, nuclear burning produces bubbles and sheets of hot ash embedded in cooler, denser fuel. The ash floats, producing Rayleigh-Taylor and Kelvin-Helmholtz instabilities and,

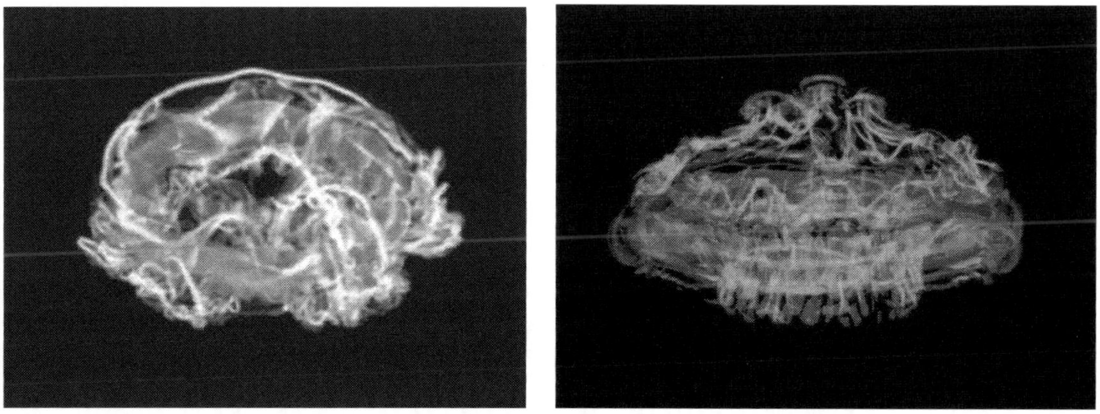

Figure 2. Vorticity in two floating, burning bubbles at a time when both have entered comparable non-linear phases. The bubble on the left was computed with SNe at nearly constant pressure and density. The flame was fully resolved at a density of 10^7 g cm^{-3} and the box size was 2 m. White indicates high vorticity and yellow coloration indicates regions of large nuclear energy generation. The bubble on the right was calculated with CASTRO using a thickened flame model for the burning at a density of $\sim 2 \times 10^9$ g cm^{-3}, gravity 1.5×10^{10} cm s^{-2}, and on a scale of 100 km, comparable to the stellar scale height. The background pressure and density decreased about 50% from bottom to top. Note the tubes of high vorticity, characteristic of turbulent flow [12], in both studies. Each calculation employed a resolution of $256\times256\times512$ zones with 2 levels of AMR and ran on 1024 CPU for about 75 hours. SNe was run on ATLAS at LLNL and CASTRO on Franklin at NERSC.

ultimately, turbulence. These instabilities, which feed back on the flame sheet, folding and distorting it, govern the rate at which the burning progresses, and, ultimately, the power and brightness of the supernova. A major effort of our Consortium has been the development of scalable codes to follow these instabilities on both large (stellar) and small (resolved flames) scales.

For small scales, we have use low-Mach number code, SNe [5, 6], developed in part with SciDAC support. This code is particularly useful for small-scale situations where the Courant condition would otherwise be prohibitive. For stellar-scale studies though, we are starting to use the new SciDAC code CASTRO. This code, developed by Almgren, Bell, and collaborators, uses compressible unsplit hydrodynamics with AMR and subcycles in time in the refined regions, resulting in considerable CPU savings. It has recently been shown to scale nearly linearly up to over 27,000 CPU at ORNL. As part of a suite of tests, we are running burning floating bubbles in planar geometry. Some results are given in Figure 2. The same generic flame and turbulence geometries appear both in calculations at low density, that well resolve the flame (SNe), and at high densities, using a thickened flame model (CASTRO). In the near future we will be moving to full star explosion models using CASTRO and a level set to track the flame.

4. Delayed Detonation

Detonation will not occur so long as the flame remains in the flamelet regime, regardless of how distorted the surface may become, but as the density declines, turbulence eventually tears the flame, stirring hot ash and cold fuel into a potentially explosive mixture. Using SNe, we have extended our previous study of this distributed regime [15] to larger domains. The distributed flame from Aspden *et al.* [6] was taken as a starting point. This flames was dominated by turbulent mixing so that the reactions occur at the turbulent nuclear time-scale. Damköhler [8]

Case	(a)	(b)	(c)	(d)	(e)
Domain width (L)	1.5×10^2	1.2×10^3	9.6×10^3	7.68×10^4	6.14×10^5
Domain height (H)	1.2×10^3	4.8×10^3	3.84×10^4	3.07×10^5	2.46×10^6
Integral length scale (l)	1.5×10^1	1.2×10^2	9.6×10^2	7.68×10^3	6.14×10^4
Turbulent intensity ($\check{u}$)	2.47×10^5	4.93×10^5	9.86×10^5	1.97×10^6	3.95×10^6
Damköhler number (Da_T)	6.4×10^{-3}	2.6×10^{-2}	1×10^{-1}	4.1×10^{-1}	1.7×10^0

Table 1. Simulation properties.

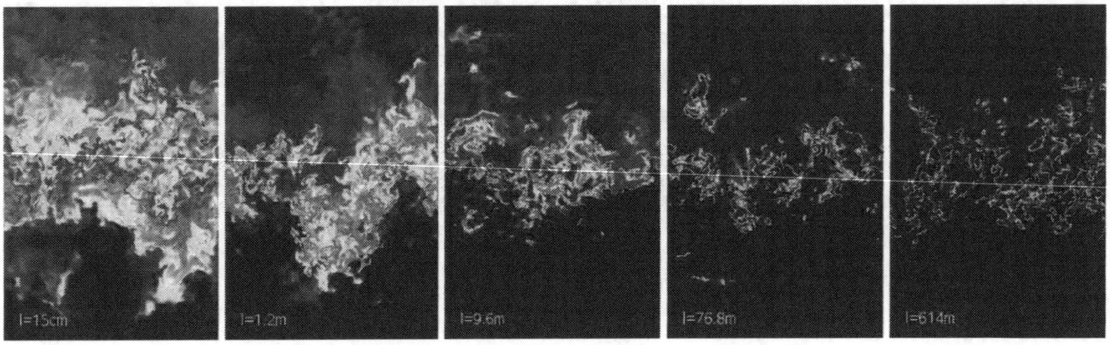

Figure 3. Instantaneous vertical slices of fuel consumption rate (intense burning in red, no burning in blue). The whole domain width is shown but the height is cropped. Each box is eight times as large as the previous one; the last one is 6 km across.

argued that in this "small-scale turbulence" regime (Ka $\gg$ 1, Da $\ll$ 1) mixing is dominated by a turbulent diffusion $D_T \sim \check{u}l$, where $\check{u}$ and l are the turbulent intensity and integral length scale, respectively. A turbulent Damköhler number can be defined as $\mathrm{Da}_T = \tau_T/\tau^T_{\mathrm{nuc}} = l/\left(\tau^T_{\mathrm{nuc}}\check{u}\right)$, where $\tau^T_{\mathrm{nuc}} = l_T/s_T$ and $\tau_T = l/\check{u}$ are the turbulent nuclear time-scale and the turbulence time-scale, respectively, and s_T and l_T are the turbulent flame speed and flame width, respectively. A turbulent Karlovitz number can then be defined as $\mathrm{Ka}^2_T = \check{u}^3 l_T/s^3_T l$, so that DaKa $\equiv$ 1. The turbulent flame speed can be shown to scale with $s_T = \check{u}\mathrm{Da}_T^{1/2}$.

This scaling breaks down when $\mathrm{Da}_T \approx 1$, corresponding to when the turbulent flame speed is equal to the turbulent intensity and the turbulent flame width is equal to the integral length scale. For larger length scales the turbulence can no longer affect the flame interior – the flame burns as a turbulently broadened flame with constant speed and thickness, which is moved around by the large scales. This limiting turbulent flame speed s_λ and width l_λ can be predicted from solely the turbulent flame speed in the original base case, $s_\lambda = \check{u}_0^2/s^0_T$ and $l_\lambda = l_u(\check{u}_0/s^0_T)^3$, where s^0_T, $\check{u}_0$ and l_0 are the turbulent flame speed, turbulent intensity and integral length scale in the base case.

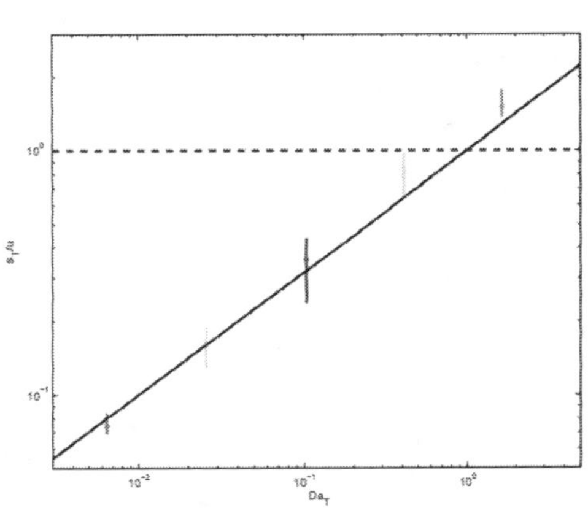

Figure 4. Turbulent flame speeds normalised by turbulent intensity as a function of turbulent Damköhler number. The linear scaling,$s_T = \check{u}Da_T^{1/2}$, is predicted to break down at about one.

Figure 3 shows slices of fuel consumption rate through the five flames. Figure 4 shows the normalised turbulent flame speed as a function of

Damköhler number. The solid black line denotes the predicted scaling $s_T = \breve{u}\mathrm{Da}_T^{1/2}$, which matches the data very well. Calculations using the Linear Eddy Model [14, 15] suggest that, for higher Damköhler numbers, the normalised flame speed levels off at a value of approximately one. The burning should also become highly irregular and potentially explosive in this regime.

An important consequence of this is that small-scale simulations can be carried out, using the three-dimensional code presented here (or using the Linear Eddy Model), to measure the turbulent flame speed under a wide range of conditions in the distributed regime (Ka $\gg$ 1). From these flame speeds, the limiting turbulent flame speed s_λ can be evaluated and used as a turbulent flame model for much larger calculations in a level set code, for example.

5. Confronting the Observations and Cosmology

In anticipation of full star hydrodynamical models soon to be available using CASTRO, we have been improving our ability to do multi-dimensional Monte Carlo studies of radiation transport. The principal code for this purpose, SEDONA, was discussed last year [9]. More recently, we have used SEDONA to postprocess a two-dimensional survey of SN Ia explosion models calclated using the Munich SN Ia code [10]. The Munich code has the necessary physics for following both the deflagration and detonation phases of the explosion including: a) a level set approach to front tracking; b) a subgrid-model for the turbulent flame interaction; and c) the necessary nuclear physics to follow recombination and electron capture in material that is in nuclear statistical equilibrium.

The initial conditions of the explosion were selected based upon insights from our studies of ignition (Sec. 2). Specifically, from 20 to 150 ignition points were considered, distributed from the center out to $\sim$ 300 km in a solid angle whose opening varied from 60° to 360°. Following the results reported last year [15] a transition to detonation was assumed when a minimal Karlovitz number and a certain range of fuel densities was achieved. Like ignition, detonation should be a stochastic process though, so five different values of Karlovitz number were considered in the range 1 to 2250 (though mostly 250 - 750).

A total of 46 hydrodynamical models was calculated and their light curves and spectra determined at all wavelengths as a function of observed angle and time. The models predict a range of ^{56}Ni yields, from 0.3 to 1.1 solar masses. The agreement with observations shown in Figure 5 is encouraging and suggests that the models have captured the essential features of the explosion, so far as light curves and spectra are concerned. They also allow us to better understand the observed *width* of the WLR and the possible dependence of its slope on the composition of the host galaxies. In this study, the width was a consequence of the high degree of asymmetry in the explosions, caused by both asymmetric ignition and detonation. Of course we are eager to see the extent to which these results remain valid in the 3D survey we plan to carry out later this year.

Acknowledgments

Research at LBNL was supported by the SciDAC Program under contract No. DE-AC02-05CH11231. A. Aspden was also supported by a Seaborg Fellowship. Work at UCSC was also supported by SciDAC under contract DE-FC02-06ER41438. At Sandia, research was supported by the Sandia Corporation, a Lockheed Martin Company, for the DOE under contract DE-AC04-94AL85000. M. Zingale was supported by a DOE/Office of Nuclear Physics Outstanding Junior Investigator award, grant No. DE-FG02-06ER41448, to Stony Brook. This research used resources provided by an INCITE award at the National Center for Computational Sciences at Oak Ridge National Laboratory, which is supported by the Office of Science of the DOE (DE-AC05-00OR22725), and the Franklin computer at the National Energy Research Scientific Computing Center. Additional computing resources were provided on the ATLAS Linux Cluster at LLNL as part of a Grand Challenge Project.

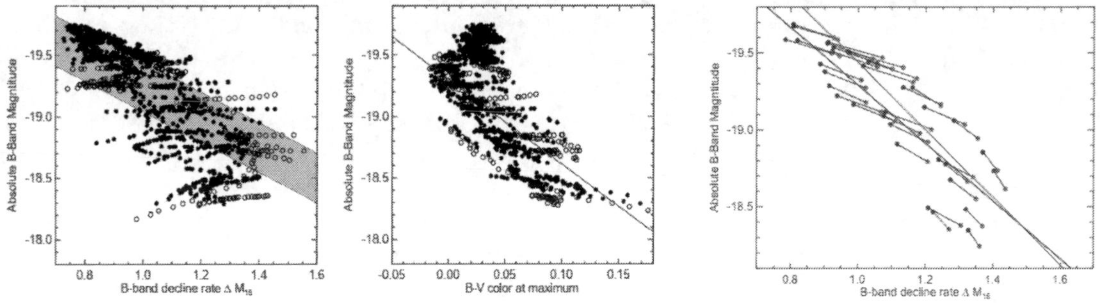

Figure 5. The width-luminosity relation (WLR) for SN Ia [10]. The shaded band in the left plot is the observed relation between peak brightness (absolute magnitude) and decline rate (change in B-band magnitude in 15 days after peak). Points are the result of a 2D survey of explosions described in the text. Most of the width is due to the variable angle at which an asymmetric explosion is observed. The middle plot gives the peak brightness as a function of the color of the supernova. The right hand plot shows how the angle-averaged WLR relation depends on metallicity for a range corresponding to 1/3 to 3 times solar. The different slope for the WLR for different metallicities may complicate the use of SN Ia for cosmology at the 1 - 2% level.

References

[1] Almgren, A. S., Bell, J. B., Rendleman, C. A., & Zingale, M. 2006a, Astrophysical Journal, 637, 922
[2] —. 2006b, Astrophysical Journal, 649, 927
[3] Almgren, A. S., Bell, J. B., Nonaka, A., & Zingale, M. 2008, Astrophysical Journal, 684, 449
[4] Arnett, W. D. 1969, Astrophysics and Space Science, 5, 180
[5] Bell, J. B., Day, M. S., Rendleman, C. A., Woosley, S. E., & Zingale, M. A. 2004, J. Comput. Phys., 195, 677
[6] Aspden A. J., Bell J. B., Day M. S., Woosley S. E. & Zingale M. 2008 Astrophysical Journal, 689, 1173
[7] Hoyle, F., & Fowler, W. A. 1960, Astrophysical Journal, 132, 565
[8] Damköhler G. 1940 *Z. Elektrochem* 46, 601 – 652
[9] Kasen, D., Thomas, R. C., Röpke, F., & Woosley, S. E. 2008, Journal of Physics Conference Series, 125, 012007
[10] Kasen, D., Röpke, F., & Woosley, S. E. 2009, submitted to Nature.
[11] Kuhlen, M., Woosley, S. E., & Glatzmaier, G. A. 2006, Astrophysical Journal, 640, 407
[12] Sreenivasan, K. R., & Antonia, R. A. 1997, Annual Review of Fluid Mechanics, 29, 435
[13] Woosley, S. E., Bell, J. B., Glatzmaier, G., Kasen, D., Nugent, P., Röpke, F., & Zingale, M. 2007, Journal of Physics Conference Series, 78, 012081
[14] Woosley, S. E., Kerstein, A., Sankaran, V., and Röpke, F. 2009, Astrophysical Journal, submitted, astroph-0811-3610.
[15] Woosley, S. E., Aspden, A. J., Bell, J. B., Kerstein, A. R., & Sankaran, V. 2008, Journal of Physics Conference Series, 125, 012012
[16] Zingale, M., Almgren, A. S., Bell, J. B., Malone, C. M., Nonaka, A. 2008, Journal of Physics Conference Series, 125, 012013
[17] Zingale, M., Almgren, A. S., Bell, J. B., Nonaka, A., & Woosley, S. E. 2009, ApJ, submitted.

Journal of Physics: Conference Series **180** (2009) 012024 doi:10.1088/1742-6596/180/1/012024

Phylogenetic molecular function annotation

Barbara E Engelhardt[1,1], Michael I Jordan[1,2], Susanna T Repo[3] and Steven E Brenner[3,4,2]

[1] EECS Department, University of California, Berkeley, CA, USA. [2] Department of Statistics, University of California, Berkeley, CA, USA. [3]Plant and Microbial Biology Department, University of California, Berkeley, CA, USA.[4] Department of Molecular and Cell Biology, University of California, Berkeley, CA, USA

E-mail: brenner@compbio.berkeley.edu

Abstract. It is now easier to discover thousands of protein sequences in a new microbial genome than it is to biochemically characterize the specific activity of a single protein of unknown function. The molecular functions of protein sequences have typically been predicted using homology-based computational methods, which rely on the principle that homologous proteins share a similar function. However, some protein families include groups of proteins with different molecular functions. A phylogenetic approach for predicting molecular function (sometimes called "phylogenomics") is an effective means to predict protein molecular function. These methods incorporate functional evidence from all members of a family that have functional characterizations using the evolutionary history of the protein family to make robust predictions for the uncharacterized proteins. However, they are often difficult to apply on a genome-wide scale because of the time-consuming step of reconstructing the phylogenies of each protein to be annotated. Our automated approach for function annotation using phylogeny, the SIFTER (Statistical Inference of Function Through Evolutionary Relationships) methodology, uses a statistical graphical model to compute the probabilities of molecular functions for unannotated proteins. Our benchmark tests showed that SIFTER provides accurate functional predictions on various protein families, outperforming other available methods.

1. Introduction

As sequencing technologies develop, sequence data is accruing at a fast rate, and the potential for medical applications of genomic data to human biology is just beginning to be realized. Sequencing also heralds unprecedented opportunities for understanding human-associated microbiota, whose genetic diversity is perhaps 100 times that of the human genome [1]. However, despite this large body of new sequence information, functional annotation remains a major challenge. Molecular functions of proteins in the human genome continue to be discovered, in large part by homology to those experimentally characterized in model organisms.

Typically, protein function annotation involves finding homologs of a protein sequence, followed by database queries and computational techniques to predict function from the annotated homologs.

[1] Current address: Computer Science Department, University of Chicago, Chicago, IL, USA

[2] Corresponding author

These methods rely on the principle that proteins from a common ancestor may share a similar function. However, most protein families have sets of proteins with different functions and therefore traditional bioinformatics approaches are unable to reliably assign the appropriate function to unannotated proteins. Currently, protein function databases have a large proportion of erroneously annotated proteins, where the incorrect annotations were either derived using an imprecise computational technique or inferred using another incorrect annotation [2-4].

We have proposed integrating available functional data using the evolutionary relationships of a protein family, and we implemented this method in the program SIFTER. This phylogenetic approach to molecular function annotation, sometimes termed "phylogenomics" [5-8], uses an explicit phylogenetic tree to make functional predictions. The basic principle of phylogenetic function annotation is that function will tend to evolve in parallel with sequence [9], and that function is more likely to change after a duplication than after a speciation event [10-12]. Thus, the traditional evolutionary based approach involves building a phylogenetic tree from homologous protein sequences, identifying the most likely location of duplication events, and propagating known functions within each clade descendant from a duplication event. This evolutionary approach is a more sophisticated method of determining protein function than using sequence similarity (BLAST), or unstructured homology data (COGs) because it is unclear how to transfer function annotation in an evolutionarily consistent way when homology is represented by clusters rather than by a phylogeny. Accelerated rates of evolution will result in closely related sequences having less significant sequence similarity scores; this means that BLAST will systematically find proteins that appear more distant in an evolutionary tree as being most similar [13]. By contrast, a phylogenetic tree reflects the evolutionary path of a set of homologous proteins (figure 1). By using phylogenetic trees, we can directly apply the assumptions about how function evolves in order to enable a consistent, meaningful method of transferring sparse and noisy functional evidence.

The use of phylogenetic function prediction to annotate individual proteins has proliferated [e.g., 6, 14-16] and provided some of the most reliable function annotations [17]. Manual phylogenetic

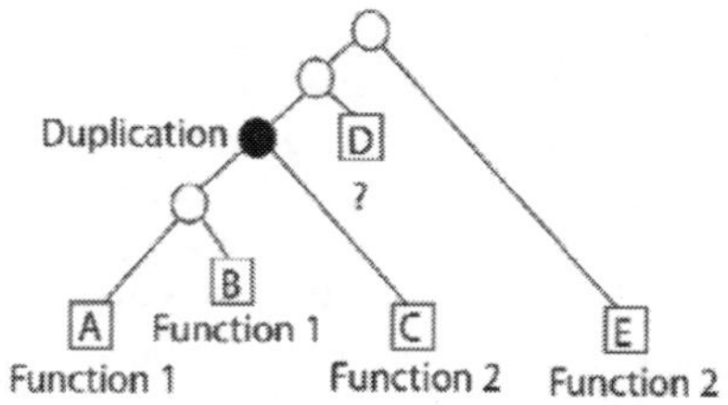

Figure 1. Function evolution is generally parsimonious; BLAST makes systematic errors in predicting function. Protein D is unannotated, while proteins A and B share a function, as do C and E. Parsimonious reconstruction dictates that D shares a function with C and E; a BLAST search gives the highest score to protein B, yielding erroneous annotations.

function prediction studies on a genome-wide level are rare, because of the time-consuming step of reconstructing and then analyzing phylogenies for each of the unannotated proteins, but they do occur [18].

2. The SIFTER approach

The SIFTER methodology [19] is based on evolutionary principles [6], using a statistical representation. Currently, SIFTER takes as input a reconciled phylogeny and a set of annotations for some of the proteins in the protein family. Given a query sequence, the appropriate Pfam [20] protein family can be used to build the reconciled phylogeny. The Pfam alignment of the query sequence and its homologs is used as input to a phylogenetic reconstruction program, such as PAUP* with maximum likelihood [21]. The phylogeny is reconciled [22] against a known species phylogeny using, for example, the Forester [23] software to determine which nodes of the tree represent protein duplication events and which represent speciation events. (This approach will treat horizontal transfer effectively like a duplication, which is appropriate in the sense that we would like to assume that function is more likely to change after a horizontal transfer event.) We include prior information about function by computing the probability of each of the candidate functions given the available evidence for the proteins in the tree with functional evidence from the GOA database [24]. The candidate molecular functions are represented as a boolean vector, where initially the probability associated with each candidate function is a function of the set of annotations for that protein and their corresponding evidence types (e.g., experimental, electronic). Finally, from this reconciled phylogeny with sparse observations, SIFTER computes the posterior probability of each molecular function for all proteins in the family using a simple statistical model of protein function evolution.

The model of protein function evolution in SIFTER allows every candidate molecular function to mutate to every other candidate function for the family, and the likelihood of a mutation is related to estimated mutability of each pair of molecular functions, branch length, and whether an internal tree node represents a speciation or a duplication event. Whereas the branch length is a (fixed) value that is input to the method through the reconciled tree, we are free to estimate the two parameters controlling function mutability and general rate of mutation along a branch. The mutability parameters in effect relate sequence change to functional change for every pair of candidate functions if you consider a constant rate of amino acid substitution along each branch. In other words, if one function has a high rate of mutability to another function, then on average it takes fewer mutations to change the first function into the second. Two different parameters reflect the relative rates of functional change after a speciation event versus after a duplication event.

The phylogenetic tree is the structure for inferring molecular function, with the phylogenetic characters replaced by the molecular function random variables. Using message passing [25-27] it is possible to propagate this information throughout the tree to infer the posterior probability of each candidate function for all nodes. We chose a probabilistic approach to protein function prediction because it is well suited to the nature of the evidence. As in other areas of computational biology [see e.g. 28], a probabilistic framework has the major advantage that it allows multiple, noisy sources of evidence to be used for a single prediction, by weighting and combining this evidence appropriately. This is a fundamental feature of SIFTER—it computes posterior probabilities for each possible function of a query protein by combining evidence from related proteins in a coherent, evolutionarily motivated way through the phylogeny. Finally, a key feature of the protein function prediction problem is the sparsity of available experimental annotations within any particular protein family. The probabilistic approach takes sparsity into account in a natural way, as posterior probabilities are lower when the supporting evidence is weak or conflicting.

The SIFTER algorithm makes predictions using the evolutionary structure of a protein family and all available functional information. It provides traceable evidence, making it straightforward to understand the posterior probability of any leaf node by looking at posterior probability of the hidden nodes throughout the phylogeny, and it provides probabilistic results for each possible function.

3. Performance of SIFTER

We tested the performance of SIFTER on two different protein families: AMP/adenosine deaminases, and aminotransferases [29]. The sequences and alignments for each family were downloaded from the Pfam database [20] and the function annotations were from the GOA database [24] and a manual literature search. Each family posed unique challenges to function prediction.

The performance of SIFTER was estimated with leave-one-out cross validation experiments, where the available annotation of each protein was removed from the training set before a SIFTER run for the protein family was performed [29], then checking whether the maximum posterior probability for the predicted protein agreed with the held-out annotation. We ran cross validation for each of the two protein families with experimental annotations, and also with a combination of experimental and electronic annotations. SIFTER's performance was compared with three other function prediction algorithms: BLAST [30], GOtcha [31] and Orthostrapper [32].

Our first dataset was composed of the Pfam AMP/adenosine deaminase family, which contained 251 proteins. This family has 33 proteins with experimental annotations, which came from the GOA database, a manual literature search, and our own characterization experiment. The challenge in assigning functions for proteins in this family is that a subset of proteins has multiple functions, where the additional function is active in a second protein domain. Thus, in our analysis we conclude that a prediction was correct if one of the two functions of a protein was assigned correctly. Cross validation on experimental annotations yielded 93.9% accuracy, while the accuracy for cross validation on experimental and electronic annotations was 96.3% [29]. In comparison, cross validation on experimental data yielded 66.7% accuracy for BLAST and GOtcha (GOtcha-exp), while Orthostrapper achieved 78.8% accuracy. GOtcha achieved 87.9% accuracy with both experimental and electronic annotation [29]. In Figure 2, we show the ROC analysis of this family, where SIFTER outperforms all the other methods on the deaminase family. This analysis looks at the percentage of false positive predictions relative to the percentage of true positive predictions as the cutoff for the posterior probability for function prediction goes from 0 to 1 (in SIFTER). This method of comparison was especially suitable for the deaminase family due to the multiple functions some of the family members have.

The aminotransferase dataset was a difficult test case for our function prediction method due to homoplasy within the family. It appears that the preference for one of the substrates might have appeared multiple times independently during evolution in this family. Despite this, SIFTER maintained good prediction accuracy (75%; 9 of 12) when only the experimental data was used in cross validation [29]. The inclusion of electronic annotations increased SIFTER prediction accuracy to 92.6% (50 of 54). BLAST, GOtcha and GOtcha-exp achieved 66.7% accuracy (8 of 12) in the cross-validation experiments with experimental data, while Orthostrapper was unable to annotate any of the proteins [29].

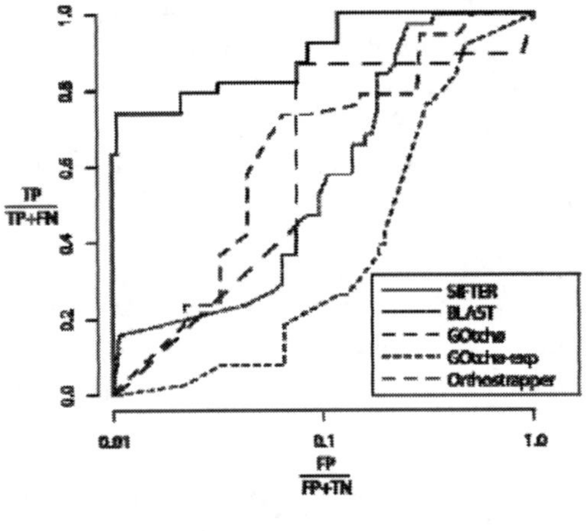

Figure 2. ROC figure comparing the five tested methods on the deaminase family using experimental annotations [29] (For GOtcha, both experimental and electronic annotation, GOtcha-exp means experimental annotation only). The X-axis is in log scale, and the key to axis labels is: TP=true positives, FP=false positives, TN=true negatives, and FN=false negatives.

4. Conclusions

The development of SIFTER is an ongoing project and we now have a new version of the program available (manuscript currently submitted). The new version includes a more general model of protein function evolution and a fast method for calculating the posterior probabilities; these improvements make SIFTER applicable on large and functionally diverse protein families and on genome-scale function annotation. Furthermore, we are validating SIFTER predictions experimentally using the extremely diverse Nudix family of hydrolases as a test bed.

References

[1] Gill SR, Pop M, Deboy RT, Eckburg PB, *et al.* 2006 *Science.* **312** 1355-9
[2] Brenner SE 1999 *Trends Genet.* **15** 132-3
[3] Galperin MY and Koonin EV 1998 *In Silico Biol.* **1** 55-67
[4] Jones CE, Brown AL and Baumann U 2007 *BMC Bioinformatics.* **8** 170
[5] Eisen JA and Hanawalt PC 1999 *Mutat Res.* **435** 171-213
[6] Eisen JA 1998 *Genome Res.* **8** 163-7
[7] Sjölander K 2004 *Bioinformatics.* **20** 170-9
[8] Brown D and Sjölander K 2006 *PLoS Comput Biol.* **2** e77
[9] Atchley WR and Fitch WM 1997 *Proc Natl Acad Sci U S A.* **94** 5172-6
[10] Ohno S 1970 *Evolution by gene duplication* (New York: Springer-Verlag)
[11] Fitch WM 1970 *Syst Zool.* **19** 99-113
[12] Lynch M and Conery JS 2000 *Science.* **290** 1151-5
[13] Koski LB and Golding GB 2001 *Journal of Molecular Evolution.* **52** 540-2
[14] Theodorides K, De Riva A, Gomez-Zurita J, Foster PG and Vogler AP 2002 *Insect Mol Biol.* **11** 467-75.
[15] Danchin A 2003 *Curr Issues Mol Biol.* **5** 37-42.
[16] Gadelle D, Filee J, Buhler C and Forterre P 2003 *Bioessays.* **25** 232-42.
[17] Eisen JA and Fraser CM 2003 *Science.* **300** 1706-7
[18] Tettelin H, Masignani V, Cieslewicz MJ, Eisen JA, *et al.* 2002 *Proc Natl Acad Sci U S A.* **99** 12391-6
[19] Engelhardt BE, Jordan MI, Muratore KE and Brenner SE 2005 *PLoS Comput Biol.* **1** e45
[20] Bateman A, Coin L, Durbin R, Finn RD, *et al.* 2004 *Nucleic Acids Res.* **32** D138-41
[21] Swofford DL 2001 *Paup*: Phylogenetic analysis using parsimony (*and other methods)* (Sunderland, Massachusetts: Sinauer Associates)
[22] Perriere G, Duret L and Gouy M 2000 *Genome Res.* **10** 379-85
[23] Zmasek CM and Eddy SR 2001 *Bioinformatics.* **17** 821-8
[24] Camon E, Magrane M, Barrell D, Lee V, *et al.* 2004 *Nucleic Acids Res.* **32** D262-6
[25] Elston RC and Stewart J 1971 *Hum Hered.* **21** 523-42
[26] Felsenstein J 1981 *J Mol Evol.* **17** 368-76
[27] Hilden J 1970 *Clin Genet.* **1** 319-48
[28] Durbin R, Eddy S, Krogh A and Mitchison G 1998 *Biological sequence analysis* (Cambridge University Press)
[29] Engelhardt BE, Jordan MI and Brenner SE 2006 *Proc 23rd Intl Conf Machine Learning.* 038.1-.8
[30] Altschul SF, Gish W, Miller W, Myers EW and Lipman DJ 1990 *J Mol Biol.* **215** 403-10
[31] Martin DM, Berriman M and Barton GJ 2004 *BMC Bioinformatics.* **5** 178
[32] Storm CE and Sonnhammer EL 2002 *Bioinformatics.* **18** 92-9

Journal of Physics: Conference Series **180** (2009) 012025 doi:10.1088/1742-6596/180/1/012025

Application of high-performance computing to the reconstruction, analysis, and optimization of genome-scale metabolic models

Christopher S Henry,[1*] Fangfang Xia,[2*] Rick Stevens[1,2]

[1]Mathematics and Computer Science Division, Argonne National Laboratory, Argonne, IL 60439, USA
[2]Department of Computer Science, University of Chicago, Chicago, IL 60637, USA
[*]These authors contributed equally to this work.

Email: chenry@mcs.anl.gov

Abstract. Over the past decade, genome-scale metabolic models have gained widespread acceptance in biology and bioengineering as a means of quantitatively predicting organism behavior based on the stoichiometry of the biochemical reactions constituting the organism metabolism. The list of applications for these models is rapidly growing; they have been used to identify essential genes, determine growth conditions, predict phenotypes, predict response to mutation, and study the impact of transcriptional regulation on organism phenotypes. This growing field of applications, combined with the rapidly growing number of available genome-scale models, is producing a significant demand for computation to analyze these models. Here we discuss how high-performance computing may be applied with various algorithms for the reconstruction, analysis, and optimization of genome-scale metabolic models. We also performed a case study to demonstrate how the algorithm for simulating gene knockouts scales when run on up to 65,536 processors on Blue Gene/P. In this case study, the knockout of every possible combination of one, two, three, and four genes was simulated in the *i*Bsu1103 genome-scale model of *B. subtilis*. In 162 minutes, 18,243,776,054 knockouts were simulated on 65,536 processors, revealing 288 essential single knockouts, 78 essential double knockouts, 99 essential triple knockouts, and only 28 essential quadruple knockouts.

Introduction

Genome-scale metabolic models represent one important end product of the genome annotation process. These models provide a means of rapidly translating detailed knowledge of thousands of enzymatic processes into quantitative predictions of whole-cell behavior. They have been applied extensively to identify essential genes and genes sets, predict organism phenotypes and growth conditions, design metabolic engineering strategies, and simulate the effects of transcriptional regulation on organism behavior [1-4].

A genome-scale metabolic model of an organism consists of three primary components: (1) a list of the reactions that take part in the organism metabolism including data on reaction stoichiometry and reversibility, (2) a set of gene-protein-reaction (GPR) mappings that capture how genes in the organism encode enzymes and how these enzymes catalyze metabolic reactions, and (3) a biomass objective function that indicates which small molecules must be produced for an organism to grow and divide [1]. All of these components are used in a method called flux balance analysis (FBA) to simulate organism metabolism in a set of specified environmental conditions.

FBA involves the use of linear optimization to define the limits on the metabolic capabilities of a model organism by assuming that the interior of the cell exists in a quasi-steady state [2-5]. This quasi-steady-state assumption is enforced by a set of linear mass balance constraints written for each metabolite included in the model.

$$\boldsymbol{N} \cdot \boldsymbol{v} = \boldsymbol{0} \tag{1}$$

In the mass balance constraints (Eq. 1), $\boldsymbol{N}$ is the m x r matrix of the stoichiometric coefficients for the r reactions and m metabolites in the genome-scale metabolic model, and $\boldsymbol{v}$ is the r x 1 vector of the steady-state fluxes through the r reactions in the model. Bounds are placed on the reaction fluxes ($\boldsymbol{v}$) depending on the reversibility of the reactions.

$$-100 \text{ mMol/gm CDW hr} \leq v_{i,reversible} \leq 100 \text{ mMol/gm CDW hr} \tag{2}$$

$$0.0 \text{ mMol/gm CDW hr} \leq v_{i,irreversible} \leq 100 \text{ mMol/gm CDW hr} \tag{3}$$

These mass balance constraints and reaction flux bounds form a set of underdetermined linear equations with many possible solutions. Because these equations are underdetermined, an optimization criterion is used to capture the most physiologically relevant region of the solution space. The optimization criteria vary depending on the application, but the most common criterion is the maximization of growth yield [5, 6]. Maximum growth yield is simulated by maximizing the flux through the biomass reaction in the model while the uptake of nutrients is fixed at a specific ratio. This is a meaningful optimization criterion because organisms have been observed to grow at the maximum predicted yield when nutrients are plentiful [7].

The FBA formulation described here forms the core of many different algorithms for the reconstruction, analysis, and optimization of genome-scale metabolic models. For this reason, software developed to run FBA in a high-performance computing (HPC) environment may be directly applied to solving many different problems. FBA is also an attractive algorithm to run with HPC because only a small amount of data is required to fully define the variables, constraints, and objective necessary to simulate even the largest genome-scale models, and once FBA has been run, the only result that must be returned is the objective and the list of reaction fluxes. Thus FBA requires minimal input and output, thereby improving scalability. Additionally, once the constraints and variables defining the FBA mass balances for a single genome-scale metabolic model have been initially loaded, many different simulations can be performed simply by adjusting the bounds on the loaded variables and repeating the optimizations. Thus, much work can be done without loading additional data, further improving scalability.

Here we describe the various FBA-based algorithms that currently exist for the reconstruction, analysis, and optimization of genome-scale metabolic models. We focus in particular on algorithms that are good candidates for processing in HPC, and we discuss the implications of running these algorithms in parallel. As a case study, we implemented the FBA algorithm for gene knockout simulation using MPI on Blue Gene/P, and we used this algorithm to simulate the 10^{10} possible quadruple gene knockouts for the *i*Bsu1103 genome-scale model of *B. subtilis*. This represents the first time that the essential combinations of up to four genes have been identified in *B. subtilis*.

1. Model reconstruction algorithms

In order to run FBA simulations in HPC for a large variety of different organism, genome-scale metabolic models must first be created for those organisms. Fortunately opportunities now exist for the use of HPC to accelerate this process. *Reconstruction* is the word used to describe the process of producing a functioning genome-scale model of an organism starting from the sequence of nucleotides in the organism's genome. This process includes gene calling, annotation, literature data mining, reaction mapping, biomass objective function assembly, error correction, and gap filling. Until recently, this process required years of manual effort to complete, and as a result, the rate of development of new genome-scale models lagged far behind the rate at which new genomes were being sequenced (figure 1) [8]. Fortunately, methods have recently emerged that expedite many of the

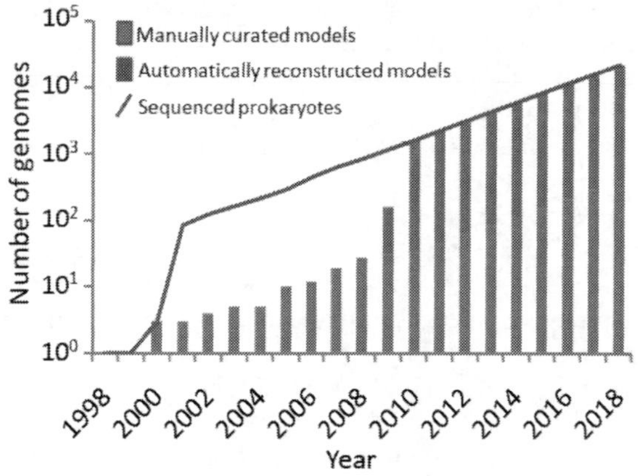

Figure 1. Past and projected rates of release of new genome sequences and new genome-scale metabolic models. Using manual reconstruction, the number of available models (blue bars) lags far behind the number of sequenced genomes (red line). However, with the introduction high-throughput reconstruction and the application of HPC to high-throughput reconstruction, model reconstruction can rapidly catch up to and keep pace with genome sequencing (red bars).

steps involved in the reconstruction process [9-16], and these methods may now be combined into a high-throughput, genome-scale reconstruction pipeline. With this high-throughput reconstruction pipeline, it will be possible to build up-to-date genome-scale models of every sequenced prokaryote on a monthly basis (figure 1). However, this lofty goal will require significant computational power to accomplish.

The rate limiting *computational* step in the high-throughput reconstruction process is an FBA-based algorithm called *auto completion*. Initial genome-scale metabolic models consisting only of reactions associated with genes, spontaneous reactions, and a biomass objective function always contain gaps that prevent them from producing all of the building blocks necessary for growth. As a result, these models are incapable of simulating growth using FBA. In auto completion, a mixed integer linear optimization problem (MILP) is solved to identify the minimal set of reactions that must be added to the initial model in order for every biomass component to be produced [16]. Often, multiple equally optimal minimal reaction sets satisfy these criteria. Recursive MILP [17] may be used to identify each of these solutions, and every solution identified can be preserved until the most physiologically realistic solution has been manually selected. Depending on the number of gaps in the initial model, run time for this algorithm can range from one minute to 24 hours. The long run time combined with the minimal input and output data size (table 1) means that scalability will not be an issue for the auto completion algorithm. The auto completion algorithm could be run in parallel to produce functioning genome-scale models for all ~800 prokaryotes with completely sequenced genomes today. Use of a parallel MILP solver [18] will make it possible to spread the work involved in producing each model across up to eight processors while retaining scalability. In this manner, the auto completion algorithm could be run on a maximum of 6,400 processors. Additionally, we anticipate the number of complete genome sequences to continue growing rapidly over the next decade (figure 1), meaning number of processors that could be consumed by the auto completion algorithm will also grow rapidly.

We note that the genome annotation and literature data-mining steps of the reconstruction process will probably always require significant manual effort to complete. However, the Rapid Annotation using Subsystems Technology (RAST) method developed within the SEED annotation framework [11, 19, 20] makes it possible for these steps to be performed for every known genome simultaneously and to be extended to new genomes with minimal effort. In RAST, most well studied biological functions have been divided into a set of subsystems (e.g., glycolysis) that are assigned to an expert annotator. That annotator performs the genome annotation and literature data-mining steps of the reconstruction process for that subsystem using a set of comparative genomics tools to project curations and annotations across all known genomes simultaneously. Whenever a new genome sequence (or set of sequences) is integrated into the SEED, annotators will curate the addition of those genomes to each of their subsystems to ensure that the subsystem is correctly propagated into the new genome. Similarly, as new literature emerges, annotators study this literature to determine how the reported results will impact the annotations in their subsystems. In this manner, annotation and literature data-mining efforts are able to keep pace with the sequencing of new genomes, making high-throughput reconstruction possible despite the need for some manual curation.

Table 1. HPC considerations model reconstruction, analysis, and optimization algorithms

Algorithm	Input	Output	Run Time per Simulation	Approximate Thread Count
Reconstruction algorithms				
Automated completion	~2 MB/thread	1 KB/thread	1-24 hours	$O(10^3)$ genomes
Optimization algorithms				
Gap filling	~2 MB/thread	1 KB/thread	1-60 min	$O(10)$ models $O(10^2)$ errors
Gap generation	~2 MB/thread	1 KB/thread	1-4 hours	$O(10)$ models $O(10^2)$ errors
Analysis algorithms				
Growth simulation	~1 MB broadcast	8 B/thread	~0.01 sec	$O(10^3)$ models $O(10^2)$ media
Gene knockout	~1 MB broadcast	8 KB/thread	~0.01 sec	$O(10^3)$ models $O(10^2)$ media $O(10^{3k})$ knockouts
Gene knockout with thermodynamic constraints	~2 MB broadcast	8 KB/thread	~3 hours	$O(10^3)$ models $O(10^2)$ media $O(10^{3k})$ knockouts
Reaction classification	~2 MB broadcast	16 KB/thread	~2 minutes	$O(10^3)$ models $O(10^2)$ media
Reaction classification with thermodynamic constraints	~2 MB broadcast	16 KB/thread	~6 hours	$O(10^3)$ models $O(10^2)$ media
Metabolic engineering	~2 MB /thread	1 KB/thread	1-4 hours	$O(10^3)$ models $O(10^2)$ media $O(10^2)$ products $O(10^2)$ raw materials

2. Model analysis algorithms

High-throughput implementation of the auto completion algorithm will enable the AGRO pipeline to produce thousands of new genome-scale metabolic models, paving the way for the use of HPC in the analysis of these models as well. Numerous FBA-based algorithms exist for analyzing genome-scale models to produce predictions of growth conditions, essential genes, metabolite concentration ranges, and metabolic engineering strategies. Application of HPC to these algorithms will produce a wealth of valuable new data.

2.1. Growth simulation

All published genome-scale metabolic models include one or more intracellular compartments separated from an extracellular compartment by a cell membrane. Only nutrients with explicit transport reactions in the model are allowed to pass into the intracellular compartment(s) from the extracellular compartment. While mass balances are maintained in the intracellular compartments, metabolites in the extracellular compartment may have a net consumption or production by the cell. This compartmental structure makes it possible to use FBA to simulate growth in a variety of environments by adjusting the set of compounds that are allowed to be consumed by the cell from the extracellular compartment. For example, when modeling rich media, all amino acids, nucleotides, and many cofactors may be consumed from the environment. When modelling minimal media, only glucose, ammonia, phosphate, sulphate, and various ions may be consumed from the environment.

The simplicity of the linear optimization algorithm means that the growth of a single model may be simulated in many different media conditions on a single processor with great speed (~70 media conditions/second). The slow step of the algorithm is the initial loading of the model into the optimization solver, so it is essential for efficiency that simulation software interacts with the optimization software through an API as this allows variable bounds and problem objectives to be adjusted on the fly without reloading the model. This simulation speed indicates that a very large number of simulations must be performed in order to justify the application of HPC.

The 324 media conditions included in the Biolog phenotyping arrays currently represent the most significant large set of media conditions that would be informative to run growth simulations on with a large set of models. Biolog phenotyping arrays are experimental devices that measure the ability of an organism to respire in hundreds of distinct media environments [21]. Each media environment in the phenotyping array contains a different nutrient, making the Biolog array useful as a means of determining whether an organism has a transporter and a catabolic pathway for many different carbon, nitrogen, phosphate, and sulphate sources. Even with thousands of models and hundreds of media conditions, growth simulation alone generates insufficient computational demand to require HPC. However, growth simulation is still a feasible application for HPC when combined with other algorithms. For example, gene knockout simulations must often be performed in a variety of media conditions.

2.2. Gene knockout simulation

Gene essentiality prediction is one of the primary uses for genome-scale metabolic models. Knowledge of essential or coessential genes in pathogenic organisms is valuable for identifying target proteins for the development of new antibacterial agents. Knowledge of essential and coessential genes in industrial organisms is also useful for ensuring that proposed metabolic engineering strategies avoid disruption of these genes.

To identify essential and coessential genes, gene knockouts are simulated in the models by using the GPR associations to identify reactions exclusively encoded by knocked out genes, setting the bounds on the flux through these reactions to zero, and maximizing the flux through the biomass reaction. If the maximum flux through the biomass reaction is zero, the knocked out gene(s) are predicted to be essential (for a single gene) or coessential (for multiple genes). The nutrients present in the environment during the knockout have a significant effect on gene essentiality, and for this reason knockouts often must be simulated in multiple media conditions. Like the growth simulation, knockout simulations can be performed very rapidly (~50 knockouts/sec), meaning any application of HPC to gene knockout simulation must involve billions of distinct simulations on thousands of processors. For example, HPC could be applied to the simulation of every possible single, double, and triple gene knockout in every sequenced pathogen on a hundred different media conditions representing distinct physiological environments for infection. With 10^2 pathogens, 10^2 media conditions, and 10^6 double knockouts, this scenario involves 10^{10} distinct simulations. With perfect scalability on 100,000 processors, the run time would be approximately 24 minutes.

2.3. Reaction classification

The gene knockout simulation algorithm provides a means of classifying the genes included in a genome-scale model. A similar FBA-based algorithm exists to classify the reactions in the model. This algorithm, called *flux variability analysis* (FVA), involves minimizing and maximizing the flux through each individual reaction in the model; model reactions are then classified based on the minimum and maximum fluxes calculated [22]. If the min and max are both positive or both negative, the reaction is essential; if they are both zero, the reaction is nonactive; if one is zero and the other is nonzero, the reaction is nonessential but active; and if one is positive and the other is negative, the reaction is nonessential but reversibly active.

The HPC scenario for the FVA algorithm is essentially identical to the scenario for single gene knockout simulation. Both algorithms have the same run time and require approximately the same number of simulations. However, it is sometimes useful to combine FVA with gene knockout simulation to study how reaction behavior is affected by gene knockout, which involves $O(10^6)$ simulations per model studied.

2.4. Thermodynamics-based metabolic flux analysis

One issue with standard FBA that can significantly impact the accuracy of the predictions generated is that thermodynamically infeasible flux profiles still satisfy the mass balance constraints and flux bounds in FBA. In thermodynamics-based metabolic flux analysis (TMFA), the mass balance constraints of FBA are supplemented by a set of thermodynamic constraints (Eq. 4) in order to eliminate thermodynamically infeasible fluxes from the feasible solution space [23]. With some additional modifications, these constraints also allow for the exploration of feasible ranges for metabolic concentrations [23].

$$\left(\sum_{j}^{m} n_{i,j} U_i\right) v_i < 0 \quad (4)$$

In the thermodynamic constraints, U_j' is the chemical potential of metabolite j, $n_{i,j}$ is the stoichiometric coefficient for compound j in reaction i, and v_i is the flux through reaction i. Because compounds shared by multiple reactants must have the same potential, these constraints prevent thermodynamically infeasible reaction sets from carrying flux. Unfortunately, these constraints are also nonlinear, forcing the problem to be formulated as either a mixed integer linear program or a quadratic program. While the algorithms for growth simulation, gene knockout, and FVA may all be supplemented with thermodynamic constraints to improve accuracy, the nonlinear formulation of TMFA decreases the rate at which simulations may be run, from 70 simulations per second to 0.25 simulations per second. Of course, this also improves the parallel scalability of these algorithms and increases the need for HPC.

2.5. Metabolic engineering

One of the major industrial applications for genome-scale metabolic models is the design of metabolic engineering strategies for inducing the production of organic chemicals such as ethanol, 1,3-PDO, and 3HP by microorganisms [24]. Often the natural metabolic network of an organism will be missing one or more of the enzymatic steps required to produce a desired product. The *pathway addition* algorithm involves identifying the minimal number of new reactions that would have to be engineered into an organism to allow for the production of a desired product from a specified raw material with a minimum yield. Mathematically this algorithm is identical to the automated completion algorithm; uptake of the selected raw material is fixed at one; production of the desired product is constrained to be greater than the minimum yield; and the number of nonnative reactions allowed to carry flux in the final solution is minimized. In the HPC scenario for this algorithm, the pathway addition algorithm could be run for 10^2 different raw materials, 10^2 different potential products, and 10^3 different organisms. In total, 10^7 simulations would be required, each with a run time ranging from one minute to one hour. Completion of this HPC run would provide the following insights: the ideal host organism to produce each desired product, the ideal raw material to produce each desired product from, the maximum yield attainable for each product from each raw material in each organism, and the number of new enzymes that would have to be added to produce each product in each organism.

The OptStrain [25] and OptKnock [26] algorithms are also used extensively to design metabolic engineering strategies. The algorithms take the problem a step further by determining how the existing metabolism of an organism must be modified to drive flux from a specified raw material towards a desired product. These algorithms are also MILP optimizations that require one minute to one hour to complete.

3. Model optimization algorithms

One of the major uses for the gene knockout and growth simulation algorithms is the validation of genome-scale metabolic models against experimentally determined gene essentiality data [27] and Biolog phenotyping array data [21]. Large-scale gene knockout and phenotyping experiments have produced a tremendous amount of experimental data that may be compared with the model predictions, and the validation that this comparison provides is important not only for assessing model accuracy but also for assessing the accuracy of the annotations on which the models are based.

Inconsistencies between model predictions and experimental data indicate that the model and possibly the underlying annotations contain an error that must be identified and fixed. The large size of genome-scale models, however, makes the manual identification of errors a monumental task. In order to assist in this task, the GrowMatch algorithm was introduced [9]. This algorithm attempts to automatically repair model errors that cause prediction inconsistencies by adding or removing reactions from the model. The GrowMatch algorithm consists of two stages, each representing a potential application for HPC: gap filling, and gap generation.

3.1. Gap filling

The gap-filling stage of GrowMatch attempts to correct prediction errors for conditions where growth is not predicted to occur *in silico* while growth is observed *in vivo* (called false negative predictions). This type of inconsistency is corrected by either adding one or more new reactions to the model or converting existing irreversible reactions into reversible reactions. The gap-filling algorithm is an MILP identical to the automated completion algorithm, but the run time is typically shorter (one minute to one hour) because of the smaller size of the gaps being filled. Also, like the auto completion algorithm, the gap-filling algorithm often produces multiple equally optimal solutions for filling each gap in the model that causes an incorrect prediction. Once the gap filling has been completed for every false negative prediction condition in the model, a reconciliation optimization must be performed [12] to select the optimal combination of all solutions that corrects as many errors as possible with minimal additions to the model. In the HPC scenario, the gap-filling algorithm would be run in parallel for all conditions with false negative predictions in all models with available data. Unoptimized models have 10^2 false negative predictions on average, and currently experimental gene essentiality or Biolog phenotyping array data exists for 10 models. Thus, up to 10^3 simultaneous simulations can be running for one minute to one hour. However, the number of organisms with available experimental data is likely to grow rapidly over time, which will result in a similar growth in the number of gap-filling simulations required by the GrowMatch process.

3.2. Gap generation

The gap generation stage of GrowMatch attempts to correct prediction errors for conditions where growth is predicted to occur *in silico* while growth is not observed *in vivo* (called false positive predictions). This type of inconsistency is corrected by either removing one or more reactions from the model or converting existing reversible reactions into irreversible reactions. Like the gap-filling algorithm, the *gap generation* algorithm is an MILP, but the problem formulation (which is described in detail elsewhere [9, 12]) is much more complex. As a result, the run time is longer, ranging from one to four hours. Like the gap-filling algorithm, the gap generation algorithm often produces multiple equally optimal solutions, and a reconciliation optimization is necessary [12] to identify the optimal combination of all alternative solutions that corrects as many errors as possible. Also like gap filling, the HPC scenario for *gap generation* involves running in parallel for all conditions with false positive predictions in all models with available data, adding up to a total of 10^3 required simulations.

4. Case study: application of HPC to gene knockout simulation

As a case study for the application of HPC to an FBA algorithm, we implemented the gene knockout simulation algorithm using MPI on the Blue Gene/P supercomputer located at Argonne National Laboratory. The Blue Gene/P consists of 40 racks each containing of 1,024 chips with four 850 MHz PowerPC 450 processors on each chip, for a total of 163,840 processors. Blue Gene/P was selected for our case study because the massive number of available processors presents a unique opportunity to demonstrate the scalability characteristics of an FBA algorithm on a truly large scale.

B. subtilis was selected for this study because it is one of the most well studied prokaryotes (second only to *E. coli*) and because knowledge of essential combinations of genes will be beneficial to continuing efforts to produce a minimal strain of *B. subtilis* [28, 29]. The knockout simulations were performed by using the largest and most up-to-date model of *B. subtilis* available, *i*Bsu1103[12]. This model captures 1,103 of the 4,105 genes in the *B. subtilis* genome, meaning that the total pool of genes considered in our gene knockout analysis will be 1,103.

The gene knockout simulations were carried out in multiple stages, with the single gene knockouts performed first, followed by the double, triple, and quadruple gene knockouts. This multistage strategy was necessary because essential gene combinations identified in earlier stages had to be filtered out of the combinations explored in later stages (table 2). For example, it is unnecessary to simulate double gene knockout combinations that include genes found to be essential during the single knockout study because double knockouts involving essential genes will always be essential as well.

Table 2. Gene knockout combinations considered at each stage of the knockout algorithm

	Single KOs	Double KOs	Triple KOs	Quadruple KO
Naïve estimate	1,103	607,753	223,045,351	61,337,471,525
Estimate based on single KO		331,705	89,892,055	18,248,087,165
Estimate based on double KO			89,870,917	18,243,796,151
Estimate based on triple KO				18,243,776,054
Lethal knockouts identified	288	78	99	28

The parallel FBA algorithm was implemented in two versions that exploit two levels of parallelism respectively: fine-grained loop-level parallelism (v1) and coarse-grained embarrassing parallelism (v2). In both approaches, a master node is used to load input files from the file system and broadcast the buffers to every slave node. In the fine-grained approach, the master node loops over the whole set of gene knockout combinations and dispatches a subset to a slave node whenever it becomes available. The granularity (G) is decided based on the total number of processors to strike a balance between minimizing interprocessor communication and balancing slave workload. The actual job sizes are randomly drawn from Uniform ($G/2$, $3G/2$) to avoid clashes of communication. In the coarse-grained approach, instead of getting their jobs from the master node, each slave determines the work that must be performed on based on the slave's rank; the slaves skip through the universal list of genes knockout combinations to find the knockouts they must simulate.

When applied to the simulation of every possible triple gene knockout, the fine-grained loop-level parallelism (v1) approach yielded a high peak performance in moderate parallel settings but scaled poorly beyond 1,024 processors because of the significant interprocessor communication involved in this approach. The coarse-grained embarrassing parallelism (v2) approach initially took a slight performance hit compared with the v1 algorithm because of the overhead involved in generating the complete knockout list on every node. However, this algorithm gave near linear scalability up to 4,096 nodes (figure 2). Because every possible combination of single, double, and triple gene knockout could be simulated in under 10 minutes on 4,096 nodes (one rack of Blue Gene/P running in virtual node mode), it is not unreasonable to expect the coarse-grained approach to scale well beyond 4,096 nodes since at this run time overhead begins to rival computation. We confirmed this hypothesis by simulating quadruple gene knockouts on 65,536 in 162 minutes. The throughput from the quadruple knockout run scaled almost linearly with all of the triple knockout runs (figure 2).

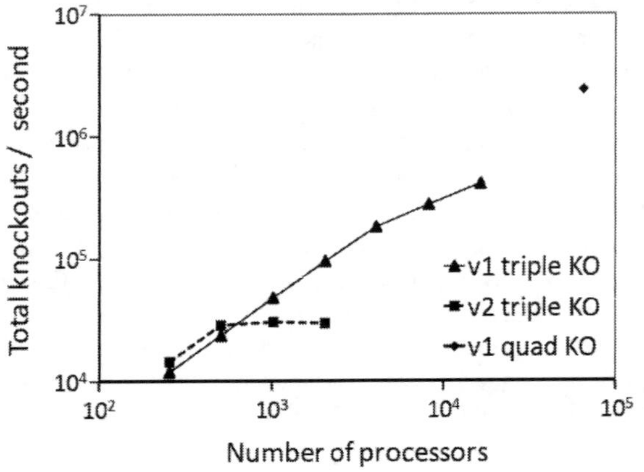

Figure 2. Scalability plot for gene knockout simulations. Here we show how the total number of knockouts simulated every second scaled with the number of processors used in the simulations for multiple parallel algorithms. Algorithm v1 scaled linearly up to 4,096 processors when simulating all possible triple knockouts and up to 65,536 processor when simulating all possible quadruple knockouts.

In the first stage of our knockout study 1,103 single gene knockouts were simulated in *B. subtilis*, resulting in the identification of 288 essential genes. In the next stage 331,705 double gene knockouts were simulated, 78 of which were found to be lethal combinations. In the third stage 89,870,917 triple gene knockouts were simulated with 99 combinations identified as lethal. In the final stage 18,243,776,054 quadruple knockouts were simulated with only 28 lethal combinations identified (table 2). The complete list of the essential gene sets in *B. subtilis 168* identified during this case study can be found here: http://bioseed.mcs.anl.gov/~chenry/BSubEssentials.txt.

5. Conclusions

Numerous opportunities exist for the application of HPC to the reconstruction, analysis, and optimization of genome-scale metabolic models. The field of biology is evolving rapidly, and high-throughput techniques for genome sequencing, gene knockout and mutation, and phenotype analysis are producing an explosion of new experimental data. HPC has an important role to play in ensuring that analytical techniques like FBA can keep up with this explosion. The analysis presented here shows that only with the computational power provided by HPC will the FBA algorithms existing today be capable of handling the explosion of biological data anticipated in the near future.

The results of the quadruple knockout simulation case study clearly show that FBA algorithms can scale well on massively parallel architectures like Blue Gene/P. The caveat to this statement is that sufficient work must be available to drown out the overhead involved in distributing and initializing problem data on a large number of processors. For standard linear programming simulations like the gene knockout algorithm, results indicate that linear scaling can be achieved if there are at least 20,000 simulations per processor. However, we expect that FBA algorithms involving MILP formulations (like TMFA, GrowMatch, or auto completion) will scale linearly with far fewer simulations per processor because of the longer run times involved.

The biological results of the quadruple knockout case study were surprising, as one might intuitively expect to see more lethal double knockouts than lethal single knockouts, more lethal triple knockouts than lethal double knockouts, and so on. Instead the simulation results indicate that *B. subtilis* is remarkably robust to multiple gene knockout. There are two explanations for this nonintuitive result: (1) a large fraction of the essential metabolic functions in *B. subtilis* are associated with essential genes, meaning they have no genetic redundancy, and (2) when a metabolic function required for growth does have genetic redundancy or biochemical redundancy, it is very redundant, requiring three or more knockouts before functionality is lost. Apparently, the increased robustness that is gained by preserving redundant copies of the genes encoding essential functions is not worth the additional cost of copying and maintaining those redundant copies. Thus it is likely that when redundancy does occur in the genome, it exists to increase flexibility in the expression of the functions related with these redundant genes, rather than as a mechanism to enhance robustness to mutation.

Acknowledgments
This work was supported in part by the U.S. Department of Energy under contract DE-ACO2-06CH11357. We thank the Argonne Leadership Computing Facility and INCITE program for compute time spent on Blue Gene/P. We thank the SEED development team for advice and assistance in using the SEED annotation system.

References
1. Edwards JS, Palsson BO: The Escherichia coli MG1655 in silico metabolic genotype: its definition, characteristics, and capabilities. *Proceedings of the National Academy of Sciences of the United States of America* 2000, 97:5528-5533.
2. Papoutsakis ET, Meyer CL: Equations and calculations of product yields and preferred pathways for butanediol and mixed-acid fermentations. *Biotechnology and Bioengineering* 1985, 27:50-66.
3. Jin YS, Jeffries TW: Stoichiometric network constraints on xylose metabolism by recombinant Saccharomyces cerevisiae. *Metab Eng* 2004, 6:229-238.
4. Varma A, Palsson BO: Stoichiometric flux balance models quantitatively predict growth and metabolic by-product secretion in wild-type Escherichia-coli W3110. *Applied and Environmental Microbiology* 1994, 60:3724-3731.
5. Varma A, Palsson BO: Metabolic capabilities of Escherichia-coli. 2. Optimal-growth patterns. *Journal of Theoretical Biology* 1993, 165:503-522.
6. Varma A, Palsson BO: Metabolic capabilities of Escherichia-coli.1. Synthesis of biosynthetic precursors and cofactors. *Journal of Theoretical Biology* 1993, 165:477-502.
7. Edwards JS, Ibarra RU, Palsson BO: In silico predictions of Escherichia coli metabolic capabilities are consistent with experimental data. *Nature Biotechnology* 2001, 19:125-130.
8. Feist AM, Herrgard MJ, Thiele I, Reed JL, Palsson BØ: Reconstruction of Biochemical Networks in Microbial Organisms. *Nat Rev Microbiol* 2009, 7:129-143.
9. Kumar VS, Maranas CD: GrowMatch: an automated method for reconciling in silico/in vivo growth predictions. *PLoS Comput Biol* 2009, 5:e1000308.
10. Kumar VS, Dasika MS, Maranas CD: Optimization based automated curation of metabolic reconstructions. *BMC Bioinformatics* 2007, 8:212.
11. Aziz RK, Bartels D, Best AA, DeJongh M, Disz T, Edwards RA, Formsma K, Gerdes S, Glass EM, Kubal M, Meyer F, Olsen GJ, Olson R, Osterman AL, Overbeek RA, McNeil LK, Paarmann D, Paczian T, Parrello B, Pusch GD, Reich C, Stevens R, Vassieva O, Vonstein V, Wilke A, Zagnitko O: The RAST Server: rapid annotations using subsystems technology. *BMC Genomics* 2008, 9:75.
12. Henry CS, Zinner J, Cohoon M, Stevens R: iBsu1103: a new genome scale metabolic model of B. subtilis based on SEED annotations. *Genome Biol* 2009:in press.
13. Kummel A, Panke S, Heinemann M: Systematic assignment of thermodynamic constraints in metabolic network models. *BMC Bioinformatics* 2006, 7:512.
14. Jankowski MD, Henry CS, Broadbelt LJ, Hatzimanikatis V: Group contribution method for thermodynamic analysis of complex metabolic networks. *Biophys J* 2008, 95:1487-1499.
15. DeJongh M, Formsma K, Boillot P, Gould J, Rycenga M, Best A: Toward the automated generation of genome-scale metabolic networks in the SEED. *BMC Bioinformatics* 2007, 8:-.
16. Suthers PF, Dasika MS, Kumar VS, Denisov G, Glass JI, Maranas CD: A genome-scale metabolic reconstruction of Mycoplasma genitalium, iPS189. *PLoS Comput Biol* 2009, 5:e1000285.
17. Lee S, Phalakornkule C, Domach MM, Grossmann IE: Recursive MILP model for finding all the alternate optima in LP models for metabolic networks. *Computers & Chemical Engineering* 2000, 24:711-716.

18. Ladányi L, Ralphs TK, Saltzman MJ: Implementing Scalable Parallel Search Algorithms for Data-intensive Applications. *The Proceedings of the International Conference on Computational Science* 2002:592.
19. Overbeek R, Disz T, Stevens R: The SEED: A peer-to-peer environment for genome annotation. *Communications of the Acm* 2004, 47:46-51.
20. Seed Viewer - Home [http://seed-viewer.theseed.org/]
21. Bochner BR: Global phenotypic characterization of bacteria. *Fems Microbiol Rev* 2009, 33:191-205.
22. Mahadevan R, Schilling CH: The effects of alternate optimal solutions in constraint-based genome-scale metabolic models. *Metab Eng* 2003, 5:264-276.
23. Henry CS, Broadbelt LJ, Hatzimanikatis V: Thermodynamics-based metabolic flux analysis. *Biophys J* 2007, 92:1792-1805.
24. Willke T, Vorlop KD: Industrial bioconversion of renewable resources as an alternative to conventional chemistry. *Applied Microbiology and Biotechnology* 2004, 66:131-142.
25. Pharkya P, Burgard AP, Maranas CD: OptStrain: a computational framework for redesign of microbial production systems. *Genome Res* 2004, 14:2367-2376.
26. Burgard AP, Pharkya P, Maranas CD: OptKnock: A bilevel programming framework for identifying gene knockout strategies for microbial strain optimization. *Biotechnology and Bioengineering* 2003, 84:647-657.
27. Gerdes S, Edwards R, Kubal M, Fonstein M, Stevens R, Osterman A: Essential genes on metabolic maps. *Curr Opin Biotechnol* 2006, 17:448-456.
28. Morimoto T, Kadoya R, Endo K, Tohata M, Sawada K, Liu S, Ozawa T, Kodama T, Kakeshita H, Kageyama Y, Manabe K, Kanaya S, Ara K, Ozaki K, Ogasawara N: Enhanced Recombinant Protein Productivity by Genome Reduction in Bacillus subtilis. *DNA Res* 2008, 15:73-81.
29. Fabret C, Ehrlich SD, Noirot P: A new mutation delivery system for genome-scale approaches in Bacillus subtilis. *Mol Microbiol* 2002, 46:25-36.

Molecular mechanism of pore creation in bacterial membranes by amyloid proteins

I F Tsigelny[1,2], Y Sharikov[1], M A Miller[1] and E Masliah[2]

[1] S an Diego Supercomputer Center, University of California at San Diego, La Jolla, CA 92093-0505, USA
[2] Department of Neurosciences, University of California at San Diego, La Jolla, CA 92093-0505, USA

Email: itsigeln@ucsd.edu

Abstract. This study explores the mechanism of pore creation in cellular membranes by MccE92 bacterial proteins. The results of this study are then compared with the mechanism of alpha-synuclein (aS)-based pore formation in mammalian cells, and its role in Parkinson's disease.

1. Introduction

The process of self-organization of proteins and peptides into long fibrils is historically called amyloidogenesis, and the proteins/peptides capable of organizing into such structures are often called *amyloids*. The process of fibril formation was found to be responsible for a number of neurodegenerative diseases, such as Parkinson's, Alzheimer's, Huntington's, prion diseases, etc. [1]. Lashuel et al. [1, 6] pointed out that the toxicity observed during amyloidogenesis does not arise from the formation of long amyloid fibrils, but instead, by intermediate ring-like oligomers. It is presumed that these amyloid proteins have a physiological role in mammalian cells, but the question of why these toxic fibrils are created by the mammalian cells is still unanswered.

Recently scientists found that bacterial cells create amyloid-like fibrils that have very specific functions in supporting bacterial survival and development. For example, the bacterium *Escherichia coli* and some other Gram-negative bacteria create functional amyloid fibers called *curli*. These curli support bacterial binding to other cells and the organization of biofilms [2]. It is noteworthy that the expression of the curli-creating proteins is controlled genetically *with the appropriate feedbacks to proceed in proper time of bacteria life*. Other bacteria use a fibril formation process to prevent growth of other bacteria in their vicinity. The protein Microcin E492 (MccE92) produced by *Klebsiella pneumoniae* self-assembles to create cytotoxic pores in neighboring cells [6], and it also self-assembles into amyloid fibrils. Bian et al. [5] showed that cytotoxic pore formation in bacteria is related to the "protofibril" stage of oligomerization, while the MccE92 protein loses most of its cytotoxicity when it oligomerizes into long fibrils [3, 4]. Thus, as noted by Lashuel et al. [7], there is a striking analogy between the bacterial and mammalian proteins that form cytotoxic protofibrils and large fibrils.

In view of the phenomenological similarity between these bacterial and mammalian proteins, the comparative study of these proteins may help us better understand the mechanism of pore formation in both mammalian and bacterial cell membranes. Accordingly, this study was undertaken to explore the mechanism of pore creation in cellular membranes by MccE92 bacterial proteins. The results of this study are then compared with the mechanism of alpha-synuclein (aS)-based pore formation in mammalian cells, and its role in Parkinson's disease [7, 14, 15].

2. Methods

Both MccE92 and aS exist in a numerous conformations in solution, and undergo rapid changes among these various forms. Annular structures were found in a number of amyloid systems [1] and these are thought to be responsible for pore formation of the cell membrane. These structures cause cytotoxicity by permitting flow of metal ions across the cell membrane. The number of events is consistent with such pore formation and we use the multi-program, multi-data modeling approach to address all of these steps. By simulating the time course of conformational changes of MccE92 and aS, it is possible to predict the tendency of these proteins to form oligomers in solution and on the surface of a phospholipid bilayer, and their tendency to aggregate into annular structures in both environments.

2.1. Homology modeling

The structure model of MccE92 was based on its 29–30% sequence homology with the redox-sensing transcriptional repressor rex (Pdb ID 1XCB_G). The homology template structure was selected by our protein structure prediction server HMM-SPECTR [8].

2.2. Molecular dynamics

We used two types of molecular dynamics (MD) simulations: first, MD of MccE92 in a water box; second, MD of the MccE92 oligomers on the POPC membrane. Once MccE92 protein is expressed by the bacteria, it travels toward the neighbor bacteria and penetrates the cell membrane toward the cytoplasm. After entering the cell, the MccE92 molecule proceeds to form amyloid-like fibrils [2]. This process goes through the stage of protofibril formation with self-organization into ring-like oligomers. These oligomers are created after various times of MccE92 penetration into the cell. Our simulations of MccE92 in water model this situation — after MccE92 exists for some time in solution, it undergoes conformational changes and then docks to the membrane.

We used periodic boundary conditions at constant pressure (1 atm) and temperature (300 K). The NAMD molecular dynamics program [9] version 2.5 was used with the CHARMM27 force-field parameters [10], cutoff 12 Å, to simulate behavior of an aS molecule in water in normal conditions. The temperature was maintained at 300 K by means of Langevin dynamics using a damping coefficient of 1/ps. A constant pressure (1 atm) was employed as implemented in the NAMD software package. Initial coordinates were taken from a previously equilibrated 500 ps system. Data for analysis were taken between 50 ps and 5 ns of the simulation. MD simulations of a single MccE92 molecule (first type of MD) were conducted in a water box with the shortest distance between the protein molecule and the box walls 30Å. The system contained more than 200,000 atoms of water. BlueGene computers at Argonne National Laboratory and San Diego Supercomputer Center were used for these simulations.

For the second type of simulation, we used 1-palmitoyl 2-oleoyl phospahtidylcholine (POPC) membrane model 180 x 180 Å. The heptamer was constructed with MccE92 molecules (3 ns conformers of single molecule MD) using minimum energy docking, taking into consideration the predicted membranephilic surfaces of proteins. Then the multimer was positioned on the surface of the membrane. The system was inserted in the water box with the following geometrical parameters: the shortest distance between the protein molecule and the box walls 25 Å, the shortest distance between the membrane molecules (on the surfaces not contacting proteins) and the box walls 10 Å. Both MD simulations were conducted in four steps: initially the system of protein and water molecules (or the system of protein, membrane and water) was minimized for 10,000 iterations; then the system was heated in 0.1° increments and equilibrated for 10 ps; then the molecular dynamics simulation was conducted; data for analysis were taken between 50 ps and 5.0 ns of the simulation.

2.3. Membrane-contacting surface of proteins prediction

During their existence in solution, different conformers dock to the cell membrane and organize the membrane-bound aggregates. To study these aggregates' formation, we first had to find which surfaces of the protein are in contact with the membrane and which are free to be in contact with other MccE92 proteins during the aggregation process. To elucidate the possible surface of MccE92 protein that will most probably contact the membrane, we used the program MAPAS [11]. The MAPAS algorithm involves deconvoluting the protein surface into a set of planes, and then selects the planes that most probably will create strong membrane contacts. This program was applied to the 3 ns MD conformer of MccE92 identified as having the best combination of membrane-contacting parameters.

The MAPAS program uses three scoring parameters: membranephilic residues score (MRS), membranephilic area score (MAS), and coefficient of membranephilic asymmetry (K_{mpha}).

MRS is calculated in the following way:

$$MRS = (\sum w(Nu_i * Su_i/Su_{imax}) + \sum((W*Nc_i)/(Nu_i + Nc_i)$$

where W is the membrane disengagement score [11] for each amino acid (called the "membranephilicity" of the residue); Nu_i is the number of uncharged residues in the plane; Nc_i is the number of charged residues in the plane; Su_i is the solvent accessible surface of the uncharged residue that is included in the plane; Su_{imax} is the maximum solvent accessible surface of the uncharged residue that is included to the plane; and Nc_i is the number of charged residues in the plane.

MAS is calculated the following way:

$$MAS = (S_{top(plane)}/S_{all(plane)})$$

where $S_{top(plane)}$ is the solvent-accessible surface of a selected number of the residues having top membranephilicity in the best putative membranephilic plane; and $S_{all(plane)}$ is the solvent-accessible surface of all residues in this plane.

K_{mpha} is calculated the following way:

$$K_{mpha} = (MAS)/(S_{top(protein)}/S_{all(protein)})$$

where $S_{top(protein)}$ is the solvent-accessible surface of a selected number of the top membranephilic residues of the entire protein, and $S_{all(protein)}$ is the solvent-accessible surface of all residues of the protein. (See more descriptions of these parameters and prediction strategy in [11] and its web supplements.)

2.4. Molecular docking

We conducted theoretical docking of MccE92 molecules during their aggregation. The MD conformers of MccE92 that have the best membrane-contacting scores were selected for docking. Docking was conducted using the program Hex [12, 15]. The final docking refinement was done with the InsightII program (Accelrys, 2007) using the 1 Å grid with the cutoff 14 Å. Only the docking complexes with the vdW energy of interaction less than –10 kcal/mole were selected for further consideration. The lowest energy two-molecular complexes where both molecules contact the same membrane surface by the predicted membrane-contacting amino acids were selected for further docking toward the final aggregate configuration.

3. Results

Microcins belong to a family of proteins produced and excreted by some enterobacteria including *E. coli* and *K. pneumoniae* [13]. One of the best studied proteins from this family, Microcin E492 (MccE92), obtained this name because it is produced by the RUC492 strain of *K. pneumoniae*. It has been shown that this protein creates pores in the bacteria cell membranes [2] and artificial lipid bilayers [13]. This protein has antibacterial activity as a result of its ability to organize into membrane-disrupting pores [2].

We used MD simulations to study the process of aggregation and membrane penetration of MccE92. Our simulations suggest that MccE92 is an unstructured protein that can have a significant number of conformational states. In our calculations we explored the set of conformations that MccE92 samples in 5 ns MD simulation runs in a water box; structures observed ranged from more globular forms with mostly helical N-terminal parts to more extended conformations. To select conformers that have the greatest probability of binding to the membrane (to start the process of aggregation and penetration), we assessed the membranephilicity score, i.e., finding which conformer would have the best membrane contact and the specific region of membrane contact on the surface of this protein.

Figure 1 shows the MccE92 conformers of the 5 ns MD run. The overall conformation of the protein changes quite significantly during the simulations. The residues creating the surface of the protein and the residues that are involved in the intramolecular interactions are changing during this process. These changes actually define the further destiny of these molecules. If the membranephilic residues prevail on the surface, the protein will most likely bind to the membrane and became a potential candidate for further aggregation. If the membranephilic residues are not exposed enough, the protein will remain in solution and will not bind to the membrane (at least until its conformation changes significantly during its life span).

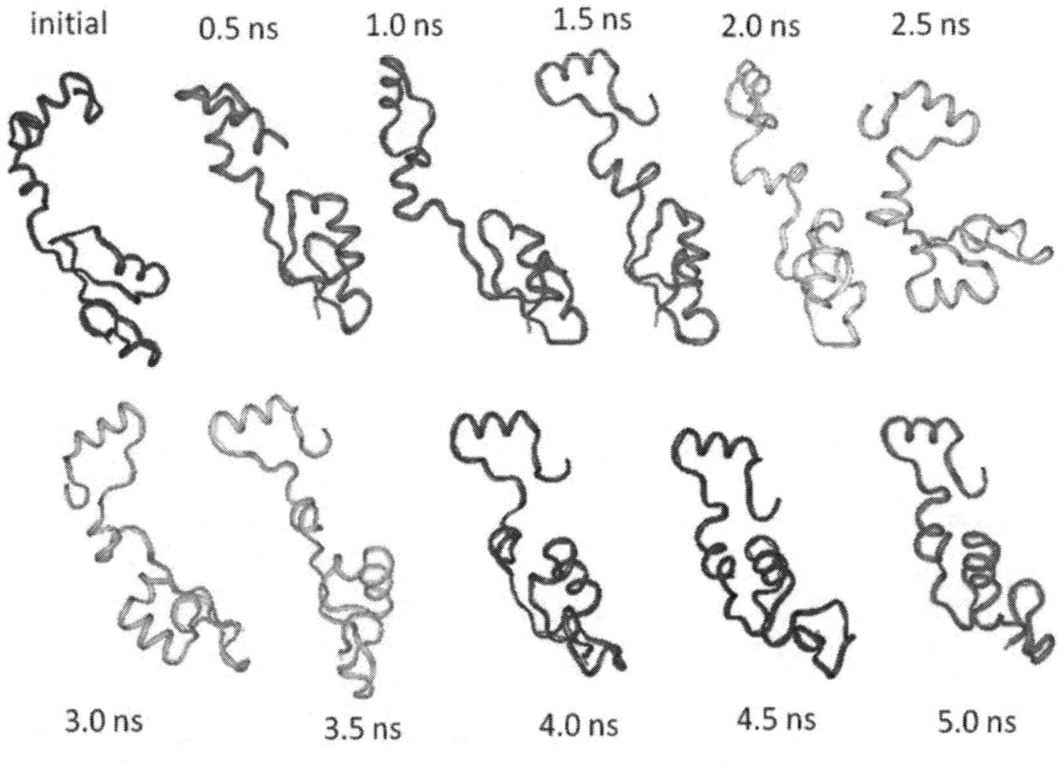

Figure 1. Conformations of Microcin E492 during 5 ns MD.

Figure 2A shows the evolution of the secondary structure of this protein. One can see that the secondary structure changes familiar to us from previous studies of alpha-synuclein also happen to the MccE92 protein during MD simulations. We see the evolution of the alpha-helical part of the secondary structure to the pi-helical structure (figures 2B and 2C). Transforming the alpha-helical parts of the molecule to pi-helical parts is a very important process. Pi-helices have been shown to be involved in a number of membrane contacts and usually serve a specific purpose in proteins. Further evolution of pi-helices is toward the unfolded turns and similar elements of the secondary structure with possible farther transformation to beta strands.

All of the conformers were then submitted to our membrane-contacting prediction program package MAPAS [11]. This program makes it possible to sort the conformers by their membranephilicity. Some of the prediction results are shown in Table 1. One can see that during MD, the membranephilic score (figure 3) actually increases at the time point of 3 ns. This observation supports the increased probability

A

```
     MELRMREISQKDLNLAFGAGETDPNTQLLNDLGNNMAWGAALGAPGGLGSAALGAAGGALQTVGQGLIDXGPVNVPIPVLIGPSW
0    CGGGTCCSHHHHHHHHHHTCCSTTCTTHHHHTTCSSCEEEEEESCCHHHHHHSCCTTTTSEEEEEEECCCTTTTTCCCSSSCCC
200  CCSSSSCHHHHHHHHHHTCCCSSSCSHHHHHHTTTSCEEESCCCTTTHHHHTTSCSSSCSSEEESCCCCCCTTTTTSCSSSSCCC
400  CCSSSSCHHHHHHHHHHSCCCCSCCHHHHHHHTTTSCEEECCBCTTTHHHHHTHHHHTTTTEEETTBCCCSTTTTTCBCSSSBCC
600  CCSSSSCHHHHHHHHHHSCCCCTTCHHHHHHHTTTSCEEEEEECTTHHHHTHHHHHHTTTTEEEEEECCSSSSSCSSBCSSSBCC
800  CCSSSSCHHHHHHHHHHTCCCCCCCHHHHHHHTCSSCEEEEECCTTTHHHHHTHHHHCSSSEEEEECCCSSSTTTTCBSSSSBCC
1000 CCSCSSCHHHHHHHHHHTCCCCCSCHHHHHHHTTTCCEEEEECCTTTIIIIIIHHHHCSSSEEEEECCCCSTTTSSSCCSSSCCC
1200 CCSCCSCHHHHHHHHHHSCCCCCSCTTHHHHHTSSSCEEEEECCTTTIIIIIHHHHTCSSSEEEEECCCCSSSSTTSBSSSSBCC
1400 CCSSCSCHHHHHHHHHHSCCCSSSCIIIIIITTTSSSCEEEEECCSSSTTTTHHHHHHTTTTEEEEECCCCSTTTSSCBCSSSBCC
1600 CCSCCSCHHHHHHHHHHSCCCCSSCHHHHHHHCSSSCCCEEECCTTIIIIIIIHHHHHTTTTCCEEECCCCSTTTTTCCSSSSCCC
1800 CCSCSSCHHHHHHHHHHSCCSCSSCHHHHHHHCSSSCCCCSBCCTTIIIIIIIHHHHCTTTTCCCCBCCCCSTTTTTCBSSSSBCC
2000 CCSSSSCHHHHHHHHHHSSCCSCSCSSHHHHHCSSSCBCCCEECSTIIIIIIIIIIIIITTTBCCCEECCCSTTTTTSBSSSSBCC
2200 CCSCSSCHHHHHHHHHHTSSCCSSCHHHHTSSCSSSCBCCCBCCTTIIIIIIIIIIIICTTTBCCSBCCCCSTTTTTSCSSSSCCC
2400 CCSSSSCHHHHHHHHHHTCCCSCCCIIIIIIICSSSCCEEEECCTTTTIIIIIIIITTCTTTSEEEECCCCSTTTTTSBSSSSBCC
2600 CCSSSSCHHHHHHHHHHSCCCCCSCHHHHHTTCSSSCEEEEECCTTTHHHHHIIIIICTTTEEEEECCCCSTTTTTCCSSSSCCC
2800 CCSSSSCHHHHHHHHHHSSCCCCCCIIIIIIICCSSCEEEEECCTTTHHHHIIIIIIIITTTEEEEECCCCSTTTSSCCSSSSCCC
3000 CCSSSSCHHHHHHHHHHSCCCCCCCHHHHTTTCSSSCEEEEECCTTTIIIIIIIIIIIIITTTEEEEEBCCSSSTTTTCBSSSSBBC
3200 CCSSSSCHHHHHHHHHHSCCCCCSCHHHHTTTCSSSCBCCCCBCIIIIIIIIIIIIIICTTTBCCCCBCCSSGGGTTCBCSSSBBC
3400 CCSSSSCHHHHHHHHHHSCCCCCSCHHHHTTTCSSSCEECCEECIIIIIIIIIIIIIICCTTTEECCEECCSSSTTCSCCSSSSCBC
3600 CCSSSSCHHHHHHHHHHSCCCCCCCHHHHTTTTSSSCEECCEECIIIIIIIIIIIIIISCTTTEECCEECCSSCCSCSSCSSSSCBC
3800 CCSSSSCHHHHHHHHHHSCCSCCSCHHHHTTTCSSSCCBCSCCCTTTIIIIIIIIISSCTTTSBCSCBCCSSCSSCSSCSSSSCBC
4000 CCSSSSCHHHHHHHHHHSSCCCCSCHHHHTTTTSSSCCEEEEECTTTHHHHTTTTSSCTTTSEEEEECCSSCCSCSCCCSSCCBC
4200 CCSSSSCHHHHHHHHHHSCCCCCSCHHHHTTTCSSSCCCCSCCCIIIIIIIIIIIIICCSSCSCSSCBCCSSSCSCSSCSSSCCBC
4400 CCSCSSCHHHHHHHHHHSCCCCCSCHHHHTTTTSSSCCSCSSBCTTTIIIIIIIIIIISSSCCCCSCBCCSSSCSSSCCCSSSCBC
4600 CCSCCSCHHHHHHHHHHSCCCCCSCHHHHTTTTSSSCCCCCEECIIIIIIIIIIIIIISSSCCCCSEECCSCTTCCSSCSSSSCBC
4800 CCSCCSCHHHHHHHHHHSCCCCCSCHHHHTTTTSSSCCCCCEECTTTHHHHIIIIIISSSCCCCSEECCSSSSSCSSCSSSSCBC
4900 CCSCCSCHHHHHHHHHHSCCCCCSCHHHHTTTTSSSCCCCCEECIIIIIIIIIIIIIISSSSSCCSEECCSSSSSCSSCSSSCCBC
        T = turn; E = extended conformation; B = isolated bridge; H = alpha helix;
                 G = 3-10 helix; I = pi helix; C = coil; S = bend.
```

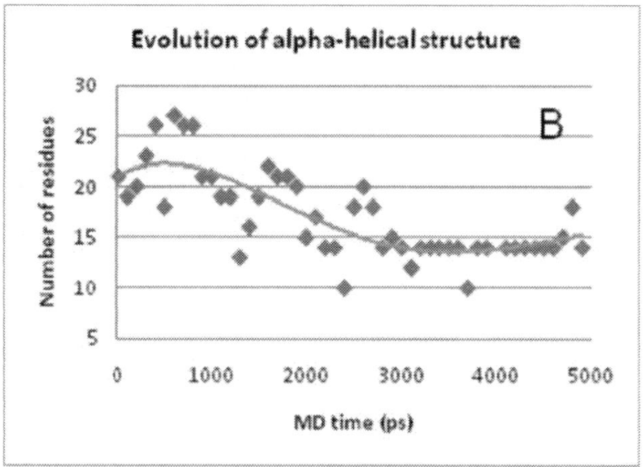

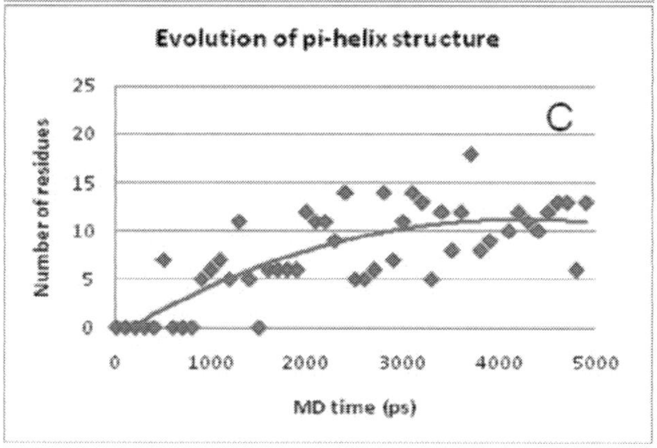

Figure 2. (A) Evolution of the secondary structure of Mcc during MD. (B) Evolution of the alpha-helical and (C) pi-helical secondary structure of Mcc during MD. Transformation of alpha-helix to pi-helix causes the increase of membrane-contacting properties of the protein and eventually supports creation of the ring-like structures of protein aggregates with further pore creation.

of the molecule becoming membrane-contacting. The study of evolution of the secondary structure of MccE92 during MD sheds light on the possible mechanism of influence of the conformational changes to the membrane-contacting properties of the protein.

Table 1. Evolution of membrane-contacting properties of the Mcc during MD.

Time (ns)	members	MRS	MAS	Kmpha
Initial	12	2.08	33.65	1.82
0.5	9	1.70	30.36	1.67
1.0	12	2.18	40.77	2.23
1.5	18	2.14	33.91	1.80
2.0	12	1.74	38.32	2.06
3.0	14	1.85	44.83	2.27
3.5	10	1.53	42.17	2.33

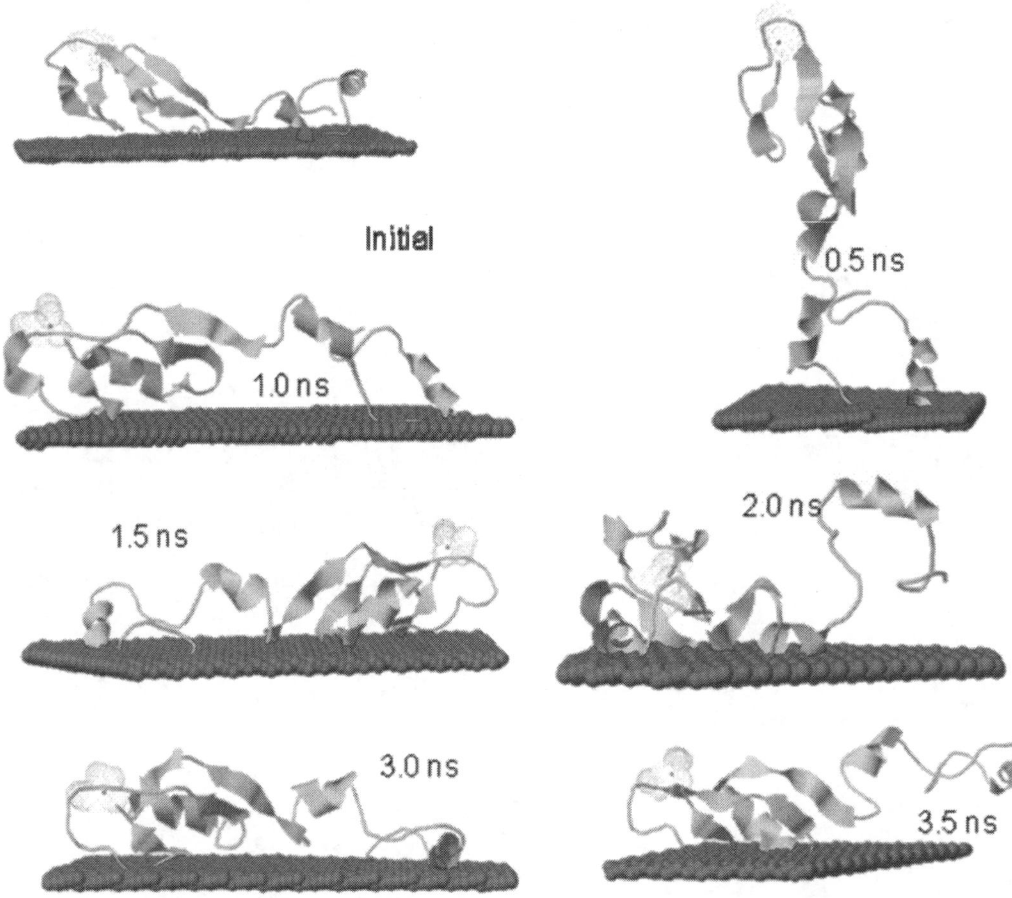

Figure 3. MAPAS selected membrane-contacting surfaces of Microcine E92 (their scoring parameters are presented in table 1). Conformer 3 ns has the best combination of the scoring parameters.

Consecutive docking steps of MccE92 were conducted until the free energy of multimer formation did not decrease further. Figure 4 shows the structure of a final heptameric aggregate of MccE92 on the membrane surface on the initial stage of simulation.

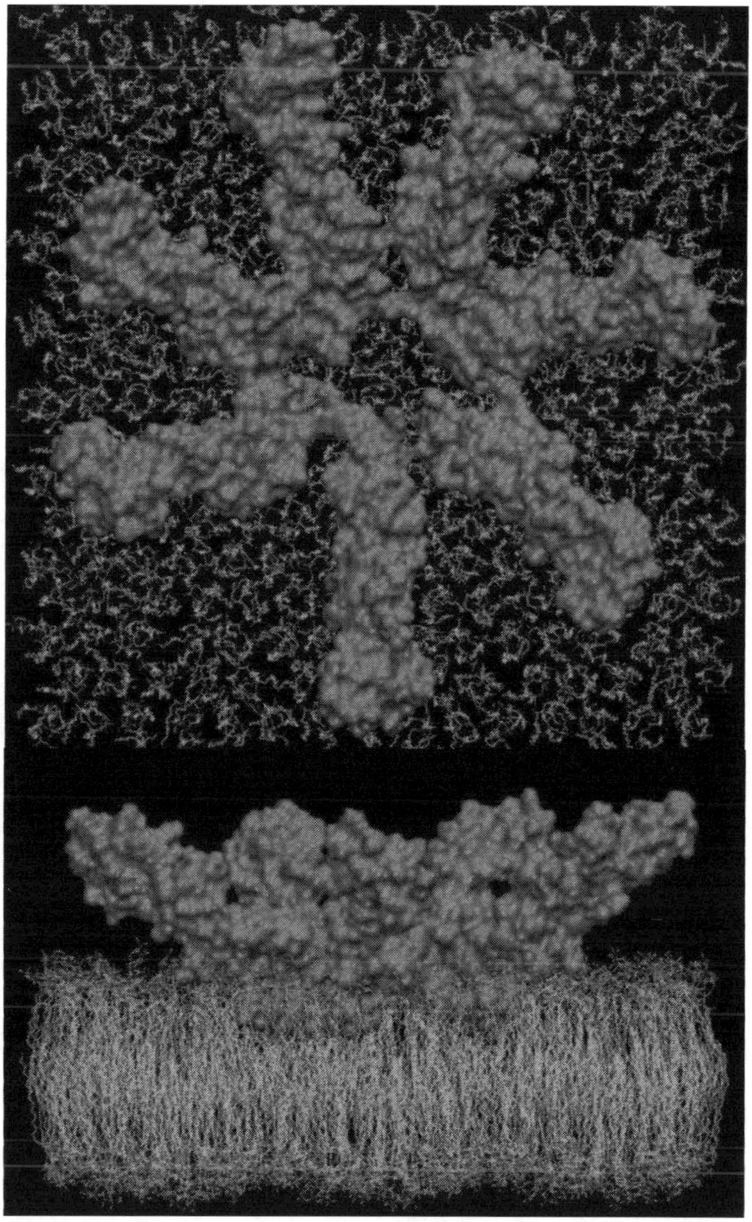

Figure 4. Position of MccE92 on the membrane at the first 200 ps of MD.

The development mechanism of these oligomeric structures is very similar to the mechanism of oligomerization of another unstructured protein, alpha-synuclein (known to be involved in Parkinson's disease). This mechanism is explicitly elucidated in [14, 15]. The final MD simulation is done with the ring-like oligomer situated on the membrane in a water box with a small additional force toward the plane of the membrane to accelerate the possible penetration.

If the conformation of a protein is not suitable for penetration, such a force would not cause any significant movement of the complex to the membrane; on the other hand, when the configuration and orientation of the complex vs the membrane are appropriate, the force will just accelerate the already occurring penetration, so we can use less computational time for modeling. Figure 5 shows penetration of the MccE92 heptamer into the membrane at the 4 ns stage of MD. These results strongly support our concept of the molecular mechanism of pore formation by bacterial proteins. MccE92 can form the ring-like oligomers that can penetrate the membrane and create the transmembrane pores. Our further experiments show that MccE92 eventually penetrates to the other side of the membrane, creating a transmembrane pore.

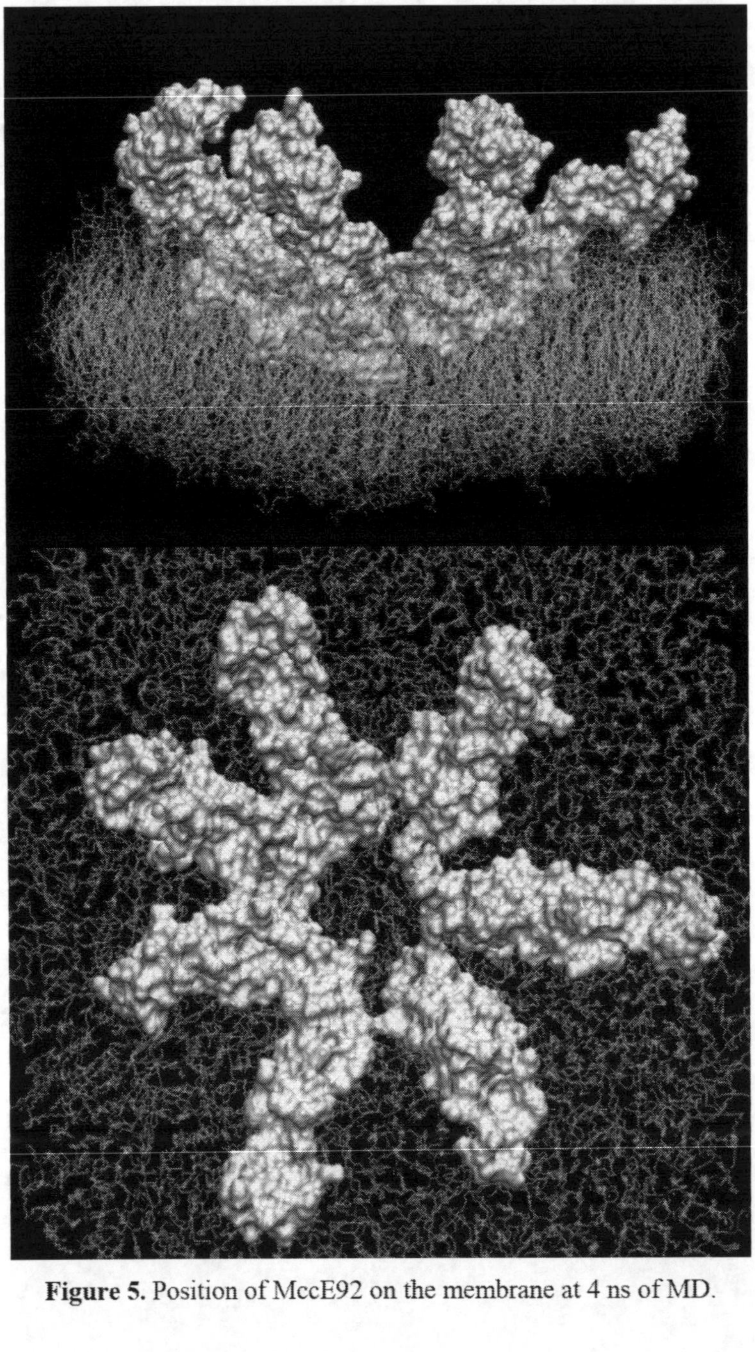

Figure 5. Position of MccE92 on the membrane at 4 ns of MD.

Acknowledgments
This work was supported by a DOE INCITE grant. The authors are also grateful for computational support on BlueGene computers at Argonne National Laboratory and at San Diego Supercomputer Center.

References

[1] Lashuel H A and Lansbuty P T 2006 Are amyloid diseases caused by protein aggregates that mimic bacterial pore-forming toxins? *Quarterly Reviews of Biophysics* **39**:1676-201

[2] Hammer N D, Wang X, McGuffie, B A, Chapman M R 2008 Amyloids: Friend or Fue *J Alzheimer's Disease* **13**:407-419

[3] de Lorenzo V, Martinez J L and Asensio C 1984 Microcin mediated interactions between Klebsiella pneumoniae and Escherichia coli strains, *J Gen Microbiol* **130:**391–400

[4] de Lorenzo V 1985 Factors affecting microcin E492 production *J Antibiot* (*Tokyo*) **38**, 340–345

[5] Bieler S, Estrada L, Lagos R, et al. 2005 Amyloid formation modulates the biological activity of a bacterial protein *J Biol Chem* **280** :26880–26885

[6] Lagos R., Wilkens M., Vergara C., Cecchi X. and Monasterio O.1993. Microcin E492 forms ion channels in phospholipid bilayer membranes. *FEBS Lett.* **321**:145–148.

[7] Lashuel H A, Petre B M, Wall J, et al. 2002 Alpha-synuclein, especially the Parkinson's disease-associated mutants, forms pore-like annular and tubular protofibrils *J Mol Bio* **322**:1089-1102

[8] Tsigelny I, Sharikov Y, Ten Eyck L F 2002 Hidden Markov models-based system (HMMSPECTR) for detecting structural homologies on the basis of sequential information Protein Eng. 2002 **15**:347-52

[9] Kalé L, Skeel R, Bhandarkar M, at al, 1999 *NAMD2:* Greater scalability for parallel molecular dynamics *J Comp Physics*, **151**:282-312

[10] Feller S, MacKerell A 2000 An improved empirical potential energy function for molecular simulations of phospholipids *J Phys Chem* B. **104**:7510-5751

[11] Sharikov Y, Walker RC, Greenberg J, Kouznetsova V, Nigam S, Miller MA, Masliah E, Tsigelny I 2008MAPAS: a tool for predicting membrane-contacting protein surfaces. *Nature|Methods*, 5:119

[12] Ritchie D W and Kemp G J L Protein docking using spherical polar Fourier correlations 2000 Proteins **39**:178-194.

[13] Barnhart M M and Chapman M R 2006 Curli biogenesis and function *Annu Rev Microbiol* **60** :131–147.

[14] Tsigelny I F, Bar-On P, Sharikov Y et al. 2007 Dynamics of alpha-synuclein aggregation and inhibition of pore-like oligomer development by beta-synuclein. *FEBS J.* **274**:1862-1877

[15] Tsigelny IF, Sharikov Y, Miller MA, Masliah E. 2008 Mechanism of alpha-synuclein oligomerization and membrane interaction: theoretical approach to unstructured proteins studies. *Nanomedicine* 2008 4:350-357

Computational chemistry at the petascale: are we there yet?

E Aprà, R J Harrison, W A Shelton, V Tipparaju and A Vázquez-Mayagoitia

Computer Science and Mathematics Division, Oak Ridge National Laboratory, Oak Ridge, TN

E-mail: aprae@ornl.gov

Abstract.
We have run computational chemistry calculations approaching the Petascale level of performance (~ 0.5 PFlops). We used the Coupled Cluster CCSD(T) module of the computational chemistry code NWChem to evaluate accurate energetics of water clusters on a 1.4 PFlops Cray XT5 computer.

1. Introduction

The field of electronic structure is struggling to get efficient parallel implementation on Petascale class hardware. One notable exception has been the achievement of Qbox, a planewave pseudopotential electronic structure code that obtained a performance of 207 TFlops on a BlueGene/L computer [1].

Qbox makes use of the message-passing MPI library for parallelization. Instead, NWChem makes use of the Global Arrays library; this allows the software developer to reach a high level of abstraction and, at the same time, to use one-sided communication to efficiently exploit the network hardware. In the remainder of the paper, we will discuss recent benchmarks and scientific results obtained with NWChem on a parallel computer whose theoretical peak performance is in excess of 1 PFlops.

2. Parallelization approaches

Multi-level parallelization has been successfully used by several authors ([2], [3] or [4]) to achieve good parallel performance at large scale. This parallelization approach is applied to simulation methods where it is possible to subdivide the calculation in a large number of independent tasks; therefore the parallelization issue remain restricted at the level of each task that typically use only a few hundred of computational cores.

As an example of this multilevel parallelization approach, our research group is currently implementing an algorithm for the calculation Raman vibrational intensities by density functional theory. Raman spectra intensities depend on the changes of the molecular polarizability due the normal mode vibrations. Specially, the frequency dependent polarizability evaluation requires demanding quantities of computational resources, and its derivatives are too complex to be evaluated analytically. Our implementation of the Raman intensities divides the computation uniformly in groups of processors by numerical derivatives, each group having a computational task whose size is consistent with the scalability limit of solution of the polarizability problem [5].

However, in the rest of the paper we will describe an computational scheme that does not resort to subdivide the computational load at multiple level, but implementation a parallelization approach that uses all the available cores together, in their entirety.
Before moving to the description of the scientific algorithm, we will briefly describe the main aspect of the parallelization framework used in NWChem.

3. The Global Arrays library on the Cray XT5

The primary component of the Global Arrays library is the distributed arrays layer. This forms an **abstraction layer** that greatly simplifies the effort of the developer of the scientific application by isolating most of the intricacies involved in parallelize an application that uses dense matrices. For example, this layer does all the necessary translation from the users shared memory styled access to the actual distributed array. It has several components that it uses: a Message Passing library, the ARMCI one-sided communication library and a memory allocator (MA library). The Distributed Arrays layer is mostly implemented on top of the ARMCI one-sided communication library.

A message passing library (MPI) is used in the Global Arrays for process management and for some collective communication . For the most of the NWChem modules (including the CCSD(T) module that we discuss below), the amount of MPI usage is minimal, most of the communication is via ARMCI one-sided communication library using the non-contiguous one-sided communication interfaces it provides.

The ARMCI communication library has interfaces for contiguous, strided and vector communication along with Read-Modify-Write operations. All of the Global Array one-sided calls are translated into ARMCI communication calls. ARMCI library, underneath, uses the fastest possible mechanism to transmit data. It uses shared memory with-in the node and, on the Cray XT5 system, uses Portals library for inter-node communication. The Global Arrays library is optimized to overlap intra-node data transfers in shared memory and inter-node data transfers in the network using non-blocking calls. On the Cray XT5, in addition to regular MPI tasks, ARMCI spawns a thread to assist with communication (particularly Read-Modify-Write and non-contiguous data transfer operations). Each of the cores on the octacore XT5 runs an MPI task or, as an alternative, one can sacrifice one core for the sever thread (discussed below) and use the remaining seven cores for the computing pool.

4. CCSD(T) implementation in NWChem

Coupled-cluster (CC) is a numerical many-body technique that studies the effect of electron correlation on the electronic structure of molecular systems. CCSD(T) is one of several CC methods that estimates the effect of electron correlations by considering single, double and triple excitations; single and double excitations are fully computed with a self-consistent approach, while the contribution of the triple excitations is computed pertubatively [6]. Here we describe the CCSD(T) module of the NWChem library, its usage of the Global Arrays and discuss how Global Arrays library is structured on the Cray XT5.

Valence only coupled cluster CCSD(T) [6] calculations have been called the "gold standard of quantum chemistry" [7] for their chemical accuracy in determining molecular energetics. The computational cost of CCSD(T) calculations formally scales as the seventh power of the number of basis functions (N^7), hence this is a method that could effectively utilize large supercomputers. A few parallel implementations of CCSD(T) are available [8–12], however, most of them have been designed either for clusters or for moderate scale parallelism.

In this work we have substantially modified and enhanced the original parallel implementation of CCSD(T) in NWChem of Kobayashi and Rendell [8] that was designed to effectively

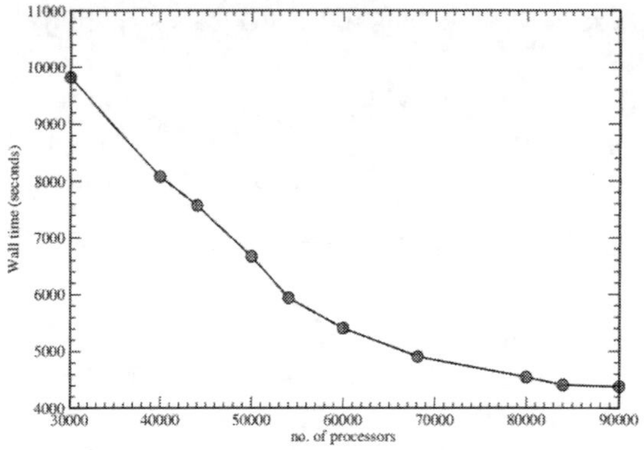

Figure 1. Walltime for the CCSD(T) calculation of total energy of $(H_2O)_{18}$ as a function of the number of processors

utilize massively parallel processors and to make minimal use of I/O resources. The focus of the performance numbers reported here is the perturbative triples correction implemented in NWChem following the "*aijkbc* algorithm" of Rendell and coworkers [13] that makes no use of I/O by storing intermediate quantities (two-electron integrals and coupled-cluster wave function amplitudes) in the global memory managed by Global Arrays.

The floating-point intensive kernel of this algorithm is a series of calls to the BLAS DGEMM [14] matrix multiplication routine. Several data parallel and one-sided Global Arrays operations are used in the CCSD(T) implementation of NWChem, most notable of these are the `ga_get()` and `ga_acc()` calls that are used to get/accumulate sections of distributed array.

Our revisions to this algorithm emphasized increased locality to reduce communication and implemented a more careful tiling of intermediates to reduce memory consumption and increase parallelism and load balance. The dynamically load balanced algorithm explicitly considers three levels of the memory hierarchy. It proceeds by tiling the full computation so that intermediate results fit in available global memory, then tiling the nested loops so that data associated with each task fits into local memory. A process access a global shared counter to determine the next task, moves data from global to local memory using the `ga_get()` operation, computes, and accumulates results into the result using the `ga_acc()` operation. The local computation employs the BLAS DGEMM [14] matrix multiplication routine to optimize for the local memory hierarchy.

5. Results

We reported performance number by using as base the parallel implementation of CCSD(T) in NWChem of Kobayashi and Rendell [8] that was designed to effectively utilize massively parallel processors and to make minimal use of I/O resources.

What distinguishes the benchmark numbers reported here is the unprecedented scale of the calculations and floating-point performance achieved. We run a series of benchmark with the 5.1 version of NWChem [15] using the water cluster $(H_2O)_{18}$ and a modified cc-pvtz basis set [16] for a total of 918 Gaussian basis functions; wall-time of the CCSD(T) runs as a function of processor number are reported in Figure 1. The last datapoint at 90000 processor reached a sustained 64-bit floating-point performance of 358 TFLOP/s.

All eight cores of the octacore Cray XT5 node were used in the performance results reported here. In order to alleviate memory usage and improved load-balancing, we had to modify the distribution of the larger Global Arrays used in this algorithm. The effect of last two distribution

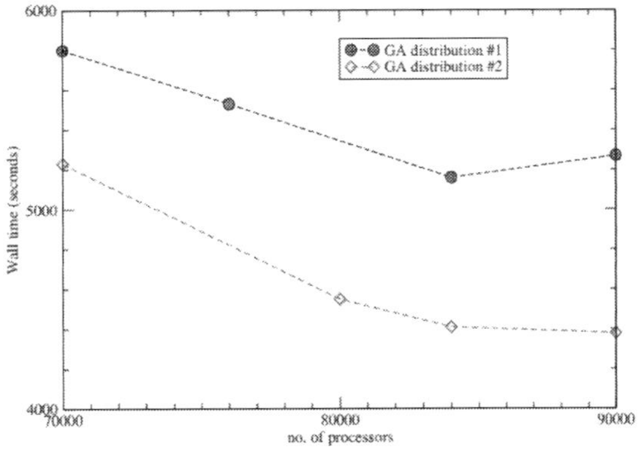

Figure 2. $(H_2O)_{18}$ benchmark: effect of different `ga_create` distributions

approaches we developed are shown in the plot[1] of Figure 2.

We modified the 5.1 NWChem source code in two main aspects by moving communications outside of inner loops (sometime at the additional cost of more use of local memory) and by modifying the distribution of the largest Global Arrays used in the triples kernel of the CCSD(T) method. The first more obvious distribution we attempted (labeled as distribution #1 in the plot) allowed us to scale at 7000 processors, but it could not scale beyond 80000 processors. This first distribution was already an improvement compare to the previous one since it drastically cut down the memory requirements (the previous one would have required 2GB or more of memory, clearly not an option on a system that has only 2GB of RAM on each core). In the next step, we refined the distribution by having the large Global Arrays being distributed over as many processors as possible; the finer granularity of this second distribution improved load-balancing and allowed us to reach better scaling beyond 80000 processors.

A further benchmark run, on the even larger cluster $(H_2O)_{20}$, we ran a coupled-cluster calculation on the ORNL's Jaguar petaflop computer that used over 100 TB of memory for a sustained performance of 487 TFLOP/s (double precision) in the triples sections on 96,000 processors, lasting for 2 hours. This floating-point performance corresponds to 55% of theoretical aggregate performance of the 96000 2.3GHz AMD Opteron cores. The number of basis functions for this run is equal to 1020 and the number of correlated orbitals is equal to 80. While the 487 TFLOP/s refers just to the triples section of the calculation, if we consider the run from beginning to end (including I/O operations such as reading the 21GB amplitudes file plus the initial step of transforming the Atomic Orbitals into Molecular orbitals), the floating point results is of 444 TFLOP/s. In figure 3 we reported the scaling curve of this benchmark.

64-bit floating precision is a necessary requirement for this kind of calculations because of the error propagation in the computation of the gaussian integrals being accumulated and of the numerical instability of high accuracy gaussian basis functions given used.

As far as scientific results are concerned, we present here (table 1) some preliminary numbers that show the relative stability of twenty water clusters in two different conformations. The binding energies computed in this study are of great value for describing the the properties of water at the molecular level.

[1] The data point at 70000 processors in the upper curve of the right plot was obtained by interpolating between the measured values at 50000 and 76000 processors

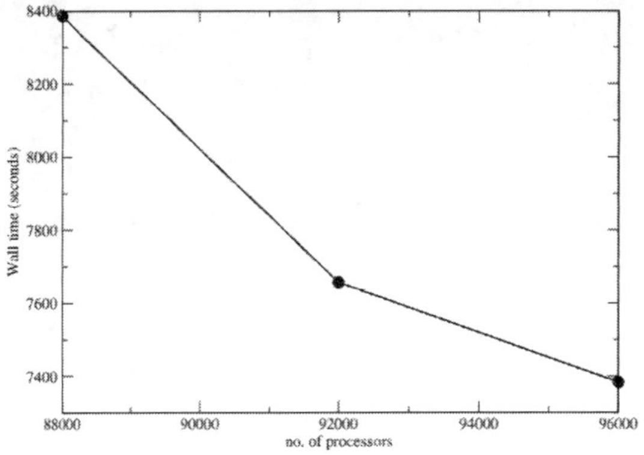

Figure 3. Walltime for the CCSD(T) calculation of total energy of $(H_2O)_{20}$ as a function of the number of processors

6. Conclusion

A unique aspect of this work is the use of the one-sided Global Arrays programming model that enabled us to scale a single tightly-coupled calculation to reach a performance approaching the petascale level. We project performance approaching 1 PFlops on larger water clusters on an upgraded computer systems.

Acknowledgments

This work was supported by the Department of Energy, offices of Basic Energy Science and Advanced Scientific Computing Research as part of the SciDAC program, and was also in part sponsored by the Laboratory Directed Research and Development Program of Oak Ridge National Laboratory (ORNL), managed by UT-Battelle, LLC for the U. S. Department of Energy under contract DE-AC05-00OR22725 with Oak Ridge National Laboratory. This research used resources of the National Center for Computational Sciences at Oak Ridge National Laboratory, which is supported by the Office of Science of the U.S. Department of Energy under Contract No. DE-AC05-00OR22725.

[1] Gygi F, Draeger E W, Schulz M, de Supinski B R, Gunnels J A, Austel V, Sexton J C, Franchetti F, Kral S, Ueberhuber C W and Lorenz J 2006 *SC '06: Proceedings of the 2006 ACM/IEEE conference on Supercomputing* (New York, NY, USA: ACM) p 45 ISBN 0-7695-2700-0

[2] Fedorov D G, Olson R M, Kitaura K, Gordon M S and Koseki S 2004 *Journal of Computational Chemistry* **25** 872–880

[3] Windus T L, Kathmann S M and Crosby L D 2008 *Journal of Physics: Conference Series* **125** 012017 (10pp) URL `http://stacks.iop.org/1742-6596/125/012017`

[4] Wang L W, Lee B, Shan H, Zhao Z, Meza J, Strohmaier E and Bailey D H 2008 *SC '08: Proceedings of*

	Edge-sharing		Dodecahedron	
	E^{tot}	B.E.	E^{tot}	B.E.
SCF	-1521.371595	152.87	-1521.363070	147.52
MP2	-1526.467996	251.02	-1526.444136	236.05
CCSD	-1526.539945	228.14	-1526.518107	214.43
CCSD(T)	-1526.709594	245.10	-1526.684946	229.64

Table 1. Total Energies (E^{tot}) in hartrees and Binding energies (B.E.) in Kcal/mol for the edge-sharing pentagonal prism and the dodecahedron $(H_2O)_{20}$ structures

the 2008 ACM/IEEE conference on Supercomputing (Piscataway, NJ, USA: IEEE Press) pp 1–10 ISBN 978-1-4244-2835-9

[5] Johnson B G and Florin J 1995 *Chemical Physics Letters* **247** 120 – 125 ISSN 0009-2614 URL `http://www.sciencedirect.com/science/article/B6TFN-3TJ5JRM-50/2/080969b3cc20c5a690f9a439f97ef378`

[6] Raghavachari K, Trucks G W, Pople J A and Head-Gordon M 1989 *Chemical Physics Letters* **157** 479 – 483 ISSN 0009-2614 URL `http://www.sciencedirect.com/science/article/B6TFN-44F1NYB-58/2/dc7ab0fbf4f6f7beddd4f5e9f7494ac6`

[7] Dunning T H, Peterson K A, Woon D E and Wilson A K 1999 *American Conference on Theoretical Chemistry* unpublished

[8] Kobayashi R and Rendell A P 1997 *Chemical Physics Letters* **265** 1 – 11 ISSN 0009-2614

[9] Hirata S 2003 *The Journal of Physical Chemistry A* **107** 9887–9897 (*Preprint* `http://pubs.acs.org/doi/pdf/10.1021/jp034596z`) URL `http://pubs.acs.org/doi/abs/10.1021/jp034596z`

[10] Bentz J L, Olson R M, Gordon M S, Schmidt M W and Kendall R A 2007 *Computer Physics Communications* **176** 589 – 600 ISSN 0010-4655 URL `http://www.sciencedirect.com/science/article/B6TJ5-4N7SBTR-1/2/bbe1d382122689b04c012b875569b7b0`

[11] Janowski T and Pulay P 2008 *Journal of Chemical Theory and Computation* **4** 1585–1592 (*Preprint* `http://pubs.acs.org/doi/pdf/10.1021/ct800142f`) URL `http://pubs.acs.org/doi/abs/10.1021/ct800142f`

[12] Lotrich V, Flocke N, Ponton M, Yau A D, Perera A, Deumens E and Bartlett R J 2008 *The Journal of Chemical Physics* **128** 194104 (pages 15) URL `http://link.aip.org/link/?JCP/128/194104/1`

[13] Rendell A P, Lee T J, Komornicki A and Wilson S 1992 *Theoretical Chemistry Accounts: Theory, Computation, and Modeling (Theoretica Chimica Acta)* **84** 271–287

[14] Dongarra J, Croz J D, Duff I and Hammarling S 1990 *ACM Transactions on Mathematical Software* **16** 1–17

[15] Bylaska E and et al 2007 *NWChem, A Computational Chemistry Package for Parallel Computers, Version 5.1*

[16] Dunning T H 1989 *The Journal of Chemical Physics* **90** 1007–1023 URL `http://link.aip.org/link/?JCP/90/1007/1`

Journal of Physics: Conference Series **180** (2009) 012028 doi:10.1088/1742-6596/180/1/012028

Hard scaling challenges for *ab initio* molecular dynamics capabilities in NWChem: Using 100,000 CPUs per second

Eric J. Bylaska[1], Kevin Glass[1], Doug Baxter[1], Scott B. Baden[2] and John H. Weare[3]

[1] Environmental Molecular Sciences Laboratory, Pacific Northwest National Laboratory, P.O Box 999, Richland, WA 99354

[2] Department of Computer Science and Engineering, University of California, San Diego
9500 Gilman Drive, # 0404, La Jolla, CA 92093-0404

[3] Department of Chemistry and Biochemistry, University of California, San Diego
9500 Gilman Drive, # 0303, La Jolla, CA 92093-0303

Email: Eric.Bylaska@pnl.gov

Abstract. An overview of the parallel algorithms for *ab initio* molecular dynamics (AIMD) used in the NWChem program package is presented, including recent developments for computing exact exchange. These algorithms make use of a two-dimensional processor geometry proposed by Gygi *et al.* for use in AIMD algorithms. Using this strategy, a highly scalable algorithm for exact exchange has been developed and incorporated into AIMD. This new algorithm for exact exchange employs an incomplete butterfly to overcome the bottleneck associated with exact exchange term, and it makes judicious use of data replication. Initial testing has shown that this algorithm can scale to over 20,000 CPUs even for a modest size simulation.

1. Introduction

The ability to predict the properties of complex materials important in toxic waste disposal, disease treatment, efficient chemical processing, and electronic device performance optimization, among others, is of great importance to DOE's efforts to address the nation's energy and environmental problems. Because the required properties are highly sensitive to complex interactions at the fundamental electronic structure level (e.g., chemical bond saturation, shell structure), reliable parameter-free simulation of their properties requires methods based directly on the solution to the electronic structure problem posed by the electronic Schrödinger equation. Development of methods at the fundamental electronic structure level also is important in light of DOE's investment in large-scale facilities such as the synchrotron light and neutron sources. These new probes are providing an unprecedented level of detail at the atomic and molecular scale. However, without appropriate theories or models, many of these new measurements cannot be readily interpreted.

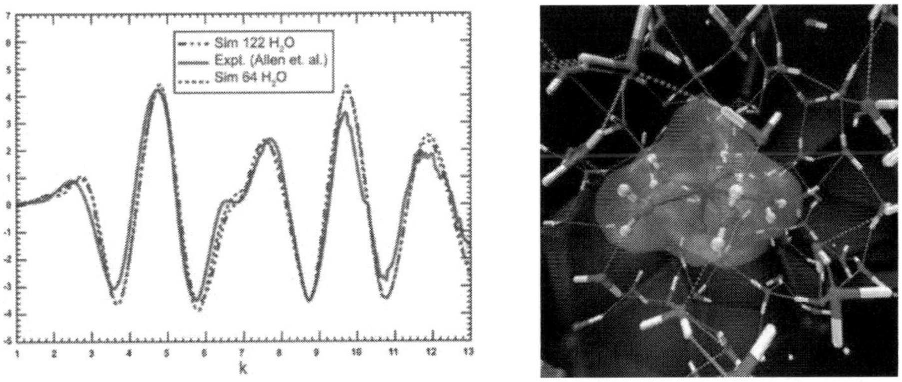

Figure 1. The experimental [1] and simulated extended x-ray absorption fine structure spectra of UO_2^{2+} in water are in perfect agreement (left). On the right, a snapshot of the inner solvation shell UO_2^{2+} is shown. The blue surface identifies the inner-coordination spheres and golden lines show the array of hydrogen bonds that are formed in the structure. The results shown are from NWChem *ab initio* molecular dynamics simulation of UO_2^{2+} and 122 H_2O molecules [2].

The method of *ab initio* molecular dynamics (AIMD) enables researchers to treat the dynamics of these systems while retaining a first-principles-based description of their interactions [3-8]. This approach is similar to classical molecular dynamics where the motions of the atoms and molecules are simulated over a period of time, but here the interactions between the atoms are calculated directly from the electronic Schrödinger equation, rather than from empirical interaction potentials or force fields. That is, at every step of a molecular dynamics simulation the locations of the electrons in the atoms and molecules are determined by solving a suitable approximation to the electronic Schrödinger equation. The instantaneous forces on the atoms are then determined by calculating their electrostatic forces from their interaction with other ions and electrons in the system.

An example of the use of AIMD simulations is for the study of hydrated radionuclides under extreme conditions. A major obstacle to the development of nuclear power is the ability to safely store highly dangerous waste materials containing uranium and other radionuclides [9-11]. Most current storage strategies are designed to store waste in saturated and unsaturated geological formations [12-15]. Stored in this way, the most likely means for uranium to migrate into the biosphere is through groundwater contact with containment canisters, resulting in a solvated UO_2^{2+} cation (or complexes) [12]. To reliably predict behavior of the radionuclide (e.g., uranium, thorium, and plutonium) waste products of nuclear power production over the range of conditions encountered in a storage facility requires a theory based at the most fundamental level on the electronic Schrödinger equation. In figure 1, results from a recent AIMD simulation of the UO_2^{2+} cation in aqueous solution are shown [2]. Even though the UO_2^{2+} cation has been studied extensively over the years using a variety of static *ab initio* (e.g. static coupled cluster calculations) and classical molecular dynamics methods, these prior simulations either have been incomplete or inaccurate. Static *ab initio* simulations have been incomplete because they were not able to take into account the motion of the water molecules in the second and apical solvent shells, and classical molecular dynamic simulations have been plagued by inaccurate force fields of the complex interactions between the UO_2^{2+} cation and water. AIMD simulations, which are able to take into account complex interactions and dynamics, are able overcome the well-known deficiencies of other molecular simulation methods. In fact, this AIMD simulation was the first molecular simulation able to reproduce the measured Extended X-ray Absorption Fine Structure (EXAFS) spectrum from experiments.

Because the electronic Schrödinger equation is solved at every step in the simulation, this type of simulation requires an enormous amount of computational power. The typical time-step in an AIMD simulation is quite small (~0.1 femtosecond=10^{-16} seconds) and the simulations needs to run at least 10 picoseconds. Many chemical processes of interest occur on the order of nanoseconds (10^{-9} seconds). Even for a 10 picosecond AIMD simulation, at least $10^{-11}/10^{-16}=10^5$ evaluations of the electronic

Schrödinger equation are needed. Compared to merely optimizing a molecular or solid-state structure, which requires at most a few hundred evaluations, this is extremely expensive. In order for this to be a practical method, the solution to the electronic Schrödinger equation in a single time-step must be able to complete within seconds. For example the computational time needed to simulate 100 picoseconds will be 11.5 days with a single time-step taking 1 second to complete, 115 days with a 10 second time-step, and ~1 year with a 30 second time-step. In general, for systems beyond a few atoms, even the least expensive approximations to the electronic Schrödinger equation are expensive to calculate.

With the advent of massively parallel computers and the development of new parallel algorithms and software, the costs of AIMD are becoming manageable. Scalable implementations of AIMD began appearing on hundreds of processors in the early to mid 1990's [16-18] and improvements continue to this day [19-22]. Notably (Gordon Bell Prize), F. Gygi *et al.* [22] have scaled a band structure calculation on 64K nodes of Blue Gene L using the Qbox FPMD code.

The most popular approximation to the electronic Schrödinger equations used for AIMD today is Density Functional Theory (DFT) based on computationally efficient approximations to the exact exchange-correlation functional (e.g. LDA and GGA). While this level of approximation is suitable for many applications, it is also becoming clear that higher levels of exchange-correlation potential that are augmented with some fraction of exact exchange (hybrid-DFT, e.g. PBE0 [23]) are needed. The lower levels of exchange correlation potentials presently used in AIMD simulation software are unable to reliably predict the properties of many materials in basic research. Examples of interest to DOE include charge localization in transition elements with tightly bound *d* electrons in oxide materials (see figure 2) [24, 25], the underestimation of reaction barriers and band gaps in solids [26-28], and accurate predictions of spin structure of solids [29] and nanoparticles. The drawback of hybrid-DFT is that it adds a significant amount of expense to an already expensive AIMD simulation. Until recently, it was infeasible to contemplate such computations. However, as we approach the Petaflop milestone, we can think about coping with the high computational costs through vastly increased parallelism.

Figure 2. Illustration of a localized electron (i.e. polaron) on the surface of hematite calculated with a higher and more expensive level (e.g. hybrid DFT) of *ab initio* molecular dynamics. Lower levels of *ab initio* molecular dynamics will predict a delocalized electron.

In this study, we present an overview of the parallel algorithms for AIMD used in the NWChem program package [30], including recent developments for computing exact exchange [31]. These algorithms make use of a two-dimensional processor geometry proposed by Gygi *et al.* for use in AIMD algorithms. Using this strategy, we have recently developed a highly scalable algorithm for exact exchange and incorporated it into an AIMD application. This new algorithm for exact exchange employs an incomplete butterfly to overcome the bottleneck associated with exact exchange term, and makes judicious use of data replication.

2. Key computations in AIMD

The bulk of the computations in *ab initio* molecular dynamics (AIMD) algorithms revolves around the solution of N_e weakly nonlinear partial differential eigenvalue equations (PDEs), $H\psi_i = \varepsilon_i\psi_i$, for the electron orbitals ψ_i, appearing as a result of the DFT approximation to the Schrödinger equation for an N_e electron system. Generally, only the N_e lowest eigenfunctions are required. Most standard AIMD algorithms use non-local pseudopotentials and plane-wave basis sets to perform the DFT calculations. In this framework, solutions are typically approached by means of a conjugate gradient algorithm or, for dynamics, a Car-Parrinello algorithm [3] that requires many evaluations of $H\psi_i$, along with maintaining orthogonality

$$\int_{\Omega} \psi_i(\mathbf{r})\psi_j(\mathbf{r})d\mathbf{r} = \delta_{i,j} \tag{1}$$

Similar FFT-based solution methods are implemented in a number of widely distributed first principles simulation software packages, e.g. NWChem [30].

For hybrid-DFT, the Hamiltonian operator H may be written as [32]

$$H\psi_i(\mathbf{r}) = \begin{pmatrix} -\frac{1}{2}\nabla^2 + V_l(\mathbf{r}) + \hat{V}_{NL} + V_H[\rho](\mathbf{r}) \\ +(1-\alpha)V_x[\rho](\mathbf{r}) + V_c[\rho](\mathbf{r}) \end{pmatrix} \psi_i(\mathbf{r}) - \alpha \sum_j K_{ij}(\mathbf{r})\psi_j(\mathbf{r}) \tag{2}$$

where the one electron density is given by

$$\rho(\mathbf{r}) = \sum_{i=1}^{N_e} |\psi_i(\mathbf{r})|^2 \tag{3}$$

The local and non-local pseudpootentials, V_l and V_{NL}, represent the electron-ion interaction. The Hartree potential V_H is given by

$$\nabla^2 V_H(\mathbf{r}) = -4\pi\rho(\mathbf{r}) \tag{4}$$

The local exchange and correlation potentials are V_x and V_c, and exact exchange kernels K_{ij} are given by

$$\nabla^2 K_{ij}(\mathbf{r}) = -4\pi\psi_j^*(\mathbf{r})\psi_i(\mathbf{r}) \tag{5}$$

During the course of a total energy minimization or AIMD simulation, the electron gradient $H\psi_i$, (equation (2)) and orthogonalization (equation (1)) are evaluated many times (i.e. >10,000 for AIMD), and hence need to be calculated as efficiently as possible. For a pseudopotential plane-wave calculation the main parameters that determine the cost of a of the electron gradient are N_g, N_e, N_a, and N_{proj}, where N_g is the size of the three-dimensional FFT grid, N_e is the number of occupied orbitals, N_a is the number of atoms, and N_{proj} is the number of projectors per atom. Summaries of the computational costs for each of the constituent parts of electron gradient (and orthogonality) are given in figure 3. The major parts of the electron gradient in order of increasing asymptotic cost are (note that conventional DFT does not compute the exact exchange term; $\alpha = 0$ in equation (2)):

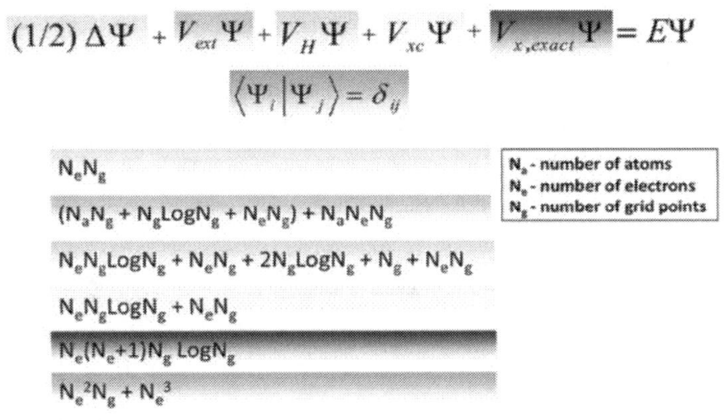

Figure 3. Operation count of Hψ in a plane-wave DFT simulation.

- The Hartree potential V_H, including the local exchange and correlation potentials V_x+V_c. The main computational kernel in these computations is the calculation of N_e three-dimensional FFTs.
- The non-local pseudopotential, V_{NL}. The major computational kernel in this computation can be expressed by the following matrix multiplications: $\mathbf{W} = \mathbf{P}^t{*}\mathbf{Y}$, and $\mathbf{Y}_2 = \mathbf{P}{*}\mathbf{W}$, where $\mathbf{P}$ is an N_g x $(N_{proj}{*}N_a)$ matrix, $\mathbf{Y}$ and $\mathbf{Y}_2$ are N_g x N_e matrices, and $\mathbf{W}$ is an $(N_{proj}{*}N_a)$ x N_e matrix. We note that for most pseudopotential plane-wave calculations $N_{proj}{*}N_a \approx N_e$.
- Enforcing orthogonality. The major computational kernels in this computation are following matrix multiplications: $\mathbf{S}=\mathbf{Y}^t{*}\mathbf{Y}$ and $\mathbf{Y}_2 = \mathbf{Y}{*}\mathbf{S}$, where $\mathbf{Y}$ and $\mathbf{Y}_2$ are N_g x N_e matrices, and $\mathbf{S}$ is an N_e x N_e matrix.

- And when exact exchange is included, the exact exchange operator $\Sigma K_{ij}\psi_j$. The major computational kernel in this computation involves the calculation of $(N_e+1)*N_e$ three-dimensional FFTs. The computation of the exact exchange operator, in which $O(N_e^2)$ independent Poisson equations must be solved, is by far the most demanding term in a pseudopotential plane-wave hybrid-DFT calculation.

3. Parallelization of AIMD (w/o exact exchange)

There are several ways to parallelize a plane-wave Hartree-Fock and DFT program [7, 17, 18, 22, 33, 34]. For many solid-state calculations, the computation can be distributed over the Brillouin zone sampling space [33]. This approach is very simple to implement, however, it cannot be used for Γ-point (**k**=0) calculations with large unit cells. Another approach is to distribute the one-electron orbitals across processors [17]. The drawback of this method is that orthogonality will involve a lot of message passing. Furthermore this method will not work for simulations with very large cutoff energy requirements (i.e., using large numbers of plane-waves to describe the one-electron orbitals) on parallel computers that have nodes with a small amount of memory, because a complete one-electron must be stored on each node. Hence this approach is not practical for Car-Parrinello simulations with large unit cells, however, this approach can work well for simulations with modest size unit cells and with small cutoff energies, when used in combination with minimization algorithms that perform orthogonalization sparingly, e.g. RMM-DIIS [6, 35].

Another straightforward way parallelize AIMD is to spatially decompose the one-electron orbitals [7, 18, 34]. This approach is versatile, easily implemented, and well suited for performing Car-Parrinello simulations with large unit cells and cutoff energies. Moreover the parallel implementation of the non-local pseudopotential and orthogonality is very easy to implement, since they can be implemented using the simple global operation reduce. The drawback of this approach is that a parallel three-dimensional fast Fourier transform (FFT) must be used, which is known not to scale beyond $\sim N_g^{1/3}$ CPUs (or processor groups), where N_g is the number of FFT grid points.

In figure 4, an example of timings versus the number of CPUs for this type of parallelization is shown. These simulations were taken from a Car-Parrinello simulation of UO_2^{2+}+122H_2O with an FFT grid of N_g=96^3 (N_e=1000) using the plane-wave DFT module (PSPW) in NWChem [30]. These calculations were performed on all four cores on the quad-core Cray-XT4 system (NERSC Franklin) composed of a 2.3 GHz single socket quad-core AMD Opteron processors (Budapest). The NWChem program was compiled using Portland Group FORTRAN 90 compiler, version 7.2.4, and linked with the Cray MPICH2 library, version 3.02, for message passing. The performance of the program is reasonable with an overall parallel efficiency of 84% on 128 CPUs dropping to 26% by 1024 CPUs. However, not every part the program scales in exactly the same way. For illustrative purposes, the timings of the FFTs, non-local pseudopotential, and orthogonality are also shown. The parallel efficiency of the FFTs is by far the worst of the three major parts of the computation. Beyond 100 CPUs no gainful work was found in the FFT computation. However, at smaller CPU sizes the inefficiency of the FFTs are damped out due to the fact that these parts of the code make up less than 5% of the overall computation, and the largest part of the

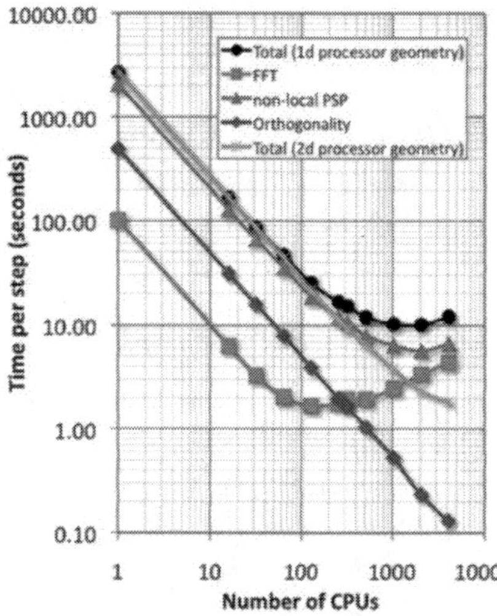

Figure 4. Overall and component timings and component from AIMD simulations of UO_2^{2+}+122H_2O using 1d processor geometry. Overall best timings also shown for 2d processor geometry. Timings from calculations on the Franklin Cray-XT4 computer system at NERSC.

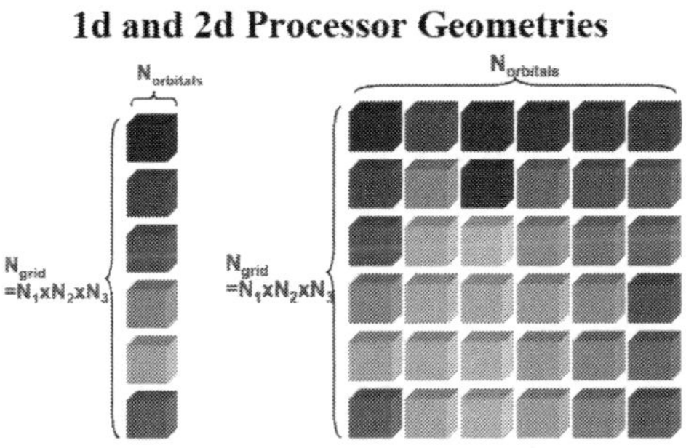

Figure 5. A parallel distribution (shown on the left), implemented in most plane-wave DFT software, each of the one-electron orbitals is identically spatially decomposed. The 2d parallel distribution suggested by Gygi *et al.* is shown on the right.

calculation is the non-local pseudopotential evaluation. Ultimately, however, the lack of scalability of the 3D FFT algorithm beyond $\sim N_g^{1/3}$ CPUs prevails and the simulation ceases to speedup. By 1000 CPUs, the non-local pseudopotential has also stalled. Interestingly, at this number of CPUs the costs of the non-local pseudopotential and the FFTs are roughly the same. Only orthogonality continues to scale beyond 1000 CPUs.

These results demonstrate an important guiding principle that is needed in the design of a parallel AIMD program: The number of CPUs that can be gainfully used in each of the major parts of the calculation is limited because they rely on global operations or all to all operations that use all CPUs in the calculation. Hence the overall parallel algorithm for AIMD should be designed to avoid global communications that span all CPUs in the calculation. For example, Gygi *et al.* distribute across orbitals as well as over space [22], resulting in a in a 2d processor geometry as shown in figure 5 (where the total number of processors, Np, can be written as Np=Npi*Npj). This decomposition reduces the cost of the global operations in the major parts of the electron gradient computation, which only need O(log Npi) or O(log Npj) communications per CPU, instead of O(log Np). For example, the FFT and non-local pseudopotential tasks only need to use global operations that span over Npi, while the orthogonality step can be broken down into a series of alternating global operations that span over either Npi or Npj, e.g. like the SUMMA algorithm [36].

The overall performance of our AIMD simulations was found to improve considerably using this new approach. With the optimal processor geometries, the running time per step took 2,699 seconds (45 minutes) for 1 CPUs down to 3.7 seconds with a 70% parallel efficiency on 1024 CPUs. The fastest running time found was 1.8 seconds with 36% parallel efficiency on 4096 CPUs. As shown in figure 6, these timings were found to be very sensitive to the layout of the two-dimensional processor geometry. For 256, 512, 1024, and 2048 CPUs, the optimal processor geometries were 64x4, 64x8, 128x8 and 128x16 processor grids, respectively. The timings of the FFTs, non-local pseudopotential, and orthogonality are also shown in figure 6. Not every part the program scaled perfectly. The parallel efficiency of several other key operations depends strongly on the shape of the processor geometry. It was found that distributing the processors over the orbitals significantly improved the efficiency of the FFTs and the non-local pseudopotential, while distributing the processors over the spatial dimensions favored the orthogonality computations.

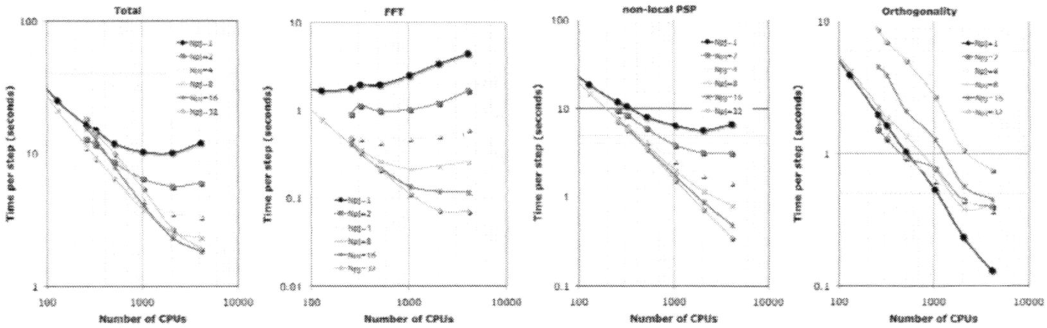

Figure 6. Overall and component timings in seconds for UO_2^{2+}+122H_2O plane-wave DFT simulations at various processor sizes (Np) and processor grids (Npj, Npi=Np/Npj). Timings were determined from calculations on the Franklin Cray-XT4 computer system at NERSC.

4. Parallelization of AIMD with exact exchange

The 2d processor geometry method can also be used to parallelize the computation of the exact exchange operator. This operator has a cost of $O(N_e^2 * N_g * \log(N_g))$, and when it is included in a plane-wave DFT calculation it is by far the most demanding term. The exchange term is well suited for this method, whereas if only the spatial or orbital dimensions are distributed then the exchange term does not scale well. When only the spatial dimensions are distributed, each of the $N_e(N_e+1)$ 3d FFTs will be computed one at a time on the entire machine. The drawback of this approach is that it significantly underutilizes the resources, since parallel efficiency of a single 3d FFT is effectively bounded to $\sim N_g^{1/3}$ processors. When only the orbital dimensions are distributed the parallelization is realized by multicasting the $O(N_e)$ orbitals to set up the $O(N_e^2)$ wave-function pairs. This multicast is followed by a multi-reduction which reverses the pattern. We note that with this type of algorithm one could readily fill a very large parallel machine by assigning each a few FFTs to each processor. However, in order to obtain reasonable performance from this algorithm it is vital to mask latency, since the interconnects between the processors will be flooded with $O(N_e)$ streams, each on long messages comprising N_g floating point words of data. In which both the spatial and orbital dimensions are distributed we only need to compute parallel three-dimension FFTs along the processor grid columns, as well as broadcast the orbitals along the processor grid rows. Compared with a multicast across all processors the benefit of this approach is to reduce latency costs, since broadcasting is done across the rows of the two-dimensional processor grid only. The basic steps involved in calculating the exchange operator with this distribution are given in the following algorithm.

Basic parallel algorithm for calculating exact exchange in a plane-wave basis using a 2d processor geometry

Input: ψ - (N_g/Npi) x (N_e/Npj) *array*
Output: $K\psi$ -(N_g/Npi) x (N_e/Npj) *array*
Work Space: Ψ – (N_g/Npi) x N_e *array*
$K\Psi$ – (N_g/Npi) x N_e *array*
N_e = *total number of orbitals*
*Np = total number of processors, where Np=Npi*Npj*
Npi/ Npj = size of column (row) processor group
taskid_i/j = rank along the column (row) processor group

```
Ψ(:,:), KΨ (:,:) = 0
s=taskid_j*(Ne/Npj); e = s+(Ne/Npj);Ψ(:,s:e) = ψ
Row_Reduction(Ψ)
counter = 0
for m=1,Ne
  for n=1,m
     if counter==taskid_j then
       ρ(:) ← Column_FFT_rc( (Ψ(:,m) *. Ψ(:,n)) )
       V(:) ← Column_FFT_cr( (f_screened(:) *. ρ(:)) )
       KΨ(:,m ) -=V(:)*.Ψ(:,n)
      if m != n then KΨ(:,n) -=V(:)*.Ψ(:,m) end if
   end if
   counter = mod(counter + 1,Npj)
end for
end for
Row_Reduction(KΨ)
Kψ=HΨ(:,s:e)
```

In this algorithm, the routines **Column_FFT_rc**, and **Column_FFT_cr** are the forward and reverse parallel three-dimensional Fast Fourier Transforms (FFTs) along the processor columns,

Journal of Physics: Conference Series **180** (2009) 012028 doi:10.1088/1742-6596/180/1/012028

Row_Reduction is a parallel reduction along the processor rows, and $f_{screened}$ is the cutoff of Coulomb kernel, given by the Fourier transform of the following real-space function,

$$f(r) = \frac{1-\left(1-\exp\left\{-(r/R)^N\right\}\right)^N}{r} \tag{6}$$

where R and N are adjustable parameters [24, 31, 37]. This algorithm is very simple to implement, and it is perfectly load balanced since each CPU only computes $(N_e*(N_e+1)/N_p)$ three-dimension FFTs. However, this simple algorithm can be improved. One problem with it is that it uses a lot of workspace. Another is that each CPU in the **Row_Reduction** subroutine receives and sends a total of $(Npj-1)*(N_g/Npi)*(N_e/Npj)\sim=N_gN_e/Npi$ amount of data, which is approximately twice as much as is necessary to compute all pairs of wave functions. We have recently developed a slightly more ambitious parallel algorithm for calculating the exchange operator, which halves the workspace and communication overhead, while maintaining a good load balance. The key idea behind this algorithm is that a specially designed broadcast and reduce routines are called that basically implement a standard asynchronous radix-2 butterfly diagram except that instead of transferring a $(Npj/2)N_gN_e/N_p$ chunk of data at the last step, they transfer only a $(N_g/Npi)(N_e/Npj)(1+(Npj-2^{**}Floor(Log_2(Npj))))\sim=N_gN_e/N_p$ chunk data.

The overall best timings for hybrid-DFT calculations of an 80 atom supercell of hematite (Fe_2O_3) with an FFT grid of $N_g=72^3$ ($N_e^{up}=272$, $N_e^{down}=272$), and a 160 atom supercell of hematite (Fe_2O_3) with an FFT grid of $N_g=144\text{x}72\text{x}72$ ($N_e^{up}=544$ and $N_e^{down}=544$) (wavefunction cutoff energy=100Ry and density cutoff energy=200Ry) and orbital occupations of $N_e^{up}=272$ and $N_e^{down}=272$ are shown in figure 7. The overall best timing per step found for the 80 atom supercell was 3.6 seconds on 9792 CPUs, and for the 160 atom supercell of hematite was 17.7 seconds on 23,936 CPUs. The timings results are somewhat uneven, since limited numbers of processor grids were tried at each processor size. However, even with this limited amount of sampling, these calculations were found to have speedups to at least 25,000 CPUs. We expect that further improvements will be obtained by trying more processor geometry layouts.

The time to compute the exchange operator using can be broken up into two parts, the time to compute the $N_e(N_e+1)$ three-dimensional FFTs (t_{fft}) and the time to perform the data transfer ($t_{butterfly}$) in the **Brdcst_Step** and **Reduce_Step** subroutines, i.e.

$$t_{exchange} = t_{fft} + t_{butterfly} \tag{7}$$

The three-dimensional parallel FFT in the NWChem program packages uses a two-dimensional Hilbert decomposition of the three-dimensional FFT grid [5]. For this type of FFT, the time to compute one three-dimensional FFT can be approximated by summing up three terms corresponding to serial computational, data transfer, and latency times.

$$t_{fft}^{1} = \frac{a*N_g Log_2(N_g)}{Npi} + \frac{2\gamma N_g}{Npi} + 2\theta\sqrt{Npi} \tag{8}$$

where N_g is the size of the FFT grid, Npi is the number of rows in the two-dimensional processor grid and a, θ, γ are adjustable parameters corresponding to the rate to compute one 3d FFT in serial, θ is the

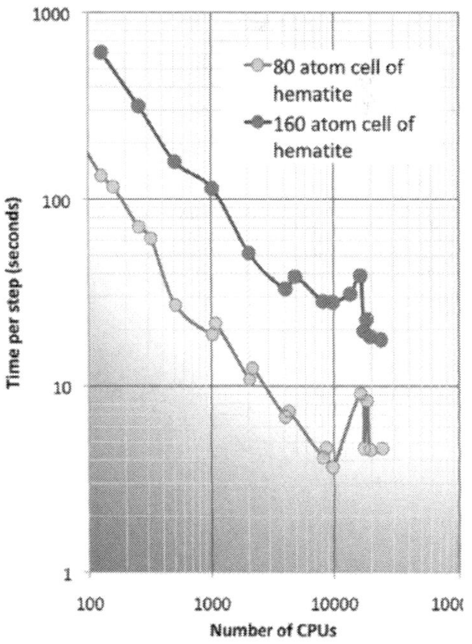

Figure 7. The overall fastest timings taken for an 80 and 160 atom Fe_2O_3 hybrid-DFT energy calculations. Timings from calculations on the Franklin Cray-XT4 computer system at NERSC.

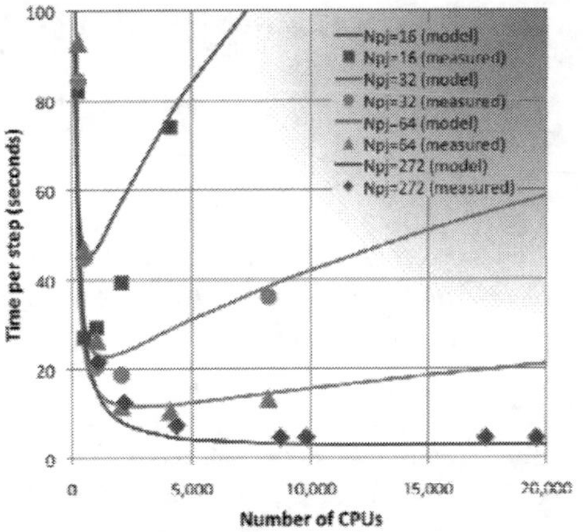

Figure 8. Comparison of performance model with measured timings for 80 atom Fe_2O_3 hybrid DFT calculation (a=*4e-9*, θ=*2.5e-4*, and γ=*5e-8*).

effective latency to start a message, and γ is the effective rate to transfer a message.

For each incomplete butterfly broadcast and reduction, the bandwidth part of the data transfer per CPU will be proportional to $(Npj/2)*(N_g/Npi)*(N_e/Npj)$ or $(1/2)*(N_eN_g/Npi)$, and the number of messages per CPU will be proportional $Log_2(Npj)$. Thus the overall timing for performing the data transfer can be written as

$$t_{butterfly} = 4\gamma(1/2)\frac{N_g N_e}{Npi} + 4\theta Log_2(Npj) \quad (9)$$

The extra factor of 4 is due to the fact that both $Log_2(Npj)$ broadcast and reductions steps are performed, and each broadcast and reduction step contains a send and a receive call.

By combining equations (7-9), we arrive at the following equation for the time to compute the exchange operator.

$$\begin{aligned} t_{exchange} = {} & \frac{N_e(N_e+1)}{Npj} t^1_{fft} \\ & + 4\gamma(1/2)\frac{N_g N_e}{Npi} \\ & + 4\theta Log_2(Npj) \end{aligned} \quad (10)$$

From an approximate fitting of the model to the 80 Fe_2O_3 simulation atom simulation, the parameters were found to be a=*4e-9*, θ=2.5e-4 and γ=5e-8 (see figure 8). As shown in figure 4, the fastest execution time will be obtained when Npj=Ne. From this relation the maximum number of CPUs can be readily estimated by solving the following equation

$$\left.\frac{dt_{exchange}}{dNpi}\right|_{Npj=Ne} = 0 \quad (11)$$

for Npi. The solution to equation (11) is then

$$\max Npi = \left(\frac{2A}{B}\right)^{2/3} \quad (12)$$

where

$$A = a(N_e+1)N_g \log_2(N_g) + 2\gamma N_g N_e\left(1 + \frac{N_e+1}{N_e}\right) \quad (13)$$

$$B = 2(N_e+1)\theta \quad (14)$$

Using this formula we find that the maximum number of CPUs that can be gainfully used in the 80 and 160 atom Fe_2O_3 Hybrid-DFT calculations to be 14,977 CPUs and 48,032 CPUs respectively and we plan to validate these results in the future.

5. Conclusion

An overview of the parallel algorithms used for AIMD in the NWChem program package is presented, including recent developments for computing exact exchange. These algorithms were based on using a two dimensional processor geometry, which allowed us to overcome the inefficiencies associated with using parallel three-dimensional FFTs. A unique aspect of our implementation of exact exchange is the development of an incomplete butterfly that halves the amount of data communicated, as well as making judicious use of data replication, in the exact exchange compared to a standard multicast algorithm.

The overall performances of our AIMD calculations were found to be fairly reasonable. For a UO_2^{2+}+122H_2O AIMD simulation the running time per step decreased from 2,699 seconds (45 minutes) for 1 CPUs down to 3.7 seconds with a 70% parallel efficiency on 1024 CPUs. The fastest running time found was 1.8 seconds with 36% parallel efficiency on 4096 CPUs. For an 80 atom Fe_2O_3 hybrid- DFT AIMD simulation, the overall parallel efficiency to 1024 processors was 99% decreasing to 53% by 9792 processors. Beyond 10,000 processors no speedups were observed for this calculation, and larger systems are required if we are to use higher levels of parallelism. Limited simulations for a larger 160 atom Fe_2O_3 hybrid-DFT AIMD simulations were also carried out. These calculations did not produce as good parallel efficiencies as the 80 atom simulations, however these calculations were found to have speedups to at least 25,000 CPUs. We believe the communications overheads are still an issue, and we are exploring latency hiding techniques via run time substrates that implement a dataflow execution model [38-40].

Significant progress has been made in terms of accuracy, efficiency, and scalability of AIMD methods in recent years. However the algorithms and implementations of these methods need to be constantly upgraded to capture the performance of emerging massively parallel computers. The projected size of the next generation supercomputers are very large (10^6-10^7 threads of computation) suggesting that current limitations in simulation times, particle sizes and levels of theory will be overcome in the coming years by brute force increases in computer size.

6. Acknowledgements

This research was partially supported by the ASCR Multiscale Mathematics program, the BES Geosciences program, and the BES Heavy element program of the U.S. Department of Energy, Office of Science ~ DE-AC06-76RLO 1830. The Pacific Northwest National Laboratory is operated by Battelle Memorial Institute. We wish to thank the Scientific Computing Staff, Office of Energy Research, and the U. S. Department of Energy for a grant of computer time at the National Energy Research Scientific Computing Center (Berkeley, CA). Some of the calculations were performed on the Chinook computing systems at the Molecular Science Computing Facility in the William R. Wiley Environmental Molecular Sciences Laboratory (EMSL) at PNNL. EMSL operations are supported by the DOE's Office of Biological and Environmental Research. EJB would like to acknowledge partial support for the development of tera/petascal parallel algorithms and the writing of this manuscript from the Extreme Scale Computing Initiative, a Laboratory Directed Research and Development Program at Pacific Northwest National Laboratory. SBB and JHW also acknowledge support by the ASCR Multiscale Mathematics program of the U.S. Department of Energy, Office of Science ~ DE-FG02-05ER25707.

References

[1] P. G. Allen *et al.*, Inorganic Chemistry **36**, 4676 (1997).
[2] P. Nichols *et al.*, Journal of Chemical Physics **128**, 124507 (2008).
[3] R. Car, and M. Parrinello, Phys. Rev. Lett. **55**, 2471 (1985).
[4] M. C. Payne *et al.*, Reviews of Modern Physics **64**, 1045 (1992).
[5] D. K. Remler, and P. A. Madden, Mol. Phys. **70**, 921 (1990).
[6] G. Kresse, and J. Furthmuller, Computational Materials Science **6**, 15 (1996).
[7] D. Marx, and J. Hutter, in *Modern Methods and Algorithms of Quantum Chemistry*, edited by J. Grotendorst (Forschungszentrum, Julich, Germany, 2000), pp. 301.
[8] M. Valiev *et al.*, in *Reviews In Modern Quantum Chemistry: A Celebration Of The Contributions Of R. G. Parr*, edited by K. D. Sen (World Scientific, Singapore, 2002).
[9] J. F. Ahearne, Physics Today **50**, 24 (1997).

[10] R. L. Garwin, Interdisciplinary Science Reviews **26**, 265 (2001).
[11] R. Ewing, Elements **2**, 331 (2006).
[12] J. Bruno, and R. Ewing, Elements **2**, 343 (2007).
[13] P. C. Burns, and A. L. Klingensmith, Elements **2**, 351 (2007).
[14] B. Gambow, Elements **2**, 357 (2006).
[15] G. R. Lumpkin, Elements **2**, 365 (2006).
[16] J. Wiggs, and H. Jonsson, Computer Physics Communications **87**, 319 (1995).
[17] J. Wiggs, and H. Jonsson, Computer Physics Communications **81**, 1 (1994).
[18] J. S. Nelson, S. J. Plimpton, and M. P. Sears, Phys. Rev. B **47**, 1765 (1993).
[19] R. A. Kendall *et al.*, Computer Physics Communications **128**, 260 (2000).
[20] E. J. Bylaska *et al.*, Comput. Phys. Comm. **143**, 11 (2002).
[21] T. L. Windus *et al.*, Computational Science - Iccs 2003, Pt Iv, Proceedings **2660**, 168 (2003).
[22] F. Gygi *et al.*, in SC '06: Proceedings of the 2006 ACM/IEEE conference on Supercomputing2006).
[23] C. Adamo, and V. Barone, J. Chem. Phys. **110**, 6158 (1997).
[24] J. C. Du *et al.*, Nuclear Instruments & Methods in Physics Research Section B-Beam Interactions with Materials and Atoms **255**, 188 (2007).
[25] M. Lundberg, and P. E. M. Siegbahn, Journal of Chemical Physics **122** (2005).
[26] M. Marsman *et al.*, Journal of Physics-Condensed Matter **20** (2008).
[27] M. Stadele *et al.*, Physical Review Letters **79**, 2089 (1997).
[28] M. Stadele *et al.*, Physical Review B **59**, 10031 (1999).
[29] F. Cinquini *et al.*, Physical Review B **74**, 165403 (2006).
[30] E. J. Bylaska *et al.*, NWChem, A Computational Chemistry Package for Parallel Computers, Version 5.1.1 (Pacific Northwest National Laboratory, Richland, Washington 99352-0999, USA, 2009).
[31] E. J. Bylaska *et al.*, Parallel Implementation of Pseudopotential Plane-Wave DFT with Exact Exchange, Journal of Computational Chemistry, (submitted) (2009).
[32] R. G. Parr, and W. Yang, *Density-functional theory of atoms and molecules* (Oxford University Press ; Clarendon Press, New York; Oxford [England], 1989).
[33] L. J. Clarke, I. Stich, and M. C. Payne, Comput. Phys. Comm. **72**, 14 (1992).
[34] E. J. Bylaska *et al.*, Computer Physics Communications **143**, 11 (2002).
[35] T. P. Hamilton, and P. Pulay, Journal of Chemical Physics **84**, 5728 (1986).
[36] van de Geign, Robert, and J. Watts, Concurrency: Practice and Experience **9**, 255 (1997).
[37] E. J. Bylaska, K. Tsemekhman, and F. Gao, Physica Scripta **T124**, 86 (2006).
[38] P. Cicotti, and S. B. Baden, in Proc. 15th IEEE International Symposium on High Performance Distributed ComputingParis, France, 2006).
[39] J. Sorensen, and S. B. Baden, in Proc. 12 SIAM Conf. Parallel Processing for Scientific ComputingSan Francisco, CA, 2006).
[40] J. Sorensen, and S. B. Baden, in *International Conference on Computational Science 2009 (ICCS 2009)*Baton Rouge, LA, 2009), p. (to appear).

Use of Bayesian inference and data to improve simulations of multi-physics climate phenomena

Charles S Jackson

Institute for Geophysics, Jackson School of Geosciences, The University of Texas at Austin

charles@ig.utexas.edu

Abstract. Bayesian inference provides a means incorporate activities usually associated with post-model development evaluation into the process of model development. Presented is a review of the factors that make this approach challenging, strategies for making the process practical for model development of complex multi-physics phenomena, and suggestions on areas requiring additional research. An analysis is presented of the strategy that was used to determine the joint probability for six non-linearly related parameters important to clouds and convection within the NCAR Community Atmosphere Model version 3.1.

1. Introduction

Research activity concerning climate phenomena, whether observational, theoretical, or simulation is directed at the goal of understanding the physical processes or balance of processes that govern nature. The challenge is that observations are limited, theories tend to address idealizations, and simulations can mimic nature for the wrong reasons. Climate model construction involves the independent development of component models, each of which seem to be ok when forced by observations, but yet can go astray when coupled to each other (e.g. [19]). The experience is much like a fishing expedition in which there is much preparation in setting up the gear, location, and timing, but with very little by way of a guarantee on the success of the catch.

Presented here are the concepts, strategies, and an example application of an alternate, Bayesian inference approach to synthesizing data constraints on choices in climate model development. It has its own set of challenges, but may ultimately provide a more robust avenue to how we use data to inform model development.

2. Use of Bayesian inference in model development

Bayesian inference can be useful for model development by synthesizing information from different sources of data, modeling results, and corresponding error information to inform difficult choices in model development such as the values assigned to multiple, non-linearly related parameters. Bayesian inference involves the stochastic sampling of sources of uncertainty, each sample of which requires a model integration to estimate the sensitivity to the input parameters, which may make it seem impractical for model development. Currently assessments of modeling uncertainty occur after climate model development is complete and is revealed through model inter-comparison projects or specially designed sensitivity experiments that tease out aspects of model predictions that are more robust (e.g. [15], [18], [23]). The problem has been that these activities mainly concern the forward propagation of errors. But unlike Bayesian inference, these more informal approaches to quantifying uncertainties are not well suited to the task of identifying the sources of uncertainty or providing guidance on choices in model development.

The best opportunity for scientific understanding to enter into the process of model development, within the Bayesian framework, is in the care that is given to the definition of the metric that quantifies model skill. Many different kinds of metrics are being developed in the context of summarizing differences that arise in multi-model projections of climate change (e.g. [3]). Another common example of the use of metrics is to provide weights to multi-model ensemble mean predictions and uncertainty (e.g. [24], [22]). The metrics that have been proposed so far do not have model development in mind. Rather they usually only quantify univariate mean statistics. We need better ways to quantify relationships if we are going to impose this goal in model development.

The other challenge facing model developers is that the volume of solution space, the space where there is a good fit to observational data, is likely a small fraction of the total volume of possibilities. This makes Bayesian inference much more similar to problems in geophysical inversion. The stochastic sampling methods developed for Bayesian inference such as Markov Chain Monte Carlo (MCMC) may not perform well if they cannot focus sampling in an efficient way. However, if we treat complex multi-physics systems as a typical inverse problem, we will only be getting a single solution for a problem where many solutions exist. Least-squares inverse, Maximum *a posteriori* (MAP), or gradient-based strategies to solving inverse problems target point estimates of the most likely model parameters, rather than full probability densities. This singular objective may be overly sensitive to assumptions concerning the value of collected data or a particular modeled phenomenon. Since, climate phenomena represent a spectrum of relationships, model development should take this into account. What maybe more helpful to the scientific aims of developing a multi-physics model is a Bayesian approach that provides a way to obtain more information about the complete statistical description of the uncertainties inherit in the data and modeling choices.

Suppose we are interested in assessing the joint probability for selecting a vector **m** of model parameter values such that the climate model g(**m**) is consistent with observations $\mathbf{d}_{obs}$. Assuming Gaussian errors, the solution posterior probability density (PPD), as it is called in Bayesian nomenclature, would be given as

$$PPD(\mathbf{m} \mid \mathbf{d}_{obs}, g(\mathbf{m})) \propto \exp\left[-\frac{1}{2}\left(g(\mathbf{m})-\mathbf{d}_{obs}\right)^T \mathbf{C}_{noise}^{-1}\left(g(\mathbf{m})-\mathbf{d}_{obs}\right)\right] \cdot prior(\mathbf{m}) \tag{1}$$

with $\mathbf{C}_{noise}^{-1}$ as an expression of the uncertainty in comparing model output to data and prior(**m**) represents prior probability for model parameters m independent of dobs. The quantity within the square brackets is typically referred to the "cost" function or metric of errors in the model relative to what is observed. Even with the Gaussian assumption for observational and modeling uncertainties, the PPD is generally not Gaussian through the nonlinearities represented in the climate model g(**m**). Because there is generally no closed-form solution to equation (1), it needs to be estimated.

It may be helpful to draw a connection between equation (1), statistical physics, and the solution that Metropolis et al. [16] proposed for estimating its solution in order to highlight the nature of the problem that is being solved. In particular, the problem is analogous to estimating the probability that a physical system have a particular property *s* (corresponding to the choices of parameter values **m**) under the constraint that the system has, on sum, an energy E (corresponding the cost function that quantities modeling errors). This probability is composed of the number of states *N*(s) available to the system with property s, out of all states of the system *Z* and is related to entropy by $\sigma = \ln(N(s))$. In order to estimate $N(s)/Z$ numerically, it is sufficient to select candidate configurations in proportion to $N(s)/Z$. Metropolis et al. [16] proposed a relatively simple procedure achieving this that has proven invaluable for simulating complex physical systems on a computer. They proved that the distribution described by $N(s)/Z$ could be approximated iteratively through by randomly stepping through configuration space with a given step size and with each step applying a set of acceptance and rejection criteria for each step. Specifically, if a step through state space reduces a system's energy then that step will always be accepted. If a step increases a system's energy then that step will only be accepted in proportion to

$$P = \exp\left(-\frac{\Delta E}{T}\right), \tag{2}$$

where P is the acceptance probability, $\Delta E = (E_2 - E_1)$ is the change in energy going from energy state E_1 to E_2, and T is the system's temperature. This transition probability may be re-expressed in terms of N(s) using the definition for temperature T

$$\frac{1}{T} = k_b\left(\frac{\partial \sigma}{\partial E}\right) \tag{3}$$

where k_b is the Boltzman constant. In discrete form

$$\frac{\Delta E}{k_b T} = \Delta\sigma . \tag{4}$$

Substituting eq. (4) into eq. (2) and redefining the temperature in eq. (2) to include k_b one finds that this transition probability is

$$P = \exp(-\Delta\sigma) = \exp(-(\sigma_2 - \sigma_1)) = \frac{N(s1)}{N(s2)} \tag{5}$$

That is, the transition probability is given by ratio of the number of states available to the system with property s1 to the number of states available to the system with property s2. So while it is very difficult to measure $N(s)/Z$ directly, the above derivation shows how one may use statistical sampling and calculations of a system's energy to derive estimates of $N(s)/Z$. This sampling strategy is an early example of a MCMC. If the solution space is well behaved, the sampling procedure is ergodic (it can visit all possible configurations). However the procedure can be sensitive to details such as the average size of the step that is taken through solution space. The following section reviews different strategies to estimating equation (1) and the various advantages or disadvantages for its applicability to climate model development.

3. Sampling strategies appropriate for climate model development

Among the challenges to use Bayesian inference in the development of a climate model are (1) the computational expense for testing choices, (2) the potentially complex structure of the response surface that describes how a cost function changes as a function of sources of uncertainty, (3) the large number of uncertain parameters, or (4) that the volume of parameter space in which viable solutions reside is may be small. Climate model development is also primarily interested in quickly defining a single working model (this does not negate the interest in finding many working models!). Therefore any method that is of practical use must make inferences with relatively few modeling experiments. These objectives and constraints do not map well onto the spectrum of strategies that have been developed for Bayesian inference, which typically assume an infinite capacity to obtain samples. Discussed below are several approaches to mitigating these challenges that, when combined with high performance computing (HPC) resources, may prove sufficiently practical for climate model development.

3.1. Multiple Very Fast Simulated Annealing (MVFSA)

MVFSA balances a number of the objectives described above and does so adaptively such that it can provide efficient estimates of complex response surfaces that may reside in small volumes of solution space [12]. The strategy uses the rules of the Metropolis sampler to statistically seek out regions of parameter space where the cost function is minimized. However over the course of the sampling, the average step size is reduced and centered on the last accepted model configuration with a Cauchy distribution the width of which is determined by the experiment number [9]. Also over time, the cost function is scaled to exaggerate the response surface through an algorithm control parameter analogous to temperature in equation (2). Any single Markov Chain is not guaranteed to find the global minimum. Rather through multiple convergence attempts, one may increase the probability of identifying regions of parameter space that allow the model to track the observational constraints. This approach works well for HPC resources because it allows one to distribute the workload across the multiple chains. Although Simulated Annealing approaches have the connotation of a technique that is useful for global optimization, not uncertainty quantification, the MVFSA strategy outlined in [12] is deployed so that it makes use of these optimization features to enhance the efficiency of estimating the PPD.

Among the shortcomings of this approach is that this strategy does not have ergodic properties. The PPD tends to have long tails from the fact that sampling starts from a random position in parameter space and may require a number of iterations before finding regions of interest [26]. This may be considered a feature for strategies that depend on statistical emulation (discussed below).

3.2. Adaptive Metropolis (AM)

A class of MCMC strategies have been developed which can adapt to solution characteristics, making them more robust and much more efficient when compared to the Metropolis/Gibbs sampler [4]. These AM methods work by making better use of samples taken to update the proposal distribution for new samples. The single component AM (SCAM) [5] and the delayed rejection AM [25] has shown sampling efficiencies comparable to MVFSA with the advantage that these approaches have been shown to be ergodic [26]. One of the surprising results of the tests of the AM strategies is the large sample autocorrelations (a measure of sample mixing) for the full component AM which would make this choice non-competitive relative to other options. Another issue that needs to be explored further is the trouble many sampling strategies have, particularly MCMC, in sampling multi-modal distributions. It would be good to know if we should expect multi-modal PPDs for the climate problem.

3.3. Statistical emulation

The point of statistical emulation is to minimize the number of expensive physics-based modeling experiments that would be required to estimate uncertainties by predicting what the output would be at untried parameter settings [14] [7]. However there has only been limited experience in developing statistical emulators of climate, almost exclusively focusing on univariate approaches with a few variables (e.g. [18], [20]). Thus as interest grows to construct more sophisticated metrics of model output that take into account relationships in space, time, and across fields such as in Smith et al. [22], statistical emulation will need to similarly evolve to model these relationships as well.

4. Application to the Community Atmosphere Model

The objective here is to evaluate the utility of MVFSA for model development based on the results of Jackson et al. [11] whose primary objective was to estimate a joint PPD for six parameters important to clouds and convection within the NCAR Community Atmosphere Model version 3.1 (CAM3.1) [1]. In particular, how many experiments are required to get useful results? How might the sampling process affect the distribution of errors for individual fields? What can we infer from the results about how appropriate MVFSA is for quantifying uncertainties and as a tool for model development? The Jackson et al. [11] exercise included over 800 11-year long model integrations from 1990 to 2001 in which MVFSA selected different combinations of parameter values and compared the model output to observations. Six independent chains were run in parallel with each experiment using 64 processors of a Intel Xeon-based linux cluster with high-speed InfiniBand interconnect. With each experiment requiring about 10 hours, all experiments were completed in two months.

An analysis of the computational costs of MVFSA for estimating the PPD was previously completed with a surrogate climate system model [12]. This analysis has been extremely useful in designing new experiments and predicting the computational requirements for a wide range of

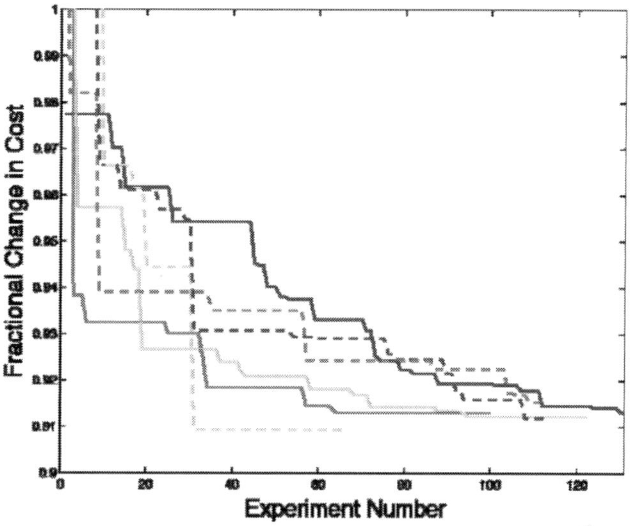

Figure 1. Fractional change in cost relative to the stand model configuration of the best performing model configuration as a function of experiment number. Shown is the results of six sampling chains from [11].

problems including systems involving 50 parameters (e.g. [13], [27]). That analysis predicts that one would need ~3000 experiments to estimate statistically stable estimates of 15-bin marginals of the PPD, thus the 800 experiments provides only a very rough estimate of these probabilities.

The cost function considered in [11] is much more typical of what we may find in model development than the version in [12] which only considers a perfect model analysis of temperature. The [11] cost function quantifies model discrepancies with instrumental, satellite, and reanalysis data including radiative balances at the top and bottom of the atmosphere, precipitation, winds, humidity, sea level pressure, and clouds. The cost function did not represent relationships among these fields. However there was an attempt to normalize the cost function by the effects of internal model variability on individual fields. The correlations among errors among multiple fields were accounted for through a renormalization process through the choice of a scaling factor 'S' [17]. Jackson et al. [11] took a more Bayesian approach to the section of 'S' by treating its selection as being condition on the choice of the other model parameters and allowing it to scale inversely by the mean bias that exists between models and predictions. The point of this step was to introduce some additional adaptivity in the acceptance/rejection of model configurations with the logic that modelers need to be more accepting of non-optimal configurations when there is significant biases. A full discussion of this feature is beyond the objectives of the present analysis. However the treatment of 'S' illustrates some of the flexibility that Bayesian inference can provide to this type of analysis that incorporate the scientific sensibilities that would be helpful to model development.

4.1. Number of experiments needed

If the goal is to identify a single model configuration, one must perform multiple sampling chains and evaluate the configuration with the lowest cost value. Figure 1 shows the path to convergence among the six chains performed in [11]. Shown is the fractional change in the cost function relative to the standard configuration of cam3.1. Any value less than 1 indicates a model configuration whose errors across all fields in the cost function sum to a number less than cost of the standard model. That does not imply that the model is better scientifically. Any of the selected model configurations may include large compensating errors. An evaluation of this point will be considered in the following section. Each sampling chain indicated in figure 1 shows the lowest cost value achieved up to a given experiment number. Six out of six chains were able to identify a model that was better than the default model configuration by experiment number 10. There were exponential reductions in the cost function of the best performing configurations, with an approximate e-folding scale of 40 experiments, a similar rate to what was found in [12]. The overall reduction in cost values amounted to 9% which was similar to that found in the perturbed physics ensemble of [23] derived from a 2578 Monte Carlo (random) sampling among six uncertain model parameters of the Hadley Center atmospheric climate model. The analysis of the model configurations after 41 experiments within each line is given by [11]. The present analysis shows that there was an additional 3% reduction in the cost function that occurred in subsequent sampling such that all models had nearly identical reductions in cost by experiment 120. Presumably one could run more lines in parallel an increase the likelihood that an optimal configuration could be found by experiment 20. There will be some additional analysis of this last point in a following section. Again, these numbers are appropriate for a six-parameter problem, but are very much in line with expectation established by [12] that the number of experiments required to get 66% of the way to convergence is by ~6 times the number of uncertain parameters. The same analysis also shows that a six-parameter problem would need ~125 experiments to fully converge, which is also in line with the present analysis.

4.2. Distribution of component errors in optimization solution

The question that is addressed here is whether the evolution of errors favors particular fields over others on the path to optimization. The answer to this question may affect the number of experiments that are required to obtain satisfactory results. If, for instance, a particular field dominates the cost function, one may expect, as I did, that progress on reducing errors among the "less important" fields would be delayed. Figure 2 shows the inter-chain average cost among the optimal configurations as a function of experiment number. Figure 2 shows that the above expectation is wrong. From the fields that dominate the cost function in panel (a) to the minor actors in panel (c), there is a similar evolution to the change in errors (with a value of zero for the standard model configuration). The reason for this is likely because these fields are not independent from one

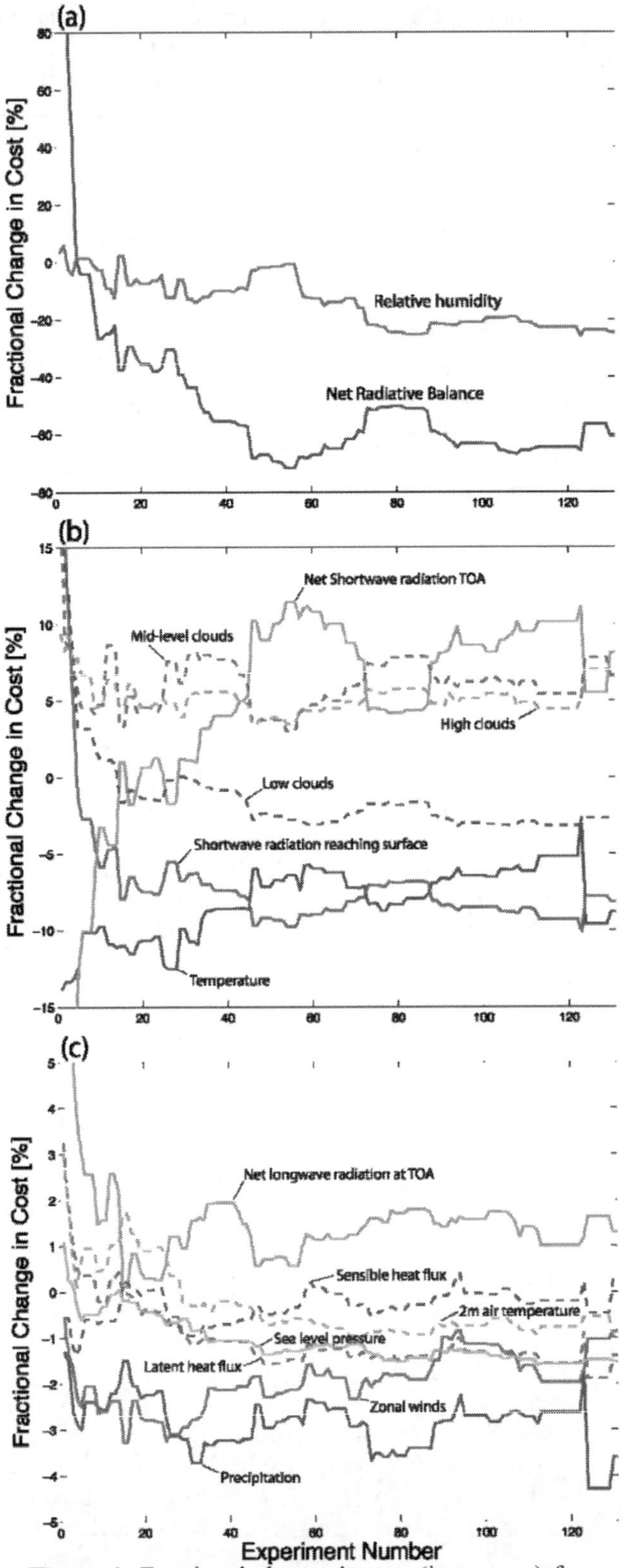

Figure 2. Fractional change in cost (in percent) for each component of the cost function shown in figure 1.

another. They are all physically linked, such that a reduction in errors in one field should give rise to a corresponding reduction of errors among related fields, if the model is an adequate description of the system. This appears to fail with the model depiction of clouds. In particular the improvements that occurred with surface variables, came at the expense of top of the atmosphere shortwave radiative fluxes and clouds.

After approximately 40 experiments, the errors that occur across many of the fields appear to evolve randomly with improvements and correspondingly sized degradations. Thus, the overall march to convergence that is apparent in figure 1 is subtle feature when viewed component by component. It may be important to realize that any of these model configurations are acceptable and the jumping around represents a fundamental uncertainty in competing choices of how to balance the multiple objectives.

4.3. Characteristics of the response surface

Presented next is two-dimensional projections of the response surface estimated as a weighted average over all accepted experiments defined as models with equal or better cost metric to the default model. The inference is binned into smaller paired subregions of parameter space p1 and p2. All experiments that fell into that subregion contribute to the weighted average through

$$\overline{\mathrm{cost}(p1,p2)} = \frac{\sum_{i=1}^{\mathrm{nexp}} \mathrm{cost}(p1,p2)_i \exp\left[-\overline{S}\cdot\left(\mathrm{cost}(p1,p2)_i - \mathrm{cost}(p1,p2)_{\min}\right)\right]}{\sum_{i=1}^{\mathrm{nexp}} \exp\left[-\overline{S}\cdot\left(\mathrm{cost}(p1,p2)_i - \mathrm{cost}(p1,p2)_{\min}\right)\right]} \qquad (6)$$

where $\mathrm{cost}_{\min}$ refers to the minimum value of the cost function over all sub-regions of parameter space, and S, as previously discussed, is determined by Bayesian inference to be the range in cost values that are deemed acceptable. As it turns out, S selects a similar range in cost values than from the $\mathrm{cost}_{\min}$ to the cost of the standard model. Thus the decision to filter experiments by only those models that have a cost value equal or less than the default model does not appreciably affect the results. Figure 3 shows the inferences of the response surface for the six model parameters which are initial cloud downdraft mass flux (alfa) and consumption rate of CAPE (tau) in panel (a) which are parameters important to convection, environmental air entrainment rate (ke) and precipitation efficiency (c0) which are important to cloud micro-physics, and low and high cloud critical relative humidity (rhminl and rhminh) which are important to cloud macro-physics. In each panel, the lowest cost values mostly appear to reside in a particular region of parameter space (the distribution does not appear to be multi-modal). The response surface becomes much more varied away from these favored regions, however the number of samples (indicated by the white and black '+' marks of the accepted and rejected configurations, respectively) become less dense precisely where the colors are more varied. Thus there are not likely enough samples to make robust inferences of the response surface.

Also presented in figure 3 are the optimal parameter choices that were selected by experiment 20. There can be considerable variation parameter values selected, however the location where the cost value is smallest also tends to be close to global optimized model configuration over all lines and experiments. Given the large variations in the cost values for each of the model components illustrated in figure 2 shows how important even small variations in parameter values have on modeling results.

4.4. Volume of solution space

The models that were deemed acceptable fill a 2% volume of the total six-dimensional volume. This is remarkably small and underscores the need to have sampling strategies that can aggressively find and sample these regions densely. This estimate of the total volume is only approximate and is dependent on how one infers regions of acceptability around the samples that have been completed. The 2% estimate comes from an accounting of the fraction of squares that have been deemed acceptable in 25 by 25 binned segments from the parameter minimum to maximum. Note that this region includes a broader range than indicated by figure 3, which focuses only on regions that contained at least a single accepted experiment. The probability is then the product of the fraction for the three pairs of variables indicted in figure 3 (0.35 for panel (a) times 0.31 for panel (b) times 0.2 for panel (c)). The estimate of the volume goes up to 8% if one subdivides parameter space into 15 bins for each parameter. The hit rate for MVFSA for sampling within the acceptable region is 70%, with a slightly lower percentage at the beginning.

5. Discussion and conclusion

The merit of using Bayesian inference in the development of numerical models of multi-physics systems such as a climate model is that it formalizes the process of model evaluation and uncertainty quantification for evaluating choices in model development. With so many degrees of freedom, such models can reproduce realistic behaviors for the wrong reasons. With Bayesian inference, the scientist can define meaningful metrics and make use of innovative sampling strategies for selecting model configurations consistent with data. The result is an ensemble of model configurations that provides a more rigorous assessment of uncertainties that remain.

The example application is an atmospheric general circulation model where features of interest could be gleaned from integrating the model for a few years with a given change in parameter values. However, some of the most challenging aspects of climate model development are in getting realistic simulations of coupled phenomena like ENSO. Coupled models take centuries to equilibrate and it may take a few centuries or millennia to fully characterize a model's ENSO characteristics [28]. One strategy for tackling this problem is to focus sampling on only a few parameters that affect the coupled phenomena and use Bayesian inference to select the appropriate combination of parameter values. A second strategy would be to make use of the results of a more elaborate quantification of uncertainties in each sub-system and use those configurations as a prior for the coupled model. One could then systematically test a limited number of coupled model

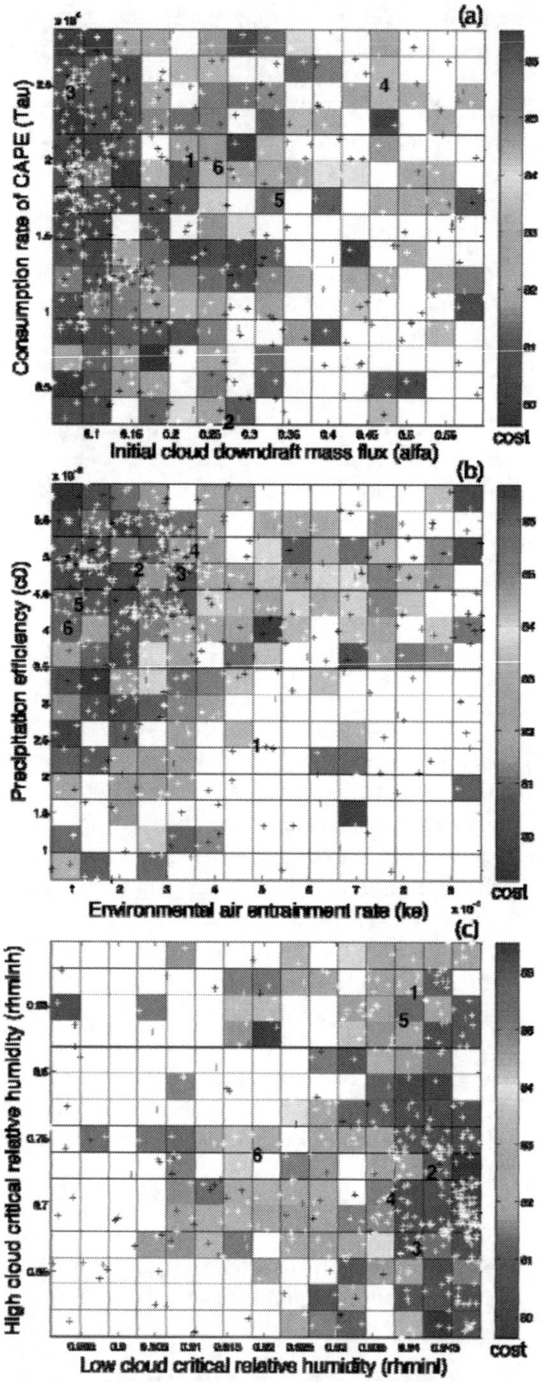

Figure 3. Estimate of cam3.1 response to six parameters important to clouds and convection. The average cost within each box follows equation (6) with the '+' marks indicating an experiment whose cost is less (white) or greater (black) than the standard model configuration. Numbers refer to the optimal model configuration after experiment 20.

configurations with longer model integrations, knowing that each component configuration has been filtered for its realism.

The number of parameters that can be explored with Bayesian inference is realistically limited to only a handful, perhaps less than 20. The limitation is that the selection process assumes all parameter choices are correlated and the number of samples required to quantify these relationships become very large. There are innovative strategies for using multi-scale techniques to reduce the apparent dimensionality of a large set of parameters [8].

The issue that would most solidify the utility of Bayesian inference for model development is whether it aids in the effort to learn the physical processes or balance of processes that govern nature. The path for this to happen is to first identify models that capture observed climate change for the right reasons and then to incorporate its key elements into a cost function so that we do not lose the ability to maintain this level of skill in future model development efforts.

References

[1] Collins, W.D., and Coauthors, 2006: The Formulation and Atmospheric Simulation of the Community Atmosphere Model Version 3 (CAM3). *J. Climate*, **19**, 2144–2161.

[2] Gleckler, J. J. Hnilo, S. M. Marlais, T. J. Phillips, G. L. Potter, B. D. Santer, K. R. Sperber, K. E. Taylor, D. N. Williams, 1999: An Overview of the Results of the Atmospheric Model Intercomparison Project (AMIP I). *Bulletin of the American Meteorological Society* 80(1), 29-56.

[3] Gleckler, P. J., K. E. Taylor, and C. Doutriaux (2008), Performance metrics for climate models, J. Geophys. Res., 113, D06104, doi:10.1029/2007JD008972.

[4] Haario, H. (2001). "An Adaptive Metropolis algorithm." Bernoulli , 7: 223–242.

[5] Haario, H., Laine, M., Lehtinen, M., Saksman, E., and Tamminen, J. (2004) Markov Chain Monte Carlo methods for high dimensional inversion in remote sensing. Journal of Royal Statistical Society , 66: 591–607.

[6] Haario, H., Saksman, E., and Tamminen, J. (1999). "Adaptive proposal distribution for random walk Metropolis algorithm." Computational Statistics, 14: 375–395.

[7] Higdon, D., M. Kennedy, J. C. Cavendish, J. A. Cafeo, and R. D. Ryne (2004) Combining field data and computer simulations for calibration and prediction, SIAM Journal on Scientific Computing, 26, 448-466.

[8] Hong, T., and M. K. Sen (2009) A new MCMC algorithm for seismic waveform inversion and corresponding uncertainty analysis, *Geophys. J. Int.* doi:10.1111/j.1365-246X.2008.04052.x

[9] Ingber, L. (1989). "Very fast simulated re-annealing." Mathematical Computational

[10] Modelling , 12: 967–973.

[11] Jackson, C. S., M. K. Sen, G. Huerta, Y. Deng, and K. P. Bowman (2008) Error Reduction and Convergence in Climate Prediction. *Journal of Climate*, **21**(24), 6698-6709. DOI: 10.1175/2008JCLI2112.1

[12] Jackson, C., M. Sen, and P. Stoffa (2004) An Efficient Stochastic Bayesian Approach to Optimal Parameter and Uncertainty Estimation for Climate Model Predictions, Journal of Climate, 17(14), 2828-2841.

[13] Jackson, C., Y. Xia, M. Sen, and P. Stoffa (2003) Optimal parameter and uncertainty estimation of a land surface model: A case example using data from Cabauw, Netherlands, Journal of Geophysical Research, 108 (D18), 4583 10.1029/2002JD002991.

[14] Kennedy, M. C., and A. O'Hagan (2001) Bayesian calibration of computer models, Journal of the Royal Statistical Society. Series B, Statistical Methodology, 63, 425-464.

[15] Meehl, G.A., G.J. Boer, C. Covey, M. Latif, R.J. Stouffer, 2000: The Coupled Model Intercomparison Project (CMIP). *Bulletin of the American Meteorological Society* 81(2), 313-318.

[16] Metropolis, N., Rosenbluth, A., Rosenbluth, M., Teller, A., and Teller, E. (1953) Equations of State Calculations by Fast Computing Machines. Journal of Chemical Physics, 21: 1087–

1091.

[17] Mu, Q., C. S. Jackson, P. L. Stoffa (2004) A multivariate empirical-orthogonal-function-based measure of climate model performance, *J. Geophys. Res.*, **109**, D15101, doi:10.1029/2004JD004584.

[18] Murphy, J. M., Sexton, D. M. H., Barnett, D. N., Jones, G. S., Webb, M. J., Collins, M. & Stainforth, D. A. 2004 Quantification of modelling uncertainties in a large ensemble of climate change simulations. Nature 430, 768–772. (doi:10.1038/nature02771)

[19] Richter, I., and S.P. Xie (2008) On the origin of equatorial Atlantic biases in coupled general circulation models. Clim. Dyn. (31) 587-598. (doi:10.1007/s00382-008-0364-z)

[20] Rougier, J. C. 2007 Probabilistic inference for future climate using an ensemble of simulator

[21] evaluations. Clim. Change 81, 247–264. (doi:10.1007/s10584-006-9156-9)

[22] Smith, R. L., C. Tebaldi, D. Nychka, and L. O. Mearns (2009) Bayesian Modeling of Uncertainty in Ensembles of Climate Models. *Journal of the American Statistical Association*, **104**(485), doi:10.1198/jasa.2009.0007

[23] Stainforth, D. A., and Coauthors, 2005: Uncertainty in predictions of the climate response to rising levels of greenhouse gases, *Nature* **433**, 403-406 (27 January 2005) doi:10.1038/nature03301.

[24] Tebaldi, C., Smith, R. L., Nychka, D., and Mearns, L. O. (2005) Quantifying Uncertainty in Projections of Regional Climate Change: A Bayesian Approach to the Analysis of Multi-model Ensembles. *Journal of Climate*, **18**, 1524-1540 (corrigendum p. 3405).

[25] Tierney, L. and Mira, A. (1999). "Some Adaptive Monte Carlo Methods for Bayesian Inference." Statistics in Medicine, 18: 2507–2515.

[26] Villagran, A., G. Huerta, C. S. Jackson, M. K. Sen (2008) Parameter Uncertainty Estimation in Climate Models. Bayesian Analysis 3(4), 823-850. DOI:10.1214/08-BA331

[27] Wang, F. (2007), Investigating ENSO sensitivity to mean climate in an intermediate model using a novel statistical technique, Geophys. Res. Lett., 34, L07705, doi:10.1029/2007GL029348.

[28] Wittenberg, A. T., 2009: Are historical records sufficient to constrain ENSO simulations? Geophys. Res. Lett., in press, May 2009

Development of a Computational Environment for the General Curvilinear Ocean Model

Mary P. Thomas[1], **Jose E. Castillo**[2]

[1] Department of Computer Science, San Diego State University, San Diego, CA
[2] Computational Sciences Research Center, San Diego State University, San Diego, CA

Email: mthomas@sciences.sdsu.edu

Abstract. The General Curvilinear Ocean Model (GCOM) differs significantly from the traditional approach, where the use of Cartesian coordinates forces the model to simulate terrain as a series of steps. GCOM utilizes a full three-dimensional curvilinear transformation, which has been shown to have greater accuracy than similar models and to achieve results more efficiently. The GCOM model has been validated for several types of water bodies, different coastlines and bottom shapes, including the Alarcon Seamount, Southern California Coastal Region, the Valencia Lake in Venezuela, and more recently the Monterey Bay. In this paper, enhancements to the GCOM model and an overview of the computational environment (GCOM-CE) are presented. Model improvements include migration from F77 to F90; approach to a component design; and initial steps towards parallelization of the model. Through the use of the component design, new models are being incorporated including biogeochemical, pollution, and sediment transport. The computational environment is designed to allow various client interactions via secure Web applications (portal, Web services, and Web 2.0 gadgets). Features include building jobs, managing and interacting with long running jobs; managing input and output files; quick visualization of results; publishing of Web services to be used by other systems such as larger climate models. The CE is based mainly on Python tools including a grid-enabled Pylons Web application Framework for Web services, pyWSRF (python-Web Services-Resource Framework), pyGlobus based web services, SciPy, and Google code tools.

1. Introduction

The science of ocean modeling and simulation is complex, involving systems spanning a large number of geographical and climate systems. Computer models have evolved to be the primary tools used to study and predict physical, chemical, and biological characteristics of the oceans; integrating simulated results and observational data from the ocean, and enhancements to the realism of ocean model simulations. Models of the ocean range in complexity from simple mathematical models to massive, realistic global ocean circulation simulators encompassing many different models, requiring

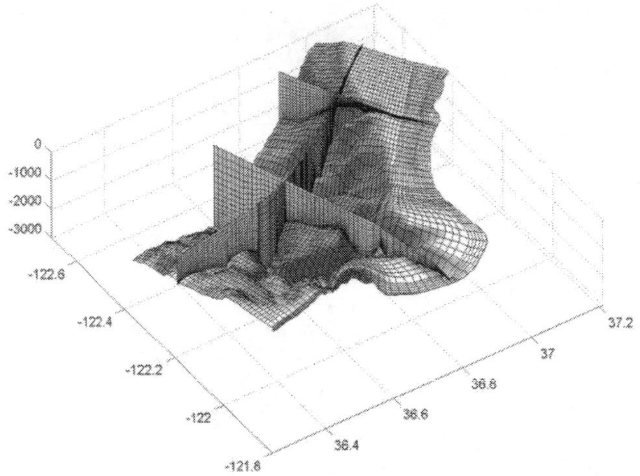

Figure 1. GCOM Mesh for the Monterey Bay, the largest bay of the West Coast of the United States. It is characterized by complex coastlines and regions of steep bathymetry. The 3D adaptive mesh used by GCOM concentrates grid points in regions where there are the most changes in the features.

thousands or millions of lines of computer code to solve, and integrating disparate and distributed data and computational resources. As a result, there are a large number of ocean models in use today, many of which address specific aspects of this complex problem such as global, coastal, regional, or basin systems. Because of increases in computer power and data resources, modelers are able to construct models where the elements of the ocean and atmosphere interact across many spatial and temporal scales. Consequently, there is a driving need for new models that can be integrated into larger systems that can also take advantage of the computational infrastructures being utilized within the ocean and computational modelling communities.

In November, 2008, the US DOE held a workshop addressing the grand challenges and limitations that exist today in the field of high-resolution climate and Earth modeling systems [1]. The outcome of this meeting is a comprehensive report with recommendations for the next generation models. Among the suggestions, the panels identified the need to develop models that can take advantage of the specialized computational environments available for us by modern computational scientists. These are needed to integrate multiple ocean models and clients with advanced compute and data resources, emerging Web and internet technologies, and cyberinfrastructure. Additionally, the workshop pointed out a need to have more precise coastal models.

In this paper, we discuss updates being made to the General Curvilinear Ocean Model (GCOM) [2] and the development of a computational environment (GCOM-CE) that allows a component-based, serial or parallel model to run on a variety of resources (local workstations, large grids such as the TeraGrid, clouds) using emerging Web and internet technologies to seamlessly access the model and its results. This paper is organized as follows: in Sections 2 and 3 we introduce the GCOM model and present a prototype system of the proposed computational environment, in Section 4, we present our conclusions and future plans.

2. The General Curvilinear Ocean Model (GCOM)

This paper does not have enough room to discuss the model in detail, so we include a brief overview of the GCOM model. The interested reader should refer to the included references of the work by Torres and Castillo. GCOM solves 3-D, time-dependent, curvilinear Navier-Stokes flow equations to

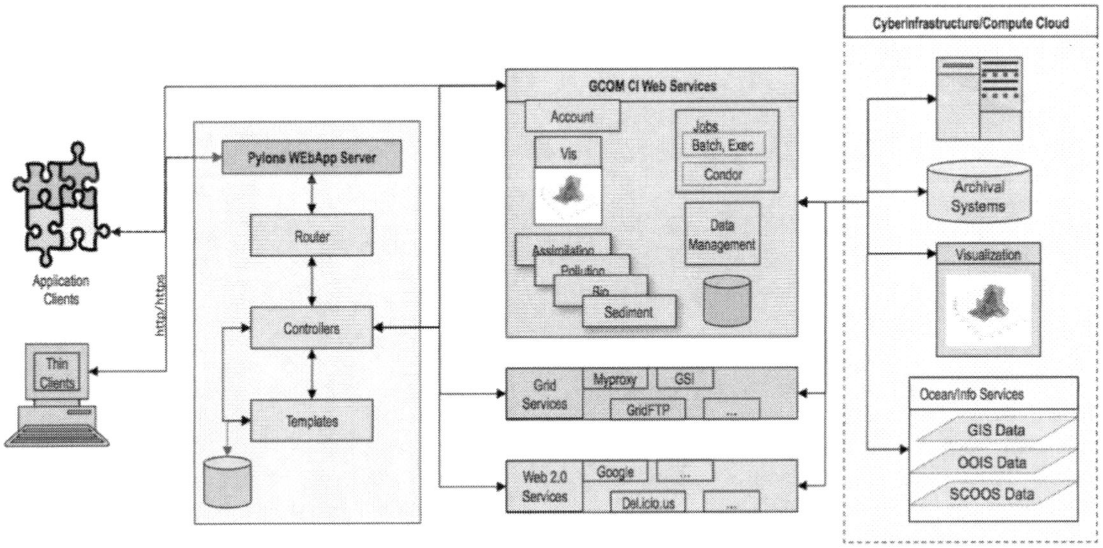

Figure 2. GCOM-CE architecture: The layers (left to right) are the user interface layer, the middleware layers (Pylons Web application framework and Web services), and the internet "cloud" of resources and services.

simulate 3D marine currents and density gradients over irregularly shaped coastlines and sea floors [3]. GCOM differs from more traditional approaches where the use of Cartesian coordinates force the model to simulate terrain as a series of steps. Instead, it uses a full 3-D curvilinear grid (see figure 1) that is adapted both to the bottom topography and to the coastline, and a numerical model that is capable of resolving accurately the dynamics of multiple boundary-layers and their interactions with the topographic features [4]. Adaptable curvilinear grids result in more efficient algorithms and increased resolution. Although the grid generation technique allows grid points to be clustered in any region of the computational domain, the model can handle both orthogonal and non-orthogonal grids. The resulting model equations are more complex than in case of sigma-coordinate models, but the boundary conditions are easily implemented and there is no need to interpolate near solid boundaries. GCOM has been validated against several types of water bodies, different coastline and bottom shapes. These include the Alarcon Seamount, Southern California Coastal Region, the Valencia Lake in Venezuela, and the Monterey Bay. In addition, GCOM has been proven to produce improved resolutions and smaller errors over other approaches [4].

2.1. Current Efforts

Originally developed in F77, the sequential model has been ported to F95, using a modular or component design, and is undergoing validation tests for the Monterey Bay region. Modifications and upgrades include the integration of new driving models: sediment transport, wind, biological, geochemical, and pollutants. Plans for parallelizing the code include the use of the Distributed Coupling Toolkit (DCT) for data distribution [5] and red-black ordering of the successive over-relaxation (SOR) sections. The structure of the code base and its' components are being used to help architect the computational environment to extract data and task parallelism, and to identify places in the code where modules can be operated in a distributed or asynchronous manner (see Section 3 below). As an example, we are developing a service to update the bathymetric grid in response to changes in sediment transport. This task can be done on any host, and uses messaging to interact with

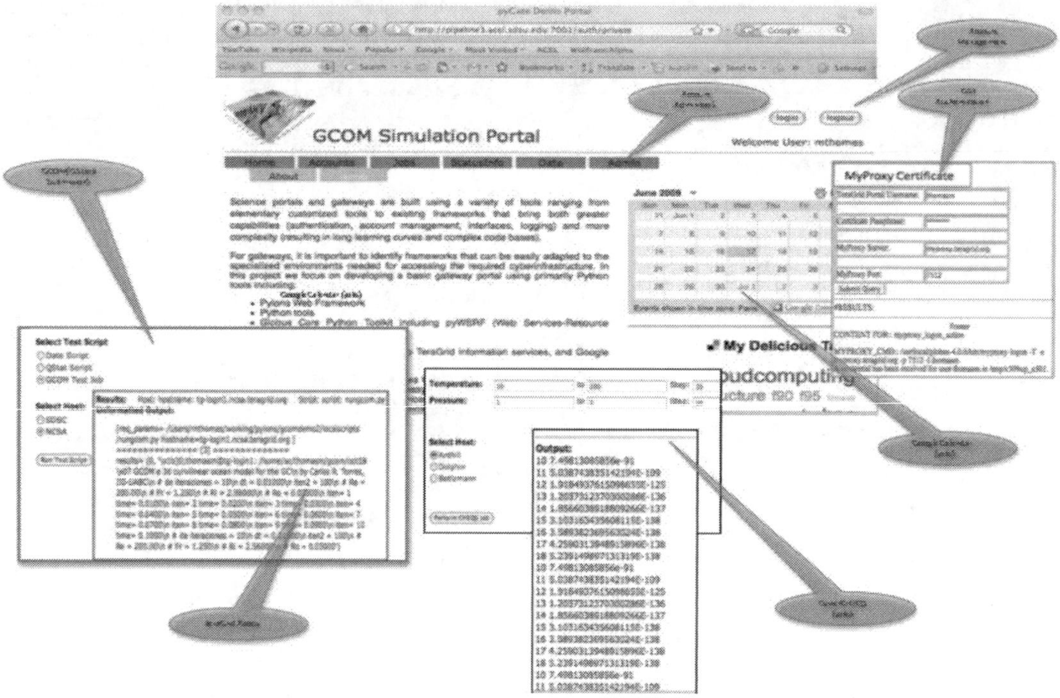

Figure 3. The Pylons GCOM-CE Demo Portal showing basic portal interface, account management, authentication and remote job execution for two applications.

other components in the system. In addition, we are designing "unit test" suites for basic grid generation, simple PDE example, timing diagnostics and input conditions.

Related Work

Relevant ocean models include the Parallel Ocean Program (POP), the Regional Oceanic Modeling System (ROMS), and the Hybrid Coordinate Ocean Model (HYCOM). These models all require that one or more of the axes must be orthogonal to surface, which results in loss of detail where there are significant changes in the bathymetry. Thus, when analyzing events along coastal areas, GCOM is capable of providing more accurate results with greater resolution.

3. GCOM Computational Environment (GCOM-CE)

In the computational science community there are significant number of efforts to develop common tools and services that can be reused at all layers of the computational environment, including the compute and data resources, as well as the distributed middleware needed to interconnect clients and systems. Based on experiences gained by Thomas in developing reusable toolkits (Open Grid Computing Environments or OGCE, the GridPort Toolkit, the Perl COG Commodity Grid Kit [6]) a community of students, developers, and scientists exist who prefer (or need) lightweight rapid prototyping capability. These requirements, and those described by the workshop on climate change, drove technology choices for this work.

3.1. Architecture

The architecture of the GCOM-CE is shown in figure 2, and includes layers for the distributed, services oriented architecture of the Web, as well as the cyberinfrastructure and middleware services being used by the GCOM CE. In this design, the front tier clients can be a human running a browser or desktop application, or an application or service exchanging messages. The backend tier includes local services, web services, other applications, and remote computing and data services.

For the middleware, we evaluated several Web 2.0 technologies, and chose to investigate how Web application frameworks would work in the CE. To keep the CE simple, we chose to work with Python because it has a strong, active open source developer community; it is object-oriented; libraries have been proved to work well for science applications (SciPy, NumPy, PyNGL PyNIO); and libraries for grid integration (pyGlobus, pyGridWare, pyGSI). The framework must have the ability to integrate emerging Web 2.0 technologies including WSGI (Web Server Gateway Interface), databases, XML/JavaScript/AJAX, Google Gadgets, social networks, Web 2.0 services and toolkits, and security. After testing several candidates, the Pylons Web Framework emerged as a viable framework [7].

Pylons is a Web application framework, hence, any number of components and libraries developed by other projects can be incorporated into the system, which facilitates customization. Pylons uses the model-view-controller (MVC) request-response architecture. The Model contains the data that the application works with; the View reads the data from the model and displays it to the user; the Controller manages the logic of the application. Pylons has several features that are beneficial to gateways and developers: Routing and "Beautiful URLs" that simplify the naming and mapping of URI's through the REST interface; multiple database interfaces; interfaces to multiple template packages; dynamic update and interactive debugger; built-in security and authentication using the AuthKit module; RESTful web services using Paste; toolkits to interact with Google code and Web 2.0 gadgets; Easy-Install installation and packaging into Python eggs (similar to jar or tar); auto deployment using Paste (similar to Ant or Maven); "virtual environment" packaging into an isolated working directory containing all necessary libraries. An extremely important feature of the Pylons framework is that the services layer is decoupled from the logic of the code behind it, allowing the same code to be hosted, for example, as a Web service, a portal page, or an iGoogle gadget.

3.2. GCOM-CE Current Status

The Pylons framework has been extended to include GSI security (using MyProxy services). In addition, a working user portal has been developed which includes authentication, job submission, unit tests, and other services. Figure 3 shows a composite of several of the functions of the demo portal. All software has been bundled into a deployable egg that can be installed on any system and can be modified for use by other applications. System admin pages are being developed to facilitate configuration and management of users and resources. Some of the portal pages are being migrated to Web services (job submission, data viewing). An interface to the Google calendar API is being developed for posting system task status and other messages.

4. Conclusions and Future Work

The migration of the GCOM code base from F77 to F90 is completed, and validation of the model is in progress. New science components such as dynamic grid generation, sediment transport, biogeochemical processes are being tested and integrated. A parallel framework, based on the DCT is being developed. The GSI-enabled Pylons framework is being used for hosting Web services, including job submission to multiple types of resources. Key features include GSI integration, the extensibility of the framework (e.g. the ease with which one can add external modules), the ability deploy copies onto other systems, the ability to dynamically reload services without restarting the server and the work being done with Web2.0 applications (Google Apps, Gadgets, etc.). A simple user portal has been developed and is being used to submit GCOM jobs to the TeraGrid.

Future plans include: further development of the framework to be included as an OGCE incubator project; adding components such as a data browser, job monitoring, simple data visualization; and exposing GCOM-CE services for use by other applications. In particular, we are using the software to develop a portal interface to the CyberCHEQS computational Web service, an NSF funded CI-TEAM project investigating flame thermochemistry properties.

Acknowledgements
This work was supported in part by the National Science Foundation (Grants #0753283 and #0721656), and with resources available on the NSF funded TeraGrid Project (NSF #12345) at the San Diego Supercomputer Center and the National Center for Supercomputing Applications, the San Diego State University Computational Sciences Research Center, and the DOE DCT project.

References

[1] Washington W. Challenges in Climate Change Science and the Role of Computing at the Extreme Scale. Proc. of the Workshop on Climate Science, Nov., 2008, Washington D.C.

[2] The General Coastal Ocean Model Project Website. Last accessed on 01-Jul-09 at http://gcom.acel.sdsu.edu.

[3] Torres C R and Castillo J E. Stratified Rotating Flow over Complex Terrain. J. Appl. Num. Math. 47 531-541, 2003.

[4] Torres, C. R., José E. Castillo, Jim Mueller, Charles Trees and Germán Larrazábal (2006Modeling 3D Coastal Circulation in Boundary-fitted Coordinates: The GCOM Project. California and the World Ocean 06 Conference. Long Beach, CA. Sept.17-20, 2006.

[5] Berbegall V, Drummond L A, Verde G, and Vidal V. A Purely Distributed Approach for Coupling Scientific and Engineering Applications. Lecture Notes in Computer Science, Vol.3984, pp. 192-201, 2006

[6] Alameda J, Christie M, Fox G, Futrelle J, Gannon D, Hategan M, Kandaswamy G, von Laszewski G, Nacar M A, Pierce M, Roberts E, Severance C, and Thomas M. The Open Grid Computing Environments collaboration: portlets and services for science gateways. Concurr. Comput. : Pract. Exper. 19, 6 (Apr. 2007), 921-942.

[7] Thomas M P, Using the Pylons Web Framework for Science Gateways. rid Computing Environments Workshop, GCE'08 12-16 Nov. 2008.

SciDAC 2009 IOP Publishing
Journal of Physics: Conference Series **180** (2009) 012031 doi:10.1088/1742-6596/180/1/012031

Cellular burning in lean premixed turbulent hydrogen-air flames: coupling experimental and computational analysis at the laboratory scale

M. S. Day[1], J. B. Bell[1], R. K. Cheng[2], S. Tachibana[3], V. E. Beckner[1], M. J. Lijewski[1]

[1]Center for Computational Science and Engineering, Lawrence Berkeley National Laboratory, Berkeley, CA, 94720

[2]Environmental Energy Technologies Division, Lawrence Berkeley National Laboratory, Berkeley, CA, 94720

[3]Environmental Energy Technologies Division, Lawrence Berkeley National Laboratory, Berkeley, CA, 94720, currently at Japan Aerospace Exploration Agency, Tokyo, Japan

E-mail: `msday@lbl.gov`

Abstract. One strategy for reducing US dependence on petroleum is to develop new combustion technologies for burning the fuel-lean mixtures of hydrogen or hydrogen-rich syngas fuels obtained from the gasification of coal and biomass. Fuel-flexible combustion systems based on lean premixed combustion have the potential for dramatically reducing pollutant emissions in transportation systems, heat and stationary power generation. However, lean premixed flames are highly susceptible to fluid-dynamical combustion instabilities making robust and reliable systems difficult to design. Low swirl burners are emerging as an important technology for meeting design requirements in terms of both reliability and emissions for next generation combustion devices. In this paper, we present simulations of a lean, premixed hydrogen flame stabilized on a laboratory-scale low swirl burner. The simulations use detailed chemistry and transport without incorporating explicit models for turbulence or turbulence/chemistry interaction. Here we discuss the overall structure of the flame and compare with experimental data. We also use the simulation data to elucidate the characteristics of the turbulent flame interaction and how this impacts the analysis of experimental measurements.

1. Introduction

Within the combustion industry there is considerable interest in developing fuel-flexible burners that can be used to stabilize lean premixed flames in a stationary turbine designed for power generation. Low-swirl burner technology, originally introduced by Bedat and Cheng [1] as a tool for studying the fundamental properties of lean, premixed turbulent flames, has the potential for meeting this need. Burners based on modifications of the original design have been used by a number of research groups [2–5], and have the potential for use in the design of next-generation, lean premixed combustion systems, including those burning lean hydrogen at both atmospheric and elevated pressures [6].

The low-swirl burner concept is extremely simple: premixed fuel exits a pipe after passing through a turbulence generation plate and an annular set of curved vanes. The vanes impart a swirl component to the flow over a narrow layer near the pipe wall, and a detached premixed flame anchors in the diverging flow above the pipe exit. Turbulence in the fuel stream wrinkles the flame, which enhances the overall rate of combustion in the device; the flames stabilize where the mean burning speed matches the axial flow velocity. Analysis of experimental velocity measurements for both lean hydrogen-air and methane-air mixtures shows that the mean burning speed of the low-swirl flames is proportional to the intensity of turbulence in the approach flow to the flame [6, 7], and that effective enhancement of the burning speed can exceed a factor of 50 or more. However, because the burner supports ultra-lean mixtures while also lacking a significant post-flame flow recirculation, nitrogen-based emissions are extremely low.

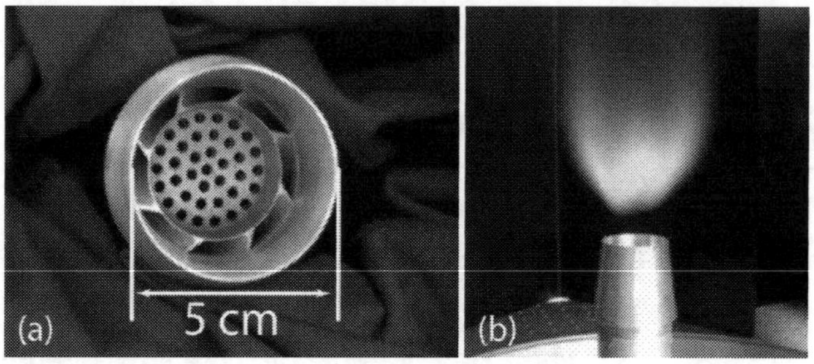

Figure 1: (a) Low-swirl nozzle, showing vanes and turbulence generation plate. (b) typical turbulent low-swirl methane flame stabilized in divergent flow above the low-swirl nozzle.

Continued development of these types of burners, particularly for alternative fuels, depends on improving our understanding of basic flame structure, stabilization mechanisms, emissions and response to changes in fuel. Numerical modeling has the potential to address some of these issues, but simulation of these types of burners has proven to be difficult because of the large range of spatial and temporal scales in the system; the bulk of the analysis to date has been experimental.

As noted in Bell et al. [8], the detailed structure of lean premixed flames becomes particularly important and difficult to simulate when burning hydrogen. Lean hydrogen-air flames burn in cellular structures—localized regions of intense burning, separated by regions of local extinction. In this regime, the flame surface is broken into discontinuous segments. This type of structure introduces severe difficulties in applying standard turbulence/chemistry interaction models, which are based on the presence of a highly wrinkled but continuous flame surface that propagates locally as an idealized laminar flame structure. The cellular burning patterns additionally make the analysis of experimental data problematic, and can lead to significant inaccuracies or misinterpretations. Analysis in Day et al.[9] examines the structure of lean premixed hydrogen flames in an multi-dimensional, time-dependent (but idealized) setting in some detail. It was found that turbulent fluctuations tend to disrupt the spontaneous development and evolution of the cellular burning structures. With increasing fluctuation instensity, the cells became larger, while the burning intensity, even in locally flat regions, is elevated on average by factors of 3 or more. In the absence of a suitably general model for the turbulent combustion of lean hydrogen-air mixtures, numerical simulations must incorporate sufficient detail in the chemical kinetics, differential species transport and turbulent fluid dynamics to capture all the important couplings in these flows.

In this paper, we focus on detailed numerical simulation of a laboratory-scale lean hydrogen flame that is stabilized on a low-swirl burner. As in our previous work, the simulation is performed with detailed chemistry and transport without explicit turbulence models. The case considered here is at a relatively high turbulence intensity. We use the low-swirl burner configuration presented in [2] that is under investigation by several groups internationally using matched nozzle hardware. This particular configuration is designed specifically to support validations with computer simulation.

2. Numerical methodology

The simulations presented here are based on a low Mach number formulation [10] of the reacting Navier-Stokes equations. The methodology treats the fluid as a mixture of perfect gases. We use a mixture-averaged model for differential species diffusion, which is critical to capturing the thermodiffusive behavior of lean hydrogen flames (see [11] for a complete discussion of this approximation). We ignore Soret, Dufour and radiative transport processes. A lean hydrogen-air inlet fuel mixture (ϕ=0.37) was modeled with the hydrogen sub-mechanism of GRI-Mech 2.11; the transport coefficients and thermodynamic relationships are obtained from EGLib [12].

The basic finite-volume numerical discretization [13] combines a symmetric operator-split treatment of chemistry and transport with a density-weighted approximate projection method to impose the evolution constraint. The resulting integration proceeds on the time scale of the relatively slow advective transport. Faster diffusion and chemistry processes are treated implicitly in time. This integration scheme is embedded in an adaptive mesh refinement algorithm based on a hierarchical system of rectangular grid patches. The data and work are apportioned over a parallel computing system using a coarse-grained load distribution strategy [14]. The complete integration algorithm is second-order accurate in space and time, and discretely conserves species mass and enthalpy.

3. Computational Setup

There are two possible approaches for how to treat the nozzle in a simulation of this sort. One possiblility is to include the flow in the nozzle directly as part of the simulation; the alternative is to use measured data to prescribe the flow at the exit to the nozzle. In a series of studies, Nogenmyr et al. [4, 15, 16], considered both options in the context of LES simulations of lean methane-air mixtures. Considerably better agreement was found between simulated and measured profiles when the flow inside the nozzle was included. In particular, the simulations correctly predicted large-scale structures emanating in their configuration from the swirl vanes inside the nozzle, and these structures ultimately played a key role in stabilizing their flame. The measured and computed flows exiting the nozzle additionally exhibited a pronounced imprint of the swirl vane structure. Mean flow rates varied by as much as 30% azimuthally in the swirling-flow region, depending on whether the measurement plane was directly above a vane or a gap. A larger scale additional azimuthal variabilities of up to 20% were observed, and attributed to basic nonuniformities in the experimental setup. These nonuniformities led to an observed global shifting of the flame from the device center.

Recent modifications to the basic low-swirl nozzle design include: airfoil-like tapering of the trailing edges of the vanes to avoid formation of the large downstream vortical structures, varying the number and length of swirl vanes and their tilt with respect to the mean flow to improve the azimuthal symmetry of the device, and adjusting the geometry and orientation of the perforated turbulence generation plates. While such details have had significant impact on the global flame shape and resulting mean flow fields, which are important from an engineering/ performance point of view, experimental evidence suggests that the local flame propagation, including the detailed interactions of the flame chemistry and fine scales of the turbulence, is largely independent of these adjustments. This independence holds particularly well in the core region of the flame inside of the swirl-induced turbulent shear layer, and where much of the fuel consumption (and emissions generation) occurs.

The simulations here incorporate an experimental characterization of the flow at the nozzle exit. Typical profiles for this purpose were provided by Petersson et al. [2], and are shown in Fig. 2 as a function of radius from the centerline of the fuel pipe. Outside the pipe, a 35 cm/s upward coflow of cold air is specified. These profiles were generated using data collected from a ϕ=0.62 methane-air flame at a mean fueling rate of 6 m/s. Stereo PIV data was collected in a set of vertical planes to infer the three components of the mean velocity field.

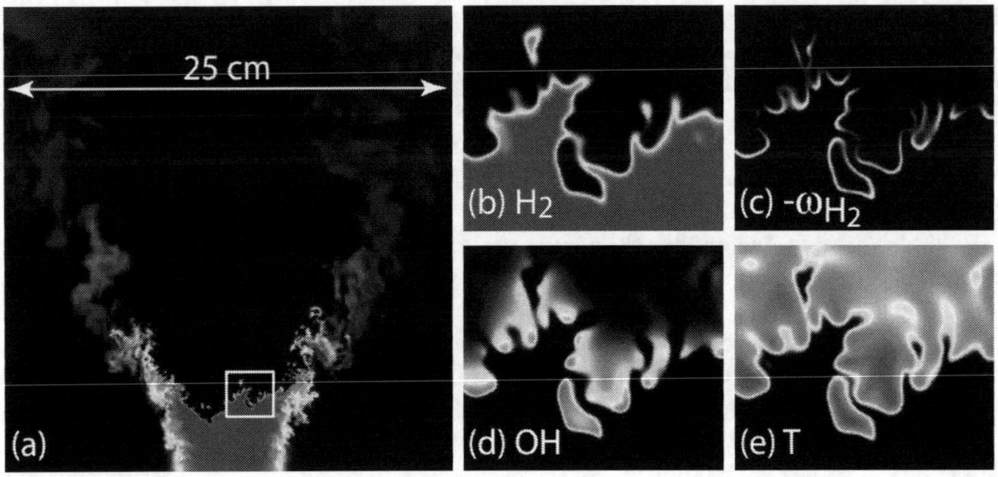

Figure 3: Typical slice from a snapshot in time of the computed flame solution. Panel represents a vertical slice centered on the symmetry axis of the nozzle. (a) The mole fraction of H_2 over a $(25\ cm)^2$. (b)-(e) Enlargements of the white-boxed region (approx. 2.5 cm wide) in (a). (b) H_2 mole fraction, (c) Destruction rate of H_2, (d) Mole fraction of OH, and (e) temperature. All scales are relative to the peak values over the domain at that instant.

These profiles were averaged azimuthally, and scaled by 250% to obtain a "ideal azimthally symmetric" laboratory-scale low-swirl nozzle operating at a mean mass flow rate of 15 m/s. Turbulent fluctuations in the nozzle were prepared in an auxiliary simulation and added to the mean flows. An initially zero-mean velocity field, representative of flow through a perforated plate, is evolved until the measured integral length scale was 4 mm. A radial scaling factor was used to shape these synthetic fluctuations to the measured data.

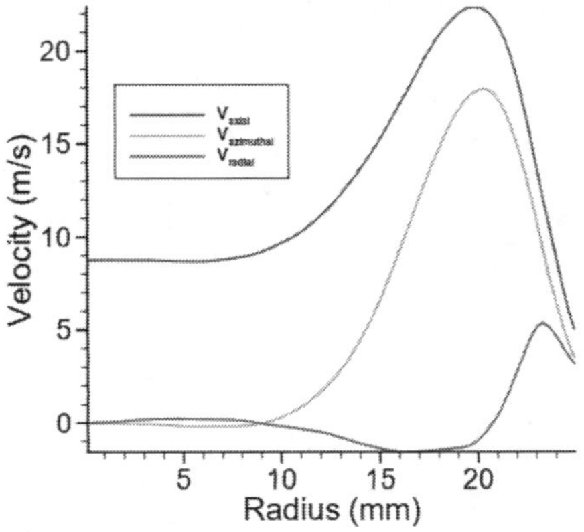

Figure 2: Inlet velocity profiles at the exit plane of the swirl nozzle, adapted from experimental data [2].

The thermal thicknesses of this hydrogen-air flame is approximately 800 μm; the full half-width maximum of the fuel consumption rate is approximately 0.5 mm. The computational domain for the reacting flow simulations measures 25 cm^3. We implicitly assume that the boundaries of this box are sufficiently far from the flame as to not significantly affect its dynamics. The base mesh for the simulation is a uniform grid of 256^3 cells and uses 3 additional levels of factor-of-two dynamic grid refinement which track regions of high vorticity (turbulence) and reactivity (combustion/flame). The flame is contained entirely within the finest level, for an effective resolution of 2048^3. We run the simulations from an arbitrary initial condition until the flame is statistically stationary. In previous work, we demonstrated that this level of resolution is adequate to capture the detailed structure of the flame including the peak fuel consumption, the thermal field and major species.

4. Simulation Results

In Figure 3, we show a cross-section through the middle of the simulation that provides a picture of the overall structure of the flame. The swirl introduced in the outer portion of the injected mixture induces a flow divergence, and a negative gradient in vertical velocity.

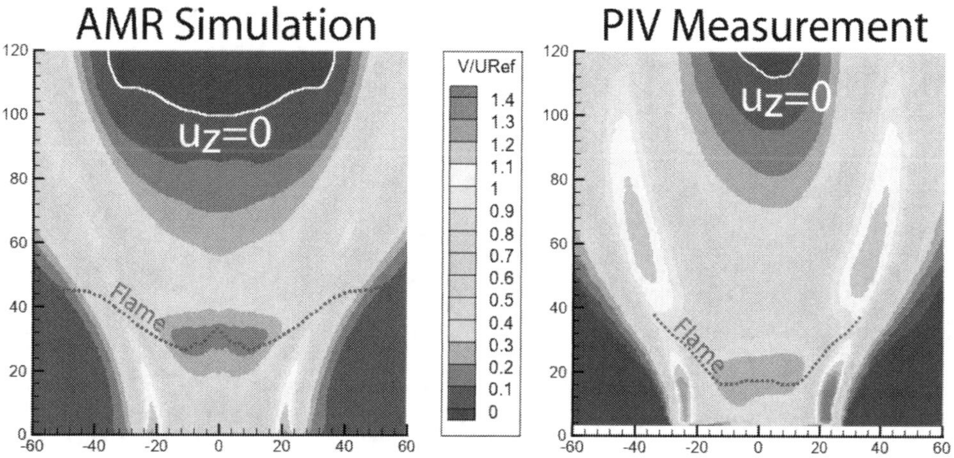

Figure 5: Comparison of the mean vertical velocity (scaled by the mean fueling rate) for the simulated ideal low-swirl nozzle, and a representative experimental profile. A number of key features are recovered in the simulations, including the weak recirculation approximately 10 cm downstream of the nozzle. The computed flame position sits somewhat downstream of the measured data, consistent with the simulation's higher mean velocity in the core.

The decreasing axial velocity provides a robust aerodynamic-based mechanism for flame stabilization. In the upper part of the simulation, one sees a low concentration of unburnt fuel (blue). This represents fuel that has been sufficiently diluted with air that it is below the flamability limit. Turbulent fluctuations from two primary sources interact with the flame (see Figure 4). Fluctuations introduced at the inlet boundary advect toward the flame, wrinkling the central core region. Also a swirl-generated shear layer forms on the outer edge of the burner nozzle, which leads to mixing of fuel and air on the sides of the burner.

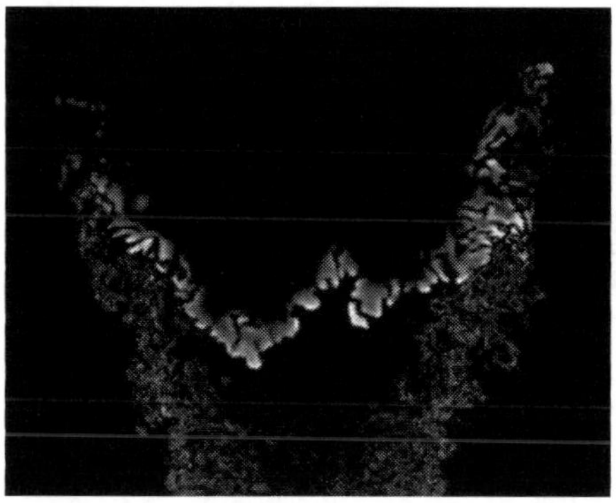

Figure 4: Composite image (cropped in near the flame/nozzle region) simultaneously shows the OH intensity (in orange) and the magnitude of vorticity (in grey). Turbulence approaches the flame from the core of the inlet nozzle at the lower domain boundary, and is generated by shear in the swirling flow region.

The idealized inlet profile we used has no experimental counterpart to date; the physical laboratory device exhibits considerable azimuthal variabilities due to the vane structure and location. Furthermore, additional analysis of experimental data suggests a Reynolds number dependence of the inflow at lower fueling rates from which the initial data from this simulation was taken. Nevertheless, Figure 5 compares the mean axial velocity profile taken from a low-swirl nozzle flame with identical fuel mixture, fueling flow rates and nozzle swirl number. Coloring in the two plots is identical. Obtaining converged statistics for the simulation is difficult because of the computational expense of long-time integrations. To reduce this requirement, we used azimuthal averages of the simulation data to compute mean profiles. This produces reasonable data for most of the domain but give poor statistics near the center of the burner. From the comparison, the experiment shows a significantly lower central velocity and somewhat higher velocity near the edge of the burner (20 mm) than the boundary data used for the simulation. The simulation predicts a higher flame location, which is consistent with the higher inflow velocity. (We note

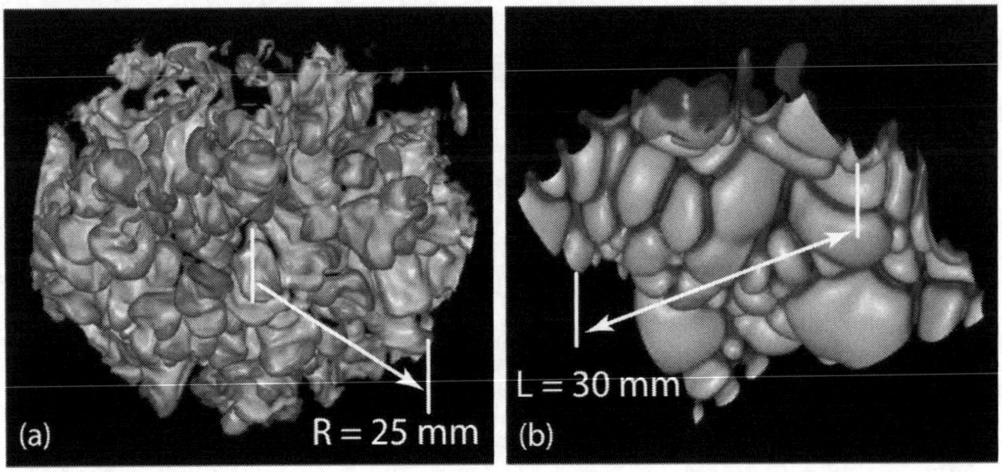

Figure 6: Flame isotherm, T= 1144 K, colored by the local rate of H_2 consumption. (a) Low-swirl flame isotherm, conditioned on $r < 25mm$, $z < 8mm$. (b) Representative isotherm of an idealized ϕ=0.37 H_2-air flame propagating freely in uniform flow.

that the appearance of a raised flame location near the origin may well be an artifact of the lack of data in this region.) Nevertheless, the simulation is consistent with a low-swirl nozzle with somewhat different flow characteristics.

Although there are qualitative differences in the global structure of the flame, we can still compare the details of the local flame structure in the experiment and the simulation. Lean hydrogen-air flames are thermodiffusively unstable and burn in cellular structures; even in the absence of turbulence, the flame will not burn as a stable continuous interface. Here, the turbulent flow in the central core of the burner serves as additional source of wrinkling of the flame. In Figure 6, we show the $T=$ 1144 K isotherm, colored by local fuel consumption in the core of the burner. For comparison, we also show the analogous image for a freely propagating hydrogen flame (taken from the study in [9]). Both images are based on the same color map, and the length scales are as indicated in the figure. Comparing these images, one can see that in the turbulent flame, there are finer structures than in the non-turbulent case. The turbulent flame shows more ridge-like structures compared to the freely propagating flame in which the features are more bulbous. Finally, the burning rate over most of the flame surface is considerably higher in the turbulent configuration, particularly where the flame is tightly folded.

One of the principal diagnostics used in the experiments is OH-PLIF (planar laser-induced fluorescence) based on imaging the flouresence of OH radicals excited by a tuned laser sheet. In Figure 7, we show a typical vertical slice of the OH concentration from the simulation alongside typical PLIF images from the experiment. Figure 7(a) shows the profile of OH over the $(25\ \text{cm})^2$ slice, while Figures 7(b) and (d) show progressive enlargements of the data corresponding to the field of view of the typical OH-PLIF data shown in Figures 7(c) and (e). The experimental data are representative images taken from the low-swirl experiment. The figure show that the simulation captures features of these flames, with remarkably similar sizes, shapes and global structure. The simulation also captures the observed variability of the OH signal (brightness on the experimental images) along the flame surface. Previous work on turbulent hydrogen flames shows that the high diffusivity of H_2 can lead to local enrichment of the fuel mixture along the front. This local enrichment can lead to intensification of local burning, which leads to increased OH that can be seen experimentally as a brigher signal in the OH PLIF (and red regions in the simulation slice data).

From analysis of the simulation data, we can obtain a detailed characterization of the

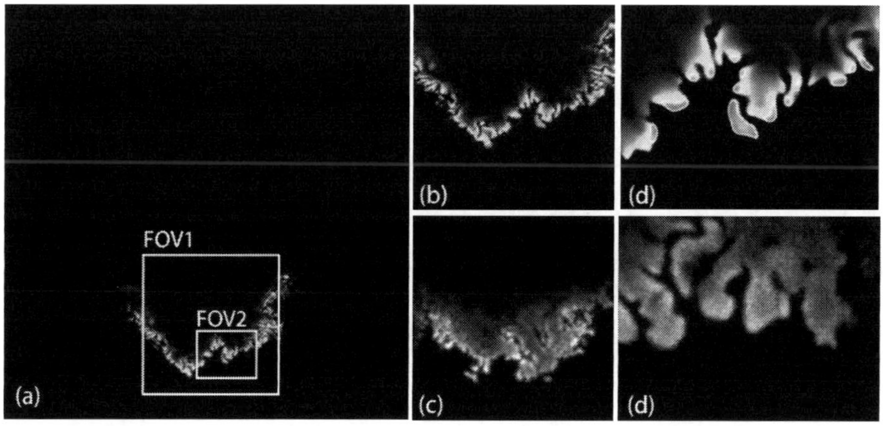

Figure 7: (a) Typical slice of OH concentration on vertical midplane from the simulation. Red indicates high relative values, and are correlated with the flame front. The white box ("FOV1") represents the field of view of OH-PLIF measurements from the representative low swirl experiment focused on the large-scale flame structure. The smaller yellow box ("FOV2") represents the experimental field-of-view OH-PLIF measurements focused more tightly on the local structure of the flame surface (width of FOV2 is approximately 26 mm). (b) Zoom of profile in (a) over FOV1, (c) typical experimental OH-PLIF image over FOV1, (d) Zoom of profile in (a) over FOV2, and (e) typical experimental OH-PLIF image over FOV2.

local flame structure that is not feasible from the experimental data. We begin with an examination of the fuel consumption rate. As with the OH-PLIF data, the fuel consumption shows considerable variability along the flame front. To quantify this behavior, we extract the T=1144 K surface (trimmed to $r<5$ cm, $z<8$ cm; outside this window, the isotherm is shredded by the swirl-induced shear layer). This temperature corresponds to the that of peak fuel consumption in the flat laminar flame at ϕ=0.37. We further trim the surface to exclude regions where the fuel consumption is less than half the laminar flame value. On the remaining sections of the flame isotherm, we can then evaluate mean and Gaussian curvatures (using approaches outlined in [9]). In Figure 8a and Figure 8b we show PDF's of of the two measures of curvature. We note that for the turbulent flame, the distribution of mean curvature is considerably broader and more symmetric than the non-turbulent flame. The Gaussian curvature is also more symmetric about zero. These observations indicate that turbulence serves to disrupt the bulbous structures characteristic of the freely propagating flame. The symmetric Gaussian curvature peaked near zero is a signature of the rib-like structures seen in Figure 6.

(a)
(b)

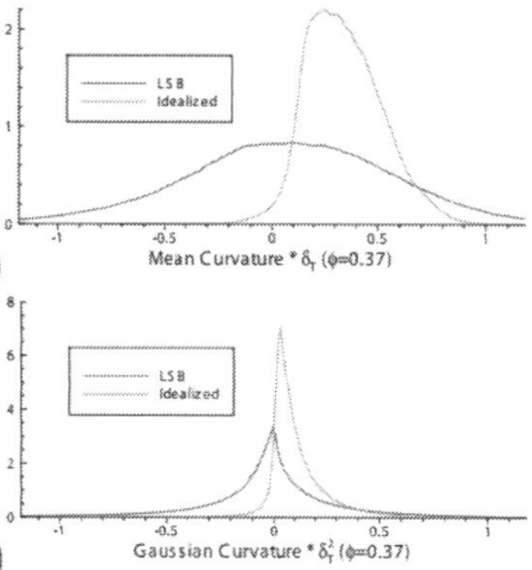

Figure 8: Distributions of mean and Gaussian curvatures of the flame isotherm (conditioned on $r<5$ cm, $z<8$ cm), scaled to the thermal thickness of the laminar H_2-air flame at ϕ=0.37.

To relate the local geometry to the local flame properties, we define a local consumption-based flame speed on the isotherm. For this construction we first define a local coordinate system on that surface as depicted in Figure 9. Associated with each triangle on the surface, we can then associate a prismatic region. We can

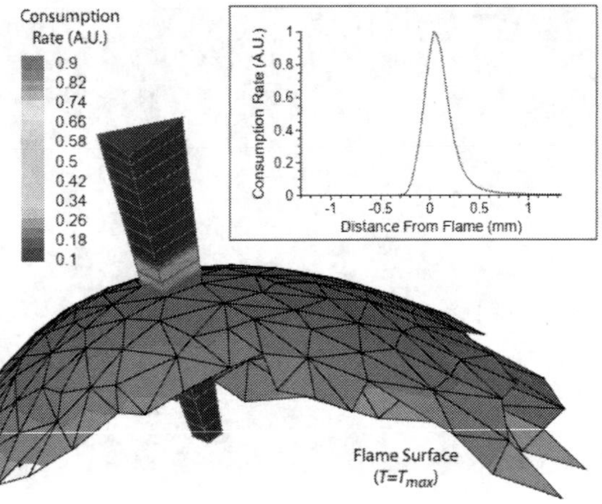

Figure 9: Prism-shaped volume, Ω, constructed using curves locally normal to the temperature isotherms (both the flame and Ω are colored by a typical profile of fuel consumption rate, ω_{H2}). The inset plot shows the variation of ω_{H2} normal to the flame surface. Ω discretizes the flame volume, and the integral curves of T that bound Ω define a set of local flame coordinates.

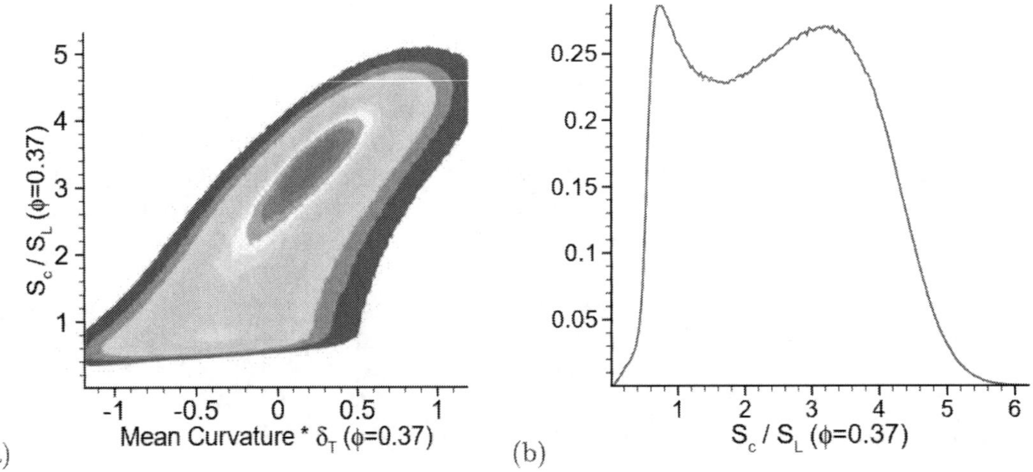

Figure 10: (a) Joint PDF of local consumption speed, s_c^ℓ, with mean curvature, (b) Distributions of s_c^ℓ over the entire conditioned flame surface. in both plots, the burning speed is normalized by the speed of the flat H_2-air flame at ϕ=0.37.

then define a local consumption speed as

$$s_c^\ell = \frac{1}{A\left(\rho Y_{\mathrm{H}_2}\right)_{\mathrm{in}}} \int_\Omega \omega_{\mathrm{H}_2} d\Omega \tag{1}$$

In Figure 10(a) we show a joint PDF of mean curvature and s_c^ℓ, normalized the laminar burning speed, s_L, of a flat steady hydrogen-air flame at ϕ=0.37. This figure shows a strong correlation between local burning speed and curvature. Regions of intense burning correspond to regions of positive curvature (positive curvature corresponds to regions where the center of curvature is on the products side of the flame interface). Also note that the dominant behavior corresponds to local burning speeds that range from 2 to 4 times the laminar flame speed. Even at zero curvature, the most probable local burning speed is three times the laminar flame speed. As discussed in [9], this enhanced burning reflects local enrichment of the fuel resulting from preferential diffusion of H_2 into burning regions. If we integrate out the curvature dependence, we can derive a PDF of local consumption speed along the flame front, shown in Figure 10(b). This PDF shows dual peaks, one corresponding to weak burning at less than the laminar flame speed and a second broader peak at about 3.3 s_l, again illustrating the structure of cellular structures.

Experimentally, it is not possible to directly measure fuel consumption. However, the similarities in the variation of fuel consumption rate and the OH concentation along the flame front suggest that there might be a correlation between them (see Figure 3). In Figure 11, we show a JPDF of s_c^ℓ versus the peak of OH mole fraction along the local coordinate normal to the flame surface. In the Figure, we have overlain the corresponding relationship for flat laminar flames at a range of equivalence ratios. The data serves to quantify the enrichment of the flame due to preferential diffusion mentioned earlier. In particular, the highest probability region in the JPDF corresponds to laminar flames burning in the range $\phi = 0.48$–0.50.

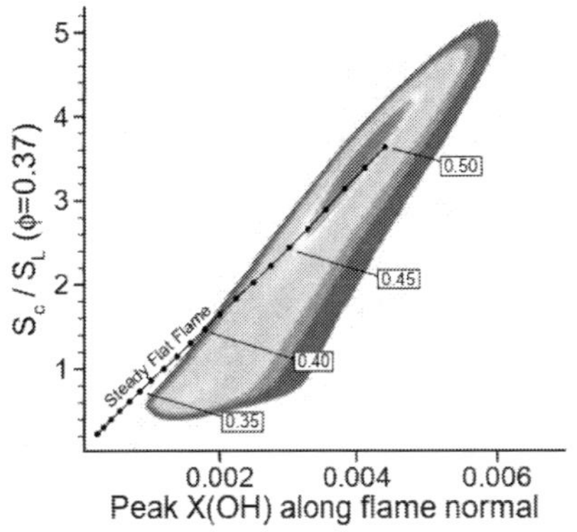

Figure 11: Joint PDF showing the strong correlation between local burning speed and the nearby peak values of the OH concentration.

There is a clear and strong correlation between peak OH and local flame speed, which is somewhat steeper than the locus of laminar flames, although the variability is somewhat broader for lower flame speeds. Nevertheless, this correlation could be used to relate measured OH signal from PLIF to a local consumption-based flame speed [17]. However, to fully develop this type of diagnostic, one would need to account for the effects of quenching. In particular, one could compute the synthetic laser intensity as in [18, 19], using the simulation data to compute quenching and develop a correlation between local flame speed and synthetic OH intensity. Of course, the viability of this type of approach would assume that image intensities can be measured with sufficient accuracy to interpret them quantitatively. It is worth noting that in this lean hydrogen-air mixture, the OH-PLIF signals will be low, making making quantitative measurement challenging to obtain.

5. Discussion and Future Work

We have presented results from a detailed numerical simulation of a laboratory-scale low-swirl turbulent lean hydrogen flame. The simulation incorporated detailed chemistry and transport without explicit turbulence models. Mean velocities and flame positions were compared with a similar laboratory configuration, and many key salient features of the flame and flowfield were reproduced. The turbulent flame surface was found to be highly corrugated by the turbulence and shows considerable local variability in instantaenous fuel consumption rates. The local and global distribution of OH compared well with experiment, and was shown to correlate well with the local speed of flame propagation. Compared to a freely propagating idealized case, the local curvature of the flame surface shifts from predominantly positive (where the center of curvature is on the products side of the flame) to symmetrically distributed around zero, with a much broader distribution. The predominant shape of the burning cells also shifts dramatically from spherical to cylindrical.

This work lays the foundation for a program of closely collaborative inverstigations between computational and experimental combustion scientists. Practical difficulties remain, including gathering suitable inlet profile specifications from the experiments, and extremely long simulation times which complicate the gathering of converged statistics of the quasi-steady flow, in spite of the highly optimized solution strategy. Nevertheless, the resulting simulations can be used to explore experimental diagnostic assumptions and to gather higher-dimensional statistics and correlations that are simply unavailable from the experiment.

Future work will involve the incorporation of emission chemistry and further characterization of the burner configurations under a variety of fuels relevant to ultra-low emission burner scenarios. Related studies currently underway are related to turbulence/chemistry interactions across a much more broad range of fluctuation intensities, and in high-pressure environments. The studies provide valuable insight that will enable the development of reduced models for turbulent flame propagation.

Acknowledgments

This work was supported by the SciDAC Program of the DOE Office of Mathematics, Information, and Computational Sciences and the experimental effort supported by the Advanced Turbine Program of the DOE Office of Fossil Energy both under under the U.S. Department of Energy under contract No. DE-AC02-05CH11231. The computations presented here were performed on Franklin, the XT4 at NERSC as part of an INCITE award. The authors would like to thank A. Dreizler and P. Petersson for providing experimental velocity measurements used to define the boundary conditions.

References

[1] Chan C K, Lau K S, Chin W K and Chang R K 1992 *Proc. Combust. Inst.* **24** 511–518
[2] Peterson P, Olofsson J, Brackman C, Seyfried H, Zetterberg J, Richter M, Alden M, Linne M, Cheng R, Nauert A, Geyer D and Dreizler A 2007 *Appl. Opt* **46** 3928–3936
[3] Cheng R K 1991 *Combust. Flame* **101** 1–14
[4] Nogenmyr K, Peterson P, Bai X S, Nauert A, Olofsson J, Brackman C, Seyfried H, Zetterberg J, Li Z S, Richter M, Dreizler A, M, Linne and Alden M 2007 *Proc. Combust. Inst* **31** 1467–1475
[5] Mansour M and Chen Y C 2007 *Experimental Thermal Fluid Sci. in press*
[6] Cheng R K, Littlejohn D, Strakey P A and Sidwell T 2009 *Proc. Combust. Inst.* **32** 21–46
[7] Cheng R K and Littlejohn D 2008 Effects of combustion geometry on the flowfields and flame properties of a low-swirl burner *Proceedings of GT2008: ASME Turbo Expo 2008* (Berlin, Germany: ASME) paper GT2008-50504
[8] Bell J B, Cheng R K, Day M S and Shepherd I G 2007 *Proc. Combust. Inst.* **31** 1309–1317
[9] Day M S, Bell J B, Bremer P T, Pascucci V and Beckner V E 2009 *Combustion and Flame* **156** 1035–1045
[10] Rehm R G and Baum H R 1978 *N. B. S. J. Res.* **83** 297–308
[11] Ern A and Giovangigli V 1994 *Multicomponent Transport Algorithms* (*Lecture Notes in Physics* vol m24) (Berlin: Springer-Verlag)
[12] Ern A and Giovangigli V 2005 *J. Comput. Phys.* **120** 105–116
[13] Day M S and Bell J B 2000 *Combust. Theory Modelling* **4** 535–556
[14] Rendleman C A, Beckner V E, Lijewski M, Crutchfield W Y and Bell J B 2000 *Computing and Visualization in Science* **3** 147–157
[15] Nogenmyr K J, Fureby C, Bai X, Petersson P, Collin R and Linne M 2008 *Combustion Flame* **155** 357
[16] Nogenmyr K J, Fureby C, Bai X S, Petersson P, Collin R and Linne M 2009 *Combustion and Flame* **156** 25–36
[17] Yamamoto K, Ozeki M, Hayashi N and Yamashita H 2009 *Proc. Combust. Inst.* **32** 1227–1235
[18] Bell J B, Day M S, Grcar J F, Bessler W G, Schultz C, Glarborg P and Jensen A D 2002 *Proc. Combust. Inst.* **29** 2195–2202
[19] Connelly B C, Bennett B A V, Smooke M D and Long M B 2009 *Proc. Combust. Inst.* **32** 879–886

Journal of Physics: Conference Series **180** (2009) 012032 doi:10.1088/1742-6596/180/1/012032

Shock-turbulence interaction: What we know and what we can learn from peta-scale simulations

Sanjiva K. Lele[1] **and Johan Larsson**[2]

[1]Professor, Dept. of Aeronautics & Astronautics, Durand Building, Stanford University, Stanford, CA 94305-4035

[2]Research Associate, Center for Turbulence Research, Stanford University, Stanford, CA 94305-3035

E-mail: lele@stanford.edu

Abstract. Many applications in engineering and physical sciences involve turbulent flows interacting with shock waves. High-speed flows around aerodynamic bodies and through propulsion systems for high-speed flight abound with interactions of shear driven turbulence with complex shock waves. Supernova explosions and implosion of a cryogenic fuel pellet for inertial confinement fusion also involve the interaction of shockwaves with turbulence and strong density variations. Numerical simulations of such physical phenomena impose conflicting demands on the numerical algorithms. Capturing broadband spatial and temporal variations in a turbulent flow suggests the use of high-bandwidth schemes with minimal dissipation and dispersion, while capturing the flow discontinuity at a shock wave requires numerical dissipation.

Results from three promising shock-capturing schemes a) high order WENO, b) nonlinear artificial diffusivity with compact finite differences, and c) a hybrid approach combining high-order central differencing with WENO near the shocks are compared using the Taylor-Green problem and compressible isotropic turbulence with eddy-shocklets. The performance of each scheme is characterized in terms of an *effective bandwidth*. The comparison highlights the damaging effect of numerical dissipation when the WENO scheme is applied everywhere. The hybrid approach is found to be best suited for studying shock-turbulence interactions.

Results from previous DNS and LES studies of the canonical shock-turbulence interaction problem, i.e. the interaction of isotropic turbulence with a (nominally) normal shock and comparison with available theory and experimental data are recalled. The principal physical effects include the amplification of turbulent kinetic energy across the shock and its anisotropy, change in turbulence length scales across the shock, departure from the common assumption of strong Reynolds analogy (used in modeling turbulence in high-speed flows), and the distortion of the shock due to its interaction with the incident turbulence. In this context new results from our SciDAC sponsored project are shown and open issues are mentioned. New DNS results achieve a significantly larger turbulence Reynolds number and allow an exploration of the nonlinear effects. While the linear interaction theory of Ribner provides useful estimates of the amplification of vorticity fluctuations across the shock, it misses the strong nonlinear dynamics of the energized and highly anisotropic vorticity downstream of the shock. It is found that previous DNS studies also underestimated this effect. The simulations show that turbulent self-stretching and tilting mechanisms bring about a relatively rapid *return to isotropy in the turbulent vorticity field*. The turbulent velocity field, however, does not show any appreciable tendency towards isotropy. It is further observed that when the turbulence interacting with the shock is sufficiently energetic the instantaneous shock structure is significantly modified; local regions of significant over-compression are found as well as regions where the mean shock compression is nearly isentropic. Estimates for the computational resources necessary for studying this fundamental shock-turbulence interaction problem at higher Reynolds number on peta-scale computing systems are given.

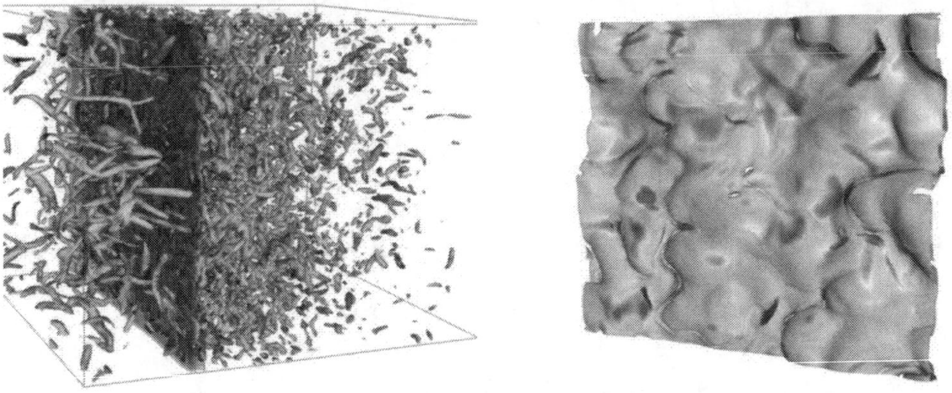

Figure 1. Shock/turbulence interaction. Left: Turbulent eddies (green structures, flowing from left to right) are compressed and amplified upon passing through a stationary shock (thin blue sheet). Right: Strongly wrinkled shock in the nonlinear regime with strong incoming turbulence, with colors indicating regions of high (red) and low (blue) streamwise velocity.

1. Introduction

Most fluid flows of scientific and engineering interest involve and are strongly affected by turbulence. For example, turbulence has a great effect on the process of mixing, which in turn has a major effect on flow separation, combustion, and other phenomena. When interacting with a shock wave, the state of turbulence changes dramatically. Thus it is clear that the interaction of turbulence with a shock wave is a key feature in many high-speed engineering systems. The most fundamental problem in this area is that of isotropic turbulence passing through a nominally normal shock. This idealized problem allows a study of the mechanics of the mutual interaction between the turbulence and the shock wave without the complications which arise from the boundary layer separation and the attendant low frequency unsteadiness (cf. Dolling [1] and Dupont *et al.* [2] for discussions of the shock wave turbulent boundary layer interaction).

This canonical problem shown in figure 1 has received considerable attention in past theoretical [3], experimental [4, 5], and computational [6–8] studies. The linear theory by Ribner [3] predicts that the velocity and vorticity variances are amplified during the shock interaction, and that there is an inviscid adjustment region behind the shock where the evanascent acoustic modes and turbulence adjusts to establish the post-shock state. This theory is formally valid in the limits of infinite Reynolds number and zero turbulent Mach number, but has thus far [6–8] only been tested at Reynolds numbers of $Re_\lambda \approx 20$. At these low Reynolds numbers, the viscous decay behind the shock is comparable to or larger than the inviscid adjustment, which makes direct comparisons to the linear theory difficult. Furthermore, the spectrum of turbulence is not truly broadband at such Reynolds numbers. The objective of the present study is to use direct numerical simulation (DNS) to produce a set of highly resolved databases at higher Reynolds number, to be used in examination of the validity of the linear theory as well as in *a priori* and *a posteriori* studies of modeling shock/turbulence interaction in the context of large eddy simulation (LES) and engineering turbulence closures.

2. Numerical method assessment for shock-turbulence interaction

A collaborative assessment of several numerical approaches to shock capturing on a suite of benchmark test problems deemed relevant for studying the interaction of shocks with turbulence was undertaken by our SciDAC project team. A comprehensive discussion of the test problems, algorithms, and detailed results are given in Ref. [9]. A short summary is included here for completeness. Results from three approaches are compared in this paper.

2.1. WENO
Weighted essentially non-oscillatory (WENO) schemes are generally considered to be state-of-the-art for shock-capturing. Here we include a standard 7th-order WENO method with Roe flux-splitting [10].

2.2. Nonlinear artificial viscosity
The nonlinear artificial viscosity/hyperviscosity approach originally proposed by Cook and Cabot [11] and further refined by Cook [12] is used. It essentially consists of adding artificial (or numerical) viscosities, diffusivities, and conductivities to the physical properties that are designed to smooth out discontinuities sufficiently such that they can be captured by a high-order central difference scheme, to mimic the energy transfer to unresolved small scales (i.e., function as a subgrid scale model in the context of large-eddy simulation), and to vanish in well resolved regions of the flow. These artificial properties take the form of high order derivatives of the resolved velocity field, thus become large around discontinuities and small elsewhere. The augmented governing equations are integrated using high-order compact finite differences. To maintain stability and minimize aliasing errors, the solution is filtered after each time step using a compact filter biased towards the smallest scales. Results shown in this paper were obtained using an in-house code by A. Bhagatwala which has very good resolution properties [13]. While the method was originally intended to always include all different artificial fluid properties, we consider two different permutations here: a minimally dissipative method with only the artificial bulk viscosity and the compact filter (the minimal requirements for capturing shock waves), and the same with added shear-viscosity (closer to the original [11, 12] model).

2.3. Hybrid approach
The hybrid method is based on the idea that broadband turbulence and discontinuities represent different physics and thus should be treated by different numerical methods. It consists of essentially three components that can be chosen rather freely: a central finite-difference scheme for the smooth regions, a shock-capturing scheme for the regions containing discontinuities, and a solution-adaptive sensor which identifies these regions. For shock-capturing, a 5th-order WENO method [10] is chosen due to its ability to capture shocks in a sharp fashion with few adjustable parameters. Away from the shocks, an 6th-order central difference scheme is used on a split form of the convective terms that improves nonlinear stability. These schemes are coupled in a conservative manner that preserves stability [14]. The shock sensor is inspired by the observation that shock waves are associated with large negative dilatation, whereas turbulence is associated with large vorticity. Comparing these quantities then yields an effective shock sensor. The primary advantage of the hybrid approach, as compared to a 'shock-capturing only' method, is that numerical dissipation is restricted to the regions flagged by the sensor, which will be shown to drastically improve predictions of turbulence spectra.

3. Effective bandwidth on nonlinear problems
Results from the Taylor-Green problem and from compressible isotropic turbulence are used to characterize the range of well-resolved scales or *effective bandwidth* of the algorithms being used. The effective bandwidth is quantified by comparing the results to those obtained from a reference calculation using much higher resolution; it was shown in Ref. [15] that the reference calculations are of sufficiently high quality to be useful as a benchmark. The effective bandwidth is defined as the wavenumber range over which the computed energy spectrum is within a specified tolerance

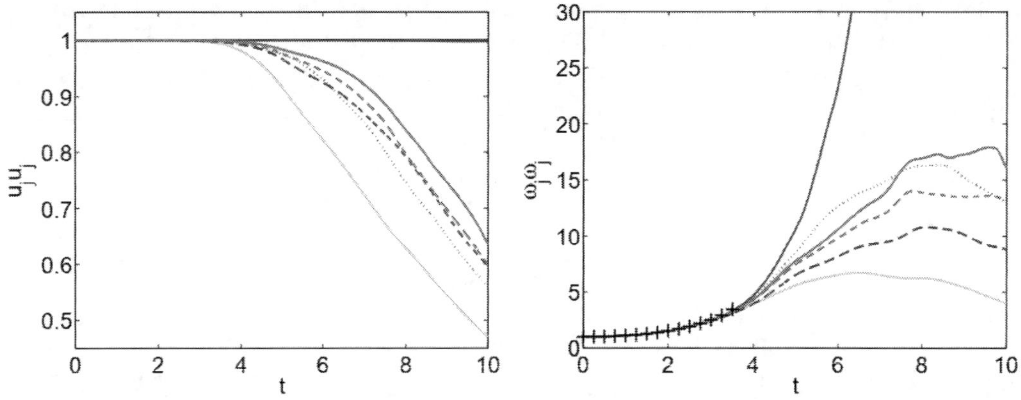

Figure 2. Evolution of normalized kinetic energy (left) and enstrophy (right) for Taylor Green problem on 64^3 grids. Different curves correspond to different algorithms. Solid blue: hybrid method; solid red: optimized compact scheme; dashed red: optimized compact scheme with shear-viscosity; dashed blue: hybrid with 8th order dissipation; dotted blue: hybrid with dynamic Smagorinsky model; green: WENO. Also plotted with + symbols is the analytic solution valid for short times.

from the 'reference energy spectrum' obtained in the reference calculation. The tolerance is set to be a deviation of $\pm 25\%$ from the reference spectrum.

3.1. Taylor-Green results

Figure 2 shows the evolution of kinetic energy and enstrophy for the Taylor-Green problem. For this inviscid problem there is no viscous dissipation and the kinetic energy should remain the same as its initial value. However beyond the non-dimensional time of 3 units there is a rapid build-up of fine-scale structures in the flow due to vortex-stretching and tilting, and numerical schemes begin to dissipate the kinetic energy at various rates. The hybrid scheme is exceptional in that it conserves kinetic energy exactly and maintains this property even for severely under-resolved flow features. Consequently the enstrophy growth predicted by the hybrid scheme is the highest amongst the methods used.

The effective bandwidth of the various schemes can be judged from figure 3 where compensated kinetic energy spectra, i.e. $E(k)/E_{ref}(k)$, are plotted at $t = 5$ (after the small-scale growth). The lines corresponding to $\pm 25\%$ deviation are also shown. The effective bandwidth for WENO is only about $1/4$ of the wavenumber range which can be represented on mesh used. The non-dissipative methods (hybrid and compact) are accurate up to almost $2/3$ of the wavenumber range, while the different forms of added dissipation lowers this to about $1/2$.

3.2. Compressible turbulence with eddy shocklets

A more stringent test case is that of decaying isotropic turbulence at high turbulent Mach number M_t. The turbulence spontaneously generates eddy shocklets which activate the various shock-capturing numerics and excessive numerical dissipation can lead to dramatic errors. We first compute a well-resolved DNS on a 256^3 grid at initial $M_t = 0.6$ and microscale Reynolds number $Re_\lambda = 100$. On this grid all methods converge to the same solution.

We then assess the different algorithms by comparing results obtained on a 64^3 grid. The decay of turbulent kinetic energy (TKE) with time is not sufficiently sensitive to judge the difference between the various schemes used here [15]; it is thus not shown here. Figure 3 shows the effective bandwidths obtained in the simulation of this nonlinear compressible turbulence

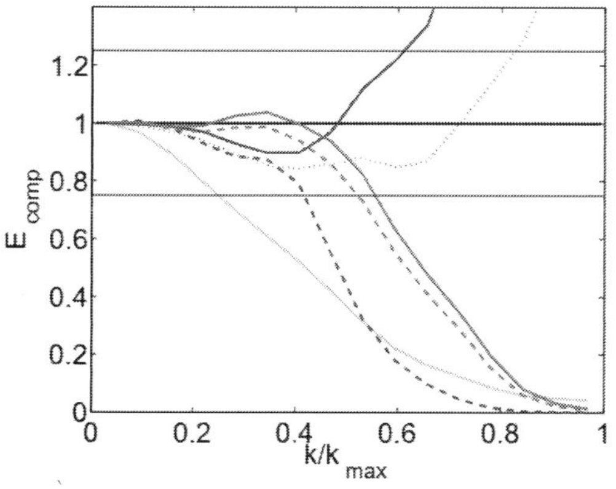

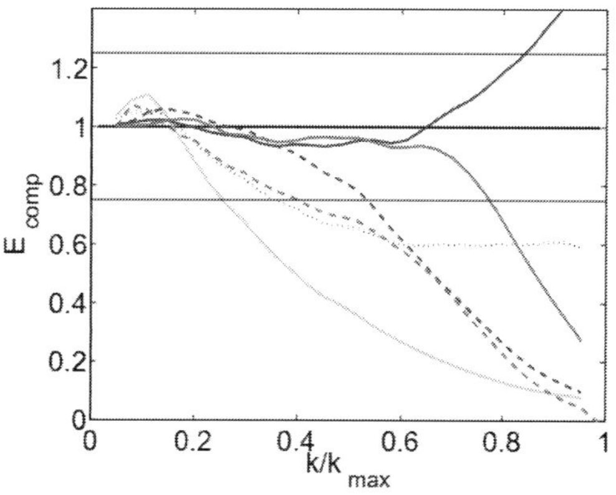

Figure 3. Compensated kinetic energy spectrum (at $t = 5$ on left panel for Taylor Green problem and at $t = 4$ (right panel) for compressible isotropic turbulence) on 64^3 grids. Different curves correspond to different algorithms. Solid black: converged reference; solid blue: hybrid method; solid red: optimized compact scheme; dashed red: optimized compact scheme with shear-viscosity; dashed blue: hybrid with 8th order dissipation; dotted blue: hybrid with dynamic Smagorinsky model; green: WENO.

problem. These results are consistent with the trends noted earlier for the Taylor-Green problem. The WENO scheme is accurate over only 1/4 of the wavenumber range. The compact scheme with artificial viscosity and hybrid scheme with high-order dissipation or dynamic SGS model all capture approximately 1/2 of the wavenumber range accurately. Only the minimally dissipative schemes capture 2/3 of the wavenumber range.

These results characterize the performance of various shock-capturing schemes on compressible turbulence problems in a fundamental way. Assessment of several additional numerical schemes was included in our collaborative work under the SciDAC project. These tests also included test cases emphasizing shock capturing ability for strong shocks, and interactions with density interfaces. A comprehensive report on these comparisons is available through the project website (see http://shocks.stanford.edu).

4. Canonical shock/turbulence interaction

The most fundamental shock/turbulence interaction problem is that of isotropic turbulence passing through a nominally normal shock wave (or, equivalently, a normal shock passing through isotropic turbulence). The essence of the problem is shown in figure 1. The incoming turbulence is isotropic, as evidenced by the random orientation of the vortex cores. The shock compresses the turbulence in the x direction, increasing the vorticity and making the post-shock turbulence axisymmetric with vortex cores predominantly oriented in the $y-z$ plane. Ribner [16] studied the problem analytically by solving the linearized Euler equations with linearized shock jump conditions for incoming purely vortical turbulence. This linear interaction analysis (LIA) assumes that the turbulence comprises a small perturbation relative to the shock and that nonlinear effects in the post-shock evolution are small (as well as the standard assumption of a difference in time scales). In addition, the Rankine-Hugoniot shock jump conditions are incorporated into LIA; one consequence is that LIA captures the generation of sound and entropy waves from incoming purely vortical turbulence.

Lee *et al.* performed a set of landmark direct numerical simulations (DNS) of canonical shock/turbulence interaction in a sequence of papers [6, 7]. The first of these papers considered shocks at Mach numbers up to 1.2 where the viscous structure of the shock was resolved; these were therefore truly direct solutions of the Navier-Stokes equations. In the second paper they verified that these "true" DNS results at Mach 1.2 could be replicated by instead capturing the shock (at considerably lower cost), provided sufficient grid resolution in the shock-normal direction at the shock. This methodology was then used to compute cases at Mach numbers up to 3. When comparing the results to LIA predictions, they found that LIA realistically represents many features, including the amplification of transverse vorticity, the amplification and post-shock evolution of the Reynolds stresses, and the decrease in transverse Taylor length scale. Mahesh *et al.* [8] considered the influence of entropy fluctuations in the upstream turbulence, and found that negatively correlated velocity and temperature fluctuations lead to enhanced amplifications of turbulence kinetic energy and vorticity. Later, Jamme *et al.* [17] resolved the viscous shock structure at Mach 1.2 and 1.5, and essentially confirmed the findings of Mahesh *et al.*. Barre *et al.* [5] studied the problem experimentally in a wind tunnel, and measured velocity variances using hot-wires and LDV.

The present study builds on these previous studies, especially those by Lee *et al.* [6, 7]. In this paper, DNS in the extended sense of capturing the shock while directly resolving all scales of turbulence is used. Highlights of the new DNS results are shown in this paper, with a more detailed account available in Larsson and Lele [18]. It is shown that a simple argument about the Kolmogorov scale implies that DNS requires a refined grid in both the shock-normal *and* the transverse directions to fully resolve the viscous scales of turbulence. This is verified by a grid convergence study, and implies that the calculations in the studies mentioned above were, most likely, under-resolved. The present DNS data is fully resolved, which leads to larger differences between the data and LIA. The Reynolds stresses are more anisotropic in the present DNS, and there are qualitative differences in the Taylor length scales. Moreover, the vorticity components return to isotropy in all cases in the present study. This raises the interesting question of whether under-resolution of the post-shock turbulence in previous DNS essentially neglects some phenomenon that is also neglected in the LIA. This will be explored below.

Throughout this paper the subscripts "u" and "d" refer to average states upstream and downstream of the shock, respectively. For quantities that evolve in the streamwise direction, these states are obtained by extrapolation to the average shock location. The mean flow and turbulent Mach numbers are defined as $M = u_{1,\mathrm{u}}/c_\mathrm{u}$ and $M_t = \sqrt{\langle u_i' u_i' \rangle_\mathrm{u}}/c_\mathrm{u}$, respectively, where u_1 is the streamwise velocity and c is the speed of sound. To minimize errors around the shock, a stretched grid is used in the streamwise direction with finer mesh spacing in the vicinity of the shock. The inflow turbulence is taken from a pre-computed database of isotropic turbulence. At the outflow, a sponge region is used to gently drive the flow towards a laminar state.

4.1. Grid resolution requirements

The most severe resolution requirement for DNS of shock-turbulence interaction come from the need to resolve the dissipative scale motions in the post-shock turbulence. In addition, since the vorticity field in the post-shock turbulence rapidly becomes isotropic, a finer grid in the post-shock region is needed in all directions. Larsson and Lele [18] estimated the post-shock Kolmogorov scale using rapid-distortion theory scaling. Comparison of this estimate with DNS data is shown in figure 4. They also conducted grid refinement studies for several shock-turbulence interactions using successively refined grids. At the relatively modest turbulence microscale Reynolds number $Re_\lambda = 40$, a grid with $1040 \times 384^2 \approx 153 \times 10^6$ grid points was found to be necessary to ensure that the estimated error in the velocity and vorticity variances was less than 2%.

4.2. Turbulence amplification and anisotropy

The variances of velocity and vorticity fluctuations are shown in figure 5 for a sequence of cases. The transverse vorticity is directly amplified at the shock due to the compression, while the

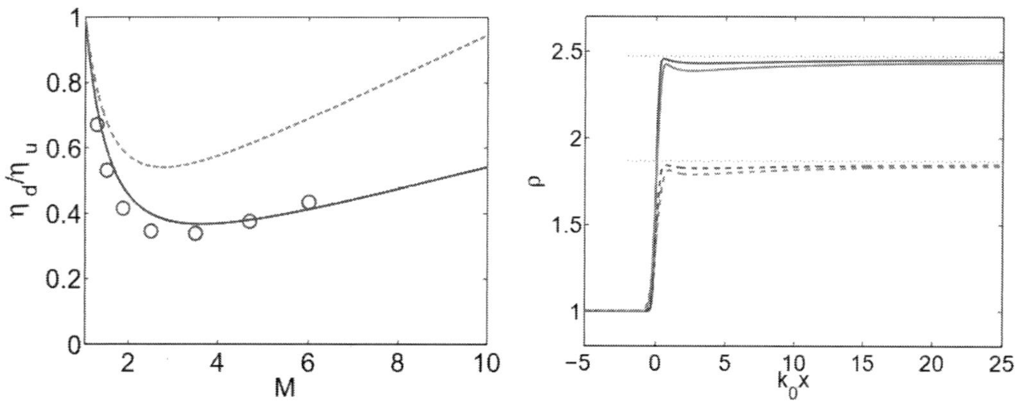

Figure 4. Left: Change in the Kolmogorov scale across the shock as a function of Mach number. Estimate from Ref. [18] in lines with $\gamma = 1.4$ (solid blue) and $\gamma = 5/3$ (dashed red). Results from DNS in symbols. Right: Mean density profiles for mean Mach number M of 1.5 (dashed lines) and 1.9 (solid lines), with turbulent Mach number M_t of 0.22 (blue) and 0.31 (red). The inviscid Rankine-Hugoniot jumps are shown in dotted black lines. Note that the mean pressure profiles are similar.

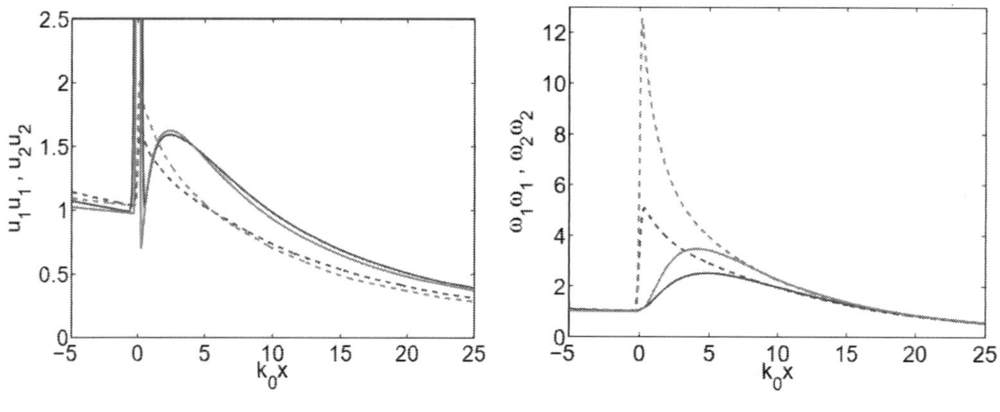

Figure 5. Canonical shock/turbulence interaction on 1040×384^2 grid. Left: variance of Reynolds stress ($\langle u'u' \rangle$ in solid lines, $\langle v'v' \rangle$ in dashed lines). Right: variance of vorticity ($\langle \omega_1 \omega_1 \rangle$ in solid lines, $\langle \omega_2 \omega_2 \rangle$ in dashed lines). Mean Mach number M of 1.9 (blue) and 3.5 (red).

streamwise vorticity is initially unchanged. Behind the shock, the out-of-equilibrium turbulence adjusts towards an isotropic state, although the Reynolds stresses never reach isotropy in these runs. One outstanding question in shock/turbulence interaction is whether the turbulence truly returns to isotropy. The low Reynolds number of the present (and all previous) runs implies a quick viscous decay behind the shock, which overwhelms any return-to-isotropy. Large-scale calculations at higher Reynolds numbers are needed to illuminate this issue.

4.3. Mean fields

The coupling between the changes in the turbulence and the mean flow can be illustrated by the profiles of mean density across the shock. Figure 4 shows such profiles for a sequence of simulations at different shock Mach number M and turbulence Mach number M_t. Note that for a fixed low value of $M_t = 0.15$ as M is increased the mean profiles show the expected tendency

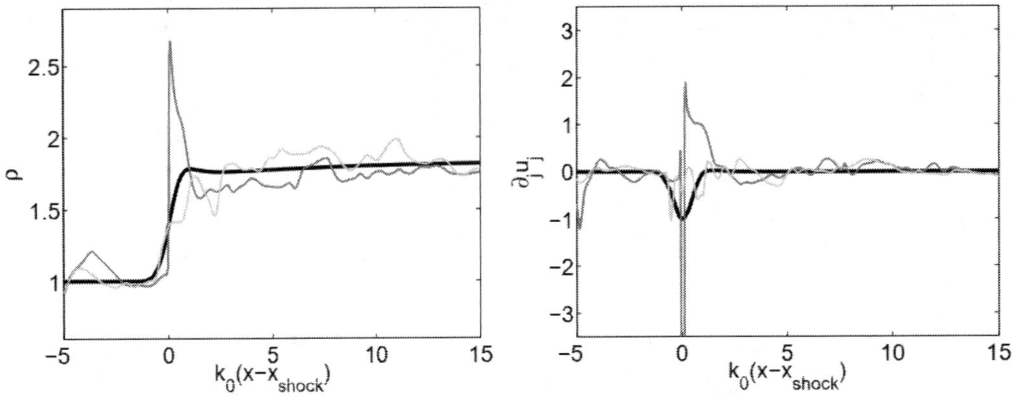

Figure 6. Instantaneous (lines in various colors) and average (thick black line) profiles of density (left) and dilatation (right) across the shock for $(M, M_t) = (1.5, 0.38)$. The instantaneous profiles are taken at different cross-stream positions but at the same time.

following the standard Rankine-Hugoniot jump conditions. However, when we consider the changes at fixed M but increasing M_t an effect associated with turbulence becomes apparent. Profiles of mean density (and pressure) display a more complex two-stage structure at higher M_t; evidently in the immediate vicinity of the shock the mean density rise is smaller than the Rankine-Hugoniot jump, with this deficit recovered in an extended region downstream of the shock. Further study of the change in mean shock structure for sufficiently vigorous incident turbulence is warranted. The present data are qualitatively consistent with an analysis of the turbulent flow shock jumps by Lele [19], who used rapid distortion theory to close the jump conditions.

4.4. Change in shock structure

At high enough turbulent Mach number M_t, the turbulent pressure fluctuations in the turbulence upstream of the shock become comparable to the pressure jump associated with the shock, which significantly alters the instantaneous shock structure. A snapshot of a strongly wrinkled instantaneous shock is shown in figure 1. It is clear that the shock is not simply planar at high M_t, but instead a rather complex object. The small "holes" in the shock are regions of nearly isentropic smooth compression. Some instantaneous profiles of density and dilatation along the x axis are shown in figure 6 for such a case. The instantaneous structure of the shock varies wildly, from being twice as strong as on average (note the significant over-compression and subsequent recovery) to being replaced by a relatively smooth compression wave. This intermittency is largely absent at lower values of M_t, and represents an outstanding fundamental question in shock/turbulence interaction. It is expected that at higher Reynolds number such intermittency may be quite strong and substantially affect the change across the shock wave.

4.5. Effect of numerical dissipation in shock-turbulence simulations

In a previous section we noted the effect of numerical dissipation in the context of the Taylor-Green and compressible isotropic turbulence problems. The 7th order WENO scheme provided an accurate representation of the energy spectrum in these flows only for 1/4 of the resolvable wavenumber range, i.e. significant dissipation was observed over the upper 3/4 range of wavenumbers. We now demonstrate the serious effect of numerical dissipation in the canonical shock-turbulence interaction problem. Results from the previous section on shock-turbulence simulations ($M = 1.5$ and $M_t = 0.22$) computed with the hybrid code are plotted against results

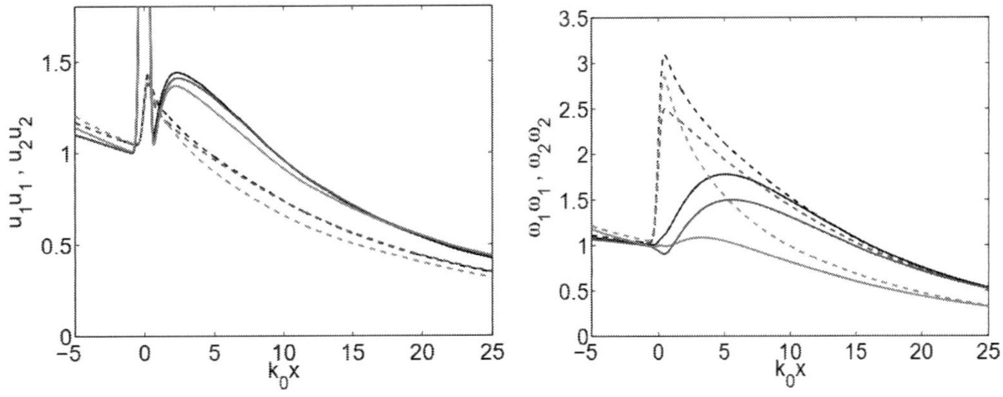

Figure 7. Effect of numerical dissipation on shock/turbulence interaction. Converged DNS on 1040×384^2 grid (black), 5th-order hybrid scheme on 347×128^2 grid (blue), and 7th-order WENO scheme on 347×128^2 grid (red). Left: variance of Reynolds stress ($\langle u'u' \rangle$ in solid lines, $\langle v'v' \rangle$ in dashed lines). Right: variance of vorticity ($\langle \omega_1 \omega_1 \rangle$ in solid lines, $\langle \omega_2 \omega_2 \rangle$ in dashed lines). Mean Mach number M of 1.5.

obtained with the 7th order WENO scheme applied everywhere in the flow. Additionally these results are compared with a finer resolution calculation using the hybrid code.

Figure 7 shows the streamwise evolution of the variances of fluctuating velocity and vorticity. The quantities plotted are normalized by the values immediately upstream of the shock allowing a quantitative comparison of the curves. Note that the hybrid results at the 'coarse' resolution of 347×128^2 are consistently close to the higher resolution 1040×384^2 results which are regarded as converged results. On the other hand the WENO results at 347×128^2 resolution show significant dissipation. In particular, the transverse component of the fluctuating velocity and all components of vorticity fluctuation show marked attenuation relative to the converged results. The nonlinear stretching/tilting of shock-normal vorticity is almost fully damped in the WENO results. These results suggest that strong caution must be exercised in using dissipative numerical algorithms in shock-turbulence interaction calculations. Even with relatively high-order dissipative schemes (in this case, 7th-order WENO), important turbulence dynamics can be numerically damped.

4.6. Cost estimates for simulation at higher Reynolds number

Well-resolved direct numerical simulation (DNS) requires adequate resolution of the viscous dissipation, which in isotropic turbulence amounts to having the maximum wavenumber $k_{\max} = \pi N_y / D \geq 1.5/\eta$, where N_y is the number of points in the transverse directions, D is the transverse domain size, and η is the Kolmogorov length scale. Consider turbulence at $M = 2$, $\gamma = 1.4$, and some Re_λ immediately upstream of the shock. The shock compression modifies the Kolmogorov scale by about a factor of $\eta_{\rm d}/\eta_{\rm u} \approx (\rho_{\rm d}/\rho_{\rm u})^{-1}(T_{\rm d}/T_{\rm u})^{3/8} \approx 0.46$ (using the simple theory presented in Ref. [18]); therefore the critical resolution is downstream of the shock. The upstream isotropic turbulence approximately satisfies $\eta_{\rm u} \approx 4.15\, L\, Re_\lambda^{-3/2}$, where L is the dissipation length scale. For the spectra used in this study we have $L \approx 1.7D/(\pi m_0)$, where m_0 is the most energetic mode. Putting this together yields the required transverse resolution of $N_y \geq 0.46\, m_0\, Re_\lambda^{3/2}$. This is a more stringent criterion than used in previous studies (since it accounts for the change in η across the shock), but is necessary to adequately resolve the post-shock dissipative motions.

To resolve the compressed post-shock motions (finer grid in x) and cover a sufficiently large streamwise domain, our present calculations use $N_x \approx 3N_y$ points in the streamwise direction. The grid required to compute a case with $Re_\lambda = 100$ and $m_0 = 4$ is therefore about $5520 \times 1840^2 \approx 19 \cdot 10^9$ points. Finally, the present calculations require about $50N_y$ time steps to fully remove the initial transients and compute statistics, so the sample calculation would require about 10^5 time steps. With the hybrid code this amounts to approximately 50 million core-hours, which shows the extreme cost of properly performed direct numerical simulation of shock/turbulence interaction.

5. Summary and outlook

The progress towards large-scale calculations of the canonical shock/turbulence interaction discussed here is part of a larger effort towards improved computational methods for complex shock/turbulence interaction problems. The canonical shock/turbulence problem described here serves two purposes in this larger effort. First, it is a useful problem for computational efficiency/accuracy assessment and improvement, since it includes all the relevant algorithmic complexities. It was used here to highlight the strong adverse effects of numerical dissipation. Even for a very high order method (7th order WENO) the numerical dissipation overwhelmed the turbulence evolution downstream of the shock. Secondly, the shock/turbulence calculations will produce large databases at different combinations of turbulent and mean flow Mach numbers, at higher Reynolds numbers than previously possible. These databases will be of use both in theoretical developments, e.g. for assessment of the validity of Ribner's linear theory at higher levels of compressibility, and in model development, e.g. for *a priori* studies or *a posteriori* comparisons in subgrid scale modeling.

Acknowledgments

Financial support has been provided by the DoE Scientific Discovery through Advanced Computing (SciDAC) program, with additional support from the Natural Sciences and Engineering Research Council of Canada to JL. The fine-grid calculations at $Re_\lambda = 40$ were run at the National Energy Research Scientific Computing Center under the ERCAP program. Higher $Re_\lambda = 60$ calculations which use up to 500M grid points are currently being conducted at ANL under the INCITE 2009 award.

References

[1] Dolling D S 2001 *AIAA J.* **39**
[2] Dupont P, Haddad C and Debieve J F 2006 *J. Fluid Mech.* **559** 255–277
[3] Ribner H S 1953 Convection of a pattern of vorticity through a shock wave NACA Report 1164
[4] Agui J H, Briassulis G and Andreopoulos Y 2005 *J. Fluid Mech.* **524** 143–195
[5] Barre S, Alem D and Bonnet J P 1996 *AIAA J.* **34** 968–974
[6] Lee S, Lele S K and Moin P 1993 *J. Fluid Mech.* **251** 533–562
[7] Lee S, Lele S K and Moin P 1997 *J. Fluid Mech.* **340** 225–247
[8] Mahesh K, Lele S K and Moin P 1997 *J. Fluid Mech.* **334** 353–379
[9] Johnsen E, Larsson J, Bhagatwala A V, Cabot W H, Moin P, Olsson B J, Rawat P S, Shankar S K, Sjögreen B, Yee H C, Zhong X and Lele S K 2009 *J. Comput. Phys.* under review
[10] Jiang G S and Shu C W 1996 *J. Comput. Phys.* **126** 202–228
[11] Cook A W and Cabot W H 2005 *J. Comput. Phys.* **203** 379–385
[12] Cook A W 2007 *Phys. Fluids* **19** 055103
[13] Lui C and Lele S K 2001 *AIAA* (2001-0291)
[14] Larsson J and Gustafsson B 2008 *J. Comput. Phys.* **227** 2886–2898
[15] Larsson J, Lele S K and Moin P 2007 *Annual Research Briefs* (Center for Turbulence Research) pp 47–57
[16] Ribner H S 1954 Shock-turbulence interaction and the generation of noise NACA Report 1233
[17] Jamme S, Cazalbou J B, Torres F and Chassaing P 2002 *Flow, Turbulence and Combustion* **68** 227–268
[18] Larsson J and Lele S K 2009 *Phys. Fluids* under review
[19] Lele S K 1992 *Phys. Fluids* **4** 2900–2905

High-fidelity simulations for clean and efficient combustion of alternative fuels

J. C. Oefelein[†], J. H. Chen[†] and R. Sankaran[‡]
[†]Reacting Flow Research Department, Sandia National Laboratories, Livermore, CA 94550
[‡]National Center for Computational Sciences, Oak Ridge National Laboratory, Oak Ridge, TN 37831

E-mail: oefelei@sandia.gov

Abstract. There is an urgent and growing demand for high-fidelity simulations that capture complex turbulence-chemistry interactions in propulsion and power systems, and in particular, that capture and discriminate the effects of fuel variability. This project addresses this demand using the Large Eddy Simulation (LES) technique (led by Oefelein) and the Direct Numerical Simulation (DNS) technique (led by Chen). In particular, we are conducting research under the INCITE program that is tightly coupled with funded projects established under the DOE Basic Energy Sciences and Energy Efficiency and Renewable Energy programs that will provide the foundational science required to develop a predictive modeling capability for design of advanced engines for transportation. Application of LES provides the formal ability to treat the full range of multidimensional time and length scales that exist in turbulent reacting flows in a computationally feasible manner and thus provides a way to simulate reacting flow phenomena in complex internal-combustion engine geometries at device relevant conditions. Application of DNS provides a way to study fundamental issues related to small-scale combustion processes in canonical configurations to understand dynamics that occur over a range of reactive-diffusive scales. Here we describe the challenges and present representative examples of the types of simulations each respective tool has been used for as part of the INCITE program. We focus on recent experiences on the Oak Ridge National Laboratory (ORNL) National Center for Computational Sciences (NCCS) Cray-XT Platform (i.e., Jaguar).

1. Introduction

Turbulent combustion processes are prevalent in a wide variety of propulsion and power systems including internal-combustion (IC) engines, gas-turbines and liquid-rockets. As such, development and rigorous validation of science-based predictive models for turbulent combustion is recognized as an important priority in research and there are a variety of challenges. Turbulent flows involving heterogeneous chemically reacting mixtures (as is the case for all propulsion and power systems) have a variety of complicating factors including highly nonlinear chemical kinetics, small-scale velocity and scalar-mixing, turbulence-chemistry interactions, compressibility effects (volumetric changes induced by changes in pressure), and variable inertia effects (volumetric changes induced by variable composition or heat addition). Coupling between processes occurs over a wide range of time and length scales, many being smaller than can be resolved in a numerically feasible manner. Further complications arise when multiple phases are present due to the introduction of dynamically evolving interface boundaries and the complex exchange processes that occur as a consequence.

At the device level, high-performance, dynamic stability, low pollutant emissions, and low soot formation must be achieved simultaneously in highly confined geometries that generate extremely complex flow and acoustic patterns. Flow and combustion processes are highly turbulent (i.e., integral-scale Reynolds numbers of $\mathcal{O}(10^5)$ or greater), and the turbulence dynamics are inherently dominated by geometry or various operating transients. In many cases operating pressures approach or exceed the thermodynamic critical pressure of the fuel or oxidizer. Operation at elevated pressures significantly increases the system Reynolds number(s) and inherently broadens the range of spatial and temporal turbulence scales over which interactions occur. The limitations and challenges associated simulating these phenomena requires that a hierarchy of approaches be taken to fully understand key processes and work toward predictive models. From this perspective, a hallmark of our combustion program is the interaction between theory, experiments, modeling, and simulations. Each approach complements the others. Here we highight two key research tools used toward this goal. The first is the solver developed by Oefelein *et al.* called RAPTOR [1–9]. The second is the solver developed by Chen *et al.* called S3D [10–15]. RAPTOR is a massively parallel flow solver designed specifically for application of the Large Eddy Simulation (LES) technique to turbulent, chemically reacting, multiphase flows. S3D is a massively parallel flow solver designed specifically for application of the Direct Numerical Simulation (DNS) technique. The combination of these techniques has enabled fundamental investigations of the fully coupled dynamic behavior of reacting flows with detailed chemistry. Following are representative results acquired from each as part of our DOE INCITE allocation.

2. Large Eddy Simulation

RAPTOR is a massively parallel flow solver designed specifically for application of LES to turbulent, chemically reacting, multiphase flows. It solves the fully coupled conservation equations of mass, momentum, total-energy, and species for a chemically reacting flow system (gas or liquid) in complex geometries. It accounts for detailed chemistry, thermodynamics, and transport processes at the molecular level and uses detailed chemical mechanisms. The code is sophisticated in its ability to handle complex geometries and a generalized subgrid-scale model framework. It is capable of treating spray combustion processes and multiphase flows using a Lagrangian-Eulerian formulation. The numerical formulation treats the compressible form of the conservation equations, but can be evaluated in the incompressible limit. The theoretical framework handles both multi-component and mixture-averaged systems. The baseline formulation employs a general treatment of the equation of state, thermodynamics, and transport properties that accommodates real gas or liquids with detailed chemistry (i.e., not constrained to ideal gas applications). Details are given by Oefelein [9].

As one example, we consider direct-injection (DI) processes for IC-engine applications. Direct injection has proven to be a promising option in Diesel and low temperature combustion engines. In conventional Diesel and homogeneous charge compression ignition (HCCI) applications, DI lowers soot and NO_x production and improves fuel economy. In hydrogen fueled engines (which are an attractive alternative to fuel cells), DI provides the appropriate energy density required for high efficiency and low NO_x emissions. To realize the full benefit of DI, however, the effect of various injection parameters such as injection timing, duration, pressure, and dilution, must be investigated and optimized under a range of engine operating conditions. In this work, we have developed a model for high-fidelity calculations of DI processes using LES and an advanced property evaluation scheme. The combination of High Performance Computing (HPC) and LES has significant potential to provide new insights into the dynamics of IC-engine flow processes, which is imperative for the development of next generation engines. The objective of this task is to fully integrate the combined merits of HPC and LES in a manner that provides some of the highest-fidelity, most detailed calculations ever performed to investigate turbulent reacting flow processes in IC-engines.

Recent calculations performed on the ORNL Cray-XT Platform (Jaguar) have been focused on detailed simulations of direct-injection (DI) processes in IC-engines, with emphasis on high-pressure processes relevant to low-temperature combustion applications. Fundamental high-fidelity representations of DI processes in IC-engines are an essential component in the development of a validated predictive modeling capability. However, detailed models are still lacking and few provide a rigorous description of transient processes that occur between the jets and the in-cylinder flow environment as fuel is injected into an engine. As part of our 2008-2009 INCITE award, we have

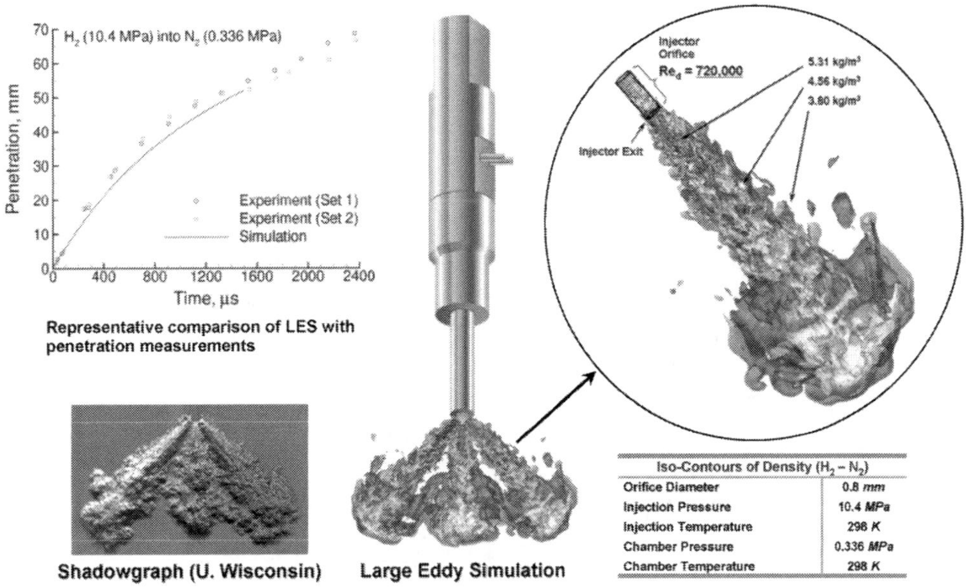

Figure 1: LES of high-pressure injection processes under actual operating conditions. To the left are penetration measurements and a corresponding shadow graph image acquired experimentally. In the center is a representative LES calculation at identical conditions showing the complete computational domain. To the right is an enlargement of one of the jets showing the turbulent structure at an instant in time.

performed one of the highest fidelity simulations of DI processes ever as a precursor to a full engine simulation (which will also be performed as part of our INCITE grant). Figure 1 shows a representative result, which has been validated using experimental data from the University of Wisconsin (see T.G. Drozda and J.C. Oefelein (2008). Large eddy simulation of direct injection processes for hydrogen and LTC engine applications. SAE World Congress, Paper 2008-01-0939). To the left of the figure are penetration measurements made experimentally along with our predicted results (which show the rate at which the fuel is injected into the engine) and a corresponding shadowgraph image (which shows the transient structure of the turbulent jets). In the center is the corresponding LES calculation showing the complete computational domain, which is an actual 3-orifice injector used in the optical engines at the Combustion Research Facility. To the right is an enlargement of one of the jets showing the turbulent structure in both the injector orifice and the ambient environment at an instant in time. The calculation was performed by gridding the entire injector, including the internal flow passages and needle valve actuator that controls the flow of fuel. A key issue in modeling flows at actual conditions is the fact that they exhibit extremely high Reynolds numbers. For the calculations performed here, the jet Reynolds number is 720,000. These studies are precursor calculations to a more detailed set of simulations that are currently being conducted in a full internal combustion engine. The objective is to identically match the operating conditions and geometry and in so doing establish a direct collaboration with our experimental efforts at the Combustion Research Facility.

3. Direct Numerical Simulation

Petascale simulations provide a way to study fundamental issues related to small-scale combustion processes in well-defined laboratory configurations, in environments where measurements are difficult or impossible to obtain. As one example, we have performed DNS of a lifted ethylene-air jet flame with emphasis placed on stabilization in a heated coflow. The objective was to understand how a lifted autoignitive flame is stabilized in a canonical configuration with relevance to diesel jet flame stabilization.

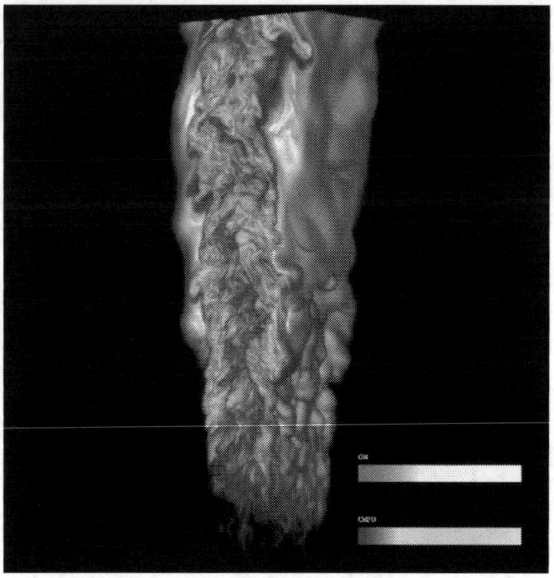

Figure 2: DNS of a lifted ethylene-air jet flame at Reynolds Number 10,000. The hydroxyl radical (red/white) denotes the lifted flame whereas formaldehyde (blue/green) denotes ignition intermediates upstream of the lifted flame.

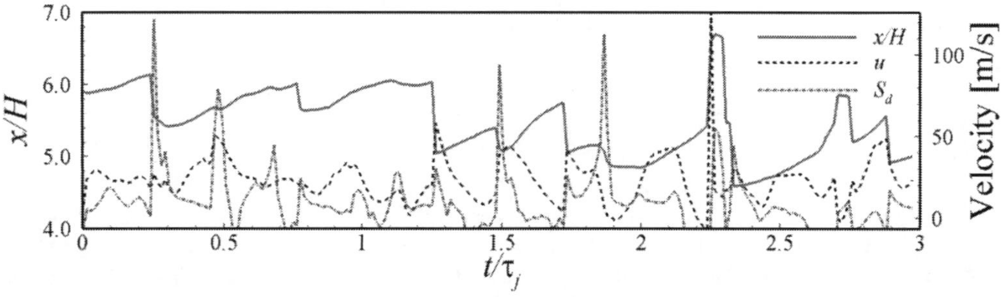

Figure 3: Time history of the axial fluctuation in the flame stabilization height, x/H, where H is the jet height, the local axial gas velocity (u) and the displacement speed of the stabilization point (S_d).

DNS of the near field of a three-dimensional spatially-developing turbulent lifted jet flame in a 1550K heated air coflow was performed with a reduced ethylene-air mechanism (22 species; 18 global steps; 167 reactions) [16] to determine the stabilization mechanism. The DNS was performed at a jet Reynolds number of 10,000 with 1.28-billion grid points. The results reveal that auto-ignition in a fuel-lean mixture immediately upstream of the flame base is the main source of stabilization of the lifted jet flame. This is verified by the presence of a key ignition precursor, HO_2, along with low levels of CH_3, OH and CH_2O upstream of where the flame is established as shown in the instantaneous image in Fig. 2. The dynamics of the fluctuations in the lifted flame stabilization point are studied by temporal tracking of the most upstream point where the hydroxyl radical is observed at a threshold of 5 percent of its peak value in the domain. The fluctuations of the flame lift-off height and the local gas and flame displacement speed are shown in Fig. 3. Large displacement speeds (i.e greater than 100 m/s) in excess of the laminar flame speed are indicative of spontaneous ignition. As well, the fast front propagation speed is accompanied by a rapid motion of the lift-off height upstream, similar to observations made in a lifted hydrogen/air jet flame [17]. Fundamental understanding of this issue, provided by detailed scalar and velocity statistics from the DNS, and many others is needed to develop robust and reliable combustion models for the combustion regimes observed under low-temperature combustion engine environments.

As a second example, we have performed DNS of ignition processes in a lean n-heptane/air mixture under HCCI conditions with temperature inhomogeneities. Homogeneous charge compression-ignition (HCCI) engines have been considered as an alternative for diesel and spark-ignition (SI) engines because HCCI engines can provide high efficiencies and very low NO_x and particulate emissions. However, under high-load HCCI operation, a high-rate of pressure rise during combustion can result in damaging engine knock. Thermal stratification induced by temperature inhomogeneities in engines has been proposed as a control measure for too rapid a pressure rise rate by spreading out the heat release through mixture preparation. To understand the effect of the thermal stratification on HCCI combustion characteristics, we performed two-dimensional DNS of lean n-heptane/air mixture ignition at constant volume with temperature inhomogeneities. As in hydrogen/air mixture ignition under HCCI conditions [18], large temperature fluctuations spread out the mean heat release rate and reduces its peak during both ignition stages. However, the overall heat release rate during the second-stage ignition is retarded as the temperature fluctuation increases. This fact will become important as we move forward and consider a variety of alternative fuels with wide variations in ignition characteristics. In addition, the effect of different temperature fluctuation length scales, larger temperature fluctuation RMS, equivalence ratio fluctuations, different turbulence time scales, and addition of EGR will be investigated. Several three-dimensional DNS studies will be performed to provide statistics for mixing and ignition model development and validation.

4. Acknowledgments

This work was supported by the U. S. Department of Energy, Office of Basic Energy Sciences, Division of Chemical Sciences, Geosciences, and Biosciences and the Office of Vehicle Technologies. Sandia National Laboratories is a multiprogram laboratory operated by Sandia Corporation, a Lockheed Martin Company, for the United States Department of Energy under contract DE-AC04-94-AL85000. The work at Oak Ridge National Laboratory (ORNL) was supported by and used resources of the National Center for Computational Sciences (NCCS) at ORNL, supported by the DOE Office of Science under contract DE-AC05-00OR22725. Computer time was granted under the Innovative and Novel Computational Impact on Theory and Experiment (INCITE) program.

References

[1] Oefelein J C and Sankaran R 2009 *Proceedings of the 21st International Conference on Parallel Computational Fluid Dynamics* Moffett Field, California
[2] Frank J H, Kaiser S A and Oefelein J C 2009 *Proceedings of the 6th Joint Meeting of the US Sections of the Combustion Institute, Paper 32D1* Ann Arbor, Michigan
[3] Sankaran V, Drozda T G and Oefelein J C 2009 *Proceedings of the Combustion Institute* **32** 1571–1578
[4] Oefelein J C 2008 *IFP Workshop on Large Eddy Simulation for Internal Combustion Flows* Ruell-Malmaison, France
[5] Drozda T G and Oefelein J C 2008 *SAE World Congress, Paper 2008-01-0939* Detroit, Michigan
[6] Drozda T G, Wang G H, Sankaran V, Mayo J R, Oefelein J C and Barlow R S 2008 *Combustion and Flame* **155** 54–69
[7] Williams T C, Schefer R W, Oefelein J C and Shaddix C R 2007 *Review of Scientific Instruments* **78** 035114–1–9
[8] Oefelein J C, Sankaran V and Drozda T G 2007 *Proceedings of the Combustion Institute* **31** 2291–2299
[9] Oefelein J C 2006 *Progress in Aerospace Sciences* **42** 2–37
[10] Hawkes E R, Sankaran R, Chen J H, Kaiser S and Frank J H 2009 *Proceedings of the Combustion Institute* **32** 1455–1463
[11] Richardson E S, Yoo C S and Chen J H 2009 *Proceedings of the Combustion Institute* **32** 1695–1703
[12] Lignell D, Chen J H and Smith P J 2008 *Combustion and Flame* **155** 316–333
[13] Sankaran R, Hawkes E R, Chen J H, Lu T and Law C K 2007 *Proceedings of the Combustion Institute* **31** 1291–1298
[14] Cook D J, Pitsch H, Chen J H and Hawkes E R 2007 *Proceedings of the Combustion Institute* **31** 2903–2911
[15] Sutherland J C, Smith P J and Chen J H 2007 *Combustion Theory and Modelling* **11** 287–303
[16] Lignell D O, Chen J H, Smith P J, Lu T F and Law C K 2007 *Combustion and Flame* **151** 2–28
[17] Yoo C S, Chen J H and Sankaran R 2009 *Journal of Fluid Mechanics* In Press
[18] Hawkes E R, Sankaran R, Pebay P P and Chen J H 2006 *Combustion and Flame* **145** 145–159

Journal of Physics: Conference Series **180** (2009) 012034 doi:10.1088/1742-6596/180/1/012034

Advanced coal gasifier designs using large-scale simulations

M Syamlal[1], C Guenther[1], A Gel[1,2] and S Pannala[3]

1 National Energy Technology Laboratory, Morgantown, WV 26505, U.S.A.
2 ALPEMI Consulting, LLC, Phoenix, AZ 85044, U.S.A.
3 Oak Ridge National Laboratory, Oak Ridge, TN 37831, U.S.A.

E-mail: madhava.syamlal@netl.doe.gov

Abstract. Porting of the legacy code MFIX to a high performance computer (HPC) and the use of high resolution simulations for the design of a coal gasifier are described here. MFIX is based on a continuum multiphase flow model that considers gas and solids to form interpenetrating continua. Low resolution simulations of a commercial scale gasifier with a validated MFIX model revealed interesting physical phenomena with implications on the gasifier design, which prompted the study reported here. To be predictive, the simulations need to model the spatiotemporal variations in gas and solids volume fractions, velocities, temperatures with any associated phase change and chemical reactions. These processes occur at various time- and length-scales requiring very high spatial resolution and large number of iterations with small time-steps. We were able to perform perhaps the largest known simulations of gas-solids reacting flows, providing detailed information about the gas-solids flow structure and the pressure, temperature and species distribution in the gasifier. One key finding is the new features of the coal jet trajectory revealed with the high spatial resolution, which provides information on the accuracy of the lower resolution simulations. Methodologies for effectively combining high and low resolution simulations for design studies must be developed. From a computational science perspective, we found that global communication has to be reduced to achieve scalability to 1000s of cores, hybrid parallelization is required to effectively utilize the multicore chips, and the wait time in the batch queue significantly increases the actual time-to-solution. From our experience, development is required in the following areas: efficient solvers for heterogeneous, massively parallel systems; data analysis tools to extract information from large data sets; and programming environments for easily porting legacy codes to HPC.

1. Introduction

Integrated Gasification Combined Cycle (IGCC) is a promising technology for meeting the growing demand for power using fossil fuel resources, while economically controlling the emission of CO_2 and other pollutants. A National Energy Technology Laboratory (NETL) study has shown that IGCC systems with carbon capture and storage (CCS) can generate power at efficiencies of 32%, compared to 25% for a conventional pulverized coal (PC) plant, and at a cost of electricity of 10.6 cents/kWh, compared to 11.9 cents/kWh for a PC plant [1]. The centerpiece of an IGCC system is the gasifier, which converts coal or other carbonaceous materials such as biomass into syngas, a mixture of CO and H_2. The syngas can be used for the production of liquid fuels and chemicals or for power generation.

In an IGCC system the syngas is shift converted into a mixture of CO_2 and H_2, CO_2 is captured and stored, and H_2 is used for power generation.

A reliable gasifier is critical for the commercial viability of IGCC. An advanced gasifier technology based on a *transport gasifier* is being developed at the Power Systems Development Facility (PSDF) in Wilsonville, Alabama, a joint project between the U.S. Department of Energy (DOE), Southern Company, and Kellogg Brown and Root (KBR). The gas-solids multiphase flow such as occurring in the *transport gasifier* is known to make the design of commercial-scale units using traditional scale-up methods unreliable [2]. Multiphase computational fluid dynamic (CFD) models are being developed at NETL to address that challenge, and this paper discusses the potential for using large-scale CFD simulations for gasifier design.

The validation of the NETL gasifier model with data from the PSDF pilot-scale facility has been reported before [3]. The results of the validation simulations were frequently communicated with the design engineers, and the acceptance of the results dramatically improved when simulations showed unexpected physical phenomena that were subsequently validated with experimental data. For example, the simulations showed that oxygen reached the upper region of the mixing section in the transport gasifier, whereas the engineers had expected that all the oxygen would be consumed in the lower region where the air is mixed with the hot, recycled char. This prediction was later verified when sampling probes were installed in the mixing section and the oxygen concentration in its upper region was measured. In another instance, the calculations showed high concentration of CO and solids near the top of the riser section of the transport gasifier. This prediction was also later validated with experimental data. These validated predictions showed that the model does not merely reproduce what is already known, but could provide information on unobserved phenomena that could be exploited for improving the gasifier design.

Another criticism that the design engineers often have is that the computational time is much larger than the physical time simulated. For example, typical validation simulations took over a million seconds of computational time (on 32-64 parallel processors) for simulating around fifteen seconds of gasifier operation. But the computational time is much smaller than the time taken to make geometric changes in the physical gasifier and the computational cost, significantly smaller. In some cases those are more useful comparisons. This was well illustrated when NETL researchers completed the simulation of the transport gasifier with certain design modifications in less than two weeks at a cost of about \$10 K, whereas the physical modification of the gasifier took around fourteen weeks at a cost of around \$6 M. The simulation predicted a significant increase in CO concentration in the syngas, which was found to agree well with data from the modified gasifier. The simulation had the added advantage that the results could be interrogated to explain why the CO concentration would increase [2]. This showed that the computations could be completed much faster (~10 times faster) than the physical modifications and required only a small fraction of the cost of modifications (< 0.2%).

The validated model was then used to study the impact of the exit geometry on syngas composition, the effect of reactor height on CO concentration, the coal trajectory within the gasifier, the gas temperature profiles in the gasifier, and the effect increasing the pressure. Later the model was used to help with the design of a Clean Coal Power Initiative (CCPI) gasifier (285 MW electric), which was scaled up from the pilot-scale gasifier (13 MW thermal), representing a factor of 50 scale-up. The multiphase CFD simulations were used to study the effect of pressure, the height to diameter ratio, coal feed rate, coal feed nozzle operation, solids circulation rate, and the effect of recycled syngas. Several sets of parametric evaluations showed an unexpected dependence of the gas species concentrations and solids temperature upon the coal jet penetration. Temperature dependency on the coal feed can impact the thermal decomposition of the coal, produce unwanted agglomeration, and could have a significant economic impact due to material degradation.

This result created interest in conducting simulations of the CCPI gasifier at greater grid resolutions. The rule-of-thumb for grid independence in gas-solids simulations is that the grid size should be of the order of 10 particle size [4]. This meant that the CCPI gasifier needed to be simulated with about 100 billion cells. This resolution would require well over three million compute cores and result in file sizes on the order of 745 GB for a single time record for each scalar tracked in the simulation. This was certainly beyond the computational capability available at NETL and, in fact, even at the fastest HPC platforms. However, the INCITE grant is allowing the group to conduct CCPI gasifier simulations at greater grid resolution than was previously possible. This paper describes how the model was improved to run on HPC platforms and presents some of the preliminary computational results.

2. Coal gasifier model

2.1. The multiphase CFD model

Gasifiers contain a reacting mixture of coal (or other solids) flowing with the gas. In the continuum modeling approach used in this study the coal particles are represented as a granular phase, which co-locates with the gas phase to form a multiphase mixture, the volume fraction ε_m giving the amount of phase m at each spatial location. The multiphase CFD model consists of the mass, momentum, energy, and species-mass balances for each phase, gas ($m = g$) or solids ($m = s$):

$$\frac{\partial}{\partial t}(\varepsilon_m \rho_m) + \nabla\cdot(\varepsilon_m \rho_m \vec{v}_m) = \sum_{l=1}^{N_m} R_{ml}$$

$$\frac{\partial}{\partial t}(\varepsilon_m \rho_m \vec{v}_m) + \nabla\cdot(\varepsilon_m \rho_m \vec{v}_m \vec{v}_m) = \nabla\cdot\overline{\overline{S}}_m + \varepsilon_m \rho_m \vec{g} + \sum_n \vec{I}_{mn}$$

$$\frac{3}{2}\varepsilon_m \rho_m \left(\frac{\partial \Theta_m}{\partial t} + \vec{v}_m \cdot \nabla\Theta_m\right) = \nabla\cdot\vec{q}_{\Theta_m} + \overline{\overline{S}}_m : \nabla\vec{v}_m - \varepsilon_m \rho_m J_m + \Pi_{\Theta_m} \quad (m \neq g)$$

$$\varepsilon_m \rho_m C_{pm}\left(\frac{\partial T_m}{\partial t} + \vec{v}_m \cdot \nabla T_m\right) = -\nabla\cdot\vec{q}_m + \sum_n \gamma_{mn}(T_n - T_m) - \Delta H_{rm}$$

$$\frac{\partial}{\partial t}(\varepsilon_m \rho_m X_{ml}) + \nabla\cdot(\varepsilon_m \rho_m X_{ml} \vec{v}_m) = R_{ml}$$

where subscript m and n stands for phases and l for a species in a phase and C_{pm} , heat capacity at constant pressure; ΔH_{rm} is the heat of reaction; $\vec{I}_{mn}$, momentum exchange between phases m and n; J_m , collisional dissipation of granular energy; $\vec{g}$, gravitational acceleration; $\vec{q}_m$, heat flux; $\vec{q}_{\Theta_m}$, granular heat flux; R_{ml} , chemical reaction rate of the lth species of the mth phase; $\overline{\overline{S}}_m$, stress tensor; T_m , temperature; $\vec{v}_m$, velocity vector; X_{ml} , mass fraction of the lth species in the mth phase; γ_{mn} , coefficient of heat transfer between phases m and n; ε_m , volume fraction; Π_{Θ_m} , dissipation of granular energy due to interaction with gas; ρ_m , density; and Θ_m , granular temperature or one-third the mean square of particle velocity fluctuations. These equations must be closed with constitutive relations for the momentum and energy exchange, and the phasic stresses, heat capacities, and heat fluxes. These equations are solved using a finite volume technique available in the open source code MFIX (http://mfix.netl.doe.gov) [5]. The gas phase pressure gradient multiplied by ε_m appears in all the momentum equations, the term accounting for a buoyancy force in the solids momentum equation. In addition, the solids momentum equation also contains a pressure term that is a function of the granular temperature and gas-phase volume fraction.

For modeling coal gasifiers, the chemical reaction rates and heats of reaction must be supplied, which is available through the module carbonaceous chemistry for continuum modeling (C_3M) developed at NETL. C_3M represents the gas phase composition with eight species (O_2, CO, CO_2, CH_4, H_2, H_2O, N_2, and tar) and the solids phase (coal) composition with four pseudo-species (fixed carbon, volatile matter, moisture, and ash). Ash does not take part in any reactions, moisture is released in an initial stage reaction, and the volatile matter produces several gas-phase species through devolitization. The gas phase reactions are tar decomposition, water gas shift reaction, and combustion reactions for H_2, CH_4, CO. The heterogeneous reactions are char combustion and gasification. All the reaction schemes are shown in Figure 1 [6]. (The sorbent reactions shown in the figure were not included in the simulations reported here.)

2.2. Computational challenges

The multiphase CFD model of coal gasifier poses enormous computational challenges, primarily stemming from the gas-solids hydrodynamics, complicated by the homogeneous and heterogeneous chemical reactions. Gas-solids flows are in general unsteady – the volume fraction, stress, and energy typically fluctuating spatially and temporally with amplitudes comparable to the mean – and no reliable, time-averaged (turbulence like) equations are available for describing their stationary state behaviour. Therefore, the multiphase CFD approach involves computing the transient solution for a sufficiently long time to reach a statistically stationary state, from which a time-averaged solution is extracted to evaluate the reactor performance. The transient calculations usually require small time-steps, necessitated by rapid changes in the physics: The solids volume fraction can vary by several orders of magnitude over short length or time scales, a variation reminiscent of a single-phase flow transitioning from *hypersonic to incompressible*. When the solids volume fraction increases beyond a certain value (~0.5), the solids-phase undergoes a phase change, similar to *glass transition*, necessitating a drastic change in the solids-phase constitutive relations. The coupling terms between the gas and the solids phases dominate the momentum and energy equations, introducing small relaxation time-scales that vary with the solids volume fraction.

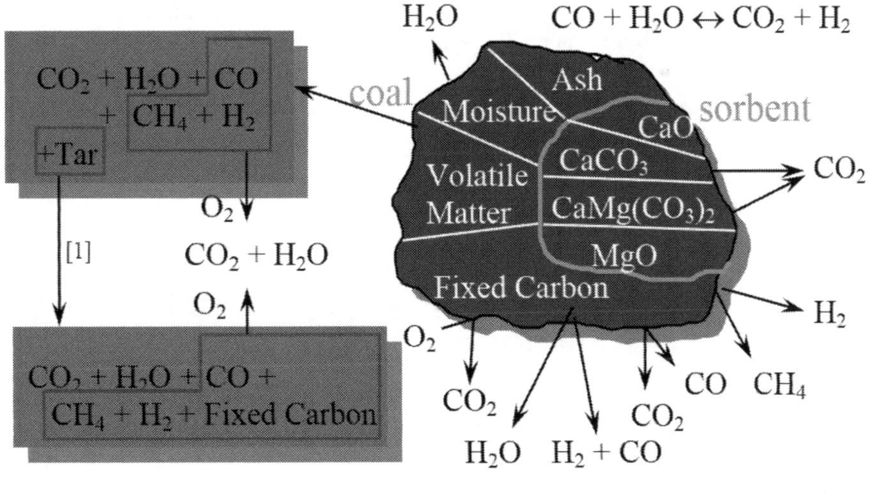

Figure 1. Schematic representation of chemical reactions in C_3M Module.

The inclusion of the coal gasification and combustion chemistry aggravates the computational challenge. The computational burden increases simply from the addition of twelve partial differential equations for tracking the species mass fractions, which more than doubles the total number of equations solved. The chemical reaction rates vary from the slow gasification reactions to very fast devolatilization and combustion reactions. Although the fast reactions are confined to small regions of the gasifier, they may control the size of the time-steps. Also the net reaction rates introduce coupling terms between the phasic continuity equations, which may reduce the rate of convergence of the non-linear iterations.

2.3. Porting the model to HPC

The model described in the previous section was developed at NETL over many years and was recently ported to HPC platforms. Figure 2 shows examples of multiphase flow reactor simulations, the computational cell count, the number of processors or cores, and the parallelization methods used over the years. In less than twenty years the cell count has grown by four orders of magnitude, necessitating the use of parallel computational platforms with a similar increase in the number of cores. The last two points on the graph labeled "CCPI gasifier" correspond to the simulations performed under the INCITE grant on the HPC platforms Cray XT4 and XT5. The results presented in this paper are from those simulations.

Extensive profiling of MFIX was conducted on HPC platforms using standard benchmark cases that scaled up to 10 million cells on 6032 cores, to determine the computational bottlenecks. Based on the profiling information and the experience in running on HPC platforms, MFIX was improved in several phases:

- Phase I: The compiler flags and MPI parameters were tuned to improve the scaling.
- Phase II: The global collective operations become very expensive as the number of processors is increased because a small amount of data is transferred between the processors and the data transfer speed becomes limited by the latency. It was found that successive over relaxation (SOR) could be used for all the variables, except the gas pressure and solids volume fraction. SOR was found to scale well as it does not require global collective operations (e.g., dot-product). Furthermore, the BiCGStab linear equation solver used for the gas pressure and solids volume fraction was improved by decreasing the number of global collective operations. One iteration of the standard BiCGStab [7] involves 6 vector-vector dot products, two of which are for residual checks. An intermediate residual check was eliminated, and the frequency of the final check was reduced. The last two dot products are independent and they were combined into one collective operation on the two vectors. This reduced the number of dot products to ~3 per iteration, thus reducing the cost associated with dot products by 50%. The change of compiler from Portland Group to PathScale also substantially contributed to the improvements obtained in Phase II.

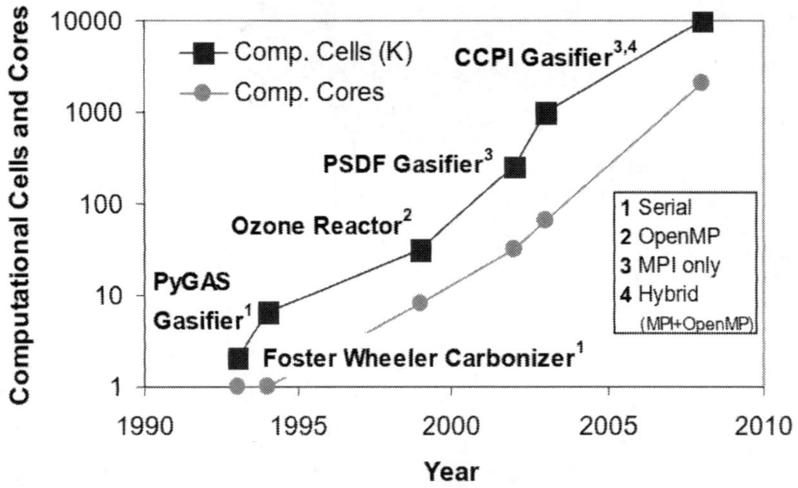

Figure 2. Size of the coal gasifier simulations over the years and the number of cores used to solve the problem.

- Phase III: Currently work is ongoing to conduct hybrid mode operation of MFIX to take advantage of multicore platforms with MPI and OpenMP. Although MFIX includes OpenMP directives for most of the loops, enabling multithreading for all the loops degrades the performance on a large number of processors. An extensive profiling study was performed to determine the loops that will most benefit from multithreading. As a result of this study, multithreading for several subroutines (e.g., linear equation solver routine, leq_msolve) were enabled. Identification of other subroutines that could improve the time-to-solution performance of MFIX is still under progress.

- Phase IV: Current MFIX I/O is not equipped to handle such high resolution runs for long transient periods and post-processing becomes a challenge. For example, when one of the 11 output files storing nearly 5 s of data reaches a size of 95 GB, the automated script for storing the data on the tape archival utility (HPSS at NCCS) fails because of the large file size. The large file size is a peculiarity of how data sets are stored in MFIX output files, all time-records of a group of variables being stored in a single file. The storage format is being improved by using netCDF/pnetCDF formats.
- Phase V: Currently we have plans to integrate MFIX with highly scalable and tuned solver library such as Trilinos.

3. Results of scaling study

Table 1 shows the results obtained from the actual production runs for the 10 million cell transport gasifier model on 2064 cores [8]. A sampling size of eight batch job run sessions were used to obtain the average simulated time per session (last column). The duration for each batch job session was 12 h of wall clock time, the maximum permitted for the requested number of cores. As seen from the results, runs employing 1032 MPI processes and 2 threads per process (in linear equation solver routines only) performed nearly 2.5 times better than those employing 2064 cores of MPI processes with a single thread. On average the simulation progressed for 0.18 s at the end of 12 hour run for 2064 cores of MPI processes only, whereas the simulation running in hybrid mode progresses 0.47 s at the end of the same duration.

Table 1. Comparison of simulated times for MPI only and hybrid MPI+ + OpenMP mode of operation on Cray XT4 at NCCS.

Mode Of Operation	Number of MPI cores	Number of Threads	Average simulated time (s) / 12 h job
MPI only	2064	1	0.18
MPI + OpenMP	1032	2	0.47

A continuous improvement cycle to monitor and improve execution performance of MFIX with the deployment of new HPC platforms is a critical component in utilizing these resources effectively for advanced coal gasifier designs and reduce time-to-solution. For this purpose a new benchmarking study to investigate the performance of MFIX on the new Cray XT5 platform at NCCS with quad cores was started. Preliminary results indicate that MFIX performance degrades if all eight cores are used for MPI processes (without any multithreading) possibly due to memory contention issues [8]. For example, for the standard 10M cell benchmarking case where time to solution is measured for integrating over 100 steps of 2.5×10^{-4} s, running a 516 processor job shows that requesting only 4 cores per node (remaining cores being idle) versus all 8 cores in a node results in 40 % reduction of time-to-solution.

The next improvement task was to investigate the effect of hybrid mode execution (i.e., MPI + OpenMP) on Cray XT5 with more than 2 threads per node. The results of the performance analysis conducted with CrayPAT tool show that boundary condition setup routine, bc_phi was the most time consuming routine after MPI and liner equation solver subroutine leq_msolve. Figure 3 shows the comparison for employing multithreading only in leq_msolve routine versus both leq_msolve and bc_phi subroutines by varying across 1 to 8 threads. Although enabling multithreading in bc_phi didn't improve time-to-solution as much as the leq_msolve subroutine, additional benchmarking needs to be conducted to replicate the runs to reduce the impact of variability due to operating system noise and to determine the optimal number of cores for MPI and OpenMP threads on Cray XT5 for the given configuration of the gasifier simulations.

Journal of Physics: Conference Series **180** (2009) 012034 doi:10.1088/1742-6596/180/1/012034

For using large-scale simulations for design, the practical issue of achieving a desirable time-to-solution must be considered. For large-scale computations on HPC, the time-to-solution is dependent upon factors beyond the computational challenges described previously. A target time-to-solution is 12 h for simulating a fully-resolved 25 MW unit, which is to be achieved by the year 2015 [2]; that is, the designers would like to get the results after an overnight simulation. To reach a stationary state in one of our large-scale simulations, it must be carried out for around 15 s (the residence time being typically 1–3 s for gas and 3–5 s for solids) and the time-average of the last 5 s are used for further analysis. Although an adaptive time-step algorithm is employed, on the average the time-step is 10^{-4} s, requiring around 150,000 time-steps for each calculation. For a 10 million cell case (285 MW unit) on 2064 processors (MPI only operation mode), the total run time on a Cray XT4 (Jaguar system at NCCS) is around 19 days. But the actual time-to-solution is significantly longer, when the factors such as wait time in the batch queue, system downtime, and job termination due to intermittent node failure are considered. Because of the long duration of the transient simulations, it is necessary to have high availability of a large number of processors for a long duration. At a shared resource center like National Center for Computational Sciences (NCCS) the runs are constrained by the throughput of the batch queue; for example, 2064-core jobs depending on the system status typically wait for 2–4 days in the queue for every 12 h compute session, which simulates 0.35~0.45 s of physical time (if employing hybrid mode of operation with MPI and OpenMP). This is equivalent to about 112 days to complete a calculation for 15 s simulation of physical time. We also need to consider the additional time required for analyzing the results, a challenge complicated by the much larger output files generated by the high resolution simulations conducted on HPC.

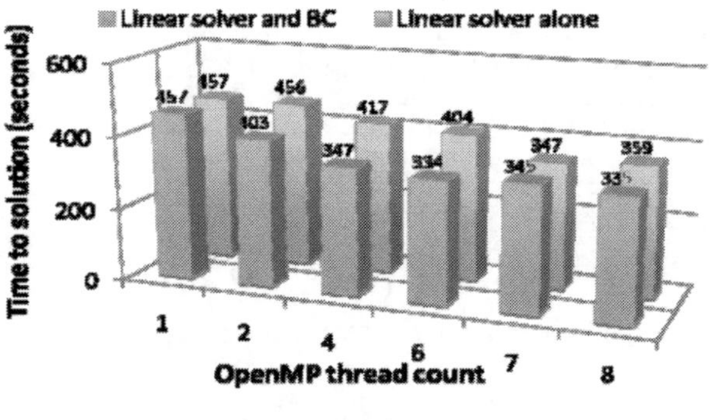

Figure 3. Comparison of time-to-solution after enabling multithreading across 1 to 8 cores on Cray XT5 for different subroutines.

4. Results of high resolution simulation

Preliminary results of a high resolution simulation (10 million cells) conducted through the INCITE award are compared with the results of a low resolution simulation (~1 million cells) previously conducted at NETL clusters (typically taking 2 weeks on 24 cores). Figures 4 and 5 show the gas temperature (in Kelvin), and, the carbon mass fraction and the volatile mass fraction in the solids-phase in the bottom section of the gasifier at a typical instance in time. They also include a horizontal clip plane, showing the mass fractions of O_2, CO, and CO_2. Not surprisingly, the higher-resolution results show greater structure in the field variable distributions. These preliminary results also show a difference in the value of the gas temperature, the high resolution simulation producing overall lower gas temperatures, especially near the coal jets (located just above the horizontal clip plane). The carbon and volatile mass fractions show that the bulk of the coal as it enters the gasifier tends to move away from the wall to a greater extent in the high resolution case than in the low resolution case. The high-resolution simulation shows (in the clip plane of the middle frame) much higher CO concentration coming up into the coal jet region from the lower portion of the gasifier. These higher levels of CO indicate better mixing of the recycled material in the lower portion of the gasifier with the burner air resulting in greater consumption of the carbon in the recycled stream entering the gasifier from below.

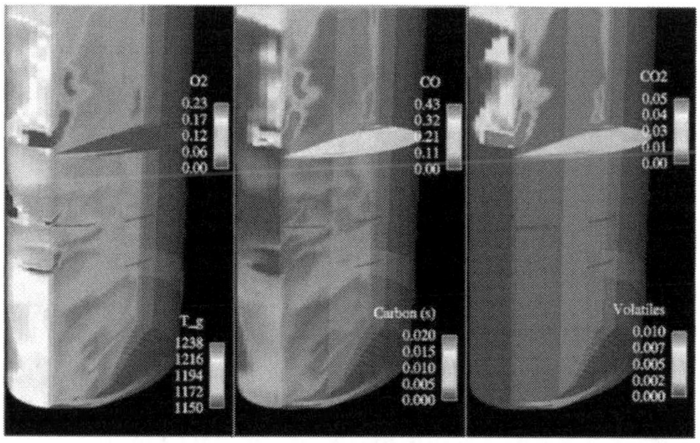

Figure 4. High-resolution simulation of CCPI gasifier.

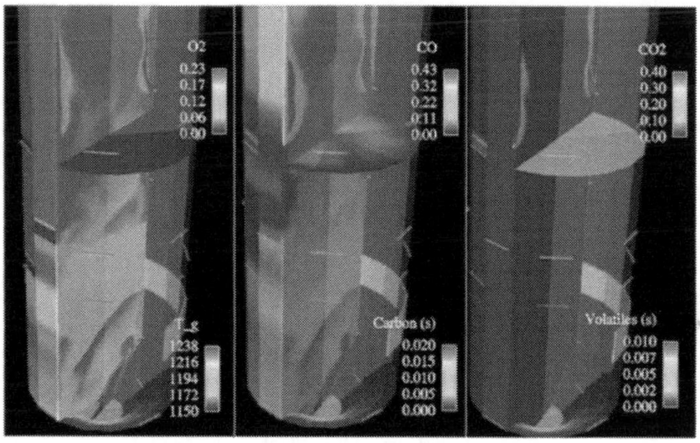

Figure 5. Low-resolution simulation of CCPI gasifier.

Figures 6 and 7 compare the high and low resolution simulations zooming into the coal jet region of the gasfier. The coal jet entering the gasifier is shown by plotting the isosurfaces for five values of the void fraction ($\varepsilon_g = 1 - \varepsilon_s$). The three horizontal clip planes are located where the coal jets enter the gasifier and show the gas temperature. The boundary conditions imposed on the coal jets set the flow of coal in a 45 degree downward direction. The high-resolution plots show that the isosurfaces go below the clip planes, predicting that the coal jets flow downward after entering the gasifier, turn around, and then move upward. Perhaps the upward turning movement causes the jet to move away from the wall. The low resolution simulation, however, shows that the coal jet quickly turns upward with little or no downward movement and takes an upward trajectory closer to the wall.

The differences in the simulation results at the two resolutions have practical implications for the gasifier design. For example, the trajectory of the coal jet could have significant effect on the gasifier reliability. A coal jet trajectory closer to the wall could lead to greater erosion of the refractory. So the low resolution simulation would predict a shorter refractory life than the high resolution simulation. Given that it is not currently feasible to conduct high resolution simulations for all the design cases, it remains to be determined how a few high resolution simulations could be combined with a large number of low resolution simulations to help with the design.

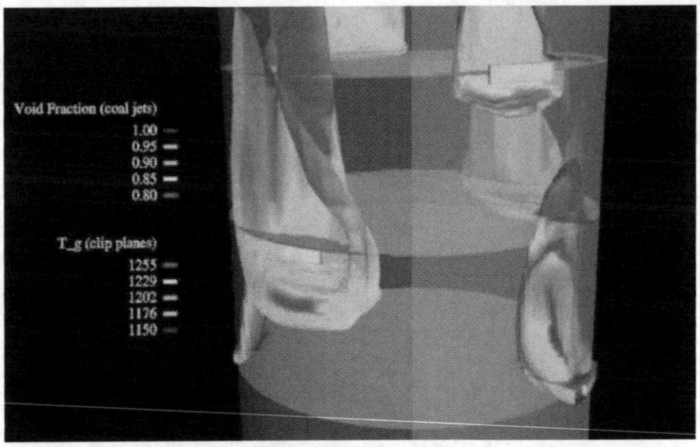

Figure 6. High resolution of coal jet region.

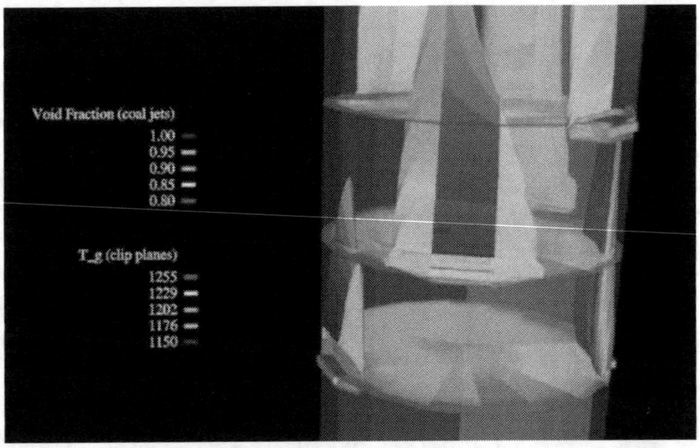

Figure 7. Low resolution coal jet region.

5. Conclusions

Computational gas-solids flows such as occurring in coal gasifiers are one of the highly non-linear processes observed in natural and man-made devices. To be predictive, the simulations need to model the spatiotemporal variations in gas and solids volume fractions, velocities, temperatures with any associated phase change and corresponding chemical reactions. These processes occur at various time- and length-scales requiring very high spatial resolution and large number of iterations with small time-steps. To solve these stiff set of non-linear partial differential equations at high spatial resolution, large-scale computer resources are required, and with the INCITE allocation, we were able to perform perhaps the largest known simulations of gas-solids reacting flows, providing detailed information about the gas-solids flow structure and the pressure, temperature and species distribution in the gasifier. One key finding is the new features of the coal jet trajectory revealed with the higher spatial resolution, which provides information on the accuracy of the lower resolution simulations. Methodologies for effectively combining high and low resolution simulations for design studies must be developed. Further analysis has to be performed to understand the causality relationships between the hydrodynamics, heat and mass transfer, and chemical reactions.

From a computational science perspective, we have found that global communication has to be reduced to achieve scalability to 1000s of cores. In addition, we have to employ hybrid parallelization to effectively utilize the multicore chips. We also find that the factors such as wait time in the batch queue could significantly increase the time-to-solution. From our experience, developments have to be made in the following areas to efficiently use large petascale resources and enable computational discovery:

- Efficient solvers for heterogeneous, massively parallel systems
- Data analysis tools to extract information from large data sets
- Programming environments for easily porting legacy codes to HPC platforms

Acknowledgements: The authors would like to acknowledge the financial support of the U.S. DOE, Fossil Energy Advanced Research Program. The submitted manuscript has been authored by several contractors for the U.S. Government under Contract Numbers DE-AM26-04NT41817.670.01.03 and DE-AC05-00OR22725. This technical effort was performed in support of the National Energy Technology Laboratory's on-going research in multiphase flows under the RDS contract DE-AC26-04NT41817. Accordingly, the U.S. Government retains a non-exclusive, royalty-free license to publish or reproduce the published form of this contribution, or allow others to do so, for U.S. Government purposes. This research used resources of the National Center for Computational Sciences at Oak Ridge National Laboratory, which is supported by the Office of Science of the U.S.DOE under Contract No. DE-AC05-00OR22725.

References

[1] Klara J M 2007 Fossil Energy Power Plant Desk Reference, DOE Report DOE/NETL-2007/1282.

[2] Report on Workshop on Multiphase Flow Research, Morgantown, WV, June 6-7, 2006, DOE/NETL-2007/1259, http://tinyurl.com/c9r7ux

[3] Guenther C, Shahnam M, Syamlal M, Longanbach J, Cicero D, and Smith P, "CFD modeling of a transport gasifier", Proceedings of the 19th Annual Pittsburgh Coal Conference, Pittsburgh, PA, September 23-27, 2002.

Guenther C, Syamlal M, Longanbach J, and Smith P, "CFD modeling of a transport gasifier Part II", Proceedings of the 20th Annual Pittsburgh Coal Conference, Pittsburgh, PA, September 15-19, 2003.

[4] Guenther C, Syamlal M, Shadle L, and Ludlow C, "A numerical investigation of an industrial scale gas-solids CFB", Proceedings of the 7th International Conference on Circulating Fluidized Beds, Niagra Falls, Ontario, Canada, May 5-8, 2002.

[5] Syamlal M, Rogers W, and O'Brien T J, "MFIX documentation: theory guide," DOE/METC-94/1004, NTIS/DE94000087, National Technical Information Service, Springfield, VA (1993).

Benyahia S, Syamlal M, and O'Brien T J, "Summary of MFIX Equations 2005-4," March 2006, http://www.mfix.org/documentation/MfixEquations2005-4-1.pdf.

Syamlal M, "MFIX Documentation: Numerical Techniques," DOE/MC-31346-5824. NTIS/DE98002029, December 1998.

[6] Syamlal M, and Bisset L, "METC gasifier advanced simulation (MGAS) model," Technical Note, DOE/METC-92/4108, NTIS/DE92001111, National Technical Information Service, Springfield, VA (1992).

[7] Barrett R, Berry M, Chan T, Demmel J, Donato J, Eijkhout V, Pozo R, Romine C, and van der Vorst H, Templates for the Solution of Linear Systems: Building Blocks for Iterative Methods, SIAM Publishing, Philadelphia, PA, 1994. www.netlib.org/templates/.

[8] Gel A, Pannala S, Sankaran R, Guenther C, Syamlal M, and O'Brien T J, "Accelerating Clean Coal Gasifier Designs with Hybrid MPI/OpenMP High Performance Computing", Proceedings of the 21st International Conference on Parallel Computational Fluid Dynamics, May 18-22, 2009, Moffett Field, CA.pp: 224-228

Massively parallel LES of azimuthal thermo-acoustic instabilities in annular gas turbines

Pierre Wolf[1], Gabriel Staffelbach[1], Laurent Gicquel[1] and Thierry Poinsot[2]

[1] CERFACS, 42 Avenue G. Coriolis, 31057 Toulouse Cedex, France
[2] Institut de Mécanique des Fluides de Toulouse, Avenue C. Soulas, 31400 Toulouse, France

E-mail: pierre.wolf@cerfacs.fr

Abstract. Most of the energy produced worldwide comes from the combustion of fossil fuels. In the context of global climate changes and dramatically decreasing resources, there is a critical need for optimizing the process of burning, especially in the field of gas turbines. Unfortunately, new designs for efficient combustion are prone to destructive thermo-acoustic instabilities. Large Eddy Simulation (LES) is a promising tool to predict turbulent reacting flows in complex industrial configurations and explore the mechanisms triggering the coupling between acoustics and combustion. In the particular field of annular combustion chambers, these instabilities usually take the form of azimuthal modes. To predict these modes, one must compute the full combustion chamber comprising all sectors, which remained out of reach until very recently and the development of massively parallel computers. A fully compressible, multi-species reactive Navier-Stokes solver is used on up to 4096 BlueGene/P CPUs for two designs of a full annular helicopter chamber. Results show evidence of self-established azimuthal modes for the two cases but with different energy containing limit-cycles. Mesh dependency is checked with grids comprising 38 and 93 million tetrahedra. The fact that the two grid predictions yield similar flow topologies and limit-cycles enforces the ability of LES to discriminate design changes.

1. Introduction

To enhance combustion for cleaner and more efficient engines implies new designs and operating ranges, usually by using lean combustion. Unfortunately, these technological choices often lead to combustion instabilities that sometimes take the form of azimuthal modes in annular gas turbines [1, 2, 3, 4]. Strong coupling of acoustics and non-linear heat release results in thermo-acoustic instabilities that can appear as both standing or rotating modes [5], though it has been shown in [1] that the rotating mode is more likely to be found when considering the limit cycle.

To predict these modes requires to understand the underlying physics governing the stability of annular gas turbines. Experimental studies require full test rigs that are complex and rare. Numerically, the study and prediction of annular chamber stability can be achieved by using one dimensional networks [1, 6] or Helmholtz solvers [7, 8]. These methods rely on the concept of the flame transfer function [9], which remains a key element and needs to be evaluated, using Large Eddy Simulation (LES) for example, or modeled [4, 9, 10]. Although LES could potentially predict combustion instabilities, until recently its use was restrained to the modeling of the flame transfer function obtained on a single sector simulation. LES of the full combustion chamber stayed out of reach until very recently through the development of massively parallel computers.

Today, full annular chambers can be computed on massively parallel machines by running LES codes on a thousand to several thousand processors, producing real engine operating conditions and their thermo-acoustic activity (limit-cycle).

A fundamental but unfortunately often neglected issue is the effect of mesh resolution on numerical results. As LES is based here on implicit filtering of the governing equations, hence by the grid size, the question of mesh dependency has been qualified as being of upmost importance by several authors [11, 12]. However, mesh dependency has been addressed mostly in academic configurations [13, 14] or for single sector configurations [15]. The computing power that is available today makes it possible to compute two grids of 38 and 93 million elements, representing a full annular helicopter chamber, and thus to compare the results in terms of mean and fluctuating fields and subgrid scale (SGS) quantities. This first comparison allows to obtain grid criteria that can then be used in an industrial context: *i.e.* to qualify two designs in terms of thermo-acoustic activity.

To start addressing such issues for industrial applications, two full annular versions of the considered helicopter engine only differing on the design of the swirlers are computed and analyzed with regard to their thermo-acoustic stability.

The LES code and the models used throughout this study are described in the next section. The target configurations are exposed before conducting an analysis of the effects of mesh resolution on the mean flow and on thermo-acoustic response of one of the two designs. LES results that help discriminating the two designs in terms of thermo-acoustic stability are then discussed.

2. Numerical tools and target configurations

The numerical tool used here is called AVBP [16]. Co-developed by CERFACS[1] and IFP[2], it solves the three-dimensional Navier-Stokes equations on unstructured and hybrid grids. Based on the LES approach [4, 11, 17], it is able to account for the influence of reacting and two-phase flow phenomena. In this investigation, the SGS stresses are modeled using the Smagorinsky approach [18] while combustion is modeled using Arhenius type reaction rates. The fuel is JP10 [15, 19], which is a surrogate for Kerosene. The interaction between the flame and turbulence is considered with the Dynamic Thickened Flame (DTF) model [20, 21, 22]. The third-order numerical scheme TTGC [23] is used. Multiple validations of the LES tool have been published [24, 25, 26]. The code has been thoroughly tested in all possible architectures available today (Power, Opteron, Xeon, Sicortex, etc..) and has exhibited excellent strong scaling performances on most super-computers (Fig. 1).

The base configuration (Fig. 2) considered throughout this study is a full annular reverse-flow helicopter gas turbine demonstrator designed by Turbomeca (Safran group). The whole chamber is computed with its casing, which helps avoiding uncertainties on the boundary conditions. Indeed, the calculated domain starts immediately after the compressor's outlet, and extends to a chocked nozzle corresponding to the throat of the high pressure distributor (Fig. 2). Fuel is supposed to be totally vaporized and only gaseous phase is computed. Air inflow feeds the combustion chamber through the swirlers, cooling films and dilution holes. Multi-perforated walls used to cool the liners are taken into account by a homogeneous boundary condition [27].

Two grids with different resolutions are compared to assess the impact of mesh resolution on LES: a light one that comprises 37.7 million tetrahedral cells and a fine one, composed of 93.1 million tetrahedral cells. Typical time steps are $5.9\ 10^{-8}$ and $3.1\ 10^{-8}$ seconds respectively.

Two variants of this annular helicopter gas turbine are computed. Version A and B only differ by the swirlers' design, the combustion chamber itself remains the same and is equipped with

[1] http://www.cerfacs.fr
[2] http://www.ifp.fr

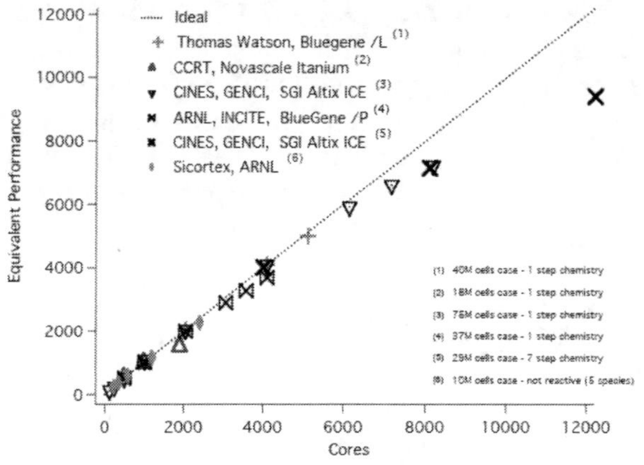

Figure 1. Strong scaling for the AVBP CFD tools on a variety of super computers.

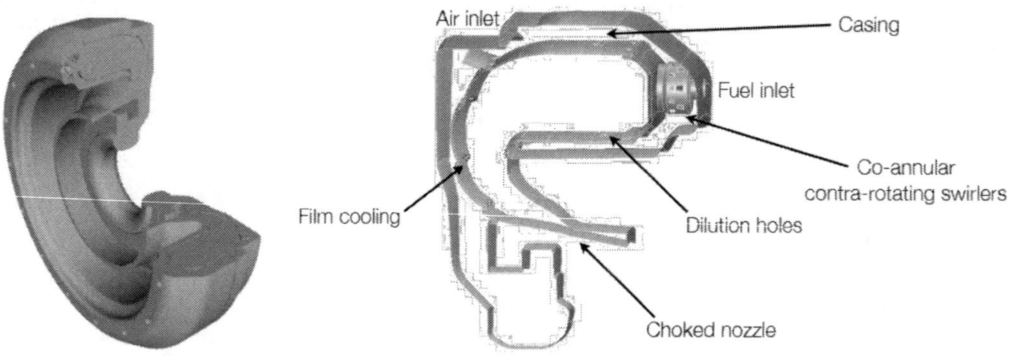

Figure 2. 3/4 view of the fully annular combustion chamber and boundary conditions shown on a single sector.

fifteen burners. Each swirler consists of two co-annular counter-rotating swirl stages. Differences between version A and B are geometrical details.

3. Results

The lighter grid has been run on 2,048 processors on a SGI Altix Ice 3[3] equipped with Intel Xeon 3GHz CPUs. LES of 0.1 seconds physical time required around 400,000 CPU hours. The fine grid case has been run on 4,096 processors on a IBM Blue Gene/P[4] and 3,000,000 CPU hours were required to perform a LES of 30 ms physical time. Those two LES reveal almost identical results. The flow topology, not presented here, is the same. Figure 3 displays the mean temperature fields on a transversal cut of a single sector and infers similar combustion regimes on both meshes. Note also that the thermo-acoustic behavior obtained from both grids draws analogous conclusions. Mesh convergence is reasonable in terms of mean and fluctuating quantities with the lighter grid resolution. The lighter mesh resolution is thus chosen for computing the two variants of the chamber (Version A and B).

[3] GENCI : http://www.genci.fr - CINES : http://www.cines.fr
[4] Argonne National Laboratory - ALCF : http://www.anl.gov

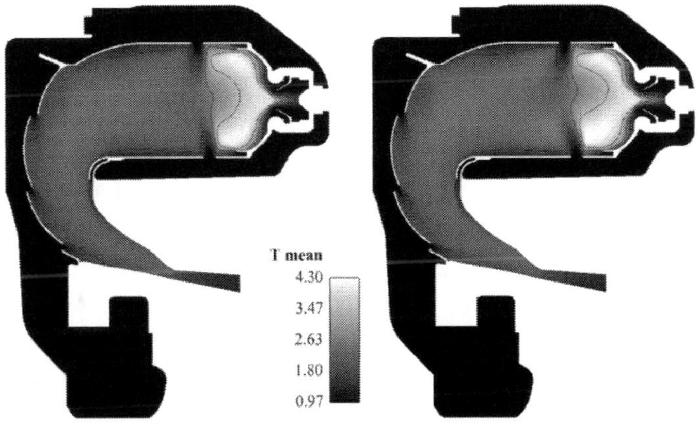

Figure 3. Mean temperature fields for the coarse (38 million cells - left) and the fine (93 million cells - right) grids. Temperatures are normalized by the inlet temperature.

When considering these two swirler designs, self-established azimuthal modes are observed. However the two cases exhibit very different limit-cycles. Version A presents intense fluctuations of heat-release and flame position, whereas version B displays weaker pressure fluctuations and better flame anchoring. Figure 4 shows the pressure RMS profiles for both versions on a developed line containing all swirlers' axis and indicates the same mode structure. However, version B demonstrates dimmer fluctuations when compared to version A, which presents strong local peaks corresponding to each swirler's exit and result in much more perturbed flame fronts. Based on such observations, version B is recommended since qualified as more stable by LES for the considered operating point.

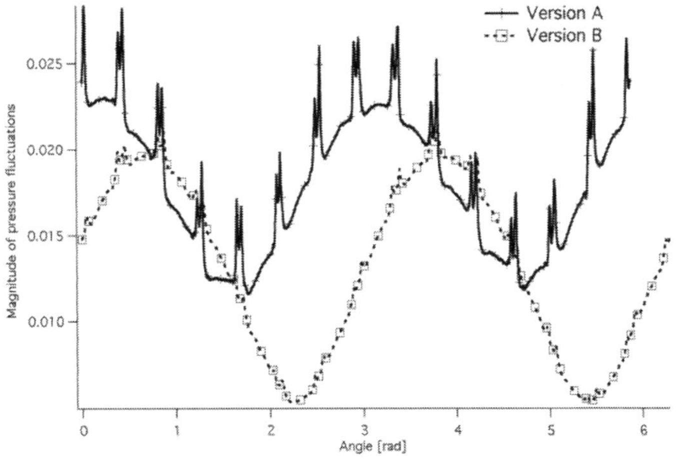

Figure 4. Normalized pressure fluctuations on a line passing by all burner's exit for version A and B of the considered combustion chamber.

4. Conclusion

Combustion instabilities, which can dramatically alter correct operation of engines, are a crucial issue. Large Eddy Simulation is a very promising path to apprehend thermo-acoustic instabilities in complex geometries. Massively parallel architectures offer the potential to compute full annular gas turbines. For the first time, a 15-sector full annular helicopter chamber has been simulated and reveals self-established azimuthal modes. Effects of grid resolution have been addressed and proved the light mesh to be adequate for the analysis. Two variants of this chamber, differing by geometrical details of the injection system, have then been gauged against each other and LES is able to discriminate the two cases in terms of azimuthal thermo-acoustic stability.

Acknowledgments

The authors thank the Argonne Leadership Computing Facility at Argonne National Laboratory, which is supported by the Office of Science of the U.S. Department of Energy under contract DE-AC02-06CH11357.

Thanks are also expressed to GENCI (Grand Equipement National de Calcul Intensif) and CINES (Centre Informatique National de l'Enseignement Supérieur).

Special credits are given to Turbomeca (C. Bérat, T. Lederlin and V. Moureau).

References

[1] Paschereit C O, Schuermans B and Monkewitz P 2006 Non-linear combustion instabilities in annular gas-turbine combustors. *44th AIAA Aerospace Sciences Meeeting and Exhibit*
[2] Candel S 1992 Combustion instabilities coupled by pressure waves and their active control. *24th Symp. (Int.) on Combustion* (The Combustion Institute, Pittsburgh) pp 1277–1296
[3] Crighton D G, Dowling A P, Williams J E F, Heckl M and Leppington F 1992 *Modern methods in analytical acoustics* Lecture Notes (New-York: Springer Verlag)
[4] Poinsot T and Veynante D 2005 *Theoretical and Numerical Combustion* (R.T. Edwards, 2nd edition.)
[5] Lieuwen T and Yang V 2005 Combustion instabilities in gas turbine engines. operational experience, fundamental mechanisms and modeling *Progress in Astronautics and AeronauticsAIAA* vol 210
[6] Stow S R and Dowling A P 2001 Thermoacoustic oscillations in an annular combustor *ASME Paper* (New Orleans, Louisiana)
[7] Nicoud F and Benoit L 2003 Global tools for thermo-acoustic instabilities in gas turbines. *APS/DFD meeting* (*Bull. Amer. Phys. Soc.* vol 48) (New York)
[8] Nicoud F, Benoit L, Sensiau C and Poinsot T 2007 *AIAA Journal* **45** 426–441
[9] Crocco L 1951 *Journal of the American Rocket Society* **21** 163–178
[10] Noiray N, Durox D, Schuller T and Candel S 2006 *Combustion and Flame* **145** 435–446
[11] Sagaut P 2002 *Large eddy simulation for incompressible flows* (Springer)
[12] Pope S B 2004 *New Journal of Physics* **6** 35
[13] Vreman B, Geurts B and Kuerten H 1996 *International Journal for Numerical Methods in Fluids* **22** 297–311
[14] Meyers J, Geurts B J and Baelmans M 2003 *Physics of Fluids* **15** 2740–2755
[15] Boudier G, Gicquel L Y M, Poinsot T, Bissières D and Bérat C 2008 *Combustion and Flame* **155** 196–214
[16] Schönfeld T and Rudgyard M 1999 *AIAA Journal* **37** 1378–1385
[17] Pitsch H 2006 *Annual Review of Fluid Mechanics* **38** 453–482
[18] Smagorinsky J 1963 *Monthly Weather Review* **91** 99–164
[19] Boudier G, Staffelbach G, Gicquel L and Poinsot T 2007 Mesh dependency of turbulent reacting large-eddy simulations in a gas turbine combustion chamber *QLES (Quality and reliability of LES) workshop* ed ERCOFTAC Leuven B
[20] Colin O, Ducros F, Veynante D and Poinsot T 2000 *Physics of Fluids* **12** 1843–1863
[21] Colin O and Rudgyard M 2000 *Journal of Computational Physics* **162** 338–371
[22] Légier J P, Poinsot T and Veynante D 2000 Dynamically thickened flame LES model for premixed and non-premixed turbulent combustion *Proceedings of the Summer Program* (Center for Turbulence Research, NASA Ames/Stanford Univ.) pp 157–168
[23] Moureau V, Lartigue G, Sommerer Y, Angelberger C, Colin O and Poinsot T 2005 *Journal of Computational Physics* **202** 710–736
[24] Roux S, Lartigue G, Poinsot T, Meier U and Bérat C 2005 *Combustion and Flame* **141** 40–54
[25] Selle L, Nicoud F and Poinsot T 2004 *AIAA Journal* **42** 958–964
[26] Prière C, Gicquel L Y M, Kaufmann A, Krebs W and Poinsot T 2004 *Journal of Turbulence* **5** 1–30
[27] Mendez S and Nicoud F 2008 *American Institute of Aeronautics and Astronautics Journal* **46** 2623–2633

SciDAC 2009 IOP Publishing
Journal of Physics: Conference Series **180** (2009) 012036 doi:10.1088/1742-6596/180/1/012036

Scaling to 150K cores: recent algorithm and performance engineering developments enabling XGC1 to run at scale

Mark F Adams
Department of Applied Physics and Applied Mathematics, Columbia University
E-mail: mark.adams@columbia.edu

Seung-Hoe Ku
Courant Institute of Mathematical Sciences, New York University
E-mail: sku@cims.nyu.edu

Patrick Worley
Computer Science and Mathematics Division, Oak Ridge National Laboratory
E-mail: worleyph@ornl.gov

Ed D'Azevedo
Computer Science and Mathematics Division, Oak Ridge National Laboratory
E-mail: dazevedoef@ornl.gov

Julian C Cummings
Center for Advanced Computing Research, California Institute of Technology
E-mail: cummings@cacr.caltech.edu

C-S Chang
Courant Institute of Mathematical Sciences, New York University
E-mail: cschang@cims.nyu.edu

Abstract. Particle-in-cell (PIC) methods have proven to be effective in discretizing the Vlasov-Maxwell system of equations describing the core of toroidal burning plasmas for many decades. Recent physical understanding of the importance of edge physics for stability and transport in tokamaks has lead to development of the first fully toroidal edge PIC code – XGC1. The edge region poses special problems in meshing for PIC methods due to the lack of closed flux surfaces, which makes field-line following meshes and coordinate systems problematic. We present a solution to this problem with a semi-field line following mesh method in a cylindrical coordinate system. Additionally, modern supercomputers require highly concurrent algorithms and implementations, with all levels of the memory hierarchy being efficiently utilized to realize optimal code performance. This paper presents a mesh and particle partitioning method, suitable to our meshing strategy, for use on highly concurrent cache-based computing platforms.

1. Introduction

The effective simulation of burning plasmas is essential in the development of commercially viable fusion power, and the physical understanding and prediction capability of the edge plasma pedestal are very important in this tokamak plasma research. Many important phenomena in tokamak plasmas require a kinetic description as given by the Vlasov equation. The 6-dimensional Vlasov equation can be approximately reduced to a 5D (3 real space + 2 velocity) gyrokinetic Vlasov equation, which models the particle gyro-motion as a charged ring.[1, 2, 3] This simplification gives us orders of magnitude increase in timestep size in the simulation and makes it possible to simulate kinetic turbulence physics in tokamaks with modern-day supercomputers.

Particle-in-cell (PIC) methods have proven to be effective in discretizing the gyrokinetic Vlasov-Maxwell system for core tokamak plasmas for many decades[4, 5]. Recent physical understanding of the importance of edge physics in tokamaks has lead to the development of the first fully toroidal gyrokinetic edge PIC code to include the magnetic separatrix – XGC1[6]. XGC1 in its present form is a full-f electrostatic gyrokinetic ion-electron particle code specifically designed for edge plasmas. The simulation region of XGC1 can be from the magnetic axis to the material wall. The equations of motion from the discretized gyrokinetic Vlasov equation are

$$\begin{aligned} \dot{\mathbf{X}} &= (1/D)[v_{\|}\hat{b} + (v_{\|}^2/B)\nabla B \times \hat{b} + \{\mathbf{B} \times (\mu\nabla B - \bar{\mathbf{E}})\}/B^2] \\ \dot{v_{\|}} &= -(1/D)(\mathbf{B} + v_{\|}\nabla B \times \hat{b}) \cdot (\mu\nabla B - \bar{\mathbf{E}}), \end{aligned} \tag{1}$$

where $\dot{\mathbf{X}}$ is the time derivative of ion gyro-center position, $v_{\|}$ is the speed of the particle parallel to the local magnetic field vector $\mathbf{B}$, $\hat{b} = \mathbf{B}/B$, $\mu = v_{\perp}^2/2B$ is the magnetic moment, $\bar{\mathbf{E}}$ is the gyro-averaged electric field, and $D = 1 + (v_{\|}/B)\hat{b} \cdot (\nabla \times \hat{b})$. The following gyrokinetic Poisson equation is solved for electrostatic potential Φ with the four-point averaging technique[3, 7]:

$$-\nabla_{\perp}\frac{\rho_i^2}{\lambda_{Di}^2}\nabla_{\perp}\Phi = e\left(1 - \nabla_{\perp}\rho_i^2\nabla_{\perp}\right)(\overline{n}_i - n_e), \tag{2}$$

where ρ_i is the ion gyroradius, λ_{Di} is the ion Debye length, and $\overline{n}_i$ is the ion gyro-center density in real space,

$$\overline{n}_i = \frac{1}{2\pi}\int f_i(\mathbf{X}, \mu, u)\delta(\mathbf{X} - \mathbf{x} + \vec{\rho}_i)d\mathbf{X}d\mu d\alpha.$$

Here $\mathbf{x}$ is the particle position vector in real space and α is the gyro-phase angle.

XGC1 solves the above equations with particles and electric field data on a spatial grid. Each timestep of an XGC1 execution includes, minimally, the following stages:

(i) Collect particle charge density on underlying grid
(ii) Solve gyrokinetic Poisson equation on grid
(iii) Compute electric field and any derivatives needed in particle equations of motion
(iv) Calculate and output diagnostic quantities
(v) Update particle positions and velocities

This is a simplified view of the algorithm in that these steps are used within a Runge-Kutta or predictor-corrector time integration method. Depending on the experiment configuration, the particles may be ions, electrons, or both. Experiments can also include collisions and other physical processes important to full-device simulations.

The core tokamak plasma has well confined single-particle orbits without a particle sink or source. In this situation, the plasma is well conserved in phase space, and the perturbed

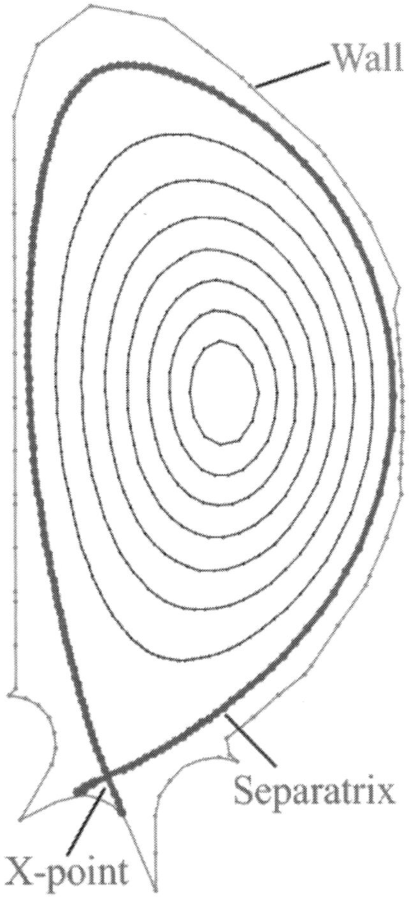

Figure 1. Magnetic flux surfaces of diverted tokamak geometry in a poloidal plane. X-point is shown on the separatrix.

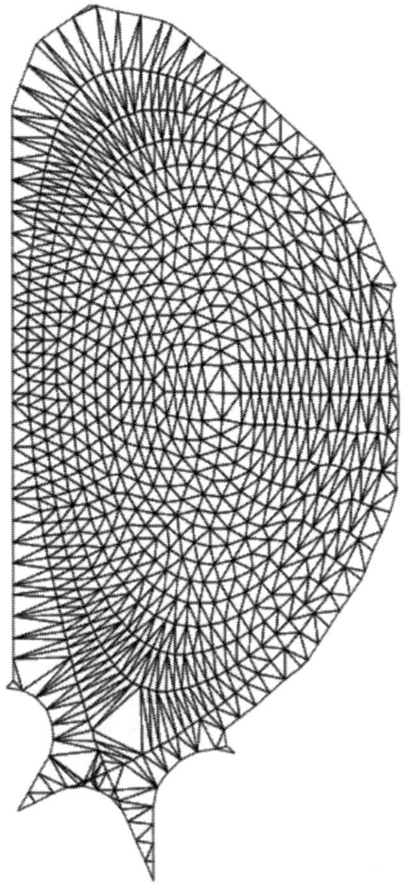

Figure 2. A sample unstructured triangular mesh with ITER magnetic field. The actual scale of the grid used in the simulations is approximately 30 times finer.

distribution function (delta-f) method has been popular. The edge tokamak plasma has some critical differences from the core region. First, the magnetic coordinate system diverges in the presence of a divertor, which all modern tokamaks have been designed to have for controlling wall damage due to escaping hot plasma. The magnetic field geometry with divertor has a separatrix and X-point (see Figure 1). The poloidal field is vanishingly small near the X-point, and the magnetic coordinate system is undefined on the separatrix. Second, the existence of the magnetic separatrix and the material wall creates unconfined single-particle orbits[8, 9]. In the scrape-off layer, the plasma is ill-confined. Third, in the edge pedestal the gradient scale length is on the same order as the radial particle excursion width, and a large self-consistent radial electric field is formed. Due to these differences, XGC1 uses the full-f PIC method and a cylindrical coordinate system.

Strong magnetic guide fields generally make the turbulence structure approximately aligned along the field. Therefore, field aligned meshes allow a significant reduction in the number of cells (via fewer grid points in the toroidal or parallel direction) in tokamak simulations. The field aligned mesh can be easily obtained from a magnetic coordinate system. As previously noted, however, the magnetic coordinate system has a singularity at the separatrix and is problematic in the diverted edge region. This paper introduces a meshing strategy that allows for semi-field aligned meshes with a cylindrical coordinate system that is both naturally axisymmetric and also effective in achieving the efficiencies of field aligned meshes.

One of the core computational kernels of a PIC code is the charge deposition process. Charge deposition requires that the interpolates of a charge at an arbitrary position in the tokamak domain be determined and the charge be deposited onto the grid that is used in the gyrokinetic Poisson equation to compute the electrostatic potential. This charge deposition can account for a significant fraction of the computation time in a PIC code. Our mesh method requires a 2D unstructured triangular mesh (*e.g.*, Figure 2) that is used for each of the 16-64 poloidal planes (toroidal cross-sections) in the simulation. These unstructured meshes pose significant challenges in terms of computational efficiency due to the relatively complex nature of unstructured data. One of the computation bottlenecks of XGC1 is the determination of these interpolates. To ameliorate this problem, we have developed an efficient search method and code to find the triangles that a particle is contained in, once the particle has first been interpolated within the toroidal domain along a local magnetic field line.

Parallelization of XGC1 is based on decompositions of both the spatial grid and the particle data. In particular, the assignment of particles to processes is based on a decomposition of the spatial domain. Minimally, both the spatial grid and the particle decompositions utilize a one-dimensional decomposition in the toroidal direction of the tokamak geometry.

All but one of the basic stages during an XGC1 timestep requires MPI communication in the parallel implementation. The charge density collection requires MPI communication between processes assigned adjacent parts of the spatial grid in neighboring poloidal planes (neighboring slices in the one-dimensional toroidal decomposition), as well as two distributed reductions. The gyrokinetic Poisson solver uses parallel numerical algorithms supplied by the PETSc library [10]. MPI communication in the diagnostic routines is limited to gathers and reductions. The update of particle positions requires reassigning some of the particles to different processes.

The remainder of this paper is organized as follows. Section 2 presents a new mesh design for edge gyrokinetic methods. Section 3 describes the triangle search method implemented in XGC1. Section 4 describes a 3D particle and grid partitioning scheme suitable for XGC1 applications. Section 5 describes performance optimizations of the parallel implementation. Section 6 presents numerical experiments on leadership class Cray computers. Finally, we conclude in Section 7.

2. Meshing and coordinate system methods

The unperturbed magnetic field in XGC1 is symmetric in the toroidal direction, which is the direction of positive or negative cylindrical angle ϕ. The magnetic field can be decomposed into toroidal field $\mathbf{B}_T = (\mathbf{B} \cdot \hat{\phi})\hat{\phi}$ and poloidal field $\mathbf{B}_P = \nabla\phi \times \nabla\psi$, where $\psi = RA_\phi$, A_ϕ is the $\hat{\phi}$ component of the vector magnetic potential $\mathbf{A}$, and R is the radial coordinate. We can define a field-line following poloidal variable

$$\lambda \equiv \int \frac{B_T}{RB_P} ds, \tag{3}$$

where ds is the infinitesimal line segment on the surface of constant ϕ and ψ[11]. The magnetic field can then be written as $\mathbf{B} = \nabla\psi \times \nabla(\lambda - \phi)$, and $\lambda - \phi$ is constant along the B-field. Unlike a magnetic coordinate system, λ is defined both inside and outside of the separatrix with a proper reference line, and it does not diverge on the separatrix except at the X-point. λ is used for the field-line following property of mesh generation, and it is calculated using the method of characteristics and an ordinary differential equation solver in a separate code from XGC1.

The mesh of XGC1 is chosen to be identical in each poloidal plane for simplicity, and the field-line following property is applied approximately. When we follow the field line, the change in toroidal angle ϕ implies a change in λ, and $\Delta\phi = \Delta\lambda$ is satisfied. The grid points are the intersection of the field line with a poloidal plane. In the mesh of XGC1, $\Delta\phi = 2\pi/n$, where n is the number of poloidal planes. Therefore, generating a grid point at every $\Delta\lambda = 2\pi/n$ on a flux surface will give a field-line following grid. Usually the grid spacing of $\Delta\lambda = 2\pi/n$ is much larger than the characteristic grid spacing except in the regions near the magnetic axis or X-point where the poloidal magnetic field vanishes. Reducing the grid distance in λ space to $2\pi/nm$ (where m is an integer) will still give the approximate field-line following property and keep the grid distance to the order of the characteristic grid spacing. In the ITER tokamak example, m is around 10 in the middle of the magnetic axis and separatrix.

In the core region, the grid should satisfy poloidal periodicity, which means that the generation of grid points should return to the original point as it makes a complete loop in the poloidal plane. In the grid of XGC1, the grid spacing $\Delta\lambda$ is set to $\lambda_{2\pi}/N$, where $\lambda_{2\pi}$ is the change in λ as it makes a loop on the poloidal plane, and N is the nearest integer to $mn\lambda_{2\pi}/2\pi$. N for the ITER grid is on the order of 100 to 1000 except in the region near the magnetic axis, and the field-line following error is less than 1%.

Even though λ is well behaved on the separatrix and outside of it, it diverges at the X-point. The field-line following grid becomes too fine near the X-point. The field-line following property of the turbulence structure cannot be satisfied to arbitrary accuracy, so we can apply a cut-off of $1/B_P$ in Eq. 3 and redefine λ as

$$\lambda \equiv \int \frac{B_T}{R \max(B_P, B_{P0})} ds. \tag{4}$$

This cut-off keeps the size of the grid near the X-point finite and reasonable without breaking the field-line following property, and the grid size is fully determined by the cut-off value B_{P0}.

The grid points generated in the above procedure are unstructured in the cylindrical coordinate system. We have used a triangle mesh to connect the grid points. The connectivity between nearest grid points in a flux surface is enforced. The TRIANGLE[1] code[12] is used for mesh generation. Figure 2 shows a sample triangle mesh for a very large grid size.

3. Mesh search algorithm

One key computational kernel is the deposition of charges by particles into this unstructured mesh. The Poisson solver computes the effective electric field using these accumulated charges as equivalent sources. The primary concern is, given a particle location (x, y), to find the enclosing triangle and corresponding barycentric coordinates. If (x_i, y_i) are the vertices of the triangle, then the barycentric coordinates (or area coordinates) of an interior point satisfy $0 \le \lambda_i \le 1$, $1 = \sum_{i=1}^{i=3} \lambda_i$, and $(x, y) = \sum_{i=1}^{i=3} \lambda_i(x_i, y_i)$. The barycentric coordinates can also be used for linear interpolation of a function $f(x, y) \approx \sum_{i=1}^{i=3} \lambda_i f_i$ where $f_i = f(x_i, y_i)$.

To speed up this search kernel, XGC1 uses geometric hashing into a background rectangular grid to locate a close initial triangle. Further testing of a small number of neighbor triangles is used to determine the target triangle. An optional preprocessing can relabel or reorder the vertices and triangles numbers by a Hilbert or Z-Morton space-filling curve to further improve the cache performance by encouraging better spatial locality.

4. Particle partitioning scheme

Modern supercomputers require highly concurrent (*i.e.*, highly parallel) algorithms and implementations, with all levels of the memory hierarchy being efficiently utilized, to realize optimal performance. PIC codes present unique computational challenges in that the equations

[1] The TRIANGLE Delaunay triangulation software is developed by Jonathan Shewchuk and available at `http://www.cs.cmu.edu/~quake/triangle.html`

to be solved are discretized with two very different methods: relatively common grid (cell) based methods such as finite elements for Maxwell's equations and particle methods for the Vlasov equation. This leads to the need to optimize two very different processes and more importantly the interaction between the two. The majority of the operations in a PIC code for tokamak plasmas, especially with the full-f methods used in XGC1, are in particle processing. That is, there are many more particles than grid points. Thus, the purely mesh-based tasks, such as the Poisson solve, smoothing and field calculations, do not produce computational bottlenecks. They can still impact performance, however, as discussed in the next section.

The particle processing is fairly straightforward, both in terms of serial and parallel performance. One of the most challenging aspects of performance in PIC codes is the interaction between particles and the grid. This is because the particles are distributed randomly throughout the mesh, and the deposition of the charge on the grid requires (some degree of) random access to an unstructured grid data structure. This random access poses significant performance challenges due to inefficient use of the cache of common modern computer architectures. Managing the cache effectively is critical to achieving high performance on our cache-based machines. A common technique for managing caches effectively is to partition or block data into pieces that can fit comfortably in as high a level of the cache as possible. (This suggests recursive or multi-level methods to accommodate the multiple levels of cache in typical processors.)

We have developed an approach for particle decomposition that is effective in providing good data locality, can be implemented efficiently, and is the first step in fully parallelizing the grid in XGC1. Our approach takes advantage of the fact that our grid points are ordered using the "natural" ordering on flux surfaces as discussed above. Grid points are ordered first by flux surface, starting from the inner surface, and within each flux surface grid points are ordered by poloidal angle. This results in an approximate *spiral* space-filling curve within the separatrix that provides good data locality, and similarly good ordering in the region outside the separatrix. Our algorithm defines a non-overlapping partitioning of the computational domain – the Voronoi diagram of the grid points (or a nearest grid point partitioning). This algorithm is practical because the triangle search method discussed in Section 3 is fast and can be used to compute the nearest grid point to a coordinate.

Our algorithm first assigns weights to each grid point by counting the number of particles in its domain. This weight is approximately a third of the area of all triangles that touch a grid point with a uniform particle distribution, but we elect to explicitly compute the weight to accommodate nonuniform particle distributions that develop as the simulation evolves. The array of grid points is segmented to keep about the same total weight in each segment or partition. Each partition is then mapped to an MPI process. After a particle is moved, in what is called the *push* phase, its new coordinate is used to find its nearest grid point; the partition or segment to which the grid point belongs determines the MPI process to which the particle belongs. The particle data is then sent, if necessary, to a new processor in what is called the *shift* phase. This particle decomposition strategy is an effective blocking strategy to improve the data locality for cache sensitive processes such as the charge deposition, and it is the first step in a fully parallelized XGC1 implementation.

5. Performance engineering

5.1. OpenMP

As mentioned in the previous section, computation solely on the mesh, such as the Poisson solve, smoothing and field calculations, is dominated by the processing of the particles. However, the runtime of phases associated with the mesh-only computation can limit performance when running with large numbers of MPI processes.

For a given experiment (and a corresponding fixed mesh), the runtime for the mesh-only computation stops decreasing when using more than a certain number of MPI processes, a number dependent on the computing system and on the problem particulars. This follows simply from there being a fixed amount of work to do. The lower bound is a function of the MPI communication overhead, the amount of redundant computation, limited parallelism bottlenecks (such as serial fraction), load imbalance, etc. It is still profitable to use more parallelism by applying it to the particle processing phases, especially when the particle count grows with the processor count in a weak scaling mode.

Problems arise when the MPI communication overhead of the mesh-only phases begins growing with the MPI process count. This clearly holds for the Poisson solve. A mitigation technique that has been effective in other PIC codes within the fusion simulation community is to exploit OpenMP parallelism within SMP compute nodes, decreasing the total number of MPI processes to the total number of processor cores divided by the number of OpenMP threads per process. We are currently utilizing this approach on computer systems with multiprocessor compute nodes that support shared memory parallel programming paradigms such as OpenMP. In our experience, decreasing the number of MPI processes also improves stability and performance in all MPI communciation phases when running on large numbers of processors, as it lessens the demand for MPI and related system resources.

Some of the particle processing is logically independent, and OpenMP can be applied to the loops over particles in these phases without code modifications. Other loops, such as those related to charge deposition, are reductions onto the mesh. In these cases, the particles are divided between the threads and thread-specific contributions to the reduction are computed. A final loop over the nodes of the mesh sums the thread-specific contributions. This loop over nodes can itself be parallelized using OpenMP. In fact, many loops over nodes or triangles of the mesh can be parallelized with OpenMP, further parallelizing the mesh-only work even when MPI overhead limits the amount of MPI parallelism that can be exploited.

This discussion glosses over a number of issues that arose during the introduction of OpenMP parallelism. In one case, the data structures being manipulated were arrays referenced via pointers and the compiler was not able to determine that these arrays were nonoverlapping, disabling any OpenMP parallelism. Passing in the arrays as subroutine arguments eliminated this particular problem. A more difficult issue arises from the memory-intensive nature of much of the particle processing work. As described earlier, a significant part of the work involves identifying the location of each particle in the mesh. Over time, the particles assigned to a process are effectively randomized, and performance is constrained by the corresponding randomized access to memory. Periodic sorting can improve the locality of memory accesses, but this in itself introduces additional cost. The decomposition of particles over MPI processes is an approximate bin sort, and improves performace at high process count. Using OpenMP to decrease the MPI process count runs counter to this, and many of the particle processing loops take less time when using MPI-only decompositions than with hybrid OpenMP/MPI parallel implementations. Currently the optimal number of MPI processes and OpenMP threads per process for a given total number of processor cores is determined empirically for a given system and problem specification.

5.2. MPI Communication Algorithms

With respect to performance, the most important MPI communication algorithms outside of those used by PETSc in the Poisson solve are those used to reassign particles between MPI processes. After the location of each particle is advanced, it is determined whether a particle has left the domain assigned to the process that currently *owns* the particle, and which process it needs to be reassigned to. The original algorithm for this communication was very efficient for small numbers of processes, but it failed, for example, on the Cray XT4 and XT5 systems when using more than 8000 processes.

The problem arose because the original algorithm posted send requests, did some local processing, then posted receive requests. Many of the messages were arriving before the receive requests were posted and were going into MPI system buffer space. Eventually there was no more space available to receive the messages, and the program failed. Note that due to the nature of the simulations, the number of processes sending particles to any given process is typically not large, but the number of particles received can be large.

An alternative algorithm was developed that is equally efficient for small numbers of processes, and has been demonstrated to work on up to 64,000 processes on the Cray XT5. This new algorithm introduces explicit flow control: handshaking messages guarantee that receive requests on the destination processes are posted before send requests on the source processes, and the total number of outstanding send and receive requests on any given process at any given time is kept below a specified threshold.

6. Numerical results

The previous sections discussed the major algorithmic innovations that have been introduced into XGC1 to enable good computer performance on leadership class computing systems. This section describes recent results in benchmarking XGC1 performance on one of the largest computers available – the Cray XT5 *jaguarpf* machine at Oak Ridge National Laboratory. This system includes 18,722 compute nodes and a custom, three-dimensional torus (25x32x24) interconnect. Each compute node contains two 2.3 GHz quad-core Opteron processors (AMD 2356 *Barcelona*) linked with dual HyperTransport connections and 16 GB of DDR2-800 memory. Thus each compute node contains 8 processor cores and 2 GB memory per core, and an aggregate of 149,776 processor cores and 299,552 TB of memory for the entire system. Memory access performance is non-uniform (NUMA) at the node level, but uniform memory access can be enforced by restricting thread placement and associated memory to a single socket in the dual socket node architecture.

To investigate parallel scalability we conducted a series of *weak scaling* or scaled speedup studies in which the number of particles per processor core is kept constant (at 900,000). These studies use one mesh, a 3 mm mesh of the ITER edge region. So the mesh part of the work is a *strong scaling* study. The grid has 893,884 grid points and 32 poloidal planes are used in all cases. Only 10 time steps are used in these simulations; this is adequate to understand the performance of the code. While there is the potential for load imbalance to develop during real physics runs (with more than 10,000 time steps), we do not address this issue here.

The left graph in Figure 3 shows the time spent in the main time stepping loop of XGC1 and in the Poisson solver within that loop for a number of different processor core counts, from 2048 to 149,248 cores. Data is presented for MPI-only experiments, for experiments with 4 OpenMP threads per MPI process, and for experiments with 8 OpenMP threads per process. The total number of computational threads and the total number of particles are fixed for a given processor core count.

From these results we see that using OpenMP does decrease the cost of the Poisson solve. While the OpenMP parallelization of the loops over particles is not as efficient as using MPI only, using 4 OpenMP threads results in an overall increase in performance compared to the MPI-only experiments. MPI-only experiments attempting to use 131,072 or higher process counts failed. The source of the failure is still under investigation, but it is repeatable and is strictly a function of the process count.

When using 4 OpenMP threads, two MPI processes are assigned to each node. In this case, each process and its 4 threads are assigned to a single quad-core processor, and the memory subsystem as seen by each process and its threads is essentially Uniform Memory Access (UMA). In contrast, when using 8 OpenMP threads, a single process and 8 threads are assigned to a compute node, and performance suffers from the Non-Uniform Memory Access (NUMA) performance characteristics of the compute node memory subsystem. Improved memory

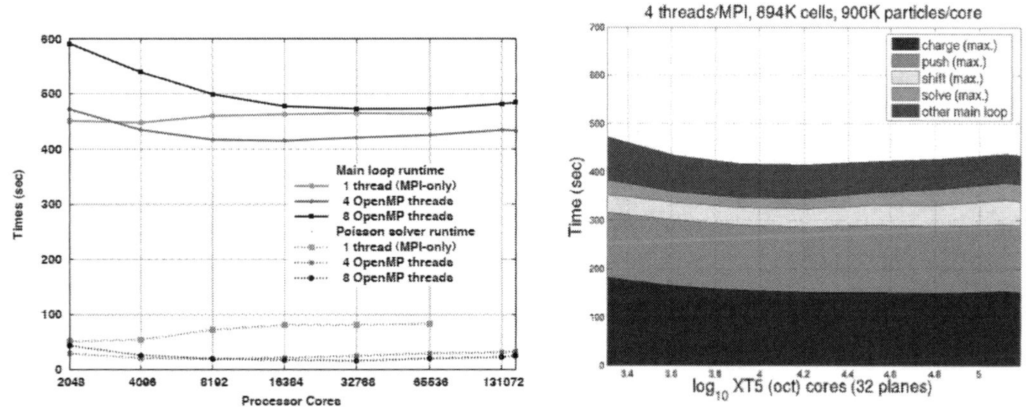

Figure 3. Run times for weak scaling with an ITER mesh and 900,000 particles per core, from 2048 to 149,248 cores

placement may address this performance problem for the 8 OpenMP thread experiments.

The right graph in Figure 3 breaks down where time is spent for the experiments with 4 OpenMP threads per process. This shows the maximum time that a processor spent in the four major phases of the code: 1) the charge deposition (*charge*), 2) the particle evolution (*push*), 3) the particle communication (*shift*) and 4) the Poisson solver (*solve*). Note that the solve phase is dominated by large all-to-all communication, which results in slightly increasing run times. These data show that the code is scaling very well with very nearly constant run time as the particle and processor count is increased.

Many of the optimizations discussed in this paper, such as the fast search routine and the particle decomposition strategy, are for serial performance and are not directly related to scalability. However, we have chosen to implement the particle decomposition strategy by placing one domain on each MPI process. Thus, as the number of processors is increased the number of particle domains increases as well, resulting in better data locality and better performance. This results in a *decrease* in memory-bound routines like the charge deposition routine shown in the right graph in Figure 3. In particular, time spent in the primary loop over particles in this phase decreases from 176 seconds with 2K cores to 130 seconds with 64K cores.

7. Conclusions

XGC1 is a new gyrokinetic particle code for tokamak plasma simulation which differs from other gyrokinetic codes in that it is a full-f code and includes the magnetic separatrix and scrape-off region, thus requiring a computationally intricate unstructured mesh. We have described here recent algorithmic advances in the XGC1 particle code that have resulted in our being able to scale to the edge of current *extreme scale* computing using about 150,000 cores. We have demonstrated new decomposition strategies for unstructured PIC methods and the effectiveness of using multi-threaded programming on a common extreme scale computer architecture of today – the Cray XT5. Future work will include continuing to parallelize the grid operations and data storage via a fully parallel or non-redundantly stored mesh.

Acknowledgments

This research was sponsored by the Office of Fusion Energy Sciences and by the Office of Advanced Scientific Computing Research, both in the Office of Science, U.S. Department of Energy. The research of D'Azevedo and Worley was performed at the Oak Ridge National

Laboratory, which is managed by UT-Battelle, LLC under Contract No. De-AC05-00OR22725. Accordingly, the U.S. Government retains a nonexclusive, royalty-free license to publish or reproduce the published form of this contribution, or allow others to do so, for U.S. Government purposes. Mark Adams was funded under DOE grant DE-FC02-06ER54862.

This research used resources (Cray XT5) of the National Center for Computational Sciences at Oak Ridge National Laboratory, which is supported by the Office of Science of the U.S. Department of Energy under Contract No. DE-AC05-00OR22725.

References

[1] Lee W W 1983 *Phys. Fluids* **26** 556
[2] Lee W W 1987 *J. Comput. Phys.* **72** 243
[3] Hahm T S 1988 *Phys. Fluids* **31** 2670
[4] Lee W W and Tang W M 1988 *Phys. Fluids* **31** 612
[5] Lin Z, Hahm T S, Lee W W, Tang W M and White R B 1998 *Science* **281** 5384
[6] Chang C S, Ku S, Diamond P H, Lin Z, Parker S, Hahm T S and Samatova N 2009 *Phys. Plasmas* **16** 056108
[7] Lin Z and Lee W W 1995 *Phys. Rev.* E **52** 5646
[8] Chang C S and Ku S 2004 *Phys. Plasmas* **11** 2649
[9] Ku S, Baek H and Chang C S 2004 *Phys. Plasmas* **11** 5626
[10] Balay S, Buschelman K, Eijkhout V, Gropp W D, Kaushik D, Knepley M G, McInnes L C, Smith B F and Zhang H 2008 PETSc users manual Tech. Rep. ANL-95/11 - Revision 3.0.0 Argonne National Laboratory
[11] Weitzner H 1981 *Phys. Fluids* **24** 2280
[12] Shewchuk J R 1996 *Applied Computational Geometry: Towards Geometric Engineering* (*Lecture Notes in Computer Science* vol 1148) ed Lin M C and Manocha D (Springer-Verlag) pp 203–222 from the First ACM Workshop on Applied Computational Geometry

Numerical linear algebra on emerging architectures: the PLASMA and MAGMA projects

Emmanuel Agullo, Jim Demmel, Jack Dongarra, Bilel Hadri, Jakub Kurzak, Julien Langou, Hatem Ltaief, Piotr Luszczek, Stanimire Tomov[1]

Department of Electrical Engineering and Computer Science, University of Tennessee, USA

E-mail: eagullo,dongarra,hadri,kurzak,ltaief,luszczek,tomov@eecs.utk.edu

E-mail: demmel@cs.berkeley.edu

E-mail: julien.langou@ucdenver.edu

Abstract. The emergence and continuing use of multi-core architectures and graphics processing units require changes in the existing software and sometimes even a redesign of the established algorithms in order to take advantage of now prevailing parallelism. Parallel Linear Algebra for Scalable Multi-core Architectures (PLASMA) and Matrix Algebra on GPU and Multics Architectures (MAGMA) are two projects that aims to achieve high performance and portability across a wide range of multi-core architectures and hybrid systems respectively. We present in this document a comparative study of PLASMA's performance against established linear algebra packages and some preliminary results of MAGMA on hybrid multi-core and GPU systems.

1 Introduction

Recent activities of major chip manufacturers, such as Intel, AMD, IBM and NVIDIA, make it more evident than ever that future designs of microprocessors and large HPC systems will be heterogeneous in nature, relying on the integration of two major types of components. On the first hand, multi/many-cores CPU technology have been recently developed and the number of cores will continue to escalate because of the desire to pack more and more components on a chip while avoiding the power wall, instruction level parallelism wall, and the memory wall [1]. And on the other hand special purpose hardware and accelerators, especially Graphics Processing Units (GPUs) are in commodity production, and have outpaced standard CPUs in floating point performance in recent years, and have become as easy, if not easier to program than multi-core CPUs.

To address the critical and highly disruptive situation that is facing the linear algebra and high performance computing (HPC) community due to the introduction of multi-core architectures and GPUs, we have developed two projects called Parallel Linear Algebra Software for Multi-core Architectures (PLASMA) [2, 3] and Matrix Algebra on GPU and Multi-core Architectures (MAGMA). PLASMA is a redesign of LAPACK [4] and ScaLAPACK [5] for shared memory computers based on multi-core processor architectures. To achieve high performance on this type of architecture, PLASMA relies on tile algorithms, which provide fine granularity parallelism. The MAGMA project aims to develop a dense linear algebra library similar to LAPACK but for heterogeneous/hybrid architectures, starting with current "Multi-core+GPU" systems.

The document is organized as follows. Section 2 gives an overview on PLASMA and its performance

[1] Research reported here was partially supported by the National Science Foundation and Microsoft Research.

against established linear algebra packages such as LAPACK, ScaLAPACK as well as equivalent commercial software offerings (Intel MKL, IBM ESSL and IBM PESSL) on two different architectures. Section 3 presents MAGMA and some preliminary results of the project on hybrid multi-core and GPU systems.

2 PLASMA

To achieve high performance on multi-core architectures, PLASMA relies on tile algorithms, which provide fine granularity parallelism. The standard linear algebra algorithms can then be represented as Directed Acyclic Graphs (DAG) [6] where nodes represent tasks and edges represent dependencies among them. Our programming model enforces asynchronous, out of order scheduling of operations. This concept is used as the basis for a scalable yet highly efficient software framework for computational linear algebra applications.

In LAPACK, parallelism is obtained though the use of multithreaded Basic Linear Algebra Subprograms (BLAS) [7]. In PLASMA, parallelism is no longer hidden inside the BLAS but is brought to the fore to yield much better performance.

PLASMA performance strongly depends on tunable execution parameters trading off utilization of different system resources. The outer block size (NB) trades off parallelization granularity and scheduling flexibility with single core utilization, while the inner block size (IB) trades off memory load with extra-flops.

We present in this paper a study of the three one sided factorization present in LAPACK: Cholesky, QR, and LU. Each of the one-sided tile factorizations presents increasingly complex challenges to parallel programming. Cholesky is the easiest. The tile algorithm is represented by a DAG with relatively little work required on the critical path. LU and QR factorizations have exactly the same dependency pattern between the nodes of the DAG. These two factorizations exhibit much more severe scheduling constraints than the Cholesky factorization. Moreover the stability of the tile LU factorization is not yet well understood.

PLASMA is currently scheduled statically with a trade off between load balancing and data reuse.

2.1 Comparison to other libraries

We perform a comparative study of PLASMA against established linear algebra packages (LAPACK and ScaLAPACK) as well as equivalent commercial software offerings (Intel MKL, IBM ESSL and IBM PESSL) for Cholesky, LU and QR factorization. We also compare with a new approach at parallel execution called TBLAS [8] (Task Based Linear Algebra Subroutines) in the Cholesky and QR cases – the TBLAS LU factorization has not yet been implemented. The experiments were conducted on two different multi-core architectures based on Intel Xeon EMT64 and IBM Power6,

PLASMA, TBLAS, LAPACK and ScaLAPACK are all linked with the optimized vendor BLAS available on the system provided within Intel MKL 10.1 and IBM ESSL 4.3 on the Intel64 and Power6 architectures, respectively. The first architecture is a quad-socket quad-core machine based on an Intel Xeon EMT64 E7340 processor operating at 2.39 GHz. Its theoretical peak is equal to 9.6 Gflop/s/ per core or 153.2 Gflop/s for the whole node (16 cores). The second architecture is a SMP node composed of 16 dual-core Power6 processors. Each dual-core Power6 processor runs at 4.7 GHz, leading to a theoretical peak of 18.8 Gflop/s per core and 601.6 Gflop/s per node (32 cores).

2.2 Methodology

Factorizations are performed in double precision. PLASMA is tuned with the pruned search method as described in [9]. TBLAS, ScaLAPACK and PESSL have been tuned with an exhaustive search (because their search space is smaller). LAPACK, MKL and ESSL have been tuned by the vendor.

Furthermore, to capture the best possible behavior of each library, we repeat the number of executions (up to 10 times) and we report the highest performance obtained. We do not flush the caches [10] before timing a factorization[2]. However, the TLB (Translation Lookaside Buffer) is flushed between two executions: the

[2] It is kernel usage, not problem size, that dictates whether one wish to flush the cache [10]. Warm (or partially warm) cache executions are plausible for dense linear factorizations. For instance, sparse linear solvers, which rely on dense kernels, intend to maximize data reuse between successive calls to dense operations.

loop over the different executions is performed in a script (rather than within the executable) and calls several times the same executable.

ScaLAPACK, PESSL and PLASMA interfaces allow the user to provide data distributed on the cores. In our shared-memory multi-core environment, because we do not flush the caches, these libraries have thus the advantage to start the factorization with part of the data distributed on the caches. This is not negligible. For instance, a 8000×8000 double precision matrix can be held distributed on the L3 caches of the 32 cores of a Power6 node.

2.3 Experimental results

We present in this section results of experiments conducted on a large number of cores (16 cores on Intel64; 32 cores on Power6). Figures 1(a), and 1(d) present the Cholesky (DPOTRF routine) performance of the different libraries. PLASMA consistently outperforms the other libraries, followed by TBLAS. These results illustrate the performance improvement brought by tile algorithms. The higher efficiency of PLASMA compared to TBLAS is essentially due to a better data reuse. Indeed, PLASMA scheduling strategy maximizes data reuse and thus benefits from a better cache effect than TBLAS whose scheduler does not take into account date reuse. PLASMA is even faster than the parallel DGEMM reference up to a matrix size

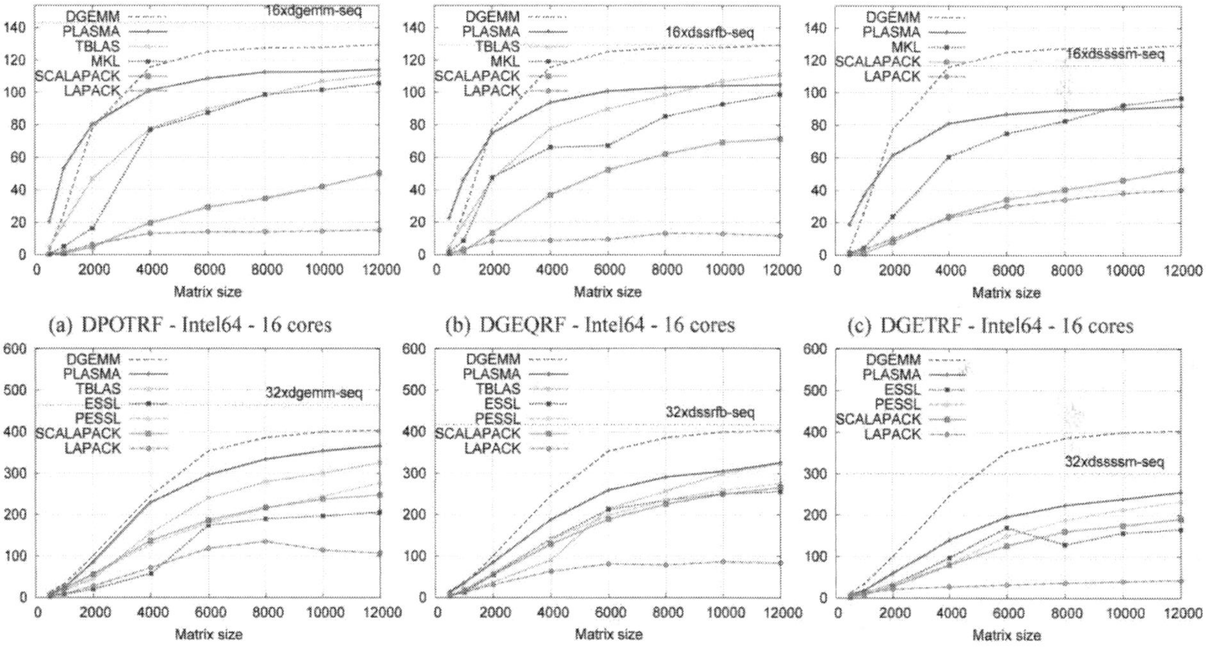

(a) DPOTRF - Intel64 - 16 cores (b) DGEQRF - Intel64 - 16 cores (c) DGETRF - Intel64 - 16 cores

(d) DPOTRF - Power6 - 32 cores (e) DGEQRF - Power6 - 32 cores (f) DGETRF - Power6 - 32 cores

Figure 1: Performance comparison on a large number of cores (Gflop/s).

$N = 2000$ on the Intel64 machine when 16 cores are used. This is not contradictory since PLASMA does not rely on the parallel version of DGEMM. Each PLASMA thread indeed uses the serial dgemm-seq. Better performance can then be achieved thanks to better scheduling. For large matrix size, the parallel DGEMM dominates all the other operations. This illustrates the fact that using a fine enough granularity (as in tile algorithms) is more critical when processing small or moderate size matrices. This also explains the major improvement brought by PLASMA compared to the other libraries on matrices of small and moderate size.

Figures 1(b) and 1(e) illustrate the performance of the QR factorization (DGEQRF routine) when all the available cores are used (16 on Intel64 or 32 on Power6). PLASMA outperforms the other libraries and

TBLAS is also very competitive. These results demonstrate the excellent scalability of tile algorithms. On the Intel64 machine, TBLAS actually has a better performance than PLASMA when 16 cores are used and when the matrix size is larger than or equal to 10,000. Indeed, when the matrices processed are large, the critical scheduling problem corresponds to maximizing a steady state throughput. The main disadvantage of a static schedule is that cores may be stalling in situations where work is available. This throughput is easier to maximize with a dynamic scheduling strategy. Approaches such as TBLAS, which do implement a dynamic scheduling, are thus likely to achieve a higher performance than approaches that implement a static scheduling (such as PLASMA currently does). All in all, these results are motivation for a dynamic scheduling mechanism that would assign priorities according to a trade off between data reuse and critical path progress.

Finally, figures 1(c), and 1(f) show that the LU factorization (DGETRF routine) has a performance behavior similar to the QR factorization and PLASMA again outperforms the other libraries. However, the lower efficiency of dssssm-seq compared to dssrfb-seq (dssssm-seq performs more extra-flops) induces a lower performance of the PLASMA LU factorization compared to the PLASMA QR one. On Intel64, this leads MKL to be slightly better than PLASMA when the matrix size is larger than or equal to 10,000 ($N \geq 10,000$). But, similarly to the QR case, moving towards a hybrid scheduling should remove the penalty due to the static scheduling strategy used in PLASMA which should improve the performance on large matrices. Furthermore, we note that an optimized implementation of the dssssm-seq kernel will improve the performance of tile algorithms.

3 MAGMA

The MAGMA research is based on the idea that, to address the complex challenges of the emerging hybrid environments, optimal software solutions will themselves have to hybridize, combining the strengths of different algorithms within a single framework. Building on this idea, we aim to design linear algebra algorithms and frameworks for hybrid manycore and GPUs systems that can enable applications to fully exploit the power that each of the hybrid components offers.

The problems and the challenges for developers in the new computational landscape of hybrid processors are daunting. Critical parts of the software infrastructure are already having a very difficult time keeping up with the pace of change: In some cases, performance is not scaling up as the number of cores grows because more and more time is spent on slow data movement rather than fast arithmetic.

Preliminary studies on a new class of "heterogeneity-aware" algorithms of "reduced communication" and "high-parallelism" confirm that this is the case. An example of the new class of communication-optimal algorithms is the RBT LP LU(NB) algorithm that we developed for Multi-core + GPU systems [11]. This algorithm aims at solving a nonsymmetric linear system of equations. Figure 2 shows its performance, comparing it with the pairwise pivoting (PwP) LU from PLASMA on current state-of-the-art multi-core systems, and the PP LU for 1 Core + 1 GPU. This algorithm underscores another change in the design space that is characteristic for many of the new techniques [12], namely that the new techniques often gain in speed for the price of relaxed accuracy. Understanding this trade-off of speed *vs* accuracy has to be further theoretically studied as it can lead to very efficient algorithms. For example here, experiments with random matrices show that LP LU(NB+64) is comparable in accuracy to PP LU, and LP LU(NB) loses only from 1 to 2 digits of accuracy to gain up to 30% in speed compared to PP LU. Another example, also related to reducing communication, is using mixed precision algorithms. Mixed precision solvers for example often achieve significant speedups on GPUs (e.g. up to $4\times$ on the GTX 280 [12]) compared to double precision solvers but the speed depends on the condition number of the matrix.

4 Conclusion and perspectives

We have shown the performance improvements brought by tile algorithms on up to 32 cores – the largest shared memory multi-core system we could access. We may expect that these results generalize somewhat to other linear algebra algorithms and even any algorithm that can be expressed by a DAG of fine-grain tasks.

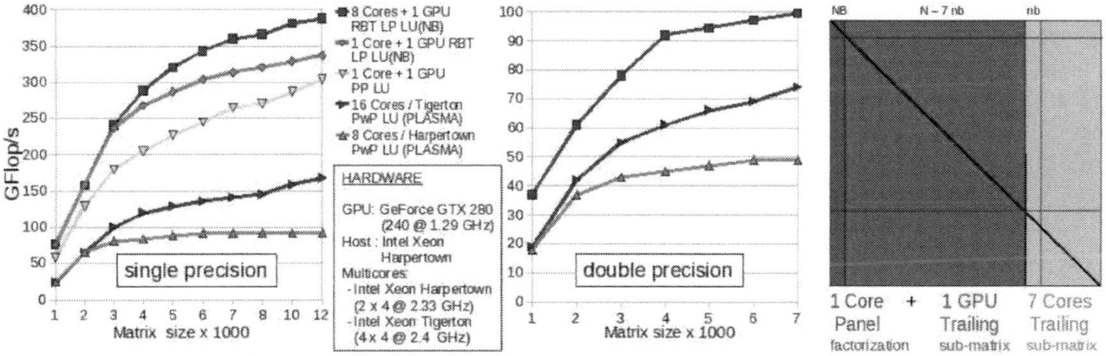

Figure 2: Performance of a communication-optimal LU for a Multi-core + GPU system in both single (left) and double (middle) precision arithmetic *vs* the PwP LU from PLASMA for multi-cores and the PP LU for 1 Core + 1 GPU system. The heterogeneity-aware work splitting used is shown on the right (for an 8 cores host).

We are also working on the interpolation of the optimum tuning parameters from a limited number of parallel executions among the range of cores and matrix sizes to the full set of possibilities. This on-going auto-tuning work should eventually be incorporated within the PLASMA software distribution. Our experiments have also shown the limits of static scheduling for the factorization of large matrices. We are currently working on the implementation of a dynamic scheduling for PLASMA.

Current work on MAGMA show that architecture trends have moved towards heterogeneous (GPU + CPU) designs of increased parallelism and communication costs, and software trends have to reflect on that. MAGMA addresses this with innovative heterogeneity-aware algorithms/techniques on extracting parallelism and reducing communication.

[1] K. Asanovic, R. Bodik, B. C. Catanzaro, J. J. Gebis, P. Husbands, K. Keutzer, D. A. Patterson, W. L. Plishker, J. Shalf, S. W. Williams, and K. A. Yelick. The landscape of parallel computing research: A view from Berkeley. Technical Report UCB/EECS-2006-183, Electrical Engineering and Computer Sciences Department, University of California at Berkeley, 2006.

[2] Buttari A., Langou J., Kurzak J., and Dongarra J. A class of parallel tiled linear algebra algorithms for multicore architectures. *Parallel Computing*, 35(1):38–53, 2009.

[3] Buttari A., Dongarra J., Kurzak J., Langou J., Luszczek P., and Tomov S. The impact of multicore on math software. *PARA 2006, Umea, Sweden*, June 2006.

[4] Anderson E., Bai Z., Bischof C., Blackford L. S., Demmel J. W., Dongarra J., Du Croz J., Greenbaum A., Hammarling S., McKenney A., and Sorensen D. *LAPACK Users' Guide*. SIAM, 1992.

[5] Blackford L. S., Choi J., Cleary A., D'Azevedo E., Demmel J., Dhillon I., Dongarra J., Hammarling S., Henry G., Petitet A., Stanley K., Walker D., and Whaley R. C. *ScaLAPACK Users' Guide*. SIAM, 1997.

[6] Christofides N. *Graph Theory: An algorithmic Approach*. 1975.

[7] BLAS: Basic linear algebra subprograms. http://www.netlib.org/blas/.

[8] Song F., YarKhan A., and Dongarra J. Dynamic task scheduling for linear algebra algorithms on distributed-memory multicore systems. Technical report, UTK CS Technical Report 638, 2009.

[9] E. Agullo, B. Hadri, H. Ltaief, and J. Dongarra. Comparative study of one-sided factorizations with multip le software packages on multi-core hardware. Technical report, 2009. Submitted to SC09.

[10] Whaley R. Clint and Castaldo Anthony M. Achieving accurate and context-sensitive timing for code optimization. *Software: Practice and Experience*, 38(15):1621–1642, 2008.

[11] Stanimire Tomov, Jack Dongarra, and Marc Baboulin. Towards dense linear algebra for hybrid GPU accelerated manycore systems. Technical Report UT-CS-08-632, University of Tennessee, 2008. Parallel Computing (submitted).

[12] M. Baboulin, J. Demmel, J. Dongarra, S. Tomov, and V. Volkov. Enhancing the performance of dense linear algebra solvers on GPUs [in the MAGMA project]. Poster at Supercomputing 08, November 18, 2008. http://www.cs.utk.edu/~tomov/SC08-poster.pdf.

Community-oriented support and research structures

Norbert Attig, Thomas Eickermann, Paul Gibbon and Thomas Lippert

Institute for Advanced Simulation, Jülich Supercomputing Centre, Forschungszentrum Jülich, 52425 Jülich, Germany

E-mail: th.lippert@fz-juelich.de

Abstract. Coordinated by the *Partnership for Advanced Computing in Europe* (PRACE) Europe is restructuring and strengthening its high-performance computing infrastructure with the aim to create a model HPC ecosystem. At the tip of the pyramid, up to six centres are envisaged that will operate systems of the highest performance class. The HPC Research Infrastructure (HPC-RI) will comprise European, national and regional centres. Science communities are integral partners, strong links will include Grid and Cloud users. The HPC-RI strives at providing scientists all over Europe, on the one hand, with unlimited and independent access to state-of-the-art computer resources in all performance classes and, on the other hand, with a world-class pan-European competence and support network.

While the hardware-oriented buildup of the infrastructure is making progress, high-quality user support and software development in the upcoming era of unprecedented parallelism and exascale on the horizon have become the imminent challenges. This has been clearly recognized by the European Commission, who will issue calls for proposals to fund petascale software development in summer 2009. Although traditional support structures are well established in Europe's major supercomputing centres, it is questionable if these structures are able to meet the challenges of the future: in general, support structures are based on cross-disciplinary computer science and mathematics teams; disciplinary computational science support usually is given in an ad-hoc, project-oriented manner.

In this paper, we describe our approach to establish a suitable support structure—Simulation Laboratories (SL). SLs are currently being established at the Jülich Supercomputing Centre of the Forschungszentrum Jülich (FZJ) and at the Steinbuch Centre for Computing (SCC) of the Karlsruhe Institute for Technology (KIT) in Germany. While SLs are community-oriented, *i.e. each SL focusses on a specific community*, they are structured in a strictly interdisciplinary manner, comprising mathematicians, computer scientists and technicians along with disciplinary scientists. SLs are led by a disciplinary scientist, and representatives of the respective disciplines give guidance to its operation. This concept is proposed as a model for and might become an integral element of a future pan-European HPC support and software research structure.

1. Background

The dramatic growth in the available computing power, including memory, data storage, and data transfer capabilities, has revolutionized many aspects of science, technology and political providence. In many areas, simulation sciences are paramount to achieving novel, high-quality results. Politicians and decision makers in industry and society all over the world ever more benefit from this innovative methodology.

1.1. Partnership for advanced computing in Europe

The European Commission is aware of the potential of the simulation sciences and considers the implementation of a pan-European HPC infrastructure as indispensable for the future competitiveness of Europe in science and technology and for the welfare of Europe's citizens [1].

In September 2006, the European Strategy Forum on Research Infrastructures (ESFRI) published a European Roadmap for Research Infrastructures [2], which—in line with the suggestions of the ad-hoc group HPCEUR [3]—is including a European High-Performance Computing (HPC) Service. The key idea of this service is the establishment of a sustainable HPC research infrastructure for the simulation sciences.

Soon after the publication of the ESFRI report in December 2006, the HPC in Europe Taskforce (HET) was formed and started discussing proposals and white papers for the definition of such an infrastructure [4]. HET recommended integrating HPC resources in a comprehensive HPC ecosystem, structured according to a performance pyramid (*cf.* Fig. 1). At the tip of this pyramid there are a few leadership-class systems of highest possible performance level and highest scalability, *i.e.* petaflop/s in 2009, mainly funded through national sources, and supported by additional European funds. The middle layer of the pyramid consists of a larger number of national and regional supercomputers. The latter are powerful supercomputers able to cope with loads below the highest performance level. The bottom of the pyramid consists of regional and local compute servers that nurture the development of a strong, pan-European competence base of computational scientists. HET advocated for a European Research Grid to provide transparent and most secure access to the systems at all levels.

It was evident that only a broad European consortium comprising all interested and capable European partners would be able to implement this idea in form of a coherent research infrastructure. Following the recommendations of HET a consortium was formed in March 2007 as an initiative under a Memorandum of Understanding named *Partnership for Advanced Computing in Europe* (PRACE) [5]. The PRACE partners defined a "preparatory phase" for the creation of the European HPC research infrastructure (HPC-RI) which extends from January 2008 to December 2009. The preparatory phase is devoted to develop concrete action plans, *i.e.* defining transparent, globally competitive organizational structures for high-performance scientific computing in Europe that transcend boundaries, coordinating the use of existing and upcoming high-end hardware, and creating and mobilizing HPC competence from all over Europe. Overarching goals are the long-term sustainability and secured funding of the HPC-RI.

In order to tackle these challenges in a coordinated manner, the PRACE consortium submitted a proposal for funding to the European Commission which was granted end of 2007. This *PRACE Project* started begin of 2008. It is running for two years. After the preparatory period, the PRACE project aims at creating a single legal entity as the distributed provider of high-performance computing resources for researchers all over Europe, the European HPC-RI. Besides preparative work on legal, organizational and administrative issues, in 2008, the PRACE project has carried out an evaluation process in order to define and implement hardware prototypes suited for petaflop/s scalability within the next two years. With begin of 2009, the PRACE project has founded a European platform, named STRATOS. The objectives of STRATOS are of long-term nature, covering technology watch, evaluation and development of next generation hardware and software in cooperation with industrial companies that are active in research and development in the European area.

1.2. Software activities under the PRACE umbrella

The required tight integration and optimal utilization of petascale supercomputers and distributed scientific expertise in Europe bears very new challenges for supercomputing. Substantial efforts are needed in order to enable computational science communities to solve

Journal of Physics: Conference Series **180** (2009) 012038 doi:10.1088/1742-6596/180/1/012038

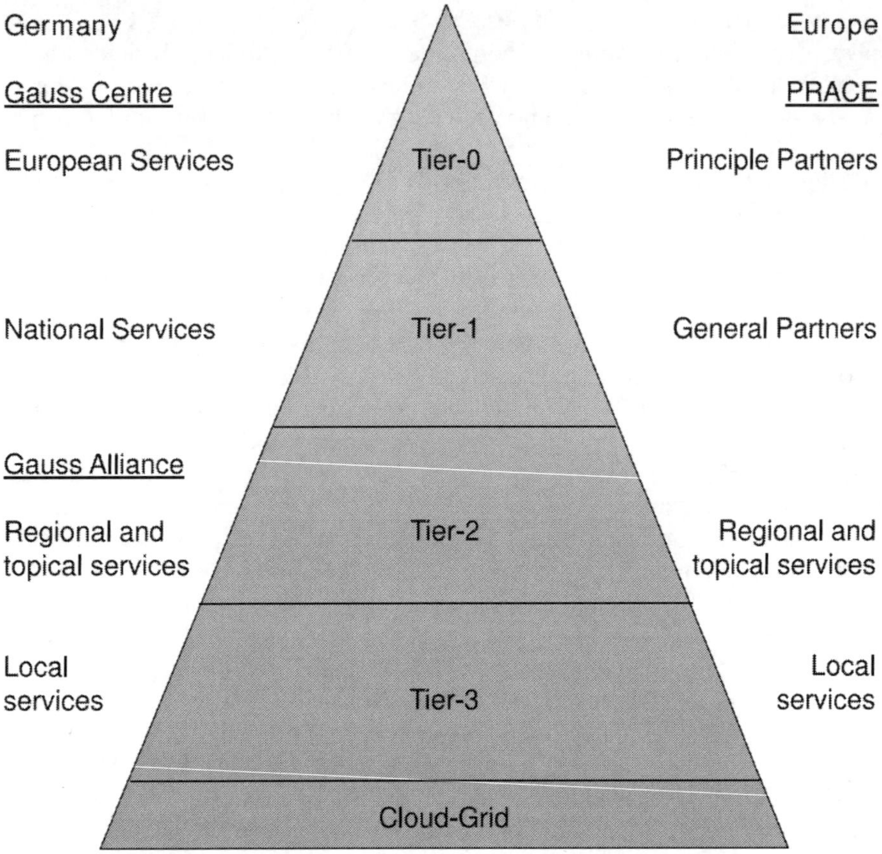

Figure 1. "Performance pyramid" in Germany and Europe. The right hand side shows at which levels the PRACE principal and general partners are positioned on the pyramid with respect to the definition of the performance levels (Tier-0 to Tier-3). The left hand side shows the German situation as a specific example for a PRACE principal partner country.

the most complex problems with highest scientific and societal impact through efficient use of high-end computational resources.

While the focus of the PRACE project lies on the preparation of an integrated sustainable HPC infrastructure including the procurement of supercomputer hardware of the highest performance class, the PRACE initiative is aware that the complementary challenge, *i.e.* enabling a most productive usage of current and future high-end systems and carrying out software development towards the exascale, so far is not at all addressed in a sufficient manner.

Fortunately, in the next round of calls for proposals scheduled for mid-summer 2009, the European Commission starts to close this gap between hardware and software. The EC will be funding the development of software in order to enable applications to scale beyond petaflop/s. The targets of the call will range from system software to parallel compilers, programming environments for accelerators and highly scalable performance analysis. A substantial part of the funds is foreseen to enable software research and development at PRACE principal and general partner sites. In so far, these efforts—being well aligned with the hardware program—will be a very important complementary part on the software side of the process defining the PRACE HPC infrastructure.

2. Simulation Laboratories

While there are good prospects for funding of petascale software research and development in Europe, it is not at all evident that our traditional support and research structures, extending from support desks to project oriented high level support and research teams, are able to cope with the new challenges to come. In general, they are based on cross-disciplinary performance and mathematics teams. Disciplinary computational science support, if provided, usually is done in an ad-hoc, project-oriented manner.

How do we guarantee that the petascale software developments, with the final goal to enable the scaling of scientific and industrial applications, really can meet the very different and complex requirements as set by the various research and technological disciplines the HPC-RI will have to serve? How can we create true interdisciplinarity, where computer scientists and technicians interact side-by-side with disciplinary scientists? How can we mobilize the tremendous potential of disciplinary research communities and trigger best cooperation of the HPC-RI and these communities as well as industry? How do we achieve long-term sustainability?

Jülich Supercomputing Centre (JSC) at the Forschungszentrum Jülich (FZJ) and the Steinbuch Centre for Computing (SCC) at the Karlsruhe Institute for Technology (KIT), joined in the "Helmholtz[1] Programme Supercomputing", are convinced that new, innovative support and research structures are needed in order to master these questions.

2.1. Structure

Research involving infrastructures is more and more carried out within large collaborations that belong to organized topical scientific communities, in particular, when unique large-scale experimental facilities are involved (LHC, Hubble telescope, *etc.*). Such instruments are made available on a competitive basis to specific experiments, carried out by larger user groups. The experimental collaborations extend far beyond national borders or organizational structures.

Today, researchers who use supercomputers of the highest performance class as their major research instruments face a similar situation: *the large-scale instruments* of computational scientists and engineers are the supercomputers together with fast networks, data centres and peripheral hardware, the classical experiments corresponds to the simulation, and very complex detectors with their periphery stand in analogy to very complex application codes and associated analysis and enabling tools, libraries, compilers and operating systems.

Our support structures must be adapted to these developments! In this spirit, the JSC and the SCC are creating *Simulation Laboratories* (SL) as specific research and support instruments[2]. The SL is a community-oriented research and support organization for a specific scientific community. It serves as an integral part of the community and strengthens its activities by collaborating with community members and assisting the scientists in performing simulations on supercomputers. The SL consists of a core group located at the supercomputer centre—or several such core groups in case of distributed SLs—and a number of associated scientists from the research community.

The SL as a whole takes care of high-level know-how on codes and algorithms of a specific scientific community. It is staffed to the level of a self-contained research team. The expertise provided to a research community by the SL thus can reach beyond the support a traditional

[1] The Helmholtz Association of German Research Centre contributes to solving major challenges facing society, science and industry with top scientific achievements in six research areas: Energy, Earth and Environment, Health, Key Technologies, Structure of Matter, Transport and Space. With 28,000 employees in 16 research centres and an annual budget of approximately 2.8 billion euros, the Helmholtz Association is Germanys largest scientific organization. Its work follows in the tradition of the great natural scientist Hermann von Helmholtz (1821-1894) [6].

[2] Simulation Laboratories bear some similarity with *Computational End Stations* as developed by the Oak Ridge National Laboratory.

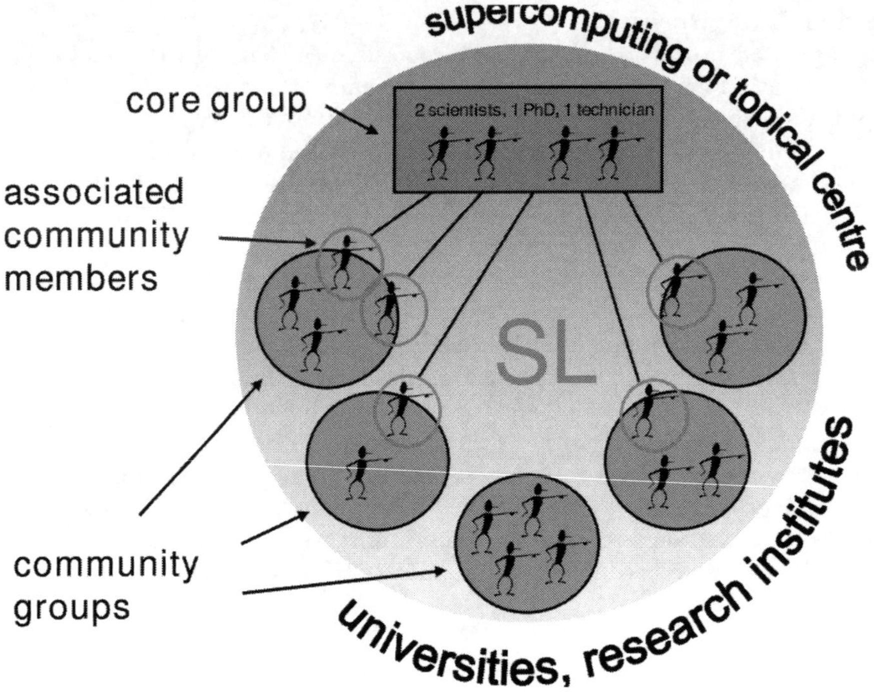

Figure 2. Structure of a simulation laboratory.

expert advisor can give.[3] With SLs, the support and research structures of supercomputing centres are transformed from being just *reactive* to become *active* support instruments for the benefit of scientific communities.

2.2. Tasks

Disciplinary Science

- Participation in research projects of the community including active collaboration in user projects and joint publications
- Participation in consortia applying for and carrying out 3rd-party-funded research
- Development and maintenance of important software for the entire community, in particular with respect to long-term provision of generic algorithms and kernel codes
- Organization of data bases and management of meta data including data curation

Supercomputing

- Development of computational methods for simulations on high-end systems
- Adaptation of software to highly scalable HPC systems and performance optimization

Education and Training

- Training in efficient computational methods and in the efficient usage of HPC systems
- Participation in education

Community Building

- Organization of a community-oriented web portal

[3] The traditional user support model can not completely be substituted by SLs, in fact it is substantially enhanced by the new structure.

- Organization of joint workshops *etc.* to trigger the exchange of experiences in the community and with the community

2.3. Organization

A typical SL consists of a core group of scientists, technicians and PhD-students located at the supercomputer centre (or at two or three centres in case of a distributed SL) and of associated members at institutes of the scientific community (*cf.* Fig. 2).

Core Group The core group comprises between four to six (or more) persons:

- An experienced scientist as the head of the Simulation Laboratory, who is recognized by the community and who is actively involved in the research of the field
- Two postdocs, who concentrate their work on methodological or complex software-related questions
- Two PhD students, who are able to address specific, thematically related development tasks
- A technician with skills in software adaptation, program optimization and data management

Associated Members The research groups of the specific community name associated members to the SL. They work at their home institutions and participate actively in the work of the SL.

Steering Committee The work plan of the SL is defined and monitored by the community. For this purpose a Steering Committee is established by the community together with the supercomputing centre.

- The steering committee is composed of representatives of the different research branches of the community. The core group is represented in this committee
- Work packages for the Simulation Laboratory are formally proposed by the community and submitted to the Steering Committee. The board evaluates the proposals, makes a selection and a final ranking. Prior to the beginning of a project, a common understanding must be reached and precisely documented of the distribution of work between the core group of the SL and the associated community members. This guarantees a close interaction of the SL with the research groups of the community.

Accompanying Measures

- Guest program in order to enable face-to-face contact with scientists of the community
- Supercomputer resources for software development and testing tasks

2.4. Operation

The work performed by a SL is integrated in the research performed by its community. It comprises work packages of two to six months duration, formally proposed by the community groups and submitted to the Steering Committee. The committee selects the proposals to be supported. This procedure guarantees a close interaction of the SL with research groups from the community and ensures the continual interest of various community groups. As an example, Fig. 3 describes the workflow of the SL Plasma Physics (PP), which is established at the JSC. Once a project (work package) has been selected, a specific work plan detailing the allocation of tasks, the goals to be achieved, dissemination, *etc.*, is agreed upon between the core group of the SL and the associated community group members, *cf.* Fig. 4. In practice, these tasks will be carried out by a temporarily assigned team, comprising one to two members of the SL and two to three members of the community group. The nature of the tasks may range from performance analysis, code improvement and I/O tuning to assembler kernel improvement and basic algorithmic development, depending on the priorities defined by the particular SL's steering committee.

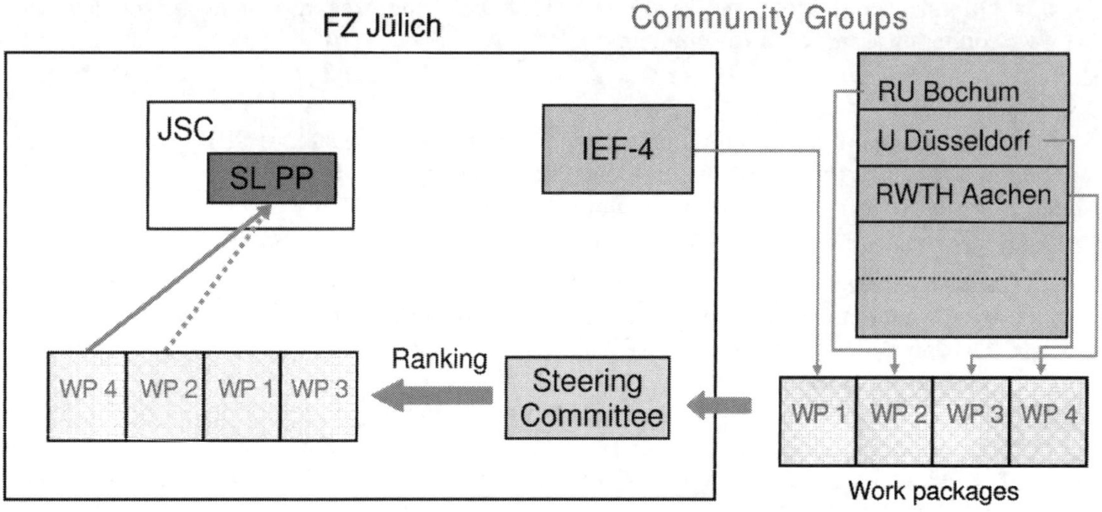

Figure 3. Work-flow diagram for a simulation laboratory.

2.5. A working simulation laboratory: "SL Plasma Physics"

In the SL Plasma Physics, illustrated in Figs. 3 and 4, first projects are currently carried out involving groups from FZJ (Plasma Physics), the University of Bochum and the Helmholtz Alliance EMMI.

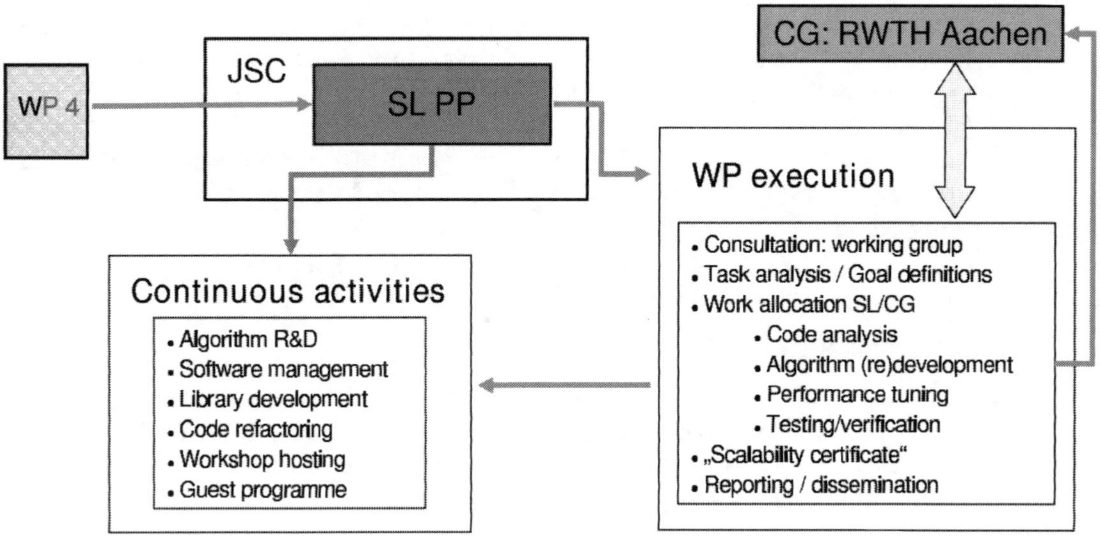

Figure 4. Execution of a work package in the SL "Plasma Physics".

To give an example, a Monte-Carlo transport code is being parallelized as part of a diploma thesis (approximately six months); as another example, the fluid turbulence code "racoon" has been ported to the Blue Gene/P system at JSC and is being further examined for potential optimization.

2.6. Cross-sectional support groups

In addition to the SLs, JSC has strengthened its Cross-Sectional Groups (CSG) to achieve an in-depth intensification of the development of HPC methods and algorithms, performance analysis tools and petabyte data repositories. The CSGs work in a horizontal manner, *i.e.* for all SLs and users of the centre. They are considered as essential to achieve a synergetic interaction of the SLs on the algorithmic and computer science level. From the work-flow point of view, SLs and CSGs form a matrix structure and complement each other in an ideal manner. The aim is to establish a first-class science-based partnership between disciplinary communities and the genuine research in applied computer science and applied mathematics carried out by the supercomputing centres.

3. SLs at FZJ and KIT

The JSC and the SCC as protagonists of the Helmholtz Programme Supercomputing have the objective to establish at least seven SLs until the end of 2010. Our recent 5-years programme evaluation attributed highest priority to the SL-concept. The international Helmholtz referee committee expressed the conviction that SLs are the most appropriate structure to meet the challenges of the petascale era and to address scalability beyond petaflop/s. SLs are seen as a guarantee that the available funds are utilized most effectively.

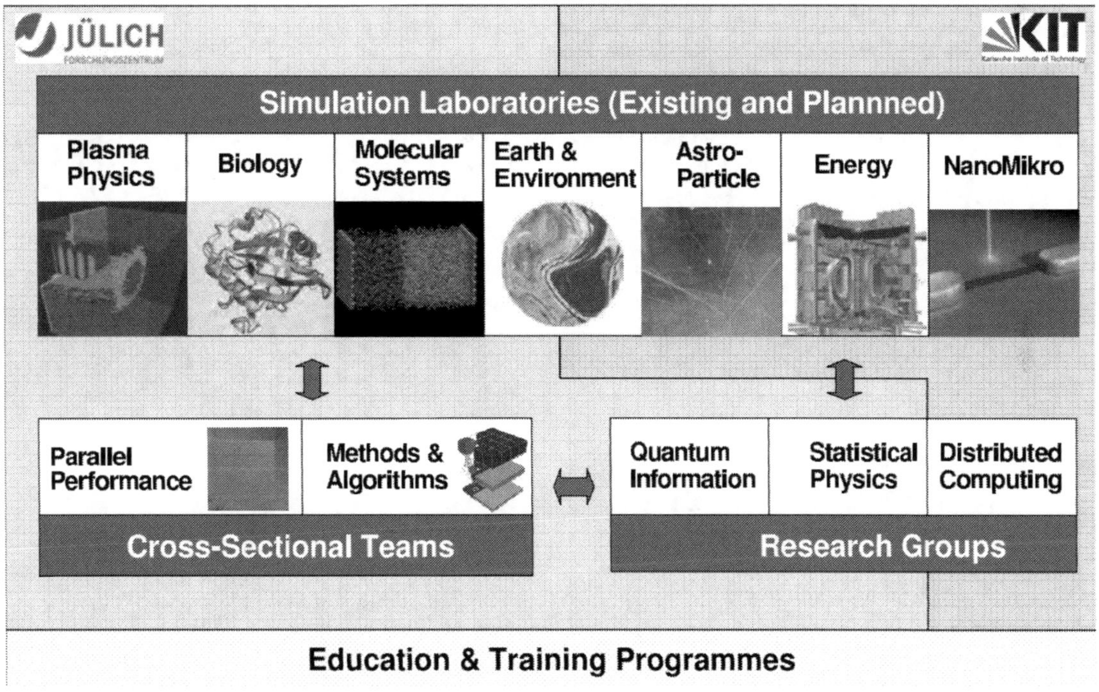

Figure 5. Pictogram of existing and planned SLs at FZJ and KIT being established within the Helmholtz Programme "Supercomputing".

In the first round until the end of 2010, the SLs *Plasma Physics, Biology and Molecular Systems* are established by the JSC at FZJ, the SLs *Astro-Particle, Energy and Nano-Mikro* will be established by the SCC at KIT, see Fig. 5. The SL *Earth & Environment* is a joint lab, run together by JSC and SCC. Further SLs are under discussion for the field of lattice quantum chromodynamics, jointly with CastorC at the Cyprus Institute, Nicosia, and in the field of engineering (CFD, mechanics) in the framework of the Jülich-Aachen Research Alliance JARA.

3.1. SL Biology (FZJ)

Research

- Protein folding & interaction
- Structure prediction
- Systems biology

Support

- Libraries
- Bio databases
- Benchmarking
- Monte Carlo
- FFT docking
- Machine learning

Codes

- PROFASI
- SMMP

Outreach

- FZJ: Biological institutes
- Helmholtz: Diverse groups
- Regional: North-Rhine Westphalia Life Science Area
- International: UC Berkeley, Michigan Tech

3.2. SL Plasma Physics (FZJ)

Research

- Kinetic methods: Particle-in-Cell, Vlasov, MD
- Fluid and MHD models
- Transport: Monte Carlo

Support

- Plasma model porting & scaling
- Code benchmarking, eg: 3D PIC

Codes

- PSC
- ILLUMINATION
- PEPC
- racoon
- EIRENE

Outreach

- FZJ: Plasma and Nuclear
- Regional: Universities Aachen, Bochum, Düsseldorf
- National: GSI (HA-EMMI), Garching, FZ-Rossendorf

3.3. SL Molecular Systems (FZJ)

Research

- Macroscopic properties from microscopic information

Journal of Physics: Conference Series **180** (2009) 012038 doi:10.1088/1742-6596/180/1/012038

- Model larger systems / longer timescales
- Integrated multi-scale approaches

Support

- Methods: electronic structure, force-field, MD
- Quantum-chemical modelling and tools
- Scalable algorithms for supercomputers

Codes

- MP2C
- P3MG
- TURBOMOLE
- COLUMBUS

Outreach

- FZJ: Plasma and Nuclear
- Helmholtz, Regional, National: NIC users
- International: cooperations with quantum chemistry developer groups

3.4. SL Earth & Environment (FZJ and KIT)

Objectives

- Development of regional & global Earth system models (coupling tools, up-/downscaling methods)
- Large-scale processing of Earth observation data (inverse modelling, assimilation)

Major Research Projects

- Air quality forecasting
- Solar effects in Earth system models
- Transport in soils and aquifers
- High impact weather

Codes

- CLaMS
- JURASSIC
- ECHAM
- SACADA
- COSMO
- WRF
- MOBILE
- MESSy

Outreach

- Helmholtz (FZJ-KIT)
- Regional, National: NIC users
- International: ESA. IPCC

3.5. SL Energy (KIT)

Research

- Power plant operation, planning and power trading
- Material behavior in nuclear facilities
- Biofuel synthesis
- Geothermal reservoirs
- Security of nuclear waste repositories

- Fuel cell design

Support

- Plasma modelling for ITER fusion facility
- Monte-Carlo simulation of radiation transport
- Numerical simulation of biofuel synthesis

Codes

- MCNP5
- Eirene
- Turbit-VOF
- DETCHEM

Cooperations

- EnBW
- SAP
- Siemens
- ABB

3.6. SL NanoMikro (KIT)

Research

- Nanoelectronics
- Nanophotonics
- Weak interactions in metal nanoclusters
- Phonons and electron-phonon coupling
- Protein structure prediction
- Interaction of actinide ions with metal surfaces
- Advanced materials for fusion and fission facilities

Support

- Molecular building Blocks
- Simulation of nanoscale materials
- Nanobiology

Codes

- TURBOMOLE
- MBPP PERT
- POEM
- FlexScreen
- CASINO

3.7. SL Elementary & Astroparticle Physics (KIT)

Research

- Astrophysical objects as laboratories
- CORSIKA—simulation of extensive air showers

Support

- Code optimization projects

Codes

- CORSIKA

Outreach

- Users from 47 countries for 50 experiments

- AUGER South - Argentina
- soon also AUGER North in USA
- KASCADE Grande - Germany
- AMANDA - Antarctica

3.8. CSG Mathematical Methods and Algorithms (FZJ)

Numerical Algorithms

- Large-scale linear systems
- Eigenvalue problems
- Error-controlled fast multipole method
- Kernels for new architectures (cell engine)

Modelling

- Pedestrian dynamics (simulation faster than real-time)
- 3D soil-root water transfer

Software & Services

- Parallel eigenvalue library
- FMM program
- Toolboxes mathematical software
- Benchmarking

Cooperations

- FZJ: research groups active in simulation
- SLs Helmholtz: diverse groups
- Regional: Universities of Wuppertal, Aachen, Bonn, Cologne, FH Aachen
- MPI-KS Dresden
- Industrial partners

3.9. CSG Parallel Performance (FZJ)

Objective

- Optimization tools for parallel codes with highest scalability

Research

- Performance analysis tool for large-scale systems
- Scalasca

Cooperative Projects

- Helmholtz: VI-HPS with international links (Oak Ridge)
- National: SILC
- European: ParMA

4. Summary

We have presented the concept and have given examples of working and planned Simulation Laboratories currently being established by the Jülich Supercomputing Centre at the Forschungszentrum Jülich (FZJ) and by the Steinbuch Centre for Computing (SCC) at the Karlsruhe Institute for Technology (KIT). Each SL puts its support and research focus on a specific community, while being of interdisciplinary nature, with mathematicians, computer scientists disciplinary scientists working together, guided by a steering committee with community representatives as members. The approach to organize the user support in terms of Simulation Laboratories for different scientific communities received strongest support during the evaluation by the Helmholtz Programme Supercomputing in spring 2009.

JSC and SCC are confident that the SLs can provide the proper organizational structure to handle the coming challenges in hardware and software development most effectively, for the benefit of computational science and engineering. The SLs will guarantee that the enablement of application software towards petaflop/s scalability and beyond will become an interactive process, bringing computer centres and user communities much closer together.

Looking forward to the formation of the European HPC infrastructure, coordinated by the *Partnership for Advanced Computing in Europe* (PRACE), Simulation Laboratories will be proposed on a European scale as an integral bottom-up element of the PRACE HPC support and software research activities, realizing a world-class pan-European competence and support network.

References

[1] Reding V 2008 ICT based infrastructures for a new scientific renaissance Web-page European Commission, DG INFSO http://www.neuroprion.org/streaming/domain18/2008/12/m683/index.html
[2] Wood J 2008 European roadmap for research infrastructures pdf-document European Strategy Forum on Research Infrastructures (ESFRI) ftp://ftp.cordis.europa.eu/pub/esfri/docs/esfri-roadmap-report-26092006_en.pdf
[3] Ramalho M 2006 High performance computing infrastructures in Europe pdf-document European Commission, DG INFSO ftp://ftp.cordis.europa.eu/pub/ist/docs/rn/changes-060511-DD-event-supercomp-report.pdf
[4] Guest M *et al.* 2006 The scientific case for a European supercomputing infrastructure pdf-document HPC in Europe Task Force
http://www.hpcineuropetaskforce.eu/files/Scientific%20case%20for%20European%20HPC%20infrastructure%20HET.pdf
[5] 2009 PRACE Web-page Partnership for Advanced Computing in Europe
[6] 2009 Web-page Helmholtz Association http://www.prace-project.eu/

SciDAC 2009 IOP Publishing
Journal of Physics: Conference Series **180** (2009) 012039 doi:10.1088/1742-6596/180/1/012039

Modeling the Office of Science ten year facilities plan: The PERI Architecture Tiger Team

Bronis R. de Supinski[1], Sadaf Alam[2], David H. Bailey[3], Laura Carrington[4], Chris Daley[5], Anshu Dubey[5], Todd Gamblin[1], Dan Gunter[3], Paul D. Hovland[6], Heike Jagode[7], Karen Karavanic[8], Gabriel Marin[2], John Mellor-Crummey[9], Shirley Moore[7], Boyana Norris[6], Leonid Oliker[3], Catherine Olschanowsky[4], Philip C. Roth[2], Martin Schulz[1], Sameer Shende[10], Allan Snavely[4], Wyatt Spear[10], Mustafa Tikir[4], Jeff Vetter[2], Pat Worley[2], and Nicholas Wright[4]

[1] Lawrence Livermore National Laboratory, Livermore, California
[2] Oak Ridge National Laboratory, Oak Ridge, Tennessee
[3] Lawrence Berkeley National Laboratory, Berkeley, California
[4] San Diego Supercomputing Center, San Diego, California
[5] University of Chicago, Chicago, Illinois
[6] Argonne National Laboratory, Argonne, Illinois
[7] University of Tennessee - Knoxville, Knoxville, Tennessee
[8] Portland State University, Portland, Oregon
[9] Rice University, Houston, Texas
[10] University of Oregon, Eugene, Oregon

E-mail: bronis@llnl.gov, alamsr@ornl.gov, dhbailey@lbl.gov, lcarring@sdsc.edu, cdaley@flash.uchicago.edu, a-dubey1@uchicago.edu, tgamblin@llnl.gov, dkgunter@lbl.gov, hovland@mcs.anl.gov, jagode@eecs.utk.edu, karavan@cs.pdx.edu, maring@ornl.gov, johnmc@cs.rice.edu, shirley@cs.utk.edu, norris@mcs.anl.gov, loliker@lbl.gov, cmills@sdsc.edu, rothpc@ornl.gov, schulzm@llnl.gov, sameer@cs.uoregon.edu, allans@sdsc.edu, wspear@cs.uoregon.edu, mtikir@sdsc.edu, vetter@ornl.gov, worleyph@ornl.gov, nwright@sdsc.edu

Abstract. The Performance Engineering Institute (PERI) originally proposed a tiger team activity as a mechanism to target significant effort optimizing key Office of Science applications, a model that was successfully realized with the assistance of two JOULE metric teams. However, the Office of Science requested a new focus beginning in 2008: assistance in forming its ten year facilities plan. To meet this request, PERI formed the Architecture Tiger Team, which is modeling the performance of key science applications on future architectures, with S3D, FLASH and GTC chosen as the first application targets. In this activity, we have measured the performance of these applications on current systems in order to understand their baseline performance and to ensure that our modeling activity focuses on the right versions and inputs of the applications. We have applied a variety of modeling techniques to anticipate the performance of these applications on a range of anticipated systems. While our initial findings predict that Office of Science applications will continue to perform well on future machines from major hardware vendors, we have also encountered several areas in which we must extend our modeling techniques in order to fulfill our mission accurately and completely. In addition, we anticipate that models of a wider range of applications will reveal critical differences between expected future systems, thus providing guidance for future Office of Science procurement decisions, and will enable DOE applications to exploit machines in future facilities fully.

1. Introduction

Sustained performance improvements are integral to the DOE Office of Science SciDAC program's mission to advance large-scale scientific modeling and simulation. Simulation is a key investigative technique for disciplines where experimentation is expensive, dangerous, or impossible. Increased performance can enable faster simulations and more timely predictions, or it can be used to increase the accuracy of existing physical models, enabling more predictive simulations. Research enabled by the SciDAC program will have far-reaching effects in fields such as basic energy, biology, environmental science, fusion energy, and high-energy physics.

The Performance Engineering Research Institute (PERI) tiger team activity targets critical SciDAC performance needs. The original intent was for each tiger team to focus the efforts of several PERI researchers on improving performance of an Office of Science application, with the application selected based on Office of Science mission objectives and application readiness for the focused effort. Thus, each tiger team was envisioned as a relatively short-term activity (six months to at most one year). In 2007, our tiger teams had a positive impact on two key DOE applications participating in the JOULE metric. We improved the performance of a turbulent combustion code (S3D [1]) on Oak Ridge's Cray XT5 Jaguar system by 13%. Similarly, we improved the performance of the Gyrokinetic Toroidal Code (GTC) [2, 3] by 10% on Jaguar and by 15% on Argonne National Laboratory's (ANL's) Intrepid Blue Gene/P system.

In 2008, the Office of Science requested that PERI provide assistance in their ten year facilities plan. In particular, they wanted PERI to provide guidance in how key applications would perform across the range of future systems expected to be offered by major vendors in that period. Thus, we redefined the scope of our tiger team activity to handle this request and started the PERI Architecture Tiger Team. This team's goal is to model the behavior of selected applications and to predict their performance on anticipated future systems, instead of to improve their performance on current systems. To fulfill the goal, we must consider a wider range of Office of Science applications and evaluate the suitability of current and future high performance computing (HPC) architectures for the applications. This broader scope has led us to include nearly all PERI researchers on the Architecture Tiger Team.

Several factors complicate the Architecture Tiger Team's charge. Large scale simulations are complex software artifacts for which the performance depends on input and frequently evolves during the course of a simulation. Further, small source code changes can lead to significant performance changes. Thus, modeling their performance across a variety of existing architectures remains a topic of research. For example, modeling at larger scales than are currently run requires changes to most existing modeling methodologies. In order to model the performance for systems that will emerge over the ten year period, we must not only overcome these challenges but also anticipate how the software, as well as the hardware, will evolve.

For these reasons, we have developed a three-part, iterative plan, with each iteration focusing our modeling effort on a different (or growing) set of applications. First, we extensively measure the performance of the applications at scale with a variety of state-of-the-art performance analysis tools. These measurements ensure that we have appropriate versions of the applications: although we are no longer focused on optimization, we still apply our expertise in this direction. Thus, this activity can also provide some benefit to the application teams.

Second, we use these measurements and other data to create predictive performance models that estimate the scaling properties of current applications on future hardware. In the first iteration of the Architecture Tiger Team, we have applied this strategy to three Office of Science early science applications: S3D, GTC, and FLASH [4, 5, 6], an astrophysical thermonuclear flash simulation. In this paper, we detail the preliminary results of this study, which indicate that these applications will perform well across the breadth of anticipated architectures.

In the third part of our process, we report findings to the Office of Science and work with them to select the applications for the next iteration. We are currently engaged in that selection process for the Architecture Tiger Team's second iteration. We are employing criteria that both reflect the importance of the applications to the Office of Science's mission and attempt to capture the breadth of characteristics of its applications. Simply put, we must ensure that the ten year facilities plan reflects the range of needs of the Office of Science's broad mission.

The rest of this report is organized as follows. We summarize key tools for our large-scale performance measurement activity in Section 2. Section 3 describes our performance modeling techniques. In Section 4, Section 5 and Section 6, we detail our initial findings with the S3D, FLASH, and GTC codes. We then state our initial conclusions and lessons learned for this on-going activity, including guidance in selecting the next set of applications, in Section 7.

2. Measurement

We have used a wide variety of performance analysis tools to characterize the behavior of S3D, GTC and FLASH on current Office of Science platforms at scale. The large volumes of performance data complicate performance measurement on systems such as Jaguar and Intrepid. Because these modern parallel applications can have dynamic behavior, understanding their performance can potentially require measuring all application processes. However, the overhead of data collection and aggregation on large machines can perturb running applications, making the measurements, and thus the models that we derive from them, inaccurate. Further, too much performance data can make analysis prohibitively expensive.

2.1. Performance analysis tools

To address these challenges, we have employed a wide variety of tools for measuing performance data of the three applications that we studied in the first iteration of the Architecture Tiger Team. We briefly describe some of our key performance tools in this section; we present results of applying them in Sections 4, 5, and 6.

2.1.1. Vampir The Vampir suite, developed at the University of Dresden - Germany, consists of VampirTrace for instrumentation, monitoring and recording and VampirServer for visualization and analysis [7, 8]. It stores event traces in the Open Trace Format (OTF) [9]. VampirTrace can examine many performance metrics, *e.g.,* MPI communication events, subroutine calls from user code, hardware performance counters, I/O events and memory allocation. VampirServer implements a client/server model with a distributed server, allowing interactive visualization of traces with over 1,000 processes and uncompressed size of up to 100 GBytes [8].

2.1.2. mpiP mpiP [10], an MPI profiling tool, measures cumulative time spent in all MPI call sites across all processes in an application. Like other profiling tools, mpiP only collects statistical information, as opposed to full trace data like Vampir. mpiP generates a single file, which is much smaller than a full trace file, but which loses timing information.

2.1.3. TAU The TAU Performance System is a portable profiling and tracing toolkit for performance analysis of parallel programs [11]. It comes with a wide selection of features to measure specific functions, code regions and user-defined events in parallel applications. The user must recompile his or her application with the TAU compilers and then run a parallel job. TAU then outputs a trace or profile as desired. TAU also provides extensive data mining and analysis tools for processing information after it has been measured and stored.

2.1.4. Libra Libra [12] is a tool for scalable load-balance analysis developed at Lawrence Livermore National Laboratory and the University of North Carolina at Chapel Hill. Unlike full trace tools, Libra uses aggressive, lossy wavelet compression to reduce the volume of load-balance data significantly before recording it. Libra can achieve 100:1 to 1000:1 compression on load-balance data and it provides a scalable client-side visualization tool for viewing recorded traces. Libra records measured code regions by call site.

2.2. Platforms

In the first iteration of the Architecture Tiger Team, we conducted extensive measurements of our target applications' performance on two leadership-class systems. The first system is Argonne National Laboratory's Blue Gene/P system, Intrepid. Intrepid contains 163,840 PowerPC 450 cores running at 850 MHz, and sustained LINPACK performance of 450 teraflops. The second system is the Cray XT4 Jaguar system at Oak Ridge National Laboratory. Jaguar contains 31,328 Opteron cores running at 2.1 GHz, and has sustained performance of over 205 teraflops. Both systems use quad-core nodes.

Jaguar and Intrepid have different network and I/O configurations. Jaguar uses a 3D mesh network for communication between nodes, whereas Intrepid uses a full 3D torus and also uses a tree network and a barrier network for collective communication. Both systems have dedicated I/O nodes that relay I/O operations between applications and the parallel filesystem. On Intrepid, I/O nodes are internal nodes in the tree network; compute nodes communicate with I/O nodes through the tree network, and the I/O nodes communicate with the parallel filesystem over Myrinet links. On Jaguar, the I/O nodes are situated along one side of the 3D mesh, and compute nodes communicate with them over the mesh network.

3. Modeling

We have developed several techniques to predict performance of DOE applications on leadership class facilities. In this section, we give a brief overview of these techniques, while we discuss the results of applying them to S3D, FLASH and GTC in Sections 4, 5 and 6.

3.1. Convolving machine profiles with application signatures

To predict the performance of applications on future architectures, we have developed an approach [13] that separates application-specific measurements from machine-specific measurements. Our approach involves two key components:

Machines Profiles that characterize the rates at which a machine can (or is expected to) carry out fundamental operations abstract from any particular application;

Application Signatures that characterize the fundamental operations that an application must execute independent of any particular machine.

Our approach enables performance predictions of applications on current systems by convolving application signatures with profiles of the existing systems, and on future systems by convolving the application profiles with profiles generated from the *expected* performance parameters of the future systems. Conceptually, a *convolution* defines an algebraic mapping of application signatures onto runtimes to arrive at a performance prediction.

Given an application profile A, and a machine profile M, we define P, a matrix of runtimes, such that $p_{ij} = \sum_{k=1}^{p} a_{ik}m_{kj}$, or:

$$\begin{bmatrix} p_{11} & p_{12} & p_{13} \\ p_{21} & p_{22} & p_{23} \\ p_{31} & p_{32} & p_{33} \end{bmatrix} = \begin{bmatrix} a_{11} & a_{12} & a_{13} & a_{14} & a_{15} \\ a_{21} & a_{22} & a_{23} & a_{24} & a_{25} \\ a_{31} & a_{32} & a_{33} & a_{34} & a_{35} \end{bmatrix} \begin{bmatrix} m_{11} & m_{12} & m_{13} \\ m_{21} & m_{22} & m_{23} \\ m_{31} & m_{32} & m_{33} \\ m_{41} & m_{42} & m_{43} \\ m_{51} & m_{52} & m_{53} \end{bmatrix}$$

$$e.g., p_{32} = a_{31}m_{12} + a_{33}m_{22} + a_{33}m_{32} + a_{34}m_{42} + a_{35}m_{53}$$

The rows of P correspond to applications while the columns correspond to systems and each p_{ij} is the expected runtime of application i on system j. The rows of A are applications, and the columns are operation counts. Any row of A is the *signature* for application i. Likewise, rows of M are bandwidths measured by some benchmark for a particular system while each column is the *profile* of a particular system.

This approach is generic, and we could apply it to measurements from any of the tools mentioned in Section 2 to produce application signatures targeted at particular types of analysis. However, in order to reflect the impact of timing considerations, we currently use traces of memory operations to characterize the fundamental operations of computational code regions and message traces similar to those produced by Vampir [14, 15]. We use cache simulation to convolve the signatures of the computational regions with characterizations of the memory system obtained with the MultiMAPS benchmark from SDSC. We then use a high-level network simulation such as Dimemas [16, 17] or SDSC's PSiNS [18] to convolve message trace signatures with simple network signatures that capture latency and bandwidth and the predicted performance of the computational regions.

3.2. Modeling assertions

An alternative modeling strategy, called *Modeling Assertions* (MA) [19], constructs symbolic models of application performance. This technique allows users to annotate their code with expressions revealing the relationships among important input parameters, computation, and communication. These annotations, in the form of pragmas or directives, capture the anticipated performance, in terms of time or other metrics such as cache misses or floating point operations. As the application runs, the MA library checks the model against application structure and key model input parameters. Symbolic performance models complement empirically derived models because the symbolic models expose sensitivities across important parameters and can be scaled to any parameter range.

3.3. Load balance modeling

In addition to models for application runtimes, we have developed models of the load balance properties of large systems. Libra's compressed representation of system-wide load balance traces uses a wavelet approximation to allow for multiscale representations of load balance properties. The structure of this approximation supports extraction of a low-resolution model of an application's load balance without recording exhaustive measurements.

This type of compact model will enable us to verify, possibly with distributed extensions of MA, that specific processes have workloads within their expected bounds. Further, our low-resolution models will eventually enable us to *predict*, within some confidence, the evolution of load on large systems and to incorporate dynamically derived load balance guidelines to application-specific load balance components. This solution will dramatically reduce the burden of large-scale measurement on the application developers, enabling them to concentrate on how best to redistribute work within their applications.

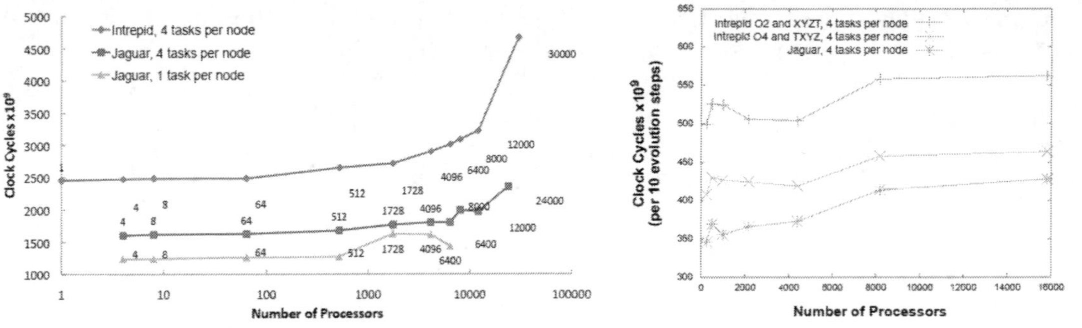

(a) S3D Turbulent Combustion (b) FLASH White Dwarf Deflagration

Figure 1. Weak Scaling for Office of Science Codes on Intrepid and Jaguar

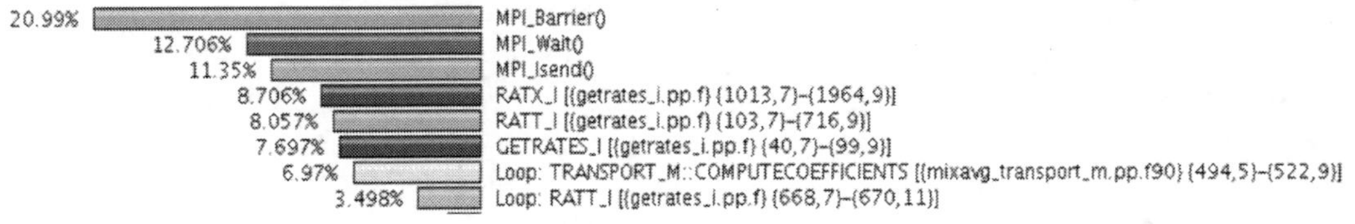

Figure 2. Most Time Consuming S3D Routines on Jaguar at Scale

4. S3D

S3D [1] is a state of the art turbulent combustion simulation. The code, which was developed at the Combustion Research Facility at Sandia National Laboratories in Livermore, California, won a 2007 INCITE award for six million hours on the XT3/4 Jaguar system at ORNL's National Center for Computational Sciences. S3D solves the compressible reacting Navier-Stokes equations by using high-fidelity numerical methods. Principal components include an eighth-order finite-difference solver, a fourth-order Runge-Kutta integrator, a hierarchy of molecular transport models and detailed chemistry. The use of direct numerical simulation (DNS) enables scientists to study the microphysics of turbulent reacting flows, as this gives full access to time-resolved fields and provides physical insight into chemistry turbulence interactions. Perhaps more importantly, S3D is critical for accurate simulations of larger systems. The detail afforded by the DNS model enables the development of reduced model descriptions that can be used in macroscale simulations of engineering-level systems.

S3D is architected for scalability. It uses a 3D domain decomposition, where each MPI process manages an equal number of grid points and has the same computational load. Interprocessor communication in this decomposition is only between nearest neighbors, and S3D uses large messages and can overlap communication and computation. All-to-all communication is required only for monitoring and synchronization ahead of I/O.

4.1. Measurement of S3D

We conducted weak scaling measurements of S3D simulating an ethylene burn, with 27,000 grid points per process. Figure 1(a) shows results for Intrepid and Jaguar, in clock cycles required for an entire run. On both systems, S3D scales almost perfectly up to 4,096 cores. After this point, runtimes begin to increase, until at 30,000 processes the runtime is twice that of the baseline, 4-core run on Intrepid. On Jaguar, our 24,000-core run took approximately 30% longer than the baseline run.

We used optimized TAU instrumentation in order to determine the cause of S3D runtime

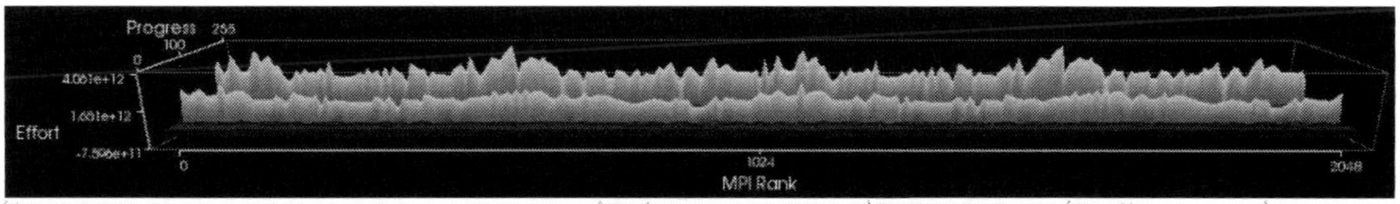

Start	TotalTime	Deviation	MeanSkew	MeanKurtosis
write_savefile(io.f90:312)	1.28868e+16 (19.98%)	7.79937e+11	3.42771	14.5355
write_savefile(io.f90:322)	8.83365e+15 (13.69%)	5.33235e+11	3.30364	13.637
write_savefile(io.f90:312)	7.97592e+15 (12.36%)	7.96813e+11	5.00885	27.2989
write_savefile(io.f90:322)	5.35586e+15 (8.30%)	5.35976e+11	5.01255	27.3279

Figure 3. Percent Total Runtime for Most Costly MPI Operationsin S3D Checkpoint on Intrepid at 16,384 Cores. Load Balance Profile of Topmost `MPI_Barrier()` Callpath is Shown.

increases at scale. Figure 2 shows the top eight entries in the profile; the first three are (`MPI_Barrier`, `MPI_Wait`, and `MPI_Isend`). S3D spends the bulk of its time in these routines at large core counts with its default I/O scheme that writes a file per MPI task. The next most time consuming routines (three subroutines — `RATX_I`, `RATT_I` and `GETRATES_I` — and two loops) are in the parallel solver.

Applying Libra to S3D reveals the underlying issue: the default IO configuration taxes the I/O systems excessively. Figure 3 shows a Libra plot of time spent in the most time-consuming `MPI_Barrier` call on 2,048 processes of a 200-timestep, 16,384 process run of S3D. MPI rank is on the x-axis, and time spent in `MPI_Barrier` per timestep is on the z-axis. Clearly, the variance in the MPI times shown in Figure 2 reflects a load imbalance caused by highly variable times to complete the I/O phase across the MPI tasks. Other I/O configurations, including one that performs writes from a subset of the MPI tasks, offer better scaling performance. Since I/O behavior is a critical component in S3D's overall performance, we are currently working with the Petascale Data Storage Institute (PDSI) to understand and to model it.

4.2. Modeling of S3D

We now detail performance models of S3D's computational regions, leaving models that include I/O for future work. The Kiviat diagram in Figure 4(a) shows anticipated memory system parameters for several hardware vendors, which we anonymize here due to NDA considerations. The four axes are memory bandwidths for L1, L2, and L3 caches and for main memory (MM). System 1 represents these parameters for Jaguar. In the diagram, the noticeable differences between current and future systems are the significant changes in L3 and main memory bandwidth. Our modeling analysis explores the impact that this difference will have on S3D.

Table 1 shows the results of convolving the machine profiles shown in Figure 4(a) with S3D memory profiles. These results indicate that the differences in memory system performance *will* impact S3D runtimes significantly. We predict that S3D's C_2H_4 problem will perform well on all expected future systems but will perform best on those systems with the most main memory bandwidth. Although not shown here, this effect is not true for all applications. For example, our predictions indicate the memory system differences will provide little benefit to WRF, a weather forecasting simulation.

An important consideration for models of S3D is that its memory behavior is scale invariant. We have compared memory traces across a range of job sizes and found that they are very consistent under weak scaling. We have previously shown that we can predict S3D performance on Jaguar as we scale the number of MPI tasks directly from traces of smaller runs. Thus, we expect the results in Table 1 to hold for much larger systems.

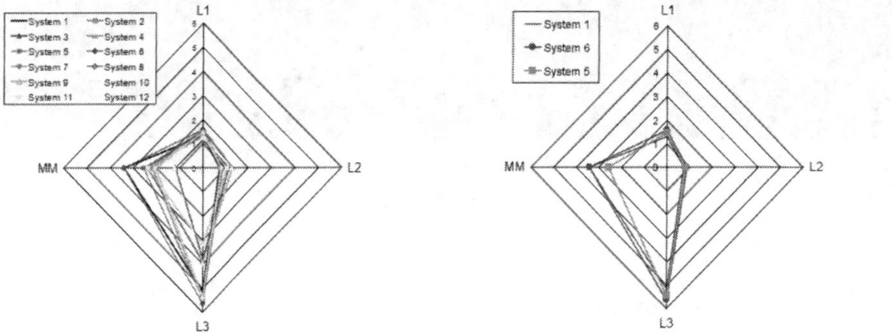

(a) Twelve Machine Profiles Used with S3D (b) Three Machine Profiles Used with FLASH

Figure 4. Memory Profiles of Systems Expected by 2012

CPUs	Sys. 1	Sys. 2	Sys. 3	Sys. 4	Sys. 5	Sys. 6	Sys. 7	Sys. 8
8	1	.72	.78	.66	.77	.81	.74	.61
64	1	.73	.78	.67	.77	.81	.73	.62
512	1	.72	.78	.66	.77	.81	.74	.62

Table 1. Prediction of S3D C_2H_4 Benchmark Performance on Systems Anticipated by 2012

5. FLASH

FLASH is a parallel, block-structured AMR code designed for compressible reactive flows [4, 5, 6]. Its capabilities span a broad range of applications, from laser-driven shock instabilities to fusion burn in type Ia supernovae. FLASH has run successfully on many leadership class systems. It is fully modular, and its components are used to create many different astrophysical applications.

5.1. Measuring FLASH

Figure 1(b) shows CPU cycles per 10 evolution steps of a FLASH white dwarf deflagration simulation, varying system size. The code scales well for on both Intrepid and Jaguar. We observe that the curves are similar, increasing slightly as the MPI task count increases. We note that using a slightly different assignment of MPI tasks to processors makes a significant performance difference on Intrepid, with the normalized performance similar to that on Jaguar, with indications that they might cross at even higher core counts.

5.2. Modeling FLASH

As with S3D, we modeled the memory behavior of the computational phases of FLASH for three anonymous future architectures for which Figure 4(b) shows memory system profiles. Our predictions in Table 2 show preliminary results for 128, 256 and 384 cores on those systems as well as Jaguar and Lonestar, the Sun Infiniband cluster at the Texas Advanced Computing Center. The reasonable accuracy on the existing systems lends confidence to the predictions on future systems. Overall, the results demonstrate that FLASH also will perform well on anticipated future systems and will benefit from improvements to main memory bandwidth.

We found that FLASH memory traces are not scale-invariant. Thus, although we expect FLASH to scale well based on our empirical measurements, we need to extend our modeling techniques to extrapolate memory traces from a set of traces gathered from smaller runs. We are currently pursuing this research direction. Initial predictions confirm that FLASH will scale reasonably well on future systems, although additional work remains to confirm this hypothesis on the very large processor counts anticipated during the Office of Science's ten year plan.

CPUs	Lonestar			Jaguar			Prediction of Systems		
	Predicted	Real	% err	Predicted	Real	% err	Sys. 1	Sys. 5	Sys. 6
128	246	227	8.2	285	258	10.3	195	188	148
256	127	131	-2.8	145	138	5.5	99	95	75
384	86	103	-16.4	99	97	2.0	67	64	51

Table 2. Performance Predictions for FLASH on Current and Future Systems

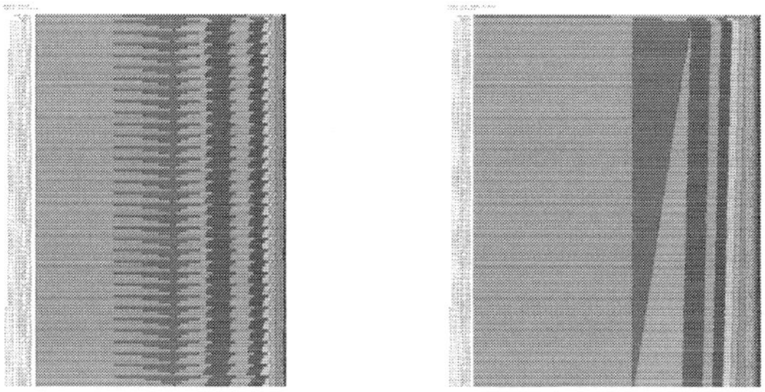

(a) Unoptimized Particle Initialization (b) Optimized Particle Initialization

Figure 5. GTC Load Balance Without and With Optimized Initialization

6. GTC

The Gyrokinetic Toroidal Code (GTC) is a particle-in-cell code to study microturbulence in magnetically confined fusion plasmas. GTC solves the gyro-averaged Vlasov equation and the gyrokinetic Poisson equation. This global code simulates the entire torus rather than just a flux tube. Written in Fortran 90/95, GTC was originally optimized for superscalar processors but is now a massively parallel code and frequently uses 1024 or more cores.

Our preliminary GTC results demonstrate how our measurement activity ensures that we use a valid version for modeling. Our initial GTC version had portions of the particle initialization commented out. Although the code executed correctly, this change led to significant load imbalance even at small scales. Figure 5(a) shows the TAU profile for GTC on 128 Jaguar cores with the unoptimized initialization. A different color in the figure represents each routine and each row, an MPI task. The figure clearly shows the load imbalance in the staggering of some routines based on the per task workload. The profile in Figure 5(b) shows that the optimized initialization corrects this imbalance. The optimizations result in the profile bars for the routines lining up much more evenly across tasks, thus improving GTC's runtime. This test demonstrates the importance of proper test code configuration for performance modeling.

7. Conclusion

We formed the PERI Architecture Tiger Team to assist the Office of Science in formulating its ten year facilities plan. Our role is to provide confidence that future leadership class systems will serve the broad range of simulations needed for the Office of Science to fulfill its mission. The first iteration of our iterative, three phase plan is nearing completion. Its results clearly demonstrate that S3D will perform well on anticipated future platforms with a preference for those that provide the highest main memory bandwidth. Our initial models for FLASH provide similar expectations although we must complete additional research that will enable scaling models for applications that do not exhibit scale-invariant memory reference behavior.

We are currently completing work on measuring and modeling GTC. Our measurement activity of both GTC and S3D demonstrated its value by ensuring that we model an appropriate version of the software. As we complete this first iteration, we are preparing for the next. We recommend to the Office of Science that we select the next set of applications with a careful eye toward those that exhibit performance differences on current platforms and that stress different aspects of memory and network performance. Thus, we anticipate focusing on at least one latency-sensitive application and one bandwidth-sensitive application.

Acknowledgments

The work of de Supinski, Gamblin and Schulz was performed under the auspices of the U.S. Department of Energy by Lawrence Livermore National Laboratory under contract DE-AC52-07NA27344 (LLNL-CONF-413427). The work of Alam, Jagode, Roth, Vetter and Worley was sponsored by the Office of Advanced Scientific Computing Research, Office of Science, U.S. Department of Energy, under contract DE-AC05-00OR22725 with UT-Batelle, LLC. The work of Bailey, Gunter and Oliker was supported by the Director, Office of Computational and Technology Research, Division of Mathematical, Information, and Computational Sciences of the U.S. Department of Energy, under contract DE-AC02-05CH11231. The work of Hovland and Norris was supported by the Office of Advanced Scientific Computing Research, Office of Science, U.S. Department of Energy, under contract DE-AC02-06CH11357. The work of Moore was supported by the U.S. Department of Energy Office of Science under contract DE-FC02-06ER25761. The work of Shende and Spear was performed under the auspices of the U.S. Department of Energy by the University of Oregon under contracts DEFG02-07ER25826 and DE-FG02-05ER25680. Accordingly, the U.S. Government retains a nonexclusive, royalty free license to publish or reproduce the published form of this contribution, or allow others to do so, for U.S. Government purposes.

This research used resources of the National Center for Computational Sciences at Oak Ridge National Laboratory, which is supported by the Office of Science of the U.S. Department of Energy under contract DE-AC05-00OR22725, and resources of the Argonne Leadership Computing Facility at Argonne National Laboratory, which is supported by the Office of Science of the U.S. Department of Energy under contract DE-AC02-06CH11357.

References

[1] Hawkes E R and Chen J H 2004 *Combustion and Flame* **138** 242–258
[2] Lee W W 1983 *Physics of Fluids* **26** 556
[3] Lee W W 1987 *Journal of Computional Physics* **72** 243
[4] Fryxell B, Olson K, Ricker P, Timmes F, Zingale M, Lamb D, MacNeice P, Rosner R, Truran J and Tufo H 2000 *Astrophysical Journal, Supplement* **131** 273–334
[5] Dubey A, Reid L and Fisher R 2008 *Physica Scripta* Special edition from Proceedings of the International Conference "Turbulent Mixing and Beyond," Trieste, Italy, August 2007.
[6] ASC Flash Center 2008 FLASH user's guide `http://flash.uchicago.edu/website/codesupport/secure/flash3_ug/`
[7] Brunst H 2008 *Integrative Concepts for Scalable Distributed Performance Analysis and Visualization of Parallel Programs* Ph.D. thesis Shaker Verlag
[8] Vampirserver user guide URL `http://www.vampir.eu`
[9] Jurenz M *VampirTrace Software and Documentation* ZIH, Technische Universität Dresden URL `http://www.tu-dresden.de/zih/vampirtrace`
[10] Vetter J and Chambreau C 2005 mpiP: Lightweight,scalable MPI profiling URL `http://www.llnl.gov/CASC/mpip`
[11] Shende S and Maloney A 2006 *International Journal of HPC Applications* **20** 287–331
[12] Gamblin T, de Supinski B R, Schulz M, Fowler R J and Reed D A 2008 Scalable load-balance measurement for SPMD codes *Supercomputing 2008 (SC'08)* (Austin, Texas) pp 46–57
[13] Snavely A, Wolter N and Carrington L 2001 Modeling application performance by convolving machine signatures with application profiles *IEEE Workshop on Workload Characterization, 2001.*
[14] Brunst H, Hoppe H C, Nagel W E and Winkler M 2001 Performance optimization for large scale computing: The scalable VAMPIR approach *Proceedings of the 2001 International Conference on Computational Science (ICCS 2001)* (San Francisco, CA) pp 751–760
[15] Brunst H, Kranzlmüller D and Nagel W 2005 *The International Series in Engineering and Computer Science, Distributed and Parallel Systems* **777** 92–102
[16] Labarta J, Girona S and Cortes T 1997 *Parallel Computing* **23** 23–34
[17] Girona S, Labarta J and Badia R M 2000 Validation of Dimemas communication model for MPI collective operations *European PVM/MPI Users' Group Meeting* pp 39–46
[18] Tikir M M, Laurenzano M, Carrington L and Snavely A 2009 PSINS: An open source event tracer and execution simulator for MPI applications *Euro-Par* (Delft, the Netherlands)
[19] Alam S R and Vetter J S 2006 A framework to develop symbolic performance models of parallel applications *IPDPS* (IEEE)

Journal of Physics: Conference Series **180** (2009) 012040 doi:10.1088/1742-6596/180/1/012040

Communication-optimal iterative methods

J. Demmel[1], M. Hoemmen[1], M. Mohiyuddin[1] and K. Yelick[1,2]

[1] Computer Science Division, University of California Berkeley, Berkeley, CA 94720-1776

[2] Lawrence Berkeley National Laboratory, One Cyclotron Road, MS 50B-4230, Berkeley, CA 94720

E-mail: {mhoemmen, demmel, marghoob, yelick}@eecs.berkeley.edu

Abstract. Data movement, both within the memory system of a single processor node and between multiple nodes in a system, limits the performance of many Krylov subspace methods that solve sparse linear systems and eigenvalue problems. Here, s iterations of algorithms such as CG, GMRES, Lanczos, and Arnoldi perform s sparse matrix-vector multiplications and $\Omega(s)$ vector reductions, resulting in a growth of $\Omega(s)$ in both single-node and network communication. By reorganizing the sparse matrix kernel to compute a set of matrix-vector products at once and reorganizing the rest of the algorithm accordingly, we can perform s iterations by sending $O(\log P)$ messages instead of $\Omega(s \cdot \log P)$ messages on a parallel machine, and reading the on-node components of the matrix A from DRAM to cache just once on a single node instead of s times. This reduces communication to the minimum possible. We discuss both algorithms and an implementation of GMRES on a single node of an 8-core Intel Clovertown. Our implementations achieve significant speedups over the conventional algorithms.

1. Introduction

The costs of arithmetic and communication continue to decrease, where "communication" means moving data, either between fast and slow memory (e.g., cache and DRAM) in the sequential case, or between processors over a network in the parallel case. However, arithmetic is getting faster much more quickly than communication, so that the primary challenge is developing algorithms to avoid communication. This is especially true in sparse matrix computations, where conventional algorithms do only a few arithmetic operations per datum, so that communication costs frequently dominate already.

Many Krylov subspace methods (KSMs) for solving $Ax = b$ and $Ax = \lambda x$ begin with a starting vector v_1, create a sequence of vectors $v_1, \ldots, v_{s+1}$ where each v_i is gotten by multiplying some linear combination of $v_0, \ldots, v_{i-1}$ by A, and then chooses the "best" (in some sense) approximate solution to $Ax = b$ resp. $Ax = \lambda x$ using the *Krylov subspace span*$\{v_1, \ldots, v_{s+1}\}$. This means that the cost in arithmetic and communication of GMRES (or any KSM) grows like $\Omega(s)$ (in fact GMRES does $\Omega(s^2)$ floating-point operations). Preconditioning only adds an additional step without changing the performance characteristics of the above two.

In [1], we show several new iterative methods that reduce the aforementioned communication cost from $\Omega(s)$ to $\Theta(1)$, which is optimal. In the parallel case, we decrease the *latency cost* (total number of messages) to $\Theta(1)$ from $\Omega(s)$, and in the sequential case, we decrease the *bandwidth cost* (total numbers of words moved) from $\Omega(s)$ (and sometimes $\Theta(s^2)$) to $\Theta(1)$. The new methods are mathematically equivalent to existing methods and converge in about the same number of iterations in practice. They include both sparse linear system solvers (with or without preconditioning) and sparse eigenvalue solvers. They are based on algorithms presented in [2, 3, 4, 5] and [6, 7]. Unlike previous authors, we have managed to reduce communication in all steps of the algorithms. Our other contributions include sparse eigenvalue solvers, new ways to improve numerical stability, entirely new algorithms that reduce the need to restart frequently, and the ability to add preconditioning to the methods while still avoiding communication. For details, see [1].

In [8], we implemented one of our iterative methods, CA-GMRES. It is mathematically equivalent to standard GMRES (the Generalized Minimal Residual Method [9]), a widely used iterative solver for sparse systems of equations $Ax = b$, but communicates a factor of $\Omega(s^2)$ less.

These new Krylov subspace methods and implementations required several algorithmic innovations. First, we developed a new computational kernel, the *matrix powers kernel.* It computes

$$W = [p_0(A) \cdot b, p_1(A) \cdot b, \ldots, p_k(A) \cdot b],$$

where $p_i(A) = \prod_{j=1}^{i}(A - \lambda_j \cdot I)$ is a degree-i polynomial in the sparse matrix A, and the λ_i are constants chosen as described in [8] and [1]. Using ideas previously introduced in [10, 11], we describe how to avoid all but $\Theta(1)$ communication while computing W. In [8], our matrix powers implementation is hierarchical: it uses the parallel algorithm on the top level (for multiple cores), and a sequential algorithm at the lower level (for off-chip data movement). Thus, the matrix is thread-blocked first for the parallel algorithm and then cache-blocked within each thread. Performance data on an 8-core 2.33 GHz Intel Clovertown shows speedups for computing W of up to 2.7× over s calls to the best optimized algorithm just for a single sparse matrix-vector multiplication $A \cdot x$) [8]. Earlier work in [11, 10] discusses parallel algorithms using distributed memory, as well as out-of-core sequential algorithms, which also achieved significant modeled and benchmarked speedups.

The columns of W span the same Krylov subspace as the vectors originally computed by GMRES, but GMRES must be reformulated to compute the best linear combination of these new basis vectors. Part of this reformulation is described in [8], where a communication-avoiding QR factorization of W is discussed; this was introduced in [12]. Performance data shows a speedup of 8 × over LAPACK's QR factorization in Intel's MKL 10.0.1.014 library, for a $10,000,000 \times 10$ matrix on an 8-core Intel Clovertown machine [8], as well as significant performance gains on multiple-node clusters [12]. TSQR also does significantly better on both shared-memory and distributed-memory platforms than the Modified Gram-Schmidt (MGS) algorithm used in many conventional iterative methods (such as Arnoldi and GMRES) [12].

Given the performance of the above two kernels, we predict significant speedups in our iterative solvers. We are nearly finished completing benchmark implementations. Future work includes developing and implementing preconditioners that work well with our iterative methods.

2. Matrix powers kernel performance results

In [8], we presented the performance results of our matrix powers kernel on an 8-core Intel Clovertown, which runs at 2.3 GHz, and has 2 MB of L2 cache per code (4 MB L2 cache shared by 2 cores). We compared it against applying SpMV s times, where we used a highly optimized SpMV implementation [13]. For speedup calculation, the time taken for the matrix powers kernel is normalized by dividing by the number of steps s and compared with the time taken for a single SpMV. Thus, the speedup is defined as: $\frac{time(\text{matrix powers kernel})/k}{time(\text{SpMV})}$. In addition to matrices whose graphs are meshes, which are expected to perform well [11], we selected sparse matrices from a variety of real applications [14]. Figure 1 shows the performance of the matrix powers kernel for different matrices.

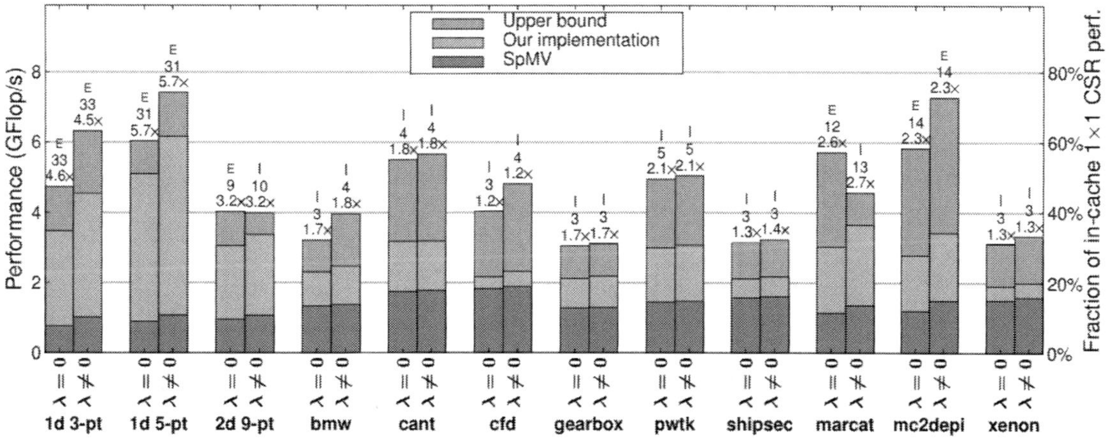

Figure 1: Matrix powers kernel performance. The yellow bars indicate the best performance over all possible s. Each bar is annotated with some of the parameters for the best performance: whether implicit or explicit cache blocking was used (indicated by 'I' or 'E' at the top), the corresponding value of s (in the middle) and the speedup with respect to the naïve implementation, which is the highly optimized SpMV code. $\lambda = 0$ indicates the case when all λ_is are 0, whereas $\lambda \neq 0$ indicates the case when all λ_is are nonzero. The 'upperbound' indicates the performance predicted by scaling the naïve performance by the change in arithmetic intensity.

3. TSQR performance results

The TSQR factorization used in these performance results is discussed in [8]. It is a hybrid of the sequential and parallel TSQR algorithms described in [12], and requires only a constant number of synchronization steps, rather than $\Omega(s)$ reductions for s columns as either the textbook Householder QR algorithm or the blocked Householder QR implementations in LAPACK and ScaLAPACK require. Figure 2 shows the performance results on a single 8-core Intel Clovertown node. We presented modeled and benchmarked performance results of TSQR in [12].

References

[1] Demmel J W and Hoemmen M in preparation Communication-optimal Krylov subspace methods Tech. rep. University of California Berkeley, Department of Electrical Engineering and Computer Science

[2] Walker H F 1988 *SIAM Journal on Scientific and Statistical Computing* **9** 152–163

[3] de Sturler E 1991 A parallel variant of GMRES(m) *Proceedings of the 13th IMACS World Congress on Computation and Applied Mathematics* ed Miller J J H and Vichnevetsky R (Dublin, Ireland: Criterion Press)

[4] Joubert W D and Carey G F 1992 *International Journal of Computer Mathematics* **44** 243–267

[5] Erhel J 1995 *Electronic Transactions on Numerical Analysis* **3** 160–176

[6] Chronopoulos A T and Gear C W 1989 *J. Comput. Appl. Math.* **25** 153–168 ISSN 0377-0427

[7] Toledo S 1995 *Quantitative performance modeling of scientific computations and creating locality in numerical algorithms* Ph.D. thesis Massachusetts Institute of Technology

[8] Mohiyuddin M, Hoemmen M, Demmel J and Yelick K 2009 Minimizing communication in sparse matrix solvers (Submitted to Supercomputing 2009)

[9] Saad Y and Schultz M H 1986 *SIAM J. Sci. Stat. Comput.* **7**

[10] Demmel J, Hoemmen M, Mohiyuddin M and Yelick K 2008 Avoiding Communication in Sparse Matrix Computations *Proceedings of the IEEE International Parallel and Distributed Processing Symposium*

[11] Demmel J, Hoemmen M, Mohiyuddin M and Yelick K 2007 Avoiding communication in computing Krylov subspaces Tech. Rep. UCB/EECS-2007-123 EECS Department, University of California, Berkeley URL \url{http://www.eecs.berkeley.edu/Pubs/TechRpts/2007/EECS-2007-123.html}

[12] Demmel J, Grigori L, Hoemmen M and Langou J 2008 Communication-optimal parallel and sequential QR

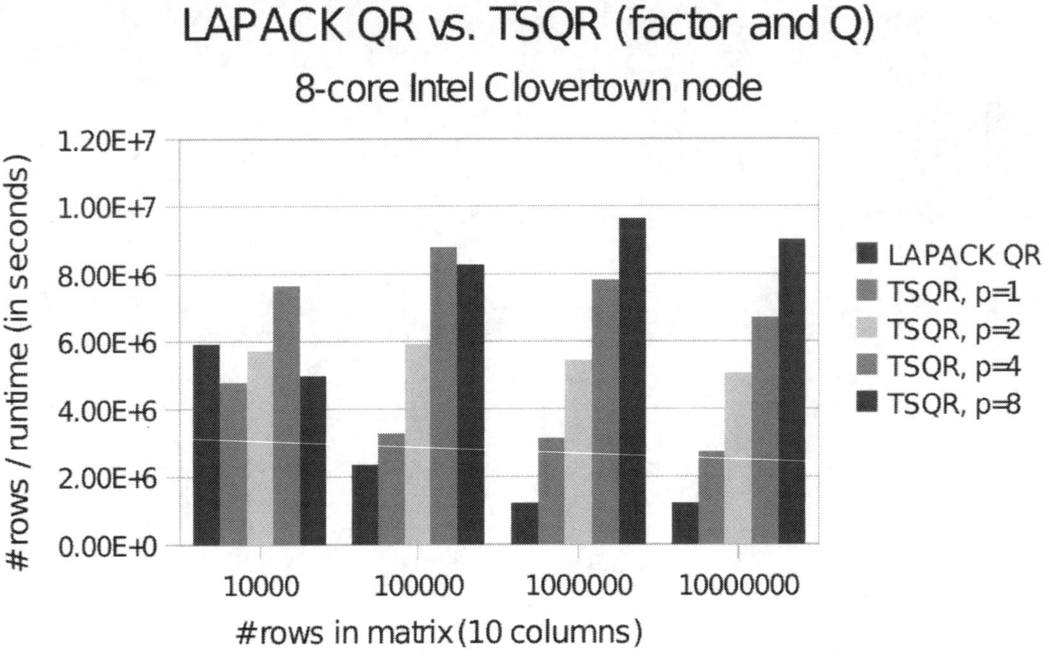

Figure 2: Performance of TSQR and LAPACK QR for $m \times 10$ matrices, where m varies from 10^4 to 10^7.

and LU factorizations Tech. Rep. UCB/EECS-2008-89 University of California EECS (Also published as LAPACK Working Note #204, and submitted to SIAM Journal on Scientific Computing)

[13] Williams S, Oliker L, Vuduc R, Shalf J, Yelick K and Demmel J 2007 Optimization of Sparse Matrix-Vector Multiplication on Emerging Multicore Platforms *Proc. of Supercomputing*

[14] Im E J, Yelick K and Vuduc R 2004 *Int. J. HPCA*

Journal of Physics: Conference Series **180** (2009) 012041 doi:10.1088/1742-6596/180/1/012041

Frontiers of performance analysis on leadership-class systems

R Fowler[1], L Adhianto[2], B de Supinski[3], M Fagan[2], T Gamblin[3], M Krentel[2], J Mellor-Crummey[2], M Schulz[3], and N Tallent[2]

[1] Renaissance Computing Institute, UNC, Chapel Hill, North Carolina
[2] Rice University, Houston, Texas
[3] Lawrence Livermore National Laboratory

E-mail: rjf@renci.org, laksono@cs.rice.edu, bronis@llnl.gov, mfagan@cs.rice.edu, tgamblin@llnl.gov, krentel@cs.rice.edu, johnmc@cs.rice.edu, schulzm@llnl.gov, tallent@cs.rice.edu

Abstract. The number of cores in high-end systems for scientific computing are employingis increasing rapidly. As a result, there is an pressing need for tools that can measure, model, and diagnose performance problems in highly-parallel runs. We describe two tools that employ complementary approaches for analysis at scale and we illustrate their use on DOE leadership-class systems.

1. Extreme Scaling and its Performance Challenges

In recent years, the emergence of multi-core computer chips has been a factor in the dramatic increase in the total number of cores, or computational elements (CEs), deployed in top-end systems. In the November, 2008 Top500 list [1], 12 systems had over 32K cores, and one had over 200K. Systems are being designed to scale to over one million. Data parallelism using an SPMD programming style has been remarkably successful at scaling applications up to thousands of cores, but computational science researchers who aspire to use very large systems efficiently must face the harsh reality of Amdahl's law: as the count of CEs goes up, every part of an application must exhibit appropriate scaling behavior on every CE or else there will be a significant loss in performance. Thus, it is vital to characterize the scaling properties of all phases of the computation. The characterization must cover all of the CEs over the entire duration of a run. Furthermore, it is necessary for tools to present specific details about performance problems that must be addressed to achieve the required efficiency at scale.

There does not yet exist a practical approach that can simultaneously achieve both the needed coverage and detail with a single set of measurements obtained with a single tool. To do so would require recording, communicating, and storing a massive quantity of data that grows with the sizes of system and application as well as with the duration of a problem run. Even if it were possible to collect the data, analysis of the data would be daunting and expensive.

We describe practical tools that address tractable and complementary aspects of the problem, together covering a large part of the problem space. These tools and methods were developed with DOE SciDAC support. We illustrate the use of these tools through examples that apply them to scientific codes that run on the DOE Office of Science leadership-class systems.

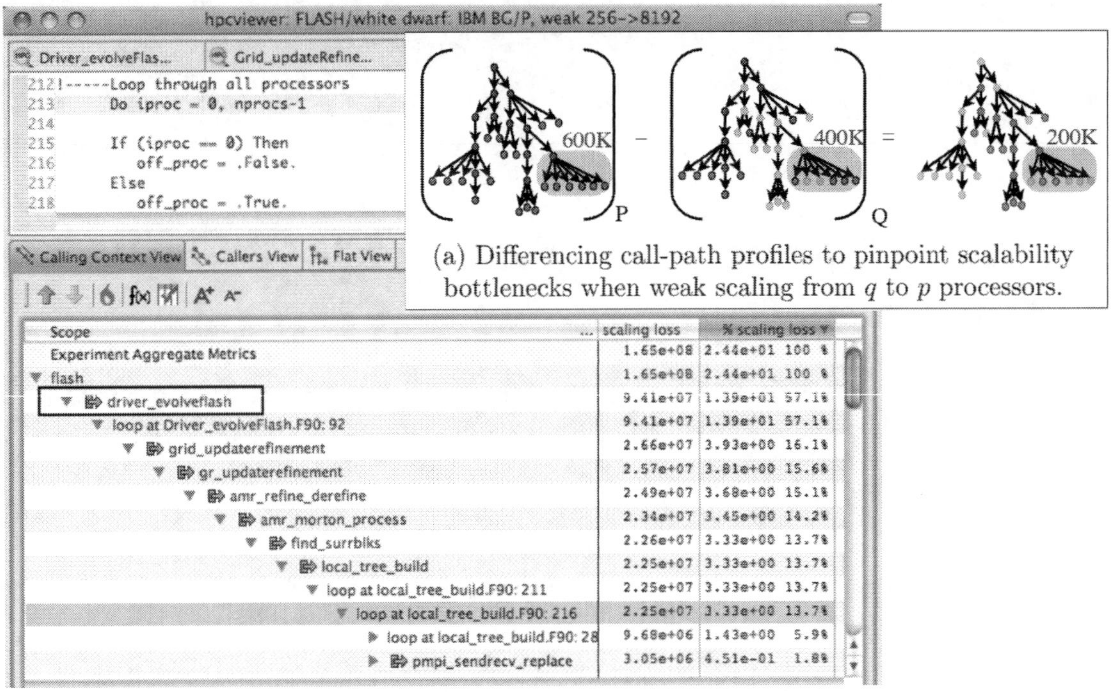

(a) Differencing call-path profiles to pinpoint scalability bottlenecks when weak scaling from q to p processors.

(b) Using HPCTOOLKIT to analyze scaling losses (microseconds) for FLASH when running at 256 and 8192 cores on an IBM BG/P.

Figure 1. Pinpointing scalability bottlenecks using differential analysis of call path profiles.

2. Measuring and Characterizing Scalability.

2.1. Differential Call-Path Profiling using HPCTOOLKIT

If an execution of a data-parallel program is sufficiently balanced and symmetric, the behavior of the program as a whole can be characterized by examining the scaling performance of any one of the CEs. If behavior is sufficiently stable over time, then an entire run can be characterized by analyzing profiles that are integrated over any appropriate time interval.

Differential profiling [2] is a strategy for analyzing multiple executions of a program by combining their execution profiles mathematically. A comparison of profiles from two executions yields information about where and how much the costs in the executions differ. While a flat profile difference may indicate that more time is spent in some procedure X in one execution relative to another, if X is called from multiple places, a flat profile difference lacks information about how the extra cost in X was incurred on behalf of different calls to X.

To understand an execution's costs in context, Rice University's HPCTOOLKIT uses call-path profiling [3]. HPCTOOLKIT's call-path profiler uses call-stack unwinding in conjunction with sampling to attribute execution costs to full calling contexts [4]. This capability is particularly important for procedures such as `MPI_Wait`, whose costs are very dependent on calling context.

To pinpoint and quantify scalability bottlenecks for parallel programs and to attribute them to calling contexts, we compare call path profiles from a pair of executions using different levels of parallelism [5]. Consider two parallel executions of an application, one executed on q processors and the second executed on p processors, where $p > q$. In a weak scaling scenario, each processor in both executions computes on an identical amount of local data. If the application exhibits perfect weak scaling, then the total cost (*e.g.*, execution time) should be the same in both executions. If every part of the application scales uniformly, then this equality should hold in *each scope* of the application. Figure 1(a) pictorially shows the analysis of weak scaling by comparing two representative calling context trees (CCTs)[1]—one tree from a process in each execution. For instance, the difference in cost incurred in the subtree highlighted in magenta in the CCTs represents parallel overhead when scaling from q to p processors. This difference in cost for the subtree can be converted into percent scalability loss by dividing by the inclusive cost of the root node in the CCT for p processors and multiplying by 100. Computing inclusive scalability losses for each node in a CCT enables one to locate scalability bottlenecks easily in a top-down fashion by tracing patterns of losses down paths from the root.

2.2. Scalable Analysis of Dynamic Load-Balance using Libra

Characterizing and tuning perfectly balanced SPMD codes, however, is only part of the problem. Many large-scale codes use irregular meshes, so partitioning of problems is not perfect. Methods such as adaptive mesh refinement (AMR) have behavior that evolves over time and the evolution induces load imbalances over time. Such behavior can be highly dependent on input data and the form of imbalances may change from problem to problem. The run-time layout of the application on the hardware can also introduce imbalances due to imperfect mappings of application topology to the system, causing differences in communication and I/O costs among the population of CEs. Finally, balance and efficiency can be impacted by run-time conditions such as contention for communication and I/O between jobs that happened to be scheduled together, or by the overhead of using error-correcting mechanisms to tolerate memory or interconnect (*e.g.*, Infiniband) problems. The challenge of dealing with these problems requires methods that can track economically the time evolution of scalable computations while still retaining enough detail to identify and diagnose anomalous behavior.

In prior work, we developed the Libra [6, 7] load-balance measurement tool. In the presence of adaptive and dynamic behavior, tools must be capable of characterizing the evolution of imbalances and of localizing the sources of imbalance to semantically meaningful parts of the application. We therefore use a model in which performance metrics are presented as *rates* normalized to a hierarchical decomposition of the application's behavior. *Progress* steps correspond to program phases that represent application-defined progress towards completing the computation. A time step in a simulation is a typical progress step. A hierarchy of *effort* steps can be nested within each of the progress steps. Effort steps represent the phases of work done to make a unit of progress. Effort steps in large parallel programs correspond to communication actions (especially collective operations), each iteration of an implicit solver, local physics computations, I/O operations, etc.

Libra uses the PMPI interface to attach measurement operations to synchronous MPI calls. These are used to denote the end points of effort regions automatically. Currently, the user must mark each progress region by manualling adding a call to a Libra library routine. We are investigating techniques to automate this process.

For each effort region, Libra generates a two-dimensional profile representing effort in that region across processes and over time. The data from these regions can be copious, so we use aggressive, parallel wavelet compression to aggregate and communicate these data to a central location. Our compression algorithm is lossy, but it preserved significant features even at very low bit rates. Figure 2 shows Libra's compression pipeline in detail.

This strategy allows us to achieve between two and three orders of magnitude of compression, typically, while still preserving outliers, load distribution, and evolutionary behavior. Our compression algorithm also exhibits good scaling behavior. Each region is compressed independently, thus permitting the work of compression to be balanced evenly across all processes

[1] A CCT is a weighted tree in which each calling context is represented by a path from the root to a node and a node's weight represents the cost attributed to its associated context.

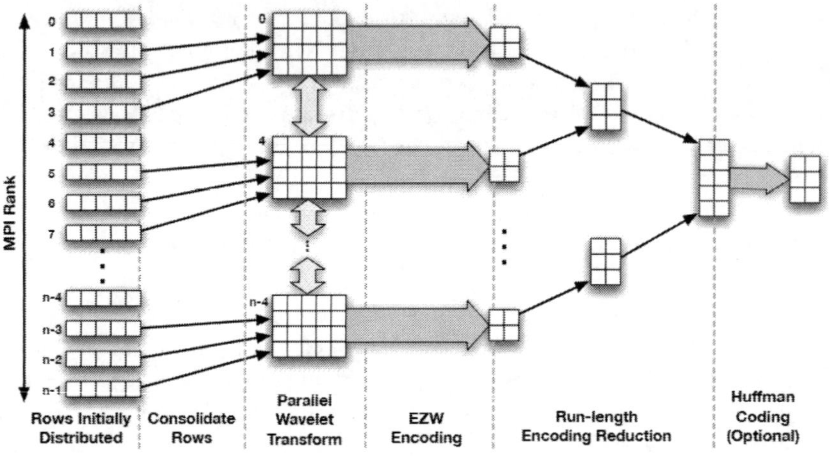

Figure 2. Libra Compression Pipeline

in the application, The time required for compression does not increase significantly as system size grows.

3. Case Studies

We present brief examples illustrating the application of these tools to real problems.

3.1. Pinpointing Scalability Bottlenecks using HPCToolkit

To show the power of HPCToolkit's differential call-path profiling analysis, we used it to pinpoint scaling bottlenecks in FLASH [8], a code that uses adaptive mesh refinement to model astrophysical thermonuclear flashes. We performed a weak scaling study of a white dwarf explosion by executing 256-core and 8192-core simulations on an IBM BlueGene/P located at the Argonne Leadership Computing Facility. For the 8192-core execution, both the input and the number of cores are 32x larger than the 256-core execution. With perfect scaling, we would expect identical run times and call-path profiles for both configurations.

Figure 1(b) shows a portion of the residual calling-context tree, annotated with two metrics: "scaling loss" and "% scaling loss". The former quantifies the scaling loss (in microseconds) while the latter expresses that loss as a percent of total execution time (shown in scientific notation). The top-most entry in each metric column gives the aggregate metric value for the whole execution. A percentage to the right of a metric value indicates the magnitude of that particular value relative to the aggregate. Thus, for this execution, there was a scaling loss of 1.65×10^8 μs, accounting for 24.4% of the execution.[2] By sorting calling contexts according to metric values, we immediately see that the evolution phase (`Driver_evolveFlash`) of the execution (highlighted with a square) has a scaling loss that accounts for 13.9% of the total execution time on 8192 cores, which represents 57.1% of the scaling losses in the execution.

To pinpoint the scalability bottleneck in FLASH's simulation, we use the "hot path" button to expand automatically the unambiguous portion of the hot call path according to this metric. Figure 1(b) shows this result. HPCToolkit identifies a loop (beginning at line 213), within its full dynamic calling context – a unique capability – that is responsible for 13.7% of the scaling losses. This loop uses a ring-based all-to-all communication pattern known as a *digital orrery* to update a processor's knowledge about neighboring blocks of the adaptive mesh. Although FLASH only

[2] A scaling loss of 24.4% means that FLASH is executing at 75.6% parallel efficiency on 8192 cores relative to its performance on 256 cores.

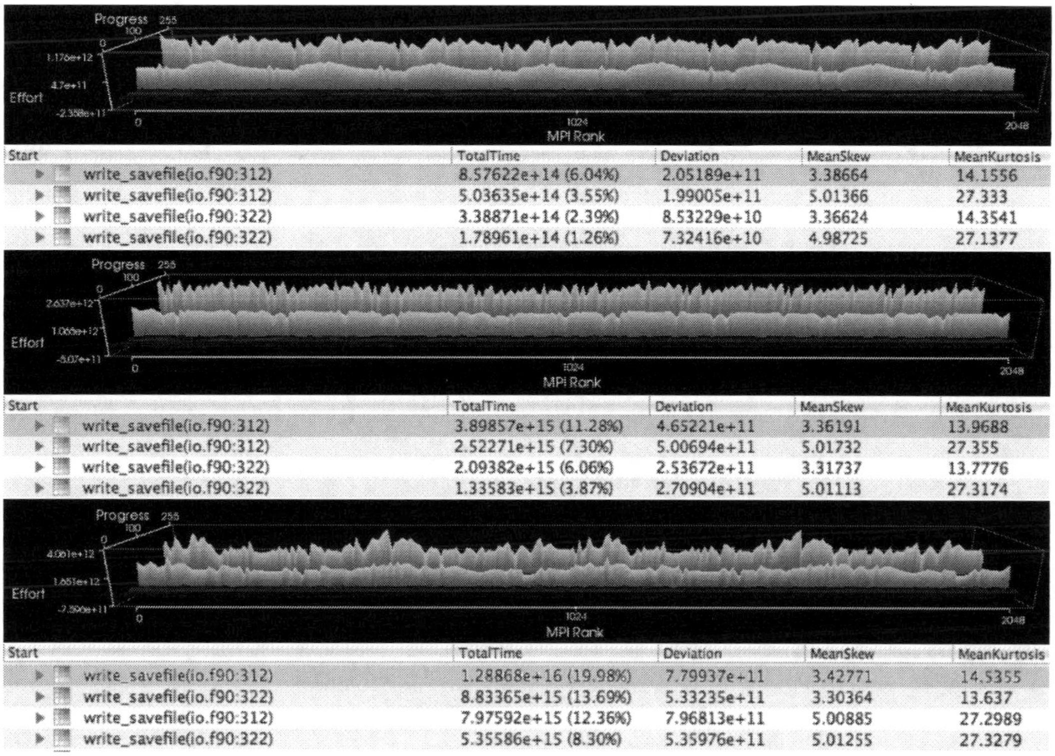

Figure 3. Load balance plots for checkpointing operations in S3D runs at 4096 (top), 8192 (middle), and 16384 (bottom) cores on an IBM BG/P.

uses the orrery pattern to set up subsequent communication to fill guard cells, looping over all processes is inherently unscalable. Other scalability bottlenecks can be identified readily by repeating the "hot path" analysis for other subtrees in the computation where scaling losses are present. The power of HPCToolkit's scalability analysis is apparent from the fact that it immediately pinpoints and quantifies the scaling loss of a key loop deep inside the application's layers of abstractions.

3.2. Examining Load Balance Using Libra

We used Libra to measure the load-balance behavior of S3D, a Direct Numerical Simulation (DNS) of turbulent combustion. Fine-grain calculations to simulate micro-turbulence with S3D can be used to generate coarser models for engineering-level simulation codes.

The computational regions of S3D exhibit almost perfect weak scaling [9]. We applied Libra to 200-time step runs of S3D on a Blue Gene/P system (Intrepid) at Argonne National Laboratory's Leadership Computing facility. For each time step, we examined the time spent in `MPI_Wait` and `MPI_Barrier`, uniquely identifying each call-site by its full call-path labeled with file names and line numbers for each frame. We confirmed that that `MPI_Wait` and `MPI_Barrier` communication routines dominated the runtime of S3D at large scales, and that most of this time is attributed to two call-sites in S3D's `write_savefile` checkpoint routine. Since `write_savefile` is called from two places in the code, four call-paths dominated execution time.

Figure 3 shows load balance profiles reported by Libra for MPI operations used in checkpointing for 4096, 8192, and 16384-process runs (top to bottom) of S3D on Intrepid. The tables below the profiles present summary statistics for each call-site. In the plots the vertical axis represents time, the depth axis represents time steps (progress), and the horizontal axis represents processes by MPI rank. In each image there are two "ridgelines" parallel to the page corresponding to two of the periodic checkpoint operations. As we increased the size of the runs, the plots and summaries reveal a large increase in total time spent in checkpointing. I/O imbalance (inter-process variation) also increases with the number of nodes. This version of S3D writes a separate checkpoint file per process. Some processes quickly write their data to disk, while others encounter contention delays in writing theirs, resulting in the sawtooth load patterns seen in the figures. We are currently investigating and measuring runs of S3D with alternative I/O strategies designed to reduce and balance contention during the checkpoint phase.

4. Conclusions and Future Directions

The two approaches that we describe are complementary. We are currently investigating the question of coordinating their usage. For example, a unified approach would have Libra dynamically enable HPCToolkit profiling to track problems in detail.

An emerging challenge is that explicit shared-memory parallelism on multi-core chips is contributing a substantial multiplicative factor to the overall performance of large systems. This adds a new dimension to the complexity of performance analysis.

Acknowledgments

This work was supported by the DOE SciDAC program under cooperative agreement numbers DE-FC02-06ER25764, DE-FC02-06ER41445, DE-FC02-07ER25800, and DE-FC02-06ER25762.

This research used resources of the Argonne Leadership Computing Facility at Argonne National Laboratory, which is supported by the DOE Office of Science under contract DE-AC02-06CH11357.

References

[1] Meuer H, Strohmaier E, Dongarra J and Horst S Top500 supercomputer sites (http://www.top500.org)
[2] McKenney P E 1998 *Software: Practice and Experience* **29** 219–234
[3] Hall R J 1992 Call path profiling *Proc. of the* 14th *Intl. Conf. on Software Engineering* (NY, NY, USA: ACM) pp 296–306
[4] Tallent N, Mellor-Crummey J and Fagan M 2009 Binary analysis for measurement and attribution of program performance *Proc. of the ACM SIGPLAN Symp. on Program Language Design and Implementation* (NY, NY, USA: ACM)
[5] Coarfa C, Mellor-Crummey J, Froyd N and Dotsenko Y 2007 Scalability analysis of SPMD codes using expectations *Proc. of the 21st Annual Intl. Conf. on Supercomputing* (NY, NY, USA: ACM) pp 13–22
[6] Gamblin T, de Supinski B R, Schulz M, Fowler R J and Reed D A 2007 Scalable load-balance measurement for SPMD codes *In Submission*
[7] Gamblin T 2009 *Scalable Performance Measurement and Analysis* Ph.D. thesis University of North Carolina at Chapel Hill
[8] Dubey A, Reid L B and Fisher R 2008 *Physica Scripta* **132** 014046
[9] de Supinski B R, Alam S, Bailey D H, Carrington L, Daley C, Dubey A, Gamblin T, Gunter D, Hovland P, Jagode H, Karavanic K, Marin G, Mellor-Crummey J, Moore S, Norris B, Oliker L, Olschanowsky C, Roth P C, Schulz M, Shende S, Snavely A, Spear W, Tikir M, Vetter J, Worley P and Wright N 2009 Modeling the Office of Science ten year facilities plan: The PERI Architecture Tiger Team *These proceedings*

LAPACKrc: Fast linear algebra kernels/solvers for FPGA accelerators

Juan Gonzalez and Rafael C. Núñez

Accelogic, 1830 Main Street, Suite 204, Weston, FL, U.S.A.

juan.gonzalez@accelogic.com

Abstract. We present LAPACKrc, a family of FPGA-based linear algebra solvers able to achieve more than 100x speedup per commodity processor on certain problems. LAPACKrc subsumes some of the LAPACK and ScaLAPACK functionalities, and it also incorporates sparse direct and iterative matrix solvers. Current LAPACKrc prototypes demonstrate between 40x-150x speedup compared against top-of-the-line hardware/software systems. A technology roadmap is in place to validate current performance of LAPACKrc in HPC applications, and to increase the computational throughput by factors of hundreds within the next few years.

1. Introduction

As one of the core supporting technologies of scientific discovery, high-performance computing (HPC) has the challenge of being ahead of computational needs in the scientific community. This is particularly true in the case of DOE applications, and a reason for the Department of Energy targeting peta- and exa-scale performance for the next few years. Superior levels of computational muscle are needed to conceive and validate the technological innovations that will advance scientific areas such as nuclear energy, physics, bioengineering, and weather modeling.

The challenge for HPC is more difficult today than in previous years. Reduced footprints and operating costs are required in addition to increased computational power, efficiency, and accuracy. The conventional approach of obtaining increased throughput by adding CPU cores is not enough if it is not accompanied by further efficiency in the use of arithmetic cores.

Field Programmable Gate Arrays (FPGAs) are emerging as an alternative to provide the energy efficiency and computational power of a new generation of HPC systems. FPGAs are semiconductor devices that can be programmed and interfaced to a conventional computer or supercomputer. As programmable devices, FPGAs can be configured as specialized computing units able to perform specific tasks several orders of magnitude faster than a traditional von Neumann CPU. FPGAs have proven successful in numerous applications such as those described in [1], [2], and [3].

Although potentially able to provide vast amounts of GFLOP/s, FPGAs are yet to be fully exploited in linear-algebra-intensive applications, which make about 70% of the HPC market worldwide. Significant risks have typically been associated with the incorporation of FPGA technology in linear-algebra-intensive supercomputing systems, which have limited the tremendous potential that FPGAs promise for those applications. First, leading-edge technology permits only a few scientists to understand the algorithm designs necessary for very high speedups, and this algorithm innovation remains a difficult design component. Second, the costs associated with the development of an FPGA application are still high compared with the costs of building traditional FORTRAN or C software applications. Third, the typical software-development project for FPGA systems still lacks robustness.

The LAPACKrc program emerges as an alternative to address these risks. LAPACKrc is a family of FPGA-based linear algebra solvers able to achieve speedups larger than 100x per commodity processor ("rc" stands for "reconfigurable computing"). LAPACKrc subsumes some of the LAPACK and ScaLAPACK functionalities for linear algebra, and incorporates sophisticated general direct and iterative sparse matrix solvers that are at least two orders of magnitude faster than traditional solvers.

In this paper we introduce the LAPACKrc library, describe its design principles, as well as current status and on-going work targeted at providing the levels of performance required by modern scientific applications.

2. Designing accelerated numerical kernels based on FPGA systems

The use of CPU/FPGA heterogeneous systems, in which different types of specialized processor architectures co-exist and collaborate to solve numerical computing problems more efficiently, are becoming ubiquitous in the HPC eco-system. As an example, Cray Inc. recently launched its hybrid CPU/FPGA supercomputer, called the Cray XT5h [4], which is fed with both CPU and FPGA blades. Cray's FPGA blades can be plugged directly into the supercomputer interconnect network to obtain all the benefits of high-speed communication and memory access that typical CPU blades have.

Three components must be considered when designing numerical methods for FPGA systems: (1) Exploiting on-chip (many-core) parallelism, (2) designing for high-speed data transfer, and (3) mixed-precision arithmetic to reduce both data transfer and time-to-solution. Exploiting many-core parallelism is perhaps the most natural technique for numerical computing experts, as it is similar to exploiting parallelism using multiple CPU cores. The challenge when designing FPGA algorithms is being able to combine this parallelism with extremely fast communication links among cores. It is also important to understand that not all the cores are designed equally – in FPGA computing, the operations performed by each core are decided and configured by the programmer according to the needs of the algorithm at hand. As an example, it is possible to create an arrangement of very efficient matrix-vector multipliers, and concurrently in the same chip, a number of general-purpose arithmetic cores. Moreover, appropriately combining FPGAs with the very efficient 64-bit arithmetic units of commercial CPUs, can bring increased computing efficiency to scientific codes.

3. The LAPACKrc family of solvers

During the last four years we have been working on the creation of extreme-speed numerical linear algebra solvers. This linear algebra technology is being called LAPACKrc, and it is defined as a family of FPGA-based linear algebra solvers able to achieve speedups larger than 100x per commodity processor [5].

Three types of solver prototypes have been produced so far: Dense linear equation solvers, sparse iterative linear equation solvers, and dense least squares solvers. The fundamentals of each of these designs are described next, followed by a short presentation of current results.

3.1. Dense linear equation solvers

The most general case of a dense solver is based on the so-called LU factorization. The LU factorization is a procedure for decomposing an $n \times n$ matrix A into a product of a lower triangular matrix L and an upper triangular matrix U, $A=LU$. When solving the matrix equation $Ax=b$, the LU factorization enables the formulation of the equation as: $Ax=(LU)x=L(Ux)=b$. This system can be solved by first solving the auxiliary system $Lz=b$ for z, and then solving $Ux=z$ for x [6].

The LU algorithm can be efficiently implemented using FPGA systems. The key concept behind our implementation is the use of a "mega-instruction" specifically built into the system architecture to exploit the super-modular structure of the LU equations. This mega-instruction is made of many (hundreds or even thousands) single arithmetic operators appropriately connected to render the exact mathematical expression of the LU factorization. Massive parallelism is enabled by the concurrent execution of all of these arithmetic operators. The complete mega-instruction, consisting of a massive interconnection of atoms is shown in Figure 1. A similar mega-instruction for Cholesky factorization is shown in Figure 2. Both LU and Cholesky mega-instructions can be combined with efficient implementations of forward and backward substitutions, so to render a complete linear solver. All of the above implementations can be combined with mixed-precision arithmetic techniques to achieve increased processing speed without loss of accuracy [7][8].

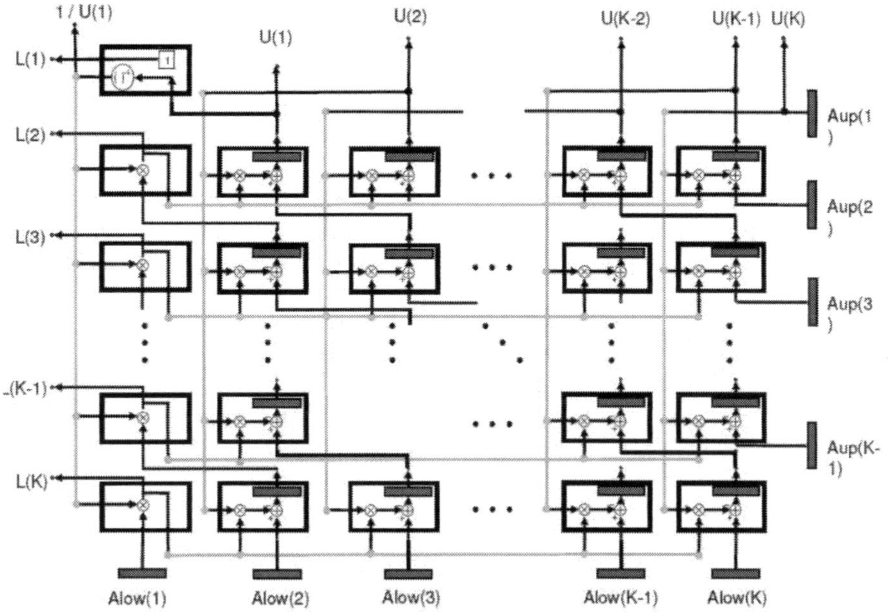

Figure 1. LAPACKrc's LU architecture.

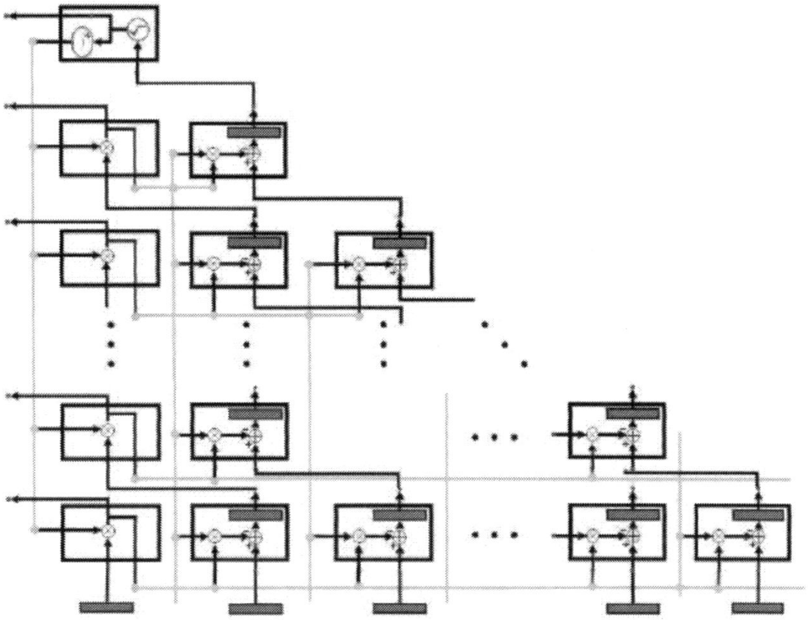

Figure 2. LAPACKrc's Cholesky architecture.

3.2. Sparse iterative linear equation solvers

The LAPACKrc library is also populated with iterative algorithms of the Krylov type [9]. The Conjuage Gradient method (CG) is the most suitable Krylov method for solving linear systems $Ax=b$, in which the coefficient matrix A is both symmetric and positive definite. Our formulation of the CG method for parallelization using FPGAs is shown in Figure 3. In the figure, A is the matrix of coefficients, $r_i=b-Ax_i$ is the vector of residuals at iteration i, and x_i is the solution of the linear system at the iteration i. The operations in the CG algorithm are repeated until convergence is achieved.

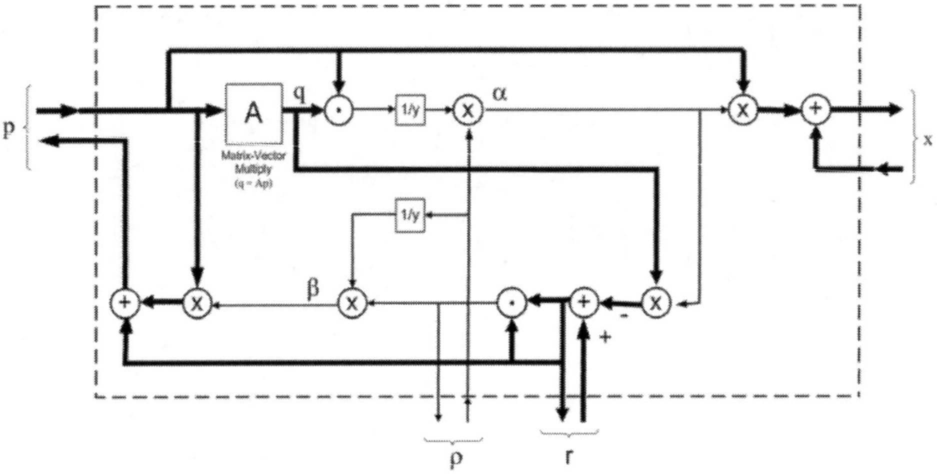

Figure 3. LAPACKrc's CG architecture.

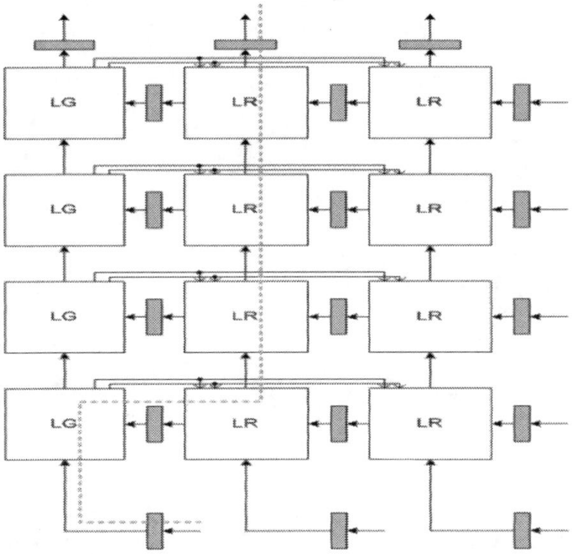

Figure 4. LAPACKrc's QR architecture.

3.3. Dense least squares solvers

The general least-squares problem can be stated as a minimization of the type: $\min_{x\in\mathbb{R}^n} \|Ax-b\|_2$, where A is a matrix of size $m\times n$ and b is a vector of m elements. In general, m and n are large and $m >> n$. An efficient method for solving this problem is based on orthogonal transformations. In particular, QR decomposition has shown to be an important and efficient method to solve least-squares problems [9].

Based on a novel scheduling algorithm that mixes Givens rotations with traditional rotations in the logarithmic domain, we have designed a super-modular architecture for QR decomposition. The basic architecture is shown in Figure 4. This architecture is based on what we call LG and LR atoms. LG and LR are Log-Givens and Log-Rotate computing units that perform the core operations of the QR decomposition with a high degree of parallelism and arithmetic optimization. This QR architecture is connected to specialized backward and forward substitutions to build a complete LS solver.

4. Performance

Our current benchmarks indicate that our LU algorithm implementation renders speedups of up to 140x when using a single commercial off-the-shelf FPGA computing platform. Figure 5 shows the speedup of Accelogic's banded LU solver compared against the Intel MKL-optimized LAPACK solver (DGETRF) running in the Xeon Woodcrest processor. The x-axis represents the bandwidth of the matrix, whereas the y-axis represents the matrix size, *N*. As it can be appreciated from the figure, our solver surpasses the performance of the state-of-the-art solver for small/medium-size problems. Figure 6 illustrates the performance of our Cholesky solver, with peak speedup factor above 40x.

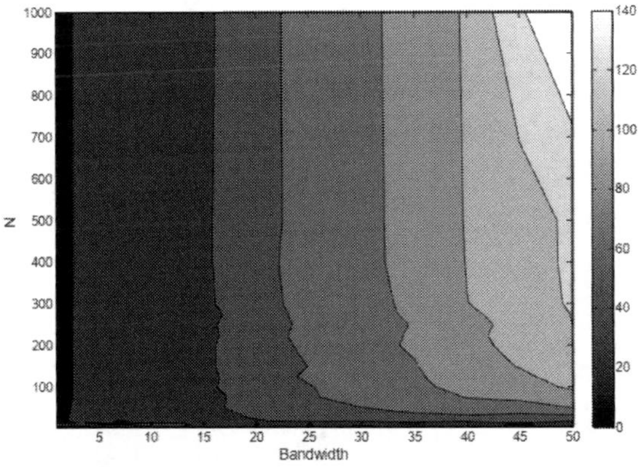

Figure 5. Performance of dense LU solver (speedup factor).

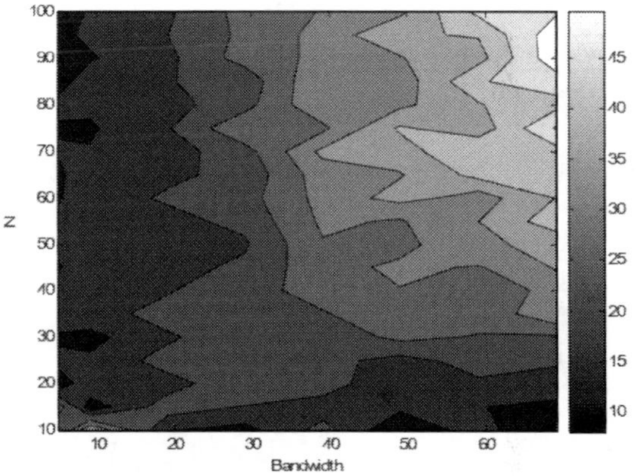

Figure 6. Performance of dense Cholesky solver (speedup factor).

The results of our iterative solver are also similar: More than 80x speedup factors are achieved for typical medium-size problems. Figure 7 shows a typical convergence curve of our iterative algorithms in time when compared against the time convergence of PETSc's CG algorithm [10], when running in the Xeon Woodcrest processor. The plotted curves show the L_2-norm time evolution of the residual of the large-scale linear equation problem "Wissgott/parabolic_fem [11]," for which the matrix is highly sparse, with only 3,674,625 non-zeros.

Figure 8 shows a performance comparison of different configurations of LAPACKrc's Least-Squares solver against an Intel-MKL optimized version of DGELS (LAPACK's solver of Least-Squares problems based on QR factorization) running in the Xeon Woodcrest processor. Although the speedup level depends on the problem size, LAPACKrc's technology consistently delivers more than 40x speedup levels. As with our linear equation solvers, the least squares solver renders 64-bit results even when lower precision arithmetic is used in particular internal operations.

5. Conclusions and future work

FPGA-based software accelerators are emerging as a strong technology for providing efficient parallelism able to feed scientific applications with the right combination of processing speed and energy efficiency. Among the advantages of FPGA computing are massive (many-core) parallelism with extremely-high data transfer speeds. Moreover, increased performance can be achieved by combining FPGAs with 64-bit CPUs using mixed-precision arithmetic methods.

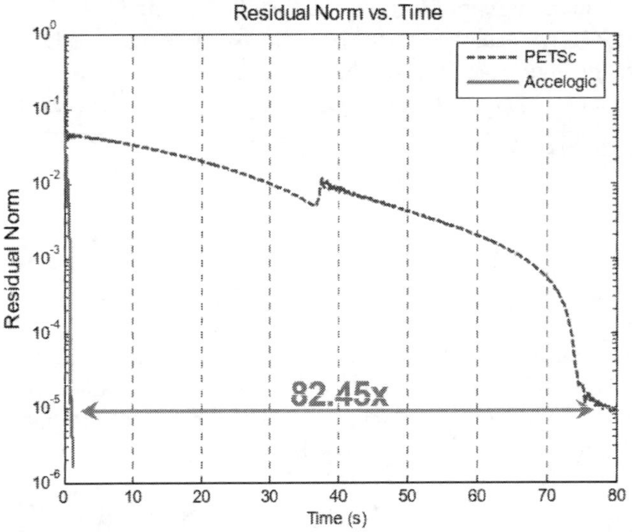

Figure 7. Performance of conjugate gradient algorithm when solving a large-scale problem.

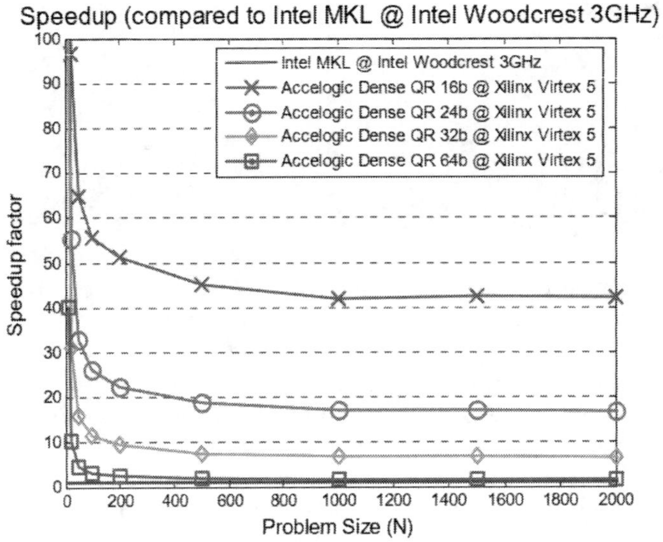

Figure 8. Performance of LAPACKrc, dense least squares solver based on QR factorization.

LAPACKrc is the crystallization of FPGA supercomputing for linear algebra applications. Designed to subsume some of the LAPACK and ScaLAPACK functionalities, as well as to incorporate accelerated sparse direct and iterative linear algebra methods, we have currently developed prototypes of both dense (LU and Cholesky) and iterative (CG) linear equation solvers, as well as dense (QR) least squares solvers. Our current results demonstrate speedups of 40x to 150x for these solvers compared with state-of-the-art software running on top-of-the-line CPUs. Future work includes, in addition to the productization of the current prototypes, the creation of additional methods for linear algebra.

Acknowledgments
This material is based upon work supported by the Department of Energy, U.S. Air Force, and NASA, under award numbers DE-FG02-07ER84728, DE-FG02-08ER85136, FA9550-07-C-0117, FA9550-09-C-0052, USAF-3446-14-SC-0001, and NNX07CA19P. We also wish to thank Jim Demmel from the University of California Berkeley, Tim Davis from the University of Florida, and Jack Dongarra from the University of Tennessee, for their consultation on this project.

Disclaimer
Several of the methods in this article are patent-pending by Accelogic, LLC. For licensing, contact info@accelogic.com. This report was prepared as an account of work sponsored by several agencies of the U.S. Government. Neither the U.S. Government nor any agency thereof, nor any of their employees, makes any warranty, express or implied, or assumes any legal liability or responsibility for the accuracy, completeness, or usefulness of any information, apparatus, product, or process disclosed, or represents that its use would not infringe privately owned rights. Reference herein to any specific commercial product, process, or service by trade name, trademark, manufacturer, or otherwise does not necessarily constitute or imply its endorsement, recommendation, or favoring by the U.S. Government or any agency thereof. The views and opinions of authors expressed herein do not necessarily state or reflect those of the U.S. Government or any agency thereof.

References

[1] Gonzalez J, Upegui A, Nunez R C, Orozco D 2006 High Performance Non-vonNeumann Algorithms for Large-Scale Optimization *31st AIAA Dayton-Cincinnati Aerospace Sciences Symposium (Dayton, OH, March 2006)*
[2] Biocceleration 2006 BioXL/S http://www.biocceleration.com/BioXLS-General.html
[3] Woods N, VanCourt T 2008 FPGA acceleration of quasi-Monte Carlo in finance *International Conf. on Field Programmable Logic and Applications, FPL 2008 (Sept.2008)* pp.335-340
[4] Cray XT5h Supercomputer. http://www.cray.com/downloads/CrayXT5hBrochure.pdf. 2008
[5] Gonzalez J, Nunez R C 2008 LAPACKrc: A portable linear algebra library for extreme-speed acceleration of HPC codes using FPGAs *DoD High Performance Computing Modernization Program, 2008 Users Group Conference (Seattle, WA, July 14-18, 2008)*
[6] Demmel J 1997 *Applied Numerical Linear Algebra* SIAM
[7] Buttari A, Dongarra J, Langou J, Luszczek P, Kurzak J 2007 Mixed Precision Iterative Refinement Techniques for the Solution of Dense Linear Systems *International Journal of High Performance Computing Applications* vol 21 issue 4 (Nov. 2007) pp 457-466
[8] Veley D, Camberos J 2007 Computational Uncertainty in Reconfigurable Computing *NATO RTO Applied Vehicle Technology Panel (AVT) Symposium (Athens, Greece, Dec 2007)*
[9] Golub G, VanLoan C 1996 *Matrix Computations* The Johns Hopkins University Press, 3 ed
[10] Balay S, Gropp W, McInnes L, Smith B 2003 *PETSc Users Manual* ANL-95/11–Rev.2.1.5.
[11] Davis T 2008 The University of Florida Sparse Matrix Collection http://www.cise.ufl.edu/research/sparse/matrices, submitted to ACM TOMS

Journal of Physics: Conference Series **180** (2009) 012043 doi:10.1088/1742-6596/180/1/012043

GPU accelerated computing—from hype to mainstream, the rebirth of vector computing

Satoshi Matsuoka[1], Takayuki Aoki[1], Toshio Endo[1], Akira Nukada[1], Toshihiro Kato[2] and Atushi Hasegawa[3]

[1]Tokyo Institute of Technology
[2]NEC Corporation
[3]NEC Informatec Systems, Ltd.

matsu@is.titech.ac.jp

Abstract. Acceleration technologies, in particular GPUs and Cell, are receiving considerable attention in modern-day HPC. Compared to classic accelerators and traditional CPUs, these devices not only exhibit higher compute density, but also sport significant memory bandwidth and vector-like capabilities to stream data at bandwidth of 100 GB/s or more. The latter qualifies such accelerators as a rebirth of vector computing. With large-scale deployments of GPUs such as Tokyo Tech's TSUBAME 1.2 supercomputer facilitating 680 GPUs in a 100-Teraflops scale supercomputer, we can demonstrate that, even under a massively parallel setting, GPUs can scale both in dense linear algebra codes as well as vector-oriented CFD codes. In both cases, however, careful algorithmic developments, especially latency hiding, are important to maximize their performance.

1. Introduction

1.1. Commoditization of HPC and its acceleration, but niche still remains

Acceleration technologies such as Cell[1], GPU[2], ClearSpeed[3], MD Grape[4] are receiving considerable attention in modern-day HPC. By all means, acceleration via specialized hardware or processors is not new — various application areas such as graphics, multimedia, networks, cryptography, etc. typically are accelerated in PCs and embedded devices. Also in the past, HPC architectures typically had vector acceleration options in CM-5, Fujitsu AP1000, etc., but all failed or faded away. However, recently renewed attention to acceleration in HPC is driven by commodity hardware intended for other uses now being used in typical HPC application scenarios and achieving great leaps in performance with regard to cost, power, size, etc. Let us first review how acceleration and commoditization interplayed in HPC. Later on in the article we will see concretization of modern acceleration in our latest TSUBAME 1.2 supercomputer at the Tokyo Institute of Technology, which is the largest open science GPU accelerated machine to date, and we will discuss its various performance characteristics.

Commoditization in HPC fundamentally changed its position from being a niche market to occupying 20% of all the server markets. The first sign of this was actually a long time ago historically, around 1980, with the availability of the 8087 acceleration option in the 8086 processor, where IEEE 754 double floating point came into actual being, leading to a series of x86 processors with increasing FP and memory performance as multimedia and other needs became more prevalent in

PCs. The first Beowulf cluster, called Wigraf, was born in 1994 at Caltech/NASA, signifying the start of commoditization of networks and the HPC software stacks to build a complete supercomputer—from then on, cluster technology has matured to the extent that we can build petascale machines embodying tens of thousands of CPU cores. Such wide proliferation of commodity technology in HPC over time has allowed its widespread proliferation and mainstream adoption, not only allowing one to save costs in building large-scale machines, but also realizing integration and smooth growth paths in the whole HPC ecosystem from simple PCs and small, few node clusters all the way up to petascale machines. It is now commonplace for someone to develop and run the same application in both his office machine and the largest supercomputer using thousands of CPU cores—with just the size of the problem being different.

However, even today various vendors including IBM, Cray, NEC, Fujitsu, etc. sell custom-made supercomputers, and more are in future development. Given the cost and other technical advantages of commodity clusters, such "dinosaurs" should have disappeared a long time ago — however, if one would observe various benchmarks such as the TOP500 or the HPC challenge, such machines dominate the top ranks. There are several reasons for this, including the ones below:

(1) High-performance, highly efficient vector processing: although vector processing capabilities in x86 processors (SSE) and others (e.g., AltiVec on PowerPC) have dramatically improved their FP performances, for single-thread applications they are still no match for dedicated vector machines such as the NEC SX. There are several reasons for this, including much wider vectors (8~16 vs. 2~4) and deeper vector pipelines, augmented by significantly greater memory bandwidth as well as random access capabilities. Although in the past, increases in both the processor and the memory clock rate helped to improved single-thread performance of commodity CPUs, such an increase has virtually come to a halt since around 2005, when multicore parallelism became the dominant methodology for increasing CPU performance. As such, expensive vector machines are still preferred for applications such as CFD where the algorithm is memory bound, or time-dependent code where there is inherent serialization in the algorithm itself.

(2) Power/installation/maintenance efficiency: even if one would overcome the inefficiency of PC clusters by utilizing more parallelism, various costs such as physical power, space, failure and maintenance may become overwhelming. Modern large supercomputers consume more than a Megawatt of power, greatly reducing their installation possibilities due to the costs involved. In such situations, a dedicated design that would reduce the extra components and thus improve the failure rates, power consumption, etc., may be desirable.

Accelerators may solve the above two problems while being merely additive to commodity clusters, thus preserving their cost advantage while addressing the niche advantages of dedicated supercomputers. Already, supercomputers as LANL's Roadrunner and Tokyo Tech's TSUBAME exist, which are clusters augmented with extensive commodity acceleration. In fact, Roadrunner, based on a combination of AMD Opteron and IBM/Sony/Toshiba PowerXCell, became the first supercomputer ever to achieve Petaflops in 2008.

The differences between the failed acceleration in HPC in the past, versus the possible success in modern times, are twofold. Firstly, modern-day accelerators are designed as part of the commodity PC ecosystem. Even customized hardware accelerators such as MDGrape and ClearSpeed are designed with standard fast I/O buses such as PCI-Express, and fit within standard high-performance cluster nodes as hosts, where the bulk of the code will still run on the PC while time-critical bottlenecks are executed on accelerators. Cell and GPUs are more fundamentally commodity, in that they had been developed with high-end embedded devices and multimedia gaming as target application areas, with an abundance of both hardware and software in the accompanying PC ecosystem. As such, leveraging this ecosystem has exactly the same characteristics as leveraging commodity clusters, with a possible similar success scenario.

Secondly, as discussed above, up until 2005 due to the continued exponential speedup of single-thread performance of CPUs, any acceleration speedups were quickly caught up and exceeded; however, this is now hitting a roadblock as shown above, and extensive acceleration would sustain its advantage over conventional CPUs in terms of resources required for a long time. For example, on a single node of TSUBAME sporting 8 Opteron 280s (2.4 Ghz x 16 cores), the best 3-D FFT performance utilizing all 16 cores has been approximately 20 Gigaflops both single and double precision using the most optimized code available. Contrastingly, a single GPU card within the Tesla 1070p records over 150 Gigaflops in single precision, and 40 Gigaflops using the NU-FFT [5] we developed. Moreover, when we parallelize across nodes, we need more than 16 nodes (= 256 cores) to recover the parallelization overhead incurred to match a single GPU performance. This difference is expected to worsen as GPU performance is expected to skyrocket, both in terms of pure flops and memory bandwidth, and most importantly, excel in cost and flops/watt performance as a result.

1.2. Commodity vector acceleration—the rebirth of vector computing

Overall, it is fair to state that GPUs and Cell are receiving the most attention as accelerators of choice, primarily because of their commodity nature, but also because of their high compute as well as memory bandwidth density. The IBM/Toshiba/Sony PowerXCell 8i sports 102.8 GigaFlops double precision peak performance as well as 25 GB/s memory bandwidth. GPU absolute performance is more dramatic, with both the latest AMD FireStream and NVIDIA Tesla GPU sporting over 1 TFlops of (albeit single precision) performance and over 100 GB/s memory bandwidth using multithreaded SIMD-vector processor array architecture, augmented by point-to-point fast memory interconnect. Such memory bandwidth effectively rivals the performance of the most expensive NEC SX series.

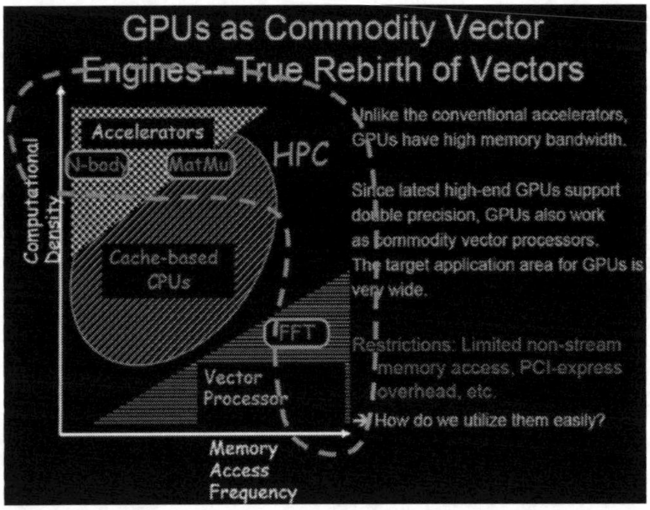

Figure 1: GPU as "rebirth of vectors" in HPC — supporting both high computational density as well as high memory access frequency.

Such a situation is depicted in figure 1. Here the Y-axis is the compute density of an application, and the X-axis is the memory access frequency. HPC apps would be naturally situated farther away from the axis, but can be categorized as (a) those with high computational density such as N-body or dense linear algebra, situated towards the upper part of the graph, and (b) those with high memory access frequency such as CFD and FFT, situated towards the right. Standard CPU performances are by and large mediocre, appropriate for applications near the center, and do not efficiently support (a) or (b). Modern accelerators such as GPUs and Cell, contrastingly, excel at both (a) and (b) because of their architectural properties as described above. As such, modern accelerators could be general HPC accelerators for various types of codes, not just specific ones as in the past.

In fact, both GPUs and Cell can be regarded as a rebirth of vector computing. Although different from traditional architectures with deep vector pipelining, nonetheless both architectures are quite amenable to where vector computers excelled in the past. In fact, modern algorithms intended to run on cache-oriented standard CPUs often do not work well due to their requirements for shorter memory access latency and locality enabled by extensive caching and associated blocking. Instead, classic algorithms intended for long vector-length "streaming" memory access work quite well. Our recent bandwidth-intensive FFT work [5] is a manifestation of this, in that we principally employ the classic vector-oriented multi-row FFT algorithm to compute the FFT for Y- and Z-axis (but not for the X-axis—for details refer to [5]).

2. Acceleration in action—TSUBAME 1.2 exemplars

As we have seen, accelerators exhibit high promise in becoming the next dominant resources in providing the majority of the compute capabilities in HPC, serving both dense computations as well as vector-friendly streaming codes. This is not to say, however, that such will occur automatically. With commodity clusters, it took years of continued research and development effort (still continuing to date) to mature to be useable in production, especially at large scale. Similarly, since development of commodity accelerators has focused on their principal market, i.e., non-HPC, single-node applications, considerable R&D as well as large-scale deployments are required for them to become truly mainstream in HPC.

Here we present a concrete example of the largest high-performance GPU deployment to date on an open science supercomputer, TSUBAME 1.2 at the Tokyo Institute of Technology, and some of the early results in scaling representative large parallel benchmark codes.

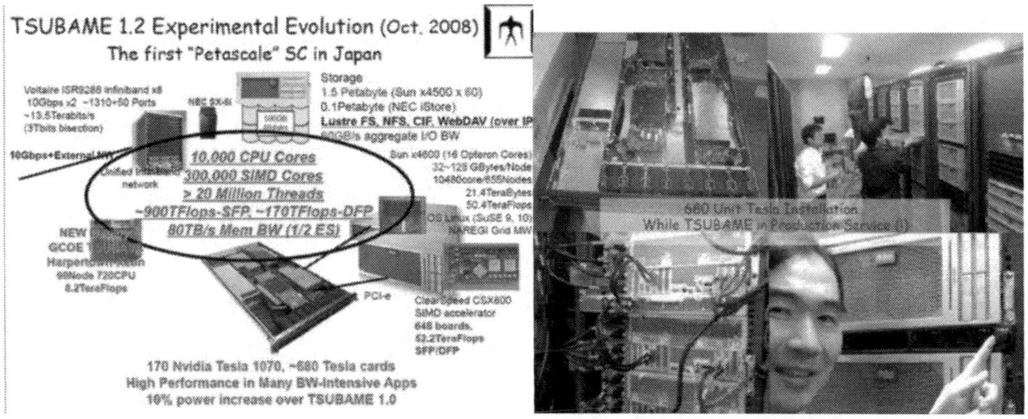

Figure 2: TSUBAME 1.2 GPU accelerated supercomputer.

The original TSUBAME 1.0 was deployed as a 10,480-CPU core cluster at the Tokyo Institute of Technology GSIC Center in the spring of 2006 [6]. In addition to the 5,120 2.4 Ghz dual-core AMD Opteron CPUs in 655 nodes (8 socket Sun x4600 nodes) providing approximately 50.4 Gigaflops of peak FP performance, it also sported 360 ClearSpeed CS600 Advance Accelerators, one card per node for about half of the nodes, providing approximately 30 Teraflops of additional compute power. Over the years, additional ClearSpeeds as well as Intel "Harpertown" Xeon nodes were added, pushing the total performance up to approximately 105 Teraflops.

TSUBAME 1.2 was implemented similarly as a follow-on upgrade to the existing TSUBAME, adding 170 NVIDIA Tesla s1070 units, each embodying four Tesla GPU cards for a total of 680 cards. This allowed the single-precision FP performance of TSUBAME to be boosted to nearly 900

Teraflops, and double-precision to approximately 160 Teraflops. The added power and thermal load on TSUBAME is only about 150 KW, which is about 1/8~1/10th of TSUBAME itself.

Speedups of various libraries and applications using GPUs have been quite astounding on a single GPU, as exemplified by the FFT speedup demonstrated in Section 2. We have utilized such capabilities, for example to greatly speed up sophisticated applications such as all-to-all 3-D protein docking, achieving equivalent speed to our earlier published result in [7] on BlueGene/L with almost four times better energy efficiency.

Here we mention our current results on two multi-node large-scale benchmarks, one dense linear algebra (Linpack) and the other sparse finite difference (multi-node Himeno) benchmark codes.

2.1. Linpack on CPU-GPU-ClearSpeed heterogeneous configuration

One of the major issues in heterogeneous supercomputers equipped with a batch of GPUs/accelerators is how users develop parallel programs that effectively use hybrid computing resources, especially tightly coupled programs. Here we take the High Performance Linpack (HPL) benchmark [8] and describe how it is heterogeneously accelerated on TSUBAME1.2. For more detailed implementation, refer to our previous paper [9].

2.1.1. High Performance Linpack (HPL)

HPL is a well known MPI based parallel benchmark used in the Top500 ranking, solving a set of dense linear equations using a direct method. It employs a block-based solver algorithm, so as to harness the highly tuned underlying Level 3 BLAS library—in particular, using a fast matrix-multiply (DGEMM) function is the key to obtain good HPL performance. Parallelization is achieved based on two-dimensional block cyclic distribution of the matrix, where each MPI process possesses a sub-matrix of the almost same size—as such HPL is designed for homogeneous environments.

Thus there are several challenging issues to port HPL onto TSUBAME1.2.

(1) All the processors, CPUs, GPUs, ClearSpeed boards should be used cooperatively for DGEMM kernel computation (intra-node heterogeneity). On TSUBAME, the ratios of theoretical peak computation performance for CPUs, GPUs, and ClearSpeed are 35%, 33%, 32%, therefore omitting any type of processor that causes heavy performance degradation.

(2) We also have to consider inter-node heterogeneity, which is that only 312 out of 648 nodes have two Tesla GPUs, and the rest have none, due to the configuration restrictions of Tesla 1070p.

(3) Finally, PCI-e/PCI-X communication overhead should not be underestimated, since the matrix data is basically allocated on host memory.

2.1.2. Kernel functions and matrix block size

In HPL, the most time consuming part in the overall benchmark is the DGEMM function calls that multiply the (M′ x B) matrix and (B x N′) matrix, where B is a tuneable block size parameter. To maintain a favorable communication-computation ratio, B should be sufficiently large.

As a CPU BLAS library, we use GotoBLAS [10]. For GPUs, we use a matrix multiply function written from scratch by our group, in order to achieve maximum efficient, and at the same time allow asynchronous double buffering to hide the data transfer latency, a feature not available in NVIDIA's native CUBLAS. Its on-board performance is about 80 GFlops, and we are able to almost completely hide the latency to effectively achieve this speed. For ClearSpeed, the CSXL BLAS library by ClearSpeed Inc. (DGEMM speed is 63 GFlops) is used. Through preliminary experiments, we found that B = 1152 would be the optimal block size.

2.1.3. Coping with heterogeneity

In TSUBAME 1.2, we are faced with intra- and inter-node heterogeneity as described above. We conduct load balancing of the DGEMM kernels running on different CPUs/accelerators based on the following strategies. By contrast, non-kernel computations, such as pivoting or row exchange, are simply done on x86 CPUs.

- *Intra-node heterogeneity*: We conceptually regard all hybrid processors on each node as a farm that provides kernel computation facility for processes. Here each process may delegate some fraction of DGEMM computation to accelerators, while other fractions may be computed by the CPU-based BLAS as in figure 3.
- *Inter-node heterogeneity*: With the above model, inter-node heterogeneity appears as imbalance of the DGEMM performances among nodes. To keep the performance per process almost identical, we configure *the number* of processes per node to reflect their compute capabilities.

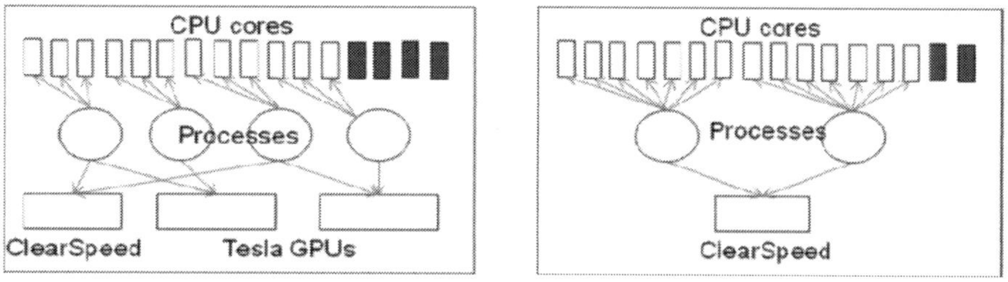

Figure 3: Mapping between processes and processors (CPUs, GPUs, ClearSpeed). The left figure shows a node with GPUs; the right one shows a node without GPU.

2.1.4 Other techniques

- We observed that PCI-e/PCI-X communication consumes considerable CPU load, and as a result conducting CPU-accelerator communication and computing DGEMM on the same cores incurs considerable performance degradation. Instead, we assign several dedicated cores solely for communication (black cores in figure 3). Although we lose considerable performance by such loss of cores (over 10 Teraflops for TSUBAME1.2), it becomes an overall win when acceleration is considered.
- In the original code, DGEMM calls are sometimes fragmented when communication and DGEMM overlap. We have eliminated such fragmentation by using Pthreads.
- We have implemented and added overlapping between pivot rows communication and DGEMM computation, which was not in the original code but is effective for accelerated machines due to the inherent parallelism between CPUs and accelerators.

2.1.5 Performance evaluation

In running the whole-machine Linpack, we used 648 of the 655 TSUBAME nodes, each of which has eight dual-core Opteron 880 processors and a ClearSpeed accelerator, and 312 nodes embodying two Tesla GPUs. Additionally, another production Xeon cluster called TSUBASA was integrated and used cooperatively—there we used 80 nodes, each of which has two quad-core Xeon E5440 processors. Combining all these hybrid resources, we have achieved 87.01 TFlops. Since the total peak speed is 163 TFlops, the efficiency is 53%, partly due to performance loss we mentioned above assigning dedicated CPU to support multiple accelerator, as well as compromises made such as the block size (B=1152). Figure 4 compares peak speed and Linpack speed for each category of processors. Interestingly, we observe here that all the processors show similar efficiency (48 to 56%), so the resulting share of runtime performance largely matches that of the theoretical peak performance. Here, GPU is the most efficient (56%) in a more careful comparison, but it is unclear whether this is inherent or due to compromise weighing in favour of GPU, and we are conducting follow-up research to clarify this. The figure also shows electrical power consumption; we observe that while accelerators provide 66% of the computation performance, their power consumption is only 15% of the entire machine.

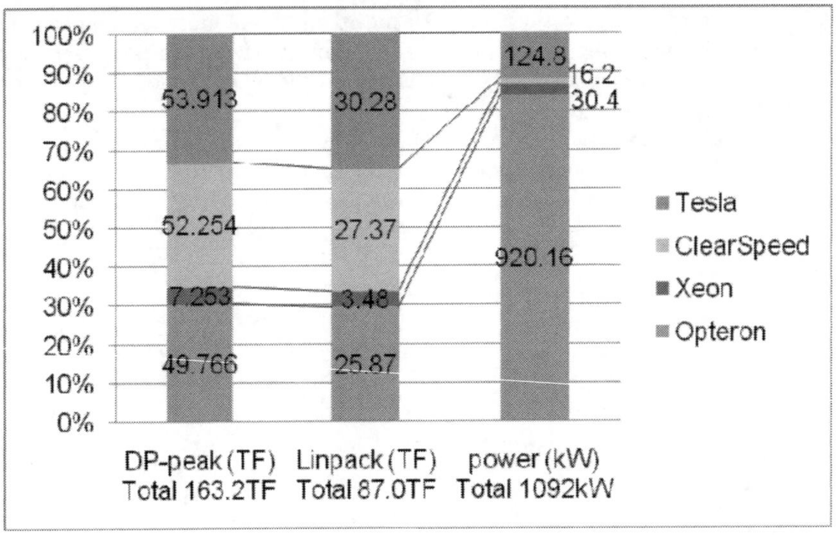

Figure 4: Performance ratio of each kind of processor. Peak speed (double-precision), contribution to Linpack speed, and electrical power consumption are shown.

	Jun06	Nov06	Jun07	Nov07	Jun08	Nov08	Jun09
Speed(TF)	38.18	47.38	48.88	56.43	67.70	77.48	87.01
Rank	7	9	14	16	24	29	?

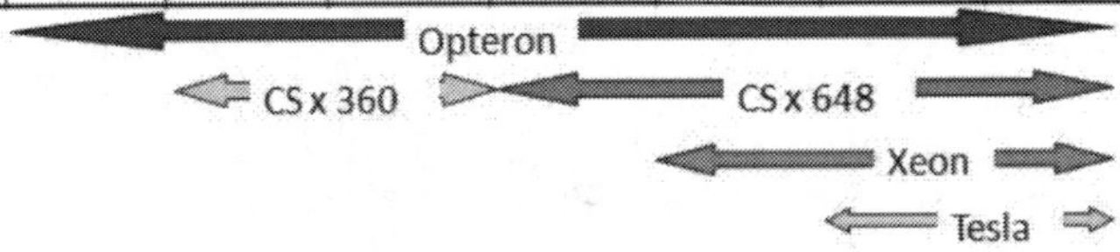

Figure 5: History of TSUBAME in the Top500.

Figure 5 shows the history of our Linpack experiments on the TSUBAME, in particular the achieved Teraflops and rank in the Top500. Ever since the original TSUBAME was installed in 2006, processor resources have been steadily increased due to system upgrade, which is represented by continuous improvement for every Top500 since its inception. Since the performance only with Opteron CPUs was 38.18 TFlops, acceleration technology improved the TSUBAME system performance by a factor of 2.3.

2.2. Himeno CFD benchmark acceleration over multiple GPU nodes

The Himeno CFD benchmark is a simple 3-D CFD code that solves a Poisson equation using finite difference Jacobi iterations. Although production-quality CFD code would be substantially more complex and employ sophisticated methods, the Himeno benchmark nonetheless is extremely bandwidth demanding and serves as a benchmark that measures the worst-case scaling scenario for bandwidth intensive codes. Himeno benchmark scores are published for a variety of architectures on the benchmark website [11]. On a standard x86 CPU, the benchmark is completely memory bound and the score would be approximately 1 GB/s. Contrastingly, a single GPU implementation on a NVIDIA CUDA GPU has seen reported scores of over 70 GB/s (figure 6) by utilizing high bandwidth device memory as well as shared memory.

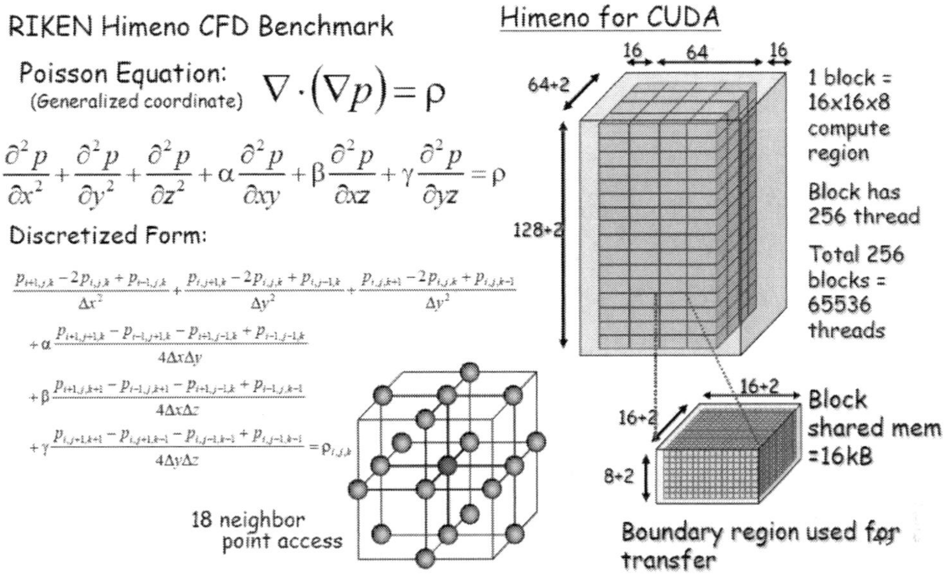

Figure 6: Riken Himeno benchmark and parallelization using CUDA GPUs.

The Himeno benchmark's main loop involves 34 FP calculations, 31 memory reads, and 1 memory write. By optimization, we can reduce this to 13 reads and 1 write—still significant and very much memory bound. There are four grid sizes, the smallest S being (65 x 65 x 129) up to the largest XL being (513 x 513 x 1025). For our benchmark, we slightly modified XL so that it would be (1025 x 513 x 513), i.e., will have a prolonged X-axis instead of Z for ease of programming.

For parallelization across GPUs, we conduct one-dimensional block distribution along the X-axis across multiple nodes. Within a GPU, we extensively parallelize and optimize the code using shared memory and coalesced memory access. Across the GPUs, we overlap MPI communication across nodes and Jacobi iteration computation within a GPU in each node.

On TSUBAME1.2, CPU ⇔ GPU communication bandwidth is greatly sacrificed by being based on PCI-E Gen1 x8 instead of the more modern Gen2 x16, and furthermore by two GPU cards sharing a single PCI-E lane—as such, in the worst case we have only $1/8^{th}$ the bandwidth of a standard Gen2 x16 implementation, and care must be taken to hide the latency effectively. Under this circumstance, communication of the boundary region on every Jacobi iteration takes approximately 8.2 ms. If the GPU compute granularity is greater than this, we would have effectively hid the latency. However, as we increase the number of GPUs while maintaining the problems size (strong scaling), the boundary area adjacent to the next GPU will not change, and only the size of the X-axis segment will get smaller, shortening the compute time.

Figure 7 (left) shows the result. We observe that latency hiding technique greatly improves the scalability of the code, achieving over 700 Gigaflops on 32 CPUs. At that point we achieve 1.1 TB/s in memory bandwidth. Reflecting upon the theoretical compute models, we observe that there is still a little bit of room for improvement, but the benchmark is approaching the hardware limits.

By doubling the problem size to XXL (figure 7 (right), 2049 x 513 x 513), we achieve considerable improvement in scalability; scaling from 16 to 32 GPUs enables 1.79 times speedup. Further details of the benchmark and the implementation will be the subject of another paper.

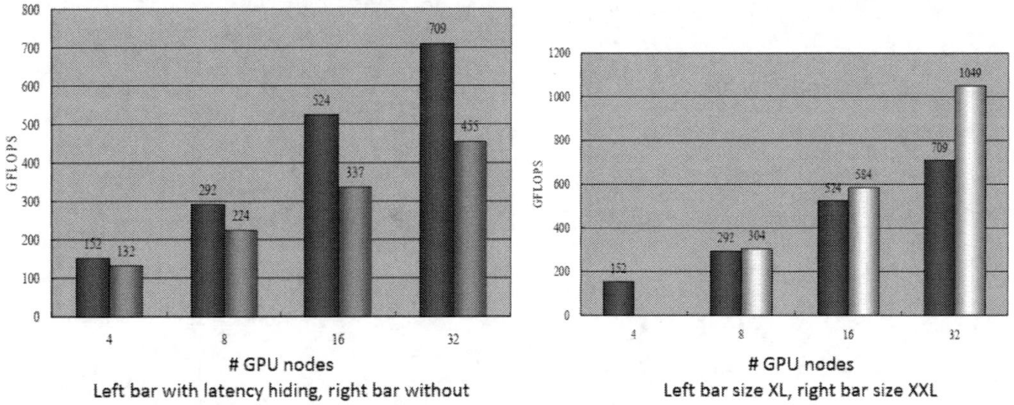

Figure 7: Riken Himeno benchmark parallelized across nodes on TSUBAME 1.2. The left graph shows the effect of latency hiding as we scale to 32 GPU nodes. The right graph shows the effect of doubling the problem size to achieve larger granularity and thus better scalability.

3. Conclusion and future work

Commodity acceleration is finally here, and in fact might dominate the computing aspects of future supercomputers, as one of its major properties is the rebirth of vector computing, which has been largely punted in commodity clusters. Already there are numbers of successful application ports with dramatic speedups reported in the literature, and large-scale multi-node applications and deployments are starting to materialize, where significant speedups and/or power savings are being reported. TSUBAME1.2, one of the first large-scale deployments of GPUs in a production supercomputer, is allowing us to understand the attractions as well as limitations of the approach, and what technical problems still lie ahead. Even as we speak, many GPU-enabled clusters are being planned/deployed, but without sufficient software support, they are reminiscent of early clusters.

As future work, we plan to conduct extensive research to help GPUs become the dominant compute resource. This research includes various systems issues, as well as algorithmic and application issues, in preparation for deployment of a petascale TSUBAME 2.0, anticipated to be deployed in 2010.

References

[1] "Cell Broadband Engine Technology and Systems", IBM Systems Journal, 51-5, May, 2007

[2] Owens, J.D. Houston, M. Luebke, D. Green, S. Stone, J.E. Phillips, J.C. "GPU Computing", Proc. IEEE, 96-5, May 2008, pp. 879-899

[3] ClearSpeed Inc. "ClearSpeed Whitepaper: CSX Processor Architecture", http://www.clearspeed.com/docs/resources/ClearSpeed_Architecture_Whitepaper_Feb07v2.pdf, Feb. 2007.

[4] Taiji M. "MDGRAPE-3 chip: a 165 Gflops Application Specific LSI for Molecular Dynamics Simulations", Proc. Hot Chips 16, IEEE Computer Society Press (CD-ROM), 2004.

[5] Akira Nukada, Yasuhiko Ogata, Toshio Endo, and Satoshi Matsuoka. "Bandwidth Intensive 3-D FFT kernel for GPUs using CUDA", Proc. ACM/IEEE Supercomputing 2008 (SC2008), Austin, Texas, the IEEE Press, Nov., 2008.

[6] Satoshi Matsuoka. *Petascale Computing Algorithms and Applications --- Chapter 14 The Road to TSUBAME and Beyond*, Chapman & Hall CRC Computational Science Series, pp.289-310, 2008

[7] Akira Nukada, Yuichiro Hourai, Akira Nishida, and Yutaka Akiyama. ""High Performance 3D Convolution for Protein Docking on IBM Blue Gene", Parallel and Distributed Processing and Applications, Springer LNCS Vol. 4742, pp. 958-969, 2007.

[8] A. Petitet, R. Whaley, J. Dongarra and A. Cleary. HPL – a portable implementation of the high-performance Linpack benchmark for distributed computers. http://www.netlib.org/benchmark/hpl

[9] Toshio Endo and Satoshi Matsuoka. "Massive Supercomputing Coping with Heterogeneity of Modern Accelerators", IEEE International Parallel & Distributed Processing Symposium (IPDPS 2008), the IEEE Press, April, 2008.
[10] K. Goto. Goto BLAS. http://www.tacc.utexas.edu/resources/software.
[11] The Riken Himeno CFD Benchmark. http://accc.riken.jp/HPC/HimenoBMT/index_e.html

DOE's Institute for Advanced Architecture and Algorithms: an application-driven approach

Richard C. Murphy

Sandia National Laboratories, PO Box 5800, MS-1319, Albuquerque, NM 87185

E-mail: rcmurph@sandia.gov

Abstract. This paper describes an application driven methodology for understanding the impact of future architecture decisions on the end of the MPP era. Fundamental transistor device limitations combined with application performance characteristics have created the switch to multicore/multithreaded architectures. Designing large-scale supercomputers to match application demands is particularly challenging since performance characteristics are highly counter-intuitive. In fact, data movement more than FLOPS dominates. This work discusses some basic performance analysis for a set of DOE applications, the limits of CMOS technology, and the impact of both on future architectures.

1. Introduction

As the MPP era draws to a close, the energy inefficiency of modern supercomputer architectures is rapidly becoming the limiting factor for architectural scalability. The MPP era's total reliance on commodity processors that are optimized for less challenging application spaces has created energy inefficiencies throughout the system. Further, fundamental transistor device physics has forced scaling in the form of increased parallelism at the chip level, rather than the exponential increase in single thread performance that characterized the MPP era. Indeed, the transition to multicore architectures is driven entirely by the aging of the CMOS fabrication process. Without a well-defined successor capable of Moore's Law scaling, the transistor limits the potential of future supercomputers deployed in energy constrained environments. As a result, changes to computer architecture are required to enable newer, more highly-scalable applications. However, choosing the right architectural path is extremely difficult. Generally, computer architects do everything possible to hide the details of the hardware implementation from the programmer: cache sizes, branch prediction policies, out-of-order execution mechanisms, and thread scheduling at the hardware level are all examples of critical architectural mechanisms and structures that are typically observed only indirectly by the programmer.

More worrisome still, each of these structures that architects use to boost single thread performance and hide the depths and complexities of modern communication hierarchies (memory or interconnect) consumes energy that could otherwise be available for the computation if the programmer could exploit simpler, lighter-weight mechanisms. It has long been argued that these mechanisms should be exposed to the programmer[9, 5, 1, 4], but the transition to multithreaded/multicore architectures leaves little choice in an energy constrained environment.

This paper presents a synthesis of application analysis results derived from simulation, analysis tools, and real hardware. The remainder of the paper is structured as follows: Section

2 discusses some of the applications studied in this work; Section 3 examines the limits of CMOS transistors; Section 4 provides results to show that performance bottlenecks tend to be latency and concurrency dominated; and Section 5 describes the impact of these trends on future architectures.

2. Applications Discussed In This Work

The application studies referenced in this work are divided into two basic application classes: physics codes, which tend to be floating point oriented; and informatics codes, which tend to be more integer oriented. The full application analysis and simulation methodology has been previously described[7, 6, 8]. The results are summarized in this paper.

2.1. Physics Applications

Broadly speaking, the physics application suite analyzed by Sandia are large-scale MPI applications, and represent real world physical simulations. They are:

- **ALEGRA** is a finite element shock physics code capable of modeling near- and far-field responses to explosions, impacts, and energy depositions.
- **CTH** is a multi-material, large deformation, strong shock wave, solid mechanics code developed at Sandia National Laboratories over the last 30 years. CTH models multi-phase, elastic viscoplastic, porous and explosive materials with multiple mesh refinement methods.
- **Cube3** Cube3 is a generic linear solver that drives the Trilinos framework for parallel linear and eigensolvers. It mimics a finite element analysis problem by creating hexagonal elements, then assembling and solving a linear system. The width, depth, and degrees of freedom (e.g., temperature, pressure, velocity, etc.) can be varied.
- **ITS** performs Monte Carlo simulations of linear time-independent coupled electron/photon radiation transport.
- **MPSalsa** is a high resolution 3d simulation of reacting flow. The simulation requires both fluid flow and chemical kinetics modeling.
- **Xyce** is a parallel circuit simulation system capable of modeling very large circuits at multiple layers of abstraction (device, analog, digital, and mixed-signal). It includes both SPICE-like models and radiation models.

This suite of applications is somewhat more narrow than those of interest to the Office of Science, but future work in the DOE Institute for Advanced Architecture will apply the same fundamental analysis techniques to specific Office of Science applications.

2.2. Informatics Applications

Recently, new large-scale applications in the area of informatics have emerged as important and substantially different from physics applications that have been the core of supercomputing over the past three decades. These applications tend to be more integer-oriented, and represent problems in discrete math. They are typically less structured (in terms of memory access patterns), and are often techniques used to analyze the output of a physical simulation, or to examine real world data sets in an attempt to find patterns or form hypotheses. Additionally, because Informatics applications tend to be less structured in nature, they are written using a variety of programming models. The application set discussed are:

- **DFS** implements a depth-first search on a large graph and forms the basis for several high-level algorithms including connected components, tree and cycle detection, solving the two-coloring problem, finding Articulation Vertices, and topological sorting.

- **Connected Components** breaks a graph into components. Two vertices are in a connected component if and only if there is a path between them.
- **Subgraph Isomorphism** determines whether or not a subgraph exists within a larger graph.
- **Full Graph Isomorphism** determines whether or not two graphs have the same shape or structure.
- **Shortest Path** computes the shortest path between two vertices using a breadth first search. Real world applications include path planning and networking and communication.
- **Graph Partitioning** is used extensively in VLSI circuit design and adaptive mesh refinement. The problem divides a graph in to k partitions while minimizing the *cut* between the partitions (or the total weight of the edges that cross from one partition to another).
- **BLAST** is the Basic Local Alignment Search Tool and is the most heavily used method for quickly search nucleotide and protein databases in biology.
- **zChaff** is a heuristic for solving the Boolean Satisfiability Problem. In propositional logic, a formula is *satisfiable* if there exists an assignment of truth values to each of its variables that make the formula true.

These applications have the potential to significantly change the methodology used in computer-driven science, and fundamentally represent a new and disruptive use of high performance computing resources.

3. CMOS Transistor Limitations

The switch to multicore is fundamentally driven by a change in the CMOS transistor that forms the implementation technology for commodity processors. Power dissipation (manifesting itself as heat which must be removed from the device) can be broken into two components: Static Power Dissipation (P_{stat}), which is dissipated continuously while the circuit is powered and Dynamic Power Dissipation (P_{dyn}) which is dissipated during switching. Although increasingly important, static power dissipation cannot be addressed by the computer architect (other than by powering off parts of the machine not in use), and is primarily a function of the fabrication process. It is defined as:

$$P_{stat} + I_{leakage} V_{dd}$$

Where $I_{leakage}$ is the leakage current of the device, and V_{dd} is the voltage. It should be noted that while the computer architect cannot do much to control leakage current, the system architect and environment in which the machine is deployed can. At the junction, leakage is generally caused by thermally generated carriers, and *increases* exponentially as temperature rises. In fact, at 85°C, leakage current increases by a factor of 60 over room temperature values. These factor and increased failure rates must be taken into account in the "hotter running" compute environments that have become a topic of recent significant interest.

The majority of power dissipation is dynamic, and is defined as as:

$$P_{dyn} = C_L V_{dd}^2 f$$

Where C_L is the capacitance of the entire circuit, V_{dd} the voltage, and f the switching frequency (proportional to the clock frequency since not all devices will switch every cycle). This equation is critically important in describing the switch to multicore.

Moore's Law as originally defined demands an exponential increase in the number of transistors, which increases C_L. Specifically, for a given area, the total capacitance is the sum of the capacitance of each device. Although capacitance per device decreases as the feature size decreases (decreasing the individual transistor's contribution to C_L), the number of devices is increasing exponentially. Similarly, until the multicore era the frequency increased substantially in each generation. To keep the total dynamic power dissipation essentially constant (so that chips could be cooled with relatively inexpensive heat syncs at roughly room temperature), the equation was balanced by *decreasing* V_{dd}, which as it approaches 1 volt is nearing a fundamental limit (in fact, the rate at which V_{dd} has decreased has already begun to flatten).

Although there are additional manufacturing benefits to moving to multicore architecture, including simplifying the verification of designs, this fundamental technology limit is the real driver. The devices are capable of higher clock rates, at higher densities, but engineering a system to cope with the power dissipation is nearly impossible. The era is characterized as an increasing bounty of transistors (Moore's original observation), but the inability to increase single thread performance (through faster clock rates). The inevitable result is more parallelism.

Of course the end of the CMOS scaling curve in the next decade or so will cause additional fundamental physical limitations, unless a solution can be found.

4. Performance Bottlenecks Tend to Be Counterintuitive

The following section discusses basic application properties, and performance bottlenecks in the network and memory system.

4.1. Real Floating Point Applications Do Less Floating Point Than Expected

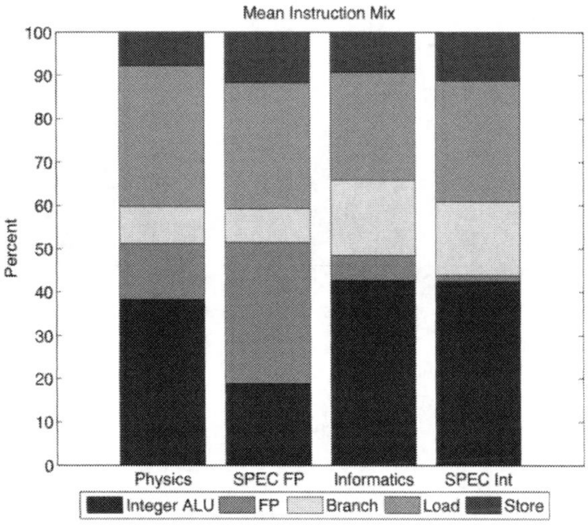

Figure 1. Instruction Mix

The instruction mix for the Physics and Informatics suites is depicted in Figure 1, and compared to the industry standard SPEC CPU 2000 suite. While SPEC FP averages approximately one floating point instruction in three, real Sandia Physics codes average only 12% (and is typically much less than 10%) because they exhibit more complex memory addressing patterns, which, in turn require more integer instructions to calculate the addresses. In fact, nearly 85% of integer instructions are calculating memory addresses, with the remaining 15% split between real integer data and branch (boolean) support.

The Informatics applications are very different from the Physics codes as well, suggesting the possibility of two different supercomputer architecture classes in the future. They perform 15% fewer memory references, and those references are much more likely to miss the cache, which stresses the memory subsystem. Additionally, the Informatics applications perform nearly twice as many branches as their Physics counterparts, leading to relatively smaller basic blocks. (This result is unsurprising due to the complex floating point calculations typically performed by the Physics codes[9].) This combination makes the Informatics codes more difficult to scale on modern superscalar processors than their physics counterparts.

4.2. Latency and Concurrency Matter More than Bandwidth

Data movement systems are typically rated by bandwidth, but most evidence points to applications being more sensitive to latency or concurrency. This can be simplistically described by Little's Law (from Economics) which says that:

$$throughput = \frac{concurrency}{latency}$$

Particularly in the case of memory systems, increasing the number of memory requests outstanding (currently limited by hardware and instruction sequences) or decreasing the latency required to receive a response has significantly more impact than changing the bandwidth.

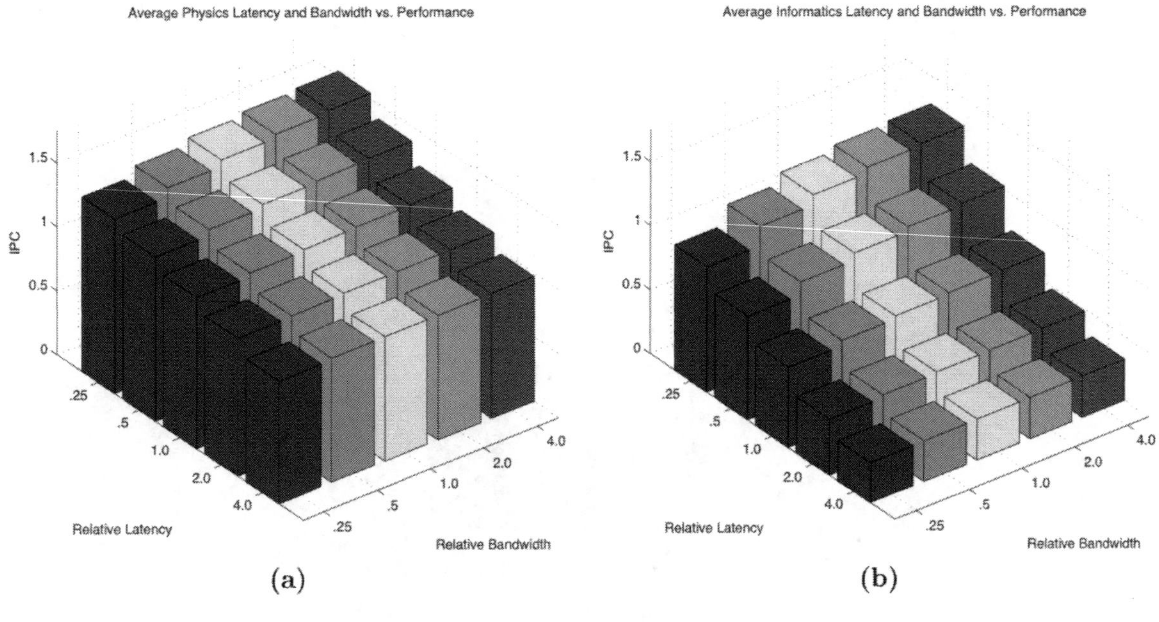

Figure 2. Latency and Bandwidth Sensitivity for (a) Physics Applications and (b) Informatics Applications

Figure 2 shows the change in Instructions Per Cycle (IPC, the computer architect's typical measure of performance) as relative latency and bandwidth are altered (the center bar being the latency and bandwidth of a typical processor's memory system. The full results and methodology have been described previously[6]. The results demonstrate that both application suites are significantly more sensitive to latency than bandwidth. A decrease in bandwidth by half on either suites affects performance by less than 5%, while doubling the memory latency halves performance[1]

[1] It should be noted that this bandwidth measure is more strict than that presented for a typical memory part, which will also include a decrease in latency in subsequent generations.

These result combined are confirmed by observations from performance counters and other real system measurements going back to ASCI Red that demonstrates that real applications have difficult saturating the memory bus on a modern architecture because address generation (the integer instructions discussed in Section 4.1) is the bottleneck more than typical memory performance.

This is further demonstrated by typical out-of-order processors that are designed to mask the multi-cycle latency of a Level 1 cache **hit**, more than the latency of a memory access (due to limitations of both the incoming instruction stream and the implementation technology).

The performance of the two application suites is also significantly different. The Physics application achieve reasonable performance on the simulated out-of-order execution unit, achieving an IPC of 1.22 for the base case. The processor is capable of retiring 4 instructions per cycle, so the achieved IPC is approximately 30% of maximum. However, any IPC greater than one is quite good for a typical application suite. In contrast, the baseline case for the Informatics suite is only 0.70, indicating that the applications are significantly more memory intensive, causing the processor additional idle time. Key graph algorithms, including both isomorphism problems and connected components show IPCs significantly less than 0.5. Aside from being less than an eighth of achievable performance, the results indicate that significant power is being wasted on high clock rates for those applications.

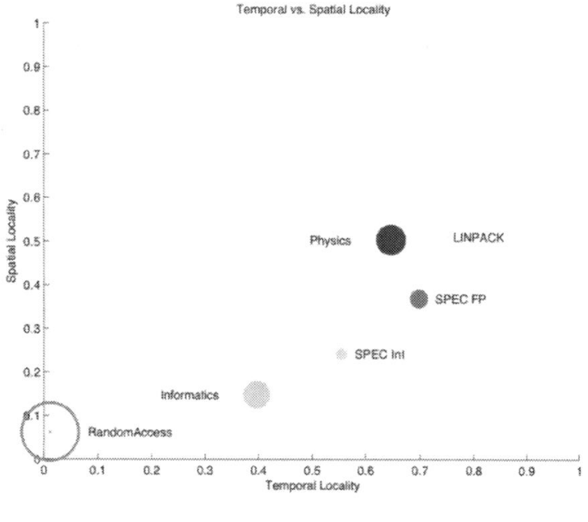

Figure 3. Application Memory Performance Properties

Figure 3 describes the fundamental memory performance properties of each benchmark suite, and has been fully described previously[7]. The spatial locality describes the probability that data near previously referenced data will be referenced. The temporal locality describes the probability that data already referenced will be reused. And the relative size of the dots represents the amount of unique data that each application suite consumes over a fixed interval of instructions. These fundamental application properties dictate the performance observed in Figure 2, with applications consuming more data, and doing so in a more random fashion (i.e., closer to the origin on the graph) showing similarly low performance in comparison to applications with smaller data sets consumed in a more structured pattern. These fundamental application properties determine the class of architecture required to provide high performance at low power more than any other issue.

4.3. Network Performance

These issues are much more difficult to measure for the network, but preliminary results and prior experience show the importance of network injection rate vs. raw bandwidth.

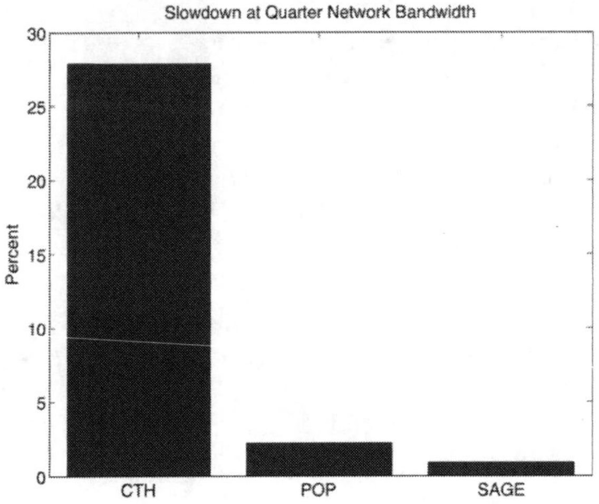

Figure 4. Measurement of Bandwidth Restricted Performance on Red Storm. This is a recast of data collected by Kurt Ferreira at Sandia, a subset of which is discussed in [3]

The trade-off between latency, bandwidth, and message rates are difficult to measure for modern networks, and the trade-offs appear to be highly application dependent. This state of affairs makes drawing firm conclusions on minimal network design difficult, which may lead to either poor application performance for a subset of applications or over-provisioned networks for some subset of applications.

Figure 4 shows the effects of running three applications on Sandia's Red Storm machine, with bandwidth degraded in the mesh by 75%, while not degrading small message latency or injection rate. The mesh is intentionally capable of approximately twice the bandwidth of the link between processor and NIC to offer more stable performance in the face of message contention, whether from non-3D communication patterns or poor application and process placement. The applications shown are CTH, SAGE, and POP. CTH and Sage were run at 2,048 nodes, and POP at 2,500. CTH and Sage were run with weak scaling data sets, and POP as a strong scaling problem.

POP is limited by a combination of message rate and latency, and it is therefore not surprising that reducing network bandwidth does not impact application performance. SAGE, like POP, appears to be immune to bandwidth changes, although it is unclear whether this is problem specific. CTH, on the other hand, shows a 27% performance degradation when mesh bandwidth is reduced to half of injection bandwidth, and is traditionally thought to be insensitive to latency and message rate. The difference between behavior of POP and CTH suggests a significant challenge for hardware design – a need to balance hardware cost, energy usage, and performance across a wide variety of applications.

In an energy optimized environment, one could envision a system that dynamically scales network bandwidth performance to meet both application and overall system demands (contention, I/O, etc.) when full topological bandwidth is not required for the current set of running applications.

5. Impact on Future Architectures: Multicore, Multithreading

Section 3 described the fundamental technology changes pushing multicore processors. The discussion of IPC in Section 4.2 shows that the processor is tremendously underutilized, especially for applications with very large, sparse data sets. Because the problem is fundamentally one of latency, there are only two basic solutions: avoiding the latency through faster devices, memory hierarchies, and caches; or tolerating the latency via mechanisms such as multiple outstanding loads, out-of-order execution, vector pipelines, or threads.

Avoiding latency via faster devices is fundamentally an economically and technologically driven process, and often provides a "one time" benefit rather than a sustained and continuous benefit over multiple generations. For example, cheap 3D integration of memory onto the processor would reduce wire lengths between the processor and memory from centimeters to microns, providing a one-time latency reduction.

Architectural techniques for avoiding latency such as caching also show diminishing returns. Caches place frequently used items closer to the processor, but become less effective proportional to cost as they increase in size. They are also large consumers of energy.

In terms of memory toleration, many of the classic techniques have reached their limits. Given the relatively aggressive out-of-order execution in modern processors and the observed IPCs from Figure 2, it is clear that there is not much room for improvement. In fact, most processors are moving towards simpler latency toleration mechanisms because the power dissipation for out-of-order execution is so high compared with the relatively small benefits. As discussed previously, most applications fail to fully utilize memory system bandwidth, meaning that the problem of generating more outstanding loads is not due to a hardware limitation, but fundamental application properties. For the Informatics suite in particular, the pointer chasing requires the results from one load to be available before another load can be generated, limiting architectural options for solving the problem.

The remaining option for power-efficient latency toleration is threads, and it can be seen in the successes of early throughput-oriented architectures such as the Sun Niagara. Threads are:

- Relatively inexpensive to implement in hardware, requiring additional memory structures to support larger register files, but not requiring complex logic as is required for out-of-order execution;
- Efficient in that switching to another thread while the currently executing thread waits on a long latency event still allows "useful" work to be performed during the idle time (in contrast with techniques such as speculative execution which simply create heat when the architecture predicts an erroneous path of execution);
- Low power, primarily because useful work is always being done and they replace more complex hardware structures;
- Capable of fully utilizing the memory system by exposing more address generation streams to the programmer (if the programmer can exploit them), which can be seen in the trend toward fewer memory controllers per core in modern processors; and,
- Potentially plentiful, in that for the desktop and server space for which the processors are optimized there typically exists sufficient parallelism (even at the task level) to allow for multithreaded processors not to starve, which is potentially a significant application problem for future supercomputers.

Consequently, the trend towards multicore/multithreaded architectures is inevitable, especially in highly energy constrained environments. The DARPA Exascale Report[2] describes dominance of data movement in future exascale architecture energy budgets.

Acknowledgments
I would like to thank Kurt Ferreira for sharing his data used in the networking discussion, as well as Arun Rodrigues, Brian Barrett, Scott Hemmert, and Jim Ang for many useful discussions over the course of preparing this paper.

References
[1] Jay B. Brockman, Peter M. Kogge, Vincent Freeh, Shannon K. Kuntz, and Thomas Sterling. Microservers: A new memory semantics for massively parallel computing. In *ICS*, 1999.

[2] Peter M. Kogge (editor). *ExaScale Computing Study: Technology Challenges in Achieving Exascale Systems.* University of Notre Dame, CSE Deptartment Technical Report TR-2008-13, September 28, 2008.

[3] Kurt Ferreira, Ron Brightwell, and Patrick Bridges. Characterizing Application Sensitivity to OS Interference Using Kernel-Level Noise Injection. In *Supercomputing 2008 (SC08), Austin TX, November 2008.*

[4] Mary Hall, Peter Kogge, Jeff Koller, Pedro Diniz, Jacqueline Chame, Jeff Draper, Jeff LaCoss, John Granacki, Apoorv Srivastava, William Athas, Jay Brockman, Vincent Freeh, Joonseok Park, and Jaewook Shin. Mapping Irregular Applications to DIVA, A PIM-based Data-Intensive Architecture. In *Supercomputing, Portland, OR*, November 1999.

[5] Shannon K. Kuntz, Richard C. Murphy, Michael T. Niemier, Jesus Izaguirre, and Peter M. Kogge. Petaflop Computing for Protein Folding. In *Proceedings of the Tenth SIAM Conference on Parallel Processing for Scientific Computing, Portsmouth, VA*, March 12-14, 2001.

[6] Richard C. Murphy. On the Effects of Memory Latency and Bandwidth on Supercomputer Application Performance. In *IEEE International Symposium on Workload Characterization 2007 (IISWC2007)*, September 27-29, 2007.

[7] Richard C. Murphy and Peter M. Kogge. On the Memory Access Patterns of Supercomputer Applications: Benchmark Selection and its Implications. *IEEE Transactions on Computers*, 56(7):937–945, July 2007.

[8] Richard C. Murphy, Arun Rodrigues, Peter Kogge, and Keith Underwood. The implications of working set analysis on supercomputing memory hierarchy design. In *The 2005 International Conference on Supercomputing*, June 20-22, 2005.

[9] Arun Rodrigues, Richard Murphy, Peter Kogge, and Keith Underwood. Characterizing a New Class of Threads in Scientific Applications for High End Supercomputers. In *Proceedings of the 18th Annual ACM International Conference on Supercomputing*, pages 164–174, June 26-July 1 2004.

Software Challenges in Extreme Scale Systems

Vivek Sarkar[1], **William Harrod**[2], **and Allan E Snavely**[3]

[1] Department of Computer Science, Rice University

[2] Information Processing Technology Office, Defense Advanced Research Projects Agency

[3] San Diego Supercomputer Center, University of California, San Diego

E-mail: vsarkar@rice.edu, william.harrod@darpa.mil, asnavely@ucsd.edu

Abstract. Computer systems anticipated in the 2015 – 2020 timeframe are referred to as *Extreme Scale* because they will be built using *massive multi-core processors* with 100's of cores per chip. The largest capability Extreme Scale system is expected to deliver *Exascale* performance of the order of 10^{18} operations per second. These systems pose new critical challenges for software in the areas of *concurrency, energy efficiency* and *resiliency*. In this paper, we discuss the implications of the concurrency and energy efficiency challenges on future software for Extreme Scale Systems. From an application viewpoint, the concurrency and energy challenges boil down to the ability to express and manage parallelism and locality by exploring a range of *strong scaling* and *new-era weak scaling* techniques. For expressing parallelism and locality, the key challenges are the ability to expose all of the intrinsic parallelism and locality in a programming model, while ensuring that this expression of parallelism and locality is *portable* across a range of systems. For managing parallelism and locality, the OS-related challenges include *parallel scalability, spatial partitioning* of OS and application functionality, *direct hardware access* for inter-processor communication, and *asynchronous* rather than interrupt-driven events, which are accompanied by runtime system challenges for *scheduling, synchronization, memory management, communication, performance monitoring,* and *power management*. We conclude by discussing the importance of *software-hardware co-design* in addressing the fundamental challenges for application enablement on Extreme Scale systems.

1. Introduction

It is widely recognized that the computer systems anticipated in the 2015 – 2020 timeframe will be qualitatively different from current and past computer systems. Specifically, they will be built using *massive multi-core processors* with 100's of cores per chip, their performance will be *driven by parallelism and constrained by energy*, and they will be subject to *frequent faults and failures*. We use the term, *Extreme Scale*, to refer to these systems. A characterization of Extreme Scale systems can be found in the recent report on "Technology Challenges in Achieving Exascale Systems" [21]. This characterization identifies three distinct classes of systems:

- **Data-center-sized Exascale systems**, capable of delivering 1 ExaFlops or 1 ExaOps *i.e.*, 1,000× the capability of currently emerging Petascale data-center-sized systems[1].
- **Departmental-sized Petascale systems** that allow the capabilities of a Petascale system to be shrunk in size and power to fit within a few racks, allowing widespread deployment.

[1] Following common usage, "ops" refers to operations per second in this paper unless otherwise specified.

- **Embedded Terascale systems** that reduce Terascale capability to a few chips and a few ten's of watts, thereby enabling deployment in a range of embedded environments.

Since the first system class listed above achieves Exascale performance, the terms *Exascale* and *Extreme Scale* are often interchangeably in the community. However, in this paper, we prefer to use *Extreme Scale* to refer to systems across all three classes and *Exascale* specifically for the largest data-center sized system class.

The focus of this paper is on *software challenges* for Extreme Scale systems. The scope of software considered spans the spectrum of operating systems; runtimes for scheduling, memory management, communication, performance monitoring, power management, and resiliency; computational libraries; compilers; programming languages; and application frameworks. Though there are significant differences in the software environments and requirements for the three classes of Extreme Scale systems, all of them share some critical challenges namely, *concurrency, energy efficiency* and *resiliency*. The scope of this paper is restricted to concurrency and energy efficiency challenges related to software and identification of opportunities for software-hardware co-design, as well as potential new interfaces between applications and system software and between system software and hardware.

The concurrency challenge is manifest in the need for software to expose at least 1000× more concurrency in applications for Extreme Scale systems, relative to current systems. It is further exacerbated by the projected memory-computation imbalances in Extreme Scale systems, with Bytes/Ops ratios that may drop to values as low as 10^{-2} where Bytes and Ops represent the main memory and computation capacities of the system respectively. These ratios will result in 100× reductions in memory per core relative to Petascale systems, with accompanying reductions in memory bandwidth per core. Thus, a significant fraction of software concurrency in Extreme Scale systems must come from exploiting more parallelism within the computation performed on a single datum *i.e.*, from strong scaling or from the "new-era" weak scaling discussed in Section 2. Strong scaling typically involves more frequent communication and synchronization than weak scaling, which in turn exacerbates the energy efficiency challenge since data movement and synchronization are major contributors to energy costs in Extreme Scale systems. Another major obstacle to achieving a large degree of concurrency arises from the communication and synchronization overheads in current system software that contribute directly to the serial (Amdahl's Law) fraction of the program's execution. A new software stack can reduce these overheads by orders of magnitude, especially with software-hardware co-design, thereby making it possible to achieve the parallel efficiency needed for Extreme Scale systems.

The energy efficiency challenge is also critical because all three classes of Extreme Scale systems will be expected to deliver their 1000× improvements in computation capability while essentially remaining within the power budgets of current systems. An aggressive hardware design for data-center-sized systems will need at least 60MW of power to achieve an Exa-op level of performance, under highly idealized zero-overhead assumptions for software [21]. When current software overheads are taken into account, it is clear that the Exascale capability cannot be achieved without a significant redesign of the system software stack.

As discussed in this paper, current software approaches will be inadequate in enabling future Grand Challenge applications (outlined in Section 2) on Extreme Scale systems. Instead, the potential for a 1000× increase in computation capability offered by each class of Extreme Scale system will only be achievable through radical re-design of the underlying system software and hardware. Current system designs won't work at Extreme Scale because of their sequential foundations and their inherent energy inefficiencies. Recent trends in High Productivity Computing Systems (HPCS) have demonstrated reductions in human effort involved in developing high-productivity software for current Petascale systems, but do not address the requirements of Extreme Scale architectures such as energy-constrained many-core parallelism and heterogeneous processors. Also, while there will be some overlap between system software requirements for Extreme Scale and those for large scale commercial data centers, there are also significant differences. Commercial system software for cloud computing is primarily focused on optimizing throughput capacity of independent jobs, whereas system software for Extreme Scale must be capable of delivering 1000× (or more) increase in parallelism to a single job.

The rest of the paper is organized as follows. Section 2 discusses selected grand challenge applications that are being developed from scratch or are being scaled up from existing Petascale applications. Section 3 discusses challenges in expressing parallelism and locality at the finest granularities possible so as to support *forward scalability*, whereas Section 4 outlines the challenges in low overhead management of this fine-grained parallelism and locality. Finally, Section 5 discusses how future Extreme Scale software stacks must be tightly integrated with new Extreme Scale hardware via software-hardware co-design, and Section 6 contains a concluding summary.

2. Challenges in Application Scaling

Application scaling remains a major challenge in the utilization of high-end parallelism. Of applications that operate at sustained Terascale performance today, only a small fraction is expected to be successful at reaching Petascale and an even smaller fraction at Exascale. Further, even for applications that reach Petascale performance today, the nature of scaling necessary to obtain Petascale performance on an Extreme Scale departmental system will raise new challenges for rewriting the application to address the concurrency and locality requirements of such systems. In this paper, we argue that the existing software "stack" is a major contributor to these scalability limitations, and that the approaches discussed in the following sections could have a significant impact in removing obstacles to scaling. We start by examining two primary ways to scale applications.

Strong scaling refers to the concept of applying more resources to the same problem size to get results faster. Unfortunately, few applications are amenable to strong scaling. As you strongly scale an application, the work at a node/processor/core decreases and the relative overhead increases. Speedup may initially equal the number of processors, but eventually the amount of overhead causes the slope of the speedup curve to flatten. At this point, adding processors does not cause the application to run faster. Eventually, it is possible that overhead grows so rapidly that adding processors actually causes the time to solution to increase and speedup to decrease. An application that demonstrated reasonable scaling over three orders of magnitude increase in the number of processors is the first principles molecular dynamics "Qbox" code that won the 2006 ACM Gordon Bell Prize for "peak performance" with over 200 Tflop/s sustained performance (56% efficiency) on the LLNL BlueGene/L [14]. More recent winners of the 2008 ACM Gordon Bell Prize further underscored the importance of algorithmic innovations that attain very high levels of spatial and temporal locality.

Weak scaling refers to the concept of adding work as an application is run on more processors. By adding work, it is possible to assure that overhead does not destroy performance. Traditionally, weak scaling has referred to adding work due to spatial scaling. Meanwhile, there are additional sources of scaling – referred here as "new-era" weak scaling — arising from new application trends in which additional work is done per datum *e.g.*, multi-scale, multi-physics, interaction analysis, and data mining. Weak scaling permits the user to look at larger or more complicated problems and use the additional processors to solve larger problems, obtain better resolution, or learn more about the phenomenon being examined.

Traditional weak scaling occurs in classical mechanics simulations, where either (1) larger problems are examined or (2) the grid size and time-step interval are reduced. Solving larger problems – e.g., modeling the airflow around an entire airplane versus modeling the airflow over a section of the wing – results in a situation when memory scales nearly proportionally with work.

When the grid size is reduced (refined) in a 3-D mechanics simulation, the time step also needs to be reduced thereby increasing the amount of work relative to the amount of memory. When scaling in three dimensions, a 3/4 power rule applies to the amount of memory required whereas a 2/3 power rule applies when scaling in only two dimensions. Thus, for these applications, the required increase in memory size for a 1000× increase in work is 180× and 100× for 3-D and 2-D applications respectively.

In summary, we do not expect most applications to get to Extreme Scale by strong scaling. Instead, applications will be scaled by using algorithmic innovations with a combination of traditional and new-era weak scaling techniques.

3. Challenges in Expressing Parallelism and Locality

The focus of this section is on the challenges in expressing parallelism and locality that are encountered by application-level programmers across the three classes of Extreme Scale systems. The task of managing the parallelism and locality is relegated to the system software discussed in Section 4. It is likely that some heroic programmers, particularly for the data-center and embedded configurations, will wish to program directly at the system level using the interfaces in Section 4. However, for the remainder, it will be critical to address the challenges outlined in this section to enable them to use the capabilities of Extreme Scale systems.

3.1. Portable Expression of Parallelism and Locality

The requirement of pervasive extreme-scale parallelism outlined earlier demands that all of the intrinsic parallelism be exposed at all levels in the application. This is a marked contrast to current practice where programmers repeatedly rewrite applications to expose incrementally more parallelism for the next generation of hardware. Instead, our goal should be to express all opportunities for parallelism, leaving the choice of what to exploit to the layers of the software stack responsible for managing parallelism and locality (Section 4). While this is likely to be a more demanding task for current programmers who have been trained in sequential programming, it is expected that expression of parallelism and locality will be simplified in future programming models that *break sequential habits of thought* [38]. In this section, we briefly summarize some of the key points made in [38] and related work.

First, a major focus in writing efficient sequential code is to *minimize the total number of operations*. In contrast, efficient parallel code needs to focus on maximizing parallelism *i.e., minimizing the number of operations on the critical path*. With modern memory hierarchies, both sequential and parallel code must also focus on *improved locality*, but parallel code offers more opportunities than sequential code to (say) perform redundant operations to reduce communication. As mentioned earlier, locality optimization will have a first-order impact on energy reduction for future Extreme Scale systems. Second, good sequential algorithms attempt to minimize space usage and often include clever tricks to reuse storage; however, parallel algorithms need to use extra space to permit temporal decoupling and to achieve larger scales of parallelism. Finally, sequential idioms often stress linear problem decomposition through sequential iteration and linear induction. On the other hand, good parallel code usually requires multi-way problem decomposition and multi-way aggregation of results. A simple example is the difference between specifying a summation as a sequential iteration vs. a Fortran 90 SUM intrinsic for arrays.

A fundamental issue in the portable expression of parallelism is the need for data structures that lend themselves naturally to *data-parallel* operations. Fortunately, the *array* or *vector* data structure, which is a cornerstone of traditional HPC applications, is very well suited to data parallelism as evidenced by programming languages such as APL [17], Fortran 90 [29], NESL [2]and Ct [11]. These languages are able to express both flat data parallelism on vectors (element-wise operations, reductions, constrained permutations) and nested data parallelism on sparse or indexed vectors. *Streams* represent another data structure that is well suited for parallelism, as exemplified in data flow languages and programming models such as Sisal [27], Synchronous Data Flow [24], Brook [4] and StreamIt [12]. However, graph and other pointer-based data structures necessary for new Extreme Scale applications pose additional challenges for expression of parallelism and locality. The notion of *abstract collections* in modern object-oriented languages can help bring some of the benefits of data parallelism from arrays and streams to pointer-based data structures. In addition, *asynchronous dynamic parallelism*, as embodied in languages such as Cilk [3], Chapel [7], Fortress [1], and X10 [6], is necessary for operating on irregular data structures. Compared to current approaches, a key challenge for Extreme Scale is the ability to express this parallelism at the finest granularity possible, while delegating to the implementation the choice of what parallelism to exploit in a locality-sensitive manner.

3.2. Portable Expression of Synchronization with Dynamic Parallelism

Writing programs using today's state-of-the-art synchronization primitives is akin to using assembly language for programming. All the burden of performance, scalability and correctness falls squarely on the shoulders of the programmer with minimal support from the programming languages, development tools, runtime systems or hardware. In order to make parallel programs robust, portable, and scalable and reduce the burden on the programmers, many innovations are required in synchronization and communication. As an example, the *phasers* construct [34, 35] extends X10's clocks [6] so as to integrate collective and point-to-point synchronization with fine-grained dynamic parallelism, while providing a *next* statement that guarantees that all synchronizations will be performed in a deadlock-free manner. Each fine-grained task has the option of registering with a phaser in *signal-only*/*wait-only* mode for producer/consumer synchronization or *signal-wait* mode for barrier synchronization. Support for dynamic parallelism dictates that it should be possible for new tasks to be dynamically added and dropped from phaser registrations, which creates a potential challenge to avoid race conditions between synchronization operations and registration add/drop requests. The fine-grain synchronization that accompanies fine-grain parallelism also presents the challenge of *phaser contraction* [36] to reduce the synchronization overhead when reducing the actual parallelism that is exploited on a given system.

One of the biggest challenges with synchronization for a programmer is the difficulty in avoiding deadlock and data races, both of which can appear non-deterministically in current programming models. Of the two, data race avoidance is more challenging than deadlock avoidance, since deadlock freedom can be enforced by well-defined programming practices and the use of deadlock-free programming constructs such as phasers. Removing non-determinism from the programming model (as in declarative and functional programming approaches) can greatly simplify the testing and debugging of parallel programs, but the key challenge there is to ensure that the resulting model is sufficiently expressive for Extreme Scale software while still being efficient enough for execution on Extreme Scale hardware.

3.3. Expressing Heterogeneity in a Portable Manner

With the advent and increasing popularity of hybrid architectures, programming systems face the challenge of how to efficiently exploit multiple levels of parallelism, often coupled with different memory systems, instruction sets, or even numerics. In current-day systems, such as the LANL Roadrunner system [20], there may be as many as three distinct types of processors with distinct memory, messaging, and performance characteristics. These elements of heterogeneity are managed explicitly by the application programmers — through the use of coroutine-style models (one for each type of heterogeneity), explicit message passing for data movement, and distinct address spaces. Other examples of heterogeneous systems might include instruction set and performance heterogeneity or simply differences in memory structure such as cache coherent shared-memory, partitioned global address space, or shared-nothing. Unfortunately, if such characteristics of hardware heterogeneity are explicitly addressed by the programmer, not only is the programming effort increased, it is likely that the software will not be functionally portable, much less performance portable to other systems.

Extreme Scale systems of the future may have both designed heterogeneity (in dimensions such as architecture, organization, instruction set that are exhibited today in hybrid systems), as well as heterogeneity that arises from manufacturing variability, configuration, or aging differences. It is critical the software built for such large-scale parallel systems address the heterogeneity of the system in a fashion that supports portability of the applications. That is, it should be possible to move applications from one machine to another — with different heterogeneous characteristics — without significant change at the application source code level. This imposes major challenges in expression of parallelism, locality, and computation so as to both enable the compiler and runtime to deliver performance on one Exascale system, but also in a form portable and flexible enough that it can enable the compiler and runtime to deliver performance on other heterogeneous Exascale systems. This is a daunting challenge, but is in our view a fundamental requirement for a technology landscape that supports Exascale computing.

4. Challenges in Managing Parallelism and Locality

Earlier in this paper, we summarized the hardware characteristics of future Extreme Scale systems as well as the challenges involved in developing applications (Section 2) and expressing parallelism and locality (Section 3) for such systems. In this section, we focus on the challenges and implications in *software management of parallelism and locality* for Extreme Scale. Current software for high-end data-center, departmental and embedded systems build on a *classical software stack* which primarily consists of operating systems, parallel runtimes, static compilers, and libraries. However, as described in the following sections, the general structure of the classical software stack has remained largely unchanged for decades, and will be highly mismatched to the requirements of all three classes of future Extreme Scale systems.

4.1. Operating System Challenges

Extreme Scale processors containing hundreds or even thousands of cores will challenge current operating system (OS) practices. Many of the fundamental assumptions that underlie current OS technology are based on design assumptions that are no longer valid for a Extreme Scale processor containing thousands of cores. In the context of Exascale system requirements, as machines grow in scale and complexity, techniques to make the most effective use of network, memory, processor, and energy resources are becoming increasingly important. A baseline challenge for the Exascale software stack is how to get the OS out of the way without compromising the need to protect hardware state from errant (or malicious) software.

Execution models that support more asynchrony will be necessary to hide *latency*. Such execution models will also require more carefully coordinated scheduling to balance resource utilization and minimize work *starvation* or resource *contention*. These execution models will also require extraordinarily *low-overhead*, fine-grained messaging. However, the attributes required by the execution model are nearly impossible to achieve when the OS intervenes for every operation that touches its privileged domain – it must intervene for inter-processor communication operations, has exclusive and privileged control of scheduling policy, and exclusive ownership of resource management policies. Over time, operating systems have evolved into multifaceted and hugely complex software implementations that have accreted a broad range of capabilities. We refer to the challenge of breaking the OS apart based on separation of concerns as "deconstructing the OS".

In its role as the gate-keeper to shared resources, operating systems have traditionally been a major bottleneck in achieving scalability on SMP's. This is especially true for the open source Linux operating system, which has historically lagged behind commercial Unix OS's such as AIX and Solaris in scalability but has now become the dominant OS of choice for high-end systems. Significant attention has been devoted by the Linux community over multiple years to bridge the scalability gap with commercial OS's, starting with efforts such as improvements to the Linux scheduler in 2001 [22]. More recent examples of scalability efforts explored and undertaken by the Linux community include large-page support, NUMA support [26], and the Read-Copy Update (RCU) API [28]. While these Linux enhancements have resulted in improvements for commercial workloads with independent requests and flow-level parallelism [40] on small-scale SMP's, the scalability requirements for even a single socket of an Extreme Scale system will be two orders of magnitude higher than what can be supported by Linux today. It is clear that this gap cannot be bridged by business-as-usual efforts; in fact, future scalability improvements in Linux are expected to be harder rather than easier to achieve, as evidenced by the RCU experience [28] and the complexities uncovered by ongoing efforts to reduce the scope of the Linux Big Kernel Lock (BKL) *e.g.,* see [19].

4.2. Runtime Challenges

Runtime support for parallel programming requires key innovations in lightweight mechanisms for communication and memory hierarchy management, and user-controllable policies for managing the system resources. Expected contributions to this area of research include:

- Lightweight runtime mechanisms to exploit the novel features of interconnection networks, including topology queries, atomic operations, remote procedure invocation, fast one-sided transfer notification used in synchronization.
- Extensions of the execution models to handle fast and slow memory associated with a single thread, and demonstration of that model on a single-chip system with software-managed local memory that replaces or augments the traditional hardware-managed cache hierarchy.
- Runtime support to virtualize the set of processors through the use of multi-threading and dynamic task migration. Programming model extensions that allow for such virtualization when needed, without enforcing it for all applications.
- Runtime support for memory system virtualization, including object caching and migration. As with processor resources, the programming model will be extended to permit runtime-managed data placement in addition to the user-managed placement already available.
- Support for multiple runtime systems for different execution models and soft real-time applications.

4.3. Compiler Challenges

The crucial role of compilers at the extreme scale is to map from language constructs that express a very high-level decomposition of an application to highly power-efficient and memory-efficient architecture-specific code and runtime layer calls. As has been proven historically, completely automatic compiler optimization from high level code will not meet the performance requirements at Extreme Scale, and in this regime, we will also encounter memory and power constraints that programmers and tools could previously ignore. Further, compiler-based approaches, such as, for example, compilers for PGAS languages, have generally focused on regular, static parallelism. As we expand the applications for Exascale platforms to encompass irregular, unstructured and dynamic algorithms, so must the compiler technology support these challenging application domains. It is important to note that the compiler requirements for

Exascale are similar to those for compilers at all scales. An article reporting on a recent NSF-sponsored workshop on the future of compiler research listed the following 6 research challenges, in addition to other guidelines on enhancing the research approach and enriching education [15].

Research challenges in optimization:

- Make parallel programming mainstream;
- Write compilers capable of self improvement (i.e., auto-tuning); and
- Develop performance models to support optimizations for parallel code.

Research challenges in correctness:

- Enable development of software as reliable as an airplane;
- Enable system software that is secure at all levels; and
- Verify the entire software stack.

Therefore, the compilers we must develop for Exascale as discussed further in this section will necessarily be very different. Compilers at the extreme scale must collaborate closely with the application programmer to derive an architecture-independent algorithm description that can be mapped to high-quality code; further, the compiler must incorporate lightweight mechanisms that interface with the runtime layer and architecture to dynamically map this code for a specific execution context to be both high performing and power efficient.

5. Software-Hardware Co-Design

We believe that software-hardware co-design will be a critical necessity for Extreme Scale systems, in addition to the interfaces outlined in the previous section. This form of co-design has been essential for *vector parallelism* [18] in current and past systems, and is also being explored for scalable approaches to mutual exclusion using *transactional memory* [23]. In this section, we discuss a few additional examples of software runtime capabilities that will be necessary for future Extreme Scale systems, and examine how they can be made more effective with software-hardware co-design.

5.1. Scheduling dynamic parallelism with fine-grained tasks

As discussed in Section 3, it is important to ensure that the intrinsic parallelism in a program can be expressed at the finest level possible *e.g.*, at the statement or expression level, and that the compiler and runtime system can then exploit the subset of parallelism that is useful for a given target machine. There have been multiple proposals for expressing fine-grained parallelism *e.g.*, statement-level *spawn* [3] or *async* [6] operations, expression-level *future* [16] operations, and operator-level data flow graphs [8, 37]. These operations for fine-grained parallelism are in stark contrast with the *bulk-synchronous parallel model* [39]. While profile-directed compile-time partitioning can be used to optimize the granularity of fine-grained tasks in certain cases [32, 33], in general the runtime system also needs to participate in the partitioning so as to best adapt to unpredictable execution times. A classic approach to runtime partitioning is *lazy task creation* [30], which has been extended into *work-stealing runtimes* for fine-grained tasks [10, 13]. A work-stealing runtime system creates a fixed number of worker threads, with one local double-ended queue (deque) per worker. Each worker repeatedly picks up work from a deque of lightweight tasks using scheduling policies that are designed to achieve good load balance while bounding the size of the deques. This approach has been shown to yield scalability that is orders-of-magnitude superior to the scalability achieved if each task were to be created as a thread at the OS level.

However, there are still significant overheads that remain in a software-only approach, that will likely prevent it from being usable at Extreme Scale. These overheads involve locking operations, and in the case of nonblocking algorithms involve spin loops on shared-memory locations with their accompanying cache consistency overheads. As mentioned earlier, these overheads are especially important because they occur on critical paths in parallel programs. *Hardware support* for shared queue data structures can result in orders-of-magnitude reductions in scheduling overheads and scalability bottlenecks, while still retaining the flexibility of task scheduling policies in software. Section 5.4 identifies other uses for hardware support for shared queues in Extreme Scale systems.

Another source of overhead in task scheduling lies in the operations that need to be performed on the fast path to save local variables, so as to ensure that the task can be resumed on a separate worker from the one that it started on (if needed). A software-only approach introduces word-at-a-time store instructions to save the local variables, and some of these stores are often redundant. In contrast, hardware support for saving and restoring local variables (as in calling conventions) can help reduce this overhead that occurs on the fast path.

5.2. Distribution and co-location of tasks and data

Another candidate for software-hardware co-design pertains to distribution and co-location of tasks and data, which is one mechanism that can be used in support of locality optimization. As observed throughout this report, it will be critical to optimize vertical locality so as to satisfy the energy constraints of Extreme Scale systems. Runtime systems for programming languages such as UPC [9] and Co-Array Fortran [31] that are based on a *Partitioned Global Address Space* (PGAS) model include the notion of virtual *home location* for each shared datum. The more recent HPCS languages extend this notion of home locations to computational tasks, as in Chapel's *locales* [7] and X10's *places* [6], so as to enable tasks to be shipped to data, data to be shipped to tasks, or any meet-in-the-middle combination thereof. The translation from global to local addresses is a major source of overhead in a software-only approach to implementing such languages, along with the communications that accompany non-local accesses. Thus, it becomes important for a compiler for such languages to perform redundancy elimination on address computations, to coalesce contiguous accesses into a single communication operation, and to overlap communication with computation [41]. Opportunities for software-hardware co-design include the use of translation buffers to accelerate virtual-to-physical address translations, and DMA-like hardware support to reduce the processor overhead of data transfers.

5.3. Collective and point-to-point synchronization with dynamic parallelism

As discussed in Section 3.2, the fine-grained parallelism intrinsic to a program may need to be accompanied by fine-grained collective and point-to-point synchronization among dynamically varying sets of fine-grained tasks. These synchronization structures may be irregular, and tasks are permitted to dynamically join or leave these structures as in the *phasers* construct [34]. Further, it is usually desirable to augment the synchronization structures with communication for reductions [35], collectives, and systolic computations. As mentioned in Section 5.1, synchronization structures represent good candidates for software-hardware co-design since a software-only approaches for synchronization suffer from unnecessary cache consistency and serialization bottlenecks. Hardware support (*e.g.*, in the form of counting semaphores) can be used to reduce the overhead of inter-core synchronization, and extensions in the form of register-level inter-core communication (*e.g.*, as in the Raw project [25]) can reduce the overhead of communication. Further, the use of a single master task to perform a reduction in software can be a scalability bottleneck, and a software-only approach to creating combining trees incurs high setup and tear-down overhead. Instead, hardware support for combining synchronization and reductions will greatly reduce the overhead of collective and point-to-point synchronization with dynamic parallelism.

5.4. *Producer-consumer parallelism*

Another common idiom in fine-grained parallel programs is that of producer-consumer parallelism. In this model, a single-writer task serves as the producer of a datum for multiple readers. To accomplish this, the writer task typically stores its result in a designated location, and the reader tasks block when they request the result (if the result is not ready). In the case of *futures* [16], the execution of the writer task may (optionally) be deferred till the datum's value is requested by one of the readers. Once again, we observe that a software-only approach suffers cache consistency and serialization bottlenecks, and hardware support can be used to reduce these bottlenecks. A classical example of hardware support in this area is the *full-empty bit*, but there may be many other variations. Also, in many cases, the location on which the producers and consumers wish to synchronize may be designated by a tag rather than an address. Hardware support to accelerate the translation of tags to addresses can be very useful. Intel's Concurrent Collections (CnC) [5] is an example of a high-level programming model that relies heavily on producer-consumer parallelism and that would benefit greatly from any hardware support.

6. Summary

There are several reasons for paying attention to software in the development of Extreme Scale systems. First, the Exascale systems that are projected for the 2015 – 2020 timeframe are dramatically different from today's Petascale systems and will require correspondingly fundamental changes in the execution model and structure of system software (both of which have remained relatively stagnant during the last two decades). Second, while there has been significant innovation at the hardware and system level for today's Petascale systems, previous approaches have not paid much attention to the co-design of multiple levels in the system software stack (OS, runtime, compiler, libraries, application frameworks) that is needed for Exascale systems. Third, while certain execution models such as Map-Reduce in cloud computing and CUDA in GPGPU data parallelism have demonstrated large degrees of concurrency, they haven't demonstrated the ability to deliver 1000× increase in parallelism to a single job with the energy efficiency and strong scaling fraction necessary for Extreme Scale systems.

To better understand the software challenges for Extreme Scale systems, we studied the challenges and implications in developing applications for Extreme Scale computing by examining multiple application classes (Section 2). From an application viewpoint, the concurrency and energy challenges boil down to the ability to express and manage parallelism and locality in the applications. This section concluded that applications can be enabled for exploiting extreme scale hardware by exploring a range of *strong scaling* and *new-era weak scaling* techniques, but only with suitable attention to efficient parallelism and locality.

Given this context, Section 3 summarized the challenges in *expressing* parallelism and locality in Extreme Scale software. One of them is the ability to expose all of the intrinsic parallelism and locality in an application, so as to make the application *forward scalable.* Another is to ensure that this expression of parallelism and locality is *portable* across vertical and horizontal dimensions. The challenges in *managing* parallelism and locality were discussed next in Section 4. OS-related challenges include *parallel scalability, spatial partitioning* of OS and application functionality, *direct hardware access* for inter-processor communication, and *asynchronous* rather than interrupt-driven events. There are additional challenges in *runtime systems* for scheduling, memory management, communication, performance monitoring, power management, and resiliency, all of which will be built atop future Extreme Scale operating systems. The section also outlined challenges in *compilers* for Extreme Scale systems. Finally, Section 5 identified opportunities for addressing the concurrency and energy challenges through *software-hardware co-design.*

Acknowledgments

This paper was derived in large part from an Exascale Software study conducted over a series of seven meetings held from June 2008 to February 2009. The goal of the study was to examine the state of the art, identify key challenges, and outline elements of a technical approach that can address the challenges without prescribing specific solutions. We would like to thank the study members and guests for their dedication to the field of parallel software and systems, and for all their hard work in contributing to the study. A detailed report on the study will be forthcoming in the near future.

The applications discussion in this paper also benefited from numerous meetings held in 2007 to investigate future Exascale computing applications and hardware/software requirements. These included three DOE Exascale Townhall Meetings held at Lawrence Berkeley National Laboratory, Oak Ridge National Laboratory and Argonne National Laboratory, the Council on Competitiveness meeting on Exascale Applications, and the Frontiers of Extreme Computing 2007/Zettaflops workshop.

This work has been supported in part by the Defense Advanced Research Projects Agency (DARPA) under AFRL contract FA8650-07-C-7724. The views, opinions, and/or findings contained in this presentation are those of the author(s) and should not be interpreted as representing the official policies, either expressed or implied, of the Defense Advanced Research Projects Agency or the Department of Defense.

References

[1] Eric Allen et al. Project Fortress: A multicore language for multicore processors. *Linux Magazine*, pages 38–43, 2007.

[2] Guy E. Blelloch and John Greiner. A provable time and space efficient implementation of NESL. In *ICFP '96: Proceedings of the first ACM SIGPLAN International Conference on Functional Programming*, pages 213–225, New York, NY, USA, 1996. ACM.

[3] R. D. Blumofe et al. CILK: An efficient multithreaded runtime system. *Proceedings of Symposium on Principles and Practice of Parallel Programming (PPoPP'95)*, pages 207–216, July 1995.

[4] Ian Buck et al. Brook for GPUs: stream computing on graphics hardware. In *SIGGRAPH '04: ACM SIGGRAPH 2004 Papers*, pages 777–786, New York, NY, USA, 2004. ACM.

[5] Zoran Budimlić et al. Multi-core Implementations of the Concurrent Collections Programming Model. In *CPC '09: 14th International Workshop on Compilers for Parallel Computers*. Springer, January 2009.

[6] Philippe Charles et al. X10: an object-oriented approach to non-uniform cluster computing. In *Proceedings of the ACM SIGPLAN conference on Object oriented programming, systems, languages, and applications*, pages 519–538, New York, NY, USA, 2005.

[7] Cray Inc. The Chapel language specification version 0.4. Technical report, Cray Inc., February 2005.

[8] Jack B. Dennis. Data Flow Supercomputers. *IEEE Computer*, 13(11):48–56, November 1980.

[9] Tarek El-Ghazawi et al. UPC Language Specification v1.1.1, October 2003.

[10] Matteo Frigo et al. The implementation of the Cilk-5 multithreaded language. In *PLDI '98: Proceedings of the ACM SIGPLAN 1998 conference on Programming language design and implementation*, pages 212–223, New York, NY, USA, 1998. ACM.

[11] Anwar Ghuloum. Ct: channelling NeSL and SISAL in C++. In *CUFP '07: Proceedings of the 4th ACM SIGPLAN workshop on Commercial users of functional programming*, pages 1–3, New York, NY, USA, 2007. ACM.

[12] Michael I. Gordon et al. A stream compiler for communication-exposed architectures. In *ASPLOS-X: Proceedings of the 10th international conference on Architectural support for programming languages and operating systems*, pages 291–303, New York, NY, USA, 2002. ACM.

[13] Yi Guo et al. Work-First and Help-First Scheduling Policies for Async-Finish Task Parallelism. In *Proceedings of the International Parallel and Distributed Processing Symposium (IPDPS)*, May 2009.

[14] Francois Gygi et al. Large-scale electronic structure calculations of high-z metals on the bluegene/l platform. In *SC '06: Proceedings of the 2006 ACM/IEEE conference on Supercomputing*, page 45, New York, NY, USA, 2006. ACM.

[15] Mary Hall et al. Compiler research: the next 50 years. *Commun. ACM*, 52(2):60–67, 2009.

[16] Robert Halstead, Jr. Multilisp: A Language for Concurrent Symbolic Computation. *ACM Transactions of Programming Languages and Systems*, 7(4):501–538, October 1985.

[17] Kenneth Iverson. *A Programming Language*. John Wiley and Sons, New York, NY, 1962.

[18] K. Kennedy and J. R. Allen. *Optimizing compilers for modern architectures: a dependence-based approach*. Morgan Kaufmann Publishers Inc., San Francisco, CA, USA, 2002.

[19] Removing the Big Kernel Lock. `http://kerneltrap.org/Linux/Removing_the_Big_Kernel_Lock`, May 2008. Viewed on September 27, 2008.

[20] Michael Kistler et al. Petascale computing with accelerators. In *PPoPP '09: Proceedings of the 14th ACM SIGPLAN symposium on Principles and practice of parallel programming*, pages 241–250, New York, NY, USA, 2009. ACM.

[21] Peter Kogge et al. ExaScale Computing Study: Technology Challenges in Achieving Exascale Systems. `http://users.ece.gatech.edu/~mrichard/ExascaleComputingStudyReports/exascale_final_report_100208.pdf`, 2008.

[22] Mike Kravetz and Hubertus Franke. Implementation of a multi-queue scheduler for linux. `http://lse.sourceforge.net/scheduling/mq1.html`, 2001.

[23] James R. Larus and Ravi Rajwar. *Transactional Memory*. Morgan & Claypool, 2006.

[24] Edward A. Lee and David G. Messerschmitt. Synchronous Data Flow. *Proceedings of the IEEE*, 75(9):1235–1245, 1987.

[25] Walter Lee et al. Space-Time Scheduling of Instruction-Level Parallelism on a Raw Machine. *Proceedings of the Eighth International Conference on Architectural Support for Programming Languages and Operating Systems (ASPLOS-VIII)*, October 1998.

[26] Tong Li et al. Efficient operating system scheduling for performance-asymmetric multi-core architectures. In *SC '07: Proceedings of the 2007 ACM/IEEE conference on Supercomputing*, pages 1–11, New York, NY, USA, 2007. ACM.

[27] J. McGraw et al. SISAL: Streams and Iteration in a Single Assignment Language Reference Manual Version 1.2. Technical report, Lawrence Livermore National Laboratory, 1985. No. M-146, Rev. 1.

[28] Paul E. McKenney and Jonathan Walpole. Introducing technology into the Linux kernel: a case study. *SIGOPS Oper. Syst. Rev.*, 42(5):4–17, 2008.

[29] M. Metcalf and J. Reid. *Fortran 90 Explained*. Oxford Science Publications, Oxford, England, 1990.

[30] E. Mohr et al. Lazy Task Creation: A Technique for Increasing the Granularity of Parallel Programs. *IEEE Transactions on Parallel and Distributed Systems*, 2(3):264–280, July 1991.

[31] Robert W. Numrich and John Reid. Co-Array Fortran for parallel programming. *ACM SIGPLAN Fortran Forum Archive*, 17:1–31, August 1998.

[32] Vivek Sarkar. *Partitioning and Scheduling Parallel Programs for Multiprocessors*. Pitman, London and The MIT Press, Cambridge, Massachusetts, 1989. In the series, Research Monographs in Parallel and Distributed Computing.

[33] Vivek Sarkar. Automatic Partitioning of a Program Dependence Graph into Parallel Tasks. *IBM Journal of Research and Development*, 35(5/6), 1991.

[34] Jun Shirako et al. Phasers: a unified deadlock-free construct for collective and point-to-point synchronization. In *ICS '08: Proceedings of the 22nd annual international conference on Supercomputing*, pages 277–288, New York, NY, USA, 2008. ACM.

[35] Jun Shirako et al. Phaser accumulators: a new reduction construct for dynamic parallelism. In *Proceedings of the International Parallel and Distributed Processing Symposium (IPDPS)*, May 2009.

[36] Jun Shirako et al. Chunking parallel loops in the presence of synchronization. In *ICS '09: Proceedings of the 23rd International Conference on Supercomputing*. ACM, June 2009.

[37] S. Skedzielewski and J. Glauert. IF1 – An Intermediate Form for Applicative Languages. Technical report, Lawrence Livermore National Laboratory, 1985. No. M-170.

[38] Guy Steele. The Future Is Parallel: What's a Programmer to Do? Breaking Sequential Habits of Thought. One-hour talk presented at the 5 March 2009 meeting of the New England Programming Languages and Systems (NEPLS) Symposium at Mitre. `http://research.sun.com/projects/plrg/Publications/NEPLSMarch2009Steele.pdf`, 2009.

[39] Leslie G. Valiant. A bridging model for parallel computation. *Commun. ACM*, 33(8):103–111, 1990.

[40] Bryan Veal and Annie Foong. Performance scalability of a multi-core web server. In *ANCS '07: Proceedings of the 3rd ACM/IEEE Symposium on Architecture for networking and communications systems*, pages 57–66, New York, NY, USA, 2007. ACM.

[41] Katherine Yelick et al. Productivity and performance using partitioned global address space languages. In *PASCO '07: Proceedings of the 2007 international workshop on Parallel symbolic computation*, pages 24–32, New York, NY, USA, 2007. ACM.

Journal of Physics: Conference Series **180** (2009) 012046 doi:10.1088/1742-6596/180/1/012046

Extreme-scale scripting: Opportunities for large task-parallel applications on petascale computers

Michael Wilde[1,2], Ioan Raicu[3], Allan Espinosa[3], Zhao Zhang[1], Ben Clifford[1], Mihael Hategan[1], Sarah Kenny[1], Kamil Iskra[1,2], Pete Beckman[1,2], Ian Foster[1,2,3]

[1]Computation Institute, University of Chicago and Argonne National Laboratory; [2]Mathematics and Computer Science Division, Argonne National Laboratory; [3]Department of Computer Science, University of Chicago.

Abstract. Parallel scripting is a loosely-coupled programming model in which applications are composed of highly parallel scripts of program invocations that process and exchange data via files. We characterize here the applications that can benefit from parallel scripting on petascale-class machines, describe the mechanisms that make this feasible on such systems, and present results achieved with parallel scripts on currently available petascale computers.

1. The parallel scripting paradigm

John Ousterhout describes scripting as *higher-level programming for the 21st Century*[3]. Scripting has revolutionized application development on the desktop and server, accelerating and simplifying programming by focusing on the composition of programs to form more powerful applications. Understanding how to scale scripting to 21st century computers should be among the priorities for researchers of next generation parallel programming models. Might scripting not provide the same benefits for extreme-scale computers?

We believe that the answer to this question is *yes*. We introduce the concept and implementation of *parallel scripting*, and describe what becomes possible when simple scripts turn "ordinary" scientific programs into petascale applications running on 100,000 cores and beyond. Scripting languages allow users to assemble sophisticated application logic quickly by composing existing codes. In parallel scripting, users apply parallel composition constructs to existing sequential or parallel programs. Using this approach, they can quickly develop highly parallel applications that can be run efficiently on a 16-core workstation, a 16,000-core cluster, or a 160K-core petascale system.

Parallel scripting is not a substitute for existing tightly coupled programming models such as MPI. Rather, it is an alternative (and higher-level) path to massive parallelism, a path particularly suitable for increasingly feasible and important problem-solving methods such as the use of parameters sweeps and ensemble studies for exploring sensitivity to parametric, structural, and initial condition uncertainty. The availability of extreme-scale computers makes such methods feasible and attractive, even in the case of complex computations. Parallel scripting allows users to apply these methods while leveraging the vast value embodied in modern application codes—both serial and parallel—that empower the scientific, engineering, and commercial computing of today and the foreseeable future.

We have been exploring such "many task" computing models[6] for several years, from the perspective of both technologies and applications. On the technology front, we have explored, in particular, a dataflow-driven parallel programming model that treats application programs as functions, and their datasets as structured objects mapped to a simple abstract data model. We have incorporated this model in a parallel scripting language, Swift, and implemented that language on large parallel computers, including a 160K-core Blue Gene/P and a 62K-core Sun Constellation. Swift programs may define hundreds of thousands (and soon millions) of tasks that read and write an even greater number of files. We have developed task and data management methods that can scale to extremely high dispatch rates and data volumes, and used them to scale applications to up to 160K cores, with high efficiency and fault tolerance.

2. Software architecture for petascale parallel scripting

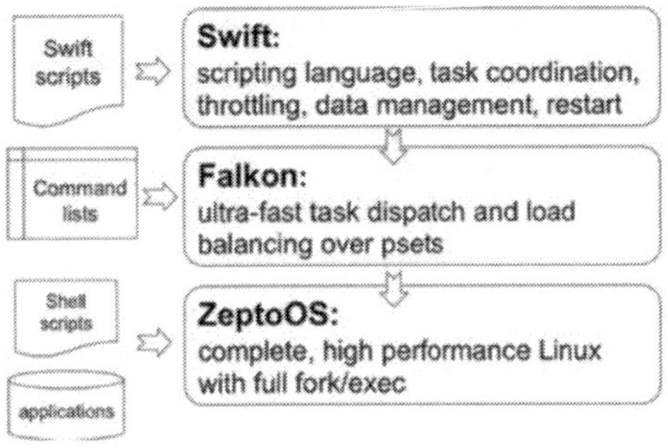

Figure 1: Architecture for petascale scripting

Our approach to high-performance parallel scripting requires three layers, which from the bottom up are: (1) a POSIX environment in which to execute the individual application tasks of a script; (2) a means to allocate compute node resources, hold them for long and varying periods of time (typically hours rather than seconds) while rapidly scheduling small independent tasks on the compute nodes – even tasks of very short duration (down to fractions of a second, but typically many seconds or minutes); and (3) a language in which to abstractly express scripts of highly parallel application invocations, their data accesses, and the data interchange between them.

Modern concepts of scripting depend on POSIX system services such as fork() and exec() in order to execute application programs from the script, and this require an operating system that supports these or similar capabilities – notably, the ability to launch a new application program and wait for it to complete. On the BG/P, the native IBM compute node kernel lacks these features, and we provide them instead through the ZeptoOS compute node kernel[4], which implements these features in a POSIX-compliant manner. On the Ranger Constellation system, we use the native compute node operating system, which provides complete POSIX support.

3. The Falkon resource provisioner and lightweight scheduler

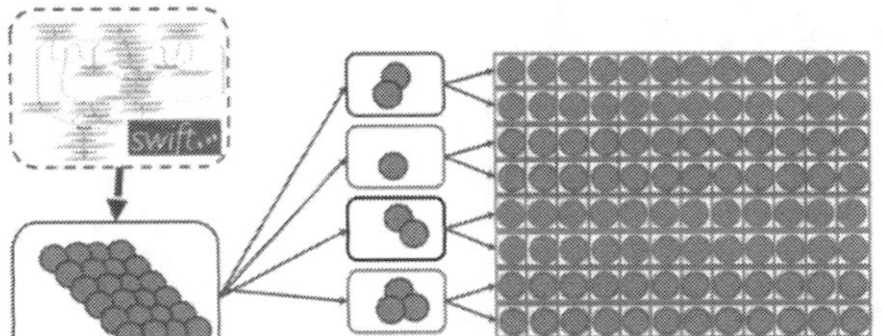

Figure 2: Falkon Provisioning and Scheduling System

The compute node resources of petascale computing systems are typically managed by traditional batch schedulers, which are designed and configured with policies for running large parallel jobs that execute the same application program on all compute nodes allocated to the job, and which run for extended periods of time. Parallel scripting, however, requires that many application programs, each with an independent set of arguments and different sets of input and output files, and having likely short and often widely varying execution times, be executable on any compute node. This far more dynamic scheduling model demands a multi-level scheduling approach, which we have implemented in a component called *Falkon*, a Fast and Light-weight tasK executiON framework[4].

Falkon allocates nodes of a compute resource in large quantities, using the native batch scheduler of the system, and runs a persistent task execution agent on each compute core that rapidly executes arbitrary and independent POSIX processes on the allocated nodes. Falkon consists of several components, shown in Figure 2: (1) a compute node agent that executes one task at a time on a compute node core; (2) a service that maintains a queue of jobs for a set of compute node resources, and which rapidly selects the next job to run on a FIFO basis; and (3) a load-balancing client that evenly distributes work to the services

Using Falkon, we have been able to meet performance requirements necessary for petascale scripting. Measurements of Falkon performance[4] indicate it can:

- execute over 3,000 tasks per second on the BG/P;
- launch, execute, and terminate 160K tasks on the BG/P at 160K-core scale in under one minute;
- execute workloads of 913K science tasks on 116K BG/P cores in 2 hours, totalling 21.4 CPU years at 99.7% efficiency and 99.6% utilization (Figure 3); and
- execute one billion trivial tasks in 18 hours in multicore stress tests.

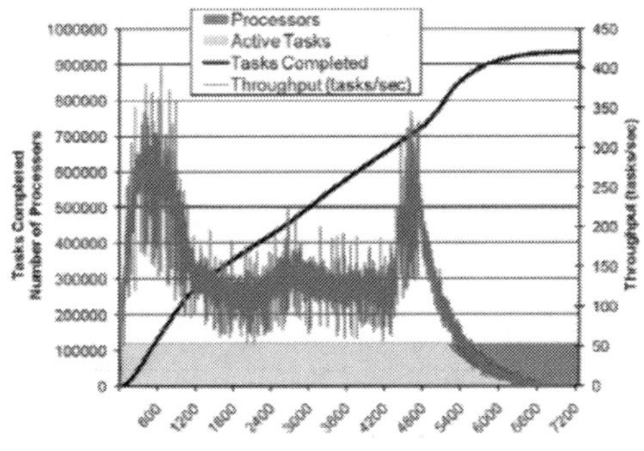

Figure 3: Falkon performance (DOCK application)

Users utilize Falkon by creating simple scripts that contain a list of tasks to execute, with arguments. Tasks are executed with FIFO scheduling and with maximum parallelism. Since tasks are sent to available agents in the order that they appear in the Falkon input script, users can sort tasks "longest first" when task duration can be estimated, thus achieving optimal utilization of a block of compute nodes. Falkon functionality has been packaged as a Swift "execution provider" to enable higher level programming (described in the next section). It has also been implemented as a newer Swift execution provider, which has the additional capability of allocating multiple "time-space" blocks of varying number of cores and wall-time duration, to best fit the current task demand.

4. The Swift parallel scripting language

Swift is a scripting language that makes it easy for a user to specify and run application programs in parallel, and to specify the data passing and dependencies between different application invocations, as well as the structure of the data (typically in terms of files and directories). It provides run time support that automates data management and enables the same Swift script to run on multi-core workstations, local clusters, remote grids, cloud resources, and petascale supercomputers. It is common to write Swift applications in three layers: a top layer which specifies the workflow, a middle layer of interface code to adapt specific applications, and the lowest layer, which defines the interfaces of the applications themselves. Studies on early versions of Swift have shown that the amount of code needed to express applications in this form is substantially lower than by *ad hoc* scripting in "shell scripts" [9]

Swift also provides a provenance recording mechanism that enables users to log how each data item was produced, query that knowledge base to locate data and methods, and retrieve the history of an object for validation, sharing, or reproduction of computational results.

```
foreach prot in protein {
  foreach sT in startT {
    foreach tUp in tUpdate {
      ItFix(prot, nsim, maxrounds, sT, tUp);
    }
  }
}
```

To show the power of the Swift language, here is a fragment of Swift code to perform protein-folding simulations using an application called "ItFix"[1]. This simple code fragment, given ten protein sequences, nsim=1000, two starting temperatures and five update intervals, will, in each of three (maxrounds) rounds of prediction, execute 10 x 1000 x 2 x 5 = 100,000 simulations. ItFix scripts been executed on up to 32,000 cores on the Argonne BG/P. Similar code can sweep across any combination of ItFix parameters. This abstract script runs without change on multiple TeraGrid clusters including the 62K-core supercomputer "Ranger".

The Swift data model provides the ability to describe nested on-disk directories as simple structures and arrays. These datasets are transparently sent to remote and parallel Swift procedures on various platforms. An operation called "mapping" translates between the simple abstract data model of Swift and the potentially messy, complex model of real-world directory structures and file naming and structuring conventions.

Swift has a C-like syntax, but many of the semantic aspects of a "functional" programming language. Procedures are expressed as functions, which are permitted to return multiple values; statements are executed in data-dependency order; variables (including array elements and structure members) are single-assignment, which makes it significantly simpler for Swift to automatically execute independent operations in parallel. Application programs are abstracted as functions, whose arguments and results are files and file-structured datasets. Layering on the ability to represent application programs as procedures, the user can define compound procedures to create libraries of higher-level processes that capture the essential protocols of an application domain's data preparation and analysis procedures.

Users interact with Swift at many levels. Scientific programmers create Swift libraries that encapsulate the execution of scientific applications, data preparation, and analysis methods. These libraries provide a stable base of functionality specific to a user community. Higher-level users write simple scripts using these libraries to perform large-scale computing tasks. The highest-level Swift users will not need to program at all: they will invoke their scripts and view their results through web interfaces.

5. Example applications

Applications to which we have applied large-scale parallel scripting [1][2][7][8] are listed in Table 1. Each is capable of consuming a large fraction, or even all, of a petascale computer. All involve doing many tasks at once, with often quite substantial amounts of communication both within each task and among tasks.

Table 1: Parallel Scripting Applications

Field	Description	Characteristics	Status
Astronomy	Creation of montages from many digital images	Many 1-core tasks, much communication, complex dependencies	E
Astronomy	Stacking of cutouts from digital sky surveys	Many 1 core tasks, much communication	E (Falkon)
Biochemistry	Analysis of mass spec data for post translational protein modifications	10,000 – 100,000 K jobs for proteomic searches using custom serial codes	D
Biochemistry	Protein folding using iterative fixing algorithm, also exploring other biomolecule interactions	100s to 1000s of 1-1000 core simulations & data analysis	O
Biochemistry	Identification of drug targets via computational screening	Up to 1M x 1 core	O (Falkon)
Bioinformatics	Metagenome modeling	1000's of 1 core integer programming problems	D
Business economics	Mining of large text corpora to study media bias	Analysis and comparison of 70M+ text files of news articles	D
Climate	Ensemble climate model runs and analysis of output data	10s to 100s of 100-1000 core simulations	E
Economics	Generation of response surfaces for various economic models	1K to 1M 1-core runs (10K typical), then data analysis	O
Neuroscience	Analysis of functional MRI datasets	Comparison of images; connectivity analysis with SEM, many tasks (100K+)	O
Radiology	Training of computer aided diagnosis algorithms	Comparison of images; many tasks, much communication	D
Radiology	Image processing and brain mapping for neurosurgical planning research	1000's of MPI application executions	O

Legend: E: evaluated, D: in development, O: in some degree of operational use

6. Conclusion

Our experience in applying Swift indicates that it enables the productive programming paradigm of knitting together existing application programs using a functional model to rapidly and productively compose new, higher level applications that can efficiently use the parallel resources of a petascale system. We are currently developing enhancements that apply "collective" data management techniques[11] to efficiently move data in and out of applications in a way that leverages the interconnect and filesystem hardware of petascale systems. We believe that, when complete, the parallel scripting model will play an indispensible role in the extreme-scale programming tool chest.

Acknowledgements.
This research is supported in part by NSF grants OCI-721939 and PHY-636265, NIH grants DC08638 and DA024304-02, U.S. Dept. of Energy under Contract DE-AC02-06CH11357, NASA Ames Research Center GSRP grant NNA06CB89H, and the UChicago/Argonne Computation Institute.

References

[1] Hocky, G., Wilde, M., Debartolo, J., Hategan, M., Foster, I., Sosnick, T.R., and Freed, K.F.). Towards petascale ab initio protein folding through parallel scripting. Argonne Technical Report ANL/MCS-P1612-0409.

[2] Kenny, S, M. Andric, M. Wilde, Michael C. M. Neale, S. Boker, SL. Small, Parallel Workflows for Data-Driven Structural Equation Modeling in Functional Neuroimaging, Argonne Technical Report ANL/MCS-P1613-0409.

[3] Ousterhout, J. Scripting: Higher Level Programming for the 21st Century IEEE Computer March 1998

[4] Raicu, R,. Zhao Zhang, Mike Wilde, Ian Foster, Pete Beckman, Kamil Iskra, Ben Clifford. "Toward Loosely Coupled Programming on Petascale Systems", IEEE/ACM Supercomputing 2008.

[5] Raicu. I. "Many-Task Computing: Bridging the Gap between High Throughput Computing and High Performance Computing", Doctorate Dissertation, Computer Science Department, University of Chicago, March 2009

[6] Raicu, I, Ian Foster, Yong Zhao. "Many-Task Computing for Grids and Supercomputers", Invited Paper, IEEE Workshop on Many-Task Computing on Grids and Supercomputers (MTAGS08), 2008, co-located with IEEE/ACM Supercomputing 2008.

[7] Small, S. L., Wilde, M., Kenny, S., Andric, M., & Hasson, U. (2009). Database-managed Grid-enabled analysis of neuroimaging data: The CNARI framework. *International Journal of Psychophysiology, in press, published online 2/20/2009*

[8] Stef-Praun, I. Foster, U. Hasson, M. Hategan, S.L. Small and M. Wilde, Accelerating medical research using the Swift Workflow System, *Paper Presented at the HealthGrid 2007, Geneva* (2007).

[9] Zhao, Y., Dobson, J., Foster, I., Moreau, L., Wilde, M., A Notation and System for Expressing and Executing Cleanly Typed Workflows on Messy Scientific Data, SIGMOD Record, September 2005.

[10] Zhao, Y., Hategan, M., Clifford, B., Foster, I., von Laszewski, G., Nefedova, V., Raicu, I., Stef-Praun, T., Wilde, M. "Swift: Fast, Reliable, Loosely Coupled Parallel Computation," *IEEE Workshop on Scientific Workflows 2007.*

[11] Zhang, Z., Allan Espinosa, Kamil Iskra, Ioan Raicu, Ian Foster, Michael Wilde, "Design and Evaluation of a Collective I/O Model for Loosely-coupled Petascale Programming", IEEE Workshop on Many-Task Computing on Grids and Supercomputers (MTAGS08), co-located with IEEE/ACM Supercomputing, 2008.

Using Hadoop as a Grid Storage Element

Brian Bockelman

Postdoc Researcher, Holland Computing Center, University of Nebraska, Lincoln, NE 68588-0150, USA

E-mail: bbockelm@cse.unl.edu

Abstract. Hadoop is an open-source data processing framework that includes a scalable, fault-tolerant distributed file system, HDFS. Although HDFS was designed to work in conjunction with Hadoop's job scheduler, we have re-purposed it to serve as a grid storage element by adding GridFTP and SRM servers. We have tested the system thoroughly in order to understand its scalability and fault tolerance. The turn-on of the Large Hadron Collider (LHC) in 2009 poses a significant data management and storage challenge; we have been working to introduce HDFS as a solution for data storage for one LHC experiment, the Compact Muon Solenoid (CMS).

1. Introduction

The Large Hadron Collider in Geneva, Switzerland is not only a unique experiment for investigations in particle physics, but also presents an exceptional computing challenge for those tasked with recording and processing huge amounts of data. When originally tasked with the problem, the four detector experiments on the LHC decided to take a decentralized processing approach through using grid computing. The grid formed to handle this task is the Worldwide LHC Computing Grid (WLCG).

At the Holland Computing Center in Nebraska, we have established a Tier-2 site for the CMS experiment. A Tier-2 site is primarily an analysis and simulation site for physicists. The site currently has about 200TB of storage and will grow to 400TB during 2009. The heaviest usage of the storage system comes from transferring data from a Tier-1 site (such as FNAL in the US) and from running I/O-intensive analysis jobs. The downloads are expected to sustain 5 Gbps and burst at 9 Gbps. The analysis jobs read at a rate of a few MB/s, but currently have a huge number of random read operations. The compute cluster is approximately 1,000 cores; as certain analysis jobs are CPU-bound, the cluster needs a few million I/O operations per minute.

Since the LHC experiments formed the WLCG, several Internet-based companies have emerged as having far larger data needs. Google grew from processing 100 TB of data a day in 2004 [1] to processing 20 PB a day [2]. As a comparison, the CMS estimated 2009 needs for total tape space are 25PB [3]. Google, Yahoo, Facebook, and others have approached the large data processing challenge with a processing model called MapReduce. If a user casts their problem into the MapReduce paradigm (loosely borrowed from functional programming), they will be able to scale the problem from a laptop to a cluster of 2,000 machines. A MapReduce implementation needs both an intelligent job scheduler and a very scalable, fault-tolerant distributed file system.

While Google's MapReduce implementation is proprietary, other large companies use an open-source implementation called Hadoop [4]. Hadoop is a product of the Apache Software

Foundation; the primary contributor is Yahoo. The largest Hadoop cluster is approximately 14PB of online disk in 4,000 machines. Hadoop is meant to work on thousands of commodity machines and therefore the Hadoop distributed file system (HDFS) has been designed to treat hardware failure as a normal occurrence that should be worked around. The reputation for scalability and fault-tolerence led us to evaluate and integrate HDFS at a large scale at our site.

The architecture of HDFS (and the GoogleFS, which HDFS is based on) is covered in [5, 6]. It is composed of many datanodes containing storage and a single namenode, which handles the namespace and its serialization. Any metadata operations goes to the namenode; all data movement goes through one of the datanodes. The I/O operations scale based on the datanode hardware capabilities. The files are broken up into 64MB chunks called "chunks" and distributed amongst the datanodes based on a given policy. The namenode is a clever design which keeps all metadata in memory, allowing it to scale well in terms of operations per second, but has the drawback that the number of files is limited by the machine's RAM.

2. Adopting Hadoop for the WLCG

Hadoop and HDFS are designed for processing and storing large volumes of data for web companies. Each Hadoop instance is usually run by the same company using it; transfers between Hadoop installs primarily happen within the same datacenter (or at least the same administrative domain). On the other hand, grid computing emphasizes the ability to transfer data between sites using common protocols; each site is under a different administrative domain and possibly using a completely different underlying storage system.

On the WLCG, the protocol used for WAN data transfers is GridFTP [7] and the protocol used for metadata operations is SRM [8]; both are necessary for interoperability between site storage and the grid. In addition, most grid users expect an efficient way for jobs to access data within the site's storage system. These three needs were addressed in the following way:

- For local access, HDFS has a binding for Linux called FUSE [9]. FUSE allows one to write a mountable file system in userspace (greatly simplifying the development of a file system and removing the need for custom kernels). Hence, worker nodes can mount HDFS and users access it in a similar manner to a NFS server. This transforms HDFS into more "normal looking" file system for users, as opposed to a more specialized data storage system. In the process of testing and deploying FUSE at a large scale, we have contributed back several bug reports and improvements. To our knowledge, we are the largest and heaviest user of FUSE, as none of the physics applications use the native Hadoop API.
- For SRM access, we have utilized the Berkeley Storage Manager (BeStMan) SRM server [10]. BeStMan's gateway mode implements all of the SRM calls utilized by CMS and ATLAS (but not all the SRM protocol functionalities). This stripped-down implementation has an advantage in that it can be used on any POSIX-like file system; other implementations of the SRM protocol are often tied to a specific storage system. The FUSE mounts which provide data access to jobs on worker nodes also provide metadata access to the SRM server. Hence, we were able to use BeStMan to provide SRM access to HDFS without any initial modifications.
- For GridFTP, we selected the Globus GridFTP server. The Globus server is in wide use elsewhere, and allows for pluggable modules to be implemented to interact with a storage system. Unlike SRM, GridFTP actually writes data in and out of HDFS. Because HDFS only supports appends, and not random writes, we had to implement in-memory data re-ordering and write an interface module between GridFTP and HDFS's C bindings.

Journal of Physics: Conference Series **180** (2009) 012047 doi:10.1088/1742-6596/180/1/012047

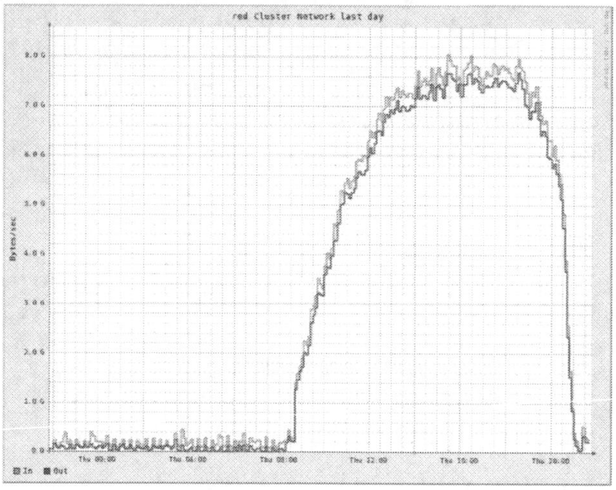

Figure 1. Network graph from a CMS user application; this graph is an aggregate for the HDFS cluster. The read rate peaks at around 8GB/s. This graph was recorded while the system was in production. The user's application had a very large readahead (10MB; the currently used one is 64KB), which causes an impressive network rate but doesn't happen for normal configurations. The readahead value is the minimum network transfer size used by the filesystem; it is an effective way to reduce latency in sequential reads, but the buffer thrashes constantly for random read workloads.

3. Performance Aspects of Hadoop

This section deals with the performance aspects of the Hadoop install at Nebraska. We currently have a dedicated Hadoop namenode (a machine with similar hardware to our worker nodes), a SRM server virtualized in a Xen machine, and 5 dedicated GridFTP servers. There are three datanode types: those having a single 80GB hard drive, those having two 1.5TB hard drives, and two Sun X4500 "Thumpers" with 48TB each. Except for the thumpers, each datanode has a 1Gbps network connection (non-blocking); the Thumpers have 4Gbps each.

Figure 1 shows the network rate (peaking at 8GB/s) from a user application reading from Hadoop. Figure 2 shows the namenode metadata operation rate we have observed in production; we have measured about 450 metadata operations / second through tests for "open" operations, and much higher for mixed workflows. Figure 3 shows the aggregate number of I/O operations per minute. This translates into 60 IOPS per hard drive (our cluster has approximately 330 hard drives), which is roughly the limit for the consumer-class harddrives that are contained in the majority of the cluster. All three measurements we show have been observed in production, but for unusual workflows. We have significant headroom with respect to performance for normal day-to-day operations, and believe we can scale further with more drives.

4. Hadoop Reliability and Manageability

As a CMS T2, Nebraska needs to provide more than just a highly-scalable Grid storage element. A storage element provides essential data to jobs; storage that can provide gigabytes a second under test condition is useless if it is not highly available under production. The SE must prove itself in terms of:

- Reliability. Even when reliable hardware is used (unless very *expensive* hardware is used), the system as a whole would be unreliable if it depended on every piece of hardware being available at all times. HDFS provides reliability by having the namenode track every block's

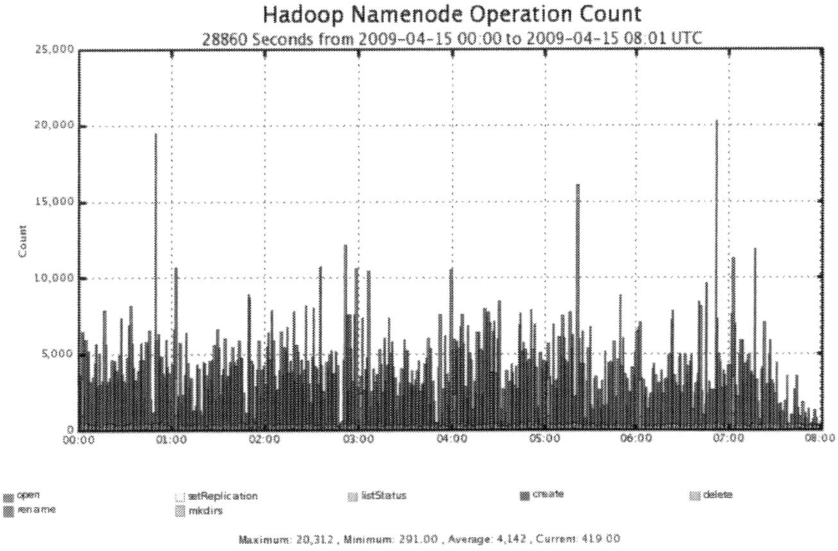

Figure 2. A graph derived from the logs of the Hadoop namenode. Each bar shows the number of times a certain operation is performed in a minute. The peak rate is 338 operations / second. During the time period shown, the load was primarily generated by a grad student working with many small files (about 64KB); it was not artificially generated. The CMS workflows use 2GB files, and do not have such a load on the metadata server.

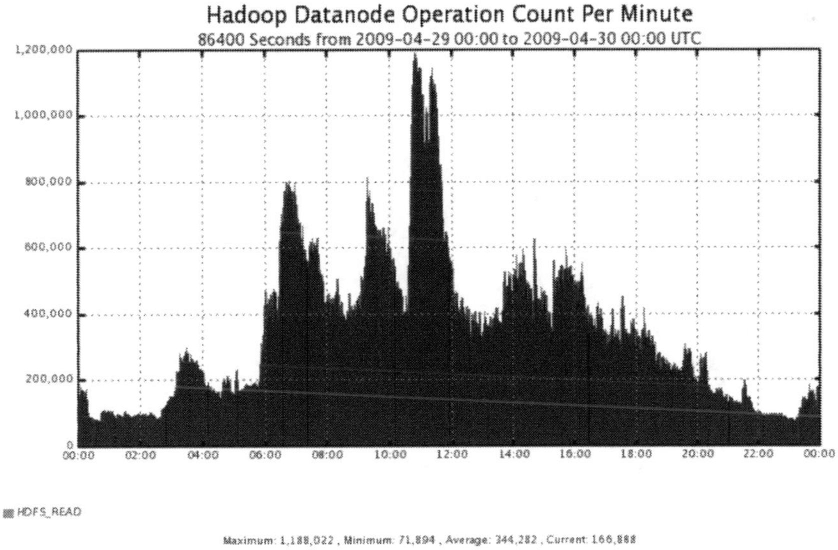

Figure 3. The number of data operations per minute delivered by the cluster. This graph shows the busiest day we've recorded during production. 1.2M I/O operations per minute is 20,000 IOPS; this is about 60 IOPS per hard drive. This is approximately the hardware limit for the class of hard drives we utilize.

location and replicate blocks as soon as they are lost. HDFS keeps a checksum of each block and performs background checksums on blocks on each datanode. The namenode is not currently highly available, but this can be mitigated by using more reliable hardware and RAIDs (this is an actively worked-on area).

- Manageability. Common management tasks must either be automated or simple. HDFS has a built-in tool to distribute a pool's data to others in order to allow it to be taken offline. It has a command-line tool that tells the admin the system is corrupt if a file block has no physical replicas. Hadoop has hooks for site monitoring tools such as Ganglia and JMX.
- Support. Other storage solutions have a close working relationship with CMS. With HDFS, CMS has no ability to influence development priorities of core developers or steer the file system's development. We feel this is mitigated by a few facts. HDFS has a large active development team which addresses bugs quickly when they are discovered; if a bug critically affects the T2, it would affect larger deploys which do employ developers. We find that HDFS covers CMS's core needs - no feature development is critically needed; we have been running the current install at the desired scale for months. Finally, if an unmet need does appear, there are companies such as Cloudera which provide paid support for Hadoop.

5. Conclusions and Future Directions

We have successfully adopted and used Hadoop's Distributed File System at our CMS site. We believe this shows HDFS is a viable alternative for a grid SE, especially in terms of scalability, reliability, and managability. While it has scaled well in our testing, there are improvements we would like to pursue. MapReduce provides capabilities CMS currently lacks; even without using the MapReduce framework, job scheduling ideas could be reused. CMS splits tasks on the submit side, and is unaware of the optimal splits on the remote cluster. Hadoop's scheduler tries to collocate the jobs and data; a MapReduce job might be able to schedule 70% or greater of its jobs so inputs are on local disk. Currently, this only happens in about 2% of the jobs. By using MapReduce - or a data-aware batch scheduler - CMS could drastically reduce its networking footprint.

There are desired improvements in HDFS and its grid adaptation which would have appreciable improvements at our site:

- Decreasing the memory footprint of GridFTP servers. As we re-order data from multiple streams in memory, we pay a significant penalty in memory per GridFTP process. We wish to explore other transport layers, such as UDT instead of TCP, which scale better and do not require multiple streams to achieve high throughput.
- Providing a strong security system. Currently, the security model is similar to NFS: it depends on anyone already inside the cluster's private network being trusted. There is no encryption or strong authentication. This is something that Hadoop hopes to address (perhaps by integration with Kerberos) in the 6-8 month timeframe.
- Implementing datacenter-aware policies. The Holland Compute Center has datacenters in two cities. A Hadoop instance is not currently optimized to understand the network topology characteristic of having two datacenters; we would like to see it optimized for data placement and access policies that understands having a segmented cluster.

Acknowledgments

Acknowledgements. This work was jointly funded by the Open Science Grid, funded jointly by DOE and NSF. This work was completed utilizing the Research Computing Facility with the associated USCMS Tier-2 site at the University of Nebraska - Lincoln. The requirements for a grid SE was provided by USCMS, and other USCMS Tier-2 sites provided valuable testing for the Hadoop installation packages.

References

[1] Jeffrey Dean and Sanjay Ghemawat. Mapreduce: Simplified data processing on large clusters. *Proceedings of USENIX OSDI'04*, pages 137–150, 2004.
[2] Jeffrey Dean and Sanjay Ghemawat. Mapreduce: simplified data processing on large clusters. *Commun. ACM*, 51(1):107–113, 2008.
[3] D Espriu. Computing resources scrutiny group. Technical Report CERN-RRB-2008-106, CERN, Geneva, Oct 2008.
[4] Apache hadoop, 2009. http://hadoop.apache.org/.
[5] Dhruba Borthakur. Hdfs architecture, 2009. http://hadoop.apache.org/core/docs/r0.20.0/hdfs_design.html.
[6] Sanjay Ghemawat, Howard Gobioff, and Shun-Tak Leung. The google file system. *19th ACM Symposium on Operating Systems Principles*, October 2003.
[7] I Mandrichenko, W Allcock, and T Perelmutov. Gridftp v2 protocol description, 2005. http://www.ggf.org/documents/GFD.47.pdf.
[8] A Sim and A Shoshani et al. The storage resource manager interface specification version 2.2, 2008. http://www.ggf.org/documents/GFD.129.pdf.
[9] Filesystem in userspace, 2009. http://fuse.sourceforge.net.
[10] A Shoshani et al. Storage resource managers: Recent international experience on requirements and multiple co-operating implementations. In *24th IEEE Conference on Mass Storage Systems and Technologies*. IEEE Computer Society, September 2007.

Scalable I/O and analytics

Alok Choudhary[1], Wei-keng Liao[1], Kui Gao[1], Arifa Nisar[1], Robert Ross[2], Rajeev Thakur[2], and Robert Latham[2]

[1] Electrical Engineering and Computer Science Department, Northwestern University
[2] Mathematics and Computer Science Division, Argonne National Laboratory

E-mail: {choudhar,wkliao,kgao,ani662}@eecs.northwestern.edu,
{rross,thakur,robl}@mcs.anl.gpv

Abstract. High-performance computing systems have already approached peta-scale with hundreds of thousands of processors/cores in many deployments. These systems promise a new level of predictive and knowledge discovery ability as researchers gain the capability to model dependencies between phenomena at scales not seen earlier. These applications are highly I/O and data intensive, leading scientists to observe that performing I/O and subsequent analyses are major bottlenecks in effectively utilizing peta-scale systems and a major hurdle in accelerating discoveries. Although significant progress has been made in performance, interfaces, and middleware runtime systems for I/O in the recent past, significantly more research and development needs to be carried out to scale the performance to the desired levels for systems containing tens to hundreds of thousands of cores. In this work we outline our recent achievements and current research for designing scalable I/O software and enabling data analytics in storage systems. We also enumerate key challenges for the I/O systems and discuss ongoing efforts that address these challenges.

1. Introduction

Research and development in science and engineering is increasingly based on simulations and/or on analysis of observational data requiring the use of high-performance computing (HPC) systems. HPC systems have already approached peta-scale in many deployments and efforts are already underway towards designing and scaling systems to exascale. As systems scale, simulations in areas such as combustion, chemistry, nanoscience, astrophysics, cosmology, fusion, climate prediction, environmental science, and biology have the potential of providing a new level of predictive and knowledge discovery, as researchers gain the capability to model dependencies between phenomena at scales. Many of these applications are expected to use systems with hundreds of thousands of processors/cores in order to achieve the resolution needed to better analyze and understand complex science problems. One common characteristic of all these applications is that they are all highly I/O and data intensive.

For example, in the area of atmospheric science and climate modeling it would be possible to use more realistic models of clouds, precipitation, and convective processes by substantially increasing the grid resolution, which would be feasible with computing power at this scale. At a mesh spacing of 1 km and higher resolution, it is possible to accurately predict tropical cyclones and other extreme weather events. It has been estimated that a global model on a uniform grid with 1 km mesh spacing would require a computer with sustained performance of 10 peta flops or more [1]. These types of applications require simulations, data assimilation,

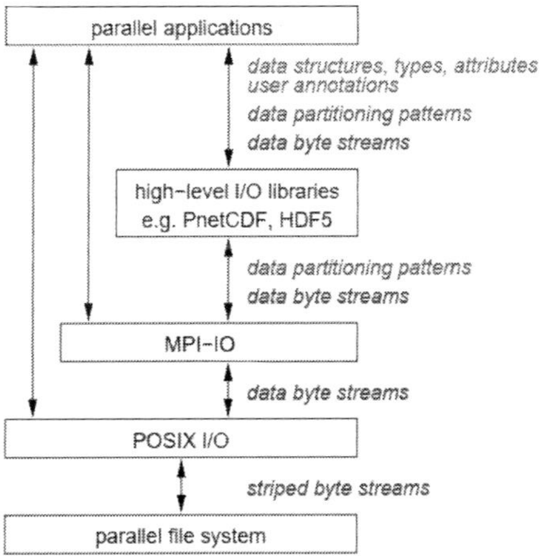

Figure 1. I/O software stack. High-level data information is lost as data reaches lower levels.

analytics, validation of simulations with observations, etc. This requirement is quite common across a range of applications in science and engineering. However, these characteristics across applications also lead the scientists to observe that performing I/O and subsequent analyses are major bottlenecks and most scientists also point that out as a major hurdle to effectively utilizing peta-scale systems and accelerating discoveries.

Although significant progress has been made in performance, interfaces, and middleware runtime systems for I/O in the recent past, significantly more research and development needs to be carried out to scale the performance to the desired levels for systems containing tens to hundreds of thousands of cores. Figure 1 illustrates a typical parallel I/O software stack that is found in typical HPC systems. We note that this is a logical view and not meant to illustrate physical connectivity or configuration of the I/O stack, rather to describe the software layers. A parallel application may directly call MPI-IO or POSIX I/O functions to access files from the parallel file system. Alternatively, a typical large-scale data-intensive application may use several layers of software as illustrated in the figure. For example, a weather modeling application may use NetCDF [2] for data representation and storage, where I/O is carried out through parallel netCDF (PnetCDF) [3] for fast performance. PnetCDF is built on top of MPI-IO [4], which may be layered on top of the Parallel Virtual File System (PVFS) [5] or some other parallel file system. The highest layer normally has semantic information associated with the data, such as data structures, types, attributes, user annotations, and data partitioning patterns, which describe the intent of a user. For example, at the high-level I/O library layer, the user may have represented a three-dimensional structure with a specific distribution of data (each element of which may be a collection of variables such as temperature, pressure, density, humidity, etc.). Some relationships can be preserved in the MPI-IO layer, such as the data partitioning patterns among a group of processes. Furthermore, the MPI-IO may use the partitioning information to enable optimizations of data caching, process coordination, and other techniques for performance optimizations [6, 7, 8, 9, 10, 11]. However, the data structures, types, attributes, and annotations will be converted to byte streams at the MPI-IO layer. Further down, at the POSIX I/O and file system layers, without proper interfaces for complex metadata communication, high-level information gets lost, and data appears just a stream of bytes.

As we move towards peta-scale systems and beyond, we have to reconsider the traditional approaches to I/O optimizations. Traditional interfaces in file systems and storage systems are designed to handle the worst-case scenarios for conflicts, synchronization, locking, coherence checks, and other issues, which adversely affect the performance. In other words, in this conventional design, the lower layers of the software stack are designed pessimistically, rather than optimistically. Of course, the reasons for those designs are the semantics and expectations of consistency, reliability, and other features needed from the storage and file systems. However, by providing and using the appropriate interface at the middleware (e.g., MPI-IO) level, understanding the access patterns, using some system resources within user application space, and performing intelligent caching, the desired scalability in performance can be achieved in such a manner that the optimizations are also portable across different systems through the use of this middleware. We need to understand the source of the problems and limitations of the current approaches, rather than blindly trying to deal with their manifestations for particular architectures and applications. In many cases, the problem is not that of having insufficient I/O capacity or bandwidth, but it is the excessive synchronization and file system control overhead at the I/O layer to account for semantics and the worst-case scenarios. This can have devastating results on overall system performance. Even though research and development for implementing efficient memory synchronization or file locking scheme (when really needed) is important, high-level information about access patterns may be the keys to eliminate or reduce such bottlenecks and are, therefore, critical to achieving real scalability. In Sections 2 and 3, we discuss our recent research on developing I/O strategies for PnetCDF and MPI-IO, respectively.

The storage systems in most of today's HPC computers employ a group of servers dedicated to I/O tasks and managed by a parallel file system. The computational capacity of one of these I/O servers is equal or even greater than that of one of the compute nodes that run the application programs. However, due to little computation required for serving I/O requests, the CPUs at the I/O servers are underutilized. The computational resources at the storage systems have been proposed as active storage to perform data filtering, a common operation in database queries. Examples are active disks that embed processing hardware into disk devices so that data filtering can be accomplished much faster. Utilizing the computational resources on the I/O servers for data processing extends the same idea of reducing the data amount transferred between application clients and the storage system. There are two advantages of enabling data processing at I/O servers over active disks. First, the I/O servers require no special hardware, unlike active disks which are still not in mass production. Second, the I/O servers are interconnected and can potentially serve as a computer cluster, running full-blown parallel programs.

Computational scientists must understand results from experimental, observational and computational simulation data to gain insights and achieve knowledge discovery. As systems approach the petascale range, problems that were unimaginable a few years ago are within reach. With the increasing volume and complexity of data produced by ultra-scale simulations and high-throughput experiments, understanding the science is largely hampered by the lack of comprehensive I/O, storage, data manipulation, analysis, and mining tools. Scientists require techniques, tools, and infrastructure to facilitate better understanding of their data, in particular the ability to effectively perform complex data analysis, statistical analysis and knowledge discovery. In Section 4, we discuss the issues of enabling the storage systems to carry out data analytics computation as part of I/O operations. Without the proper interfaces and standardized protocols, data structures and their access information at the higer levels of the I/O stack cannot be passed to the lower leverls systematically.

2. Optimizations in parallel netCDF

Dataset storage, exchange, and access play a critical role in scientific applications. For such purposes, netCDF serves as a software library and self-describing machine-independent data format that supports the creation, access, and sharing of array-oriented scientific data [2]. NetCDF stores metadata in a file header that describes the structures of the arrays and their layout in a netCDF file. Additional information, such as user annotations, can also be saved as

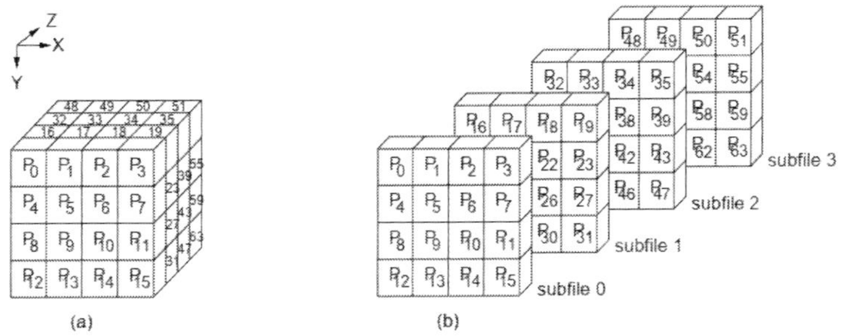

Figure 2. A 3D array partitioned among 64 processes and its mapping to 4 subfiles.

attributes. Built on top of MPI-IO for portability, Parallel netCDF (PnetCDF) was developed to support parallel I/O operations while retaining the netCDF file format for compatibility with existing netCDF datasets and tools [3]. PnetCDF I/O functions take an additional argument of an MPI communicator to indicate the processes participating in the shared-file I/O operations. PnetCDF internally constructs MPI derived data types to describe complex access patterns, such as reading a subarray of a multidimensional variable. MPI hints supplied by users are passed to the underlying MPI-IO library, so that PnetCDF can take advantage of the I/O optimizations available in MPI-IO.

PnetCDF encourages a shared-file I/O approach, one of the two parallel I/O programming styles commonly used by today's parallel applications, where all application processes store and access related data in single, shared files. The other style is termed as unique I/O, also known as the unique-file-per-process style, in which each process accesses files that are unique to the process. Unique I/O can create a management nightmare for file systems when a parallel job running on thousands of processes produces hundreds of thousands or millions of files. Furthermore, accessing millions of files can be a daunting task for post-run data analysis.

The shared-file I/O programming style avoids such problems. It can also be used to save global partitioned data structures in the canonical order. However, shared-file I/O performance often suffers from significant file system overheads, at large scale, due to conflicted I/O requests among processes [12, 11]. Such overheads include the system's data consistency control for cache coherency and I/O atomicity. Standard file system atomicity rules require that the results of an individual write call are either entirely visible or completely invisible to any successive read call from the same or different process [13]. Similarly, any caches must be kept synchronized so that a consistent view of the file is preserved. In modern parallel file systems, data consistency control is usually implemented using a file locking mechanism. As the number of concurrent accesses to a shared file increases, the overhead due to lock conflicts can increase dramatically.

As a middle ground between the two extremes of small files and very large files, we have developed a subfiling scheme for PnetCDF that allows a large multi-dimensional global array to be split into a number of smaller subarrays, each saved in a separate file. The subfiling scheme reduces the file system control overhead by decreasing the number of processes concurrently accessing a shared file. By saving the subfiling information in the netCDF files, the structure of global arrays can be portably reconstructed. In PnetCDF subfiling, the multi-dimensional array partitioning is always along the most significant axis, and subarrays are stored in the canonical order. Figure 2 shows an example of a three-dimensional array partitioned among 64 processes and the mapping of the subarrays to four subfiles. Through the same PnetCDF I/O functions, the partitioned global array still appears to users as a single netCDF array and access to it is kept the same as without subfiling. In order to achieve this goal, the subfiling implementation automatically generates the MPI fileviews for each subfile for a given I/O request. Mapping information of a global array to subfiles is stored as netCDF attributes which are duplicated

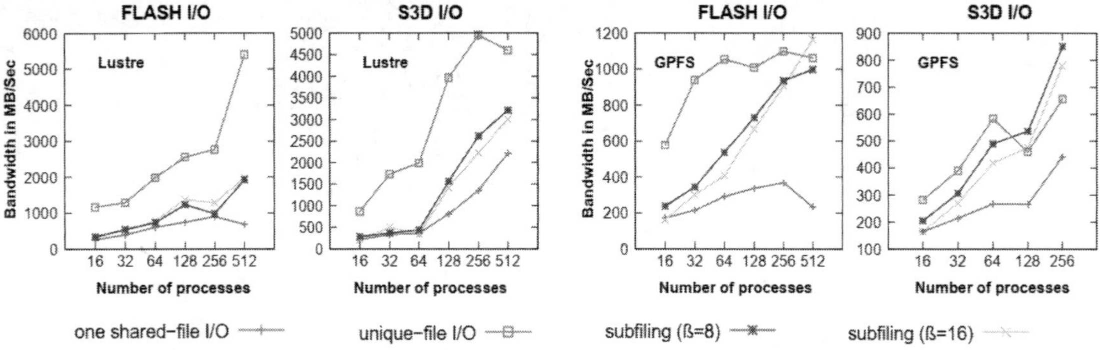

Figure 3. PnetCDF subfiling performance results for FLASH and S3D I/O kernels.

in all subfiles. Each netCDF subfile is self-contained with sufficient information to describe the layout of the local array mapped to the global array, including the names of all subfiles and partitioning of each subfile. Therefore, accessing the metadata from one of the subfiles is enough to understand the subfiling structure of the global array.

We evaluate this subfiling scheme using two scientific production application I/O kernels, FLASH-IO and S3D-IO. The performance is compared with the approaches of unique I/O and single shared-file I/O using two parallel file systems: Lustre on Franklin, the Cray XT4 at NERSC, and GPFS on Mercury, a TeraGrid machine at NCSA. Figure 3 summarizes the write bandwidths where β = nproc/nsubfile is the ratio of the number of processes sharing a subfile. In most of the cases, subfiling performance falls between the two I/O approaches. On GPFS, subfiling even outperforms the unique-file I/O approach when 256 or 512 numbers of processes are used. This is because the large number of concurrent file open operations overwhelms the file system metadata server, causing significant increase in the file open cost.

3. Scalable MPI-IO

The storage systems on high-performance parallel computers usually consist of a small number of I/O servers to serve requests from a much larger number of compute nodes that run user applications. As the size of modern computer systems increases, it is obvious that the storage system is becoming a performance bottleneck. Envisioning this problem, the IBM BlueGene systems dedicate a set of nodes as intermediate I/O nodes that forward the requests from application nodes to the I/O servers. These I/O nodes become the actual I/O clients to the underlying file system, where one I/O node is responsible for the requests from a group of compute nodes [14]. For example, in BG/L, there is usually an I/O node for every 8 to 64 processors [15, 16]. This design decreases the chances of possible I/O server contention, when multiple compute nodes are concurrently performing I/O. Imagine, if thousands of process clients on a HPC system perform I/O operations simultaneously, then the high volume of I/O requests can easily choke the available network bandwidth at the storage system.

Based on a similar concept, we have developed a software solution, the I/O Delegate Cache system (IODC), that uses an additional set of MPI processes as I/O delegate nodes (IOD) dedicated to handle I/O requests from application client processes [17]. The IODC system is built inside ROMIO [18], an MPI-IO implementation developed at Argonne National Laboratory. In the IODC system, all I/O requests initiated by a user application running on application nodes are intercepted and redirected to IOD nodes. This is achieved when parallel applications pass their I/O requests to the MPI-IO layer directly or through a high-level I/O library, e.g., HDF5 [19] or PnetCDF. Like the I/O nodes in a Blue Gene, only IOD nodes access the underlying parallel file system directly. In addition to forwarding requests, a collaborative distributed

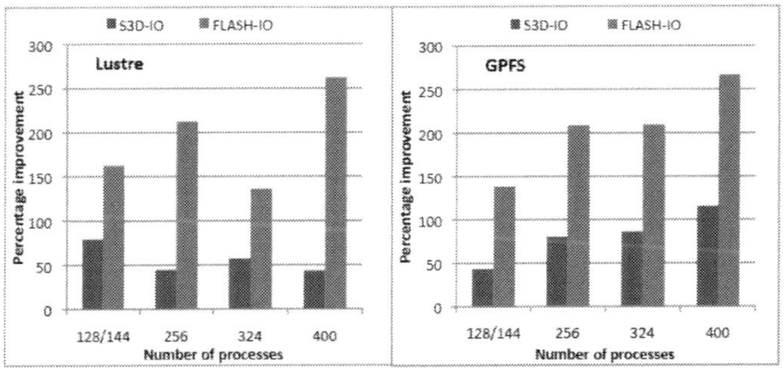

Figure 4. Performance results of using I/O delegation for FLASH and S3D I/O kernels.

cache is maintained across the IOD nodes for further I/O optimization. Features such as request aggregation, lock request alignment, cache page migration, and atomicity control are incorporated. By exploiting processing power and memory of IOD nodes for performing I/O caching and data aggregation, we achieve significant I/O bandwidth improvements.

We conducted our experiments using the FLASH and S3D I/O kernels using Lustre on Tungsten and GPFS on Mercury. Both Tungsten and Mercury are TeraGrid machines at NCSA. Experimental results are given in Figure 4. By allocating only 2-3% of additional computer nodes as I/O delegates, we achieved the improvement percentages ranging from 25% to 260%. Such significant performance improvement demonstrates the potential for using a separate group of MPI processes for I/O tasks to reduce the contention at the I/O servers, a technique that is applicable to both shared-file and unique-file I/O patterns.

3.1. Adaptive file domain assignment for collective I/O

Although the shared-file I/O method provides a way to preserve the canonical order of structured data and ease the file management for post-run data analysis and visualization, this method often performs poorly when the I/O requests are not well coordinated. To address such performance issue, MPI collective I/O is designed to provide an opportunity to induce collaboration among the requesting processes for better I/O performance. Well-known examples of using such a collaboration are two-phase I/O [20] and disk directed I/O [21]. These process collaboration strategies have demonstrated significant performance improvements over uncoordinated I/O. However, even with these improvements, the shared-file can I/O still perform far worse than the unique-file-per-process approach. One of the main reasons is that concurrently accessing shared files incurs higher file system locking overhead from data consistency control. The same overhead can never happen if a file is only accessed by a single process. Two of most important data consistency issues that POSIX-compliant file systems must enforce are I/O atomicity and cache coherence. Most file systems rely on a locking mechanism for these tasks. Due to the nature of file striping, the lock granularity is usually set to the file block size or stripe size, instead of a byte. If two I/O requests simultaneously access the same file block and at least one of them is a write, they must be carried out serially, even if they do not overlap in bytes.

ROMIO's collective I/O implementation is based on the two-phase I/O strategy proposed in [20], which includes a request redistribution phase and an I/O phase. The two-phase strategy first calculates the aggregate access file region and then partitions it among the I/O aggregators into approximately equal-sized contiguous segments, named **file domains**. This partitioning strategy is referred as the even partitioning method. The I/O aggregators are a subset of the processes that act as I/O proxies for the rest of the processes. In the data redistribution phase, all processes exchange data with the aggregators based on the calculated file domains. In the I/O phase, aggregators access the shared file within the assigned file domains. Two-phase I/O can

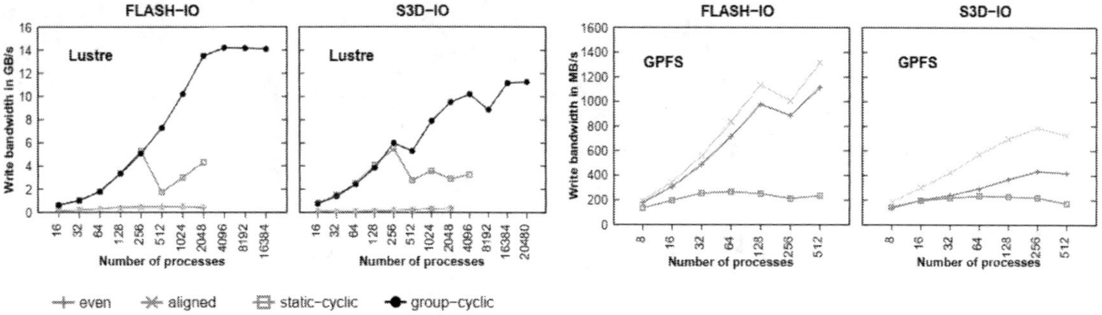

Figure 5. Results for various file domain partitioning methods for FLASH and S3D I/O kernels.

combine multiple small non-contiguous requests into large contiguous ones and has demonstrated to be very successful, as modern file systems handle large contiguous requests more efficiently. However, the even partitioning method does not necessarily produce the best I/O performance on all file systems due to the lock contentions. Since the lock granularity on most parallel file systems is set to the file stripe size, the file domain partitioning must be carefully aligned with the lock boundaries in order to minimize the lock conflicts.

We have developed a few file domain partitioning methods [22] as replacements for the even partitioning method used by ROMIO. The first method aligns the partitioning with the file system's lock boundaries. The second method, the *static-cyclic* method, partitions a file into fixed-size blocks based on the lock granularity and statically assigns the blocks in a round-robin fashion among the I/O aggregators. The third method, the *group-cyclic* method, divides the I/O aggregators into groups, each being of size equal to the number of I/O servers. Within each group, the static-cyclic partitioning method is used. This method is particularly designed for the situation that the number of I/O aggregators is much larger than the I/O servers. We evaluate these methods using four I/O benchmarks on two parallel file systems, Lustre and GPFS. Due to the different file locking mechanism implemented in Lustre and GPFS, these partitioning methods result in significant performance differences on the two file systems.

Figure 5 presents the write bandwidth results of the FLASH-IO and S3D-IO benchmarks on Jaguar, a Cray XT4 at ORNL, and Mercury at NCSA. These results indicate that the group-static and lock-boundary aligned methods give the best write performance on Lustre and GPFS, respectively. The lessons learned from this work can be helpful for the MPI-IO implementation to adapt the best partitioning method based on the underlying file system configuration. On file systems that implement server-based locking protocols, such as the Lustre, the group-cyclic file domain partitioning method is the best choice for collective write operations, as it minimizes the lock acquisition cost. For token-based locking protocols, such as the one used by GPFS, the method that aligns the partitioning to the lock boundaries produces the least number of file lock conflicts. These observations support developing ad hoc file domain partitioning methods based on the underlying file system characteristics, such as file locking and striping.

4. Data analytics in storage systems

It is known that most large-scale applications carry out their I/O tasks in bursts. That is, periodically a significant time of I/O inactivity is followed by bursts of I/O. Figure 6 illustrates the I/O activities of the CTH application in one 3-hour run. CTH is a family of codes developed at Sandia National Laboratories for modeling complex multi-dimensional, multi-material problems that are characterized by large deformations and/or strong shocks [23, 24]. Through our collaboration with the Sandia National Laboratory, we are studying the I/O traces produced from production runs [25]. Data checkpointing was enabled for every 25 minutes and

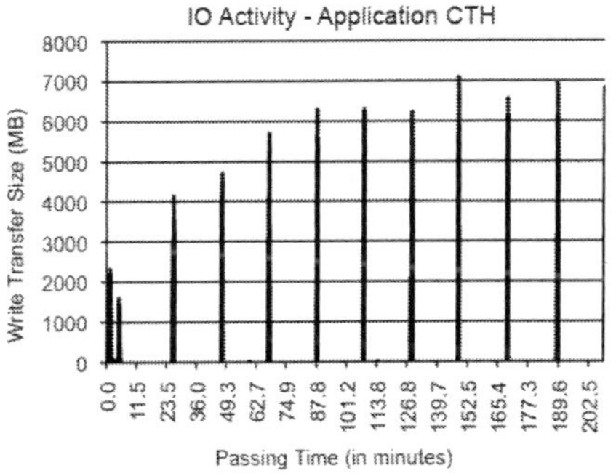

Figure 6. The bursty nature I/O activity of application CTH. Checkpoint writes occurs at a regular interval. This I/O pattern is very common in large-scale applications.

hence results in bursts of I/O activities on the file system. Such I/O pattern explains a low utilization of I/O servers commonly observed in today's parallel file systems.

Active storage has been proposed since late 1990s and prototyped in hard disk drives [26, 27, 28]. The majority of active storage applications are database operations, which perform mostly data filtering operations. The concept of active storage has been extended from the disk devices to the entire storage system, particularly, the I/O servers [29, 30]. Given the computational capabilities of the I/O servers in a parallel file system, their idle time can potentially be used to share the burden of computational nodes. Data intensive operations, such as data analytics and mining, can benefit from such as an active storage model on parallel file systems. However, offloading computational tasks from application clients to file system servers has several obstacles on today's file systems.

As illustrated in Figure 1, high-level information is lost when the data arrives at the file system. Information such as data types and structures are critical for data analytics programs to operate correctly. On the POSIX-compliant parallel file systems, the file metadata that can be transferred between client applications and file systems is limited to the attributes describing the file status. To enable computation offloading from application clients to file systems, the file system must adapt object-based protocols to allow user-defined attributes and operations be associated with the data objects. Object-based concepts have been incorporated into some parallel file systems, such as PVFS, Lustre, and Panasas, but without standardized metadata protocols, the potential optimization from metadata availability at the file level is still difficult to achieve.

Another challenge is the collaboration of multiple file system servers in order to carry out a single data analytics task in parallel. As files stored on a parallel file system can be striped across multiple servers, these servers must be able to communicate with each other as part of a parallel data analytics task. However, the I/O servers in modern parallel file systems are not programmed to communicate with each other directly. Enabling active storage on parallel file systems will soon change this phenomenon. Offloading the analytics tasks from application clients to the I/O servers will require identifying the group of servers holding parts of the data and informing the server group of data partitioning patterns. This is analogous to defining an MPI communicator in an MPI program, carried out at run time.

We have started the active storage design for parallel data analytics by extending the PnetCDF library and the PVFS file system. The well defined metadata in PnetCDF is borrowed in order to construct the interfaces that allow users to pass customized metadata along with the data to the PVFS file system. PVFS servers collaborate to execute data analytics operations as parallel programs. Our initial work also uses the I/O delegation discussed in Section 3 as another option for operation offloading. Since I/O delegation runs at user space, close to the applications, it can retain more high-level metadata details for better performance. In our design, the results of analytics are represented and saved in the netCDF format, so that both client and server components have a consistent and metadata-rich data presentation. Results are forthcoming.

5. Conclusions

Hardware and software improvements are constantly being made in the area of HPC I/O to address the needs of efficient and effective storage system in computational science. However, there are still challenges for achieving scalable performance as HPC systems enter the peta-scale era. The most significant one is obsolete standards that cannot address large-scale I/O requirements. From the performance point of view, abiding by the traditional semantics for data consistency has been identified as a major obstacle. Such semantics were designed for small-scale, loosely-coupled distributed environments and the problems addressed by these semantics seldom appear in parallel I/O operations. New standards for interfaces and semantics must be adopted to enable performance at scale. From the productivity point of view, the high-level I/O information has no proper communication channels to be delivered to other components in the I/O stack. Many aspects of such information can be used to bypass unnecessary system controls and enable optimizations for better performance. Our work demonstrated the importance of reducing file system locking overhead in collective MPI-IO operations. Such optimization relies on the awareness of application I/O patterns as well as the underlying system configurations. Extensions to POSIX I/O, the ANSI T-10 OSD, and pNFS standards are all important steps in the direction of high-performance, standards-based I/O systems.

To increase the functionality of I/O systems for computational science, flexible interfaces are important for exploiting the computational potential in storage systems. We have highlighted design issues for enabling data analytics in parallel file system servers and pointed out the deficiency of current standards. As the I/O systems and software are still evolving to meet the challenges of existing and future HPC systems, adapting informative application programming interfaces in file systems will elevate the I/O systems to a more versatile component from its traditional role of simple file access.

Acknowledgments

This work was supported in part by U.S. Department of Energy (DOE) SCIDAC-2: Scientific Data Management Center for Enabling Technologies grant DE-FC02-07ER25808, DOE FASTOS award number DE-FG02-08ER25848, the DOE Office of Advanced Scientific Computing Research, Office of Science, under contract DE-AC02-06CH11357 and DE-FG02-08ER25835. This work was also supported in part by the National Science Foundation (NSF) under HECURA CCF-0621443, NSF SDCI OCI-0724599, and NSF ST-HEC CCF-0444405. We thank Lee Ward and Ruth Klundt at Sandia National Laboratories for providing the I/O trace data and valuable suggestions toward the application I/O characteristic analysis on HPC systems. This research was supported in part by NSF through TeraGrid resources provided by NCSA, under TeraGrid Projects TG-CCR060017T, TG-CCR080019T, and TGASC080050N. This research also used resources of the National Center for Computational Sciences at Oak Ridge National Laboratory, which is supported by the Office of Science of DOE under Contract No. DE-AC05-00OR22725.

References

[1] Committee 2008 *National Research Council Report, National Academies Press: Washington, D.C.*
[2] netCDF *Network Common Data Form* The UniData program center, University Corporation for Atmospheric Research http://www.unidata.ucar.edu/software/netcdf/
[3] Li J, Liao W, Choudhary A, Ross R, Thakur R, Gropp W, Latham R, Siegel A, Gallagher B and Zingale M 2003 Parallel netCDF: A Scientific High-Performance I/O Interface *the Supercomputing Conference*

[4] Message Passing Interface Forum 1997 *MPI-2: Extensions to the Message Passing Interface* http://www.mpi-forum.org/docs/docs.html
[5] Carns P, Ligon W, Ross R and Thakur R 2000 PVFS: A Parallel File System for Linux Clusters *the Third Annual Linux Showcase and Conference* pp 317–327
[6] Ching A, Choudhary A, Coloma K, Liao W, Ross R and Gropp W 2003 Noncontiguous I/O Accesses Through MPI-IO *IEEE/ACM International Symposium on Cluster Computing and the Grid*
[7] Ma X, Winslett M, Lee J and Yu S 2003 Improving MPI-IO Output Performance with Active Buffering Plus Threads *the International Parallel and Distributed Processing Symposium*
[8] Coloma K, Choudhary A, Liao W, Lee W, Russell E and Pundit N 2004 Scalable High-level Caching for Parallel I/O *the International Parallel and Distributed Processing Symposium*
[9] Coloma K, Choudhary A, Liao W, Lee W and Tideman S 2005 DAChe: Direct Access Cache System for Parallel I/O *the 20th International Supercomputer Conference*
[10] Coloma K, Ching A, Choudhary A, Liao W, Ross R, Thakur R and Ward L 2006 A new flexible MPI collective I/O implementation *the IEEE Conference on Cluster Computing*
[11] Liao W, Ching A, Coloma K, Choudhary A and Ward L 2007 An Implementation and Evaluation of Client-Side File Caching for MPI-IO *the International Parallel and Distributed Processing Symposium*
[12] Ross R, Latham R, Gropp W, Thakur R and Toonen B 2005 Implementing MPI-IO Atomic Mode Without File System Support *the 5th IEEE/ACM International Symposium on Cluster Computing and the Grid*
[13] IEEE/ANSI Std 10031 1996 *Portable Operating System Interface (POSIX)-Part 1: System Application Program Interface (API) [C Language]*
[14] Yu H, Sahoo R, Howson C, Almasi G, Castanos J, Gupta M, Moreira J, Parker J, Engelsiepen T, Ross R, Thakur R, Latham R and Gropp W D 2006 High Performance File I/O for the BlueGene/L Supercomputer *the 12th International Symposium on High-Performance Computer Architecture (HPCA-12)*
[15] Almasi G, Archer C, Castanos J, Erway C, Heidelberger P, Martorell X, Moreira J, Pinnow K, Ratterman J, Smeds N, Steinmacher-burow B, Gropp W and Toonen B 2004 Implementing MPI on the BlueGene/L Supercomputer *Euro-Par Conference on Parallel Processing*
[16] Loft R D 2005 Blue Gene/L Experiences at NCAR *IBM System Scientific User Group meeting (SCICOMP11)*
[17] Nisar A, Liao W and Choudhary A 2008 Scaling Parallel I/O Performance Through I/O Delegate and Caching System *International Conference for High Performance Computing, Networking, Storage and Analysis*
[18] Thakur R, Gropp W and Lusk E 1997 *Users Guide for ROMIO: A High-Performance, Portable MPI-IO Implementation* Technical Report ANL/MCS-TM-234, Mathematics and Computer Science Division, Argonne National Laboratory
[19] HDF Group *Hierarchical Data Format, Version 5* The National Center for Supercomputing Applications http://hdf.ncsa.uiuc.edu/HDF5
[20] del Rosario J, Brodawekar R and Choudhary A 1993 Improved Parallel I/O via a Two-Phase Run-time Access Strategy *the Workshop on I/O in Parallel Computer Systems at IPPS '93* pp 56–70
[21] Kotz D 1997 *ACM Transactions on Computer Systems* **15** 41–74
[22] Liao W and Choudhary A 2008 Dynamically Adapting File Domain Partitioning Methods for Collective I/O Based on Underlying Parallel File System Locking Protocols *International Conference for High Performance Computing, Networking, Storage and Analysis*
[23] Hertel E, Bell R, Elrick M, Farnsworth A, Kerley G, Mcglaun J, Petney S, Silling S, Taylor P and Yarrington L 1993 CTH: A Software Family for Multi-Dimensional Shock Physics Analysis *the 19th International Symposium on Shock Waves* pp 377–382
[24] McGlaun J, Thompson S and Elrick M 1989 CTH: A three-dimensional shock wave physics code *the Hypervelocity Impact Symposium* pp 12–14
[25] Klundt R, Weston M and Ward L 2008 I/O Tracing on Catamount Tech. Rep. SAND2008-3684 Sandia National Laboratories Albuquerque, New Mexico
[26] Riedel E, Gibson G and Faloutsos C 1998 Active storage for large-scale data mining and multimedia *International Conference on Very Large Data Bases* (San Francisco, CA, USA: Morgan Kaufmann Publishers Inc.) pp 62–73 ISBN 1-55860-566-5
[27] Keeton K, Patterson D and Hellerstein J 1998 *SIGMOD Record* **27** 42–52
[28] Acharya A, Uysal M and Saltz J 1998 Active disks: programming model, algorithms and evaluation *International Conference on Architectural support for programming languages and operating systems* (New York, NY, USA: ACM) pp 81–91 ISBN 1-58113-107-0
[29] Lim H, Kapoor V, Wighe C and Du D 2001 *Mass Storage Systems, IEEE Symposium on* **0** 101 ISSN 1051-9173
[30] Felix E, Fox K, Regimbal K and Nieplocha J 2006 Active storage processing in a parallel file system *International Conference on Linux Clusters: The HPC Revolution*

SciDAC 2009 IOP Publishing
Journal of Physics: Conference Series **180** (2009) 012049 doi:10.1088/1742-6596/180/1/012049

Data Management and Its Role in Delivering Science at DOE BES User Facilities – Past, Present, and Future

Stephen D. Miller, Kenneth W. Herwig, Shelly Ren, Sudharshan S. Vazhkudai

Oak Ridge National Laboratory

Pete R. Jemian

Argonne National Laboratory

Steffen Luitz, Andrei A. Salnikov, Igor Gaponenko

SLAC National Accelerator Laboratory

Thomas Proffen, Paul Lewis

Los Alamos National Laboratory

Mark L. Green

Tech-X Corporation

millersd@ornl.gov

Abstract. The primary mission of user facilities operated by Basic Energy Sciences under the Department of Energy is to produce data for users in support of open science and basic research [1]. We trace back almost 30 years of history across selected user facilities illustrating the evolution of facility data management practices and how these practices have related to performing scientific research. The facilities cover multiple techniques such as X-ray and neutron scattering, imaging and tomography sciences. Over time, detector and data acquisition technologies have dramatically increased the ability to produce prolific volumes of data challenging the traditional paradigm of users taking data home upon completion of their experiments to process and publish their results. During this time, computing capacity has also increased dramatically, though the size of the data has grown significantly faster than the capacity of one's laptop to manage and process this new facility produced data. Trends indicate that this will continue to be the case for yet some time. Thus users face a quandary for how to manage today's data complexity and size as these may exceed the computing resources users have available to themselves. This same quandary can also stifle collaboration and sharing. Realizing this, some facilities are already providing web portal access to data and computing thereby providing users access to resources they need [2]. Portal based computing is now driving researchers to think about how to use the data collected at multiple facilities in an integrated way to perform their research, and also how to collaborate and share data. In the future, inter-facility data management systems will enable next tier cross-instrument-cross facility scientific research fuelled by smart applications residing upon user computer resources. We can learn from the medical imaging community that has been working since the early 1990's to integrate data from across multiple modalities to achieve better diagnoses [3] – similarly, data fusion across BES facilities will lead to new scientific discoveries.

1. Introduction

The US Department of Energy Office of Basic Energy Sciences operates nine synchrotron radiation and high-flux neutron sources. In addition, it supports five nanoscale science research centers and another three electron beam centers [1]. In 2008, these user facilities enabled over 10,000 researchers to perform experiments which in turn engaged additional thousands of collaborators to also analyze data from these experiments. Common to these facilities is the production of experiment data, and the need for users to access, process, and analyze these data, resulting in peer-reviewed science publications.

This paper examines data management methodologies and their relationship to producing scientific results across selected DOE BES user facilities including neutron scattering facilities Intense Pulsed Neutron Source (IPNS - now closed [4]), High-Flux Isotope Reactor (HFIR) [5], Spallation Neutron Source (SNS) [6], and Los Alamos Neutron Science Center (LANSCE) [7], and X-ray facilities Advanced Photon Source (APS) [8] and Linac Coherent Light Source (LCLS) [9]. This combination accounts for about 67% of the total annual BES operating budget for user facilities [10]. The maturity of these facilities is also significant as it varies from retired, mature, emerging, and new user facilities. Examining these facilities provides a glimpse into the past and present, and a look to the future for the role of data management in accomplishing their mission. Facilities are examined individually, trends are identified, and a model for the future is put forward. Of particular interest are the volume of data produced by typical experiments, user access to data and computing resources, and the role of the facility in enabling scientific research by providing intelligent data management.

The premise of this paper is that facility produced data are valuable. User facilities with beam lines are rate limited as to the number of experiments that can be performed annually according to a number of factors. The unique scientific capabilities of these facilities are a scarce resource with demand exceeding availability typically by factors of two to three. Access is obtained via a peer-reviewed process unique to each facility but which is uniformly extremely competitive. Those who are successful in obtaining beam time greatly value the resource and often depend upon it for their individual science programs. We can also calculate a metric for the value of the data produced. In round numbers, one can take the facility operating budget divided by the number of beam lines and the number of operating days to determine a $/instrument/day operating cost. Knowing the average data rate per instrument per day, one can then determine the $/TB to produce data. This cost varies widely from thousands to millions of dollars per Terabyte of data produced by many of the instruments at these facilities. Given the high cost of producing this data, it is reasonable to consider the investment in a reasonable data management strategy.

Currently much of this data particularly at X-ray sources are ultimately discarded based upon facility data retention practices which are often, understandably, based upon available storage resources for storing data. When a facility discards data, the burden falls upon users to manage (store and maintain its provenance), process, and analyze their data. Add in the complexity for doing so with data obtained at multiple instruments across multiple facilities over many visits and one can quickly see that users face significant data management and access bottlenecks along the road to scientific discovery.

Steps to change this process have been taken by the data management team for HFIR and SNS which has produced a web based portal that manages experiment data on behalf of the users and also provides users access to computing resources. Thus far, all experiment data produced for SNS have been archived and are available via the portal as well as secure shell access. Eventually the composite size of the repository will grow beyond what the facility can maintain itself, and this point may only be only a few years away even though the facility is expected to have a 40 year lifespan. The challenge increases significantly for X-ray facilities which typically produce multi-TB data sets per experiment. Yet even these data volumes look quite tractable when compared to the data volumes produced by the High Energy Physics (HEP) community which has data volumes approaching the Exascale [21]. To facilitate international research, the HEP community has developed an extensive data storage and analysis infrastructure including over 140 computing centers in 34 countries [23].

Another premise of this paper is to enable user access to data on the same material obtained with multiple techniques (at multiple facilities) will facilitate a new mode of scientific discovery. This leads to the desire for a common approach to data management across these user facilities. There are already a significant number of cross-over users between X-ray and neutron scattering in the fields of powder and single crystal diffraction, small angle scattering, reflectometry, and imaging and tomography. Ultimately, the success of this approach also depends on development of more sophisticated analysis methods which allow easy integration of these multiple data sets collected in a variety of modalities but all seeking to address a common scientific goal.

To this end, work is already underway between APS and SNS to facilitate inter-facility data caching where selected data are to be made available via a storage lease for a period of time. This work is being done among the facility data management personnel working in conjunction with the CEDPS SciDAC team [22]. There are many challenges to face in providing such a resource including user authentication among users and user facilities, cyber security, and cyber infrastructure development, and human labor to develop the necessary software and systems. It is hoped that demonstrating the benefits and how they outweigh the costs will help bring more resources to this work perhaps leveraging more of the SciDAC resources and teams.

2. Neutron Scattering User Facilities

Neutron scattering user facilities generally fall into two broad categories – reactor-based providing a continuous source of neutrons, or accelerator-based generating pulses of neutrons. HFIR is reactor-based, and IPNS, SNS, and LANSCE are pulsed sources. Neutron scattering techniques first utilized continuous neutron sources and were found useful for uncovering both the structure and dynamics at the atomic and molecular levels. The Nobel Prize for physics was awarded to Clifford G. Shull and Bertram N. Brockhouse in 1994 "for pioneering contributions to the development of neutron scattering techniques for studies of condensed matter" [11].

Utilizing an event data format is taking root at the neutron scattering facilities where detected neutrons are stored in a list where the individual event detection records the detected position along with the timing information for when an event is detected. Additional metadata (sample conditions, dynamic instrument parameters) are also archived in a time-stamped list and can be correlated with the data post-collection. The event list can then be (re)histogrammed as desired allowing researchers to sift through the raw data to construct data sets of interest. For example, parametric studies over temperature, pressure, and/or magnetic field changes etc. can be rebinned over desired parameter spans in order to reach a desired statistical data quality or desired "error bar" range. Thus the event data format is ideal for supporting pump-probe type experiments, which is also of interest for X-ray experiments – significant new scientific discoveries are forecast utilizing the pump-probe technique.

The neutron scattering community is quite fortunate in that up to this point, facilities have managed to collect and store all of the data that has been produced. Though this is the case now, the future holds challenges to continue this pace. Several new instruments will have the capability to produce data sets of similar size as those more typical of X-ray instruments. Thus, a new strategy will be needed for neutron scattering facility based data management.

2.1. IPNS

As of January 2008, IPNS began to undertake decommissioning and decontamination activities which included ramping down and terminating the user program [4]. IPNS had been operational for more than 26 years and had collected approximately 520GB of data across 14 instruments over the lifetime of the facility. Users typically went away with their experiment data or had anonymous ftp access to the data. Metadata were sparse, thus finding data via ftp could be challenging.

2.2. HFIR

The HFIR facility has been refurbished in 2006-07 and currently has 15 instruments operating via an 85 MW reactor neutron source. During the lifetime of the facility, the reactor has had over 420 run-cycles to produce neutrons where a run cycle is approximately 3 weeks in duration. Since 2007, HFIR data for 5 of the instruments has been archived totalling around 2GB of storage space. Experiment and processed data are generally manageable in size and instrument users typically leave taking their experiment data with them. However, HFIR has plans to operate a tomography beamline which will produce hundreds of GB to TB size experiment data sets, causing data management for this instrument to be rethought.

2.3. LANSCE

The Lujan Neutron Scattering Center within LANSCE utilizes a pulsed source and has a complement of 15 instruments. It maintains a data archive of approximately 4 TB that includes all neutron scattering data collected since its 1986 inception. The size of experimental data sets has grown significantly over the past few years. Raw data collected for a typical user experiment is on the order of 2 GB, but can vary widely depending on the instrument and the experiment particulars. Generally, initial data reduction is performed on site and users depart with a reduced data set. Access to both the raw and reduced data is available through the instrument scientist. Ever tightening computer security regulations over the past several years has made it difficult to provide users with direct offsite data access.

An initial proof of concept effort has been undertaken to include data from the Lujan NPDF instrument into the Neutron Science Web Portal in which approximately 2.5GB of data have been made available via the portal. To access this data, users are required to get an ORNL external computing user (XCAMS) ID and request portal access.

2.4. SNS

The SNS facility took its first data in April 28, 2006 with instruments collecting data via the user program in September of 2007 with first user data collected on the BASIS backscattering spectrometer. Currently there are 7 instruments in the user program with another 4 to enter commissioning soon in 2009. These instruments are located in target station 1 which will have a complement of approximately 23 instruments when fully built out, and an additional target station 2 is being planned to house a similar complement of instruments. Thus far data from every experiment has been archived for the SNS instruments providing excellent opportunities to examine data statistics. The total SNS archive consists of approximately 2.8TB with another 1TB of user contributed data (shared) data. Experiment data rates vary from around 200MB/ day to around 4.7GB/day per instrument with an average of around 1.3GB/day/instrument with a standard deviation of around 1.6GB/day. Thus on average assuming an operating budget of approximately $180M and about 200 operating days over a full complement of 23 instruments the value of data at about $32M/TB. Compare this with around $2K/TB in a high performance data storage system, which is over 4 orders of magnitude difference.

Within the next year, SNS will begin commissioning a diffractometer which will have the ability to produce around 200GB/hour resulting in tens of TB per experiment. In addition, both SNS and HFIR have plans to support tomography instruments which could routinely produce hundreds of GB to tens of TB per experiment. Techniques such as inelastic single crystal experiments will produce multi-TB sized data sets as well. In addition, pump-probe experiments utilizing an event data format have the potential to produce very large data sets perhaps approaching 1TB per experiment in addition to the raw experiment data.

3. X-ray User Facilities

BES supports 6 synchrotron radiation light sources with techniques for spectroscopy, scattering, and imaging. Here we focus on APS which is a mature, flagship facility supporting thousands of users producing hundreds of publication per year, and LCLS which is a new highly-anticipated facility expecting first beam in September of 2009. APS provides a full suite of instruments addressing all of the above techniques where LCLS produces ultrafast pulses of X-rays enabling imaging of single molecules and stop-motion of atoms and molecules in motion. The LCLS coherent hard X-ray beam will be delivered in a form of 120 Hz pulses (30 Hz in Fall 2009) with 100 fsec timing. LCLS instruments will support imaging, spectroscopy, and pump-probe measurement techniques for both imaging and spectroscopy.

3.1. APS

APS has almost 60 beamlines with 1-3 instruments per beamline and rapidly produces copious amounts of data. Typical experiments will produce between a few KB to 100GB, while imaging experiments (such as tomography) produce more, on the order of 1TB to 10TB of data per experiment. Data rates for some instruments, such as X-ray Photon Correlation Spectroscopy and 3D X-ray Diffraction Microscopy will approach 300 MB/s on a continuous basis. Data management across beam lines varies considerably but can be best described as central to each instrument. A variety of data formats are in use, each chosen by the instrument support team based on the needs of the particular community. Some beam lines utilize facility-defined formats while other beam lines, based upon expediency, just utilize company proprietary data formats as provided by the detector or data acquisition system. Adoption of international standard formats such as NeXus is slow due to many factors. Data storage and archiving policies are local to each instrument. For instruments with small data volumes, data is retained in perpetuity. For high-volume instruments such as imaging or tomography, the amount of local storage is constrained by cost in dollars, time, and physical storage space for the media. For these instruments, on-site storage resources are limited and thus users are responsible for storing and managing their own data. The facility will provide users a two week window of opportunity to copy their data before it is deleted from its servers.

APS is beginning to address the need for a facility approach to the coordination of metadata associated with the primary data from each experiment. Collection of metadata can facilitate many aspects of downstream data processing (reduction, analysis, access, and archival) and provide the possibility of data fusion. The metadata collection process begins at the facility with the user's experiment proposal and continues through publication.

Real-time access to high-performance computing is necessary to handle the data processing for advanced instruments such as the XPCS noted above. Communication of the data through the network and access to the data and computing capabilities must be built upon standardized security infrastructure and high-volume data communication protocols that spans beyond the APS to include the breadth of the DOE facilities and user communities. Standardized infrastructure to manage the HPC capabilities will enable the user community to focus on the production of science from the mass of accumulated data.

3.2. LCLS

There will be 6 sets of versatile high data bandwidth instruments installed in two hatches of the LCLS experimental area. Some instruments will be capable of producing up to tens of Gbytes/sec of data in peak. In the final implementation of the system up to two of those instruments can be used simultaneously. This will results in multi-terabyte data volumes to be handled on daily basis. The data rate will ramp up over the next 3 years. The first experiments begin their data collection in September 2009 and will produce up to 1 TB of data per day with nearly 100 TB of data to be stored before the end of the year 2009. In three years the amount of data to be stored per day will increase up to 15 TB from only one of the instruments, and that would correspond to nearly 2-3 PB of data per year. All stored data will be preserved at SLAC Computer Center for the duration of at least 1 year. All stored data will have to be delivered to the corresponding experimental groups for further analysis.

Unprecedented amounts and complexity of data expected at LCLS which poses a number of challenging problems to be addressed:

- Implementing a high-bandwidth data delivery pipeline for sending hundreds of MB/sec of data from the detectors to the long-term storage and archival system

- Providing a reliable and secure storage mechanism for many petabytes of data for a period of at least one year (most likely – much longer)
- Building an efficient data access mechanism to the stored data. Such mechanism should allow parallel access to the data from an analysis farm of thousands of computing cores.
- Building a data exportation system to send terabytes of data to off SLAC consumers. The system would need to cope with the same aggregate data rate the data are delivered from instruments.

The architecture of the data management system which is currently under development is based on iRODS [17] – the next generation data management infrastructure introducing the highly customizable rule-based management model. The main file format for stored data will be HDF5 [18]. In addition there will be a number of specialized relational databases augmenting the bulk data.

Users will interact with the data storage system primarily via the Web portal which is currently under development. The portal will be based on contemporary dynamic Web 2.0 technologies in order to deliver enhanced user experience with the system.

The off-site data exportation will employ high-performance multi-stream data transfer protocols such as BBCP [19] which would allow utilizing 30 Gbit/sec SLAC WAN connection with ESNet [20]. Building a distributed data exportation system involving a number of regional centers is also planned for later stages of the LCLS experiments. Such a distributed architecture will facilitate efficient data delivery to remote clusters of users. The development will require substantial cooperation and coordination efforts with other BES laboratories and with international counterparts.

4. Future Cross-Cutting Science Opportunities

In looking at methods for scientific discovery, today's facilities appear as islands with the instrument appearing as isles among these islands. As a result, scientific research is performed in relative isolation. To illustrate, today there are a plethora of data file formats among these various instruments, and this alone is a significant inhibitor of software re-use and interoperability, and collaboration. Another serious inhibitor is the challenge to simply move data between facilities. Even more basic is the lack of robust authentication systems to support facility users.

Assume for the moment that we have overcome these obstacles and let us look at what we could stand to achieve.

- Data mining spanning across both neutron and X-ray data sets facilitating research for both experiment teams and non-experimentalist researchers.
- Creation of new cross-technique reference databases for researchers.
- Common software to work with data for visualization and basic analysis.
- Definition of experiments spanning across facilities such as advanced pump-probe experiments utilizing both neutron and X-ray sources – novel scientific discoveries await this capability.
- Seamless co-refinement of data from X-ray and neutron single crystal and powder diffraction instruments.
- Better enable the study of biological materials via integrated SANS and SAX techniques.
- Combination of imaging and tomography utilizing neutrons and X-rays in order to better exploit the measurement properties of each technique.
- Incorporation of high performance computing at the beamline to provide better sample alignment, experiment steering, and experiment simulations.
- Development of a common support infrastructure across the user facilities better enabling cross-facility support and integration.
- The ability to readily collaborate with academic and international partners.
- A common or federated to approach user authentication and data access controls which comply with federally-mandated computer security.

5. The Future of Facility Data Management

The BES user facilities are well poised to take advantage of software and systems developed via the SciDAC program [12]. With the advent of user facilities producing "bigger data", the need for data movement and data management tools grows larger. However it appears that software produced by the SciDAC Collaboratories may be applicable for helping to address some of the facility data management issues[13]. There is also good intersection between facility needs and the resources provided by a number of SciDAC Institutes [14] and Centers [15] as well such as visualization, data storage, scientific data management, networking, computing, and simulation.

The user facilities have several significant additional requirements necessary to establish dynamic and versatile scientific research communities. These can be roughly divided into two categories: hardware infrastructure, and software systems – yet these both must be combined to perform synergistically. At the foundation is needed the infrastructure to provide data storage capacity at some level serving the user facilities. This includes the ability to capture metadata to support data cataloguing incorporated into a "facility node" (or "network node") which provide some level of computing, data storage, and networking capacity independent of the operation of the facility. Then a user facility network we are calling the User Facility Network (UFNet) needs to be established to enable networking across the facilities interfacing to the network nodes at each facility. UFNet would operate synergistically and leverage ESNet to provide inter-facility networking capacity and capabilities with federated user authentication to support scientific community growth. These *network nodes* would provide the abstraction layer between facility operations and the infrastructure supporting the user to access data and computing resources. Facilities need to have what we are calling "*nodeware*" to interface their facility data collection processes, archival systems and computing resources with their *network node* and UFNet will provide the middleware infrastructure connecting the nodes together.

At the upper level is the need for the users to work smarter providing them the ability to easily leverage the above described infrastructure. Users need not be concerned about computing-related issues in order to pursue their research. Work is already underway via a company called the Tech-X Corporation to via SBIR efforts to produce an "*Orbiter*" thin- and thick client application [16] residing on a user computer which provides programmatic access to remote data and computing resources provided by a common Service Oriented Architecture (SOA) hosted within the HFIR/SNS network. Users are also provided the opportunity to work autonomously or to define their own research teams via the definition of *ad hoc* virtual organizations.

There are significant benefits to this approach:

1. Facilities provide a uniform interface to data and computing resources thus better enabling their reach into research communities.
2. *Significant* return on investment for the SciDAC project
3. Extensible to incorporate scientific research communities external to DOE
4. The user can perform scientific research holistically by creating data portfolios from among the user facilities as necessary
5. Users can work better both collaboratively and autonomously.
6. As we learn from medical research, advancement requires multi-technique data integration. Enabling integrated scientific research *is* the next tier of scientific discovery.

References

[1] http://www.sc.doe.gov/bes/BESfacilities.htm

[2] "The SNS/HFIR Web Portal System – How Can it Help Me?" Stephen D. Miller Invited Talk I3.1 2009 International Conference on Neutron Scattering. http://www.mrs.org/s_mrs/bin.asp?CID=19203&DID=239713&DOC=FILE.PDF

[3] Medical Imaging, Analysis of Multimodality 2D/3D Images Kuhn, M. And Beolchi, L. January 1, 1995. **ISBN-13:** 978-9051992106

[4] IPNS: http://www.pns.anl.gov/

[5] HFIR: http://neutrons.ornl.gov/hfir/hfir_facilities.shtml
[6] SNS: http://neutrons.ornl.gov/facilities/facilities_sns.shtml
[7] LANSCE: http://lansce.lanl.gov/
[8] APS: http://www.aps.anl.gov/
[9] LCLS: http://lcls.slac.stanford.edu/
[10] http://www.science.doe.gov/bes/archives/budget/BES_FY2009budget.pdf
[11] http://nobelprize.org/nobel_prizes/physics/laureates/1994/press.html
[12] SciDAC: http://www.scidac.gov/
[13] SciDAC Collaboratories: http://www.scidac.gov/collabs.html
[14] SciDAC Institutes: http://www.scidac.gov/institutes.html
[15] SciDAC Centers: http://www.scidac.gov/aboutET.html
[16] "Doing Your Science While You're in Orbit". Mark L. Green, Poster I4.7 2009 International Conference on Neutron Scattering. http://www.mrs.org/s_mrs/bin.asp?CID=19203&DID=239713&DOC=FILE.PDF
[17] "IRODS – Rule-Oriented Data Systems": https://www.irods.org/
[18] HDF5: http://www.hdfgroup.org/HDF5/
[19] BBCP – multi-stream file transfer protocol for WAN: http://www.slac.stanford.edu/~abh/bbcp/
[20] ESNet – Energy Sciences Network: http://www.es.net/
[21] http://www.science.doe.gov/sbir/solicitations/FY%202009/34.HEP.Computing.htm
[22] http://www.cedps-scidac.org/
[23] http://lcg.web.cern.ch/LCG/

Building a parallel file system simulator

E Molina-Estolano[1], C Maltzahn[1], J Bent[2] and S A Brandt[1]

[1] UCSC Systems Research Lab, University of California, Santa Cruz, CA 95064, USA
[2] Los Alamos National Lab, Los Alamos, NM 87545, USA

E-mail: `{carlosm,eestolan,scott}@cs.ucsc.edu, johnbent@lanl.gov`

Abstract. Parallel file systems are gaining in popularity in high-end computing centers as well as commercial data centers. High-end computing systems are expected to scale exponentially and to pose new challenges to their storage scalability in terms of cost and power. To address these challenges scientists and file system designers will need a thorough understanding of the design space of parallel file systems. Yet there exist few systematic studies of parallel file system behavior at petabyte- and exabyte scale. An important reason is the significant cost of getting access to large-scale hardware to test parallel file systems. To contribute to this understanding we are building a parallel file system simulator that can simulate parallel file systems at very large scale. Our goal is to simulate petabyte-scale parallel file systems on a small cluster or even a single machine in reasonable time and fidelity. With this simulator, file system experts will be able to tune existing file systems for specific workloads, scientists and file system deployment engineers will be able to better communicate workload requirements, file system designers and researchers will be able to try out design alternatives and innovations at scale, and instructors will be able to study very large-scale parallel file system behavior in the class room. In this paper we describe our approach and provide preliminary results that are encouraging both in terms of fidelity and simulation scalability.

1. Introduction

We develop a novel approach to building a simulator for parallel file systems. Parallel file systems using object-based storage devices have been the storage architecture of choice as the scale of scientific high-end computing systems increases exponentially [1]. There are a handful of parallel file systems in use today, namely Lustre [2], PanFS [3], and PVFS [4], that are all based on the object-based storage paradigm [5] popularized by NASD [6]. New variations of parallel file system architectures have been proposed or are in development: Ceph [7], zFS [8], Sorrento, Ursa Minor [9], Kybos [10]. There is tremendous pressure to improve the performance of these systems. Yet, presently the performance of these systems cannot be easily tested without access to large clusters and expensive compute time. Existing large-scale systems are generally reserved for production use, running scientific codes 24/7, with very limited opportunities for other uses. Analytical models and simulators that are carefully validated by the high-end performance community would allow testing of file system designs at scale using a tiny fraction of the resources required by real system tests.

In this paper we introduce IMPIOUS (Imprecisely Modeling Parallel I/O is Usually Successful), a trace-driven parallel file system simulator using an *abstract simulation model* of object-based parallel file systems that accounts for just enough detail to reproduce important effects observed in real systems. IMPIOUS will (1) enable file system designers and researchers to try out innovative data placement strategies and other novel subsystems at scale, (2) facilitate file system deployment by providing a low-cost platform for "what-if" workload and file system tuning scenarios, (3) empower scientists to quickly tune existing file systems for specific workloads, and (4) aid instructors by providing a platform for in-classroom experiments.

The benefit of abstract simulation models is their simplicity, which allows users to simulate very large systems on much smaller systems within an acceptable amount of time. Our central hypothesis is that an abstract model with a few simple components suffices to describe important aggregate behaviors of parallel filesystem without having to model detailed behavior. For example, a simple model of disks works as well as detailed models as used in DiskSim [11].

The only parallel file system simulator we are aware of is the one reported in [12]. That simulator uses a detailed model of PVFS and Linux using the OMNeT++ simulation framework [13], focuses on network communications among OSDs by using an accurate simulation of TCP/IP over Ethernet, and is tuned to the particular infrastructure of the author's institution. Thus, the authors have chosen a what we call "bottom-up" strategy in that they first strive to model the behavior of the underlying systems with high fidelity before combining these simulation models to simulate a parallel file system. The disadvantage of this bottom-up approach is that it introduces a large amount of detail into the simulation which lengthens simulation run time, makes it more difficult to separate behavior due to a specific file system environment from behavior due to fundamental file system design decisions, and requires greater effort to change and experiment with alternative simulation models and therefore discourages rapid hypothesis testing.

In this project we hope to achieve the following contributions: (1) *a simple, scalable, and general simulation model* – the model supports file system-specific plugins and consists of about 10 general model parameters such as number of clients, number of OSDs, page size, and network latency; (2) *a framework for rapid hypothesis testing* – the simulation model can be refined easily to include a hypothesis for a particular behavior of a parallel file system for a particular workload; (3) *a simulator* – we are planning to make IMPIOUS available to the open source community. It currently consists of about 10k lines of Python code.

2. An abstract model of parallel file systems

Parallel file systems are complex systems, and the details of the implementations of commonly used systems differ significantly. Yet, they consist of the same subsystems: *object-based storage devices (OSDs)* which run their own local file system and export a flat name space of objects; *metadata management* which implements the file name space and all metadata operations; and *clients*, which provide the (typically POSIX) file system interface to applications. Clients communicate with metadata management to stat/open/close files and manipulate the name space, and directly access OSDs to read and write data.

The key characteristics distinguishing particular parallel file systems are the *data placement strategy*, which maps a request to the client to requests to one or more OSDs using a particular striping strategy and a redundancy scheme; the *resource locking protocol*, which determines which disks or OSDs are locked when for a particular client (e.g. locking the entire RAID group when accessing a stripe in that group); the *redundancy strategy*, which determines the kind of redundancy and which subsystem is responsible for maintaining it (e.g. clients, OSDs, or disk controllers); and the *client buffer cache*, which specifies management of communication between clients and OSDs (e.g. asynchronous vs. synchronous and arbitrary extents vs. pages).

3. The IMPIOUS simulator

The simulator uses a plugin for the desired parallel filesystem, implementing the above components. In a simulation run, a set of simulated clients is driven by a parallel file I/O trace or by a workload generator. (We currently have generators for N-N and N-1 checkpointing workloads.) The clients produce I/O requests to send to the OSDs, following the plugin behavior. Each simulated OSD consists of a simple internal filesystem, which we call NaiveFS; and a simulated disk, with the option of using either DiskSim or a simple disk model.

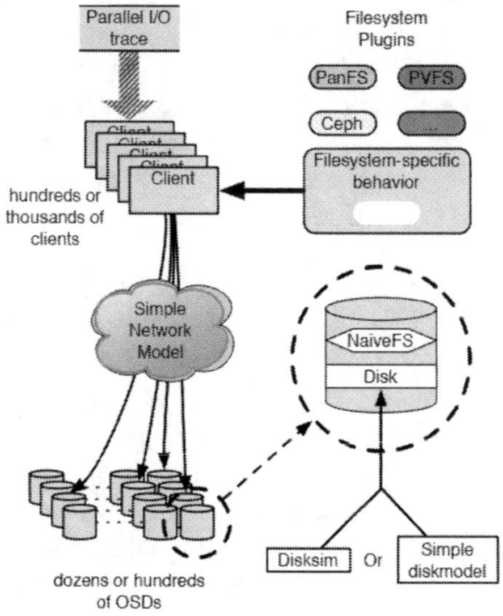

Figure 1: **The major components of IMPIOUS.** In the simulator, clients are driven by parallel I/O traces. Filesystem-specific plugins determine certain fundamental behavior: the data placement strategy, replication strategy, locking discipline, and client cache strategy. Clients communicate with OSDs to perform I/O. Each simulated OSD consists of an instance of NaiveFS, an idealized per-OSD filesystem; and a disk simulation, which can be either DiskSim or a simple disk model.

We have implemented the models for PVFS, PanFS, and Ceph; and we are in the process of implementing Lustre. We chose these file systems because they are commonly used (except Ceph), we have access to their placement strategies, and their placement strategies are sufficiently different. Our preliminary validation results are based on PanFS [3].

PanFS is the filesystem used in the Panasas parallel storage product. Its data placement uses per-file RAID 5 groups, with a fixed per-filesystem stripe width of 8 to 11 OSDs. For each file, a randomized algorithm (biased towards emptier nodes) selects the OSDs for the first RAID-5 stripe set; data is written repeatedly across the stripe set, in chunks of 64KB per OSD for each stripe. Up to 2,000 such stripes are written across the stripe set; once that limit has been reached, another stripe set, disjoint from the first one, is selected. If the file grows sufficiently that there are not enough OSDs to create another disjoint stripe set, and all stripe sets have been written across 2000 times, writes continue through the existing write sets in a round-robin fashion, again with 2000 writes per stripe set. To keep the parity consistent, a client must acquire an exclusive lock *per stripe* in order to write to it. This degrades performance for multiple-writer workloads with small interleaved writes. To handle node failure, PanFS uses *virtual spares* in conjunction with RAID-5. The stripe length for the filesystem is selected so that the maximum number of stripe sets per file will leave a small number of spare OSDs. Thus, if a node fails, each file with a stripe set on that node can replace that position in the stripe set with a spare node, and reconstruct the data from parity.

4. Preliminary validation results

In order to validate IMPIOUS, we use traces of important simulation checkpointing patterns [14]. The first experiment attempts to reproduce the effect of alignment on write bandwidth for N-1 checkpoint patterns; we use experimental N-1 checkpointing results collected using the PatternIO benchmark [15] running on 512 processes on LANL's Roadrunner test cluster, which includes a PanFS storage system consisting of 56 OSD's. These results are shown in Figure 2a: the three lines in this graph illustrate the

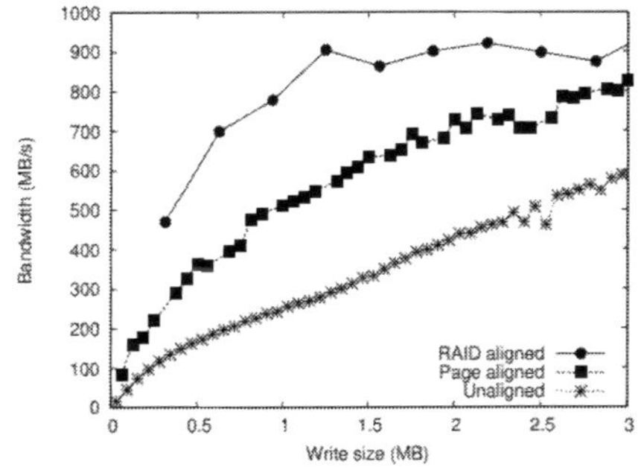

(a) Experimental Results

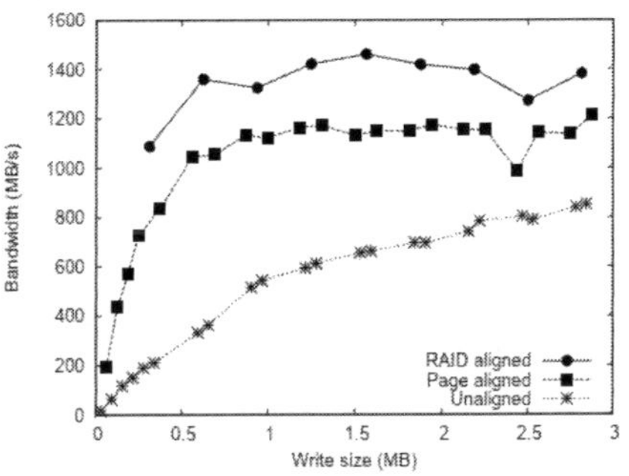

(b) IMPIOUS Simulated Results (DiskSim)

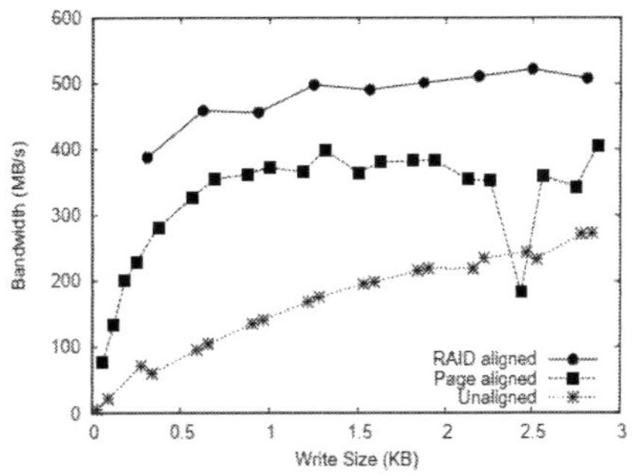

(c) IMPIOUS Simulated Results (simple disk model)

Figure 2: **N-1 Write Alignment** Graph (a) shows how checkpoint write bandwidth for an N-1 workload is affected by the alignment of writes. The three lines show writes that are aligned with stripe lengths, writes that are not aligned with stripe length but are page-aligned, and unaligned writes. We run IMPIOUS with both DiskSim (b) and the simple disk model (c). The graphs show that IMPIOUS is capable of approximating the performance differences among these different sets of write sizes (we are still investigating the pronounced dip at the 2496 KB chunk size under the simple disk model).

effect of write size on write bandwidth. All three lines improve as the writes get larger.

The top line shows the bandwidth achieved when the size of the writes is a multiple of the full stripe length (10 OSDs or 640K in this case); this achieves the highest bandwidth. The middle line shows the bandwidth for page aligned write sizes (4K in this case) which are not a multiple of the full stripe length. Writing partial stripes is slower for two reasons. First, each partial write is a read-update-write cycle, since existing data from the stripe is needed to compute the parity. Second, since the stripe is locked during each write, any other writers trying to write to the same stripe are blocked. Finally, the bottom line shows the bandwidth for write sizes that are not a multiple of the 4K page size.

These unaligned writes achieve the least bandwidth for two reasons. First, the PanFS client synchronously flushes dirty data that is smaller than a page. For example, compare a 32K write to a 31K write, both beginning at the start of a page. PanFS will buffer the 32K write until memory pressure forces it to be flushed, at which point it will write the entire 32K at once. Conversely, it will split the 31K write into 28K and 3K writes. Thus, the 31K write will incur an extra round trip to the OSDs, and an extra parity update. Second, an idiosyncrasy of the experimental run improves the performance of the aligned writes, but not the unaligned writes. The 512 processes were placed in groups of eight consecutive processes per compute node, with each group sharing a single PanFS client and buffer cache. This allowed each group's writes to be aggregated; for instance, the workload of 512 processes performing 128K writes was effectively a workload of 64 clients performing 1024K writes. This aggregation could not happen for unaligned write sizes, because of the synchronous flushing behavior described above.

Figure 2b illustrates how IMPIOUS accurately models these trends. It does not match the absolute throughput values: we used DiskSim with an available recent disk model to model the OSDs' disks, and made no attempt to match the OSDs' precise performance characteristics. IMPIOUS correctly reproduces the relative results, capturing the performance differences among the different types of write sizes.

Even though IMPIOUS does not reproduce the absolute throughput values, the relative results are still very useful for a user wondering what sort of change they might expect if they were able to manipulate either their application or the underlying storage configuration to match the size of their application's write to the full stripe size of their storage system. Except for DiskSim these results are obtained with very simple simulation models. In fact, the bulk of the simulation's runtime is spent in DiskSim. To make IMPIOUS more scalable, we are developing a simple disk model, intended to give accurate *aggregate* results without precisely modeling the disk's internals. Figure 2c shows results with IMPIOUS using the simple disk model instead of DiskSim. With the exception of the pronounced dip with a chunk size of 2496K, which we are still investigating, using the simple disk model in IMPIOUS instead of DiskSim produces similar qualitative results, showing the three trend lines, while reducing the simulation's runtime by an order of magnitude. However, the absolute throughputs are considerably lower. We are still refining the simple disk model; we hope ultimately to use DiskSim parameters in the simple disk model, and match the aggregate behavior of DiskSim disks without the large runtime.

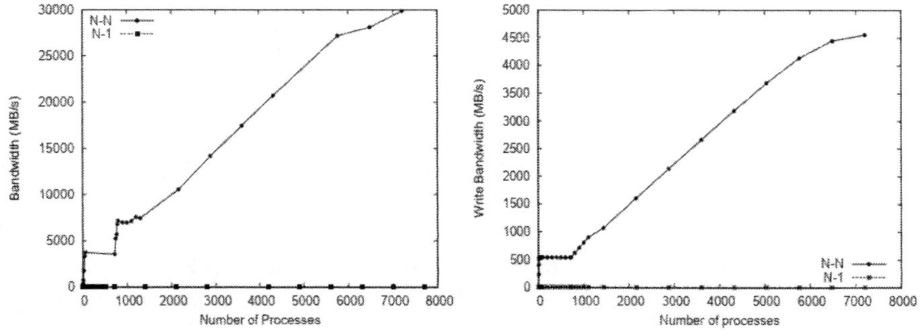

(a) Experimental Results (b) Simulated Results

Figure 3: **Checkpoint Types.** These two graphs depict the challenge to a parallel file system caused by an N-1 checkpoint pattern. Checkpoint write bandwidth is shown on the y-axis as a function of the number of processes. Due to locking, N-1 checkpoints have been demonstrated across multiple HPC parallel file systems to perform much more poorly than N-N. The graph on the left contains experimental data confirming this from the PanFS storage system attached to LANL's Roadrunner supercomputer. IMPIOUS is able to correctly predict the relative ordering of these two lines, and the linear growth of the N-1 throughput.

We also simulate the detrimental performance effects of an N-1 checkpoint pattern. The locking present in an N-1 checkpoint pattern, from multiple clients writing to the same file, significantly degrades performance compared to an otherwise similar N-N workload. We simulated a checkpointing experiment that was run on LANL's Roadrunner supercomputer, using a PanFS storage system of 512 OSDs. In this workload, a number of clients generate an N-1 strided workload in 47 KB writes.

Figure 3b shows the experimental Roadrunner results and our simulated results. IMPIOUS correctly captures the performance degradation caused by the N-1 workload. Each data point was produced by a simulation of 512 OSDs running on an 2 GHz Opteron, using only one core. With the aforementioned simple disk model, the simulation runtime varied from 12 minutes to 3 hours, scaling approximately linearly with the total number of I/O operations. Thus, the entire set of simulations producing this graph can be run in under a day on a single machine, or in a few hours on a modestly sized compute cluster by running simulations concurrently.

However, the simulated absolute throughput values for the N-N workload are much lower than the Roadrunner results, and the bandwidth saturates at a low number of processes. According to our current simulation model implementation each 47KB write is treated separately, which increases communication overhead and incurs a large number of extra seeks. We hypothesize that PanFS contains an optimization where a client that is the only writer to a file coalesces multiple unaligned requests before writing them. We are in the process of implementing client level request coalescing in IMPIOUS ; as an approximation, we ran an N-N workload with the stripe-aligned write size of 576KB, which increased the throughput considerably.

The linear curve of the experimental N-N performance is due to an idiosyncrasy of the real system's architecture: The Roadrunner cluster is partitioned into sub-clusters (*CU*s), each supporting up to 720 processes, and each with an equal fraction of the total I/O bandwidth. This leads to a stair step function where each time the increasing number of processes on the x-axis spills over to a new CU, the total I/O bandwidth increases by roughly the fraction of the I/O bandwidth available to the new CU (Figure 3a appears to show a linear curve but at sufficient resolution the entire curve would reveal its stair step shape). We implemented the I/O bandwidth partitioning in IMPIOUS to capture this effect; without it, the throughput would rise asymptotically.

5. Conclusions and future work

Our preliminary simulation results are encouraging. Given two different sets of experimental results from checkpointing workloads on PanFS, we were able to simulate the important performance effects of those workloads with IMPIOUS . The wall-clock runtime IMPIOUS is also encouraging. Simulation runs required far fewer computing resources than running the original workload: for experimental runs consuming 2 minutes across a 512-node storage cluster, the slowest simulated runs took three hours on a single core.

We will continue to improve and validate IMPIOUS . To do so, we will acquire experimental results from different workloads. We wish to demonstrate IMPIOUS reproducing results from considerably different workloads on the same filesystem; and show it reproducing the performance differences of a single workload running on different filesystems.

References

[1] Top500 2008 Performance development URL `http://top500.org/lists/2008/11/performance_development`
[2] Schwan P 2003 *Proceedings of the 2003 Linux Symposium*
[3] Welch B, Unangst M, Abbasi Z, Gibson G, Mueller B, Small J, Zelenka J and Zhou B 2008 *Proceedings of the 6th USENIX Conference on File and Storage Technologies (FAST '08)* pp 17–33
[4] Carns P H, Ligon W B, Ross R B and Thakur R 2000 *Proceedings of the 4th Annual Linux Showcase and Conference* (Atlanta, GA) pp 317–327
[5] Azagury A, Dreizin V, Factor M, Henis E, Naor D, Rinetzky N, Rodeh O, Satran J, Tavory A and Yerushalmi L 2003 *Proceedings of the 20th IEEE / 11th NASA Goddard Conference on Mass Storage Systems and Technologies* pp 165–176
[6] Gobioff H, Gibson G and Tygar D 1997 Security for network attached storage devices Tech. Rep. TR CMU-CS-97-185 Carniege Mellon
[7] Weil S A, Brandt S A, Miller E L, Long D D E and Maltzahn C 2006 *Proceedings of the 7th Symposium on Operating Systems Design and Implementation (OSDI)* (Seattle, WA)
[8] Rodeh O and Teperman A 2003 *Proceedings of the 20th IEEE / 11th NASA Goddard Conference on Mass Storage Systems and Technologies* pp 207–218
[9] Abd-El-Malek M, II W V C, Cranor C, Ganger G R, Hendricks J, Klosterman A J, Mesnier M, Prasad M, Salmon B, Sambasivan R R, Sinnamohideen S, Strunk J D, Thereska E, Wachs M and Wylie J J 2005 *Proceedings of the 3nd USENIX Conference on File and Storage Technologies (FAST'05)* (San Francisco, CA)
[10] Wong T M, Golding R A, Glider J S, beth Borowsky E, Becker-Szendy R A, Fleiner C, mana Hosekote D R K and

Zaki O A 2005 Kybos: self-management for distributed brick-based storage Research Report RJ 10356 IBM Almaden Research Center

[11] Bucy J S, Schindler J, Schlosser S W, Ganger G R and Contributors 2008 The disksim simulation environment version 4.0 reference manual Tech. Rep. CMU-PDL-08-101 Parallel Data Laboratory, Carnegie Mellon University Pittsburgh, PA

[12] Carns P H, Settlemyer B W and Ligon W B 2008 *SC '08: Proceedings of the 2008 ACM/IEEE conference on Supercomputing* (Piscataway, NJ, USA: IEEE Press) pp 1–8 ISBN 978-1-4244-2835-9

[13] Varga A 2001 *Proceedings of the European Simulation Multiconference*

[14] Bent J 2009 PLFS Maps http://www.institutes.lanl.gov/plfs/maps

[15] National Energy Research Scientific Computing Center I/O Patterns from NERSC Applications https://outreach.scidac.gov/hdf/NERSC_User_IOcases.pdf

Eucalyptus: an open-source cloud computing infrastructure

Daniel Nurmi, Rich Wolski, Chris Grzegorczyk, Graziano Obertelli, Sunil Soman, Lamia Youseff and Dmitrii Zagorodnov

Computer Science Department, University of California, Santa Barbara, CA 93106; and Eucalyptus Systems Inc., 130 Castilian Dr., Goleta, CA 93117

E-mail: rich@cs.ucsb.edu

Abstract. Utility computing, elastic computing, and cloud computing are all terms that refer to the concept of dynamically provisioning processing time and storage space from a ubiquitous "cloud" of computational resources. Such systems allow users to acquire and release the resources on demand and provide ready access to data from processing elements, while relegating the physical location and exact parameters of the resources. Over the past few years, such systems have become increasingly popular, but nearly all current cloud computing offerings are either proprietary or depend upon software infrastructure that is invisible to the research community. In this work, we present Eucalyptus, an open-source software implementation of cloud computing that utilizes compute resources that are typically available to researchers, such as clusters and workstation farms. In order to foster community research exploration of cloud computing systems, the design of Eucalyptus emphasizes modularity, allowing researchers to experiment with their own security, scalability, scheduling, and interface implementations. In this paper, we outline the design of Eucalyptus, describe our own implementations of the modular system components, and provide results from experiments that measure performance and scalability of a Eucalyptus installation currently deployed for public use. The main contribution of our work is the presentation of the first research-oriented open-source cloud computing system focused on enabling methodical investigations into the programming, administration, and deployment of systems exploring this novel distributed computing model.

1. Introduction

Scalable Internet services [1, 2, 3, 4] deliver massive amounts of computing power (in aggregate) on demand to large, internationally distributed user communities through well-defined software interfaces. Until recently, however, access to these services has been restricted to human-oriented and simple query-style application programming interfaces (APIs). With few exceptions, an application programmer wishing to incorporate such a service as a software component had little ability to direct and control computation inside the service explicitly.

Cloud computing [5, 6] has emerged as a new paradigm for providing programmatic access to scalable Internet service venues.[1] While significant debate continues with regard to the "optimal" level

[1] The term "cloud computing" is considered by some to be synonymous with the terms "elastic computing," "utility computing," and occasionally "grid computing." For the purposes of this paper, we will use the term "cloud computing" to refer to cloud, elastic, or utility computing but not to grid computing. The difference is explained in Section 4.

of abstraction that such programmatic interfaces should support (c.f., software-as-a-service versus platform-as-a-service versus infrastructure-as-a-service [7, 8, 9, 10]), the general goal is to provide users with the ability to program resources within a verylarge-scale resource "cloud" so that they can take advantage of the potential performance, cost, and reliability benefits that access to scale makes possible.

In short, the model is to provide a large user base with the ability to program some specified fraction of the resources hosted by a scalable service provider (e.g., Google [3], Amazon [2], SalesForce [4], 3Tera [1], etc.) through one or more well defined service interfaces. However, while the interfaces are public, the infrastructure maintained by the various service providers is almost exclusively proprietary. Thus it is not possible (or at least not easy) for researchers to build, deploy, modify, instrument, or experiment with a cloud infrastructure under their own control.

In this paper, we describe the design and implementation of Eucalyptus — an open-source software infrastructure architected specifically to support cloud computing research and infrastructure development. The design of Eucalyptus is distinctive in that it

- must be able to deploy and execute in hardware and software environments not under the control of its designers, and
- must be modularized to allow component-wise modification or replacement,

while achieving the greatest degree of scalability possible. This work describes the system architectural tradeoffs imposed upon the design by these two requirements, the way in which they have been addressed by the current version of Eucalyptus that is currently available and in use, and the degree to which these trade-offs impact the functionality and performance of the overall system.

The motivation for Eucalyptus is an exploratory one. Cloud computing as an emerging concept has great potential, but the speed of commercial engineering leaves fundamental questions either not fully defined or unanswered. Thus, while cloud systems are providing users a valuable service, the closed nature of the software has created a situation where researchers interested in cloud computing topics are finding it difficult to formulate experiments due to the lack of a common, flexible framework in which they can work.

1.1. Open-source infrastructure as a service

Although most existing cloud computing implementations share the common high-level notion of flexible, scalable, and dynamic computational "provisioning," there is significant variation in exactly how that power is presented to the end user. Some systems, such as Amazon's Elastic Compute Cloud (EC2) [11] and Enomalism [12], allow users to allocate entire virtual machines (VMs) on demand, thus providing what is commonly referred to as Infrastructure as a Service (IaaS). Here, the user is responsible for providing the operating system kernel, base OS software, and any user level software and applications they wish to run and the IaaS system provisions physical resources and instantiates the user's VMs.

Eucalyptus implements IaaS, with the key differentiations being that it is specifically designed to be easy to install and maintain in a research setting, *and* that it is easy to modify, instrument, and extend. Specifically, commercial cloud infrastructures take advantage of the ability to control the local resource configuration (hardware versioning, O.S. versioning, network and storage policies, etc.) and access to large collections of potentially expensive resources (e.g., publicly visible and routable Internet addresses). In a research setting, it is unlikely that the cloud infrastructure can mandate a specific configuration for all hardware and software it manages, nor is it possible to predicate functionality on the availability of very large resource sets.

Further, because IaaS systems each typically target a specific installation, they are not engineered with extensibility or portability as a primary concern, nor is the need for ease of system administration given primacy in the design. The difficulties are compounded by the need to be able to incorporate multiple compute clusters into a single resource pool from which cloud allocations are to be drawn. Few open-source software packages of any kind are designed to install and deploy on multiple compute clusters that then operate together as an ensemble. Thus, Eucalyptus is a relatively unique example of IaaS and also a harbinger of future multi-cluster open-source design experiences. The way in which it frames and then addresses the challenges that arise as a result, forms the basis of the contribution this paper makes.

Specifically, we describe

- a simple open architecture for implementing cloud functionality at the IaaS level,
- experiences with implementing this architecture using open-source Web-service software as the intrinsic technology, and
- performance results demonstrating the viability of the resulting cloud computing system.

IaaS, however, is not the only approach to implementing cloud computing that the commercial sector is currently pursuing. Amazon and Google also both provide Data as a Service (DaaS) capabilities, through the Simple Storage Service (S3) [13] and parts of App Engine [14] respectively, where users can both store and access massive amounts of data from the provided computational resources. In addition, Google's App Engine [14], also provides a language-level abstraction, making it generically categorizable as Platform as a Service (PaaS), where access to computational power and storage is gained through language-specific APIs and libraries. Finally, companies such as Salesforce.com [4] provide a number of high-level software service packages (e.g., Web-accessible Customer Relationship Management, Enterprise Resource Planning, Inventory Control, Payroll, etc.). This higher-level approach is often described as Software as a Service (SaaS).

We have chosen to focus Eucalyptus at the IaaS level for two reasons. First, Amazon.com's EC2 is perhaps the most commercially successful cloud computing endeavor to date and it implements IaaS. Eucalyptus is interface-compatible with EC2, making it possible to test its functionality against one of the most mature commercial examples of cloud computing. This availability of a "gold standard" greatly influenced the design since it is possible to gauge immediately how closely our open-source rendition of the functionality matches its exemplar. Second, higher-level cloud computing abstractions all seem to depend on similar IaaS functionality, at least conceptually. We do not claim that all cloud computing infrastructures include an IaaS layer in their software architecture. However, for the purposes of further research and open-source development, we speculate that self-contained IaaS functionality that can be layered upon will prove both foundational and beneficial.

2. Eucalyptus design

The Eucalyptus design is primarily motivated by two engineering goals: extensibility and non-intrusiveness. Eucalyptus is extensible as a result of its simple organization and modular design. Further, we have implemented Eucalyptus using open-source Web-service technologies, which serve to illuminate its internals. As a collection of Web services, Eucalyptus components have well defined interfaces (described by WSDL documents), support secure communication (using WS-Security policies), and rely upon industry-standard Web-services software packages (Axis2, Apache, and Rampart). This choice of implementation technology also supports the second design goal — that of nonintrusive or "overlay" deployment. We do not assume that researchers interested in Eucalyptus are necessarily willing to dedicate entire collections of machines to Eucalyptus alone (although this model of operation is also supported), nor do we assume that they are willing to allow Eucalyptus to modify the local software configuration in potentially disruptive ways. Intrusiveness is admittedly a subjective metric. For the purposes of our work, we assume that a site wishing to use Eucalyptus is willing to support virtualized execution through Xen [15] and to host Web services. With these two requirements fulfilled, Eucalyptus can be deployed and executed without modification to the underlying infrastructure.

2.1. Architectural overview

Academic research groups have access to a number of resources; for instance, small clusters, pools of workstations, and various server/desktop machines. Since public IP addresses are usually scarce, and the security ramifications of allowing complete access from the public Internet can be daunting, system administrators commonly deploy clusters as pools of "worker" machines on private, unroutable networks with a single "head node" responsible for routing traffic between the worker pool and a public network. Although this configuration provides security while using a minimum of publicly routable addresses, it usually means that, while most machines can initiate connections to external hosts, external hosts cannot typically connect to machines running within each cluster.

For example, an administrator might configure two small Linux clusters, a small server pool, and a collection of computer lab workstations. The clusters each have a single front-end machine with a publicly accessible IP address, while the nodes are connected via a private network such that they can only contact each other and their respective front-ends. The server and workstation machines have public IP addresses, but the workstations are behind a firewall and can not be contacted from the outside world. In this scenario, it is clear that it is not possible to install a fully connected system, since many of the machines can only initiate connections to external hosts or are entirely isolated from external networks. In addition, the two sets of cluster nodes may even have overlapping IP addresses since their networks are fully private and unroutable. In order to make all of these types of resources part of a single cloud, we reflect the hierarchical nature of this typical configuration in the architecture of Eucalyptus, as depicted in figure 1, where the three hierarchical levels are shown. These hierarchical components are sufficiently general to accommodate installation on common network hierarchies found within many institutions, an example of which is depicted in figure 2.

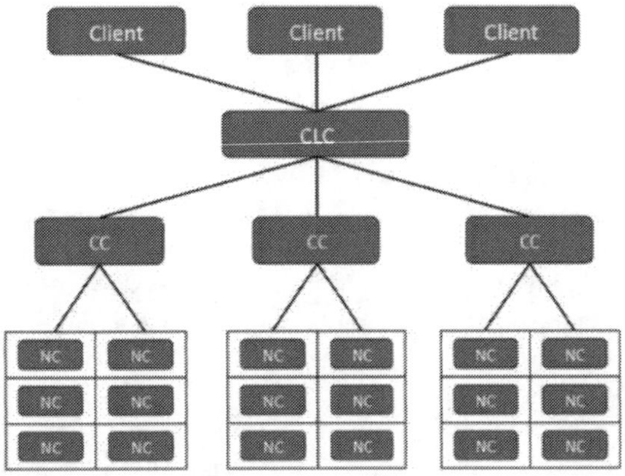

Figure 1. Eucalyptus employs a hierarchical design to reflect underlying resource topologies.

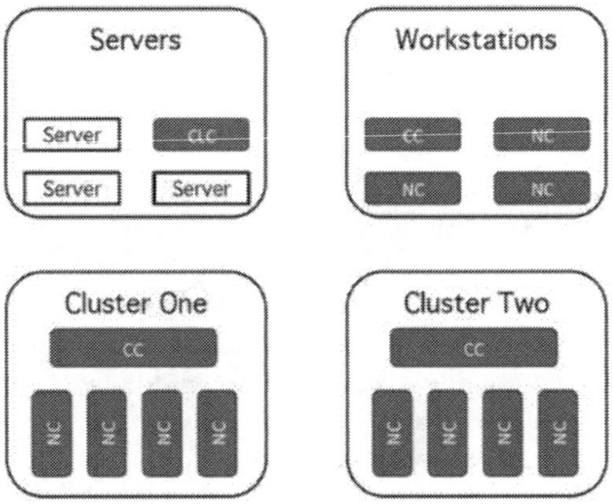

Figure 2. Example location of CLC, CC and NC components running within a typical resource environment.

2.1.1. Node Controller

The Node Controller (NC) is the component that executes on the physical resources that host VM instances and is responsible for instance start up, inspection, shutdown, and cleanup. There are typically many NCs in a Eucalyptus installation, but only one NC needs to execute per physical machine, since a single NC can manage multiple virtual machine instances on a single machine. The NC interface is described via a WSDL document that defines the instance data structure and instance control operations that the NC supports (*runInstance, describeInstance, terminateInstance, describeResource* and *startNetwork*). The run, describe, and terminate operations on an instance perform minimal system setup, followed by calls to the underlying hypervisor (Xen in the current implementation) to control and inspect running instances. The *describeResource* operation reports current physical resource characteristics (compute cores, memory, and disk capacity) to the caller and the *startNetwork* operation sets up and configures the virtual networking described in more detail in Section 2.2.

2.1.2. Cluster Controller

A collection of NCs that logically belong together report to a single Cluster Controller (CC) that typically executes on a cluster head node or server that has access to both private and public networks. The CC is responsible for gathering state information from its collection of NCs, scheduling incoming VM instance execution requests to individual NCs, and managing the configuration of public and private instance networks. The WSDL that describes the CC interface is similar to the NC interface, except that each operation is plural instead of singular (*runInstances, describeInstances, terminateInstances, describeResources*). The describe and terminate instance control operations are merely pass-thru operations to the relevant NC module. When a CC receives a *runInstances* request, it performs a simple scheduling task of determining which NCs can support the incoming instance by querying each NC through *describeResource* and choosing the first NC that has enough free resources. The CC also implements a *describeResources* operation, however, instead of reporting actual physical resources available, this operation takes as input a description of resources that a single instance could occupy, and returns the number of instances of that type can be simultaneously executed on the NCs.

2.1.3. Cloud Controller

Each Eucalyptus installation includes a single Cloud Controller (CLC) that is the user-visible entry point and global decision-making component of an Eucalyptus installation. The CLC is responsible for processing incoming user-initiated or administrative requests, making high-level VM instance scheduling decisions, processing service-level agreements (SLAs) and maintaining persistent system and user metadata.

The CLC itself is composed of a collection of services (figure 3) that handle user requests and authentication, persistent system and user metadata (e.g., VM images and ssh key pairs), and the management and monitoring of VM instances. The services are configured and managed by an enterprise service bus (ESB) [16] that publishes services and mediates handling of user requests while decoupling the service implementation from message routing and transport details. Our design emphasizes transparency and simplicity in order to foster experimentation and extension of Eucalyptus, particularly with respect to cloud behavior. To achieve extensibility at this level of granularity, the architectural components of the CLC (including, but not limited to the VM scheduler, SLA engine, and user/administrative interfaces) are mutually isolated behind well-defined internal interfaces where ESB configuration controls their orchestration. With this as a foundation, our CLC implementation can function as an Amazon EC2 work-alike by interoperating with the EC2 client tools using bothWeb-services and Query interfaces (Amazon publishes specification documents

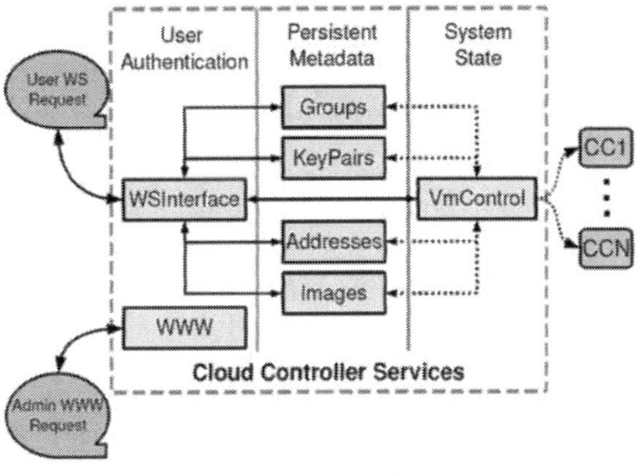

Figure 3. Overview of services that comprise the Cloud Controller. Lines indicate the flow of messages where the dashed lines correspond to internal service messages.

describing these interfaces). We chose EC2 because it is relatively mature, has a large existing user community, and because it implements a well-defined IaaS functionality. However, the interface parsing is modularized so that Eucalyptus can support different interfaces, either as a way of emulating other infrastructures or to allow interface customization.

2.1.4. Client interface

The CLC's client interface service essentially acts as a translator between the internal Eucalyptus system interfaces (i.e., the NC and CC instance control interfaces) and some defined external client interface. For example, Amazon provides a WSDL document that describes a Web-service SOAP-based client interface to their service as well as a document describing an HTTP Query-based interface, both of which can be translated by the CLC user interface service into Eucalyptus internal objects. We use JiBX [17] binding tool to specify a mapping of XML elements onto instances of Java objects, which we have used to create bindings that map the body of EC2 SOAP messages onto internal Eucalyptus objects.

The Query interface does not lend itself to this model however. First, there is no XML document to consume. Second, the authentication mechanism is different and in conflict with the WS-Security policy enforced. Third, conflicts exist between the structure of SOAP requests and Query requests for the same field of the same kind of request.

The solution stems from the observation that the Query interface for EC2 is a strict subset of the SOAP interface. As a result, we have developed a simple binding framework that maps HTTP Parameter names onto object fields guided by annotations. We then rely on annotations of the target object to aid in de-obfuscating inconsistencies such as elided lists and unwrapped complex types (i.e., field names of a child class). Ultimately, JiBX is used to marshal the bound object using the namespace for the EC2 SOAP interface. The result is two-fold: First, JiBX will validate the object that is actually a legal SOAP interface request, hence, a legal EC2 client request. Second, the marshalled XML document can be supplied as the SOAP body to allow further processing to continue along the exact same path it would have taken if the message had been SOAP to begin with.

2.1.5. Administrative interface

In addition to supporting primary tasks, such as starting and stopping instances, a cloud infrastructure must support administrative tasks, such as adding and removing users and disk images. Eucalyptus supports such tasks though aWebbased interface, implemented by the cloud controller, and command-line tools. Unlike the client interface, however, the administrative interface is unique to Eucalyptus. That is, while cloud purveyors do publish their client interfaces they do not generally publish administrators' interfaces. Thus, we have defined one for the system that is independent of any specific client interface or intrinsic IaaS functionality.

Users are added to a Eucalyptus installation either through the action of an administrator or by filling out an on-line form that is sent to the administrator for approval. Control over account creation thus rests in the hands of a human being, which we found necessary given the absence of automated approval methods, such as credit card verification used by Amazon. It is up to the cloud administrator to try to ensure that a new account will not be misused. By forcing new users to confirm their interest in the account by clicking on a link received in an email message, Eucalyptus maps the identity of a user to the their email address. If that is not sufficient, the administrator may choose to verify the identity of the applicant with the help of other information on the sign-up form. Once added, a user account can be temporarily disabled or permanently removed by an administrator. At any point, the administrator can find out which instances a user is executing and terminate them.

Currently, disk images in Eucalyptus can be added to the system only by an administrator. An image consists of a Xen-compatible guest OS kernel, a root file system image, and, optionally, a RAM disk image. Adding an image constitutes uploading these three components into the system and naming the image. After a image is added, any user can run instances of that image. Administrators may temporarily disable or permanently remove the image. Finally, the administrator is in charge of adding and removing nodes from cluster controller's configuration.

2.1.6. Instance control

Creation of virtual machine instance metadata in Eucalyptus is managed by a component of the CLC named the VmControl service. VmControl continuously maintains a simple local representation of the state of underlying resources (i.e., number of instances each CC could potentially create). When instance creation events are initiated, it coordinates with the other services in the CLC to resolve user request references to image, keypairs, networks, and security groups. Allocation then consists of validating references to metadata, application of an allocation strategy producing a "pre-allocation," meaning that as far as the VmControl component is concerned, the resources have been locally reserved. Messages are then disseminated to the CCs involved in the allocation. Each such CC will schedule the instance request to its locally controlled NCs which, finally, create the virtual machine instance itself and respond accordingly.

2.1.7. SLA implementation and management

Service-level agreements (SLAs) are implemented as extensions to the message handling service which can inspect, modify, and reject the message, as well as the state stored by VmControl. Ultimately, the VmControl rationally arbitrates access to resources and enforces system-wide or user-specific service-level agreements. These decisions require data about the state of resources that is captured in a system model and the result of update events (i.e., either a change to the model or information about a failure). We have implemented an extensible SLA scheme, which couples the state model with event handling to support further work in quantitative study of service level agreements.

The VmControl relies on a local model for decision-making purposes. To keep the model up to date, each CC is passively polled to obtain the state of its instance availability, allocations, virtual network, and registered images. Information gathered via polling is treated as *ground truth* and user requests are handled in transactions that commit only when they are reflected on the resources.

Nonetheless, the model may become inconsistent, causing the system to agree to an SLA with a user that is unsatisfiable. This can happen when when messages are lost (e.g., due to network partition) and the state of resources changes (the period between polling events can be thought of as a network partition). However, loss of messages can be identified (polling is semi-synchronous) and times when the model is in an invalid state can, ultimately, also be detected (after the system recovers and ground truth can be inspected). Consequently, the likelihood that the model will be incorrect at a given moment can be computed.

We have implemented a simple yet powerful initial SLA that allows users to control the high-level network topology of their instances. While resource providers typically think of collections of machines in terms of "clusters" or "pools", we have adopted the more general concept of "zones" that is currently used by Amazon EC2. Within EC2, a "zone" is correlated to a vague geographic location such as "east coast U.S." or "west coast U.S.", while we use the term to refer to a logical collection of machines that has several NC components and a single CC component. Eucalyptus allows users to specify a zone configuration upon instance execution, which allows an instance set to reside within a single cluster or potentially across clusters. Each configuration offers different administrative and network performance characteristics, which we explore in more detail in Section 2.2. In addition, Eucalyptus further co-opts the notion of zone, extending it to support different SLAs with respect to tradeoffs between the number of resources acquired and their relative topology. In the current implementation, the default set of zones supplied allows users to request a specific cluster, the emptiest cluster, any single cluster unless no cluster provides the minimum requested, and multiple clusters.

2.2. Virtual networking

Perhaps one of the most interesting challenges in the design of a cloud computing infrastructure is that of VM instance interconnectivity. One of the most attractive characteristics of cloud systems stems from the fact that although the underlying physical machines may have complex and restrictive networking topologies, a simpler, more configurable VM interconnection topology can be presented to the user through virtualization. When designing Eucalyptus, we recognized that the VM instance network solution must address *connectivity, isolation,* and *performance.*

First and foremost, every virtual machine that Eucalyptus controls must have network connectivity to each other, and at least partially to the public Internet (we use the word "partially" to denote that at least one VM instance in a "set" of instances must be exposed externally so that the instance set owner can log in and interact with their instances). Because users are granted super-user access to their provisioned VMs, they may have super-user access to the underlying network interfaces. This ability can cause security concerns, in that, without care, a VM instance user may have the ability to acquire system IP or MAC addresses and cause interference on the system network. In addition, if two instances are running on one physical machine, a user of one VM may have the ability to snoop and influence network packets belonging to another. Thus, in a cloud shared by different users, VMs belonging to a single cloud allocation must be able to communicate, but VMs belonging to separate allocations must be isolated. Note that current hypervisor offerings do not support this notion of grouping directly. Finally, one of the primary reasons that virtualization technologies are just now gaining such popularity is that the performance overhead of virtualization has diminished significantly over the past few years, including the cost of virtualized network interfaces. Our design attempts to maintain inter-VM network performance as close to native as possible.

Each instance controlled by Eucalyptus is given two virtual network interfaces; one is referred to as "public" while the other is termed "private". The public interface is assigned the role of handling communication outside of a given set of VM instances, or between instances within the same availability zone as defined by the SLA. For example, in an environment that has available public IP addresses, they may be assigned to VM instances at instance boot time, allowing communication both to and from the instance. In environments where instances are connected to a private network with a router that supports external communication through network address translation (NAT), the public interface may be assigned a valid private address giving it access to systems outside the local network through the NAT-enabled router. The instance's private interface, however, is used only for inter-VM communication across zones, handling the situation where two VM instances are running inside separate private networks (zones) but need to communicate with one another.

Within Eucalyptus, the cluster controller currently handles the set up and tear down of instance virtual network interfaces. The CC can be configured to set up the public interface network in three ways corresponding to three common environments we currently support. The first configuration instructs Eucalyptus to attach the VM's public interface directly to a software Ethernet bridge connected to the real physical machine's network, allowing the administrator to handle VM network DHCP requests the same way they handle regular DHCP requests. The second configuration allows the administrator to define a dynamic pool of IP addresses that will be assigned via a DHCP server that is executed by the CC. In this configuration, the administrator defines a network, an interface on the CC that is connected to that network, and a range of IP addresses that are dynamically assigned as instances are started. Finally, we support a configuration that allows an administrator to define static Media Access Control (MAC) and IP address tuples. In this mode, each new instance created by the system is assigned a free MAC/IP tuple, which is released when the instance is terminated.

Another requirement of the virtual network is that it supports instance network traffic isolation. We require that if two instances, owned by separate users, are running on the same host or on different hosts connected to the same physical Ethernet, they do not have the ability to inspect or modify each other's network traffic. To meet this requirement, each set of instances owned by a particular user is assigned a tag that is then used as a virtual local area network (VLAN) identifier assigned to that user's instances. Eucalyptussupports a VLAN enabled networking mode (called MANAGED mode) in which the system assigns VLANs dynamically to user requests. Each VM gets a private network address and "Elastic IPs" (public addresses that can persist across user requests) are supported via IP-tables at the cluster head-node.

3. Experiment

To illustrate the performance characteristics of Eucalyptus as well as to observe its functionality under user load not generated by the development team, we have installed Eucalyptus on a small research Linux cluster at our home institution and made it available for general use to the wider community as a "public cloud." The hardware configuration comprises 7 compute nodes and one head-node. The compute nodes are on an isolated network, while the frontend is publicly accessible. Each system has two Intel Xeon 3.2GHz processors, 3GB of RAM and approximately 40GB of available disk (single SCSI drive). We are running a single CLC on the front-end, a single CC on the front-end, and one NC per compute node.

Users request access to the Eucalyptus Public Cloud (EPC) by requesting credentials from the CLC through the user signup web page. Subsequent cloud allocation requests are limited to 4 instances which will be terminated automatically after 6 hours. A reverse firewall prevents EPC hosted instances from making network connections to external network addresses (public Linux distribution sites are excepted to allow instance configuration) to avoid inadvertent "spambot" hosting. Only a local EPC zone is available as an externally accessible SLA.

All experiments detailed in this section have been conducted using the EPC in the presence of ambient induced load. That is, unless otherwise indicated, we measured the performance of the EPC in the presence of load being generated by its users (i.e. in a non-dedicated mode).

3.1. Instance throughput

The first experiment we perform is designed to measure the performance of VM instance control operations. Because Eucalyptus is interface compatible with Amazon's EC2, we are able to perform the same experiments on both Eucalyptus and EC2 without customization. The primary purpose in doing so is to verify that the EC2 functionality is, indeed, fully replicated by Eucalyptus. Less rigorously, the quantitative comparison serves as a high-level test for whether our implementation is pathologically inefficient. In fact, during early phases of Eucalyptus design we discovered a number of performance "bugs" through comparisons with EC2.

Because one of the primary functions of Eucalyptus is to control the execution of VM instances on a collection of resources, we perform an "instance throughput" experiment where we measure the time from when a user wishes to execute a collection of instances to the time the instances are booted and available for use on the network. For this experiment, we measure the total time between an instance execution request to the point when we can first detect that the instance is running. In order to measure the instance state, we rely on the Amazon EC2 command-line tool "ec2- describe-instances", which simply queries the cloud server for information about a user's instances and prints the information to the user's terminal. To gather a single data point, we first take a timestamp followed immediately by a launch of an instance or set of instances using the client tool "ec2-run-instances". Then, we repeatedly poll the server using "ec2-describe-instances" until our initiated instance enters a "running" state, at which point we take another timestamp. The difference between the two timestamps constitutes a single data point that represents the number of seconds between a user instance creation request and the user becoming aware that the instance(s) are available for use. Each trial is characterized by four variables and timings are reported in seconds. The first variable is the VM type requested, where the VM type is defined as the number of cores, amount of RAM, and allocated disk space. The second variable is the instance image itself, which we control by loading identical copies inside both EC2 and the EPC. For this experiment we run trials for a "small" VM type with a corresponding image called "ttylinux" [18]; a compact Linux image that boots very quickly and offers a minimal networked Linux installation when fully booted. The third variable of interest is the number of instances simultaneously requested, which we vary from one to eight instances. The final variable is of course the system used; either Eucalyptus or Amazon EC2.

In figure 4 we show the results of the instance throughput experiment. The figure shows two empirical cumulative distribution functions that allow us to examine both the magnitude and variance of time taken to create instances in EC2 and the EPC. Each data point represents the percentage (Y axis) of instance creation trials that took at least the number of seconds denoted at the point's corresponding position on the X axis.

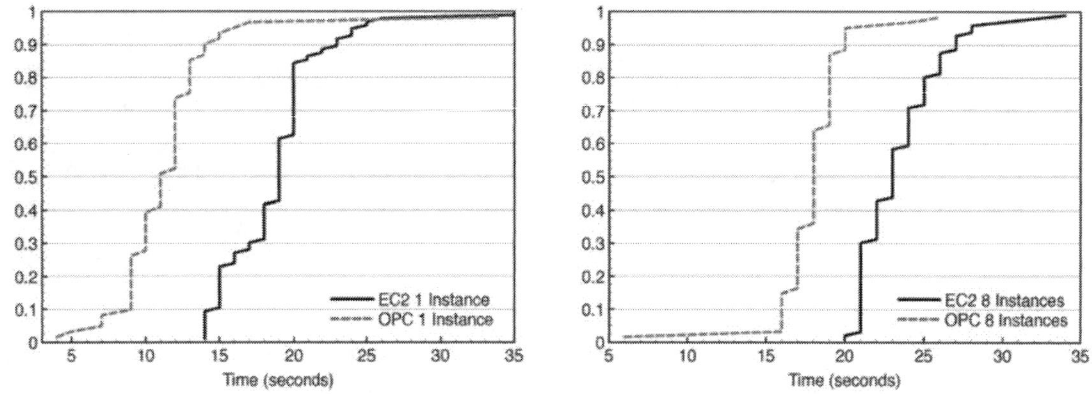

Figure 4. Empirical CDFs comparing number of seconds taken to start one and eight VM instances within EC2 and Eucalyptus.

Notice that for both cases (one and eight concurrent instance creation trials), though the range of creation times overlap, all empirical quantiles in the EPC case are lower than those of EC2. For example, 98 percent of the eight concurrent instance creation trials completed in less than 24 seconds within EPC, while only 75 percent completed in less than 24 seconds within EC2. For the one instance case, the difference is even more striking, with 98 percent of the trials completing in less than 17 seconds within the EPC and only 32 percent completing in less than 17 seconds within EC2.

This result, we believe, indicates that the Eucalyptus implementation is relatively efficient given its target environment. However it does not indicate that the EPC is outperforming EC2. The actual EC2 harnesses a vast resource pool and, as such, should almost certainly incur a measurable performance overhead over an Eucalyptus implementation running on a small cluster. At the same time, the implementation of the EPC does seem to compare well with that of the system it emulates indicating that its implementation is, at least, relatively high performance. This supposition is further supported by the (somewhat surprising) similarity in the shapes of two distribution plots. Both are unimodal with relatively similar tail weights. At present we are unable to go beyond this observation and to make a direct inference (say from confidence bounds on the variance) about the similarity of the two performance profiles however doing so is something we hope to achieve in the near future.

4. Related work

Cloud computing stems from recent innovations in operating system virtualization and scalable Internet services. It also shares intellectual underpinning with grid computing, although the precise nature of this sharing is a matter of some debate.

Machine virtualization projects producing VM hypervisor software [15, 19, 20, 21] have enabled new mechanisms for providing resources to users. In particular, these efforts have influenced hardware design [22, 23, 24] to support transparent operating system hosting. The "right" virtualization architecture remains an open field of study [25]): analyzing, optimizing, and understanding the performance of virtualized systems [26, 27, 28, 29, 30] is an active area of research. Eucalyptus implements a cloud computing "operating system" using Xen-based virtualization as its initial target hypervisor and this work, particularly with respect to performance benchmarking, serves as a starting point for studying the overheads introduced by Eucalyptus.

Thanks in part to the new facilities provided by virtualization platforms, a large number of systems have been built using these technologies for providing scalable Internet services [1, 31, 32, 33, 34, 3, 4], that share in common many system characteristics: they must be able to rapidly scale up and down as workload fluctuates, support a large number of users requiring resources "on-demand", and provide stable access to provided resources over the public Internet. While the details of the underlying resource architectures on which these systems operate are not commonly published, Eucalyptus is almost certainly shares some architectural features with these systems due to shared objectives and design goals.

In addition to the commercial cloud computing offerings mentioned above (Amazon EC2/S3, Google AppEngine, Salesforce.com, etc.), which maintain a proprietary infrastructure with open interfaces, there are open-source projects aimed at resource provisioning with the help of virtualization. Usher [35] is a modular open-source virtual machine management framework from academia. Enomalism [12] is an open-source cloud software infrastructure from a start-up company. Virtual Workspaces [36] is a Globus-based [37] system for provisioning workspaces (i.e., VMs), which leverages several pre-existing solutions developed in the grid computing arena. The Cluster-on-demand [38] project focuses on the provisioning of virtual machines for scientific computing applications. oVirt [39] is a Web-based virtual machine management console.

While these projects produced software artifacts that are similar to Eucalyptus, there are several differences. First, Eucalyptus was designed from the ground up to be as easy to install and as non-intrusive as possible, without requiring sites to dedicate resources to it exclusively (one can even install it on a laptop for experimentation). Second, the Eucalyptus software framework is highly modular, with industry-standard, language-agnostic communication mechanisms, which we hope will encourage third-party extensions to the system and community development. Third, the external interface to Eucalyptus is based on an already popular API developed by Amazon. Finally, Eucalyptus is unique among the open-source offerings in providing a virtual network overlay that both isolates network traffic of different users and allows two or more clusters to appear to belong to the same Local Area Network (LAN).

Grid computing must also be acknowledged as an intellectual sibling of, if not ancestor to, cloud computing [40, 41, 42, 43]. The original metaphor for a computational utility, in fact, gives grid computing its name. While grid computing and cloud computing share a services oriented approach [44, 45] and may appeal to some of the same users (e.g., researchers and analysts performing loosely-coupled parallel computations), they differ in two key ways. First, grid systems are architected so that individual user requests can (and should) consume large fractions of the total resource pool [46]. Cloud systems often limit the size of an individual request to be tiny fraction of the total available capacity [47] and, instead, focus on scaling to support large numbers of users.

A second key difference concerns federation. From its inception, grid computing took a middleware-based approach as a way of promoting resource federation among cooperating, but separate, administrative domains [37]. Cloud service venues, to date, are unfederated. That is, a cloud system is typically operated by a single (potentially large) entity with the administrative authority to mandate uniform configuration, scheduling policies, etc. Eucalyptus conforms to the design constraints governing cloud systems.

Several research projects and white papers in the last few years have studied the performance ramifications of deploying specific workloads (often scientific ones) in today's commercial clouds. For example, Palankar et al. [48] benchmarked Amazon's S3 cloud storage solution for scientific applications, pointing out several current characteristics of the system that need to be addressed before it is appropriate for access to scientific data. Garfinkel [49] analyzed Amazon's EC2 management, performance and security facilities and reported on their experience with moving large scale research application to the cloud. This work, while valuable by itself, could be significantly augmented through experimentation with Eucalyptus, both in terms of experimental verification and by allowing the researchers of these works to more precisely understand the measured resource performance response through system instrumentation. In addition, the performance results presented in this paper are directly relevant to these other benchmarking efforts.

Overall, we find that there are a great number of cloud computing systems in design and operation today that expose interfaces to proprietary and closed software and resources, a smaller number of open-source cloud computing offerings that typically require substantial effort and/or dedication of resources in order to use, and no system antecedent to Eucalyptus that has been designed specifically with support academic exploration and community involvement as fundamental design goals.

5. Conclusion and future work

In this work, we have presented the Eucalyptus open-source cloud computing software framework. We have shown that Eucalyptus is distinctive among other cloud computing IaaS systems in that it supports an industry standard interface (Amazon EC2), deploys as an overlay atop existing commonly encountered resource configurations (small clusters, workstation pools, etc), and has been designed as a modular system where components may be replaced or enhanced in order to foster future cloud computing research efforts. The entire Eucalyptus system is available for download and has been successfully installed both on clusters and numerous personal computing environments.

Benchmarking Eucalyptus against EC2 reveals that it is relatively efficient. While it outperforms EC2 in absolute terms, it does so in an environment with significantly fewer resources. Only when a process-level virtual network overlay is employed is performance substantially degraded. However, by adapting the concept of availability zone from EC2, Eucalyptus allows users to trade network performance for scalability explicitly though a default set of SLAs supplied with the system. This adaptation supports the claim that Eucalyptus allows new cloud computing techniques and policies to be developed. Thus we conclude that Eucalyptus is not inherently inefficient and provides facilities for cloud computing research that are otherwise unavailable.

In addition to constantly supporting new features, we are particularly interested in using Eucalyptus as a platform for experimenting with novel cloud computing concepts such as dynamic SLA generation, new virtual networking topologies for floating static IP addresses across clouds, investigations on how to implement a truly secure cloud infrastructure, and investigating novel user and administrative cloud interfaces.

References

[1] 3Tera home page `http://www.3tera.com/`

[2] Amazon.com home page `http://www.amazon.com/`

[3] Google – `http://www.google.com/`

[4] Salesforce Customer Relationships Management (CRM) system `http://www.salesforce.com/`

[5] Buyya R, Yeo C S and Venugopa S Market-oriented cloud computing: Vision, hype, and reality for delivering it services as computing utilities *Proceedings of the 10th IEEE International Conference on High Performance Computing and Communications (HPCC-08, IEEE CS Press, Los Alamitos, CA, USA) 2008*

[6] Skillicorn D 2002 *Parallel and Distributed Processing Symposium., Proceedings International, IPDPS 2002, Abstracts and CD-ROM* 247–251

[7] Chang M, He J and Castro-Leon E 2006 Service-orientation in the computing infrastructure *SOSE '06: Proceedings of the Second IEEE International Symposium on Service-Oriented System Engineering* (Washington, DC, USA: IEEE Computer Society) pp 27–33, ISBN 0-7695-2726-4

[8] Greschler D and Mangan T 2002 *Int. J. Netw. Manag.* **12** 317–321 ISSN 1099-1190

[9] Greschler D and Mangan T 2002 *Int. J. Netw. Manag.* **12** 339–345 ISSN 1099-1190

[10] Laplante P, Zhang J and Voas J 2008 *IT Professional* **10** 46–50 ISSN 1520-9202

[11] Amazon elastic compute cloud – `http://aws.amazon.com/ec2/`

[12] Enomalism elastic computing infrastructure `http://www.enomaly.com`

[13] Amazon simple storage service – `http://aws.amazon.com/s3/`
[14] Google appengine – `http://code.google.com/appengine/`
[15] Barham P, Dragovic B, Fraser K, Hand S, Harris T, Ho A, Neugebauer R, Pratt I andWarfield A 2003 Xen and the art of virtualization *SOSP '03: Proceedings of the nineteenth ACM symposium on Operating systems principles* (New York, NY, USA: ACM) pp 164–177 ISBN 1-58113-757-5
[16] Schmidt M, Hutchison B, Lambros P and Phippen R 2005 *IBM Systems Journal* **44** 781–797
[17] JiBX home page `http://jibx.sourceforge.net/`
[18] Ttylinux home page – `http://www.minimalinux.org/ttylinux/`
[19] Bellard F 2005 *Proceedings of the USENIX Annual Technical Conference, FREENIX Track* 41–46
[20] Hyper-v home page – `http://www.microsoft.com/hyperv`
[21] Vmware home page – `http://www.vmware.com`
[22] Advanced Micro Devices, AMD Inc 2005
[23] Hiremane R 2007 *Technology@Intel Magazine* 4
[24] Intel 2005 *Intel Solutions White Paper*
[25] Adams K and Agesen O 2006 A comparison of software and hardware techniques for x86 virtualization *ASPLOS-XII: Proceedings of the 12th international conference on Architectural support for programming languages and operating systems* (New York, NY, USA: ACM) pp 2–13 ISBN 1-59593-451-0
[26] Huang W, Koop M, Gao Q and Panda D Virtual machine aware communication libraries for high performance computing *Proceedings of Supercomputing 2007*
[27] Huang W, Liu J, Abali B and Panda D K 2006 A case for high performance computing with virtual machines *ICS '06: Proceedings of the 20th annual international conference on Supercomputing* (New York, NY, USA: ACM) pp 125–134 ISBN 1-59593-282-8
[28] Menon A, Cox A and Zwaenepoel W 2006 *Proc. USENIX Annual Technical Conference (USENIX 2006)* 15–28
[29] Mergen M F, Uhlig V, Krieger O and Xenidis J 2006 *SIGOPS Oper. Syst. Rev.* **40** 8–11 ISSN 0163-5980
[30] Youseff L, Seymour K, You H, Dongarra J andWolski R 2008 The impact of paravirtualized memory hierarchy on linear algebra computational kernels and software. *HPDC* (ACM) pp 141–152
[31] Amazon Web Services home page `http://aws.amazon.com/`
[32] Chang F, Dean J, Ghemawat S, Hsieh W, Wallach D, Burrows M, Chandra T, Fikes A and Gruber R 2006 *Proceedings of 7th Symposium on Operating System Design and Implementation (OSDI)* 205218
[33] Dean J and Ghemawat S 2004 *Proceedings of 6th Symposium on Operating System Design and Implementation (OSDI)* 137–150
[34] DeCandia G, Hastorun D, Jampani M, Kakulapati G, Lakshman A, Pilchin A, Sivasubramanian S, Vosshall P and VogelsW2007 *Proceedings of twenty-first ACM SIGOPS symposium on Operating systems principles* 205–220
[35] McNett M, Gupta D, Vahdat A and Voelker G M 2007 Usher: An Extensible Framework for Managing Clusters of Virtual Machines *Proceedings of the 21st Large Installation System Administration Conference (LISA)*
[36] Keahey K, Foster I, Freeman T and Zhang X 2005 *Sci. Program.* **13** 265–275 ISSN 1058-9244
[37] Foster I and Kesselman C 1997 *International Journal of Supercomputer Applications*
[38] Chase J, Irwin D, Grit L, Moore J and Sprenkle S 2003 *High Performance Distributed Computing, 2003. Proceedings. 12th IEEE International Symposium on* 90–100
[39] oVirt home page `http://ovirt.org/`
[40] Berman F, Fox G and Hey T 2003 *Grid Computing: Making the Global Infrastructure a Reality* (Wiley and Sons)

[41] Foster I and Kesselman C (eds) 1998 *The Grid – Blueprint for a New Computing Infrastructure* (Morgan Kaufmann)
[42] NSF TeraGrid Project `http://www.teragrid.org/`
[43] Tannenbaum T and Litzkow M 1995 *Dr. Dobbs Journal*
[44] Foster I, Kesselman C, Nick J and Tuecke S 2002 The physiology of the grid: An open grid services architecture for distributed systems integration URL `citeseer.ist.psu.edu/foster02physiology.html`
[45] Foster I, Kesselman C and Tuecke S 2001 *Int. J. High Perform. Comput. Appl.* **15** 200–222
[46] Ostriker J P and NormanML 1997 *Commun. ACM* **40** 84–94 ISSN 0001-0782
[47] Amazon Elastic Compute Cloud (Amazon EC2) `http://aws.amazon.com/ec2/`
[48] Palankar M R, Iamnitchi A, Ripeanu M and Garfinkel S 2008 Amazon s3 for science grids: a viable solution? *DADC '08: Proceedings of the 2008 international workshop on Data-aware distributed computing* (New York, NY, USA: ACM) pp 55–64 ISBN 978-1-60558-154-5
[49] Garnkel S L 2007 An evaluation of amazons grid computing services: Ec2, s3 and sqs Tech. Rep. TR-08-07 Center for Research on Computation and Society, School for Engineering and Applied Sciences, Harvard University 12

Building and testing a production quality grid software distribution for the Open Science Grid

A Roy on behalf of the OSG consortium

University of Wisconsin-Madison, Department of Computer Sciences, Madison, WI, USA

Email: roy@cs.wisc.edu

Abstract. We describe how we integrate, build, and test the Open Science Grid (OSG) software stack, which is used to provide a production quality infrastructure for grid sites and users across OSG to run their grid jobs. The software stack is sufficiently complex that building and testing the software stack is non-trivial and requires a variety of types of testing including internal integration testing as well as end-user testing.

1. Introduction

The Open Science Grid (OSG) [1] is a national, distributed computing grid for data-intensive research. It brings together resources from campuses and research communities around the nation and the world into a common, shared grid infrastructure via a common set of software. OSG is a production facility that provides millions of CPU hours per week to scientific researchers.

In order to provide this production quality facility, the OSG provides a reference software distribution to all sites and users. (A *site* accepts computational jobs and/or data to be stored, while a *user* accesses various sites.) While sites and users can choose to provide their own software, the OSG provides testing and support for the reference distribution, and most people in OSG choose to use it.

The OSG software stack is divided into two pieces. The first is the Virtual Data Toolkit (VDT). [2] This contains nearly the entire OSG software stack. (The name "Virtual Data" is historical and derives from the goals of a precursor project, iVDGL.) An important goal of the VDT is to be a grid software distribution that is grid-agnostic. That is, it does not contain specific configuration required to be part of a specific grid. In fact, portions of the VDT are used by a variety of other grids, such as LCG [3], TeraGrid [4], and campus/regional grids such as NYSGrid [5].

The second portion of the OSG software stack contains configuration information specific to OSG. For example, there is a configuration program that knows how to adjust the settings for various software components to point at central OSG services, such as the GRATIA accounting service.

The effort to create the OSG software stack is primarily an integration effort: we write almost none of the software that is in the software stack, but we instead build, test, and package a set of software that we know works together reliably.

This paper describes how we build, test, and distribute the OSG software stack.

1.1. Goals of the OSG software stack

There are three major goals for the OSG software stack:

- The software should include all of the grid software needed by the OSG stakeholders. However, we do not distribute the science applications themselves—just the grid software.
- It should be easy to install and configure the software.
- It should support as many operating system/CPU architecture combinations (a.k.a. *platforms*) as possible. We do not mandate a single platform because sites in OSG are autonomous: they do not receive funding to join OSG, and they make their own choices about the kinds of hardware and software they deploy. Also, it is our goal for users to be able to access as many resources as possible, which requires us to support a variety of platforms. In VDT 2.0.0 (the latest release as of this writing), the VDT supports 22 different platforms [6].

1.2. Design choices in the OSG software stack

We have made several design decisions about the software stack.

- *Binary distributions*: Similar to most OS distributions, but different from some such as FreeBSD Ports, we distribute binaries, not source. In our experience, building all of the different software is both complex and slow, so we prefer to do this on behalf of users. For example, software components require many other pieces of software in order to build (such as GNU make, autoconf, or Ant), and often have very specific requirements on the versions of this underlying software. While all of this can be accommodated on a user's computer, it is complex and error-prone.
- *Configuration*: We provide tools for configuring the software to work together in specific scenarios. For example, we provide configuration to install specific web applications into Tomcat, to connect Tomcat with Apache, and to configure specific security settings across Tomcat and Apache.
- *Pacman*: The majority of our software distribution is done with Pacman. [7] Pacman is not a widely used tool, but it has several significant advantages. In particular, users can easily install as any user on the system, not just the root user (like the usual usage of RPM), and this enables easy client tool installations. Also, users can easily install the software simultaneously in multiple locations, which can be very useful for using and testing multiple versions of the software. We do distribute some software as native packages (RPMs and debs) and will increase support in the future: see Section 5 for more details.
- *Focus on integration*: We focus on integrating software that we know works together. In some cases, we will ship software that users might expect to get from their OS (such as MySQL) because we require specific versions that are known to work together reliably.

1.3. An example software integration

The OSG software stack includes a wide variety of grid software such as Condor [8], Globus [9], dCache [10], VOMS [11] the OSG Matchmaker [12] and much more. It also includes non-grid support software such as Apache [13], MySQL [14] and more. As an example of how the OSG software stack integrates these different software components, consider VOMS.

VOMS is software that maintains membership lists for Virtual Organizations (VOs), and issues X509 attribute certificates that can be used as to authorize users. VOMS includes a back-end daemon and a front-end web interface, which itself relies on Tomcat, and relies on a specific version of MySQL. There is also a plug-in for Globus (which accepts job submissions) that compiles against the VOMS libraries and uses the VOMS attribute certificate for authorization. All of these components need to be built, packaged and tested against each other, and configuration scripts were created to link them together easily for users.

2. Related work

Linux distributions face a similar problem to the OSG software stack, though they focus on a single operating system. Also they include software that is lower-level (such as the kernel or basic Linux commands) than are included in the OSG software stack.

Other grid efforts also build and test software stacks. A notable example is the gLite software distribution [15] which is part of the European EGEE project. There are many interesting comparisons and contrasts to draw which we are unable to include due to space constraints.

3. Building the OSG software stack

A principle of our software distribution is that we must be able to build all components that we ship from their source code. While there are a few exceptions to this rule (which we are working to correct), the VDT currently includes more than forty software builds. Building the software can be complicated because we build each one of the software components on many platforms. For example, in VDT 2.0.0 (the latest release at the time of writing), we built most software on 14 to 16 platforms, where "platform" is defined as the combination of a specific OS version (such as RedHat Enterprise Linux 5.x) and CPU architecture (such as x86). In some cases, we do not support a software component on all platforms, or we re-use software on multiple platforms.

Building forty software components on sixteen platforms requires careful management. We rely on the NMI Build and Test Facility [16] at the University of Wisconsin-Madison, which uses the Metronome software [17] to manage builds. We have built a thin layer on top of Metronome that allows us to extract build information from our source code repository, submit a build to Metronome, and stage the binaries from the build facility into the VDT software distribution. While builds often have to be carefully crafted in order to run successfully on multiple platforms, once they have been written we can easily manage the builds and use the resulting software.

To support our goal of smoothly integrating software for users, we sometimes patch the software to provide bug fixes or features needed by OSG stakeholders.

4. Testing the OSG software stack

When supporting the wide variety of platforms we support in the complex software environment we have in OSG, testing is absolutely critical to our success. We have developed a software testing process that works quite well and helps us iron out many of our bugs. An important aspect of this process is that we not only test the software integration, but we also test that the software installation process works correctly.

Our software testing process has three steps: internal daily automated testing by the VDT team, validation on a small testing grid in OSG (the validation testbed), and end-user testing on a larger testing grid in OSG. Each one of these steps is crucial to our testing process.

4.1. VDT internal testing

During development of a new VDT release, we perform daily, automated, integration tests of most of the software in the VDT. By *integration tests*, we try to differentiate these tests from the unit and functional tests that software developers often think about, though there are similarities. We expect that developers will do unit and functional testing and we only perform those tests lightly, or to ensure that old bugs are not re-introduced. Instead our integration tests verify that we have built the software correctly and it works together as expected. Recall our example of how we distribute VOMS in Section 1.3. In our testing, we will probe the entire system: can we create a VOMS attribute certificate and use it to authorize a job submission? This will test the breadth of our installation.

We do not label a platform as "supported" unless we run our tests on that platform. In some cases, we support platforms that are very similar to each other; for example we support CentOS 5 which is a freely-available rebuild of Red Hat Enterprise Linux 5. Because one of our major OSG stakeholders uses CentOS 5, we test on it in addition to Red Hat, to be absolutely certain that the software works on both platforms.

Our testing framework installs the software just as a user would install it, so we are not only testing the software, but the installation process. This is particularly important because we are testing an integrated entire software distribution.

We do not stop testing software once it is released. Because we are testing the installation process and this involves downloading the software from our web site, we want to ensure that nothing has been accidentally changed that would break a user's installation.

We normally run the tests continuously, but we can also run the tests on demand for immediate developer feedback.

Results of tests are examined in two ways. First, there is a daily email at 8:00 am, so the VDT developers can look over the tests from the previous night for any problems that need to be addressed. Second, the results of the daily builds are stored in a database so that we can look at past results to compare them.

Finally, we leave the software installed for a day or two, so test failures can be easily examined. Recall that the VDT software can be installed multiple times on a single computer using Pacman. We take advantage of this and keep many recent installations in place. The reports for our test results include the location on disk of the installation, and a developer can often quickly debug a failed test because the installation is still there. Installations are automatically rotated to ensure that we leave sufficient disk space for new tests.

4.2. Validation Testbed

When the VDT team produces a new VDT release or pre-release, it is given to the OSG Validation Testbed (VTB). [18] The VTB team creates the OSG software stack; recall that this is a small addition to the VDT to provide OSG-specific configuration. Once it is created, three to five grid sites install the software. The testers ensure that the installation process goes smoothly with no errors. They then run a small set of "smoke tests", such as a running a simple grid job or transferring a file. While the tests are not extensive, they provide two important kinds of feedback.

First, they are external testers not connected with the VDT development, so they can provide feedback on all aspects of the software installation and usage that is not colored by too intimate a knowledge of the VDT.

Second, their environments are not identical to the ones tested by the VDT team. There is a wide variety of software that they have pre-installed: different Linux kernels, different revisions of the OS, different batch systems, different file systems, and more. Although it cannot reproduce the full range of conditions that will be experienced by end-users, it does provide an important variety that helps eliminate bugs.

The VTB provides feedback and bug reports and these are then are addressed by the VDT team and result in a new version of the VDT and OSG software stacks.

The VTB testing is usually relatively quick: 2–4 weeks. Because it is a small group of testers, interaction with the VDT team tends to be intimate and quick, and ideally there is a quick turnaround on problems.

4.3. Integration Testbed

No matter how good a testing process is, it is important to include end-user testing in scenarios that are as close to production as possible. The OSG Integration Testbed (ITB) [19] is a grid of 10–20 sites. Each site donates a small number of computers and storage to provide a production-like environment to run grid jobs. Unlike the VTB, the testing is not done by the sites, but is done by the end-users in OSG. They are members of the various *Virtual Organizations* (VOs) that have scientific applications that need to be run. They dedicate some time to the ITB testing process and verify that their scientific workflows operate correctly with the new version of the OSG software stack. With a combination of the scientists verifying their workflows and the site administrators watching their ITB sites, we have a good assurance that a new software release works as advertised.

Usually the ITB process does discover a number of problems, which are resolved, provided as software updates and re-tested.

OSG has about 15 active VOs at any given time, but not all of them are able to dedicate time to ITB testing. Therefore, our testing is unable to be 100% comprehensive. However, we find that this process helps fix problems before the software is released to production, gives confidence to our users that the software really works, and assures VOs and production grid sites that it is safe to upgrade to the new version. This involvement from the community in testing the software has been invaluable.

4.4. Production

After the software has been tested, it is released as a new production release. In our experience, it takes a long time (many months) for grid sites to upgrade to new versions of the software, even if it is generally agreed to work and be a useful upgrade. Site administrators are busy with a wide variety of projects and need to carefully schedule downtime at their sites. Therefore OSG always supports both the current release and the previous release in tandem. As much as possible, we maintain backwards compatibility between the current release and previous release.

4.5. Security fixes or minor updates

The process described above is used for major upgrades that require significant testing. However, we often have minor bug fixes, and these do not require the heavyweight process described above. Similarly, security fixes must sometimes be delivered much more quickly than this process allows.

In these cases, we rely on either just the VDT testing, or perhaps also some VTB testing to validate that the software update will not change behavior. While there is a risk of introducing new bugs or incompatibilities, the reduced time to ship important updates is usually worth the risk.

5. Future plans and conclusions

As any software developer knows, testing is never thorough enough or complete enough. There is always more testing that can be accomplished. In the future, we plan to expand the thoroughness of the VDT internal testing, and add some automated test suites that VTB and ITB grid sites can use to verify the basic functionality of the software.

As successful as Pacman has been in meeting many of our packaging needs, it is often unsatisfying to system administrators who prefer packages that are installed just like the software they receive from their OS vendor. Therefore we have begun producing native packages (RPMs and Debian packages) to better meet the needs of system administrators, and we expect to improve our support for native packages in the coming months.

Even with more work to be done, we have found this three-step process (internal testing, external validation, end-user validation) to be enormously useful in providing a stable, production-ready grid software distribution.

Acknowledgements

Although the author list for this paper is short, many people have contributed to the work described here. While it is too difficult to list everyone, a few groups and people must be acknowledged. The current VDT team includes Tim Cartwright, Alan De Smet, Scot Kronenfeld, Tanya Levshina, and Neha Sharma. Earlier members of the VDT Team Kirill Kireyev and Nate Mueller were instrumental in the initial development of the VDT testing framework. The OSG Integration Testbed effort led by Rob Gardner and the Validation Testbed effort led by Suchandra Thapa have provided not only tireless testing, but also many refinements on how we should test. The OSG Operations Group, currently led by Rob Quick, has helped us move from testing to production many times. The OSG management, particularly Miron Livny and Ruth Pordes, have provided excellent technical guidance and have helped us keep our sights on the big picture. And many more people have provided daily input, feedback, and guidance: a great thanks to you all.

References

[1] The Open Science Grid Web Site, http://www.opensciencegrid.org
Avery, Paul 2007 Open Science Grid: Building and sustaining general cyberinfrastructure using a collaborative approach *First Monday* **12**
[2] The Virtual Data Toolkit web site http://vdt.cs.wisc.edu
[3] Worldwide LHC Computing Grid web site, http://lcg.web.cern.ch/LCG/
[4] TeraGrid web site, http://www.teragrid.org/
[5] NYSGrid web site, http://www.nysgrid.org
[6] VDT 2.0.0 supported platforms, http://vdt.cs.wisc.edu/releases/2.0.0/requirements.html
[7] Pacman web site, http://atlas.bu.edu/~youssef/pacman/
[8] Litzkow M, Livny M, and Mutka M 1988 Condor - A Hunter of Idle Workstations, *Proceedings of the 8th International Conference of Distributed Computing Systems*, 104-11
Thain D, Tannenbaum T, and Livny M, Distributed Computing in Practice: The Condor Experience *Concurrency and Computation: Practice and Experience*, **17** 323-356
[9] Foster I, Kesselman C 1997 Globus: A Metacomputing Infrastructure Toolkit, *International Journal of Supercomputing Applications* **11** 115-28
Foster I 2006 Globus Toolkit Version 4: Software for Service-Oriented Systems, *IFIP International Conference on Network and Parallel Computing* 2-13
[10] Ernst M, Fuhrmann P, Gasthuber M, Mkrtchyan T, Waldman C, 2001 dCache: a distributed storage data caching system, *Computing in High-Energy Physics*
[11] Alfieri R, Cecchini R, Ciaschini V, dell'Agnello L, Frohner Á, Gianoli A, Lörenty K, and Spataro F 2003 VOMS, an authorization system for virtual organizations *European Across Grids Conference* 33-40
[12] OSG Matchmaker web site, http://osgmm.sourceforge.net/
[13] Apache httpd web site, http://httpd.apache.org/
[14] MySQL web site, http://www.mysql.com/
[15] gLite web site, http://glite.web.cern.ch/glite/
[16] NMI build and test web site, http://nmi.cs.wisc.edu/
[17] Pavlo A, Couvares P, Gietzel R, Karp A, Alderman I, Livny M, and Bacon C. 2006 The NMI Build and Test Laboratory: Continuous Integration Framework for Distributed Computing Software, *Proceedings of LISA '06: Twentieth Systems Administration Conference* 263-73
[18] OSG validation testbed web site, https://twiki.grid.iu.edu/bin/view/Integration/ValidationTestbed
[19] OSG integration testbed web site, https://twiki.grid.iu.edu/bin/view/Integration/WebHome

Journal of Physics: Conference Series **180** (2009) 012053 doi:10.1088/1742-6596/180/1/012053

FastBit: interactively searching massive data

K Wu[1], S Ahern[2], E W Bethel[1,3], J Chen[4], H Childs[5], E Cormier-Michel[1], C Geddes[1], J Gu[1], H Hagen[3,6], B Hamann[1,3], W Koegler[4], J Lauret[7], J Meredith[2], P Messmer[8], E Otoo[1], V Perevoztchikov[7], A Poskanzer[1], Prabhat[1], O Rübel[1,3,6], A Shoshani[1], A Sim[1], K Stockinger[1,10], G Weber[1,3] and W-M Zhang[9]

[1] Lawrence Berkeley National Laboratory, Berkeley, CA 94720, USA.
[2] Oak Ridge National Laboratory, Oak Ridge, TN 37831, USA.
[3] University of California - Davis, Davis, CA 95616, USA.
[4] Sandia National Laboratories, Livermore, CA 94551, USA.
[5] Lawrence Livermore National Laboratory, Livermore, CA 94550, USA.
[6] Technische Universität Kaiserslautern, Erwin-Schrödinger-Straße, Kaiserslautern, Germany.
[7] Brookhaven National Laboratory, Upton, NY 11973, USA.
[8] Tech-X Corporation, Boulder, CO 80303, USA.
[9] Kent State University, Kent, OH 44240, USA.
[10] Current address: Credit Suisse, Zurich, Switzerland.

Abstract. As scientific instruments and computer simulations produce more and more data, the task of locating the essential information to gain insight becomes increasingly difficult. FastBit is an efficient software tool to address this challenge. In this article, we present a summary of the key underlying technologies, namely bitmap compression, encoding, and binning. Together these techniques enable FastBit to answer structured (SQL) queries orders of magnitude faster than popular database systems. To illustrate how FastBit is used in applications, we present three examples involving a high-energy physics experiment, a combustion simulation, and an accelerator simulation. In each case, FastBit significantly reduces the response time and enables interactive exploration on terabytes of data.

1. Introduction

Current generations of scientific instruments [1, 2] and computer simulations [3, 4] are capable of producing massive amounts of data. However, a relatively small percentage of records often hold the key to the insight sought by scientists. For example, the high-energy physics experiment STAR [5] collects billions of events of colliding particles, however, only a few hundred of them might be the rare events having the unique signature of Quark-Gluon Plasma – finding evidence of Quark-Gluon Plasma is one of the key scientific objectives of STAR. Locating a relatively small number of data records distributed throughout a huge collection is a common challenge in many scientific fields. Database Management Systems (DBMS) are designed to handle such tasks, however, existing DBMS are not well-suited for scientific data explorations.

The existing DBMS systems are designed for transactional applications where data records are frequently modified such as modifying one's bank account during an ATM withdrawal. In contrast, scientific applications typically do not perform such transactions. Most interactions with a scientific data set are to select a subset of records for further analyses, for example, selecting a number of collision events with certain properties and computing a histogram of their

RID	**A**	bitmap index =0	=1	=2	=3
1	0	1	0	0	0
2	1	0	1	0	0
3	2	0	0	1	0
4	2	0	0	1	0
5	3	0	0	0	1
6	3	0	0	0	1
7	1	0	1	0	0
8	3	0	0	0	1
		b_1	b_2	b_3	b_4

Figure 1. An illustration of the basic bitmap index for a column **A** that can only take on four distinct values from 0 to 3. Note RID is the short-hand for "row identifiers".

energy distribution, or selecting regions with a high temperature in a combustion simulation and computing the amount of heat generated from each region. Such analysis tasks are generally supported by a data warehousing system and require a different data access technology [6–8].

A familiar search technology that routinely handles a large amount of data is the web search engine [9, 10]. The key underlying technology is the inverted index designed for text data. Since the bulk of scientific data are numerical values, this indexing technology is not applicable. However, other technologies employed by search engines such as the parallel iterators can be used effectively for scientific applications.

We have been working on a set of techniques specifically designed for scientific data. They are based on the bitmap index [11, 12]. The basic bitmap index is only effective on low-cardinality variables that have a small number of distinct values, such as the gender of a customer or the state of a customer's address. However, scientific data usually contains variables with very high cardinalities, such as floating-point values that hardly ever repeat. In Section 2 of this paper, we review key techniques to make bitmap indexes efficient on high-cardinality variables.

The software containing our bitmap indexes is called FastBit[1]. Since its release in 2007, it has been downloaded thousands of times, received an R&D 100 award, contributed to a number of PhD theses [13, 14]. It has also been integrated with a number of popular data management and analysis systems, such as HDF5 [15, 16] and VisIt [17, 18]. Many such work are performed without direct involvement from any FastBit developer [19, 20]. For example, a group in Brazil used FastBit to develop an indexing method for geographical data warehouses that was 10 - 20 times faster than existing techniques [20]. A group in Germany developed a drug screening software that was 100s of times faster than competing versions [19]. In Section 3 of this paper, we review three use cases that do involve the developers. These applications use very distinct features of FastBit to locate individual records satisfying specified conditions (Sec. 3.1), to find connected records (Sec. 3.2), and to compute conditional histograms (Sec. 3.3).

2. Core Bitmap Indexing Techniques

In this section we review the core bitmap indexing techniques used in FastBit. We start with a brief description of the basic bitmap index.

2.1. Basic bitmap index

Figure 1 shows a logical view of the basic bitmap index. Conceptually, this index contains the same information as a B-tree [12, 21]. The key difference is that a B-tree would store a list of

[1] FastBit software is available from `https://codeforge.lbl.gov/projects/fastbit/`.

Row IDentifiers (RIDs) for each distinct value of column **A**, whereas a bitmap index represents the same information as sequences of bits, which we call *bitmaps*. In this basic bitmap index, each bitmap corresponds to a particular value of the column. A bit with value 1 indicates that a particular row has the value represented by the bitmap.

The bitmap indexes are particularly useful for query-intensive applications, such as data warehousing and on-line analytical processing. One of the key reasons is that queries can be answered with bitwise logical operations on the bitmaps. In the example shown in Figure 1, a query "$\mathbf{A} < 2$" can be answered by performing bitwise OR on b_1 and b_2 (i.e., $b_1 \mid b_2$). Since most computer hardware support such bitwise logical operations efficiently, the queries can be answered quickly in general. Another key reason is that answers from different bitmap indexes can be easily combined. This is because the answer from each bitmap index is a bitmap and combining the different answers simply requires additional bitwise logical operations. Since combining answers from different indexes swiftly is such an important consideration, a number of DBMS that do not support bitmap indexes, such as PostgreSQL and MS SQL Server, even convert intermediate solutions to bitmaps so they can be combined them more effectively.

The biggest weakness of the basic bitmap index is that its size grows linearly with the number of distinct values. Next we review different strategies to control the index sizes and improve the query response time, namely, compression, encoding and binning.

2.2. Bitmap Compression

Each bitmap in a bitmap index usually contains a large number of 0s, making it easy to compress. To answer queries efficiently, researchers have developed specialized compression methods that offer faster bitwise logical operations and consequently faster query response time [22–24]. The most widely used bitmap compression method is Byte-aligned Bitmap Code (BBC) [25]. We developed a new bitmap compression method called Word-Aligned Hybrid (WAH) code and showed it to perform bitwise logical operations more than 10 times faster than BBC [26, 27].

WAH gains its speed partly from its simplicity. For long sequences of 0s or 1s, it uses run-length encoding to represent them and for relatively short sequences of mixed 0s and 1s, it represents the bits literally. Hence, it is a hybrid of two methods. Another key feature that enables it to achieve performance is that the compressed data is word-aligned. More specifically, WAH compressed data contains two types of words; *literal* words and *fill* words.

As illustrated in Figure 2(a), a *literal* word contains one bit to indicate its type and uses the remaining bits to store the bitmap literally. A *fill* word similarly needs 1 bit to indicate its type. It uses another bit to indicate whether the bits are all 0s or all 1s, and the remaining bits are used to store the number of bits in a bitmap it represents. The number of bits represented by a WAH *fill* word is always a multiple of the number of bits stored in a literal word.

Figure 2(b) illustrates the effects of compression on overall query response time. In this test, the commercial implementation of the bitmap index (marked as "DBMS bitmap index") uses BBC compression, while "FastBit index" uses WAH compression. The query response time reported are average time values over thousands of ad hoc range queries that produce the same number of hits. Overall, the WAH compressed indexes answer queries about 14 times faster than the commercial bitmap indexes.

2.3. Bitmap Encoding

Bitmap encoding techniques manipulate the bitmaps produced by the basic bitmap index to reduce either the number of bitmaps in an index or the number of bitmaps needed to answer a query. For example, to answer a one-sided range query of the form "$\mathbf{A} < 3$" on the data shown in Figure 1, one needs to OR three bitmaps b_1, b_2 and b_3. It is possible to store C bitmaps that correspond to $\mathbf{A} \leq a_i$ for each of the C distinct values of $\mathbf{A}$, where C is called the *column cardinality* of **A**. Such a bitmap index would have the same number of bitmaps as the basic

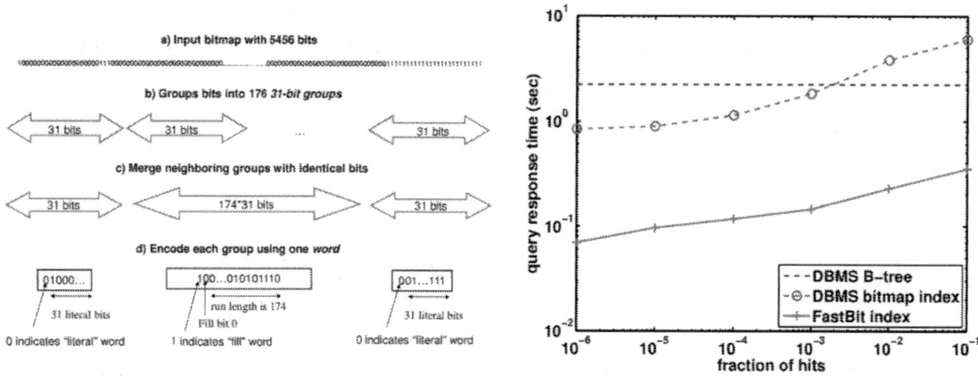

(a) an illustration of WAH (b) time (s) to answer queries

Figure 2. The effects of compression on query response time. The faster WAH compression used in FastBit reduces the query response time by an order of magnitude (Timing measurements collected with a 400 MHz UltraSPARC II CPU [27, 28]).

bitmap index, but can answer each one-sided range query by reading just one bitmap. This is the *range encoding* [11]. Along with the *interval encoding* and the *equality encoding* (used in the basic bitmap index), there are three basic bitmap encoding methods. Next, we briefly explore strategies to compose more advanced encodings using these basic methods.

The two common strategies of composing bitmap encodings are multi-component encoding [11] and multi-level encoding [29, 30]. The central idea of multi-component encoding is to break the key value corresponding to a bitmap into multiple components in the same way an integer number is broken into multiple digits in a decimal or binary representation. In general, using more components reduces the number of bitmaps needed, which may reduce the index size. Carrying this to the extreme, we can make all base sizes as small as possible (where the minimum is 2). This particular multi-component encoding can be optimized to be the binary encoding. This encoding produces the minimum number of bitmaps and the corresponding index size is the smallest without compression.

The new class of encoding method we proposed was the multi-level index composed of a hierarchy of nested bins on the base data. Each level in such an index is a complete index on its own, and can answer queries on its own or in combination with other levels. This flexibility can be exploited to achieve better performance [31]. Figure 3 shows some timing measurements of five multi-level and multi-component indexes. These timing measurements indicate that two-level encodings, especially the range-equality encoding RE and the interval-equality encoding IE, can be as much as ten times faster than the basic bitmap index (marked E1). On average, the two-level encoded indexes are about 3 to 5 times faster than the basic bitmap index and the binary encoded index (BN).

2.4. Binning

Scientific data are often computed or collected with high precision. For example, the temperature and pressure in a combustion simulation are computed with 64-bit floating-point values having 16 decimal digits of precision. To capture the full precision of the data in a bitmap index, one typically needs millions of bitmaps in the basic bitmap index. Such indexes can be large and slow to work with, even with the best compression and bitmap encoding techniques. However, we observe that an exploratory process does not usually query data at the full precision. For

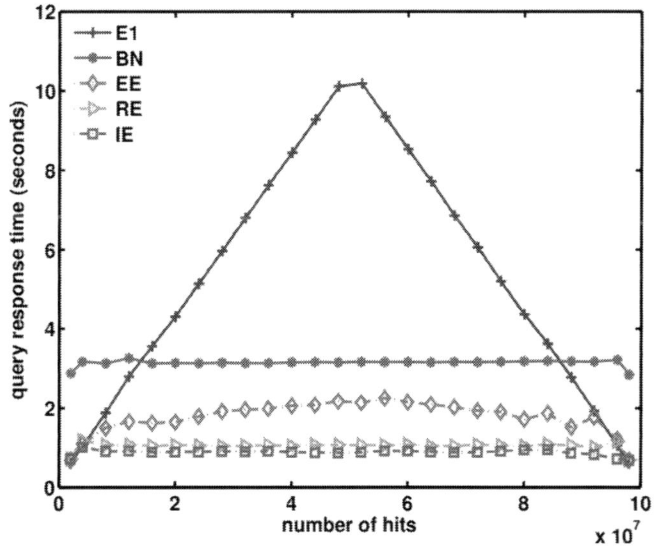

Figure 3. The average query response time of five different bitmap encoding methods with WAH compression (BN: binary encoding, E1: the basic one-component equality encoding, EE: two-level equality-equality encoding, RE: two-level range-equality encoding, IE: two-level interval-equality encoding; timing measurements collected with a 2.8 GHz Pentium 4 [31]).

example, a typical query involving pressure is of the form "pressure $> 2 \times 10^7$ Pascal." In this case, the constant in the query expression has only one significant digit. To answer all such queries with one-significant-digit constants, we only need to make sure all the values are binned to one significant digit, which only requires 9 bins for each distinct exponent in the data. Even if the user requires more than one significant digit, typically, 2 or 3 significant digits are enough. The number of bins needed are still modest.

Binning can reduce the number of bitmaps in an index and improve the query response time as well. However, for some queries, we have to go back to the base data in order to answer the queries accurately. For example, if we have 99 equal-width bins for pressure between 0 and 10^9 (with 100 bin boundaries at $i \times 10^7, i = 0, \ldots, 100$), then the query condition "pressure $> 2.5 \times 10^7$" cannot be resolved with the index only. More specifically, bins 0 and 1 contain records that satisfy the query condition, and bins 3 and onwards contain records that do not satisfy the condition, but records in bin 2 (representing $2 \times 10^7 \leq$ pressure $< 3 \times 10^7$) need to be examined further. In this case, we say the records in bin 2 are *candidates* of the query. The process of determining which candidate actually satisfies the condition is called a *candidate check*. When a candidate check is needed, it often dominates the total query response time.

We have studied different approaches to minimize the impact of candidate checks. One is to reorder the expression being evaluated to minimize the overall cost of candidate checks for a query [33]. Another approach is to place the bin boundaries to minimize the cost of evaluating a fixed set of queries [34, 35]. More recently, we proposed a new approach to more directly reduce the candidate check cost by providing a clustered copy of the base data named Order-preserving Bin-based Clustering (OrBiC) [32]. This approach organizes all values of each bin contiguously on disk to reduce the I/O time needed for candidate checks. Figure 4 shows some performance numbers to illustrate the advantages of this new approach. Figure 4(a) shows the binned index with OrBiC outperforms the one without OrBiC for all query condition tested, and Figure 4(b) shows the average time used by a binned index with OrBiC to be less than that used by the

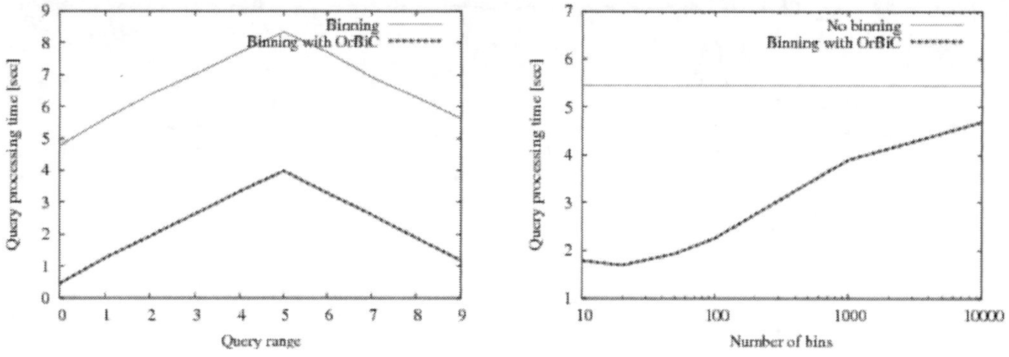

(a) individual query (b) average time

Figure 4. Time needed to process range queries using indexes with and without Order-preserving Bin-based Clustering (OrBiC) (Timing measurements collected on a 2.8 GHz Pentium 4 [32]).

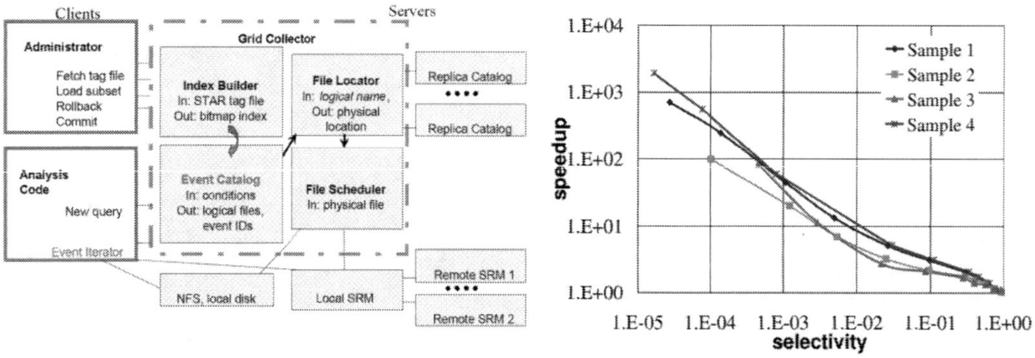

(a) Schematics of Grid Collector (b) Speedup using Grid Collector

Figure 5. Using Grid Collector can significantly speed up analysis of STAR data (Timing measurements collected on a cluster with 2.8 GHz Intel Xeon processors [36, 37]).

precise index (with no binning).

3. Extended Technologies to Support Applications

After reviewing the core techniques behind the FastBit searching capability, we now show how this basic searching capability enables more complex searching operations.

3.1. Smart Parallel Iterator

A large data set is often organized as many files each containing a number of data records, and the files are usually stored on massive storage systems. To perform an analysis, one needs a cluster of computers to hold the necessary files. The analysis system needs to retrieve and replicate the files on the cluster. Managing the file replication and extracting the desired records can be challenging. The mapReduce systems from google and Yahoo! automate some of these operations [38]. The core concept in such a system is the parallel iterator that coordinates the

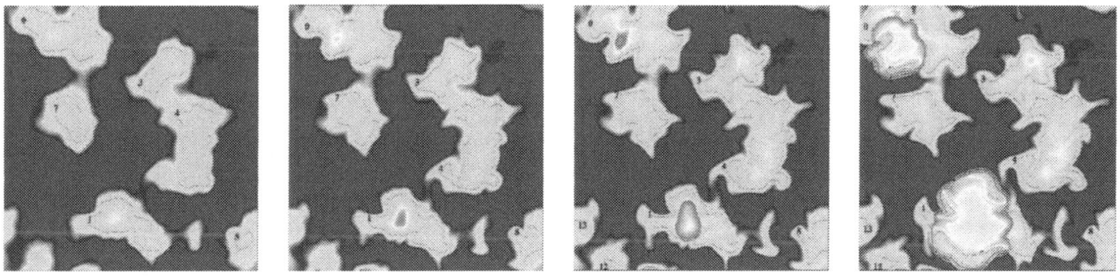

Figure 6. Tracking the evolution of ignition kernels in a combustion simulation [39].

accesses to files and records. In this example, we demonstrate how FastBit fits into an efficient parallel iterator system.

The version of parallel iterator system we built was called Grid Collector. The prototype was used in the STAR experiment and integrated into the STAR Data Analysis Framework [5, 37]. The parallel iterator extended the functionality of the Event Iterator in the STAR analysis framework, and could be used transparently with existing analysis programs. Figure 5(a) shows the schematics of Grid Collector. FastBit is responsible for building the indexes for the Event Catalog and resolving the selection criteria specified by the users to identify files and events. The Grid Collector server also scheduled the file movements and cached files with the storage resource managers.

Figure 5(b) shows the observed speedup values versus the selectivity of the analysis task, where the baseline is the time needed to run the same analysis without Grid Collector. The selectivity is the fraction of events in these files that are selected for the analysis task. The Grid Collector speeds up analysis tasks primarily by reducing the amount of disk pages accessed. When one out of 1000 events is selected, the speedup values are observed to be between 20 and 50. When one in ten events is used in an analysis job, the observed speedup is more than 2-fold.

3.2. Tracking spatial features

The previous example contains independent data records; however, many others have correlated records and the search operations depend on this correlation. For example, the records produced by a combustion simulation are related to each other by a mesh, to find and track ignition kernels, we need the connectivity information of this underlying mesh (see Figure 6) [39, 40]. FastBit indexes are uniquely able to take advantage of the underlying correlations among the data records because they preserve the ordering of data provided by the user.

We divide the task of tracking ignition kernels into three steps: 1 finding points (records) satisfying the user provided conditions for ignition kernels, 2 grouping the points into connected components, and 3 tracking the evolution of the components by computing the overlap among them. On a casual examination, FastBit indexes seem to only offer help in the first step. Indeed, many database indexes can only accelerate the first step. However, FastBit indexes and the underlying compressed bitmap data structure can speed up all three steps.

In step 2, we are able to take advantage of the compressed answers generated from step 1 to provide a compact representation of the points in the regions of interest. Instead of recording each data point separately, we store *line segments* consisting of multiple consecutive points [39]. We further provide an efficient union-find data structure to accelerate step 2 [41]. At the end of step 2, we again represent each connected component as a compressed bitmap, which allows step 3 to be completed quickly with our efficient bitwise AND operations.

In general, the first two steps can find any regions of interest. We observed that the time needed is a linear function of the number of line segments [39]. Since the number of line segments

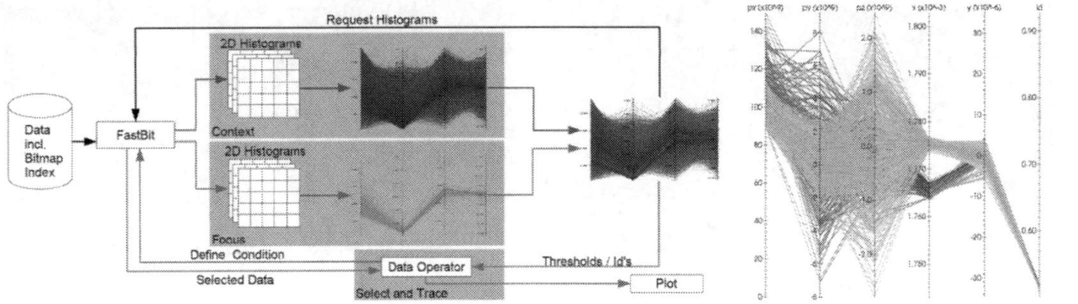

(a) system schematics (b) parallel coordinate

Figure 7. The schematics of the parallel coordinate system used for selecting and displaying particles from a Laser Wake-Field Accelerator simulation (a) and a parallel coordinate display for a time step (b) [18].

is bounded by the number of points on the surface of the regions, our algorithm for finding regions of interest is bounded by a linear function of the number of points on the surface, instead of the number of points in the regions. For well-defined regions, where the number of point inside is much larger than the number of points on the surface, our approach has a significant advantage.

3.3. Dynamic conditional histograms

Aside from its searching capability, FastBit also has a set of efficient functions for computing conditional histograms [18, 42, 43]. In this example of visual exploration of Laser Wake-Field Accelerator (LWFA) simulation data, we illustrate how this capability is used to produce parallel coordinate plots and facilitate the tracking and understanding of particle bundles in LWFA[2]. The new visual analysis system reduced the turn-around time from hours to seconds.

Figure 7(a) illustrates the use of FastBit to produce a parallel coordinate plot in this application. FastBit is used to select records according to user specified conditions and to compute the histograms on the selected records. For example, the context in Figure 7(b) (shown in blue) was selected with "$px > 8.872 \times 10^{10}$." The focus (shown in green) was selected with an additional condition of "$x > 1.77 \times 10^{-3}$." The parallel coordinate plots contain only particles with a high forward momentum px (shown as the first axis on the left) $> 8.872 \times 10^{10}$. These particles form two well-separated groups on the x-axis (shown as the 4th axis from the left), an important feature for understanding the particle bundles in LWFA.

These parallel coordinate plots are generated using 2-D histograms involving neighboring axes. FastBit produced these histograms with the data records satisfying the user specified conditions and removed the need for user (or the visualization system) to read the whole data set before perform the filtering. This reduced the volume of data accessed and improved the interactivity of the visual exploration [18].

In the same application, FastBit was also used to track particles through time and select data for volume rendering and other computations. For example, after a localized group of accelerated particles was found in one time step, scientists might track its history to understand how the group was formed. To accomplish this task, we first extracted the identifiers of the particles, then found particles with these identifiers in other time steps. This search operation involved many thousands of particle identifiers over millions of particles in the base data. The ability to handle this operation efficiently is another feature of FastBit.

[2] Note that the user data is stored in HDF5 files [15, 16] and visualization is performed with VisIt [17].

4. Summary and Future Plans

FastBit employs a number of advanced bitmap indexing techniques to significantly reduce the query response time on large scientific data. Its advantages have not only been observed in timing measurements, but also proven through theoretical computational complexity analyses [23, 24]. Furthermore, we have extended FastBit to include functions for handling simple mesh information and computing conditional histograms. All these are useful features in analyzing massive data sets. In addition, FastBit software is available open source so that anyone can derive from it or extend it for their own work.

A number of new features are currently planned for FastBit, such as handling of queries that join multiple tables. To support more applications, we are working on integrating information about irregular and semi-regular meshes for queries involving spatial data. We are also studying ways to perform text searches using compressed bitmap indexes.

FastBit shares a number of shortcomings of all other bitmap indexes; the most important of which is that they are slow to update after modifications of a few existing records. This prevents it from being used effectively in applications where data records change frequently. However, if the changes are limited to relatively new records, FastBit could work effectively by avoiding indexing those new records.

Acknowledgments

This work was supported in part by the Office of Science of the U.S. Department of Energy under Contract No. DE-AC02-05CH11231, the Scientific Discovery through Advanced Computing (SciDAC) program's Scientific Data Management (SDM) center, Visualization and Analytics Center for Enabling Technologies (VACET) and the COMPASS project. This research used resources of the National Energy Research Scientific Computing Center (NERSC), which is supported by the Office of Science of the U.S. Department of Energy under Contract No. DE-AC02-05CH11231. The authors also gratefully acknowledge FastBit users for their efforts to test and debug the software package, and to find new and creative ways of using the software.

References

[1] Anticic T, Barroso V, Carena F, Carena W, Chapeland S, Cobanoglu O, Denes E, Divia R, Fuchs U, Kiss T, Makhlyueva I, Ozok F, Roukoutakis F, Schossmaier K, Soos C, Vyvre P V and Vergara S 2008 Commissioning of the ALICE data acquisition system *CHEP'07* (IOP Publishing)
[2] Gameiro S, Crone G, Ferrari R, Francis D, Gorini B, Gruwe M, Joos M, Lehmann G, Mapelli L, Misiejuk A, Pasqualucci E, Petersen J, Spiwoks R, Tremblet L, Unel G, Vandelli W and Yasu Y 2006 *IEEE Transactions on Nuclear Science* **53** 907–911
[3] Chen J H, Choudhary A, Supinsky D D, DeVries M, Hawkes E R, Klasky S, Liao W K, Ma K L, Mellor-Crummey J, Podhorszki N, Sankaran R, Shende S and Yoo C S 2009 *Computational Science & Discovery* **2** 015001–015031
[4] Geddes C, Toth C, van Tilborg J, Esarey E, Schroeder C, Bruhwiler D, Nieter C, Cary J and Leemans W 2004 *Nature* **438** 538–541
[5] Harris J and STAR collaboration 1994 *Nuclear Physics A* **566** 277 – 285
[6] Chaudhuri S and Dayal U 1997 *ACM SIGMOD Record* **26** 65–74
[7] Devlin B A and Murphy P T 1988 *IBM Systems Journal* **27** 60–80
[8] Inmon W H 1996 *Communications of the ACM* **39** 49–50
[9] Brin S and Page L 1998 *Computer Networks and ISDN Systems* **30** 107–117
[10] Zobel J and Moffat A 2006 *ACM Comput. Surv.* **38** 6
[11] Chan C Y and Ioannidis Y E 1998 Bitmap index design and evaluation *Proceedings of the 1998 ACM SIGMOD* (New York, NY, USA: ACM press) pp 355–366
[12] O'Neil P 1987 Model 204 architecture and performance *2nd International Workshop in High Performance Transaction Systems, Asilomar, CA* (Springer-Verlag) pp 40–59
[13] Reiss F R 2007 *Data Triage* Ph.D. thesis EECS Department, University of California, Berkeley URL `http://www.eecs.berkeley.edu/Pubs/TechRpts/2007/EECS-2007-79.html`
[14] Sinha R 2007 *Indexing Scientific Data* Ph.D. thesis UIUC
[15] The HDF Group 2007 HDF5 user guide URL `http://hdf.ncsa.uiuc.edu/HDF5/doc/H5.user.html`

[16] Gosink L, Shalf J, Stockinger K, Wu K and Bethel W 2006 HDF5-FastQuery: Accelerating complex queries on HDF datasets using fast bitmap indices *SSDBM 2006* (IEEE Computer Society Press.)
[17] Childs H, Brugger E S, Bonnell K S, Meredith J S, Miller M, Whitlock B J and Max N 2005 A contract-based system for large data visualization *IEEE Visualization 2005* pp 190–198 software available at http://wci.llnl.gov/codes/visit/
[18] Rübel O, Prabhat, Wu K, Childs H, Meredith J, Geddes C G R, Cormier-Michel E, Ahern S, weber G H, Messmer P, Hagen H, Hamann B and Bethel E W 2008 High performance multivariate visual data exploration for extemely large data *SuperComputing 2008* (Austin, Texas, USA) p 51
[19] Schlosser J and Rarey M 2009 *Journal of Chemical Information and Modeling* **49** 800–809
[20] Siqueira T L L, Ciferri R R, Times V C and de Aguiar Ciferri C D 2009 A spatial bitmap-based index for geographical data warehouses *SAC '09* (New York, NY, USA: ACM) pp 1336–1342
[21] Comer D 1979 *Computing Surveys* **11** 121–137
[22] Johnson T 1999 Performance measurements of compressed bitmap indices *VLDB'99* (San Francisco: Morgan Kaufmann) pp 278–289
[23] Wu K, Otoo E and Shoshani A 2004 On the performance of bitmap indices for high cardinality attributes *VLDB* pp 24–35
[24] Wu K, Otoo E and Shoshani A 2006 *ACM Transactions on Database Systems* **31** 1–38
[25] Antoshenkov G and Ziauddin M 1996 *VLDB Journal* **5** 229–237
[26] Wu K, Otoo E and Shoshani A 2001 A performance comparison of bitmap indexes *CIKM* pp 559–561
[27] Wu K, Otoo E and Shoshani A 2002 Compressing bitmap indexes for faster search operations *Proceedings of SSDBM'02* (Edinburgh, Scotland) pp 99–108
[28] Stockinger K and Wu K 2006 *Bitmap Indices for Data Warehouses* (Idea Group, Inc.) pp 179–202
[29] Wu K, Otoo E and Shoshani A 2001 Compressed bitmap indices for efficient query processing Tech. Rep. LBNL-47807 Lawrence Berkeley National Laboratory Berkeley, CA
[30] Sinha R R and Winslett M 2007 *ACM Trans. Database Syst.* **32** 16
[31] Wu K, Stockinger K and Shoshani A 2007 Performance of multi-level and multi-component compressed bitmap indexes Tech. Rep. LBNL-60891 Lawrence Berkeley National Laboratory Berkeley, CA
[32] Wu K, Stockinger K and Shosani A 2008 Breaking the curse of cardinality on bitmap indexes *SSDBM'08* pp 348–365
[33] Stockinger K, Wu K and Shoshani A 2004 Evaluation strategies for bitmap indices with binning *International Conference on Database and Expert Systems Applications (DEXA 2004), Zaragoza, Spain* (Springer-Verlag)
[34] Koudas N 2000 Space efficient bitmap indexing *Proceedings of the ninth international conference on Information knowledge management CIKM 2000 November 6 - 11, 2000, McLean, VA USA* (ACM) pp 194–201
[35] Rotem D, Stockinger K and Wu K 2005 Optimizing candidate check costs for bitmap indices *CIKM*
[36] Wu K, Gu J, Lauret J, Poskanzer A M, Shoshani A, Sim A and Zhang W M 2005 Grid collector: Facilitating efficient selective access from data grids *Proceedings of International Supercomputer Conference In Heidelberg, 2005*
[37] Wu K, Zhang W M, Perevoztchikov V, Lauret J and Shoshani A 2004 The grid collector: Using an event catalog to speed up user analysis in distributed environment *Proceedings of Computing in High Energy and Nuclear Physics (CHEP) 2004*
[38] Dean J and Ghemawat S 2004 MapReduce: Simplified data processing on large clusters *OSDI'04*
[39] Wu K, Koegler W, Chen J and Shoshani A 2003 Using bitmap index for interactive exploration of large datasets *Proceedings of SSDBM 2003* (Cambridge, MA, USA) pp 65–74
[40] Stockinger K, Shalf J, Bethel W and Wu K 2005 Query-driven visualization of large data sets *IEEE Visualization 2005, Minneapolis, MN, October 23-28, 2005*
[41] Wu K, Otoo E and Suzuk K 2009 *Pattern Analysis & Applications* **12** 117–135
[42] Stockinger K, Bethel E W, Campbell S, Dart E and Wu K 2006 Detecting distributed scans using high-performance query-driven visualization *SC '06* (New York, NY, USA: ACM) p 82
[43] Rübel O, Prabhat, Wu K, Childs H, Meredith J, Geddes C G R, Cormier-Michel E, Ahern S, Weber G H, Messmer P, Hagen H, Hamann B and Bethel E W 2008 Application of high-performance visual analysis methods to laser wakefield particle acceleration data *IEEE Visualization 2008* (Columbus, Ohio, USA)

Advances in simulation of wave interactions with extended MHD phenomena

D Batchelor[1], G Abla[6], E D'Azevedo[1], G Bateman[2], D E Bernholdt[1], L Berry[1], P Bonoli[3], R Bramley[4], J Breslau[5], M Chance[5], J Chen[5], M Choi[6], W Elwasif[1], S Foley[4], G Fu[5], R Harvey[7], E Jaeger[1], S Jardin[5], T Jenkins[10], D Keyes[8], S Klasky[1], S Kruger[9], L Ku[5], V Lynch[1], D McCune[5], J Ramos[3], D Schissel[6], D Schnack[10] and J Wright[3]

[1] Oak Ridge National Laboratory; [2] Lehigh University; [3] Plasma Science and Fusion Center, Massachusetts Institute of Technology; [4] Indiana University; [5] Princeton Plasma Physics Laboratory; [6] General Atomics; [7] CompX International; [8] Columbia University; [9] Tech-X Corporation; [10] University of Wisconsin

E-mail: batchelordb@ornl.gov

Abtract: The Integrated Plasma Simulator (IPS) provides a framework within which some of the most advanced, massively-parallel fusion modeling codes can be interoperated to provide a detailed picture of the multi-physics processes involved in fusion experiments. The presentation will cover four topics: 1) recent improvements to the IPS, 2) application of the IPS for very high resolution simulations of ITER scenarios, 3) studies of resistive and ideal MHD stability in tokamk discharges using IPS facilities, and 4) the application of RF power in the electron cyclotron range of frequencies to control slowly growing MHD modes in tokamaks and initial evaluations of optimized location for RF power deposition.

1. Introduction

The Center for Simulation of Wave Interactions with Magnetohydrodynamics (SWIM) has the scientific objectives of: improving our understanding of interactions that both RF wave and particle sources have on extended-MHD phenomena, improving our capability for predicting and optimizing the performance of burning plasmas, developing an integrated computational system for treating multi-physics phenomena with the required flexibility and extensibility to serve as a prototype for the Fusion Simulation Project, addressing mathematics issues related to the multi-scale, coupled physics of RF waves and extended MHD, and optimizing the integrated system on high performance computers.

The SWIM Center consists of three elements: (1) Development of a computational platform referred to as the Integrated Plasma Simulator (IPS) that will allow efficient coupling of the full range of required fusion codes or modules and carrying out a program of research on as optimizing burning plasma scenarios, (2) A physics campaign addressing long timescale discharge evolution in the presence of sporadic fast MHD events. This involves interfacing the IPS to both linear and 3D non-linear extended MHD codes and improving the understanding of how RF can be employed to achieve long-time MHD stable discharges and control sawtooth events, and (3) A physics campaign for modeling the direct interaction of RF and extended MHD for slowly growing modes. The primary physics focus of this campaign is to improve the understanding of how RF can be employed to control neoclassical tearing modes. This requires development of new approaches to closure for the fluid equations and the interfacing of RF modules directly with the extended MHD codes and with code modules that implement the fluid closures.

2. New developments in the Integrated Plasma Simulator

The computational core of the SWIM project is the Integrated Plasma Simulator (IPS), a software framework designed and developed not only to support the coupled simulations required by the SWIM science plan, but also to explore issues associated with integrated fusion simulation at an even larger scale for future efforts such as the proposed Fusion Simulation Project (FSP). The design of the IPS is based on concepts from the Common Component Architecture, expressed in a simplified form that satisfies the project's needs without introducing unnecessary complexity [1, 2]. For example, the IPS itself, as well as computational driver components, are written in Python, which is readily modified and extended in response to new scientific needs. The physics components of the IPS are generally wrappers around standalone physics codes. The physics codes used by the IPS are routinely used as stand-alone models, and are under continual development. So the ability to use them unmodified in the IPS for integrated simulations is a major advantage. Likewise, with the help of a Plasma State component as a central store for key simulation data, file-based data exchange remains adequate for current IPS applications.

A major focus of recent development in the IPS has been support for the execution of multiple components concurrently, where individual components may be parallel [3]. There are several motivations for this "multiple-component multiple-data" (MCMD) execution model, which can ultimately be traced back to the fact that in this field, existing codes vary widely in their level of parallel scalability. One illustrative use for the MCMD-style of execution arises in the slow MHD campaign, where a scalable long-running MHD code such as NIMROD [15] requires updates to certain quantities as it evolves the system through time. In this case, updates, which couple only loosely to the MHD, are done by the GENRAY code [14], which has only a small degree of parallelism. Stopping a large-scale MHD task in order to run a small-scale GENRAY task would leave the majority of compute nodes allocated to the job idle during this phase of the simulation. The MCMD execution model allows the small GENRAY task to be run on a small number of additional processors while the MHD code keeps iterating, resulting in a much more efficient use of resources. Only a small number of nodes are idled between GENRAY executions. Another example of MCMD execution comes from the fast MHD campaign. In this case (see figure 1) each iteration of the simulation consists of a phase in which one or two scalable simulation codes run (in this case the AORSA full wave ICRF code [8] and the NUBEAM neutral beam code [5, 12]), followed by a phase in which several less scalable codes must be run. Fortunately, their data dependencies allow several of the non-scalable codes to be run concurrently – in this case CQL3D [6], and TSC [9, 10], which in turn may require several GLF23 evaluations [16]. Additionally, many of the stability analyses of the plasma from the just-completed time step can easily be run concurrently over the many toroidal modes of the plasma (DCON, PEST-II, ELITE, and BALLOON in this example; see for example [4]). The IPS MCMD capability even allows for several distinct simulations to be run simultaneously as a single job, interleaved so that scalable components in one simulation run concurrently with unscalable components of the other. With judicious allocations of resources, designed to produce similar execution times for the scalable and unscalable components, it is possible to achieve a more efficient overall utilization of the computing resources than would be possible running either simulation individually.

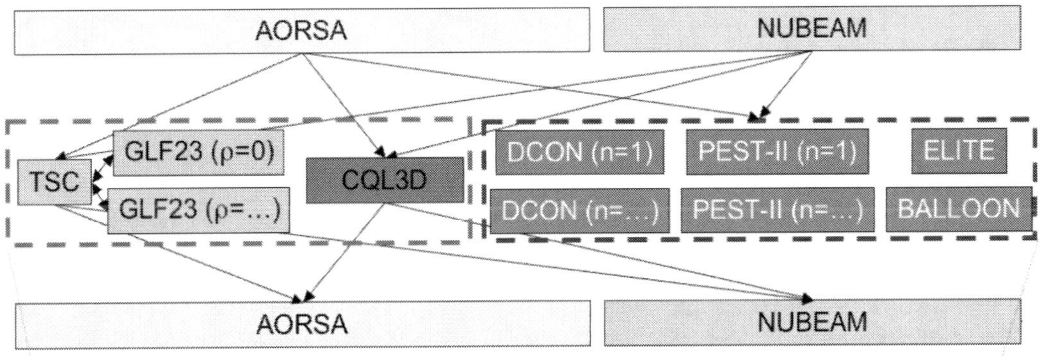

Figure 1. Sample MCMD concurrent execution use case based on SWIM fast MHD research campaign. [3].

3. Applications of the Integrated Plasma Simulator to ITER

Integrated modeling of projected ITER discharges already plays a very important role in the ITER program, both for the purpose of finalizing the design of auxiliary systems and for planning the research program. Presently a single run of an integrated model can take a month or more of continuous running time if the most powerful physics models are employed. The number of such runs that will be needed is quite large, particularly when surveys of various discharge parameter values are required such as for optimization studies.

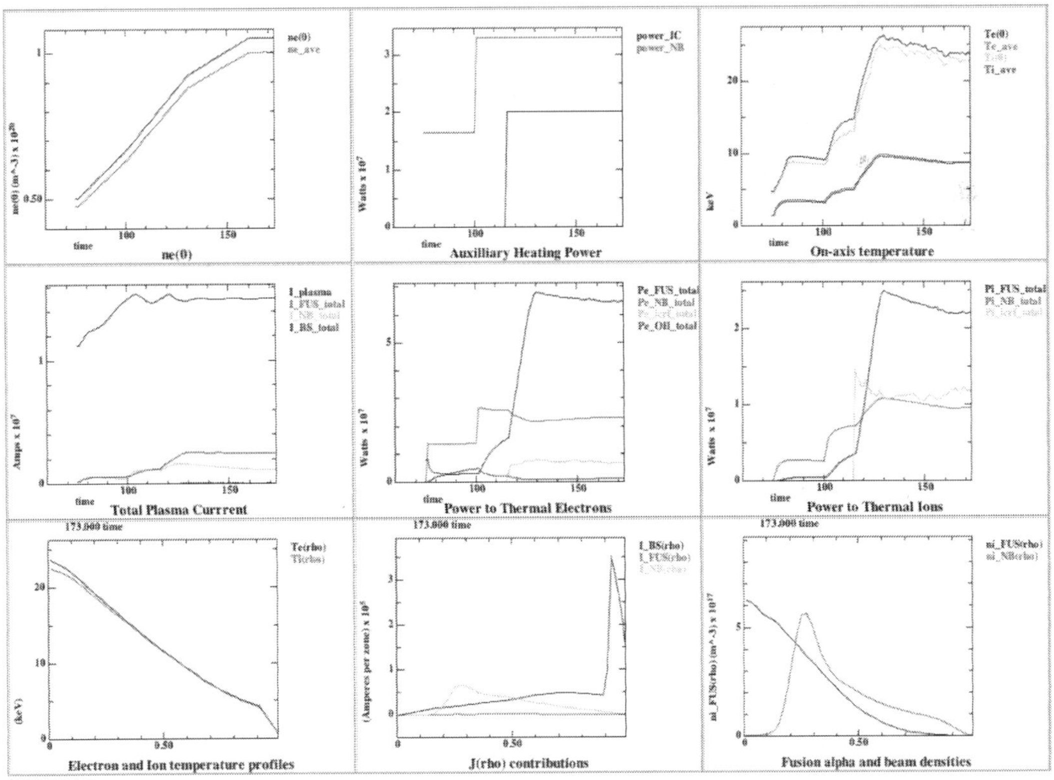

Figure 2. IPS results for an ITER simulation. Power sources include two neutral beams and ion cyclotron heating. In this case the plasma heating is dominated by fusion alpha particles.

The SWIM project has adopted a near term goal of demonstrating the capability to use massively-parallel computers to greatly accelerate the integrated simulation process for ITER, while also improving the level of physics fidelity of the simulations. To this end we are using the SWIM IPS with TSC/GLF23 coupled to the AORSA full-wave ICRF code and the NUBEAM code. Figure 2 shows IPS results for an ITER H-mode scenario that has been previously studied using the TSC/PTRANSP code with serial versions of the heating and transport modules. For this run the ICRF power deposition was calculated with AORSA using 2056 processors and the neutral beams and fusion physics was calculated with parallel NUBEAM running on 512 processors on the Franklin computer at NERSC. The number of Monte Carlo particles used by NUBEAM was 10^6 for neutral beam injection and10^6 for fusion products, while the AORSA calculations used 256×256 Fourier modes for the radial and poloidal angle grids. These are much higher resolution calculations than were feasible previously. We expect soon to do similar studies with the parallel TORIC code for ICRF [17], to begin the process of benchmarking against the PTRANSP simulation, and to work on optimizing the running time using the MCMD capability.

4. Progress in fast MHD campaign

We are evaluating the ideal and resistive MHD stability of a CMOD discharge simulated using the IPS. The stability codes read the equilibrium from the plasma state as the simulation proceeds. Shown in figure 3 is the linear phase of a n=1 resistive instability as calculated by the new toroidal MHD stability code M3D-C1. We are able to use realistic values for the resistivity and other parameters by packing the mesh points around the magnetic flux surface where the magnetic island forms (using routines provide by the RPI SCORE Center). Shown in the figure is the perturbed part of the toroidal current density for a resistive instability in CMOD with magnetic Lundquist number $S=10^7$. The configuration is ideally stable.

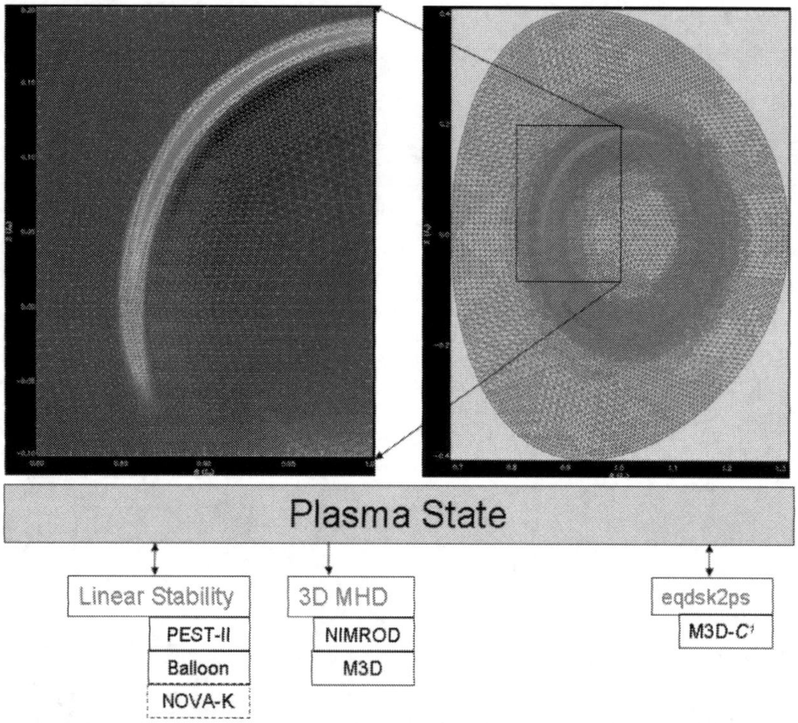

Figure 3. Perturbed toroidal current density for a resistive instability in CMOD with Lundquist number $S=10^7$. Equilibrium file is read from the plasma state and analyzed by M3D-C^1.

5. Progress in slow MHD campaign

Tokamaks have achieved their high performance by relying on strong magnetic fields to create a nested set of surfaces on which the magnetic field is confined. Near rational surfaces, where the magnetic field lines close on themselves rather than cover the surface, the plasma is susceptible to instabilities which can break the magnetic field into island structures. The break in topology allows particles and energy to escape the plasma more readily and reduces the plasma performance. The most promising technique to control this process uses localized radiofrequency (RF) current drive to stabilize the islands. One of the key control parameters of the stabilization experiments is the location of the RF current deposition relative to the location of the rational surface. Numerical experiments have begun to understand the influence of this parameter.

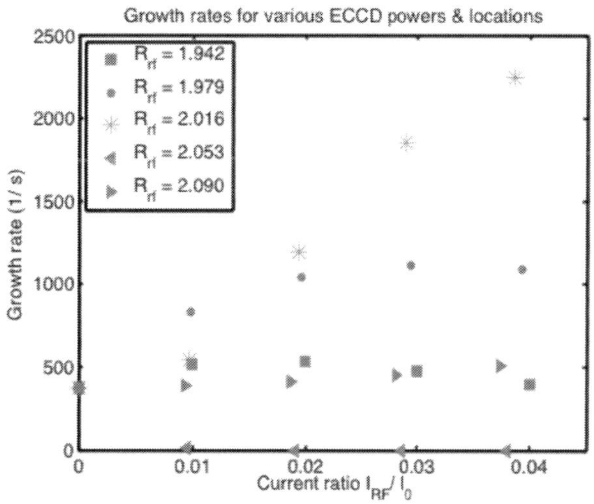

Figure 4. The stabilization or destabilization of the tearing mode, in response to different deposition locations for RF-induced current, is clearly seen in the numerical results from NIMROD.

Initial numerical simulations with the NIMROD code have reproduced many of the qualitative features seen in experiments, including the stabilization or destabilization of the instability. Stabilization is sensitive to the location of the RF current deposition is changed as seen in figure 4. Nonlinear effects complicate attempts to deposit current directly at the rational surface, as the position of this surface shifts in response to the deposited current, figure 5. Simulations suggest that the goal of inducing current at the island center (enabling suppression) is best achieved by depositing current at a point initially just outside the island, and that robust, safe approaches to island stabilization will need to take the motion of rational surfaces into account (especially at high RF powers). These preliminary calculations provide useful guidance as we move to more quantitative predictions.

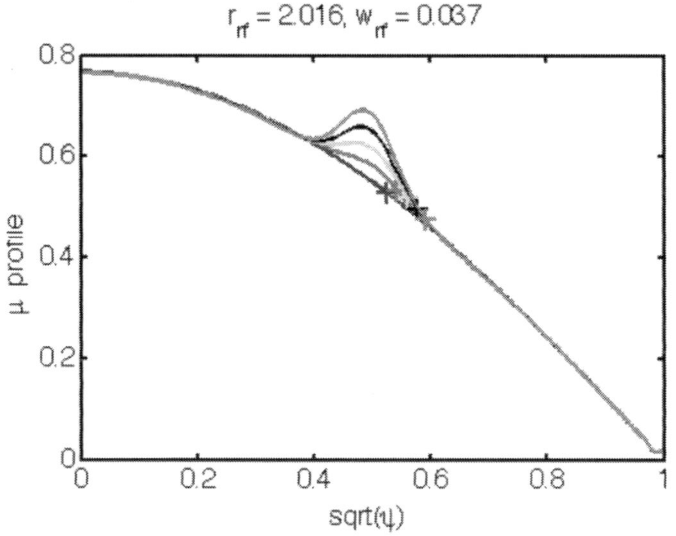

Figure 5. The rational surface position, denoted by the cross, shifts increasingly outward as higher RF powers induce larger localized plasma currents near the original rational surface position.

To make more quantitative predictions, future work will include coupling the NIMROD and GENRAY codes. The mathematical formulation for this coupling has been developed under the SWIM project [7, 13] and work on completing this coupling is under way. The coupling makes use of the Plasma State component to transmit the appropriate quantities from NIMROD to GENRAY without modification of the GENRAY component (which was developed under the fast MHD campaign) and utilizes the recently developed MCMD capabilities described in Section 2. Once the full coupling is complete, truly quantitative predictions of RF feedback stabilization of magnetic islands will be possible.

References

[1] D. Batchelor, C. Alba, G. Bateman, D. Bernholdt, L. Berry, P. Bonoli, R. Bramley, J. Breslau, M. Chance, J. Chen, M. Choi, W. Elwasif, G. Fu, R. Harvey, E. Jaeger, S. Jardin, T. Jenkins, D. Keyes, S. Klasky, S. Kruger, L. Ku, V. Lynch, D. McCune, J. Ramos, D. Schissel, D. Schnack, and J. Wright, Simulation of Wave Interations with MHD, in Rick Stevens, editor, *SciDAC 2008, 14-17 July 2008, Washington, USA*, volume 125 of *Journal of Physics: Conference Series*, page 012039, Institute of Physics, 2008.

[2] Wael R. Elwasif, David E. Bernholdt, Lee A. Berry, and Don B. Batchelor, Component Framework for Coupled Integrated Fusion Plasma Simulation, in *HPC-GECO/CompFrame 2007, 21-22 October, Montreal, Quebec, Canada*, 2007

[3] Samantha S. Foley, Wael R. Elwasif, Aniruddha G. Shet, David E. Bernholdt, and Randall Bramley, Incorporating Concurrent Component Execution in Loosely Coupled Integrated Fusion Plasma Simulation, in *Component-Based High-Performance Computing (CBHPC) 2008.*

[4] A. H. Glasser, American Physical Society, 45th Annual Meeting of the Division of Plasma Physics, October 27-31, 2003, Albuquerque, New Mexico, MEETING ID: DPP03, abstract QP1.109

[5] R.J. Goldston et al., "New Techniques for Calculating Heat and Particle Source Rates due to Neutral Beam Injection in Axisymmetric Tokamaks", J. Comp. Phys. 43 (1981) 61

[6] R. W. Harvey and M. G. McCoy, The CQL3D Code, Proc. IAEA TCM on Thermonuclear Plasmas, Montreal (1992), p. 527, available through USDOC/NTIS No. DE93002962.

[7] C. C. Hegna and J. D. Callen, to be published, Phys. Plasmas 2009

[8] E. F. Jaeger, L. A. Berry, E. D'Azevedo et al., *Physics of Plasmas* **9**, 1873 (2002).

[9] S. C. Jardin, N. Pomphrey, and J. Delucia, "Dynamic Modeling of Transport and Positional Control of Tokamaks", J. Comput. Phys. 66 481-507 (1986)

[10] S. C. Jardin, M. G. Bell, and N. Pomphrey, "TSC Simulation of Ohmic Discharges in TFTR", Nucl. Fus. 33 3710382 (1993)

[11] "Fusion Simulation Project Workshop", Journal of Fusion Energy, Eds. A. Kritz and D. Keyes, 28 (1) 1-59 (2009).

[12] A. Pankin, D. McCune, R. Andre et al., "The Tokamak Monte Carlo Fast Ion Module NUBEAM in the National Transport Code Collaboration Library", Computer Physics Communications Vol. 159, No. 3 (2004) 157-184

[13] J. J. Ramos, Phys. Plasmas **15**, 082106 (2008)

[14] A. P. Smirnov and R. W. Harvey, Bull. Am. Phys. Society **40**, 1837 (1995); "The GENRAY Ray Tracing Code", CompX Report CompX-2000-01 (2001).

[15] C.R. Sovinec, A.H. Glasser, D.C. Barnes, T.A. Gianakon, R.A. Nebel, S.E. Kruger, D.D. Schnack, S.J. Plimpton, A. Tarditi, M.S. Chu and the NIMROD Team, "Nonlinear Magnetohydrodynamics with High-order Finite Elements," Journal of Computational Physics, **195**, 355 (2004)

[16] R. E. Waltz, G. M. Staebler, W. Dorland, G. W. Hammett, and M. Kotschenreuther, *Phys. Plasmas* **4**, 2482 (1997).

[17] J. C. Wright, P. T. Bonoli, M. Brambilla, F. Meo, E. D'Azevedo, D. B. Batchelor, E. F. Jaeger, L. A. Berry, C. K. Phillips, and A. Pletzer, "Calculations of fast wave mode conversion and lower hybrid propagation in tokamaks", Phys. Plasmas, 11(5), May 2004

SciDAC 2009 IOP Publishing
Journal of Physics: Conference Series **180** (2009) 012055 doi:10.1088/1742-6596/180/1/012055

Advances in petascale kinetic plasma simulation with VPIC and Roadrunner

K J Bowers[1,4], B J Albright[1], L Yin[1], W Daughton[1], V Roytershteyn[2], B Bergen[3], and T J T Kwan[1]

[1] Applied Physics Division / Plasma Theory and Applications Group (X-1-PTA)
[2] Theoretical Division / Applied Mathematics and Plasma Physics Group (T-5)
[3] Computer, Computational, and Statistical Sciences Division / Computational Physics and Methods Group (CCS-2), Los Alamos National Laboratory, PO Box 1663, Los Alamos, NM, 87545

E-mail: `kevin.j.bowers@gmail.com`

Abstract. VPIC [1], a first-principles 3d electromagnetic charge-conserving relativistic kinetic particle-in-cell code, was recently adapted to run on Los Alamos's Roadrunner [2], the first supercomputer to break a petaflop (10^{15} floating point operations per second) in the TOP500 supercomputer performance rankings. [3] We summarize VPIC's modeling capabilities, VPIC's optimization techniques and Roadrunner's computational characteristics. We then discuss three applications enabled by VPIC's unprecedented performance on Roadrunner: modeling laser plasma interaction in upcoming inertial confinement fusion experiments at the National Ignition Facility, modeling short-pulse laser GeV ion acceleration and modeling reconnection in space and laboratory plasmas.

1. Introduction

First-principles *particle-in-cell (PIC)* simulations [4,5] are the only viable tool for understanding many plasma physics phenomena. PIC divides the simulation volume into a voxel (cell) mesh and represents plasma as a many particles. These particles are "pushed" by mesh-sampled electric and magnetic fields. The fields in turn are updated using electrical currents "accumulated" from the particle motions. Though this is computationally expensive and, if implemented conventionally, inefficient on modern supercomputers, it gives insight into numerous phenomena impossible to measure directly or understand with more approximate methods.

Employing millions of voxels and billions of particles, large PIC simulations require large supercomputing resources. Due to physical limitations, moving data between and even within a supercomputer's microprocessors is more time consuming than performing even basic computations. Unfortunately, traditional PIC implementations, evolved from experience on vector supercomputers, require more data motion per computation than other methods often used in supercomputing (e.g., dense matrix, molecular dynamics N-body and Monte-Carlo calculations) and cannot utilize recently developed petascale supercomputers efficiently. Unlike traditional implementations, VPIC [1], a first-principles 3d electromagnetic charge-conserving

[4] Guest Scientist. Present adddress: D. E. Shaw Research LLC, 120 West 45th Street, 39th Floor, New York, NY 10036

relativistic kinetic PIC code, was designed from the ground up to minimize data motion. As a result, VPIC can more fully exploit petascale supercomputer potential, enabling simulations of numerous plasma physics phenomena at unmatched fidelity and scale—using over a trillion particles, a billion voxels and/or a hundred thousand processing cores. VPIC running on Los Alamos's Roadrunner supercomputer is a prime example. In a parameter study of particle trapping physics within laser-driven hohlraums of inertial confinement fusion experiments [6], we measured end-to-end sustained performance exceeding 0.374 Pflop/s (s.p.) on 122,240 processing cores (17 of Roadrunner's 18 connected units).

This paper is structured as follows: We give a brief overview VPIC's modeleling capabilities and optimization techniques and Roadrunner's computational characteristics (ref. [1] and ref. [2] present these in greater detail). We then discuss three applications enabled by VPIC's performance on Roadrunner: modeling *laser-plasma interaction (LPI)* in upcoming *inertial confinement fusion (ICF)* experiments at the *National Ignition Facility (NIF)* [7–9], modeling short-pulse laser GeV ion acceleration [10–13] and modeling magnetic reconnection in space and laboratory plasmas [14–16].

2. VPIC on Roadrunner

2.1. VPIC overview

VPIC integrates the relativistic Maxwell-Boltzmann equations in a linear background medium. To avoid a prohibitively costly discretization, VPIC samples each species' 6d position-momentum phase-space with particles that follow the Boltzmann equation's characteristics. As a result, though computational particles typically represent many physical particles, their phase-space distribution obeys the Boltzmann equation outside of discrete particle collisional effects, which must be handled separately.

The Maxwell equations and the particle equations of motion are integrated in time with a quasi-symplectic (e.g., phase-space volume conserving, when using lossless background materials and boundary conditions) Trotter factorization based scheme. The resulting timestep is an explicit-implicit mixture of leap-frog, Verlet, exponential differencing and Boris magnetic field rotations [17], modified to use a sixth-order rotation angle approximation [18]. The simulation domain is divided into a regular mesh of identical rectangular voxels with potentially irregular but mesh-aligned boundaries, with tens to thousands of particles per voxel. The electric and magnetic fields are sampled staggered on this mesh and interpolated to the particles with an "energy-conserving" scheme (i.e., consistent with a discretized energy principle). The fields' curls are computed with second-order accurate finite differencing. The current density is sampled on the same mesh and extrapolated from the particles by the method of Villasenor and Buneman [19], making VPIC "charge-conserving" (i.e., in exact arithmetic, the fields satisfy an implied discretized charge continuity relation). VPIC also infrequently applies Marder passes [20] tuned specifically to clean finite precision arithmetic error induced charge conservation violations. Relativistic particle motion can generate non-physical Cherenkov radiation in the discretized Maxwell equations; VPIC uses the charge-conserving transverse current adjustment method [21] to combat this.

Achieving high performance on modern supercomputers requires minimizing the number of internode communications, number of memory accesses, amount of data in internode communications, amount of memory accessed and number of arithmetic operations performed, in roughly that order. Fortunately, due to the finite c in relativistic simulations, internode communications are naturally optimized by VPIC's spatial domain decomposition. Thus, as there is typically more field and particle data per node than can fit in any cache, DRAM access optimization is main technical challenge in VPIC.

Key to this is heavy single precision use. This allows better *single-instruction-multiple-data (SIMD)* utilization (e.g., four single precision operations can be processed simultaneously versus two double precision operations on typical processors) while reducing data motion by half over double precision. For typical time steps, voxel dimensions and particles per voxel, model discretization error exceeds single precision arithmetic error and thus is acceptable provided it does not introduce other non-physical artifacts (e.g., large violations of otherwise conserved quantities). To help ensure this, VPIC is stabilized against single precision arithmetic error. For example, particle coordinates are stored as a voxel index plus a normalized offset from the center of that voxel; this preserves various symmetries better and gives orders of magnitude improvement in the worst case absolute position resolution over more conventional representations.

Also key is heavy streaming memory access pattern use. For example, VPIC updates the majority of particles each timestep in a single pass. This, and storing the particle data contiguously and organized for efficient SIMD, means the particle update streams through the particles once using large aligned memory transfers. Likewise, to avoid randomly accessing field data each timestep, field interpolation coefficients for each voxel are precomputed before the particle update and saved in a contiguous SIMD compatible layout. Additionally, the particles are periodically sorted [22] such that all particles in a given voxel are processed approximately sequentially. The necessary field data for the particle update is thus effectively streamed sequentially from memory into cache with large aligned memory transfers.

In the method of Villensor and Buneman, a particle makes a current contribution in each voxel through which it passes. To utilize SIMD here, VPIC exploits that particles do not cross voxel boundaries often due to the simulation speed-of-light and stability conditions. VPIC updates 4 particles simultaneously by assuming no boundary crossings occur. Particles that do cross are detected and are processed individually subsequently. Like the interpolation coefficients, current contributions in a voxel are made to a contiguous SIMD compatible set of partial currents which is postprocessed into the current density after the particle update. If a particle hits an "exceptional" boundary during current accumulation (e.g., needs to be sent to a neighboring node), the index and remaining particle displacement are saved to an exception list for later processing. This avoids additional passes through the particles to find exceptions.

2.2. Roadrunner overview

Roadrunner is a first-of-its-kind heterogeneous cluster-of-clusters at Los Alamos National Laboratory, utilizing 6,480, 1.8 GHz, dual-core AMD Opteron CPUs and 12,960, 3.2 GHz, IBM Cell eDP CPUs. Each Cell contains one general purpose *power processing element (PPE)* core and eight single precision 4-vector SIMD optimized *synergistic processing element (SPE)* cores. [23] Roadrunner's computational horsepower primarily resides in the SPEs, with an aggregate SPE single precision theoretical peak performance of 2.7 Pflop/s. Unlike conventional cores, the SPEs cannot access memory directly. Rather, each SPE can only access 256 KiB of high performance *local store (LS)* specific to it. The LS is analogous to cache, but, unlike conventional cache, the SPE must explicitly manage DRAM transfers to and from the LS.

VPIC's performance is asymptotically limited by the rate particles can be streamed to and from DRAM. On Roadrunner, the SPE to Cell local DRAM bandwidth greatly exceeds the SPE to Opteron local DRAM bandwidth. Accordingly, on Roadrunner, all VPIC simulation state resides within Cell DRAM, the Cells are responsible for all computations and the Opterons manage communications among Cells. Initial porting effort focused on the dominant particle update. Each SPE processes a fraction of the particles and triple-buffers its particles in blocks of 512 (i.e., while a block is processed, the next is loaded and the previous is stored). To utilize SPE resources more fully, the SIMD loop used on conventional cores was hand unrolled by 4; effectively, each SPE processes 16 particles simultaneously. A custom 512 line software managed cache of voxel data handles random memory access. The cache has a fully-associative least-recently-used policy (i.e., a line can hold any voxel's data and the least recently used line is replaced when caching new data). As the last 512 requests are guaranteed to be cached, before processing a block, all needed voxel data is requested. Because particles are approximately sorted, most requests are already cached but the cache is sufficient to accomodate every particle in a different voxel. Core-for-core and clock-for-clock, Roadrunner's SPE cores are faster than its Opteron cores and, in aggregate, the SPEs can process particles 15.6 times faster than the Opterons. [2]

2.3. Recent VPIC developments

To support ICF experiment LPI and thermonuclear burn modeling and space and laboratory plasma reconnection modeling, VPIC provides four new capabilities. First, the method of Ref. [24] is available to model elastic and inelastic short-range ion-ion collisions. In this, nearby particle pairs are interacted to give a Monte-Carlo treatment of the Fokker-Planck collision operator in the Boltzmann equation. Second, the method of Ref. [25] is available to model ion-electron collisions. This uses a computationally inexpensive fluid model for the electrons, coupled to the ions by a stochastic Langevin process. Third, this fluid electron model was refined to incorporate electron-radiation coupling via ion-electron Bremsstrahlung, electron-electron Bremsstrahlung and Compton scattering processes. Lastly, additional particle boundary conditions appropriate for plasma inflows and outflows in magnetic reconnection are supported. [26]. This imposes a zero derivative condition on the time averaged particle moments up through the pressure tensor on plasma outflows and an absorbing Higdon [27] boundary condition on the fields.

Owing to the SPE and PPE relative performance, even minor kernels that do not exploit the SPEs are potential Amdahl bottlenecks. As such, recent computational efforts have focused on porting additional kernels to the SPEs. We first converted VPIC to use "overlays", allowing multiple kernels to share the limited SPE local store. Then, as performance is a balance between particle spatial organization (which improves the SPE's particle update software cache performance) and the time spent organizing particles (the second most expensive operation overall), we ported VPIC's particle sorting to the SPEs.

The obvious implementation is the out-of-place counting sort described in Ref. [22], parallelized over the SPE. In this, each SPE histograms a fraction of the particles by voxel. These histograms are converted into the sorted particle's partitioning by voxel and, using this partitioning, the SPEs make a copy of the particles in sorted order. For this method to scale, the conversion must also be SPE-parallel. Unfortunately, this requires large all-to-all exchanges among the SPEs within a Cell before and after the conversion. To avoid this, instead, each SPE is first assigned a fraction of voxels and then particles are sorted by the SPE assigned the voxel containing the particle with a SPE-parallel out-of-place counting sort (given there are orders of magnitude fewer SPE than voxels, the conversion step involves a miniscule data exchange and need not be SPE-parallel). Next, noting each SPE can now sort the particles in its assigned voxels completely independent, each SPE does a serial out-of-place counting sort of particles in its assigned voxels by voxel. We probe a few times more particles than SPE to estimate the optimal assignment of voxels to SPE to accomodate particle non-uniformity. Overall, this requires $O(N_{particle}/N_{spe}) + O(N_{voxel}/N_{spe})$ operations, $O(N_{particle}) + O(N_{voxel})$ auxiliary storage.

We are currently SPE-parallelizing the field update. These kernels use a custom software managed cache to handle strided memory accesses. Many have already been SPE-parallelized and, currently, the largest Amdahl bottlenecks often reside in one-off user-provided in-situ simulation diagnostics.

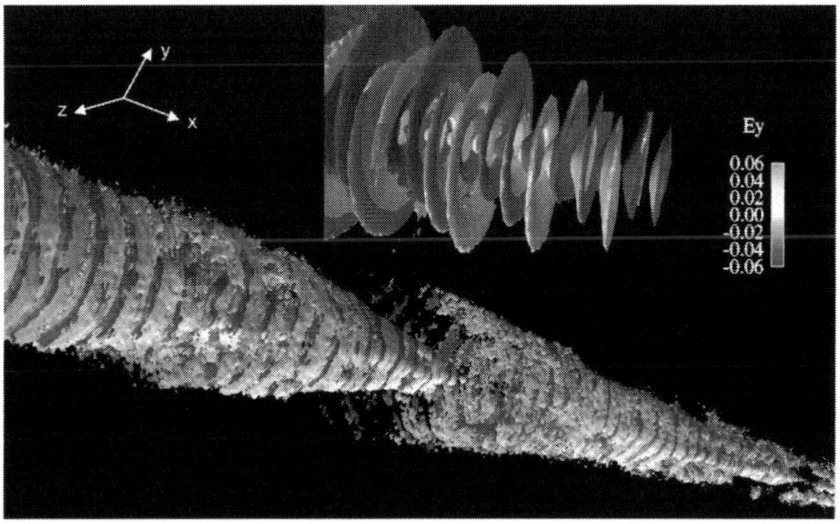

Figure 1. 3d VPIC LPI simulation under NIF plasma conditions. (Top) At SRS saturation, the EPW excited by a f/4 laser speckle is bowing, causing the source for SRS laser backscatter to become incoherent. (Bottom) Owing to the larger speckle volume, the EPW produced in a f/8 speckle is also self-focusing and filamenting, further reducing backscatter. Isosurfaces of constant EPW electric field, colored by the laser electric field during two SRS bursts is shown (color bar indicates the laser field strength in simulation units).

3. LPI modeling of NIF experiments

In 2010, ICF experiments commence at the NIF. In these, over a 10^6 Joules of laser energy are focused onto the inner walls of a gas-filled hohlraum, a hollow cylinder made of high-Z material such as gold or uranium. The hohlraum walls absorb the laser energy and re-radiate it as x-rays. These x-rays are absorbed by a spherical capsule at the hohlraum's center, causing the capsule—including deuterium-tritium fuel in the capsule—to compress, bringing the fuel to the high temperatures and pressures required for fusion reactions.

To prevent the hohlraum walls from ablating during the $\sim$ 10 ns laser drive, a low-Z fill gas (H or He) is used. As the laser propagates, it ionizes this gas and laser-plasma instabilities (i.e., stimulated laser light scattering off plasma density fluctuations) may arise. LPI affect ICF experiments in three ways: they scatter laser light out of the hohlraum, reducing the energy delivered to the capsule; they scatter laser light to other places in the hohlraum, degrading capsule implosion symmetry; and, they pre-heat the fuel with hot electrons, making compression harder.

Stimulated Raman scattering (SRS), the resonant amplification of electron density fluctuations by a laser, is the greatest LPI concern in ICF. In ICF experiments, a roughly uniform laser intensity is maintained across the beam with random phase plates that break the beam into a laser speckle ensemble. In a speckle, SRS manifests as the amplification of a forward directed *electron plasma wave (EPW)* and the backward scattering of laser light. While the linear growth of SRS is well understood [28], single speckle experiments at the LANL Trident Laser facility evince nonlinear behavior: at a threshold intensity, SRS reflectivity rises abruptly to a level orders of magnitude higher than linear theory predicts. Above this threshold, reflectivity is laser drive intensity independent [29].

Recent work [8, 9] using VPIC has identified the essential nonlinear physics governing SRS under NIF-hohlraum conditions. Versus time, the scattering manifests as a series of pulses, each of which passes through 4 distinct phases: (1) SRS grows linearly from density fluctuations. (2) Electrons trapped by the EPW reduce the EPW frequency and phase velocity by an amount that scales with EPW amplitude. (3) Near the speckle center, where the amplitude is highest, the EPW phase velocity is lower than at the speckle's edge; EPW phase front bending ensues. (4) The EPW wave amplitude exceeds the electron's *trapped particle modulation instability (TPMI)* threshhold [30, 31]. TPMI generates waves off-axis from the laser direction and leads to EPW filamentation, self-focusing, and phase front breakup. Moreover, filamentation leads to increased trapped-electron transverse-loss, rapidly increasing EPW damping. As the source for SRS backscatter is proportional the product of the EPW and laser electric fields integrated over the speckle cross section [32], combined these terminate the pulse, fundamentally limiting speckle SRS backscatter.

Figure 1 (top) shows the effect of bowing. This snapshot was taken from a 3d VPIC LPI simulation at SRS pulse saturation of a f/4 beam with intensity $I_0 = 2.5 \times 10^{16}$ W/cm^2 under NIF plasma conditions in the trapping regime ($k\lambda_{De} = 0.34$, where k is the initial EPW wavenumber and λ_D is the Debye length) [5] As bowing causes the EPW electric field's phase (but not amplitude) to change across the speckle, iso-surfaces of EPW electric field depict this bowing. The iso-surfaces are colored by the laser electric field intensity, graphically showing the source for SRS backscatter has become incoherent.

Figure 1 (bottom) shows self-focusing and filamentation effects in two bursts of SRS. This snapshot was taken from a 3d VPIC LPI simulation of a f/8 beam with intesity $I_0 = 1.0 \times 10^{16}$ W/cm^2 and wave number $k\lambda_{De} = 0.34$. The speckle volume is 16 times larger than the previous simulation (325 billion particles on a $8256 \times 512 \times 512$ mesh were used), permitting more transverse self-focusing modes to develop. This leads to chaotic phase variation across the speckle and the EPW filamentation (this process does not occur in a f/4 speckle as angular symmetry is approximately maintained). This further reduces SRS source coherence and increases wave damping, quenching the SRS pulse.

Current research is investigating whether neighboring speckles can interact via hot electron or wave exchanges to produce higher backscatter than they would individually. Earlier work [33] showed that hot electrons from one speckle can destabilize a neighboring speckle and lower the SRS nonlinear threshold under some conditions. However, this work employed 2d simulations, over-predicting hot electron communication. 3d petascale *ab initio* VPIC simulations on Roadrunner are necessary to examine this problem quantitatively.

4. GeV ion acceleration using short-pulse lasers and nano-scale targets

This past decade has seen an explosion of interest in charged particle acceleration via the interaction of ultra-intense ($I > 10^{18}$ W/cm^2) short-pulse (~ps) lasers with plasma. When such a laser strikes a ~ 10s of micron thick solid-density target, hot electrons are produced where the laser is absorbed, travel through the target and create a sheath on the rear target surface with very large (~TV/m) electric fields. These fields ionize and accelerate ions on the rear surface in the *target-normal sheath acceleration (TNSA)* process [34, 35]. The resulting ion beam current is orders of magnitude higher than conventional accelerators and transverse emittance is orders of magnitude lower. With careful target preparation, the TNSA mechanism has been shown in experiments [36], in VPIC simulations and in analytic theory [10] to generate quasi-monoenergetic ion beam spectra.

Ultra-intense lasers with contrast exceeding 10^{10} permit nano-scale target use, without the laser pre-pulse destroying the target. In such, VPIC simulations [11, 13] showed the target can become relativistically transparent to the laser. As the laser penetrates the target, a brief period

[5] Though average laser intensity in initial NIF experiments is lower, SRS is dominated by the fraction of speckles that are by chance above non-linear SRS threshold due to the random phase plates. This effect will be increasingly important as intensity is increased in subsequent NIF experiments.

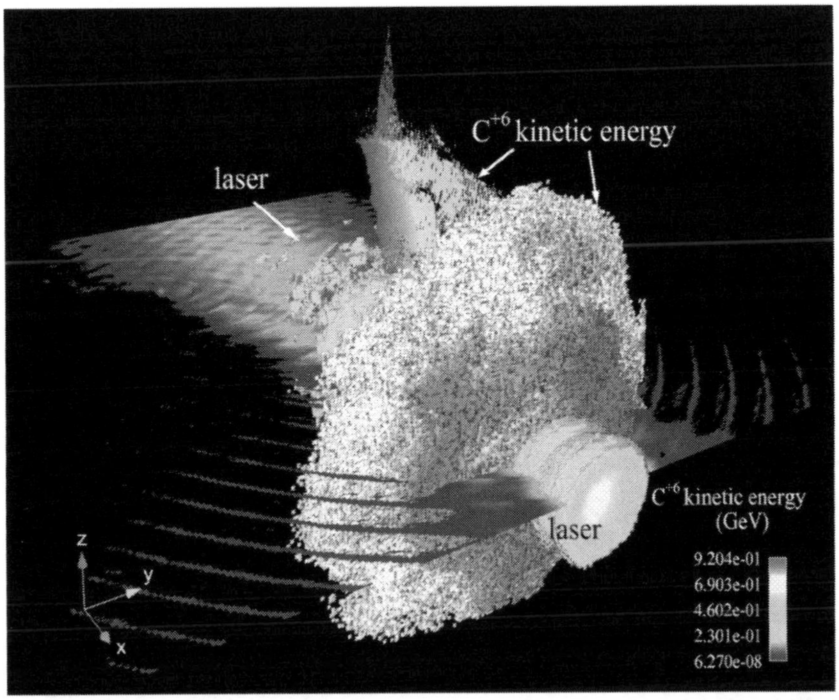

Figure 2. 3d VPIC simulation of ion acceleration using circular polarized laser light at intensity $I_0 \simeq 10^{21}$ W/cm^2 incident on a 30 nm C^{+6} target. The resulting peak ion energy is approaching a GeV, as indicated by the red-yellow isosurface.

of enhanced acceleration, the *Break-out Afterburner (BOA)*, ensues, possibly due to a relativistic Buneman instability [12]. This leads to enhanced ion beam energy and laser-to-ion-beam energy conversion efficiency. Trident was recently upgraded, allowing experimental confirmation of this. [37].

In addition to the BOA, which uses linearly polarized lasers, others [38] have examined ultra-high contrast circularly polarized lasers. In this regime, radiation pressure was thought to accelerate the ions instead of a plasma instability and 3d VPIC simulations on Roadrunner confirmed this analysis. Figure 2 shows ion acceleration using a 30 nm C^{+6} target at solid density $n_e/n_{cr} = 660$, where n_{cr} is the critical density for laser absorption, and a 312 fs laser pulse of peak intensity of $I_0 \simeq 10^{21}$ W/cm^2. C^{+6} ion acceleration is shown at when the laser is penetrating the target and the maximum ion energy is approaching a GeV. This simulation used 14×10^9 voxels and 21×10^9 particles; solid density target simulations over such extended domains are not possible without petascale resources like VPIC on Roadrunner.

As high-energy ion beam generation using ultra-thin targets and two different ion acceleration mechanisms (laser polarizations) at moderately-high laser intensity yielded promising results, current investigations are exploring even higher laser intensities and optimizing the laser pulse length, target composition and thickness for >GeV proton beam generation.

5. Magnetic reconnection

In magnetic reconnection, magnetic field energy is rapidly converted into kinetic energy, including high speed particle flows, ion and electron thermal heating and highly energetic non-thermal tails. Reconnection is ubiquitous in plasma physics, occuring in tokamak devices, planetary magnetospheres, solar flares, extragalactic jets and many other settings, yet poorly

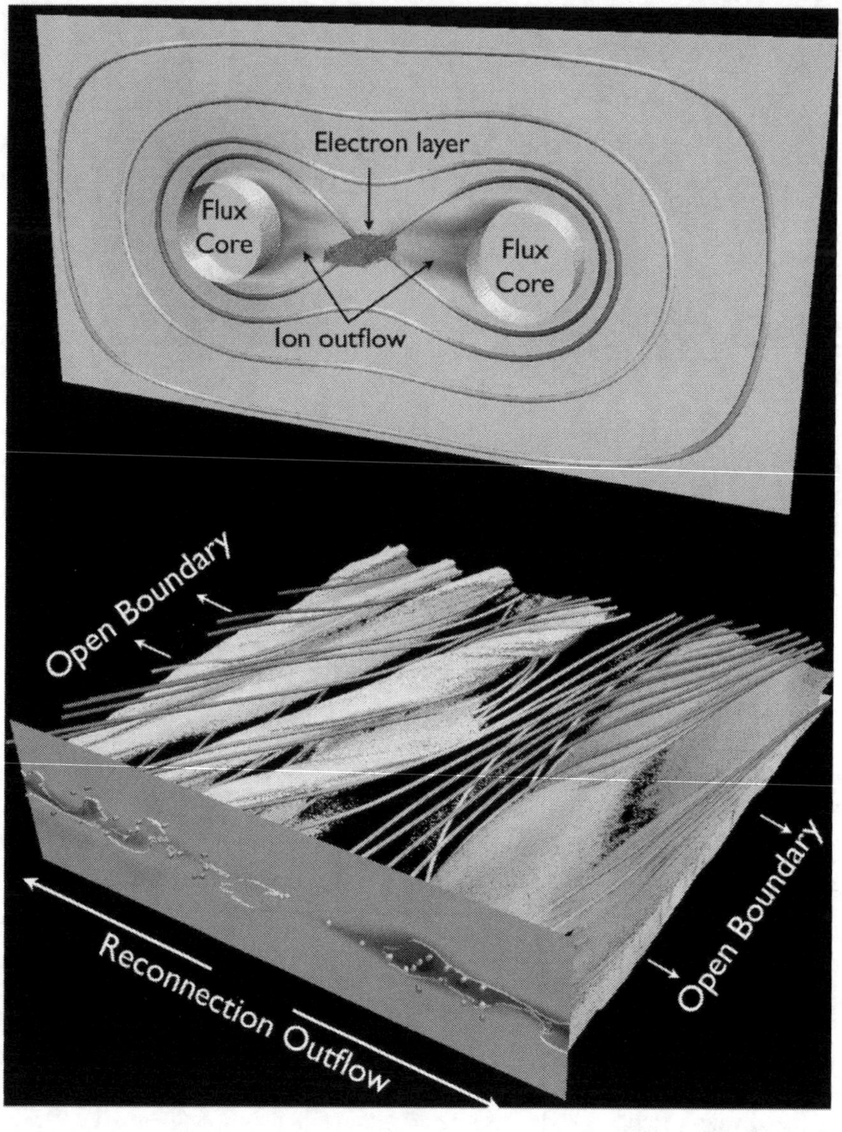

Figure 3. (Top)A 3d VPIC collisional reconnection simulation of MRX uses two flux cores; like the experiment, reconnection is driven by reducing the flux core currents. (Bottom) A 3d VPIC simulation of reconnection layers relevant Earth's magnetosphere. Open boundary conditions allow plasma and magnetic flux to enter and exit the system. Shown is a density isosurface colored by the ion outflow velocity along magnetic field lines.

understood. In collisional regimes, *magnetohydrodynamics (MHD)* gives a reasonable model, but, in weakly collisional regimes, theoretical models and fluid simulations have been unsatisfactory. VPIC's new collisional modeling capabilities permit a direct solution of the full plasma kinetic equation, the theoretical basis for reduced models like MHD. This approach allows rigorous study of reconnection at arbitrary collisionality [15, 16], covering both the fluid and kinetic regimes.

Reconnection simulations are challenging due to the widely separated space and time scales involved. High temperature plasmas are good conductors, implying they trap magnetic flux for very long times. During reconnection, this frozen-flux condition is rapidly broken within very small diffusion regions. Our present research aims to understand diffusion region dynamics and the coupling between diffusion regions and macroscopic scales. To that end, VPIC simulations on Roadrunner are (1) modeling a well-diagnosed laboratory reconnection experiment in a semi-collisional regime and (2) modeling large-scale collisionless reconnection layers relevant to Earth's magnetosphere.

The *Magnetic Reconnection eXperiment (MRX)* recently reported detailed diffusion region measurements, including the electron layer's kinetic structure [39]. MRX's reconnection is driven by reducing the current within two axi-symmetric flux cores. This was recently modeled using a 2d collisionless PIC code with boundary conditions appropriate to the actual experiment [40]. Although this demonstrated a number of striking similarities for ion scale structures, the experimentally measured electron layer thickness was $3-5\times$ larger than observed in the 2d simulation [41]. Two possibilities for this discrepancy, weak Coulomb collisions and 3d effects (e.g., plasma instabilities within the electron layer), are being investigated with VPIC. Collisional 2d VPIC simulations yield significantly broader electron layers and allow uncertain experimental parameters, such as the electron temperature, to be estimated under the assumption Coulomb collisions are sole source of layer broadening. Preliminary 3d simulations in the collisionless limit have also demonstrated plasma instabilities within the electron layer. Figure 3 (top), taken from one of these 3d simulations, shows the flux cores, magnetic field structure, ion outflow and electron current sheet.

Modeling large-scale reconnection in Earth's magnetosphere presents additional challenges: the plasma is highly collisionless and the relative separation of the ion kinetic and overall system scale is nearly $100\times$ larger than MRX. But simulations focusing about a reconnection site are still within reach. These simulations employ the open boundary model described above with plasma inflows at a density and temperature appropriate for the solar wind. This allows both magnetic flux and plasma to enter and exit the system—avoiding unphysical recirculation effects that quickly dominate periodic simulations. Figure 3 (bottom) shows magnetic island formation along the resonant surfaces across the initial current sheet. In 3d, these islands form extended flux ropes at oblique angles above and below the reconnection layer. Over longer time scales, these flux ropes coalesce and intertwine in complex ways to produce new highly elongated current sheets, which are also unstable to new magnetic island (flux rope) formation.

6. Conclusion

In less than a year from VPIC's first petascale runs on Roadrunner, many exciting developments, both scientifically and computationally, have occurred. And, as additional modeling capabilities are added, more kernels are SPE-parallelized and users become increasingly adept at tuning their simulations for petascale resources, the range of simulations for which VPIC is both capable and efficient will continue to grow.

Acknowledgments

This work was performed under the auspices of the United States Department of Energy by the Los Alamos National Security LLC Los Alamos National Laboratory under Contract No. DE-AC52-06NA25396. Work supported in part by the Laboratory Directed Research and Development (LDRD) Program.

References

[1] Bowers K J, Albright B J, Yin L, Bergen B and Kwan T J T 2008 *Phys. Plasmas* **15** 055703

[2] Bowers K J, Albright B J, Bergen B, Yin L, Barker K J and Kerbyson D J 2008 0.374 pflop/s trillion-particle kinetic modeling of laser plasma interaction on roadrunner *SC '08: Proceedings of the 2008 ACM/IEEE conference on Supercomputing* (Piscataway, NJ, USA: IEEE Press) pp 1–11

[3] Dongarra J Top500 supercomputers, *http://www.top500.org* URL `www.top500.org`

[4] Birdsall C K and Langdon A B 1985 *Plasma Physics via Computer Simulation* (New York, NY: McGraw-Hill Inc.)
[5] Hockney R W and Eastwood J W 1988 *Computer Simulation Using Particles* (Philadelphia, PA: Institute of Physics Publishing)
[6] Albright B J, Bowers K J, Yin L, Bergen B, Daughton W, Kwan T J T, Snell C, Hegelich B M, Flippo K and Fernández J C 2008 High performance modeling of laser-plasma interaction in the era of petascale computing *Proceedings of the 13th Advanced Accelerator Concepts Workshop* (AIP)
[7] Yin L, Daughton W, Albright B J, Bowers K J, Montgomery D S, Kline J L, Fernández J C and Roper Q 2006 *Phys. Plasmas* **13** 072701
[8] Yin L, Albright B J, Bowers K J, Daughton W and Rose H A 2007 *Phys. Rev. Lett.* **99** 265004
[9] Yin L, Albright B J, Bowers K J, Daughton W and Rose H A 2008 *Phys. Plasmas* **15** 013109
[10] Albright B J, Yin L, Hegelich B M, Bowers K J, Kwan T J T and Fernández J C 2006 *Phys. Rev. Lett.* **97** 115002
[11] Yin L, Albright B J, Hegelich B M and Fernández J C 2006 *Laser and Particle Beams* **24** 291–298
[12] Albright B J, Yin L, Bowers K J, Hegelich B M, Flippo K A, Kwan T J T and Fernández J C 2007 *Phys. Plasmas* **14** 094502
[13] Yin L, Albright B J, Hegelich B M, Bowers K J, Flippo K A, Kwan T J T and Fernández J C 2007 *Phys. Plasmas* **14** 056506
[14] Yin L, Daughton W, Karimabadi H, Albright B J, Bowers K J and Margulies J 2008 *Phys. Rev. Lett.* **101** 125001
[15] Daughton W, Roytershteyn V, Albright B J, Karimabadi H, Yin L and Bowers K J 2009 *Phys. Plasmas* **in press**
[16] Daughton W, Roytershteyn V, Albright B J, Karimabadi H, Yin L and Bowers K J 2009 *Phys. Rev. Lett.* **submitted**
[17] Boris J P 1970 Relativistic plasma simulation—optimization of a hybrid code *Proceedings of the 4th Conference on Numerical Simulation of Plasmas* ed Boris J P and Shanny R A (Washington D. C.: Naval Research Laboratory) pp 3–67
[18] Blahovec J D, Bowers L A, Luginsland J W, Sasser G E and Watrous J J 2000 *IEEE Trans. Plasma Sci.* **28** 821–829
[19] Villasenor J and Buneman O 1992 *Comp. Phys. Comm.* **69** 306–316
[20] Marder B 1987 *J. Comp. Phys.* **68** 48–55
[21] Eastwood J W, Arter W, Brealey N J and Hockney R W 1995 *Comp. Phys. Comm.* **87** 155–178
[22] Bowers K J 2001 *J. Comp. Phys.* **173** 393–411
[23] Kahle J A, Day M N, Hofstee H P, Johns C R, Maeurer T R and Shippy D 2005 *IBM J. Res. Dev.* **49**
[24] Takizuka T and Abe H 1977 *J. Comp. Phys.* **25** 205
[25] Lemons D S, Winske D, Daughton W and Albright B J 2009 *J. Comp. Phys.* **228** 1391–1403
[26] Daughton W, Scudder J and Karimabadi H 2006 *Phys. Plasmas* **13** 072101
[27] Higdon R L 1986 *Mathematics of Computation* **47** 437–459
[28] Drake J F, Kaw P K, Lee Y C, Schmid G, Liu C S and Rosenbluth M N 1974 *Phys. Fluids* **17** 778
[29] Montgomery D S, Cobble J A, Fernández J C, Focia R J, Johnson R P, Renard-LeGalloudec N, Rose H A and Russell D A 2002 *Phys. Plasmas* **9** 2311–2320
[30] Rose H A 2005 *Phys. Plasmas* **12** 012318
[31] Rose H A and Yin L 2008 *Phys. Plasmas* **15** 042311
[32] Russell D A, DuBois D F and Rose H A 1999 *Phys. Plasmas* **6** 1294–1317
[33] Albright B J, Daughton W, Yin L, Bowers K J, Kline J L, Montgomery D S and Fernández J C 2006 *J. Phys. IV* **133** 253–257
[34] Hatchett S P, Brown C G, Cowan T E, Henry E A, Johnson J S, Key M H, Koch J A, Langdon A B, Lasinski B F, Lee R W, Mackinnon A J, Pennington D M, Perry M D, Phillips T W, Roth M, Sangster T C, Singh M S, Snavely R A, Stoyer M A, Wilks S C and Yasuike K 2000 *Phys. Plasmas* **7** 2076–2082
[35] Maksimchuk A, Gu S, Flippo K, Umstadter D and Bychenkov V Y 2000 *Phys. Rev. Lett.* **84** 4108–4111
[36] Hegelich B M, Albright B J, Cobble J, Flippo K, Letzring S, Paffett M, Ruhl H, Schreiber J, Schulze R K and Fernández J C 2006 *Nature* **439** 441–444
[37] Henig A, Kiefer D, Markey K, Gautier D C, Flippo K A, Letzring S, Johnson R P, Shimada T, Yin L, Albright B J, Bowers K J, Fernández J C, Rykovanov S G, Wu H C, Zepf M, Jung D, Liechtenstein V K, Schreiber J, Habs D and Hegelich B M 2009 *Phys. Rev. Lett.* **in press**
[38] Robinson A P L, Zeph M, Kar S, Evans R G and Bellei C 2008 *New Journal of Physics* **10** 013021
[39] Ren Y, Yamada M, Ji H, Dorfman S, Gerhardt S P and Kulsrud R 2008 *Phys. Plasmas* **15** 082133
[40] Dorfman S, Daughton W, Roytershteyn V, Ji H, Ren Y and Yamada M 2008 *Phys. Plasmas* **15** 102107
[41] Ji H, Ren Y, Yamada M, Dorfman S, Daughton W and Gerhardt S P 2008 *Geophys. Res. Lett.* **35**

Concurrent, Parallel, Multiphysics Coupling in the FACETS Project

J R Cary[1], J Candy[2], J Cobb[3], R H Cohen[4], T Epperly[4], D J Estep[5], S Krasheninnikov[6], A D Malony[7], D C McCune[8], L McInnes[9], A Pankin[10], S Balay[9], J A Carlsson[1], M R Fahey[3], R J Groebner[2], A H Hakim[1], S E Kruger[1], M Miah[1], A Pletzer[1], S Shasharina[1], S Vadlamani[1], D Wade-Stein[1], T D Rognlien[4], A Morris[7], S Shende[7], G W Hammett[8], K Indireshkumar[7], A Yu Pigarov[6], H Zhang[9]

[1] Tech-X Corporation, 5621 Arapahoe Avenue, Suite A, Boulder, CO 80303
[2] General Atomics
[3] Oak Ridge National Laboratory
[4] Lawrence Livermore National Laboratory
[5] Colorado State University
[6] University of California at San Diego
[7] ParaTools, Inc.
[8] Princeton Plasma Physics Laboratory
[9] Argonne National Laboratory
[10] Lehigh University

E-mail: cary@txcorp.com

Abstract. FACETS (Framework Application for Core-Edge Transport Simulations), is now in its third year. The FACETS team has developed a framework for concurrent coupling of parallel computational physics for use on Leadership Class Facilities (LCFs). In the course of the last year, FACETS has tackled many of the difficult problems of moving to parallel, integrated modeling by developing algorithms for coupled systems, extracting legacy applications as components, modifying them to run on LCFs, and improving the performance of all components. The development of FACETS abides by rigorous engineering standards, including cross platform build and test systems, with the latter covering regression, performance, and visualization. In addition, FACETS has demonstrated the ability to incorporate full turbulence computations for the highest fidelity transport computations. Early indications are that the framework, using such computations, scales to multiple tens of thousands of processors. These accomplishments were a result of an interdisciplinary collaboration among computational physics, computer scientists and applied mathematicians on the team.

1. Introduction

The FACETS (Framework Application for Core-Edge Transport Simulations) project [1] has the goal of providing whole-tokamak modeling through coupling separate components for each of the core region, edge region, and wall, with fully realistic sources. This is a complex problem, as each component is parallel in its own right, and each can be parallelized in a distinct manner.

Direct simulation of the entire system is not possible due to the range of scales. The spatial scales vary from the electron gyroradius (≈ 0.01 mm in the edge to ≈ 0.1 mm in the core) to the system size (of order several meters), i.e., by a factor 3×10^5. The time scales vary from the electron gyroperiod (20 ps) to the discharge duration ($\approx$ 1000 s), i.e., by a factor of 6×10^{13}. Thus, a full simulation would require the integration of 3×10^{16} (spatial resolution lengths)^3 for 6×10^{13} temporal resolution periods, for a product of 2×10^{30}. With the need for of 10^{6-12} degrees of freedom per spatial resolution volume (100 per length for a modest fluid model, easily larger by 100 to resolve velocity space as well), and 10^2 floating point operations per update of a degree of freedom for on temporal resolution period, such a fundamental simulation will require $2 \times 10^{38-44}$ floating point operations, which even on petascale platforms, would require $2 \times 10^{23-29}$ s, exceeding the age of the universe by a factor of 10^{6-12}.

Given the large disparity between what is possible and what is needed, progress can only be achieved by separating physics into different parts, such that for each part, valid, reduced approximations exist. For example, in the core of the plasma, the rapid transport along field lines assures that the plasma parameters such as density and temperature are, over long time scales, constant on toroidally nested flux surfaces. This reduces the transport equation to one dimension for evolution on the discharge time scale. As another, in the plasma edge, though simulations must be global, they are nevertheless over a narrow region, and one can use averaging methods to reduce the time scale disparity.

The above naturally translates to a software component approach, which is the approach FACETS is taking. It is bringing together successively more accurate and, hence, computationally demanding components to model the complete plasma device. It is being constructed to run on Leadership Class Facilities (LCFs) to be able to use the most computationally demanding components, while at the same time usable on laptops for less demanding models. To do this FACETS is constructing a C++ framework for incorporating the best software packages and physics components. In this paper we discuss the FACETS progress of the last year.

2. Converting legacy applications to components suitable for Leadership Class Facilities

Transforming legacy fusion applications/libraries to FACETS components suitable for Leadership Class Facilities (LCFs) requires *glue code* to connect the legacy application to the FACETS interface and a cross-compile build environment to produce a statically linked executable. We chose to target a static executable to make FACETS portable to the widest possible collection of current and future LCFs. Glue code translates FACETS interface calls to legacy application calls and performs tasks like language interoperability, unit conversions, and calculating aggregate quantities from mesh data. In the case of UEDGE, our fluid edge component, we had to replace Python-based glue code with a new approach because Python typically uses dynamically loadable libraries — not a static executable.

Our approach to replace UEDGE's Python-based glue code involved making an extension to Forthon [2] and rewriting functions implemented in Python. UEDGE uses Forthon, a tool for generating Python wrappers for Fortran parameters, subroutines, and functions, to create a Python interface to hundreds of variables and tens of subroutines and functions. We extended Forthon to generate Babel SIDL [3] files and implementation files to provide a C++ interface to UEDGE's Fortran code. This approach leveraged all the preexisting Forthon interface description files (`.v` files); although, we had to insert additional information in the `.v` files in the form of comments. Thus, one set of `.v` files supports traditional UEDGE and LCF UEDGE. We wrote C++ glue code to replace code previously implemented in Python. The largest part of this work involved writing the routines to input and output UEDGE data structures to/from HDF5 files.

For a few of the legacy applications including UEDGE, we provided new autoconf-based configuration and build systems. These systems were designed to support multiple types of builds, for example serial, parallel, and LCF. These systems required some adjustments to perform correctly on LCF machines where the front-end nodes are different than the compute nodes. Although mechnical in nature, making these kinds of changes is time consuming.

3. Improving performance of components through algorithmic modifications

Recent work has focused on incorporating robust and scalable parallel nonlinear solvers from the PETSc library into UEDGE to solve the nonlinear system $f(u) = 0$, where u represents the vector of unknowns. We implemented complete functionality for fully implicit, parallel matrix-free Newton-Krylov solvers. The use of PETSc has allowed us to overcome a major bottleneck in the parallel implementation of UEDGE. This multispecies code evolves density and temperature profiles of hydrogen and impurity plasma ions and neutral components of the same. Strong nonlinearities exist owing to ionization and recombination, and the equation set must be well preconditioned to use the efficient Newton-Krylov solver technique. Previously, we could not simultaneously advance plasma and neutrals in parallel because of a very limited block Jacobi algorithm for the parallel preconditioner. Implementation of PETSc has opened an array of possible preconditioners ranging from Jacobi, additive Schwarz, and finally full LU, which now allows us to advance both ions and neutrals together. Furthermore, the PETSc "coloring" scheme for efficiently computing the full finite-difference preconditioning Jacobian in parallel gives further substantial savings.

4. Embedding parallel turbulence computations in FACETS

Turbulence contributes a significant fraction of total transport in the core plasma. Our recent efforts have focused on extending the FACETS core solver to incorporate fluxes from the five-dimensional gyrokinetic continuum turbulence code GYRO. Embedding turbulence calculations creates special software and algorithmic challenges. In particular, as these calculations are very expensive, special domain decomposition strategies need to be developed. Further, efficient algorithms need to be developed to minimize the time it takes to achieve statistical steady-state of the computed fluxes and ensure that the Jacobian needed to advance the transport equations are computed with minimum function calls.

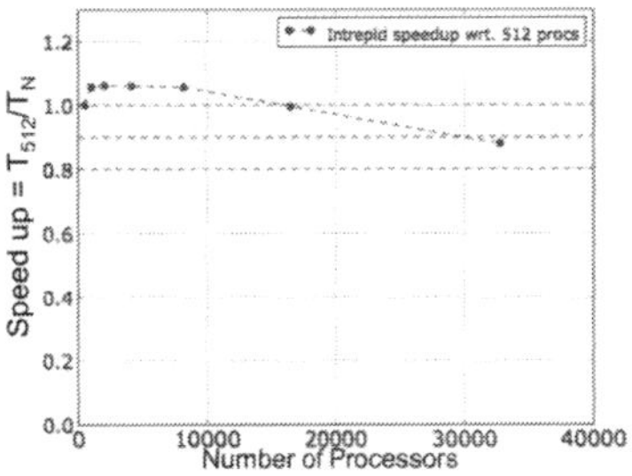

Figure 1. Weak scaling on the Intrepid LCF of embedded turbulent flux calculation using GYRO. There is a 10% drop in efficiency at high processor counts due to decomposition/network toplogy mismatch.

For performing the domain decomposition we have developed a set of C++ classes to split a given set of processors into different communicators, each running an instance of GYRO in parallel. For example, given 5120 processors and ten flux surfaces we create ten worker communicators, each running GYRO on 512 processors. For data transfer with the core solver an additional communicator of all rank 0 processors of the worker communicators is created. A distributed array is created on this messaging communicator and is used to transfer gradients and values to the GYRO instances and fluxes back to the core solver. This infrastructure was tested on the Intrepid LCF for parallel scaling efficiency by running GYRO on 512 processors per flux surface with increasing the number of flux surfaces from 4 to 64. Timing studies show that our infrastructure adds a 3 second overhead per 2 hours of embedded GYRO calculations. Further, the infrastructure scales almost linearly, showing a 10% loss in efficiencies going from 32 flux surfaces (16,384 PEs) to 62 flux surfaces (32,768 PEs). This loss is attributed for failing to take into account the network topology of the supercomputer. We are exploring ways to improve this by instrumenting FACETS with the TAU profiling tool.

Our current effort is focused on incorporating the fluxes from the GYRO calculations into the core solver so that we can advance the transport equations in time using turbulent fluxes. For this we are developing algorithms to compute running statistical averages of fluxes and the ability to stop a GYRO instance once fluxes are in statistical steady-state. Strategies to restart GYRO from perturbed previous solutions as initial conditions are being developed. These two algorithmic advances will allow us to couple the turbulence fluxes into FACET in the core solver and create a dynamic embedded-turbulence transport solver.

5. Coupling realistic particle and energy sources

The parallelized PPPL Monte Carlo package (NUBEAM) is used for computing the core sources from neutral beam injection. Progress in incorporating NUBEAM into FACETS has been made on two fronts: (i) design and development of a Plasma State physics component interface in Fortran and a mechanism to access this interface from a C/C++ driver and (ii) development of a parallelized driver for the NUBEAM component.

The strategy for coupling NUBEAM to a C/C++ driver leverages the Plasma State Component interface to NUBEAM that was already under development for the SWIM SCIDAC. The Plasma State (PS) is a fortran-90 object incorporating a time slice representation of a tokamak equilibrium, plasma profiles such as temperatures, densities, and flow velocities, and core plasma particle, momentum, and energy sources. It also incorporates a machine description section containing time invariant quantities such as neutral beam and RF antenna geometries. The PS implementing code along with the C/C++ wrapper is generated from a state specification file using a Python script.

A PPPL/Tech-X collaboration [8] developed a reliable, portable method for control of instantiation of fortran-90 Plasma State objects from C++ code, and set/get access to all state elements. This "opaque handle" method identifies state instances using a short array of integers; the C/C++ routines thus access fortran-90 routines through an interface involving fortran-77 primitive data types only. The majority of inputs to NUBEAM-including all physical quantities that would be required for any neutral beam model-are handled through the Plasma State, using this method. The remainder of inputs-NUBEAM specific numerical controls such as the sizes of Monte Carlo particle lists to use in a given simulation-are handled through two C/C++ compatible sequenced "shallow" fortran-90 data types (one for initialization, and a smaller one containing parameters that are adjustable in time). These control structures, containing no allocatable elements, are directly mapped to C structs.

A new parallelized driver for the NUBEAM has been developed. This has allowed testing, scaling studies, and served as a template for the C++ driver in FACETS. NUBEAM is the first component in FACETS using volumetric coupling with core transport equations.

6. Plasma-wall interaction module

The edge plasma and material wall are strongly coupled primarily via particle recycling, impurity production, and plasma power exhaust. The handling of transient and static peak power loads, core plasma contamination with impurities, plasma-facing components lifetime, and hydrogen retention in walls are the critical issues affecting the design of next-step fusion devices (like ITER, CTF, DEMO). To address these issues and to model self-consistently the plasma-wall coupling in FACETS, the Wall and Plasma-Surface Interactions (WALLPSI) module was developed [4].

The work on WALLPSI verification and validation against vast experimental data is in progress. Several laboratory experiments had showed clear saturation of retained deuterium in beryllium and in graphite (e.g. pyrolitic graphite). In [4], the results of simulations of static deuterium retention in graphite and beryllium at room temperatures with WALLPSI were presented showing good agreement with experimental data (see Fig. 2). Here, deuterium is retained via collisional production of broken bond traps followed by population of these traps by deuterium atoms. The modeled saturated dose for deuterium in graphite is consistent with [D]/[C]=0.4, the measured concentration of statically retained deuterium averaged over the projectile range. WALLPSI verification includes solving of simple diffusive 1-D transport problems for hydrogen in wall material.

The coupled plasma-wall modeling scheme with WALLPSI was tested by: (i) calculating the inventory build-up of mobile, chemically bonded, adsorbed and trapped hydrogen in the wall as well as the nonlinear variation of hydrogen recycling coefficient and impurity production rate in response to the incident plasma particle and energy fluxes, and (ii) simulating the spatiotemporal evolution of plasma parameters and hydrogen inventory in the wall with the coupled WALLPSI and plasma transport code EDGE1D for a range of plasma conditions [4].

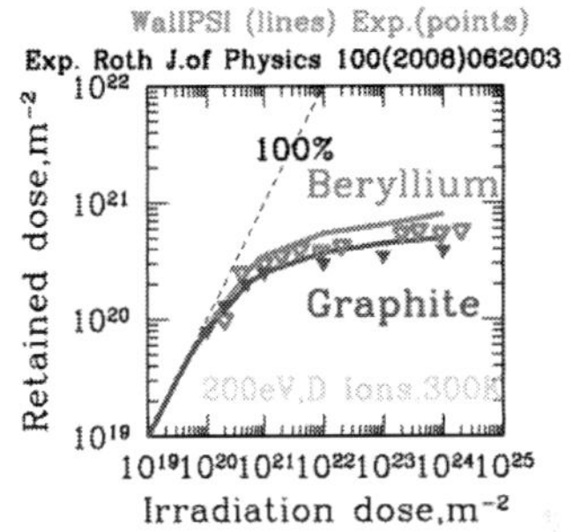

Figure 2. Example of WALLPSI validation against experimental retention data

7. Development of multicomponent visualization capabilities

Visualization is extremely valuable in providing better understanding of scientific data generated by fusion simulations. Various models need to be compared with each other and validated against experiments. That is why FACETS developed a set of standards and tools facilitating FACETS data visualization. All FACETS components use HDF5 data format for the output and comply to a particular standard in organizing HDF5 so that the data that is supposed to be visualized is easily found and interpreted. We call this standard VizSchema [5], [6]. VizSchema is a self-contained vocabulary for describing gridded data and so allows third party applications to visualize and post-process FACETS data. For example, the following pseudo-code snippet shows the metadata indicating thet the dataset (electron temperature in the scape-off-layer) needs to be visualized, needs a mesh called solmesh, should be interpolated to a zone and is a part of a composite variable called `tes`:

```
Dataset "tesSol" {
  Att vsType = "variable"
  Att vsMesh = "solmesh"
  Att vsCentering = "zonal"
  Att vsMD = "tes"
}
```

Mesh's metadata describes its kind and provides appropriate for the kind information. For example, this snippet describes a structured mesh which is a part of a bigger mesh called `sol`:

```
Group "solmesh" {
  Att vsType = "mesh"
  Att vsKind = "structured"
  Att vsMD = "sol"
}
```

The vsMD attributes shown above indicate that all tes variables will be combined into one variable which will live on a mesh combined from the `sol` meshes.

Based on the VizSchema standard, we developed a plugin for the VisIt [7] visualization tool. This plugin allows visualization of all FACETS data. An example of multicomponent visualization (core and three edge resions) is shown in Fig. 3.

8. First physics studies

The turbulence in the core region has been the focus of intense research for over 10 years, resulting in a reasonable understanding of the involved processes. As discussed above, FACETS leverages these efforts by integrating embedded gyrokinetic codes for calculating core transport fluxes. A key part of the goal for FACETS in the coming year is to extend the solver techniques to enable full time-dependent simulations using embedded gyrokinetic coupled to Monte Carlo calculation of neutral beam sources.

The embedded turbulence technique relies on the turbulence correlation lengths being much smaller than the system scale lengths. This not true for the plasma near the edge of the tokamak, and as a result the a complete understanding of the transport in the edge region is more primitive (although progress is being made in separately-funded projects). Despite the limitations of the fluid edge model used in the inital phase of FACETS, the model provides accurate models for the parallel transport and sources from the wall. For the first core-edge physics study, we exploit these strengths to study the pedestal build-up of tokamak plasmas.

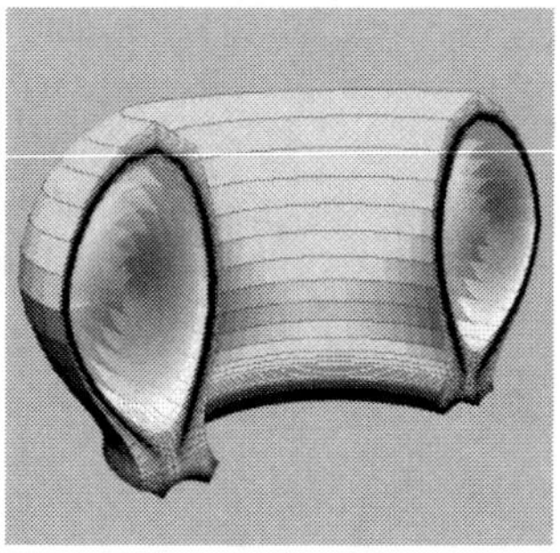

Figure 3. Three-dimensional visualization of an integrated core and edge simulation.

The highest-performing tokamak plasmas are characterized by steep edge plasma parameters known as the pedestal. Eventually, these steep profiles become so steep that they drive instabilities which destroy them. The details of how the pedestal forms has many unknowns because it is known to depend not only on the edge turbulence, but also the amount of power leaving the core and the way in which the plasma interacts with the wall [Maingi09]. We have begun simulations of pedestal buildup of the DIII-D experimental discharges utilizing an interpretive mode for UEDGE where plasma profiles are taken from experimental measurements and given sources of particles and energy, the transport coefficients are computed. On the closed magnetic field lines, the interpretive procedure assumes plasma profiles are approximately constant on flux surfaces (verifiable by direct 2D UEDGE simulation for spatially varying transport coefficients). The resulting 1D flux-surface averaged transport equation then treats the plasma fluxes as the unknowns, given the experimental profiles, UEDGE-computed neutral particle source, and input edge power and particle flux from neutral beam fueling. From the computed radial plasmas fluxes (density and ion/electron energy), transport diffusivities can be determined (as gradients are known).

What is unknown experimentally is the magnitude and detailed 2D distribution of the neutral gas source from gas puffing and wall/divertor recycling of ions into neutrals. The physics question to be answered by this procedure is how the pedestal region is fueled from such neutral sources in a setting where the transport coefficients are constrained and if this source is consistent with core density build-up using theory-based transport there.

9. Workflow for core-edge analysis

Computational workflow (in this context) refers to the process by which one starts with the initial set of data, runs the simulations to produce a final set of data, analyzes the data, and produces scientific results. Even for standalone, single-purpose codes, computational workflow issues can be cumbersome and difficult for new users who have to learn how to modify the input parameters, how to run the code on remote systems, and how to analyze the data. As FACETS encapsulates more codes into its framework, these workflow issues increase in difficulty rapidly, and a constant re-evaluation of workflow issues is necessary to ensure that users are able to use FACETS.

As described in the previous section, we are utilizing experimental data to help constrain the simulations. By constraining what we know, we are able to use simulations to help understand the role of other difficult to measure quantities. The type of analysis in this simulation is unique and has required development of new tools. The general procedure is one of using a combination of python and bash scripts to setup the simulation, and python scripts to analyze the output. The python setup scripts are able to handle the disparate data sources from the experimental analysis (multiple because most of the data is generally written to be easily used by core-only codes). The output scripts must collect the output from each component, and then present a unified visualization for analysis. This is done using either matplotlib for routine visualization, or as seen in Fig. 3, visit for higher-end visualization and analysis. Initial studies are underway and the workflow is continually improved as the result of feedback from users and as the simulations progress.

10. Summary

The FACETS project has made steady progress towards providing whole device modeling for the fusion community. Physics components have been refurbished, a framework for parallel components is undergoing continuing development, data analysis and visualization tools have been adapted, and the project is now embarking on its first physics studies. The framework has made tens of packages available on platforms from laptops to LCFs.

Acknowledgments

Work supported by the U.S. Department of Energy (DOE) Office of Science under grants and contracts including DE-FC02-07ER54907 at Tech-X, DE-AC02-06CH11357 at ANL, DE-FC02-07ER54909 at CSU, DE-AC52-07NA27344 at LLNL, DE-AC05-00OR22725 at ORNL, DE-FC02-07ER54910 at ParaTools, DE-AC02-76CH03073 at PPPL, DE-FC0207-ER54908 and DE-FG0204-ER54739 at UCSD. This work used the resources of the National Energy Research Scientific Computing Center, which is supported by the DOE Office of Science under Contract No. DE-AC02-05CH11231,and of the National Center for Computational Sciences at ORNL, which is supported by the DOE Office of Science under Contract No. DE-AC05-00OR22725, and of the Argonne Leadership Computing Facility at Argonne National Laboratory, which is supported by the Office of Science of the U.S. Department of Energy under contract DE-AC02-06CH11357.

References

[1] J R Cary, J Candy, R H Cohen, S Krasheninnikov, D C McCune, D J Estep, J Larson, A D Malony, A. Pankin *et al* 2008 First results from core-edge parallel composition in the FACETS project *J. Physics: Conf. Series* **125** 012040

[2] D P Grote 2009 Forthon, Lawrence Berkeley National Laboratory, http://hifweb.lbl.gov/Forthon (last viewed June 4, 2009).

[3] Tamara Dahlgren, Thomas Epperly, Gary Kumfert, and James Leek. *Babel User's Guide.* CASC, Lawrence Livermore National Laboratory, Livermore, CA, 2004.

[4] A.Yu. Pigarov and S.I. Krasheninnikov, Coupled plasma-wall modeling, *Journal Nuclear Materials* **390-391** (2009) 192.

[5] VizSchema, https://ice.txcorp.com/trac/vizschema/wiki/WikiStart.

[6] S Shasharina, J R Cary, S Veitzer, P Hamill, S Kruger, M Durant, and D Alexander Vizschema - visualization interface for scientific data, to be published in *proceedings of Computer Graphics, Visualization, Computer Vision and Image Processing 2009, Algarve, Portugal,* June 20-22, 2009.

[7] H. Childs, E. S. Brugger, K. S. Bonnell, J. S. Meredith, M. Miller, B, J Whitlock and N. Max A Contract-Based System for Large Data Visualization *Proceedings of IEEE Visualization* 2005, pp 190-198, Minneapolis, Minnesota, October 23–25, 2005.

[8] A Pletzer, D McCune, S Muszala, S Vadlamani and S Kruger, Exposing Fortran Derived Types to C and Other Languages, *Computing in Science and Engineering* **10** (2008) 86.

Journal of Physics: Conference Series **180** (2009) 012057 doi:10.1088/1742-6596/180/1/012057

Whole-volume integrated gyrokinetic simulation of plasma turbulence in realistic diverted-tokamak geometry

C S Chang[1], S Ku[1], P Diamond[2], M Adams[3], R Barreto[4], Y Chen[5], J Cummings[6], E D'Azevedo[4], G Dif-Pradalier[2], S Ethier[7], L Greengard[1], T S Hahm[7], F Hinton[8], D Keyes[3], S Klasky[4], Z Lin[9], J Lofstead[10], G Park[1], S Parker[5], N Podhorszki[4], K Schwan[10], A Shoshani[11], D Silver[12], M Wolf[10], P Worley[4], H Weitzner[1], E Yoon[7] and D Zorin[1]

[1] Courant Institute of Mathematical Sciences, New York University, NY 10012, USA
[2] University of California at San Diego, La Jolla, CA 92093, USA
[3] Columbia University, New York, NY 10027, USA
[4] Oak Ridge National Laboratory, Oak Ridge, TN 37831, USA
[5] University of Colorado at Boulder, Boulder, CO 80309, USA
[6] California Institute of Technology, Pasadena, CA 91125, USA
[7] Princeton Plasma Physics Laboratory, Princeton, NJ 08543, USA
[8] Hinton Associates, Escondido, CA 92029, USA
[9] University of California at Irvine, Irvine, CA 92697, USA
[10] Georgia Institute of Technology, Atlanta, GA 30332, USA
[11] Lawrence Berkeley National Laboratory, Berkeley, CA 94720, USA
[12] Rutgers University, New Brunswick, NJ 08901, USA

E-mail: `cschang@cims.nyu.edu`

Abstract. Performance prediction for ITER is based upon the ubiquitous experimental observation that the plasma energy confinement in the device core is strongly coupled to the edge confinement for an unknown reason. The coupling time-scale is much shorter than the plasma transport time-scale. In order to understand this critical observation, a multi-scale turbulence-neoclassical simulation of integrated edge-core plasma in a realistic diverted geometry is a necessity, but has been a formidable task. Thanks to the recent development in high performance computing, we have succeeded in the integrated multiscale gyrokinetic simulation of the ion-temperature-gradient driven turbulence in realistic diverted tokamak geometry for the first time. It is found that modification of the self-organized criticality in the core plasma by nonlocal core-edge coupling of ITG turbulence can be responsible for the core-edge confinement coupling.

1. Introduction

In order to achieve the design goal of Q=10 (fusion output energy/input energy), ITER[2] relies upon the so called high-mode (H-mode) plasma operation in which an edge temperature pedestal is formed as the core heating power is increased. For an unknown reason, transition into H-mode is accompanied with an instantaneous improvement of the core turbulence level and ion temperature. The time scale of this core-edge coupling is much faster than the core-edge energy transport time scale. Moreover, the overall ion temperature T_i gradient becomes stiff, raising the core T_i as the edge pedestal T_i value grows while keeping it low in the scrape-off layer. Since the scrape-off layer plasma touches the material wall, its temperature must be kept cool in order to minimize the damage to the material wall. Without the edge pedestal effect in H-mode operation, it is difficult to keep the scrape-off layer plasma cool while raising the core temperature. These H-mode phenomena (formation of edge pedestal and the simultaneous improvement of the core plasma temperature) are crucial ubiquitous experimental findings which have been making the thermonuclear fusion program brighter. However, a satisfactory theoretical understanding of these critical core-edge coupled phenomena is still absent, and awaits an integrated micro-turbulence and neoclassical simulation of core-edge plasmas in a realistic device geometry.

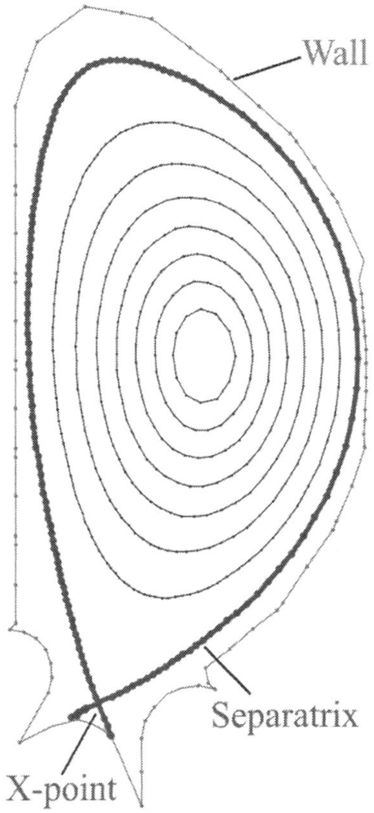

Figure 1. Magnetic flux surfaces of diverted tokamak geometry in a poloidal plane. X-point is shown on the separatrix. ITER is used as an example.

Although the necessity has thereby been recognized for a long time, it is a formidable challenge to capture both edge and core regions in an integrated global simulation. The difficulty of this problem is further exacerbated by the geometric complexity introduced in the edge region by the presence of a divertor and, thus, the magnetic separatrix surface (see Fig. 1). The magnetic separatrix surface is a singular surface to the conventional "flux" coordinate system used in the existing gyrokinetic codes. Due to this reason, the existing gyrokinetic codes do not approach the magnetic separatrix surface. At the same time, the existence of the magnetic X-point on the separatrix surface and an arbitrary wall shape in a tokamak device makes the conventional tokamak gyrokinetic mesh inadequate and requires a specially designed, field-line following unstructured mesh (see Fig. 2). This poses extra difficulties in the gyrokinetic particle simulation of diverted edge plasma. Details of the mesh property used in XGC1 can be found in the paper by M. Adams, S. Ku et al., presented at this conference [1]. In the present work such a geometrically difficult simulation challenge is combined with the formidable non-local, nonlinear, multi-scale physics challenges on a) a first-principles global gyrokinetic simulation to capture the basic physics of the nonlocal turbulence coupling and associated transport; b) a full distribution function representation (full-f) implemented to properly take into account the loss of single particle confinement across the magnetic separatrix surface and the formation of a strong edge ExB shearing layer thereby; and c) a multi-scale integrated modeling approach which includes the neoclassical physics together with the turbulence physics as demanded by a proper self-consistent evolution of the global core-edge background profile in the presence of nonlinear turbulence propagation and turbulence-driven transport. The full-f equation described here is different from the conventional full-f representation used in other fields in that the distribution function is reduced from the conventional six dimensional phase space to five dimensional phase space based upon the fast gyromotion perpendicular to the strong background magnetic field (leading to 5D gyrokinetic equation [17]). Eventhough it is a reduced equation from the conventional full-f equation, such a "full-f" global 5D gyrokinetic simulation in realistic magnetic geometry represents a major technical challenge that demands extreme scale computing resources.

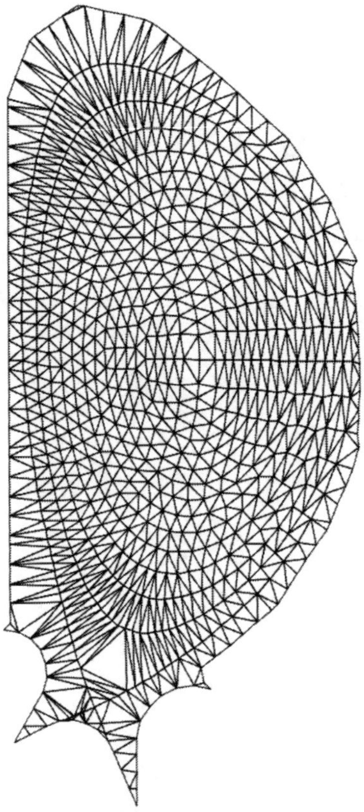

Figure 2. A sample unstructured triangular mesh with diverted magnetic field. The actual scale of the grid used in the simulations is approximately 30 times finer. ITER is used as an example.

Representative micro-turbulence phenomena in tokamak plasma are driven by the ion temperature gradient (ITG), electron temperature gradient (ETG), trapped electron drift resonance (TEM), magnetic field gradient (Ballooning Modes), and other instabilities. Among them the electrostatic ITG turbulence is the most fundamental one, setting the baseline turbulence transport at the longer wavelength scale. Other turbulence drivers change or add the micro-fluctuation and transport properties. In the present work, we include the ITG turbulence only, as a first step toward the formidable challenge. Other turbulence modes are to be added in the future as the computing power and the numerical technique improve further.

The "full-f" gyrokinetic code (XGC1) in the CPES project has been successfully developed targeting the difficulties noted in diverted edge geometry and the integrated nonlocal physics between turbulence and neoclassical physics (see Ref. [7], Invited Paper, 2008 American Physical Society, Division of Plasma Physics). Thanks to the recent petascale capability of ORNL's *Jaguar* computer and the generous award in the 2009 INCITE and OASCR/OFES Joule programs, XGC1 computing has been pushed to the extreme limit where no other gyrokinetic code has ever approached. Ion temperature gradient (ITG) driven turbulence physics has been simulated in integration with the background neoclassical physics using full-f gyrokinetic ions and adiabatic electrons. The simulation volume has been extended to the entire toroidal plasma volume (corresponding to the DIII-D tokamak facility at General Atomics) – from the magnetic axis to the material wall and across the mathematically complex magnetic separatrix region. Central plasma heating and conserving Coulomb collisions are also taken into account to make the simulation more realistic. One production run in DIII-D geometry consumes about 1.5M MPP hours, which is equivalent to the operation of 150,000 *Jaguarpf* XT5 cores for 10 hours in the well-scaled XGC1 simulation. The excellent scalability of XGC1 to petascale computing is, as reported in Ref. [1], shown in Fig. 3.

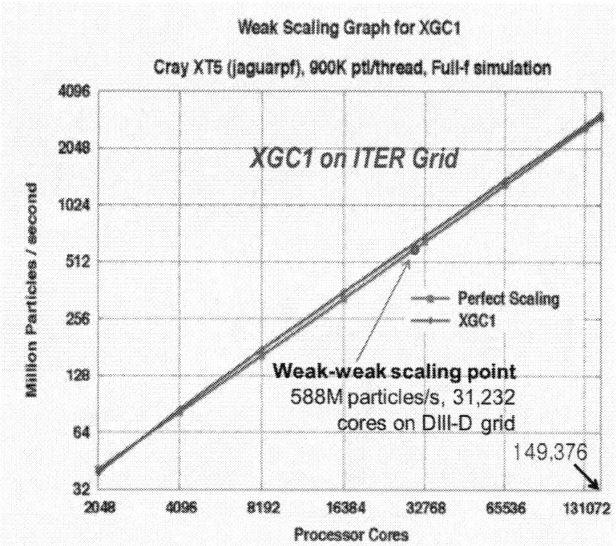

Figure 3. Two different weak scalings on fixed ITER grid (weak-strong) and on two different device grids (weak-weak, ITER vs DIII-D). Both of them show excellent scalability.

The purpose of this paper is to report such a high capacity simulation using the XGC1 code and the resulting new physics understanding of how H-mode plasma operation is maintained through nonlocal nonlinear coupling between core and edge plasmas. The paper is organized as follows. After this introduction section, the XGC1 code is described in Section 2. The new physics discovery is presented in Section 3, followed by conclusion and discussion in Section 4.

2. The full-f gyrokinetic code XGC1

XGC1 is a full-f gyrokinetic particle-in-cell code [6, 7] which can simulate the whole plasma volume including the magnetic axis, the magnetic separatrix and the scrape-off layer bounded by a biased material wall, by reading in a g-eqdsk data file generated from experimental equilibrium reconstruction [16]. XGC1 in its present form is an electrostatic code. Full-f marker ions and adiabatic electrons are used in the present study of ITG turbulence. A heat source is placed in the central core plasma to induce a heat flux into the simulation region, achieved by raising the particle kinetic energy in the source region by a small fraction while keeping the pitch angle invariant. The heat is then lost to the materal wall by self-consistent transport. A heat source is necessary to maintain the plasma in a self-organized global profile. A particle-, momentum- and energy-conserving linear Monte Carlo Coulomb collision operator is built into the particle motion [5, 23, 25, 10, 18]. Monte Carlo neutral particles can also be simulated together in the full-f XGC family codes [5]. In the present short-time simulation, however, the neutral particle routine is not used.

Marker particles follow the electrostatic Lagrangian equation of motion which conserves the mass, the canonical angular momentum and the energy [24, 4, 19].

$$\begin{aligned} \dot{\mathbf{X}} &= (1/D)[v_\parallel \hat{b} + (v_\parallel^2/B)\nabla B \times \hat{b} + \{\mathbf{B} \times (\mu \nabla B - \mathbf{E})\}/B^2] \\ \dot{v}_\parallel &= -(1/D)(\mathbf{B} + v_\parallel \nabla B \times \hat{b}) \cdot (\mu \nabla B - \mathbf{E}) \\ D &= 1 + (v_\parallel/B)\, \hat{b} \cdot (\nabla \times \hat{b}), \end{aligned} \tag{1}$$

where $v_\parallel$ is the speed of the particle parallel to the local magnetic field vector $\mathbf{B}$, $\hat{b} = \mathbf{B}/B$, $\mu = v_\perp^2/2B$ is the magnetic moment, and $\mathbf{E}$ is the gyro-averaged electric field. XGC1 uses a cylindrical coordinate system for the particle advance, which allows particle motion in true arbitrarily shaped flux surfaces including the magnetic separatrix and X-point. In a conventional magnetic flux coordinate system, the equation of motion encounters a mathematical singularity on the magnetic separatrix surface, and the error in the particle motion grows rapidly as it approaches the magnetic separatrix.

The following gyrokinetic Poisson equation is solved on a mesh with the four-point averaging technique [21],

$$-\nabla_\perp \frac{\rho_i^2}{\lambda_{Di}^2} \nabla_\perp \Phi = e\left(1 - \nabla_\perp \rho_i^2 \nabla_\perp\right)(\overline{n}_i - n_e), \tag{2}$$

where $\rho_{\rm i}$ is the ion gyroradius vector, λ_{Di} is the ion Debye length, and $\overline{n}_i$ is the ion density in real space [17],

$$\overline{n}_i = \frac{1}{2\pi} \int f_i(\mathbf{X}, \mu, u) \delta(\mathbf{X} - \mathbf{x} + \rho_{\rm i}) d\mathbf{X} d\mu d\alpha,$$

$\mathbf{x}$ is the particle position vector, and α is the gyro-phase angle. The above gyrokinetic Poisson equation is valid for steep gradient plasma, too, as long as the plasma gradient scale length is much greater than the ion gyroradius.

In order to take advantage of the slowly varying nature of the electric potential (both neoclassical and turbulent) along the magnetic field lines, the electrostatic potential is solved on an approximately field-line following mesh as is usually done in the conventional turbulence codes. Meshes on all the ploidal planes are identical (axisymmetric) in XGC1. When a mesh node is mapped along the magnetic field line, it approximately meets another mesh node on the adjacent poloidal plane. The mapping is only approximate because a magnetic field line is not closed unless it is on a mode rational surface. Due to the existence of the magnetic X-point, large flux volume expansion along the field lines, and an arbitrary wall shape, an unstructured triangular mesh system is used in the radial-poloidal plane (and a regular mesh is used in the toroidal direction). Use of the cylindrical coordinate system for particle advances and the unstructured mesh for the field-line following Poisson solver is necessary in the diverted geometry simulation, but adds complexity and demands an extreme high performance computing (HPC) resource. Details on the innovative numerical approach to these issues can be found in Ref. [1].

The macroscopic physical quantities associated with the guiding center particles in a volume ΔV are defined as follows. f_i, n_i, m_i, $q_{i\psi}$, χ_i are the distribution function, the density, the mass, the heat flux in ψ space, and the thermal conductivity of the guiding center ions, respectively. Radial particle flux is zero in the present adiabatic electron model. The volume can be defined as a flux shell for flux surface averaged quantites, or the local cell for local quantities.

$$f_i(X, v_\parallel, \mu) = (1/\pi\Delta V) \sum_{X_j \in \Delta V} w_j \delta(X - X_j)\delta(v_\parallel - v_{\parallel j}\delta(\mu - \mu_j) \quad (3)$$

$$\begin{aligned} W &= \sum_{X_j \in \Delta V} w_j \\ n_i &= = W/\Delta V \\ u_\parallel &= \sum_{X_j \in \Delta V} v_{\parallel j} w_j / W \\ T_i &= \frac{1}{3} \sum_{x_j \in \Delta V} w_j m_i \left(2B\mu + (v_{\parallel j} - u_\parallel)^2\right)/W \\ q_{i\psi N} &= (1/2) \sum_{X_j \in \Delta V} m_i \left(2B\mu + (v_{\parallel j} - u_\parallel)^2\right) w_j \dot{\vec{X}}_j \cdot \nabla\psi / W \\ \chi_i &= q_{i\psi}/(|\nabla\psi|^2 \frac{\partial T}{\partial \psi}) \end{aligned} \quad (4)$$

Advantages in a full-f particle simulation, in comparison with the conventional delta-f particle simulation [11, 22], are in the ability to calculate the large dynamical variations of the mean and the perturbed parts of the plasma together without scale separation, in the facilitation of sources and sinks, and in the non-growth of the marker particle weights. A full-f particle simulation solves the $df/dt = C(f) + S$ equation directly in a gyrokinetic form, where C is the collision operator and S represents the sources and sinks, by following the guiding center particle motion (Eq. 1) in 5D phase space in the electric field given by Eq. (2). Unlike a delta-f gyrokinetic equation, which requires the conservation of phase-space volume, a full-f simulation can accept a source (sink) model on the right-hand side. There is no scale separation in the dynamical self-organization between the background and the perturbed plasma dynamics, and there is no need to use a fixed Maxwellian temperature profile for a free energy source, as assumed in a conventional delta-f simulation. On the other hand, the simulation is more expensive computationally. A full-f gyrokinetic particle-in-cell code requires many more marker particles, roughly by a factor $\sim (n/\delta n)^2$, than a delta-f gyrokinetic code. Fortunately, the simulation can be carried out without suffering from the monotonic growth of the particle weights as in a delta-f code. The actual number of full-f marker particles used in this study will be discussed in a later section.

XGC1 has been thoroughly verified for the accuracy of the single particle motions by examining the conservation properties, the solver capability in steady state using a simplified manufactured solution method, the collisional neoclassical physics by comparing with known analytic solutions, and the delta-f GAM and linear/nonlinear ITG physics by comparing with other global codes. A method of manufactured solution approach for the turbulence solution verification is under development by our mathematical and physics team members not only for XGC1 but also for other turbulence codes.

The large amount of physics data I/O from full-f simulation in extreme scale HPC is another challenge XGC1 encounters. A new adaptive I/O library called *ADIOS* has been created to meet this challenge [20]. A factor of 30 improvement in the write speed has been achieved in XGC1, allowing for more frequent write-out of the simulation data for code restart and the real-time physics analysis. Further improvement of *ADIOS* is being made as XGC1 pushes the HPC computing to a higher dimension. The simulation data is moved to the *eSimMon* Dashboard [3], created by our team, using the Kepler workflow engine in our EFFIS (End-to-end Framework for Fusion Integrated Simulation) framework for physics analysis including vector graphics and movie viewing.

3. Full-f simulation of whole-device ITG turbulence

We load the marker particles initially as a local Maxwellian, with flux-function density and temperature profiles (the magnetic flux-surface label, such as the enclosed poloidal magnetic flux ψ, is usually used as a radial coordinate). Instead of using the marker particle density to describe the radial profile of the plasma number density, we use the particle weight distribution as a function of radius. To be more specific, the particle weight distribution w_j in Eq. (3) is initialized with ψ_N dependence, where ψ_N is the poloidal magnetic flux normalized to be zero on the magnetic axis and unity at the separatrix surface. This method can resolve the enhanced particle noise problem at the low density region of the plasma profile. The particle weight in the simulation is normalized to unity at the magnetic axis, decreasing in minor radii according to the initial density profile (variation in the particle weight is normally within factor of 10). The experimental particle density is then represented by another normalization.

We note here that the local Maxwellian loading in the laboratory frame contains some unspecified initial toroidal rotation in an inhomogeneous plasma, which is physically insignificant compared to other free energy sources (such as the radial pressure gradient). Thus, the effect of initially large toroidal rotation on turbulence cannot be studied in the present simulations. A neutral beam type of external momentum source is not considered in the present work either, even though this capability is available in the XGC family code, XGC0.

We could also initially load the marker particles as a global canonical Maxwellian, which may yield more rapid reduction of the initial GAM activities and faster self-organization to the neoclassical solution [15]; hence, an easier growth of turbulence. It is well-known that unless the initial GAM activities are subdued, ITG turbulence does not grow in a full-f simulation [8]. However, the canonical Maxwellian loading has a disadvantage in that it is difficult to customize the radial temperature and density profiles. For this reason, we choose the local Maxwellian loading to take advantage of the easier profile customization. We get compensation for the disadvantage by obtaining a neoclassical radial electric field solution in a pre-conditioning simulation and using it in the turbulence simulation as the initial electric field. Since the particle loading after the preconditioning run is still a local Maxwellian, this process does not eliminate the initial GAM oscillations completely, but helps reduce the amplitude and the relaxation time.

The boundary condition we use on the elliptic Poisson equation is that the electrostatic potential on the material wall surface vanishes. Any particle hitting the wall surface is absorbed to the wall and removed from the simulation. Initial plasma profiles are shown in Fig. 4. The initial ion temperature T_i at the magnetic axis is chosen to be 4.5 keV, dropping linearly in $\sqrt{\psi_N}$ ($\propto$ real radius, approximately) to about 1.3 keV at the temperature pedestal top $\psi_N \simeq 0.83$ before following a tanh profile to form the edge pedestal. The relation between the real distance along the outside midplane and ψ_N is given in Fig. 5, together with the safety factor q profile (which is stopped short of the magnetic separatrix surface $\psi_N = 1$ where it becomes infinite). This T_i profile is not far off from experimental observations, and hence yields a realistic ratio

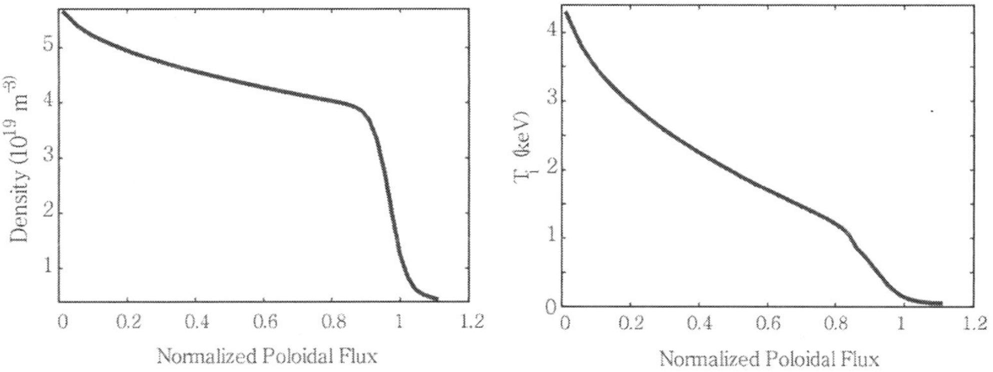

Figure 4. Early-time plasma density and temperature profiles. Electron temperature is assumed to be equal to ion temperature. Notice that the temperature pedestal knee is located at a somewhat smaller minor radius than the density pedestal knee, making the relative temperature slope high between the two knees.

of the ion larmor radius (or the shortest turbulence wave length considered) to the device size ($a/\rho_i \simeq 200$ where $\rho_i \simeq 3$ mm is the average ion gyroradius and $a \simeq 60$ cm is the minor radius along the outside midplane at the material wall of DIII-D). Of course, the initial T_i profile changes with heat transport as the simulation progresses. For a representative value of $v_i = (T_i/m_i)^{1/2}$, where m_i is the deuteron mass, we take $T_i = 4$ keV.

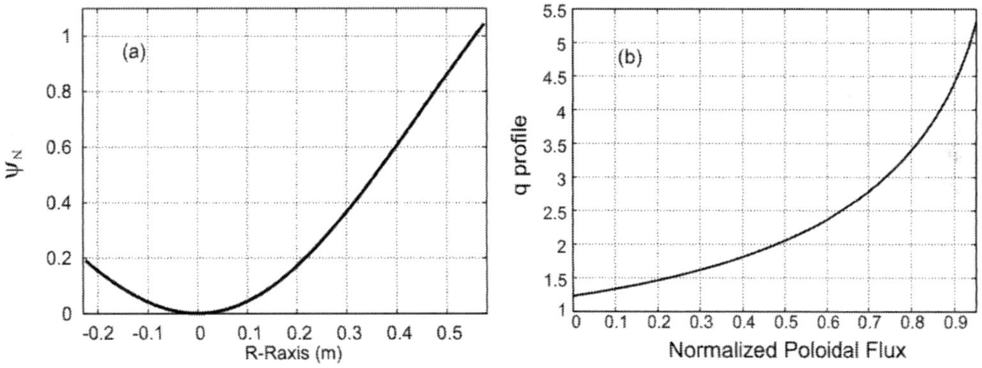

Figure 5. (a) Relationship between the normalized poloidal flux ψ_N and real distance in meters from the magnetic axis ($R_{axis} = R_o$) to the flux surface (R) along the midplane. (b) Radial profile of the safety factor q.

Plasma density n at the magnetic axis is chosen to be $5.5 \times 10^{19} m^{-3}$, dropping to about $4 \times 10^{19} m^{-3}$ at the density pedestal top $\psi_N \simeq 0.9$ with linear dependence in $\sqrt{\psi_N}$, followed by another tanh profile to form a density pedestal. Since we use adiabatic electrons, the plasma density profile does not change in time.

Stability of ion temperature gradient (ITG) modes in the core plasma is usually described in terms of the ratio between the major radius at the magnetic axis R_0 (or R_{axis}) and radial T_i gradient scale length, $R_0/L_T = R_0|\partial \log T_i/\partial r|$. The initial profile of R_0/L_T is as shown in Fig. 6. The horizontal dashed line is the nonlinear stability criterion of a core plasma from

Ref. [9], which is invalid in the edge pedestal [6]. As can be seen in Fig. 6, R_0/L_T is below the nonlinear stability limit in the significant portion of the core plasma $\psi_N < 0.5$ (r/a¡0.6), but rises steeply in the edge pedestal for a stronger ITG drive. As will be shown shortly, this does not mean that the ITG turbulence is strong in the edge pedestal, proving the invalidity of the R_0/L_T criteron there.

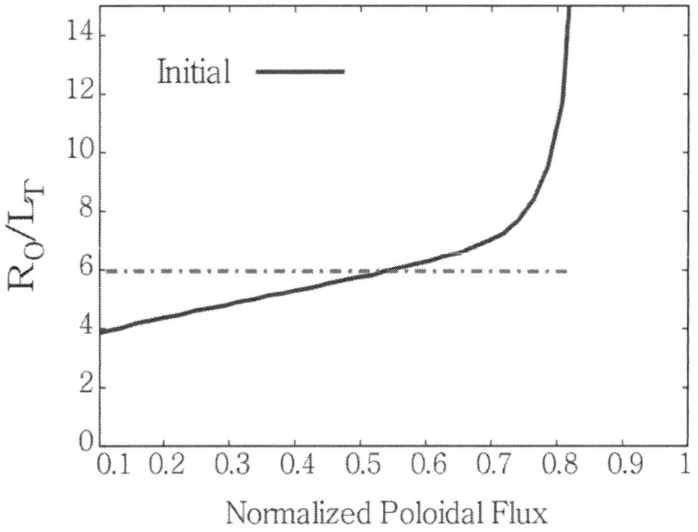

Figure 6. Initial profile of $R_0/L_T = R_0|\partial \log T_i/\partial r|$. The horizontal dashed line is the nonlinear stability criterion of a core plasma, from Ref. [9].

In an H-mode plasma, a strong radial electric field layer forms just inside the magnetic separatrix surface, consistent with a steep density pedestal as observed in experiments in the radial force balance relation. Our previous simulation shows that the formation of the strong radial electric field layer is from the X-point effect (loss of single particle ion confinement) [6]. In the present ITG simulation in H-mode type pedestals, we assume the existence of a reasonably strong edge density pedestal profile. The radial electric field and the sheared flows are then evaluated in the code self-consistently with the neoclassical and turbulence physics. Since our turbulence model uses adiabatic electrons, the plasma density profile does not evolve. Only the ion temperature T_i profile evolves. The electron temperature profile is assumed to be equal to the ion temperature profile.

The temperature and density profiles are within a realistic range of experimental values in DIII-D. However, in order to save the computing time, we artificially enhance the collision frequency by a factor of 10 (yielding the average ion collision time to be $\tau_{ic} \simeq 140\ v_i/R$). The simulation is then performed to a few collision times. Not only the neoclassical relaxations, but also the GAM and zonal flow relaxation become faster in proportion to collision frequency. A factor of 10 enhancement of the collision frequency still keeps the plasma in the banana regime all the way out to the pedestal top ($\nu_* = 0.7$ at $\psi_N = 0.83$), keeping the integrity of the particle orbit dynamics.

Convergence tests in marker particle number have been performed up to 12 billion particles and have revealed that a total of 5 billion marker particles (average $\sim$ 2,000 marker particles per grid node) is minimal for an adequate study of the global ITG turbulence. This yields a noise driven χ_i at the level $\simeq 0.05\ m^2/s$, which is identified from the simulations performed without ITG or collisions and is much lower than the χ_i level ($\gtrsim 1m^2/s$) obtained in the actual simulation with ITG turbulence and enhanced collisions. Convergence tests in the grid size show that we need $\simeq 3\ mm$ average grid size at the outside midplane, which is about the average ρ_i size. We have used various numbers of *Jaguarpf* XT5 cores (up to 149,760) for benchmarking and scaling studies. The actual simulation results presented here are from 7.5 billion marker particles (corresponding to an average of $\sim$ 3,000 marker particles per grid node) run for 88 hours on 14,976 CRAY XT5 cores to obtain a relatively steady ion heat flux or turbulence level (corresponding to a total of 2.4 ms in real time). This corresponds to about 10 hours with 149,760 XT5 cores (the full capacity of *Jaguarpf*), taking into account some performance degradation. A weak scaling study of XGC1 up to 149,760 cores is presented in Ref. [1].

The time step of the simulation ($= 0.04\ R_0/v_i$) is chosen from a convergence test. The number of total time steps taken in the simulation presented here is 10,000.

4.5MW of total heat is added to the ions around the magnetic axis ($\psi_N \leq 0.04 \simeq 10cm$) to force a heat flux into the turbulence region. The heating is achieved by raising the particle energy uniformly in the heating region by a small fraction of the kinetic energy while keeping the pitch angle invariant. This rate of heating in a small central volume is too large for the local ITG and neoclassical transports to efficiently transport the heat radially outward. As a matter of fact, the full-f plasma even shows resistance to the local heat flux by developing a strong E × B shearing rate in the heat source region. In order to ensure a smooth outward heat flux from the localized heating, we raise further the Coulomb collision rate by another factor of 10 (thus raising the neoclassical transport level) in the central core ($\psi_N \leq 0.08 \simeq 13cm$) in a somewhat greater volume than the heat source volume. This artifical collision is part of the central heating model in the code.

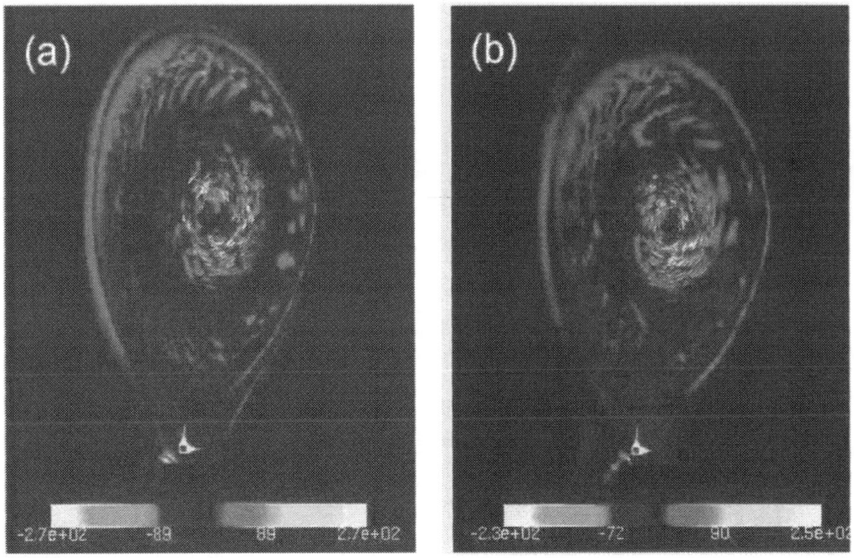

Figure 7. Snapshot figures of turbulent electrostatic potential on a poloidal (constant toroidal angle) plane at an (a) early and (b) later stage

Figure 7 shows two snapshots of fluctating electrostatic potential contour on the poloidal cross-section in the earlier and later phases of the ITG turbulence. It can be clearly seen that the turbulence starts at two places: in the edge where R_0/L_T is large and in the central core where the heating source exists. A dark ring just inside the separatrix surface indicates weak turbulence intensity there, even though R_0/L_T is very large there, which is a characteristic of an H-mode plasma. The edge-generated turbulence propagates radially inward. On the other hand, the heating-generated central turbulence does not propagate outward. It can be seen that the central core turbulence is under a strong (sheared) poloidal rotation, indicating a self-containment of the heating-generated turbulence. The inward propagation stops when the edge-originated turbulence meets the strongly-sheared central turbulence at $t \simeq 150R/v_i$ ($\simeq 0.6$ ms).

Another remarkable observation made from the present simulation is the self-organizing modification of the background temperature profile by the incoming turbulence, as can be seen in Figure 8. Before the arrival of the edge-originated turbulence in the core plasma, the ion temperature gradient was below the nonlinear ITG criticality level (dotted horizontal line, originated from Ref. [9]). However, the arrival of the edge-generated turbulence raises the local temperature gradient above the nonlinear criticality level. The ion thermal conductivity is self-regulated by the self-generated ExB shearing. In other words, the turbulence criticality is nonlocally self-organized by the edge turbulence source. This state is maintained by the outward-flowing heat flux. The combination of the low thermal conductivity near the magnetic separatrix surface and the large heat flux from the core maintains the T_i pedestal and keeps the R_0/L_T around the T_i knee well above the nonlinear criticality level, continuously supplying the ITG turbulence energy to maintain the new self-organized criticality. Without the strong heat flux from the core, the R_0/L_T value around the intitial T_i knee location would collapse, and the driver for the elevated SOC state would be lost. We have examined a few different heating power levels. It is found that the η_i profile shown here is "stiff" with respect to the change of heating power, which is consistent with the experimental findings.

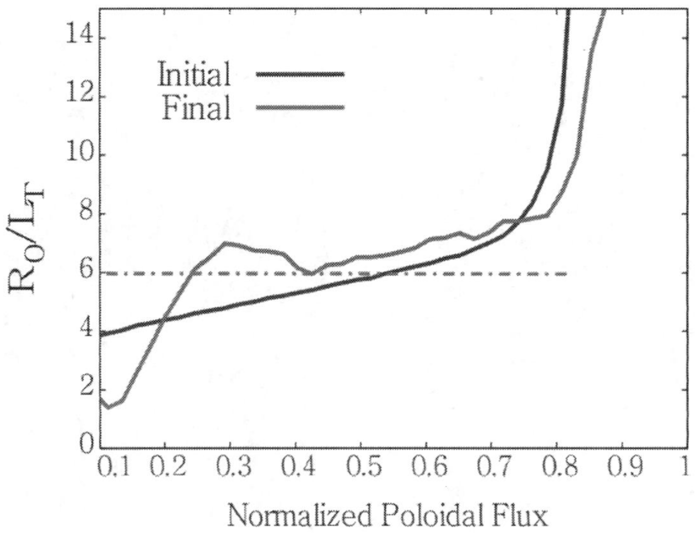

Figure 8. Self-organizing modification of the background temperature profile by the incoming turbulence. The horizontal line represents the nonlinear criticality in the core plasma [9].

Maintaing the T_i knee with a strong heat flux H can be understood in a simple one-dimensional argument as follows: As long as the heat diffusion is the dominant mechanism in determining the temperature gradient (over other energy dissipation or off-diagonal transport mechanisms), the continuity of heat fluxes between two adjacent positions x_1 and x_2 (with $x_1 < x_2$) demands $H = -\chi_1 \partial T_i(x_1)/\partial x = -\chi_2 \partial T_i(x_2)/\partial x$. If χ_2 is forced to stay low for some reason at a high heat flux, $\chi_1 > \chi_2$ results in a strong knee in the T_i profile somewhere between

x_1 and x_2. In the present problem, the strong edge effects (the magnetic separatrix with X-point, strong ExB shearing, and steep density gradient) keep the ITG driven χ_2 low. In the meantime, the high heat flux from the core and the strong ITG maintains a high χ_1 value.

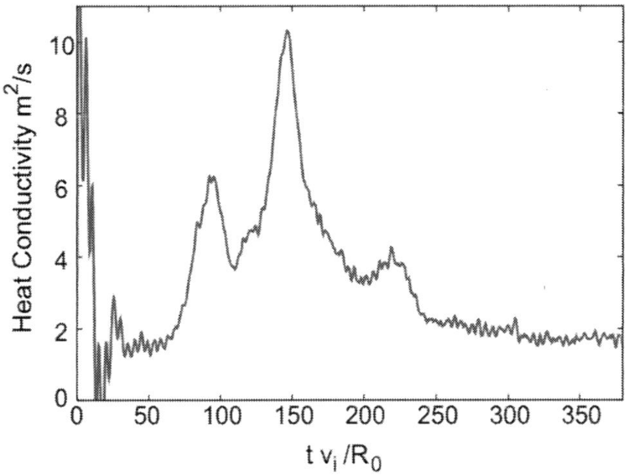

Figure 9. Time behavior of the effective ion thermal conductivity (thermal flux divided by local T_i gradient) from the start of the simulation across $\psi_N = 0.64$ which corresponds to $r = 42$ cm on the outside midplane. The self-organizing process is bursty.

Figure 9 is the time behavior of the effective ion thermal conductivity (thermal flux divided by local T_i gradient) from the start of the simulation across $\psi_N = 0.64$ which corresponds to $r = 42$ cm on the outside midplane. The short initial jittering is driven by GAM oscillations during the self-organization of the radial electric field, ion distribution function, plasma profile, and toroidal rotation toward neoclassical equilibrium in the initial local Maxwellian loading, and is subdued at about 30 v_i/R. At about 60 v_i/R, ITG modes start to grow. As mentioned in the previous section, it is well-known that unless the jittering from the initial GAM activities is subdued, ITG turbulence does not grow in a full-f simulation [8]. The total simulation time is about twice the collision time.

Figure 9 also shows that after the arrival of the turbulence front, there is a distinctive bursty type of heat flux behavior in the initial stage of nonlinear turbulent transport until about 240 v_i/R, as the initial neoclassical ion temperature profile and turbulence are self-organizing, followed by a relatively steady level of heat flux behavior. The inter-burst period is much greater than the initial GAM period. The self-organization process from one global state to another is of interest (such as the transition process from a low to a high mode in a tokamak plasma). The case studied here corresponds to a global transition from a neoclassical state to a turbulent state. The radial speed of the ballistic motion of heat burst is about $V_r \simeq (1/5)\rho_i v_i/R \simeq (1/30)\ \rho_i v_i/L_T)$, which is similar to the analytic estimate of the turbulence intensity propagation speed reported in Refs. [14, 13, 12]. Distinction between the $\rho_i v_i/R$ and $\rho_i v_i/L_T$ scalings in V_r is difficult to determine in the present simulation since R/L_T quickly self-organizes to a fixed value $\simeq 6.5$.

The inward propagation of the bursty turbulence intensity interplays with the local temperature gradient and ExB shearing dynamics. Arrival of the turbulence front at a location is first noticed by the steepening of the local temperature gradient, followed by the increase of heat flux ($\propto$ turbulence intensity) which weakens the temperature gradient, and then followed by increase in the local ExB shearing rate. There is about $\gtrsim 30 v_i/R_0$ time delay between three local events. Increase of the local ExB shearing rate then suppresses the turbulence and the outward heat flux, ending the burst. The local temperature steepens again, and the bursty cycle continues until a quasi steady avalanche turbulence is reached at the end, where the turbulence shows a 1/f power law.

In the central core, strong heat source maintains strong $E \times B$ shearing (see Fig. 7). As a result, the bursty heat flux does not appear there. It can also be noticed that the ion temperature gradient remains subcritical even though there is a strong turbulence there, indicating that the turbulence is driven by heating, not by the ion temperature gradient. This phenomenon has not been understood, yet. This subject may have an interesting relation to another uniquitous experimental observation that auxilary heating degrades tokamak core confinement. Possible relation of the local heating-driven strong ExB shearing is another interesting topic to be studied in the future, especially in a reversed-q magnetic equilibrium.

Energy conservation has been investigated within the volume $0.3 \le \psi_N \le 0.7$. Amongst the energy flowing into the volume across the inner surface $\psi_N = 0.3$, the sum of the particle energy change, the field energy created, and the outward-flowing energy across the outer surface $\psi_N = 0.7$, the total energy conservation showed about a 2% error. The energy conservation error in a full-f code should not grow unless numerical errors grow.

4. Conclusions and discussions

A special full-f gyrokinetic particle code XGC1 has been developed for a formidable integrated simulation, for the first time, of the core-edge and the neoclassical-turbulence physics in a realistic tokamak geometry including the magnetic separatrix. Turbulence studied in this report is from the electrostatic ITG mode using adiabatic electrons. Central heating is used to drive radial heat flux in the turbulence. A particle number, momentum and energy conserving Monte-Carlo collision operator is used to produce proper collision effects in neoclassical and turbulence physics.

Thanks to the capability computing enabled by the DOE Leadership Class computing systems, new understandings in the ITG turbulence physics are obtained on a) how the core ITG turbulence and ion temperature are instantaneously influenced by the edge plasma condition in a time scale much shorter than the energy confinement time scale and b) how a transition from a neoclassical state to a turbulent state is achieved.

The present numerical study shows strong evidence that the so-called "tail-wagging-dog" phenomenon, as ubiquitously observed in the experiments, can be explained from fast propagation of ITG turbulence energy from the pedestal top to the core and its modification of the core turbulence criticality through nonlocal, nonlinear self-organization. In order to validate and confirm this theoretical finding, an experimental study of the inward ITG turbulence propagation at the L-H transition from the edge pedestal top to the core is strongly suggested.

We note here again that the simulation presented here is only the first step toward understanding the L-H confinement physics in a tokamak. The ITG turbulence study reported here is only electrostatic without particle transport, in fixed magnetic equilibrium, and without turbulence interaction with plasma current. Addition of kinetic electrons and electromagnetic effects, including tearing modes, to the XGC1 turbulence study is the nearest term improvement goal. A routine to evolve magnetic equilibrium and q profile along with the background plasma profile evolution is already incorporated in the sister code XGC0 and will be used in XGC1. Other routines readily available from XGC0 include toroidal torque source and Monte Carlo neutral transport and recycling. Analytic model magnetic ripple capability is already incorporated into XGC1.

Acknowledgments

The XGC1 full-f gyrokinetic code has been developed collaboratively in the SciDAC FSP Prototype Center for Plasma Edge Simulation (CPES), which has been funded by the U.S.

Department of Energy, jointly between the Office of Fusion Energy Science and the Office of Advanced Scientific Computing Research.

This research used resources of the National Center for Computational Sciences at Oak Ridge National Laboratory, which is supported by the Office of Science of the U.S. Department of Energy under Contract No. DE-AC05-00OR22725. The extreme computing on *Jaguarpf* presented here was possible through the 2009 OASCR Joule Milestone program. The authors are indebted to Dr. D. Kothe, K. Roche, R. Kendall and other staff members at NCCS for their help in getting large-size Joule runs on *Jaguarpf*. This research also utilized *Franklin* at NERSC througn ERCAP allocation, also supported by US DOE Office of Science.

The research of D'Azevedo, Klasky, Podhorszki, and Worley was performed at the Oak Ridge National Laboratory, which is managed by UT-Battelle, LLC under Contract No. De-AC05-00OR22725. Accordingly, the U.S. Government retains a nonexclusive, royalty-free license to publish or reproduce the published form of this contribution, or allow others to do so, for U.S. Government purposes.

References

[1] M. Adams, S. Ku, P. Worley, E. D'Azevedo, J. Cummings, and C.S. Chang. *SciDAC2009, submited to J. Phys: Conf. Series.*
[2] R. Aymar, V.A. Chuyanov, M. Huguet, Y. Shimomura, ITER Joint Central Team, and ITER Home Teams. *Nucl. Fusion*, 41:1301, 2001.
[3] R. Barreto, S. Klasky, P. Podhorszki, P. Mouallem, and M. Vouk. *2009 International Symposium on Collaborative Technologies and Systems (CTS 2009).*
[4] A.H. Boozer. *Phys. Fluids*, 27:2441, 1984.
[5] C. S. Chang and S. Ku. *Phys. Plasmas*, 11:2649, 2004.
[6] C. S. Chang and S. Ku. *Phys. Plasmas*, 15:062510, 2008.
[7] C. S. Chang, S. Ku, P. Diamond, Z. Lin, S.Parker, T.S. Hahm, and N. Samatova. *Phys. Plasmas*, 16:056108, 2009.
[8] G. Dif-pradalier, V. Grandgirard, Y. Sarazin, X. Garbet, Ph. Ghendrih, and P. Angelino. *Phys. Plasma*, 15:042314, 2008.
[9] A. M. Dimits, G. Bateman, M. A. Beer, B. I. Cohen, W. Dorland, G. W. Hammett, C. Kim, J. E. Kinsey, M. Kotschenreuther, A. H. Kritz, L. L. Lao, J. Mandrekas, W. M. Nevins, S. E. Parker, A. J. Redd, D. E. Shumaker, R. Sydora, and J. Weiland. *Phys. Plasmas*, 7:969, 2000.
[10] A. M. Dimits and B. I. Cohen. *Phys. Rev. E*, 49, 1994.
[11] A. M. Dimits and W. W. Lee. *J. Comput. Phys.*
[12] X. Garbet, Y. Sarazin, F. Imbeaux, P. Ghendrih, O. D. Gurcan C. Bourdelle, and P. H. Diamond. *Phys. Plasmas*, 14:122305, 2007.
[13] D. Gurcan, P. H. Diamond, and T. S. Hahm. *Phys. Plasmas*, 14:055902, 2007.
[14] T.S. Hahm, P.H. Diamond, and Z. Lin. *Phys. Plasmas*, 12:090903, 2005.
[15] Y. Idomura, S. Tokuda, and Y. Kishimoto. *Nucl. Fusion*, 43:234, 2003.
[16] L. L. Lao, H. St. John, R. D. Stambaugh, A.G. Kellman, and W. Pheiffer. *Nucl. Fusion*, 25:1611, 1985.
[17] W. W. Lee. *Phys. Fluids*, 26:556, 1983.
[18] Z. Lin, W. M. Tang, and W. W. Lee. *Phys. Plasmas*, 2, 1995.
[19] R.G. Littlejohn. *Phys. Fluids*, 28:2015, 1985.
[20] J. Lofstead, F. Zheng, S. Klasky, and K. Schwan. *IPDPS 2009, IEEE Computer Society Press*, 2009.
[21] Y. Nishimura, Z. Lin, J.L.V. Lewandowski, and S. Ethier. *J. Comp. Phys.*, 214:657, 2006.
[22] S. E. Parker and W. W. Lee. *Phys. Plasmas*, B5:77, 1993.
[23] W. X. Wang, N. Nakajima, M. Okamoto, and S. Murakami. *Plasma Phys. Controlled Fusion*, 41, 1999.
[24] R. White. *Phys. Fluid B*, 2:845, 1990.
[25] X. Q. Xu and M. N. Rosenbluth. *Phys. Fluids B*, 3, 1991.

Exploration of finite ion orbit effects in the ion cyclotron range of frequencies

D L Green[1], E F Jaeger[1], L A Berry[1] and the rf-SciDAC team.
[1]Oak Ridge National Laboratory, P.O. Box 2008, Oak Ridge, TN, 37831-6169
E-mail: greendl1@ornl.gov

Abstract. The rf-SciDAC collaboration is developing computer simulations to predict the damping of radio frequency (rf) waves in fusion plasmas. Here we extend self-consistent quasi-linear calculations of ion cyclotron resonant heating to include the finite drift of ions from magnetic flux surfaces and rf induced spatial transport. The all-orders spectral wave solver AORSA is iteratively coupled with a particle based update of the plasma distribution function.

1. Introduction

The next decade will see the construction of ITER[1] and mark a significant step towards nuclear fusion as a power source. In part, the ITER design relies on radio frequency (rf) power as a means of auxiliary plasma heating and current drive[2]. The rf-SciDAC group[3] is developing computer simulations to predict how the ITER rf system will perform while benchmarking simulation results against present experiments. Here we report progress in developing tools to utilize leadership class computing facilities to investigate ion cyclotron resonant heating (ICRH) in tokamak fusion plasmas. Specifically, a self-consistent quasi-linear (QL) approach is presented to calculate power absorption with no assumption on the size of the Larmor radius, harmonic number or ion orbit deviation from a magnetic flux surface.

Recent work by Jaeger et al.[4] has demonstrated a QL iterative calculation of ICRH. The All-ORders Spectral Algorithm (AORSA) is used to solve for the rf wave fields assuming a Maxwellian electron distribution and a combination of thermal and non-thermal ion species. The non-thermal ion species distribution is provided as a flux surface bounce-averaged quantity by the finite difference Fokker-Planck code CQL3D [5]. CQL3D utilizes the simplifying assumption of zero ion orbit deviation from a flux surface and takes as input from AORSA a 3D, orbit averaged velocity space diffusion tensor that describes the rf heating. For the non-Maxwellian species AORSA evaluates the susceptibility directly from the ion distribution. An iterative coupling of these codes on the slow heating timescale (ms) has allowed self-consistent calculation of the power absorption for minority heating scenarios in ITER and the Alcator C-Mod tokamak[6] and also the high harmonic heating of neutral beam injected ions in NSTX[7] and DIII-D[8] [4, 3]. However, in many cases a significant fraction of ions have finite orbit widths such that coupling AORSA to alternative Fokker-Planck code that include these effects is desirable.

Several authors[9, 10, 11, 12] have investigated the importance of including finite ion orbits in ICRH calculations. The SELFO code[13] combined the LION full-wave solver[14] with the Monte-Carlo code FIDO[15]. FIDO calculates the orbit averaged ion distribution function in terms of three invariants of the unperturbed motion using Monte-Carlo collision and rf operators. Murakami et al.[16] have also performed self-consistent iterations using the TASK/WM full-wave code[17] and GNET[18] which solves the linearized drift kinetic equation. Such studies have shown that including finite orbit effects in ICRH simulations gives; (i) a broader collisional power transfer profile since power absorbed at resonance is transfered to the background plasma across the entire orbit width; (ii) additional resonant surface crossings due to orbits deviating from a flux surface; and (iii) rf-induced spatial transport associated with the ion slowing down time on electrons being shorter on the outer part of the orbit or an asymmetric toroidal antenna spectrum.

Previous calculations with finite orbits have utilised a finite Larmor radius assumption in the wave solve. Here we include finite orbit effects in calculations valid to all-orders in the Larmor radius by coupling a Monte-Carlo particle code to AORSA. Section 2 details the construction of a 4D ion distribution from a particle list. In section 3 a simple Monte-Carlo code with rf operator (sMC+rf) is iterated with AORSA for a high power ICRH minority heating scenario on the C-Mod tokamak.

2. Coupling particle and continuum codes

For non-thermal ion species AORSA requires a 4D $f_i\left(R, z, v_\perp, v_\parallel\right)$ distribution function with resolution sufficient to capture the 4D phase space gradients. Furthermore, calculation of the plasma dielectric depends on velocity space derivatives according to Eq. 5 of ref.[4]. Both requirements have implications for the number of particles required to resolve important features in 4D. Fortunately, spatial gradients of the distribution have a larger scale size than the rf wave fields such that a lower spatial resolution is required for f_i relative to that required to solve the wave fields. However a weak sensitivity to noise in the velocity space derivatives act to increase the required particle number. For a typical ICRH calculation a suitable distribution function grid resolution is 32 x 32 x 64 x 128. At this grid resolution, 10^6 test particles produce a sufficiently accurate representation of the density. However, the velocity space derivatives are somewhat noisy which may lead to spurious values of plasma conductivity and possibly contaminate the wave solution.

Hellsten et al.[13] solved the problem of noise in the velocity space derivatives by using the 'hat' function for the particle shape with derivatives reducing to derivatives of the base hat functions. Our approach here is similar. We choose a Gaussian of arbitrary width such that the final distribution is differentiable by numerical means. The Gaussian width is chosen on a per run basis, being a balance between derivative stability and velocity space resolution. In cylindrical coordinates the distribution is a sum over particles according to

$$f\left(R^i, z^j, v_\perp^k, v_\parallel^l\right) = \sum_p w_p \delta\left(R - R^i\right)\delta\left(z - z^j\right) f_p \tag{1}$$

$$f_p\left(v_\perp, v_\parallel, v_\perp^k, v_\parallel^l\right) = 2\pi \mathrm{I}_0\left(\frac{v_\perp v_\perp^k}{\sigma^2}\right)\exp\left(\frac{-\left(v_\perp^2 + \left(v_\parallel - v_\parallel^l\right)^2 + v_\perp^{k2}\right)}{2\sigma^2}\right) \tag{2}$$

where $R, z, v_\perp, v_\parallel$ are the particle locations, i, j, k, l the grid coordinates, I_0 is the Bessel function, w_p the particle weight and σ the particle width. Since derivatives are only required in the velocity directions the Dirac delta function is sufficient for the spatial dimensions.

For further improvement in resolution, or an equivalent relaxation in the required number of particles, we can take advantage of the separation of particle orbit and rf heating time scales. Each particle orbit can be traced such that it traverses many phase space grid points. The particle weight may be distributed along its orbit. However, with 10^6 particles this is rarely required and integrating guiding center trajectories in combination with a 4D or even 2D particle shape is very computationally intensive.

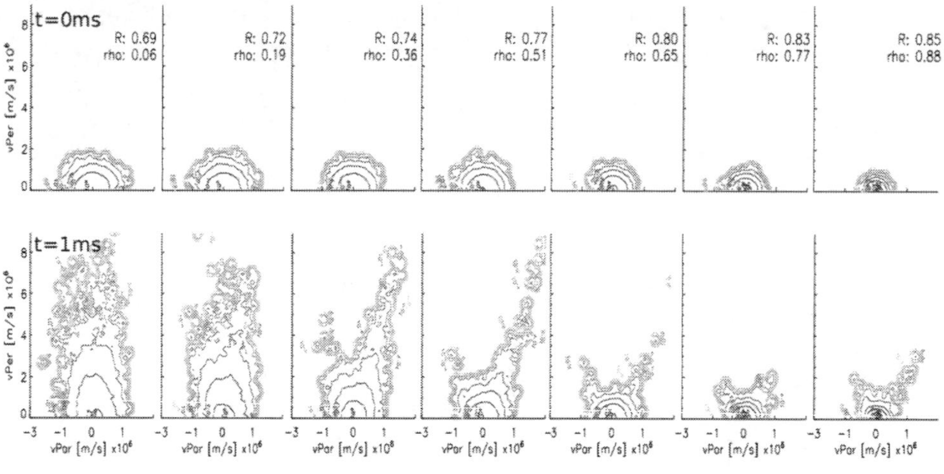

Figure 1. H distribution $f_H\left(v_\perp, v_\|\right)$ at $z = 0$.

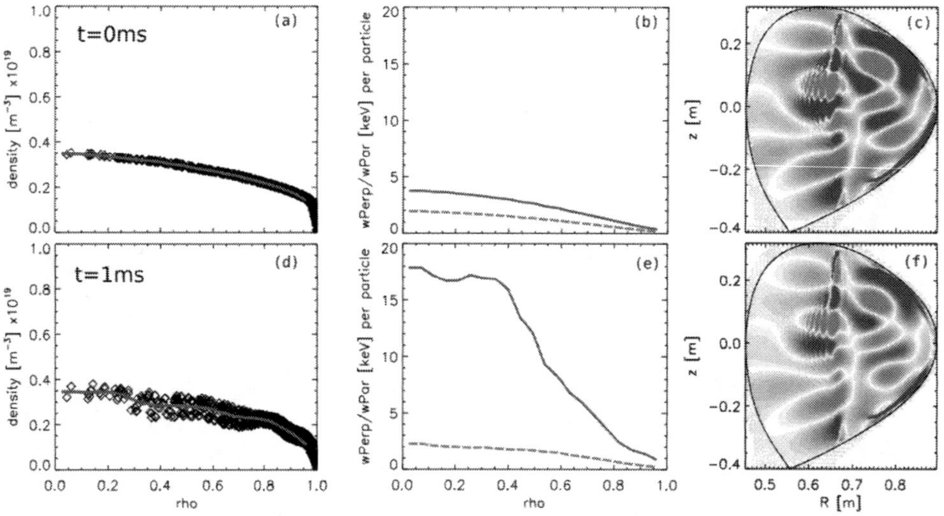

Figure 2. H density, perpendicular (black/solid) and parallel (red/dashed) energy per particle and E^+.

3. Initial results using a simple Monte-Carlo code

Prior to coupling with more comprehensive particle codes such as ORBIT-rf [19], here we use a simple Monte-Carlo code, sMC+rf, which integrates the ion guiding center equations of motion in 5D (cylindrical coordinates in space R, ϕ, z and velocity $v_\perp, v_\|$) under an axisymetric background magnetic equilibrium produced by the EFIT code[20]. We assume coulomb collisions with a Maxwellian background by including the pitch and energy scattering operators presented by Boozer et al.[21] and the usual diffusive rf scattering operator[9]. Our rf operator includes only the component of the wave electric field rotating with the gyro motion of the particle (E^+), assumes a parallel wavenumber $k_\| = n_\phi/R$ where n_ϕ is the toroidal mode number of the launched wave, and the perpendicular wavenumber is derived from the cold plasma dispersion relation. While most other Monte-Carlo approaches to ICRH make similar assumptions they are in general complicated by coordinate systems that are either constants of the particle motion or field aligned such that larger time steps reduce computational requirements. To avoid the

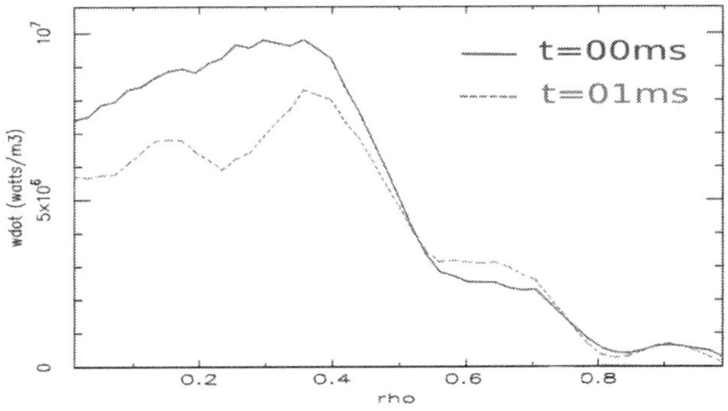

Figure 3. Flux surface averaged H power absorption calculated according to Eq. 17 of ref.[4]

complications with flux based coordinate systems, resolve interactions along a particle orbit and easily switch the effects of finite orbits on or off we retain a simple coordinate system. However, this requires we take advantage of the Jaguar Cray XT4/5 system at Oak Ridge National Laboratory and Franklin, the XT4 system at the National Energy Research Scientific Computing Center (NERSC). Furthermore, sMC+rf tracks particles below the thermal background energy such that the minority heating problem in tokamak plasmas is accessible.

A suitable scenario where the assumptions in the rf operator can be studied is fundamental minority heating of a D-H (97-3%) plasma in C-Mod. We use the EFIT equilibrium for discharge #1051206002 and launch 2.4MW of 80MHz rf as a magnetosonic fast wave with n_ϕ=10. The equilibrium is characterised by a toroidal magnetic field of B_0=5.4T and plasma current of 0.63MA with the fundamental H resonance located near the magnetic axis at R=0.69m. The background D and electron species are assumed Maxwellian with profiles described by $T(\rho) = T_{\rm lim} + (T_0 - T_{\rm lim})\left(1-\rho^\beta\right)^\alpha$ where ρ is the square root of normalised poloidal flux; $T^e_{\rm lim}$=0.05keV, T^e_0=3.6keV, $\alpha_{\rm Te}$=1.3, $\beta_{\rm Te}$=1.9, $T^D_{\rm lim}$=0.05keV, T^D_0=3.6keV, $\alpha_{\rm TD}$=1.3, $\beta_{\rm TD}$=1.9. Similarly, the electron density is described by $n^e_{\rm lim}=3.6\times10^{19}m^{-3}$, $n^e_0=12\times10^{19}m^{-3}$, $\alpha_{\rm ne}$=0.6, $\beta_{\rm ne}$=1.4. These profiles are fixed throughout the simulation.

The initial H minority distribution is chosen as a Maxwellian with temperature and density profiles according to $T^H_{\rm lim}$=0.05keV, T^H_0=3.6keV, $\alpha_{\rm TH}$=1.3, $\beta_{\rm TH}$=1.9, $n^H_{\rm lim}=10.8\times10^{17}m^{-3}$, $n^H_0=3.6\times10^{18}m^{-3}$, $\alpha_{\rm nH}$=0.6, $\beta_{\rm nH}$=1.4. An initial particle distribution is created to match the desired H temperature and density profiles. Fig. 1 shows a horizontal cut of the H ion distribution $f_H\left(R, z=0, v_\perp, v_\parallel\right)$ from the magnetic axis to the plasma edge at t=0 and 1ms. The top row shows the initial Maxwellian used as input to AORSA to solve for the wave fields. The real E^+ component of the rf electric field calculated by AORSA is shown Fig. 2c. Also shown in Fig. 2 is the density, as well as perpendicular and parallel energy profiles. Fig. 3 shows the H power absorption. The AORSA wave field is passed to sMC+rf to advance the H distribution 1ms after which the rf wave fields and power absorption are recalculated. Fig. 1 clearly shows a non-Maxwellian tail forming near the plasma core. Fig. 2e shows an accompanying increase in the average perpendicular energy. After the first iteration E^+ in Fig. 2f shows a reduced field magnitude on the magnetic axis with a corresponding drop in the on axis power absorption seen in Fig. 3.

4. Summary and future work

We have presented the first coupling of an all-orders full-wave solver (AORSA) with a particle based update of the plasma distribution to achieve quasi-linear self-consistency in ICRH calculations for fusion plasmas. The coupling of AORSA with the simple Monte-Carlo code sMC+rf will extend present simulation capabilities to include the effects of both rf induced spatial transport and finite orbits. Calculations for a minority heating scenario on the C-Mod tokamak were presented with the results showing a significant reduction in the core power absorption after iterating with the particle code relative to the initial Maxwellian. These results must be benchmarked against other code combinations including AORSA/CQL3D and other particle codes such as ORBIT-rf.

An important step for future work is to remove the limitations on the rf heating operator. By utilising the 4D quasi-linear rf diffusion tensor calculated by AORSA the assumptions of a specific $k_\parallel$ and cold plasma $k_\perp$ will be removed. sMC+rf was developed specifically with this in mind. However, the AORSA QL operator contains 4 diffusion coefficients at each point in 4D phase space on the rf wave field grid such that it requires tens of GBytes at present resolutions and grow to hundreds of GBytes at anticipated resolutions. Recent modifications to AORSA and sMC+rf utilise the parallel netCDF library[22] to allow efficient reading and writing of this data. However, using this large data set within a particle based update to the distribution function is difficult. The size of the QL operator requires the data to be read and stored across multiple processors such that preventing a communication bottleneck will be a challenge.

Acknowledgments
This work was sponsored by SciDAC and the Oak Ridge National Laboratory, managed by UT-Battelle, LLC for the US DOE under contract DE-AC-5-00OR22725.

References

[1] Shimomura Y 2004 *Journal of Nuclear Materials* **329-333** 5–11
[2] Aymar R 2002 *Journal of Nuclear Materials* **307-311** 1–9
[3] Jaeger E F, Berry L A, D'azevedo E F, Barrett R F, Ahern S D, Swain D W, Batchelor D B, Harvey R W, Myra J R, D'ippolito D A, Phillips C K, Valeo E, Smithe D N, Bonoli P T, Wright J C and Choi M 2008 *Physics of Plasmas* **15** 072513
[4] Jaeger E, Harvey R, Berry L, Myra J, Dumont R, Phillips C, Smithe D, Barrett R, Batchelor D, Bonoli P, Carter M, Dazevedo E, Dippolito D, Moore R and Wright J 2006 *Nuclear Fusion* **46** S397–S408
[5] Harvey R and McCoy M 1992 *Proc. of IAEA Technical Committe Meeting on on Numerical Modeling of Plasmas, Montreal, Canada (Vienna: IAEA) available through USDOC/NTIS No. DE93002962*
[6] Tang V, Parker R, Bonoli P and Wright J 2007 *Plasma Physics and Controlled Fusion* **49** 873–904
[7] Rosenberg A, Menard J, Wilson J and Medley S 2004 *Physics of Plasmas* **11** 2441–2452
[8] Pinsker R, Porkolab M, Heidbrink W and Luo Y 2006 *Nucl. Fusion* **46** S416–S424
[9] Kovanen M, Core W and Hellsten T 1992 *Nuclear Fusion* **32** 787–799
[10] Eriksson L and Hellsten T 1995 *Physica Scripta* **52** 70–79
[11] Hedin J, Carlsson J, Hellsten T and Jaun A 1998 *Plasma Physics and Controlled Fusion* **40** 1085–1095
[12] Hedin J, Hellsten T, Eriksson L and Johnson T 2002 *Nuclear Fusion* **42** 527–540
[13] Hellsten T, Carlsson J, Eriksson L G, Hedin J and Källbäck J 1998 *Proc. of 17th Int. Conf. on Fusion Energy, Yokohama, Japan* IAEA–CN–69 THP2/36
[14] Villard L, Brunner S and Vaclavik J 1995 *Nuclear Fusion* **35** 1173–1190
[15] Carlsson J, Eriksson L and Hellsten T 1994 *Proc. of Theory of Fusion Plasmas, Varenna* 351
[16] Murakami S, Fukuyama A, Akutsu T, Nakajima N, Chan V, Choi M, Chiu S, Lao L, Kasilov V, Mutoh T, Kumazawa R, Seki T, Saito K, Watari T, Isobe M, Saida T, Osakabe M, Sasao M and the LHD Experimental Group 2006 *Nucl. Fusion* **46** S425–S432
[17] Fukuyama A, Yokota E and Akutsu T 2000 *Proc. 18th IAEA Conf. on Fusion Energy, Sorrento, Italy* THP2–26
[18] Murakami S, Gasparino U, Idei H and Kubo S 2000 *Nuclear Fusion* **40** 693–700
[19] Choi M, Chan V, Pinsker R, Chiu S and Heidbrink W 2005 *Physics of Plasmas* **12** 072505
[20] Lao L, StJohn H, Peng Q, Ferron J and Strait E 2005 *Fusion and Science Technology* **48** 968–977
[21] Boozer A and KuoPetravic G 1981 *Physics of Fluids* **24** 851–859
[22] Li J, Liao W, Choudhary A, Ross R and Thakur R 2003 *Arxiv preprint arXiv:cs/0306048v1*

Advanced simulation of electron heat transport in fusion plasmas

Z Lin[1], Y Xiao[1], I Holod[1], W Zhang[1], W Deng[1], S Klasky[2], J Lofstead[3], C Kamath[4], and N Wichmann[5]

[1]Department of Physics and Astronomy, University of California, Irvine, CA 92697
[2]Oak Ridge National Laboratory, Oak Ridge, TN 37831
[3]Georgia Institute of Technology, College of Computing, Atlanta, GA 30332
[4]Lawrence Livermore National Laboratory, Livermore, CA 94550
[5]Cray Inc., 411 First Avenue S., Suite 600, Seattle, WA 98104
Email: zhihongl@uci.edu

Abstract. Electron transport in burning plasmas is more important since fusion products first heat electrons. First-principles simulations of electron turbulence are much more challenging due to the multi-scale dynamics of the electron turbulence, and have been made possible by close collaborations between plasma physicists and computational scientists. The GTC simulations of collisionless trapped electron mode (CTEM) turbulence show that the electron heat transport exhibits a gradual transition from Bohm to gyroBohm scaling when the device size is increased. The deviation from the gyroBohm scaling can be induced by large turbulence eddies, turbulence spreading, and non-diffusive transport processes. Analysis of radial correlation function shows that CTEM turbulence eddies are predominantly microscopic but with a significant tail in the mesoscale. A comprehensive analysis of kinetic and fluid time scales shows that zonal flow shearing is the dominant decorrelation mechanism. The mesoscale eddies result from a dynamical process of linear streamers breaking by zonal flows and merging of microscopic eddies. The radial profile of the electron heat conductivity only follows the profile of fluctuation intensity on a global scale, whereas the ion transport tracks more sensitively the local fluctuation intensity. This suggests the existence of a nondiffusive component in the electron heat flux, which arises from the ballistic radial E X B drift of trapped electrons due to a combination of the presence of mesoscale eddies and the weak de-tuning of the toroidal precessional resonance that drives the CTEM instability. On the other hand, the ion radial excursion is not affected by the mesoscale eddies due to a parallel decorrelation, which is not operational for the trapped electrons because of a bounce averaging process associated with the electron fast motion along magnetic field lines. The presence of the nondiffusive component raises question on the applicability of the usual quasilinear theory for the CTEM electron transport. This is in contrast to the good agreement between the quasilinear transport theory and simulation results of the electron heat transport in electron temperature gradient (ETG) turbulence, which is regulated by a wave-particle decorrelation. Therefore, the transport in the CTEM turbulence is a fluid-like eddy mixing process even though the linear CTEM instability is driven by a kinetic resonance. In contrast, a kinetic process dominates the transport in the ETG turbulence, which is characterized by macroscopic streamers.

1. Introduction

Understanding the physics of the turbulent transport is one of the most important scientific challenges facing magnetically confined, burning plasma experiments, since ignition is sustained by a balance between the heat loss due to the turbulent transport and the self-heating by fusion products. Current reactor design relies on the extrapolation of the transport level from present-day fusion experiments to much larger future devices, such as the International Thermonuclear Experimental Reactor (ITER)[1]. An important goal of predictive simulations is to replace such empirical scalings by first-principles numerical simulations[2].

Turbulence of the fusion plasma in a tokamak reactor is often excited by pressure-gradient-driven, electrostatic, drift-wave instabilities possessing anisotropic mode structures. Turbulence eddies rotate along surfaces of constant electrostatic potentials in the direction perpendicular to both the electric and magnetic fields (so called $E \times B$ drift). In the prevailing fluid picture[3], transport is understood as arising from the eddy mixing of a random walk process with a correlation length of the eddy size and a decorrelation time of the eddy turnover time[4,5]. The mixing length argument[6] conjectures that the transport increases significantly with a larger eddy size along the radial direction of pressure gradients. The concept of turbulent transport via the eddy mixing is valid when the collisional mean-free-path of constituent particles is much shorter than the eddy correlation length so that particles and eddies move together as fluid elements. However, in high temperature fusion plasmas, the mean-free-path of charged particles along the magnetic field lines is much longer than the typical eddy length in the parallel direction. In this nearly-collisionless, wave-dominated plasma turbulence, transport in the radial direction is carried by the random $E \times B$ motions of charged particles. If the charged particles de-couple from the turbulence eddies due to kinetic effects before the eddies can execute a complete rotation, the transport process is regulated by the kinetic wave-particle decorrelation rather than by the fluid eddy mixing. The different transport processes could lead to different transport scaling.

In this study, we examine the relevance of the fluid and kinetic processes in electron heat transport from largest ever fusion simulations resolving the electron dynamics with a real electron-to-ion mass ratio using the gyrokinetic toroidal code (GTC)[5]. Electron transport in fusion plasmas has not been studied as much as the ion transport, but is more important in a burning plasma experiment ITER since the energetic fusion products (α-particles) heat electrons first. Primary candidates for driving electron heat transport in fusion plasmas are the collisionless trapped electron mode (CTEM)[7] and electron temperature gradient (ETG)[8] turbulence. Our first-principles simulations clarify device size scaling, non-diffusive, and non-local turbulent transport. A comparative study between the CTEM and ETG turbulence provides important insights on the dynamics of nonlinear wave-particle interactions and the fluid eddy mixing in regulating the electron heat transport. In particular, the CTEM turbulence is characterized by the co-existence of micro- and meso-scale eddies. The electron heat transport is a fluid-like eddy mixing process even though the linear CTEM instability is driven by a kinetic resonance. In contrast, a kinetic process dominates the transport in the ETG turbulence, which is characterized by macroscopic streamers.

Numerical simulations of the electron turbulence are more challenging due to the small electron-to-ion mass ratio (thus large electron thermal velocity), which introduces smaller spatial scales and faster time scales as compared to the ion turbulence. Global GTC simulation is needed for the multi-scale electron turbulence, which becomes feasible thanks to the implementation of an advanced electron model[9] based on the expansion of electron response using the mass ratio as the small parameter, an efficient global field aligned mesh[10], and a highly scalable algorithm using multi-level parallelisms[11] that enables GTC to efficiently scale up to more than 100,000 cores. These breakthrough simulations with the electron dynamics in the global geometry were carried out in close collaborations with computational scientists in the areas of I/O, parallelization and optimization, workflow, and data analysis. GTC is currently an early application code for the peta-flop supercomputer at ORNL, a benchmark code for CRAY, part of the benchmark suite at NERSC and part of the code suite for the critical study of applications and architectures designated by DOE ASCR. The scientific importance of GTC simulations for ITER burning plasma experiments is also evident from the selection[12] of GTC simulations by ASCR as the "Top Breakthroughs in Computational Science".

The state-of-the-art GTC simulation of the CTEM turbulence used 28,000 cores for 42 hours and produced 60 TB of data in a dedicated run on jaguar. Simulation results show that the electron heat transport exhibits a gradual transition from Bohm to gyroBohm scaling when the device size is increased[13]. The deviation from the gyroBohm scaling can be induced by large turbulence eddies, turbulence spreading, and non-diffusive transport processes. Analysis of radial correlation function shows that CTEM turbulence eddies are predominantly microscopic but with a significant component in the mesoscale. The macroscopic, linear streamers are mostly destroyed by zonal flow shearing, which is further confirmed by a comprehensive analysis of kinetic and fluid time scales showing that zonal flow shearing is the dominant decorrelation mechanism. The time evolution of the turbulence shows that the mesoscale eddies result from a dynamical process of linear streamers breaking by zonal flows and merging of microscopic eddies. The radial profile of the electron heat conductivity only follows the profile of fluctuation intensity on a global scale, whereas the ion transport tracks more sensitively the local fluctuation intensity. This suggests the existence of a nondiffusive component in the electron heat flux, which arises from the ballistic radial $E X B$ drift of trapped electrons due to a combination of the presence of mesoscale eddies and the weak detuning of the toroidal precessional resonance that drives the CTEM instability. On the other hand, the ion radial excursion is not affected by the mesoscale eddies due to a parallel decorrelation, which is not operational for the trapped electrons because of a bounce averaging process associated with the electron fast motion along the magnetic field lines. The presence of nondiffusive component raises questions on the applicability of the usual quasilinear theory for the CTEM electron transport. This is in contrast to the good agreement between the quasilinear transport theory and simulation results of the electron heat transport in the ETG turbulence. In the ETG turbulence, wave-particle decorrelation is the dominant mechanism responsible for the electron heat transport[14]. ETG turbulence is dominated by radially elongated eddies (streamers). Various spatial and temporal scales of the ETG turbulence are calculated through comprehensive analysis of the massive data produced from simulations. We find that ETG transport is proportional to the local fluctuation intensity, since phase-space island overlap leads to a diffusive process with a time scale comparable to the wave-particle decorrelation time determined by the parallel spectral width. The kinetic time scale is much shorter than the fluid time scale of eddy mixing.

Our results have important implications on the extrapolation of transport properties from present-day tokamaks to future large reactors and on the choice of time scales in transport models. The transition from Bohm to gyroBohm scaling of the electron heat transport in the CTEM turbulence is favorable for the large fusion reactor ITER. The nonlocal transport properties require a global simulation (instead of a local, flux-tube simulation). The non-diffusive transport in the CTEM turbulence cannot be accurately described by the quasilinear theory underlying most of the existing transport models. The kinetic picture of wave-particle decorrelation for ETG transport is inconsistent with the prevailing picture of fluid eddy mixing. Since the ratio of fluid-to-kinetic time scales increases with the device size, the extrapolation of the transport level from present-day experiments to future larger reactors could be overly pessimistic, if the simplistic mixing length argument with the streamer length as the spatial step size and the fluid time scale as the time step size is invoked.

This paper is organized as follows. The GTC physics and computational models are summarized in Sec. 2. Sec. 3 presents GTC simulation results of the CTEM and ETG instability saturation and fluctuation characteristics. Transport processes in the CTEM and ETG turbulence are discussed in Sec. 4. Sec. 5 is the conclusions.

2. Gyrokinetic Toroidal Code (GTC)

The simulations in this work were carried out using the Gyrokinetic Toroidal Code (GTC), which is a fully global 3D particle-in-cell code developed for first-principle simulations of turbulence and transport in magnetically confined plasmas. We emphasize the global aspect of the simulation in contrast to the local simulation of only a small region, or flux tube, surrounding a chosen magnetic field line.

2.1. Simulation model

We are studying low frequency drift waves characterized by a short wavelength in the direction perpendicular to the magnetic field lines but a very long wavelength in the direction parallel to the field. In the gyrokinetic simulation[15], the phase attribute of the fast gyration (or cyclotron) motion of the charged particles around the magnetic field lines is averaged away, reducing the dimensionality of the system from 6D to 5D. This method[16] removes the fast cyclotron motion, which has a much higher frequency than the characteristic waves of plasma microturbulence.

The particle-in-cell method consists of moving particles along the characteristics of the gyrokinetic equation. The electrostatic potential and field are obtained by solving the Poisson equation on a spatial mesh after gathering the charge density on the grids. The electrostatic forces are subsequently scattered back to the particle positions for advancing the particle orbits. The use of spatial grids and the procedure of gyroaveraging[15] reduce the intensity of small scale fluctuations (particle noise). Particle collisions can be recovered as a "subgrid" phenomenon via Monte Carlo methods. The particle noise is further reduced by using a perturbative simulation method[17] where only the perturbed distribution function is calculated in simulation. Numerical properties of the electron dynamics are improved by an electrostatic fluid-kinetic hybrid electron model[9] based on an expansion of the electron response using the electron–ion mass ratio as a small parameter. Electron response is adiabatic in the lowest order and nonadiabatic response is taken into account in the higher order equations[18].

GTC employs the magnetic coordinates, which provide the most general coordinate system for any magnetic configuration possessing nested surfaces. General geometry with strong shaping has been implemented in GTC using a Poisson solver[19] in real space and a spline fit of the equilibrium data from an MHD code such as EFIT[20]. The property of straight field lines in the magnetic coordinates is most suitable for describing the instability with field aligned eigenmodes and enables the implementation of an efficient global field aligned mesh for the quasi-2D structure of the plasma turbulence in the toroidal geometry[10]. The global field-aligned mesh provides the highest possible computational efficiency without any simplification in terms of physics models or simulation geometry. The magnetic coordinates are also desirable for efficiently integrating the particle orbits, which move predominantly along the magnetic field line direction. The equation of motion can be derived from a Hamiltonian formulation[21] which conserves phase space volume and is best for integrating particle orbits for a long period.

2.2. Parallelization model

GTC employs three levels of parallelism. The original parallel scheme implemented in GTC[5] is a 1D domain decomposition in the symmetric, toroidal direction (long way around the torus) using Message Passing Interface (MPI). Each MPI process is in charge of a toroidal domain with both particles and fields. Particles move from one domain to another while they travel around the torus. All communications are one-way traffic to avoid congestion. A second level of parallelism was later implemented[11] to increase the concurrency. Within each toroidal domain, we now divide the particles between several MPI processes, but each process keeps a copy of all the fields on a single toroidal plane. A "particle-domain" communicator links the MPI processes within a toroidal domain of the original 1D domain decomposition, while a "toroidal-domain" communicator links in a ring-like fashion all the MPI processes with the same intra-domain rank. To take advantage of the shared memory capability of multi-core nodes, a third level of parallelism is implemented[11] at the loop level using OpenMP compiler directives. These three levels of parallelism using mixed-mode MPI-OpenMP enables GTC to scale to a very large number of processors and use a very large number of particles, which results in a very high phase space resolution and a low statistical noise[22]. Shown in **Fig. 1** is the weak scaling of the GTC computing power, which is almost a linear function of the number of cores up to 100,000 of cores on Cray XT5 supercomputer. GTC is portable and optimized for various scalar and vector supercomputers[11].

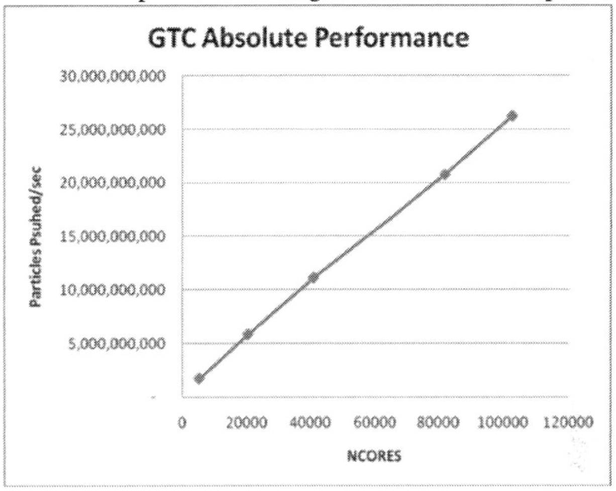

Fig. 1. *Absolute scaling of GTC computing power as a function of number of cores.*

When GTC uses tens of thousands of nodes, having each node create an individual netCDF[23] restart file causes unacceptable delay in the Lustre[24] file system due to the large number of simultaneous file creation requests. To remove this bottleneck, the IO was rewritten to use HDF-5 collectives[25]. To avoid any further changes to the IO layer, we developed an abstraction layer called ADIOS (ADapatable IO System)[26,27] in the collaboration among SciDAC fusion projects GPS-TTBP, CPES, and GSEP[28]. ADIOS provides a simple API that can select automatically the best techniques for each different grouping of data as specified by an entry in an external XML configuration file. To achieve the best performance with any chosen IO method, we have worked with experts on each method, such as synchronous collective MPI-IO[29], HDF-5, and parallel netCDF, to create an optimal implementation. We also have two asynchronous methods of the DART[30] system and DataTap[31]. ADIOS also simplifies the IO routines through an automatic code generation system that can be easily included. By modifying the XML file, all required changes to the IO routines to properly perform the IO can be generated as part of the recompilation without touching the science part of the code. For performance, with the synchronous collective MPI-IO method, we consistently measure a data transfer rate for restarts at 20 GB/s aggregate across the entire system run. Using the DART asynchronous method, it takes 12 seconds to generate 1.2 TB of data.

This concept can be readily generalized for processing data. For example, we can create a method to represent a visualization engine and add it to ADIOS. This method's implementation would incorporate all of the calls, data formatting, setup, and other bits of code necessary for connecting GTC with the visualization engine. The user could then turn on or off the in situ visualization by merely changing the method used for writing a particular grouping of data. For example, the particle data in GTC is a prime candidate for using such a method. If offline visualization is preferred or required for a particular run, the method for particle writes could be changed to something else, for example, MPI-IO. Other advantages of ADIOS include directly integrating workflow through the DataTap and DART systems and even Kepler[32]. While Kepler integration provides a natural interface with existing workflows, these largely rely on files be written to disk in order for the workflow to process. The asynchronous DataTap and DART systems have a similar workflow concept using dynamic code deployment and in stream message passing.

3. Instability saturation and fluctuation characteristics of CTEM and ETG turbulence

The collisionless trapped electron mode (CTEM) instability with a characteristic poloidal wavelength on the order of the ion gyroradius (ρ_i) is driven by the toroidal precessional resonance between trapped electrons and driftwaves. GTC simulations[13] show that this linear CTEM instability is saturated by the shearing effects of the nonlinearly-generated zonal flows. This is further confirmed by a comprehensive analysis of kinetic and fluid time scales showing that zonal flow shearing is the dominant decorrelation mechanism. As a result, the macroscopic, linear, radial streamers are mostly broken by the zonal flows. As shown in the top panel of **Fig. 2** for the contour plot of the electrostatic potential ϕ in a tokamak with a minor radius a=250ρ_i, the steady state turbulence is dominated by microscopic eddies with a scale length of 5ρ_i, but with also a significant component of mesoscale (5-50ρ_i) streamers. The mesoscale eddies result from a dynamical process of zonal flow breaking of linear streamers and spontaneous merging of microscopic eddies. For the following analysis of space and time scales, we use the fluctuation data from a simulation of tokamak size of a=500ρ_i. To quantify the perpendicular structures, we calculate the two-point correlation function, $C_{r\zeta}(\Delta r,\Delta\zeta)=\dfrac{\langle\phi(r+\Delta r,\zeta+\Delta\zeta)\phi(r,\zeta)\rangle}{\sqrt{\langle\phi^2(r+\Delta r,\zeta+\Delta\zeta)\rangle\langle\phi^2(r,\zeta)\rangle}}$, where $\langle\cdots\rangle$ represents averaging over toroidal ζ and radial r directions at a poloidal angle $\theta=0$. Radial correlation function $C_r(\Delta r)$ is then calculated by taking the maximal value along the ridge of $C_{r\zeta}(\Delta r, \Delta\zeta)$. The function $C_r(\Delta r)$ decays exponentially for the small radial separation but possesses a significant mesoscale tail in the range of $5\text{-}50\rho_i$. An exponential fit of the initial drop of $C_r(\Delta r)$ gives the short length scale as $L_r\sim 5\rho_i$. This is the average radial size for the microscopic eddies. The perpendicular structure of the CTEM turbulence is thus multi-scale.

For comparison, the instability saturation and fluctuation characteristics of the electron temperature gradient (ETG) turbulence are quite different. The ETG instability with a much shorter characteristic poloidal wavelength on the order of the electron gyroradius (ρ_e) is driven by electron parallel resonance. The nonlinear saturation of the ETG instability is primarily due to a nonlinear toroidal mode coupling[33]. In this process, the spectral energy successively flows toward longer wavelengths, eventually down to damped modes. The effects of zonal flows in the ETG turbulence are very weak. Therefore, the ETG turbulence after saturation is dominated by macroscopic radial streamers as shown in the lower panel of **Fig. 2** for the contour plot of the electrostatic potential ϕ in a much smaller tokamak with a minor radius of 500ρ_e. The perpendicular structure of the ETG turbulence is thus strongly anisotropic. In the steady state, the ETG turbulence is also mediated by a radial spreading[34] where small scale turbulence eddies are generated in the unstable region (around $r=0.5a$) and flow along the nonlinearly generated radial streamers to the stable region (near $r=0$ and $r=a$). Meanwhile, the streamers can break and reconnect, resulting in a very complex dynamical evolution.

The 2D structures on a poloidal plane (perpendicular to the magnetic field lines) shown in **Fig. 2** is extended along the magnetic field lines (in the toroidal direction), resulting in a flux-tube like eddy structures. The parallel wavevector of both CTEM and ETG is $k_{\|}\sim 1/qR$, where R is the major radius and q is the safety factor of the tokamak. The radial variations of q cause the eddy structures to twist around in the toroidal direction.

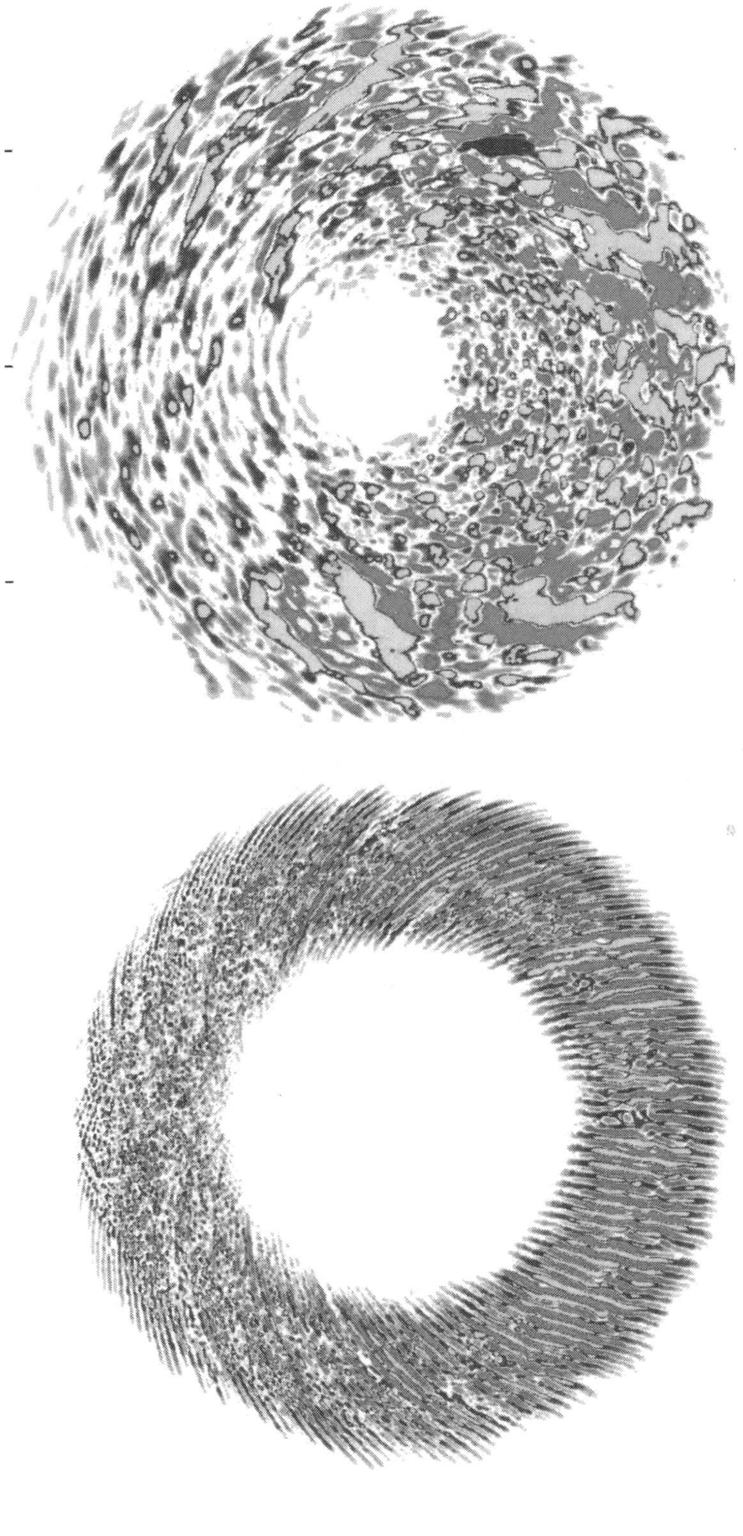

Fig. 2. *Poloidal contour plots of the electrostatic potential ϕ in the CTEM turbulence (top panel) and the ETG turbulence (lower panel).*

4. Transport processes of CTEM and ETG turbulence

To study the transport mechanism, we examine the correlation between heat conductivity to the local turbulence intensity and compare various characteristic time and spatial scales. In the CTEM turbulence[13], the time and flux-surface-averaged ion heat flux $\boldsymbol{Q_i}$, electron heat flux $\boldsymbol{Q_e}$, and radial ExB drift intensity I during the nonlinear stage is shown in the lower right panel of **Fig. 3**. The radial profile of the fluctuation intensity I contains a global envelope plus an oscillating part with a spatial period of $\sim 10\rho_i$. The correlation between the radial profiles of the fluctuation intensity I and the ion heat flux $\boldsymbol{Q_i}$ suggests that the ion heat transport is driven by the local $E\ X\ B$ drift intensity I. On the other hand, the radial profile of the electron heat flux $\boldsymbol{Q_e}$ is much smoother and lacks the small scale oscillations. Nonetheless, the global profile is quite similar to the global envelope of the fluctuation intensity I. This suggests that the electron heat transport is close to diffusive on the global scale, but not on the microscopic and meso-scale. The remarkable similarity between the ion heat flux $\boldsymbol{Q_i}$ (top left panel) and the fluctuation intensity I (top right panel) in both radial structure and time evolution further confirms that the ion heat transport is driven by the local fluctuation intensity. However, the electron transport demonstrates a ballistic propagation in the radial direction (lower left panel). Therefore, the electron heat transport follows the global structure of of $E\ X\ B$ profile but contains a non-diffusive component on the mesoscale.

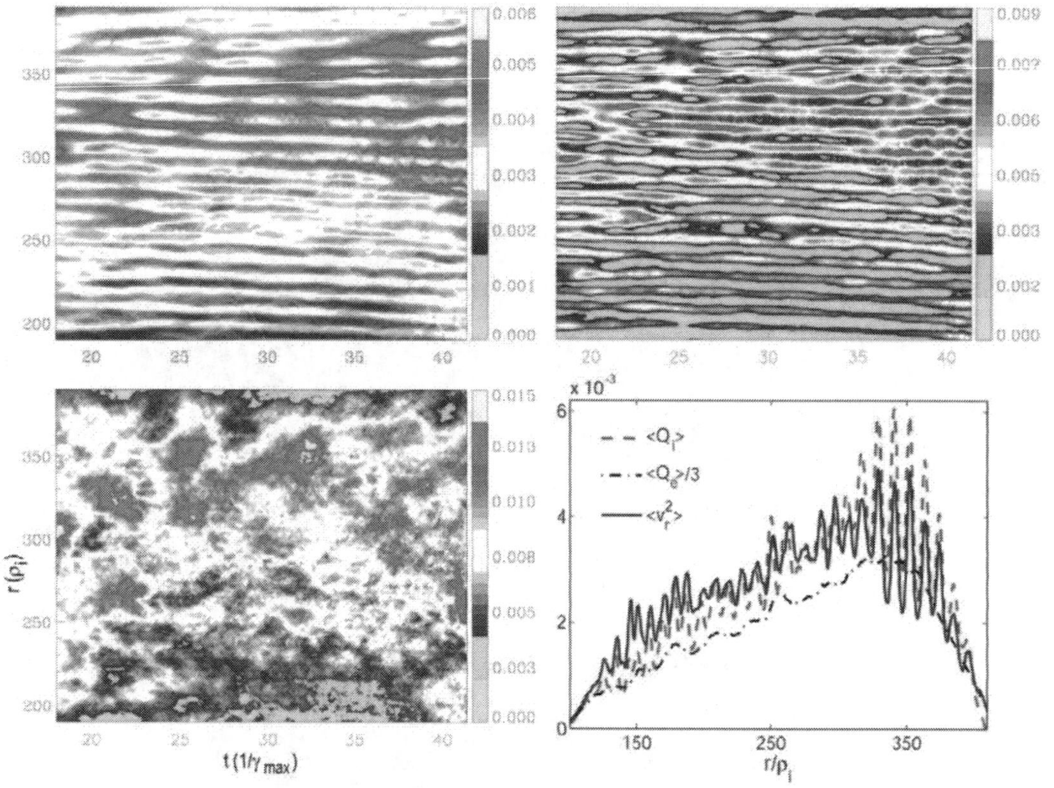

Fig. 3 *Time-radial contour plot for CTEM turbulence intensity (I, top right), ion heat flux (Q_i, top left), electron heat flux (Q_e, lower left), and the radial profiles of the time-averaged I, Q_i, and Q_e (Lower right panel)*

We can now define an effective wave-particle decorrelation time τ_{wp} since the electron heat transport is proportional to the fluctuation intensity on the global scale, i.e., $\tau_{wp}=2D/I$. Because the particle diffusivity is much smaller than the thermal conductivity, a test particle diffusivity D can be related to the thermal conductivity χ_e by $D = 2\chi_e/3$. We find $\tau_{wp}=0.61L_{ne}/v_i$ for the trapped electrons. This characteristic time reflects the physical process relevant to the transport mechanism[14]. To evaluate various time scales, we first calculate the two-time-two-point correlation function $C_{t\zeta}(\Delta t,\Delta\zeta)=\frac{\langle\phi(t+\Delta t,\zeta+\Delta\zeta)\phi(t,\zeta)\rangle}{\sqrt{\langle\phi^2(t+\Delta t,\zeta+\Delta\zeta)\rangle\langle\phi^2(t,\zeta)\rangle}}$. Then we calculate the Lagrangian time correlation function $C_t(\Delta t)$ by taking the maximal value along the ridge of $C_{t\zeta}(\Delta t, \Delta\zeta)$, i.e., moving with the wave phase velocity. This Lagrangian time correlation function $C_t(\Delta t)$ decays exponentially with an autocorrelation time $\tau_{auto}=11L_{ne}/v_i$. The eddy turnover time τ_{eddy}, which describes how fast the eddy rotates due to the $E\ X\ B$ drift without the zonal flow shearing, is $\tau_{eddy}=L_r/<\delta v_r>=1.6L_{ne}/v_i$ for microscopic eddies. Another fluid time scale that relates to the dynamics of the turbulence eddies is the zonal flow shearing time, $\tau_s=0.66L_{ne}/v_i$.

The kinetic time scales related to the linear CTEM eigenmode include the wave-particle decorrelation time scales for the trapped electrons to cross the turbulence eddies in the parallel and perpendicular directions. Because of the fast bounce motion, which averages out the parallel electric field, the trapped electrons cannot decorrelate from the wave in the parallel direction. In the spectral range of interest, the CTEM frequency is proportional to the toroidal mode number (i.e., non-dispersive). Thus the resonant electrons cannot decorrelate from the wave in the toroidal direction. On the other hand, the resonant electrons can decorrelate from the wave in the radial direction due to the radial dependence of the precessional frequency. However, this dependence is very weak (on the device size scale). Therefore, the radial de-tuning time is also very long. Trapped electrons thus remain resonant with the wave until they are diffused nonlinearly across the turbulence eddies. This nonlinear kinetic time scale is the resonance broadening time τ_{rb}, which is calculated as $\tau_{rb}=5.1L_{ne}/v_i$. Therefore, all kinetic time scales are much longer than the fluid time scales, i.e., CTEM transport is a fluid-like process.

The physical picture of the device size scaling of the electron heat transport in the CTEM turbulence thus emerges from the combination of the turbulence structures and the fluid-like nature of the electron transport process. The resonant trapped electrons can be convected by the $E\ X\ B$ drift across the large number of mesoscale eddies. This mesoscale ballistic process then drives a non-diffusive component in the electron heat transport and smooths small radial scale structure of the turbulence intensity. The mesoscale ballistic electron heat flux, together with the turbulence spreading, leads to the deviation from the gyroBohm transport scaling for the small devices. The electron heat transport thus exhibits a device size scaling characterized by a gradual transition from Bohm to gyroBohm scaling, which is good news for the large fusion reactor ITER.

In contrast, we find that stochastic wave-particle decorrelation[14] is the dominant mechanism responsible for the electron heat transport driven by the ETG turbulence with extended radial streamers. The phase-space island overlap due to the interactions of many toroidal modes leads to a diffusive transport process with a time scale comparable to the wave-particle decorrelation time, determined by the parallel spectral width. This kinetic time scale of the wave-particle decorrelation is much shorter than the fluid time scales of the eddy mixing. Consistently, the transport is proportional to the local fluctuation intensity, and quasilinear calculation of the electron heat conductivity using measured spectra agrees well with the simulation value[14].

5. Conclusions

Large scale simulations of electron heat transport in fusion plasmas find that the transport in the collisionless trapped electron mode (CTEM) turbulence is a fluid-like eddy mixing process even though the linear CTEM instability is driven by a kinetic resonance. In contrast, a kinetic process dominates the transport in the electron temperature gradient (ETG) turbulence, which is characterized by macroscopic streamers. The GTC simulations have identified, for the first time, the nonlinear transport processes that determine the level and scaling of the electron transport in fusion plasmas.

Acknowledgments: This work is supported by SciDAC GPS-TTBP, GSEP, and CPES centers. We acknowledge fruitful discussions with C. S. Chang, L. Chen, P. H. Diamond, S. Ethier, T. S. Hahm, and F. Zonca. Simulations used supercomputers at ORNL and NERSC.

References

[1] http://www.iter.org/.
[2] W. M. Tang, *Journal of Physics: Conference Series* **125,** 012047(2008).
[3] A. Hasegawa and K. Mima, *Phys. Fluids* **21**, 87 (1978).
[4] H. Biglari, P. H. Diamond, and P. W. Terry, *Phys. Fluids* B **2**, 1 (1990).
[5] Z. Lin *et al*, *Science* **281**, 1835 (1998).
[6] B. B. Kadomtsev, *Plasma Turbulence* (Academic, London, 1965).
[7] J. Adam, W. Tang, and P. Rutherford, Phys. Fluids **19**, 561 (1976).
[8] W. Horton, B. G. Hong, and W. M. Tang, *Phys. Fluids* **31**, 2971 (1988).
[9] Z. Lin and L. Chen, Phys. Plasmas **8**, 1447 (2001).
[10] Z. Lin *et al*, *Phys. Rev. Lett.* **88**, 195004 (2002).
[11] S. Ethier, W. M. Tang, and Z. Lin, *Journal of Physics: Conference Series* **16**, 1 (2005).
[12] SciDAC Review, special issue 2009, http://www.scidacreview.org/0901/.
[13] Y. Xiao and Z. Lin, submitted to *PRL*.
[14] Z. Lin *et al*, *Phys. Rev. Lett.* **99**, 265003 (2007).
[15] W. W. Lee, *Phys. Fluids* **26**, 556 (1983); *J. Comput. Phys*. **72**, 243 (1987).
[16] A. J. Brizard and T. S. Hahm, *Rev. Mod. Phys.* **79**, 421 (2007).
[17] S. E. Parker and W. W. Lee, *Phys. Fluids* B **5**, 77 (1993).
[18] I. Manuilskiy and W. W. Lee, *Phys. Plasmas* **7**, 1381 (2000).
[19] Z. Lin and W. W. Lee, *Phys. Rev. E* **52**, 5646 (1995).
[20] L. L. Lao *et al*, *Nuclear Fusion* **25**, 1611 (1985).
[21] R. B. White and M. S. Chance, *Phys. Fluids* **27**, 2455 (1984).
[22] I. Holod and Z. Lin, *Phys. Plasmas* **14**, 032306 (2007).
[23] http://www.unidata.ucar.edu/software/netcdf/.
[24] http://www.lustre.org/.
[25] http://hdf.ncsa.uiuc.edu/products/hdf5/.
[26] J. Lofstead, S. Klasky, H. Abbasi, and K. Schwan, in Supercomputing'07, Reno, 2007.
[27] J. Lofstead, F. Zheng, S. Klasky, and K. Schwan, "*Adaptable, Metadata Rich IO Methods for Portable High Performance IO*", in Proceedings of IPDPS'09, Rome, 2009.
[28] http://www.scidac.gov/fusion/fusion.html.
[29] http://www.llnl.gov/icc/lc/siop/mpiio.html.
[30] C. Docan, M. Parashar, and S. Klasky, "*High Speed Asynchronous Data Transfers on the Cray XT3*". Technical report, Rutgers, The State University of New Jersey, 2007.
[31] K. Schwan, H. Abbasi, M. Wolf, "*Live data workspace: A flexible, dynamic and extensible platform for petascale applications*". In ClusterComputing, Austin, 2007.
[32] http://kepler-project.org/.
[33] Z. Lin, L. Chen, and F. Zonca, *Phys. Plasmas* **12**, 056125 (2005).
[34] Z. Lin and T. S. Hahm, *Phys. Plasmas* **11**, 1099 (2004).

Magnetic X-points, stochasticity, and Edge Localized Modes

Linda E. Sugiyama[1] **and the M3D Team**
[1]Laboratory for Nuclear Science
Massachusetts Institute of Technology, Cambridge MA 2139-4307 U.S.A.
E-mail: sugiyama@mit.edu

Abstract. The confinement and stability properties of the edge of a magnetically confined fusion plasma have long resisted theoretical explanation. Numerical simulation with the M3D extended MHD code shows that large periodic Edge Localized Modes (ELMs), associated with otherwise good confinement, represent a new type of nonlinear MHD plasma instability, in which a coherent plasma instability couples to the favorable part of a chaotic magnetic field. The magnetic surface bounding the plasma typically contains at least one X-point. Under small perturbations, such a Hamiltonian surface splits into two different asymptotic limits. A ballooning-type plasma instability can couple to the "unstable" field manifold that forms bulges around the original magnetic surface and then grow to penetrate deep into the plasma. The magnetic field becomes chaotic over the affected region, allowing plasma loss from the core to the outside. Eventually the instability saturates and the plasma relaxes back towards its original axisymmetric shape. The interaction between the plasma and the homoclinic field perturbation drives a multi-stage process that reproduces many experimental observations. This type of instability may help to explain the wide range of ELM and ELM-free behavior observed in fusion plasmas with X-points, as well as properties of plasma confinement.

1. Plasma boundaries with X-points

Magnetically confined toroidal plasmas routinely achieve high temperatures in a configuration with one or two magnetic X-points on the bounding magnetic surface, where the field becomes purely toroidal. X-points allow good control of the plasma shape against instabilities. They support a steep edge pressure gradient with good plasma edge confinement (H-mode), favorable for fusion reactivity. The outer legs of the X are intended to channel plasma losses along the field lines, to divertors designed to handle them. In practice, the steep edge pressure gradient and associated plasma current can drive large periodic instabilities (Edge Localized Modes or ELMs) that erode the edge and dump plasma and energy onto the surrounding walls. The losses can be severe in fusion burning plasmas and are one of the major constraints on the design of next generation experiments such as ITER.

Despite its long history[1], the H-mode plasma edge has defied solid theoretical explanation. The size and onset of ELMs are still difficult to predict. Plasma stability theory almost universally assumes that the magnetic field responds to small perturbations in a coherent, unitary fashion. This may be justified in the interior of toroidal plasmas, where the magnetic flux surfaces often behave as surprisingly persistent KAM surfaces of the Hamiltonian system that describes the magnetic field.

The magnetic boundary surface with an X-point behaves differently[2]. The X-point is equivalent to a Hamiltonian hyperbolic saddle point. Under small perturbations, an X-point surface splits into two, defined by different asymptotic limits, depending on the direction in which the field line is traversed (eg, [4]). The "stable" limiting surface or manifold, on which the field line moves away from the X-point, remains close to the original surface. The "unstable" surface forms loops or bulges transverse to the stable surface, as the field line approaches the X-point. The asymptotic intersection points, called homoclinic points, become more closely spaced as the X-point is approached. Because the Hamiltonian is area preserving in a 2D surface of section, successive loops of the unstable manifold have equal areas and their excursions around the stable surface become longer and longer, becoming infinite as the X-point is approached.

Homoclinic intersections cause the magnetic field to become chaotic[4]. While a number of theoretical properties can be proven, they are primarily asymptotic and qualitative. No criterion exists for the limiting perturbation size that allows the theoretically predicted behavior. The manifold shapes and intersection points are determined by the driving mechanism, which may be far from Hamiltonian. Since in physical systems, infinite excursions are impossible because of natural length scales and propagation delays, the plasma magnetic field differs from the theoretical homoclinic tangle, although it may try to approach it.

ELMs have recently become accessible to numerical simulation with extended MHD models at meaningful parameters. Previous, more approximate simulations of ELMs[5, 6] saw field mixing, but did not consider field splitting. The present results show that ELMs in fact constitute a new type of plasma instability that grows in the presence of a split, chaotic magnetic field. This paper addresses certain aspects of the field splitting. Detailed results on ELMs are reported elsewhere[7].

2. Numerical model

The extended MHD code M3D[8] simulates a free boundary plasma surrounded by a "vacuum" region with open field lines that is bounded by a rigid, ideally conducting wall. The code solves the compressible resistive MHD equations, including separate density and temperature, over the entire region. The resistive MHD vacuum is characterized by low density, zero temperature, and high resistivity.

Initial conditions were taken from a DIII-D discharge with large ELMs, the ITER-similar shot 119690 at $t = 2656$ ms[3], chosen because it was relatively collisional, $\nu_e = 0.7$, and had a very steep, narrow edge pressure gradient strongly unstable to MHD ballooning modes. The plasma boundary had one lower X-point. A second, upper X-point existed a short distance above and inboard from the top of the plasma. The plasma had toroidal magnetic field $B_T = 1.60$ T at major radius $R_o = 1.76$ m, current $I_p = 1.042$ MA, and neutral beam heating power $P_{NB} = 4,8$ MW. The central density was $n_{eo} = 1.2 \times 10^{20} \mathrm{m}^{-3}$, temperature $T_{eo} \simeq T_{io} = 1.83$ keV, and normalized beta $\beta_N = 1.81$. The edge magnetic safety factor was $q_{95} = 3.75$ and central q_o just above unity. The simulations reported here neglect the equilibrium toroidal rotation.

The ELM is partly resistive. The model resistivity varied as $(T/T_o)^{-3/2}$ inside the plasma, up to the large, constant vacuum value. A relaxed-parameter case with large central plasma resistivity, Lundquist number $S = 10^6$, vacuum $S_{vac} = 10^3$, normalized kinematic ion viscosity $\mu_i/\rho = 10^{-5}$, and "vacuum" density $n_{vac}/n_o = 0.4$ clearly demonstrates the instability. Cases at more realistic parameters, including the actual resistivity $S = 3.3 \times 10^7$, with $\mu = 6 \times 10^6$ and $n_{vac}/n_o = 0.1$, show an analogous evolution, at reduced scale[7].

The parallel thermal conductivity, using the artificial sound wave method[8], was equivalent to a diffusion coefficient $\kappa_{\parallel} = 3.53(R_o/a_o)$. Perpendicular thermal and particle diffusivities were $\kappa_{\perp} = D_n = 10^{-5}$. The dissipation, other than resistivity, was uniform except very near the wall, where it became large, 10^{-2}. Length is normalized to $a_o = 100$ cm, not the plasma radius 55.7 cm. Time is normalized to the Alfvén time $\tau_A = R_o/v_A$, based on the plasma vacuum field and initial density at the magnetic axis R_o. The reference time is $\tau_A = 0.78$ μsec.

The grid in each poloidal plane was packed around the plasma edge, more tightly on the outboard side. The smaller S case used 66 radial by 400 poloidal points, for 12871 total points, for linear triangular finite elements. Other cases used 72 by 400 points, or 15769 total. FFTs were used in the toroidal direction. Typical toroidal mode numbers ranged over $|n| \leq 22$, or 72 poloidal planes with full dealiasing. Runs with $|n| \leq 48$ modes show that the primary nonlinear instability falls within the lower range.

3. Results

The initial configuration was taken from DIII-D experimental reconstructions that include the large bootstrap current in the plasma edge. The ELM was started with a small random perturbation of multiple toroidal harmonics at time $t = 0$.

The time evolution at large resistivity is shown in Figure 1. The top row shows temperature contours. Black represents the vacuum region and its border the vacuum vessel. The toroidal axis $R = 0$ lies to the left (inboard) side.

The early nonlinear instability at $t = 66\tau_A$ barely perturbs the outboard edge (large R) of the plasma, as seen in the small fuzzy region and uneven point spacing on the outermost magnetic surfaces. The perturbed magnetic region is a little wider at the bottom and top. By $t = 89$, the nonlinear instability is strongly developed. The midplane oscillations broaden poloidally and large plasma fingers grow above and below the midplane, around ± 60 degrees. In contrast, the standard ballooning eigenmode decreases away from the midplane. The growth rate also drops from initial levels.

As time progresses to $t = 133$, the low density fingers grow inward, penetrating deeply into the plasma. The protruding high density fingers disperse shortly outside the original magnetic boundary, where the plasma shears off in "blobs." Plasma spreads outside, over the outboard region and then into the outboard divertor, while the core density, and to a lesser extent the temperature, is reduced. (A central magnetic island grows as q_o drops below unity, then decays with little effect on the main instability.) At $t = 178$, smoothing of the edge field is visible. In this case, the instability is unphysically strong and the loss of interior plasma reduces the driving force before the later ELM stages can develop.

At more realistic parameters[7], including the actual resistivity, the growth rate and instability magnitude are reduced. The finger variation is smaller and outgoing fingers shorter, but the field still mixes over much of the plasma. Relatively early in the ELM, some of the outer magnetic field lines partially relocalize in the edge region, enough to support a new, partial edge density gradient. At a later stage, plasma moves into the inboard side and into the divertors at top and bottom. As the original instability saturates, the plasma relaxes towards the original equilibrium shape, with small changes in the profiles.

The chaotic field region mirrors the inward growth of the plasma fingers. A central core is mostly preserved, inside some limiting flux surface. The behavior agrees with the theoretical expectation[4] that the entire region containing homoclinic points becomes chaotic. This behavior differs from interior plasma instabilities, where relatively narrow magnetic island chains, interspersed with good, closely spaced KAM surfaces, often persist.

The magnetic boundary surface clearly splits. The $+\phi$-direction puncture plots clearly show lobes near the bottom X-point on the outboard side, as expected for the "unstable" manifold. The surfaces are relatively smooth on the inboard side of the lower X-point and the outboard side of the upper X-point, where this direction should coincide with the "stable" manifolds. Since the stable manifolds of two different X-points cannot intersect, the $+\phi$ points connected to the lower X-point cannot extend much above the top of the original plasma. The $-\phi$ points closely shadow the lobes of the "unstable" manifold of the upper X-point and form large lobes near the top of the plasma. They also show a smoother "stable" response on the outboard side

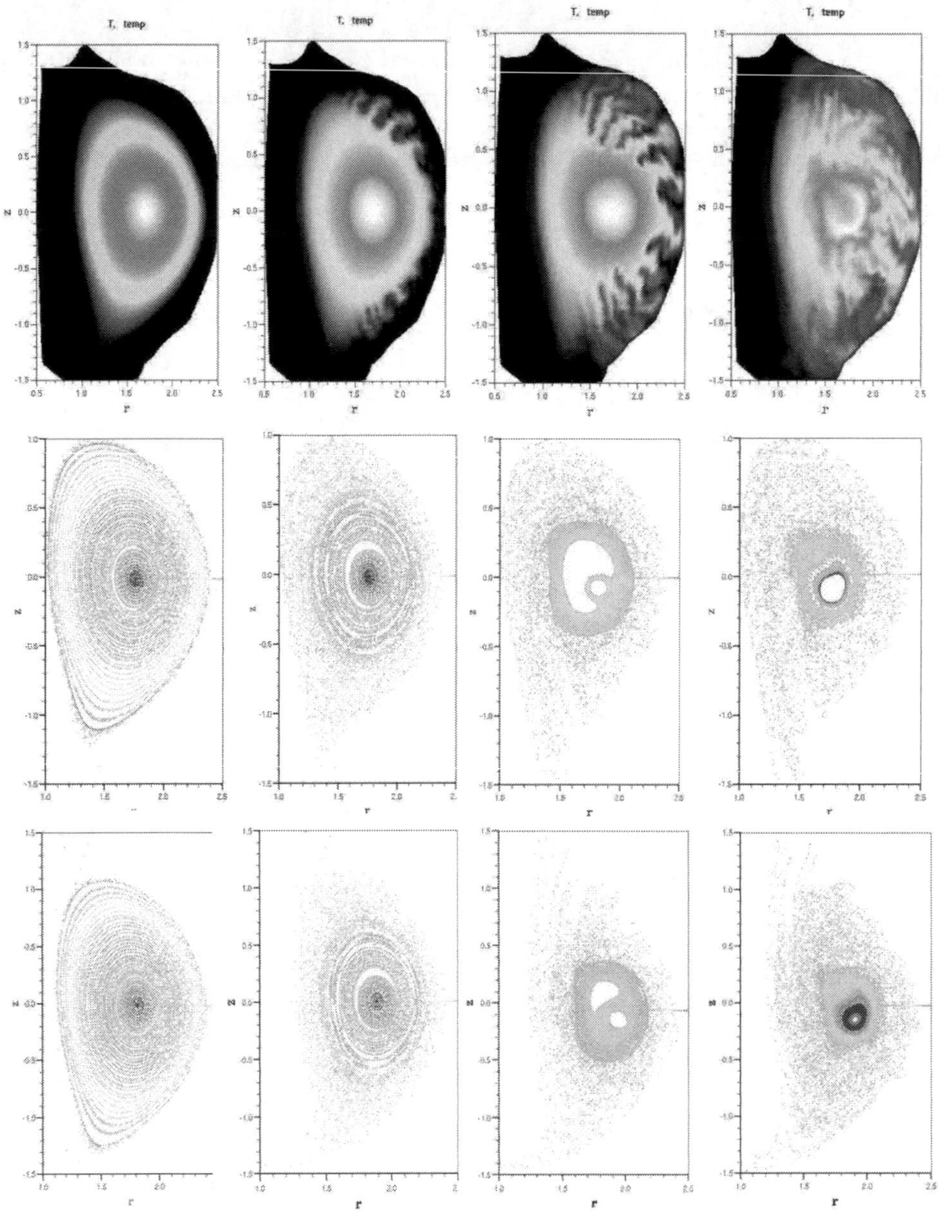

Figure 1. Time evolution of an ELM at large resistivity, shown at $t = 67$, 89, 133, and $178\tau_A$ for a perturbation started at $t = 0$. Temperature contours (top row), magnetic puncture plots traced in $+\phi$-direction (middle), and in the $-\phi$-direction (bottom). Axis labels are arbitary units.

of the lower X-point. The small wiggles on the inboard side are consistent with the "unstable" manifold of the lower X-point, which is less strongly unstable than on the outboard side.

In 3D, the field lines and the plasma perturbations have helical shapes that roughly follow the equilibrium field lines. Figure 3 shows a surface of constant temperature, colored according to values of the poloidal magnetic flux. A single nearby field line, traced in the $+\phi$-direction, shows how the overall "chaotic" field lines approximately coincide with the temperature contours for

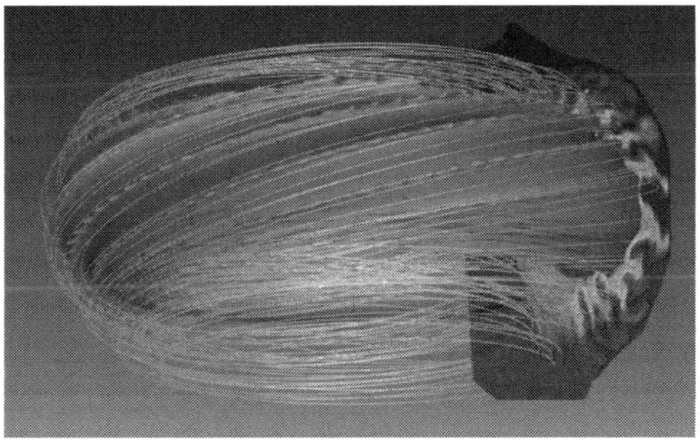

Figure 2. Helical structure at $t = 133$, showing one temperature contour, colored according to the poloidal magnetic flux, and one $+\phi$ field line, eventually lost to the bottom. Tilted upward to show the lower plasma fingers.

a significant number of toroidal transits, before (often) being lost at the X-point. The coupling of field and plasma is much tighter at the smaller resistivity of a real ELM[7], where the plasma slips through the field at a reduced rate.

The mismatch in shape between the idealized homoclinic tangle and the the plasma interchange-type instability, together with the fact that only part of the field participates constructively in the instability, reduces the growth rate over the case restricted field splitting[7].

4. Summary

Large scale MHD simulations show that Edge Localized Modes (ELMs) in high temperature, toroidal fusion plasmas constitute a new class of nonlinear plasma instability. A coherent plasma structure is able to couple dynamically to only part of a chaotic magnetic field, which then grows with the plasma instability. The magnetic X-point on the plasma boundary surface causes it to split into two asymptotic limits under perturbation. The field lines become chaotic, but retain their helical, torus-wrapping character. Instabilities driven by the strong edge pressure gradient, preferentially aligned along the equilibrium field, can proceed by a ballooning-like flux-tube interchange that couples nonlinearly to the "unstable" loops of the split field. Low density loops grow inwards by transferring plasma to the outward growing, higher density loops. The magnetic field approximates a near-Hamiltonian homoclinic tangle, modified by the plasma response. Over time, the plasma heals back to a nearly axisymmetric state. The resulting multi-stage ELM has many features similar to experiment.

The demonstration that a chaotic magnetic field with only asymptotically definable structure can couple effectively to the plasma motion may help to explain the large range of ELM and ELM-free behavior observed in H-mode fusion plasmas, as well as properties of plasma confinement.

Acknowledgments

Work supported by the U.S. Department of Energy OFES and SciDAC programs. Computation and visualization performed at NERSC. Thanks to T. Evans and L. Lao for DIII-D data.

References

[1] Wagner F, et al., 1982 *Phys. Rev. Lett.* **49** 1408
[2] Roeder R K W, Rapoport B I, and Evans T W, *Phys. Plasmas* 2003 **10** 2796-99; Evans T E, in *Chaos, Complexity and Transport: Theory and Applications* 2008, ed. Chandre C, Leoncini X, and Zaslavsky G (Singapore, World Scientific Publishing Co.) pp 147-176
[3] Evans T E, et al., *Nucl. Fusion* 2005 **45** 595-607; Evans T E, et al., 2004 *Phys. Rev. Lett* **92** 235003
[4] Lichtenberg A J and Lieberman M A 1992 *Regular and Chaotic Dynamics 2nd ed.*, (New York, Springer-Verlag); Most texts on Hamiltonian dynamics discuss these topics.
[5] Mizuguchi N, et al. 2007 *Nucl. Fusion* **47** 579-585; Huysmans G T A and Czarny O, 2007 *Nucl. Fusion* **47** 659-666; Snyder P B, Wilson H R, and Xu X Q, 2005 *Phys. Plasmas* **12** 056115
[6] Strauss H R, Sugiyama L, et al. 2009 *Nucl. Fusion* **49** 055025
[7] Sugiyama L E, and Strauss H R, in preparation.
[8] Park W, et al., 1999 *Phys. Plasmas* **6** 1796-1803; Sugiyama L E and W Park 2000 *Phys. Plasmas* **7** 4644-4658

ALPS: A framework for parallel adaptive PDE solution

Carsten Burstedde[⋆] **Martin Burtscher**[⋆] **Omar Ghattas**[⋆,†]
Georg Stadler[⋆] **Tiankai Tu**[⋆] **Lucas C. Wilcox**[⋆]

[⋆]Institute for Computational Engineering and Sciences (ICES), The University of Texas at Austin, 1 University Station C0200, Austin TX 78712, USA

[†]Jackson School of Geosciences and Department of Mechanical Engineering, The University of Texas at Austin, 1 University Station C0200, Austin TX 78712, USA

E-mail: {carsten,burtscher,omar,georgst,ttu,lucasw}@ices.utexas.edu

Abstract. Adaptive mesh refinement and coarsening (AMR) is essential for the numerical solution of partial differential equations (PDEs) that exhibit behavior over a wide range of length and time scales. Because of the complex dynamic data structures and communication patterns and frequent data exchange and redistribution, scaling dynamic AMR to tens of thousands of processors has long been considered a challenge. We are developing *ALPS*, a library for dynamic mesh adaptation of PDEs that is designed to scale to hundreds of thousands of compute cores. Our approach uses parallel forest-of-octree-based hexahedral finite element meshes and dynamic load balancing based on space-filling curves. *ALPS* supports arbitrary-order accurate continuous and discontinuous finite element/spectral element discretizations on general geometries. We present scalability and performance results for two applications from geophysics: seismic wave propagation and mantle convection.

1. Introduction

The advent of the age of petascale computing brings unprecedented opportunities for breakthroughs in scientific understanding and engineering innovation. However, the raw performance made available by petascale systems is by itself not sufficient to solve many challenging modeling and simulation problems. For example, the complexity of solving evolutionary partial differential equations often scales as $n^{\frac{4}{3}}$, where n is the number of unknowns.[1] Thus, the three-orders-of-magnitude improvement in peak speed of supercomputers over the past dozen years has meant just a factor of 5.6 improvement in spatio-temporal resolution—not even three successive refinements of mesh size. For many problems of scientific and engineering interest, there is a desire to increase the resolution of current simulations by several orders of magnitude. Thus, although supercomputing performance has outpaced Moore's Law over the past several decades due to increased concurrency [1], the curse of dimensionality imposes much slower scientific returns.

The work requirements of scientific simulations typically scale as n^{α}. The power α can be reduced through the use of optimal solvers such as multigrid for PDEs and fast multipole for integral equations and N-body problems. Once α has been reduced as much as possible, further

[1] Optimal solvers require $\mathcal{O}(n)$ work per time step, and time accurate integration often implies $\mathcal{O}(n^{\frac{1}{3}})$ time steps.

reductions in work can be achieved only by reducing n itself. This can be accomplished in two ways: through the use of adaptive mesh refinement/coarsening (AMR) techniques, and the use of higher order accurate discretizations (in space and time). AMR places mesh points only where needed to resolve features of the solution, whereas high-order approximations reduce the number of mesh points necessary to achieve a given accuracy.

Fortunately, many problems have local multiscale character, i.e., resolution is needed only in localized (possibly dynamically evolving) regions, such as near fronts, discontinuities, material interfaces, reentrant corners, boundary and interior layers, and so on. In this case, AMR methods can deliver orders-of-magnitude reductions in the number of mesh points. However, AMR methods can also impose significant overhead, in particular on highly parallel computing systems, due to their need for frequent re-adaptation and load-balancing of the mesh over the course of the simulation. Because of the complex data structures and communication patterns and frequent data exchange and redistribution, scaling dynamic AMR to tens of thousands of processors has long been considered a challenge. Space constraints preclude a proper review of existing methods and software for parallel AMR, but see the discussion and references in [2, 3].

We have developed the *ALPS* (**A**daptive **L**arge-scale **P**arallel **S**imulations) framework for parallel adaptive solution of PDEs [3]. *ALPS* includes the *octor* [4] and *p4est* libraries for parallel dynamic mesh adaptivity on single-octree-based and forest-of-octree-based geometries, respectively, and the *mangll* library for arbitrary-order hexahedral continuous and discontinuous finite/spectral element discretizations on general multi-octree geometries. *ALPS* has been shown to scale well weakly and strongly to over 60,000 processor cores [3]. In this paper, we describe several applications of the *ALPS* framework to two continuum mechanics problems—the propagation of elastic waves (with application to global seismology) and the buoyancy-driven creeping (Stokes) flow of a non-Newtonian incompressible fluid (with application to global mantle convection)—and provide sample performance results on Ranger, the 579 teraflops, 62,976-core Sun system at the Texas Advanced Computing Center (TACC).

2. Elastic wave propagation: Global seismology

The elastic wave equation models the propagation of longitudinal and shear waves in a linearly elastic medium. For isotropic media, the material stiffness tensor involves only the Lamé parameters λ and μ. In mixed velocity–stress form, the elastic wave equation reduces to

$$\rho \frac{\partial \boldsymbol{v}}{\partial t} = \nabla \cdot \boldsymbol{S} + \boldsymbol{f}, \tag{1a}$$

$$\frac{\partial \boldsymbol{S}}{\partial t} = \mu \left(\nabla \boldsymbol{v} + \nabla \boldsymbol{v}^{\top} \right) + \lambda \left(\nabla \cdot \boldsymbol{v} \right) \boldsymbol{I}, \tag{1b}$$

where $\boldsymbol{v}$ and $\boldsymbol{S}$ are the unknown velocity vector and Cauchy stress tensor of the medium, and ρ, t, f, and $\boldsymbol{I}$ are the mass density, time, body force, and identity tensor, respectively.

We have built an elastic wave propagation code based on the *mangll* and *p4est* libraries from *ALPS*. Velocity and stress fields are discretized using the discontinuous Galerkin (dG) method with a Godunov flux [5] in space and a five-stage fourth-order explicit Runge-Kutta (RK) method in time. The elements are the same as those used in the spectral element method [6], having nodes at the tensor product Legendre-Gauss-Lobatto (LGL) points. The dG method duplicates degrees of freedom between adjacent element faces. On nonconforming interfaces between elements (arising due to 2:1 subdivision between adjacent hexahedra), the numerical flux is computed by introducing a face integration mesh that incorporates the contributions from each smaller face individually using the two-dimensional tensor LGL quadrature based on the nodes of the smaller face. The LGL quadrature reduces the block diagonal dG mass matrix to a diagonal matrix, permitting faster application of its inverse.

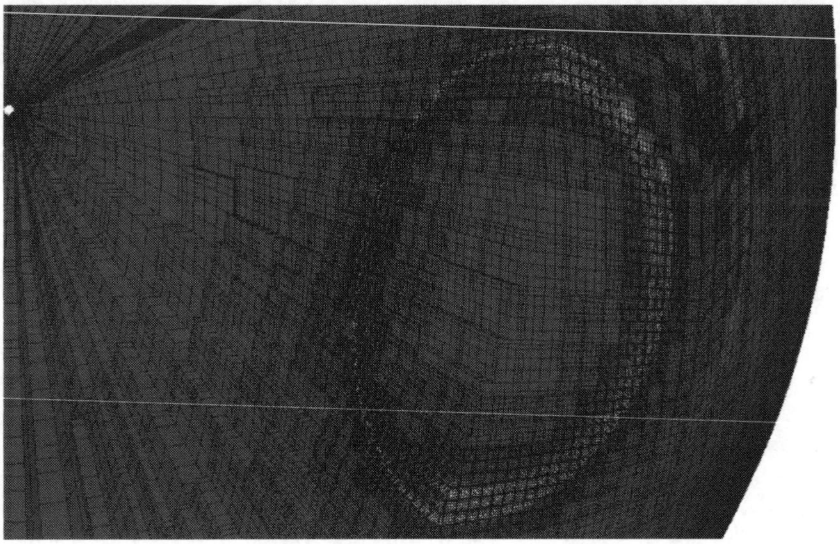

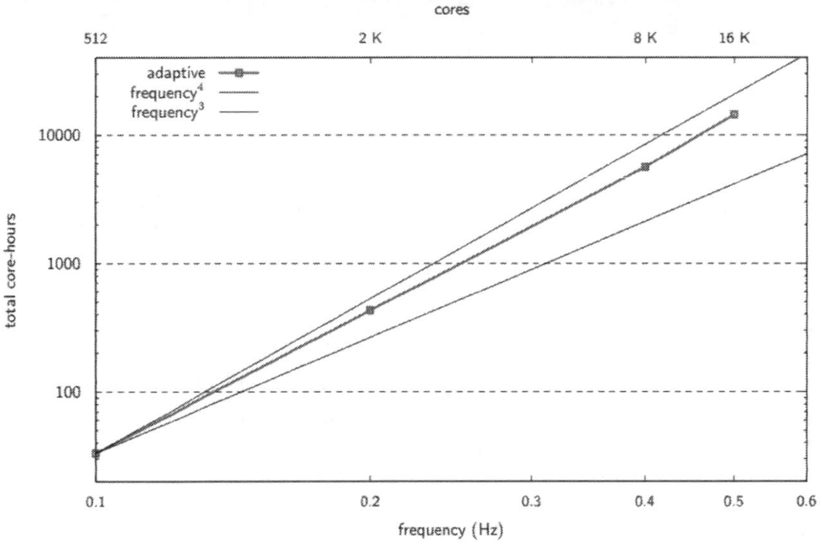

Figure 1. *Top:* cut of wavefront-adapted spherical mesh for a double-couple source. *Bottom:* scaling of run time with the number of cores for dynamic-adaptive wave propagation solver.

Figure 1 presents results from the solution of an elastic wave propagation problem in a spherical geometry. The material parameters approximate those of the Earth and the source is a simple double-couple. The top image depicts a snapshot of a mesh that has been adapted to propagating wavefronts. The significant reduction of mesh points can be seen in the image. The bottom plot presents scaling results to 16K cores, showing the growth in total processor-hours as a function of maximum source frequency. A perfectly scalable static mesh explicit wave propagation solver would have a total solution time that scales as the fourth power of frequency. The scaling is shown as the top black line (with 512 cores as the base case). Dynamic adaptivity has the potential to reduce the complexity to be closer to the third power of frequency (shown as the bottom black line), since adapting to moving wavefronts effectively eliminates one dimension from the mesh. The cubic scaling cannot be achieved for problems on finite domains or with interfaces, due to reflections and generation of new waves (the number of wavefronts

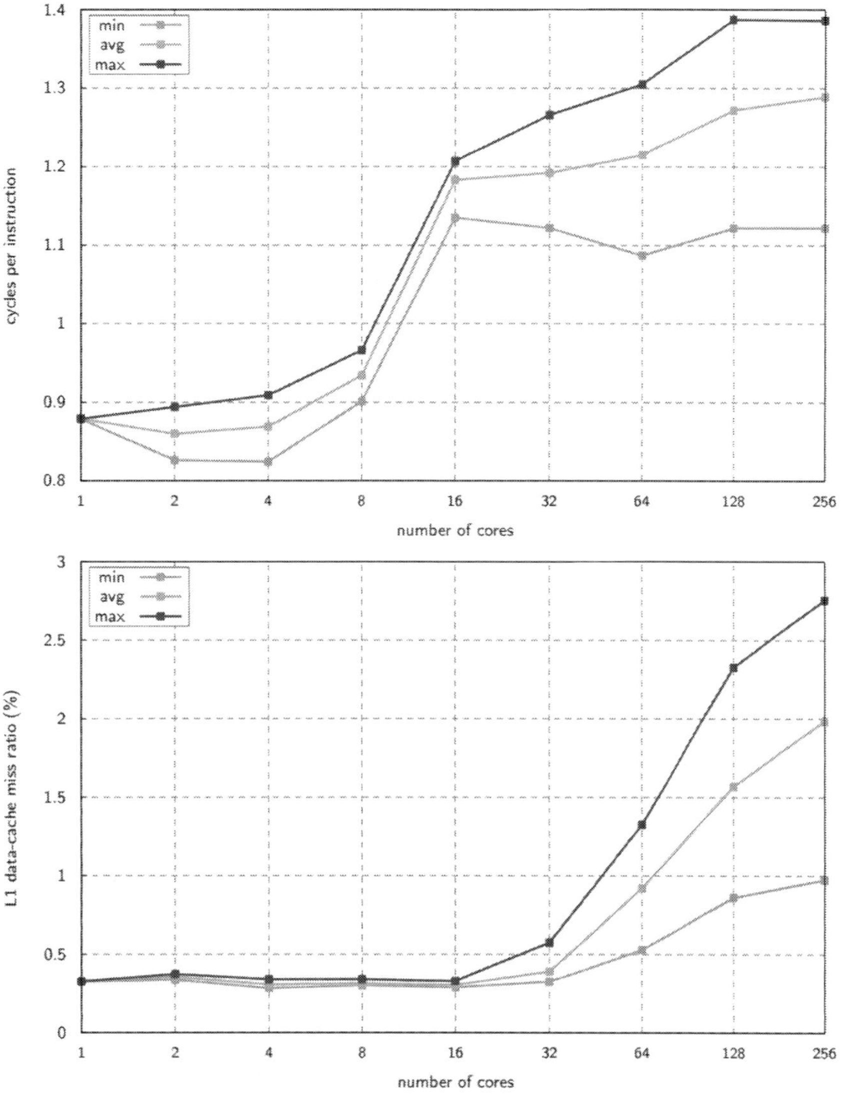

Figure 2. CPU performance for wave propagation code with 10K total elements of polynomial degree 6. *Top:* cycles per instruction. *Bottom:* L1 data-cache miss ratio in percent.

that must be tracked does not remain constant). The actual observed scaling is plotted as the red line, showing that, even in a layered Earth model, the "optimal" fourth power scaling can be improved using adaptivity.

We also studied the efficiency of the implementation using performance counter measurements from the *PAPI* library [7]; the results are summarized in Figure 2. Up to eight cores, more than one instruction is executed per cycle on average, indicating that there is significant instruction-level parallelism. As the compute nodes become saturated when using all 16 cores per node, we observe a jump in the cycles per instruction (CPI). Yet, even above 16 cores, the CPI is low because we carefully tuned important sections of the source code so that the compiler can vectorize them [8]. Beyond 16 cores, the CPI grows due to an increase in cache misses, which is caused by an increase in data that need to be exchanged between compute nodes, which results in less regular memory access. Nevertheless, the L1 data-cache miss ratio is surprisingly low for

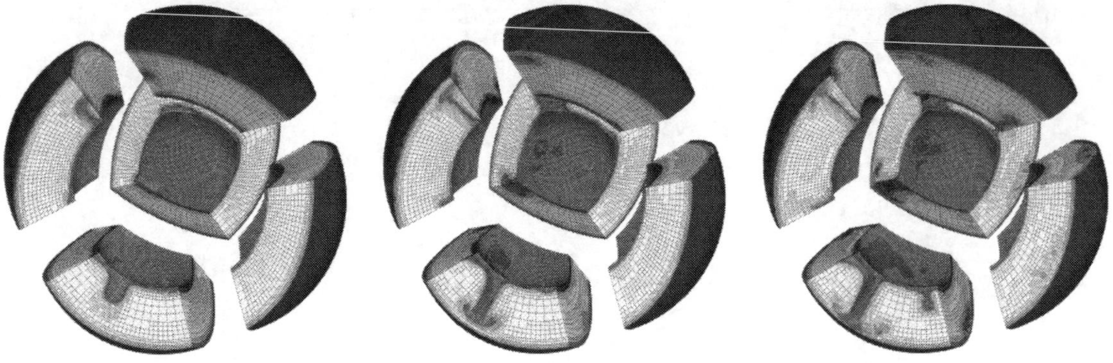

Figure 3. Snapshots of three time steps from a mantle convection simulation using *Rhea*, showing dynamically-adapted refinement near rising thermal plumes and thermal boundary layers.

code that processes about 1.75, 1.12 and 0.96 million degrees of freedom per second per core on 1, 32 and 256 cores, respectively. This low ratio is due to data being accessed in mostly regular patterns, which allows data to be prefetched effectively.

3. Non-Newtonian Stokes flow: Mantle convection

Our second example is flow of a viscous incompressible creeping non-Newtonian fluid, which is governed by equations for mass, momentum, and energy balance,

$$\nabla \cdot \boldsymbol{v} = 0, \tag{2a}$$

$$-\nabla \cdot \left[\eta \left(\nabla \boldsymbol{v} + \nabla \boldsymbol{v}^{\top}\right) - \boldsymbol{I} p\right] = \rho \boldsymbol{g}, \tag{2b}$$

$$\rho c_p \left(\frac{\partial T}{\partial t} + \boldsymbol{v} \cdot \nabla T\right) - \nabla \cdot (k \nabla T) = \rho H, \tag{2c}$$

where $\boldsymbol{v}$, p, and T are the unknown velocity, pressure, and temperature; and $\eta = \eta(T, \boldsymbol{v})$, ρ, $\boldsymbol{g}$, c_p, k and H are the viscosity, density, gravitational acceleration, specific heat, thermal conductivity, and internal heat generation rate. We are interested in convection in Earth's mantle, where the flow is driven by buoyancy. Thus we invoke the Boussinesq approximation, which replaces the density in the gravitational force term in (2b) by $\rho = \rho_0 [1 - \alpha(T - T_0)]$, where ρ_0 and T_0 denote reference temperature and density and α is the coefficient of thermal expansion. Various constitutive laws are used for the mantle, but in general the viscosity depends nonlinearly on both temperature and second invariant of the deviatoric strain rate tensor.

We have built a parallel AMR mantle convection code, *Rhea*, that solves (2a)–(2b) with appropriate boundary conditions and temperature- and strain-rate-dependent viscosity [3] using *ALPS* components. The images in Figure 3 illustrate dynamic mesh adaptivity in resolving thermal upwellings and boundary layers. The first version of *Rhea* discretizes the velocity, pressure, temperature fields with mapped trilinear hexahedral finite elements. Mantle flows are strongly advection-dominated, and therefore we employ the streamline-upwind Petrov-Galerkin (SUPG) scheme to stabilize the energy equation (2c). The equal-order discretization of the Stokes equations (2a), (2b) is stabilized with pressure projection [9]. Explicit integration of the energy equation decouples the temperature update from the nonlinear Stokes solve; the latter is carried out at each time step using the updated temperature via a lagged-viscosity (Picard) iteration. Each Picard iteration requires a variable-viscosity Stokes solve, which is discussed below.

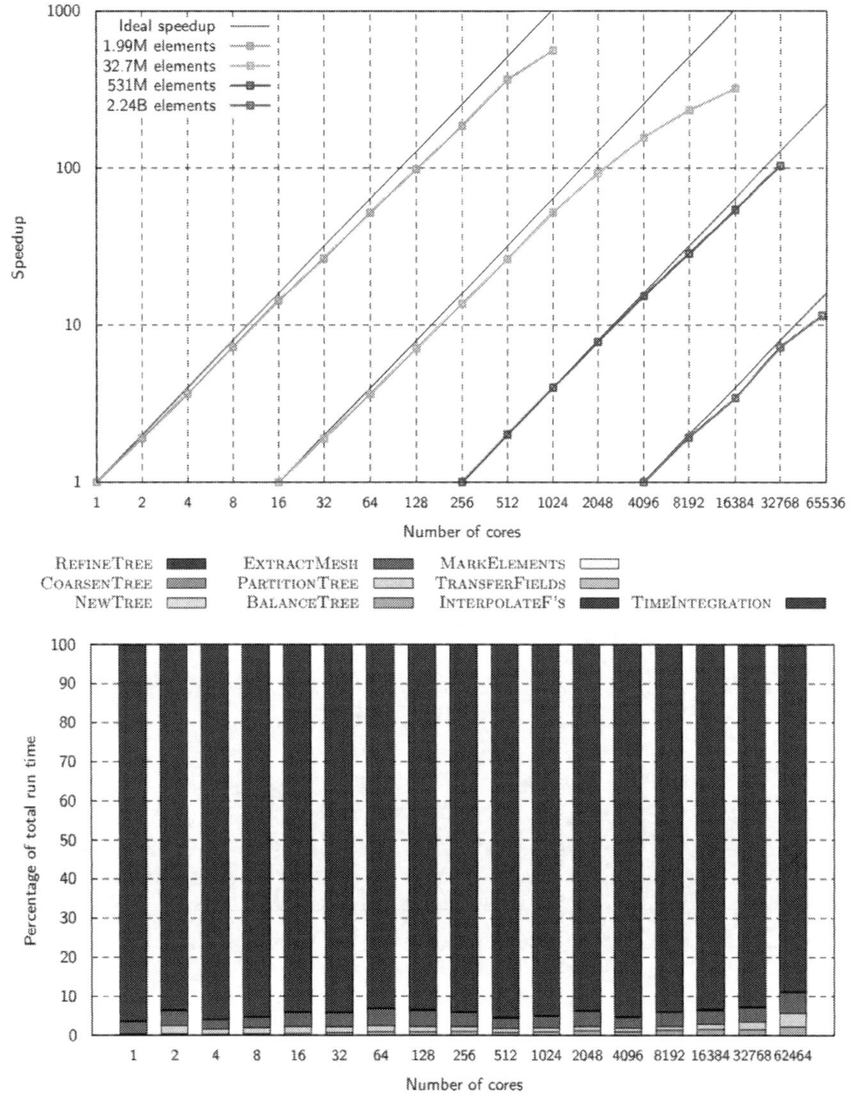

Figure 4. *Top:* fixed-size (strong) scalability for adaptive solution of energy equation for four problem sizes on cubic domain. Mesh is adapted every 32 time steps. *Bottom:* weak scalability for adaptive solution of energy equation. Total run time is broken down into numerical PDE solution (blue) and AMR functions (all other colors). Problem size increases isogranularly at 131,000 elements per core; largest problem has approximately 7.9 billion elements.

First, we study the scalability of AMR for solution of the energy equation (2c). This low-order-discretized, explicitly-solved, scalar, linear equation is a severe test of AMR, since there is very little numerical work over which to amortize AMR. As can be seen from the top plot in Figure 4, the fixed-size scaling speedups are nearly optimal over a wide range of core counts. For instance, solving the problem with 531 million elements (blue line) exhibits a speedup of 101 on 32,768 cores over the same problem running on 256 cores (128 would be optimal). Moreover, scaling weakly from 1 to 62,464 cores (bottom of Figure 4), we see that all mesh adaptation functions—including coarsening, refinement, interpolation, rebalancing and repartitioning of the mesh—together impose little overhead on the PDE solver. Only for 62K cores does the total

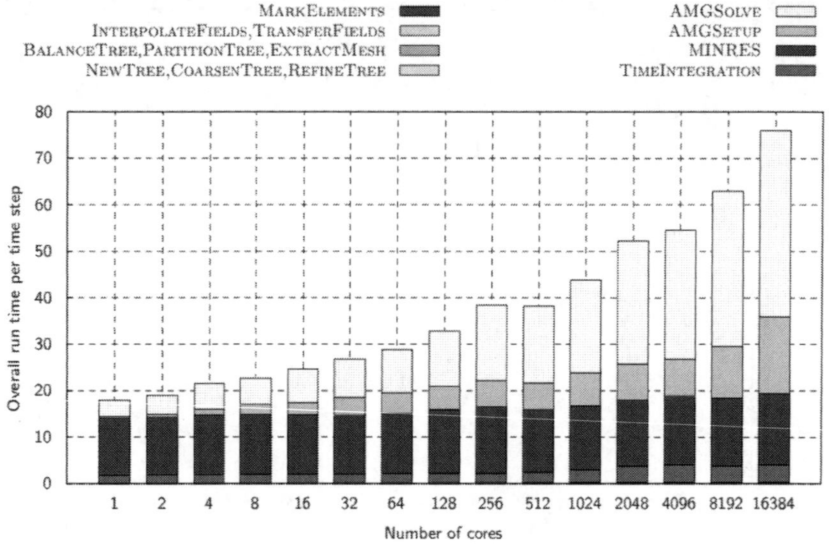

Figure 5. Breakdown of mantle convection run time per time step into AMG setup (gray), AMG V-cycle (yellow), MINRES iteration (blue), and time integration of the energy equation (red) using *BoomerAMG* preconditioning. AMR operations (all other colors) have negligible cost. Mesh is adapted every 16 time steps, and problem size increases isogranularly at 50K elements per core (largest problem has 815 million elements).

cost of AMR exceed 10% of the end-to-end run time, and even then just barely.

Next, we study the algorithmic and parallel scalability of the variable-viscosity Stokes solver, which is invoked at each Picard iteration within each time step. We use MINRES to solve the symmetric indefinite Stokes system; the preconditioner is a block-diagonal matrix, which invokes one V-cycle of an algebraic multigrid (AMG) method for the viscous (1,1) block, and an inverse-viscosity-scaled mass matrix for the pressure Schur-complement (2,2) block (see [10] for details). For Cartesian geometry, the viscous vector operator can be approximated by a scalar, variable-coefficient Poisson operator associated with each of the three velocity components (provided viscosity gradients are not too large). The Poisson operators are approximately inverted by one V-cycle of *BoomerAMG* from the *hypre* library [11]. The bar chart in Figure 5 shows the breakdown of different components of an adaptive variable-viscosity Stokes solve using *BoomerAMG* as we scale weakly from 1 to 16K cores. Spherical geometry, on the other hand, induces additional coupling among the velocity components that cannot be effectively approximated by the decomposition into scalar Poisson solves. Thus we employ one V-cycle of the adaptive smoothed aggregation AMG solver *ML* from *Trilinos* [12] on the entire viscous vector block. Table 1 presents weak scalings of a single variable-viscosity Stokes solve on the spherical domain using *ML*. We draw several conclusions from Figure 5 and Table 1. First, for the full AMR mantle convection simulations with *Rhea* (i.e. including energy and nonlinear Stokes), the cost of parallel AMR is in the noise. Second, using a V-cycle of either *BoomerAMG* or *ML* as a preconditioner for the viscous block, in conjunction with a scaled mass matrix approximation of the Schur complement, results in excellent algorithmic scalability, as evidenced by the near-insensitivity of MINRES iterations to a 16,000-fold increase in problem size and number of cores. Third, while both AMG implementations exhibit excellent algorithmic scalability, there is an opportunity for improvement of the parallel scalability of the AMG setup and V-cycle components at $O(10^4)$ cores, based on our tests on a cluster system and viscous operators with several-orders-of-magnitude variation in coefficients. Finally, the examples demonstrate that

#cores	#dofs	MINRES #iterations	AMG setup (s)	AMG V-cycle (s)	MINRES matvec (s)	ζ_A	ζ_I	ζ_V	ζ
8	1.89M	77	7.12	41.3	130.8	1.00	1.00	1.00	1.00
64	12.8M	76	8.47	38.9	114.9	1.01	1.12	1.05	1.10
512	99.7M	78	10.2	48.1	129.9	0.99	1.02	0.87	0.95
4096	672M	109	36.5	189.8	208.7	0.70	0.89	0.31	0.41
16384	2.43B	109	66.0	211.3	216.5	0.70	0.86	0.28	0.36

Table 1. Weak scaling of Stokes solve on spherical shell domain using *ML*, with 10^3 viscosity contrast, on a 3 times locally refined mesh. The numbers of cores, degrees of freedom, and MINRES iterations are tabulated, along with the time for AMG setup, AMG V-cycle, and MINRES matvecs. Also shown are algorithmic efficiency ζ_A (based on MINRES iterations), implementation efficiency ζ_I (based on MINRES timing, excluding V-cycle), V-cycle efficiency ζ_V (based on *ML* timing), and overall efficiency ζ (based on overall wall clock).

our framework for parallel forest-of-octrees AMR with high-order-accurate discretization and general geometries scales very well to at least $O(10^4)$ processor cores.

Acknowledgement

This work was partially supported by NSF (OCI-0749334, CCF-0427985, CNS-0540372, CNS-0619838, DMS-0724746) and DOE (06ER25782, 08ER25860). Computing resources at TACC were provided under TeraGrid award MCA04N026. We acknowledge many helpful discussions with Rob Falgout, Ulrike Yang, Rahul Sampath, George Biros, and Tim Warburton. We thank TACC for their outstanding support, in particular Bill Barth, Jay Boisseau, Tommy Minyard, Romy Schneider, and Karl Schulz.

References

[1] Colella P, Dunning T H, Gropp W D and Keyes D E 2004 A Science-Based Case for Large-Scale Simulation, Volume 2 Office of Science, U.S. Department of Energy URL `http://www.pnl.gov/scales/`

[2] Diachin L F, Hornung R, Plassmann P and Wissink A 2006 *Parallel Adaptive Mesh Refinement* (SIAM) chap 8 (*Software, Environments, and Tools* no 20)

[3] Burstedde C, Ghattas O, Gurnis M, Tan E, Tu T, Stadler G, Wilcox L C and Zhong S 2008 Scalable adaptive mantle convection simulation on petascale supercomputers *Proceedings of SC08* (IEEE/ACM)

[4] Tu T, O'Hallaron D R and Ghattas O 2005 Scalable parallel octree meshing for terascale applications *Proceedings of SC2005*

[5] Hesthaven J S and Warburton T 2008 *Nodal Discontinuous Galerkin Methods: Algorithms, Analysis, and Applications* (*Texts in Applied Mathematics* vol 54) (Springer)

[6] Deville M, Fischer P and Mund E 2002 *High-Order Methods for Incompressible Fluid Flow* (*Cambridge Monographs on Applied and Computational Mathematics* vol 9) (Cambridge University Press)

[7] Performance applications programming interface (PAPI) URL `http://icl.cs.utk.edu/papi/`

[8] Diamond J, Kim D, Burtscher M, Keckler S, Pingali K and Browne J 2009 Multicore optimization for Ranger *Proceedings of Teragrid '09*

[9] Dohrmann C and Bochev P 2004 *International Journal for Numerical Methods in Fluids* **46** 183–201

[10] Burstedde C, Ghattas O, Stadler G, Tu T and Wilcox L C 2009 *Computer Methods in Applied Mechanics and Engineering* **198** 1691–1700

[11] Center for Applied Scientific Computing, Lawrence Livermore National Laboratory 2007 hypre. *High Performance Preconditioners, User Manual* `https://computation.llnl.gov/casc/linear_solvers/`

[12] Gee M, Siefert C, Hu J, Tuminaro R and Sala M 2006 ML 5.0 smoothed aggregation user's guide Tech. Rep. SAND2006-2649 Sandia National Laboratories

High performance simulations for transformational earthquake risk assessments

D B McCallen[1,2] and S C Larsen[3]

[2] Program Leader for Nuclear Energy, Earth Sciences Division, Lawrence Berkeley National Laboratory, University of California, 1 Cyclotron Road, Berkeley, Ca, 94720
[3] Geophysicist and Computational Scientist, Lawrence Livermore National Laboratory, Lawrence Livermore National Security LLC, 7000 East Avenue, Livermore, Ca, 94550

E-mail: dbmccallen@lbl.gov

Abstract. Earthquakes occurring around the world are responsible for extensive loss of life and infrastructure damage. On average, 1100 earthquakes with significant damage potential occur world-wide per year, and a major societal challenge is to design a human environment that contains appropriate earthquake resistance. Design of critical infrastructure such as large buildings, bridges, industrial facilities and nuclear power plants in seismically active regions is a significant scientific and engineering challenge that encompasses the multiple disciplines of geophysics, geotechnical and structural engineering. Because of the great complexities in earthquake physical processes, traditional approaches to seismic hazard assessment have relied heavily on historical earthquake observations. In this approach, observational data from many locations is homogenized into an empirical assessment of earthquake hazard at any specific site of interest. With major advancements in high performance computing platforms and algorithms, it is now possible to utilize physics-based predictive models to gain enhanced insight about site-specific earthquake ground motions and infrastructure response. This paper discusses recent advancements in geophysics and infrastructure simulations and future challenges in implementing advanced simulations for both earthquake hazard (future ground motions) and earthquake risk (infrastructure response and damage) assessments.

1. The earthquake response problem

The complete end-to-end (e.g. fault rupture-to-structural response) physical processes associated with earthquakes and infrastructure response are very complex. The processes can be divided into four segments as indicated in figure 1: (1) earthquake fault rupture; (2) seismic wave propagation through rock; (3) response of the near-surface sedimentary soil deposits; and (4) the dynamic response of the particular structure under consideration. Through extensive scientific study, much has been learned in recent years about earthquake fault rupture processes, and observations of earthquake ground motions have provided important insights into the manner that sub-surface geologic structure can influence the observed earthquake ground motions at a particular site. In the engineering community, analysis of structural system performance in actual earthquakes and structural component and system testing on mechanical earthquake simulator shake tables has provided important quantitative data on the nonlinear response of structures to extreme earthquake motions. Drawing upon these engineering and

[1] To whom any correspondence should be addressed

scientific advancements, and the tremendous advancements in computational hardware and algorithms, there is a timely opportunity for advanced computational simulation to play a much larger role, and in fact transform the manner that we predict future earthquake motions and design major infrastructure.

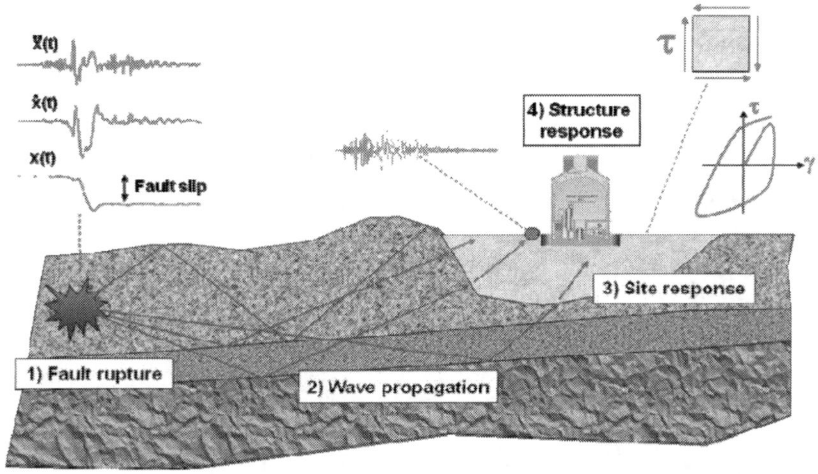

Figure 1. Four segments of the source-to-structure earthquake process.

The historical approach to estimation of earthquake ground motions has been based upon *Probabilistic Seismic Hazard Assessment* (PSHA) through which historical information from earthquakes at many observed geographic locations is synthesized to develop a probabilistic estimate of future ground motions at a specific site. For a particular site of interest, PSHA identifies near-by active faults, interprets geologic data to infer frequency of earthquake occurrence on each fault, and utilizes generic attenuation relationships that describe the decay of ground motion intensity as a function of distance from a causative fault, all combined to yield a probabilistic representation of the potential future ground motions at the site. The defined ground motion hazard, which is used to design critical structures, is typically associated with an annual mean probability of exceedance (e.g. a 1×10^{-4} mean annual frequency of exceedance).

This empirically based approach provides a standardized methodology with embedded statistical information, but it may not capture all of the site-specific complexities of a particular site under consideration. As more data from actual earthquakes have been obtained, the variability and complexities of earthquake ground motions has come into clearer focus. For example, ground motions measured in the 1991 Landers California earthquake (see figure 2) clearly show the influence of fault rupture directivity in which two strong motion instrument sites, GSC and PFO, have substantially different observed motions despite their almost identical distance from the causative fault. Site GSC is subjected to much larger motions as the fault rupture propagates in the direction of GSC and directs more wave energy toward that site. These data are indicative of the profound influence that the physical process can have on observed ground motions.

A realistic high performance computational model capable of representing source-to-structure earthquake processes would be a tremendously powerful tool for performing site specific evaluations of ground motion levels and structural response, and for evaluating the sensitivity of structural response to specific geologic parameters and processes. Such a model could augment, or replace, the current empirically based elements of the PSHA process.

Recently, significant progress has been made towards achieving the individual components of such an ambitious end-to-end modeling goal. With the highest-end computers, three-dimensional simulations of fault rupture and seismic wave propagation are now being solved for regional scale models at low frequency. Structural response simulations, which realistically capture the behavior of steel and concrete structures subjected to large ground motions, and which include significant nonlinear behavior due to cracking, yielding or other damage in the structure, are being rigorously solved with numerically robust and efficient iterative, time-stepping solution algorithms.

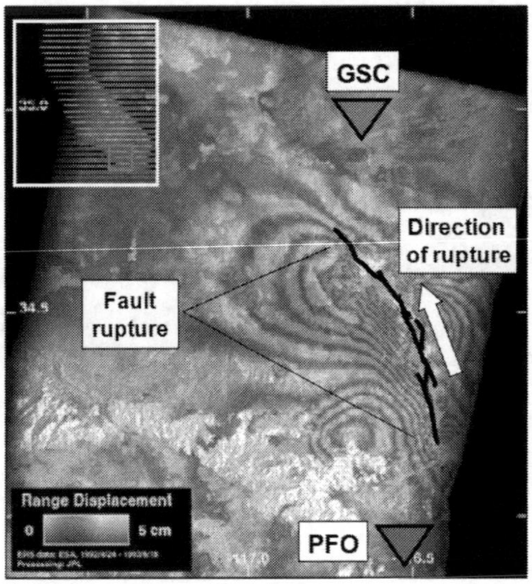

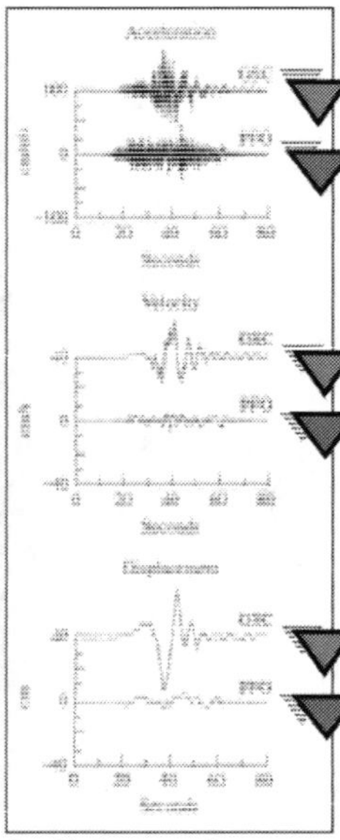

Figure 2. The observed effect of fault rupture directivity on ground motions (1992 Landers CA).

2. Simulating earthquake ground motions at regional scale

Segments 1, 2 and 3 of the earthquake response problem (figure 1) include representation of the fault rupture process and seismic wave propagation through bedrock and alluvial deposits to the surface of the ground. Computer programs that have been developed over the last few years to address this source rupture and wave propagation problem are typically finite difference based, and limited to linear elastic and viscoelastic response behavior characterized by the three dimensional equations of elastodynamics. The time dependent solutions are typically achieved through explicit time integration, and staggered grids are often applied to the finite difference stencil [1]. The three dimensional character of these programs, and the large computational domains necessary to model major earthquakes, demands the most advanced computing resources. Numerical resolution of the higher frequencies of interest (5-15 Hz depending on the structure under consideration) represents a significant computational challenge.

An example of the level of computational capabilities associated with high performance three dimensional simulation is illustrated in figure 3, where a fully three dimensional simulation of the great 1906 San Francisco earthquake (M=7.8) and a postulated Hayward fault California earthquake (M=7.1) are illustrated. For the 1906 simulation, a significant portion of northern California was included for the very extensive earthquake fault rupture. The 1906 simulation, computed with a 11.1 billion node finite difference grid, required 1 TB of memory and 16 hours of wall clock time on a 1024-cpu Linux cluster (1.1 TFlops of sustained performance) to compute 200 seconds of simulated earthquake ground motions. This simulation resolved ground motion waveforms to only 1 Hz frequency. The computational requirements to extend the simulations to 3 or 4 Hz, which would be desirable for most evaluations of typical building structures, would demand significantly more computational resources as indicated in Table 1. For example, a model with 5 Hz frequency resolution for the 1906 earthquake simulation would require a petascale system. Such a simulation could be

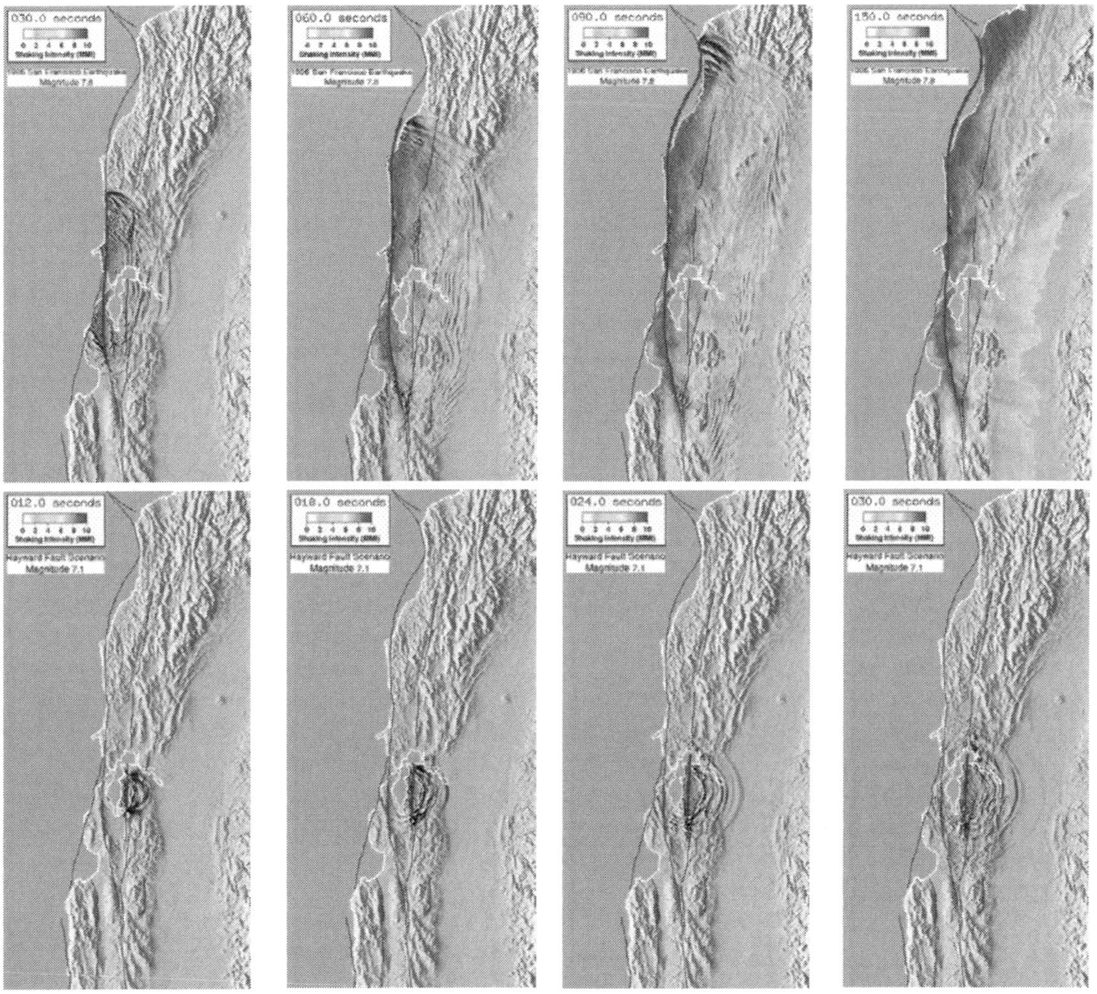

Figure 3. Snapshots of seismic wave propagation and estimated Modified Mercalli Intensity of ground shaking from a fault rupture scenario of the M=7.8 1906 San Francisco earthquake and a postulated M=7.1 Hayward fault earthquake simulation.

Table 1. Computational requirements for regional earthquake simulations (metrics based on a 128 node, 1024 CPU Linux cluster with ~ 1.1 Tflop performance).

Earthquake Scenario	Required Number of Grid Nodes	Memory / Solution Time (1 Hz Resolution)	Memory / Solution Time (3 Hz Resolution)	Memory / Solution Time (5 Hz Resolution)
M=7.8 San Francisco	11.1 Billion (1 Hz)	1 TB / 16 hrs	27 TB / 54 days	125 TB / 416 days
M=7.1 Hayward	1.49 Billion (1 Hz)	0.15 TB / 1.2 hrs	4 TB / 4 days	20 TB / 31 days

performed in a reasonable wall-clock time (e.g. 24 hours) on a commodity-base Linux cluster with 500,000 compute core and a sustained performance of about 1 PFlop. The memory required would not be excessive since large problems are more constrained by run-time. The total memory requirements would be approximately 4 GBytes for a 16 core node.

These 1906 simulations were utilized in seismological studies of the great 1906 San Francisco Earthquake [2,3,4] and provided new insight into the ground motion distribution from that event as well as insight into the earthquake hazard associated with potential future San Andreas fault earthquakes. Because strong earthquake ground motion instrumentation was not available in 1906, quantitative data on the actual observed motions (i.e. measured ground motion waveforms) are not available, and the forensic studies of this event have relied on estimates of human observations based on the Modified Mercalli Intensity scale to compare observed and simulated results. The intensity levels estimated from the massively parallel simulations were in good agreement with documented eyewitness accounts from the great 1906 earthquake [4]. The simulation plots clearly illustrate the significant difference in scale of a great M=7.8 earthquake versus a major M=7.1 earthquake.

3. Simulating nonlinear structural response to strong ground motions

Finite element based structural mechanics simulations of the response of infrastructure to extreme earthquake ground motions has made tremendous progress in the last ten to fifteen years both in the research community and in transferring the capabilities into the hands of practicing engineers. Due to the relatively long duration of earthquake excitations (e.g. 10-50 seconds), the stiff mechanical characteristics of steel and concrete structures, and the physically small dimensions of structural finite elements, the time history solutions of structural response have been carried out with implicit time integration schemes that are unconditionally stable and allow the utilization of large time steps, i.e. the time step size is not constrained by a Courant stability limit. Under extreme earthquake ground motions, most structures are designed so that some level of nonlinear response behavior will occur due to local yielding or member cracking. In fact, controlled and limited nonlinear behavior is desirable from an energy dissipation point of view during extreme earthquakes. Nonlinear behavior is currently permitted as an appropriate design feature in many engineering design codes [5]. To realistically represent nonlinear behavior of the structural system, accurate constitutive models of nonlinear element behavior (both steel and concrete) and robust solution algorithms with equilibrium iterations in each time step have been the focus of successful intense research.

Computational structural mechanics technologies have progressed to the point where realistic, nonlinear models can be computed for a significant range of structural types. Figure 4 illustrates a very detailed inelastic finite element model of a steel frame system of a high-rise building. In this model, each beam and column element is discretized with a fiber-based finite element with elasto-plastic material behavior. In the fiber model, the cross-section of each structural member is divided into a number of zones, or fibers, as shown in figure 4, and the nonlinear stress-strain behavior is invoked for each fiber of the cross section. The fiber modeling approach rigorously accounts for interaction between flexure (bending moment) and tension / compression (axial force) in each element, and also captures the spread and evolution of inelastic behavior through the element cross section. The nonlinear model also precisely represents the effects of structural displacements and the additional bending moments introduced by vertical gravity loads when the building sways significantly to one side.

This type of computational model has recently been used extensively to study the nonlinear seismic response of steel frame buildings to strong, near-fault earthquake ground motions [6]. A comparison of linear and fully nonlinear simulations of an actual building response is shown in figure 4. Under extreme ground motions, the nonlinear effects can be very pronounced, resulting in significant permanent deformation of the building and a significantly different distribution of member forces than what would be predicted by a simple linear elastic model. Figure 4 for example illustrates the significant influence of nonlinearities in a simulation of a forty story steel frame building response to an actual strong earthquake motion recording made in Taiwan during the 1999 Chi Chi earthquake. In this analysis, an implicit Newmark-Beta time integration was utilized with full-Newton based equilibrium iterations within each time step to drive the residual from nonlinear behavior to zero within each time step. The linear mode of the building is grossly in error for this particular strong ground motion as the building is driven well into the nonlinear regime.

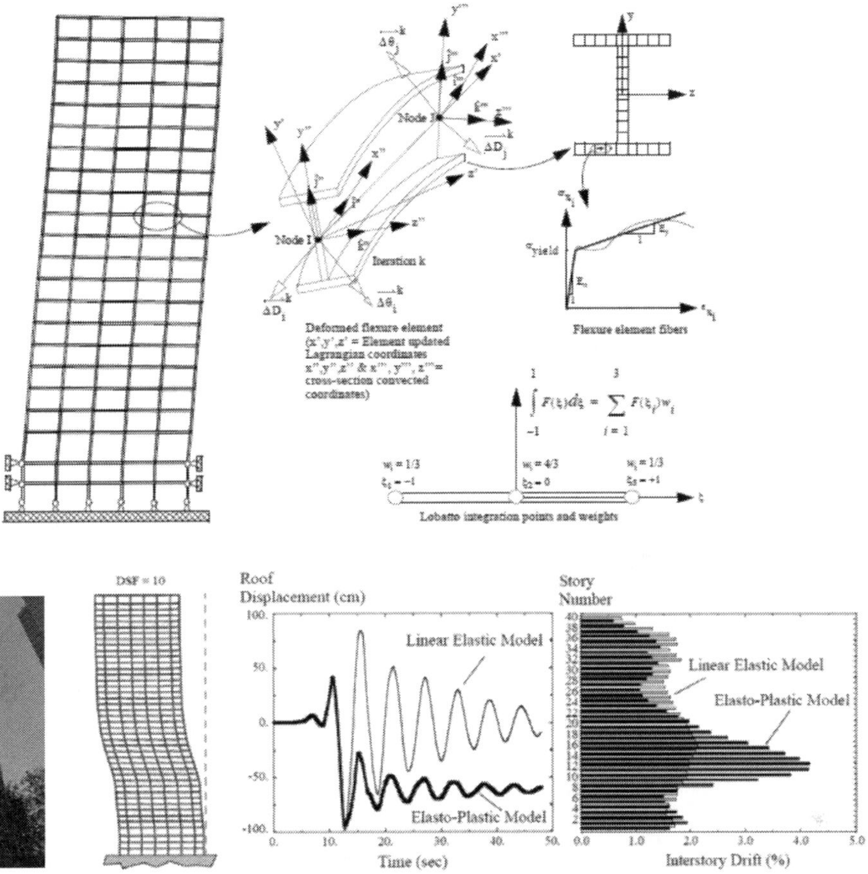

Figure 4. Nonlinear fiber element based model for steel buildings; fiber model for each beam cross section, linear and nonlinear response of a 40 story steel frame building (measured 1999 Taiwan earthquake ground motion record as input motion).

4. End-to-end simulations in hazard and risk assessment

Realistically addressing the full suite of complex elements in the earthquake processes illustrated in figure 1 will require additional computational developments, but the individual computational elements of a full end-to-end simulation process are well along in development. Initial thrusts at performing fully end-to-end simulations have been completed by the authors and coworkers in evaluating the seismic hazard and risk to major long-span bridges in the San Francisco Bay area of California. As a result of limitations in existing strong motion instruments, there were major concerns about how this class of flexible, low-frequency structures will respond to long period components of ground motions that are observed in the near-field (within 20km or so) of major earthquake faults. This concern was compounded by the fact that traditional strong motion instrumentation and companion signal processing algorithms were not designed to measure long period components of motion (waveforms with periods greater than 5 seconds) and thus there is very limited quantitative data on near-field motions.

To investigate the response of long-span bridges to near-field motions, a comprehensive geophysics-engineering simulation effort was initiated by the University of California and DOE National Laboratories to investigate the response of long-span bridges. The effort revolved around a case study for the Oakland – San Francisco Bay Bridge Suspension spans across the San Francisco

Bay (figure 5). This bridge structure is representative of long period bridge systems and is located in close proximity to the Hayward fault. In this effort, a high performance regional scale geophysics model was utilized to simulate the long-period ground motions at the bridge site, and a nonlinear structural mechanics model was utilized to evaluate the bridge response to the simulated earthquake ground motions. This effort represented the first complete end-to-end simulation of the earthquake process.

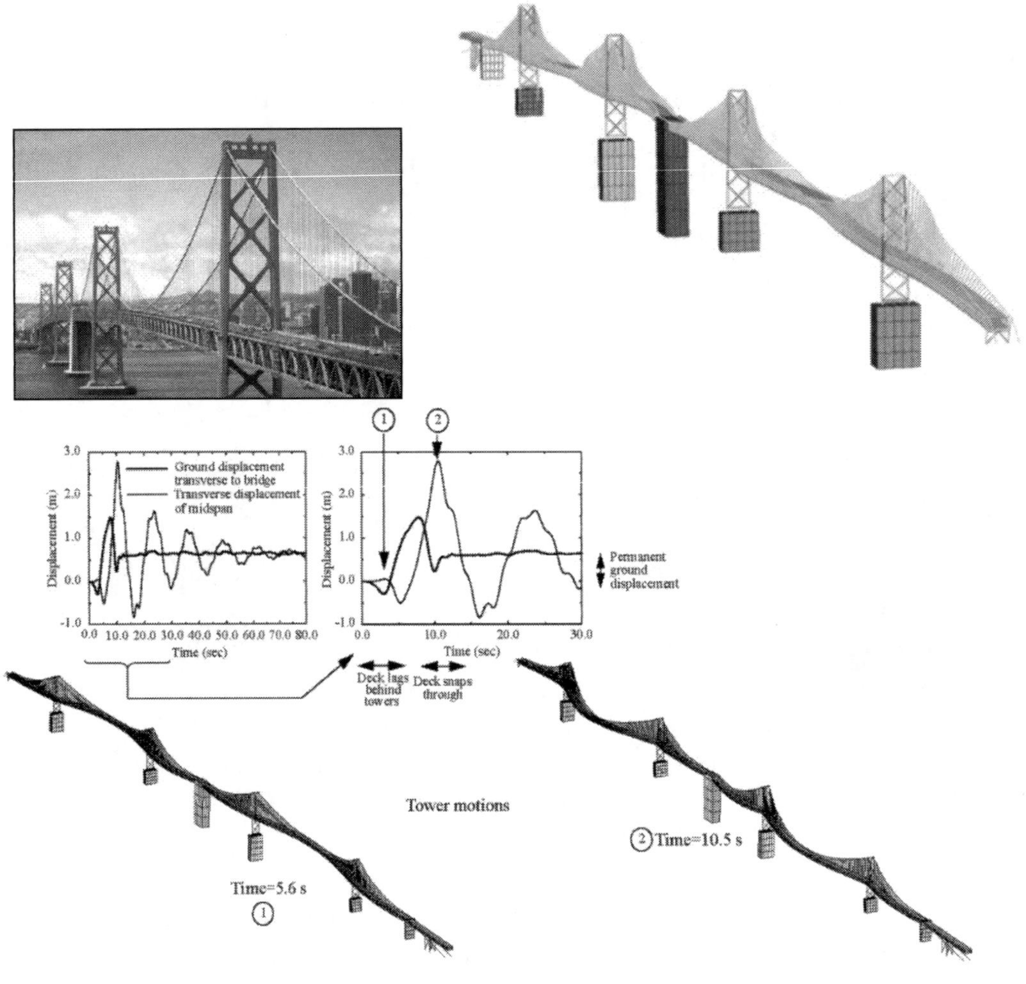

Figure 5. Nonlinear finite element model of a long-span steel suspension bridge; bridge response to long period, near-fault earthquake ground motions.

The simulated ground motions at the Bay Bridge site for a Hayward fault earthquake (see figure 3) contained both permanent ground displacement associated with tectonic fault slip, and low-frequency ground displacement pulses associated with directivity and wave superposition effects (see ground displacement from the geophysics simulation shown in figure 5 where these features are evident in the ground motion waveforms). The earthquake ground motions developed from the geophysics simulations provided direct input to the bridge computational model. It was discovered in these simulations that the long-period components of ground motion can have a very important effect on the response of this type of bridge system, and that the bridge system forces and displacements can be significantly underestimated if these long period effects are not appropriately considered [7,8,9]. Subsequently, these observations, which were based solely on end-to-end computational simulations, have been confirmed through the evaluation of bridge response using newly obtained measured broad-band ground motion records [9].

5. Moving towards full end-to-end simulations

Advanced simulation tools have and will continue to revolutionize the earthquake design and analysis of major man-made structures. The goal of performing realistic end-to-end simulations that can simulate both ground motions and subsequent structural response, and develop correlations between structural response and earthquake geophysical parameters would be an extremely powerful tool. The precise manner in which such a tool would be fully implemented in design practice remains to be established. In order to simulate waveforms at the higher frequencies of engineering interest, the geophysics computational models must be populated with representative geologic structure and material parameters at fine spatial scales. To gather general insight about processes and phenomenon, it would be sufficient to simply populate a model with realistic sub-surface parameters that are not necessarily representative of any particular actual site. Much insight can be gained about earthquake processes by studying virtual sites with realistic, but "made-up" geologic structures. This is likely the most compelling near-term application of an end-to-end simulation model. To adapt advanced simulations to hazard and risk assessments at specific actual sites, it is necessary to develop a schema for generating realistic geologic models surrounding the site. There are generally two approaches to this, one includes developing a suite of stochastically generated geologic structure models that define the unknown subsurface geology in a statistical sense, and the other is to utilize measured data from frequently occurring small earthquakes in an inversion process to further constrain the subsurface geology.

To move forward, a number of computational issues must be addressed:

- Development of the most efficient computational tool for simulation of ground motions. Resolution of higher frequencies of engineering interest will be a very significant computational challenge when including features such as topography of the free surface, geologic energy attenuation in the equations of motion, nonlinearity of shallow soils, variable density grids optimized around the significant property differences between deep and shallow geologic structures, and incorporation of detailed geologic subsurface features.
- Implementation of an algorithmic framework for numerically coupling geophysics models and structural models for cases in which interaction between the supporting soil and the structure is important (the soil-structure-interaction for a massive nuclear power plant being the classic example). In light of the fact that geophysics simulations and structural simulations are typically performed in fundamentally different algorithmic manners, i.e. implicit time integration versus explicit time integration, this will complicate rigorous coupling of these models.
- Establishment of procedures for characterizing the geologic structure parameters at fine enough scale to support the numerical computation of waveforms at high frequency. The simulation of frequencies of engineering interest will require characterization of geology on finer scales than is typically available from traditional subsurface exploration sources. Procedures must be implemented for numerical implementation of the adopted approach, which can be a significant challenge when dealing with models with billions of zones.
- Development and implementation of stochastic representations of fault rupture processes in the geophysics simulations. For a given earthquake fault, geophysical information is often available on the magnitude and frequency of occurrence of potential earthquakes. However, it is unknown precisely how the fault will rupture in future earthquakes, and data have shown that the manner in which the fault ruptures can have significant influence on realized ground motions (e.g. see figure 2). It is therefore necessary to characterize a suite of potential rupture scenarios that encompass the hazard potential of a given fault. Much research has been done

on developing this suite, and this work must be implemented into a form readily implementable by geophysics codes.

End-to-end simulations offer tremendous potential for increasing insight and understanding about earthquake processes. The ability to computationally investigate cause and effect between geophysical processes, geophysical features and structural response will provide a tool for new understanding and for evaluation of the accuracy of existing simplifications and idealizations that are employed in engineering practice (e.g. the accuracy of the assumption that ground motions consist purely of vertically propagating shear waves). Building upon the major computational research advancements in each respective area of the problem, the development of a fully end-to-end capability is within our grasp.

References

[1] McCallen D and Larsen S 2003 NEVADA – A simulation environment for regional estimation of ground motion and structural response *Lawrence Livermore National Laboratory Report* UCRL-ID-152115

[2] Larsen S, Dreger D and Dolenc D 2006 Simulations of the 1906 San Francisco earthquake using high performance computing *Seism. Res Lett.* **77** 275

[3] Aagaard B, Brocher T, Dolenc D, Dreger D, Frankel A, Graves R, Harmsen S, Hartzell S, Larsen S and Zoback M 2008 Ground motion modeling for the 1906 San Francisco earthquake I: validation using the 1989 Loma Prieta Earthquake *Bull. Seism. Soc. Am.* **92** 989-1011

[4] Aagard B, Brocher T, Dreger D, Graves R, Frankel A, Harmsen S, Hartzell S, Larsen S, McCandless K, Nilsson S, Petersson N, Rogers A, Sjogreen B, Tkalcic H, and Zobak M 2008 Ground motion modeling of the 1906 San Francisco earthquake II: ground motion estimates for the 1906 and scenario events *Bull. Seism. Soc. Am.* **92** 1012-1046

[5] ASCE Standard ASCE/SEI 43-05 2005 Seismic design criteria for structures, systems, and components in nuclear facilities, *Am. Soc. of Civil Eng.*

[6] McCallen D, Astaneh A and Larsen S 2009 Near field earthquake motions and the response of buildings *Lawrence Berkeley National Laboratory Report* (in press)

[7] McCallen D and Astaneh A 2000 An explicit integration framework for nonlinear analysis of suspension bridges *J. Seis. and Earth. Eng.* **2** 4

[8] McCallen D, Astaneh-Asl A, Larsen S and Hutchings L 2006 Dynamic response of the suspension spans of the San Francisco Bay Bridge *Proceedings of the 100th Anniversary Conference Commemorating the 1906 San Francisco Earthquake* San Francisco Ca Paper 952

[9] McCallen D, Astaneh-Asl A, Larsen S and Hutchings L 2009 The response of long-span bridges to low frequency, near-fault earthquake ground motions *Proceedings of the 2009 Technical Conference on Lifeline Earthquake Engineering Am. Soc. of Civil Eng* Oakland Ca (to appear June 2009)

Journal of Physics: Conference Series **180** (2009) 012062 doi:10.1088/1742-6596/180/1/012062

Modeling Subsurface Reactive Flows Using Leadership-Class Computing

Richard Tran Mills[1], Glenn E Hammond[2], Peter C Lichtner[3], Vamsi Sripathi[4], G (Kumar) Mahinthakumar[5], Barry F Smith[6]

[1]Computational Earth Sciences Group, Computer Science and Mathematics Division, Oak Ridge National Laboratory, Oak Ridge, TN 37831-6015

[2]Hydrology Group, Environmental Technology Division, Pacific Northwest National Laboratory, Richland, WA 99352

[3]Hydrology, Geochemistry, and Geology Group, Earth and Environmental Sciences Division, Los Alamos National Laboratory, Los Alamos, NM 87545

[4]Department of Computer Science, North Carolina State University, Raleigh, NC 27695-8206

[5]Department of Civil, Construction, and Environmental Engineering, North Carolina State University, Raleigh, NC 27695-7908

[6]Mathematics and Computer Science Division, Argonne National Laboratory, Argonne, IL 60439-4844

E-mail: `rmills@ornl.gov`, `glenn.hammond@pnl.gov`, `lichtner@lanl.gov`, `vamsi_s@ncsu.edu`, `gmkumar@ncsu.edu`, `bsmith@mcs.anl.gov`

Abstract. We describe our experiences running PFLOTRAN—a code for simulation of coupled hydro-thermal-chemical processes in variably saturated, non-isothermal, porous media—on leadership-class supercomputers, including initial experiences running on the petaflop incarnation of Jaguar, the Cray XT5 at the National Center for Computational Sciences at Oak Ridge National Laboratory. PFLOTRAN utilizes fully implicit time-stepping and is built on top of the Portable, Extensible Toolkit for Scientific Computation (PETSc). We discuss some of the hurdles to "at scale" performance with PFLOTRAN and the progress we have made in overcoming them on leadership-class computer architectures.

1. Introduction

Over the past several decades, subsurface (groundwater) flow and transport models have become vital tools for the U.S. Department of Energy (DOE) in its environmental stewardship mission. These models have been employed to evaluate the impact of alternative energy sources and the efficacy of proposed remediation strategies for legacy waste sites. For years, traditional models—simulating single-phase groundwater flow and single-component solute transport within a single continuum, with basic chemical reactions such as aqueous complexing, mineral precipitation/dissolution and linear sorption to rock/soil surfaces and including radioactive decay—have been employed. Although these simplified groundwater models are still in wide use, advances in subsurface science have enabled the development of more sophisticated models that employ multiple fluid phases and chemical components coupled through a suite of biological and geochemical reactions at multiple scales. With this increased complexity, however, comes the need for more computational power, typically far beyond that of the average desktop computer. This is especially true when applying these sophisticated multiphase flow and multicomponent reactive transport models in three-dimensional problem domains. We contend, in fact, that in-depth understanding of many subsurface flow and transport problems will require simulations that demand petascale computing capabilities. In this paper, we briefly describe PFLOTRAN—a recently developed and highly-scalable code for simulations of reactive flows in geologic media—and discuss some of the challenges encountered and the progress made in scaling PFLOTRAN simulations towards the petascale on leadership class computer architectures.

2. Problem Formulation and Discretization

PFLOTRAN solves a coupled system of continuum scale mass and energy conservation equations in porous media for a number of phases, including air, water, and supercritical CO_2, and for multiple chemical components.

The general form of the multiphase partial differential equations solved in the flow module of PFLOTRAN for mass and energy conservation can be summarized as [1]:

$$\frac{\partial}{\partial t}\left(\phi \sum_{\alpha} s_\alpha \rho_\alpha X_i^\alpha\right) + \boldsymbol{\nabla} \cdot \sum_{\alpha} \left[\boldsymbol{q}_\alpha \rho_\alpha X_i^\alpha - \phi s_\alpha D_\alpha \rho_\alpha \boldsymbol{\nabla} X_i^\alpha\right] = Q_i, \tag{1a}$$

and

$$\frac{\partial}{\partial t}\left(\phi \sum_{\alpha} s_\alpha \rho_\alpha U_\alpha + (1-\phi)\rho_r c_r T\right) + \boldsymbol{\nabla} \cdot \left[\boldsymbol{q}_\alpha \rho_\alpha H_\alpha - \kappa \boldsymbol{\nabla} T\right] = Q_e. \tag{1b}$$

In these equations, α designates a phase (e.g. H_2O, supercritical CO_2), species are designated by the subscript i (e.g. $w = H_2O$, $c = CO_2$), ϕ denotes porosity of the geologic formation, s_α denotes the saturation state of the phase, X_i^α denotes the mole fraction of species i; ρ_α, H_α, U_α refer to the molar density, enthalpy, and internal energy of each fluid phase, respectively; $\boldsymbol{q}_\alpha$ denotes the Darcy flow rate defined by

$$\boldsymbol{q}_\alpha = -\frac{k k_\alpha}{\mu_\alpha} \boldsymbol{\nabla}(p_\alpha - W_\alpha \rho_\alpha g z), \tag{2}$$

where k refers to the water saturated permeability, k_α denotes the relative permeability, μ_α denotes the fluid viscosity, W_α denotes the formula weight, and g denotes the acceleration of gravity. The source/sink terms, Q_i^α and Q_e, describe injection and extraction of mass and heat at wells, respectively.

The multicomponent reactive transport equations solved by PFLOTRAN have the form [1]:

$$\frac{\partial}{\partial t}\left(\phi \sum_{\alpha} s_\alpha \Psi_j^\alpha\right) + \boldsymbol{\nabla} \cdot \sum_{\alpha} \boldsymbol{\Omega}_j^\alpha = -\sum_{m} \nu_{jm} I_m, \tag{3}$$

for the jth primary species, and

$$\frac{\partial \phi_m}{\partial t} = \overline{V}_m I_m, \tag{4}$$

for the mth mineral. The quantities Ψ_j^α, $\boldsymbol{\Omega}_j^\alpha$ denote the total concentration and flux of the jth primary species in phase α (see [1] for more details). The mineral precipitation/dissolution reaction rate I_m is determined using a transition state rate law, and the quantities ν_{jm} designate the stoichiometric reaction coefficients. These equations are coupled to the flow and energy conservation equations through the variable p, T, s_α, and $\boldsymbol{q}_\alpha$. In turn, chemical reactions may alter the material properties of the porous medium, such as porosity and permeability, which leads to coupling between the transport equations and flow and energy equations, albeit on a much slower time scale compared to the reverse coupling.

Although support for advanced discretizations is planned, PFLOTRAN currently utilizes a first-order finite volume spatial discretization combined with backward-Euler (fully implicit) time stepping. Upwinding is used for the advective term in the transport equations. The system of nonlinear equations arising from the discretization is solved using inexact Newton-Krylov methods. Within the flow and transport modules the equations are solved fully

Journal of Physics: Conference Series **180** (2009) 012062 doi:10.1088/1742-6596/180/1/012062

implicitly, but because transport generally requires much smaller time steps than flow, we couple these modules sequentially. A linear interpolation is used to obtain flow field variables within the transport solve. To account for changes in porosity and permeability due to mineral reactions, the transport solver is used to calculate an updated porosity over a flow time step and the revised porosity is passed back to the flow solver. Future implementations may explore independent grid hierarchies for flow and transport, as well as fully coupled schemes.

3. Architecture and Parallel Implementation

PFLOTRAN is has been written from the ground up with parallel scalability in mind, and can run on machines ranging from laptops to the largest massively parallel computer architectures. Through judicious use of Fortran 90/95 features, the code employs a highly modular, object-oriented design that can hide many of the details of the parallel implementation from the user, if desired. PFLOTRAN is built on top of the PETSc framework [2–4] and uses numerous features from PETSc including nonlinear solvers, linear solvers, sparse matrix data structures (both blocked and non-blocked matrices), vectors, constructs for the parallelism of PDEs on structured grids, options database (runtime control of solver options), and binary I/O.

PFLOTRAN's parallel paradigm is based on domain decomposition: each MPI process is assigned a subdomain of the problem and a parallel solve is implemented over all processors. Message passing (3D "halo exchange") is required at the subdomain boundaries with adjacent MPI processes to fill ghost points in order to compute flux terms. A number of different preconditioners from PETSc or other packages (PETSc provides interfaces to several) can be employed, but currently we usually use single-level domain decomposition preconditioners employed inside of a global inexact Newton-Krylov solver. Within the Krylov solver, gather/scatter operations are needed to handle off-processor vector elements in matrix-vector multiplies, and numerous *MPI_Allreduce()* operations are required to compute vector inner products and norms, making communication highly latency-bound.

4. Achieving Scalability on Leadership-Class Machines

In this section, we examine performance in a strong scaling context using a benchmark problem from a model of a hypothetical uranium plume at the Hanford 300 Area in southeastern Washington state, described in [5]. The computational domain of the problem measures $1350 \times 2500 \times 20$ meters (x,y,z) and utilizes complex stratigraphy (Figure 1) mapped from the Hanford EarthVision database [6], with material properties provided by [7]. The stratigraphy must be read from a large HDF5 file at initialization time. We examine aspects of both computation and I/O in this section.

In the benchmark runs here, we utilize an inexact Newton method with a fixed tolerance for the inner, linear solve. The linear (Jacobian) solves employ the `BCGS` stabilized bi-conjugate gradient solver (BiCGstab) in PETSc, with a block-Jacobi preconditioner that applies an incomplete LU-decomposition solver with zero fill-in (ILU(0)) on each block. (For linear solves with block-structured matrices, a block-ILU(0) solve is employed.) Our choice of preconditioner is a very simple one, but we have been surprised at how well it has worked at scale on large machines such as Jaguar. We have experimented with more sophisticated pre-

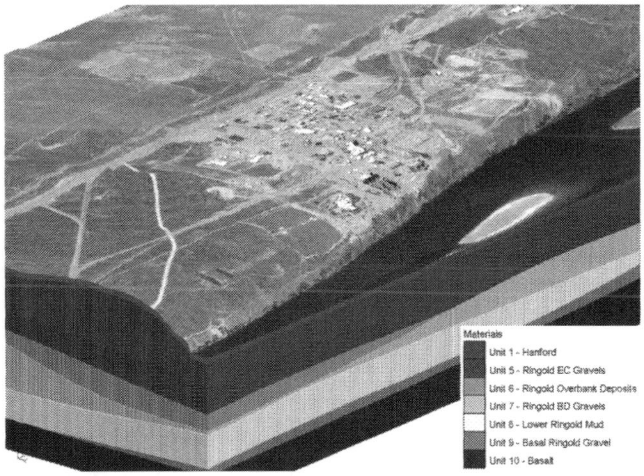

Figure 1. Hanford 300 Area stratigraphy (z scale = 20x, z axis ranges 70–130 meters).

conditioners such as the Hypre/BoomerAMG and Trilinos/ML algebraic multigrid solvers, but we have yet to identify a preconditioner that is as robust and as scalable at high processor counts as block-Jacobi. Solvers such as BoomerAMG display excellent convergence behavior, but show unacceptable growth in set up time above 1000 cores. (Determining how to make good use of multi-level solvers is an area of active research for us.)

4.1. Computational Phase

4.1.1. Strong-scaling, flow, 270M DoF Figure 2 depicts the strong scaling behavior of the flow solver for a $1350 \times 2500 \times 80$ cell flow-only version of the problem (270 million total degrees of freedom) run on the Cray XT4 at ORNL/NCCS. We ran the problem for 50 time steps. The scaling behavior is good all the way out to 27580 processor cores (near the limit of the number of cores available on the machine with 800 MHz DDR2 memory—we did not wish to introduce load imbalance by mixing memory speeds), although some departure from linear scaling is seen. Because we employ a single-level domain decomposition preconditioner (block-Jacobi), it might be supposed that the loss of effectiveness of the precondition as the number of processor cores grows (and hence the size of each local subdomain shrinks) might be responsible for a significant portion of the departure from linear scaling; the growth in the total number of BiCGstab iterations is actually very modest, however, with only approximately 10% more iterations required at 27580 cores than at 1024 cores.

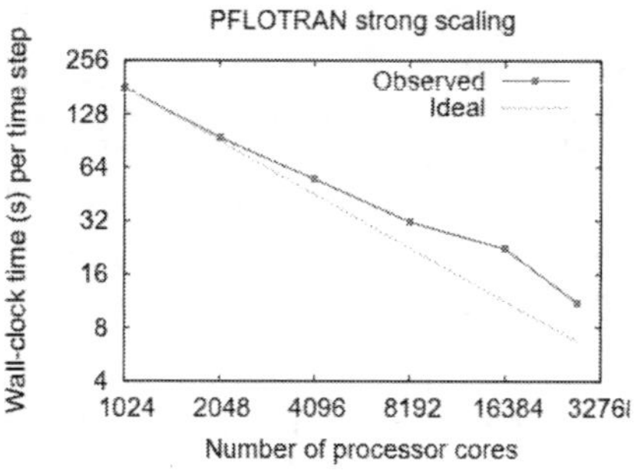

Figure 2. Strong-scaling performance on the Cray XT4 of the computation phase (no I/O) of PFLOTRAN for a variably-saturated flow problem with 270 million total degrees of freedom.

The departure from linear scaling is almost entirely due to the large growth in the proportion of time spent in the computation of vector inner products (dot products) and norms (which is dominated by the cost of `MPI_Allreduce()` operations): 10% of execution time is spent at 1024 cores, but this grows to 58% at 16384 cores. In fact, if we examine the strong-scaling behavior of the cost of vector dot products and norms versus the cost of all other function calls (Figure 3), we see that the strong scalability of the dot products and norms is quite poor, while the scalability of everything else is essentially linear (and even slightly superlinear at 16384 cores).

Because the time in the vector dot product and norm calculations is dominated by time spent in `MPI_Allreduce()` calls, the time spent in these calls was further investigated using CrayPAT, which breaks the time into two components: *MPI_Allreduce* and *MPI_Allreduce (Sync)*.

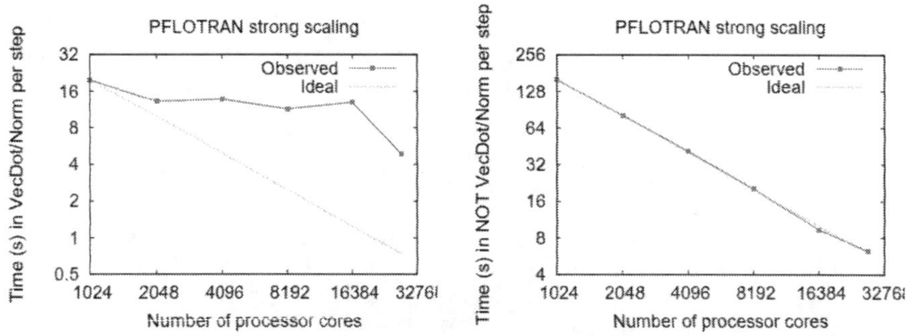

Figure 3. The left plot depicts the combined cost of all vector dot product or norm calculations for the strong-scaling case depicted in Figure 2. The right plot depicts the combined cost all routines that are not vector dot product or norm calculations.

According to Cray, *MPI_Allreduce (Sync)* is due to the implicit synchronization operation in the `MPI_Allreduce()` call and is likely due to load imbalance. However, further analysis of the code indicated that only 5% of the processors incur any significant application load imbalance. To investigate this further, we developed a perfectly load balanced dot product kernel that performs approximately the same number of `MPI_Allreduce()` calls as PFLOTRAN and takes approximately the same execution time (accomplished via redundant local dot products) as PFLOTRAN. This kernel yielded nearly the same amount of *MPI_Allreduce (Sync)* time, indicating that the *MPI_Allreduce (Sync)* experienced by the application is not due to application load imbalance but likely due to system load imbalance (i.e., performance variability among cores).

As the `MPI_Allreduce()` calls performed during the BiCGstab solve dominate the overall execution time at large core counts on the XT systems, we implemented an alternative, "improved" version of the algorithm [8] in a new `KSP` solver ("`IBCGS`") in PETSc. The standard BiCGstab method requires four `MPI_Allreduce()` calls per iteration (including the vector norm computation to check convergence). The restructured algorithm used in `IBCGS` requires only two `MPI_Allreduce()` calls, at the expense of a considerably more complicated algorithm, more local computation (additional vector operations), and computation of a transpose matrix-vector product at the beginning of the solve. By lagging the calculation of the residual norm, it is possible to further reduce communication and wrap everything into one `MPI_Allreduce` iteration, at the cost of doing one additional iteration. Despite the additional computation involved, the restructured solver is generally faster than the standard BiCGstab solver on the XT4/5 when using more than a few hundred processor cores. Table 1 displays the breakdown (as reported by CrayPAT) of time spent in *MPI_SYNC*, *MPI*, and *User* time when using the traditional ("BCGS") and restructured ("IBCGS") algorithms for a 30 time step run (including initialization time but no disk output) of the benchmark problem on 16384 cores of the XT4 version of Jaguar. The amount of time spent in *User* computations is larger in the IBCGS case, but the time spent in *MPI_SYNC* and *MPI* is far lower.

Group	Traditional	Restructured
MPI_SYNC	196	120
MPI	150	79
User	177	200

Table 1. Comparison of *MPI_SYNC*, *MPI*, and *User* time for the traditional and restructured BiCGstab implementations.

4.1.2. Strong-scaling, transport, 2B DoF We have recently made some preliminary benchmark runs on the petaflop incarnation of Jaguar, the Cray XT5 at ORNL/NCCS, with a coupled flow and transport problem from the Hanford 300 Area. The problem consists of 850×1000×160 cells and includes 15 chemical components, resulting 136 million total degrees of freedom in the flow solve and 2.04 billion total degrees of freedom in the transport solve. This problem size is at the limit of what we can do with 32-bit array indexing. (Although PFLOTRAN and PETSc can support 64-bit indices, some work will be required to get our code working with some of the libraries we link with.) The problem was run for 25 time steps of both flow and transport. In many simulations, we would use a longer time step for flow than for transport, but in this case there is a time varying boundary condition (the stage of the adjacent Columbia River) that limits the size of the flow time step (to a maximum of one hour), and higher resolution for the transport step is not needed.

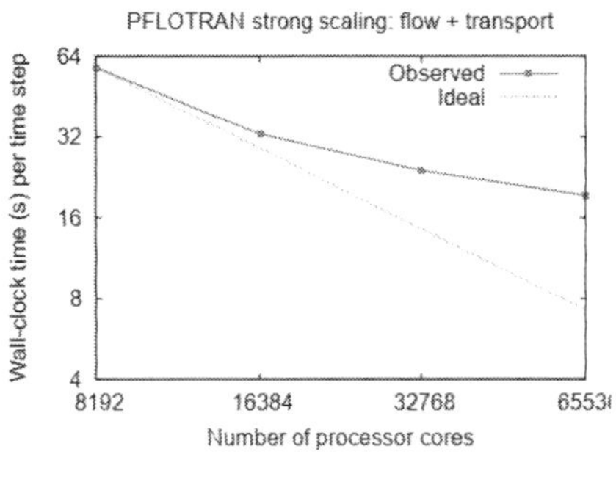

Figure 4. Combined strong-scaling performance of flow and transport solves for a coupled flow and reactive transport simulation on an $850 \times 1000 \times 160$ cell grid with 15 chemical components.

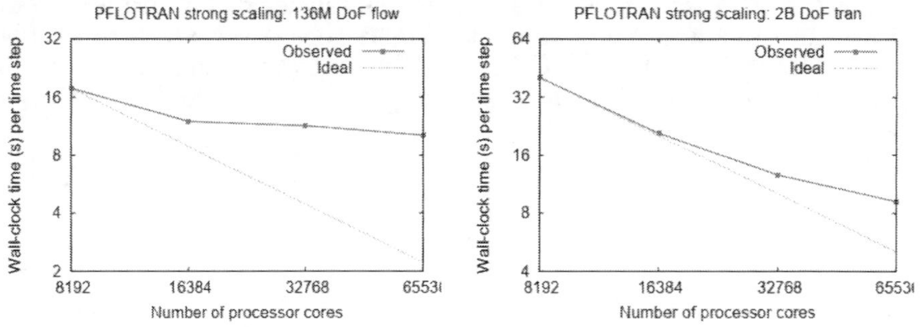

Figure 5. Strong-scaling performance of the flow solve component (136 million degrees of freedom, left) and the transport solve component (2.04 billion degrees of freedom, right) of the strong-scaling study depicted in figure 4.

Figure 4 depicts the preliminary scaling behavior we have observed on up to 65536 processor cores on Jaguar. Figure 5 breaks the performance down into time spent in the flow and transport solves, respectively. It is not surprising that the scalability of the flow problem is poor: at only 136 million degrees of freedom, this is a very small problem to be running on so many processor cores. We employ the large number of processor cores for the benefit of the much larger (15 times) and computationally more expensive transport problem. We note that the coupled flow and reactive transport problems that we might typically run with PFLOTRAN will generally involve transport problems considerably larger than their flow counterparts, perhaps going as high as 20 chemical components. If kinetic rate laws are used for sorption, then the size of the transport problem becomes even higher because a kinetic sorption equation (without flux terms) must be added for each kinetic reaction. It is possible that at some point it may make sense to solve the flow problem redundantly on disjoint groups of MPI processes, much as parallel multi-level solvers do for coarse-grid problems. This may not be necessary, however, as the cost of the transport solves may be dominant, and in many cases it is possible to run the flow solves with a much larger time step size than the transport solve.

We have not yet had the opportunity to conduct further experiments to understand and improve the performance of this benchmark problem. We believe that running it with the restructured BiCGstab solver in PETSc will be beneficial, but we first need to add support in PETSc for the transpose matrix-vector product for blocked sparse matrices with large block size. The data we collected during the run indicate that there are some significant sources of load imbalance that do not derive from the partitioning of work among the processors. (We believe that the partitioning is good, as this problem employs a regular grid and the ratio of maximum to minimum times spent in some key routines is not high even at 65536 cores; for example, it is 1.2 inside the Jacobian evaluation routines.) More detailed evaluation (running experiments with `MPI_Barrier()` calls inserted before collective communications, for example) is needed to identify the sources of load imbalance.

4.2. I/O Phase

Because PFLOTRAN spends relatively little time in I/O, serialized I/O through MPI process 0 is actually workable to about 8000 processor cores. To scale beyond this, we have added parallel I/O in the form of 1) routines employing parallel HDF5 for reading input files and writing out simulation output files, and 2) employing direct MPI-IO calls in a PETSc Viewer backend (for checkpointing). We found that straightforward use of collective HDF5 and MPI-IO calls greatly increased the scalability of the code, but initialization costs still became unacceptable when using more than about 16000 cores (roughly half the size of the Jaguar Cray XT4 system). Here we detail how we have mitigated this problem, focusing on performance on the Cray XT architecture, although the code changes that we describe actually improve performance on a wide range of architectures.

The Cray XT architecture uses the Lustre file system (lfs) [9] to facilitate high bandwidth I/O for applications. Lustre is an object based parallel file system that has 3 main components namely Object Storage Servers (OSS's), Meta Data Servers (MDS) and File system clients. Each OSS can serve multiple Object Storage Targets (OST's) which act as I/O servers, the MDS manage the names and directories for the file system and the file system clients are the Cray XT compute nodes. The Cray XT5 system has 672 OST's, 1 MDS and 37,544 file system clients or compute nodes. Files are broken into chunks and stored as objects on the OST's in a round robin fashion. The size of the object stored on a single OST is defined as File Stripe Size and the number of OST's along which the file is striped is defined as File Stripe Count. The default value for stripe size is 1MiB and stripe count is 4 OST's. We have used the default lfs settings for our runs but these parameters can be modified by the user with the help of lfs commands. Depending upon the file access pattern, the lfs settings will have an impact on I/O performance.

From the scaling studies performed on Cray XT, it was observed that the initialization phase of PFLOTRAN does not scale well with increase in processor count. The first curve in Figure 6 shows the timing of the initialization phase at different processor counts on Cray XT5. The initialization phase almost entirely consists of HDF5 routines which read from a single large HDF5 input file. All processes participate in a parallel read operation which involves a contiguous region of HDF5 datasets and there are two different read patterns i.e., every process reads the dataset entirely and every process reads a chunk of the dataset starting from its own offset within the dataset. For this kind of read pattern the lfs settings (stripe size and stripe count) will have little impact on performance because most of the processors will be accessing a single OST at any given point of time during execution.

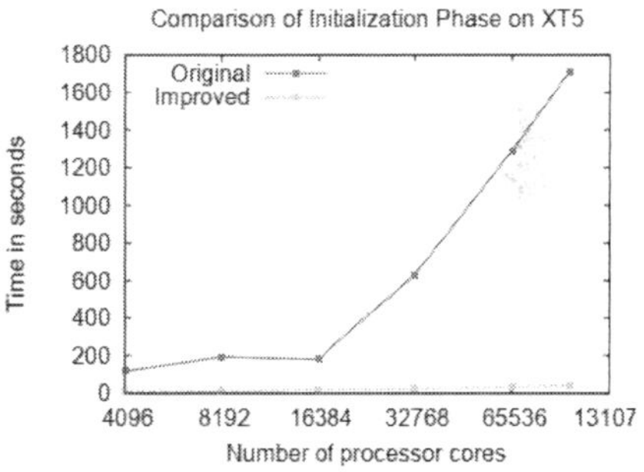

Figure 6. Performance comparison of the original and improved initialization routines on the 270 million degrees of freedom flow problem on the Cray XT5. The curve labeled "original" depicts the performance when all processes open and read from the HDF5 file, and the "improved" curve depicts performance when only a subset of processes read the data and then broadcast to the other members of their I/O group.

To investigate the reasons for the poor performance of the initialization phase at scale, we used CrayPAT to profile PFLOTRAN on up to 27572 cores of the Cray XT4. Figure 7 shows the percentage contribution of different routines to overall execution time. The profiling was done for the 270M DoF test problem with 30 time-steps and with no output. From the figure, it is evident that the MPI routines `MPI_File_open()`, `MPI_File_close()` and `MPI_File_read_at()` contribute significantly at higher processor count. It must be noted here that for the entire file system there is only 1 Meta data server (MDS) and every time a process needs to open/close a file it needs to poll the MDS. As the number of clients increases the file open/close become a much costlier operation and this setup would become a bottleneck for I/O performance. This has been identified as a problem for system scalability by others [10; 11].

To avoid this performance penalty at higher processor counts we have modified the read mechanism within PFLOTRAN. Instead of having every process participate in the read operations we implemented a mechanism where only a subset of processes execute the HDF5 routines and are involved in the read operations. Temporary buffers will have to be allocated at the reader processes to hold the data for their group. The reader processes collect the offset values and chunk sizes from the processes in their group by using `MPI_Gather()`. After the designated processes have completed reading the data from the file they would distribute the data using `MPI_Bcast()` or `MPI_Scatterv()` depending upon the offset values and chunk sizes to the rest of the processes in their respective groups.

By careful selection of reader processes, we ensure that there is no network congestion at this point. Even though we did not face out-of-memory problems with our runs, we can avoid this situation in the future, if it arises, by using a smaller group size or by using multiple `MPI_Bcast()/MPI_Scatterv()`.

With these modifications, we were able to reduce the time spent in initialization phase from 1700 to 40 seconds at 98304 cores of Cray XT5. The second curve in Figure 6 shows the timing of the initialization phase with the improved I/O method. For the runs up to 16384, we have used a group size of 1024 and for runs above 16384 it is 8192.

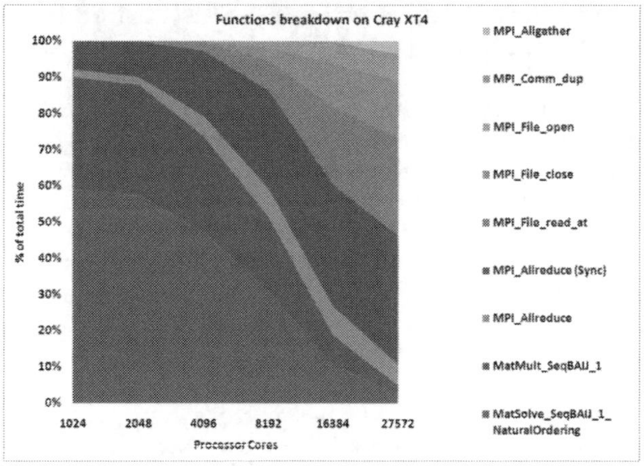

Figure 7. Breakdown of contributions of different routines to overall execution time of the 270 million degrees of freedom flow problem run on the Cray XT5 for 30 time steps, with no disk output.

5. High-fidelity Simulations of the Hanford 300 Area

PFLOTRAN is being used simulate a variety of phenomena, including geologic sequestration of CO_2 and radionuclide migration at several field sites. In one example of ongoing work, PFLOTRAN is being used to conduct high fidelity simulations of geochemical transport of U(VI) at the Hanford 300 Area, which lies in close proximity to the Columbia River in southeastern Washington State. From 1943–1975, a variety of waste from Manhattan Project and Cold War era activities was disposed of in process ponds within the site. Despite removal of highly contaminated pond bottom sediments, a U(VI) plume with concentrations well in excess of the EPA's maximum contaminant level for uranium exists in the area [12]. Previous work predicted that most of the uranium would be dissipated with a 10-15 year time period, which has proved not to be the case. PFLOTRAN is being used to understand the mobilization and transport of U(VI) at the 300 Area, which presents many modeling challenges. Aside from the many conceptual uncertainties in the problem such as the choice of initial conditions, rapid fluctuations in the Columbia River stage, which occur on an hourly basis with several meter variations, can have a dramatic impact on the size of the U(VI) plume, its migration direction, and the rate at which it migrates to the river. Due to the immense size of the physical domain needed to include the indispensable river boundary condition, the grid resolution required to preserve accuracy, and the large number of chemical components required to simulate U(VI) geochemistry, 3D simulation of the Hanford 300 Area is impractical without high-performance computing.

The simulations we are conducting are distinguished from other attempts to model U(VI) migration at the Hanford 300 Area by use of a high resolution 3D model, and a new formulation for the initial conditions imposed on the system in which the Hanford sediments are initially equilibrated with U(VI). Thus far, we have utilized a gridded domain measuring $850 \times 1000 \times 20$ meters (x, y, z) with 5 meter horizontal by 0.5 meter vertical resolution (1,360,000 grid cells, which yields 20,400,000 degrees of freedom when 15 primary aqueous species are employed). Initial conditions are generated by first running a non-sorbing simulation of U(VI) transport for several years to generate an initial U(VI) plume that reaches to the river. Using a snapshot of the flow and transport solution at that time, sorbed phase U(VI) concentrations are equilibrated with the aqueous phase to generate the initial conditions. Preliminary simulations (Figure 8) suggest that the slow release of U(VI) to the river is caused primarily by the rapid fluctuations in the river stage over time as well as slow leaching of uranium from the Hanford sediments. As river water infiltrates inland, U(VI) near the river bank is desorbed, but is strongly retarded, retaining U(VI) near the river.

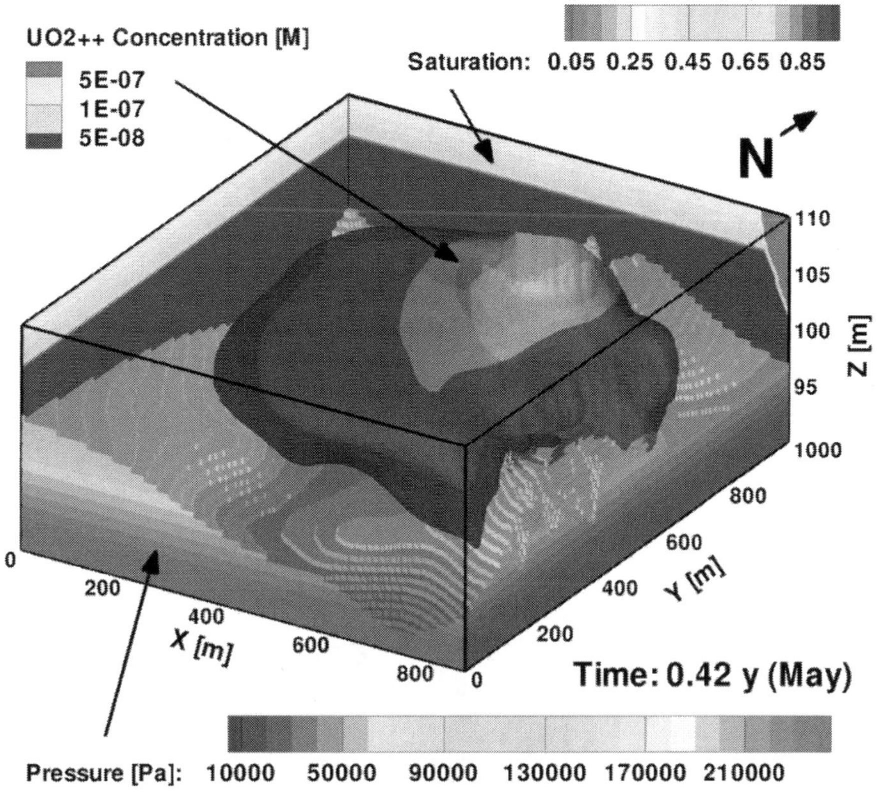

Figure 8. Isosurfaces of total aqueous U(VI) concentration in the May timeframe, when groundwater flow is primarily toward the river.

6. Conclusions and Future Work

PFLOTRAN has only been under active development for a few years, but has already become a modular, object-oriented, extensible, and scalable code. We have demonstrated that it is currently able to make effective use of a significant portion of a petascale machine. It does not do so with anywhere near the efficiency that we would like, but it is possible to use the code to solve leadership-class computing problems right now.

Further development of PFLOTRAN continues at a rapid pace. One focus is on developing better solvers/preconditioners to further improve scalability. We are exploring several types of multi-level approaches, which we believe are ultimately necessary to achieve good weak scalability. These approaches will probably be combined with physics based ones [13] that will allow multi-level methods to be employed on linear systems to which they are well-suited. Though we did not discuss it in this paper, we are also making progress on adding structured adaptive mesh refinement capabilities to the code, which will greatly decrease the memory and computational requirements for simulations that must resolve certain fine-scale features. We are also adding support for unstructured meshes, which will allow static refinement of the grid around complex geologic features, around wells, etc. Support for multiple-continuum (subgrid) models—necessary to more accurately model phenomena such as leaching of U(VI) in the source zone at the Hanford 300 Area—will also be added to the code. Incidentally, this will dramatically increase the work associated with each grid cell and result in more efficient use of leadership-class machines, as much of the subgrid work can proceed in embarrassingly parallel fashion.

Acknowledgements
This research was partially sponsored by the Climate and Environmental Sciences Division (CESD) of the Office of Biological and Environmental Research (BER) and the Computational Science Research and Partnerships (SciDAC) Division of the Office of Advanced Scientific Computing Research (ASCR) within the U.S. Department of Energy's Office of Science (SC). This research used resources of the National Center for Computational Sciences (NCCS) at Oak Ridge National Laboratory (ORNL), which is managed by UT-Battelle, LLC, for the U.S. Department of Energy under Contract No. DE-AC05-00OR22725. Pacific Northwest National Laboratory is managed for the U.S. Department of Energy by Battelle Memorial Institute under Contract No. DE-AC06-76RL01830. Argonne National Laboratory is managed by UChicago Argonne, LLC, for the U.S. Department of Energy under Contract No. DE-AC02-06CH11357. Los Alamos National Laboratory is managed is operated by Los Alamos National Security, LLC, for the National Nuclear Security Administration of the U.S. Department of Energy under contract DE-AC52-06NA25396.

References

[1] Lichtner P 1996 Continuum formulation of multicomponent-multiphase reactive transport *Reactive Transport in Porous Media* (*Reviews in Mineralogy* vol 34) ed Lichtner P, Steefel C and Oelkers E (Mineralogical Society of America) pp 1–81

[2] Balay S, Buschelman K, Gropp W D, Kaushik D, Knepley M G, McInnes L C, Smith B F and Zhang H 2009 PETSc Web page http://www.mcs.anl.gov/petsc

[3] Balay S, Buschelman K, Eijkhout V, Gropp W D, Kaushik D, Knepley M G, McInnes L C, Smith B F and Zhang H 2009 PETSc users manual Tech. Rep. ANL-95/11 - Revision 3.0.0 Argonne National Laboratory

[4] Balay S, Gropp W D, McInnes L C and Smith B F 1997 Efficient management of parallelism in object oriented numerical software libraries *Modern Software Tools in Scientific Computing* ed Arge E, Bruaset A M and Langtangen H P (Birkhäuser Press) pp 163–202

[5] Hammond G E, Lichtner P C, Mills R T and Lu C 2008 Towards petascale computing in geosciences: application to the hanford 300 area *SciDAC 2008 Scientific Discovery through Advanced Computing* (*Journal of Physics: Conference Series* vol 125) ed Keyes D (Seattle, Washington: IOP Publishing) p 012051

[6] Thorne P D, Bergeron M P and Freedman V L 2006 Groundwater data package for hanford assessments Report PNNL-14753 Rev. 1 Pacific Northwest National Laboratory, Richland, WA

[7] Williams M D, Rockhold M L, Thorne P D and Chen Y 2008 Three-dimensional groundwater models of the 300 area at the hanford site, washington state Report PNNL-17708 Pacific Northwest National Laboratory, Richland, WA

[8] Yand L T and Brent R 2002 The improved BiCGStab method for large and sparse unsymmetric linear systems on parallel distributed memory architectures *Proceedings of the Fifth International Conference on Algorithms and Architectures for Parallel Processing* (IEEE)

[9] Lustre file system Web page http://www.lustre.org

[10] Fahey M R, Larkin J M and Adams J 2008 I/o performance on a massively parallel cray XT3/XT4 *22nd IEEE International Symposium on Parallel and Distributed Processing, IPDPS 2008* (Miami, Florida) pp 1–12

[11] Yu W, Vetter J S and Oral S 2008 Performance characterization and optimization of parallel i/o on the cray XT *22nd IEEE International Symposium on Parallel and Distributed Processing, IPDPS 2008* (Miami, Florida) pp 1–11

[12] Zachara J M, Davis J A, Liu C, McKinley J P, Qafoku N, Wellman D M and Yabusaki S 2005 Uranimum geochemistry in vadose zone and aquifer sediments from the 300 Area uranium plume Report PNNL-15121 Pacific Northwest National Laboratory, Richland, WA

[13] Hammond G E, Valocchi A J and Lichtner P C 2005 *Advances in Water Resources* **28** 359–376

Massively parallel electrical conductivity imaging of the subsurface: applications to hydrocarbon exploration

Gregory A Newman and Michael Commer

Lawrence Berkeley National Laboratory, MS-90-1116, Berkeley CA, 94720

GANewman@lbl.gov

Abstract. Three-dimensional (3D) geophysical imaging is now receiving considerable attention for electrical conductivity mapping of potential offshore oil and gas reservoirs. The imaging technology employs controlled source electromagnetic (CSEM) and magnetotelluric (MT) fields and treats geological media exhibiting transverse anisotropy. Moreover when combined with established seismic methods, direct imaging of reservoir fluids is possible. Because of the size of the 3D conductivity imaging problem, strategies are required exploiting computational parallelism and optimal meshing. The algorithm thus developed has been shown to scale to tens of thousands of processors. In one imaging experiment, 32,768 tasks/processors on the IBM Watson Research Blue Gene/L supercomputer were successfully utilized. Over a 24 hour period we were able to image a large scale field data set that previously required over four months of processing time on distributed clusters based on Intel or AMD processors utilizing 1024 tasks on an InfiniBand fabric. Electrical conductivity imaging using massively parallel computational resources produces results that cannot be obtained otherwise and are consistent with timeframes required for practical exploration problems.

1. Introduction

Seismic imaging methods have a long and established history in hydrocarbon reservoir exploration. Yet the technology has encountered difficulty in discriminating different types of reservoir fluids, such as brines, oil, and gas. This limitation has led to the development of new geophysical technologies, specifically the use of low-frequency electromagnetic energy to complement seismic methods. In contrast to seismic data, electromagnetic measurements have shown to be highly sensitive to changes in fluid types and hence the location of hydrocarbons. Among such measurement techniques, the key emergent EM technology with respect to hydrocarbon exploration is controlled source electromagnetics (CSEM). CSEM utilizes low-frequency EM energy (less than 1 Hz) to map variations in the subsurface electrical conductivity, σ (in units of Siemens per meter, (S/m)) or its reciprocal ($1/\sigma$ in units of ohm-meters, (Ω.m)), called resistivity [5, 6, 4]. CSEM measurements are carried out using a deep-towed transmitter to excite low-frequency electromagnetic signals (< 1 Hz) that are sensitive to subsurface resistivity variations. The signals are measured on the sea floor by electric and magnetic field sensors/detectors with the largest transmitter-detector separations exceeding 15 km (figure 1). The method can interrogate down to 4 km depth, below the seabed. Another relevant EM technology employs natural field emissions below 0.1 Hz that arise from the interaction of the solar wind with the Earth's magnetosphere. Previously considered a source of noise when measuring CSEM fields, magnetotelluric (MT) fields complement CSEM measurements. Though not sensitive to oil bearing formations, MT fields are sensitive to electrical resistivity variations on a gross scale and interrogate much greater depths than CSEM data. A consensus is now emerging that MT data can significantly help in reducing uncertainty and ambiguity in interpreting CSEM data [2, 8].

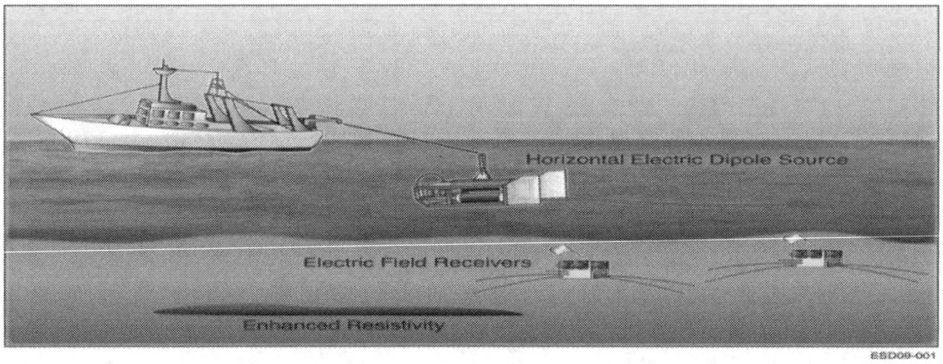

Figure 1. The CSEM technique senses regions of enhanced resistivity that can be associated with oil or gas deposits. This technique interrogates down to reservoir depths (as deep as 4 km beneath the ocean floor with the current technology). Sea bottom receivers are also used to collect natural field (MT) data.

Successfully extracting and processing the information from electromagnetic CSEM and MT data has proved up to now to be a formidable problem. The problem is especially significant in the search for hydrocarbon energy in highly complex *offshore* geological environments, where many of the world's oil and gas deposits remain to be found. Such offshore hydrocarbon exploration is an especially arduous task because reservoirs generally reside in highly complex geological environments, often beneath miles of ocean. Deep-water reservoirs are exceedingly difficult to successfully locate without recourse to imaging them and the background geology in three spatial dimensions (3D). To provide a maximally consistent electromagnetic data interpretation to geologists, such imaging requires large-scale modeling, spatially exhaustive survey coverage, and multicomponent data volumes. Moreover it is now recognized that the 3D conductivity imaging of marine CSEM data is strongly influenced by electrical anisotropy of geological media [9, 1, 7]. Failure to treat such effects in the imaging processes can produce misleading results. To effectively deal with the problem it is necessary to incorporate anisotropy within the 3D imaging framework. Here we discuss one such approach that treats transverse anisotropy, which appears to be relevant for many exploration scenarios.

The 3D imaging problem, which in this paper is also referred to as the inversion problem, usually has large computational demands, due to the computationally expensive solution of the forward modeling problem of EM field simulation on a given 3D finite-difference (FD) grid. It is described by a sparse linear system of equations that is solved using iterative Krylov methods. Such methods readily parallelize and are straight forward to implement. However, large data volumes require many forward solutions in an iterative inversion scheme. Therefore, we have developed an imaging algorithm that utilizes two levels of parallelization, one applied to the modeling, or imaging, volume and the other applied to the data volume. The algorithm is designed for arbitrarily large datasets, allowing for an arbitrarily large number of parallel tasks, while the computationally idle message passing is minimized. We have further incorporated an optimal meshing scheme that allows us to separate the imaging or modeling mesh from the simulation mesh [2]. This provides for significant acceleration of the 3D EM field simulation, having a direct impact on the time to solution for the 3D imaging process. Much of the details of the 3D imaging approach adopted in this paper have been published elsewhere for the isotropic media [2, 10]. Extension to treat media exhibiting transverse anisotropy is straightforward [9].

2. Minimization procedure

We seek to minimize the error functional

$$\phi = \tfrac{1}{2}\,\{\mathbf{D}_{CSEM}(\mathbf{d}^{p} - \mathbf{d}^{obs})\}^{\mathbf{T}*}\{\mathbf{D}_{CSEM}\,(\mathbf{d}^{p} - \mathbf{d}^{obs})\}(\alpha) + \tfrac{1}{2}\,\{\mathbf{D}_{MT}(\mathbf{Z}^{p} - \mathbf{Z}^{obs})\}^{\mathbf{T}*}\{\mathbf{D}_{MT}\,(\mathbf{Z}^{p} - \mathbf{Z}^{obs})\}(\beta)$$
$$+ \tfrac{1}{2}\,\lambda_h\,\{\mathbf{W}\mathbf{m}_h\}^{T}\{\mathbf{W}\mathbf{m}_h\} + \tfrac{1}{2}\,\lambda_v\,\{\mathbf{W}\mathbf{m}_v\}^{T}\{\mathbf{W}\mathbf{m}_v \qquad (1)$$

$\mathbf{T}^*$ denotes the transpose-conjugation operator and $\mathbf{d}^{obs}$ and $\mathbf{d}^{p}$ the observed and predicted CSEM data, consisting of the complex valued electric and magnetic fields measured at the detectors. MT predicted and observed data are in the form of impedances, $\mathbf{Z}^{p}$ and $\mathbf{Z}^{obs}$. Diagonal weighting matrices, $\mathbf{D}_{CSEM}$ and $\mathbf{D}_{MT}$ are incorporated into the error functional to help compensate for noisy measurements. A successful outcome for imaging joint data also requires careful weighting between the CSEM and MT data types, otherwise one data type can predominate in the imaging process. Here we introduce the weighting coefficients α and β so that each data type makes a meaningful contribution in the imaging process. Our strategy for selecting α is based upon testing several trial values. Only a few inversion iterations are necessary to determine if the data sets are appropriately balanced. Other approaches to this problem can be found in [2]. Stabilization terms also appear in (1) and are designed to treat media exhibiting transverse electrical anisotropy. Parameterization of anisotropic conductivity is made on a Cartesian mesh, where horizontal and vertical values are assigned to cell in the mesh. Stabilization is achieved by reducing the model curvature in three dimensions in the minimization process. To do this we employ a FD approximation to the Laplacian (∇^2) producing a roughening matrix **W**. **W** acts on both the horizontal and vertical conductivity values $\mathbf{m}_h$ and $\mathbf{m}_v$, which are bounded using log or hyperbolic transformations. The regularization parameters, λ_h and λ_h control the amount of smoothing admitted into the model. Choice of the regularization parameters is dictated by the data noise and is optimally carried out using a cooling approach. We refer the reader to [10] for additional details. Minimization of (1) is carried out using a non-linear conjugate gradient scheme, ideal for large scale data and imaging volumes.

3. Imaging examples

The Campos basin, located off shore of Brazil is a known oil and gas province with ongoing production. In 2004, a 3D CSEM survey was carried out to better quantify the hydrocarbon potential over part of the basin. The sail lines on a 40 x 40 km^2 grid used for subsurface conductivity mapping are shown at the top of Figure 2. Data was collected from nearly 1 million transmitter sites along the sail lines using 23 sea bottom detectors. Obviously, this amount of data cannot be treated with the current inversion methodology, even with a massively parallel implementation. Every source treated by the imaging algorithm requires a forward simulation, an adjoint computation, as well as two or more additional simulations for step control for each nonlinear inversion update. To efficiently deal with this data volume, we use a general reciprocity principle that involves the interchange of transmitter and receiver points. Hence, the positions of the actual CSEM transmitter along the sail line become the computational receiver profiles, and the actual CSEM detectors on the seafloor become computational sources. The equivalent reciprocal problem involves 951,423 data points and 207 effective sources, since there are 23 source locations with three polarizations and each operating at the three discrete excitation frequencies of 0.125 Hz, 0.25 Hz, and 0.5 Hz. Each effective transmitter is polarized according to the antenna orientation of its corresponding detector.

Analysis of the Campos Basin data without taking anisotropy into account produced serious image artifacts [3], where the data were analyzed using 32,768 tasks/processors on the IBM Watson Research Blue Gene/L supercomputer over a 24 hour period. Even though the imaging result was deemed an initial failure it pointed to the correct approach to analyzing the data. Subsequent data analysis by [1] incorporating formation anisotropy produced interpretable results (figure 2). It is also important to stress the need for fast processing times in imaging CSEM data in three dimensions. Initial imaging experiments required over four months of processing time on distributed clusters based on Intel or AMD processors utilizing 1024 tasks on an InfiniBand fabric.

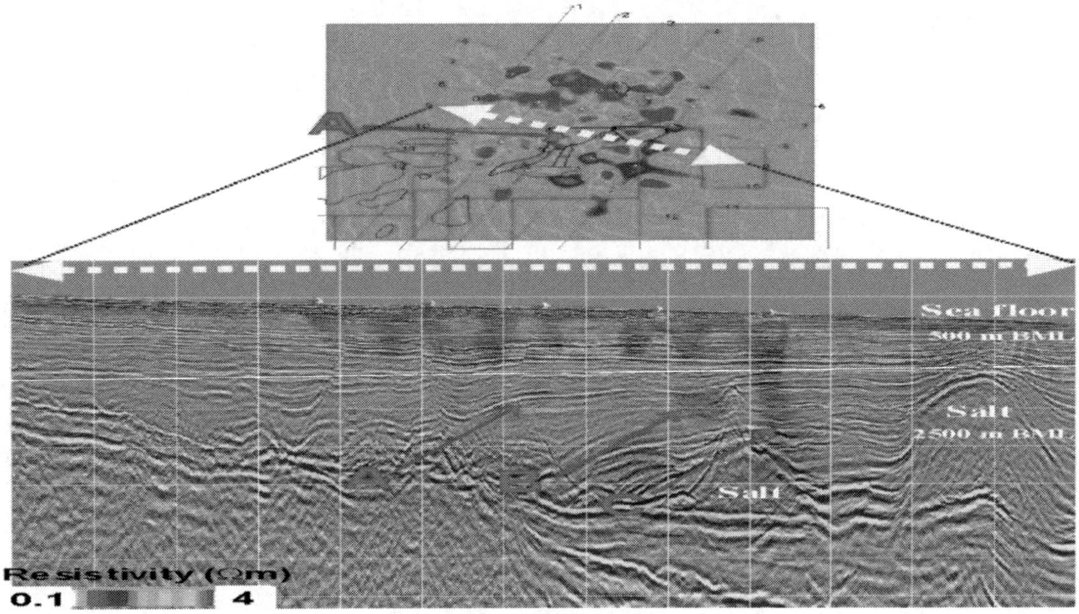

Figure 2. Rendered at the top is the average vertical resistivity map from 500 to 2500 m below the seafloor, superimposed with sail lines used to acquire the Campos Basin data. The cross-section at the bottom shows the vertical resistivity image along the indicated transect. The Campos Basin experiment demonstrated the necessity to incorporate electrical anisotropy into the imaging processes for accurate results. The CSEM image is shown together with seismic reflection horizons. Anomaly A is related to enhanced resistivity due to a known oil field. Anomaly B is enhanced resistivity and may indicate a possible hydrocarbon trap above a large salt body. Enhanced conductivity at C is likely to be related to conductive brines originating from salt below. Results presented by [1].

An important exploration problem to demonstrate the advantages of joint conductivity inversion of CSEM and MT data is the imaging of oil bearing horizons in the presence of sub salt structures [2]. The geometries of the reservoirs and salt structures are exceedingly difficult to map without recourse to 3D imaging. Such structures are encountered in the Gulf of Mexico, where seismic imaging beneath salt can be a formidable task. For oil bearing horizons above salt, the situation is better, but we will show that such structures can be identified much better under a joint CSEM/MT imaging framework.

In this synthetic imaging example, the data consists of 143 MT stations spread over a 25×25 km^2 grid with a 2.5 km sampling interval. Each of the two impedance tensor elements Z_{xy} and Z_{yx} is measured for 13 MT frequencies, ranging (logarithmically) from $5 \cdot 10^{-4} - 0.125$ Hz. The grid of CSEM detectors is a sub grid of the MT station grid, with 63 locations comprising an area of 20×15 km^2. As mentioned previously, it is common to treat marine CSEM data in a reciprocal way, owing to the enormous data volumes generated with a continuously moving transmitter towed by the vessel. Thus with two CSEM frequencies, 0.25 and 0.75 Hz, we simulate a total of 126 sources, each with a length of 200 m. A total of 6468 in-line receivers, with a length of 100 m, are evenly spread over the actual CSEM sail lines. These lines are 50 m above the sea floor. Counting in-phase and quadrature components of the complex data, the total number of CSEM and MT data points are 12,936 and 7436, respectively. More details regarding the data acquisition geometry and model can be found in [2].

The image results of the separate CSEM and MT inversions are shown in figure 3 and clearly illustrate the different degrees of resolution achieved by either method. While the oil bearing horizon is indicated by the CSEM image, yet without a clear delineation of the true shape, the MT data is only sensitive to the large salt bodies. However the MT image shows salt body conductivities which are generally above the true values. A great improvement is achieved by the joint inversion (d), both in terms of delineation of the reservoir and salt bodies, as well as in reproducing the true conductivities.

Note also that the depth of the salt bodies is reproduced to a fairly good degree. To image the different scenarios, 7,875 tasks were employed on Franklin (Cray XT4) system at National Energy Research Scientific Computing Center (NERSC). Processing time for the different imaging experiments varied between 5 and 9 hours.

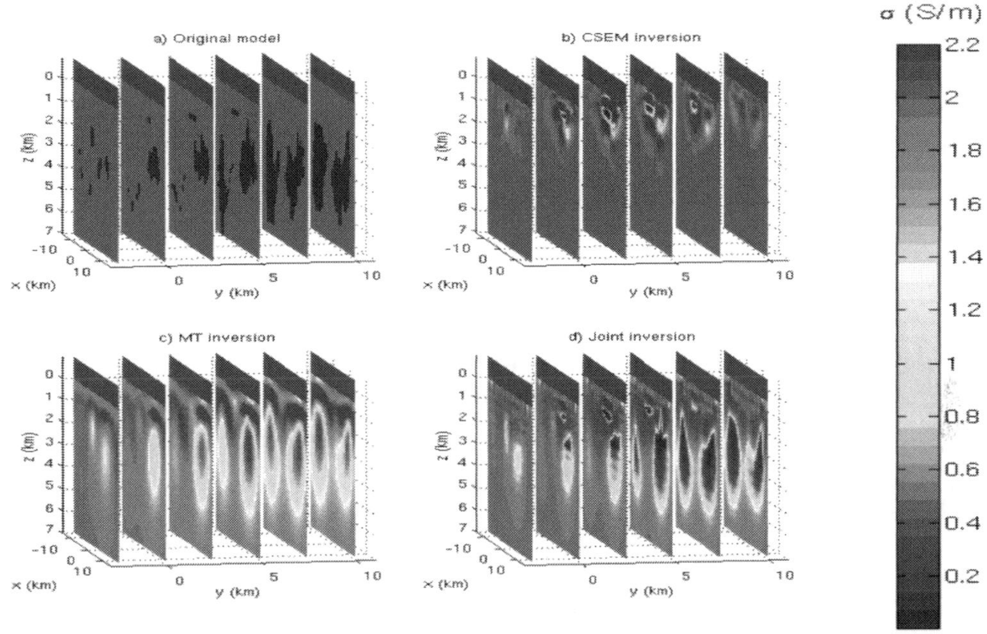

Figure 3. Marine prospecting study on joint CSEM and MT imaging showing original model (a), CSEM inversion (b), MT inversion (c), and joint inversion (d). Images are rendered in electrical conductivity.

Acknowledgements
This work was carried out at Lawrence Berkeley National Laboratory, with base funding provided by the United States Department of Energy, Office of Basic Energy Sciences, under contract DE-AC02-05CH11231. Additional funding and support was provided by the ExxonMobil Corporation.

References

[1] Carazzone J. J., Dickens T. A., Green, K. E., Jing, C., Wahrmund, L. A., Willen, D. E., Commer M., and Newman G. A., 2008, Inversion Study of a Large Marine CSEM Survey: Society of Exploration Geophysicists Expanded Abstracts, **27**, 644-647.

[2] Commer M., and Newman, G. A., 2009, Three-dimensional controlled-source electromagnetic and magnetotelluric joint inversion: Geophysical Journal International, in press.

[3] Commer M., Newman G. A., Carazzone J. J., Dickens T. A., Green K. E., Wahrmund L. A., Willen, D. E., Shiu J., 2008, IBM J. Res & Dev NO. ½, January/March 2008.

[4] Constable, S., 2006, Marine electromagnetic methods – A new tool for offshore exploration: The Leading Edge **25**, 438-444.

[5] Eidesmo T., Ellingsrud S., MacGregor L. M., Constable S., Sinha M. C, Johansen, S., Kong, F. N., Westerdahl, 2002, Sea Bed Logging (SBL), a new method for remote and direct identification of hydrocarbon filled layers in deepwater areas, First Break, 20.3, 144-152.

[6] Ellingsrud S., Eidesmo T., Johansen S., Sinha M. C., MacGregor, L. M, Constable, S., 2002, Remore sensing of hydrocarbon layers by seabed logging (SBL): Results from a cruise offshore Angola: The Leading Edge, **21**, 972-982.

[7] Jing, C., K. E. Green, and D. Willen, 2008, CSEM inversion: Impact of anisotropy, data

coverage, and initial models: 78th Annual International Meeting, SEG, Expanded Abstracts, 604-607.

[8] Mackie, R. M., Watts, M. D. & Rodi, W., 2007. Joint 3D inversion of marine CSEM and MT data, *SEG Techn. Prg. Exp. Abstr.* **26**, 574-578.

[9] Newman, G. A. and Commer, M., 2009, Imaging CSEM data in the presence of electrical anisotropy: submitted Geophysics.

[10] Newman, G. A., and Boggs, P. T., 2005, Solution accelerators for large-scale three-dimensional electromagnetic inverse problems: Inverse Problems, **20**, S151-S170.

Developing a component-based framework for subsurface simulation using the Common Component Architecture

Bruce Palmer[1], Yilin Fang[1], Vidhya Gurumoorthi[1], Glenn Hammond[1], James Fort[1] and Tim Scheibe[1]

Pacific Northwest National Laboratory,
PO Box 999 Richland WA 99352

E-mail: bruce.palmer@pnl.gov

Abstract. This paper will provide a brief overview of two frameworks being developed for subsurface simulations. The first is based on the Smoothed Particle Hydrodynamics algorithm and is designed to model flow and transport at the pore scale. The second is based on the Subsurface Transport Over Multiple Phases subsurface continuum code (STOMP) and is designed to simulate flow at the Darcy scale. Both frameworks have been built using the Common Component Architecture toolkit. These frameworks will eventually be combined into a single framework for performing hybrid multiscale simulations that seamlessly integrate both the particle and continuum simulations together.

1. Introduction

The increasing complexity of computer simulations, coupled with continuing changes in libraries, algorithms, computer architectures, and operating systems, is making the process of developing and maintaining code to perform high-end computations increasingly difficult. The traditional model of monolithic code development, in which all aspects of programming are handled by a relatively small and homogeneous group of programmers, is breaking down in the face of increasingly complex simulation requirements related to communication, I/O, linear algebra, grid management, visualization, solvers, etc. This paper will describe the decomposition of two codes into components using the Common Component Architecture toolkit[1]. This includes a Smoothed Particle Hydrodynamics (SPH) code, based on a Lagrangian formulation of the hydrodynamic equations [2], that is used for pore scale simulations of fluid flow in topologically complex configurations and a Darcy-scale continuum code used to simulate subsurface flow using finite volume numerical techniques[3,4]. The long term goal of this project is to couple the Lagrangian and continuum approaches into a single hybrid multiscale simulation of subsurface flow.

The components themselves are patterned after principles developed in object-oriented programming. Individual components represent a collection of subroutines or functions along with some internal data representing state. Data is communicated between components using explicit calls, no data is shared between components by sharing common data structures. The CCA toolkit itself supports this encapsulation, as well as providing other user services such as runtime configurability and language interoperability[5].

2. Reducing dependencies between components

A key feature of our strategy for creating components with minimal dependencies between them is to use a data dictionary and associated metadata to allow individual components to identify necessary data from a common data pool without explicit exchanges of information. At the start of the simulation, all components that generate data arrays that might be used by other components allocate memory for these arrays and deposit pointers to them in a data manager, along with associated meta-data. The meta-data consists of attributes that allow other components to determine whether the data is something that they can use. The combination of a data pointer and associated meta-data comprises an entry into the data dictionary. All entries in the data dictionary are interpreted in the same way by components in the application. For example, a data element representing the x velocities of particles located on a processor will be interpreted as such by all components. Some components, such as the component responsible for performing the time integration of particle coordinates, may recognize the x velocity as something it particularly needs to execute its function, but other components, such as the component for exporting a system snapshot, may treat the x velocity as just another particle property that should be included in the output.

A second mechanism for reducing dependencies is to provide input to each component directly from the user, instead of having all the input directives read by a single "driver" program and then delivering them to other components by passing them through the call stack. Alternatively, the driver can store the directives in globally accessible variables, but both methods can lead to dependencies between otherwise unrelated components. Having each component read its input directly from an external source eliminates these dependencies, although it may be cumbersome to require multiple independent files to run a single simulation. To simplify the input and allow users to include all input directives in a single file, a "read input" component was created that reads an input file that has been divided up into blocks with each block directed at a single component. The individual blocks have the form

```
input component1
   ....
   ....
end
```

Component 1 can retrieve the input data between the delimiters "`input`" and "`end`" by making a request to the input component for the block of input text corresponding to "`component1`". This text block can then be parsed using whatever mechanisms the author of component 1 wishes to use. The input supplies a number of functions that can be useful for parsing input, but if other code already exists for reading the input, it can be used instead. The input component itself is fairly neutral with respect to input formats and, apart from requiring the delimiters `input` and `end`, supports any kind of input (e.g. keyword-value, formatted, etc.).

The elimination of dependencies using these two concepts is best illustrated by looking at the remapping component in the SPH framework. The remapping component starts with data associated with particles, which may be randomly distributed in space, and interpolates the data onto a regular grid so that it can be exported in a format used by standard visualization software. To execute the remapping function, the component needs the x, y, z coordinates of all particles and it gets these from the data manager by looking for entries called '`xcoord`', '`ycoord`', '`zcoord`'. These arrays are *required* by the remapping component in order for it to work so it looks for them specifically in the data manager. Everything else that the remapping component uses is specified by the user. The user may specify in the remap input block that they want to visualize the velocity vector with components '`xvel`', '`yvel`', and '`zvel`' as well as the concentration field '`conc`'. After parsing the input block and extracting these names, the remap component will then check the data manager to see if there are fields with these names available. The remap component then caches pointers to these fields and will write the corresponding data to a visualization file whenever called upon to do so.

Most components contain an initialization method, which is called at calculation startup and allows each component to find the data it will use in the remaining simulation. Most components also contain a "do something" method that causes the component to execute its function whenever it is called. In the case of the remap component, this method causes the remap component to interpolate the user specified data onto a regular grid and then write the values out to a file for subsequent analysis and visualization.

3. The Smoothed Particle Hydrodynamics framework

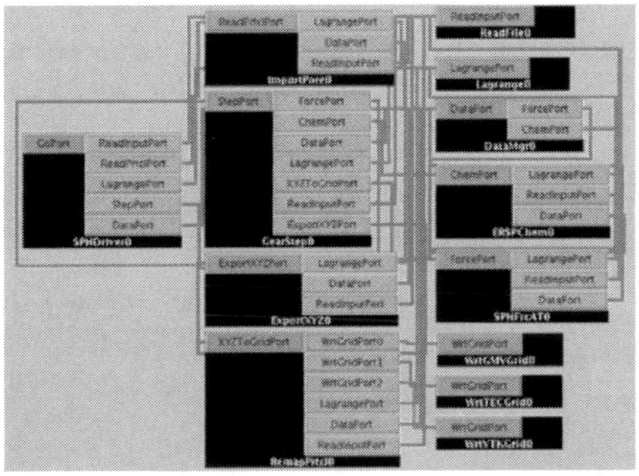

Figure 1. Schematic diagram of SPH application built from components.

The SPH framework is fairly well-developed and supports considerable functionality. It is also mature enough that new physics and chemistry can be added to it by focussing on modifying only a few routines. The remainder of the components in the framework can remain untouched. A "wiring diagram" of an SPH application is shown in Figure 1.

The diagram shows a driver component at the left that is responsible for controlling overall program flow, a set of components in the middle that are responsible for reading and writing particle configurations and integrating the particle equations of motion and a variety of components on the right that provide support functions and define the model physics and chemistry. The Lagrange component is the communication layer used to parallelize the SPH application. It is built with the Global Arrays communication library[6] and is used by a large number of the other components, the ReadFile and DataMgr components supply the input and data manager capabilities described above, and the ERSPChem and the SPHFrcAt components are responsible for implementing the chemistry and physics models that the application is simulating. Adding new models to the framework should only require modifications or new versions of the force and chemistry components.

An example of a simulation performed using the SPH framework is shown in Figure 2. This simulation contains approximately 14 million particles and represents transport of a non-reactive contaminant through a porous medium generated from a Monte Carlo simulation of 250 hard spheres. The figure shows an isosurface of concentration after the contaminant has traversed approximately halfway through the simulation cell (flow is in the z-direction). These simulations are being used to determine the effective transport constants for contaminant transport in subsurface simulations. Additional models that are under development include a simulation of uranium transport in porous media that includes the effects of intragranular diffusion.

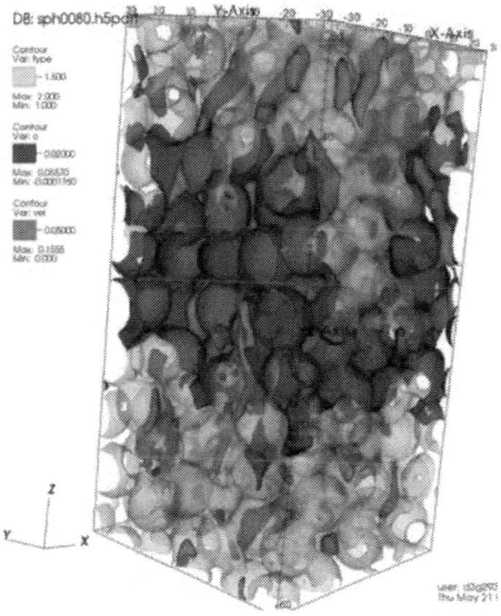

Figure 2. Simulation of non-reactive contaminant transport using SPH framework. Porous media surface is in green, isosurfaces of velocity are red and isosurfaces of concentration are blue. The visualization is done using the VisIt parallel software package.

4. The STOMP framework

Work has also been underway to develop a component version of the Subsurface Transport Over Multiple Phases (STOMP) subsurface simulation code[3,4]. This is a continuum code designed to perform simulations at the Darcy scale. The degree of componentization is not nearly as advanced as for the SPH framework, but a major objective has been achieved by splitting out the grid from the rest of the STOMP code as a separate component. At present, the STOMP framework contains only two components. One is the grid component and the second is a physics component that contains the rest of the STOMP code.

The grid component contains much of the input associated with problem setup and is responsible for creating the grid and adding all the fields required by the simulation to the grid. The grid component supports an unstructured syntax, although currently it is only implemented for a structured grid. The data mapping block refers to the necessity of mapping from a structured grid to an unstructured syntax. At some point in the future, if the grid component is reimplemented using an unstructured grid library, it should be possible to plug this into the existing framework with only modest modifications to the remaining components. Several additional components in the remaining STOMP code have been identified and work is underway to separate these from the main code. They include components for physics, chemistry, solvers, input, and output. The existing STOMP framework with possible future components is illustrated schematically in Figure 3.

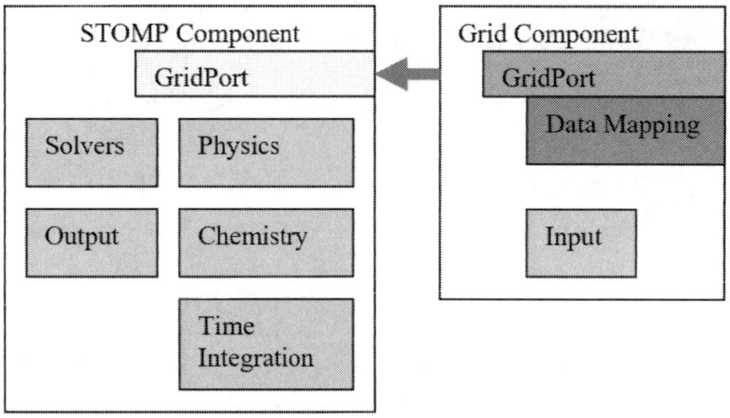

Figure 3. Schematic diagram of STOMP framework. Green boxes indicate future components that are currently contained within existing components.

A simulation of contaminant transport at the Hanford IFRC site performed with the existing STOMP framework is shown in Figure 4. The figure shows the migration of a plume through an array of monitoring wells.

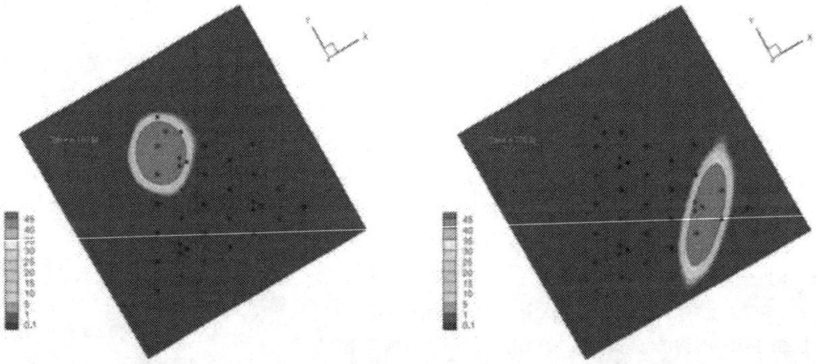

Figure 4. STOMP simulation of contaminant transport through an array of monitoring wells at the Hanford IFRC site.

Although the existing STOMP framework does not make use of an explicit data manager, the grid component can act in a similar way and many of the data concepts outlined above have been incorporated into it.

5. Summary

The frameworks described above demonstrate the feasibility of creating component-based applications that have a high degree of functional encapsulation, minimal dependencies between components, and still maintain high performance. An example of how this can increase programmer productivity is the development in the SPH framework of multiple force and input components that can be interchanged with each other without modifications to other components. The frameworks were implemented using the CCA toolkit, which allows users to configure the application at runtime and supports language interoperability between components. The SPH framework has exploited this feature by having components written in F90, C, and C++.

The componentization of the individual SPH and STOMP frameworks is expected to support the next stage of this project, which is to integrate these two simulation approaches into a single hybrid multiscale simulation that couples a pore scale description using the SPH algorithm with a Darcy scale simulation using STOMP. This will require extensions to the existing data model and a more robust version of the data manager, as well as development of appropriate coupling components that will link the two models together into a single application.

Acknowledgments

The authors are indebted to the CCA development team for its support and help on issues related to building and using the CCA toolkit as well as to Karen Schuchardt for useful discussions. This research is supported by the U. S. Department of Energy's Office of Science under the Scientific Discovery through Advanced Computing (SciDAC) program. A portion of this research was performed using the Molecular Science Computing Facility at EMSL, a national scientific user facility sponsored by the Department of Energy's Office of Biological and Environmental Research and located at Pacific Northwest National Laboratory.

References

[1] Allan BA, R Armstrong, DE Bernholdt, F Bertrand, K Chiu, TL Dahlgren, K Damevski, WR Elwasif, TGW Epperly, M Govindaraju, DS Katz, JA Kohl, M Krishnan, JW Larson, S Lefantzi, MJ Lewis, AD Malony, LC McInnes, J Nieplocha, B Norris, SG Parker, J Ray, S Shende, TL Windus, SJ Zhou. (2006) A Component Architecture for High-Performance Scientific Computing *Int. J. High Perf. Comput. Applications* **20** 163-202.

[2] Monaghan, JJ. (2005) *Smoothed particle hydrodynamics Rep. Prog. Phys.* **68** 1703-1759.

[3] White, M. D., and M. Oostrom (2000), STOMP Subsurface Transport Over Multiple Phases Version 2.0 Theory Guide, PNNL-12030, 235 pp., Pacific Northwest Natl. Lab., Richland, Wash.

[4] White, M. D., and M. Oostrom (2006), STOMP Subsurface Transport Over Multiple Phases Version 4.0 User's Guide, PNNL-15782, 120 pp., Pacific Northwest Natl. Lab., Richland, Wash.

[5] Kohn, S, G Kumfert, J Painter, C Ribbens, Divorcing Language Dependencies from a Scientific Software Library, 10th SIAM Conference on Parallel Processing, Portsmouth, VA, March 12-14, 2001.

[6] Nieplocha, J, B Palmer, V Tipparaju, M Krishnan, H Trease, E Apra. (2006) Advances, Applications and Performance of the Global Arrays Shared Memory Programming Toolkit *Int. J. High Perf. Comput. Applications* **20** 203-231.

Journal of Physics: Conference Series **180** (2009) 012065 doi:10.1088/1742-6596/180/1/012065

Application of the SALSSA framework to the validation of smoothed particle hydrodynamics simulations of low Reynolds number flows

Karen Schuchardt[1], Jared Chase[1], Jeff Daily[1], Todd Elsethagen[1], Bruce Palmer[1] and Tim Scheibe[1]

[1]Pacific Northwest National Laboratory, PO Box 999 Richland WA 99352

E-mail: karen.schuchardt@pnl.gov

Abstract. The Support Architecture for Large-Scale Subsurface Analysis (SALSSA) provides an extensible framework, sophisticated graphical user interface (GUI), and underlying data management system that simplifies the process of running subsurface models, tracking provenance information, and analyzing the model results. The SALSSA software framework is currently being applied to validating the Smoothed Particle Hydrodynamics (SPH) model. SPH is a three-dimensional model of flow and transport in porous media at the pore scale. Because fluid flow in porous media at velocities common in natural porous media occur at low Reynolds numbers, it is important to verify that the SPH model is producing accurate flow solutions in this regime. Validating SPH requires performing a series of simulations and comparing these simulation flow solutions to analytical results or numerical results using other methods. This validation study has been greatly aided by the application of the SALSSA framework, which provides capabilities to setup, execute, analyze, and administer these SPH simulations.

1. Introduction

Many subsurface flow and transport problems of importance today involve coupled non-linear flow, transport, and reaction in media exhibiting complex heterogeneity. In particular, problems involving biological mediation of reactions and/or mineral precipitation/dissolution fall into this class of problems. We are developing a hybrid multi-scale modeling framework that integrates models with diverse representations of physics, chemistry, and biology across different scales (sub-pore, pore and continuum). The modeling framework takes advantage of advanced computational technologies including parallel code components such as parallel solvers and gridding. Components of the modeling framework are being constructed using the Common Component Architecture (CCA) toolkit [1]. As part of that effort, we have developed a three-dimensional model of flow and transport in porous media at the pore scale — that is, where the geometry of solid grains and pore spaces is explicitly represented. The numerical method used to solve the flow and transport equations is the mesh-free particle method called Smoothed Particle Hydrodynamics (SPH)[2][3]. The code can accommodate large three-dimensional (3D) simulations with arbitrarily complex pore geometry. However, the complexity of these systems presents challenges for code validation and verification. Therefore, a series of simpler 3D simulations, designed for verification of the SPH solution of the Navier-Stokes equations at low Reynolds numbers typical of porous media flow is currently under way. The *Support Architecture for Large-scale Subsurface Simulation and Analysis* (SALSSA) software is the second component of this effort, addressing the macro modeling processes such as running simulations and analyzing results. SALSSA integrates workflow, data management, and visualization in support of model development and validation activities. In this paper, we describe the application of the SALSSA software to the SPH validation study and present the features and capabilities of SALSSA in this context.

2. Validation of low Reynolds number flows

Fluid flow in porous media at velocities common in natural porous media (e.g., groundwater flow) occur at low Reynolds numbers and therefore it is important to verify that the SPH model is producing accurate flow solutions in this regime. The SPH model contains parameters that do not appear in the corresponding incompressible Navier-Stokes equations and it is important to identify value ranges for these parameters that lead to good solutions for the flow profiles in topologically complex media. These parameters include a speed of sound c, which manifests itself by introducing a finite compressibility into the flow, and a weighting function diameter h that is roughly analogous to the grid spacing in grid-based numerical methods. Ideally, solutions should be independent of these parameters in some limit.

The approach taken in this study is to perform a series of simulations using relatively simple geometries where the flow solutions can be compared against analytical results or numerical results obtained using other methods, such as finite elements. The two flow geometries are slit flow between two parallel walls and flow in a cubic array of periodic spheres. The slit flow problem leads to a parabolic flow profile for which there is an analytic solution and the periodic sphere problem can be modeled using standard grid-based numerical methods. Both of the configurations will be simulated using SPH over a range of parameters and compared to validated solutions obtained by alternative means. The SPH solutions will be required to accurately reproduce the flow field and to introduce minimal density fluctuations. The second requirement is imposed to confirm that solutions correspond to relatively incompressible flow.

3. SALSSA workflow environment

For the SPH validation study considered here, users begin with an initial input file, launch the SPH code on a remote compute machine, and examine the output. Based on their analysis of the output, the input file of their previous run is edited, the model rerun, and the new outputs are analyzed. This process repeats until a conclusion is derived. The SALSSA workflow environment is composed of four primary components that support and facilitate this process: an Organizer tool component; a simulation execution component; a data management component; and an analysis and visualization component. Each of these components is described in more detail in the following sections, including how they specifically contribute to the SPH validation study.

3.1. Organizer

The main user interaction tool, the "Organizer", provides access to the capabilities needed to conduct the validation study. This includes invocation of tools and applications used to manage and analyze simulation files, access to a workflow engine for simulation execution, provenance graph as well as spreadsheet views of tasks as they are conducted, and tabular views of simulation input and output data. It is structured around the concept of a "study" where a single study is a collection of setup, execution and analysis tasks. Figure 1 shows the Organizer tool with a provenance-graph representation of an SPH slit flow validation study. Each icon in the graph represents a task performed by the user. All information regarding these tasks, including inputs and outputs, time stamps, job information for simulation tasks, and an extensible set of additional metadata, is automatically recorded by the SALSSA data management component and presented in the Organizer tool. A text file-based registry system allows for the recognition and integration of additional desktop or remote applications. The Organizer emphasizes the use of existing software such as text editors, spreadsheets, and visualization tools though we provide a Python based programming model to add tools with custom interfaces and have developed a few including a grid visualizer and parameter study tools.

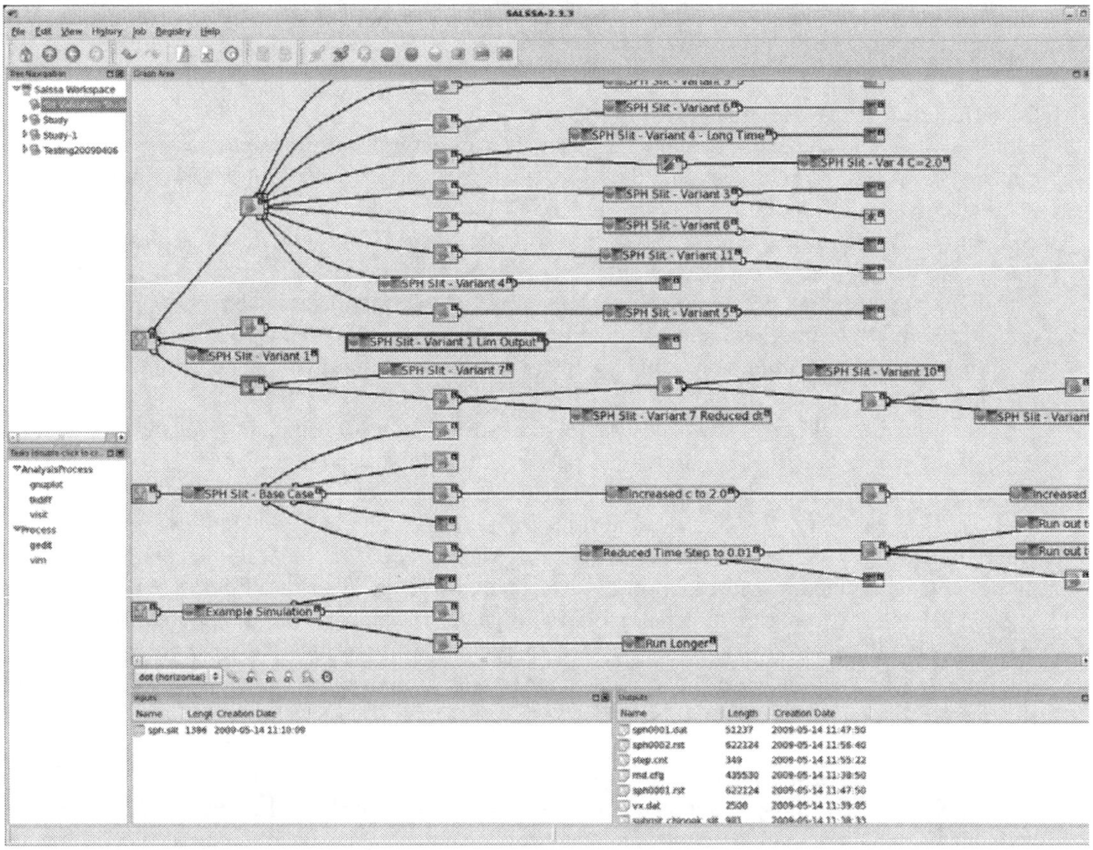

Figure 1. Screen snapshot of the Organizer and its use in the validation study. The main window contains an interactive graph that shows the provenance and relationships between individual tasks. The primary tasks used in this study were editors to adjust inputs, SPH simulations, and analysis sessions. Metadata, automatically extracted from data files, is available for each task along with a listing of all input and output files. The Organizer also supports spreadsheet style views of simulations and graph decluttering methods.

3.2. Simulation execution

The Simulation execution component includes a job Launcher tool, a customizable machine registration and job script generation mechanism, job execution workflows, job monitoring, and the ability to terminate individual jobs. The Launcher tool, invoked on simulations in the Organizer, collects user job submission inputs such as number of nodes and processors, and time, memory, and disk space limits (depending on the requirements of the target machine). This information, along with references to job input files is sent to the a workflow engine that stages files, submits the job requests, monitors the status of the jobs, and manages the disposition of outputs as directed by the user. Changes to the job states and data files are automatically sent to the Organizer. The Organizer also provides capabilities to monitor a running job through viewing the output file while it is being generated and terminate the job if necessary.

The simulation execution engine was implemented using the Kepler [4] open-source workflow system. Kepler has been applied to several scientific domains, and significant effort has gone into building components tailored for scientific workflow. By developing a custom Kepler workflow and augmenting it with a machine registry, we are able to submit and monitor arbitrary simulations run on a variety of machines ranging from Linux workstations, clusters, and supercomputers. For this study, simulations were run both at the Environmental and Molecular Sciences Laboratory (EMSL) and the National Energy Research Scientific Computing Center (NERSC). Our workflow is capable of launching multiple jobs to multiple machines while performing basic load balancing. This enables the automatic execution of a series of jobs such as parameter sensitivity studies.

SPH simulations may generate many very large files, which are typically left in place for analysis and often eventually moved to an archive. The user can specify the disposition of output files from within the Launcher tool. Later, these files can be staged to/from archives automatically as needed or under explicit user control. This capability is implemented as an additional workflow. The workflow system is also used to perform remote analysis tasks on data that resides on compute servers.

3.3. Data management

The data management component automatically tracks provenance (relationships between tasks) and inputs and outputs of each task, performs metadata extraction on data files to support search, browse, and graphical display, and performs MIME type detection. An extended version of the Open Provenance Model (OPM) [5] is used to represent provenance information. As a user executes tasks in the Organizer, relationships between data and processes are recorded using the OPM model, providing a complete record of a user's activities. The provenance information is used by the Organizer to create the interactive graph of tasks, as shown in figure 1. The provenance is stored in an open-source Resource Description Framework (RDF) store called Sesame [6]. RDF is a directed, labeled graph data format, which provides a flexible platform to describe complex relationships. The provenance graph and metadata is queried using the RDF query language, SPARQL [7]. Data artifacts associated with each task are stored in a content management system based on Alfresco [8]. The data model used to track inputs and outputs also supports references to files in the form of a URI. These are used for large simulation outputs that may be archived or left on the compute server.

In addition to the provenance, additional metadata is extracted from input and output files as they're introduced into the SALSSA environment. This capability relies on a registry system that provides rules for first establishing the MIME type of a file based on its name or, if needed, by examining the contents of the file. Once the type is known, the registry provides rules, in the form of regular expressions, to capture contextual metadata from the files. Both of these capabilities are readily extensible, by end users, through small Python scripts.

3.4. Analysis and visualization

SALSSA enables the analysis and visualization of the SPH validation runs through its registration of standard and custom applications. To date, we have integrated a range of analysis tools including TecPlot, General Mesh Viewer (GMV), Gnuplot, and Microsoft Excel, with current work focused on integrating the VisIt scientific visualization package. Depending on the analysis tool being used, the

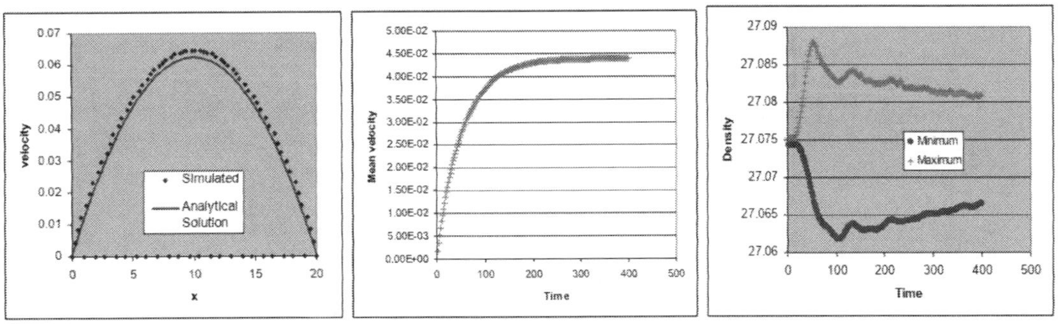

Figure 2. Results of a selected SPH validation simulation run. Left: comparison of a simulated velocity profile along a cross-section through the slit; Center: Time variation of simulated mean velocity as steady-state flow is approached; Right: Density variations within the simulated flow field as a function of time. All variables are presented in non-dimensional units.

SPH output may need to be translated to a different format. These translation scripts are also registered as tools. In this case, a custom Perl script parses the model output file and generates a time step based column-formatted text file. This file can be used within Microsoft Excel or Gnuplot to generate plots or perform further numerical analysis. Figure 2 shows examples of the plots generated from Microsoft Excel. The Gnuplot integration with SALSSA includes a custom interface that supports interactive scripting. Particle-based output from the SPH runs can be viewed directly, such as with GMV or VisIt [9]. Critical due to the size of the data, these analysis tasks can be accomplished from the user's desktop with the data residing where it was originally computed without having to stage it to a local workstation.

4. Summary

Validation of the SPH model is currently under way with tens of parameter configurations already executed and analyzed within the SALSSA workflow environment. SALSSA automates many tasks associated with running the simulation while transparently tracking data location and the provenance associated with tasks as they are executed. It emphasizes the use of familiar desktop tools while introducing new analysis tools specifically designed to handle large-scale simulation data. This case study has identified opportunities for ongoing research including improving the graphical presentation of large and complex workflows, automatic re-execution of a sequence of tasks, and automatic staging of large outputs from archival storage for analysis.

Acknowledgements

This research is supported by the U. S. Department of Energy's Office of Science under the Scientific Discovery through Advanced Computing (SciDAC) program. A portion of this research was performed using the Molecular Science Computing Facility at EMSL, a national scientific user facility sponsored by the Department of Energy's Office of Biological and Environmental Research and located at Pacific Northwest National Laboratory. The work was conducted at the Pacific Northwest National Laboratory, a multiprogram national laboratory operated by Battelle for the U.S. Department of Energy under contract DE-AC05-76RLO1830.

References

[1] Allan B A *et al.* 2006 A component architecture for high-performance scientific computing *The International Journal of High Performance Computing Applications* **20** 163-202
[2] Tartakovsky A M, Meakin P, Scheibe T D and Eichler West R M 2007 Simulations of reactive transport and precipitation with smoothed particle hydrodynamics *Journal of Computational Physics*. **222** 654-72
[3] Tartakovsky A M, Meakin P, Scheibe T D and Wood B D 2007 A smoothed particle hydrodynamics model for reactive transport and mineral precipitation in porous and fractured porous media *Water Resources Research.* **43**
[4] Ludäscher B, Altintas I, Berkley C, Higgins D, Jaeger E, Jones M, Lee MA, Tao J, Zhao Y. Scientific Workflow and the Kepler System. Concurrency and Computation: Practice and Experience. Volume 18, Issue 10, Pages 1039 – 1065.
[5] Moreau L, J Freire, J Myers, J Futrelle, and PR Paulson. 2007. "The Open Provenance Model." Presented by Luc Moreau at Workshop on Principles of Provenance, Edinburgh, Scotland on November 20, 2007
[6] The Sesame open-source RDF Framework. http://www.openrdf.org/
[7] The SPARQL Query Language for RDF. http://www.w3.org/TR/rdf-sparql-query/
[8] The Alfresco Enterprise Content Management System. http://www.alfresco.com/
[9] The VisIt Visualization System. https://wci.llnl.gov/codes/visit/

High energy physics from high performance computing

T. Blum
Physics Department, University of Connecticut, Storrs, CT 06269-3046, USA
E-mail: tblum@phys.uconn.edu

Abstract. We discuss Quantum Chromodynamics calculations using the lattice regulator. The theory of the strong force is a cornerstone of the Standard Model of particle physics. We present USQCD collaboration results obtained on Argonne National Lab's Intrepid supercomputer that deepen our understanding of these fundamental theories of Nature and provide critical support to frontier particle physics experiments and phenomenology.

1. Introduction

Thanks to recent advances in computer power, software, and algorithms, lattice gauge theory calculations of Quantum Chromodynamics (QCD), or Lattice QCD for short, have come a long way since Wilson proposed the method[1] 35 years ago. In this talk, we discuss results from lattice QCD and members of the USQCD collaboration that impact frontier particle physics research.

QCD is the fundamental theory of the strong force responsible for the material world. Quarks and gluons, interacting via QCD bind together to form hadrons like the ubiquitous protons and neutrons and less well known, less massive π and K mesons, in some ways like electrons and nuclei form atoms by exchanging the electromagnetic force carriers, the photons. QCD is different from Quantum Electrodynamics (QED) in some important ways. For one, the coupling constant, or strength of the QCD interaction, is large, unlike the electromagnetic charge e. Further, QCD is said to confine: quarks and gluons which carry the "color" charge of QCD are not free; they only exist in bound states that are neutral with respect to color charge. These eigenstates of the theory are the hadrons just mentioned. Finally, there are eight force carriers in QCD, called gluons, compared to one photon in QED, and they interact with each other as well as the quarks. For these reasons, QCD at the relatively low energy scale of a typical hadron mass, say 1 GeV, can not be studied in a perturbative framework. Being a relativistic quantum field theory, QCD must be regularized, or rigorously defined. This can be done using a discrete hypercubic lattice of points to describe space-time which we come back to in the next section.

Discretizing the continuum theory leads naturally to numerical simulations. The non-abelian nature of QCD and the fact that an almost exact chiral symmetry of the quarks exists in Nature make it difficult to solve, even with the largest supercomputers. In this talk we describe the state-of-the-art simulations being done on some of the largest supercomputers in the world, the algorithms used for them, and the theoretical physics results gained from them.

2. Dynamical fermion simulations

The heart of any lattice gauge theory (see, *e.g.*, [2]) calculation is its ensemble of gauge configurations, representing the gauge fields to be averaged over in the path integral for the quantum expectation value of some observable. For lattice QCD these configurations of "link" variables represent the gluons, for QED the photons. Feynman's path integral is to be taken over all possible configurations, or paths of the system, in configuration space. Since a configuration is the specification of the fields at every point in space-time, and since the (components of) fields take all possible values between $\pm\infty$, the un-regulated path integral diverges. It is made finite by discretizing the continuum Lagrangian to produce one that exists on a four dimensional hyper-cubic lattice with spacing a. The continuum limit, $a \to 0$, is taken practically by computing at several values of a and then extrapolating to 0. In practice the 4-volume is finite too, $VT = (N_s a)^3 \times N_t a$ ($N_{s,t}$ are the number of lattice sites in the space and time directions), so the infinite volume limit is taken in a similar way.

There are several (quark) degrees of freedom at each lattice site and several (gluon) degrees of freedom on each link connecting neighboring sites in a typical simulation, and $N_s \sim 16-64$, $N_t \sim 32-200$, so enumerating *all* possible configurations is impossible. We are therefore lead to Monte-Carlo simulations using importance sampling to estimate the path integral[3]. The configurations are chosen randomly, but with a weight equal to the exponential of their action. A typical configuration is generated from a Markov chain of configurations by performing an "update", or change, on the previous configuration. As long as the update algorithm satisfies basic constraints like detailed balance and ergodicity, in the limit of many updates, one is guaranteed to generate configurations with the correct weight, even if the starting configuration is not part of the equilibrium distribution [3]. Of course, there are many update algorithms that satisfy the basic requirements, so the trick is to find one that is optimal, that is, one that samples the configuration space as efficiently, but as cheaply, as possible. The USQCD collaboration has been using a combination of the hybrid monte-carlo[4, 5] (HMC) and rational HMC[6] algorithms. The former is used for an even number of degenerate fermions, while the latter must be used for an odd number. The RHMC can also be used for even numbers of fermions. Here hybrid refers to the two steps in the update algorithm. First the fields are evolved through phase space by adding a fictitious conjugate momentum for the gauge field and numerically integrating the coupled Hamilton's equations of motion for the system in "monte-carlo time". The trajectory length is defined as the number of integration steps times Δt, the integration time-step. Typical trajectory lengths have $O(1)$ time-units. This first step is often referred to as the molecular dynamics update. Since the integration is inexact (an error of order Δt^2 accrues in the hamiltonian after each trajectory), a second metropolis accept/reject step is performed after each trajectory which makes the algorithm exact, *i.e.*, the distribution of gauge configurations is the correct one in the limit that the number of updates is large.

State-of-the-art calculations employ 2+1 flavors of *dynamical* light-mass quarks, meaning two degenerate up (u) and down (d) quarks, and one heavier strange quark (s). In Nature, the u and d quarks are nearly massless ($(m_u + m_d)/2 \sim 3-5$ MeV), while the s quark is somewhat heavier ($m_s \sim 90-100$ MeV). Dynamical means the virtual pair creation/annihilation of the quarks, or vacuum polarization, effects on the gauge fields are included in the gauge configuration generation. In correlation functions, or expectation values, valence quarks appear, including heavier flavors of quarks. In Nature there are three, charm (c), bottom (b), and top (t). These quarks are so heavy ($\sim 1-175$ GeV) that their vacuum polarization effects are ignored compared to the dominant statistical and systematic uncertainties in present lattice calculations. However, lattice results are nearing the one-percent precision level for some quantities, so groups are starting to consider including the lightest of the heavies, the c quark, in future configuration generation. Of the heavies, only charm can be simulated directly on present day lattices ($a > 1/3$ GeV^{-1}) because the dimensionless mass should satisfy $ma \ll 1$ to avoid lattice spacing artifacts.

There are ways to treat the heavy quarks so that even b quark observables may be computed accurately. Some of these are described later. The t quark is so massive that its decays can be treated accurately in perturbation theory.

Table 1. Asqtad gauge configurations generated in 2008 by the MILC collaboration on the ANL BG/P (Intrepid). "spc" is the lattice spacing in fm, and "traj" is the number of trajectories for the entire Markov chain.

spc	m_l/m_s	$m_\pi L$	size	traj
0.06	0.10	4.3	$64^3 \times 144$	4956
0.06	0.15	4.4	$56^3 \times 144$	4800
0.045	0.20	4.6	$64^3 \times 192$	5166
0.09	0.05	4.8	$64^3 \times 96$	4500

Table 2. Same as Table 1, but for (5d) DWF, generated by LHPC, RBC, and UKQCD collaborations.

spc.	m_l/m_s	$m_\pi L$	size	traj
0.08	0.15	4.1	$32^3 \times 64 \times 16$	6000
0.08	0.22	4.9	$32^3 \times 64 \times 16$	6000
0.08	0.29	5.5	$32^3 \times 64 \times 16$	6000
0.08	0.11	4.6	$48^3 \times 64 \times 16$	400

The quarks (fermions) always appear as the inverse of a large, sparse matrix known as the propagator because it describes how the quarks move through space-time. The matrix (D) is the finite lattice representation of the so-called Dirac operator which appears in the quantum mechanical equation of motion for a relativistic spin 1/2 particle. In the continuum the Dirac operator is proportional to a first order 4d covariant partial derivative. They only enter through D because the quantum fields themselves are anti-comuting, complex numbers known as Grassman variables which we do not know how to simulate on a computer. Since they appear quadratically in the action, sandwiching the Dirac operator, their contribution to the path integral can be computed analytically, leaving only factors of the propagator and a square root of the determinant of D for each flavor of quark in the action (see [7], for example). Evaluating the factors involving D dominates any calculation with fermions and boils down to inverting D, even for the determinant factors. This is because the $\det D$ factors can be re-exponentiated to the action by introducing fictitious bosons called pseudofermions. Their presence is felt through a force term in the equations of motion during the molecular dynamics update[4]. The cost due to the gauge part of the action is almost negligible by comparison, and most correlation functions of interest involve fermions.

There is much freedom in defining D; it is only required to equal the continuum operator in the limit $a \to 0$ and should also be gauge invariant to avoid unwanted mixing with unphysical operators. Thus, several discretizations are employed which are equivalent up to small errors proportional to positve powers of a (usually two). Within USQCD three types of fermion operators are being used in gauge field generation: a^2-tadpole improved (Asqtad), domain wall and a^2 improved Wilson, or "clover" fermions. All three formally differ from the continuum theory by order a^2 errors, and they each have strengths and weaknesses. We discuss the first two, the latter is discussed in the talk by Edwards.

2.1. Asqtad fermions

Asqtad fermions are improved staggered fermions[8, 9, 10] which ameliorate the flavor symmetry breaking $O(a^2)$ errors of un-improved staggered fermions. Staggered fermions reduce the number of fermion degrees of freedom due to lattice fermion doubling[11] by diagonalizing the naive discretization of the Dirac operator and then eliminating 3 out of 4 of the spin components on each lattice site. In the limit $a \to 0$, the remaining spin degrees of freedom correspond to four degenerate Dirac fermions. The weight for a single fermion is then given by $(\det D)^{1/4}$. If $a \neq 0$, the flavor symmetry (called taste-symmetry to distinguish it from a continuum flavor symmetry) is broken. It is this symmetry that is partially restored by the Asqtad formulation. There has been vigorous debate in the literature over the validity of the "rooting" procedure (whether it yields a continuum Dirac fermion, or a sick cousin when $a \to 0$)[12, 13, 14, 15]). The primary advantages of staggered fermions are that they are relatively cheap to simulate and they have an exact axial symmetry for $a \neq 0$ left over from the full $SU(3)_L \times SU(3)_R$ chiral symmetry of continuum QCD.

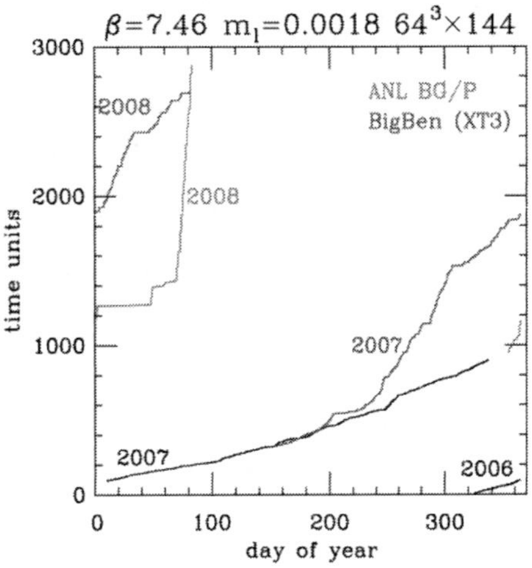

Figure 1. Time history of Asqtad fermion gauge configuration generation. The ensemble is described in the first row of Table 1 and corresponds to the lightest, most difficult quark mass.

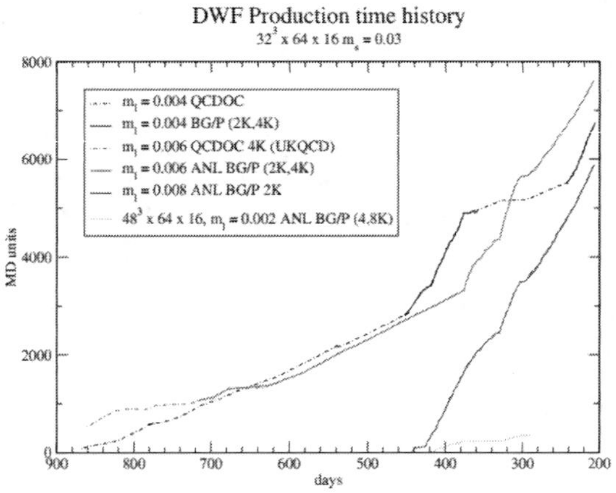

Figure 2. Time history of DWF fermion gauge configuration generation. The ensembles are described in Table 2.

2.2. Domain wall fermions

Domain wall fermions (DWF) were first formulated in [16] as a non-perturbative method for defining and simulating chiral gauge theories, but were quickly reformulated for numerical efficiency [17]. In [17] it was also recognized that DWF could be used to simulate vector gauge theories like QCD with exact chiral symmetry even when $a \neq 0$. This is the principle advantage of DWF: chiral symmetry protects the theory from unwanted lattice artifacts and renders the lattice theory continuum-like even when $a \neq 0$. The symmetry only becomes exact when the size of an extra 5th dimension, which provides the chiral symmetry to low-energy, 4 dimensional fermions stuck to the boundaries of the extra dimension, is taken to ∞. For finite size the chiral symmetry is softly broken, inducing a small shift in the quark mass in addition to the bare mass appearing in the lagrangian. This shift, however, can be computed accurately from the simulation and its effects eliminated, or subsumed into the quark mass. The addition of an extra dimension, of course, increases the computational burden. The expectation is that the extra burden is partially offset by the continuum-like character of DWF since a may be relatively larger when continuum symmetries are intact.

2.3. Early Science Program at ALCF

In 2008, the USQCD collaboration was awarded Early Science Program (ESP) time on the Blue Gene/P Intrepid supercomputer at the Argonne (National Lab) Leadership Computing

Facility (ALCF) under our three year INCITE allocation. The ESP time was primarily used for gauge field ensemble generation. The stability of Intrepid combined with USQCD's readiness to compute efficiently at the start of operation lead to spectacular gains in ensemble generation, advancing our science goals approximately two years ahead of schedule. The configurations are summarized in Tables 1 and 2 and the generation dramatized in Figs. 1 and 2.

In the chiral limit of the light quarks, $m_l \to 0$, D becomes singular and can not be inverted. Even though the physical u and d quarks are most likely massive, their masses are sufficiently small that present simulations are run with heavier masses, and expectation values are obtained by extrapolating to the physical point. Even though calculations could be done for lighter masses, even physical ones, the volumes would be too small to accommodate such light particles. A rule of thumb for QCD is that the mass of the pion (m_π) times the linear size of the box should be greater than 4 to avoid significant finite volume effects (see Tables 1 and 2). Recent work suggests this may be an underestimate of the needed size for some quantities [18].

Besides several values of the light quark mass for each type, the various ensembles of configurations summarized in Tables 1 and 2 reflect the exigencies of the two discretizations. The Asqtad configurations cover three lattice spacings (known as fine, super fine, and ultra fine) to bring the taste-breaking artifacts under control and together with existing coarser lattice spacing ensembles to take the continuum limit. The values of a and m_l are such that very large lattice sizes are required to simulate sufficiently large physical volume. For the DWF configurations, the lattice spacing is not as small, but is significantly smaller than a companion ensemble with $a = 0.114$ fm which together with these is being used to take the continuum limit[19] (recall the leading artifacts for both go like a^2). The size of the extra dimension is 16 lattice sites for each. The last row in Table 2 refers to a very demanding large volume, small mass simulation that was begun on Intrepid. However, the added size of the lattice and decreased mass have pushed the update algorithm beyond present capabilities, despite significant gains from tuning which we turn to next. Completion must await the next generation of leadership class computers.

2.4. Tuning the rational hybrid monte carlo for domain wall fermions

Improvements in algorithms have played a large role in lattice QCD simulations in the last few years. Most of the time used in generating lattices is related to including the full effects of light quarks in the Feynman path integral. The quarks enter through the determinant of the Dirac operator, which is a function of all the gluon fields on the lattice, and which also prohibits two quarks from occupying the same state, as required by the Pauli exclusion principle. The determinant is a positive definite quantity, so it can be added to an importance sampling measure. The standard way to do this, by writing the determinant as a bosonic path integral, requires all eigenvalues be positive. This is not the case here. However, the eigenvalues come in related pairs, which taken together are positive.

To handle a single quark correctly, the determinant of the product $D\,D^\dagger$ is used, which is manifestly positive eigenvalue-by-eigenvalue. On can show that $\det(D\,D^\dagger) = \det(D)\det(D)$, so that the simulation of a single quark requires that $(\det(D\,D^\dagger)^{1/2}$ be known. Using a rational approximation for $(\det(D\,D^\dagger)^{1/2}$ yields an accurate way to include single quarks into simulations, which is vital for the 2+1 flavor simulations that most accurately represent the physical world, where the up and down quarks are taken as light and degenerate in mass, and the strange quark is much heavier [6].

The RHMC algorithm, as currently implemented, uses a rational approximation for the effects of a single quark species, along with many other improvements, to allow the molecular dynamics steps to be as large as possible. We would like to detail a few of the recent improvements used in the RHMC algorithm.

In every step of the molecular dynamics part of the RHMC algorithm, the propagation of quarks in the gluon background must be determined. This is the most numerically expensive part of the calculation and is generally done by a Krylov space solver. As quark masses get smaller (the physically interesting region) the solutions are ever more expensive. One feature added to current RHMC implementations is to minimize the force from light quarks via the insertion of Hasenbush preconditioning masses [20]. The light quark forces are still expensive to evaluate, but they are not evaluated as often, since they are small. This allows one to use an integrator with multiple time scales: expensive to calculate forces are made small and only evaluated on a relatively large time scale, while large, cheaper forces are evaluated often. Use of multiple time scale integrators has allowed us to target our computing time most effectively.

The improvements mentioned above have been in place for the last 2-3 years. For the quark masses and voluems we were simulating with 2-3 years ago, speed-ups of a factor of 5 or more were seen. This was a sufficiently large gain that we then moved to smaller quark masses which were not possible without the improved RHMC algorithm. In the last year, with the availability of the large BG/L and BG/P installations, we have been doing simulations at much smaller quark masses, and on much larger volumes. In our our most demanding simulation to date (last row, Table 2), we observed that it was taking over 3 times as long as expected. This DWF simulation has a pion mass of roughly 240 MeV which is small by current standards, but still well above the physical value of 140 MeV.

It is important to know whether we had reached a fundamental limit in the capability of the RHMC, related to physics at such a light quark mass, or whether the particular choice of RHMC parameters was not optimal. Fortunately, we have found the latter to be the case. For this light quark mass, we introduced an additional Hasenbush preconditioning mass and an additional time step in our multi-time step integrator. These steps should have helped to keep the forces from the light quarks small and allowed us to use a large time step for them. Initially these improvements did not help with the acceptance. However, we then decreased the stopping condition for our Krylov solvers by a factor of 100, to 3×10^{-8}. With this increased accuracy, we were able to increase the molecular dynamics step size and regain the factor of three we had lost for this simulation. Our system has $O(10^9)$ degrees of freedom and, for a good acceptance, the change in the energy at the end of a molecular dynamics trajectory must be an $O(1)$ quantity. Thus it is not surprising that such a small residual is required in our Krylov solvers.

For our current 2+1 flavor simulations, we have found it optimal to rewrite the path integral as

$$Z = \int [dA] \left(\frac{\det \mathcal{D}(m_l)}{\det \mathcal{D}(m_x)}\right) \left(\frac{\det \mathcal{D}(m_x)}{\det \mathcal{D}(m_s)}\right) \left[\left(\frac{\det \mathcal{D}(m_s)}{\det \mathcal{D}(1)]}\right)^{1/2}\right]^3 \exp\left\{-S_g(A)\right\} \qquad (1)$$

where $\mathcal{D}(m) = D^\dagger(m_l)D(m_l)$ and $m_l = m_u = m_d$. Each ratio of determinants involves a separate pseudofermion field and the fermions are integrated on three different time scales. A represents the gluon field, and S_g the action of its kinetic term.

3. Results

3.1. Improved staggered fermions

The MILC Asqtad configurations are being used by several groups within USQCD to compute fundamental properties of QCD and the SM. The gluon topological charge susceptibility from MILC is shown in Fig. 3. When plotted with the non-Goldstone pion mass $m_{\pi 1}$, it tends to vanish in the combined chiral and continuum limits, as it should. The pion leptonic decay constant is shown in Fig. 4, accompanied by a multi-parameter fit of the data to chiral perturbation theory. The data are fit simultaneously with corresponding pion mass measurements and extrapolated in quark mass to the point where the pion mass agrees with its value from experiment. This fixes the value of the light quark mass at the "physical point". Not only does this allow all other calculated quantities to be evaluated at the physical point, but the quark mass determined this way, once renormalized, is a fundamental parameter of the SM.

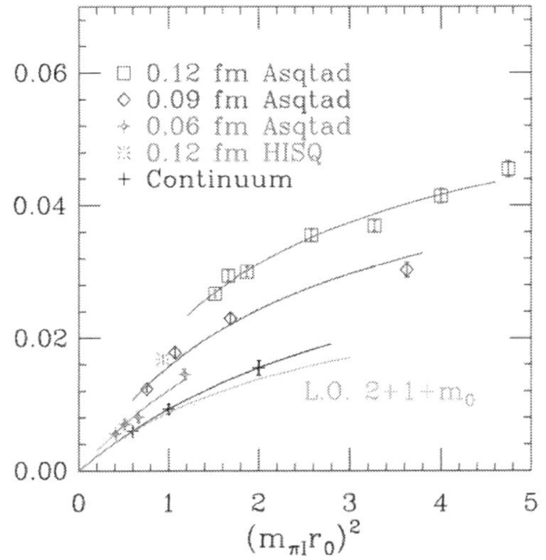

Figure 3. Gluon topological charge susceptibility (MILC Collaboration). r_0 is the Sommer parameter.

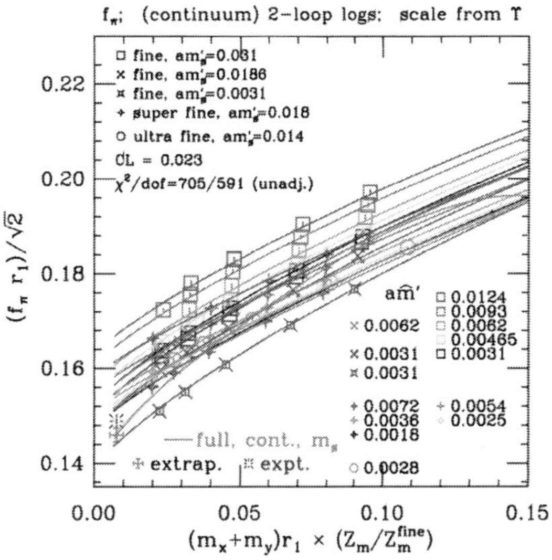

Figure 4. Pion decay constant (MILC Collaboration).

A huge experimental effort is currently devoted to the study of the decays and mixings of mesons with one heavy and one light quark with the ambitious aim of observing phenomena beyond the SM. The experimental measurements must be complemented, however, by precise lattice QCD calculations of the corresponding weak matrix elements.

The HPQCD calculation of the D_s meson decay constant is shown in Fig. 5. The new calculations on the super and ultra fine lattices are quite consistent with their earlier prediction which is 2.3 standard deviations lower than the 2009 measurement from the CLEO experiment.

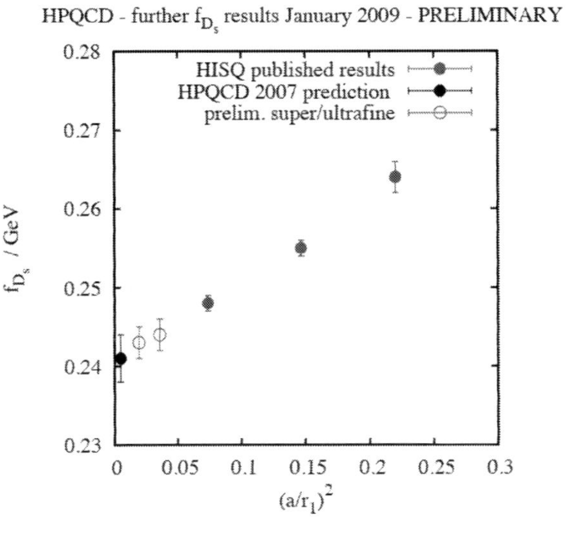

Figure 5. D_s meson decay constant (HPQCD Collaboration).

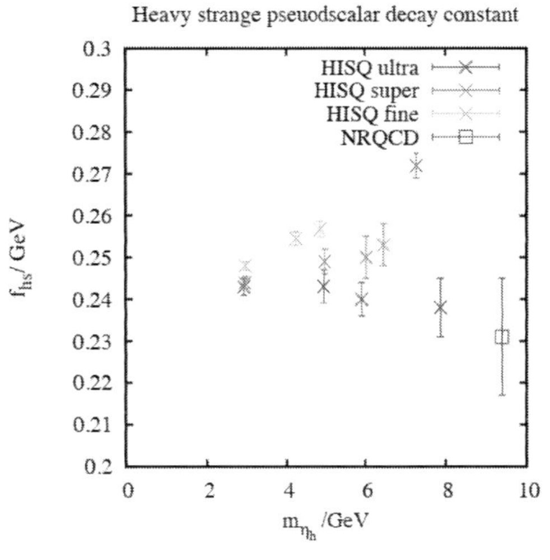

Figure 6. Heavy-strange meson decay constant. Relativistic (HISQ) heavy quark computed with several lattice spacings (HPQCD Collaboration).

The Fermilab/MILC group has been calculating heavy/light matrix elements on the Asqtad lattices, using Asqtad light valence quarks and clover improved Fermilab heavy valence quarks. They recently calculated the B meson decays $\bar{B} \to D^* l \nu_l$ [21] and $B \to \pi l \nu_l$ [22] to determine the Cabibbo-Kobayashi-Maskawa (CKM) matrix elements V_{cb} and V_{ub}, important SM parameters in the search for new physics. These matrix elements will be calculated on the ultra fine ensemble, reducing the uncertainties due to the heavy quark discretization, the largest errors in this set of calculations, by a factor of two.

With the advent of Leadership Class computers which can handle small a, large size lattices, HPQCD is investigating simulating a relativistic b quark ($m_b \sim 4$ GeV) directly on the lattice. Figure 6 shows the result for the heavy-strange meson decay constant computed with highly improved staggered quarks (HISQ) for both the heavy and s valence quarks. As a decreases, the decay constant trends toward the non-relativistic heavy quark (NRQCD) result. While inspiring, calculations at the b quark mass will require the next generation of computers. HPQCD has also used HISQ valence quarks on the super fine and ultra fine lattices to compute the c quark mass.

3.2. Domain wall fermions

The LHPC/RBC/UKQCD DWF gauge configurations listed in Table 2, along with a previously generated coarser ensemble with $a = 0.114$ fm, are being used to study the light-strange K mesons (kaons) [19, 23, 24] and to investigate nucleon structure from QCD[25]. The former is an important compliment to the heavy-light meson investigation in the search for CP violation and new physics while the latter addresses an important aspect of QCD and is discussed in the talk on nuclear physics and lattice QCD by Edwards.

Figure 7 depicts the most accurate calculation to date of the kaon B parameter B_K which also provides a constraint on the vertex of the unitary triangle (geometrical representation of the CKM matrix) through the experimental measurement of K meson mixing since CP violation in the SM arises only from a single complex phase in the CKM matrix. The figure reveals how tricky it is to extrapolate to the physical point. Heavier light-quark points exhibit markedly linear behavior which belies the true behavior in the chiral limit. The current precision on B_K is roughly 6 % from all sources, but a large fraction is due to the lattice spacing error which is about 4%. With the new fine DWF lattices, this error can be reduced significantly, constraining the SM even further. Preliminary results for B_K computed on the fine lattices are shown in Fig. 8. The RBC and UKQCD collaborations are currently working to use both calculations to establish the continuum limit of B_K. Indicating its importance to the particle physics community, two other groups in USQCD are computing B_K in the "mixed-action" approach. Sharpe, *et al.*, use the Asqtad lattices with HYP-smeared staggered valence quarks [26] while Laiho, *et al.*, use the Asqtad lattices but with DWF valence quarks [27].

Another important precision test of the SM comes from unitarity of the CKM matrix, for example $|V_{ud}|^2+|V_{us}|^2+|V_{ub}|^2 = 1$ must hold; any deviation signals new physics. To extract $|V_{us}|$ from meson decays whose underlying quark decay process is $s \to u$, a non-perturbative form factor $f_0(0)$ must be determined. Figure 9 shows $f_0(q^2)$ computed with 2+1 flavors of DWF[23]. The original calculation extrapolated to $q^2 = 0$ as required by the kinematics imposed by the lattice size and periodic boundary conditions. New points, computed using twisted boundary conditions and allowing simulations (nearly) at $q^2 = 0$, show excellent agreement. As with B_K, analysis is underway using the fine lattices to take the continuum limit of $f_0(0)$.

Like the Asqtad program, a main goal of the DWF work is to investigate QCD near the chiral limit and confront experiment. On the coarse DWF ensemble, an important finding was that SU(3) chiral perturbation theory for 2+1 flavors of quarks does not work well when the s quark mass is near its physical value of roughly 100 MeV and only terms in the next-to-leading-order in the expansion are kept. It was much more accurate to use SU(2) chiral perturbation

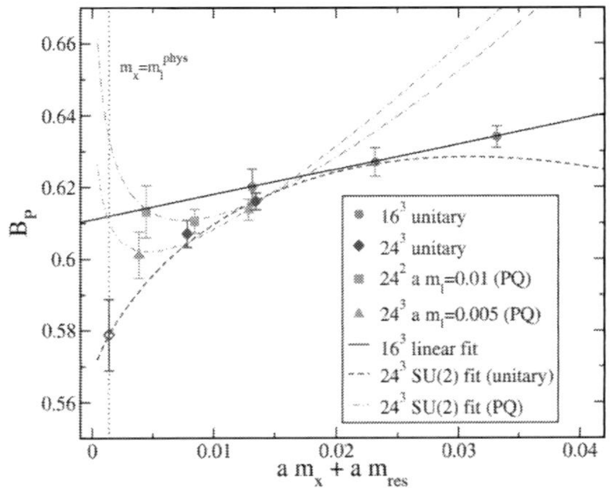

Figure 7. The kaon B parameter computed on the coarse DWF ensembles [24]. The (extrapolated) value at the physical point is shown at the lower left of the figure.

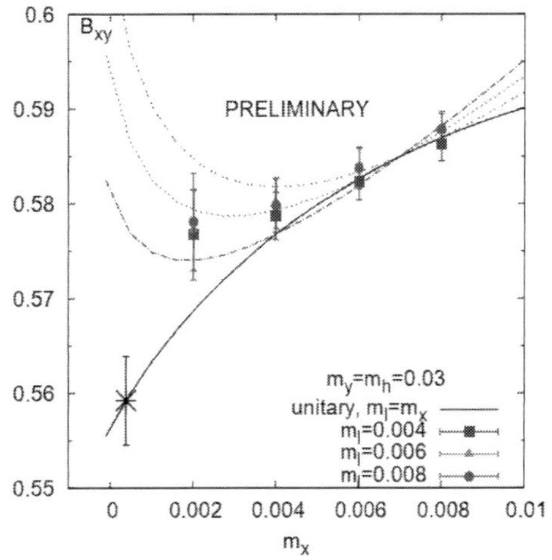

Figure 8. The kaon B parameter computed on the fine DWF ensembles (RBC and UKQCD collaborations).

theory to extrapolate the light quarks to their physical point [19]. In fact, this was done for the calculation of B_K discussed above. Confirming this behavior on the fine lattices is underway. It also suggests running simulations at even smaller quark masses. We have started such a run where the pion mass ~ 240 MeV (last row in Table 2). However even with the prodigious power of Intrepid, the simulation is out of reach and must await the next generation of Leadership Class machines. In the meantime, one can back off of a, lower the quark mass, and still have a suitably large volume. The difficulty is that for DWF, increasing a also increases the unwanted chiral symmetry breaking effects from finite 5th dimension.

To tame the chiral symmetry breaking effects, we add an irrelevant term to the action, or equivalently, an auxiliary determinant (AuxDet for short) to the weight in the path integral [28, 29, 30]. The AuxDet allows for larger a without the extra chiral symmetry breaking effects. DWF simulations with $m_\pi = 180$ and 240 MeV are being run now at the ALCF. Besides strengthening the investigation of the chiral limit of QCD begun on the conventional DWF lattices, these large volume, small mass simulations open up an interesting new challenge: simulating decays of K mesons to two-pion final states. Important outcomes of such calculations are the direct CP violation parameter ϵ' and the long standing $\Delta I = 1/2$ rule for K decays. Figure 10 shows preliminary results computed on the 16^3, coarse DWF ensemble for the isospin 0, two π meson scattering correlation function. Such calculations are extremely demanding since they involve so-called "disconnected" diagrams where the quark propagators are not connected with each other. The disconnected diagrams require huge statistics to resolve. Hence, the $K \to \pi\pi$ decay project is to run on Intrepid in 2009 as the AuxDet lattices become available.

4. Outlook

Enabled by large machines like Intrepid, the gains achieved in lattice QCD simulations were not even dreamt of a few years ago. As the continuum and chiral limits are approached, lattice QCD calculations are living up to the promise of solving the incredibly rich, complex, and beautiful theory of the strong interactions, a fundamental theory that is a cornerstone of particle and nuclear physics. Still there is work to be done to reach these difficult limits, and we can't help but look to the next generation of computers. In the next five years, with sufficient but realistic resources ($O(100)$'s of Teraflop-years), QCD calculations may be done at the physical point with

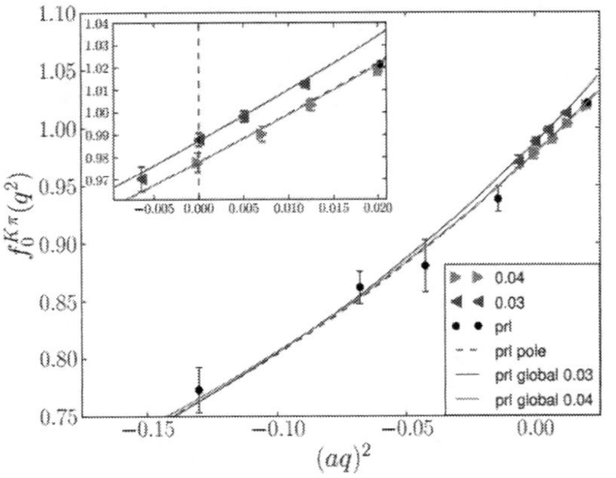

Figure 9. Form factor $f_0(q^2)$ needed to extract $|V_{us}|$ from experimental measurement of the K meson decay $K \to \pi l \nu$, computed on the coarse ensemble[23].

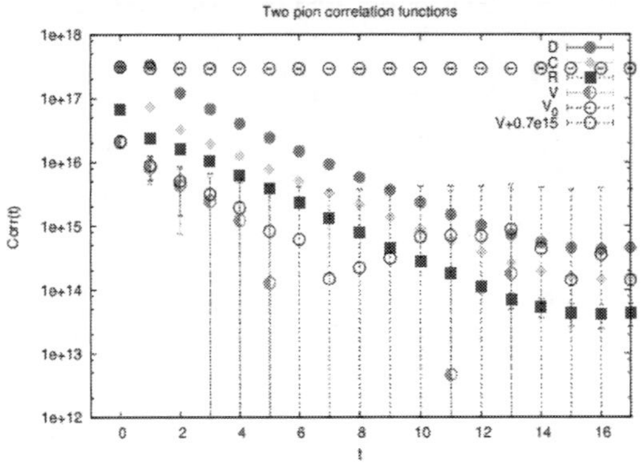

Figure 10. Correlation function for $\pi\pi$ scattering. Different curves correspond to four distinct quark contractions in the correlation function (RBC and UKQCD collaborations).

several lattice spacings and sizes to take the continuum and infinite volume limits. The Large Hadron Collider (LHC) Era is set to begin. Electro-weak symmetry breaking, supersymmetry, and beyond the SM phenomenology offer new challenges for lattice gauge theory.

5. References

[1] Wilson K G 1974 *Phys. Rev.* **D10** 2445–2459
[2] Creutz M Cambridge, Uk: Univ. Pr. (1983) 169 P. (Cambridge Monographs On Mathematical Physics)
[3] Toussaint D 1989 *Comput. Phys. Commun.* **56** 69–92
[4] Gottlieb S A, Liu W, Toussaint D, Renken R L and Sugar R L 1987 *Phys. Rev.* **D35** 2531–2542
[5] Duane S, Kennedy A D, Pendleton B J and Roweth D 1987 *Phys. Lett.* **B195** 216–222
[6] Clark M A and Kennedy A D 2007 *Phys. Rev. Lett.* **98** 051601 (*Preprint* hep-lat/0608015)
[7] Peskin M E and Schroeder D V Reading, USA: Addison-Wesley (1995) 842 p
[8] Lepage G P 1999 *Phys. Rev.* **D59** 074502 (*Preprint* hep-lat/9809157)
[9] Orginos K and Toussaint D (MILC) 1999 *Phys. Rev.* **D59** 014501 (*Preprint* hep-lat/9805009)
[10] Orginos K, Toussaint D and Sugar R L (MILC) 1999 *Phys. Rev.* **D60** 054503 (*Preprint* hep-lat/9903032)
[11] Nielsen H B and Ninomiya M 1981 *Phys. Lett.* **B105** 219
[12] Creutz M 2007 *Phys. Lett.* **B649** 230–234 (*Preprint* hep-lat/0701018)
[13] Bernard C, Golterman M, Shamir Y and Sharpe S R 2008 *Phys. Rev.* **D77** 114504 (*Preprint* 0711.0696)
[14] Bernard C, Golterman M and Shamir Y 2008 *Phys. Rev.* **D77** 074505 (*Preprint* 0712.2560)
[15] Creutz M 2008 *Phys. Rev.* **D78** 078501 (*Preprint* 0805.1350)
[16] Kaplan D B 1992 *Phys. Lett.* **B288** 342–347 (*Preprint* hep-lat/9206013)
[17] Shamir Y 1993 *Nucl. Phys.* **B406** 90–106 (*Preprint* hep-lat/9303005)
[18] Yamazaki T *et al.* (RBC+UKQCD) 2008 *Phys. Rev. Lett.* **100** 171602 (*Preprint* 0801.4016)
[19] Allton C *et al.* (RBC-UKQCD) 2008 *Phys. Rev.* **D78** 114509 (*Preprint* 0804.0473)
[20] Hasenbusch M 2001 *Phys. Lett.* **B519** 177–182 (*Preprint* hep-lat/0107019)
[21] Bernard C *et al.* 2009 *Phys. Rev.* **D79** 014506 (*Preprint* 0808.2519)
[22] Bailey J A *et al.* 2008 (*Preprint* 0811.3640)
[23] Boyle P A *et al.* 2008 *Phys. Rev. Lett.* **100** 141601 (*Preprint* 0710.5136)
[24] Antonio D J *et al.* (RBC) 2008 *Phys. Rev. Lett.* **100** 032001 (*Preprint* hep-ph/0702042)
[25] Syritsyn S N *et al.* 2008 *PoS* **LATTICE2008** 169 (*Preprint* 0903.3063)
[26] Bae T *et al.* 2008 *PoS* **LATTICE2008** 275 (*Preprint* 0809.1220)
[27] Aubin C, Laiho J and Van de Water R S 2009 (*Preprint* 0905.3947)
[28] Vranas P M 1999 (*Preprint* hep-lat/0001006)
[29] Fukaya H *et al.* (JLQCD) 2006 *Phys. Rev.* **D74** 094505 (*Preprint* hep-lat/0607020)
[30] Renfrew D, Blum T, Christ N, Mawhinney R and Vranas P 2009 (*Preprint* 0902.2587)

SciDAC 2009 IOP Publishing
Journal of Physics: Conference Series **180** (2009) 012067 doi:10.1088/1742-6596/180/1/012067

Exploring the spectrum of QCD using the lattice

***Hadron Spectrum Collaboration*: John Bulava[1], Saul Cohen[2], Jozef Dudek[2], Robert Edwards[2], Eric Engelson[3], Justin Foley[1], Balint Joo[2], Jimmy Juge[4], Huey-Wen Lin[2], Nilmani Mathur[5], Colin Morningstar[1], Mike Peardon[6], †David Richards[2], Sinead Ryan[6], Christopher Thomas[2], Anthony Thomas[2], Stephen Wallace[3]**

[1] Department of Physics, Carnegie Mellon University, Pittsburgh, PA 15213, USA
[2] Thomas Jefferson National Accelerator Facility, Newport News, VA 23606, USA
[3] Department of Physics, University of Maryland, College Park, MD 20742, USA
[4] Department of Physics, University of the Pacific, Stockton, CA 95211, USA
[5] Department of Theoretical Physics, Tata Institute of Fundamental Research, Mumbai 400005, India
[6] School of Mathematics, Trinity College, Dublin 2, Ireland

E-mail: †`dgr@jlab.org`

Abstract. The calculation of the spectrum of QCD is key to an understanding of the strong interactions, and vital if we are to capitalize on the experimental study of the spectrum. In this paper, we describe progress towards understanding the spectrum of resonances of both mesons and baryons from lattice QCD, focusing in particular on the resonances of the $I = 1/2$ nucleon states, and of charmonium mesons composed of the heavy charmed quarks.

1. Introduction

The strong interaction, one of the forces of the Standard Model of particle and nuclear physics, is responsible for binding both protons and neutrons into nuclei, and also the primordial gluons and the lightest quarks into pions, protons, neutrons and other so-called hadrons. The interaction arises from a quantum field theory known as Quantum Chromodynamics (QCD).

In order to really understand QCD and hence test whether it is the complete theory of the strong interaction, we must know the spectrum of mesons and baryons that it implies and test those spectra against high quality data. The complete combined analysis of available experimental data on the photoproduction of nucleon resonances is the 2009 milestone in Hadronic Physics (HP), and the measurement of the electromagnetic properties of the low-lying baryons is an HP 2012 milestone. The observed spectrum of QCD provides little direct evidence of the presence of the gluons. However, QCD admits the possibility of exotic mesonic states of matter in which the gluonic degrees of freedom are explicitly exhibited. The search for such states will be an important component of the upgraded JLab@12GeV.

Given the intense experimental efforts in hadron spectroscopy, the need to predict and understand the hadron spectrum from first principles calculations in QCD is clear. Hence, an important goal of the effort of the USQCD Collaboration is the study of the resonance spectrum of QCD. The remainder of this paper is laid out as follows. In the next section, we introduce lattice QCD, and outline the methodology for looking at resonances on the lattice. We then

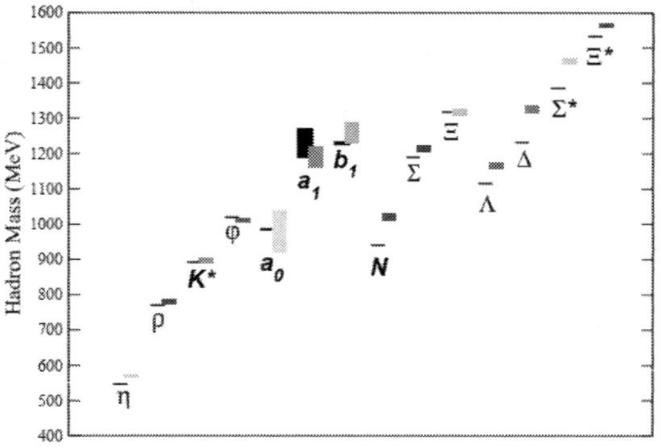

Figure 1. A comparison between the calculated values of the low-lying hadron masses and their physical values, obtained in a calculation with two flavors of light (u/d) quarks and a strange quark[1]. The black and colored bands indicate the experimental and lattice values, respectively; the width of the band indicates the uncertainty.

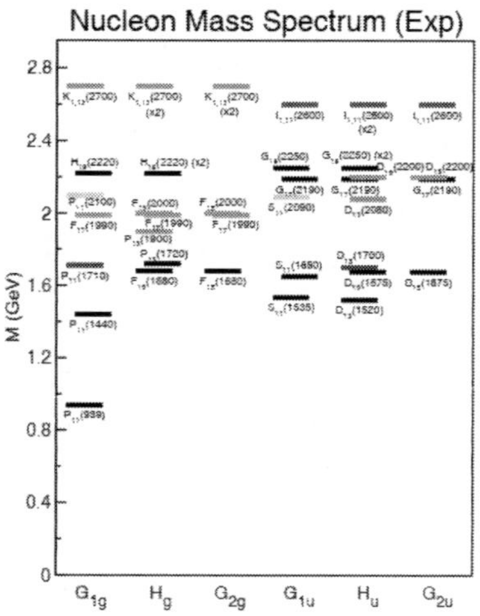

Figure 2. The figure shows the experimentally determined nucleon spectrum classified according to the transformation properties under the cubic group of the lattice; the lowest-lying state in the G_{1g} channel is the proton[2].

describe recent progress at understanding the resonance spectrum, focusing first on nucleon resonances and then on the meson sector. We conclude with prospects for future calculations.

2. Lattice QCD

The interactions of the quarks and gluons at very short distances, such as those probed at the LHC, are weak, a property known as asymptotic freedom, for which Gross, Politzer, and Wilczek won the 2004 Nobel Prize. This enables the interactions to be expanded in terms of a small coupling constant. At longer distances, typical of the binding of the quarks and gluons into hadrons, the coupling becomes strong, and the theory highly non-linear. Here, the only means of solving, as opposed to modelling, QCD is through numerical calculations on the lattice.

Lattice gauge calculations solve QCD on a four-dimensional lattice, or grid, of points in Euclidean space. The quarks reside on the points of the grid, whilst the gluons are associated with the links joining those points. Lattice calculations proceed through a Monte Carlo method, in which ensembles of gauge configurations are generated with a probability distribution prescribed by the Euclidean QCD action. Lattice QCD calculations have always been at the leading edge of exploiting the most powerful supercomputing resources available, helped by the highly regular nature of the problem.

The calculation of the ground-state spectrum has been a benchmark calculation of lattice QCD since its inception. Figure 1 shows a summary of the low-lying light-hadron masses compared with their experimental values, measured on anisotropic clover lattices designed for the study of resonance spectroscopy[1]; details of the computational methodology are given in the poster of Balint Joo.

A comprehensive picture of resonances requires that we go beyond a knowledge of the ground state mass in each channel, and obtain the masses of the lowest few states of a given quantum number. This we can accomplish through the use of the variational method[3, 4]; we calculate a matrix of correlator functions

$$C_{ij}(t) = \sum_{\vec{x}} \langle O_i(\vec{x}, t) O_j^\dagger(\vec{0}, 0) \rangle,$$

where $\{O_i; i = 1, \ldots, N\}$ are a basis of interpolating operators with given quantum numbers. We then solve the generalized eigenvalue equation

$$C(t)u = \lambda(t, t_0)C(t_0)u$$

to obtain a set of real (ordered) eigenvalues $\lambda_n(t, t_0)$, where $\lambda_0 \geq \lambda_1 \geq \ldots \geq \lambda_{N-1}$. At large Euclidean times, these eigenvalues then delineate between the different masses

$$\lambda_n(t, t_0) \longrightarrow e^{-M_n(t-t_0)} + O(e^{-\Delta M_n(t-t_0)}),$$

where $\Delta M_n = \min\{| M_n - M_i | : i \neq n\}$. The eigenvectors u are orthogonal with metric $C(t_0)$, and a knowledge of the eigenvectors can yield information about the partonic structure of the states. Crucial to the application of the variational method is the use of a basis of interpolating operators that have a good overlap with the low-lying states of interest. The cubic lattice employed in our calculations does not admit the full rotational symmetry of the continuum, but rather the more restricted symmetry of the octahedral group. Thus states at rest are classified according to the irreducible representation (irreps) of the cubic group, and for spectroscopy calculations, interpolating operators must be constructed that transform irreducibly under the cubic group; this task has been the prerequisite for our study both of baryons[5, 6], and of mesons.

3. Nucleon Resonance Spectrum

Baryons, containing three quarks, are emblematic of the non-Abelian nature of QCD, and of the three colors of the theory. An important goal in exploring baryons is attempting to discern the effective degrees of freedom that describe the spectrum; the search for so-called "missing resonances" focuses on whether the spectrum can be well described by a quark model, or whether an effective theory with fewer degrees of freedom, such as a quark-diquark picture, provides a more faithful description of the baryon spectrum.

There are three double-valued irreducible representations of the cubic group, denoted $G_{1u/g}(2)$, $H_{u/g}(4)$ and $G_{2u/g}(2)$, where g and u refer to positive and negative parity, respectively, and the brackets contain the dimension of the irrep. G_1 contains continuum spins $1/2, 7/2, \ldots$, H_g spins $3/2, 5/2, \ldots$ and G_2 spins $5/2, 7/2, \ldots$. Thus, at any fixed lattice spacing a, a state corresponding to spin-5/2 has four degrees of freedom in H, and two in G_2, with degeneracies between the energies in the two irreps emerging in the continuum limit. Figure 2 shows the experimental nucleon spectrum as seen on the lattice; the limited number of irreps requires that for each channel we be able to isolate as many energy levels as possible.

The nucleon spectrum has been analysed in a calculation with two flavors of light Wilson fermions, at two values of the pion mass[7], building on an earlier calculation in the quenched approximation to QCD[8]. The data are shown in Figure 3. For the first time in a lattice calculation, we can identify a spin-5/2 state, but the multi-hadron states that should be seen in the spectrum appear elusive; the calculation of correlation functions that are expected to be sensitive to these multi-hadron contributions is an important goal for the collaboration.

4. Meson Resonances

The new Hall D of the JLab 12GeV upgrade centers on the study of meson states produced in photoproduction reactions in the GlueX detector. Photoproduction has been proposed, within

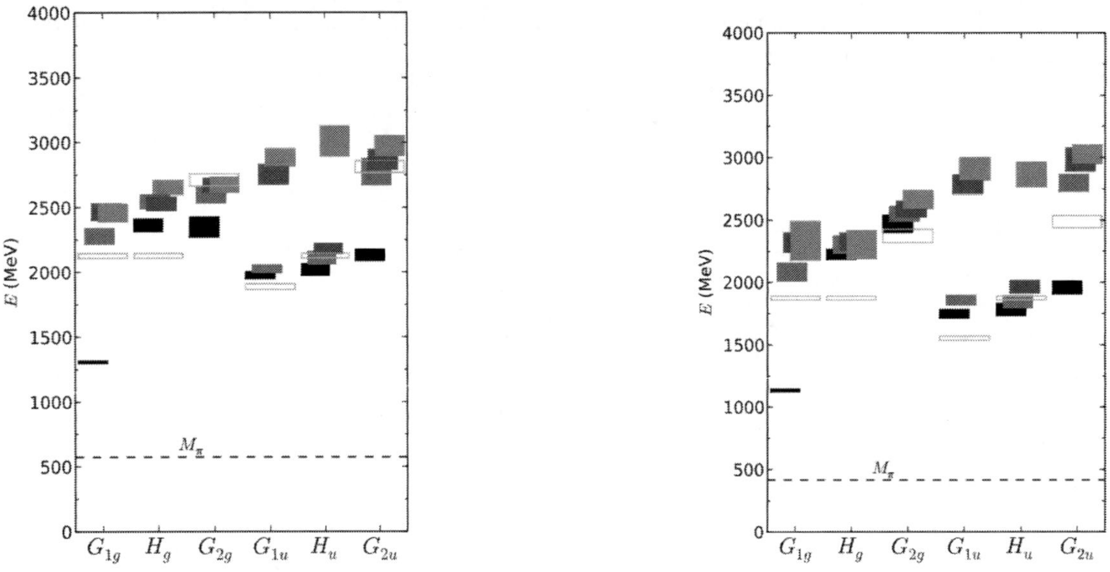

Figure 3. The left- and right-hand panels show the spectrum of $I = 1/2$ baryon resonance, indicated by the solid boxes, obtained on $N_f = 2$ Wilson fermion lattices at $m_\pi = 578$ and 416 MeV respectively[7]; the errors are indicated by the vertical width of the box. The open boxes show the expected thresholds for multiparticle states.

QCD-motivated models, as a favorable method for the production of "exotic" hybrid mesons, those mesons having J^{PC} outside the set allowed to a fermion-antifermion pair. The hybrid hypothesis is that an excited gluonic field in addition to a quark-antiquark pair can give rise to these quantum numbers.

As a theater for developing our methodology, we have studied charmonium, composed of the heavier charm quark c and its antiquark; the system is particularly attractive, in that it is both computationally far less demanding than systems composed of the light (u, d, s) quarks, and because there is a wealth of high-precision experimental data. For the study of the spectrum, performed in the quenched approximation to QCD, we used the known continuum behavior of our operators to enable the *spins* of the states in the different lattice irreps to be identified[9], illustrated for the $J^{PC} = J^{--}$ and J^{++} channels in Figure 4. Lattice QCD can further enable the electromagnetic properties of the states to be investigated. Thus the radiative transitions between some low-lying, non-exotic states were studied[10], and recently this was extended to include the radiative transition form factors for some of the higher-lying states in the spectrum, including those with exotic quantum numbers[11]. Most importantly, it was shown that radiative transitions to states with exotic quantum numbers were calculable, and furthermore the partial width for the exotic decay $\Gamma(\eta_{c1} \to J/\psi\gamma)$ was shown to be comparable to the non-exotic width. The extension of this calculation to mesons composed of light quarks, using the $N_f = 2 \oplus 1$ anisotropic clover lattices[1], is in progress.

5. Conclusions

The calculation of the masses and properties of resonances is essential to fully capitalize on the investment in experimental facilities, and to provide reliable calculations to confront the experimental results. Thus the generation and analysis of lattices designed for spectroscopy is a central goal of the USQCD Collaboration's program. The calculations outlined above are but the first stage in this program, with quark masses considerably higher than those realized in nature. An immediate challenge arises as we enter the regime where resonances are unstable under the

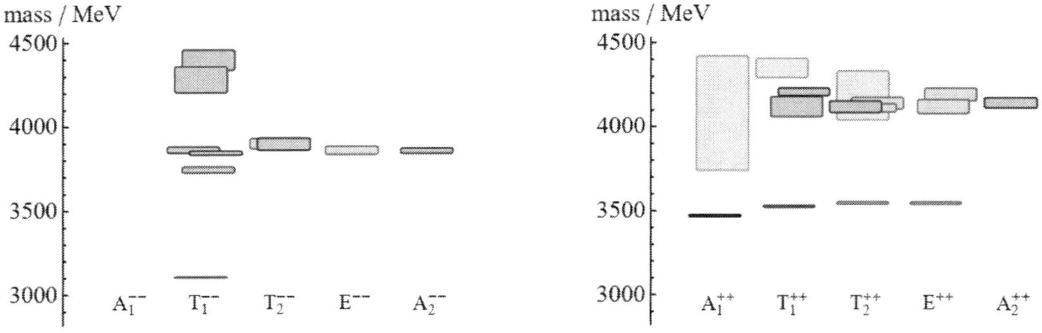

Figure 4. The left- and right-hand panels show the masses calculated in the $J^{PC} = J^{--}$ and J^{++} channels, respectively, categorized according to the lattice irreps, with the color coding indicating their estimated continuum spin assignments where known: Black ($J = 0$), Red ($J = 1$), Green ($J = 2$), Blue ($J = 3$), and Orange (unidentified).

strong interactions, and multi-hadron energies in the spectrum are expected to be important; a novel means of efficiently computing correlation functions in this regime has recently been outlined[12]. The advent of petascale computing is enabling calculations to be performed at progressively lighter values of the light-quark masses, and calculations with all masses fixed to their physical values are now, finally, within reach.

Acknowledgments
This research used resources of the National Center for Computational Sciences at Oak Ridge National Laboratory, which is supported by the Office of Science of the Department of Energy under Contract DE-AC05-00OR22725. This research was supported in part by the National Science Foundation through TeraGrid resources provided by Pittsburgh Supercomputing Center (PSC), San Diego Supercomputing Center (SDSC) and the Texas Advanced Computing Center (TACC), and under the USQCD SciDAC-2 Grant. This work used cluster resources at Jefferson Laboratory provided under the auspicies of the USQCD Collaboration.

MP and SR are supported by Science Foundation Ireland under research grants 07/RFP/PHYF168 and 06/RFP/PHY061, respectively. JB, JF and CM are supported by National Science Foundation grant numbers NSF-PHY-0653315 and NSF-PHY-0510020. KJJ is supported by grant number NSF-PHY-0704171. Authored by Jefferson Science Associates, LLC under US DOE Contract No. DE-AC05-06OR23177. The U.S. Government retains a non-exclusive, paid-up, irrevocable, world-wide license to publish or reproduce this manuscript for U.S. Government purposes.

References
[1] Lin H W *et al.* (Hadron Spectrum) 2009 *Phys. Rev.* **D79** 034502 (*Preprint* 0810.3588)
[2] Lichtl A C 2006 (*Preprint* hep-lat/0609019)
[3] Michael C 1985 *Nucl. Phys.* **B259** 58
[4] Luscher M and Wolff U 1990 *Nucl. Phys.* **B339** 222–252
[5] Basak S *et al.* (Lattice Hadron Physics (LHPC)) 2005 *Phys. Rev.* **D72** 074501 (*Preprint* hep-lat/0508018)
[6] Basak S *et al.* 2005 *Phys. Rev.* **D72** 094506 (*Preprint* hep-lat/0506029)
[7] Bulava J M *et al.* 2009 *Phys. Rev.* **D79** 034505 (*Preprint* 0901.0027)
[8] Basak S *et al.* 2007 *Phys. Rev.* **D76** 074504 (*Preprint* 0709.0008)
[9] Dudek J J, Edwards R G, Mathur N and Richards D G 2008 *Phys. Rev.* **D77** 034501 (*Preprint* 0707.4162)
[10] Dudek J J, Edwards R G and Richards D G 2006 *Phys. Rev.* **D73** 074507 (*Preprint* hep-ph/0601137)
[11] Dudek J J, Edwards R and Thomas C E 2009 *Phys. Rev.* 094504 (*Preprint* 0902.2241)
[12] Peardon M *et al.* 2009 (*Preprint* 0905.2160)

Visualization as a tool for understanding QCD evolution algorithms

Massimo Di Pierro[1], Michael Clark[2], Chulwoo Jung[3], James Osborn[4], John Negele[5], Richard Brower[2], Steven Gottlieb[6], Yaoqian Zhong[1]

[1] School of Computing, Depaul University; [2] Physics Department, Boston University; [3] Physics Department, Brookhaven National Laboratory; [4] Argonne National Laboratory; [5] Center for Theoretical Physics, MIT; [6] Indiana Univeristy

Abstract. In this paper we present a project of visualizing topological charge in Lattice QCD in order to understand its evolution under the Markov Chain Monte Carlo algorithms and how to maximize its equilibration. Lattice QCD is a very computationally expensive technology for computing properties of composite particles from a few fundamental parameters such as quark masses. Our research plays an important role in validating these computations by showing that the autocorrelation length is small and it can eventually lead to even more efficient algorithms.

1. Introduction and Motivations

The Standard Model of Particle Physics provides a strikingly successful description of the structure of the matter down to scales of 10^{-18}m. The basic degrees of freedom are elementary particles including quarks and leptons which interact with each other according to strong and electro-weak interactions. Quarks play a special role in this model because their interacting are mediated by the exchange of gluons; thus, they create bound states such as the proton, the neutron and the majority of particles that constitute the richness of structures that we observe in the physical world. The part of the Standard Model that describes quarks and gluons is called Quantum Chromo Dynamics (QCD). The ability to perform computations in QCD is of critical importance in Particle and Nuclear Physics. It enables us to determine fundamental parameters of the Standard Model from experiments, compare experimental results with model predictions, and eventually find discrepancies that can shed some light into yet unknown smaller structures and laws of physics at higher energy. It enables us to calculate the equation of state, transport coefficients, and other properties of hadronic matter and the quark gluon plasma explored in relativistic heavy ion collisions at RHIC the LHC. It also enables us to calculate the spectrum, structure, and interactions of mesons and baryons from first principles and lies at the heart of explorations at Jefferson Lab, RHIC spin, and a future electron-ion collider.

Journal of Physics: Conference Series **180** (2009) 012068 doi:10.1088/1742-6596/180/1/012068

Since quarks are subject to strong interactions, standard perturbative techniques are inadequate to perform many interesting computations. For computing the spectrum of the model and computing the lifetime of composite particles, the purely numerical approach known as Lattice QCD that has been the most productive[2].

Lattice QCD consists of taking a small portion of space (a cube of $\sim 10^{-15}$m of size) for a brief time interval ($\sim 10^{-23}$secs), discretizing it on a four dimensional lattice, and simulating numerically the dynamics of QCD inside. The spatial fields do not evolve over time as in ordinary classical systems. Instead, the fields, that are defined on the four-dimensional Euclidean space-time, evolve over a fictitious MCMC time (τ). Lattice QCD computations consist of measuring correlation functions over the possible four dimensional configurations of the fields, each weighted by a number that encodes the dynamics of QCD[1]. Quantities like particles masses and lifetimes can be determined from these correlations.

In practice, the four dimensional configurations are created using Markov Chain Monte Carlo (MCMC) algorithms. Examples are the Molecular Dynamics algorithm or the Hybrid Monte Carlo algorithm. Each configuration is generated from the previous one using a transition in probability that encodes the laws of QCD dynamics.

In this paper, we describe how visualization techniques can be useful to provide information about these algorithms, about the data that is being generated.

The quark fields can be integrated out and, while they play a critical role in the inner workings of the algorithms, they are not part of the actual configurations. The latter only contain gauge degrees of freedom representing gluons, hence they are usually referred to as gauge configurations. Typical gauge configurations have sizes that range from 32×24^3 to 192×64^3 points; which correspond to a size ranging from 256 MBytes to 15 GBytes each (in single precision). A typical Lattice QCD computation generates and utilizes about 1000 of these gauge configurations, thus requiring multiple TBytes of storage. Algorithms are always parallel because of the size of the problem of Lattice QCD.

A typical problem consists in determining whether this small subset of the infinite space of all possible gauge configuration is sufficiently representative and whether the statistical uncertainty determined from the averages can be trusted. From a physical perspective the gauge configurations can be grouped into equivalence classes characterized by their topological charge density.

Our goal is to visualize this topological charge density and measure how fast/slow it evolved under the MCMC for different algorithms and different simulation parameters (the dynamic quark mass in particular). This would provide us with a proof that current Lattice QCD computation are adequately sampling the space of all possible configurations as opposed to getting stuck in one topological sector, and provide a effective tool for developing algorithms that maximize equilibration of the topological charge.

Journal of Physics: Conference Series **180** (2009) 012068 doi:10.1088/1742-6596/180/1/012068

2. Visualization

Typical production gauge configurations appear to be separated by enough MCMC steps and there is no visible continuity in the topological charge. In order to measure autocorrelation in the topological charge we generated *ad hoc* gauge configurations with small trajectories for two different algorithms and different dynamical quark masses.

- **Wilson** Two sets of $24^3 \times 32$ dynamical Wilson gauge configurations with $\beta = 5.6$ and quark masses of $m_1 = 66$MeV and $m_2 = 33$MeV respectively. Each set consists of 1000 trajectories with a 0.1 molecular dynamics time units, saved every 2 time units ($dt = 0.2$). To integrate Hamilton's equations a multi-step 2nd order symmetric symplectic integrator was used (Omelyan) [3].
- **DWF**: One set of $24^3 \times 64$ dynamical gauge configurations generated with Domain Wall Fermions (two degenerate light ones, $m_u = 14$MeV, and one strange, $m_s = 74$MeV) and Iwasaki gauge action at $\beta = 2.13$, saved every 1/6 of a trajectory ($dt = 0.1\bar{6}$). Further details can be found in [4] and the references therein.

Each gauge configuration consists of a set of $SU(3)$ matrices $U_{x,y}$ for every pair of neighbor sites x, y on the lattice. Given a quark field q_y, the $U_{x,y}$ matrix represent the change in the quark field under parallel transport from $y \to x$, *i.e.*, $q_y \to q_x = U_{x,y} q_y$. Therefore $U_{y,x} = U_{x,y}^H$.

The topological charge is computed in three steps according to the definition:

- The gauge field was smoothed by n-iterations of a "cooling step"

$$U_{x,y} \to U_{x,y} = P_{SU(3)}(\alpha U_{x,y} + (1-\alpha) S_{x,y}) \quad (1)$$

  where $P_{SU(3)}$ is a projection operator into SU(3) and $S_{x,y}$ is the average of all products of three consecutive links connecting x with y.
- We computed the chromo-electro-magnetic field defined by

$$F_{x,\mu,\nu} = \frac{1}{2}\mathrm{Re}(A_{x,\mu,\nu} - A^H_{x,\mu,\nu}) \quad (2)$$

  where $A_{x,\mu,\nu}$ is the average of the product of four consecutive links in the μ, ν plane starting and ending at point x.
- We computed the topological charge density, defined by

$$Q_x = \frac{1}{32\pi^2} \sum_{\mu\nu\rho\sigma} \epsilon_{\mu\nu\rho\sigma} \mathrm{Tr}(F_{x,\mu,\nu} F_{x,\rho,\sigma}) \quad (3)$$

In fig. 1 we show three images of iso-surfaces of topological charge density for one gauge configuration for different numbers of cooling steps (0,10,15). In fig. 2 we show three images of iso-surfaces of topological charge density for the same number of cooling steps (20) for different consecutive gauge configurations separated by 6 trajectories.

3. Autocorrelation and Hurst Exponent

In addition to the visualization of the topological charge density we also computed the average autocorreleation $R(\tau)$ and average Hurst exponent H for each independent Markov Chain. First we computed the autocorrelation and the Hurst exponent over the MCMC time (τ) for each lattice point, then we averaged the results over all lattice points. The Hurst exponent is computed using the standard R/S technique described in ref. [5]. In fractal geometry, the Hurst exponent is also known as "index of dependence", it measures the relative tendency of a time series to either strongly regress to the mean or cluster in a direction. H is 0.5 for white noise. H is related to the auto-correlation length via $R(\tau) \simeq \tau^{2H-2}$ and $2-H$ is known as the "fractal dimension" of the problem.

Our results are summarized in the following table.

Action	quark masses	H
Wilson	66MeV	0.966
Wilson	33MeV	0.973
DWF	69,17MeV	0.908

The autocorrelation is plotted is fig. 3

Notice how the autocorrelation becomes negative at around 100 MCMC steps. This may indicates an alternating pattern in the data, but a larger MCMC chain will be required to validate this hypothesis.

4. Conclusion

Our analysis demonstrates that the autocorrelation length of a topological charge is very small compared with the trajectory length of typical production grade Lattice QCD computations. Our work is still in progress and, as more data is analyzed, we hope to derive an empirical interpolating formula for the autocorrelation length as function of the simulation parameters.

Acknowledgments

This research is funded by the Department of Energy grant DE-FC02-06ER41441. We thank Argonne National Laboratory computing resources, the DOE Advanced Simulation and Computing Initiative (ASCI) for the VisIt software, and Brian Fox at DePaul Univ.

Bibliography

[1] Pierro M D 2008 *Modern Pysics A* **125:012067**
[2] Kronfeld A 2006 *J. Phys. Conf. Ser.* **21-3**
[3] Takaishi T and de Forcrand P 2006 *Phys. Rev.* **E73** 036706 (*Preprint* `hep-lat/0505020`)
[4] Jung C 2007 *POS(LATTICE 2007)* **37** (*Preprint* `arXiv:0710.5337`)
[5] Hurst E 1951 *Trans. Am. Soc. Civil Eng.* **116** 770

Figure 1. Iso-surfaces for a 3D slice of a topological charge density for one gauge configuration for different numbers of cooling steps (0,10, 15).

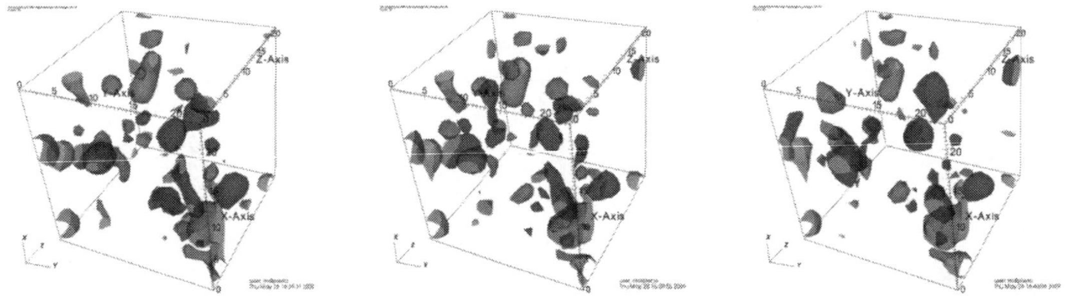

Figure 2. Iso-surfaces for a 3D slice of topological charge density for the same number of cooling steps (20) for different consecutive gauge configurations separated by 6 trajectories.

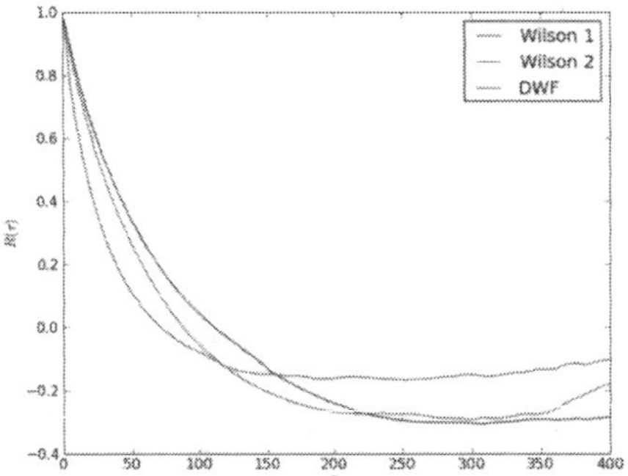

Figure 3. Autocorrelation length for the three sets of gauge configurations considered in the paper.

Nuclear physics using Lattice QCD in the SciDAC era

Robert G. Edwards, for the USQCD Collaboration
Thomas Jefferson National Accelerator Facility, Newport News, VA 23606, USA
E-mail: edwards@jlab.org

Abstract. Lattice QCD currently provides our only means of *solving* QCD (Quantum Chromo Dynamics) – the theory of the strong nuclear force – in the low-energy regime, and thus of crucial importance for theoretical and experimental research programs in High Energy and Nuclear Physics. Under the SciDAC program, a software infrastructure has been developed for lattice QCD that effectively utilize the capabilities of the INCITE facilities. These developments have enabled a new generation of Nuclear Physics calculations investigating the spectrum and structure of matter, such as the origin of mass and spin. This software infrastructure is described and recent results are reviewed.

1. Introduction

Quantum Chromo Dynamics (QCD) is the name for the theory of the strong nuclear force that binds quarks to produce hadrons, for examples the familiar proton and neutron of the atomic nucleus, which are the building blocks of all nuclear matter. Lattice QCD currently provides our only means of *solving* QCD in the low-energy regime, and thus of crucial importance for theoretical and experimental investigations in High Energy and Nuclear Physics. This proceedings contribution will describe QCD and how Lattice QCD is formulated. The software infrastructure developed for such calculations under SciDAC is outlined. These software efforts have enabled recent calculations in Nuclear Physics that determine the spin content of the proton and neutron as well as the spectrum of exotic matter. These calculations are of great relevance to the experimental program at Jefferson Lab, and address milestones within the DOE Nuclear Physics and Hadronic Physics programs.

2. Formulation

The theory put forward as Quantum Chromo Dynamics (QCD) follows a very similar form as Quantum Electrodynamics (QED) - the theory of the electromagnetic interactions (for example, light) with matter. In QED, electrons carry an electric charge and respond to electric and magnetic fields. Special Relativity tells us that we should consider these kind of interactions in a four-dimensional covariant form; namely, through the Dirac operator

$$D(A,m) = -i\sum_{\nu=1}^{4}\gamma_\nu\left(\frac{\partial}{\partial x_\nu} + ieA_\nu(x)\right) + m \tag{1}$$

where m is the mass of the particle, and the electric charge e multiplies the vector potential $A_\nu(x)$ that is a function of the four-dimensional site x_ν. This vector potential – just a real

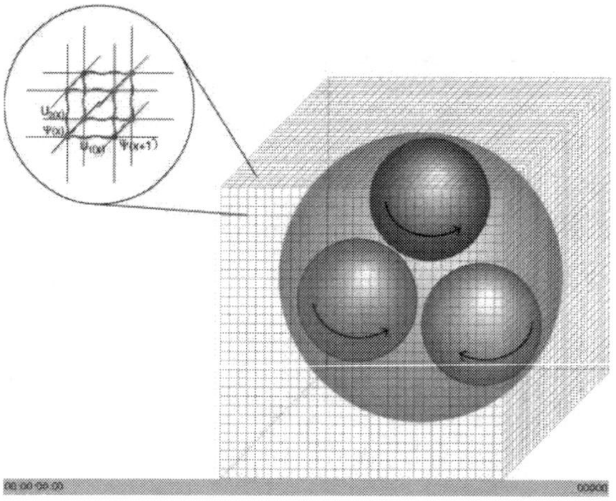

Figure 1. Representation of QCD on a lattice. The gluons are placed on links and the quarks are on the sites. The lattice spacing is chosen small enough to provide adequate resolution of a particle, such as three quarks – each carrying a "color" charge – that bind to form a proton.

number in QED – is related to ordinary electric and magnetic fields. The γ_ν are 4×4 matrices that act on spin components of electrons (spin $\frac{1}{2}$ particles).

What is so special about QCD is the idea that quarks carry a strong "charge" – called a "color" charge. The vector potentials become 3×3 traceless hermitian matrices and are called "gluons" which have some coupling g. The 3 correponds to three color charges which add together to give a neutral color-less object in analogy to a positive and negative charge adding to zero in QED. The traceless condition on the vector potentials implies there are 8 gluons in QCD while there is only 1 photon in QED. Dirac predicted that spin $\frac{1}{2}$ particles that propagate according to the Dirac operator D can come in matter and anti-matter variants - electrons and positrons. This prediction carries over to QCD; namely, quarks and anti-quarks, can be created within the vacuum (empty space).

To account for this seething vacuum, QCD states that configurations of gluons and quarks can occur with some probability

$$d\mu = e^{-S(A)}\det(D(A, m_{up}))\det(D(A, m_{down}))\det(D(A, m_{strange})) \quad (2)$$

where the $\det(D)$ is the determinant of the Dirac operator D, and an overall normalization is suppressed. There are three small mass quarks that occur and have the whimsical names *up*, *down* and *strange*. The exponent $S(A)$ is basically the energy, or rather "action", for some gluon configuration. An "observable" $\mathcal{O}$ is then averaged over these configurations (snapshots) of quarks and gluons. Since each point in space-time is in the sum, we use integrals to denote the sums

$$\langle \mathcal{O} \rangle = \int d\mu\, \mathcal{O}(A) \quad (3)$$

3. Lattice QCD

While calculations in QED are amenable to perturbation theory (in the electric charge), QCD calculations are not. The only *a priori* method known, that is systematically improveable, for such calculations is Lattice QCD. Here, space-time is discretized into a four-dimensional grid.

Journal of Physics: Conference Series **180** (2009) 012069 doi:10.1088/1742-6596/180/1/012069

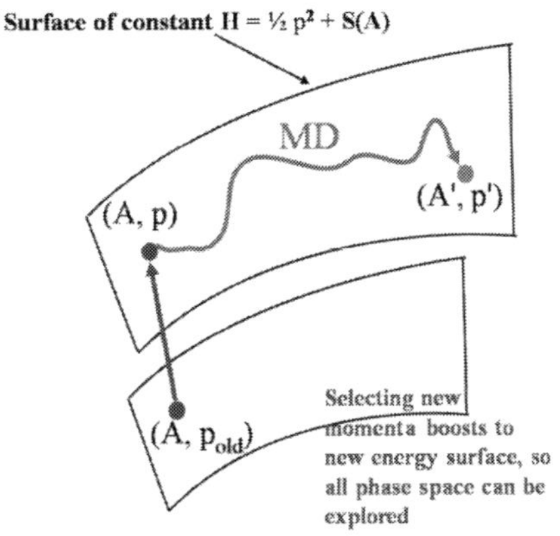

Figure 2. Lattice QCD calculations involve averages over configurations ("snapshots") of gluon fields. To produce a new configuration, molecular dynamics is use to evolve from some set of momenta and gluon field, to some new values. The time for evolution of gauge fields is the bulk of the cost of lattice calculations and requires use of INCITE facilities.

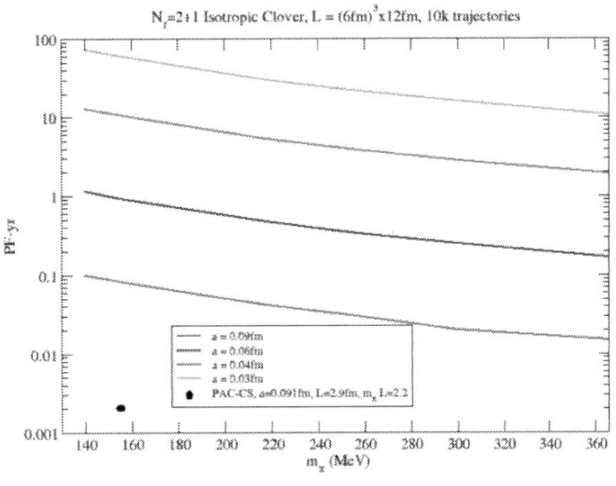

Figure 3. Scaling of cost for gauge field production as a function of pion mass and at a fixed physical volume $(6\text{fm})^4$. A recent calculation is shown as a black dot.

A representation is shown in Fig. 1. The quarks live on the sites, and the gluons are placed on the links. There is some lattice spacing between the sites, and the overall box size should be large to account for particles propagating. The continuum limit is achieved for very large box sizes and arbitrarily small lattice spacing.

The average of an observable in Eq. 3 in lattice QCD is still an integral over all possible gluon configurations as in Eq. 3. However, to make this integral amenable to numerical calculations, the integral is replaced by a sum using importance sampling. Here, one starts from some configuration of gluons, and new configurations are generated that sample the most important (largest) weights in the measure Eq. 2.

The method of choice for this "generation" of gauge fields is to use some variant of the Hybrid Monte Carlo (HMC) algorithm [1]. Namely, construct a *Hamiltonian* $H = \frac{1}{2}p^2 + S(A)$ where the momenta p describe movement in some fictitious (computer) time. Figure 2 shows a representation of the method. We propose some new gauge field and accept/reject it with some probability. Hamilton's equations (first order coupled differential equations in the gauge fields A and the momenta p) are used to update the gauge field. The new gauge field is accepted/rejected according to the change of the total energy of H. The goal then is to keep the overall energy changes small, so an accurate integration of the differential equations is essential.

The determinant $\det(D)$ is the major obstacle, whose solution is the subject of intense algorithm and software investigations. It turns out the $\det(D)$ can be recast into a solution of the Dirac equation at each step of the integration of Hamilton's equations shown in Fig. 2. Thus, the bulk of the time for these calculations is the solution of large sparse linear systems of equations whose solution involves iterative methods. These calculations require the large scale computers available through the INCITE program.

The cost for the generation of these gauge fields is governed by several factors, for example the lattice grid size and the quark's mass. As the quark mass becomes smaller, the linear system solution of the Dirac equation becomes more singular, and more iterations are needed for convergence. This quark mass is characterized by the pion mass – the lightest particle mass found in QCD.

There have been several important recent advances in the algorithms for gauge generation (reviews in Refs. [2, 3]). The cost is dependent upon the discretization methods used, but a generic estimate is that the time to produce one new configuration after integration of Hamilton's equations scales like

$$\mathrm{Cost}_{\mathrm{traj}} = C\left(\frac{\mathrm{fm}}{a}\right)^6 \cdot \left[\frac{V}{\mathrm{fm}}\right]^{5/4} \cdot \left(\frac{135\mathrm{MeV}}{m_\pi}\right)^2 \tag{4}$$

Where the volume V and lattice spacing a are in Fermi ($= 10^{-15}$ meters), and the pion mass is in units of MeV. The physical limit (the real world) is at $m_\pi = 135\mathrm{MeV}$. For comparison, the size of a proton is about 1 fm. A reasonable lattice spacing is $a \sim 0.1\mathrm{fm}$, and physical box sizes of about 6 fm are desired. So, lattice grid sizes of something like 64^4 are needed. In Fig. 3 is shown this cost scaling, and for comparison, a recent light pion mass calculation by the PACS-CS collaboration [3] is displayed. One can see that for large box sizes and small quark masses, petaflop-years are needed.

4. Development under SciDAC

4.1. Software infrastructure

A major goal for the USQCD collaboration's project under SciDAC is to develop and provide the software infrastructure for Lattice QCD calculations as well as to investigate new algorithmic and hardware techniques that accelerate these calculations [4, 5]. To address these needs, USQCD has developed a layered software infrastructure whose schematic representation is shown in Fig. 4. At Level 1, an abstraction is made of the basic communication patterns in QCD (mostly nearest neighbor send-receives) called *QMP*, as well as abstractions for the linear algebra (QLA) and threading or multi-core usage (called *QMT*). The implementation of each interface provided at this level are architectural specific, and appropriate libraries are provided. At Level 2, a data-parallel abstraction is made called *QDP*. The implementation of this level hides the architectural implementations below it, and provides a generic interface in either `C` or `C++`, along with appropriate libraries.

Level 2 provides a convenient and powerful abstraction for the manipulation of lattice objects that live on rectangular grids. This grid is partitioned over an underlying grid of processors. All communications in "shift" operations are handled below this level, and involve communicating faces between processors.

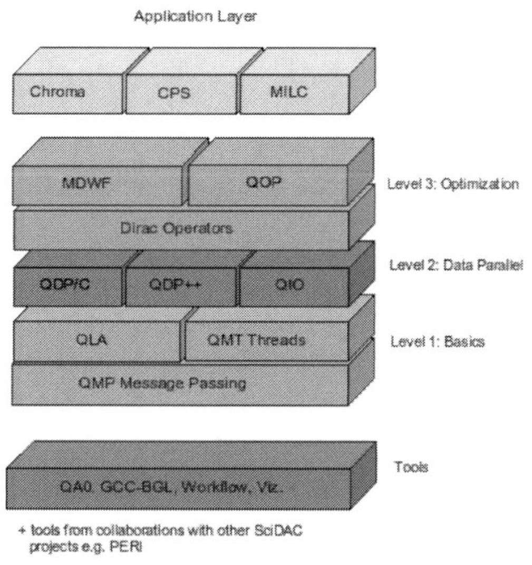

Figure 4. Schematic of the software infrastructure developed for lattice QCD under SciDAC. Each Level is a module providing a uniform programming interface and a library whose implementation is optimized for the architecture. The Q prefix is for QCD. At Level 1 are modules for Linear Algebra (*QLA*), thread support (*QMT*), and message passing (*QMP*). At level 2 is a data parallel interface called *QDP* providing an architecturally neutral view of lattice field manipulations, such as shifts combined with linear algebra operations. Level 3 provides optimized versions of critical codes, like large sparse linear-system solvers for the Dirac operators. These codes have been optimized for a range of systems, such as Bluegenes, Cray XT, QCDOC, and clusters. High level application codes utilize these levels.

It is recognized, however, that to fully utilize the underlying hardware, a further abstraction is developed – called Level 3 – that can group multiple data-parallel operations. This level provides full implementations of the solution to the Dirac equation via iterative methods such as variants on Conjugate Gradient. This level is highly architecturally dependent and developed for the Bluegene series, the Cray XT series, clusters of workstations as well as the QCDOC. More information as well as documentation can be found in Refs. [4, 5]. Scaling on Cray and Bluegene systems is shown in Fig. 5.

High level code libraries and applications (for example, *Chroma*, *CPS* and *MILC*) have been developed that utilized these libraries to implement various algorithms within Lattice QCD. Major application suites within these packages have been developed to support the HMC algorithm to produce gauge fields, as well as applications for measurements involving these configurations. The *MILC* and *CPS* codes have been developed prior to SciDAC, and have subsequently evolved to utilize the various Levels. The *Chroma* code [6] is a new application suite developed under SciDAC that is built directly over the Levels 2 and 3 described above.

Each of these application suites have tended to support particular physic calculations relevant to High Energy or Nuclear Physics. For example, the bulk of the gauge generation relevant to the spectroscopy program at Jefferson Lab have utilized Cray XT systems. Further description of those recent experiences are described by B. Joo in Ref. [7].

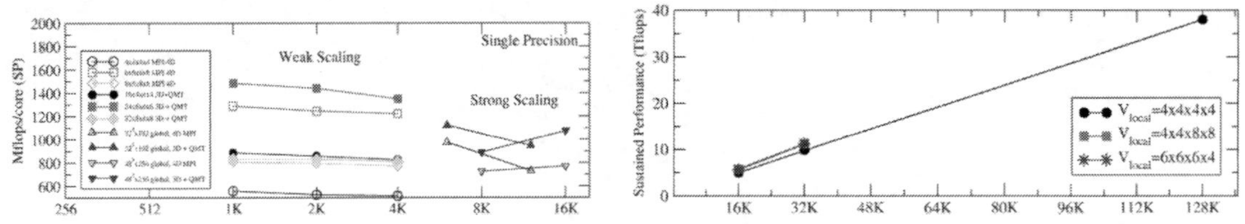

Figure 5. Left panel shows weak and strong scaling on the ORNL Cray XT4. The solid symbols indicate performance after the switch to a hybrid programming model: QMT (threading) between cores and MPI between nodes. Right panel: weak scaling on the ANL Bluegene/P.

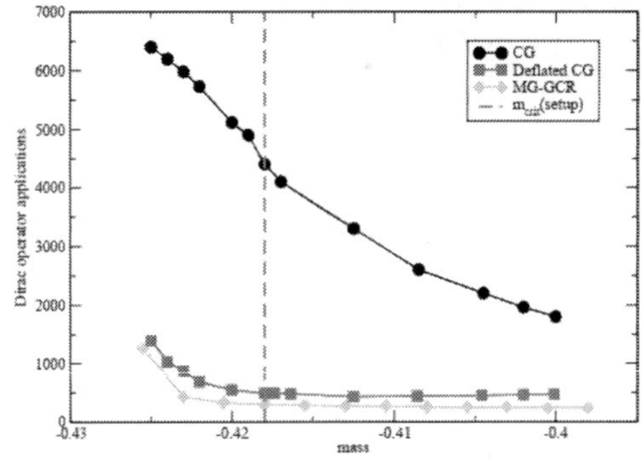

Figure 6. Comparison of the accelerated iterative linear system solvers `EigCG` (deflation from Ref. [8]) and `MG-GCR` (amg from Ref. [11]) with a non-accelerated CG method. A substantial reduction in the number of iterations is observed as the quark mass is reduced in the accelerated methods.

4.2. Acceleration techniques

Besides gauge field production, an increasingly important and time consuming part of lattice QCD calculations is involved with measurements, or rather the computations on a given gauge field. Again, the solution of large sparse linear systems via the Dirac operator is dominant here. Frequently, solutions corresponding to multiple right-hand sides are needed. Deflation methods as well as Multigrid have become important in accelerating these calculations [8, 9, 10, 11]. Deflation techniques are, in some cases, also being used to accelerate gauge generation [10]. In Fig. 6 is shown the performance of the `EigCG` inverter [8] and a variant of Algebraic Multigrid [11]. Each method shows a substantial reduction in the number of iterations versus standard iterative techniques.

The `EigCG` inverter has been used in production for more than a year in spectroscopy calculations at Jefferson Lab [12], and has resulted in more than a factor of three speed up in the calculations. Further investigations are proceeding, and work is underway to employ these techniques in other parts of the USQCD computing program.

5. Results in nuclear physics

Within the USQCD project, there are at least three major questions and avenues of research that aim to explain the origin of matter and spin. What is the structure of hadrons, what is their

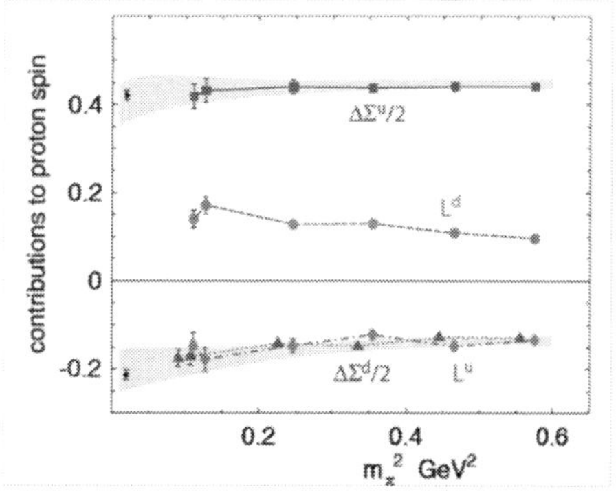

Figure 7. What is the origin of spin? The contribution of quark spin and orbital angular momentum to nucleon spin Ref. [14]. The experimental determination of the quark spin content from HERMES [15, 16] is also shown. These results indicate that the total quark orbital angular momentum within a proton and neutron is small, while the orbital angular momentum of each quark is large. The spin of the nucleon is thus carried mostly by the gluons binding the quarks.

spectroscopy, and how do hadrons and nuclear matter interact? These questions are the subject of intense theoretical and experimental investigations. A new generation of experiments are underway, such as the 12GeV upgrade to the U.S. Dept. of Energy Jefferson Lab, the upgrades to the RHIC-Spin project at Brookhaven National Lab, and GSI in Germany, that will have significant impact on our understanding.

With the recent advances in algorithm and computing capabilities, Lattice QCD is well suited to provide answers to these questions, and is a major focus of the USQCD collaboration. The next sections outline recent investigations into hadron structure and spectroscopy. The contribution by T. Luu [13] will describe results in Nuclear Structure using lattice QCD.

6. Nucleon structure

Within hadron structure, there are fundamental questions as to the origin of matter and spin that have been posed:

- What is the size, shape, distribution of charge and current in hadrons?
- How are quarks and gluons distributed?
- How does nucleon spin arise from quarks and gluons?
- What role do strange quarks play in nucleon structure?

These questions are part of the long range plan of the Nuclear Science Advisory Committee (NSAC) which advises DOE and NSF as to long term priorities within nuclear physics.

Recent calculations utilizing INCITE resources (ANL) are providing insight as to the origin of nucleon spin. In Fig. 7 is shown a recent calculation [14] of the orbital angular momentum and the spin contribution of quarks within a nucleon, as well as a recent experimental determination of the spin contribution of the quarks. The pion masses have not yet reached the physical limit, so (chiral) extrapolations are used. However, it is clear that there is reasonable agreement between the lattice calculations and the experimental determination of the spin contributions. What is new insight, though, is the breakdown. The spin contributions amount to $\sim 41\%$ of the nucleon spin, and the orbital terms are about 0%. The latter are individually large, but cancel in the sum. Thus, the rest of the nucleon spin comes from strange quarks (small contribution given other experimental observations) or more likely gluons . Thus, it appears most of the nucleon spin and mass is not from the quarks, but the gluons.

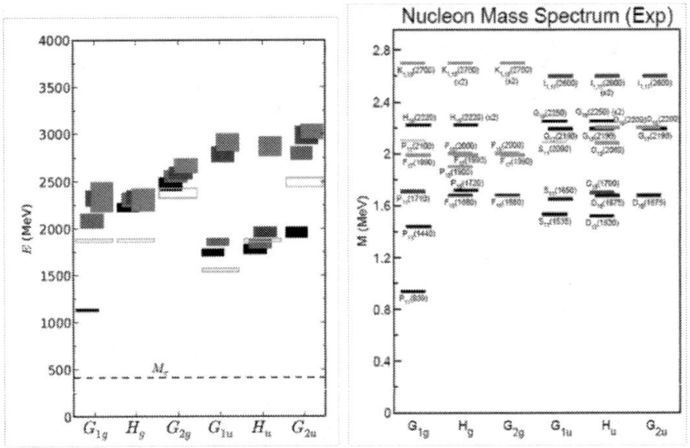

Figure 8. Left panel: the lattice determination of the excited state energies of the nucleon at a pion mass of 400MeV. Each channel corresponds to different spin and parity combinations. Right panel: the experimental measurements where lighter color indicates less well determined states. Since the lattice calculation is not at the physical value, there is some shift in energy. Empty boxes show thresholds for possible decay thresholds. Patterns of states are similar between the two panels. In particular, isolated spin $\frac{5}{2}$ states in (G_{2u} are apparent. These are the first ever lattice calculations of such high lying excited states, and represent significant advances in the methodology needed for such calculations.

Qualifications are in order. There are contributions that have been neglected due to their significant complexity; namely, disconnected terms. These arise from the probe peering into the nucleon (photons) interacting with the virtual quarks in the vacuum. There are some reasons to suspect these contributions are small, but nonetheless, the calculation is not complete until they are included. Current efforts are aimed at improving these calculations and extending the investigations to other matter to gain insight as to the origin of spin.

7. Spectroscopy

A new generation of experiments devoted to hadron spectroscopy, including GlueX at Jefferson Lab, PANDA at GSI/FAIR and BES III intend to make measurements with unprecedented precision and in previously unexplored mass ranges and quantum-number sectors. A major goal of the USQCD collaboration is to extract from realistic lattice simulations predictions for masses, decay properties and relevant matrix elements in these domains.

Spectroscopy reveals fundamental aspects of hadronic physics:

- What are the essential degrees of freedom describing the low-energy spectrum of QCD?
- Are there gluonic excitations in mesons - exotic states beyond quark-antiquark states of quark model?
- Are there exotic baryon states beyond three quark states of the quark model - tetraquark, pentaquark, or molecular states?

The determination of the low lying hadrons is a milestone within the Hadron Physics part of the NSAC long range plan.

Excited states in QCD are notoriously difficult to compute. The problem is the signal for such states dies out quickly when measured across the lattice. Making the lattice spacing very small is a solution, but very costly. A pragmatic solution has been to adopt asymmetric lattices whereby

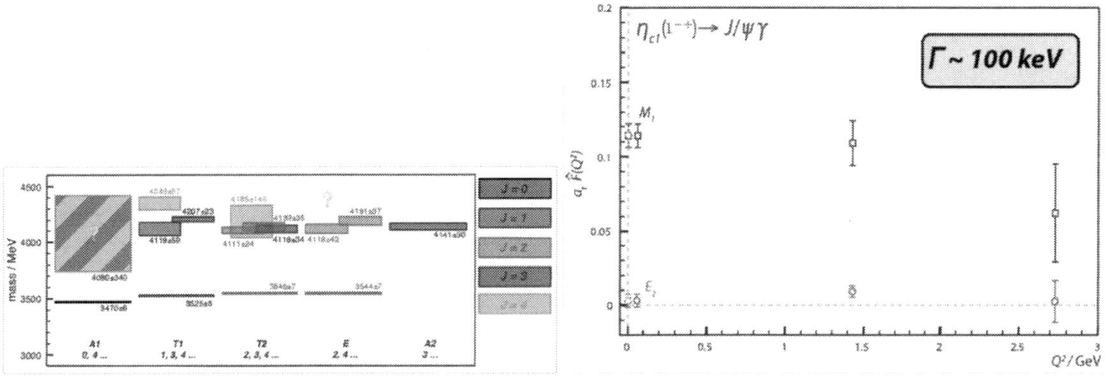

Figure 9. Left panel: high lying spectrum of charmonium J^{++} states from Ref. [19, 20]. The horizontal axis is the spin decomposition of the state, and the colors indicate the final spin J assignment. The lattice determination of the decays of these states has yielded new information as to their quark structure. Right panel: photon decay of exotic charmonium $1^{-+} \to\to 1^{--}\gamma$, whose rate is comparable to other non-exotic states indicating reasonable chance of measurement. The search for these exotic states is a major part of the new Hall D program at Jefferson Lab as part of its 12GeV upgrade.

gauge fields are generated with a temporal resolution much finer than in spatial directions [12]. These calculations have use the bulk of Lattice QCD time at ORNL. A recent account of these activities is in Ref. [7].

7.1. Nucleon excited states

Using these asymmetric lattices, recent calculations demonstrate clearly, for the first time, the ability to extract the nucleon excited state spectrum [17]. In Fig. 8 is shown the determination of the highly excited energy states of the nucleon, at a 400MeV pion mass, as well as a comparison to the experimentally known spectrum. The darker the symbol, the more well known is the experimental state.

The ground state nucleon is a spin $\frac{1}{2}$ particle. However, when excited, it jumps to higher spin states at higher energies. Many of these states are poorly determined experimentally which is where Lattice QCD can shine. For example, is there a different quark content for these excited states? What does stand-out in the comparison to the experimental spectrum is how patterns of states are repeating. Evidence is seen for a spin $\frac{5}{2}$ which has not been before in lattice QCD.

These are early stage calculations – more work is needed. The excited states will decay which is what will give us the answers to the quark content and structure. New techniques have been developed under SciDAC that enable these calculations which are underway [18].

7.2. Exotic states of matter

A crucial component of the 12GeV upgrade to Jefferson Lab is a new experimental hall that produce a very high intensity polarized photon beam that will probe matter in a novel manner. One possible outcome is that exotic states of matter will be produced where the quarks are bound with excited gluon fields. Meson particles can have a tell-tale signature of such exotic content. A crucial question is – do these exotic states exist, and can they be produced at such a rate they will be measurable experimentally?

Recent calculations [19, 20] claim that at least in a heavy quark mass limit called "charm", these exotic states likely exist, and production rates are indeed comparable with other states.

This quark mass region is the subject of intense experimental searches, for example the CLEO-c experiment at Cornell. An important result of these lattice calculations is the most extensive determination of the spectrum in the charm region to date as illustrated in Fig. 9. Shown is one part part of the spectrum at high spin. Techniques for determining the spin content of the state were developed, and used successfully in the determination of the the decay rate of exotic states as shown in the right panel. On going calculations will determine the exotic spectrum and decay rates for light quark mass regions.

Acknowledgments

This research used resources of the National Center for Computational Sciences at Oak Ridge National Laboratory, which is supported by the Office of Science of the Department of Energy under Contract DE-AC05-00OR22725. In particular, we made use of the Jaguar Cray XT facility, using time allocated through the US DOE INCITE program. This research was supported in part by the National Science Foundation through TeraGrid resources provided by Pittsburgh Supercomputing Center (PSC) as well as allocations on the Kraken Cray XT5 facility at the Univ. of Tennessee. Authored by Jefferson Science Associates, LLC under U.S. DOE Contract No. DE-AC05-06OR23177. The U.S. Government retains a non-exclusive, paid-up, irrevocable, world-wide license to publish or reproduce this manuscript for U.S. Government purposes.

References

[1] Duane S, Kennedy A D, Pendleton B J and Roweth D 1987 *Phys. Lett.* **B195** 216–222
[2] Clark M A 2006 *PoS* **LAT2006** 004 (*Preprint* `hep-lat/0610048`)
[3] Ishikawa K I 2008 (*Preprint* `0811.1661`)
[4] collaboration U http://www.usqcd.org
[5] software U http://www.usqcd.org/usqcd-software
[6] Edwards R G and Joo B (SciDAC Collaboration) 2005 *Nucl. Phys. B. Proc. Suppl.* **140** 832 (*Preprint* `hep-lat/0409003`)
[7] Joo B 2009 These proceedings
[8] Stathopoulos A and Orginos K 2007 (*Preprint* `arXiv:0707.0131`)
[9] Darnell D, Morgan R B and Wilcox W 2008 *Linear Algebra Appl.* **429** 2415–2434 (*Preprint* `0707.0502`)
[10] Luscher M 2007 *JHEP* **12** 011 (*Preprint* `0710.5417`)
[11] Clark M A *et al.* 2008 *PoS* **LATTICE2008** 035 (*Preprint* `0811.4331`)
[12] Lin H W *et al.* (Hadron Spectrum) 2009 *Phys. Rev.* **D79** 034502 (*Preprint* `0810.3588`)
[13] Luu T 2009 These proceedings
[14] Hagler P *et al.* (LHPC) 2008 *Phys. Rev.* **D77** 094502 (*Preprint* `0705.4295`)
[15] Airapetian A *et al.* (HERMES) 2007 *Phys. Rev.* **D75** 012007 (*Preprint* `hep-ex/0609039`)
[16] Ackerstaff K *et al.* (HERMES) 1999 *Phys. Lett.* **B464** 123–134 (*Preprint* `hep-ex/9906035`)
[17] Bulava J M *et al.* 2009 *Phys. Rev.* **D79** 034505 (*Preprint* `arXiv:0901.0027`)
[18] Peardon M *et al.* 2009 (*Preprint* `0905.2160`)
[19] Dudek J J, Edwards R G, Mathur N and Richards D G 2008 *Phys. Rev.* **D77** 034501 (*Preprint* `0707.4162`)
[20] Dudek J J, Edwards R and Thomas C E 2009 (*Preprint* `0902.2241`)

Lattice QCD simulations on big cats, sea monsters and clock towers

Bálint Joó on behalf of the USQCD Collaboration and the Hadron Spectrum Collaboration
Jefferson Lab, 12000 Jefferson Avenue, Newport News, VA 23606
E-mail: bjoo@jlab.org

Abstract. We present details of lattice QCD computations we are performing on the Cray XT series of computers, from BigBen – an XT3 hosted at the Pittsburgh Supercomputing Center (PSC) – through Jaguar (XT4) and Kraken (XT5) – which are hosted at the National Center for Computational Science (NCCS) and the National Institute of Computational Science (NICS), respectively, at Oak Ridge National Laboratory (ORNL). We discuss algorithmic tuning to make the computation more efficient and present some recent results.

1. Introduction

Lattice QCD is currently the only known model-independent, non-perturbative method for carrying out calculations of Quantum Chromodynamics (QCD), which are relevant to both Nuclear and High Energy physics. In this poster we report on progress on calculations which will enable the calculation of the excited state baryon spectrum, the photo-couplings of Hybrid Mesons and the investigation of multi-particle states. These calculations form the basis of DOE milestones in Hadronic Physics, and are of great relevance to the exprimental program at the Jefferson Lab.

In this contribution, we focus on the initial phase of the calculation: gauge configuration generation. The extraction of physical results requires that these configurations be subjected to sophisticated analysis campaigns. We will not discuss the analysis and results further here, but will rather refer the reader to other contributions in this volume, in particular those of Edwards [1], Richards [2] and Luu [3].

2. Hybrid Monte Carlo for gauge generation

Discretizing QCD onto a lattice proceeds by ascribing fermionic degrees of freedom to the sites of a space-time lattice, and the gluonic degrees of freedom to the lattice links. The gluon fields are represented by $SU(3)$ matrices on the lattice links, which correspond to *parallel transport operators* betwen points connected by the links. It is common to abuse notation and denote as U the collection link variables $U_\mu(x)$ for all the links on a given lattice.

Lattice QCD calculations generically involve evaluating the path integral

$$\langle \mathcal{O} \rangle = \frac{1}{\mathcal{Z}} \int \mathcal{D}U \; \mathcal{O}(U) \; \prod_{i=1}^{N_f} \det(M_i(U)) \; e^{-S(U)} \; , \tag{1}$$

where $\mathcal{O}$ is the observable of interest and $S(U)$ is the gauge action. The path integral has partition function $\mathcal{Z}$ and the measure $\mathcal{D}U$ is shorthand for integrating over every configuration U. The fermionic degrees of freedom are integrated out, and their effect amounts to a weight in the path integral of $\det(M_i)$ where M_i is the *fermionic kernel* for the i-th flavor of quark. We note that the work described in this article was carried out using anisotropic clover fermions.

An efficient way to evaluate the path integral is by a Monte Carlo method using importance sampling. This method produces lattice configurations U with a probability $P(U)$ so that

$$P(U) \propto \prod_{i=1}^{N_f} \det(M_i(U))\ e^{-S(U)} \tag{2}$$

and the expectation value reduces to a simple average: $\bar{\mathcal{O}} = \frac{1}{N}\sum_{\{U\}} \mathcal{O}(U)$. The workhorse of the gauge generation process is the Hybrid Monte Carlo [4] algorithm and its variants [5]. Treating the U link variables as canonical coordinates, these methods extend the configuration space to a phase space by assigning to each link a canonically conjugate momentum $\pi_\mu(x)$. One can then define a Hamiltonian $H = \frac{1}{2}\sum_{x,\mu} \pi(x,\mu)^2 + S(U)$. The algorithms iterate the following scheme: start from some initial gauge configuration U, generate fresh canonical momenta π and perform Hamiltonian Molecular Dynamics (MD) from this phase space state to a trial state (U', π'). The new state is subjected to a Metropolis acceptance test

$$P_{\text{acc}} = \min\left(1, e^{-H(U',\pi')+H(U,\pi)}\right) . \tag{3}$$

If the state is accepted the new starting state will be (U', π'), otherwise it will be the old state (U, π). We note that the MD algorithm should be reversible and area preserving, so that the Monte Carlo process satisfies the condition of detailed balance, which is sufficient for the generated Markov Chain to converge to the desired equilibrium distribution. To satisfy this condition symmetric, symplectic integration schemes are used. The acceptance rate of the algorithm depends on the energy change $\delta H = H(U', \pi') - H(U, \pi)$ along an MD trajectory. This can be controlled by varying the step-size(s) of the underlying integration scheme. Finally, we note that the primary cost of the algorithm comes from the fact that in order to evaluate the contribution from the quarks one must solve the system

$$M^\dagger M x = b \tag{4}$$

at least once per MD step (x and b are vectors). This system has a dimension of $O(V)$ where V is the lattice volume, and has condition number that varies inversely with the quark mass. Dimensions and condition numbers, each of $O(10^7)$, are common. Hence, typical avenues of optimization include: developing faster implementations of M, developing solver algorithms which converge fast and finding integration schemes which reduce δH on average so that one may take fewer, longer steps in the MD, thus requiring fewer solves.

3. Tuning the molecular dynamics integration

In our simulations, we employ a three level nested integration scheme where on each level we use the MD integrator due to Omelyan [6]. As described in [7], the leading order error term on a given level is given by

$$\left(\frac{6\lambda^2 - 6\lambda + 1}{12}A + \frac{1-6\lambda}{24}B\right)\delta\tau^2 \tag{5}$$

where λ is a tunable constant, $\delta\tau$ is the integration step-size, and A and B are Poisson Brackets (PB) of coordinate (gauge) and momentum update operators. In [7], minimising the variance of

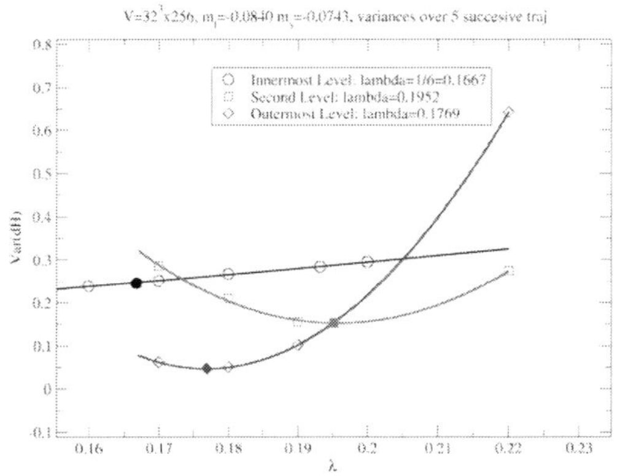

Figure 1. Tuning a 3 level MD integration scheme. We show the parameter λ from eq. (5) on the horizontal axis, and the variance of δH on the vertical axis. The three lines correspond to the innermost (circles), middle (squares) and outer (diamonds) integrators.

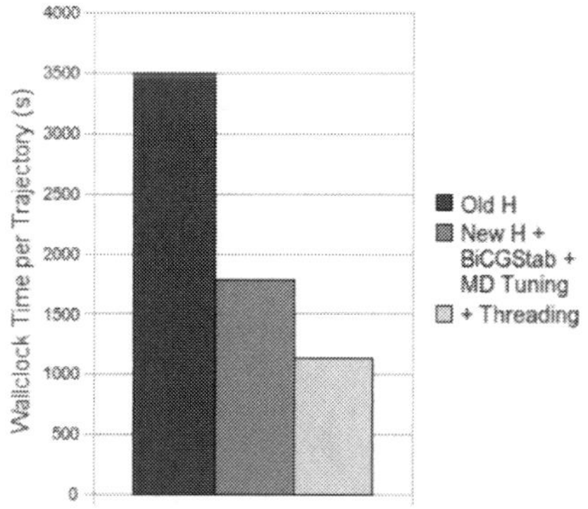

Figure 2. The effect of improvements

Table 1. Our target large volume data sets. We show the number of trajectories generated at each set of parameters at the time of writing.

	$m_\pi = 230MeV$	$m_\pi = 380MeV$
$V = 24^3 \times 128$	5380	12985
$V = 32^3 \times 256$	280	978

δH over an ensemble is advocated as the best way to tune an integration scheme. The optimal λ on each level can be determined in an experimental fashion, by varying it and measuring the δH on several trajectories to get an estimate of its variance. We have carried out the latter procedure on Kraken-XT at NICS and show our results in fig. 1, where filled symbols show our final interpolated minima. Our tuning was very rough: we carried out only 5 trajectories for each value of λ for each level of the integrator. Thus, in terms of trajectories the cost of the tuning is about 80 trajectories in total. While this is a small percentage of a full run which comprises several 1000 trajectories, it is a noticeable amount of computer time. It can be seen that we have reduced the variance of δH from ≈ 0.3 to $\approx$ 0.05, or by a factor of 6. This could in principle correspond to a factor of 6 increase in step-size, however in practice this is not attained as for such large outer steps, the integration would go unstable.

Apart from tuning our step-size, we have made several other improvements to our code. Whereas we used to solve eq. (4) with the Conjugate Gradients (CG) method, we have now switched over to a two step stabilized Bi-Conjugate Gradients (BiCGStab) procedure. Further, we have added threading to our linear operator M, our BLAS-like operations and most lattice site loops. The overall effect of these improvements is shown in fig. 2. Switching to BiCGStab and tuning have gained us a factor of 2 in wallclock time, with a small additional gain ($\approx 30\%$) from threading.

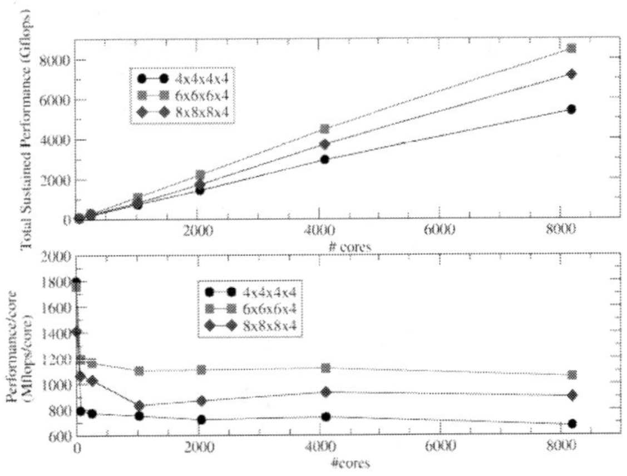

Figure 3. Weak scaling of the performance of our Conjugate Gradients code on Jaguar prior to quad core and CNL upgrades.

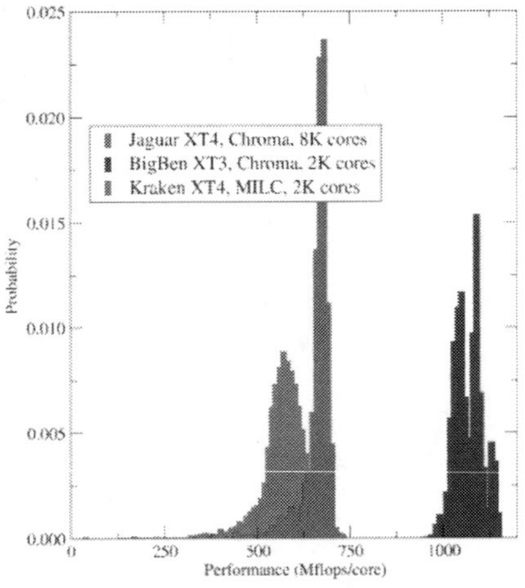

Figure 4. Performence Histograms from the Chroma Code on Jaguar XT4 (red) and BigBen (blue). We show also the performance of the MILC code on Kraken XT4 (green).

4. Cray experiences

Since 2007, we have used a variety of Cray XT platforms, starting with Jaguar at NCCS and BigBen at PSC. Both machines were then Cray XT3-s running the Catamount operating system. We have followed Jaguar through its various upgrades to dual, then quad cores; from XT3 to XT4 interconnects, and from Catamount to CNL operating systems. Currently we are running on Jaguar XT4, Jaguar XT5, Kraken XT5 at NICS and we are still active on BigBen at PSC through INCITE, NCCS discretionary and NSF Teragrid awards. Our primary code is the Chroma code [8] which was developed as part of the QCD Infrastructure project of SciDAC [9].

Our XT experiences are summarized in figs. 3 and 4. On BigBen we run jobs of size 2K cores, in single precision. On the XT3 platform we experienced good weak scaling as can be seen in fig 3 where we show the scaling of our code to 8192 cores on Jaguar after its dual core, but prior to its CNL upgrade. Indeed, the scaling is excellent. However, it can be seen that there is a distinct sweet spot in terms of local problem size (red line). Much larger local volumes fall out of cache (blue line) while much smaller problems are hit by surface to volume effects (black line).

In figure 4 we show normalized histograms of performances reported by the CG solvers on both BigBen, and on Jaguar XT4 post the quad core and CNL upgrade. While see some structure in the histogram for BigBen, essentially performances fall between 9.35-1.13 Gflops per core. On Jaguar XT4 the performances are substantially lower, around 450-700 Mflops/core. This may be due to the fact that on Jaguar we are working with a different global problem size, and that we are using 8192 cores, whereas on BigBen the partition size is 2048 cores. Nonetheless, the Jaguar histogram appears wider than the BigBen one with a shoulder extending toward lower performance numbers, with some as low or lower than 250Mflops/core. To ensure that this is not just a feature of our code or the large partitions, we show some data from running the MILC code on 2048 cores of Kraken XT4. This histogram is narrower than that of the Chroma code on Jaguar, but also exhibits a shoulder towards lower performances and is not immune from the occasional very low outlier.

5. Progress

We have made substantial progress in generating ensembles. Some initial results are presented in [10, 11]. We show our progress in table 1. We are concentrating on two volumes: $V = 24^3 \times 128$ and $V = 32^3 \times 256$ lattice sites respectively; and at two values of the quark mass which correspond to pion masses of $m_\pi = 230$ MeV and $m_\pi = 380$ MeV. The two volumes are required to ensure control over finite volume effects, and also because investigation of decays requires a finite size scaling analysis. With the INCITE 2008 resources, we have completed the run with $V = 24^3 \times 128$ and $m_\pi = 380$ MeV and with our 2008 NSF TeraGrid allocation we have begun the run at $V = 24^3 \times 128$ with a quark mass of $m_\pi = 230$ MeV which we are continuing on BigBen this year.

Using our INCITE 2009 and NSF Teragrid allocation, we have targeted the larger volume lattices using Kraken XT5 and Jaguar. We have not made as much progress as we intended for several reasons: Due to the large volumes used we found that single precision was no longer sufficient for the MD part of our simulations, so we moved to double precision. Secondly, we have scaled our running up to partitions of size 16384 cores and found a reduction in the performance of our code at this partition size. We have managed to compensate for some of this by the improvements in algorithm and software discussed earlier but progress remains slower than we'd like. We are currently in the process of investigating this issue.

6. Summary

We have discussed the program of anisotropic lattice gauge configuration generation and our experiences on the Cray XT platforms to date. In particular we have presented some algorithmic tuning results and touched on additional software improvements to our code. We discussed the performance of our code on the Cray XT architecture through its various incarnations.

7. Acknowledgements

We wish to thank Mike Clark for many helpful discussions on the tuning Molecular Dynamics, and Steve Gottlieb for supplying performance data from the MILC code. This research used resources of the National Center for Computational Sciences at Oak Ridge National Laboratory, which is supported by the Office of Science of the U.S. Department of Energy Under contract DE-AC05-00OR22725; and was supported in part by the National Science Foundation, through TeraGrid allocations at the Kraken Cray XT5 facility at the National Institute for Computational Sciences, The University of Tennessee and Oak Ridge National Laboratory and on the BigBen Cray XT3 facility at the Pittsburgh Supercomputing Center. Authored by Jefferson Science Associates, LLC under U.S. DOE Contract No. DE-AC05-06OR23177. The U.S. Government retains a non-exclusive, paid-up, irrevocable, world-wide license to publish or reproduce this manuscript for U.S. Government purposes.

8. References

[1] Edwards R G these proceedings
[2] Richards D G these proceedings
[3] Luu T C these proceedings
[4] Duane S, Kennedy A D, Pendleton B J and Roweth D 1987 *Phys. Lett.* **B195** 216–222
[5] Clark M A, Kennedy A D and Sroczynski Z 2005 *Nucl. Phys. Proc. Suppl.* **140** 835–837 (*Preprint* `hep-lat/0409133`)
[6] I P Omelyan I M M and Folk R 2003 *Comput. Phys. Commun.* **151** 272
[7] Clark M A, Kennedy A D and Silva P J 2008 Tuning HMC using Poisson brackets (*Preprint* `0810.1315`)
[8] Edwards R G and Joo B (SciDAC) 2005 *Nucl. Phys. Proc. Suppl.* **140** 832 (*Preprint* `hep-lat/0409003`)
[9] Joo B 2007 *J. Phys. Conf. Ser.* **78** 012034
[10] Bulava J M *et al.* 2009 *Phys. Rev.* **D79** 034505 (*Preprint* `0901.0027`)
[11] Lin H W *et al.* (Hadron Spectrum) 2009 *Phys. Rev.* **D79** 034502 (*Preprint* `0810.3588`)

Journal of Physics: Conference Series **180** (2009) 012071 doi:10.1088/1742-6596/180/1/012071

Simulating the quark-gluon plasma

L. Levkova,
HotQCD Collaboration[1]

Physics Department, University of Utah, Salt Lake City, Utah 84112, USA

E-mail: `ludmila@physics.utah.edu`

Abstract. We present results for the equation of state of quark-gluon plasma in 2+1 flavor QCD with physical strange quark mass and close to the physical light quark masses. We also study the chiral condensate and susceptibility which are quantities sensitive to the chiral symmetry restoration aspects of the transition between ordinary matter and the quark-gluon plasma.

1. Introduction

The quark-gluon plasma is the state in which the matter of the early universe existed microseconds after the Big Bang. As the universe cooled down to temperatures about five orders of magnitude higher than that of the core of the Sun, the quark-gluon plasma underwent a transition into the ordinary matter we observe today. Currently, the primordial conditions which allow for the quark-gluon plasma to form are recreated in the heavy-ion collision experiments conducted at RHIC (The Relativistic Heavy-Ion Collider at the Brookhaven National Laboratory) and elsewhere. To interpret the experimental data, knowledge of the quark-gluon equation of state (*i.e.* the dependence of energy density and pressure on temperature) is essential. We present here the results for the equation of state obtained by the HotQCD collaboration using the only nonperturbative, first-principles method currently available – lattice QCD, the discretized version of the fundamental theory of the strong interactions. We simulate the QCD thermal statistical ensemble on a discrete four dimensional lattice representing the space-time. To approximate the physical world as closely as possible, we need to perform our computations on fine enough lattices that the discretization errors are small and well controlled. To gain a deeper understanding of the discretization effects we used two different lattice actions, the p4 and the asqtad actions [1], to perform two independent calculations of all the physical quantities of interest. The discretization effects for the two actions are not the same, but the results obtained with them should show consistency if the these effects are small. With both actions we simulate a quark-gluon system with two light degenerate sea quarks and one heavy sea quark whose mass, m_s, is tuned to the physical strange quark mass. The light quark masses, m_l, are one-tenth of m_s. To check for discretization errors, we also compare the equation of state obtained with two different temporal extents of

[1] A. Bazavov, T. Bhattacharya, M. Cheng, N.H. Christ, C. DeTar, S. Ejiri, Steven Gottlieb, R. Gupta, U.M. Heller, K. Huebner, C. Jung, F. Karsch, E. Laermann, L. Levkova, C. Miao, R.D. Mawhinney, P. Petreczky, C. Schmidt, R.A. Soltz, W. Soeldner, R. Sugar, D. Toussaint and P. Vranas.

the lattice, $N_\tau = 8$ and 6. For both actions the discretization effects are smaller as N_τ becomes larger. Larger N_τ also increases the expense of the calculation in terms of computer power.

2. The QCD equation of state

We employ the so called integral method of the determination of the equation of state of the quark-gluon plasma. At the basis of this method is the calculation of the energy-momentum tensor trace anomaly $\Theta_{\mu\mu}$:

$$\frac{\Theta^{\mu\mu}(T)}{T^4} \equiv \frac{\epsilon - 3p}{T^4} = T\frac{\partial}{\partial T}(p/T^4) \ , \tag{1}$$

where ϵ is the energy density, p is the pressure and T is the temperature. From the above it follows that the pressure can be obtained by integrating the trace anomaly:

$$\frac{p(T)}{T^4} - \frac{p(T_0)}{T_0^4} = \int_{T_0}^{T} \mathrm{d}T' \frac{1}{T'^5}\Theta^{\mu\mu}(T') \ . \tag{2}$$

The integration is along lines of constant physics [1]. The lower integration limit T_0 is chosen such that $p(T_0)$ is small. The energy density can be obtained as a simple linear combination:

$$\frac{\epsilon(T)}{T^4} = \frac{3p(T)}{T^4} + \frac{\Theta^{\mu\mu}(T)}{T^4} \ . \tag{3}$$

Figure 1 shows our results for the trace anomaly. From the data we can conclude that away

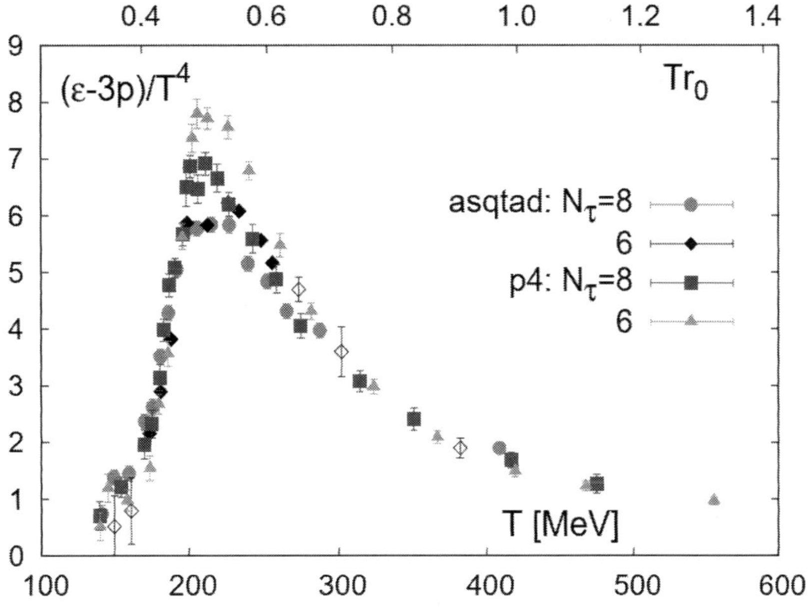

Figure 1. The trace anomaly *vs* the temperature in MeV (lower x-axis) and in units of the scale parameter $r_0 = 0.318$ fm (upper x-axis), determined form the static quark potential [1].

from the peak region, we have a good agreement between the two actions and the two N_τ's. In

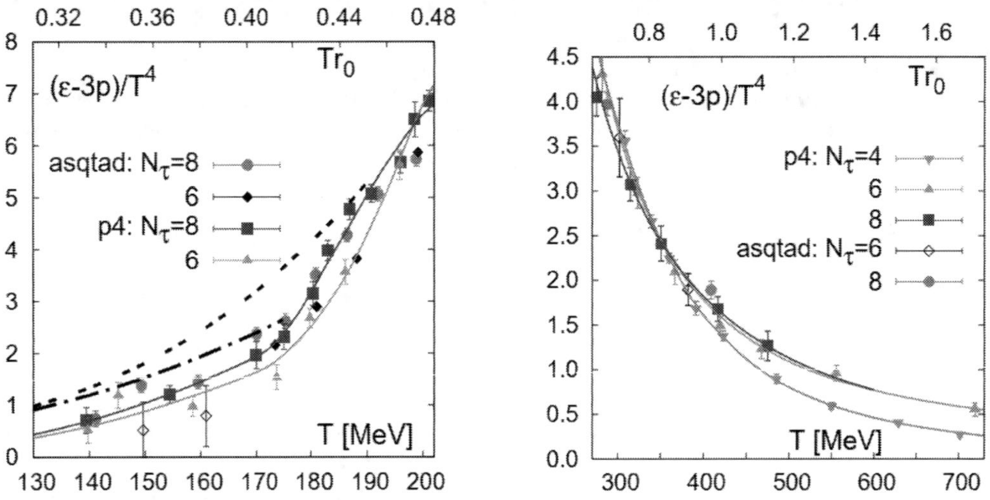

Figure 2. The trace anomaly at low (left panel) and high (right panel) temperatures. The dashed-dotted and the dashed curves in the left panel give the trace anomaly in a hadron resonance gas model with two different cutoffs for the heaviest resonance: 1.5 and 2.5 GeV, respectively. For more details on the fits to the lattice data see [1].

the peak region the two action show some differences. The p4 action gives a higher peak due to somewhat larger discretization effects. Figure 2 shows in more detail the trace anomaly in the high- and low-temperature regions. The behavior of the trace anomaly at low temperatures is commonly approximated with a hadron resonance gas model. In the left panel of Fig. 2 we compare it with our data at two resonance cutoffs of 1.5 and 2.5 GeV. In both cases the lattice results stay below the theoretical prediction. We attribute this discrepancy to the fact that both actions distort the hadronic spectrum at low temperature where the discretization error is larger than at higher T. The high-temperature behavior can be studied from the right panel of Fig. 2, where we can see that the two actions and the $N_\tau = 6$ and 8 cases show a good consistency. Also shown there are results with the p4 action at $N_\tau = 4$, which disagree with the rest of the data due to the large discretization errors at such a short temporal extent.

The energy density and pressure results are presented in Fig. 3. Both of these bulk thermodynamics quantities have a similar behavior: within the region of the transition they show a rapid growth which slows down as the temperature increases further. The rapid growth in the transition region is due to deconfinement which leads to the proliferation of new colored degrees of freedom. Still, even at temperatures of about 550 MeV the pressure and energy density do not reach the Stefan-Boltzmann ideal gas values, which means that strong interactions persist in the plasma above the transition temperature. The two actions give results which differ by no more than 10% at about 200 MeV and less than 5% at higher temperatures.

Of special interest to the hydrodynamics models of heavy-ion collisions is the speed of sound in the quark-gluon plasma. We extract it from the ratio of the pressure and energy density p/ϵ, which is obtained from the ratio of interpolating curves for the trace anomaly and pressure. The speed of sound, c_s, is given by:

$$c_s^2 = \frac{\mathrm{d}p}{\mathrm{d}\epsilon} = \epsilon\frac{\mathrm{d}(p/\epsilon)}{\mathrm{d}\epsilon} + \frac{p}{\epsilon} \ . \tag{4}$$

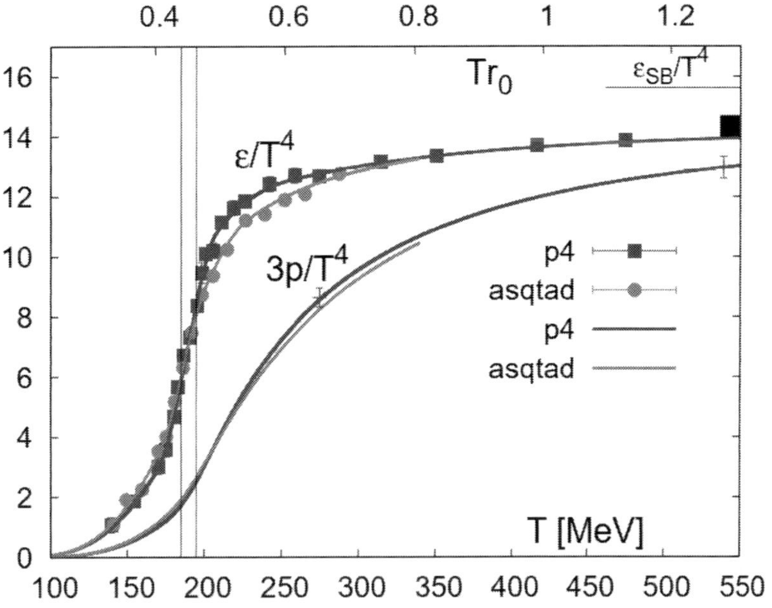

Figure 3. Energy density and three times the pressure with the p4 and asqtad actions at $N_\tau = 8$. The error bars on the pressure shown at two temperatures indicate the systematic errors arising from the particular integration schemes used. The black filled bar shows the systematic shift to the data needed to match with hadron resonance gas at low temperature. For more details refer to [1].

Figure 4 presents our results for p/ϵ and c_s^2, where the former is also compared with the hadron resonance gas prediction with a cutoff of 2.5 GeV. The determination of the speed of sound is sensitive to the details of the interpolating techniques and the spread of the curves can be interpreted as an estimate of the systematic errors involved in the calculation.

3. Chiral symmetry restoration

QCD has a "chiral" symmetry that is spontaneously broken at low T and restored at high T, when the quarks are massless. The transition between ordinary matter and a quark-gluon plasma combines features of both deconfinement and chiral symmetry restoration. In the previous section we presented our results for thermodynamic quantities, such as energy density and pressure, which are sensitive to the deconfining properties of the transition. Here we discuss observables which are related to the chiral symmetry restoration, namely the chiral condensate ($\langle\bar{\psi}\psi\rangle_q$, $q = l, s$ denotes quark flavor) and the chiral susceptibility (χ_q). The chiral condensate is an order parameter for the spontaneous symmetry breaking in the case of massless quarks. In our calculations (as in the real world) the quarks are not massless, which means that the chiral condensate is at most an approximate order parameter. To avoid ultraviolet singularities and a multiplicative renormalization in the chiral condensate, on the lattice we study the observable $\Delta_{l,s}$ constructed as:

$$\Delta_{l,s}(T) = \frac{\langle\bar{\psi}\psi\rangle_{l,T} - \frac{m_l}{m_s}\langle\bar{\psi}\psi\rangle_{s,T}}{\langle\bar{\psi}\psi\rangle_{l,0} - \frac{m_l}{m_s}\langle\bar{\psi}\psi\rangle_{s,0}} \ . \tag{5}$$

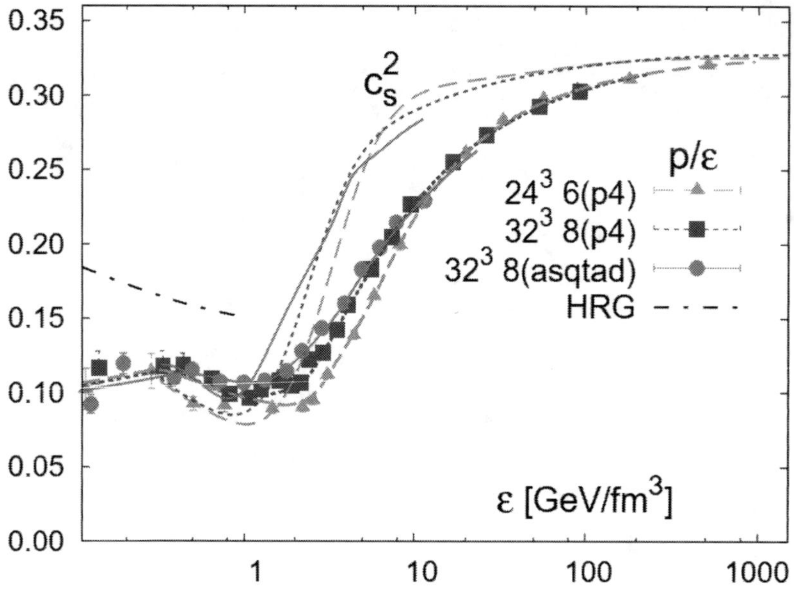

Figure 4. The ratio of pressure and energy density, p/ϵ and the square of the speed of sound, c_s^2, obtained at $N_\tau = 6$ and 8. The curves with no data points represent the square of the speed of sound. The hadron resonance gas prediction for p/ϵ with a cutoff of 2.5 GeV is given by the dash-dotted curve.

In the left panel of Fig. 5 we show the results for $\Delta_{l,s}$. The data shows some small differences between the two actions and the two N_τ's. The vertical band denotes the temperature region where we would place the chiral transition. This band has been determined from the peak in the chiral susceptibility shown on the right panel of Fig. 5. The chiral susceptibility measures the fluctuations in the chiral condensate, which become infinite at the transition temperature in the chiral theory. In our case of massive quarks, the peak has a finite height and it shows a shoulder-like shape which complicates the definition of a single value for the transition temperature. Instead we choose to refer to the transition as happening in a range of temperatures: 180 MeV $\lesssim T \lesssim$ 200 MeV. We expect this region to shift downward in temperature by about 10 MeV when the full continuum and the physical quark mass limits are taken.

References

[1] Bazavov A. *et al.* [HotQCD Collaboration] 2009 Equation of state and QCD transition at finite temperature `arXiv:0903.4379`

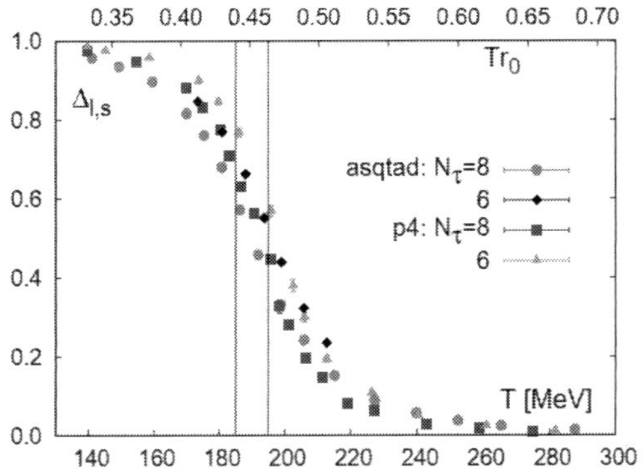

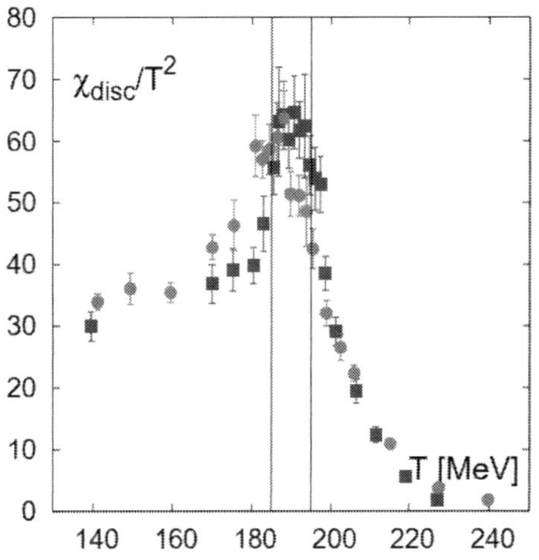

Figure 5. The subtracted and normalized chiral condensate $\Delta_{l,s}$ (left panel) and the disconnected chiral susceptibility $\chi_{\rm disc}/T^2$ (right panel).

An optimization toolkit for the Blue Gene and subsequent generations of SciDAC computers

A V Pochinsky

Laboratory for Nuclear Science, Massachusetts Institute of Technology,
77 Masssachusetts Avenue, Cambrdige, MA 02139, USA

E-mail: `avp@mit.edu`

Abstract. A tool for writing platform-independent optimized computational routines, QA_0, is described. Application to optimizing SciDAC Lattice QCD codes is presented.

1. Introduction

Large scale computation relies heavily on floating point operations. Most of SciDAC applications are numerical codes that run on the largest computers in the world. Making efficient use of compute power is of the utmost importance.

The Top 500 List [1] contains four different architectures within the first one hundred fastest computers. While some architectures will move out of the list, the situation will likely remain fundamentally the same as new players enter the field. This observation makes it attractive to be able to run an application on may different architectures. However, there is a catch. While the most portable C code could be run unmodified on a new machine, in practice one can not expect to achieve good performance without an extra effort. Code tuning for a new machine is expected, sometimes at a considerable cost. Code tuning is time consuming, breaks portability, and is prone to subtle bugs. If, however, the benefits of higher performance outweigh these problems, one has to choose between several extremes. Writing critical parts of the code in a machine language is one extreme. Another approach is to use compiler-specific pragmas and language extensions to improve code generation. In addition to breaking portability, this method exposes dark corners of the compiler implementations where bugs often lurk.

In this work we explore another approach. An abstraction layer below the high level language and machine code is designed and implemented that allows the programmer to express a limited set of computation in a portable way that however lends itself to compilation into highly optimized code.

2. Application domains

A large and important class of applications could be expressed in terms of a massively data parallel model, e.g. highly successful USQCD application suite makes the data parallel model central to its application domain libraries [2]. Operations on arrays are performed element-wise with well localized regular communications in form of shifts and reductions. On modern parallel systems one usually does not write the data parallel code explicitly. Instead, the program is converted by hand into an MPI program assuming fat nodes and reliable communication between the nodes. In addition, physical problems often exhibit data locality: to make progress computation only needs to access a limited set of neighbors of a given element. In addition, presently lattice quantum chromodynamics (LQCD) algorithms experience fast evolution, thus making the area a good testbed for tool development.

2.1. Quantum chromodynamics

The QCD code spends more than 90% of its time a computation of the following kind:

$$\forall x : \phi^{a\alpha}[x] = \sum_{\mu} \left((U_\mu[x])^a_b(\delta^\alpha_\beta + \gamma^\alpha_\beta[\mu])\varphi^{b\beta}[x+\hat{\mu}] + (U^\dagger_\mu[x-\hat{\mu}])^a_b(\delta^\alpha_\beta - \gamma^\alpha_\beta[\mu])\varphi^{b\beta}[x-\hat{\mu}] \right).$$

Computations like this one lend themselves very well to parallel execution with plenty of opportunities to overlap computation and communication. Here we concentrate on the computation part, since communication is to the nearest neighbors and standard tricks of overlapping computation and communication are simple to apply by hand.

When matrix operations are unrolled on a site, the resulting basic blocks become long enough to benefit from instruction scheduling. There is no need to guess which loops should be unrolled. In addition, LQCD algorithms are insensitive to the details of floating point operations: for code transformation purposes one does not need to worry about round off details: floating point operations may be presumed exact and associative. Overall, the LQCD code presents more optimization opportunities than the standard semantics of a programming language would allow.

3. QA_0 abstraction layer

Instead of trying to twist the compiler into doing optimization we want, or making a programmer into unrolling the loops and making code transformations by hand, we have designed an abstraction layer, QA_0 that does exactly that. It takes as an input a high level language procedure and slavishly does a fixed set of loop unrolling and constant foldings specific to the application domain. The resulting register transfer representation of the procedure is converted into a form suitable for a particular target machine.

The QA_0 input language has following features:

- All memory accesses are explicit. There are only two operations on memory: `load` and `store`. All other operations take inputs from registers and place results into registers.
- Non-memory operations have no effects other than producing results.
- Control flow is explicit. There are loops, forward branches and procedure returns in the present version. Only leaf procedures are supported.
- In addition to integer, pointer and floating point registers, there are "registers" for common QCD data: gauge matrices, Dirac fermions, staggered fermions, projected fermions and complex numbers.
- Both single and double precision is supported.
- Instruction set is extensible with user level macros to simplify coding.
- Like HLL, the same input language is used to generate code for different backends.

The language also includes a hygenic macro system that was found invaluable in writing long procedures.

3.1. Target application domains

Currently QA_0 supports only LQCD as an application domain. Other areas can be supported in a straightforward way. One only needs to define application specific types, operations and a set of transformation rules to reduce the operations on objects of these types into a simple form.

3.2. Supported targets
Currently QA_0 supports the following targets

C99–32 Standard C with built-in complex types on 32-bit machines.

C99–64 Standard C with built-in complex types on 64-bit machines.

Cee–32 Traditional C (complex operations expanded into floats) on 32-bit machines.

Cee–64 Traditional C (complex operations expanded into floats) on 64-bit machines.

xlc–BG/X BG/L and BG/P with the Double Hummer support using IBM XLC as a backend.

Target **Cee–32** is used extensively for development and testing and serves as a reference for verification of other back ends.

C99–32 and **C99–64** are comparison targets to cross check performance of other back ends. At present code generated for complex arithmetics in C compilers is rather uneven: small changes in a procedure source result in drastic changes in performance numbers.

For Blue Gene computers all complex arithmetic is expanded into XLC's Double Hummer intrinsics. This way we benefit from instruction scheduling and code generation while avoiding hand coding. XLC also take care of the C API, thus simplifying the QA_0 back end.

3.3. QA_0 Example
As an example of the QA_0 procedure, coding of

$$\forall x : \psi^{a\mu}[x] \leftarrow s\zeta^{a\mu}[x] + U^a_b[x]\phi^{b\mu}[x]$$

looks as follows:

```
(procedure sample ([stem "sam"] count-flops)
            ([psi     pointer "psi"   "struct Fermion *"]
             [s       double  "s"     "double"]
             [zeta    pointer "zeta"  "const struct Fermion *"]
             [U       pointer "U"     "const struct Gauge *"]
             [phi     pointer "phi"   "const struct Fermion *"]
             [len     int     "len"   "int"])
   (loop () ([i (const 0) (reg len)])
      (load qcd-fermion () r-zeta ([reg zeta]))
      (load qcd-fermion () r-phi ([reg phi]))
      (load qcd-qcd-su-n () r-U ([reg U]))
      (op qcd-mulf () (r-U*phi) ([reg r-U] [reg r-phi]))
      (op qcd-madd-fermion () (r-psi) ([reg s] [reg r-zeta] [reg r-U*phi]))
      (store qcd-fermion () ([reg psi]) [reg r-psi])
      (op pointer-add () (psi) ([reg psi] [const (size-of Fermion)]))
      (op pointer-add () (zeta) ([reg zeta] [const (size-of Fermion)]))
      (op pointer-add () (phi) ([reg phi] [const (size-of Fermion)]))
      (op pointer-add () (U) ([reg U] [const (size-of SU-n)]))))
```

The S-expression surface syntax is presently used because the tool is still in flux; once semantics requirements to the system are better understood a more traditional C-like syntax will be provided as well.

Table 1. BG/P strong scaling behavior.

#cpu	float, MFlops/cpu	double, MFlops/cpu
256	814	626
512	808	620
1024	755	561
2048	802	609
4096	813	622

Table 2. SiCortex strong scaling behavior.

#cpu	float, MFlops/cpu	double, MFlops/cpu
128	195	181
256	193	176
512	191	173

4. Results

We compare performance of one of the QCD inverters on different parallel systems. In the case of Blue Gene /P [3], IBM's XLC compiler was used as a back-end for QA_0. Though it does not produce the most optimal code, it is sufficiently good for our purposes. In all cases QMP over MPI was used for communication. For the SiCortex machine [4] we also used the vendor's compiler. In both cases it is possible to generate a better code for QCD than the current compilers are capable of, however, the required effort may be justified if significant time is available on a respective machine.

4.1. Strong scaling

For a given problem size strong scaling shows how far a job could be stratched accross multiple processors. We used our production lattice size, $32^3 \times 64 \times 16$, in this study. Tables 1 and 2 show per processing element performance on BG/P and SiCortex computers respectively.

4.2. Weak scaling

For weak scaling studies we keep the local problem size fixed and change the machine size. A single node performance shows how well the processor is utilized, while various machine sizes show effects of the network on the perfomance.

The BG/P performance is shown on figure 1. One can see that communications–computation separation model really works: for all local sizes performance scales linearly with the machine size. (One node data point serves as a reference, four nodes case is peculiar to the BG/P—it does not involve the torus fabric at all.)

On SiCortex, performance is also remarkably flat, see figure 2. Relatively low share of peak achived is caused by a lack of hardware prefetch capabilities in the processor.

5. Conclusions

QA_0 is a valuable tool for writing performance critical routines. By combining platform independence and exposing a low level processor abstraction to the application programmer,

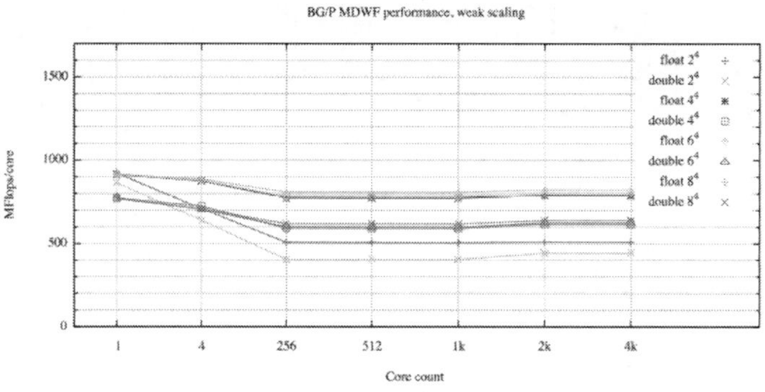

Figure 1. BG/P weak scaling.

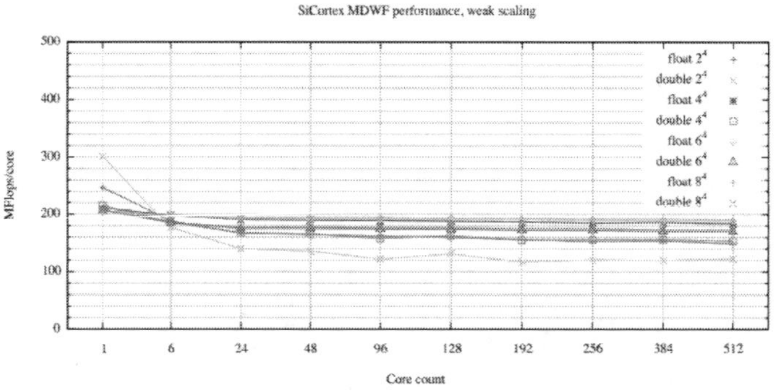

Figure 2. SiCortex weak scaling.

QA_0 makes it possible to achieve performance of hand crafted code without writing in the assembly language.

Overall design of QA_0 makes is suitable for other application domains. This direction is being exploited as well.

The tool is available under an open source license at the SciDAC Outreach LQCD page [5].

Acknowledgments

This work was supported in part by the DOE Office of Nuclear Physics under grant DEFG02-94ER40818. This research used resources under the INCITE and ESP programs of the Argonne Leadership Computing Facility at Argonne National Laboratory, which is supported by the Office of Science of the U.S. Department of Energy under contract DE-AC02-06CH11357, resources provided by the DOE through its support of the MIT Blue Gene/L. IBM has provided early access to BG/P. Sicortex Inc. has graciously provided access to its hardware for scaling studies.

References

[1] `http://www.top500.org/`
[2] `http://www.usqcd.org/software.html`
[3] `http://www.ibm.com/`
[4] `http://www.sicortex.com/`
[5] `https://outreach.scidac.gov/frs/download.php/125/qa0-1.4.0.tar.bz2`

Deflation for inversion with multiple right-hand sides in QCD

A Stathopoulos[1], **A M Abdel-Rehim**[1] **and K Orginos**[2]

[1] Department of Computer Science, College of William and Mary, Williamsburg, VA 23187
[2] Department of Physics, College of William and Mary, Williamsburg, VA 23187

E-mail: andreas@cs.wm.edu,amrehim@cs.wm.edu,kostas@wm.edu

Abstract. Most calculations in lattice Quantum Chromodynamics (QCD) involve the solution of a series of linear systems of equations with exceedingly large matrices and a large number of right hand sides. Iterative methods for these problems can be sped up significantly if we deflate approximations of appropriate invariant spaces from the initial guesses. Recently we have developed *eigCG*, a modification of the Conjugate Gradient (CG) method, which while solving a linear system can reuse a window of the CG vectors to compute eigenvectors almost as accurately as the Lanczos method. The number of approximate eigenvectors can increase as more systems are solved. In this paper we review some of the characteristics of eigCG and show how it helps remove the critical slowdown in QCD calculations. Moreover, we study scaling with lattice volume and an extension of the technique to nonsymmetric problems.

1. Introduction

Lattice QCD is one of the most computationally demanding applications. The heart of the computation is the solution of the lattice-Dirac equation in a Euclidean space-time lattice [1], which translates to a linear system of equations $Mx = b$, often for a large number of right hand sides [2]. Because of large matrix size, iterative methods are the only way to solve the problem. The Dirac operator M is γ_5-Hermitian, or $\gamma_5 M = M^H \gamma_5$, for some γ_5 matrix, but it is often preferable to solve the normal equations $A = M^H M$ problem. Also, $M = m_q I - D$, where m_q is a parameter related to the quark mass and D is an operator. Beyond the very large dimension and number of right hand sides, M becomes increasingly ill-conditioned as $m_q \rightarrow m_{critical}$. In lattice QCD this is known as critical slowdown and is a limiting computational factor.

The solution of linear systems with many right hand sides is a challenging problem. Iterative methods that solve each system one by one tend to regenerate search directions within previously explored subspaces, thus wasting iterations. Seed and block methods are often used to share information between systems [3, 4, 5]. For Hermitian matrices, the most useful information lies in invariant subspaces near zero. Deflating these eigenpairs significantly improves conditioning of iterative methods. Such ideas have been tried in QCD [6], but effective deflation methods have only appeared recently. Moreover, one does not want to precompute the required eigenpairs through an eigensolver but to reuse the information built by the linear solvers.

For non-Hermitian matrices, the GMRESDR method [7, 8] computes approximate eigenspace information during GMRES(m). In GMRESDR, when the GMRES basis reaches m vectors, it is restarted not only with the residual of the approximate solution of the linear system but also with the $nev < m$ smallest magnitude harmonic Ritz pairs. Therefore, being equivalent to a version of the Implicitly Restarted Arnoldi, GMRESDR solves the linear system and at the same time converges to the nev smallest magnitude eigenpairs.

The GMRESDR approach transfers also to the Hermitian case. For Hermitian systems, however, GMRESDR is much more expensive per step than CG and, more importantly, because it restarts every m steps, it impairs both the optimal convergence of CG and the eigenvalue convergence of unrestarted Lanczos. Recently, a Recycled MINRES method was developed in [9] which reuses a window of m iteration vectors of the MINRES to approximate nev eigenvectors. When m vectors have been reached the window is restarted similarly to GMRESDR, but the MINRES itself is not restarted. Our independently developed eigCG method similarly keeps a window of m vectors without restarting CG but restarts the window in a locally optimal way, thus achieving nearly optimal convergence for the eigenvalues [10].

2. EigCG and incremental EigCG

It is well known that the residuals produced by the CG iteration are co-linear with the Lanczos iteration vectors, and that the Lanczos tridiagonal projection matrix can be obtained through the CG coefficients [11]. Therefore, if we kept all the CG vectors, V, we could solve the Lanczos projected eigenvalue problem V^TAV and obtain approximate eigenpairs that satisfy a certain optimality. This however would imply unlimited storage for V or it would require a second CG pass to recompute V on the fly. Moreover, spurious eigenvalues would have to be dealt with.

The idea in our eigCG method is to use a limited memory window V with CG residuals as an eigenvector search space to keep track of the lowest nev Ritz vectors of the matrix A. When V grows to $m > 2nev$ vectors, we restart it as follows: We combine $u_i^{(m)}$, $i = 1, nev$, lowest Ritz vectors at the m-th step (thick restarting) with their nev locally optimal CG recurrence vectors ($u_i^{(m-1)}$ from $m-1$ step) and restart the basis V with an orthonormal basis for this set [12]:

$$\left[u_1^{(m)}, u_2^{(m)}, \ldots, u_{nev}^{(m)}, u_1^{(m-1)}, \ldots, u_{nev}^{(m-1)}\right]. \tag{1}$$

In contrast to a restarted eigensolver, we do not use V to determine the next search direction. Instead, we leverage a window of the following $m - 2nev$ residuals computed by the unrestarted CG to extend the search space. The linear system convergence is unaffected. The proposed algorithm, eigCG, can be implemented on top of CG in an efficient way that does not require extra matrix-vector multiplications or long vector orthogonalizations, and that is fully parallelizable. For details the reader is referred to [10].

Once a set of approximate eigenvectors U have been computed, we can "deflate" subsequent CG runs by using the init-CG approach [13]. Given an initial guess $\tilde{x}_0$, we consider another guess whose U components are removed through the following Galerkin oblique projection:

$$x_0 = \tilde{x}_0 + U(U^HAU)^{-1}U^H(b - A\tilde{x}_0). \tag{2}$$

Then x_0 is passed as initial guess to CG. The init-CG converges similarly to a deflated system until the linear system error approaches the eigenvalue error in U. At that point, we restart init-CG, re-applying the initial deflation. One or two restarts are sufficient to obtain linear convergence, equaling the convergence of a linear system deflated with the accurate eigenvalues.

Solving one linear system would rarely yield enough eigenvector approximations to effectively deflate subsequent systems. Our Incremental eigCG employs a simple outer scheme that calls eigCG for $n_1 < s$ of the s right hand sides and accumulates the resulting nev approximate Ritz vectors from each run into a larger set, U. While solving each new system, this incremental scheme computes additional Ritz vectors and also improves the ones that have not sufficiently converged in previous systems.

In all our experiments and with modest values of nev, eigCG has demonstrated the surprising property of converging to the nev targeted eigenvectors with a convergence rate identical to

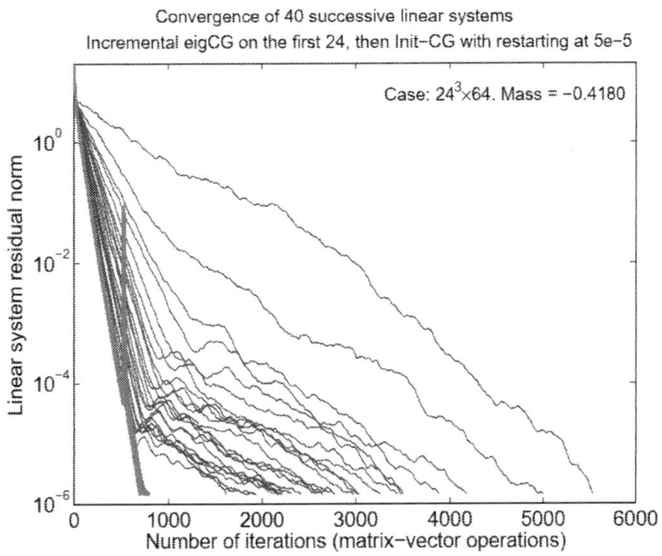

Figure 1. Improving convergence for the first 24 linear systems with Incremental eigCG(10,100) (blue). After the first 24 systems, we use the 240 vectors in U to deflate init-CG for another 16 vectors (red). Restarting is not used during the Incremental part, hence the knee in convergence curves. Restarting once at 5E-5 restores linear convergence. This is the 10 million matrix with a quark mass very close to the critical mass which causes the ill conditioning. The majority of the systems are solved with a speedup of 8.

that of unrestarted Lanczos. For example, a parameter $nev = 8$ has been sufficient for this behavior, while increasing m (the window size) offers only computational, not convergence, benefits. Several experiments in [10] study various aspects of this behavior.

3. Speeding up linear systems in QCD

We have implemented our algorithms in C and interfaced with Chroma, a lattice QCD C++ software base developed at Jefferson National Lab [14]. We have run both on an 8 node dual socket, dual core cluster with 4GB of memory per node, and on 256 processors of the Cray XT4 at NERSC, Lawerence Berkeley National Lab. Our computations involve all-to-all propagators which require the solution of several hundreds of right hand sides. Thus, we can afford to run Incremental eigCG on the first 24-32 systems accumulating about 250 vectors in U. For the rest of the systems, we use this U to deflate the init-CG as we described earlier.

The Dirac matrices used in this paper come from two ensembles of an anisotropic, two flavor dynamical Wilson fermion calculation. In both cases the sea pion mass is about 400(36) MeV and the lattice spacing is 0.108(7)fm. The anisotropy factor of the time direction is about 3 and the critical mass was determined to be -0.4188 in all cases. We consider three typical lattice sizes: one $16^3 \times 64$ and with the 12 variables per point the matrix size is $N = 3,145,728$; one $24^3 \times 64$ for a matrix size of $N = 10,616,832$; one $32^3 \times 96$ for a matrix size of $N = 37,748,736$.

Figure 1 shows a typical convergence behavior for Incremental eigCG for the 10 million matrix and a quark mass close to critical. The more ill-conditioned a matrix is, the more iterations it requires initially, so eigCG has enough time to identify more eigenvalues and more accurately.

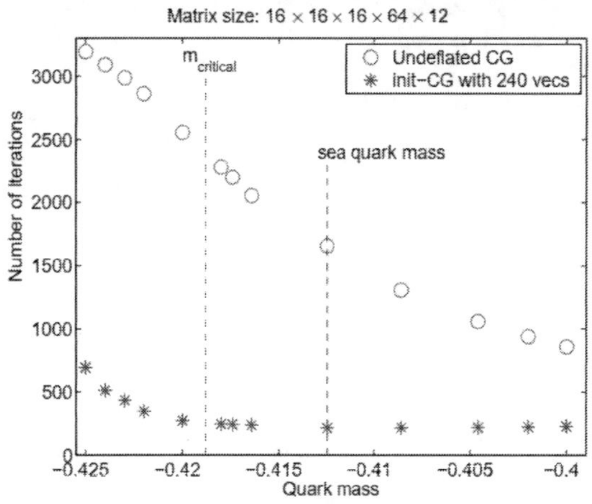

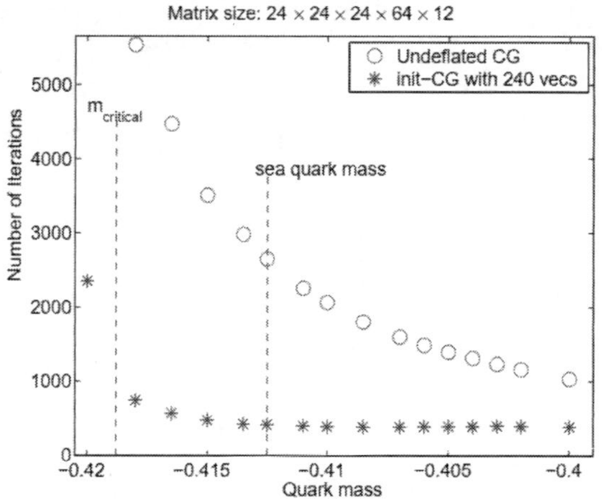

Figure 2. For each quark mass and its corresponding matrix we plot the average number of iterations required by init-CG to solve the rest 24 systems. For comparison the number of iterations of the non-deflated CG is reported. Critical slowdown is removed.

3.1. Removing critical slowdown in QCD

As the quark mass approaches $m_{critical}$, conditioning of the system rapidly deteriorates. However, shifting cannot change the density of the eigenvalues. Ill conditioning is simply due to a small number of eigenvalues approaching zero. Therefore if this small number of eigenvalues is captured and deflated, the conditioning of the matrix should be independent of the shift (quark mass). It is for this reason that our Incremental eigCG successfully removes the critical slowdown that has long plagued calculations in this field. Figure 2 shows the effects of the critical slowdown in the original, undeflated CG and how our method takes about the same number of iterations for any shift up to very close to the critical mass.

3.2. Scaling with volume

For the same configuration, increasing the lattice volume increases the matrix condition number. This volume scaling problem is much harder to address with deflation methods than critical slowdown because the eigenvalue density increases. To achieve a constant condition number, we need to capture the same portion of the spectrum which means computing more eigenvalues. Figure 3 shows the effect of 250 vector deflation on CG convergence for the three different size lattices. Note that deflated CG takes about 2.9 times more iterations on the largest lattice than on the smallest one, even though the volume increased by a factor of 12. By running Incremental eigCG with more systems we can remove this volume dependency at the expense of storing more vectors. Figure 4 shows the effect of deflating with 256, 512 and 1024 vectors on the largest lattice. We can achieve the same number of CG iterations on the large lattice as by deflating with about 5 times more vectors. However, there are diminishing returns in time.

4. EigBICG: extending to the nonsymmetric case

Recently, we extended the eigCG idea to eigBICG which is based on the nonsymmetric Lanczos. The main difference is that we need to keep two windows, W and V, corresponding to the left and right residuals built by BICG. We restart these windows similarly to (1) using left and right Ritz vectors. An Incremental eigBICG outer method accumulates the $2nev$ vectors to improve accuracy of the eigenspace, which is then used to deflated the BICGSTAB method. Figure 5 shows the deflation improvements on two smaller lattices in Matlab. Although eigBICG

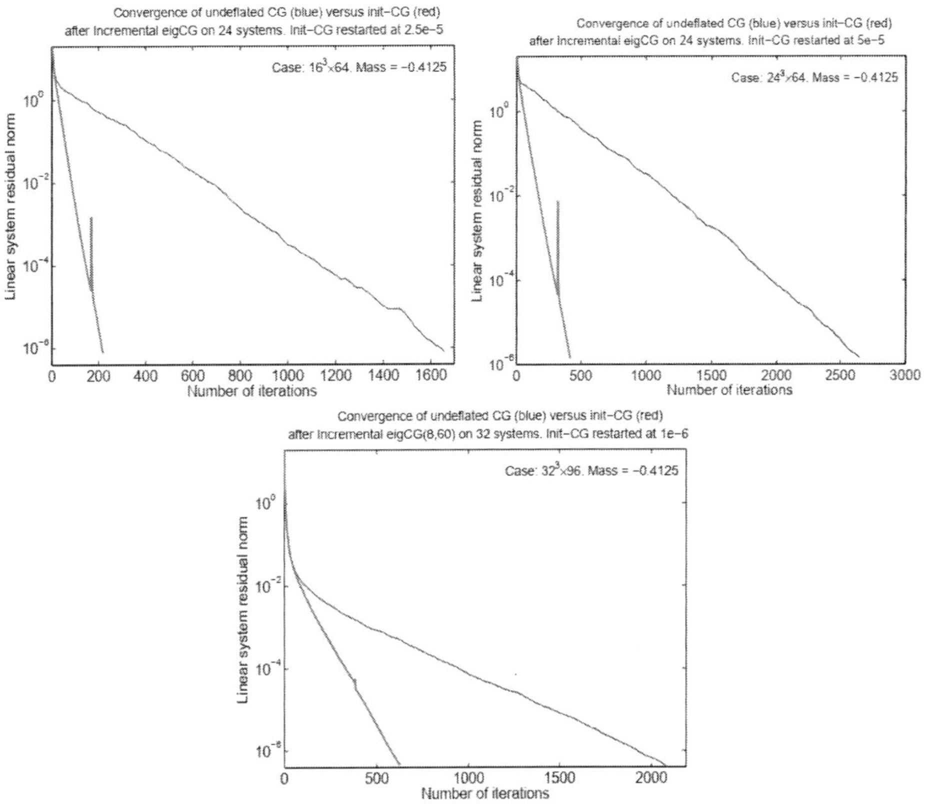

Figure 3. CG convergence improvement by deflating 240 vectors obtained through Incremental eigCG, for three lattices with increasing volumes. Volume slowdown is reduced but not removed.

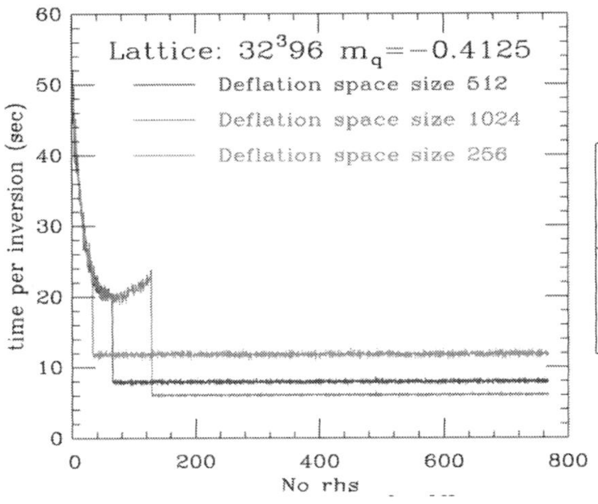

	CG iterations				
	undeflated	vectors deflated			
Case		240	256	512	1024
3M	1628	215			
10M	1907	337			
37M	2135	-	624	415	278

Figure 4. Left graph shows the time to solve the 780 linear systems after solving the first 32, 64, 128 systems with Incremental eigCG. Right table shows the CG iterations each case took.

converges similarly to unrestarted Lanczos, the Incremental part is less effective than in the symmetric case. Still, the nonsymmetric version is useful if the experiments involve much heavier quark masses as shown in the figure.

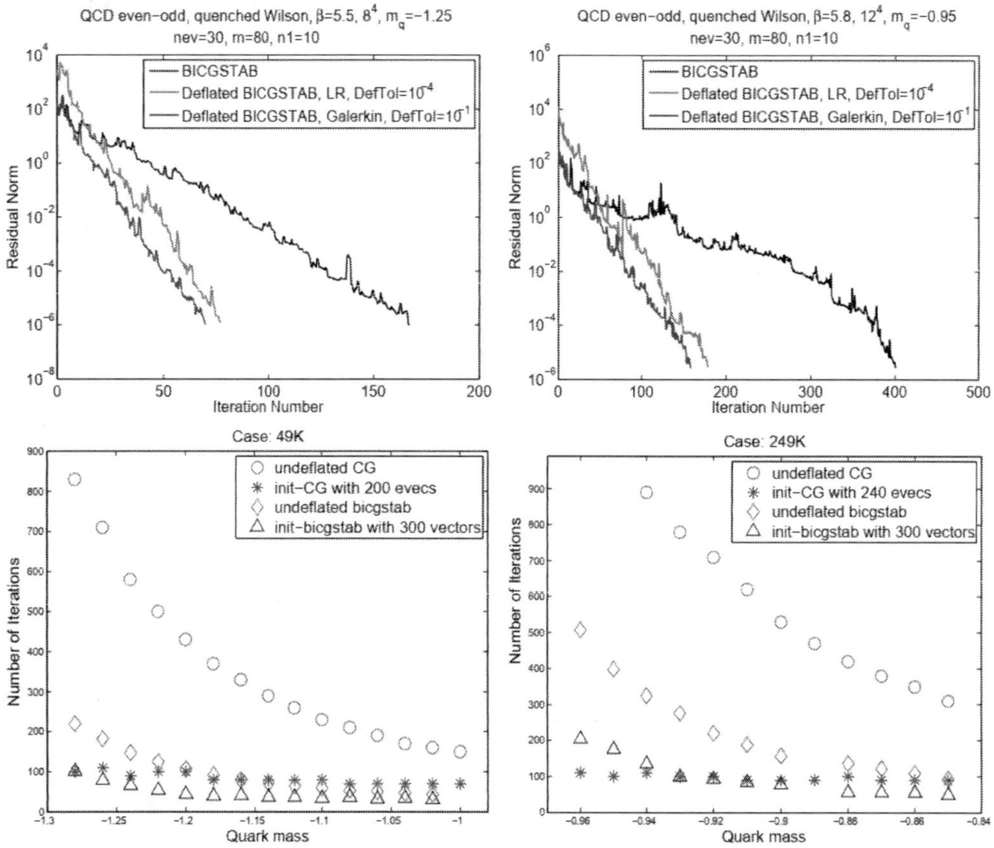

Figure 5. Deflated BICGSTAB using eigenspace from eigBICG and comparison with eigCG.

5. Conclusions

We have outlined a successful method, eigCG, that has helped us speedup dramatically the linear system solves in QCD and remove the critical slowdown. Volume slowdown is reduced but we have not yet reached lattice sizes where it will become the bottleneck. We have also presented a promising extension to the nonsymmetric case.

Acknowledgments

This work was supported by the NSF grant CCF-0728915, the DOE Jefferson Lab and the Jeffress Memorial Trust grant J-813.

References

[1] Wilson K G 1974 *Phys. Rev.* **D10** 2445–2459
[2] Foley J, Juge K J, O'Cais A, Peardon M, Ryan S and Skullerud J I 2005 *Comput. Phys. Commun.* **172** 145–162 (*Preprint* `hep-lat/0505023`)
[3] Simoncini V and Gallopoulos E 1995 *SIAM J. Sci. Comput.* **16** 917–933
[4] Chan T F and Wan W 1997 *SIAM J. Sci. Comput.* **18** 1698–1721
[5] O'Leary D P 1980 *Lin. Alg. Appl.* **29** 293–322

[6] de Forcrand P 1996 *Nucl. Phys. B (Proc. Suppl.)* **47** 228–235
[7] Morgan R B 2002 *SIAM J. Sci. Comput.* **24** 20–37
[8] Morgan R and Wilcox W 2004 Deflated iterative methods for linear equations with multiple right-hand sides Tech. Rep. BU-HEPP-04-01 Baylor University
[9] Wang S, de Sturler E and Paulino G H 2007 *International Journal for Numerical Methods in Engineering* **69** 2441–2468
[10] Stathopoulos A and Orginos K June 12, 2008 (original version July 1, 2007) Computing and deflating eigenvalues while solving multiple right hand side linear systems with an application to quantum chromodynamics Tech. Rep. arXiv.org 0707.0131v2 http://arxiv.org/pdf/0707.0131v2, Revised version under review in SIAM J. Sci. Comput.
[11] Saad Y 2003 *Iterative methods for sparse linear systems* (Philadelphia, PA, USA: SIAM)
[12] Stathopoulos A 2007 *SIAM Journal on Scientific Computing* **29** 481–514
[13] Saad Y, Yeung M, Erhel J and Guyomarc'h F 2000 *SIAM J. Sci. Comput.* **21** 1909–1926
[14] Edwards R G and Joo B (SciDAC) 2005 *Nucl. Phys. Proc. Suppl.* **140** 832 (*Preprint* `hep-lat/0409003`)

SciDAC 2009 IOP Publishing
Journal of Physics: Conference Series **180** (2009) 012074 doi:10.1088/1742-6596/180/1/012074

Practical algorithms to facilitate large-scale first-principles molecular dynamics

François Gygi[1], Ivan Duchemin[1], Davide Donadio[2], and Giulia Galli[2]

[1] Department of Applied Science, University of California Davis, Davis, CA 95616, USA
[2] Department of Chemistry, University of California Davis, Davis, CA 95616, USA

E-mail: fgygi@ucdavis.edu

Abstract. Running First-Principles Molecular Dynamics (FPMD) simulations on large parallel platforms presents a number of practical challenges related to the large size of the datasets generated during simulations and to the lack of flexibility of parallel FPMD codes for the implementation of new algorithms. In this paper, we present two approaches implemented in the Qbox code, that alleviate these problems and facilitate large-scale FPMD simulations. We first describe a parallel I/O strategy based on MPI-IO that fully exploits the performance of parallel file systems. We also describe the implementation of a client-server interface that enables coupled quantum-classical computations in which the quantum simulation run by Qbox is "driven" by another application. This feature can be used to couple FPMD simulations with other simulation methods such as e.g. Path Integral Monte Carlo, thermodynamic integration or replica exchange dynamics.

1. Introduction

Considerable efforts have been made in the past years to extend first-principles simulations to larger problems. The complexity of the quantum-mechanical part of such simulations–the electronic structure problem–is limiting the size of feasible simulations. Conventional methods exhibit unfavorable scaling, with a computational cost of $O(N^3)$ where N is the number of electrons included in the system[1]. Newer algorithms have shown the possibility of reducing this cost to $O(N)$ [2], although at the price of reduced versatility and robustness.

In the context of the plane-wave, pseudopotential method and Density Functional Theory (DFT) [1], parallel algorithms and implementations have been developed [3, 4, 5, 6, 7] to exploit the power of large platforms and reduce the time to solution in large simulations. The scalability of the plane-wave, pseudopotential approach is due to the fact that most computations in a simulation can be separated in groups of processors that operate on groups of electronic states in a weakly coupled manner [8]. Scalability up to 64k nodes was demonstrated in simulations of large systems [9] using the Qbox code [7].

Although the performance of the numerical kernels of FPMD codes is the subject of continuing research, a number of other bottlenecks severely limit the efficiency and usability of FPMD codes for large simulations. The first important limitation comes from the size of the simulation data that must be stored between runs (the "restart file"). In FPMD simulations, the data describing the electronic structure of the system must be stored at the end of each molecular dynamics run to avoid a costly recalculation of the system's electronic structure upon restart. The amount of data involved is proportional to N^2 in a dense system. For a mid-size simulation (e.g. 256 H_2O

molecules) the size of the electronic structure data reaches 10-20 GB. For larger simulations, in particular in nanoscience applications (see e.g. Ref.[10]), this size can reach 200-300 GB. Such large datasets directly affect productivity since (i) they require a significant time to be read and written, and (ii) they make it impractical to move simulation data from platform to platform when simulations must be migrated, or if post-processing of data must be performed on a separate platform.

Another factor limiting large-scale FPMD simulations is the difficulty of coupling DFT codes to other simulation codes. Such couplings are desirable in situations where the accuracy of DFT energies is needed in combination with advanced algorithms for updating atomic configurations. Examples of such hybrid simulation methods include adiabatic switching [11], reversible scaling [12], replica exchange dynamics (or parallel tempering) [13] and path integral Monte Carlo simulations [14]. The difficulty in integrating new algorithms into DFT codes is partly due to the complexity of parallel DFT codes. In particular, DFT codes are mostly designed to simulate a single instance of a physical system, whereas many of the above advanced simulation techniques involve the simultaneous evolution of multiple copies of the system.

In this paper, we describe the implementation of two features of the Qbox code that help alleviate the difficulties discussed above. First, we describe a parallel I/O strategy that greatly reduces the time needed to read and write restart files on large parallel platforms. Second, we describe a client-server interface that enables arbitrary coupling between Qbox and other simulation codes using advanced algorithms to update atomic configurations. Portability and performance measurements of the parallel I/O implementation are discussed. Finally, an application to coupled simulations using adiabatic switching in nitrogen is used to illustrate the flexibility of the client-server model. All the features described are implemented in the current publicly available distribution of Qbox [7].

2. Parallel I/O strategy in Qbox

The Qbox code [7] adopts the www.quantum-simulation.org (QSO) [15] format to save all the information needed to restart a simulation at a later time. The QSO format is based on the XML markup language [16], which ensures portability and facilitates the addition of new information into the data format as new features are needed while preserving backward compatibility. The QSO format is formally defined using the XML Schema Language [17]. XML Schema definitions are documents that define the allowable contents of restart files in a concise way and allow for automatic validation of data using validating parsers such as e.g. the Apache Xerces-C parser [18]. The Xerces-C parser is integrated in Qbox and is used to parse XML input documents such as pseudopotential information encoded in the QSO format, or QSO sample files. In order to ensure portability of QSO documents, binary formats are avoided. Numerical data is instead represented using base64 encoding and a little-endian byte ordering on all platforms.

In large-scale simulations, the growing size of restart files leads to the challenge of processing large XML documents. In the following sections, we describe the approach used in Qbox to read and write the XML restart files efficiently.

2.1. Parallel read of restart files

All currently available XML parsers process documents serially, which limits the rate at which large XML documents can be processed. Given the rate at which a file can be read from a single processor (typically less than 50 MB/s), it is critical to parallelize this process and take full advantage of the performance of parallel file systems on large parallel computers. Qbox decomposes the parsing process into three phases: 1) parallel reading of the entire XML document into memory, 2) parallel preprocessing of the largest part of the numerical data, and 3) in-memory parsing of a reduced XML document on a single node. Phase 1) consists of reading the entire XML document by calling the `fread` function from each node. This results in

Table 1. Aggregate write rates measured using Qbox on NERSC's Franklin platform (Cray-XT4) and on ANL's BlueGene/P platform.

platform	N_{tasks}	rate (MB/s)
Cray-XT4	2048	778
Cray-XT4	4096	715
Cray-XT4	8192	687
BG/P	2048	814

each node holding a character string that contains a fragment of the XML document. The I/O performance achieved in this phase is high. On NERSC's Cray-XT4 Franklin platform (which uses a Lustre parallel file system [19]), a 254 GB file is read into memory in 28 seconds using 4200 tasks on 2100 nodes, achieving an aggregate read rate of 8793 MB/s. In phase 2), the XML contents are processed by extracting the largest elements (<grid_function>) that contain the values of electronic wavefunctions on rectangular grids, and by transferring them to their appropriate destination using MPI calls. Finally, in phase 3), the remaining XML contents (i.e. omitting the <grid_function> elements) are reduced to a single string which is gathered on the master node and parsed using the Apache Xerces-C parser. This approach has proved reliable and portable on a number of large platforms including Cray-XT4, Sun Constellation, BlueGene/L and BlueGene/P. Portability is ensured by relying only on the `fread` function.

2.2. Parallel write of restart files

Writing large datasets also constitutes a challenge since the standard C functions such as `fwrite` do not support writing simultaneously in different parts of a file. MPI-IO [20, 21] in principle provides an efficient alternative, with several functions designed to write files in parallel. Qbox first prepares the document to be written as a character string residing in memory. The data must then be written in order of increasing MPI rank by each node. The MPI function `MPI_File_write_shared`, which implements this functionality is however not supported by all file systems. We circumvent this difficulty by implementing a C++ shared file pointer object which keeps track of file offsets and can be used with the better supported `MPI_File_write_at_all` function. The aggregate I/O rate achieved with this approach is in the range of 600-800 MB/s. Table 1 summarizes the write rates observed on various platforms. On NERSC's Franklin platform, a 254 GB restart file is written in 296 seconds, reducing the time spent in I/O operations to an acceptable level. We note that the move to large scale problems and the need for parallel I/O raise the issue of the compatibility of I/O libraries (such as MPI-IO) across file systems. In this case, a limited subset of MPI-IO functions had to be used in order to insure portability to different parallel file systems[1].

The efficiency of large-scale simulations is also affected by the fact that large datasets are difficult to archive and to move between platforms. This issue can be mitigated using compression algorithms that incur some data loss. Qbox implements a form of data compression by allowing for on-the-fly resizing of the plane wave basis set used to describe wavefunctions. This resizing operation can be performed before saving a restart file, and readily leads to a large reduction in data size. It requires however a reoptimization of the wavefunctions when restarting a simulation, since the compressed data has lower resolution than the original. Recently, another approach to compression of electronic structure data has been developed based on a recursive bisection algorithm [22] and will be integrated into future releases of Qbox.

[1] We are grateful to R. Latham for several discussions about MPI-IO functions.

3. A client-server interface for Qbox

The implementation of new features in parallel simulation codes is challenging due to the relatively complex structure of these codes. A number of specialized algorithms for updating atomic positions have been proposed [11, 12, 13, 14] to compute e.g. free energies or optimal transition paths. From the standpoint of software engineering, the proliferation of such methods poses a challenge. Integrating all such algorithms in an FPMD code would rapidly lead to a maintenance problem. Furthermore, the release of these new features would not likely keep up with the changes desired by the community of users. A client-server interface was implemented in Qbox in order to open the possibility of coupling Qbox with other simulation codes, thus removing the dependencies between the software implementation of new algorithms and Qbox itself.

Qbox normally functions as a command line interpreter [23], reading commands from standard input and writing output on standard output. This makes it possible to use Qbox interactively or in batch mode. In "server mode", we use the C++ IOStream library's ability to redirect output streams to make Qbox read commands from a file and write output to another file. In this case, Qbox (acting as a server) repeatedly opens a file, executes all commands it contains and closes the file. Qbox output is written to another file which is closed at the end of the current sequence of commands. This allows for coupled operation with a "driver" program (or client) that repeatedly creates Qbox input files and reads Qbox output files. Synchronization between client and server is implemented through the creation and destruction of a lock file that signals that Qbox is ready to execute a new input file. The output files written by Qbox are valid XML documents, a property that can be used by the client program to efficiently extract specific elements from the Qbox output.

Using files for interprocess communication is admittedly a primitive approach. However, it allows for client-server simulations in which the client and the server run on separate computers that share a file system. For example, the Qbox server can run on a large parallel platform while the driver program runs on the platform's front-end node. The overhead of creating and closing files was found to be negligible in real applications.

The client-server model also allows for simulations including an arbitrary number of copies of a physical system as is the case, e.g. in Path Integral Monte Carlo (PIMC) or Nudged Elastic Band (NEB) simulations. This can be implemented by running separate instances of Qbox in client-server mode, while a single driver controls the client processes through separate input and output files.

The client-server interface was used in calculations of the melting temperature of nitrogen at high-pressure. The simulations required calculations of free energies of various phases relative to a reference system represented by a classical intermolecular potential. The driver program implements the molecular dynamics simulation of the reference system and creates Qbox commands to displace atoms appropriately. Qbox repeatedly computes DFT energies, forces and stress for these atomic configurations and writes them on an output file that is read by the driver program. Atomic displacements are sent to Qbox using the `move` command [23]. The calculation of the DFT energy in Qbox is efficient since atomic displacements are usually small. The electronic wavefunctions are kept in memory between moves, and only need a minor reoptimization after each move.

This model minimizes software dependencies between Qbox and any client program. Ensuring backward compatibility of the syntax of Qbox commands is a reasonable goal and does not impose constraints on the details of the Qbox implementation. It also allows for more flexibility in the choice of the programming language used to implement the driver. Since the only required feature is the ability to manipulate text files, driver programs can be written in any language, including scripting languages. In the example discussed above, the driver program was written in Fortran90.

4. Conclusions

We have described two features of the Qbox FPMD code that facilitate large-scale FPMD simulations on parallel computers. First, a parallel I/O strategy based on MPI-IO demonstrates the possibility of using large XML files to save the information describing the electronic structure of large systems. A combination of parallel read/write and parallel XML preprocessing is used to reduce the time spent in I/O operations. Second, a client-server interface effectively decouples the software implementation of the electronic structure calculation from algorithms used to update atomic positions. These algorithmic enhancements, while not directly connected to the numerical kernels used in FPMD simulations, are important ingredients that make large scale FPMD simulations more tractable.

Acknowledgments

Support from DOE through grant SciDAC DE-FC02-06ER25794 and from NSF PetaApps OCI-0749217 is gratefully acknowledged. We also thank the NSF TeraGrid for allocation of computer time.

References

[1] Martin R M 2004 *Electronic Structure. Basic Theory and Practical Methods* (Cambridge University Press)
[2] Goedecker S 1999 *Rev. Mod. Phys.* **71** 1085–1123
[3] `http://www.cpmd.org`
[4] `http://www.quantum-espresso.org`
[5] `http://cms.mpi.univie.ac.at/vasp/`
[6] `http://www.emsl.pnl.gov/docs/nwchem/nwchem.html`
[7] The qbox first-principles molecular dynamics code `http://eslab.ucdavis.edu/software/qbox`
[8] Gygi F 2006 *J. Phys.: Conference Series* **46** 268–277
[9] Gygi F, Draeger E W, Schulz M, de Supinski B R, Gunnels J A, Austel V, Sexton J C, Franchetti F, Kral S, Ueberhuber C W and Lorenz J 2006 Large-scale electronic structure calculations of high-z metals on the bluegene/l platform *SC '06: Proceedings of the 2006 ACM/IEEE conference on Supercomputing* (New York, NY, USA: ACM) p 45 ISBN 0-7695-2700-0
[10] Li Y, Galli G and Gygi F 2008 *ACS Nano* **2** 1896–1902
[11] Watanabe M and Reinhardt W P 1990 *Phys. Rev. Lett.* **65** 3301–3304
[12] de Koning M, Antonelli A and Yip S 1999 *Phys. Rev. Lett.* **83** 3973–3977
[13] Hansmann U H E 1997 *Chem. Phys. Lett.* **281** 140 – 150
[14] Ceperley D M 1995 *Rev. Mod. Phys.* **67** 279–355
[15] Web standards for quantum simulations `http://www.quantum-simulation.org`
[16] Extensible markup language (xml) `http://www.w3.org/XML`
[17] Xml schema language `http://www.w3.org/XML/Schema`
[18] Xerces-c++ xml parser `http://xerces.apache.org/xerces-c`
[19] `http://www.lustre.org`
[20] `http://www.mpi-forum.org`
[21] Gropp W, Lusk E, Doss N and Skjellum A 1996 *Parallel Comput.* **22** 789–828 ISSN 0167-8191
[22] Gygi F 2009 *Phys. Rev. Lett.* **102** 166406
[23] *The Qbox User Guide* `http://eslab.ucdavis.edu/software/qbox/QboxUserGuide.pdf`

Advancing nanoelectronic device modeling through peta-scale computing and deployment on nanoHUB

Benjamin P. Haley[1,2], Sunhee Lee[2], Mathieu Luisier[1,2], Hoon Ryu[2], Faisal Saied[3,4], Steve Clark[3], Hansang Bae[2] and Gerhard Klimeck[1,2]

1. Network for Computational Nanotechnology, Purdue University
2. School of Electrical and Computer Engineering, Purdue University
3. Rosen Center for Advanced Computing, Purdue University
4. Computing Research Institute, Purdue University

bhaley@purdue.edu

Abstract. Recent improvements to existing HPC codes NEMO 3-D and OMEN, combined with access to peta-scale computing resources, have enabled realistic device engineering simulations that were previously infeasible. NEMO 3-D can now simulate 1 billion atom systems, and, using 3D spatial decomposition, scale to 32768 cores. Simulation time for the band structure of an experimental P doped Si quantum computing device fell from 40 minutes to 1 minute. OMEN can perform fully quantum mechanical transport calculations for real-word UTB FETs on 147,456 cores in roughly 5 minutes. Both of these tools power simulation engines on the nanoHUB, giving the community access to previously unavailable research capabilities.

1. Introduction

The rapid progress in nanofabrication technologies has led to the emergence of new classes of nano-devices, in which the quantum nature of charge carriers dominates the device properties and performance. The need for atomistic-level modeling is particularly clear in studies of quantum dots. Quantum dots are solid-state structures capable of trapping charge carriers so that their wave functions become fully spatially localized, and their energy spectra consist of well-separated, discrete levels. Existing nanofabrication techniques make it possible to manufacture quantum dots in a variety of types and sizes [1]. Among them, semiconductor quantum dots grown by self-assembly (SADs), trapping electrons as well as holes, are of particular importance in quantum optics, since they can be used as detectors of infrared radiation [2], optical memories [3], and single photon sources [4]. Arrays of quantum-mechanically coupled SADs can also be used as optically active regions in high-efficiency, room-temperature lasers [5].

Atomic scale modeling is also critical for studying concepts such as silicon-on-insulator ultra-thin-body (UTB) devices [6], multi-gate nanowires [7-8], and band-to-band tunneling field-effect transistors (TFETs) [9-10]. In these regimes, classical simulation approaches, such as the drift-diffusion [11] and effective mass [12] models, do not correctly capture the strong quantization effects, which influence device performance.

The goal of this paper is to present new capabilities that have been added to our Nanoelectronic Modeling tool set. We are in the process of developing, in OMEN, a single code that will include the functionality of the well-established NEMO 3-D code for strain and electronic structure computations as well as a new capability to solve the challenging 3D quantum transport problem. OMEN is designed to run efficiently on large NSF Track 2 systems like Ranger, at TACC, and Kraken, at NICS. We believe that OMEN will be one of the premier simulation tools for the design and analysis of realistic nanoelectronic devices, and, therefore, it will be an extremely useful tool for the Network for Computational Nanotechnology (NCN) community.

The recent advances in OMEN and NEMO 3-D include algorithmic refinements, performance analysis to identify the best computational strategies, and porting to state of the art HPC architectures, including Ranger, the BlueGene, and Cray XT5, as well as a Woodcrest Linux cluster. One important consequence of these enhancements is the ability to run 3D quantum transport computations. Performing such calculations for realistic devices is a very challenging computational problem, and it represents a new capability for which resources of the scale of Ranger and Kraken are essential. We present initial results for the transport problem on both machines. We also present results for the electronic structure computations based on both the NEMO 3-D code, and the new implementation in OMEN. From an algorithmic point of view, a key challenge for electronic structure is the extraction of interior, degenerate eigenvectors at this scale. These calculations have been carried out on up to 8K cores on the BlueGene and the Cray XT3, and for 32K cores, and 1 billion atoms, on Kraken (XT5).

The rest of the paper is structured as follows. In section 2 we review the physical models underlying NEMO 1-D, NEMO3-D, and OMEN and highlight some of they insights gained from these models; in section 3, we describe the approach used for the parallelization of the computations. In section 4, we briefly describe the algorithms implemented in the packages. Section 5 has the performance results, including benchmark results for 3D transport up to 65K cores, and up to 32K cores in the electronic structure phase. Section 6 describes the deployment of these codes on the nanoHUB.

2. Physical models

Figure 1 shows the evolution from the NEMO 1-D and NEMO 3-D codes to OMEN. NEMO 1-D is a mature code that computes quantum transport for a 1D system such as the resonant tunneling diode shown in Fig. 1a. A critical modeling capability is the ability to model an *open* quantum mechanical system where carriers are injected on one side of the device and extracted at the other side. This constitutes a quantum statistical mechanics problem in non-equilibrium. The non-equilibrium Green function formalism (NEGF) is well established [53, 54] to solve this problem where a Hermitian system Hamiltonian is augmented by open boundary conditions, creating a non-Hermitian system. A set of Green functions is needed to compute the states in the system as well as the occupancy and availability of states and the transfer of carriers from one state to the next. For an overview of the NEGF formalism we refer the reader to a web-page on nanoHUB.org which is authored and maintained by Prof. Supriyo Datta.

There were three key insights in the modeling of resonant tunneling diodes. One was the need to model extended devices through novel boundary conditions [55] which includes quantum charge self-consistency and strong ineleastic scattering or quasi equilibrium in the contacts. The second key insight is that inelastic scattering from acoustic and polar optical phonons, alloy disorder, and interface roughness in the central device region can be modeled quantitatively in NEGF [56-59] and compare well to the valley currents in experimental data at low temperatures. At room temperature, however, a very different physics explains the typical high performance, high current density resonant tunneling diode valley current: It is thermionic emission through excited states. That was the third critical insight that was enabled by the atomistic, layer by layer modeling through empirical tight binding [60,61]. The room temperature calculations require the solution of a double integral in transverse momentum k and total energy E over strongly spiked transmission coefficients (Fig 1b) to result in the overlap of theory with experimental current voltage curves (Fig 1c). Since each bias point is in principle decoupled from the others one can expose a natural multilevel parallelism in voltage, momentum and energy (Fig 1d) [47].

NEMO 3-D can compute strain and electronic structure (but not transport) for 3D devices where the devices are represented atom-by-atom (Fig 1e) [62-64]. The main NEMO 3-D calculation is the strain minimization [62, 16] in a valence force field method and the computation of electronic structure eigenstates, such as that one of an artificial atom / quantum dot depicted in Figure 1f. From these eigenstates one can devise a plethora of properties such as absorption coefficients as symbolized in Fig 1g [67], single impurity characteristics in FinFETs [68], disordered Si quantum wells on SiGe [69], even some general transport characteristics of core-shell nanowires [70] and alloyed nanowires [71]. The first multi-million atom calculations of alloy-disordered systems were demonstrated in 2002 [62,65]. NEMO 3-D is using a 1D layer decomposition as a nearest neighbor CPU communication pattern and was demonstrated to scale well on even the earliest Beowulf cluster computers [62,66]. Later we demonstrated scaling on thin and long nanowire like structures up to 8,192 cores [72]. The NEMO 3-D section below highlights our effort to parallelize NEMO 3-D on more than a single level through 3-D decomposition and parameter sweeps.

Our new OMEN embraces the quantum transport concepts of NEMO 1-D and the 3-D spatial representation of NEMO 3-D. Three dimensional atom arrangements of nanowires realistically gated with dielectrics (Fig 1i) are computed to obtain energy resolved current (Fig 1j) and current voltage characteristics (Fig. 1k). The dramatic work-load increase makes it a perfect candidate for the use of peta-scale computing systems with 4 levels of natural parallelism (Fig. 1l). The code has already been used to model ultra-scaled InAs High-Electron-Mobility Field Effect Transistors (HEMTs) [73], band-to-band tunneling transistors with 2D and 3D device configurations [74], and graphene transistors [75]. An overview of the evolution of NEMO 1-D and NEMO 3-D to OMEN is also presented in reference [76]. Computational results for these capabilities, including scaling to to over 147,000 processors, are presented in section 5.

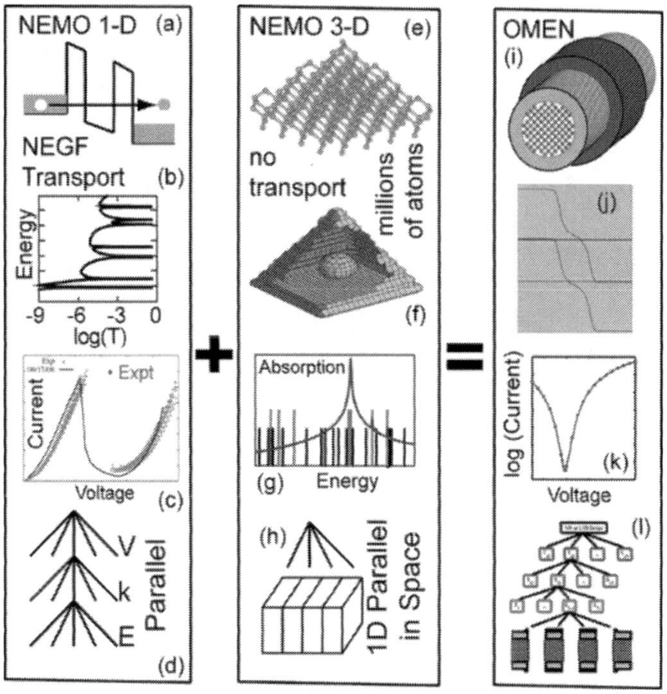

Figure 1. The NEMO 1-D code can do 1D NEGF quantum transport, while the NEMO 3-D is designed to do strain and electronic calculations. Our new code OMEN incorporates NEMO 3-D functionality and adds a 3D quantum transport capability, and includes features that make the code peta-scale capable.

2.1. NEMO 3D
We summarize here some of the key features of the model and the NEMO 3-D package, which consists of three main parts:

- The geometry constructor (setup phase)
- The strain computation
- The electronic structure computation

Geometry constructor and setup. The geometry constructor's purpose is to represent the nanostructure in atomistic detail in the memory of the computer. Each atom is represented by its coordinates, its type (atomic number), whether the atom is on the surface or in the interior of the sample (important later on in electronic calculations), in what kind of computation it will participate (strain only or strain and electronic), and its nearest neighbors. This part of NEMO 3-D consumes only a small fraction of total computational time and memory, and will not be discussed further.

The strain computation. The materials making up a quantum dot nanostructure differ in their lattice constants. This lattice mismatch leads to strain; atoms throughout the sample are displaced from their bulk positions. Knowledge of equilibrium atomic positions is crucial for the subsequent calculation of the quantum dot's electronic properties, which makes the computation of strain a necessary step in realistic simulations of these nanostructures. NEMO 3-D computes strain using an atomistic valence force field method (VFF) [16]. In this approach, the total elastic energy of the sample is computed as a sum of bond-stretching and bond-bending contributions from each atom. The equilibrium atomic positions are found by minimizing the total elastic energy of the system.

The electronic structure computation. The single-particle energies and wave functions are calculated using an empirical first-nearest-neighbor tight-binding model. The underlying idea of this approach is the selection of a basis consisting of atomic orbitals (such as s, p, d, and s*) centered on each atom. These orbitals are treated as a basis set for the Hamiltonian operator and lead to a discrete Hamiltonian matrix that has a block tri-diagonal structure due to the restriction to nearest neighbor interactions. The parameterization of bonds using analytical forms of directional cosines is given in Ref. [17]. NEMO 3-D uses the parameterization for the $sp^3d^5s^*$ model, in which each atom is assigned 20 spin-degenerate orbitals [18].

The 20-band nearest-neighbor tight-binding model is parameterized by 34 energy constants, which need to be established by fitting the computed electronic properties of materials to those measured experimentally. This is done by considering bulk semiconductor crystals (such as GaAs or InAs) under strain. A genetic algorithm is used to search for optimal parameters, described in detail in Refs. [18-20]. Once a parameterization it is known for each material constituting the quantum dot, a full atomistic calculation of the single-particle energy spectrum is carried out on samples composed of millions of atoms. Calculating the electronic structure is the most computationally intensive phase of the code. The eigenvalue problem is solved in parallel using one of the algorithms discussed in Section 4.

2.2. OMEN
OMEN is a one-, two-, and three-dimensional quantum transport simulator based on the nearest-neighbor tight-binding method and dedicated to next generation transistors like ultra-thin body (UTB), nanowire, and band-to-band tunneling field-effect transistors. It contains four levels of parallelism, and has already demonstrated scaling up to 65,536 cores with a performance of 208 TFlop/s [21].

OMEN solves the Schrödinger equation with open boundary conditions in an orbital basis. The computational burden directly depends on the complexity of this basis. As one uses more orbitals to describe an atom (from 1 to 20), the size of the Hamiltonian matrices increase, as does the work required to solve the Schrödinger equation. As an alternative to the popular Non-Equilibrium Green's Function (NEGF) formalism, OMEN uses a Wave Function (WF) approach that is physically identical to NEGF, but computationally less expensive for ballistic transport [15]. In the WF formalism the Schrödinger equation takes the form of a sparse linear system of equations "Ax=b" with a block tri-diagonal matrix "A" of size $N = N_A \times t_b$, equal to the number of atoms in the structure (N_A) times the number of orbitals of the tight-binding model (t_b). For example, t_b =10 in the $sp^3d^5s^*$ model without spin-orbit coupling. The first and the last diagonal blocks of the matrix A contain the open-boundary conditions, the multiple right-hand-side vector "b" the injection mechanism. Typical device structures are composed of 10,000 to 200,000 atoms, resulting in matrices of size 50,000< N <1,000,000, depending on the value of t_b.

The resulting charge and current densities are then self-consistently coupled to the Poisson equation expressed in a finite-element basis. Hence, ballistic transport, source-to-drain tunneling, alloy disorder, strain, and interface roughness can be treated on an atomistic level in n- or p-doped FETs with a cross section up to about 22 nm^2 for circular or square nanowire and a width of 8 nm for UTB. Transistor structures made of several semiconductor materials, such as Si, Ge, SiGe, GaAs, or InAs and with any channel orientation (<100>, <110>, <111>, or <112>) can be investigated. As an example a double-gate UTB FET with transport along the <100> axis is considered in Figure 2 [13-15].

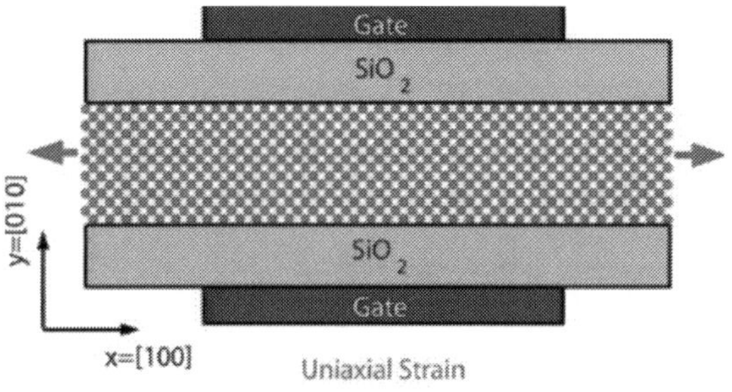

Figure 2. Schematic view of a double-gate ultra-thin body field effect transistor with transport along the <100> crystal axis and confinement along <010>.

3. Parallelization

3.1. NEMO 3-D

To extend NEMO 3-D performance, the new features of NEMO 3-D include 3D domain decomposition parallelism (Figure 3). Recent benchmarks show that 3D domain decomposition scheme can be utilized exceeding 32,000 cores on realistic electronic structures comprised of one billion atoms. The spatial decomposition can be placed in the lowest level of parallelization for solving eigenstates and charge of general electronic structures.

Additionally, NEMO 3-D is prepared with programmable multiple-level parallelization (Figure 3) which is similar to the tri-level parallelism in NEMO 1-D (bias-energy-momentum) [22] and four-level parallelism in OMEN (bias-energy-momentum-1D spatial decomposition) [23-24]. However the parallelization scheme is different in a sense that the variables involved in each level can determined depending on the application. For example, if we are interested in self-consistent band structure of a highly doped wire with respect to temperature and magnetic field of 20 and 10 samples, respectively, the branch immediately spreads out into 200 independent tasks. With the addition of 10 momentum groups and 50 processors for spatial decomposition, then 100,000 cores can be immediately employed for simulation with nearly ideal scalability.

3.2. OMEN

Quantum transport calculations involve solving many linear system of equations (LSE); when hundred of thousands of LSEs have to be solved to simulate the properties of a given device, the solution time becomes an issue. In effect a LSE "Ax=b" must be treated for each energy (typically 500 to 10,000) and momentum (1 to 50) used to inject states into the simulation domain and for each applied voltage (15 to 100 drain-to-source and gate-to-source bias combinations). In the most advantageous case this represents 7,500 LSE, but 50 million in the worst case.

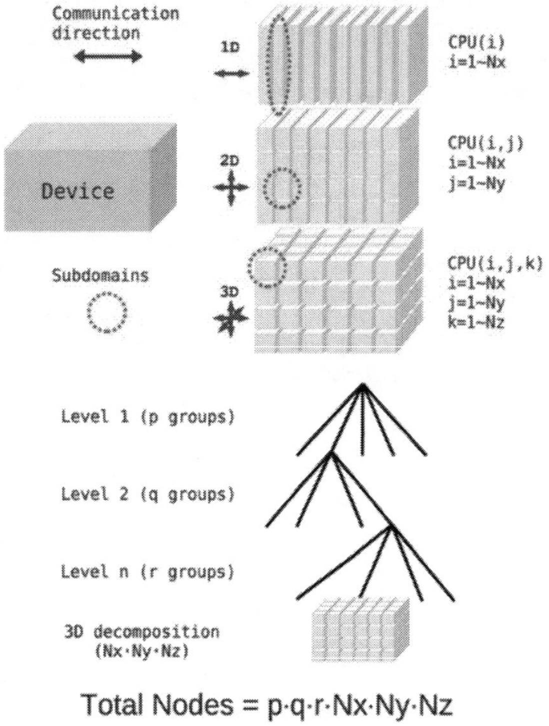

Figure 3. Multiple-level parallelism in NEMO 3-D. For example, if we want to simulate the self-consistent band structure with respect to temperature and external magnetic field, we can set the first group as temperature values(T), the second group as magnetic field(B) and the last group as k points. Therefore, each final branch (3-D decomposition) will be an independent group (T,B,k) that solves self-consistent band structure.

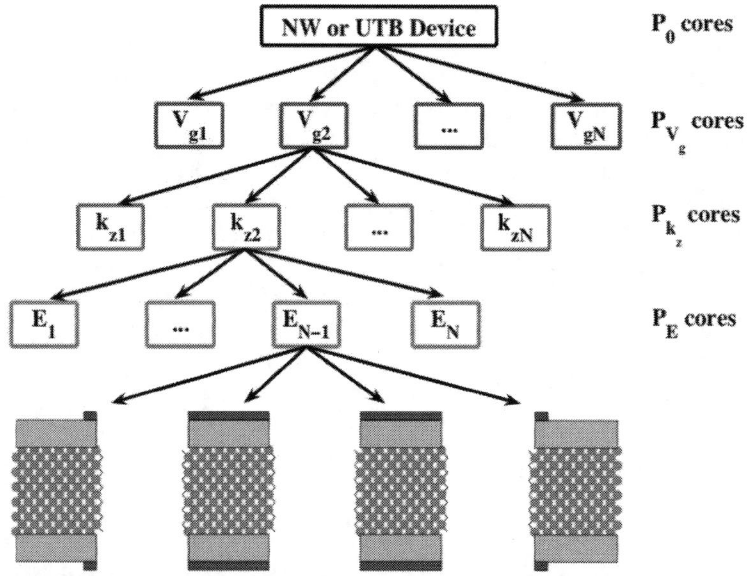

Figure 4. Parallel scheme of OMEN. Bias points, momentum points, and energy points are embarrassingly parallelized and form the three highest levels of parallelism. From the total number of cores, P_0, there are P_{V_g} cores operating per bias point, P_{kz} per momentum point, and P_E per energy point. If $P_E>1$(here4) the simulation domain is spatially decomposed (fourth level of parallelism).

To accelerate the computation of all the LSE three natural levels of parallelism have been identified in OMEN: energy, momentum, and bias points. The two first levels (energy and momentum) are almost embarrassingly parallel, they just call the function "MPI Allreduce" at the end of the calculation to sum over the contributions from all the processors. The bias points are completely independent. These three levels are controlled by MPI instructions. The parallelization scheme of OMEN is shown in Figure 4.

The distribution of cores to each task is decided by the user at the beginning of the simulation. This works fine with the bias and the momentum points since their respective number is known at the beginning of the simulation. However, the number of energy points is different for each momentum, not known at the beginning of the simulation, and requires an automatic redistribution of the number of cores per momentum point to equalize the work load as much as possible [21].

The fourth level of parallelism in OMEN, labeled "spatial domain decomposition", is directly related to the calculation of the open boundary conditions (OBCs), the construction of the matrix "A" and of the vector "b", and the solution of the sparse linear system of equations "Ax=b". The open boundary conditions are parallelized by treating each open contact on a different CPU. Since most of the time a device has only two open contacts the calculation of the OBCs does not scale beyond two cores.

4. Algorithms

4.1. NEMO 3-D

To find the energy levels and wave functions of the electronic structure, in a quantum dot, for example, we need to solve a large eigenvalue problem arising from the discretization of the time-independent Schrödinger equation $H\vec{\psi}_i = E_i\vec{\psi}_i$, where H is the n by n Hamiltonian matrix, $\vec{\psi}_i$ is the eigenvector representing the wave function associated with the eigenvalue E_i representing the i-th energy level. We have implemented the Lanczos algorithm [25] in NEMO 3-D and in OMEN. For a standard symmetric eigenvalue problem, this algorithm builds an orthonormal basis V_k for a Krylov subspace associated with the Hamiltonian, and a k by k tridiagonal matrix T_k such that the following relationship (the Lanczos factorization) is satisfied: $HV_k = V_kT_k + t_{k+1,k}\vec{v}_k\vec{e}_k^T$

It has been shown [26] that the eigenvalues of T_k tend to eigenvalues of H as k tends to n. Moreover, some eigenvalues of T_k converge to eigenvalues of H very quickly. The problem of finding eigenvalues E_i of H is then reduced to the much easier problem of finding eigenvalues of a small tridiagonal matrix T_k and the eigenvectors $\vec{\psi}_i$ of H are found to be $[\vec{\psi}_1 \quad \ldots \quad \vec{\psi}_k] = V_kW_k$, where W_k are the eigenvectors of T_k. The advantage of Lanczos algorithm is that it is fast. However, it cannot compute repeated eigenvalues. This is addressed by our implementation of block Lanczos, which we do not discuss here. We have also implemented Trace Minimization [27-28] and Rayleigh-Ritz eigenvalue solvers, and interfaced the PARPACK library [29] to NEMO 3-D.

4.2. OMEN

The LSE "Ax=b" are solved in parallel using direct sparse linear solver libraries like MUMPS [30], SuperLU$_{dist}$ [31], Pardiso [32], or a customized block cyclic reduction (BCR) of the matrix "A" [33]. Some of the algorithms work with MPI in a distributed memory way (MUMPS, SuperLU$_{dist}$, BCR), others (Pardiso, BCR again) with OpenMP on shared memory machines. In 3D nanowire structures the BCR outperforms the other solvers by a factor of 2 if spin-orbit coupling is not included [34]. Another advantage of the BCR algorithm is that it allows an interleaved calculation of the open boundary conditions and of the solution of "Ax=b". Standard packages like MUMPS, SuperLU$_{dis}$, or Pardiso need the completely assembled matrix "A" at the beginning of the factorization phase. The

BCR algorithm can start working on a matrix "A" not including its first and last diagonal blocks since they are only used at the last step of the reduction of "A". Consequently, the calculation of the OBCs can scale beyond 2 cores and better computational performance is obtained compared to a sequential approach [21].

Once all the LSE are solved the carrier density and the electrostatic potential in the device are calculated from the wave function coefficients stored in the vector "x" of "Ax=b". The solution of Schrödinger equation is self-consistently coupled to the Poisson equation described on a three-dimensional finite element mesh [35-36]. To reduce the computational burden Poisson equation is solved in parallel up to 512 cores using Aztec [37], an iterative sparse linear solver. On multi-core supercomputers OMEN has the ability to work either with four levels of MPI parallelization as described above or with 2.5 levels of MPI parallelization (bias, momentum, and energy points across the nodes) and 1.5 levels of OpenMP parallelization (energies within one node and spatial domain decomposition) [21]. The hybrid OMEN requires nested parallelization [38] which is not possible on all the machines due to compiler issues. The MPI-OpenMP hybrid version of OMEN was tested up to 60,000 cores on Ranger at TACC, scaled almost perfectly there, and helped reduce the start up time of the simulations.

5. Results

5.1. NEMO 1-D

NEMO 1-D was designed to model high-performance Resonant Tunneling Diodes (RTDs) [41,42]. In order to do this accurately, NEMO 1-D considers resonant states in the contacts, which modify the injection of carriers into the central device structure. NEMO 1-D also employs a Hartree self consistent quantum mechanical charge calculation in both the contacts and the central body of an RTD. Realistic band structures, with non-parabolic regions and band warping, are also included in NEMO 1-D, along with low temperature scattering due to optical phonons, acoustical phonons, and interface roughness [43-45]. Incorporating this level of accuracy is demanding, and this mature code is the standard to which the others are compared. The original parallelism on three levels based on bias points, momentum integral, and energy integral was developed for a small cluster system to enable the most flexible efficiency of parallelism [46]. Figure 5 shows the scaling results of a hole RTD simulation using NEMO 1-D, in a sp3d5s* basis, 180 bias points, 200 momentum points, and an adaptive energy grid to handle sharp resonant peaks. This simulation scaled to 23,000 cores on Jaguar at NICS, reducing the time for a simulation which would have required 100 days on a serial machine to 10 minutes.

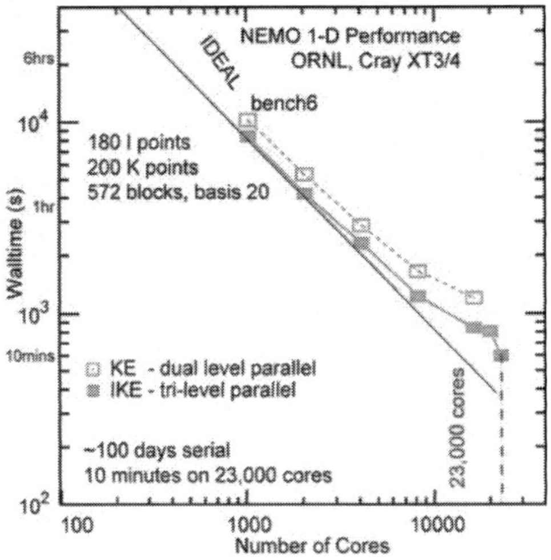

Figure 5. NEMO 1-D scaling with 2 and 3 levels of parallelism on Jaguar at NICS.

5.2. NEMO 3-D

The new 3D spatial decomposition algorithm in NEMO 3-D opens new avenues of exploration. Previously, the 1D decomposition method used in NEMO 3-D limited the range of simulations to artificially elongated domains. Many systems of interest, e.g. quantum dots, are cubic. Figure 6 shows a comparison between 1D, 2D, and 3D spatial decomposition for the calculation of band structure for a 4 million Si atom structure. The 1D decomposition cannot utilize more than 80 cores, while the 3D decomposition can use up to 2048 cores, reducing the simulation time by a factor of 25. Given the

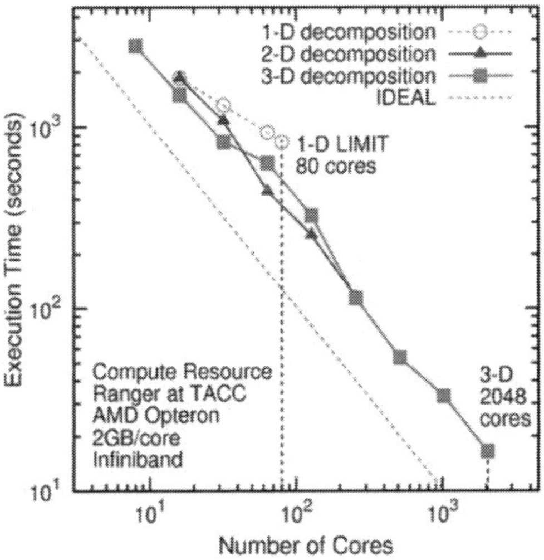

Figure 6. Comparison between different domain decomposition algorithms. Tested with 44 x 44 x 44 (nm^3) silicon structure that has 4 million atoms. Ranger@TACC computational resources were used.

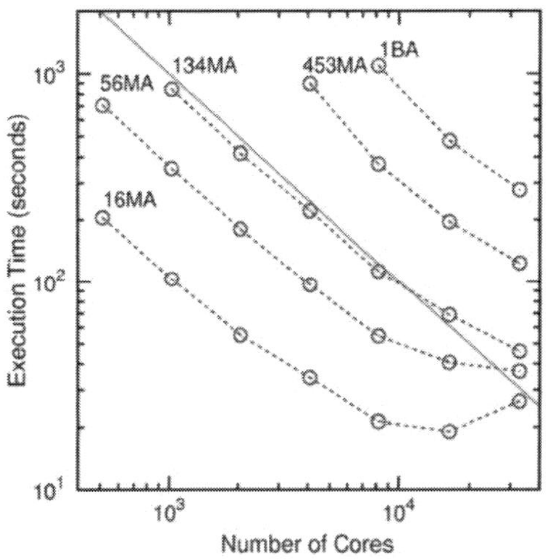

Figure 7. Strong scaling plots with NEMO 3-D for different number of atoms. Tested on Kraken@NICS (XT5).

availability of HPC resources, this capability brings previously unrealistic simulations into the realm of practical consideration.

The 3D spatial decomposition also enables much larger simulations. Figure 7 shows simulations of cubic structures with up to 1 billion atoms, using up to 32768 cores on Kraken XT5 at NICS. NEMO 3-D can now make very efficient use of HPC resources to perform studies which would otherwise be impossible.

One such system of interest is a thin highly doped phosphorous region in a Si substrate, which is of interest to experimentalists and theoreticians for potential use in quantum computing systems [39-40]. Figure 8 shows the scaling of an end-to-end band structure calculation of a 2D Si-P bar 1.1 x 1.1 x 22 nm^3. Spatial decomposition and momentum space parallelism reduce the simulation time for this system from 40 minutes to 1 minute.

5.3. OMEN

OMEN has enabled fully quantum transport simulations of realistic devices. For example, Figure 9 shows the scaling for simulations of a 2D Si double gate ultra-thin-body (UTB) transistor, with a thickness of 5 nm, gate length 22 nm, and total length 42 nm. Four levels of parallelism are employed: 16 applied bias points, 16 momentum points, 800 to 1400 energy points, and spatial decomposition on 2 cores. These simulations scaled to 65536 cores on Kraken at NICS and Ranger at TACC. The availability of the peta-scale resources makes the total simulation time for a fully quantum mechanical calculation very manageable.

Very recently we were given access to the Leadership Computing Facility at Oak Ridge with its Cray XT5 Jaguar. On that resource we demonstrated scaling to 147,456 cores. At that level of parallelism the code delivered around 504 TFLOP/s at 37% of peak for each CPU. We note here that no special tuning was performed on this platform to obtain this high peak performance.

6. Deployment on nanoHUB.org

nanoHUB.org is a community web site intended to deliver on-line simulation and more to the nanotechnology community at large. Over 150 tools are now on line which have delivered over 400,000 simulations to over 7,000 users in the 12 months leading up to May 2009. All the simulation

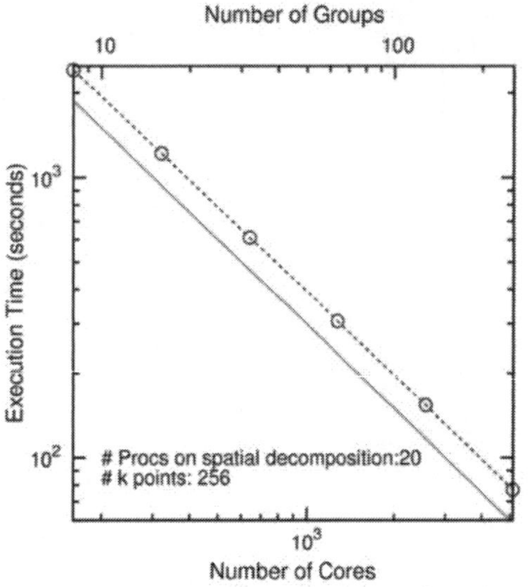

Figure 8. The performance of the multi-level parallelism. End-to-end band structure calculation is performed on a small periodic structure with 256 k points considered.

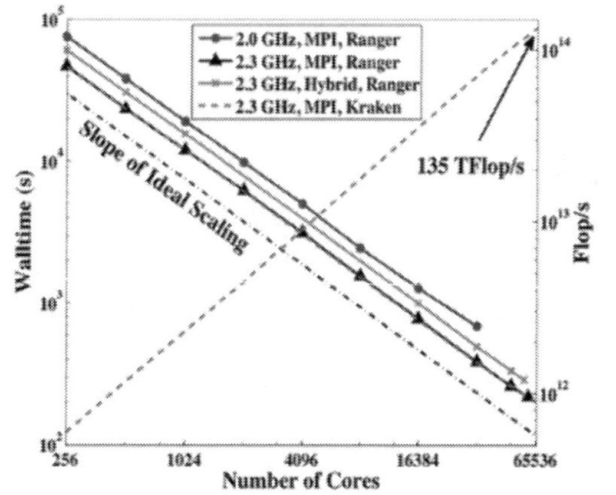

Figure 9: Scaling performances of OMEN up to 65,536 cores for a 2-D DG UTB transistor using four levels of parallelism (16 bias points, 16 momentum points, 800 to1400 energy points, spatial decomposition on 2 cores).

tools are fully interactive in that users can easily set-up their numerical experiments and "what if?" questions and compare their simulation results after the computation. No installation of software is needed on the user end, compute cycles, the graphical user interface, tools, and instructional materials are delivered remotely. Tools of various computational intensity are hosted, many of them executing in a few minutes, few of them running in parallel on computational resources like the TeraGrid. Over 430 citations in the literature are evidence of use in peer-reviewed research — 52% of these citations are unaffiliated with the host of the nanoHUB: The Network for Computational Nanotechnology. nanoHUB has also been used in over 290 classes in over 90 different institutions for educational purposes. More information can be found on http://nanoHUB.org/about and http://nanoHUB.org/usage as well as several peer reviewed articles [47-52].

Both NEMO 3-D and OMEN drive computational tools on nanoHUB.org. OMEN is the underlying engine for the tools OMEN Nanowire and Band Structure Lab, while NEMO 3-D powers Quantum Dot Lab. Figure 10 shows the usage statistics for these simulation tools. Together NEMO 3-D and OMEN have driven thousands of simulations on the nanoHUB, providing cutting-edge research capability for a worldwide user base. We plan to replace the backend of several more tools on the nanoHUB with OMEN or components of OMEN; this will greatly simplify maintenance of those tools and consolidate several competing

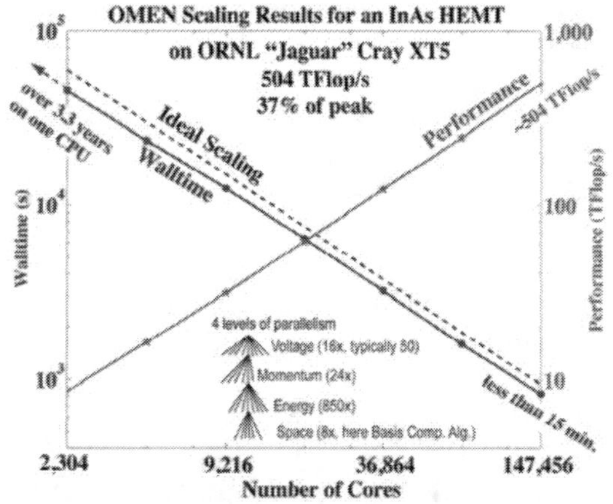

Figure 10. Scaling performances of OMEN up to 147,456 cores for a 2-D DG UTB transistor using four levels of parallelism (16 bias points, 24 momentum points, 850 energy points, spatial decomposition on 8 cores).

implementations into one code base. For example, currently several tools implement a Poisson solver using different methods. In many cases, no one maintains the code. By replacing these with the OMEN solver, we reduce the number of Poisson implementations to one, which is a highly optimized current research tool. The number of simulations run in these OMEN and NEMO3D powered tools are tabulated in Table 1 and the global users are depicted in Figure 11.

Table 1. Usage statistics for OMEN- and NEMO 3-D-powered tools on the nanoHUB.

Tool	Powered by	Users	Simulations	Mean CPU time	Release date	Months released
Quantum Dot Lab	NEMO 3-D	1888	18157	8.09 seconds	Nov 2005	42
Band Structure Lab	OMEN	2307	18251	7.61 minutes	May 2006	36
OMEN Nanowire	OMEN	104	729	4.33 hours	Sept 2008	9

Figure 11. World map of NEMO3D and OMEN-powered tool users.

7. Conclusions

We have made substantial enhancements to the codes NEMO 3-D and OMEN. The enhancements allow the codes to make full use of peta-scale computing resources and opened previously inaccessible realms of device engineering studies. NEMO 3-D can handle billion atom systems, scaling to 32768 cores, using three dimensional spatial decomposition. By making use of large scale computing resources, NEMO 3-D can reduce the calculation time for the band structure of a cutting-edge quantum computing device from 40 minutes to 1 minute. OMEN can model fully quantum mechanical transport in realistic UTB FETs on 147,456 cores in about 15 minutes. These compute engines drive tools on the nanoHUB, presenting users with state-of-the-art research capabilities in their web browsers.

References

[1] For reviews and references see, e.g., Jacak, L., Hawrylak, P., and Wojs, A, "Quantum dots", Springer-Verlag, Berlin, 1998.
[2] Aslan, B., Liu, H.C., Korkusinski, M., Cheng, S.-J., and Hawrylak, P., Appl. Phys. Lett., 82,

630, 2003.
[3] Petroff, P.M., in "Single Quantum Dots: Fundamentals, Applications, and New Concepts", Peter Michler, Ed., Springer, Berlin, 2003.
[4] Michler, P., et al., Science, 290, 2282, 2000; Moreau, E., et al., Phys. Rev. Lett., 87, 183601, 2001.
[5] Arakawa, Y., and Sasaki, H., Appl. Phys. Lett., 40, 939, 1982; Fafard, S., et al., Science, 22, 1350, 1996; Maximov, M.V., et al., J. Appl. Phys., 83, 5561, 1998.
[6] B. Doris et al., "Extreme scaling with ultra-thin Si channel MOSFETs", IEDM Tech. Dig. 2002, 267-270 (2002).
[7] Y. Cui, L. J. Lauhon, M. S. Gudiksen, J. Wang, and C. M. Lieber, "Diameter-controlled synthesis of single-crystal silicon nanowires", Appl. Phys. Lett. 78, 2214 (2001).
[8] S. D. Suk et al., "Investigation of nanowire size dependency on TSNWFET", IEDM Tech. Dig. 2007, 891-894 (2007).
[9] J. Appenzeller, Y.-M. Lin, J. Knoch, and Ph. Avouris, "Band-to-Band Tunneling in Carbon Nanotube Field-Effect Transistors", Phys. Rev. Lett. 93, 196805 (2004).
[10] W. Y. Choi, B.-G. Park, J. D. Lee, and T.-J. King Liu, "Tunneling Field-Effect Transistors (TFETs) With Subthreshold Swing (SS) Less Than 60 mV/dec", IEEE Elec. Dev. Lett. 28, 743-745 (2007).
[11] W. Fichtner, D. J. Rose, and R. E. Bank, "Semiconductor-Device Simulation", IEEE Trans. on Elec. Dev. 30, 1018-1030 (1983).
[12] J. Wang, E. Polizzi, and M. S. Lundstrom, "A three-dimensional quantum simulation of silicon nanowire transistors with the effective-mass approximation", J. Appl. Phys. 96, 2192-2203 (2004).
[13] M. Luisier, A. Schenk, and W. Fichtner, "Three-Dimensional Full-Band Simulations of Si Nanowire Transistors", IEDM Tech. Digest 2006, 811 (2006).
[14] M. Luisier, A. Schenk, and W. Fichtner, "Atomistic Treatment of interface roughness in Si nanowire transistors with different channel orientations", Appl. Phys. Lett. 90, 102103 (2007).
[15] M. Luisier, G. Klimeck, A. Schenk, and W. Fichtner, "Atomistic Simulation of Nanowires in the sp3d5s* Tight-Binding Formalism: from Boundary Conditions to Strain Calculations", Phys. Rev. B 74, 205323 (2006).
[16] F. Oyafuso, G. Klimeck, P. von Allmen, T. Boykin, and R. C. Bowen, "Strain Effects in large-scale atomistic quantum dot simulations", Phys. Stat. Sol. (b), Vol. 239, p 71-79 (2003).
[17] J. C. Slater and G. F. Koster, "Simplified LCAO Method for the Periodic Potential Problem", Phys. Rev. 94, 1498-1524 (1954).
[18] Klimeck, G., Oyafuso, F., Boykin, T.B., Bowen, R.C., and von Allmen, P., "Development of a Nanoelectronic 3-D (NEMO 3-D) Simulator for Multimillion Atom Simulations and Its Application to Alloyed Quantum Dots", Computer Modeling in Engineering and Science, 3, 601, 2002.
[19] G. Klimeck, F. Oyafuso, R. C. Bowen, T. B. Boykin, T. A. Cwik, E. Huang, E. S. Vinyard .3-D atomistic nanoelectronic modeling on high performance clusters: multimillion atom simulations., Superlattices and Microstructures, Vol. 31, Nos 2–4, 2002.
[20] F. Oyafuso, G. Klimeck, R. C. Bowen, T. B. Boykin, and P. von Allmen. "Disorder Induced Broadening in Multimillion Atom Alloyed Quantum Dot Systems", Phys. Stat. Sol. (c), vol 0004, pg 1149-1152 (2003).
[21] M. Luisier and G. Klimeck, "Numerical Strategies towards Peta-Scale Simulations of Nanoelectronics Devices", submitted to SC09 (2009).
[22] Gerhard Klimeck, "Parallelization of the Nanoelectronic Modeling Tool (NEMO 1-D) on a Beowulf Cluster," Journal of Computational Electronics, Vol 1, Issue 1-2, 2002
[23] Mathieu Luisier and Gerhard Klimeck, "A multi-level parallel simulation approach to electron transport in nano-scale transistors," Supercomputing 2008, Austin TX, Nov. 15-21 2008.

Regular paper - 59 accepted papers, 277 submissions (21%)

[24] Hansang Bae, Steve Clark, Ben Haley, Hoon Ryu, Gerhard Klimeck, Sunhee Lee, Mathieu Luisier and Faisal Saied, "A Nano-electronics Simulator for Petascale Computing: From NEMO to OMEN," TeraGrid 2008, Jun 9-13, 2008, Las Vegas

[25] C. Lanczos, "An Iteration Method for the Solution of the Eigenvalue Problem of Linear Differential and Integral Operators", Journal of Research of the National Bureau of Standards, Vol. 45, No 4, 1950

[26] R.R. Underwood, "An iterative block Lanczos method for the solution of large sparse symmetric eigenproblems", Ph.D. Dissertation, Stanford University, Stanford, CA, 1975

[27] A. H. Sameh and J. Wisniewski. A Trace Minimization Algorithm for the Generalized Eigenvalue Problem . SIAM Journal on Numerical Analysis , Vol. 19, No. 6, pp. 1243-1259, 1982.

[28] A. Sameh and Z. Tong, The trace minimization method for the symmetric generalized eigenvalue problem. *J. Comput. Appl. Math.* 123, 155-175, 2000.

[29] K. Maschhoff and D. Sorensen, "A portable implementation of ARPACK for distributed memory parallel architectures", Copper Mountain Conference on Iterative Methods, 1996

[30] P. R. Amestoy, I. S. Duff, and J.-Y. L'Excellent, "Multifrontal parallel distributed symmetric and unsymmetric solvers" Comput. Methods in Appl. Mech. Eng. 184, 501 (2000).

[31] X. S. Li and J. W. Demmel "SuperLU DIST: A Scalable Distributed Memory Sparse Direct Solver for Unsymmetric Linear Systems", ACM Trans. on Math. Software 29, 110 (2003).

[32] O. Schenk and K. Gärtner, "Solving Unsymmetric Sparse Systems of Linear Equations with PARDISO", Journal of Future Generation Computer Systems 20, 475 (2004).

[33] T. B. Boykin, M. Luisier, and G. Klimeck, "Multi-band transmission calculations for nanowires using an optimized renormalization method", Phys. Rev. B 77, 165318 (2008)

[34] M. Luisier, A. Schenk, W. Fichtner, T. B. Boykin, and G. Klimeck, "A parallel sparse linear solver for nearest-neighbor tight-binding problems", Proc. of the 14th international Euro-Par conference on Parallel Processing, 790-800 (2008).

[35] P. M. Gresho and R. L. Sani, "Incompressible Flow and the Finite Element Method: Isothermal Laminar Flow", John Wiley and Sons, New York (2000).

[36] R. E. Bank, D. J. Rose, and W. Fichtner, "Numerical Methods for Semiconductor Device Simulation", IEEE Trans. Electron Dev. 30, 1031 (1983).

[37] R. S. Tuminaro, M. Heroux, S. A. Hutchinson, and J. N. Shadid, "Official Aztec User's Guide: Version 2.1" (1999).

[38] A. Duran, M. Gonzalez, and J. Corbalan, "Automatic thread distribution for nested parallelism in OpenMP", Proceedings of the 19th annual international conference on Supercomputing, 121-130 (2005).

[39] T. C. Shen, J. Y. Ji, M. A. Zudov, R. R. Du, J. S. Kline, J. R. Tucker, "Ultradense phosphorus delta layers grown into silicon from PH3 molecular precursors," Appl. Phys. Lett. 80, 9, 2002.

[40] Hoon Ryu, Sunhee Lee and Gerhard Klimeck, "A Study of Temperature-dependent properties of n-type delta-doped Si bandstrucstures in equilibrium," to appear in IEEE proceedings of the 13th International Workshop on Computational Electronics, Tsinghua University, Beijing, May 27-29 2009.

[41] Gerhard Klimeck, Roger K. Lake, R. Chris Bowen, William R. Frensley and Ted Moise, "Quantum Device Simulation with a Generalized Tunneling Formula", Appl. Phys. Lett., Vol. 67, p.2539 (1995).

[42] R. Chris Bowen, Gerhard Klimeck, Roger Lake, William R. Frensley and Ted Moise, "Quantitative Resonant Tunneling Diode Simulation", J. of Appl. Phys. 81, 3207 (1997).

[43] Gerhard Klimeck, Roger Lake and Daniel K. Blanks,"Role of interface roughness scattering in self-consistent resonant tunneling diode simulation", Phys. Rev. B, Vol. 58, 7279 (1998).

[44] Gerhard Klimeck, Roger Lake and Daniel K. Blanks,"Numerical Approximations to the

Treatment of Interface Roughness Scattering in Resonant Tunneling Diodes", Semicond. Sci. Technology, Vol 13, pg. A165 (1998).

[45] Roger Lake, Gerhard Klimeck and Daniel K. Blanks, "Interface Roughness and Polar Optical Phonon Scattering in InGaAs/AlAs/InAs RTDs", Semicond. Sci. Technology, Vol 13, pg. A163 (1998).

[46] Gerhard Klimeck, "Parallelization of the Nanoelectronic Modeling Tool (NEMO 1-D) on a Beowulf Cluster", Gerhard Klimeck, Journal of Computational Electronics, Vol 1, Issue 1-2, pp 75-79 (2002).

[47] Gerhard Klimeck, Michael McLennan, Sean B. Brophy, George B. Adams III, Mark S. Lundstrom, "nanoHUB.org: Advancing Education and Research in Nanotechnology", IEEE Computers in Engineering and Science (CISE), Vol. 10, pg. 17-23 (2008).

[48] Nancy Wilkins-Diehr, Dennis Gannon, Gerhard Klimeck, Scott Oster, Sudhakar Pamidighantam, "TeraGrid Science Gateways, Virtual Organizations and Their Impact on Science", IEEE Computer, Vol. 41, Issue 11, Nov. 2008, Page(s):32 – 41

[49] Benjamin P. Haley, Gerhard Klimeck, Mathieu Luisier, Dragica Vasileska, Abhijeet Paul, Swaroop Shivarajapura, Diane L. Beaudoin, "Computational nanoelectronics research and education at nanoHUB.org", accepted (June 26) for publication to Journal of Computational Electronics (2009).

[50] Shaikh Ahmed, Gerhard Klimeck, Derrick Kearney, Michael McLennan, and M. P. Anantram, "Quantum Simulations of Dual Gate MOSFET Devices: Building and Deploying Community Nanotechnology Software Tools on nanoHUB.org", Journal of High Speed Electronics and Systems, Vol. 17, No 3, pp. 485-494 (2007).

[51] Gerhard Klimeck and Dragica Vasileska, "ABACUS and AQME: Semiconductor Device and Quantum Mechanics Education on nanoHUB.org", IEEE proceedings of the 13th International Workshop on Computational Electronics, Tsinghua University, Beijing, May 27-29 2009, DOI:10.1109/IWCE.2009.5091083.

[52] Gerhard Klimeck, Michael McLennan, Mark S. Lundstrom, George B. Adams III. "nanoHUB.org - online simulation and more materials for semiconductors and nanoelectronics in education and research", 8th IEEE Conference on Nanotechnology, 2008. NANO '08. Aug. 18-21 2008 Page(s):401-404, DOI:10.1109/NANO.2008.124

[53] S. Datta, "The NEGF Approach to Nano-Device Simulation", http://nanohub.org/topics/negf

[54] S. Datta, "Nanoscale Device Simulation: The Green's Function Method," Superlattices and Microstructures, 28, 253-278 (2000).

[55] Gerhard Klimeck, Roger K. Lake, R. Chris Bowen, William R. Frensley and Ted Moise, "Quantum Device Simulation with a Generalized Tunneling Formula", Appl. Phys. Lett., Vol. 67, p.2539 (1995)

[56] Roger K. Lake, Gerhard Klimeck, R. Chris Bowen, Chenjing L. Fernando, Manhua Leng, Ted Moise and Y. C. Kao, "Interface Roughness and Polar Optical Phonon Scattering And the Valley Current in Resonant Tunneling Devices", Superlattices and Microstructures, Vol. 20, p.279 (1996).

[57] Gerhard Klimeck, Roger Lake, Dan Blanks, Chenjing L. Fernando, R. Chris Bowen, Ted Moise, and Y. C. Kao, "The Effects of Electron Screening Length and Emitter Quasi-Bound States on the Polar-Optical Phonon Scattering in Resonant Tunneling Diodes", Physica Status Solidi (b), Vol. 204, 408 (1997).

[58] Gerhard Klimeck, Roger Lake and Daniel K. Blanks, "Role of interface roughness scattering in self-consistent resonant tunneling diode simulation", Phys. Rev. B, Vol. 58, 7279 (1998).

[59] Roger Lake, Gerhard Klimeck, R. Chris Bowen and Dejan Jovanovic, "Single and multiband modeling of quantum electron transport through layered semiconductor devices", J. of Appl. Phys. 81, 7845 (1997)

[60] R. Chris Bowen, Gerhard Klimeck, Roger Lake, William R. Frensley and Ted Moise, "Quantitative Resonant Tunneling Diode Simulation", J. of Appl. Phys. 81, 3207 (1997).

[61] Gerhard Klimeck, Timothy B. Boykin, R. Chris Bowen, Roger Lake, Dan Blanks, Ted Moise, Y. C. Kao, and William R. Frensley, "Quantitative Simulation of Strained InP-Based Resonant Tunneling Diodes", in IEEE Proceedings of the 1997 55th IEEE Device Research Conference Digest, IEEE, NJ, p. 92 (1997).

[62] Gerhard Klimeck, Fabiano Oyafuso, Timothy B. Boykin R. Chris Bowen, and, Paul von Allmen, "Development of a Nanoelectronic 3-D (NEMO 3-D) Simulator for Multimillion Atom Simulations and Its Application to Alloyed Quantum Dots (INVITED)", Computer Modeling in Engineering and Science (CMES) Volume 3, No. 5, pp 601-642 (2002), ISSN: 1526-1492.

[63] Gerhard Klimeck, Shaikh Ahmed, Neerav Kharche, Marek Korkusinski, Muhammad Usman, Marta Prada, and Timothy B. Boykin, "Atomistic Simulation of Realistically Sized Nanodevices Using NEMO 3-D: Part II - Applications", Special Issue on Nanoelectronic Device Modeling in IEEE Transactions on Electron Devices, Vol. 54, Issue 9, pg. 2090 - 2099 (2007).

[64] Gerhard Klimeck, Shaikh Ahmed, Hansang Bae, Neerav Kharche, Steve Clark, Benjamin Haley, Sunhee Lee, Maxim Naumov, Hoon Ryu, Faisal Saied, Marta Prada, Marek Korkusinski, and Timothy B. Boykin, "Atomistic Simulation of Realistically Sized Nanodevices Using NEMO 3-D: Part I - Models and Benchmarks", Special Issue on Nanoelectronic Device Modeling in IEEE Transactions on Electron Devices, Vol. 54, Issue 9, pg. 2079 - 2089 (2007).

[65] Fabiano Oyafuso, Gerhard Klimeck, R. Chris Bowen, and Timothy B. Boykin, "Atomistic Electronic Structure Calculations of Unstrained Alloyed Systems Consisting of a Million Atoms", Journal of Computational electronics, Vol. 1. Issue 3, pp. 317-321 (2002).

[66] Gerhard Klimeck, R. Chris Bowen, and Timothy B. Boykin, "Atomistic simulation of quantum dots including strain and bandstructure and full band simulation of hole transport in 1-D heterostructures", in IEEE proceedings of 7th International Workshop on Computational Electronics (IWCE). Book of Abstract, Glasgow 22-25 May 2000, pp 6-7, (2000)

[67] Muhammad Usman, Hoon Ryu, Insoo Woo, David S. Ebert, and Gerhard Klimeck, "Moving towards nano-TCAD through multi-million atom quantum dot simulations matching experimental data", IEEE Transactions on Nanotechnology, Vol. 8, Issue 3, pg. 330-344 (2009), DOI: 10.1109/TNANO.2008.2011900 (2009).

[68] G.P. Lansbergen, R. Rahman, C.J. Wellard, P.E. Rutten, J. Caro, N. Collaert, S. Biesemans, I. Woo, G. Klimeck, L.C.L. Hollenberg, and S. Rogge, "Gate induced quantum confinement transition of a single dopant atom in a Si FinFET", Nature Physics, Vol. 4, pg. 656 (2008).

[69] Neerav Kharche, Marta Prada, Timothy B. Boykin, and Gerhard Klimeck, "Valley-splitting in strained Silicon quantum wells modeled with 2 degree miscuts, step disorder, and alloy disorder", Applied Phys. Lett. Vol. 90, 092109 (2007).

[70] Gengchiau Liang, Jie Xiang, Neerav Kharche, Gerhard Klimeck, Charles M. Lieber, and Mark Lundstrom, "Performance Analysis of a Ge/Si Core/Shell Nanowire Field Effect Transistor", cond-mat/0611226, Nano letters, Vol. 7, 642-646 (2007).

[71] Timothy B. Boykin, Mathieu Luisier, Andreas Schenk, Neerav Kharche, Gerhard Klimeck, "Atomistic, random-alloy calculations of the electronic structure and transmission characteristics of AlGaAs nanowires", IEEE Transactions on Nanotechnology, Vol. 6, No. 1, 43-7 (2007).

[72] Hansang Bae, Steve Clark , Ben Haley, Ryu Hoon, Gerhard Klimeck, Sunhee Lee, Mathieu Luisier, and Faisal Saied, "A Nano-electronics Simulator for Petascale Computing: From NEMO to OMEN", Proceedings of TeraGrid 2008, June 9-13, 2008, Reno, NV.

[73] Mathieu Luisier, Neophytos Neophytou, Neerav Kharche, and Gerhard Klimeck, "Full-Band and Atomistic Simulation of Realistic 40 nm InAs HEMT", IEEE IEDM, San Francisco, USA, Dec. 15-17, 2008, DOI : 10.1109/IEDM.2008.4796842.

[74] Mathieu Luisier, and Gerhard Klimeck, "Atomistic, Full-Band Design Study of InAs Band-to-

Band Tunneling Field-Effect Transistors", IEEE Electron Device Letters, Vol. 30, pp. 602-604 (2009), DOI: 10.1109/LED.2009.2020442 (2009).

[75] Mathieu Luisier, and Gerhard Klimeck, "Performance analysis of statistical samples of graphene nanoribbon tunneling transistors with line edge roughness", Applied Physics Letters, Vol. 94, 223505 (2009), DOI:10.1063/1.3140505

[76] Gerhard Klimeck and Mathieu Luisier, "From NEMO1D and NEMO3D to OMEN: moving towards atomistic 3-D quantum transport in nano-scale semiconductors", IEEE IEDM, San Francisco, USA, Dec. 15-17, 2008, DOI: 10.1109/IEDM.2008.4796647.

First-principles computational discovery of materials for hydrogen storage

V Ozolins[1,4], A R Akbarzadeh[1], H Gunaydin[1], K Michel[1], C Wolverton[2] and E H Majzoub[3]

[1]Department of Materials Science and Engineering, University of California, Los Angeles, California 90095-1595, USA
[2]Department of Materials Science and Engineering, Northwestern University, Evanston, Illinois 60208, USA
[3]Department of Physics and Astronomy, University of Missouri, St. Louis, Missouri 63121, USA

E-mail: vidvuds@ucla.edu

Abstract. Hydrogen-fuelled vehicles require a cost-effective, lightweight material with precisely targeted thermodynamics and fast kinetics of hydrogen release. Since none of the conventional metal hydrides satisfy the multitude of requirements for a practical H_2 storage system, recent research efforts have turned to advanced multicomponent systems based on complex hydrides. We show that first-principles density-functional theory (DFT) calculations have become a valuable tool for understanding and predicting novel hydrogen storage materials and understanding the atomic-scale kinetics of hydrogen release. Recent studies have used DFT calculations to (i) predict crystal structures of new solid-state hydrides, (ii) determine phase diagrams and thermodynamically favoured reaction pathways in multinary hydrides, and (iii) study microscopic kinetics of diffusion, phase transformations, and hydrogen release.

1. Introduction

General adoption of hydrogen as a vehicular fuel depends on the ability to store hydrogen at high volumetric and gravimetric densities, as well as on the ability to extract it at sufficiently rapid rates. Since practical requirements for on-board hydrogen storage systems are very challenging and cannot be met with existing technologies, significant effort has been invested in searching for new materials that could efficiently store hydrogen at near ambient conditions [1]. Among the various storage options (such as compressed gas, sorbents and metal-organic frameworks [2,3], chemical hydrides [4]), solid-state hydrogen storage in complex hydrides has attracted particular attention due to very high volumetric storage densities and favorable thermodynamics [5-7]. Interest in complex hydrides was sparked by the discovery [8] that sodium alanate, $NaAlH_4$, reversibly releases approximately 4.5 wt. % hydrogen at relatively low temperatures (around 120 °C) when doped with small amounts of titanium:

$$NaAlH_4 \leftrightarrow \frac{1}{3}Na_3AlH_6 + \frac{2}{3}Al + H_2 . \quad (1)$$

[4] To whom any correspondence should be addressed.

The discovery of reaction (1) instantly doubled the reversible capacity of inexpensive hydrides and pointed to the intriguing possibility that many other complex hydrides, such as alanates, borohydrides and amides, may be used for energy storage. Subsequently, high-density hydrogen storage has been demonstrated in several complex hydride systems, including amides [9,10], borohydrides [11], and others [12]. Nevertheless, achieving the right combination of structural, thermodynamic, kinetic and economic factors remains elusive. Trial-and-error experimentation or high-throughput searches for new materials are often hampered by experimental difficulties, such as small sample sizes and slow (de)hydrogenation kinetics. Thus, it is highly desirable to accurately estimate the thermodynamic properties of new materials before substantial resources are expended on their synthesis and characterization. First-principles calculations of the total crystal binding energies using the density-functional theory (DFT) have been shown to yield sufficiently accurate thermodynamic properties to be useful in predicting new hydrogen storage reactions [13,14]. Since these calculations are parameter-free and require only the atomic numbers and crystal structures as input, they represent an attractive alternative to expensive and time-consuming experimentation.

We review here our recent work towards a systematic, rational approach to designing novel hydrogen storage materials using state-of-the-art scientific computing. We show how the accurate predictive power of first-principles DFT modeling can be used to understand the thermodynamic and kinetic factors involved in hydrogen release and uptake and design new materials systems with improved properties. Our review is organized into three main areas: (1) predicting crystal structures of new materials, (2) predicting thermodynamics of hydrogen storage reactions and identifying favored reaction pathways, and (3) studies of microscopic kinetic mechanisms, rate-limiting steps and activation energies for hydrogen release and absorption reactions.

2. Computational methods

For all calculations presented here, the first-principles density-functional theory (DFT) method of electronic structure calculations has been used to calculate the total binding energies and vibrational free energies. Structural energies and vibrational frequencies were obtained using the *ab-initio* total-energy and molecular dynamics code VASP [15,16]. Electronic exchange and correlation were treated within the generalized gradient approximation (GGA) of Perdew and Wang [17], and the projected augmented wave (PAW) approach [18,19] was used to represent the interactions between the nuclei and valence electrons. The phonon frequencies were determined using the dynamical matrix approach where the symmetry-inequivalent rows of the dynamical matrix were calculated by displacing ions for a few steps around their equilibrium positions (further details of our implementation are given in Ref. [13]). The vibrational entropies and enthalpies were obtained by summing over the directly calculated normal mode frequencies.

The free energy of the hydrogen molecule is given by:

$$F[H_2] = E_0[H_2] + \frac{7}{2}k_B T - TS_{H_2} \tag{2}$$

where k_B is the Boltzmann constant and $E_0(H_2)$ is the static DFT energy of an isolated H_2 molecule at T = 0 K; it is obtained by placing H_2 in a periodic cell with edge lengths of 10 Å. S_{H2} represents the well-known expression for the entropy of an ideal diatomic gas in the high-temperature limit:

$$S_{H_2} = k_B\left[\frac{7}{2}\ln(T) - \ln(p) + C_0\right], \tag{3}$$

where p is the pressure and C_0 = -4.22 is a dimensionless constant chosen to reproduce experimental data [20]. Equations (2) and (3) account for corrections to the free energy of gaseous H_2, which has to be treated separately from static and vibrational free energy contributions.

3. Predicting crystal structures

Determining the crystal structures of new compounds is the first step required by any first-principles materials discovery effort. However, predicting the ground state crystal structure of a compound with a given chemical formula remains one of the most challenging problems in the field of computational materials science. Perhaps the most popular method is to choose "reasonable" crystal structure types (so-called "usual suspects") and calculate their total energies with first-principles techniques. The structure with the lowest total energy (or free energy in finite-temperature studies) is then declared to be the ground state, or at least one that can provide a reasonable approximation to the thermodynamics of the true ground state. This approach has been widely used in studies of intermetallic compounds [21] and pressure-induced phase transformations [22]. With the increase in the available computer power, very large sets of trial structures have become practical, and the use of the so-called *database searching* approach, has exploded in popularity.

In database searching, candidate crystal structures are obtained from chemically similar compounds in published databases, such as the Inorganic Crystal Structures Database (ICSD) [23]. Next, DFT calculations are performed for all database matches to find the lowest-energy structure. Phonon calculations are then used to verify the dynamic stability of the structure. If the compound has imaginary phonon frequencies, *ab initio* molecular dynamics (AIMD) simulations or frozen phonon calculations for finite displacements along the unstable mode eigenvectors are used to search for stable geometries with lower energies [24]. For common stoichiometries, ICSD usually has a large set of candidate structures (from many tens to hundreds or even thousands), which constitute a large trial pool to sample the configuration space of competing crystal structure types. The database searching approach has been used with considerable success in numerous studies of complex hydrides [24-29]. The data mining technique introduced by Curtarolo *et al.* [30] offers an efficient way to narrow down database searching by using probabilistic concepts that guide the choice of the likely candidate ground states. Data mining offers significant efficiencies at common compositions where the number of trial structures can be very large.

All database methods rely heavily on the existence of an extensive catalog of "good" trial structures and have limited use if crystallographic information on chemically similar compounds is not available, as is the case for many novel hydrides. Furthermore, we will show below that even for some common chemical formulae the database searching approach may fail to predict energetically competitive structures because the database entries are too dissimilar in their bonding properties from the target hydride. This shows that *there is a strong need for methods that can predict entirely new ground state structure types, i.e. those that are not isostructural with compounds already observed in other systems*. There are few first-principles methods that can generate new ground state crystal structures "from scratch" (e.g., genetic algorithms [31] or simulated annealing [32]), but they presently require large computational effort and are efficient only for relatively small unit cells. In this section, we describe two methods that take advantage of special complex hydride bonding properties to generate entirely new crystal structure types with only a moderate computational cost.

3.1. Lattice algebra enumeration

In many materials it has been difficult to determine exactly the location of the hydrogen atoms, while the positions of the heavier ions are well known or can be guessed with a high degree of certainty. For instance, when hydrogen is covalently bonded to anions such as N or O there is often considerable freedom in the bond orientation, making the determination of the structure difficult. To predict the crystal structures in such cases of partial information, we have developed a computational search method, which provides the means for a systematic enumeration of the possible configurations for the unresolved degrees of freedom.

A typical case is presented by the structure of lithium imide, Li_2NH, which is a decomposition product in the famous lithium amide – lithium hydride reaction [9]:

$$LiNH_2 + LiH \leftrightarrow Li_2NH + H_2. \tag{4}$$

All experimental investigations of Li_2NH consistently point to a room temperature structure with Li and N ions occupying the positions of an antifluorite structure (see figure 1). However, there are conflicting reports on the H positions. Some reports suggest that hydrogen is distributed in a disordered manner across various Wyckoff positions within the fluorite framework [33-35], while recent studies by Herbst and Hector [36] and Balogh *et al.* [37] propose an ordered arrangement of H positions, resulting in an orthorhombic structure of Ima2 symmetry.

Another interesting case is presented by the crystal structures of mixed Li-Mg imides, such as $Li_2Mg(NH)_2$ and $Li_4Mg(NH)_3$ [38], which may form as the decomposition products in thermodynamically optimized reactions between lithium amide and magnesium hydride [10]:

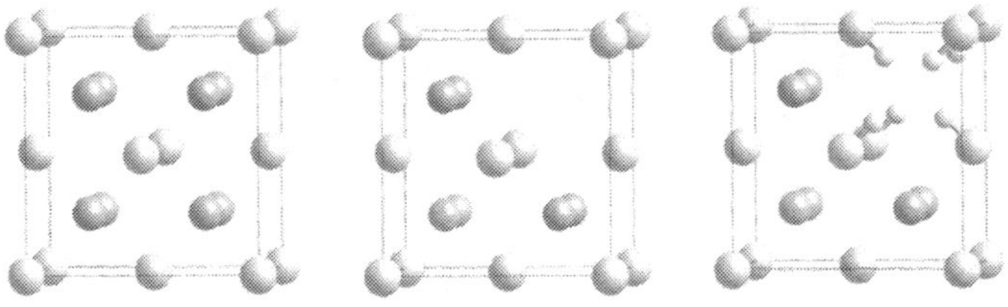

Figure 1. Enumeration of candidate $Li_2Mg(NH)_2$ structures starting from the cubic antifluorite structure of Li_2NH. *Left:* Antifluorite Li_2NH without hydrogen atoms (for clarity). *Middle:* Introducing vacancies and Mg ions into Li_2NH; each Mg substitution on Li sites is accompanied by a vacancy. *Right:* Addition of hydrogen atoms with N-H bonds pointing towards the vacant sites.

$$2LiNH_2 + MgH_2 \rightarrow 2LiH + Mg(NH_2)_2 \leftrightarrow Li_2Mg(NH)_2 + 2H_2. \quad (5)$$

The ion exchange in the first step of reaction (5) is exothermic and non-reversible, while the second step is endothermic and reversible. In comparison with the original lithium amide – lithium hydride reaction (4), addition of magnesium hydride in reaction (5) lowers the dehydrogenation enthalpy and hydrogen release temperature due to the formation of the strongly bound $Li_2Mg(NH)_2$ compound [39]. Experimentally, the structure of $Li_2Mg(NH)_2$ is known to belong to the Iba2 space group [40] and contain substitutional disorder on cation (Li and Mg) sites. Site-disordered phases present a challenge to conventional DFT calculations since they require the use of elaborate alloy-theoretical tools, such as the cluster expansion [41] to evaluate finite-temperature energetics. Screening for favorable hydrogen storage thermodynamics usually allows uncertainties of 5-10 kJ/(mol H_2) in the calculated reaction enthalpies, comparable to the physical accuracy of the DFT itself [13,14]. Within these accuracy requirements, it is usually sufficient to find the T=0 K ground state with an ordered arrangement of ions and then use its free energy to evaluate the thermodynamic properties of finite-temperature configurationally disordered phases.

The enumeration method is based on the observation that the N-H orientational ordering problem may be mapped onto a multicomponent alloy problem, which allows us to utilize well-established lattice algebra methods developed in the alloy theory community. For Li imide, we consider small unit cell input structures where all N–H dimers are aligned along one of the principle orientations (e.g., [100], [110], [111]). For a given orientational axis, the problem of listing different N–H dimer orientations can be viewed as formally identical to a binary alloy problem on a fixed fcc lattice: by placing Li and N atoms on the sites of the cubic fluorite lattice, N forms a fcc sublattice (see figure 1). We then associate different N–H alignments with 'atoms' of certain type and use symmetry to reduce the number of distinct trial structures, followed by DFT total energy calculations to relax all structural degrees of freedom and find the lowest-energy structures [42]. For mixed Li-Mg imides, we generate candidate structures with ordered arrangements of Li^+ and Mg^{2+} ions on the cation sites of the antifluorite lattice, leaving some of the sites vacant to satisfy charge neutrality. The procedure of forming a mixed imide from the cubic antifluorite structure is illustrated in figure 1 for the case of $Li_2Mg(NH)_2$. Li^+ and Mg^{2+} ions are placed on the cation sites of Li_2NH and the excess charge due to magnesium ions is accommodated through the introduction of vacancies on the lithium sites. The chemical composition in this family of compounds is given by $Li_{4-2x}Mg_x\square_x(NH)_2$, where $\square_x$ represents the vacant cation sites. To satisfy steric constraints, N-H bonds are oriented towards the neighboring vacant sites whenever possible. Our procedure not only enumerates all possible orientations of N-H dimers and orderings of Li^+ and Mg^{2+} within a chosen shape of the unit cell, but also searches through all possible choices of lattice vectors with a given number of atoms in the unit cell. Since the number of distinct symmetry-inequivalent structures increases geometrically with the number of atoms in the

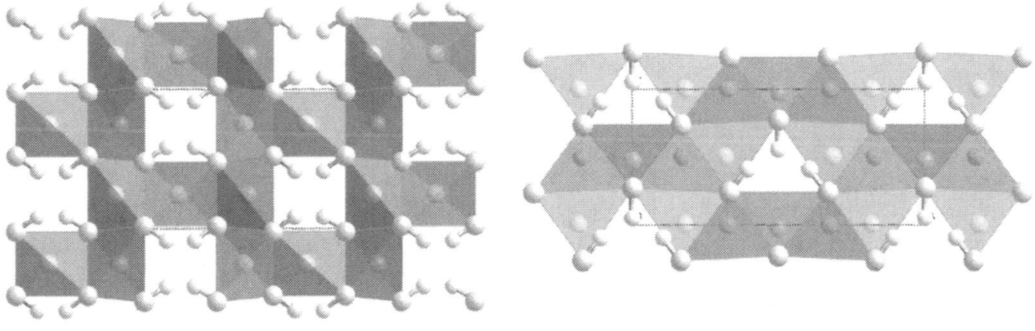

Figure 2. Proposed $Pca2_1$ (left) and Imm2 (right) structures of $Li_2Mg(NH)_2$ and $Li_4Mg(NH)_3$, respectively, shown along the direction of the vacancy rows. With respect to the fluorite lattice, vacancy rows run along the [100] ($Pca2_1$) and [110] (Imm2) directions. N atoms are in blue, Li in green, Mg in orange, and H in pink. For visual clarity, we show the tetrahedral coordination of Li and Mg cations with N anions.

unit cell, the enumeration method is limited to relatively small unit cells, e.g. one formula unit in the case of $Li_4Mg(NH)_3$.

For Li_2NH, this method yields several new low-energy ordered structures that were found to be lower in energy than all previously proposed ones [42]. An even more sophisticated alloy-theory approach, the so-called continuous cluster expansion, was used by Mueller and Ceder to model the energy of Li_2NH in terms of the relative N-H bond orientations and suggest a ground state structure of orthorhombic Pbca symmetry [43]. For the mixed-cation imide, $Li_2Mg(NH)_2$, we find a new, previously unsuspected structure of $Pca2_1$ symmetry (see figure 2), which is 4.4 kJ/mol lower in energy than the site-ordered Iba2 phase proposed in Ref. [44]. We also predict the existence of a $Li_4Mg(NH)_3$ phase, for which we predict an orthorhombic structure of Imm2 symmetry, shown in figure 2. With respect to the antifluorite structure from which these are derived, the rows of vacancies occur in the [100] direction for $Li_2Mg(NH)_2$ and in the [110] direction for $Li_4Mg(NH)_3$. In both compounds, all N-H bonds are oriented towards the vacancies, pointing to the important role that the hydrogen size constraints play in stabilizing these mixed-cation ordered-vacancy structures.

3.2. Prototype electrostatic ground state search

In many cases, the database searching method fails to predict stable structure types either because the database contains too few entries or the material under consideration is too different from the prototype compounds in the ICSD. For instance, in the case of $Mg(BH_4)_2$, the lowest-energy structure predicted by the database searching method using as many as 93 ICSD entries is dynamically unstable [24,45]. We have developed a method, which can predict entirely new, unsuspected ground state structure types for complex ionic crystals [46]. Physically, this approach is based on the observation that complex hydrides are characterized by an arrangement of positively charged cations (e.g., Mg^{2+}) and negatively charged complex anions (e.g., BH_4^-), with strong covalent bonds within each complex anion and predominantly electrostatic interactions between the cations and the anions [47,48]. These considerations suggest that the choice of the ground state crystal structure is predominantly influenced by the electrostatic energy, with secondary bonding interactions paying a minor role. Accordingly, the PEGS method is based on the assumption that the topology and symmetry (but not the total cohesive energy) of the ground state crystal structure can be determined by minimizing the total electrostatic energy for a set of charged complex anions and cations with soft-sphere repulsion between the ions. The resulting structures, henceforth referred to as prototype electrostatic ground states (PEGS), are then fully relaxed using first-principles DFT calculations to obtain accurate total energies and structural parameters. The PEGS Hamiltonian includes electrostatic and soft-core interionic repulsion terms:

$$H_{\text{PEGS}} = \sum_{i<j} \left(\frac{Z_i Z_j}{r_{ij}} + \frac{1}{r_{ij}^{12}} \right), \tag{6}$$

where the sums run over all ions, r_{ij} is the separation between ions i and j, Z_i is the ionic charge, and the second term (soft-core repulsion) is zero for non-overlapping ions. Monte Carlo simulated annealing in conjunction with distance scaling potential energy surface smoothing techniques [49] are used to minimize the electrostatic energy given by equation (6). For each compound, many runs are executed in parallel for several different values of a starting random number seed, which initializes the cell contents. The resulting candidate ground states are then used as input to DFT calculations to relax all structural parameters and obtain accurate total energies. The success of our approach does not depend on the precise values of the ionic radii and charges, since the PEGS can be determined for a range of parameters, spanning the parametric regions near the compound under consideration.

We have applied the PEGS method to predict the crystal structure of $Mg(BH_4)_2$ [24]. The PEGS search easily outperforms database searching [45] and simulated annealing AIMD methods and unexpectedly finds a new tetragonal ground state structure of $I\bar{4}m2$ symmetry (shown in figure 3), which is 5 kJ/mol lower than the $P6_1$ low-temperature (LT) structure observed in solution-grown crystals [50,51]. The primitive unit cell of the $I\bar{4}m2$ structure contains only two formula units (22 atoms), compared with 30 formula units (330 atoms) for the experimental $P6_1$ structure. Interestingly, both the predicted and experimental structures have very similar near-neighbor environments, but dramatically different densities, suggesting that coordination with solvent molecules and the associated steric factors may play an important role in promoting the growth of the higher-energy LT crystal structure. To the best of our knowledge, the $I\bar{4}m2$ structure has not been observed in any other compounds and represents a completely new type of crystal structure. This demonstrates the usefulness of our approach for rapidly generating new crystal structure types and accurately predicting thermodynamic properties of new materials. Using the derived structure, we obtain a room-temperature decomposition enthalpy of 29.5 kJ/(mol H_2), corresponding to hydrogen release at T=20 °C and p=1 bar (see table 1). Our results indicate that magnesium borohydride might be one of the most promising candidates among all known complex hydrides for high-capacity, near-ambient storage, and that the experimentally observed high H_2 release temperatures and lack of reversibility are due to poor kinetics.

Recently, we used the PEGS search to predict the unknown ground state crystal structures of $X_nB_{12}H_{12}$ compounds involving the icosahedral $[B_{12}H_{12}]^{2-}$ complex, which have been suggested as products during hydrogen release from borohydrides by recent experiments [52-54]. We conducted ground state crystal structure searches using one- and two-formula unit primitive cells for $Li_2B_{12}H_{12}$, $MgB_{12}H_{12}$, and $CaB_{12}H_{12}$. The lowest-energy structures found in this search are shown graphically in figure 4. For $Li_2B_{12}H_{12}$, our structure is ~7 kJ/mol (or ~70 meV per formula unit) lower than the $P2_1/n$ structure proposed by Ohba *et al.* [55], which was obtained by database searching using candidate structures from the ICSD. Interestingly, it is also 7 kJ/mol lower than the $Pa\bar{3}$ structure in solution-grown samples obtained in Ref. [56]. There are no existing theoretical predictions or experimental data for the crystal structures of $MgB_{12}H_{12}$ and $CaB_{12}H_{12}$, and our work constitutes the first available study of the crystal structures and thermodynamics of these compounds. All the monoclinic structures found by PEGS have no known prototypes in the ICSD and represent completely new structure types that could not have been predicted using methods based on database searching. It is also worth noting that many borohydrides exhibit structural polymorphism and high-temperature entropically stabilized phases. PEGS results should provide highly accurate reaction

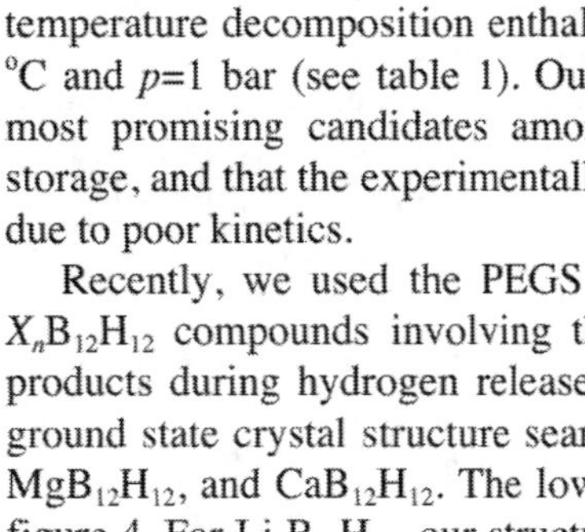

Figure 3. Tetragonal $I\bar{4}m2$ structure of $Mg(BH_4)_2$ predicted using the PEGS search.

enthalpies even in these cases, since the predicted energetics are expected to be within a few kJ/mol to the energies of the structural polymorphs. For instance, using the predicted PEGS $I\bar{4}m2$ ground state structure instead of the solution-synthesized LT phase of $Mg(BH_4)_2$ results in a negligible effect on the reaction enthalpy (less than 2 kJ/mol H_2) [24]. The PEGS method has also been successfully used to predict the ground state structures of $LiSc(BH_4)_4$, $Sc(BH_4)_3$, and $Sc_2(B_{12}H_{12})_3$ [57] and high-temperature polymorphs of $Ca(BH_4)_2$ [58,59].

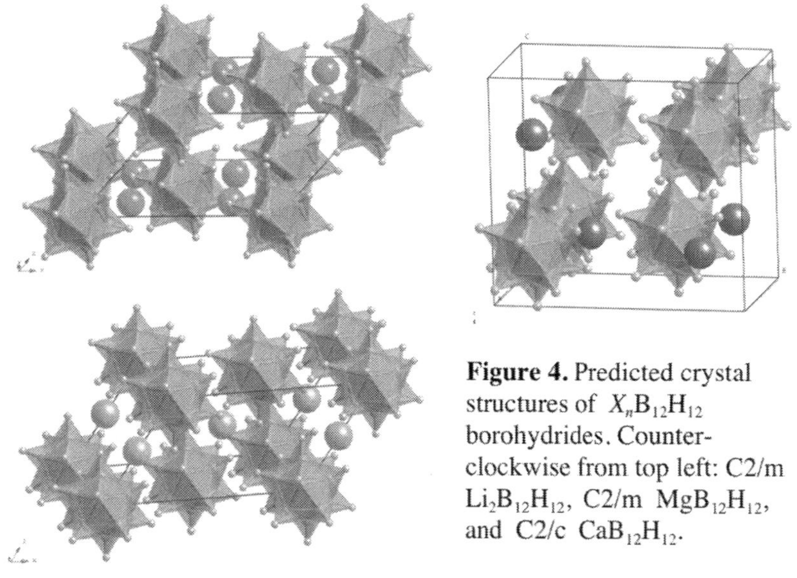

Figure 4. Predicted crystal structures of $X_nB_{12}H_{12}$ borohydrides. Counter-clockwise from top left: C2/m $Li_2B_{12}H_{12}$, C2/m $MgB_{12}H_{12}$, and C2/c $CaB_{12}H_{12}$.

4. Predicting thermodynamics and favoured hydrogen storage reactions

4.1. Basic thermodynamic considerations

Hydrogen uptake/release from materials is governed by a combination of thermodynamic and kinetic factors. The temperature at which the desorption of H_2 becomes thermodynamically possible is determined by the famous van't Hoff equation,

$$\ln(p) = -\frac{\Delta H}{RT} + \frac{\Delta S}{R}, \tag{7}$$

where ΔH and ΔS are the enthalpy and entropy of the dehydriding reaction, respectively. ΔH is related to the bond strength between H and the storage material, while ΔS is dominated by the gas-phase entropy of H_2; $\Delta S^0 = 130.68$ J/(K mol H_2) at T=20 °C and p=1 bar [20]. The practical constraints of on-board storage place limitations on the accessible temperatures and pressures, and hence may be translated into constraints on the desired values of ΔH. Assuming that ΔS is dominated by the entropy of H_2 gas and requiring hydrogen release at ambient pressures below the operating temperature of proton exchange membrane (PEM) fuel cells (80 °C) leads to the upper enthalpy limit of ~45 kJ/(mol H_2), while requiring rehydrogenation at ambient temperatures and pressures of a few hundred bar or lower gives the lower enthalpy limit of ~20 kJ/(mol H_2).

Hence, the search for a high-density H_2-storage material with ΔH between 20 and 45 kJ/(mol H_2) has attracted worldwide scientific interest. Most of the recently discovered hydrogen storage reactions with suitable thermodynamics are based on the enthalpy destabilization ideas of Reilly and Wiswall [60,61]. A canonical example of a destabilized reaction is the lithium borohydride – magnesium hydride reaction demonstrated by Vajo *et al.* [11]: $2LiBH_4 + MgH_2 \leftrightarrow MgB_2 + 2LiH + 4H_2$. Due to the formation of a strongly bound MgB_2 compound, the enthalpy of the destabilized reaction, ΔH=41 kJ/(mol H_2), is much lower than the enthalpy for the "pure" $LiBH_4 \rightarrow LiH + B + 3/2H_2$ reaction, ΔH=60 kJ/(mol H_2).

Table 1. Predicted decomposition reactions involving $Mg(BH_4)_2$. Calculated T=300 K enthalpies (ΔH) and entropies (ΔS) are given in kJ/(mol H_2) and J/(K mol H_2), respectively. T_c gives the calculated temperatures of hydrogen release at p=1 bar of hydrogen pressure.

Reaction	Wt. % (kg H_2 /kg)	Vol. density (g H_2/l)	ΔH	ΔS	T_c (°C)
$6Mg(BH_4)_2 \rightarrow MgB_{12}H_{12}+5MgH_2+13H_2$	8.1	63	29.3	99.8	20
$5Mg(BH_4)_2+2LiBH_4 \rightarrow Li_2B_{12}H_{12}+5MgH_2+13H_2$	8.4	46	24.4	100.0	-29
$5Mg(BH_4)_2+CaBH_4 \rightarrow CaB_{12}H_{12}+5MgH_2+13H_2$	7.7	51	25.7	100.5	-18

4.2. Grand-canonical linear programming

The grand canonical linear programming (GCLP) method provides a robust framework for automatically predicting new storage reactions in multicomponent systems, where the preferred reaction pathways are often complex and difficult to intuit [62,63]. GCLP was introduced in Ref. [64], where it was used to determine thermodynamically reversible hydrogen storage reactions in the Li-Mg-N-H system. Consider a multi-phase solid containing a certain fixed ratio of non-hydrogen species, which can exchange hydrogen with a reservoir of H_2 gas at a chemical potential $\mu_{H_2}(p,T)$; the latter is a function of temperature T and pressure p. The grand-canonical Gibbs free energy is given by the following expression:

$$\Xi(p,T)=\sum_i x_i F_i(T)-\frac{\mu_{H_2}(p,T)}{2}\sum_i x_i n_i^{H}, \tag{8}$$

where $F_i(T)$ is the free energy of phase "i" (we neglect the pressure dependence of the free energies of solid phases), n_i^H is the number of hydrogen ions in one formula unit of phase "i", and x_i are the unknown variable molar fractions of phases coexisting at a given composition, temperature, and pressure. The molar fractions are determined by minimizing Eq. (8), subject to the mass-conservation constraints for non-hydrogen species:

$$f_s=\sum_i x_i n_i^s=\text{const}, \tag{9}$$

where n_i^s is the number of ions of type "s" in one formula unit of phase "i", and f_s represent given molar ratios of the non-hydrogen species (e.g., Li, Mg, and N). Equations (8) and (9) constitute a linear programming problem, where the unknown variables are molar fractions of the possible phases, x_i. It is solved for a decreasing sequence of hydrogen chemical potentials $\mu_{H_2}(p,T)$, starting from T=0 K. Reactions are identified by comparing the computed molar fractions at two successive values of the chemical potential; the reactants and reaction products are found as the difference between the two sets of x_i.

4.3. Entropy-optimized reactions in the Li-Mg-Ca-B-H system

Armed with the first-principles thermodynamics of the predicted PEGS crystal structures of $Mg(BH_4)_2$ and $X_nB_{12}H_{12}$ phases, we have used the GCLP approach to identify all thermodynamically favored reactions in the quinary Li-Mg-Ca-B-H system [65]. We have discovered two new, previously unsuspected reactions belonging to a qualitatively new class of entropically optimized reactions that use fully hydrogenated high-entropy borohydrides as reactants and lead to the formation of low-entropy $X_nB_{12}H_{12}$ phases as end products. These reactions are given in table 1 and the corresponding van't Hoff curves are shown in figure 5, which demonstrates that they both pass through the target window for on-board storage. Two of these reactions exhibit nearly ideal thermodynamics. The first

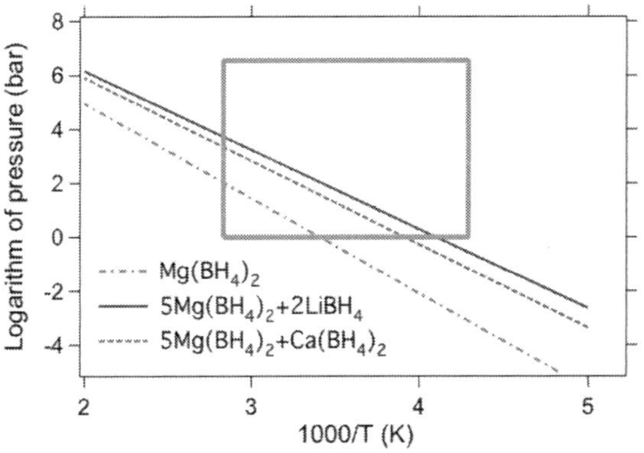

Figure 5. Calculated van't Hoff plot for reactions listed in table 1. The region within the rectangular box corresponds to the desirable temperature-pressure window for on-board hydrogen storage: from −40 to +80 °C and from 1 to 700 bar, respectively.

one uses a 5:2 molar mixture of Mg and Li borohydrides, leading to the formation of $Li_2B_{12}H_{12}$ and MgH_2, accompanied by the release of 8.4 wt. % H_2 at T=−29 °C and p=1 bar. The second reaction involves a 5:1 molar mixture of Mg and Ca borohydrides, leading to the formation $CaB_{12}H_{12}$ and MgH_2, releasing 7.7 wt. % H_2 at T=−18 °C. All the reactions in table 1 have low enthalpies [~25 kJ/(mol H_2)], which should significantly simplify heat management during on-board recharging.

Even though the enthalpy values in table 1 are at the lower limit of the commonly accepted 20-45 kJ/(mol H_2) range, they are compatible with excellent thermodynamics due to the low vibrational entropies of the $Li_2B_{12}H_{12}$ and $CaB_{12}H_{12}$ product phases.

Indeed, the calculated reaction entropies are near 100 J/(K mol H_2), approximately 25% below the standard state entropy of gaseous hydrogen. Therefore, hydrogen release occurs at temperatures that are substantially higher than those expected by considering reaction enthalpies only. Low reaction *entropies* can have a highly beneficial effect since they allow the use of materials with low reaction *enthalpies* while maintaining the same equilibrium H_2 pressure. Other things being equal, lower enthalpy reactions are always better due to smaller amount of heat released during exothermic regeneration of these hydrides. Formation of the new $X_nB_{12}H_{12}$ compounds is found to be particularly interesting, since the entropy of the tightly bound $[B_{12}H_{12}]^{2-}$ anion is significantly lower than the entropy of the $[BH_4]^-$ anion due to the smaller number of low-energy rotational and translational modes on a per-atom basis. A rule of thumb for designing new complex hydride reactions with *low entropies* suggests using reactants that have complex anions constituted by a small number of atoms (such as NH_2, BH_4, etc., resulting in a high number of low frequency phonon modes per atom) and products that are either tightly bound bulk phases or materials with large complex anions (e.g., $B_{12}H_{12}$), exhibiting few low-frequency phonon branches [65].

These results demonstrate the power of the combined theoretical framework for structure prediction and GCLP-based reaction prediction, showing that previously unsuspected reactions can be found even in extensively studied materials systems. Practical usefulness of these and other [66-68] theoretically predicted reactions will be determined not only by their thermodynamics, but also by the rate of hydrogen release, i.e., by their *kinetics*. Indeed, the dehydrogenation kinetics of $Mg(BH_4)_2$ have been measured: hydrogen desorption does not occur until very high temperatures, and the material has not been successfully rehydrided [54]. One needs to differentiate this type of thermodynamically preferred, but kinetically limited reaction from the type that is thermodynamically limited. The former is an excellent candidate for enhanced reactivity via catalysis, whereas the latter is a dead-end and should be abandoned in the search for reversible solid-state storage systems. It is interesting to note here that even though the reversibility of $Mg(BH_4)_2$ has not been observed, $Ca(BH_4)_2$ is reversible and an understanding of its reaction pathways and kinetics may lead to progress with other borohydrides.

5. Kinetics

Preceding discussion shows that recent experimental and theoretical studies have identified promising complex hydride systems with thermodynamic properties and hydrogen storage capacities approaching practical levels. However, most of the predicted reactions suffer from unacceptably poor kinetics, so that achieving fast reversible (de)hydrogenation should now be considered the central challenge in the field. Addressing this challenge requires a sustained focus on gaining a fundamental

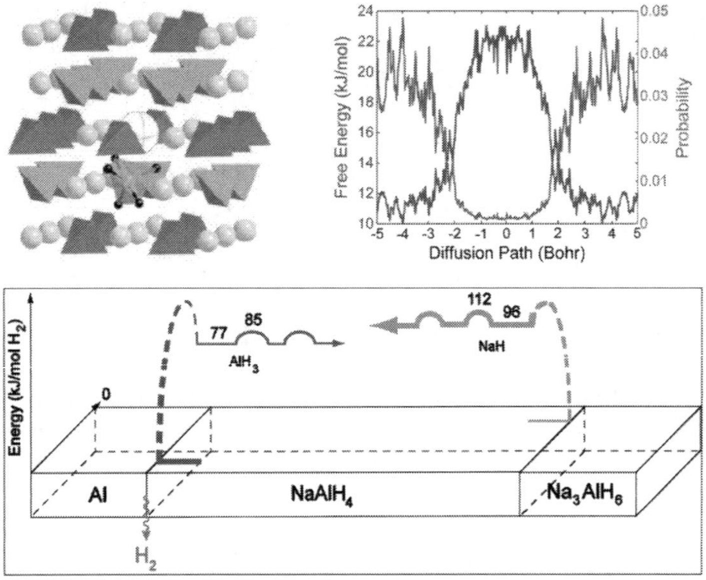

Figure 6. *Top left:* Relaxed structure of a charge-neutral AlH_3 vacancy in $NaAlH_4$. AlH_4 complexes are shown in blue, Na+ ions in yellow. Empty sphere denotes the vacant site, and the structure of the AlH_5 complex is shown using a semi-transparent polyhedron. *Top right:* Calculated free energy profile for the bulk diffusion of AlH_3 vacancies in $NaAlH_4$. *Lower left:* Proposed energy profile for the dehydrogenation pathway of $NaAlH_4$ via AlH_3 vacancies (left curve). The competing pathway via NaH vacancies (right curve) requires higher activation energy.

understanding of the detailed atomic-scale kinetic mechanisms involved in the hydrogen release and uptake reactions. First-principles DFT calculations have only recently started to address these important issues, and a complete picture of the hydrogen release process does not exist even for the most extensively studied materials, such as MgH_2 or $NaAlH_4$. Researchers have studied select aspects of the complete chain of the microscopic kinetic events that are involved in the hydrogen release process, such as hydrogen dissociation and diffusion [69-71], and direct hydrogen release [72]. Below we give an example relating to mass transport in hydrogen storage materials, which is emerging as the main candidate for the role of the rate-limiting step in the complex hydride reaction kinetics [73,74].

In the prototypical reversible complex hydride, Ti-doped sodium alanate $NaAlH_4$, hydrogen release according to reaction (1) is accompanied by long-range diffusion of metal species (Na or Al). Recently, hydrogen-deuterium scrambling experiments suggested that mass transport of the metal species might be the rate limiting step or the dehydrogenation of $NaAlH_4$ [75]. In Ref. [74], we carried out a first-principles molecular dynamics study of the kinetics of bulk diffusion in $NaAlH_4$ using the umbrella sampling weighted histogram analysis method [76] to obtain the activation free energies for the bulk diffusion of neutral vacancies in pure $NaAlH_4$. This activation energy provides the lower bound on the total activation energy of all bulk diffusion assisted decomposition mechanisms, unaffected by Ti doping in view of the fact that significant bulk substitution of Ti in $NaAlH_4$ is highly unfavorable [77] and has not been observed [47,78].

We begin by considering the structure of the simplest low-energy charge-neutral defects in $NaAlH_4$. Figure 6 shows the calculated structure of neutral AlH_3 and NaH vacancy complexes. It is seen that the introduction of a metal ion vacancy leads to a rearrangement of the coordination of the neighboring AlH_4^- complex anions. Introducing a neutral AlH_3 vacancy results in the formation of an AlH_5^{2-} moiety in $NaAlH_4$, bound to the vacant site by Coulomb attraction due to local excess of Na^+ ions. Existence of a strong Columbic attraction between the AlH_5^{2-} complex and the vacancy implies that these defects must diffuse together, which requires a mechanism of changing the orientation of the vacancy-AlH_5^{2-} pair. According to our AIMD simulations, the extra hydrogen of the AlH_5^{2-} moiety is very mobile and gets exchanged with the AlH_4^- units surrounding the AlH_3 vacancy about once in every 1 picosecond of simulation at 400 K. This observation implies that the hydrogen ion easily diffuses around the vacancy site, facilitating the re-orientation mechanism required for long-range diffusion of this complex. For the case of a NaH vacancy, two neighboring AlH_4^- tetrahedra start to share a vertex, forming an $Al_2H_7^-$ complex to compensate for the missing H ion (see figure 1 in Ref. [74]). Again, there is a Coulombic attraction between the $Al_2H_7^-$ complex and the Na^+ vacancy, requiring that a dynamic process allowing for the shared H ion to travel with the vacancy must accompany the vacancy-mediated diffusion of the Na+ species. The umbrella sampling technique [76] was used to obtain the free energy profiles for the diffusion of both types of neutral defects. The free energy profile for the diffusion of an AlH_3 vacancy is shown in figure 6. We find that the free energy barrier for AlH_3 vacancy diffusion (12 ± 2 kJ/mol) is in excellent agreement with the anelastic energy loss results of Ref. [79]. The total activation energy for Al mass transport via neutral AlH_3 vacancies, which is a sum of vacancy formation and migration energies, is found to be Q=85 kJ/mol H_2 [74], in excellent agreement with the experimentally measured activation energies in Ti-catalyzed $NaAlH_4$ of around 80 kJ/(mol H_2) [80]. The activation energy for the bulk diffusion via NaH vacancies is found to be significantly higher, Q=112 kJ/(mol H_2), and therefore not favored. Our results suggest that bulk diffusion of Al species is the rate-limiting step in the dehydrogenation of Ti-doped samples of $NaAlH_4$. The calculated diffusion coefficient and vacancy concentration at typical decomposition temperatures ($D=4\times10^{-4}$ cm^2/s and C=0.01% at T=400 K) are consistent with the observed dehydrogenation times of minutes to hours in ball-milled Ti-doped powders consisting of micron-sized grains. This study shows that Ti doping is not needed to obtain bulk mass transport at rates that are consistent with the observed dehydrogenation kinetics, and therefore the catalytic effect of Ti is not related to increased concentration of bulk defects and accelerated bulk diffusion [70].

6. Summary

We have demonstrated how a comprehensive first-principles framework involving DFT calculations of total energies and vibrational thermodynamics, efficient algorithms for crystal structure prediction, and thermodynamics-based determination of preferred reaction pathways can be used to discover new materials with superior hydrogen storage thermodynamics. In contrast to earlier studies, which only use information on known compounds [66-68], our methods allow to include new phases with crystal structures that are not yet experimentally measured. This dramatically expands the range of these studies by extending the realm of predictions from reactions with known compounds to reactions with yet-to-be-synthesized ones. Our framework, possibly complemented with other methods for structure prediction (data mining [30], genetic algorithms [31], etc.) should prove very useful for predicting new materials for other methods of energy storage (e.g., electrochemical) that are based on phase transformations between different solid state phases. Kinetics is emerging as the key area where further advances are needed both in methodology and applications. In spite of considerable progress, we do not yet have the ability to determine the microscopic mechanisms and rate-limiting steps in an unbiased and reliable manner without significant input from both the experiment and physical intuition.

Acknowledgments

Authors gratefully acknowledge financial support from the US Department of Energy under grants DE-FG02-05ER46253, DE-FG02-07ER46433, and DE-AC04-94AL85000, as well as from the National Science Foundation under grants CBET-0730044 and CBET-0730929.

References

[1] Satyapal S, Petrovic J, Read C, Thomas G and Ordaz G 2007 *Catal. Today* **120** 246
[2] Yaghi O M, O'Keeffe M, Ockwig N W, Chae H K, Eddaoudi M and Kim J 2003 *Nature* **423** 705
[3] Chae H K, Siberio-Perez D Y, Kim J, Go Y, Eddaoudi M, Matzger A J, O'Keeffe M and Yaghi O M 2004 *Nature* **427** 523
[4] Xiong Z T, Yong C K, Wu G. T, Chen P, Shaw W, Karkamkar A, Autrey T, Jones M O, Johnson S R, Edwards P P and David W I F 2008 *Nat. Mater.* **7** 138
[5] Sandrock G 1999 *J. Alloy Compound* **293** 877
[6] Sandrock G and Thomas G 2001 *Appl. Phys. A* **72** 153
[7] Grochala W and Edwards P P, 2004 *Chem. Rev.* **104** 1283

[8] Bogdanovic B and Schwickardi M 1997 *J. Alloy Compound* **253** 1
[9] Chen P, Xiong Z, Luo J, Lin J and Tan K L 2002 *Nature* **420** 302
[10] Luo W 2004 *J. Alloy Compound* **381** 284
[11] Vajo J J, Skeith S L, Mertens F 2005 *J. Phys. Chem. B* **109** 3719
[12] Pinkerton F E, Meisner G P, Meyer M S, Balogh M P and Kundrat M D, 2005 *J. Phys. Chem. B* **109** 6
[13] Wolverton C, Ozolins V and Asta M 2004 *Phys. Rev. B* **69** 144109
[14] Wolverton C, Siegel D J, Akbarzadeh A R and Ozolins V 2008 *J. Phys.: Condens. Matter* **20** 064228
[15] Kresse G and Furthmüller J 1996 *Comput. Mat. Sci.* **6** 15
[16] Kresse G and Furthmüller J 1996 *Phys. Rev. B* **54** 11169
[17] Perdew J P and Wang Y 1992 *Phys. Rev. B* **45** 13244
[18] Blochl P E 1994 *Phys. Rev. B* **50** 17953
[19] Kresse G and Joubert D 1999 *Phys. Rev. B* **59** 1758
[20] Chase, M W, Davies C A, Downey J R, Frurip D J, McDonald R A and Syverud A N 1985 *J. Phys. Chem. Ref. Data* **14**, Suppl. No. 1
[21] Zhong Y, Wolverton C, Chang Y A and Liu Z-K 2004 *Acta Mater.* **52** 2739
[22] Mujica A, Rubio A, Munoz A and Needs R J 2003 *Rev. Mod. Phys.* **75** 863
[23] Hellenbrandt M 2004 *Cryst. Rev.* **10** 17
[24] Ozolins V, Majzoub E H and Wolverton C 2008 *Phys. Rev. Lett.* **100** 135501
[25] Vajeeston P, Ravindran P, Kjekshus A and Fjellvag H 2004 *J. Alloy Compound* **363** L7
[26] Lovvik O M and Swang O 2004 *Europhys. Lett.* **67** 607
[27] Ke X Z, Kuwabara A and Tanaka I 2005 *Phys. Rev. B* **71** 184107
[28] Wolverton C and Ozolins V 2007 *Phys. Rev. B* **75** 064101
[29] Akbarzadeh A R, Wolverton C and Ozolins V 2009 *Phys. Rev. B* **79** 184102
[30] Curtarolo S, Morgan D, Persson K, Rodgers J and Ceder G 2003 *Phys. Rev. Lett.* **91** 135503
[31] Oganov A R and Glass C W 2006 J. Chem. Phys. **124** 244704
[32] Martonak R, Laio A and Parrinello M 2003 Phys. Rev. Lett. **90** 075503
[33] Juza R and Opp K 1951 *Z. Anorg. Allg. Chem.* **266** 325
[34] Noritake T, Nozaki H, Aoki M, Towata S, Kitahara G, Nakamori Y and Orimo S 2005 *J. Alloys Compounds* **393** 264
[35] Ohoyama K, Nakamori Y, Orimo S and Yamada K 2005 *J. Phys. Soc. Japan.* **74** 483
[36] Herbst J F and Hector L G Jr 2005 *Phys. Rev. B* **72** 125120
[37] Balogh M P, Jones C Y, Herbst J F, Hector L G and Kundrat M 2006 *J. Alloy Compound* **420** 326
[38] Michel K, Akbarzadeh A R and Ozolins V 2009 to appear in *J. Phys. Chem. C*
[39] Yang J, Sudik A and Wolverton C 2007 *J. Alloy Compound* **430** 334
[40] Rijssenbeek J, Gao Y, Hanson J, Huang Q, Jones C and Toby B 2008 *J. Alloy Compound* **454** 223
[41] Asta M, Ozolins V and Woodward C 2001 *JOM* **53**(9) 16
[42] Magyari-Kope B, Ozolins V and Wolverton C 2006 *Phys. Rev. B* **73** 220101
[43] Mueller T and Ceder G 2006 *Phys. Rev. B* **74** 134104
[44] Wang Y and Chou M Y 2007 *Phys. Rev. B* **76** 014116
[45] Vajeeston P, Ravindran P., Kjekshus A and Fjellvag H 2006 *Appl. Phys. Lett.* **89** 071906
[46] Majzoub E H and Ozolins V 2008 *Phys. Rev. B* **77** 104115
[47] Ozolins V, Majzoub E H and Udovic T 2004 *J. Alloy Compound* **375** 1
[48] Majzoub E H, McCarty K F and Ozolins V 2005 *Phys. Rev. B* **71** 024118
[49] Wawak R J, Pillardy J, Liwo A, Gibson K D and Scheraga H A 1998 *J. Phys. Chem. A* **102** 2904
[50] Cerny R, Filinchuk Y, Hagemann H and Yvon K 2007 *Angew. Chem.-Int. Edit.* **46** 5765
[51] Her J-H, Stephens P W, Gao Y, Soloveichik G L, Rijssenbeek J, Andrus M and Zhao J C 2007

Acta Crystallogr. B **63** 561
[52] Li H-W, Kikuchi K, Nakamori Y, Ohba N, Miwa K, Towata S and Orimoa S 2008 *Acta Materialia* **56** 1342
[53] Hwang S-J, Bowman R C, Reiter J W, Rijssenbeek J, Soloveichik G L, Zhao JC, Kabbour H, Ahn C C 2008 *J. Phys. Chem. C* **112** 3164
[54] Soloveichik G L, Gao Y, Rijssenbeek J, Andrus M, Kniajanski S, Bowman R C, Hwang S-J and Zhao J C 2009 *Int. J. Hydrogen Energy* **34** 916
[55] Ohba N, Miwa K, Aoki M, Noritake T, Towata S, Nakamori Y, Orimo S and Züttel A 2006 *Phys. Rev. B* **74** 075110
[56] Her J-H, Yousufuddin M, Zhou W, Jalisatgi S S, Kulleck J G, Zan, J A, Hwang S-J, Bowman R C and Udovic T J 2008 *Inorg. Chem.* **47** 9757
[57] Kim C, Hwang S-J, Bowman R C, Reiter J W, Zan J A, Kulleck J G, Kabbour H, Majzoub E H and Ozolins V 2009 *J. Phys. Chem. C* **113** 9956
[58] Majzoub E H and Ronnebro E C 2009, *J. Phys. Chem. C* **113** 3352
[59] Lee Y S, Kim Y, Cho Y W, Shapiro D, Wolverton C and Ozolins V 2009, Phys. Rev. B **79** 104107
[60] Reilly J J and Wiswall R H 1967 *Inorg. Chem.* **6** 2220
[61] Reilly J J and Wiswall R H 1968 *Inorg. Chem.* **7** 2254
[62] Siegel D J, Wolverton C and Ozolins V 2007 *Phys. Rev. B* **75** 014101
[63] Siegel D J, Wolverton C and Ozolins V 2007 *Phys. Rev. B* **76** 134102
[64] Akbarzadeh A, Ozolins V and Wolverton C 2007 *Adv. Mater.* **19** 3233
[65] Ozolins V, Majzoub E H and Wolverton C 2009 *J. Am. Chem. Soc.* **131** 230
[66] Alapati S V, Johnson J K and Sholl D S 2006 *J. Phys. Chem. B* **110** 8769
[67] Alapati S V, Johnson J K and Sholl D S 2007 *J. Alloys. Comp.* **446** 23
[68] Alapati S V, Johnson J K and Sholl D S 2008 *J. Phys. Chem. C* **112** 5258
[69] Hao S and Sholl D S 2008 *Appl. Phys. Lett.* **93** 251901
[70] Peles A and Van de Walle C G 2007 *Phys. Rev. B* **76** 214101
[71] Gunaydin H, Barabash S V, Houk K N and Ozolins V 2008 *Phys. Rev. Lett.* **101** 075901
[72] Vegge T 2006 *Phys. Chem. Chem. Phys.* **8** 4853
[73] David W I F, Jones M O, Gregory D H, Jewell C M, Johnson S R, Walton A and Edwards P P 2007 *J. Am. Chem. Soc.* **129** 1594
[74] Gunaydin H, Houk K N and Ozolins V 2008 *Proc. Natl. Acad. Sci. USA* **105** 3673
[75] Bellosta von Colbe J M, Schmidt W, Felderhoff M, Bogdanovic B and Schüth F 2006 *Angew. Chem. Int. Ed.* **45** 3663
[76] Roux B 1995 *Comput. Phys. Commun.* **91** 275
[77] Lovvik O M and Opalka S M 2005 *Phys. Rev. B* **71** 054103
[78] Brinks H W, Sulic M, Jensen C M and Hauback B C 2006 *J. Phys. Chem. B* **110** 2740
[79] Palumbo O, Paolone A, Cantelli R, Jensen C M and Ayabe R 2006 *Mater. Sci. Eng. A* **442** 75
[80] Sandrock G, Gross K and Thomas G 2002 *J. Alloys Compd.* **339** 299

DCA++: a case for science driven application development for leadership computing platforms

Michael S Summers, Gonzalo Alvarez, Jeremy Meredith and Thomas A Maier

Computer Science and Mathematics Division, Oak Ridge National Laboratory, P. O. Box 2008, Mail Stop 6164, Oak Ridge, TN 37831

Thomas C. Schulthess

Swiss National Supercomputer Center and Institute for Theoretical Physics, ETH Zurich, CSCS MAN E 133, Galeria 2, CH-9628 Manno

schulthess@cscs.ch

Abstract. The DCA++ code was one of the early science applications that ran on jaguar at the National Center for Computational Sciences, and the first application code to sustain a petaflop/s under production conditions on a general-purpose supercomputer. The code implements a quantum cluster method with a Quantum Monte Carlo kernel to solve the 2D Hubbard model for high-temperature superconductivity. It is implemented in C++, making heavy use of the generic programming model. In this paper, we discuss how this code was developed, reaching scalability and high efficiency on the world's fastest supercomputer in only a few years. We show how the use of generic concepts combined with systematic refactoring of codes is a better strategy for computational sciences than a comprehensive upfront design.

1. Introduction

The development of leadership computing platforms in recent years has been very remarkable. The introduction in 2002 of the Earth Simulator in Japan, resulted in a revitalization of scientific supercomputing in the United States, and the performance of the largest supercomputers available to open science increased by almost three orders of magnitude, from a few teraflop/s in 2002 to a petaflop/s in 2008. But the architectural changes of the computers were, from an application perspective, at least as challenging as the performance improvements. In November 2002, the largest system that was available to open science besides the Earth Simulator sustained the Linpack benchmark with 4.46 teraflop/s and consisted of 3,016 processor cores. Six years later, in November 2008, the largest system available to open science sustained Linpack with 1.06 petaflop/s and consisted of 150,152 processor cores. Put in simple terms, scientific simulations would only benefit from the factor 240 increase in performance if they could scale to 50 times the number of processor cores.

Weak scaling, where the required number of processors increases in proportion to the problem size, is an inherent property of many scientific simulation problems. Codes currently exhibiting this property are the natural choice to run on the next generation of supercomputers. However, limiting application selection to such codes would eventually lead to extinction of supercomputing in science since there would be fewer and fewer scalable applications to run on each generation of supercomputer. Early science applications for *Jaguar*, the first petaflop/s scale supercomputer at the National Center for Computational Sciences (NCCS), were thus selected according to their potential scientific impact and necessary investments were made to bring these applications to readiness.

The DCA++ code was one of these early science applications. It was selected because the study of models of high temperature superconductivity seemed to be a very promising endeavor for simulations, and advances in understanding the mechanism of high temperature superconductivity could have a big impact on the development of new materials for electric power transmission. But at the time of its selection, in the fall of 2006, it was not at all clear how these simulations would benefit from more than a factor 10 increase in number of cores on a NCCS leadership class system. In the present paper, we discuss how the DCA++ code was developed, and how it became the first scientific application to sustain a petaflop/s under production conditions.

2. Modelling superconductivity and the DCA/QMC method

Soon after high-temperature superconductors were discovered in 1986 [1], it became clear that the physics of superconductivity in these materials should be describable with a simple model, the single band 2 dimensional (2D) Hubbard model [2]. In this model, carriers (electrons or holes) are fermions with spin 1/2. They are able to move between sites on a square lattice with a hopping amplitude t between near neighbor sites, and they repel each other with a Coulomb interaction U, when two carriers with opposite spin occupy the same site. Despite its simplicity and the massive amount of attention paid to it by the condensed matter physics community, the problem of describing superconductivity with this model has eluded a solution for nearly two decades. The problem does not have an analytic solution and any numerical solution suffers from the curse of dimensionality inherent to quantum many-body problems. The complexity of the 2D Hubbard model scales with 4^N, where N is the number of sites. Since superconductivity is a macroscopic quantum effect, nominally N would be of order 10^{23}. At the same time, numerous experiments indicated that quantum correlations at atomic and nanometer lengths scales play an important role as well. Because of this, an approximate numerical solution of the Hubbard model capable of describing superconductivity, has to address a multi-scale problem that stretches from atomic to macroscopic length scales.

The development of quantum cluster methods [3] in the late 1990s, laid the methodological foundation for solving the 2D Hubbard model. The basic idea is to solve the quantum many-body problem accurately on a cluster that is large enough to account for the nano-meter scale quantum correlations, and to coherently embed the cluster into a mean-field that accounts for the macroscopic length scales. The resulting approximate method has the name "Dynamical Cluster Approximation" (DCA). Numerically, the cluster many-body problem is solved with a variant of the Hirsch Fye Quantum Monte Carlo method (QMC), that has been proposed by Ed D'Azevedo [4] in order to perform well on superscalar processors used in the Cray XT line of supercomputers.

In 2005 [2] it was possible to systematically show for the first time that the 2D Hubbard model describes a superconducting transition with a critical temperature that is comparable to the typical values observed in the cuprates. A series of subsequent simulations revealed details of the pairing mechanism in the model [6].

Meanwhile, a series of remarkable experimental studies were published, in which a relationship between nano-scale inhomogeneities and pair formation was found [7,8]. In particular, the paper from the Princeton group [8] showed that regions exist in the superconductor, in which the superconducting pair-formation persists at temperatures that are significantly higher than the macroscopically observed transition temperature. This indicates that understanding the role of inhomogeneities and disorder effects in the pair formation will be crucial in understanding the cuprate materials. However, these studies would require a new code base, in which variants of the Hubbard model with disorder and inhomogeneities could be quickly introduced, tested, and studied with simulations at scale. Since supercomputers with the necessary performance were on the horizon, it was now time to prepare the DCA/QMC code base for these new science problems that would be studied on petaflop/s scale systems – a decision was made to develop the DCA++ code.

3. DCA++ Code requirements

The DCA++ code is a multi-scale quantum field theory application. At the macroscopic scale the "bath" (see Figure 1) is represented with a self-energy function. At the atomic scale it uses a very large set of disorder configurations to model the many impurities that would be found in the cuprate sample. Each disorder configuration (the cluster of Figure 1) is represented by a particular set of (random) parameters of the single-band 2D Hubbard model. The self-energy of the bath is used to initialize the cluster and the many-body Green's functions (a complex valued function similar to a many-body correlation function) computed from the cluster can be used to compute a new self-energy. The code iterates until a self-consistent self-energy and many-body Green's function are found. The computation of each cluster's many-body Green's function involves a Monte Carlo integration (represented by dice in Figure 2).

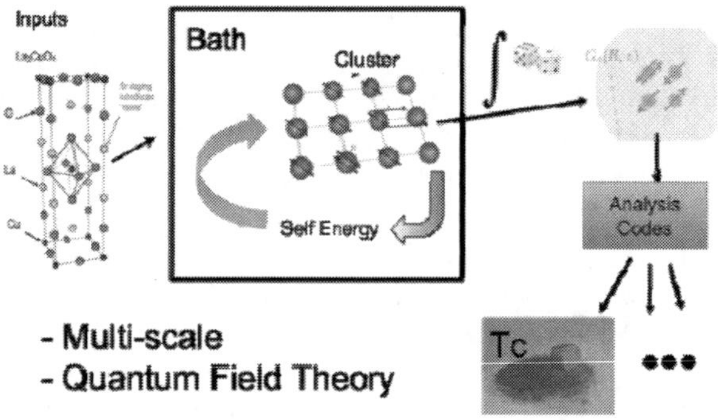

Figure 1. The DCA++ code computes the many-body Green's function from a model description of the material crystal. Materials properties such at the superconducting transition temperature can be calculated from these many-body Green's functions.

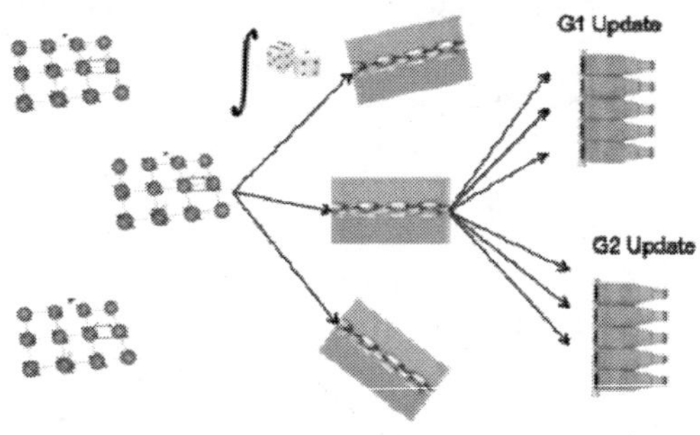

Figure 2. DCA++ has a natural parallelism structure that can make use of a very large number of processors.

The structure of this problem has at least three levels of hierarchical parallelism (see Figure 2). The available processors are first divided up into teams, one for each disorder configuration. Each disorder team can then be divided into Monte Carlo integration sub-teams. Each integration sub-team uses a separate Markov chain to generate a sequence of integrand values. The processing of each Markov chain involves an expensive update procedure (the modified Hirsch-Fye algorithm with delayed updates [4]) that can only be handled by a sub-sub-team of processors. The update procedure solves a linear algebra problem that is currently the computational bottleneck of the simulation – the code spends about 95% of the time in this procedure. The sub-sub-teams can be further divided to parallelize the updates or the calculation of observables.

4. Generic software development in stages: systematic refactoring

The key aspects of generic programming and its relationship to systematic refactoring are described by Douglas Gregor, et. al. in an OOPSLA paper entitled "Concepts: Linguistic Support for Generic Programming in C++" [10]:

- "Generic programming is a systematic approach to designing and organizing software that focuses on finding the most general (or abstract) formulation of an algorithm together with an efficient implementation. [9]
- The generic programming process derives generic algorithms from families of concrete (non-generic) implementations that exhibit some commonality. We lift away unnecessary requirements on types from an implementation, thereby raising the level of abstraction". [10]
- "Once many algorithms within a given problem domain have been lifted, we start to see patterns among the requirements. ... When this occurs, each set of requirements is bundled

> into a concept. … When the Generic Programming process is carefully followed, the concepts that emerge tend to describe the abstractions within the problem domain in some logical way. … Thus, the output of the Generic Programming process is not just a generic, reusable implementation, but a better understanding of the problem domain." [11]

For computational scientists this approach simultaneously provides multiple benefits. It permits them to organize their software into concepts that are natural to the physical problems being studied and concepts that are natural to the computational techniques they use to solve these problems. This makes the code easier to verify and extend. The lifting process provides a natural methodology for re-organizing their code so that it continually reflects the results of their research. In our development of DCA++ we have demonstrated that it is possible to do this without sacrificing efficiency and scalability.

Investigations of complex systems by means of simulations are reminiscent of Cold War era research into complex hybrid AI/numerical applications. These projects involved a mix of computer science, applied mathematics, physics, and software engineering. Software development had to be nimble, because no direct path to a solution was obvious in the beginning and thus software had to be developed and re-developed many times. In order to manage complexity and cost, these researchers needed the ability to define software abstractions, and they developed languages and tools to this end. However, it was not the languages and tools that ultimately reduced the complexity of these codes. It was the identification and use of correct/efficient concepts and their refinements that managed complexity both in the minds of researchers and in the computer code. The structure of the code had to reflect the most efficient way to think about the problem and the best way to think of the problem had to be developed while the code was being developed.

This is the view we took in the DCA++ code, namely, that it is not possible to avoid future rewriting of the code by comprehensive upfront design. Rather, the understanding of what needs to be coded, what the best abstractions are, and how these can be implemented in the most efficient way, is wisdom that has to be allowed to develop and improve. Hence the DCA++ project is designed to allow frequent refactoring of the software, and there is no expectation that there will ever be a final design of the software until the project is completed.

The basic idea of refactoring is presented in Figure 3. As every code is developed, it increases in both functionality and complexity. As the complexity of the code grows, it becomes harder to change until it reaches a ceiling where adding new capabilities is too costly. While it may appear that the code is at an impasse at this point, it usually is not. Much of the complexity of the code is not intrinsic to the problem the code is solving. During the refactoring process, code is rewritten (often just a portion of it) so that (1) the non-intrinsic complexity is greatly reduced and (2) the new abstractions used in the code change the rate at which complexity grows with additional functionality. This is illustrated in Figure 3 by the reduced slope after refactoring.

Trying to avoid refactoring by comprehensive upfront design is usually not a good strategy in computational science research codes. There are several reasons for this:

- The design of the program abstractions involves many trade-offs. Experience with the actual system is needed before good design decisions can be made.
- While the code is being developed, researchers learn about the problem they are trying to solve/explore leading to changes in requirements.
- Algorithm development is usually part of the research project, and the best algorithms for a problem on a particular architecture are only discovered during code development.
- It is efficient to identify refactoring opportunities by looking for redundant code.
- Upfront designs of a system of abstractions/concepts are usually too general, and developers end up writing more code than necessary, leaving much code untested.

5. The first sustained petaflop/s: a practical example of refactoring process

In November 2008, DCA++ was the first application code to sustain a petaflop/s under production conditions on a general-purpose supercomputer. The two-year development time that lead up to this point provides an interesting illustration of the refactoring process.

The DCA++ code was constructed in two stages. The first stage, Version 1 was developed to demonstrate that the team's approach could be implemented efficiently, correctly, and in a scalable manner to run on 50 thousand or more processors. While Version 1 was written in C++ and made some use of object orientation and generic programming, it was not written in a way that would minimize the cost of adding future extensions. The second stage, Version 2, represented a complete rewrite to meet all of Version 1's requirements, and to significantly reduce the cost of developing the many envisioned extensions.

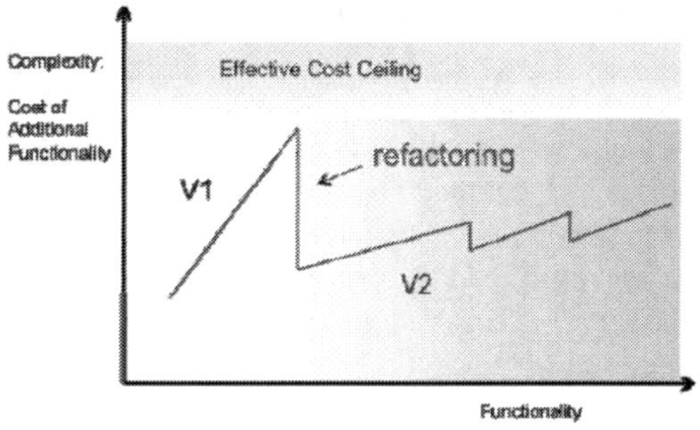

Figure 3. Refactoring enables us to afford more functionality.

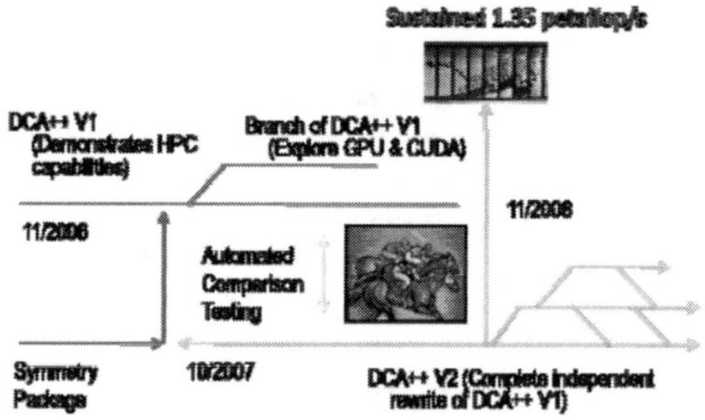

Figure 4. DCA++ development time line.

Figure 4 presents the two-year development timeframe prior to the first petaflop/s scale runs. Version 1 was developed primarily by Gonzalo Alvarez, a computational physicist with extensive background in numerical approaches similar to the ones used in the DCA/QMC method. Initially, the work on Version 2 focused on developing a symmetry package that would be available for use in Version 1 as well. The actual development of the DCA++ application Version 2 started in Fall 2007. Its primary developer was Mike Summers, a mathematician by training with vast software development experience, who was required to become familiar with the basic aspects of quantum field theory used in the DCA/QMC method. About at the same time, work was started by Jeremy Meredith, a computer scientists with experience in computer graphics and visualization, to experiment with Version 1 on graphic processing units (GPU).

The work with DCA++ on GPU showed that promising performance gains could be made on graphic accelerators, and most importantly, that the Hirsch-Fye QMC solver with delayed updates produces satisfactory results with only single precision arithmetic. This was in April 2008, at about the same time when Version 1 demonstrated scalability to the full Cray XT4 system with 8,000 quad-core processors (32 thousand cores in total). Since using a single precision QMC solver would accelerate the DCA++ code on regular CPUs as well, it was decided to introduce some of the changes of the GPU experiment back into Version 1. Making a reliable mixed precision version of the code, where the self-energy evaluation would remain in double precision and the cluster solver would be allowed to run in single precision, was difficult and the production runs with Version 1 were continued with double precision. However, the design requirement to support mixed precision capability was added to Version 2.

In the meantime a very detailed automated testing system was developed to verify that Version 2 produced exactly the same results as Version 1. This lead to a rather swift debugging and validation phase of the code and by fall 2008, Version 2 was ready for production when the Cray XT5 petaflop/s scale supercomputer was ready for initial tests with applications under production conditions. The refactoring had already paid off in two ways:

- Running on 150 thousand processors versus 50 thousand earlier in the summer revealed a weakness in the MPI collective communication approach that was taken in both versions of the DCA++ code. However, Version 2 required changes to only a few lines of code in one inherited method and near perfect scaling was reached up to the full size of the supercomputer. The required changes in Version 1 were more involved.
- The change to a mixed precision version of the DCA/QMC method was fairly trivial in Version 2 – an afternoon's worth of coding and debugging work – since the possibility to alter precision in different parts of the code had been added to the specification. This change accelerated the time to solution by a factor 1.9, resulting in 1.35 petaflop/s sustained performance of the code under production conditions.

Figure 5 summarizes the basic structure and size of the resulting DCA++ Version 2 code. For basic I/O, the code uses a Javascript Object Notation (JSON) parser. Ideally, a code like DCA++ would be based on XML, for its high-level I/O. However, maintaining the necessary parser system for XML on bleeding edge supercomputers was not practical, and a much lighter JSON parser that we wrote in C++ was used instead. Many of the elementary data structures and algorithms for condensed matter physics simulations have already been abstracted and collected in the PSIMAG library. The symmetry package, which contains all the necessary low-level machinery to apply the DCA method to arbitrary space groups in 2D and 3D, has been added to the PSIMAG library as well, as it will be used by many other condensed matter and materials science codes (this package alone consists of about 200,000 lines of code). Like most other computational science codes, DCA++ relies on libraries such as BLAS and LAPACK for dense linear algebra, MPI for message passing, and CUDA or pthreads for processor dependent fine-grained parallelism. With this structure, the actual DCA++ code ends up being relatively small – only about 25 thousand lines of code of which almost 90% are contained in generic classes.

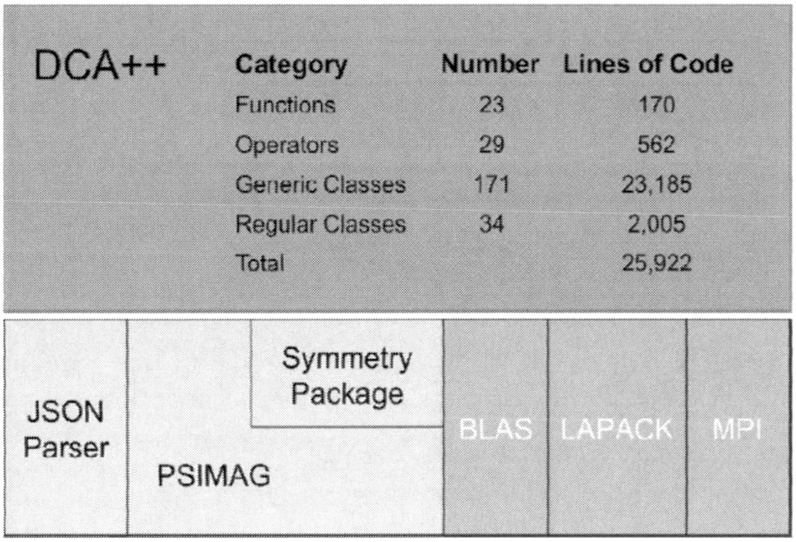

Figure 5. DCA++ code structure: 97% is Object Oriented, 89% is generic

6. Summary

The main characteristics of computational science application development is that several issues have to be addressed concomitantly: model-development, the methods to solve these models, and the algorithms to optimally map resulting computing codes onto high-performance computing hardware.

Implementing software for such applications has to follow a non-conventional path, with frequent and systematic refactoring steps built into an iterative design process. Relying on rigorous up-front design of abstractions turns out not to be practical. The DCA++ code development successfully followed such an approach, and after a two-year development process resulted in a production code that implements bleeding edge algorithms to study superconductivity in models of the cuprate high-temperature superconductors at scale. The team that developed DCA++ has a mixed background from many disciplines, bringing together experience in condensed matter and computational physics, applied mathematics, and computer science. It was the first code to sustain a petaflop/s under production conditions on a general-purpose supercomputer.

Acknowledgement
This work was supported by the Laboratory Directed Research and Development fund of Oak Ridge National Laboratory and the Center for Nanophase Material Sciences, which is sponsored by the Division for Scientific User Facilities Division of the Office of Basic Energy Sciences, as well as a grant from the National Center for Computational Sciences, which is funded by the Facilities Division of the Office of Advanced Scientific Computing Research (ASCR).

References
[1] J. G. Bednorz and K. A. Müller, "Possible high-Tc superconductivity in the Ba-La-Cu-O system", Zeitschrift f. Physik B-Cond. Matt. **64**, 189-193 (1986)
[2] F. C. Zhang and T. M. Rice, "Effective Hamiltonian for the superconducting Cu oxides", Phys. Rev. B **41**, 7243 (1988).
[3] T. A. Maier, M. Jarrell, T. Pruschke, and M. H. Hettler, "Quantum Cluster Theories", Rev. Mod. Phys. **77**, 1027 (2005).
[4] Gonzalo Alvarez, Michael S. Summers, Don E. Maxwell, Markus Eisenbach, Jeremy S. Meredith, Jeffrey M. Larkin, John M. Levesque, Thomas A. Maier, Paul R. C. Kent, Eduardo F. D'Azevedo, Thomas C. Schulthess, Proceedings of the ACM/IEEE Conference on High Performance Computing, SC 2008, November 15-21, 2008, Austin, Texas, USA. IEEE/ACM 2008, ISBN 978-1-4244-2835-9
[5] T. A. Maier, M. Jarrell, T. C. Schulthess, P. R. C. Kent, and J. B. White, "Systematic study of d-wave superconductivity in the 2D repulsive Hubbard model", Phys. Rev. Lett. **95**, 237001 (2005).
[6] T. A. Maier, M. Jarrell, and D. J. Scalapino, "Strucutre of the pairing interactions in the two-dimensional Hubbard model", Phys. Rev. Lett. **96**, 047005 (2006).
[7] K. McElroy, Jinho Lee, J. A. Slezak, D.-H. Lee, S. Uchida, and J. C. Davis, "Atomic-scale sources and mechanism of the nanoscale electronic disorder in $Be_2Sr_2CaCu_2O_{8+\delta}$", Science **309**, 1048-1052 (2005).
[8] K. K. Gomes, A. N. Pasupathy, A. Pushp, S. Ono, Y. Ando, and A. Yazdani, "Visualizing pair formation on the atomic scale in the high-Tc superconductor $Bi_2Sr_2CaCu_2O_{8+\delta}$", Nature **447**, 569 (2007).
[9] Mehdi Jazayeri, Rudiger Loos, David Musser, and Alexander Stepanov. Generic Programming. In Report of the Dagstuhl Seminar on Generic Programming, Schloss Dagstuhl, Germany, April 1998.Compiler Construction, pages 86–101, London, UK, 2001. Springer-Verlag.
[10] Douglas Gregor, Jaakko Jarvi, Jeremy Siek,Bjarne Stroustrup, Gabriel Dos Reis,Andrew Lumsdaine, OOPSLA'06 October 22-26, 2006, Portland Oregon, USA.
[11] http://www.generic-programming.org/

Journal of Physics: Conference Series **180** (2009) 012078 doi:10.1088/1742-6596/180/1/012078

Fully-coupled engineering and mesoscale simulations of thermal conductivity in UO_2 fuel using an implicit multiscale approach

M R Tonks, G Hansen, D Gaston, C Permann, P Millett and D Wolf

Idaho National Laboratory, P.O. Box 1625, Idaho Falls, ID 83415.

Abstract. Though the thermal conductivity of solid UO_2 is well characterized, its value is sensitive to microstructure changes. In this study, we propose a two-way coupling of a mesoscale phase field irradiation model to an engineering scale, finite element calculation to capture the microstructure dependence of the conductivity. To achieve this, the engineering scale thermomechanics system is solved in a parallel, fully-coupled, fully-implicit manner using the preconditioned Jacobian-free Newton Krylov (JFNK) method. Within the JFNK function evaluation phase of the calculation, the microstructure-influenced thermal conductivity is calculated by the mesoscale model and passed back to the engineering scale calculation. Initial results illustrate quadratic nonlinear convergence and good parallel scalability.

1. Introduction

Thermal conductivity of UO_2 fuel in a light water reactor strongly impacts both the design of the fuel and the operation of the reactor system employing it. As the fuel undergoes irradiation, the temperature and neutron flux gradients cause fission products and porosity to form within the pellet. Microstructure evolution results in different local thermal conductivities throughout the fuel pellet. Rokkam et al.[1] have developed a mesoscale phase field model that predicts the nucleation and growth of voids due to irradiation. The microstructure predicted by the method can then be used to calculate an effective thermal conductivity [2]. However, the computational expense of modeling the entire fuel pellet with such a mesoscale model would be prohibitively large.

To investigate the impact of a microstructurally-dependent thermal conductivity on engineering scale thermomechanics, this study proposes an approach to couple an engineering scale calculation with a mesoscale phase field calculation. Fully coupled multiscale calculations are challenging due to operator splitting and extreme diversity in both spatial and temporal scales. To address both of these considerations, the nonlinear engineering scale calculation is consistently coupled with the mesoscale solution and then solved using a preconditioned Jacobian-free Newton Krylov (JFNK) [3] method. In this paper, we describe our initial approach to a multiscale model of a UO_2 fuel pellet. Section 2 briefly summarizes the engineering scale and mesoscale models and discusses the scale bridging method. In section 3, initial results from the multiscale model are presented.

2. The multiscale model

Typically, thermal conductivity impacts heat conduction within the fuel according to

$$\begin{aligned} \rho C_p \frac{\partial T}{\partial t} + \nabla\cdot\mathbf{q} - Q &= 0 && T\in\Omega, \\ \mathbf{n}\cdot\mathbf{q} &= q(T) && T\in\Gamma^C, \\ \mathbf{n}\cdot\mathbf{q} &= 0 && T\in\Gamma^T\cup\Gamma^B, \\ T(t=0) &= T_0 && T\in\Omega, \end{aligned} \tag{1}$$

where T, ρ and C_p are temperature, density and specific heat of UO_2, respectively, and we assume fission reactions generate heat at a uniformly distributed constant rate Q. In this expression, Γ^T denotes the top, Γ^B denotes the bottom and Γ^C denotes the outer circumferential boundaries of the fuel pellet such that $\Gamma = \Gamma^T\cup\Gamma^B\cup\Gamma^C$, and $q(T)$ is an expression of the heat rejected to the cladding. The heat flux $\mathbf{q}$ within the UO_2 domain Ω may be written as $\mathbf{q} = -k\nabla T$, where k is the thermal conductivity of UO_2, assuming that the transfer of mass due to species diffusion is small.

In a fuel pellet undergoing irradiation within a reactor, the values of k, ρ, and C_p depend on the microstructure of the material, including voids and fission products, though in this work we only consider the microstructure effect on k. In our multiscale simulations, the heat flux problem is solved by the engineering scale model, while the value of k is determined by the mesoscale model.

2.1. Engineering (coarse) scale discretization

The engineering scale model is a fully-coupled finite element implementation of Eq. (1) with a linear mechanics model of a UO_2 fuel pellet [4]. The coarse test space V is approximated by V^h using linear Lagrange finite elements such that $V^h \subset V$ and $V^h = \mathrm{span}\{\phi_i\}_{i=1}^n$. The trial space hosting the engineering scale solution is similarly approximated, where $T\in V^h$ is

$$T(t,\mathbf{x}) = \sum_{j=1}^{n} T_j(t)\phi_j(\mathbf{x}). \tag{2}$$

The model is implemented in Idaho National Laboratory's `BISON` fuel performance code. `BISON` is built upon INL's `MOOSE`: a parallel, nonlinear, computational framework [5].

In our calculations, we employ Eq. (1) in weak form from Newman et al.[4]. The JFNK solution method used in `BISON` begins with casting this weak form solution as a residual function, *i.e.*

$$\mathbf{F}(\mathbf{x}) = \mathbf{0} \tag{3}$$

of length N, where N is the number of unknowns in the discrete problem. The Jacobian of this system is a $N\times N$ sparse matrix,

$$\mathcal{J}(\mathbf{x}) = \frac{\partial\mathbf{F}(\mathbf{x})}{\partial\mathbf{x}}, \tag{4}$$

where the components of $\mathbf{F}(\mathbf{x})$ are taken directly from the weak form of the problem. Given the Jacobian and the outline in Newman et al.[4], one can express Newton's method in terms of the solution of a series of linear systems. In right preconditioned form, these linear systems may be written as

$$\mathcal{J}(\mathbf{x}^{(k)})M^{-1}(M\delta\mathbf{x}^{(k)}) = -\mathbf{F}(\mathbf{x}^{(k)}), \tag{5}$$

where M^{-1} symbolically represents the preconditioning process.

Table 1. Values for the mesoscale model parameters. The diffusivity is in arbitrary units (a.u.). See Rokkam et al.[1] for the meaning of each parameter.

D_v^0	c_v^0	E_v^m	E_v^f	L_v	A	B	κ_v	κ_η	P_{casc}	c_{max}
1643.0 a.u.	1.0	1.0 eV	1.0 eV	1	1 eV	1 eV	0.5 eV/nm^2	1.0 eV/nm^2	0.3%	1.0

2.2. *Mesoscale (fine) scale simulation*

To simulate the microstructure evolution due to irradiation in the UO_2 fuel, we employ the phase field model developed in Rokkam et al.[1]. The model for this preliminary study represents a simplified description of the fuel, since it only considers vacancies and assumes the material is comprised of a single component. In addition, we evolve the microstructure within a 2D domain.

In the model, the vacancies are defined by a conserved concentration field $c_v(\mathbf{r},t)$ and the void phase is defined by the continuous order parameter $0 < \eta(\mathbf{r},t) < 1$, where $\eta = 0$ in a solid material and $\eta = 1$ in a void. The void and vacancy behavior is defined by a free energy function with two energy wells defining two stable phases; *i.e.*, the matrix phase ($\eta = 0$ and $c_v = c_v^{eq}$) and the void phase ($\eta = 1$ and $c_v = 1$), where $c_v^{eq} = c_v^0 \exp(-E_v^f/k_bT)$ is the vacancy equilibrium concentration with Boltzmann constant k_b.

The evolution of c_v and η is defined by

$$\frac{\partial c_v}{\partial t} = \nabla \cdot M_v \nabla \frac{1}{N} \frac{\partial F(c_v,\eta)}{\partial c_v} \tag{6}$$

$$\frac{\partial \eta}{\partial t} = -\frac{1}{N} L_v \frac{\partial F(c_v,\eta)}{\partial \eta}, \tag{7}$$

where $M_v = D_v/(k_bT)$ is the vacancy mobility with the diffusivity $D_v = D_v^0 \exp(E_v^m/k_bT)$, and L_v is the order parameter mobility. To model the irradiation, there is a probability P_{casc} that c_v will increase by a random amount up to the maximum value c_{max}.

For this work, Eqs. (6) and (7) are uniformly discretized in space and solved using an explicit finite-difference approach with forward Euler time-stepping. Once the microstructure has evolve for a specified amount of time at a uniform temperature, the spatially-dependent mesoscale thermal conductivity is defined as $k'(\mathbf{r}) = k_{bulk}$ if $\eta \leq 0.8$ and $k'(\mathbf{r}) = k_{void}$ if $\eta > 0.8$. We assume the void conductivity to be that of He gas, $k_{void} = 0.152$ W/m K, and we use the temperature-dependent expression for k_{bulk} in UO_2 from Lucuta et al.[6]. Table 1 summarizes the parameter values for the mesoscale model used in this work.

2.3. *Scale bridging*

There is a disparity between the spatial and time scales in the mesoscale and engineering scale models. To couple the two models, we must bridge the scales.

While the engineering scale model may take time steps of hours or days, the mesoscale model requires much smaller time steps. We bridge the time scale in a similar manner to accelerated experimentation techniques, *i.e.* a much higher irradiation rate is considered at the mesoscale. Therefore, the amount of time required to reach a specified burnup is different at the two scales. This approach does introduce some error into the calculation, since the diffusivity is equal in both time scales.

The spatial bridging method employed here is similar to the method proposed by Wagner and Liu [7]. We define coarse and fine calculational spaces, where the coarse space spans the complete problem geometry and the fine space exists only at the Gauss points on the engineering mesh. The mesoscale model calculates void formation over the fine space, resulting in the spatially-dependent mesoscale thermal conductivity $k'(\mathbf{r})$ defined in the previous section. However, the coarse space does not resolve this mesoscale spatial dependence. We determine the coarse space thermal conductivity k by solving a steady-state conductivity equation [2],

$$\nabla \cdot (k'(\mathbf{r})\nabla T) = 0, \tag{8}$$

in the fine space using the evolved microstructure. Equation (8) is discretized and solved using the same grid as that used for the phase field simulations. A constant-temperature boundary condition of $T_l = 800$ K is applied on the left boundary and a constant heat flux of $q' = 50$ MW/m^2 is applied across the right boundary; periodic boundary conditions are applied in the y-direction (note that the values of T_l and q' are arbitrarily selected and have no effect on the value of the calculated thermal conductivity). Assuming $q = q'$, k is determined with $q = -k\Delta T$, where ΔT is the difference between the temperature on the left boundary of the fine space ($T_l = 800$ K) and the average temperature on the right boundary.

The microstructure predicted by the mesoscale model is random, due to the random irradiation term. However, the engineering scale should not resolve this mesoscale randomness. To determine a mesoscale grid size for which the random variation of k is sufficiently small, we calculate k at 20 temperatures within the narrow range $999.9 \leq T \leq 1000.1$. We repeat the calculation for increasing grid sizes ranging from 128×128 to 1024×1024 and calculate the standard deviation of the random scatter (see Fig. 1(a) for the standard deviations from various grid sizes). The standard deviation decreases with increasing grid size, but even at a grid size of 1024×1024, some random variation is still apparent.

The mesoscale grid size required to eliminate the randomness from the macroscale k would be computationally prohibitive in the multiscale model. To allow for small grid sizes, we employ a dynamic curve fitting method to bridge the length scale. Using a least mean squares algorithm, we fit all the values of k calculated by the mesoscale model with a fourth-order polynomial. The thermal conductivity used by the engineering scale model is taken from the curve fit, while the values calculated by the mesoscale model are used to refine the fit. The mesoscale model does not calculate a conductivity when the current temperature is within a certain tolerance of a temperature at which k was previously calculated, and the fit is only recalculated at the start of each residual calculation.

3. Results

Here, we calculate the temperature distribution in a UO_2 fuel pellet. The purpose of this simulation is a proof-of-concept of our multiscale model, not to accurately predict the thermal behavior in a UO_2 pellet. Therefore, we make several assumptions to simplify the calculation.

We consider a dished UO_2 pellet of radius $r = 4.13$ mm and height $h = 6.75$ mm. The pellet has a constant temperature boundary condition $T_c = 700$ K on the outer circumference and a heat source of $Q = 400$ MW/m^3 applied to the pellet. Also, steady-state heat conduction is assumed and the pellet is discretized with 720 linear hexhedral elements. In the mesoscale calculations, we consider a 2D square of UO_2 with sides $l = 1.28$ μm discretized with a 128×128 finite difference grid. The parameters from Table 1 are used and the sample is irradiated for 120 arbitrary time units.

The multiscale calculation converged quadratically (see Fig. 1(b)) and it exhibits excellent parallel scalability (Fig. 1(c)). Figures 2(a) and 2(b) show the temperature distribution within the pellet and the dependence of the thermal conductivity k with temperature, respectively. From plots of the mesoscale vacancy concentration at three positions in the pellet (see Figs. 2(c) – 2(e)), we observe that large voids, surrounded by vacancy-depleted zones, have formed in the center of the pellet. At mid-radius, a larger number of small voids have formed. On the outer circumference, no voids are present.

4. Conclusions

Our multiscale model calculates the temperature distribution in a fuel pellet, considering the microstructure-dependence of the thermal conductivity. The JFNK solution converges quadratically and has excellent parallel scalability. The final microstructure shows strong spatial-dependence.

This work represents a proof-of-concept of our multiscale model. Significant work is still needed to accurately predict the thermomechanical behavior in a UO_2 fuel pellet. In future studies, more physics

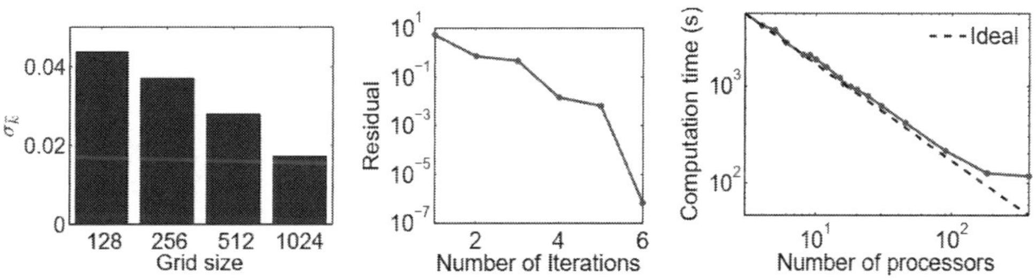

(a) (b) (c)

Figure 1. (a) The standard deviation of the random scatter in the thermal conductivity at increasing grid sizes, (b) the residual value at each of the six JFNK solution iterations and (c) the computation time from different numbers of parallel processors, showing excellent parallel scalability.

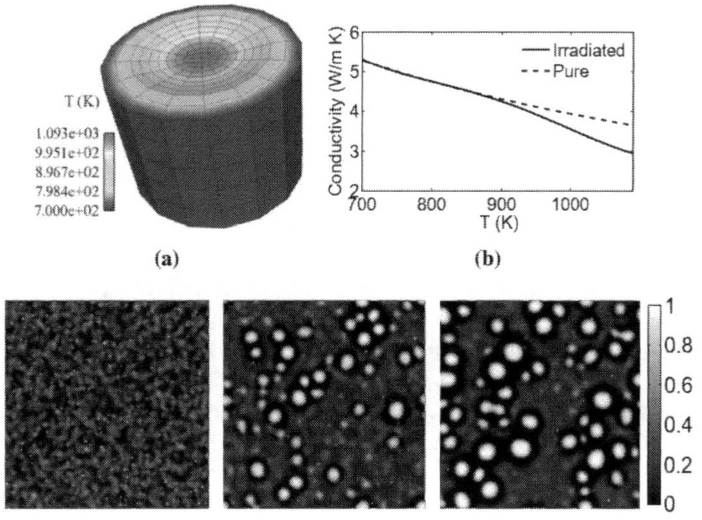

(a) (b)

(c) $r = 4.13$ mm (d) $r = 2.07$ mm (e) $r = 0.0$ mm

Figure 2. (a) The temperature distribution in the fuel pellet, (b) conductivity vs. temperature from the mesoscale model, (c) - (e) plots of the final c_v at $r = 4.13$ mm, $r = 2.07$ mm and $r = 0.0$ mm, respectively.

will be incorporated in the mesoscale model (fission gas, multiple components, self-interstitials) which will be evolved on a 3D domain. Furthermore, the engineering scale simulation will be time-dependent rather than steady-state.

References

[1] Rokkam S, Millett P, Wolf D and El-Azab A 2008 *Multiscale Materials Modeling Proc.* 405–408

[2] Millett P, Wolf D, Desai T, Rokkam S and El-Azab A 2008 *Journal of Applied Physics* **104** 033512

[3] Knoll D A and Keyes D E 2004 *J. Comput. Phys.* **193** 357–397

[4] Newman C, Hansen G and Gaston D 2009 *Journal of Nuclear Materials* In press, http://dx.doi.org/10.1016/j.jnucmat.2009.03.035

[5] Gaston D, Newman C, Hansen G and Lebrun-Grandié D 2008 *Nucl. Engrg. Design* Submitted

[6] Lucuta P, Matzke H and Hastings I 1996 *Journal of Nuclear Materials* **232** 166–180

[7] Wagner G and Liu W 2003 *J. Comput. Phys.* **190** 249–274

Journal of Physics: Conference Series **180** (2009) 012079 doi:10.1088/1742-6596/180/1/012079

The linearly scaling 3D fragment method for large scale electronic structure calculations

Zhengji Zhao[1], Juan Meza[2], Byounghak Lee[3], Hongzhang Shan[2], Erich Strohmaier[2], David Bailey[2], and Lin-Wang Wang[2]

[1] National Energy Research Scientific Computing Center (NERSC)

[2] Computational Research Division, Lawrence Berkeley National Laboratory

[3] Physics Department, Texas State University

ZZhao@lbl.gov

Abstract. The linearly scaling three-dimensional fragment (LS3DF) method is an *O(N) ab initio* electronic structure method for large-scale nano material simulations. It is a divide-and-conquer approach with a novel patching scheme that effectively cancels out the artificial boundary effects, which exist in all divide-and-conquer schemes. This method has made *ab initio* simulations of thousand-atom nanosystems feasible in a couple of hours, while retaining essentially the same accuracy as the direct calculation methods. The LS3DF method won the 2008 ACM Gordon Bell Prize for algorithm innovation. Our code has reached 442 Tflop/s running on 147,456 processors on the Cray XT5 (Jaguar) at OLCF, and has been run on 163,840 processors on the Blue Gene/P (Intrepid) at ALCF, and has been applied to a system containing 36,000 atoms. In this paper, we will present the recent parallel performance results of this code, and will apply the method to asymmetric CdSe/CdS core/shell nanorods, which have potential applications in electronic devices and solar cells.

1. Introduction

Nanostructures have wide applications in biological imaging, light emitting diodes, solar cells, and other electronic devices. The sizes of the nanostructures are so small that they have very different electronic and optical properties from their bulk counter parts, due to quantum confinement effects. Nevertheless, to study the properties of nanostructures, one needs to do *ab initio* calculations on systems containing 1,000 to 100,000 atoms, which are too large for direct *ab initio* methods. Direct methods have been applied to systems with 2,000 atoms at most [1]. This is because even the simplest *ab initio* methods, e.g., the density functional theory (DFT) methods under the local density approximation (LDA), are computationally expensive, scaling as $O(N^3)$, where N is the size of the system. In addition, due to communication bottlenecks the parallelization of direct LDA methods is limited to roughly 10,000 processors [1]. In fact, the most widely used direct LDA code, VASP, is difficult to scale to even several thousand processors. Therefore, both the computational costs and the limit on parallelization call for a new approach.

Over the past decade, many $O(N)$ methods have been developed [2]. These approaches can be classified into three main categories: the local orbital methods [3], the truncated D-matrix methods [4], and divide-and-conquer methods [5]. While these methods have successfully reduced the computational cost and have been applied to large systems, several fundamental technical issues remain that are difficult to overcome. For example, in the commonly used local orbital methods, there exist extraneous local minima in the total energy functional, which make the total energy minimization difficult (convergence problem). This is due to constraining the wavefunctions on the local orbital manifold. Moreover, the overlap between neighboring local orbitals makes it difficult to scale these methods to large numbers of processors. In short, for the previous *O(N)* methods, especially the local orbital methods and the truncated D-matrix methods, on top of some fundamental technical issues, a main challenge is to scale the codes to tens of thousands of computer processors while preserving the *ab initio* accuracy.

Recently we have developed a new *O(N)* method, the linearly scaling three-dimensional fragment (LS3DF) method [6,7]. It is a divide-and-conquer approach that scales to tens of thousands of computer processors quite easily, and yields essentially the same results as direct LDA methods. The LS3DF method won the 2008 ACM Gordon Bell Prize for algorithm innovation [8]. In this paper, we will briefly describe the LS3DF method, focusing on its recent parallel performance results, then describe how we have used LS3DF to study the electronic structures of asymmetric CdSe/CdS core/shell nanorods, which have potential applications in solar energy conversions.

2. The LS3DF method

The LS3DF method is based on the short-ranged nature of quantum mechanical effects. The total energy of a system can be split into the classical electrostatic energy and the quantum mechanical energy (kinetic energy and exchange correlation energy). The electrostatic interaction is long-ranged, therefore the electrostatic energy must be calculated by solving a Poisson equation globally. But the quantum mechanical effects are short-ranged, therefore they can be solved locally, and the total quantum mechanical energy for the system can be obtained by combining the locally calculated quantum energies. In our LS3DF method, we divide a large system into small pieces (fragments), and independently calculate a solution on each fragment. We then patch the local solutions together to obtain the total energy and the total charge density for the whole system. The heart of the LS3DF method is a novel patching scheme that effectively cancels out the artificial boundary effects. Figures 1 and 2 illustrate our division and patching scheme using a 2D example for simplicity. In figure 1, a periodic super cell is divided into 4x4 pieces. At each grid point *(i,j)*, we introduce 4 fragments (along the right-upper direction) with different sizes, 1x1, 2x1, 1x2 and 2x2. All fragments at all grid points *(i,j)* *(i=1,...,4; j=1,...,4)* will be calculated independently using a direct LDA method, e.g., PEtot [9], a planewave pseudo potential LDA code. The fragments will then be summed up according to the patching scheme illustrated in figure 2. Where the 1x1 (red) and 2x2 (blue) fragments are positive fragments, and the other two, the 1x2 (yellow) and the 2x1 (green) are negative fragments.

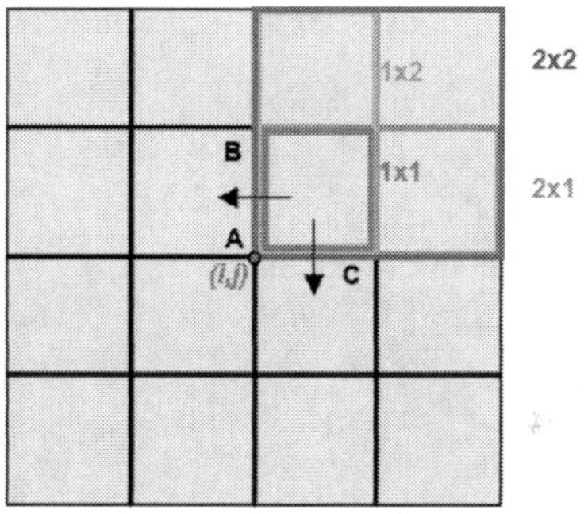

Figure 1. A schematic view of the division of a system into small fragments. This 2D periodic super cell is divided into 4x4 fragment grids. At each fragment grid point *(i,j)*, 4 fragments with different sizes are introduced. Where the red, green, yellow, and blue rectangles represent the fragments of size 1x1, 2x1, 1x2 and 2x2, respectively.

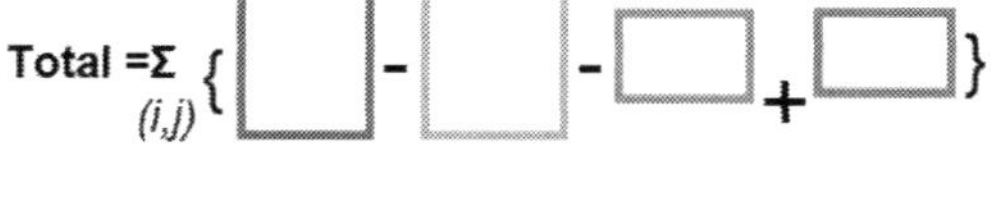

Figure 2. The schematic view of the fragment patching scheme in the LS3DF method for 2D systems. Here the yellow (1x2) and the green (2x1) fragments are negative fragments, and blue (2x2) and red (1x1) are positive fragments.

We can illustrate how the patching scheme yields the original system from the fragments. Let's consider the area covered by the red square in figure 1. For convenience, we denote the fragment 1x1 introduced at the grid point *(i,j)* as $F_{11}(i,j)$. By counting how many positive and negative fragments cover this area, one can easily see whether this area is described properly after all the fragments are added up. This area is covered by 5 positive fragments, they are $F_{11}(i,j)$, $F_{22}(i\text{-}1,j\text{-}1)$, $F_{22}(i,j\text{-}1)$, $F_{22}(i,j)$ and $F_{22}(i\text{-}1,j)$. And this area is also covered by four negative fragments, which are $F_{21}(i,j)$, $F_{21}(i\text{-}1,j)$, $F_{12}(i,j)$ and $F_{12}(i,j\text{-}1)$. When these fragments are summed up using the patching scheme in figure 2, the red square area is covered only once after 4 positive and 4 negative fragments cancel out. We can also show the artificial boundary is removed in this patching scheme. Let's consider the left boundary of the red square (edge AB). We can define a direction (outward) for this boundary as shown with a left arrow in figure 1. We can count how many fragments contain this boundary. They are three positive fragments, $F_{11}(i,j)$, $F_{22}(i,j)$, and $F_{22}(i,j\text{-}1)$ and three negative fragments, $F_{12}(i,j)$, $F_{12}(i,j\text{-}1)$, and $F_{21}(i,j)$. When these six fragments are summed up, the edges from the three negative fragments cancel out the edges from the other three positive fragments. Similarly we can see the artificial corners (e.g., the corner BAC, outward direction) cancel.

The patching scheme for 2D systems can be generalized to 3D systems straightforwardly. In 3D cases, at each grid point *(i,j,k)*, eight fragments (along the right-upper direction) with different sizes are introduced, they are four positive fragments, 2x2x2, 2x1x1, 1x2x1 and 1x1x2, and four negative fragments, 2x2x1, 2x1x2, 1x2x2, and 1x1x1. The patching scheme is

$$\text{Total} = \sum_{i,j,k} \{F_{222} + F_{211} + F_{121} + F_{112} - F_{221} - F_{212} - F_{122} - F_{111}\}$$

The boundary effect cancellation assumes that the fragments that share a given boundary have very similar charge densities near that boundary. Our tests have shown that this assumption always holds as long as the smallest fragment (1x1x1) is not too small and the fragments have a band gap (after some artificial boundary passivations). When a typical 8-atom unit cell is chosen as the 1x1x1 fragments, the errors of the total energy, charge density, atomic force and the dipole moment are well below the typical errors introduced by other approximations of the numerical calculations (e.g., pseudopotentials, planewave basis cut offs). Thus the LS3DF method gives essentially the same results as the direct methods. In contrast to the convergence issues that is common in other *O(N)* methods, the self consistent (SC) iterations in LS3DF converges at a comparable rate as the direct methods. For a system containing a few thousands atoms, the total energy typically converges to 10^{-6} a.u. within 50-60 SC charge density mixing iterations. The LS3DF method outperforms the direct LDA methods when the system contains more than 550 atoms. For nanostructures with 10,000 atoms, the LS3DF would be faster by 3 orders of magnitudes, even presuming the direct LDA methods can scale up to thousands of processors. For more details, see [6,7].

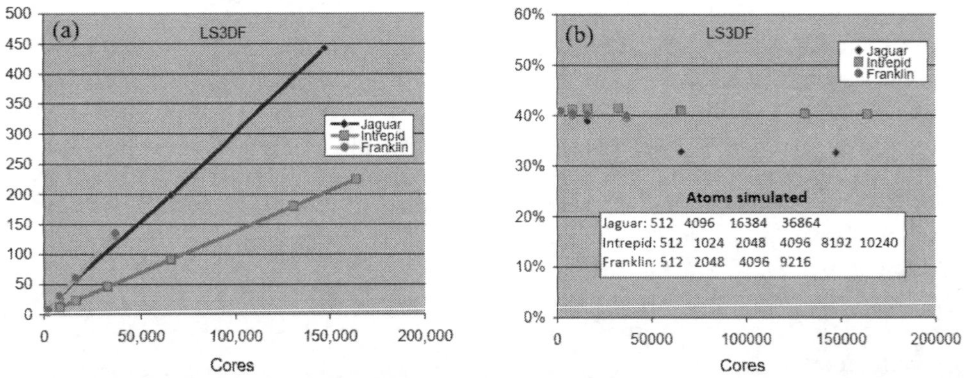

Figure 3. Weak scaling floating point operation rates (a) and the computational efficiency (b) of the LS3DF method on different machines. The systems used were $ZnTe_{1\text{-}x}O_x$ alloy (x=8%) with various number of atoms. The planewave cut off energy was 60 Ry. The flops were measured for one SCF iteration step. The floating point operations were measured in double precision, using the profiling tool Craypat 4.1.

One of the great advantages of the LS3DF method is its excellent parallel scaling. We have run our code on the Cray XT4 (Franklin) at NERSC, the Cray XT5 (Jaguar) at OLCF, and the Blue Gene/P (Intrepid) at ALCF. Figure 3 shows the weak scaling results on these three machines, indicating that LS3DF exhibits nearly linear scaling up to the maximum available processor cores on all three systems. It achieved 135 Tflop/s on 36,864 cores on Franklin at 40% efficiency; 224 Tflop/s on 163,840 cores on Intrepid at 40% efficiency; 442 Tflop/s on 147,456 cores on Jaguar at 33% efficiency. Additional LS3DF performance results are given in [8].

3. Electronic structure calculations for asymmetric CdSe/CdS core/shell nanorods

With the advance of synthetic methods, numerous nanostructures with different shapes have been synthesized in research laboratories. Recently Carbone and his colleagues have synthesized asymmetric core/shell structures, using the newly developed seed growth method [10]. In the asymmetric core/shell nanorods, a CdSe core is embedded at one end of a CdS quantum rod shells. By changing the sizes of the core and the shell (both diameter and length), one can manipulate the electronic structures inside the nanorods. Hence these nanostructures appear to be particularly interesting for solar cell applications. In addition, these asymmetric core/shell structures provide a system on which one can study the quantum confinement effect, the band alignment, the strain (due to the core/shell lattice constant mismatch), and the surface effects.

We have applied the LS3DF method to the asymmetric CdSe/CdS core/shell nanorods with both the Cd atoms terminated and with the Cd+S atoms terminated surfaces, as well as to their counterparts: the pure CdS nanorods, to study how the core and the surface affect the electronic structures inside these nanorods. The LS3DF calculations give us the total charge density and the total potential. Using the resulting potential, we have calculated the valence band maximum (VBM, hole) and the conduction band minimum (CBM, electron) states and the band gaps (Table 1), utilizing the folded spectrum method [11]. We have also calculated the dipole moments (Table 2) of these nanorods. One can see that in the nanorods with the Cd terminated surface, the introduction of the CdSe core results in a significant change in the band gaps (0.24 eV) and dipole moment (-9.65 a.u.), while in the nanorods with the Cd+S terminated surface, the role of the CdSe core seems to be surprisingly small, both on the band gap and the dipole moment. Figure 4 shows the charge density isosurface of the hole and the electron states of the four nanorods. One can see that in the nanorods with the Cd terminated surface, the introduction of the CdSe core significantly changes the localization of the hole states, while in the nanorods with the Cd+S terminated surface, the hole state localization seems not be affected by the presence of the CdSe core, indicating the Cd+S terminated surface has the dominant effects to the hole localizations. Further study on the surface passivation effect is under way [12].

Table 1. The calculated band gaps of the four nanorods. The band gap changes due to the different surfaces (column 4) and the CdSe core (row 4) are also shown in the table.

	Cd termin. (eV)	Cd+S termin. (eV)	Band gap change (eV)
Pure CdS nanorods	2.655	2.498	-0.156
CdSe/CdS Core/shell	2.415	2.403	-0.011
Band gap change	-0.240	-0.095	

Table 2. The calculated dipole moments along the c-axis of the four CdS nanorods (the components of the dipole moments in the other two directions are small, not shown here). The dipole moment changes due to the different surfaces (column 4), and the CdSe cores (row 4) are also shown in the table.

	Cd termin. (a.u.)	Cd+S termin. (a.u.)	Dipole mom. change (a.u.)
Pure CdS nanorods	-15.623	-28.415	-12.792
CdSe/CdS core/shell	-25.277	-28.108	-2.830
Dipole mom. change (a.u.)	-9.654	0.307	

4. Conclusions

We have briefly described the LS3DF method for *ab initio* electronic structure calculations. This scheme features $O(N)$ scaling in terms of system size, and also scales very well in parallel performance. We presented parallel scaling results, summarized the accuracy and the SCF convergence rate, and then applied this method to analyze asymmetric CdSe/CdS core/shell nanorods. We demonstrated that the scheme scales linearly to hundreds of thousands of computer processors; it yields essentially the same results as the direct LDA methods; it can calculate a nanosystem with thousands of atoms self-consistently in a couple of hours. We expect wide applications of LS3DF method in nanostructure calculations.

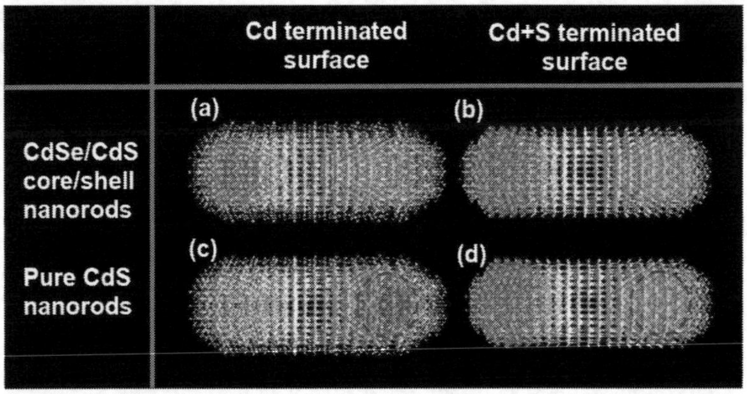

Figure 4. Isosurface of the charge densities of the conduction band minimum (CBM, green) and the valance band maximum (VBM, red) states of the four CdS nanorods with/without CdSe core. Where (a) and (b) are for the pure CdS nanorods with the Cd terminated and the Cd+S terminated surfaces, while (c) and (d) are for the corresponding CdSe/CdS core/shell nanorods, respectively. The isovalue larger than 0.001 e/bhor^3 was shown for both VBM and CBM states. The blue dashed circle shows the CdSe core area. These nanorods are constructed as a wurzite structure, and have 2.8 nm in diameter, and 8.4 nm in length (c-axis). The diameter of the CdSe core is 2.1 nm. There are 3063 and 2298 atoms in the nanorods with the Cd terminated and the Cd+S terminated surfaces, respectively. Where the magenta, yellow, and blue dots represent the Cd, S, and Se atoms, respectively, and the white dots represent pseudo H atoms which passivate the surface dangling bonds.

5. Acknowledgments

This work was supported by the ASCR and the BES offices in the DOE, Office of Science, under LAB03-17 initiative, contract No. DE-AC02-05CH11231. It was supported by the DOE INCITE project. It used the resources of National Energy Research Scientific Computing Center (NERSC), Oak Ridge Leadership Computing Facility, and Argonne Leadership Computing Facility (ALCF). The authors would like to thank Jeff Larkin at Cray Inc., and Katherine Riley at ALCF for their help with running the LS3DF code on Jaguar and Intrepid, respectively. They would also thank Ying Luo at Beijing Normal University for providing the VFF relaxed core/shell nanostructures.

References

[1] F. Gygi, *et al*, Proceedings of the 2005 ACM/IEEE Conference on Supercomputing, (2005).
[2] G. Goedecker, Rev. Mod. Phys, 71:1085, 1999.
[3] G. Galli, *et al*, Phys. Rev. Lett. **69**, 3547(1992); J.-L.Fattebert, *et al*, Phys.Rev.B **73**,115124 (2006)
[4] X.P. Li, *et al*, Phys. Rev. B **47**, 10891 (1993); D.R. Bowler, *et al*, phys. stat. sol. b **243**, 989 (2006).
[5] W. Yang, Phys. Rev. Lett. **66**, 1438 (1991).
[6] L. W. Wang, Z. Zhao, and J. Meza, Phys. Rev. B **77**, 165113 (2008).
[7] Z. Zhao, J. Meza, and L. W. Wang, J. Phys.: Condens. Matter **20** (2008) 294203.
[8] L. W. Wang, *et al*, ACM Gordon Bell submission, Supercomputing (2008).
[9] http://hpcrd.lbl.gov/ linwang/PEtot/PEtot.html
[10] L. Carbone, *et al*, Nano Lett., **7(10)** 2942-2950 (2007)
[11] L. W. Wang and A. Zunger, J. Chem. Phys. **100**, 2394 (1994).
[12] Z. Zhao, and L. W. Wang, *Application of the LS3DF method in CdSe/CdS core/shell nanorods*, to be submitted.

Journal of Physics: Conference Series **180** (2009) 012067 doi:10.1088/1742-6596/180/1/012067

Exploring the spectrum of QCD using the lattice

***Hadron Spectrum Collaboration*: John Bulava[1], Saul Cohen[2], Jozef Dudek[2], Robert Edwards[2], Eric Engelson[3], Justin Foley[1], Balint Joo[2], Jimmy Juge[4], Huey-Wen Lin[2], Nilmani Mathur[5], Colin Morningstar[1], Mike Peardon[6], †David Richards[2], Sinead Ryan[6], Christopher Thomas[2], Anthony Thomas[2], Stephen Wallace[3]**

[1] Department of Physics, Carnegie Mellon University, Pittsburgh, PA 15213, USA
[2] Thomas Jefferson National Accelerator Facility, Newport News, VA 23606, USA
[3] Department of Physics, University of Maryland, College Park, MD 20742, USA
[4] Department of Physics, University of the Pacific, Stockton, CA 95211, USA
[5] Department of Theoretical Physics, Tata Institute of Fundamental Research, Mumbai 400005, India
[6] School of Mathematics, Trinity College, Dublin 2, Ireland

E-mail: †`dgr@jlab.org`

Abstract. The calculation of the spectrum of QCD is key to an understanding of the strong interactions, and vital if we are to capitalize on the experimental study of the spectrum. In this paper, we describe progress towards understanding the spectrum of resonances of both mesons and baryons from lattice QCD, focusing in particular on the resonances of the $I = 1/2$ nucleon states, and of charmonium mesons composed of the heavy charmed quarks.

1. Introduction

The strong interaction, one of the forces of the Standard Model of particle and nuclear physics, is responsible for binding both protons and neutrons into nuclei, and also the primordial gluons and the lightest quarks into pions, protons, neutrons and other so-called hadrons. The interaction arises from a quantum field theory known as Quantum Chromodynamics (QCD).

In order to really understand QCD and hence test whether it is the complete theory of the strong interaction, we must know the spectrum of mesons and baryons that it implies and test those spectra against high quality data. The complete combined analysis of available experimental data on the photoproduction of nucleon resonances is the 2009 milestone in Hadronic Physics (HP), and the measurement of the electromagnetic properties of the low-lying baryons is an HP 2012 milestone. The observed spectrum of QCD provides little direct evidence of the presence of the gluons. However, QCD admits the possibility of exotic mesonic states of matter in which the gluonic degrees of freedom are explicitly exhibited. The search for such states will be an important component of the upgraded JLab@12GeV.

Given the intense experimental efforts in hadron spectroscopy, the need to predict and understand the hadron spectrum from first principles calculations in QCD is clear. Hence, an inportant goal of the effort of the USQCD Collaboration is the study of the resonance spectrum of QCD. The remainder of this paper is laid out as follows. In the next section, we introduce lattice QCD, and outline the methodology for looking at resonances on the lattice. We then

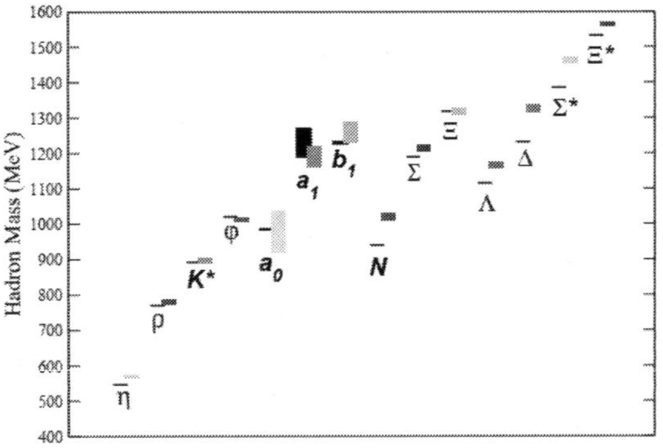

Figure 1. A comparison between the calculated values of the low-lying hadron masses and their physical values, obtained in a calculation with two flavors of light (u/d) quarks and a strange quark[1]. The black and colored bands indicate the experimental and lattice values, respectively; the width of the band indicates the uncertainty.

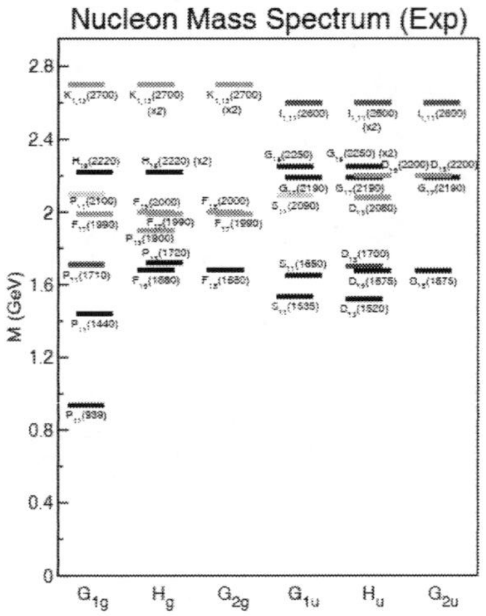

Figure 2. The figure shows the experimentally determined nucleon spectrum classified according to the transformation properties under the cubic group of the lattice; the lowest-lying state in the G_{1g} channel is the proton[2].

describe recent progress at understanding the resonance spectrum, focusing first on nucleon resonances and then on the meson sector. We conclude with prospects for future calculations.

2. Lattice QCD

The interactions of the quarks and gluons at very short distances, such as those probed at the LHC, are weak, a property known as asymptotic freedom, for which Gross, Politzer, and Wilczek won the 2004 Nobel Prize. This enables the interactions to be expanded in terms of a small coupling constant. At longer distances, typical of the binding of the quarks and gluons into hadrons, the coupling becomes strong, and the theory highly non-linear. Here, the only means of solving, as opposed to modelling, QCD is through numerical calculations on the lattice.

Lattice gauge calculations solve QCD on a four-dimensional lattice, or grid, of points in Euclidean space. The quarks reside on the points of the grid, whilst the gluons are associated with the links joining those points. Lattice calculations proceed through a Monte Carlo method, in which ensembles of gauge configurations are generated with a probability distribution prescribed by the Euclidean QCD action. Lattice QCD calculations have always been at the leading edge of exploiting the most powerful supercomputing resources available, helped by the highly regular nature of the problem.

The calculation of the ground-state spectrum has been a benchmark calculation of lattice QCD since its inception. Figure 1 shows a summary of the low-lying light-hadron masses compared with their experimental values, measured on anisotropic clover lattices designed for the study of resonance spectroscopy[1]; details of the computational methodology are given in the poster of Balint Joo.

A comprehensive picture of resonances requires that we go beyond a knowledge of the ground state mass in each channel, and obtain the masses of the lowest few states of a given quantum

Journal of Physics: Conference Series **180** (2009) 012067 doi:10.1088/1742-6596/180/1/012067

number. This we can accomplish through the use of the variational method[3, 4]; we calculate a matrix of correlator functions

$$C_{ij}(t) = \sum_{\vec{x}} \langle O_i(\vec{x}, t) O_j^{\dagger}(\vec{0}, 0) \rangle,$$

where $\{O_i; i = 1, \ldots, N\}$ are a basis of interpolating operators with given quantum numbers. We then solve the generalized eigenvalue equation

$$C(t)u = \lambda(t, t_0) C(t_0) u$$

to obtain a set of real (ordered) eigenvalues $\lambda_n(t, t_0)$, where $\lambda_0 \geq \lambda_1 \geq \ldots \geq \lambda_{N-1}$. At large Euclidean times, these eigenvalues then delineate between the different masses

$$\lambda_n(t, t_0) \longrightarrow e^{-M_n(t-t_0)} + O(e^{-\Delta M_n(t-t_0)}),$$

where $\Delta M_n = \min\{| M_n - M_i | : i \neq n\}$. The eigenvectors u are orthogonal with metric $C(t_0)$, and a knowledge of the eigenvectors can yield information about the partonic structure of the states. Crucial to the application of the variational method is the use of a basis of interpolating operators that have a good overlap with the low-lying states of interest. The cubic lattice employed in our calculations does not admit the full rotational symmetry of the continuum, but rather the more restricted symmetry of the octahedral group. Thus states at rest are classified according to the irreducible representation (irreps) of the cubic group, and for spectroscopy calculations, interpolating operators must be constructed that transform irreducibly under the cubic group; this task has been the prerequisite for our study both of baryons[5, 6], and of mesons.

3. Nucleon Resonance Spectrum

Baryons, containing three quarks, are emblematic of the non-Abelian nature of QCD, and of the three colors of the theory. An important goal in exploring baryons is attempting to discern the effective degrees of freedom that describe the spectrum; the search for so-called "missing resonances" focuses on whether the spectrum can be well described by a quark model, or whether an effective theory with fewer degrees of freedom, such as a quark-diquark picture, provides a more faithful description of the baryon spectrum.

There are three double-valued irreducible representations of the cubic group, denoted $G_{1u/g}(2)$, $H_{u/g}(4)$ and $G_{2u/g}(2)$, where g and u refer to positive and negative parity, respectively, and the brackets contain the dimension of the irrep. G_1 contains continuum spins $1/2, 7/2, \ldots$, H_g spins $3/2, 5/2, \ldots$ and G_2 spins $5/2, 7/2, \ldots$. Thus, at any fixed lattice spacing a, a state corresponding to spin-5/2 has four degrees of freedom in H, and two in G_2, with degeneracies between the energies in the two irreps emerging in the continuum limit. Figure 2 shows the experimental nucleon spectrum as seen on the lattice; the limited number of irreps requires that for each channel we be able to isolate as many energy levels as possible.

The nucleon spectrum has been analysed in a calculation with two flavors of light Wilson fermions, at two values of the pion mass[7], building on an earlier calculation in the quenched approximation to QCD[8]. The data are shown in Figure 3. For the first time in a lattice calculation, we can identify a spin-5/2 state, but the multi-hadron states that should be seen in the spectrum appear elusive; the calculation of correlation functions that are expected to be sensitive to these multi-hadron contributions is an important goal for the collaboration.

4. Meson Resonances

The new Hall D of the JLab 12GeV upgrade centers on the study of meson states produced in photoproduction reactions in the GlueX detector. Photoproduction has been proposed, within

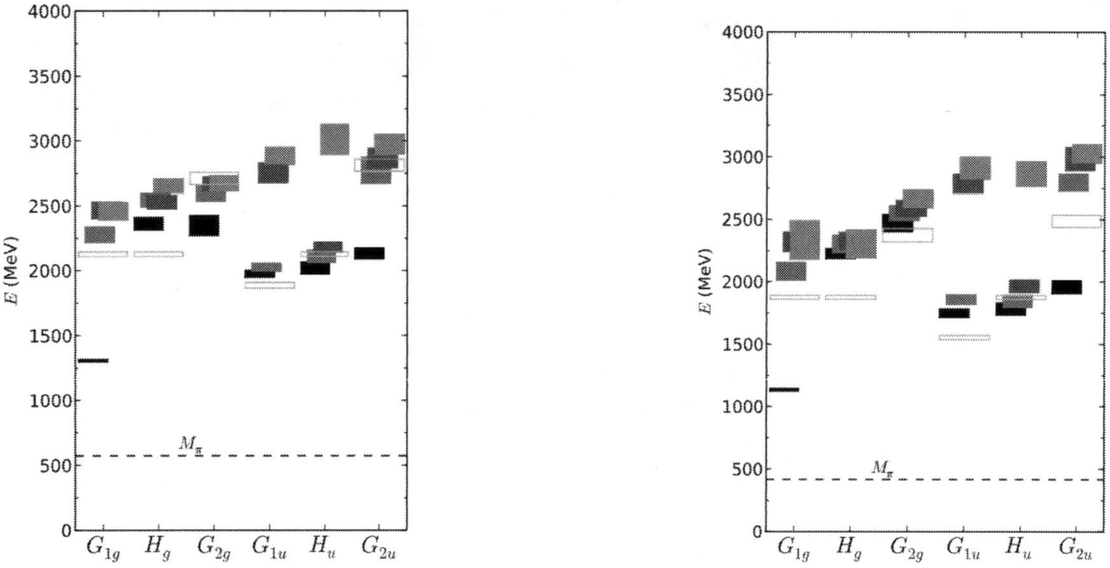

Figure 3. The left- and right-hand panels show the spectrum of $I = 1/2$ baryon resonance, indicated by the solid boxes, obtained on $N_f = 2$ Wilson fermion lattices at $m_\pi = 578$ and 416 MeV respectively[7]; the errors are indicated by the vertical width of the box. The open boxes show the expected thresholds for multiparticle states.

QCD-motivated models, as a favorable method for the production of "exotic" hybrid mesons, those mesons having J^{PC} outside the set allowed to a fermion-antifermion pair. The hybrid hypothesis is that an excited gluonic field in addition to a quark-antiquark pair can give rise to these quantum numbers.

As a theater for developing our methodology, we have studied charmonium, composed of the heavier charm quark c and its antiquark; the system is particularly attractive, in that it is both computationally far less demanding than systems composed of the light (u, d, s) quarks, and because there is a wealth of high-precision experimental data. For the study of the spectrum, performed in the quenched approximation to QCD, we used the known continuum behavior of our operators to enable the *spins* of the states in the different lattice irreps to be identified[9], illustrated for the $J^{PC} = J^{--}$ and J^{++} channels in Figure 4. Lattice QCD can further enable the electromagnetic properties of the states to be investigated. Thus the radiative transitions between some low-lying, non-exotic states were studied[10], and recently this was extended to include the radiative transition form factors for some of the higher-lying states in the spectrum, including those with exotic quantum numbers[11]. Most importantly, it was shown that radiative transitions to states with exotic quantum numbers were calculable, and furthermore the partial width for the exotic decay $\Gamma(\eta_{c1} \rightarrow J/\psi\gamma)$ was shown to be comparable to the non-exotic width. The extension of this calculation to mesons composed of light quarks, using the $N_f = 2 \oplus 1$ anisotropic clover lattices[1], is in progress.

5. Conclusions

The calculation of the masses and properties of resonances is essential to fully capitalize on the investment in experimental facilities, and to provide reliable calculations to confront the experimental results. Thus the generation and analysis of lattices designed for spectroscopy is a central goal of the USQCD Collaboration's program. The calculations outlined above are but the first stage in this program, with quark masses considerably higher than those realized in nature. An immediate challenge arises as we enter the regime where resonances are unstable under the

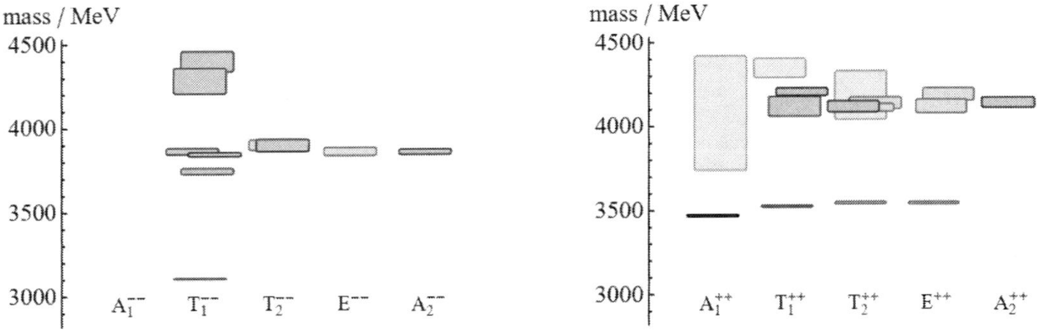

Figure 4. The left- and right-hand panels show the masses calculated in the $J^{PC} = J^{--}$ and J^{++} channels, respectively, categorized according to the lattice irreps, with the color coding indicating their estimated continuum spin assignments where known: Black ($J = 0$), Red ($J = 1$), Green ($J = 2$), Blue ($J = 3$), and Orange (unidentified).

strong interactions, and multi-hadron energies in the spectrum are expected to be important; a novel means of efficiently computing correlation functions in this regime has recently been outlined[12]. The advent of petascale computing is enabling calculations to be performed at progressively lighter values of the light-quark masses, and calculations with all masses fixed to their physical values are now, finally, within reach.

Acknowledgments

This research used resources of the National Center for Computational Sciences at Oak Ridge National Laboratory, which is supported by the Office of Science of the Department of Energy under Contract DE-AC05-00OR22725. This research was supported in part by the National Science Foundation through TeraGrid resources provided by Pittsburgh Supercomputing Center (PSC), San Diego Supercomputing Center (SDSC) and the Texas Advanced Computing Center (TACC), and under the USQCD SciDAC-2 Grant. This work used cluster resources at Jefferson Laboratory provided under the auspicies of the USQCD Collaboration.

MP and SR are supported by Science Foundation Ireland under research grants 07/RFP/PHYF168 and 06/RFP/PHY061, respectively. JB, JF and CM are supported by National Science Foundation grant numbers NSF-PHY-0653315 and NSF-PHY-0510020. KJJ is supported by grant number NSF-PHY-0704171. Authored by Jefferson Science Associates, LLC under US DOE Contract No. DE-AC05-06OR23177. The U.S. Government retains a non-exclusive, paid-up, irrevocable, world-wide license to publish or reproduce this manuscript for U.S. Government purposes.

References

[1] Lin H W *et al.* (Hadron Spectrum) 2009 *Phys. Rev.* **D79** 034502 (*Preprint* `0810.3588`)
[2] Lichtl A C 2006 (*Preprint* `hep-lat/0609019`)
[3] Michael C 1985 *Nucl. Phys.* **B259** 58
[4] Luscher M and Wolff U 1990 *Nucl. Phys.* **B339** 222–252
[5] Basak S *et al.* (Lattice Hadron Physics (LHPC)) 2005 *Phys. Rev.* **D72** 074501 (*Preprint* `hep-lat/0508018`)
[6] Basak S *et al.* 2005 *Phys. Rev.* **D72** 094506 (*Preprint* `hep-lat/0506029`)
[7] Bulava J M *et al.* 2009 *Phys. Rev.* **D79** 034505 (*Preprint* `0901.0027`)
[8] Basak S *et al.* 2007 *Phys. Rev.* **D76** 074504 (*Preprint* `0709.0008`)
[9] Dudek J J, Edwards R G, Mathur N and Richards D G 2008 *Phys. Rev.* **D77** 034501 (*Preprint* `0707.4162`)
[10] Dudek J J, Edwards R G and Richards D G 2006 *Phys. Rev.* **D73** 074507 (*Preprint* `hep-ph/0601137`)
[11] Dudek J J, Edwards R and Thomas C E 2009 *Phys. Rev.* 094504 (*Preprint* `0902.2241`)
[12] Peardon M *et al.* 2009 (*Preprint* `0905.2160`)

SciDAC 2009 IOP Publishing
Journal of Physics: Conference Series **180** (2009) 012080 doi:10.1088/1742-6596/180/1/012080

Fast multiresolution methods for density functional theory in nuclear physics

G. I. Fann[1], J. Pei[5,6], R. J. Harrison[1,4], J. Jia[1], J. Hill[1], M. Ou[2], W. Nazarewicz[5,6], W. A. Shelton[1] and N. Schunck[5,6]

[1] Computer Science and Mathematics Division, Oak Ridge National Laboratory, Oak Ridge, TN 37830

[2] UT Joint Institute for Computational Sciences, Oak Ridge, TN 37831

[4] Department of Chemistry, University of Tennessee, Knoxville, TN 37996

[5] Department of Physics, University of Tennessee, Knoxville, TN 37996

[6] Physics Division, Oak Ridge National Laboratory, Oak Ridge, TN 37831

E-mail: fanngi@ornl.gov

Abstract.
We describe a fast real-analysis based $O(N)$ algorithm based on multiresolution analysis and low separation rank approximation of functions and operators for solving the Schrödinger and Lippman-Schwinger equations in 3-D with spin-orbit potential to high precision for bound states. Each of the operators and wavefunctions has its own structure of refinement to achieve and guarantee the desired finite precision. To our knowledge, this is the first time such adaptive methods have been used in computational physics, even in 1-D. Accurate solutions for each of the wavefunctions are obtained for a sample test problem. Spin orbit potentials commonly occur in the simulations of semiconductors, quantum chemistry, molecular electronics and nuclear physics. We compare our results with those obtained by direct diagonalization using the Hermite basis and the spline basis with an example from nuclear structure theory.

1. Introduction

Solving the Schrödinger and the Lippman-Schwinger equations are important for understanding the chemistry and the nuclear structures of atoms and molecules. In this paper, we present a fast real-analysis based algorithm using an adaptive multiwavelet basis for solving the Schrödinger and Lippman-Schwinger equations with spin-orbit interactions for a non-spherically symmetric potential in 3-D.

It has been known for some time that multiresolution analysis (MRA) and multiwavelets should lead to useful numerical algorithms. However, straightforward generalizations of such an approach from one and two spatial dimensions to higher dimensions yield algorithms that are too costly for practical applications. The development of efficient and robust algorithms in three and higher dimensions has recently been successful. These algorithms rely on low-separation rank (LSR) approximations of weakly singular kernels coupled with MRA[3, 5, 7]

In this paper we describe an extension of these ideas to solving the Schrödinger equation and the Lippman-Schwinger equation with spin-orbit interactions. The spin-orbit term is an important part of many quantum-mechanical Hamiltonians. For example, in chemistry, spin-orbit interaction models the shifts in an electron's atomic energy levels, e.g. the splitting of spectral lines. It also occurs in models of the effects of electric field or magnetic field on an electron's spin and it is an important component of the strong forces acting between nucleons.

The spin-orbit interaction introduces rotational terms into the Schrödinger equation. In particular, terms involving the gradient of the potential and the gradient of the wave functions are introduced into the equation. In practice, physicists and chemists have been able to perform this computation, but it is difficult to obtain accurate eigenvalues and eigenfunctions beyond four to five digits in the general case.

The problem can be described as follows. Let $\mathcal{H}$ denote the Hamiltonian with eigenvalues E and eigenfunctions ϕ. The Schrödinger equation, with spin-orbit, is written as

$$\mathcal{H}\phi = E\phi$$

where $\mathcal{H} = \left(-\frac{\hbar^2}{2m}\Delta + V\right)\mathbf{I} + V_{SO}$ with $\mathbf{I}$ being the 2×2 identity matrix, Δ is the Laplacian operator, $\hbar$ is Planck's constant, m is the mass of the particle and V is a L_2 integrable compact scalar potential function.

The spin-orbit potential term, V_{SO}, assuming reflection symmetry with respect to the y-axis, is an application of a 2 by 2 matrix operator on a vector-valued complex valued wavefunction $\phi = (u, v)^T$, with $u, v : R^3 \rightarrow C$. The operator V_{SO} is given by:

$$V_{SO} := -i\lambda \begin{bmatrix} (\partial_y V)\partial_x - (\partial_x V)\partial_y & (\partial_z V)\partial_\perp - (\partial_\perp V)\partial_z \\ (\partial_z V)\partial_\perp^* - (\partial_\perp^* V)\partial_z & -(\partial_y V)\partial_x + (\partial_x V)\partial_y \end{bmatrix}$$

where $\partial_\perp := i\partial_x + \partial_y$, $i := \sqrt{-1}$, λ is a constant, and A^* denotes complex conjugation of A. Each wave-function satisfies the zero Dirichlet boundary condition for a sufficiently large box or asymptotically.

A common spectral method for solving this eigensystem problem is to expand the wavefunctions and its first and second derivatives in a basis (e.g., harmonic oscillators, splines, Gaussians) of fixed length, form a linear eigensystem problem and solve for the eigenvalues and the eigenvectors. The coefficients of the wavefunctions form the entries of the unknown eigenvectors. Another common method is to numerically discretize the Hamiltonian operator. In both cases a dense or a sparse generalized eigensystem problem $Hx = \lambda Sx$ is solved using a standard eigensolver, LAPACK for the dense case [2], and typically Davidson or Lanczos eigensolvers for the sparse case [4, 6].

Formally, we rewrite the Schrödinger equation as a Lippman-Schwinger equation:

$$(-\frac{\hbar^2}{2m}\Delta - E)\mathbf{I}\phi = (-V\mathbf{I} - V_{SO})\phi.$$

The formal inverse $G = (-\frac{\hbar^2}{2m}\Delta - E)^{-1}$ is the convolution operator with the bound-state Helmholtz kernel. We describe the application and approximation with this Green's function below.

2. Multiwavelets and low-separation rank representations

The representation of functions and operators is based on an adaptive pseudo-spectral representation of functions and operators using the discontinuous Alpert's multiwavelets[1] and low-separation rank approximation of weakly singular kernels. The use of these representations for solving Schrödinger equation without spin-orbit potential in computational chemistry and density functional theory were presented in [7, 8, 10]. To represent functions in 3-D, we use tensor products of 1-D multiwavelets.

The kernel for the bound state Helmholtz convolution operator e^{-kr}/r is approximated by LSR using Gaussians

$$\frac{e^{-kr}}{r} = \sum_{m=1}^{M} w_m e^{-p_{1,m}x_1^2} e^{-p_{2,m}x_2^2} e^{-p_{3,m}x_3^2} + O(\frac{\epsilon}{r}).$$

The number of terms, M in the sum, is nearly optimal and depends logarithmically on precision ϵ. In [3], we showed the compatibility of the fast application of low-separation rank approximation of the Helmholtz operators with the multiwavelet approximation of functions. The approximation of the Helmholtz kernel is tabulated and is constructed for each wavefunction. The multiwavelet approximation of each wavefunction has its own adaptive resolution to the desired precision.

3. Algorithm

In [7, 5], we described algorithms for solving Schrödinger equation without the spin-orbit term. We proceeded in two steps. First, we represented the approximate wave functions and the Hamiltonian operator in the multiwavelet basis in a large enough domain so that the tails of the wavefunctions are 0, and diagonalized the eigensystem using LAPACK's Hermitian eigensolver to obtain estimates for the eigenvalues and orthogonal eigenfunctions. This is fast for a small number of eigenvalues and eigenvectors. Once the orthogonal eigenvectors have been obtained, we refined the approximation for the eigenvectors, constructed and applied the bound-state Helmholtz Green's functions for each estimated eigenvalues and iteratively solved the Lippman-Schwinger equation associated to the Schrödinger equation.

Our algorithm based on direct diagonalization with fixed-point iteration is given below. If V_{SO} is not included, the Green's function iteration summarized above [7] converges quickly to high accuracy.

Our algorithm for the solving the spin-orbit Hamiltonian is

- Initialize all guess wave-functions ϕ_i^0 for all i, $n = 0$
- Loop on n until converged
  1. Form matrices $\mathbf{H} =< \phi_i^n, \mathcal{H}\phi_j^n >$, $\mathbf{S} =< \phi_i^n, \phi_j^n >$
  2. Solve the orthogonal generalized eigenproblem for the eigenpair (E_i^n, ϕ_i^n)
  3. Refine the multiwavelt expansion of ϕ_i^n
  4. Construct the LSR approximation of the Green's function G_i
  5. Apply a fixed-point convolution iteration[1], $\phi_i^{n+1} = -G_i \star (VI + V_{SO})\phi_i^n$
  6. Compute the relative norm of the change in eigenvectors ϕ_i
  7. If the relative norm is less than norm, label ϕ_i converged
  8. Update counter, $n = n + 1$

where $G_i(r) = \frac{e^{-k_i r}}{r}$, $r = \sqrt{x^2 + y^2 + z^2}$, $k_i = \sqrt{-\frac{2m}{\hbar^2}E_i}$. In practice one can show that if the norm of $G \star V_{so}\phi$ is small relative to $G \star V\phi$ then the iteration converges. Formal proof will be presented elsewhere.

4. Test problem

The above algorithm has been applied to solve the Schrödinger equation for a non-spherically symmetric potential from nuclear density function theory using the MADNESS code. The potential V is the two-center cosh potential

$$V(x, y, z) = \frac{V_1}{1 + e^{-R_1/a_1}\cosh(r_+/a_1)} + \frac{V_2}{1 + e^{-R_2/a_2}\cosh(r_-/a_2)}$$

[1] The symbol $\star$ denotes the convolution operation.

eval	spin	π	HO	spline	MRA
0	1/2	+	-22.23984	-22.2400	-22.240107
1	1/2	-	-22.23949	-22.2399	-22.239980
2	1/2	+	-9.43509	-9.4365	-9.436627
3	3/2	+	-9.42925	-9.4319	-9.432028
4	1/2	+	-9.42911	-9.4310	-9.430805
5	3/2	-	-9.42490	-9.4278	-9.427881
6	1/2	-	-8.77589	-8.7782	-8.778282
7	1/2	-	-8.77013	-8.7737	-8.773836
8	1/2	+	-1.71593	-1.7239	-1.725169
9	1/2	+	-1.51149	-1.5251	-1.526936

Figure 1. The solutions from using different bases for solving the spin-orbit problem are tabulated.

with $V_1 = V_2 = -50$ MeV; $a_1 = a_2 = 1.0$ fm; $R_1 = R_2 = 2.0$ fm. The potential wells are at $r_- = (0, 0, -\zeta)$ and $r_+ = (0, 0, \zeta)$ with $\zeta = 7.5$ fm. The spin-orbit term is describe above as V_{SO} and is also given by Eq. (22) in [9]

A comparison of the results from basis expansions using the harmonic oscillator basis and also the spline basis are included for comparison in Figure 1. The computation with the spline basis uses a 2-D tensor product expansion with up to degree 11 elements in each dimension. For the calculation using the harmonic oscillator basis, a 3-D tensor product of up to 25 shells and with degree up to 22 in each dimension is used. Due to memory usage both of these boxes were only able to use a domain $[-40., 40]$, in fermies, in each spatial direction. The MRA computation uses a tensor product of degrees 0 to degree 7 multiwavelets in each dimension. The precision of the multiwavelet calculation is $1e - 5$ using a domain $[-120., 120.]^3$. Some of our results are reported in [9].

Three dimensional plots of the absolute value of the real part of the 6-th and the 8-th wavefunctions, $log(|u|)$, are displayed in Figure 2. Various isosurfaces of the wavefunctions show the complicated structures as well as the symmetries of the wavefunctions.

5. Acknowledgements

This work was supported by the Scientific Discovery through Advanced Computing (SciDAC) program of the U.S. Department of Energy, Office of Science, the Division of Basic Energy Sciences, and the Office of Advanced Scientific Computing Research, Program in Applied Mathematics and through the SciDAC Scientific Application Pilot Program under contract number DE-AC05-00OR22725 with UT-Battelle, LLC (Oak Ridge National Laboratory). It is also supported under contracts DE-FG02-96ER40963 (University of Tennessee), DE-FG05-87ER40361 (Joint Institute for Heavy Ion Research), and DE-FC02-07ER41457 with UNEDF SciDAC Collaboration. Computational resources were provided by the National Center for Computational Sciences at Oak Ridge and the National Energy Research Scientific Computing Facility.

References

[1] B. Alpert. A class of bases in L^2 for the sparse representation of integral operators. *SIAM J. Math. Anal*, 24(1):246–262, 1993.

[2] E. Anderson, Z. Bai, C. Bischof, S. Blackford, J. Demmel, J. Du Croz, A. Greenbaum, S. Hammarling,

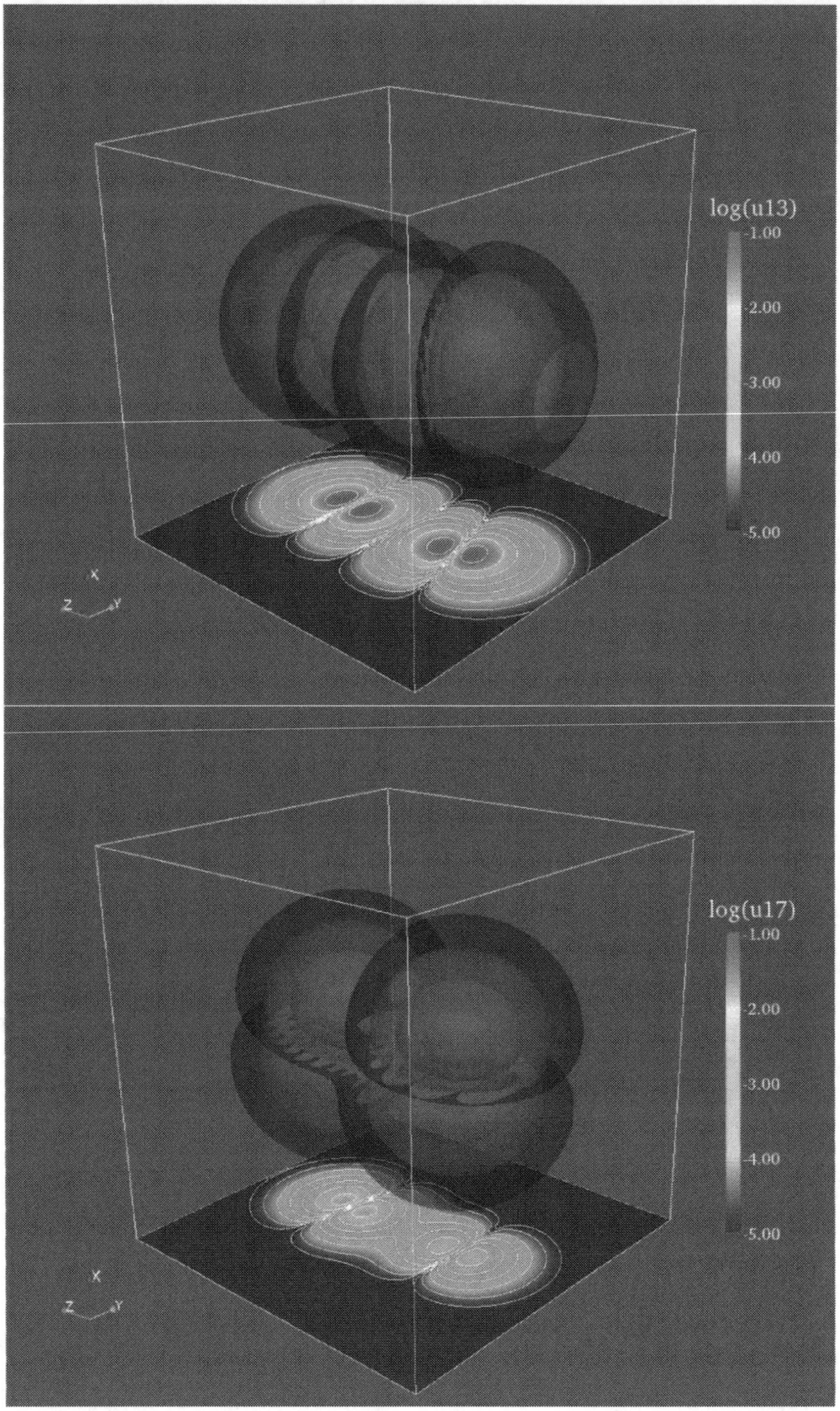

Figure 2. The logarithm of the absolute value of the real part of the wavefunctions corresponding to eigenvalue number 6 (labeled log(u13)) and number 8 (labeled log(u17)) with various isosurfaces are plotted with domain $[-20, 20]^3$.

A. McKenney, and D. Sorenson, editors. *LAPACK User's Guide, Third Edition.* SIAM, Philadelphia, PA, 1999.

[3] G. Beylkin, R. Cramer, G.I. Fann, and R.J. Harrison. Multiresolution separated representations of singular and weakly singular operators. *Appl. Comput. Harmon. Anal.*, 23(2):235–253, 2007.

[4] Ernest R. Davidson. Super-matrix methods. *Comp. Phys. Comm.*, 53:49–60, 1989.

[5] G.I. Fann, G. Beylkin, R.J. Harrison, and K. Jordan. Singular operators in multiwavelet bases. *IBM Journal of Research and Development*, 48(2):161–171, March 2004.

[6] G. Golub and C. Van Loan. *Matrix Computations.* Johns Hopkins University Press, 3rd edition, 1996.

[7] R.J. Harrison, G.I. Fann, T. Yanai, and G. Beylkin. Multiresolution quantum chemistry in multiwavelet bases. In P.M.A. Sloot et. al., editor, *Lecture Notes in Computer Science. Computational Science-ICCS 2003*, volume 2660, pages 103–110. Springer, 2003.

[8] R.J. Harrison, G.I. Fann, T. Yanai, Z. Gan, and G. Beylkin. Multiresolution quantum chemistry: basic theory and initial applications. *J. Chem. Phys.*, 121(23):11587–11598, 2004.

[9] J.C. Pei, M.V. Stoitsov, G.I. Fann, W. Nazarewicz, N. Schunck, and F.R. Xu. Deformed coordinate-space Hartree-Fock-Bogoliubov approach to weakly bound nuclei and large deformations. *Physics Review C*, 78, 064306, 2008.

[10] T. Yanai, G.I. Fann, Z. Gan, R.J. Harrison, and G. Beylkin. Multiresolution quantum chemistry: Analytic derivatives for Hartree-Fock and density functional theory. *J. Chem. Phys.*, 121(7):2866–2876, 2004.

Lattice QCD simulations on big cats, sea monsters and clock towers

Bálint Joó on behalf of the USQCD Collaboration and the Hadron Spectrum Collaboration
Jefferson Lab, 12000 Jefferson Avenue, Newport News, VA 23606
E-mail: bjoo@jlab.org

Abstract. We present details of lattice QCD computations we are performing on the Cray XT series of computers, from BigBen – an XT3 hosted at the Pittsburgh Supercomputing Center (PSC) – through Jaguar (XT4) and Kraken (XT5) – which are hosted at the National Center for Computational Science (NCCS) and the National Institute of Computational Science (NICS), respectively, at Oak Ridge National Laboratory (ORNL). We discuss algorithmic tuning to make the computation more efficient and present some recent results.

1. Introduction

Lattice QCD is currently the only known model-independent, non-perturbative method for carrying out calculations of Quantum Chromodynamics (QCD), which are relevant to both Nuclear and High Energy physics. In this poster we report on progress on calculations which will enable the calculation of the excited state baryon spectrum, the photo-couplings of Hybrid Mesons and the investigation of multi-particle states. These calculations form the basis of DOE milestones in Hadronic Physics, and are of great relevance to the exprimental program at the Jefferson Lab.

In this contribution, we focus on the initial phase of the calculation: gauge configuration generation. The extraction of physical results requires that these configurations be subjected to sophisticated analysis campaigns. We will not discuss the analysis and results further here, but will rather refer the reader to other contributions in this volume, in particular those of Edwards [1], Richards [2] and Luu [3].

2. Hybrid Monte Carlo for gauge generation

Discretizing QCD onto a lattice proceeds by ascribing fermionic degrees of freedom to the sites of a space-time lattice, and the gluonic degrees of freedom to the lattice links. The gluon fields are represented by $SU(3)$ matrices on the lattice links, which correspond to *parallel transport operators* betwen points connected by the links. It is common to abuse notation and denote as U the collection link variables $U_\mu(x)$ for all the links on a given lattice.

Lattice QCD calculations generically involve evaluating the path integral

$$\langle \mathcal{O} \rangle = \frac{1}{\mathcal{Z}} \int \mathcal{D}U \; \mathcal{O}(U) \; \prod_{i=1}^{N_f} \det(M_i(U)) \; e^{-S(U)} \; , \tag{1}$$

where $\mathcal{O}$ is the observable of interest and $S(U)$ is the gauge action. The path integral has partition function $\mathcal{Z}$ and the measure $\mathcal{D}U$ is shorthand for integrating over every configuration U. The fermionic degrees of freedom are integrated out, and their effect amounts to a weight in the path integral of $\det(M_i)$ where M_i is the *fermionic kernel* for the i-th flavor of quark. We note that the work described in this article was carried out using anisotropic clover fermions.

An efficient way to evaluate the path integral is by a Monte Carlo method using importance sampling. This method produces lattice configurations U with a probability $P(U)$ so that

$$P(U) \propto \prod_{i=1}^{N_f} \det(M_i(U))\ e^{-S(U)} \tag{2}$$

and the expectation value reduces to a simple average: $\bar{\mathcal{O}} = \frac{1}{N}\sum_{\{U\}} \mathcal{O}(U)$. The workhorse of the gauge generation process is the Hybrid Monte Carlo [4] algorithm and its variants [5]. Treating the U link variables as canonical coordinates, these methods extend the configuration space to a phase space by assigning to each link a canonically conjugate momentum $\pi_\mu(x)$. One can then define a Hamiltonian $H = \frac{1}{2}\sum_{x,\mu}\pi(x,\mu)^2 + S(U)$. The algorithms iterate the following scheme: start from some initial gauge configuration U, generate fresh canonical momenta π and perform Hamiltonian Molecular Dynamics (MD) from this phase space state to a trial state (U', π'). The new state is subjected to a Metropolis acceptance test

$$P_{\text{acc}} = \min\left(1, e^{-H(U',\pi')+H(U,\pi)}\right) . \tag{3}$$

If the state is accepted the new starting state will be (U', π'), otherwise it will be the old state (U, π). We note that the MD algorithm should be reversible and area preserving, so that the Monte Carlo process satisfies the condition of detailed balance, which is sufficient for the generated Markov Chain to converge to the desired equilibrium distribution. To satisfy this condition symmetric, symplectic integration schemes are used. The acceptance rate of the algorithm depends on the energy change $\delta H = H(U', \pi') - H(U, \pi)$ along an MD trajectory. This can be controlled by varying the step-size(s) of the underlying integration scheme. Finally, we note that the primary cost of the algorithm comes from the fact that in order to evaluate the contribution from the quarks one must solve the system

$$M^\dagger M x = b \tag{4}$$

at least once per MD step (x and b are vectors). This system has a dimension of $O(V)$ where V is the lattice volume, and has condition number that varies inversely with the quark mass. Dimensions and condition numbers, each of $O(10^7)$, are common. Hence, typical avenues of optimization include: developing faster implementations of M, developing solver algorithms which converge fast and finding integration schemes which reduce δH on average so that one may take fewer, longer steps in the MD, thus requiring fewer solves.

3. Tuning the molecular dynamics integration

In our simulations, we employ a three level nested integration scheme where on each level we use the MD integrator due to Omelyan [6]. As described in [7], the leading order error term on a given level is given by

$$\left(\frac{6\lambda^2 - 6\lambda + 1}{12} A + \frac{1 - 6\lambda}{24} B\right) \delta\tau^2 \tag{5}$$

where λ is a tunable constant, $\delta\tau$ is the integration step-size, and A and B are Poisson Brackets (PB) of coordinate (gauge) and momentum update operators. In [7], minimising the variance of

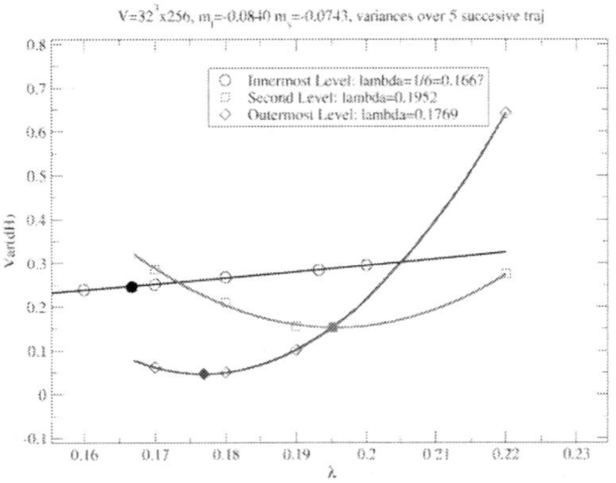

Figure 1. Tuning a 3 level MD integration scheme. We show the parameter λ from eq. (5) on the horizontal axis, and the variance of δH on the vertical axis. The three lines correspond to the innermost (circles), middle (squares) and outer (diamonds) integrators.

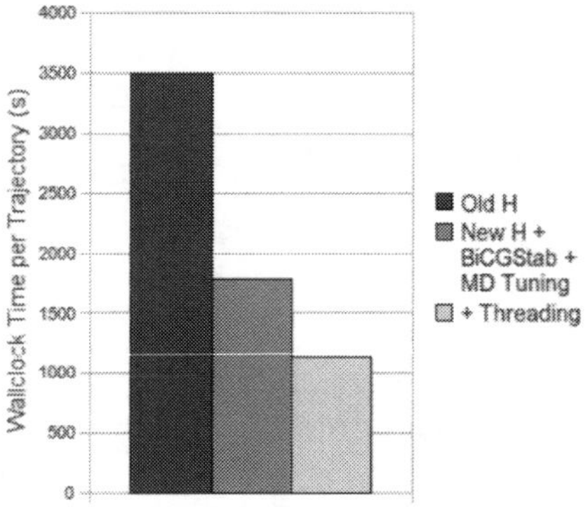

Figure 2. The effect of improvements

Table 1. Our target large volume data sets. We show the number of trajectories generated at each set of parameters at the time of writing.

	$m_\pi = 230MeV$	$m_\pi = 380MeV$
$V = 24^3 \times 128$	5380	12985
$V = 32^3 \times 256$	280	978

δH over an ensemble is advocated as the best way to tune an integration scheme. The optimal λ on each level can be determined in an experimental fashion, by varying it and measuring the δH on several trajectories to get an estimate of its variance. We have carried out the latter procedure on Kraken-XT at NICS and show our results in fig. 1, where filled symbols show our final interpolated minima. Our tuning was very rough: we carried out only 5 trajectories for each value of λ for each level of the integrator. Thus, in terms of trajectories the cost of the tuning is about 80 trajectories in total. While this is a small percentage of a full run which comprises several 1000 trajectories, it is a noticeable amount of computer time. It can be seen that we have reduced the variance of δH from ≈ 0.3 to $\approx$ 0.05, or by a factor of 6. This could in principle correspond to a factor of 6 increase in step-size, however in practice this is not attained as for such large outer steps, the integration would go unstable.

Apart from tuning our step-size, we have made several other improvements to our code. Whereas we used to solve eq. (4) with the Conjugate Gradients (CG) method, we have now switched over to a two step stabilized Bi-Conjugate Gradients (BiCGStab) procedure. Further, we have added threading to our linear operator M, our BLAS-like operations and most lattice site loops. The overall effect of these improvements is shown in fig. 2. Switching to BiCGStab and tuning have gained us a factor of 2 in wallclock time, with a small additional gain ($\approx 30\%$) from threading.

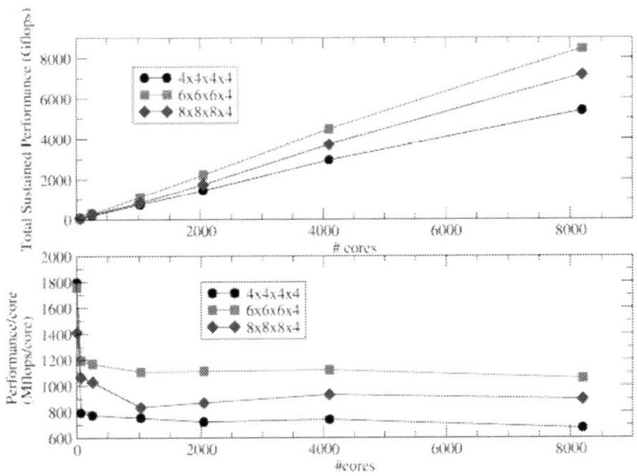

Figure 3. Weak scaling of the performance of our Conjugate Gradients code on Jaguar prior to quad core and CNL upgrades.

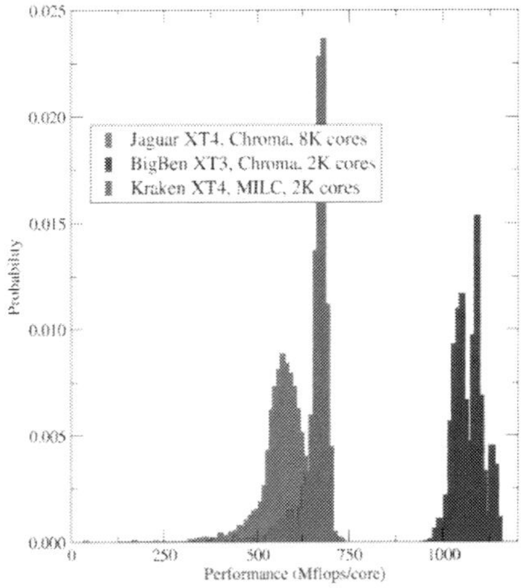

Figure 4. Performence Histograms from the Chroma Code on Jaguar XT4 (red) and BigBen (blue). We show also the performance of the MILC code on Kraken XT4 (green).

4. Cray experiences

Since 2007, we have used a variety of Cray XT platforms, starting with Jaguar at NCCS and BigBen at PSC. Both machines were then Cray XT3-s running the Catamount operating system. We have followed Jaguar through its various upgrades to dual, then quad cores; from XT3 to XT4 interconnects, and from Catamount to CNL operating systems. Currently we are running on Jaguar XT4, Jaguar XT5, Kraken XT5 at NICS and we are still active on BigBen at PSC through INCITE, NCCS discretionary and NSF Teragrid awards. Our primary code is the Chroma code [8] which was developed as part of the QCD Infrastructure project of SciDAC [9].

Our XT experiences are summarized in figs. 3 and 4. On BigBen we run jobs of size 2K cores, in single precision. On the XT3 platform we experienced good weak scaling as can be seen in fig 3 where we show the scaling of our code to 8192 cores on Jaguar after its dual core, but prior to its CNL upgrade. Indeed, the scaling is excellent. However, it can be seen that there is a distinct sweet spot in terms of local problem size (red line). Much larger local volumes fall out of cache (blue line) while much smaller problems are hit by surface to volume effects (black line).

In figure 4 we show normalized histograms of performances reported by the CG solvers on both BigBen, and on Jaguar XT4 post the quad core and CNL upgrade. While see some structure in the histogram for BigBen, essentially performances fall between 9.35-1.13 Gflops per core. On Jaguar XT4 the performances are substantially lower, around 450-700 Mflops/core. This may be due to the fact that on Jaguar we are working with a different global problem size, and that we are using 8192 cores, whereas on BigBen the partition size is 2048 cores. Nonetheless, the Jaguar histogram appears wider than the BigBen one with a shoulder extending toward lower performance numbers, with some as low or lower than 250Mflops/core. To ensure that this is not just a feature of our code or the large partitions, we show some data from running the MILC code on 2048 cores of Kraken XT4. This histogram is narrower than that of the Chroma code on Jaguar, but also exhibits a shoulder towards lower performances and is not immune from the occasional very low outlier.

5. Progress

We have made substantial progress in generating ensembles. Some initial results are presented in [10, 11]. We show our progress in table 1. We are concentrating on two volumes: $V = 24^3 \times 128$ and $V = 32^3 \times 256$ lattice sites respectively; and at two values of the quark mass which correspond to pion masses of $m_\pi = 230$ MeV and $m_\pi = 380$ MeV. The two volumes are required to ensure control over finite volume effects, and also because investigation of decays requires a finite size scaling analysis. With the INCITE 2008 resources, we have completed the run with $V = 24^3 \times 128$ and $m_\pi = 380$ MeV and with our 2008 NSF TeraGrid allocation we have begun the run at $V = 24^3 \times 128$ with a quark mass of $m_\pi = 230$ MeV which we are continuing on BigBen this year.

Using our INCITE 2009 and NSF Teragrid allocation, we have targeted the larger volume lattices using Kraken XT5 and Jaguar. We have not made as much progress as we intended for several reasons: Due to the large volumes used we found that single precision was no longer sufficient for the MD part of our simulations, so we moved to double precision. Secondly, we have scaled our running up to partitions of size 16384 cores and found a reduction in the performance of our code at this partition size. We have managed to compensate for some of this by the improvements in algorithm and software discussed earlier but progress remains slower than we'd like. We are currently in the process of investigating this issue.

6. Summary

We have discussed the program of anisotropic lattice gauge configuration generation and our experiences on the Cray XT platforms to date. In particular we have presented some algorithmic tuning results and touched on additional software improvements to our code. We discussed the performance of our code on the Cray XT architecture through its various incarnations.

7. Acknowledgements

We wish to thank Mike Clark for many helpful discussions on the tuning Molecular Dynamics, and Steve Gottlieb for supplying performance data from the MILC code. This research used resources of the National Center for Computational Sciences at Oak Ridge National Laboratory, which is supported by the Office of Science of the U.S. Department of Energy Under contract DE-AC05-00OR22725; and was supported in part by the National Science Foundation, through TeraGrid allocations at the Kraken Cray XT5 facility at the National Institute for Computational Sciences, The University of Tennessee and Oak Ridge National Laboratory and on the BigBen Cray XT3 facility at the Pittsburgh Supercomputing Center. Authored by Jefferson Science Associates, LLC under U.S. DOE Contract No. DE-AC05-06OR23177. The U.S. Government retains a non-exclusive, paid-up, irrevocable, world-wide license to publish or reproduce this manuscript for U.S. Government purposes.

8. References

[1] Edwards R G these proceedings
[2] Richards D G these proceedings
[3] Luu T C these proceedings
[4] Duane S, Kennedy A D, Pendleton B J and Roweth D 1987 *Phys. Lett.* **B195** 216–222
[5] Clark M A, Kennedy A D and Sroczynski Z 2005 *Nucl. Phys. Proc. Suppl.* **140** 835–837 (*Preprint* hep-lat/0409133)
[6] I P Omelyan I M M and Folk R 2003 *Comput. Phys. Commun.* **151** 272
[7] Clark M A, Kennedy A D and Silva P J 2008 Tuning HMC using Poisson brackets (*Preprint* 0810.1315)
[8] Edwards R G and Joo B (SciDAC) 2005 *Nucl. Phys. Proc. Suppl.* **140** 832 (*Preprint* hep-lat/0409003)
[9] Joo B 2007 *J. Phys. Conf. Ser.* **78** 012034
[10] Bulava J M *et al.* 2009 *Phys. Rev.* **D79** 034505 (*Preprint* 0901.0027)
[11] Lin H W *et al.* (Hadron Spectrum) 2009 *Phys. Rev.* **D79** 034502 (*Preprint* 0810.3588)

Journal of Physics: Conference Series **180** (2009) 012081 doi:10.1088/1742-6596/180/1/012081

Nuclear Forces and High-Performance Computing: The Perfect Match

T Luu[1], **A Walker-Loud**[2]

[1] N-division, Lawrence Livermore National Laboratory, Livermore, CA 94551 USA

[2] Physics Department, College of William and Mary, Williamsburg, VA 23187-8795 USA

E-mail: tluu@llnl.gov

Abstract. High-performance computing is now enabling the calculation of certain hadronic interaction parameters directly from Quantum Chromodynamics, the quantum field theory that governs the behavior of quarks and gluons and is ultimately responsible for the nuclear strong force. In this paper we briefly describe the state of the field and show how other aspects of hadronic interactions will be ascertained in the near future. We give estimates of computational requirements needed to obtain these goals, and outline a procedure for incorporating these results into the broader nuclear physics community.

1. Introduction

High-performance computing is fast becoming an integral part of nuclear physics, acting as a third pillar of scientific research standing next to theory and experiment. This integration has brought with it a new understanding of how certain aspects of hadronic (nuclear) forces comes from Quantum Chromodynamics (QCD), the field theory that describes the interactions of a set of nature's fundamental particles, quarks and gluons.

Lattice QCD (LQCD), the numerical implementation of QCD on a discretized space-time lattice, is now emerging as a powerful calculational tool of nuclear physics observables. Even at this early stage, the results are impacting the broader nuclear physics community. With ever increasing high-performance computing, longstanding nuclear physics questions, such as the nature of the nuclear spin-orbit coupling and the three-body interaction, will be tackled directly by LQCD. Further, the predictive capability of LQCD will allow scientists to understand nuclear phenomena in extreme environments not accessible by experiment like those found in the birth and death throes of a star or in the crusts of neutron stars.

Current computer limitations have restricted lattice QCD calculations to a regime where the light quark masses are equal (isospin limit) and unnaturally large. This equates to pion masses m_π typically ≥ 300 MeV.[1] Even though this does not represent the physical world, calculations in this regime have resulted in a deeper understanding of nuclear physics. We highlight a few examples below. These examples are by no means exhaustive.

[1] With the ever increasing high-performance computing capabilities, this scenario is rapidly changing. Within the last year lattice gauge configurations have been generated with $m_\pi \simeq 190$ MeV.

1.1. Nuclear physics is perturbative at large m_π

The study of hadron interactions with lattice QCD is different from scattering experiments in a conceptually important way; lattice calculations are limited to probing multi-particle interactions with a finite, Euclidean volume. Therefore, the particles can not be asymptotically separated from each other, nor can the in/out states be brought onto their mass shell. Consequently, one can not perform "scattering experiments" with lattice calculations, except at kinematic thresholds, since Euclidean Green's functions are not simply related to the physical S-matrix elements [1]. However, it has been shown that the energy levels of particles in a finite volume can be related to the infinite volume scattering parameters [2], and in particular for two particles to the scattering phase shift. This formalism was extended by Lüscher to interacting relativistic field theories [3, 4], with the degenerate, two-particle phase shift given by

$$\frac{p\cot\delta(p)}{m} = \frac{1}{\pi m L} S\left(\frac{pL}{2\pi}\right), \tag{1}$$

where $S(x)$ is a known function dependent upon the geometry of the spatial volume and the interaction momenta is defined by

$$\Delta E_{HH} \equiv E_{HH} - 2m = 2\sqrt{p^2 + m^2} - 2m, \tag{2}$$

and $\delta(p)$ is the infinite volume scattering phase shift[2] and m is the mass of either particle. For low momentum interactions, one has

$$\frac{p\cot\delta(p)}{m} = -\frac{1}{ma} + \frac{1}{2}(mr)\left(\frac{p}{m}\right)^2 + \cdots \tag{3}$$

where a is the scattering length, r is the effective range and the ellipses denote further suppressed terms in the effective-range expansion. It is important to note the range of validity of this formalism, which has come to be known as Lüscher's method. Equation (1) is only valid below the inelastic threshold. Further, it has been assumed the effective range of the interaction is much smaller than the spatial volume, $r \ll L$. However, there are no restrictions on the scattering length, which is important for nuclear physics, since at the physical pion mass, we know the $^1S_0$ and $^3S_1$ scattering lengths are unnaturally large, with $a(^1S_0) \simeq -23.7$ `fm` and $a(^3S_1) \simeq 5.4$ `fm`, while typical lattice calculations today have $L \simeq 2.5$–3 `fm`. For $a \ll L$, the interaction energy can be expressed as

$$\Delta E = \frac{4\pi a}{mL^3} - \frac{4a^2}{mL^4}\mathcal{I} + \frac{4a^3}{\pi^2 m L^5}\left[\mathcal{I}^2 - \mathcal{J}\right] + O(L^{-6}) \tag{4}$$

where $\mathcal{I}$ and $\mathcal{J}$ are known geometrical constants whose numerical values are given in tab. 1 for a square, periodic spatial volume. The relation in the converse limit, $a \gg L$, relevant for NN interactions near the physical point was derived in Ref. [10].

The first dynamical calculation of nucleon-nucleon scattering with lattice QCD was only recently carried out at three values of the pion mass with $m_\pi \geq 350$ `MeV` [11]. It was found that at these pion masses, the interaction was weakly repulsive in both the $^1S_0$ and $^3S_1$ channels, see Fig. 1. This is in contrast to the same system at the physical pion mass, where we know that, for example, the deuteron is (weakly) bound and has an unnaturally large scattering length, $a/r \gg 1$. *Nature's fine tuning of the NN system disappears for large m_π.*

[2] It has been asserted in the literature that lattice QCD can be used to directly calculate the nuclear potential [5, 6]. However, as pointed out in [7, 8, 9], this assertion is false since in an interacting field theory, like QCD, except for infinitely heavy particles, there is no unique definition of a potential as it is not an observable quantity. The only things which can be rigorously computed are n-point Green's functions, which can then be related to infinite volume scattering parameters.

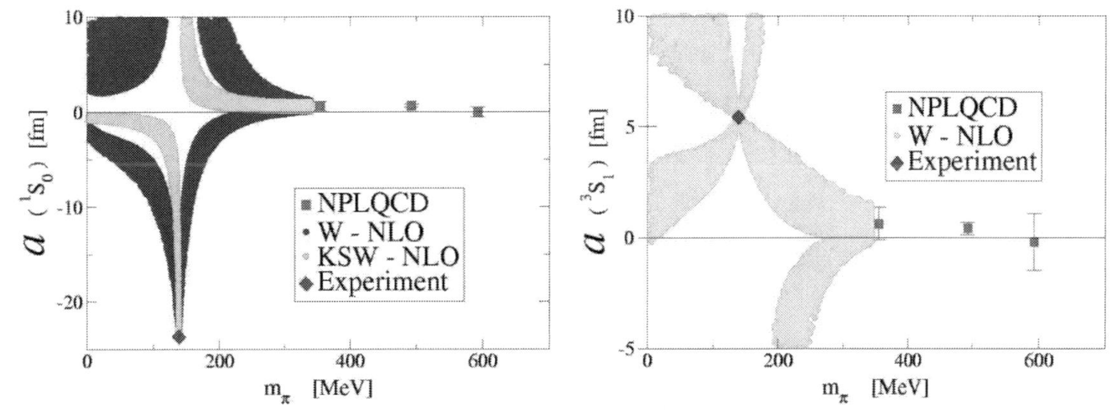

Figure 1. *NPLQCD calculation of NN scattering lengths [11].*

Beane et al.'s results suggest that standard perturbation theory applies to hadronic systems in the large m_π regime. We can parametrize the interaction between hadrons by partial waves using short-ranged repulsive forces that have interaction terms of 'natural size',

$$V_0(\vec{p}',\vec{p}) = \frac{4\pi a_0}{m}\left[1+\frac{a_0 r_0}{2}\left(\frac{p'^2+p^2}{2}\right)+\ldots\right] \qquad \text{(s-wave)} \tag{5}$$

$$V_1(\vec{p}',\vec{p}) = \frac{12\pi a_1}{m}\vec{p}'\cdot\vec{p}\left[1+\frac{a_1 r_1}{2}\left(\frac{p'^2+p^2}{2}\right)+\ldots\right] \qquad \text{(p-wave)} \tag{6}$$

$$V_2(\vec{p}',\vec{p}) = \frac{10\pi a_2}{m}\left(3(\vec{p}'\cdot\vec{p})^2-p'^2\,p^2\right)\left[1+\frac{a_2 r_2}{2}\left(\frac{p'^2+p^2}{2}\right)+\ldots\right] \qquad \text{(d-wave)}, \tag{7}$$

and so on. The parameters a_l and r_l constitute part of the effective range parameters from scattering theory. Depending on the partial wave, these parameters have different dimensions. Perturbation theory remains valid as long as $a_0/L \ll 1$ and $r_0/L \ll 1$, as well as $a_1/L^3 \ll 1$ and $r_1 L \ll 1$, and so forth.

In most of the sections below we assume that perturbation theory holds for hadronic systems, since to date LQCD calculations are still in a regime where m_π is large. In sect. 4 we briefly outline the methods needed to go beyond perturbation theory.

1.2. ***Weinberg's leading order prediction works very well for mesons***

The simplest system of interacting hadrons, both theoretically and numerically is that of two pions in the isospin $I = 2$ channel.[3] The simplicity of this system follows from the fact that the pions are the pseudo-Goldstone bosons associated with the spontaneous breaking of the approximate chiral symmetry of QCD. At leading order (LO) in chiral perturbation theory (χ-PT), the $\pi\pi$ interactions are uniquely predicted, first determined by Weinberg in 1966 [12], with the scattering lengths given by

$$m_\pi a_{\pi\pi}^{I=0} = -\frac{7m_\pi^2}{16\pi f_\pi^2}, \qquad\qquad m_\pi a_{\pi\pi}^{I=2} = \frac{m_\pi^2}{8\pi f_\pi^2}. \tag{8}$$

Subleading orders in the chiral expansion give rise to perturbatively small corrections containing both non-analytic contributions as well as analytic terms whose coefficients are not determined

[3] The $I = 0$ channel is complicated by the presence of the scalar resonance, σ. Additionally, numerical computations of scalar scattering channels are complicated by the necessity of including the numerically-expensive disconnected quark propagators.

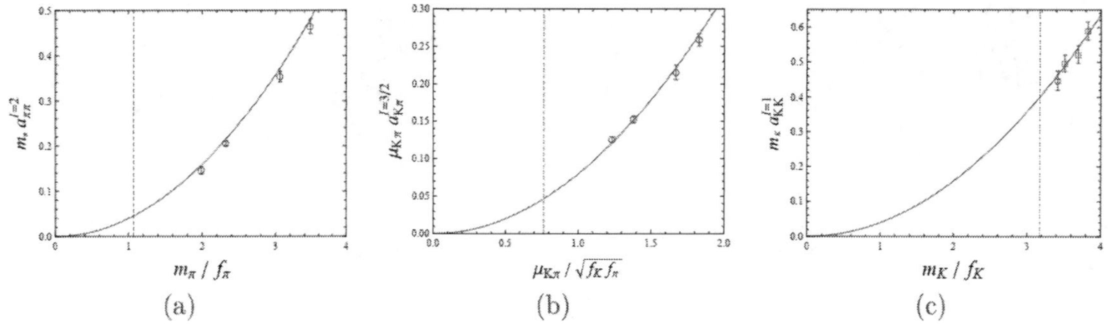

Figure 2. *NPLQCD calculated values of the $I = 2$ $\pi\pi$ (a), $I = 3/2$ $K\pi$ (b) and $I = 1$ KK (c) scattering lengths. In all cases, the numerical-data are plotted vs. the lattice-physcal masses and decay constants, as calculated for each ensemble. The dashed lines represent the physical point and the curves are Weinberg's LO prediction. The reduced mass is $\mu_{K\pi} = m_K m_\pi/(m_K + m_\pi)$. The LO predictions for the $K\pi$ and KK system at the physical pion mass are $m_K a_{KK}^{I=1} = \frac{m_K^2}{8\pi f_K^2}$ and $\mu_{K\pi} a_{K\pi}^{I=3/2} = \frac{\mu_{K\pi}^2}{4\pi f_\pi f_K}$.*

Table 1. Numerical values of geometrical sums appearing in this document.

$\mathcal{I}$	$\mathcal{J}$	$\mathcal{K}$	$\mathcal{L}$	$\mathcal{M}$
-8.9136	16.5323	8.4019	-6.3748	18.3

by chiral symmetry alone [13]. These coefficients must be determined either with a comparison to experiment or to numerical results from lattice QCD.

In 2005, the NPLQCD Collaboration performed the first lattice calculation of $I = 2$ $\pi\pi$ scattering containing both light and strange dynamical quarks [14]. Because the low energy pion interactions are strongly constrained by chiral symmetry, the lattice calculation was performed with a mixed action using domain-wall valence quarks [15, 16, 17] (which retain a lattice chiral symmetry), with the `asqtad` improved [18, 19] staggered MILC gauge ensembles [20]. In 2007, the calculation was updated with a significant increase in statistics, as well as the use of the relevant mixed action χ-PT expression to perform the extrapolation to the physical point [21, 22, 23, 24, 25, 26, 27]. The result was a 1% prediction (including systematic errors) of the $I = 2$ $\pi\pi$ scattering length, expressed at the value of the charged pion mass [28],

$$m_\pi a_{\pi\pi}^{I=2} = 0.04330 \pm 0.00042\,, \tag{9}$$

a value competitive with the best theoretical predictions [29, 30]. *The precision of these LQCD measurements surpasses that of experiment.*

With the same mixed action, the NPLQCD Collaboration has similarly computed the values of the $I = 3/2$ $K\pi$ [31] and $I = 1$ KK [32] scattering lengths as well as f_K/f_π [33]. The high precision attained in the lattice calculation of these quantities, combined with the $I = 2$ $\pi\pi$ results, can be used to provide stringent checks of low energy dynamics predicted by $SU(2)$ and $SU(3)$ χ-PT.

One interesting phenomenon which has been observed is that the scattering lengths, for all values of the light and strange quark masses used, are remarkably well described by the LO predictions of Weinberg, see Fig.2, for which there is presently no theoretical understanding.

1.3. *Lattice QCD can probe systems that are not accessible experimently*

The most powerful aspect of LQCD is its predictive capability. With sufficient computer resources, systems not accessible by experiment can be calculated on a lattice. In most cases, these systems have impact in many areas of nuclear physics, such as nuclear astrophysics. A recent example of this predictive capability came from LQCD simulations of multiple pions and kaons in a box. Because of chiral symmetry, the perturbative nature of these systems is well established and general finite-volume formulas can be derived that relate the ground state interacting energies of these systems (measured by LQCD) to interaction parameters as a function of the number of mesons n [34, 35],

$$E_{A_1}(n) = \frac{4\pi a_0}{mL^3}\binom{n}{2}\left\{1 - \frac{a_0\mathcal{I}}{\pi L} + \left(\frac{a_0}{\pi L}\right)^2 [\mathcal{I}^2 + (2n-5)\mathcal{J}] - \left(\frac{a_0}{\pi L}\right)^3 [\mathcal{I}^3 + (2n-7)\mathcal{I}\mathcal{J} + (5n^2 - 41n + 63)\mathcal{K}]\right\} + \binom{n}{2}\frac{8\pi^2 a_0^2 r_0}{mL^6} + \binom{n}{3}\frac{\eta_3}{L^6} + O(L^{-7}) \,. \qquad (10)$$

At order $1/\mathrm{L}^6$, a dressed pure three-body interaction is needed to enforce cutoff independence. Detmold et al. used LQCD-measured interaction energies of different numbers of pions [36, 37] to extract this force, finding that it is repulsive with a size consistent with naive dimensional analysis. A subsequent work determined the equivalent three body interaction for charged kaons [38] finding an interaction strength consistent with zero. *For the first time, LQCD is allowing definitive statements of hadronic three-body forces to be made.*

1.4. *Three Baryon States with lattice QCD*

Recently a high-satistics exploration of a three-baryon signal with the same quantum numbers of the $\Xi^0\Xi^0 n$ system was undertaken, finding on a single ensemble with $m_\pi \simeq 390$ MeV, $E_{\Xi\Xi N} = 3877.9 \pm 6.9 \pm 9.2 \pm 3.3$ MeV, where the first uncertainty is statistical and the second and third uncertainties are estimates of fitting systematics [39]. Despite the remarkable level of precision in this calculation, no extraction of interaction parameters was performed. Lattice QCD calculations of multi-baryon systems are emblematic of the stochastic nature of these simulations, suffering poorer signal-to-noise ratios compared to their mesonic counterparts. Extracting a pure three-baryon interaction will require sufficiently more computational resources than available today, as well as improved algorithms. In Refs. [39, 40], the NPLQCD Collaboration undertook a significant, high-statistics scaling study of the extraction of (multi) baryon energy levels from lattice QCD calculations. This provides the basis for our resource estimates required for these types of calculations in sect. 5.

2. Going beyond A_1 (s-wave) cubic symmetry

Lattice calculations typically involve simulations inside a periodic box. System wavefunctions with angular momentum for integer and odd-half states are therefore characterized by the irreducible representations (irreps) of the cubic (octahedral) group O and its cover, O_h, respectively. For zero center of mass (CM) systems, integer (spatial) states fall into one of five representations: A_1, A_2, T_1, T_2, and E. Odd-half states belong to either G_1, G_2, or H representations. There is extensive literature on how these states are comprised of infinite-volume, continuum angular momentum states of SO(3) (*e.g.* [41]). For example, ground-state A_1 systems represent s-wave systems in the infinite-volume limit, whereas ground-state T_1 systems represent p-wave systems in this limit.

To date all calculations have been on systems with spatial A_1 symmetry. Projection onto this symmetry is simplest, as it represents the most 'symmetric' and least time-consuming operation.

With future high-performance computing, and the need to look beyond two-baryon systems, other cubic symmetries will be probed, giving new information about the interaction between hadrons.

For example, with two identical baryons, projection onto the negative parity T_1 spatial symmetry will force ground-states to be perturbitively connected to the non-interacting state with kinetic energy $4\pi^2/mL^2$, or the first cubic shell. The simplest system to consider consists of a neutron and proton with spatial T_1 and spin A_1 coupled to total angular momentum T_1, *i.e.* $|(T_{1u}A_1)T_{1u}>$. The energy of this system to order $1/\mathrm{L}^5$ is

$$E_{(T_{1u}A_1)T_{1u}} = \frac{4\pi^2}{mL^2} + \frac{96\pi^3 a_1}{mL^5} + O(L^{-7}) \ . \tag{11}$$

Beyond the kinetic energy term, $4\pi^2/mL^2$, the leading order term of this system is only sensitive to the "scattering volume", a_1. In the infinite volume, continuum limit this state maps to the $^1P_1$ scattering state.

At $1/\mathrm{L}^5$ we can also extract information about non-central forces by looking at, for example, two neutrons with spatial T_1 symmetry coupled to spin T_1 symmetry. In the infinite volume limit, this system can belong to three possible scattering states: $^3P_0$, $^3P_1$, and $^3P_2$. In a box, these states are split up into four possibilities, $|(T_{1u}T_1)A_{1u}>$, $|(T_{1u}T_1)T_{1u}>$, $|(T_{1u}T_1)T_{2u}>$, and $|(T_{1u}T_1)E_u>$. If life consisted of only central forces these states would be degenerate. Non-central forces, on the other hand, will break this degeneracy. The strength and order of the broken degeneracies give us information about the nature of the non-central force. For example, if we identify a source of the non-central force as coming from virtual one-pion exchange (OPE),

$$V_\pi(\vec{q}) = -\left(\frac{g_A}{\sqrt{2}f_\pi}\right)^2 (\tau_1\cdot\tau_2)\frac{(\vec{\sigma}_1\cdot\vec{q})\,(\vec{\sigma}_2\cdot\vec{q})}{q^2+m_\pi^2} \ , \tag{12}$$

where $\vec{q}$ is the momentum transfer carried by the virtual pion, g_A is the axial-vector coupling, and f_π is the pion decay constant, we find that the degeneracies are broken in the following manner:

$$E_{(T_{1u}T_1)A_{1u}} = \frac{4\pi^2}{mL^2} + \frac{96\pi^3 a_1}{mL^5} + \frac{8\pi^2 g_A^2}{3f_\pi^2 m_\pi^2 L^5} - \frac{80\pi^2 g_A^2}{3f_\pi^2 m_\pi^2 L^5} + O\left(L^{-7}\right) \tag{13}$$

$$E_{(T_{1u}T_1)T_{1u}} = \frac{4\pi^2}{mL^2} + \frac{96\pi^3 a_1}{mL^5} + \frac{8\pi^2 g_A^2}{3f_\pi^2 m_\pi^2 L^5} + \frac{40\pi^2 g_A^2}{3f_\pi^2 m_\pi^2 L^5} + O\left(L^{-7}\right) \tag{14}$$

$$E_{(T_{1u}T_1)T_{2u}} = \frac{4\pi^2}{mL^2} + \frac{96\pi^3 a_1}{mL^5} + \frac{8\pi^2 g_A^2}{3f_\pi^2 m_\pi^2 L^5} + O\left(L^{-7}\right) \tag{15}$$

$$E_{(T_{1u}T_1)E_u} = \frac{4\pi^2}{mL^2} + \frac{96\pi^3 a_1}{mL^5} + \frac{8\pi^2 g_A^2}{3f_\pi^2 m_\pi^2 L^5} + O\left(L^{-7}\right) \ , \tag{16}$$

where the terms in red (blue) come from the central (non-central) part of OPE. At this order, the scattering volume can be re-defined to absorb the central part of OPE, $a_1 \rightarrow a_1 + \frac{g_A^2 m}{36\pi f_\pi^2 m_\pi^2}$ [4]. But the non-central parts clearly break the degeneracy for the $|(T_{1u}T_1)A_{1u}>$ and $|(T_{1u}T_1)T_{1u}>$ states. LQCD calculations of these energies will therefore tell us of the existence and strength of the tensor interaction.

Interaction terms of higher partial waves can be probed with appropriate projections onto other spatial symmetry groups. For example, the infinite volume $^1D_2$ scattering state will have

[4] The central part of OPE also contributes at order $1/\mathrm{L}^5$ in eq.(11) but can be absorbed into the scattering volume with the following re-definition: $a_1 \rightarrow a_1 + \frac{g_A^2 m}{4\pi f_\pi^2 m_\pi^2}$.

the following ground state energy in a box,

$$E_{(E_g A_1)E_g} = \frac{4\pi^2}{mL^2} + \frac{240\pi^5 a_2}{mL^7} + O(L^{-8}) . \quad (17)$$

In general higher partial wave interaction terms will be suppressed by larger powers of L, and because of the finite dimensionality of the cubic group, operator mixing between different partial waves will occur.

Similar results can be obtained in three-baryon systems. In [42] perturbative results were given for a three-neutron system. Here we concentrate on the "perturbative" triton system whose energy is

$$E_{(A_{1g}G_1)G_{1g}} = 0 + \frac{12\pi}{mL^3}\left(\frac{a_0^s + a_0^t}{2}\right) - \frac{12}{mL^4}\left(\frac{(a_0^s)^2 + (a_0^t)^2}{2}\right)\mathcal{I}$$
$$+ \frac{12}{\pi mL^5}\left(\frac{(a_0^s)^3 + (a_0^t)^3}{2}\right)\mathcal{I}^2 + \frac{12}{\pi mL^5}\left(\frac{a_0^s + a_0^t}{2}\right)\left[(a_0^s)^2 + (a_0^t)^2\right]\mathcal{J} + O(L^{-6}) . \quad (18)$$

Here a_0^s and a_0^t are the scattering lengths for the spin-singlet and spin-triplet channels, respectively. The negative parity triton has two states that are perturbatively connected to the first cubic shell,

$$E_{(T_{1u}G_1)G_{1u}} = \frac{4\pi^2}{mL^2} + \frac{12\pi}{mL^3}\left(\frac{a_0^s + a_0^t}{2}\right) - \frac{12}{mL^4}\left(\frac{(a_0^s)^2 + (a_0^t)^2}{2}\right)\left[\mathcal{L} + \frac{1}{4}\right] + \frac{9a_0^s a_0^t}{mL^4}$$
$$+ \frac{12}{\pi mL^5}\left(\frac{(a_0^s)^3 + (a_0^t)^3}{2}\right)\left[\mathcal{L}^2 + \frac{1}{2}(\mathcal{L} - \mathcal{M}) - \frac{1}{16}\right] - \frac{(a_0^s)^2 a_0^t + (a_0^t)^2 a_0^s}{\pi mL^5}\left[9\mathcal{L} + 3\mathcal{M} - \frac{15}{8}\right]$$
$$+ \frac{6\pi^3}{mL^5}\left(\frac{(a_0^s)^2 r_0^s + (a_0^t)^2 r_0^t}{2}\right) + \frac{108\pi^3}{mL^5}\left(\frac{a_1^s + a_1^t}{2}\right) + O(L^{-6}) , \quad (19)$$

and

$$E_{(T_{1u}G_1)H_{1u}} = \frac{4\pi^2}{mL^2} + \frac{144\pi^3}{mL^5}\left(\frac{a_1^s + a_1^t}{2}\right) + O(L^{-6}) . \quad (20)$$

r_0^s (a_1^s) and r_0^t (a_1^t) are the effective ranges (scattering volumes) in the spin-singlet and spin-triplet channels, respectively. In the infinite volume limit these states correspond to $|J^\pi> = |\frac{1}{2}^-> $ and $|\frac{3}{2}^->$.

3. LQCD offers an ideal place for performing nuclear physics "experiments"

Though not utilized to date, LQCD calculations of nuclear systems can be performed with non-zero CM motion. Because of the box boundary conditions, one can take advantage of this option to further probe aspects of hadronic interactions. Angular momentum states are not part of the full cubic group, however, as only cubic operations that preserve the CM momentum vector are allowed. As such, the classification of angular momentum states falls into subgroups, or little groups, of the cubic group [43, 44]. A consequence of this is that the mapping of infinite volume angular momentum states to point-group irreps can be different for non-zero CM. For example, ground state p-wave systems can be part of the A_1 symmetry for states with non-zero CM, as opposed to T_1 symmetry for zero CM.

For two baryons, the simplest way to enforce non-zero CM motion is to give one of the baryons the smallest momentum allowed by the box, $2\pi/L$. Angular momentum states must fall into the irreps of the C_{4v} group. For example, two neutrons with spins oppositely aligned have their ground state energy as

$$E_{A_1} = \frac{2\pi^2}{mL^2} + \frac{8\pi a_0}{mL^3} - \frac{8a_0^2}{mL^4}\mathcal{L} + \frac{8a_0^3}{\pi mL^5}\left[\mathcal{L}^2 - \mathcal{M}\right] + \frac{4\pi^3 a_0^2 r_0}{mL^5} + O(L^{-6}) . \quad (21)$$

Compare this to the same energy with zero CM motion given in eq.(4). Because eq.(21) is not perturbatively connected to zero, but to $2\pi^2/mL^2$, expressions involving the effective range are not derivatively suppressed and show up at $1/L^5$.

On the other hand, if the two neutrons have their spins aligned their ground state energy is

$$E_{A_1} = \frac{2\pi^2}{mL^2} + \frac{24\pi^3 a_1}{mL^5} + O(L^{-6}) \ . \tag{22}$$

In this case the energy shift is sensitive to a_1. This system also belongs to the A_1 irrep of C_{4v}, even though it would correspond to a p-wave system in the infinite volume limit. Further, it is perturbatively connected to $2\pi^2/mL^2$, as is eq.(21). One does not have to go to a higher cubic shell to probe the scattering volume, as was done in the previous section; one need only project onto the correct spin state.

In addition to working with non-zero CM motion, calculations can also be done on lattices with asymmetric spatial volumes. Angular momentum states, in this case, will also be restricted to subgroups of O in a similar (but not exact) manner as the examples above. This gives another handle on probing various aspects of the interaction between hadrons. In this sense, LQCD offers a unique and powerful "experimental" platform for learning the nature of hadronic forces.

4. As $m_\pi \to 140$ MeV, life *will* become non-perturbative

Self-bound, non-perturbative states of multi-nucleons exist in nature (i.e. nuclei). As lattice calculations approach the physical pion mass these states are expected to form and the perturbative expressions of the previous sections will become inadequate. Fortunately for the two nucleon case eq. (1) still holds. In principle, EFT prescriptions for constructing potentials from these results can be used, giving (schematically)

$$V_0(\vec{p}',\vec{p}) = f_0^R(p'^2,p^2) \qquad \text{(s-wave)} \tag{23}$$

$$V_1(\vec{p}',\vec{p}) = (\vec{p}'\cdot\vec{p})\, f_1^R(p'^2,p^2) \qquad \text{(p-wave)} \tag{24}$$

$$V_2(\vec{p}',\vec{p}) = \left(3(\vec{p}'\cdot\vec{p})^2 - p'^2\, p^2\right) f_2^R(p'^2,p^2) \quad \text{(d-wave)}\ , \tag{25}$$

and so on. Here f_i^R, being functions of only the magnitude of the initial and final momenta, are chosen to reproduce, among other things, the effective range parameters. The superscript R refers to the particular regularization scheme employed.

For three-baryon systems there is no analogue of eq. (1). Recently there has been work on understanding the three-fermion system at the unitary limit ($a_0 \to \infty$) within a box [45], as well as three-bosons [46], but a general non-perturbative method utilizing box-boundary conditions that includes other interaction terms has not been developed. Here we outline a method that solves the secular equation,

$$H_{eff}(E)P|\Psi> = \left\{H + V\frac{1}{E-QH}QV\right\} P|\Psi> = E\ P|\Psi>\ , \tag{26}$$

using Faddeev techniques. The term V represents the interaction between baryons (*e.g.* Eqs.(23)-(25)) which can include a pure three-baryon term. The projection operators Q and P are chosen such that $Q + P = 1$ and they commute with the kinetic energy operator, $[Q,T] = [P,T] = 0$. The state $|\Psi>$ represents the (as yet unknown) eigenstate of the interacting three-body system and E represents the exact (to be determined) eigenvalue. The effective Hamiltonian is only applicable in the 'model' space defined by P and is energy-dependent. The space defined by P usually represents the lowest modes of the system, and in this case, will be defined by non-interacting, fully anti-symmetric states of definite cubic symmetry that reside in the first few cubic shells of the box.

As described in [47], part of eq.(26) can be recast into integral form by using Faddeev decompositions,

$$|\psi_i> = \mathcal{T}^Q|\Omega_i> + \mathcal{T}^Q G_0(E)\Pi|\psi_i> \; , \tag{27}$$

where $\Pi = 1 + P_{12}P_{13} + P_{12}P_{23}$ is a sum of permutation operators, $G_0(E) = 1/(E-T)$ is the non-interacting three-body propagator, and

$$\mathcal{T}^Q = V + VG_0(E)Q\mathcal{T}^Q \tag{28}$$

is the 'excluded-space' T-matrix. The state $|\Omega_i>$ resides in the model space, *i.e.* $|\Omega_i>\in P$, and $|\psi_i>$ must be numerically determined. There are standard techniques for solving the integral equation in eq.(27).

Given $|\psi_i>$, model space matrix elements of H_{eff} are related by

$$<\Omega_j|H_{eff}(E)|\Omega_i> = <\Omega_j|T|\Omega_i> + <\Omega_j|\psi_i> \; . \tag{29}$$

As the energy E is not known *a priori* (it is what we are trying to determine), in practice a starting guess for E is used to construct these matrix elements, which in turn are diagonalized to obtain a new E. This value is used to contruct new matrix elements, which are then diagonalized, and so on. The process repeats itself until the input energy is the same as the energy after diagonalization. In [42] we describe how to construct the non-interacting, anti-symmetric states $|\Omega_i>$. This procedure has been applied to self-bound nucleons in the infinite volume limit [47] and to fermions bound in an optical trap [48]

Having a non-perturbative, three-baryon formalism with box-boundary conditions is an essential step for extracting any pure three-baryon force. In a similar fashion described in sect. 1.3, information about the 4-, 5-, and higher-baryon systems can be used to also extract the three-baryon force, but due to complexities inherent to baryon systems this procedure will be tedious and require enormous computing resources. On the other hand, one can investigate the three-baryon system (including its excited states) using non-zero CM motion, or within asymetric spatial volumes, in an analagous manner described in the previous sections. These systems will also give insight into the three-baryon force.

5. What does it take to get there?

The work described in sects. 1.1–1.4 resulted from teraflop-sized calculations using lattice gauge ensembles with large pion masses ($m_\pi > 290$ MeV), relatively course lattice spacings ($a \sim 0.1$ fm), and limited spatial volumes ($L = 3.5 \sim 4$ fm). Ensembles at the physical pion mass, using fine lattice spacings $a < 0.1$ fm, and within large volumes $L > 7$ fm will require full petaflop resources or more. Further, measurements performed on these ensembles that are relevant to nuclear physics will require commensurate resources. Extraction of interaction parameters within two-baryon systems at a precision that competes with experiment will also require dedicated petaflop resources (at least). In the meson sector requirements are less restrictive. The superior signal-to-noise behavior in this sector means that interaction parameters of higher-partial waves should be accessible in the very near future, requiring 10X to 100X teraflop resources. Probing the three-nucleon force should coincide with measuring the triton with adequate precision–something that is expected to take 10X to 100X petaflop resources. Alpha particle precision measurements will take full exascale resources, and only at this scale can investigations away from the isospin limit be made. An understanding of isospin breaking forces, as well as non-leptonic parity violating nuclear forces, will require exascale resources and beyond.

These rough estimates are made using current experience and knowledge about lattice calculations. It is highly probable that disruptive technologies and algorithms will accelerate the timescale for achieving these goals.

6. Conclusion

High-performance computing is bringing nuclear physics into a new era where predictive capability will become the norm. In this process, answers to longstanding questions in fundamental physics will be attainable, and the connection between different subfields of nuclear physics will become stronger. Nowhere is this more realizable than in the area of nuclear forces, where high-performance computing is now allowing aspects of hadronic interactions to be ascertained from true *ab intio*, first principles LQCD calculations. The impact of these calculations will be far reaching, affecting the fields of nuclear structure and nuclear astrophysics, for example.

In these proceedings we described how future LQCD calculations will be used to extract hadronic interaction parameters and outlined a method for analyzing non-perturbative three-body systems that utilize box boundary conditions. We give estimates on computational resources needed to obtain important milestones. As we argued in these proceedings, LQCD promises to be calculational tool for nuclear physics of unprecedented power and scope. High-performance computing will make this promise come true.

Acknowledgments

We thank members of NPLQCD for their inciteful critiques of this manuscript. Luu thanks M. Hruška for useful discussions. Walker-Loud was supported under OJI Grant DE-FG02-07ER41527. Luu performed this work under the auspices of the U.S. Department of Energy by Lawrence Livermore National Laboratory under Contract DE-AC52-07NA27344.

References

[1] Maiani L and Testa M 1990 *Phys. Lett.* **B245** 585–590
[2] Huang K and Yang C N 1957 *Phys. Rev.* **105** 767–775
[3] Luscher M 1986 *Commun. Math. Phys.* **105** 153–188
[4] Luscher M 1991 *Nucl. Phys.* **B354** 531–578
[5] Ishii N, Aoki S and Hatsuda T 2007 *Phys. Rev. Lett.* **99** 022001 (*Preprint* nucl-th/0611096)
[6] Nemura H, Ishii N, Aoki S and Hatsuda T 2009 *Phys. Lett.* **B673** 136–141 (*Preprint* 0806.1094)
[7] Detmold W, Orginos K and Savage M J 2007 *Phys. Rev.* **D76** 114503 (*Preprint* hep-lat/0703009)
[8] Beane S R, Orginos K and Savage M J 2008 *Int. J. Mod. Phys.* **E17** 1157–1218 (*Preprint* 0805.4629)
[9] Beane S R 2008 (*Preprint* 0812.1236)
[10] Beane S R, Bedaque P F, Parreno A and Savage M J 2004 *Phys. Lett.* **B585** 106–114 (*Preprint* hep-lat/0312004)
[11] Beane S R, Bedaque P F, Orginos K and Savage M J 2006 *Phys. Rev. Lett.* **97** 012001 (*Preprint* hep-lat/0602010)
[12] Weinberg S 1966 *Phys. Rev. Lett.* **17** 616–621
[13] Gasser J and Leutwyler H 1984 *Ann. Phys.* **158** 142
[14] Beane S R, Bedaque P F, Orginos K and Savage M J (NPLQCD) 2006 *Phys. Rev.* **D73** 054503 (*Preprint* hep-lat/0506013)
[15] Kaplan D B 1992 *Phys. Lett.* **B288** 342–347 (*Preprint* hep-lat/9206013)
[16] Shamir Y 1993 *Nucl. Phys.* **B406** 90–106 (*Preprint* hep-lat/9303005)
[17] Furman V and Shamir Y 1995 *Nucl. Phys.* **B439** 54–78 (*Preprint* hep-lat/9405004)
[18] Orginos K and Toussaint D (MILC) 1999 *Phys. Rev.* **D59** 014501 (*Preprint* hep-lat/9805009)
[19] Orginos K, Toussaint D and Sugar R L (MILC) 1999 *Phys. Rev.* **D60** 054503 (*Preprint* hep-lat/9903032)
[20] Bernard C W *et al.* 2001 *Phys. Rev.* **D64** 054506 (*Preprint* hep-lat/0104002)
[21] Bar O, Bernard C, Rupak G and Shoresh N 2005 *Phys. Rev.* **D72** 054502 (*Preprint* hep-lat/0503009)
[22] Chen J W, O'Connell D, Van de Water R S and Walker-Loud A 2006 *Phys. Rev.* **D73** 074510 (*Preprint* hep-lat/0510024)
[23] Bedaque P F, Sato I and Walker-Loud A 2006 *Phys. Rev.* **D73** 074501 (*Preprint* hep-lat/0601033)
[24] Chen J W, O'Connell D and Walker-Loud A 2007 *Phys. Rev.* **D75** 054501 (*Preprint* hep-lat/0611003)
[25] Orginos K and Walker-Loud A 2008 *Phys. Rev.* **D77** 094505 (*Preprint* 0705.0572)
[26] Chen J W, O'Connell D and Walker-Loud A 2009 *JHEP* **04** 090 (*Preprint* 0706.0035)
[27] Chen J W, Golterman M, O'Connell D and Walker-Loud A 2009 (*Preprint* 0905.2566)
[28] Beane S R *et al.* 2008 *Phys. Rev.* **D77** 014505 (*Preprint* 0706.3026)

[29] Colangelo G, Gasser J and Leutwyler H 2001 *Nucl. Phys.* **B603** 125–179 (*Preprint* hep-ph/0103088)
[30] Leutwyler H 2008 (*Preprint* 0808.2825)
[31] Beane S R *et al.* 2006 *Phys. Rev.* **D74** 114503 (*Preprint* hep-lat/0607036)
[32] Beane S R *et al.* (NPLQCD) 2008 *Phys. Rev.* **D77** 094507 (*Preprint* 0709.1169)
[33] Beane S R, Bedaque P F, Orginos K and Savage M J 2007 *Phys. Rev.* **D75** 094501 (*Preprint* hep-lat/0606023)
[34] Beane S R, Detmold W and Savage M J 2007 *Phys. Rev.* **D76** 074507 (*Preprint* 0707.1670)
[35] Detmold W and Savage M J 2008 *Phys. Rev.* **D77** 057502 (*Preprint* 0801.0763)
[36] Beane S R *et al.* 2008 *Phys. Rev. Lett.* **100** 082004 (*Preprint* 0710.1827)
[37] Detmold W *et al.* 2008 *Phys. Rev.* **D78** 014507 (*Preprint* 0803.2728)
[38] Detmold W, Orginos K, Savage M J and Walker-Loud A 2008 *Phys. Rev.* **D78** 054514 (*Preprint* 0807.1856)
[39] Beane S R *et al.* 2009 (*Preprint* 0905.0466)
[40] Beane S R *et al.* 2009 (*Preprint* 0903.2990)
[41] Johnson R C 1982 *Phys. Lett.* **B114** 147
[42] Luu T 2008 (*Preprint* 0810.2331)
[43] Moore D C and Fleming G T 2006 *Phys. Rev.* **D73** 014504 (*Preprint* hep-lat/0507018)
[44] Gottlieb S A and Rummukainen K 1996 *Nucl. Phys. Proc. Suppl.* **47** 819–822 (*Preprint* hep-lat/9509088)
[45] Pricoupenko L and Castin Y 2007 *Journ. Phys. A* **40** 12863 (*Preprint* 0705.1502)
[46] Kreuzer S and Hammer H W 2009 *Phys. Lett.* **B673** 260–263 (*Preprint* 0811.0159)
[47] Luu T C, Bogner S, Haxton W C and Navratil P 2004 *Phys. Rev.* **C70** 014316 (*Preprint* nucl-th/0404028)
[48] Luu T and Schwenk A 2007 *Phys. Rev. Lett.* **98** 103202 (*Preprint* cond-mat/0606069)

Journal of Physics: Conference Series **180** (2009) 012082 doi:10.1088/1742-6596/180/1/012082

Towards the universal nuclear energy density functional

M Stoitsov[1−3], J Moré[4], W Nazarewicz[1,2,5], J C Pei[1,2], J Sarich[4], N Schunck[1,2], A Staszczak[1,2,3] and S Wild[4]

[1] Department of Physics and Astronomy, University of Tennessee Knoxville, TN 37996, USA
[2] Oak Ridge National Laboratory, P.O. Box 2008, Oak Ridge, TN 37831, USA
[3] INRNE, Bulgarian Academy of Sciences, Sofia-1784, Bulgaria
[4] Argonne National Laboratory, Argonne, IL 60439, USA
[5] Institute of Theoretical Physics, Warsaw University, ul.Hoża 69, PL-00681 Warsaw, Poland
[6] Institute of Physics, Maria Curie-Skłodowska University, pl. M. Curie-Skłodowskiej 1, 20-031 Lublin, Poland

E-mail: `stoitsovmv@ornl.gov`

Abstract. The UNEDF SciDAC project to develop and optimize the energy density functional for atomic nuclei using state-of-the art computational infrastructure is briefly described. The ultimate goal is to replace current phenomenological models of the nucleus with a well-founded microscopic theory with minimal uncertainties, capable of describing nuclear data and extrapolating to unknown regions.

1. Introduction

An understanding of the atomic nucleus is crucial for a complete nuclear theory for element formation, for properties of stars, and for various societal applications. The "Universal Nuclear Energy Density Functional" (UNEDF) SciDAC-2 collaboration [1] carries out a comprehensive study of all nuclei based on the most reliable theoretical approaches, the most advanced algorithms, and extensive computational resources, with a view of scaling to the petaflop platforms and beyond. We seek to replace current phenomenological models of nuclear structure and reactions with a well-founded microscopic theory that delivers maximum predictive power with well-quantified uncertainties.

In order to ensure close alignment of the necessary applied mathematics and computer science research with the necessary physics research, partnerships have been formed consisting of computer scientists and applied mathematicians linked with specific physicists. In this short overview, we describe one research task under UNEDF aiming at development of the optimized nuclear energy density functional and the related computational infrastructure to solve self-consistent, nonlinear equations of the nuclear energy Density Functional Theory (DFT). Our goal is to combine the best knowledge of the nuclear Hamiltonian with newly developed many-body techniques to construct a novel class of energy functionals. These functionals, validated by using the relevant data, will then be applied to predict nuclear properties and describe nuclear reactions.

Currently used energy density functionals, such as popular zero-range Skyrme functionals, are characterized by 10-15 coupling constants adjusted to selected experimental data. In spite of their simplicity, they offer a remarkably effective and robust description of atomic nuclei. Over the past few years, however, it has become clear that commonly used functionals suffer from various deficiencies that make it impossible to obtain a consistent and highly quantitative description of nuclear properties. From the point of view of DFT practitioners, the challenges by the UNEDF roadmap are, therefore, threefold: (i) How can we compute the properties of *all* atomic nuclei within a reasonable timescale? (ii) Given a mathematical form of the functional, how can we optimize its parameters to a set of experimental data? (iii) What mathematical form should the energy functional take?

2. Nuclear energy density functional

Within the DFT, the total binding energy of the nucleus is given by

$$E = \int d^3\boldsymbol{r}\mathcal{H}(\boldsymbol{r}), \tag{1}$$

where the energy density $\mathcal{H}(\boldsymbol{r})$ is a functional of the nucleonic intrinsic density matrices and their derivatives. A representative example is the commonly used nuclear Skyrme energy functional:

$$\mathcal{H}(\mathbf{r}) = \frac{\hbar^2}{2m}\tau_0 + \sum_{t=0,1}\Big(U_t^{\rho\rho}\rho_t^2 + U_t^{\rho\tau}\rho_t\tau_t + U_t^{J^2}\mathbf{J}_t^2 + U_t^{\rho\Delta\rho}\rho_t\Delta\rho_t + U_t^{\nabla\rho\nabla\rho}(\nabla\rho_t)^2 + U_t^{\rho\nabla J}\rho_t\nabla\cdot\mathbf{J}_t + U_t^{J\nabla\rho}\mathbf{J}_t\cdot\nabla\rho_t\Big) + \sum_{t=0,1}U_t^{\tilde\rho\tilde\rho}\tilde\rho_t^2, \tag{2}$$

where the isospin index t labels iso-scalar ($t = 0$) and iso-vector ($t = 1$) particle ρ_t, pairing $\tilde\rho_t$, kinetic τ_t, and spin-current $\mathbf{J}_t$ local densities, respectively. In the most general case, the amplitudes U_t^m with $m = \{\rho\rho, \tilde\rho\tilde\rho, \rho\tau, J^2, \rho\Delta\rho, \nabla\rho\nabla\rho, \rho\nabla J, J\nabla\rho\}$ are functions of τ_t, ρ_t, $\tilde\rho_t$, $\mathbf{J}_t$, and possibly their derivatives.

The requirement that the total energy (1) be minimal under a variation of the particle and pairing densities leads to the Hartree-Fock-Bogoliubov (HFB; or Bogoliubov-de Gennes) equations. They form a self-consistent set of non-linear integro-differential equations which has to be solved iteratively.

3. DFT solvers

The computational cost of HFB iterative convergence schemes can become very expensive, especially when the size of the model space, largely determined by the imposed symmetries or the number of many-body configurations processed simultaneously, is huge. Of particular interest is the coordinate-space formulation of HFB, which has an advantage to treat arbitrarily complex intrinsic shapes, including those seen in fission or heavy-ion fusion. A number of coordinate-space DFT solvers have been developed over the years, and their performance strongly depends on the size and symmetries of the spatial mesh employed. Of particular interest is the recently developed 2D solver HFBAX [2] based on B-splines that contain a number of enhancements resulting in a significant numerical speedup of the code. Figure 1 shows the neutron density distribution of a deformed exotic nucleus ^{110}Zr calculated with HFBAX in a 2D spacial box of dimension **AxB** fm. Under UNEDF, efforts are under way to develop a new-generation 3D DFT solver based on the multi-resolution wavelet analysis under the MADNESS framework [3].

The configuration-space DFT solvers are based on the expansion of HFB eigenstates in a discrete basis, such as the harmonic oscillator basis. The codes that use the basis expansion technique can be very versatile. For example, the 3D DFT solver HFODD [4] is capable of treating arbitrary shapes, which is of particular importance when studying complex configurations such as high-spin states or nuclear fission. In many problems of physical interest, however, symmetry-restricted solvers are sufficient. The HFBTHO solver [5] is a highly optimized equivalent of HFODD that is restricted to axially-symmetric shapes. It is particularly adapted to large-scale nuclear mass calculations, since the vast majority of atomic nuclei are axially-deformed in their ground-states.

4. Large scale nuclear mass table

Calculating the entire nuclear mass table is a challenge of its own. Around 10,000 nuclei are expected to exist between the proton and neutron limits of stability, representing about 600,000 many-body configurations for which a DFT solution needs to be found. They are either spherical (quadrupole deformation $\beta = 0$), prolate ($\beta > 0$), or oblate ($\beta < 0$). In order to determine the ground-state equilibrium deformation, constrained calculations are first carried out for every nucleus on a β-deformation grid. The resulting energy curves define the regions within which the local minima are found. The ground-state configuration is assigned to the lowest-energy minimum.

In the nuclear DFT, the ground-state of an odd-mass nucleus is represented by the blocking of a one-quasi-particle excitation out of a paired HFB vacuum representing a neighboring even-even system [6]. The quantum numbers of the quasi-particle excitation that yield the lowest total energy are not known beforehand; hence, one must select several blocking candidates in a given energy window around the Fermi level. In HFBTHO, the quasi-particle calculations are carried out for each configuration within the so-called equal-filling approximation, which preserves the time-reversal symmetry. Odd-odd nuclei are handled by a double-blocking procedure: for each blocked one-quasi-neutron state, we solve the DFT equations by considering all possible one-quasi-proton states within the energy cutoff range, and the lowest-energy final state is assigned to the ground-state of the odd-odd system. The total number of different configurations to be taken into account across the mass table is of the order of one million. In spite of the high efficiency of HFBTHO, computing the entire mass-table would take almost 12 CPU years on a single processor. However, with the help of massively parallel computers such as the Cray XT5 Jaguar and Kraken, the entire calculation is accomplished in a single 14-CPU-hour run involving 9,000 processors. As an illustrative example, Figure 2 shows the result of a large-scale HFBTHO mass-table run with the SLy4 Skyrme functional [7] and a zero-range pairing force covering all even-even, odd-even and odd-odd nuclei with neutron number $2 \leq N \leq 190$ and proton number $2 \leq Z \leq 110$. For some recent applications of mass-table results, see Refs. [8, 9].

5. DFT solvers enhancements

The complexity of the large-scale nuclear mass table problem cannot be resolved by a trivial increase in the number of processors used. To find converged DFT results for the multitude of nuclear configurations is a challenging task that requires the development of efficient DFT solvers, optimized for leadership-class architectures.

One of the crucial enhancements, proposed and analyzed in detail in Ref. [10], is the use of the modified Broyden mixing for the numerical solution of nonlinear equations in many variables to nuclear self-consistent calculations. (The size of the Broyden vector in our calculations ranges from ~300,000 in HFBTHO to ~3,000,000 in HFODD.) This method provides impressive performance improvements against the conventional linear mixing procedure. For example, using linear mixing, one is left with almost 30% unconverged cases assuming the standard accuracy requirements. With the Broyden method, all nuclei in the mass table of figure 2 are resolved. Other recent enhancements include the application of the augmented Lagrangian method for constrained DFT calculations [11] and parallelization of HFODD on the Jaguar Cray XT5.

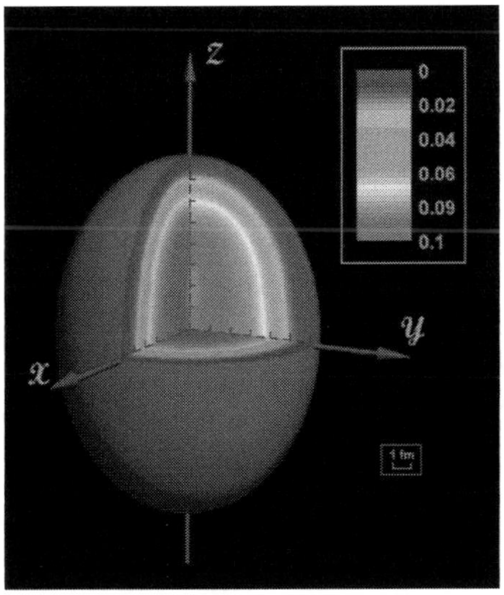

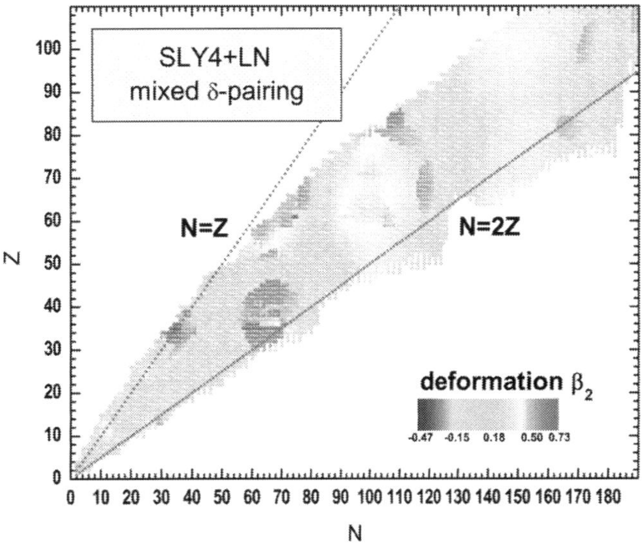

Figure 1. Neutron local density distribution in the exotic neutron-rich nucleus ^{110}Zr obtained in coordinate space calculations using the HFBAX DFT solver with the SLy4 Skyrme functional and a zero-range pairing force.

Figure 2. Ground-state shape deformation of even-even, odd-even, and odd-odd atomic nuclei across the entire nuclear chart obtained in a singe mass-table run on the Cray XT5 Jaguar supercomputer. The calculations were carried out with the HFBTHO DFT solver using the SLy4 energy density functional augmented by density-dependent zero-range pairing force.

6. Optimization of the functional

Recent progress in code development and hardware architectures make it possible to undertake a global parameter search and statistical analysis of energy functionals.

For values of p parameters $\boldsymbol{\theta}$, our simulators compute T observables $\{s_{t,n}(\boldsymbol{\theta})\}_{t=1}^{T}$ for nucleus n. Given an experimental data set $\{d_{t,n}\}_{t,n=1}^{T,N}$, corresponding to different nuclear properties (such as binding energies, radii, and moments), our objective is to find the parameters $\boldsymbol{\theta}$ so that the theoretical output approximates the experimental data. To this end, we solve the nonlinear least squares problem $\min\left\{f(\boldsymbol{\theta}) = \sum_{t,n}\left(d_{t,n} - s_{t,n}(\boldsymbol{\theta})\right)^2/\sigma_{t,n}^2 : \boldsymbol{\theta} \in \Re^p\right\}$, where $\sigma_{t,n} > 0$ reflects the relative accuracy of the experimental data. The optimization procedure is based on updating a quadratic model $q(\boldsymbol{\theta}) = f(\hat{\boldsymbol{\theta}}) + (\boldsymbol{\theta} - \hat{\boldsymbol{\theta}})^T g + \frac{1}{2}(\boldsymbol{\theta} - \hat{\boldsymbol{\theta}})H(\boldsymbol{\theta} - \hat{\boldsymbol{\theta}})$ approximating $f(\boldsymbol{\theta})$ near the calculated solution $\hat{\boldsymbol{\theta}}$.

Since the Hessian H is positive semi-definite at a solution, we can conduct sensitivity and correlation analysis at $\hat{\boldsymbol{\theta}}$ as a byproduct of the optimization. Under certain statistical assumptions, H can be used to approximate the covariance $Cov(\boldsymbol{\theta}) \approx H^{-1}$, and hence the correlation matrix $R_{ij} = Cov(\boldsymbol{\theta})_{ij}/\sqrt{Cov(\boldsymbol{\theta})_{ii}Cov(\boldsymbol{\theta})_{jj}}$. Such an analysis is a standard approach in other domains of physics, but could not be addressed in nuclear structure until very recently due to the extremely high cost of the function evaluations.

7. New generation nuclear energy density functionals

A straightforward empirical generalization of the standard Skyrme functional is obtained by replacing all coupling constants U_t^m in (2) with density-dependent functions. In this way, the standard set of parameters is extended, thus increasing the flexibility of the functional. Another extension of the DFT currently pursued, which retains the general functional structure (2) but offers novel density dependence, comes from the Density Matrix Expansion method applied to the long-range part of chiral nucleon-nucleon interactions [12].

8. Visualization tools

One of the bottlenecks in large-scale calculations is the massive amount of data produced in mass-table runs. Specialized tools must be developed to analyze and exploit this rich theoretical output. To this end, a web server has been dedicated to collecting, sorting, and visualizing the data produced by UNEDF mass-table runs. It can be freely accessed at *http://massexplorer.org* . This web site contains links to the raw data files, and more useful data will become available as the UNEDF project progresses.

9. Conclusions

Finding the nuclear energy density functional for atomic nuclei capable of describing stable and short-lived nuclei and their reactions requires the unique combination of innovative theoretical tools that can link microscopic interactions to energy functionals, significant code development with special emphasis on massively parallel algorithms, and state-of-the-art optimization techniques that can efficiently cope with the high computational cost of function evaluations. The UNEDF SciDAC project will develop key computational codes and algorithms for reaching the goal of solving the nuclear quantum many-body problem, thus paving the road to the comprehensive model of the nucleus.

10. Acknowledgments

Our DFT/Optimization UNEDF team includes M. Kortelainen, T. Lesinski, J. McDonnell, W. Nazarewicz, N. Nikolov, J.C. Pei, N. Schunck, and M. Stoitsov from ORNL/UT; G.I. Fann, H.A. Nam, and W.A. Shelton from ORNL; J. Moré, J. Sarich, and S. Wild from ANL; D. Furnstahl (OSU); S.K. Bogner and Biruk Gebremariam (NSCL/MSU); J. Dobaczewski (Warsaw/Jvyäskylä), and T. Duguet (Saclay).

The UNEDF SciDAC Collaboration is supported by the U.S. Department of Energy (DOE) under grant No. DOE-FC02-09ER41583. This work was also supported by DOE Contract Nos. DE-FG02-96ER40963 (University of Tennessee), DE-AC05-00OR22725 with UT-Battelle, LLC (Oak Ridge National Laboratory), DE-FG05-87ER40361 (Joint Institute for Heavy Ion Research), and DE-FG03-03NA00083 (Stewardship Science Academic Alliances).

References

[1] G.F. Bertsch, D.J. Dean, and W. Nazarewicz, SciDAC Review, Winter 2007, p. 42; http://www.unedf.org.
[2] J.C. Pei, M.V. Stoitsov, G.I. Fann, W. Nazarewicz, N. Schunck, and F.R.Xu, Phys. Rev. C **78**, 064306 (2008).
[3] T. Yanai, G.I. Fann, Z. Gan, R.J. Harrison and G. Beylkin, J. Chem. Phys. **121**, 2866 (2004).
[4] J. Dobaczewski and P. Olbratowski, Comput. Phys. Commun. **158**, 158 (2004); J. Dobaczewski *et al.* arXiv:0903.1020 (2009).
[5] M.V. Stoitsov, J. Dobaczewski, W. Nazarewicz, and P. Ring, Comput. Phys. Commun. **167**, 43 (2005).
[6] G. Bertsch, J. Dobaczewski, W. Nazarewicz, and J. Pei, Phys. Rev. A **79**, 043602 (2009).
[7] E. Chabanat, P. Bonche, P. Haensel, J. Meyer, and R. Schaeffer, Nucl. Phys. A **635**, 231 (1998).
[8] G.F. Bertsch, C.A. Bertulani, W. Nazarewicz, N. Schunck, and M.V. Stoitsov, Phys. Rev. C **79**, 034306 (2009).
[9] M. Stoitsov, W. Nazarewicz, and N. Schunck, Int. J. Mod. Phys. E **18**, (2009).
[10] A. Baran, A. Bulgac, M. Forbes, G. Hagen, W. Nazarewicz, N. Schunck, and M.V. Stoitsov, Phys. Rev. C **78**, 014318 (2008).
[11] J. Nocedal and S.J. Wright, *Numerical Optimization* (Springer Series in Operations Research and Financial Engineering, 2006).
[12] B. Gebremariam, S.K. Bogner, and T. Duguet, in preparation.

Journal of Physics: Conference Series **180** (2009) 012083 doi:10.1088/1742-6596/180/1/012083

Ab initio nuclear structure - the large sparse matrix eigenvalue problem

James P. Vary[1], Pieter Maris[1], Esmond Ng[2], Chao Yang[2], Masha Sosonkina[3]

[1]Department of Physics, Iowa State University, Ames, IA, 50011 USA

[2]Computational Research Division, Lawrence Berkeley National Laboratory, Berkeley, CA 94720, USA

[3]Scalable Computing Laboratory, Ames Laboratory, Iowa State University, Ames, IA, 50011, USA

E-mail: jvary@iastate.edu

Abstract. The structure and reactions of light nuclei represent fundamental and formidable challenges for microscopic theory based on realistic strong interaction potentials. Several *ab initio* methods have now emerged that provide nearly exact solutions for some nuclear properties. The *ab initio* no core shell model (NCSM) and the no core full configuration (NCFC) method, frame this quantum many-particle problem as a large sparse matrix eigenvalue problem where one evaluates the Hamiltonian matrix in a basis space consisting of many-fermion Slater determinants and then solves for a set of the lowest eigenvalues and their associated eigenvectors. The resulting eigenvectors are employed to evaluate a set of experimental quantities to test the underlying potential. For fundamental problems of interest, the matrix dimension often exceeds 10^{10} and the number of nonzero matrix elements may saturate available storage on present-day leadership class facilities. We survey recent results and advances in solving this large sparse matrix eigenvalue problem. We also outline the challenges that lie ahead for achieving further breakthroughs in fundamental nuclear theory using these *ab initio* approaches.

1. Introduction

The structure of the atomic nucleus and its interactions with matter and radiation have long been the foci of intense theoretical research aimed at a quantitative understanding based on the underlying strong inter-nucleon potentials. Once validated, a successful approach promises predictive power for key properties of short-lived nuclei that are present in stellar interiors and in other nuclear astrophysical settings. Moreover, new medical diagnostic and therapeutic applications may emerge as exotic nuclei are predicted and produced in the laboratory. Fine tuning nuclear reactor designs to reduce cost and increase both safety and efficiency are also possible outcomes with a high precision theory.

Solving for nuclear properties with the best available nucleon-nucleon (NN) potentials, supplemented by 3-body (NNN) potentials as needed, using a quantum many-particle framework that respects all the known symmetries of the potentials is referred to as an "*ab initio*" problem and is recognized to be computationally hard. Among the few *ab initio* methods available for light nuclei beyond atomic number $A = 10$, the no core shell model (NCSM) [1] and the no core full configuration (NCFC) [2] methods frame the problem as a large sparse matrix eigenvalue problem one of the focii of the SciDAC-UNEDF program.

2. Nuclear physics goals

Many fundamental properties of nuclei are poorly understood today. By gaining insight into these properties we may better utilize these dense quantum many-body systems to our advantage. A short list of outstanding unsolved problems usually includes the following questions.

- What controls nuclear saturation - the property that the central density is nearly the same for all nuclei?
- How does a successful shell model emerge from the underlying theory, including predictions for shell and sub-shell closures?
- What are the properties of neutron-rich and proton-rich nuclei, which will be explored at the future FRIB (Facility for Rare Isotope Beams)?
- How precise can we predict the properties of nuclei - will it be feasible to use nuclei as laboratories for tests of fundamental symmetries in nature?

The past few years have seen substantial progress but the complete answers are yet to be achieved. We highlight some of the recent physics achievements of the NCSM that encourage us to believe that those major goals may be achievable in the next 10 years. In many cases, we showed the critical role played by 3-body forces and the need for large basis spaces to obtain agreement with experiment.

Since 3-body potentials enlarge the computational requirements by one to two orders of magnitude in both memory and CPU time, we have also worked to demonstrate that similar results may be achieved by suitable adjustments of the undetermined features of the nucleon-nucleon (NN) potential, known as the off-shell properties of the potential. The adjustments were constructed so as to preserve the original high precision description of the NN scattering data. A partial list of the NCSM and NCFC achievements is included below.

- Described the anomaly of the nearly vanishing quadrupole moment of ^{6}Li [3];
- Established need for 3-body potentials to explain, among other properties, neutrino-^{12}C cross sections [4];
- Found quenching of Gamow-Teller (GT) transitions in light nuclei due to effects of 3-body potential [5] (or off-shell modifications of NN potential) plus configuration mixing [6];
- Obtained successful description of $A = 10$ to 13 low-lying spectra with chiral NN + NNN interactions [7];
- Explained ground state spin of ^{10}B by including chiral NNN interaction [7].

We have identified several major near-term physics goals on the path to the larger goals listed above. These include:

- Evaluate the properties of the Hoyle state in ^{12}C and the isoscalar quadrupole strength function in ^{12}C to compare with inelastic α scattering data;
- Explain the lifetime of ^{14}C (5730 years) which depends sensitively on the nuclear wavefunction;
- Calculate the properties of the Oxygen isotopes out to the neutron drip line;
- Solve for nuclei surrounding shell closure in order to understand the origins shell closures.

Preliminary results for these goals are encouraging and will be reported when we obtain additional results closer to convergence.

3. Recent NCFC results

Within the NCFC method, we adopt the harmonic oscillator (HO) single particle basis which involves two parameters and we seek results independent of those parameters either directly in a sufficiently large basis or via extrapolation to the infinite basis limit. The first parameter $\hbar\Omega$ specifies the HO energy, the spacing between major shells. Each shell is labeled uniquely by the quanta of its orbits $N = 2n + l$ (orbits are specified by quantum numbers n, l, j, m_j) which begins with 0 for the lowest shell and increments in steps of unity. Each unique arrangement of fermions (neutrons and protons) within the HO orbits that is consistent with the Pauli principle, constitutes a many-body basis state. Many-body basis states satisfying chosen symmetries are employed in evaluating the Hamiltonian H in that basis. The second parameter is $N_{\rm max}$ which limits the total number of oscillator quanta allowed in the many-body basis states and thus limits the dimension D of the Hamiltonian matrix. $N_{\rm max}$ is defined to count the total quanta above the minimum for the specific nucleus needed to satisfy the Pauli principle.

3.1. Quadrupole moment of ^{6}Li

Experimentally, ^{6}Li has a surprisingly small quadrupole moment, $Q = -0.083\,e\,{\rm fm}^2$. This means that there are significant cancellations between contributions to the quadrupole moment from different parts of the wave function. In Ref. [3] the NCSM was shown to agree reasonably well with the data using a nonlocal NN interaction. Here we employ the realistic JISP16 NN potential from inverse scattering that provides a high-quality description of the NN data [8]. This NN potential is sufficiently well-behaved that we may obtain NCFC results for light nuclei [2]. Indeed, the left panel of Fig. 1 shows smooth uniform convergence of the ground state energy of ^{6}Li. The quadrupole moment does not exhibit the same smooth and uniform convergence. Nevertheless, the results for 15 MeV$< \hbar\Omega <$ 25 MeV where the ground state energy is closest to convergence, strongly suggest a quadrupole moment in agreement with experiment.

3.2. Beryllium isotopes

Next, we use the Be isotopes to portray features of the physics results and illustrate some of the computational challenges, again with the JISP16 nonlocal NN potential [8]. Fig. 2 displays the NCFC results for the lowest states of natural and unnatural parity respectively in a range of Be isotopes along with the experimental ground state energy. Since our approach is guaranteed to provide an upper bound to the exact ground state energy for the chosen Hamiltonian, each point displayed for a particular value of $N_{\rm max}$ represents the minimum value as a function of $\hbar\Omega$

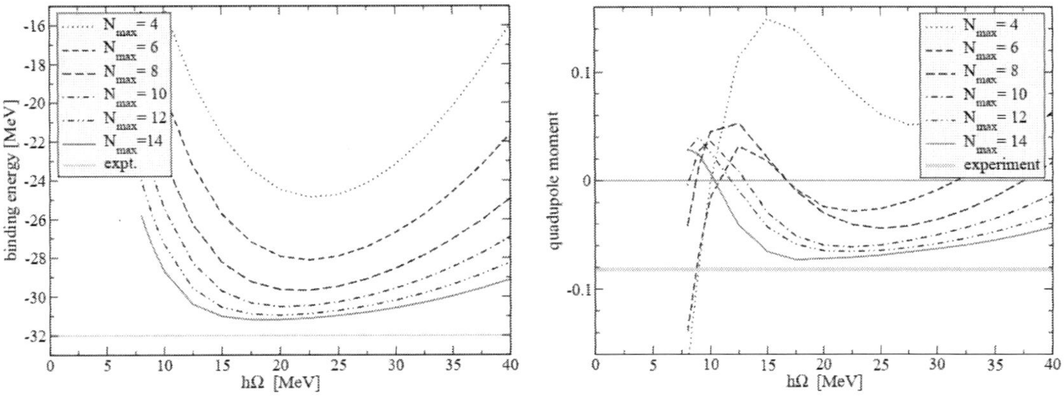

Figure 1. Ground state energy of ^{6}Li (left) and quadrupole moment (right) obtained with the JISP16 NN potential as function of $\hbar\Omega$ for several values of $N_{\rm max}$.

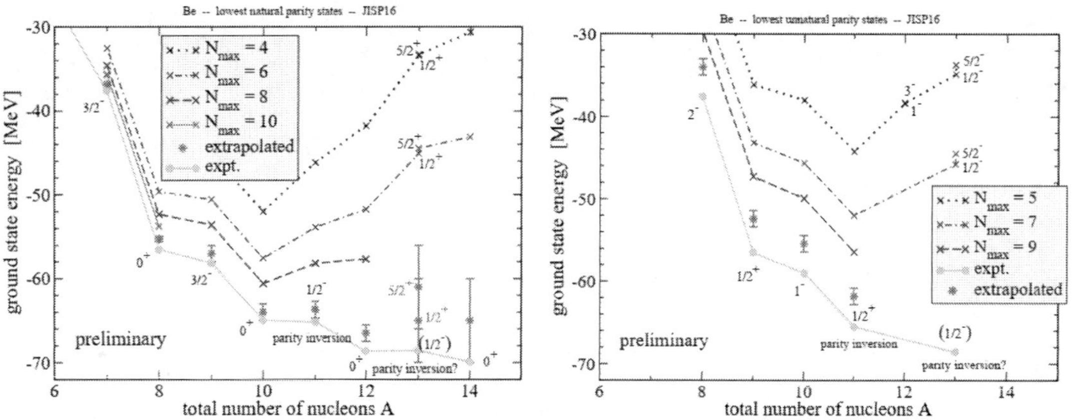

Figure 2. Lowest natural (left) and unnatural (right) states of Be isotopes obtained with the JISP16 NN potential. Extrapolated results and their assessed uncertainties are presented as red points with error bars.

obtained in that basis space.

In small model spaces, $N_{\max} = 4$ and 5, the calculations suggest a "U-shaped" curve for the binding energy as function of A, the number of nucleons, whereas experimentally, the ground state energy keeps decreasing with A, at least over the range in A shown in Fig. 2. However, as one increases $N_{\max}$, the calculated results get closer to the experimental values. In order to obtain the binding energies in the full configuration space (NCFC), we can use an exponential fit to a set of results at 3 or 4 subsequent $N_{\max}$ values [2]. After performing such an extrapolation to infinite model space, our calculated results follow the data very closely as function of A.

In both panels of Fig. 2, the dimensions increase dramatically as one increases $N_{\max}$ or increases atomic number A. Detailed trends will be presented in the next section. The consequence is that results terminate at lower A as one increases $N_{\max}$. Additional calculations are in process in order to reduce the uncertainties in the extrapolations.

We note that the ground state spins are predicted correctly in all cases except $A = 11$ (and possibly at $A = 13$, where the spin-parity of the ground state is not well known experimentally) where there is a reputed "parity inversion". In this case a state of "unnatural parity", one involving at least one particle moving up one HO shell from its lowest location, unexpectedly becomes the ground state. While Fig. 2 indicates we do not reproduce this parity inversion, we do notice, however, that the lowest states of natural and unnatural parity lie very close together. This indicates that small corrections arising from neglected effects, such as three-body (NNN) potentials and/or additional contributions from larger basis spaces could play a role here. We are continuing to investigate these improvements.

4. Computer Science and Applied Math challenges

We outline the main factors underlying the need for improved approaches to solving the nuclear quantum many-body problem as a large sparse matrix eigenvalue problem. Our goals require results as close to convergence as possible in order to minimize our extrapolation uncertainties. According to current experience, this requires the evaluation of the Hamiltonian in as large a basis as possible, preferably with $N_{\max} \geq 10$, in order to converge the ground state wavefunction.

The dimensions of the Hamiltonian matrix grow combinatorially with increasing $N_{\max}$ and with increasing atomic number A. To gain a feeling for that explosive growth, we plot in Fig. 3 the matrix dimension (D) for a wide range of Oxygen isotopes. In each case, we select the "natural parity" basis space, the parity that coincides with the lowest HO configuration for that

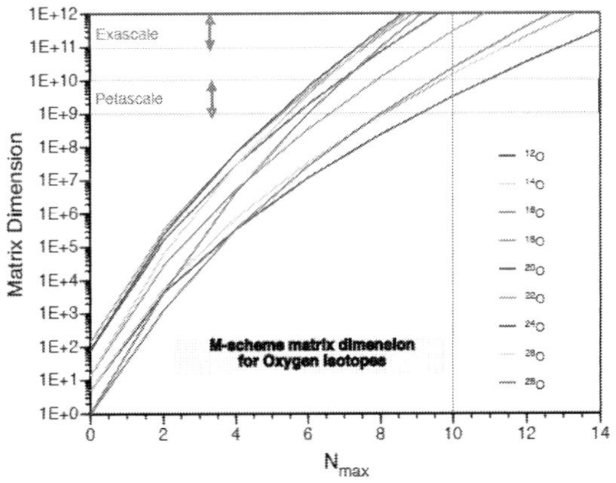

Figure 3. Matrix dimension versus $N_{\max}$ for stable and unstable Oxygen isotopes. The vertical red line signals the boundary where reasonable convergence emerges on the right of it.

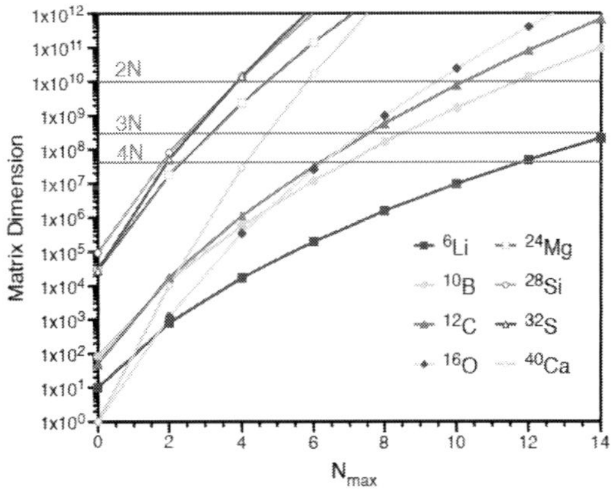

Figure 4. Matrix dimension versus $N_{\max}$ for a sample set of stable N=Z nuclei up to $A = 40$. Horizontal lines show expected limits of Petascale machines for specific ranks of the potentials ("2N"=NN, "3N"=NNN, etc).

nucleus. The heavier of these nuclei have been the subject of intense experimental investigation and it is now believed that ^{28}O is not a particle-stable nucleus even though it is expected to have a doubly-closed shell structure according to the phenomenological shell model. It would be very valuable to have converged *ab initio* NCFC results for ^{28}O to probe whether realistic potentials are capable of predicting its particle-unstable character.

We also include in Fig. 3 the estimated range that computer facilities of a given scale can produce results with our current algorithms. As a result of these curves, we anticipate well converged NCFC results for the first three isotopes of Oxygen will be achieved with Petascale facilities since their curves fall near or below the upper limit of Petascale at $N_{\max} = 10$.

Dimensions of the natural parity basis spaces for another set of nuclei ranging up to $A = 40$ are shown in Fig. 4. In addition, we include estimates of the upper limits reachable with Petascale facilities depending on the rank of the potential. It is important to note that theoretically derived 4N interactions are expected to be available in the near future. Though relatively less important than 2N and 3N potentials, their contributions are expected to grow dramatically with increasing A.

A significant measure of the computational burden is presented in Figs. 5 and 6 where we display the number of non-zero many-body matrix elements as a function of the matrix dimension (D). These results are for representative cases and show a useful scaling property. For Hamiltonians with NN potentials, we find a useful fit $F(D)$ for the non-zero matrix elements with the function

$$F(D) = D + D^{1+\frac{12}{14+\ln D}}. \tag{1}$$

The heavier systems displayed tend to be slightly below the fit while the lighter systems are slightly above the fit. The horizontal red line indicates the expected limit of the Jaguar facility (150,000 cores) running one of these applications assuming all matrix elements and indices are stored in core. By way of contrast, we portray the more memory-intensive situation with NNN potentials in Fig. 6, where we retain the fitted curve of Fig. 5 for reference. The horizontal red line indicates the same limit shown in Fig. 5.

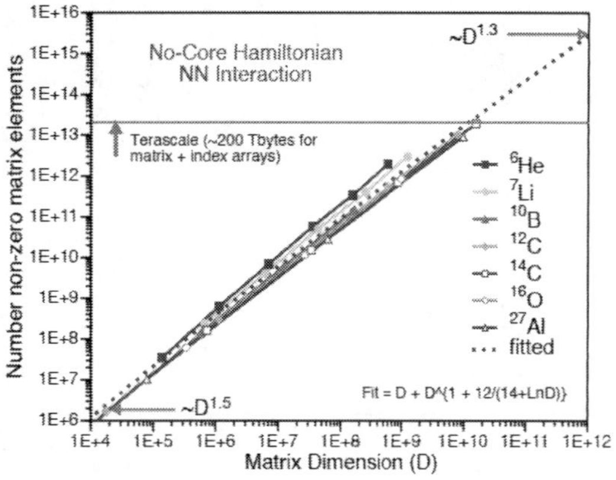

Figure 5. Number of nonzero matrix elements versus matrix dimension (D) for an NN potential and for a selection of light nuclei. The dotted line depicts a function describing the calculated points.

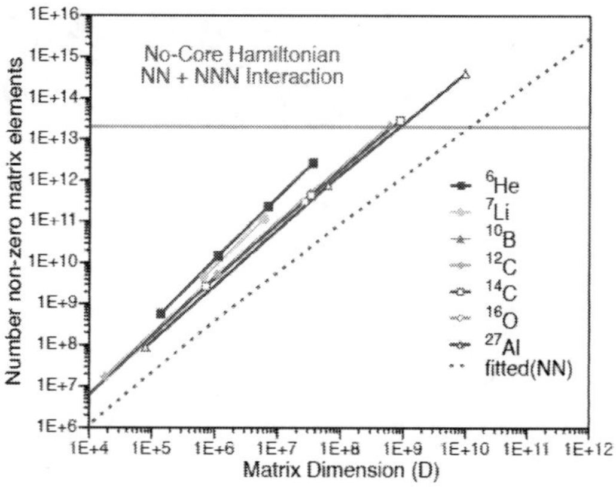

Figure 6. Same as Fig. 5 but for an NNN potential. The NN dotted line of Fig. 5 is repeated as a reference. The NNN slope and magnitude are larger than the NN case. The NNN case is larger by a factor of 10-100.

nucleus	$N_{\max}$	dimension	2-body	3-body	4-body
^{6}Li	12	$4.9 \cdot 10^{6}$	0.6 GB	33 TB	590 TB
^{12}C	8	$6.0 \cdot 10^{8}$	4 TB	180 TB	4 PB
^{12}C	10	$7.8 \cdot 10^{9}$	80 TB	5 PB	140 PB
^{16}O	8	$9.9 \cdot 10^{8}$	5 TB	300 TB	5 PB
^{16}O	10	$2.4 \cdot 10^{10}$	230 TB	12 PB	350 PB
^{8}He	12	$4.3 \cdot 10^{8}$	7 TB	300 TB	7 PB
^{11}Li	10	$9.3 \cdot 10^{8}$	11 TB	390 TB	10 PB
^{14}Be	8	$2.8 \cdot 10^{9}$	32 TB	1100 TB	28 PB
^{20}C	8	$2 \cdot 10^{11}$	2 PB	150 PB	6 EB
^{28}O	8	$1 \cdot 10^{11}$	1 PB	56 PB	2 EB

Table 1. Table of storage requirements of current version of MFDn for a range of applications. Roughly speaking, entries up to 400TB imply Petascale while entries above 1PB imply Exascale facilities will likely be required.

Looking forward to the advent of exascale facilities and the development of 4N potentials, we present in Table 1 the matrix dimensions and storage requirements for a set of light nuclei for potentials ranging from 2N through 4N.

5. Algorithms of MFDn

Our present approach embodied in the code Many Fermion Dynamics - nuclear (MFDn) [9, 10, 11] involves several key steps:

(i) enumerate the single-particle basis states for the neutrons and the protons separately with good total angular momentum projection m_j,

(ii) enumerate and store the many-body basis states subject to user-defined constraints and symmetries such as $N_{\max}$, parity and total angular momentum projection M,

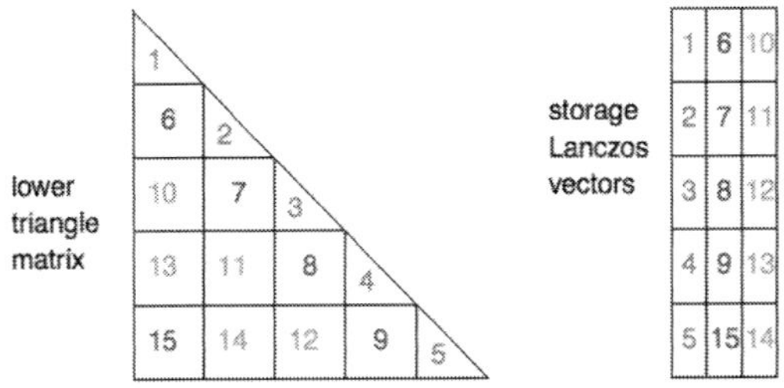

Figure 7. MFDn scheme for distributing the symmetric Hamiltonian matrix over $n(n+1)/2$ PE's and partitioning the Lanczos vectors. We use $n = 5$ "diagonal" PE's in this illustration.

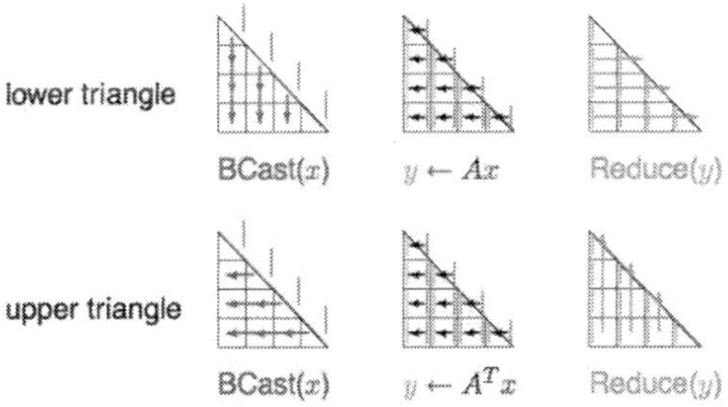

Figure 8. MFDn scheme for performing the Lanczos iterations based on the Hamiltonian matrix stored in $n(n+1)/2$ PE's. We use $n = 4$ "diagonal" PE's in this illustration.

(iii) evaluate and store the many-nucleon Hamiltonian matrix in this many-body basis using input files for the NN (and NNN interaction if elected),

(iv) obtain the lowest converged eigenvalues and eigenvectors using the Lanczos algorithm with a distributed re-orthonormalization scheme,

(v) transform the converged eigenvectors from the Lanczos vector space to the original basis and store them in a file,

(vi) evaluate a suite of experimental observables using the stored eigenvectors.

All these steps, except the first, which does not require any significant amount of CPU time, are parallelized using MPI. Fig. 7 presents the allocation of processors for computing and storing the many-body Hamiltonian on $n(n+1)/2$ processors. (We illustrate the allocation with 15 processors so the pattern is clear for an arbitrary value of n.) We assume a symmetric Hamiltonian matrix and compute/store only the lower triangle.

In step two, the many-body basis states are round-robin distributed over the n diagonal processors. Each vector is also distributed over n processors, so we can (in principle) deal with arbitrary large vectors. Because of the round-robin distribution of the basis states, we obtain excellent load-balancing; however, the price we pay is that any structure in the sparsity pattern of the matrix is lost. We have implemented multi-level blocking procedures to enable an efficient determination of the non-zero matrix elements to be evaluated. These procedures have been presented recently in Ref. [11].

After the determination and evaluation of the non-zero matrix elements, we solve for the lowest eigenvalues and eigenvectors using an iterative Lanczos algorithm, step four. The matvec operation, and its communication patterns, for each Lanczos iteration is shown in Fig. 8 where two sets of operations account for the storage of only the lower triangle of the matrix.

MFDn also employs a distributed re-orthonormalization technique developed specifically for the parallel distribution shown in Fig. 7. After the completion of the matvec, the new elements of the tridiagonal Lanczos matrix are calculated on the diagonals. The (distributed) normalized Lanczos vector is then broadcast to all processors for further processing. Each previously computed Lanczos vector is stored on one of $(n+1)/2$ groups of n processors. Each group receives the new Lanczos vector, computes the overlaps with all previously stored vectors within that group, and constructs a subtraction vector, that is, the vector that needs to be subtracted from the Lanczos vector in order to orthogonalize it with respect to the Lanczos vectors stored in that group. These subtraction vectors are accumulated on the n diagonal processors, where they are subtracted from the current Lanczos vector. Finally, the orthogonalized Lanczos vector is re-normalized and broadcast to all processors for initiating the next matvec. One group is designated to also store that Lanczos vector for future re-orthonormalization operations.

The results so far with this distributed re-orthonormalization appear stable with respect to the number of processors over which the Lanczos vectors are stored. We can keep a considerable number of Lanczos vectors in core, since all processors are involved in the storage, and each part of a Lanczos vector is stored on one and only one processor. For example, we have run 5600 Lanczos iterations on a matrix with dimension 595 million and performed the distributed re-orthonormalization using 80 groups of stored Lanczos vectors. We converged about 450 eigenvalues and eigenvectors. Tests of the converged states, such as measuring a set of their symmetries, showed they were reliable. No converged duplicate eigenstates were generated. Test cases with dimensions of ten to forty million have been run on various numbers of processors to verify the strategy produces stable results independent of the number of processors. A key element for the stability of this procedure is that the overlaps and normalization are calculated in double precision, even though the vectors (and the matrix) are stored in single precision. Furthermore, there is of course a dependence on the choice of initial pivot vector; we favor a random initial pivot vector.

6. Accelerating convergence

We are constantly seeking new ideas for accelerating convergence as well as saving memory and CPU time. The situation is complicated by the wide variety of strong interaction Hamiltonians that we employ and we must simultaneously investigate theoretical and numerical methods for "renormalizing" (softening) the potentials. Softening an interaction reduces the magnitude of many-body matrix elements between very different basis states such as those with very different total numbers of HO quanta. However, all methods to date soften the interactions at the price of increasing their complexity. For example, starting with a strong NN interaction, one renormalizes it to improve the convergence of the many-body problem with the softened NN potential [12] but, in the process, one induces a new NNN potential that is required to keep the final many-body results invariant under renormalization. In fact, 4N potentials and higher are also induced[7]. Given the computational cost of these higher-body potentials in the many-body problem as described above, it is not clear how much one gains from renormalization - for some NN interactions it may be more efficient to attempt larger basis spaces and avoid the renormalization step. This is a current area of intense theoretical research.

In light of these features of strongly interacting nucleons and the goal to proceed to heavier nuclei, it is important to investigate methods for improving convergence. A number of promising methods have been proposed and each merits a significant research effort to assess their value over a range of nuclear applications. A sample list includes:

- Symplectic no core shell model (SpNCSM) - extend basis spaces beyond their current $N_{\max}$ limits by adding relatively few physically-motivated basis states of symplectic symmetry which is a favored symmetry in light nuclei [13]
- Realistic single-particle basis spaces [14]
- Importance truncated basis spaces [15]
- Monte Carlo sampling of basis spaces [16, 17]
- Coupled total angular momentum J basis space

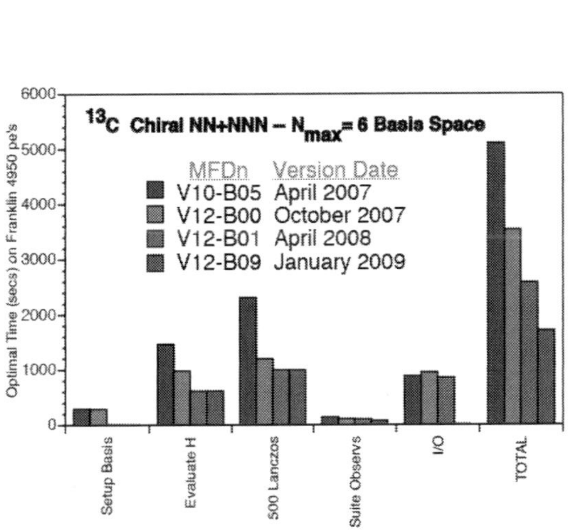

Figure 9. Performance improvement of MFDn over a 2 year span displayed for major sections of the code.

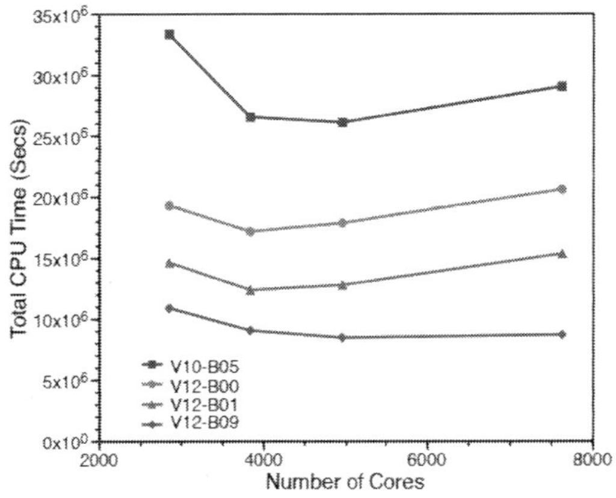

Figure 10. Total CPU time for the same test cases evaluated in Fig. 9 running on various numbers of Franklin cores.

Several of these methods can easily be explored with MFDn. For example, while MFDn has some components that are specific for the HO basis, it is straightforward to switch to another single-particle basis with the same symmetries but with different radial wavefunctions. Alternative truncation schemes such as Full Configuration Interaction (FCI) basis spaces are also implemented and easily switched on by appropriate input data values. These flexible features have proven convenient for addressing a wide range of physics problems.

There are additional suggestions from colleagues that are worth exploring as well such as the use of wavelets for basis states. These suggestions will clearly involve significant research efforts to implement and assess.

7. Performance of MFDn

We present measures of MFDn performance in Figs. 9 and 10. Fig. 9 displays the performance characteristics of MFDn from the beginning of the SciDAC-UNEDF program to the present, roughly a 2 year time span. Many of the increments along the path of improvements have been documented [11, 18, 19].

Fig. 9 displays the timings for major sections of the code as well as the total time for a test case in ^{13}C using a NNN potential on 4950 Franklin processors. The timings are specified for MFDn versions that have emerged during the SciDAC-UNEDF program. The initial version tested, V10-B05, represents a reasonable approximation to the version prior to SciDAC-UNEDF. We note that the overall speed has improved by about a factor of 3. Additional effort is needed to further reduce the time to evaluate the many-body Hamiltonian H and to perform the matvec operations as the other sections of the code have undergone dramatic time reductions.

We present in Fig. 10 one view of the strong scaling performance of MFDn on Franklin. Here the test case is held fixed at the case shown in Fig. 9 and the number of processors is varied from 2850 to 7626. Due to memory limitations associated with the very large input NNN data file (3 GB), scaling is better than ideal at (relatively) low processor numbers. At the low end, there is significant redundancy of calculations since the entire NNN file cannot fit in core and must be processed in sections. As one increases the number of processors, more memory becomes available for the NNN data and less redundancy is incurred. At the high end, the scaling begins to show signs of deteriorating due to increased communications time relative to computations.

The net result is the dip in the middle which we can think of as the "sweet spot" or the ideal number of processors for this application. Clearly, this implies significant preparation work for large production runs to insure maximal efficiency in the use of limited computational resources. For the largest feasible applications, this also implies there is high value to finding a solution for the redundant calculations brought about by the large input files. We are currently working on a hybrid OpenMP/MPI version of MFDn, which would significantly reduce this redundancy. However, at the next model space, the size of the input file increases to 33 GB, which we need to process in sections on Franklin.

8. Conclusions and Outlook

Microscopic nuclear structure/reaction physics is enjoying a resurgence due, in large part, to the development and application of *ab initio* quantum many-body methods with realistic strong interactions. True predictive power is emerging for solving long-standing fundamental problems and for influencing future experimental programs.

We have focused on the no core shell model (NCSM) and the no core full configuration (NCFC) approaches and outlined our recent progress. Large scale applications are currently underway using leadership-class facilities and we expect important progress on the path to detailed understanding of complex nuclear phenomena.

We have also outlined the challenges that stand in the way of further progress. More efficient use of memory, faster I/O and faster eigensolvers will greatly aid our progress.

Acknowledgments

This work was supported in part by DOE grants DE-FC02-09ER41582 and DE-FG02-87ER40371. Computational resources are provided by DOE through the National Energy Research Supercomputer Center (NERSC) and through an INCITE award (David Dean, PI) at Oak Ridge National Laboratory and Argonne National Laboratory.

References

[1] P. Navrátil, J. P. Vary and B. R. Barrett, Phys. Rev. Lett. **84**, 5728 (2000); Phys. Rev. C **62**, 054311 (2000).
[2] P. Maris, J. P. Vary and A. M. Shirokov, Phys. Rev. C. **79**, 014308(2009), arXiv 0808.3420.
[3] P. Navrátil, J.P. Vary, W.E. Ormand, B.R. Barrett, Phys. Rev. Lett. **87**, 172502(2001)
[4] E. Caurier, P. Navrátil, W.E. Ormand and J.P. Vary, Phys. Rev. C **66**, 024314(2002); A. C. Hayes, P. Navrátil and J. P. Vary, Phys. Rev. Lett. **91**, 012502 (2003).
[5] A. Negret, et al, Phys. Rev. Lett. **97**, 062502 (2006).
[6] S. Vaintraub, N. Barnea and D. Gazit, nucl-th 0903.1048
[7] P. Navrátil, V. G. Gueorguiev, J. P. Vary, W. E. Ormand and A. Nogga, Phys. Rev. Lett. **99**, 042501(2007); nucl-th 0701038.
[8] A. M. Shirokov, J. P. Vary, A. I. Mazur and T. A. Weber, Phys. Letts. B **644**, 33(2007); nucl-th/0512105.
[9] J.P. Vary, The many-fermion dynamics shell-model code, 1992, unpublished.
[10] J.P. Vary and D.C. Zheng, The many-fermion dynamics shell-model code, ibid. 1994, unpublished.
[11] P. Sternberg, E.G. Ng, C. Yang, P. Maris, J.P. Vary, M. Sosonkina and H. V. Le, Accelerating Configuration Interaction Calculations for Nuclear Structure, Conference on High Performance Networking and Computing, IEEE Press, Piscataway, NJ, 1-12. DOI= http://doi.acm.org/10.1145/1413370.1413386
[12] S. K. Bogner, R. J. Furnstahl, P. Maris, R. J. Perry, A. Schwenk and J. P. Vary, Nucl. Phys. A **801**, 21 (2008) [arXiv:0708.3754 [nucl-th].
[13] T. Dytrych, K.D. Sviratcheva, C. Bahri, J.P. Draayer and J.P. Vary, Phys. Rev. Lett. **98**, 162503 (2007); Phys. Rev. C **76**, 014315 (2007); J. Phys. G. **35**, 123101 (2008); and references therein.
[14] A. Negoita, to be published.
[15] R. Roth and P. Navrátil, Phys. Rev. Lett. **99**, 092501 (2007).
[16] W.E. Ormand, Nucl. Phys. A **570**, 257(1994).
[17] M. Honma, T. Mizusaki and T. Otsuka, Phys. Rev. Lett. **77**, 3315 (1996). [arXiv:nucl-th/9609043].
[18] M. Sosonkina, A. Sharda, A. Negoita and J.P. Vary, Lecture Notes in Computer Science, **5101**, Bubak, M.; Albada, G.D.v.; Dongarra, J.; Sloot, P.M.A. (Eds.) pp. 833 842 (2008).
[19] N. Laghave, M. Sosonkina, P. Maris and J.P. Vary, paper presented at ICCS 2009, to be published.

Visualization as a tool for understanding QCD evolution algorithms

Massimo Di Pierro[1], Michael Clark[2], Chulwoo Jung[3], James Osborn[4], John Negele[5], Richard Brower[2], Steven Gottlieb[6], Yaoqian Zhong[1]

[1] School of Computing, Depaul University; [2] Physics Department, Boston University; [3] Physics Department, Brookhaven National Laboratory; [4] Argonne National Laboratory; [5] Center for Theoretical Physics, MIT; [6] Indiana Univeristy

Abstract. In this paper we present a project of visualizing topological charge in Lattice QCD in order to understand its evolution under the Markov Chain Monte Carlo algorithms and how to maximize its equilibration. Lattice QCD is a very computationally expensive technology for computing properties of composite particles from a few fundamental parameters such as quark masses. Our research plays an important role in validating these computations by showing that the autocorrelation length is small and it can eventually lead to even more efficient algorithms.

1. Introduction and Motivations

The Standard Model of Particle Physics provides a strikingly successful description of the structure of the matter down to scales of 10^{-18}m. The basic degrees of freedom are elementary particles including quarks and leptons which interact with each other according to strong and electro-weak interactions. Quarks play a special role in this model because their interacting are mediated by the exchange of gluons; thus, they create bound states such as the proton, the neutron and the majority of particles that constitute the richness of structures that we observe in the physical world. The part of the Standard Model that describes quarks and gluons is called Quantum Chromo Dynamics (QCD). The ability to perform computations in QCD is of critical importance in Particle and Nuclear Physics. It enables us to determine fundamental parameters of the Standard Model from experiments, compare experimental results with model predictions, and eventually find discrepancies that can shed some light into yet unknown smaller structures and laws of physics at higher energy. It enables us to calculate the equation of state, transport coefficients, and other properties of hadronic matter and the quark gluon plasma explored in relativistic heavy ion collisions at RHIC the LHC. It also enables us to calculate the spectrum, structure, and interactions of mesons and baryons from first principles and lies at the heart of explorations at Jefferson Lab, RHIC spin, and a future electron-ion collider.

Since quarks are subject to strong interactions, standard perturbative techniques are inadequate to perform many interesting computations. For computing the spectrum of the model and computing the lifetime of composite particles, the purely numerical approach known as Lattice QCD that has been the most productive[2].

Lattice QCD consists of taking a small portion of space (a cube of $\sim 10^{-15}$m of size) for a brief time interval ($\sim 10^{-23}$secs), discretizing it on a four dimensional lattice, and simulating numerically the dynamics of QCD inside. The spatial fields do not evolve over time as in ordinary classical systems. Instead, the fields, that are defined on the four-dimensional Euclidean space-time, evolve over a fictitious MCMC time (τ). Lattice QCD computations consist of measuring correlation functions over the possible four dimensional configurations of the fields, each weighted by a number that encodes the dynamics of QCD[1]. Quantities like particles masses and lifetimes can be determined from these correlations.

In practice, the four dimensional configurations are created using Markov Chain Monte Carlo (MCMC) algorithms. Examples are the Molecular Dynamics algorithm or the Hybrid Monte Carlo algorithm. Each configuration is generated from the previous one using a transition in probability that encodes the laws of QCD dynamics.

In this paper, we describe how visualization techniques can be useful to provide information about these algorithms, about the data that is being generated.

The quark fields can be integrated out and, while they play a critical role in the inner workings of the algorithms, they are not part of the actual configurations. The latter only contain gauge degrees of freedom representing gluons, hence they are usually referred to as gauge configurations. Typical gauge configurations have sizes that range from 32×24^3 to 192×64^3 points; which correspond to a size ranging from 256 MBytes to 15 GBytes each (in single precision). A typical Lattice QCD computation generates and utilizes about 1000 of these gauge configurations, thus requiring multiple TBytes of storage. Algorithms are always parallel because of the size of the problem of Lattice QCD.

A typical problem consists in determining whether this small subset of the infinite space of all possible gauge configuration is sufficiently representative and whether the statistical uncertainty determined from the averages can be trusted. From a physical perspective the gauge configurations can be grouped into equivalence classes characterized by their topological charge density.

Our goal is to visualize this topological charge density and measure how fast/slow it evolved under the MCMC for different algorithms and different simulation parameters (the dynamic quark mass in particular). This would provide us with a proof that current Lattice QCD computation are adequately sampling the space of all possible configurations as opposed to getting stuck in one topological sector, and provide a effective tool for developing algorithms that maximize equilibration of the topological charge.

2. Visualization

Typical production gauge configurations appear to be separated by enough MCMC steps and there is no visible continuity in the topological charge. In order to measure autocorrelation in the topological charge we generated *ad hoc* gauge configurations with small trajectories for two different algorithms and different dynamical quark masses.

- **Wilson** Two sets of $24^3 \times 32$ dynamical Wilson gauge configurations with $\beta = 5.6$ and quark masses of $m_1 = 66$MeV and $m_2 = 33$MeV respectively. Each set consists of 1000 trajectories with a 0.1 molecular dynamics time units, saved every 2 time units ($dt = 0.2$). To integrate Hamilton's equations a multi-step 2nd order symmetric symplectic integrator was used (Omelyan) [3].
- **DWF**: One set of $24^3 \times 64$ dynamical gauge configurations generated with Domain Wall Fermions (two degenerate light ones, $m_u = 14$MeV, and one strange, $m_s = 74$MeV) and Iwasaki gauge action at $\beta = 2.13$, saved every 1/6 of a trajectory ($dt = 0.1\bar{6}$). Further details can be found in [4] and the references therein.

Each gauge configuration consists of a set of $SU(3)$ matrices $U_{x,y}$ for every pair of neighbor sites x, y on the lattice. Given a quark field q_y, the $U_{x,y}$ matrix represent the change in the quark field under parallel transport from $y \to x$, *i.e.*, $q_y \to q_x = U_{x,y}q_y$. Therefore $U_{y,x} = U^H_{x,y}$.

The topological charge is computed in three steps according to the definition:

- The gauge field was smoothed by n-iterations of a "cooling step"

$$U_{x,y} \to U_{x,y} = P_{SU(3)}(\alpha U_{x,y} + (1-\alpha)S_{x,y}) \tag{1}$$

  where $P_{SU(3)}$ is a projection operator into SU(3) and $S_{x,y}$ is the average of all products of three consecutive links connecting x with y.
- We computed the chromo-electro-magnetic field defined by

$$F_{x,\mu,\nu} = \frac{1}{2}\mathrm{Re}(A_{x,\mu,\nu} - A^H_{x,\mu,\nu}) \tag{2}$$

  where $A_{x,\mu,\nu}$ is the average of the product of four consecutive links in the μ, ν plane starting and ending at point x.
- We computed the topological charge density, defined by

$$Q_x = \frac{1}{32\pi^2}\sum_{\mu\nu\rho\sigma}\epsilon_{\mu\nu\rho\sigma}\mathrm{Tr}(F_{x,\mu,\nu}F_{x,\rho,\sigma}) \tag{3}$$

In fig. 1 we show three images of iso-surfaces of topological charge density for one gauge configuration for different numbers of cooling steps (0,10,15). In fig. 2 we show three images of iso-surfaces of topological charge density for the same number of cooling steps (20) for different consecutive gauge configurations separated by 6 trajectories.

3. Autocorrelation and Hurst Exponent

In addition to the visualization of the topological charge density we also computed the average autocorreleation $R(\tau)$ and average Hurst exponent H for each independent Markov Chain. First we computed the autocorrelation and the Hurst exponent over the MCMC time (τ) for each lattice point, then we averaged the results over all lattice points. The Hurst exponent is computed using the standard R/S technique described in ref. [5]. In fractal geometry, the Hurst exponent is also known as "index of dependence", it measures the relative tendency of a time series to either strongly regress to the mean or cluster in a direction. H is 0.5 for white noise. H is related to the auto-correlation length via $R(\tau) \simeq \tau^{2H-2}$ and $2 - H$ is known as the "fractal dimension" of the problem.

Our results are summarized in the following table.

Action	quark masses	H
Wilson	66MeV	0.966
Wilson	33MeV	0.973
DWF	69,17MeV	0.908

The autocorrelation is plotted is fig. 3

Notice how the autocorrelation becomes negative at around 100 MCMC steps. This may indicates an alternating pattern in the data, but a larger MCMC chain will be required to validate this hypothesis.

4. Conclusion

Our analysis demonstrates that the autocorrelation length of a topological charge is very small compared with the trajectory length of typical production grade Lattice QCD computations. Our work is still in progress and, as more data is analyzed, we hope to derive an empirical interpolating formula for the autocorrelation length as function of the simulation parameters.

Acknowledgments

This research is funded by the Department of Energy grant DE-FC02-06ER41441. We thank Argonne National Laboratory computing resources, the DOE Advanced Simulation and Computing Initiative (ASCI) for the VisIt software, and Brian Fox at DePaul Univ.

Bibliography

[1] Pierro M D 2008 *Modern Pysics A* **125:012067**
[2] Kronfeld A 2006 *J. Phys. Conf. Ser.* **21-3**
[3] Takaishi T and de Forcrand P 2006 *Phys. Rev.* **E73** 036706 (*Preprint* `hep-lat/0505020`)
[4] Jung C 2007 *POS(LATTICE 2007)* **37** (*Preprint* `arXiv:0710.5337`)
[5] Hurst E 1951 *Trans. Am. Soc. Civil Eng.* **116** 770

Journal of Physics: Conference Series **180** (2009) 012068 doi:10.1088/1742-6596/180/1/012068

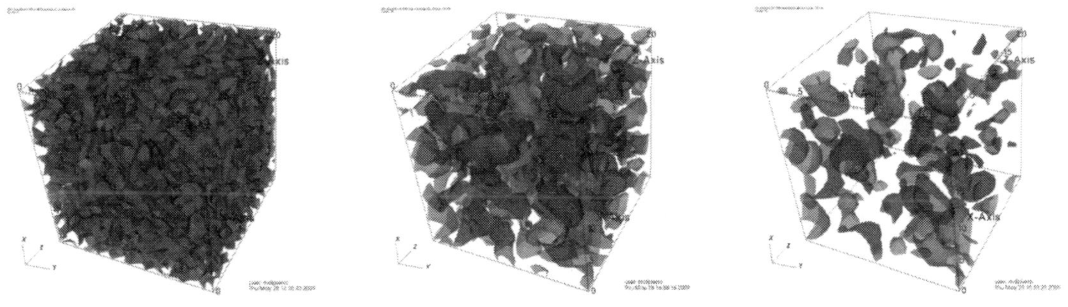

Figure 1. Iso-surfaces for a 3D slice of a topological charge density for one gauge configuration for different numbers of cooling steps (0,10, 15).

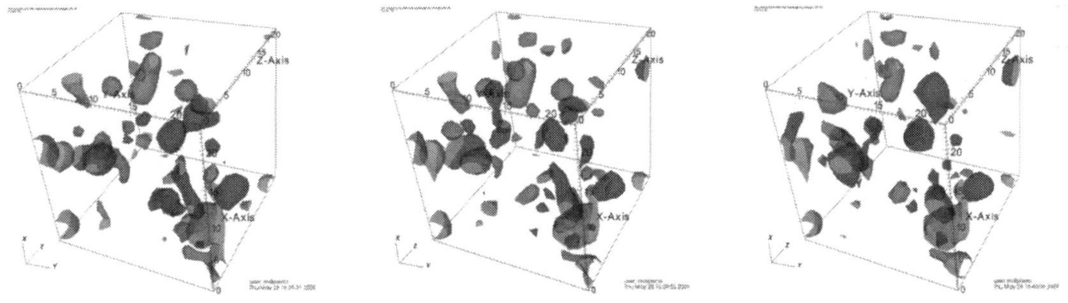

Figure 2. Iso-surfaces for a 3D slice of topological charge density for the same number of cooling steps (20) for different consecutive gauge configurations separated by 6 trajectories.

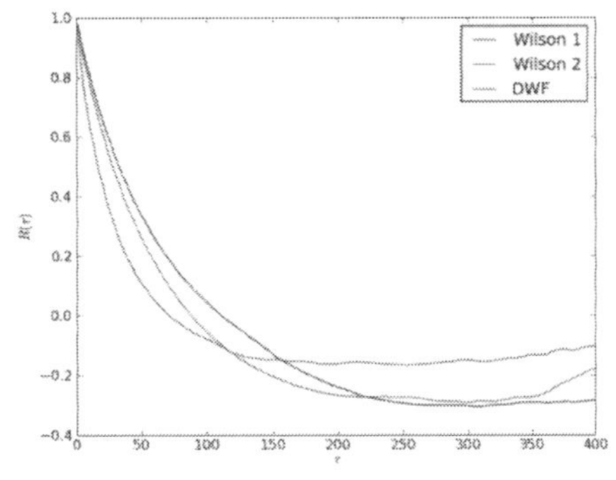

Figure 3. Autocorrelation length for the three sets of gauge configurations considered in the paper.

Occam's razor and petascale visual data analysis

E. W. Bethel[1], C. Johnson[2], S. Ahern[3], J. Bell[1], P.-T. Bremer[5], H. Childs[1], E. Cormier-Michel[1], M. Day[1], E. Deines[4], T. Fogal[2], C. Garth[4], C. G. R. Geddes[1], H. Hagen[7], B. Hamann[4], C. Hansen[2], J. Jacobsen[1], K. Joy[4], J. Krüger[2], J. Meredith[3], P. Messmer[6], G. Ostrouchov[3], V. Pascucci[2], K. Potter[2], Prabhat[1], D. Pugmire [3], O. Rübel[1], A. Sanderson[2], C. Silva[2], D. Ushizima[1], G. Weber[1,4], B. Whitlock[5], K. Wu[1]

[1]Lawrence Berkeley National Laboratory
[2]University of Utah
[3]Oak Ridge National Laboratory
[4]University of California, Davis
[5]Lawrence Livermore National Laboratory
[6]Tech-X Corporation
[7]Technische Universität Kaiserslautern, Germany

E-mail: ewbethel@lbl.gov

Abstract. One of the central challenges facing visualization research is how to effectively enable knowledge discovery. An effective approach will likely combine application architectures that are capable of running on today's largest platforms to address the challenges posed by large data with visual data analysis techniques that help find, represent, and effectively convey scientifically interesting features and phenomena.

1. Introduction

The principle of *Occam's Razor*, which is attributed to the 14th century English logician and Franciscian Friar William of Ockham, is roughly translated from Latin as "entities must not be multiplied beyond necessity"[1]. The principle suggests that when there are multiple competing hypotheses, the simplest one is usually the correct explanation. Another interpretation, and one we focus on here, is that "less is more." Within the context of modern computational science and visual data exploration and analysis, applying this principle yields a fruitful course of research and development: simulations and experiments produce larger and more complex data, yet the sheer size and complexity of these results confounds attempts to gain insight; insights are drawn from simple yet elusive observations.

Our colleagues in the computational and experimental sciences are using increasingly powerful computational technology to their advantage. The advantage for science is that there is the opportunity to model scientific phenomena with increasing physical realism and to collect and store data at increasingly finer spatiotemporal resolution. Data from both simulations and experiments exhibits increasing spatial resolution: climate models, for example, are moving towards sub-kilometer resolution, resulting in data files that are unwieldy at best and that at

[1] `http://en.wikipedia.org/wiki/Occam's_razor`

worst confound existing analysis and visualization algorithms and implementations. The grid or sample values are evolving from simple scalars or vectors to more complex types like tensors, functions, and distributions. For some of these data types, there do not exist any satisfactory method for visual data exploration. Large and distributed systems enable larger ensemble runs, which produce many more datasets. The process of studying the relationships between many datasets and input parameters grows increasingly difficult due to the sheer volume of data and the potential complexity of the analysis. More powerful platforms enable codes to model phenomena for a longer duration in simulation time, thereby increasing temporal complexity.

From a visual data analysis perspective, the resulting challenges are numerous and profound and provide the basis for much of the current research in our community. For example, [1] concludes that our ability to generate data far exceeds our ability to gain insight from such data using existing techniques, even if scaled up to run on large, parallel platforms. This theme restates the findings of an earlier work [2] pointing out that the sheer size and complexity of scientific data is itself an impediment to scientific insight.

Insight comes in many forms and via many different paths. From a visualization perspective, there are three broad visualization use modalities [3]. "Presentation visualization" is where you know what is there and want to show it to someone else. "Analytical visualization" is where you know what you are looking for. "Discovery visualization" occurs when you have no idea what you're looking for. Discovery visualization is characterized as an "undirected search," or "unconstrained navigation" through visualization or rendering parameter space. One might expect that most "unexpected" discoveries would occur in discovery and analytical visualization, with these unexpected results clearly communicated to others through presentation visualization techniques.

One central thesis of this article is that focusing visualization research on techniques for discovery and analytical visualization with a special emphasis on the challenges posed by science in the petascale era will yield the greatest impact, or potential for enabling scientific knowledge discovery. One conclusion we hope to convey here is that simply making higher resolution images of higher resolution data exacerbates the problems pointed out in earlier reports, e.g., [2], where the scientist is burdened with more work. Instead, our approach is to realize an effective implementation of Occam's Razor, where petascale visual data analysis produces less work for the scientist, thereby increasing the likelihood of scientific insight.

We discuss this theme in several different dimensions in this article. First, we, like many other scientific fields, seek to leverage increasingly powerful computational platforms to tackle ever larger problems. We present results in Section 2 where we show effective use, from a visual data analysis perspective, of today's largest platforms on problem sizes of unprecedented scale. While this result is noteworthy in its own right, Section 3 takes a different approach by combining advances in visualization and data management to tackle a hero-sized problem in accelerator modeling to replace a manual process that previously required hours with yesterday's "small" datasets with one that takes seconds on today's "hero-sized" datasets. Next in Section 4, we present work that uses advances in computational geometry to discover elusive and difficult to quantify features in scientific data to produce insights never before possible using traditional visualization or analysis technologies. Finally, Section 4 presents advances in visualization technology that help to convey deeper understanding of flow-based data.

2. Petascale visual data exploration and analysis

2.1. Introduction

As the by-product of advances in technology is "more and more data," one issue facing the visualization and analysis community is the feasibility of using today's largest computational platforms for knowledge discovery. To gain better insight into this issue, VACET researchers recently conducted a series of experiments aimed at fostering a better understanding of

Journal of Physics: Conference Series **180** (2009) 012084 doi:10.1088/1742-6596/180/1/012084

functional and performance limits that might be encountered when running a production-quality visualization application at extreme levels of concurrency on datasets of unprecedented size. The results, which we discuss in this section, suggest this approach is viable and that visualization research and development efforts have produced technology that is today capable of ingesting and processing tomorrow's datasets.

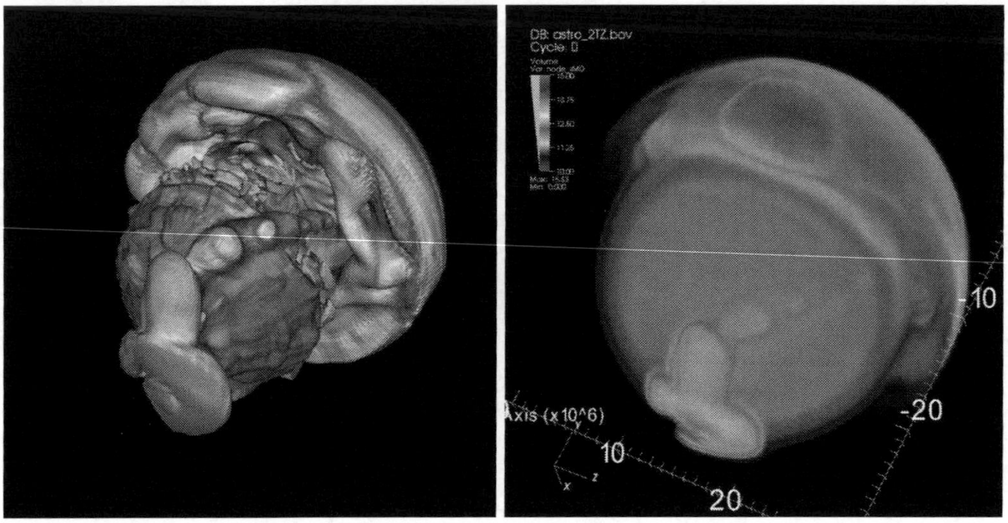

(a) Isocontouring of two trillion zones on 32,000 Opteron cores of JaguarPF, a Cray XT5 at OLCF/ORNL.

(b) Volume rendering of two trillion zones on 32,000 Opteron cores of Franklin, a Cray XT4 at NERSC/LBNL.

Figure 1. Our functional performance experiments consist of loading extremely large datasets and executing visualization algorithms at extreme levels of concurrency producing images of isocontouring (left) and volume rendering (right).

Another purpose of these runs was to prepare for establishing VisIt's credentials as a "Joule code," or a code that has demonstrated scalability at a large number of cores. VisIt is the first and only visual data analysis code that is part of the ASCR Joule metric, which aims to track code performance (scalability) over a period of time.

The team's experiments consisted of running the VisIt software application [4] on several of the nation's largest computing platforms and on dataset sizes ranging from 500 billion (two terabytes per scalar) to 2 trillion cells (eight terabytes per scalar) and at concurrency levels ranging from 8000 to 32,000 cores. Each experiment consisted of running VisIt in parallel: loading in data, performing two common visualization tasks (isosurfacing and volume rendering), and producing an image (Figure 1).

2.2. Parallel processing of data

As there is some variation in how tools process data in parallel, we will describe VisIt's approach for parallel processing. VisIt employs a client-server model. The VisIt server is parallelized using an MPI-based communication model and can run on a large, parallel platform. The VisIt client, which runs on the user's workstation, communicates with the VisIt server over an IP-based connection. To begin, the end user sets up the visualization or analysis task from the VisIt client-side interface. The client then communicates this information to MPI rank 0 of the server. MPI rank 0, in turn, communicates this information to the other processors through a broadcast command. Each processor then sets up a data flow network, which consists of modules to read data, process data, and ultimately to render data to images. VisIt's parallelization approach uses domain decomposition so that each MPI process, which runs data I/O, processing, and rendering, operates only on its data subset. The partitioning occurs by operating on meta-data, for example by partitioning pre-defined domains.

Once the data flow networks are set up on each MPI process, execution begins. First, each process sets up its reader to load only the data for its subset of the larger data set. Next, each process executes one or more algorithms (e.g., isocontouring) with user-specified parameters (e.g., isocontouring level). For processing sequences that includes rendering, each processor renders its portion of the problem into an image, then the resulting images from all processes are assembled into a final image. The resulting image is then transferred to the client and displayed to the user. The processing and rendering for volume rendering is somewhat different; the volume rendering algorithm is not embarrassingly parallel. Our approach is described in more detail in [5].

2.3. Experiment description

One goal of this experiment was to demonstrate the viability of these techniques across diverse supercomputing environments, in terms of operating system, I/O performance, FLOPs, and network bandwidth. We performed these tests on on Crays (OLCF/ORNL's Jaguar & NERSC/LBNL's Franklin), a Sun Linux machine (TACC's Ranger), a CHAOS Linux machine (LLNL's Juno), and an AIX machine (LLNL's Purple).

Because each machine has a different number of cores, we performed a weak scaling study. We started with a single data set, which we then upsampled to an appropriate resolution. This data set was from a core-collapse supernova simulation done by the CHIMERA code, on a curvilinear mesh of more than three and one half million cells[2]. We chose the upsampling approach since we are not aware of any current datasets containing 2T cells and since the primary objective for our studies is to better understand the performance and functional limits of parallel visual data analysis software. These objectives can be achieved using upscaled data.

The upsampling process involved interpolating a scalar field onto a high resolution rectilinear mesh and then writing the data out as compressed binary data (gzipped). There were ten files for every core used, and every file contained 6.25 million data points, for a total of 62.5 million data points per core. In our experience, the visualization tool often has one tenth (or less) of the resources (e.g., cores) as the simulation code and simulations codes often write out one file per core. Hence, having multiple files per core was our best approximation at emulating these common real world conditions.

We ran with 16000 cores on each machine visualizing one trillion cells[3]. On the Jaguar and Franklin machines, we ran additional, larger-sized problems consisting of two trillion cells on 32000 cores. Although times varied from machine to machine, I/O was the dominant factor (taking two or more minutes at 16000 cores), with contouring taking approximately ten seconds and rendering taking one to ten seconds.

2.4. Issues discovered during scaling study

Although the majority of VisIt's infrastructure scaled well to a large number of cores, we ran into several obstacles along the way:

- VisIt's MPI rank 0 was collecting status information from all other processors through point-to-point communications, which is non-scalable. We worked around this issue for our study and plan on implementing a complete solution for a future release of VisIt.

[2] Sample data courtesy of Tony Mezzacappa (ORNL), Bronson Messer (ORNL), Steve Bruenn (Florida Atlantic University) and Reuben Budjiara (University of Tennessee).

[3] We ran with only 8000 cores and one half trillion cells on Purple, because the full machine has only 12208 cores, and only 8000 are easily obtainable for large jobs.

- VisIt's volume rendering algorithm was attempting an optimization for sample point communication that required a buffer that had an $O(nProcs \times nProcs)$ space requirement. This "optimization," while appropriate for low levels of concurrency, proved to cause problems at high levels of concurrency. The workaround for these tests was to remove this "optimization." We are evaluating the best course of action for a solution for a future release of VisIt.
- We observed that the loading of shared libraries took quite a long time at scale (as much as five minutes) and VisIt's plugin model may need to adapt for this case, likely by switching to a static binary with precompiled plugins.

2.5. Lessons learned

The primary objective for these experiments was to gain a better understanding of functional and performance limits when running visual data analysis applications at extreme levels of concurrency and and problem sizes. We encountered a couple of minor problems that will be rectified and appear in a future public VisIt release.

From a visual data analysis perspective, these problem sizes and concurrency levels are a "first." The successful completion of these functional and performance tests show progress towards petascale computing by demonstrating that today's technology is capable of ingesting and processing tomorrow's datasets.

The performance data the team collected during the experiments reveals insights into potential bottlenecks and opportunities for performance optimization on different machine architectures at high levels of concurrency and ultrascale datasets. Future work will include a more detailed, end-to-end performance study of several different visualization algorithms to better understand performance limits and opportunities for VisIt, a production-quality visual data analysis software application.

3. Accelerator modeling

3.1. Background

Laser wakefield simulations model the behavior of individual particles as well as the behavior of the plasma electric and magnetic fields. Output from these simulations can become quite large: today's datasets, such as the ones we study here, can grow to be on the order of 200GB per timestep, with the simulation producing ≈ 100 timesteps. The scientific challenge we help address in this study is first to quickly find particles that have undergone wakefield acceleration, then trace them through time to understand acceleration dynamics, and perform both visual and quantitative analysis on the set of accelerated particles. This approach better serves science needs than attempting to visualize and analyze *all* particles at once.

To solve this problem, we focused on combining and extending two different but complementary technologies aimed at enabling rapid, interactive visual data exploration and analysis of contemporary scientific data [6]. To support highly effective visual data exploration, knowledge discovery and hypothesis testing, we have adapted and extended the concept of parallel coordinates, in particular binned or histogram-based parallel coordinates, for use with high performance query-driven visualization of very large data. In the context of visual data exploration and hypothesis testing, the parallel coordinates display and interaction mechanism serves multiple purposes. First, it acts as a vehicle for visual information display. Second, it serves as the basis for the interactive construction of compound Boolean data range queries. These queries form the basis for subsequent "drill down" or data mining actions. To accelerate data mining, we leverage state-of-the-art index/query technology to quickly mine for data of interest as well as to quickly generate multiresolution histograms used as the basis for the visual display of information. This combination provides the ability for rapid, multiresolution visual data exploration, and implements an easy-to-use interface for composing multivariate and multidimensional queries in a high performance, query-driven visualization and analysis system [7].

We apply this new technique to large, complex scientific data created by a numerical simulation of a laser wakefield particle accelerator. In laser wakefield accelerators, particles are accelerated to relativistic speeds upon being "trapped" by the electric fields of plasma density waves generated by the radiation pressure of an intense laser pulse fired into the plasma. These devices are of interest because they are able to achieve very high particle energies within a relatively short amount of distance when compared to traditional electromagnetic accelerators. The VORPAL [8] simulation code is used to model experiments, such as those performed at the LOASIS facility at LBNL headed by W. Leemans [9], and is useful in helping to gain deeper understanding of phenomena observed in experiments [10], as well as to help formulate and optimize the methodology for future experiments.

One scientific impact of our work is that we have vastly reduced the duty cycle in visual data exploration and mining. In the past, accelerator scientists would perform the "trace backwards" step using serial scripts that performed a search at each timestep for a set of particles. Runtimes for this operation were on the order of hours. Using our implementation, those runtimes are reduced from hours to seconds.

In our implementation, a user will see an "overview" of all data in the multidimensional dataset, then select subsets for subsequent visual data exploration and analysis. The multidimensional presentation of data – both the "overview" as well as subset selections – is a statistical display known as a *parallel coordinates* plot. Examples of such visual data exploration and analysis are shown in Figure 2, and are described in more detail in [6].

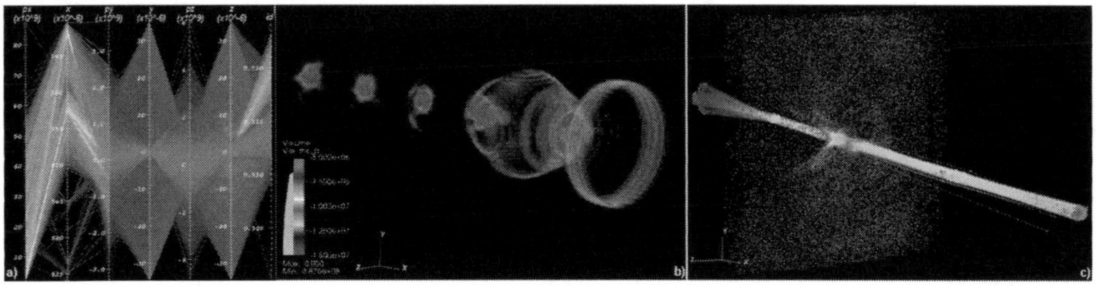

Figure 2. The interface on the left, which shows a multivariate view of particle distribution across all simulation variables, is a vehicle for selecting particles having high acceleration and spatial coherency. All particles are shown in gray, and selected particles shown in red. This selection interface is linked with other forms of visual data exploration, so that a physical view of selected particles (middle, red particles) helps the scientist to quickly gain insight into the relationship between statistical-space and physical-space features. Paths of "interesting" particles over a range of timesteps, shown on the right, are colored to indicate particle momentum. That image shows particles are constantly accelerated over this range of time. This new capability, developed as part of VACET's research and development portfolio, is now part of widely distributed, production-quality, parallel capable visual data exploration software infrastructure (VisIt). Image, courtesy O. Rübel et al., LBNL, generated from VORPAL simulation results, courtesy C. Geddes et al., LBNL.

3.2. Multidimensional data display: histogram-based parallel coordinates

Parallel coordinates, believed to be introduced in 1885[4], provide an effective interface for displaying multivariate data and for defining multi-dimensional queries (range selections) based on thresholding. The idea of combining parallel coordinates displays with physical space views of subset selections is not new. Applications range from multidimensional geometry [11, 12], GIS/remote sensing [13], data mining [14], currency and economic analysis [15]. These concepts have had success in the visual data exploration and analysis of particle-based data in life sciences data [16–18], and multiple teams concurrently and explored applying these concepts to fusion simulation data [19].

Using sliders attached to each axis of the parallel coordinates plot, a user defines range thresholds in each displayed dimension. By rendering the user-selected data subset (the focus view) in front of a parallel coordinates plot created from the entire data set (or in many cases a subset of it) (the context view), the user receives immediate feedback about general properties of the selection. Data outliers stand out visually as single or small groups of lines diverging from the main data trends. Data trends appear as dense groups of lines (bright colored bins in our case). A quick visual comparison of the focus and context views helps to convey understanding about similarities and differences between the two.

In practice, parallel coordinates have disadvantages when applied to very large datasets. First, each data record is represented with a single polyline that connects each of the parallel coordinates axes. As data size increases, the plot becomes more cluttered and difficult to interpret. Second, data records drawn later will occlude information provided by data records drawn earlier. Finally and worst of all is the fact this approach has computational and rendering complexity that is proportional to the size of the dataset. As data sizes grow ever larger, these problems become intractable.

To address these problems, which stem from the attempt to use an existing visualization technique on very large datasets, we developed an efficient rendering technique based on two-dimensional (2D) histograms. Rather than viewing the parallel coordinates plot as a collection of polylines, one per data record, we approach rendering by considering instead the relationships of all data records between pairs of parallel coordinate axes. That relationship can be discretized as a 2D histogram and then later rendered. This idea was introduced in earlier work [20], and we extended the basic idea to accomodate very large datasets through a combination of adaptive histograms, novel rendering, and rapid histogram computation.

Complementary to the histogram-based rendering technique is the need to quickly compute histograms from large data and user selections. Our approach is to leverage FastBit to quickly generate conditional histograms. This approach has proven useful in the past in a "hero-sized" cybersecurity visual analytics and data mining application [21]. The key here is to provide an implementation that runs sufficiently quickly as to enable *interactive* visual data exploration of large particle-based datasets. Our implementation, described in more detail in [6], achieves such performance levels through extensions to FastBit and use on parallel computing platforms to accelerate I/O and histogram computation.

3.3. Particle paths and visual data analysis

Once "interesting" particles are located, the next step in the knowledge discovery process is to learn something about the temporal characteristics of these particles. In our earlier work [6], we iteratively formulate complex multidimensional queries to isolate interesting particles and display them. Typically, the multidimensional query is applied at a timestep after particles have begun to undergo acceleration. From that query results a list of particles; we then use the FastBit machinery to find those named particles across all simulation timesteps and display

[4] See `http://en.wikipedia.org/wiki/Parallel_coordinates`

them.

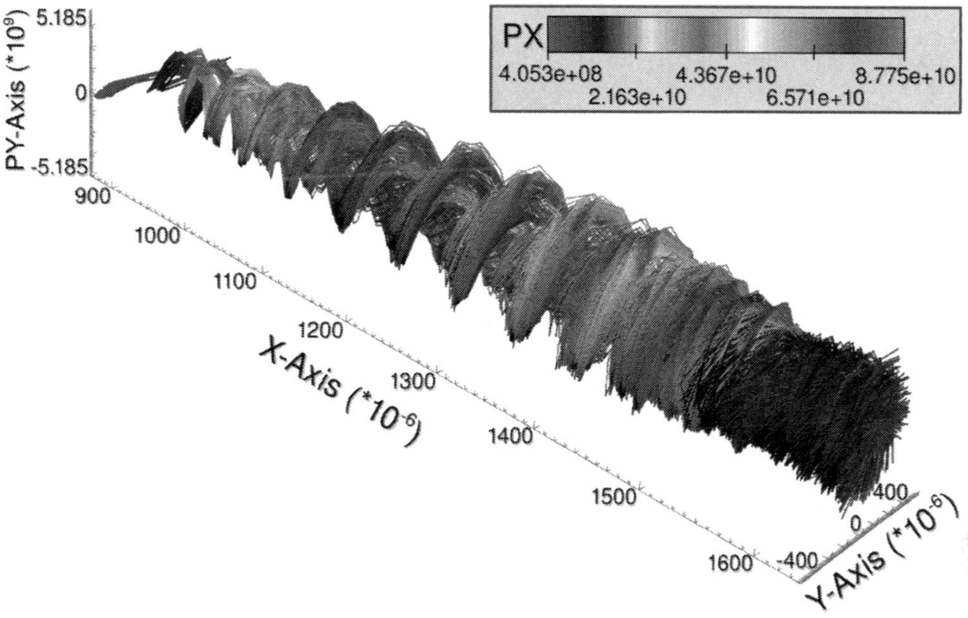

Figure 3. Particle paths colored by x-momentum so that color indicates the level of acceleration. The spiral structure of the paths illustrates the particles' oscillating motion over time. Image courtesy O. Rübel et al., LBNL.

Figure 3 shows recent results where we employ visual data analysis techniques to help reveal scientifically interesting information. Figure 3 shows a subset of high-energy particles traced through time. The paths are shown in (x, y, py) space (x and y are positions in space, while py is a particle's momentum in the y direction) colored by px, a particle's momentum in the x direction, so that color indicates the level of acceleration of particles. The particles move along the paths from left to right. The spiral structure of the paths illustrates the oscillating motion of the particles in y. Along any given particle pathline, color transitions reveal interesting phenomena. First, particles exhibit low momentum in x when they first enter the simulation (blue at the left). Second, after being injected, the particle are constantly accelerated (transition from blue to red). Finally, later in the simulation – at $x \approx 1250\mu m$ – the particles "outrun" the wave and decelerate (transition from red to blue). As they accelerate, the tracks show the particles oscillate transversely in the focusing field. Figure 3 shows the paths only until the particles have completed the deceleration process.

3.4. Knowledge discovery: traditional analysis

Complementary to high performance visual data exploration technology, which helps with discovering characteristics of particles undergoing acceleration, are automated methods for finding and analyzing such particles of interest. Recent work [22] combines signal processing and machine learning techniques to locate particles undergoing a high degree of acceleration and that are spatiotemporally compact. Once individual particles of interest have been identified, they are grouped into compact "bunches" using unsupervised fuzzy clustering. Concurrently, the pathlines of the bunches are derived using a graph-based algorithm. This approach proved sufficiently robust to successfully identify high-energy beams in multiple datasets that exhibit markedly different characteristics and behavior. This type of approach holds promise for helping to accelerate scientific knowledge discovery by finding and analyzing "features" that are difficult to define and that vary from dataset to dataset.

4. Topologically based feature detection and tracking

4.1. Introduction: topological data analysis

In recent years, topological structures have proven to be very useful for building very general "feature" definitions [23] in both visualization and quantitative data analysis. The success of these feature definitions stems from the fact that topological structures are based on fundamental properties of isosurfaces. An isosurface [24, 25] connects all locations in a domain where the scalar field assumes a specified isovalue and is an extremely versatile and ubiquitous visualization and analysis component. By varying the isovalue and observing changes to the corresponding isosurface, one gains considerable insights into properties of that scalar. Isosurfaces arise naturally in data analysis since distinct isosurfaces often have a useful physical interpretation directly related to the application domain. For example, in premixed combustion simulations, an isotherm (a surface of constant temperature) of an appropriate temperature is often associated with the location of the flame.

In the context of scalar field analysis, topological properties provide a quantifiable characterization of fundamental isosurface properties: the number of connected isosurface components and the genus of an isosurface (i.e., the number of independent tunnels or "holes" in the surface). Morse theory [26] provides insight into the topological evolution of isosurfaces of scalar functions: isosurface topology changes at distinct isolated *critical points* in the scalar field that can be identified and analyzed analytically or combinatorially.

Critical points identified by the Morse theory constructions provide hints about the location of interesting behavior in a scalar field. However, as a set of individual disconnected points they are only of limited use to decipher global structures in a data set. Topological structures relate critical points to each other and provide a means for ranking topological features and simplifying the global topological structure [27, 28]. This simplification supports further "boiling down" of the data to essential features and extracting a simpler, overall coarser structure.

The *Reeb graph* [29] is one of these structures; it expresses the evolution of individual contours as a graph that is defined by these critical points and their relationships. The Reeb graph describes the evolution of contours, i.e., individual connected components of a level set, as the value for which we extract the level set changes. Essentially, the Reeb graph is a skeleton of the manifold. Edges of the Reeb graph correspond to families of topologically equivalent contours, and nodes correspond to critical points where the number of contours change. For simply connected domains, the Reeb graph is always a tree structure, called a *contour tree* [30].

The *Morse complex* and the *Morse-Smale complex* [31–33] comprise alternate means of relating critical points to each other. Assuming the function on the manifold is differentiable, and thus the gradient is defined at each location, it is possible to start at any location in the domain and follow a gradient line either to its origin or its destination, i.e., follow a line of steepest descent or ascent from that point. This line of steepest ascent/descent will end in a local maximum/minimum. The Morse complex is the segmentation of the domain into ascending or descending manifolds of the critical points. An ascending/descending manifold of a maximum/minimum is the union of all locations in the domain for which the path of steepest ascent/descent ends at that maximum/minimum. Superimposing ascending and descending manifolds results in the Morse-Smale complex.

Using contour tree simplification, it is possible to manipulate isosurfaces in terms of fewer, important connected components [27]. Furthermore, one can represent the contour tree in a hierarchical layout that makes it possible to explore the global structure of a data set and subsequently obtain a more detailed topological representation in regions of interest [34]. Finally, it is possible to represent the topological information contained in the contour tree in a more intuitive fashion [35]. Use of a multiresolution-representation of the Morse-Smale complex [28, 33] has also proven useful in analyzing scientific simulation data. For example, it is possible to specify features, such as bubbles, that are difficult to characterize using traditional feature detection methods, and use this definition for analyzing turbulent mixing simulations [36]. More recent work used the Morse-Smale complex for calculating clean distance fields in porous solids hit by a projectile [37] and used the results to calculate measures like filament density. Other topological structures, such as Jacobi sets, allow relating critical points between different functions/simulations [38]. Recently, Bremer et al. [23] gave an overview of topology-based feature definition and feature tracking.

4.2. Science application: analysis of turbulent, premixed hydrogen flames

Understanding combustion processes over a broad range of operational regimes is of great interest for a variety of applications like engine or power plant design. To this end, there has been considerable recent interest in developing premixed burners capable of stably burning ultra-lean hydrogen-air fuel mixtures. Such burners could, for example, be used as one component of a clean-coal power plant utilizing hydrogen extracted from coal gasification. Lean premixed systems are subject to a variety of hydrodynamic and combustion instabilities that render practical flame stabilization, and traditional approaches to flame analysis, extremely difficult. The flames burn in a cellular mode that is highly nonuniform, time-dependent, and difficult to characterize [39].

To study this combustion process, researchers at the Center for Computational Sciences and Engineering at Lawrence Berkeley National Laboratory performed numerical simulations of lean premixed hydrogen flames at three different levels of turbulence: no turbulence, weak turbulence, and strong turbulence. Each flame burns in a volume periodic in the x- and y-coordinate with pre-mixed fuel being injected through the plane $z = 0$. The flame simulations were carried out with a low Mach number model that incorporates a detailed description of the combustion kinetics and differential species transport. For more details of the model and its implementation, see [40]; for example analysis of lean hydrogen flames in this regime, see [41]. Each solution analyzed here is represented as a sequence of three-dimensional snapshots, taken at uniform intervals in time, where each snapshot consists of cell-centered data on a uniform Cartesian grid describing the temperature, chemical composition, and effective fuel consumption rate. The "flame" is represented as an isotherm extracted from the datasets using standard techniques. The local fuel consumption rate, interpolated to these isosurfaces, is used to divide the surface into burning "cells" separated by non-burning regions, as defined by a threshold in the consumption rate. To gain new insights into the combustion process, scientists are interested in two types of analysis: time-aggregate statistics and detailed analysis of cell evolution.

4.3. Characterizing influence of turbulence

The top row of Figure 4 shows isotherms of combustion simulations performed at the three turbulence levels under consideration. Mapping the fuel consumption rate to isotherm color, the cellular structure of the burning process becomes immediately obvious. Essentially we are interested in the influence of turbulence level on the size and evolution of these burning cells.

It is possible to use the Morse complex to study the evolution of the cellular burning patterns observed in lean premixed hydrogen-air flames. One can construct a quantitative analysis through segmentation of this flame isotherm based on threshold values of the local burning rate. This thresholding divides the isotherm contour into individual burning cells that evolve in time. Such a definition allows a quantitative analysis of the influence of turbulence on the evolution of these flame structures, for example. However, the network of these burning cells (and its dual, the network of extinction regions) may depend strongly on the threshold that determines the segmentation, and there is no universally accepted threshold value. It is crucial then to

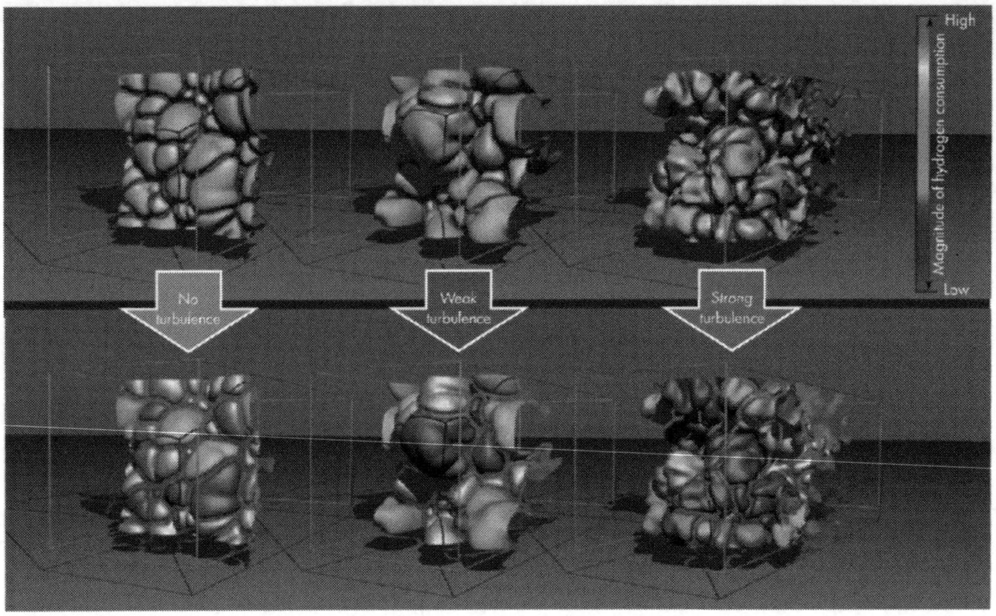

Figure 4. The top 3D panels show flame surfaces of a lean premixed hydrogen flame at different levels of turbulence colored by the local fuel consumption. In the bottom 3D panels a small set of burning cells is randomly colored to show the irregularity of the more turbulent cells.

examine the statistical characterization of these flame cells as a function of the threshold value to ensure that the conclusions drawn from the analysis do not depend on the detailed values of this rather arbitrary parameter [39]. The Morse complex [42] provides a framework to represent segmentations for all possible thresholds and thus for evaluating the segmentation for a range of thresholds and determining the stability of these arbitrary parameters. Furthermore, it supports simplifying the topological structure of burning regions, merging very small burning regions into larger ones, thus simplifying analysis.

Once an appropriate choice of parameters has been determined we track the corresponding cells over time. A tracking graph can be constructed efficiently from a three-dimensional hypersurface (embedded in four-dimensional space-time) of the evolution of the flame isotherm over time. Based on this hypersurface and a segmentation in each time step, boundaries of burning regions as they evolve over time can be constructed. The Reeb graph of this system (Figure 5) can efficiently represent a complete tracking graph of the evolution of the burning regions over time, including their creation, destruction, splitting and merging behavior, and provides a basis for quantitative analysis of their evolution.

4.4. Lessons learned

Our methods allow, for the first time, a quantitative analysis of the cellular burning structures (Figure 6), and yield important scientific insights: Contrary to common intuition, higher turbulence levels, which generally lead to increased energy in the finer length scales of the approach flow to the flame, apparently lead to *larger* cellular burning structures. Moreover, these larger cells tend to burn more intensely than would be expected from simple theories of flame propagation. The combination of these two effects dramatically increases the global propagation speed of the turbulent flames over the steady flat flame case.

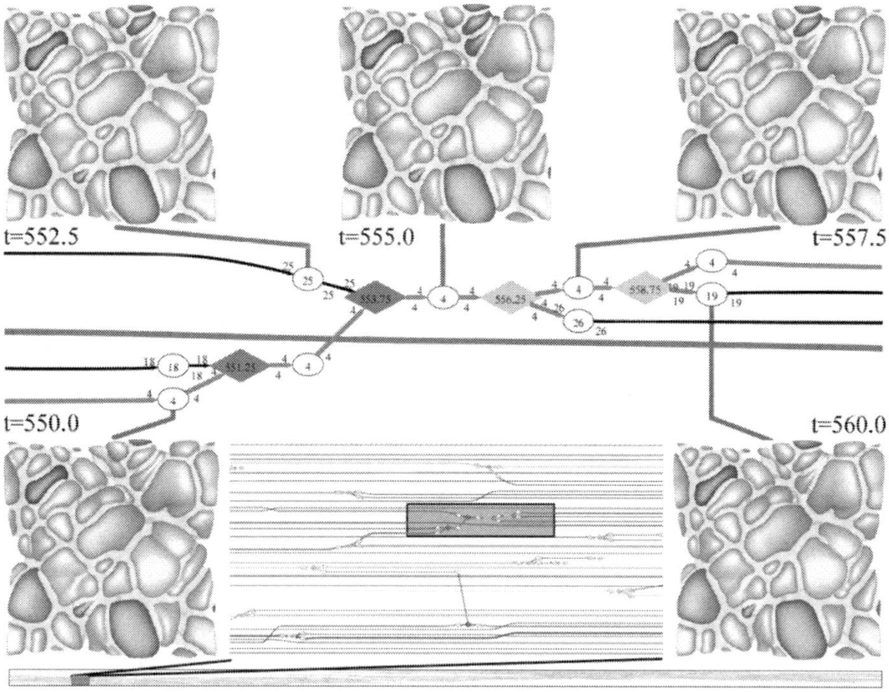

Figure 5. Subsection of the tracking graph of a combustion simulation of the "no turbulence" case that is compressed to only show two merge and two split events. Arc color corresponds to region color in corresponding segmentations. Round nodes correspond to cells explicitly segmented by the Morse complex, diamonds to topological events between time steps. Red signifies a merge, green a split, and turquoise a birth/death event. The figure shows three zoom levels of the graph: graph for the entire simulation (bottom graph), a portion corresponding to several subsequent time steps (middle graph) and a zoom into the two merge and split events annotated with colored segmentations of the isotherm.

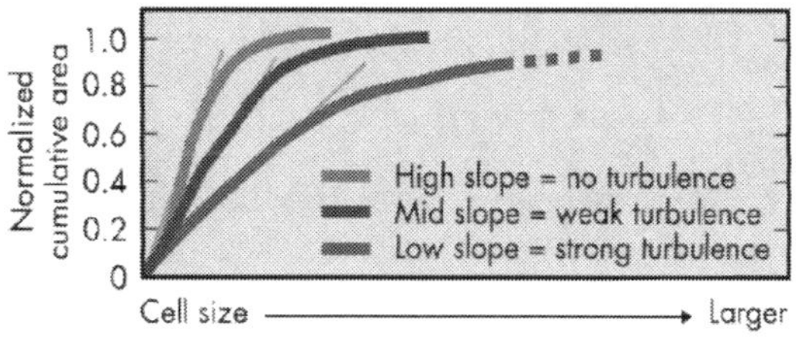

Figure 6. Cumulative density function of cell area distributions. The graph reveals that more turbulence creates larger cells with a wider distribution of normalized surface areas indicating a more intense burning process.

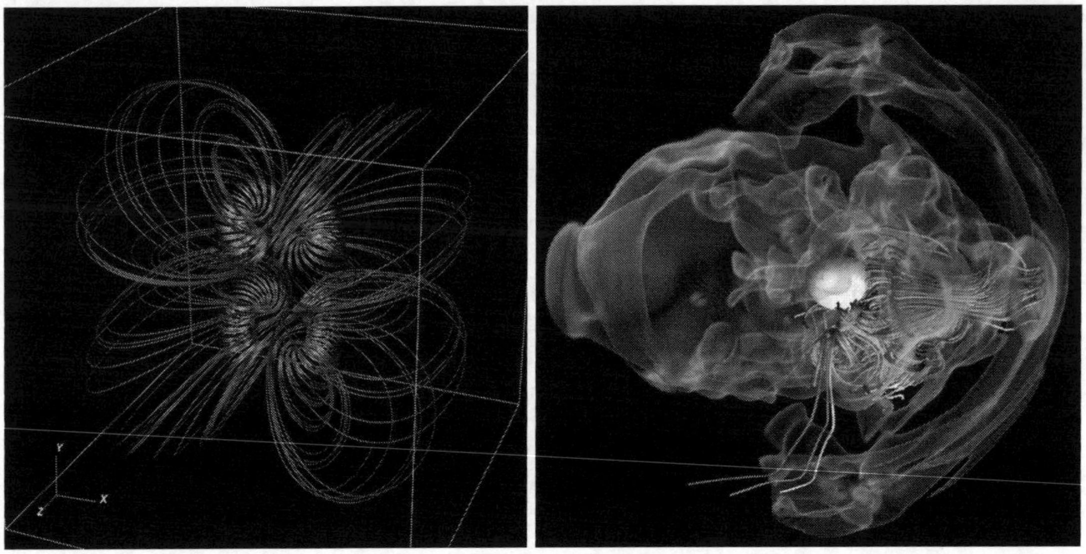

Figure 7. Integral curves illustrate the paths of massless particles in a vector field. The left image is generated from the numerical simulation of vortex merging (simulation courtesy ANAG, LLNL [44]). On the right, streamlines illustrate the magnetic field of a collapsing stellar core (simulation from E. Endeve et al., ORNL).

5. High performance, parallel, and accurate computation of particle trajectories, integral curves and integral surfaces

5.1. Introduction: integration-based visualization

Modern methods for the analysis and visualization of vector fields, which play a key role in many application domains like flow simulation, astrophysics and fusion, are built on empirically studying the behavior of massless particles advected with the vector field. These so-called integration-based or Lagrangian methods offer unparalleled insight into vector field structures especially for time dependent vector fields [43].

One simple yet effective visualization approach for vector fields is the straightforward depiction of particle trajectories, as described by vector field integral curves. For a vector field $v(t, x)$, an integral curve started at a point x_0 at time t_0 is defined by the ordinary differential equation

$$\dot{\phi}(t) = v(t, \phi(t)), \quad \text{with initial condition } \phi(t_0) = x_0,$$

and is readily approximated using numerical integration schemes. The resulting curves can then be directly depicted graphically to allow an identification of general vector field behavior (see Figure 7).

5.2. Integral surfaces

Recently, VACET scientists have developed a novel approach of grouping trajectories of particles emanating from a common curve seeding into so-called integral surfaces. Technically, an integral surface S corresponding to a seeding curve C is defined by

$$\frac{\partial}{\partial t} S(s, t) = v(t, S(s, t)) \quad \text{where } S(s, t_0) = C(s).$$

The resulting visualizations are a significant improvement over visualizing individual curves due to the fact they offer the possibility for greatly improved visual understanding of complex 3D flow (see Figure 8).

To approximate integral surfaces, the continuum of integral curves over s is reduced to a finite number of such curves that represent a skeleton of the integral surface. However, due to the strong deformation by the flow a fixed discretization does not result in a good approximation of the surface; rather, the newly developed algorithm adapts the discretization to the observed distortion of the surface as it progresses through the vector field [45, 46]. This allows the generation of high-quality surfaces exhibiting fine detail using a vastly reduced number of integral curve computations in comparison to a naive approach. Moreover, care is taken to preserve temporal locality during the surface propagation, permitting computation of surfaces in large time-dependent vector fields in an out-of-core, streaming fashion.

5.3. Lagrangian coherent structures

The recently introduced notion of Lagrangian Coherent Structures (LCS) and the associated Finite-Time Lyapunov Exponents (FTLE) measure (e.g. [47, 48]) have garnered much attention in both visualization and application science communities for their ability to visualize and analyze time-varying vector fields in terms of the convergence and divergence of neighboring particles. FTLEs are a finite-time analogue of Lyapunov exponents in dynamical systems; they measure the average exponential separation rate between neighboring particles over a time interval Δt via

$$\sigma := \frac{1}{|\Delta t|} \log \|D\phi\|_2,$$

where $\|D\phi\|_2$ denotes the spectral norm of the spatial derivative of the particle advection over Δt. LCS are then identified as ridge lines in the resulting scalar measure σ. It can be shown that these LCS represent particle trajectories of maximal hyperbolicity and hence form a "dynamic skeleton" of a vector field. Through these structures, it is possible to accurately visualize the fully dynamic nature of vector fields, as exemplified in Figure 9. In contrast, earlier techniques such as topological analysis are unable to easily capture time-varying structures [49]. Furthermore, FTLEs enable the representation of the structural dynamics in terms of scalar fields, opening up vector field structural analysis to a host of existing methods developed in the context of scalar fields. VACET scientists have delivered visualization tools that enable fast computation and interactive visualization of vector fields using these techniques. Furthermore, they continue to examine the value of FTLEs and closely related measures can be generalized to other applications domains beyond fluid flow such as dynamical systems and medical visualization (see Figure 10).

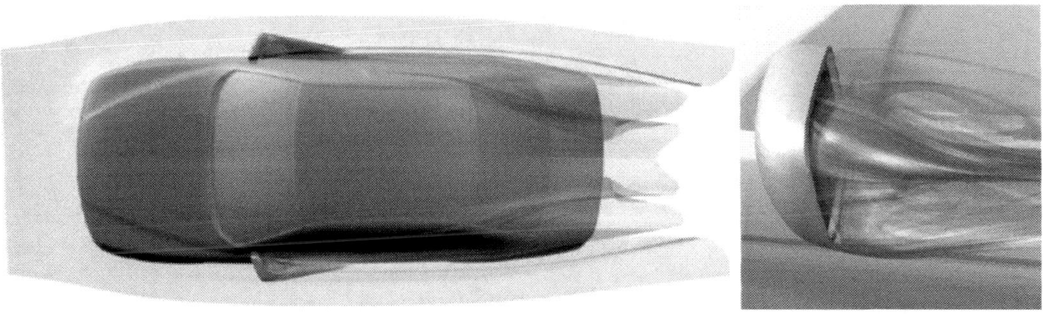

Figure 8. A path surface illustrates the flow of air around a car. The surface represents a dense sheet of massless particles, and the surface color coding provides insight into the lifetime of individual particles. The most notable features identified in this visualization are the vortices behind the rear-view mirror, where the particles are drawn into a vortex (close-up on the right). The path surface is constructed from several thousand integral curves (simulation courtesy M. Rütten, German Aerospace Research Center).

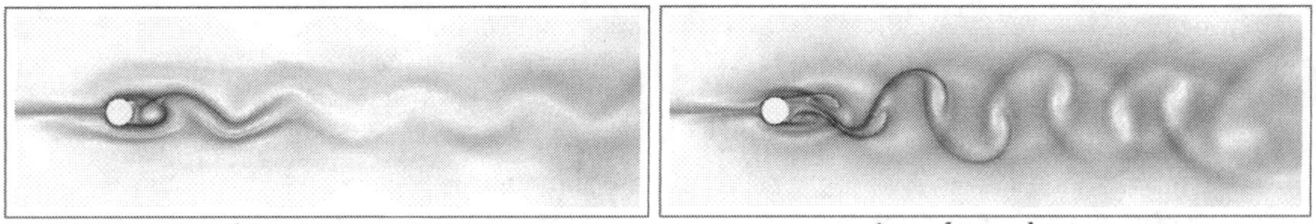

instantaneous time-dependent

Figure 9. The two-dimensional flow behind a cylinder is visualized using a direct color coding of FTLE measures for forward (red) and backward (blue) advection. The left image shows an analysis of the instantaneous flow, whereas the right image is computed from the fully time-dependent flow field. In comparison, it is apparent that the instantaneous depiction does not adequately reveal the flow dynamics. The time-varying FTLE analysis reveals strongly localized ridge-like LCS that capture the characteristic vortex street downstream of the cylinder (cf. [50]).

Unfortunately, the computation of FTLE measures entails computing a huge number of particle traces that densely cover the domain of interest. Even for medium-sized datasets, it is not uncommon that application of Lagrangian visualization techniques requires computing millions of trace particles. VACET researchers have proposed a variety of methods to address this problem. For smaller data, they leveraged the massive parallel nature of GPUs to allows a direct FTLE computation [50] and corresponding visualization with a 30× to 50× speedup compared to a CPU implementation, resulting in interactive computation and visualization. For large to very large data, an approach was presented that evaluates FTLEs adaptively by varying the density of seeded particles in dependence of the spatial variation of the FTLE measure. Depending on the vector field, this can reduce the number of required integral curve computations to between 5% and 20% of the naive, full implementation [52].

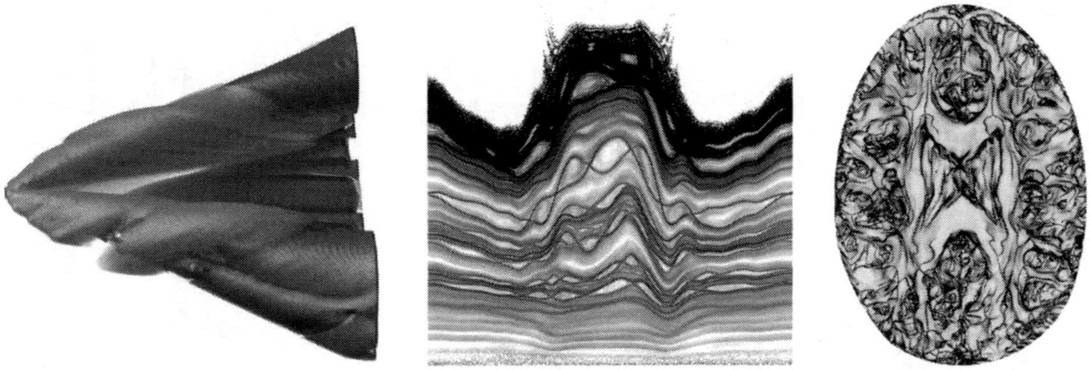

Figure 10. Lagrangian Visualization using FTLE measures is applicable over a wide variety of application domains. The right image visualizes the LCS over a Delta wing using a direct volume rendering of FTLE measures for forward (red) and backward (blue) advection (simulation courtesy M. Rütten, German Aerospace Research Center). The center image depicts an FTLE-like measure that captures coherent structures in a Poincaré section from a Tokamak fusion simulation (data from NIMROD project [51]). The rightmost image depicts a FTLE-like coherency measure for fiber bundles in medical DT-MRI data.

5.4. High-performance, scalable integral curves

For all integration-based visualization techniques, very large datasets represent a significant challenge. The non-local, non-linear nature of integral curve computation defies established parallelization paradigms and requires novel approaches to distribution of both data and computation. Furthermore, the sheer number of integral curves that must be computed to apply these techniques poses a significant challenge. Even though VACET researchers have successfully worked to reduce this number using adaptive approximation and other techniques, the computational effort required to apply Lagrangian methods to very large, petascale datasets is still significant. This problem will be addressed by the development of novel parallelization techniques that will allow integral curve computation to scale to large supercomputers. These new capabilities, together with the Lagrangian visualization techniques that are build on them, will be implemented in production-quality, parallel-capable visual data exploration software that runs on virtually all modern platforms, ranging from desktop-class machines to DOEs petascale computer systems. This work represent a major new capability for domain scientists concerned with vector field analysis, as it will enable broad use of modern visualization methods and allow the treatment of petascale datasets when run on large parallel computing platforms.

Acknowledgments

This work was supported by the Director, Office of Science, Office and Advanced Scientific Computing Research, of the U.S. Department of Energy under Contract No. DE-AC02-05CH11231 through the Scientific Discovery through Advanced Computing (SciDAC) program's Visualization and Analytics Center for Enabling Technologies (VACET). This research used resources of the National Energy Research Scientific Computing Center (NERSC), which is supported by the Office of Science of the U.S. Department of Energy under Contract No. DE-AC02-05CH11231.

References

[1] 2007 Visualization and Knowledge Discovery: Report from the DOE/ASCR Workshop on Visual Analysis and Exploration at the Extreme Scale `http://www.sci.utah.edu/vaw2007/`

[2] Mount R 2004 The office of science data-management challenge. report from the doe office of science data-management workshops Tech. Rep. SLAC-R-782 Stanford Linear Accelerator Center URL `http://www.slac.stanford.edu/cgi-wrap/getdoc/slac-r-782.pdf`

[3] Butler D M, Almond J C, Bergeron R D, Brodlie K W and Haber R B 1993 Visualization Reference Models *VIS '93: Proceedings of the 4th conference on Visualization '93* pp 337–342 ISBN 0-8186-3940-7 (PAPER)

[4] Childs H, Brugger E S, Bonnell K S, Meredith J S, Miller M, Whitlock B J and Max N 2005 A contract-based system for large data visualization *Proceedings of IEEE Visualization 2005* pp 190–198

[5] Childs H, Duchaineau M A and Ma K L 2006 A scalable, hybrid scheme for volume rendering massive data sets *Eurographics Symposium on Parallel Graphics and Visualization* pp 153–162

[6] Rübel O, Prabhat, Wu K, Childs H, Meredith J, Geddes C G R, Cormier-Michel E, Ahern S, weber G H, Messmer P, Hagen H, Hamann B and Bethel E W 2008 High performance multivariate visual data exploration for extemely large data *SuperComputing 2008 (SC08)* (Austin, Texas, USA) lBNL-716E (to appear)

[7] Stockinger K, Shalf J, Wu K and Bethel E W 2005 Query-Driven Visualization of Large Data Sets *Proceedings of IEEE Visualization 2005* (IEEE Computer Society Press) pp 167–174 lBNL-57511

[8] Nieter C and Cary J R 2004 *J. Comput. Phys.* **196** 448–473 ISSN 0021-9991

[9] Geddes C, Toth C, van Tilborg J, Esarey E, Schroeder C, Bruhwiler D, Nieter C, Cary J and Leemans W 2004 *Nature* **438** 538–541 lBNL-55732

[10] Geddes C G R, Bruhwiler D L, Cary J R, Mori W B, Vay J L, Martins S F, Katsouleas T, Cormier-Michel E, Fawley W M, Huang C, Wang X, Cowan B, Decyk V K, Esarey E, Fonseca R A, Lu W, Messmer P, Mullowney P, Nakamura K, Paul K, Plateau G R, Schroeder C B, Silva L O, Toth C, Tsung F S, Tzoufras M, Antonsen T, Vieira J and Leemans W P 2008 *Journal of Physics: Conference Series* **15**

[11] Inselberg A and Dimsdale B 1990 Parallel coordinates: a tool for visualizing multi-dimensional geometry *VIS '90: Proceedings of the 1st conference on Visualization '90* (Los Alamitos, CA, USA: IEEE Computer Society Press) pp 361–378 ISBN 0-8186-2083-8

[12] Inselberg A 2008 *Parallel Coordinates Visual Multidimensional Geometry and Its Applications* (Springer-Verlag)

[13] Inselberg A 1984 Parallel coordinates for multidimensional displays *Spatial Information Technologies for Remote Sensing Today and Tomorrow, The Ninth William T. Pecora Memorial Remote Sensing Symposium* (IEEE Computer Society Press) pp 312–324

[14] Inselberg A, Hauser H, Ward M and Yang L 2006 Modern parallel coordinates: from relational information to clear patterns, tutorial *IEEE Visualization*

[15] Inselberg A 2009 *Parallel Coordinates: Interactive Visualization for High Dimensions* (Springer-Verlag) Advanced Information and Knowledge Processing

[16] Rübel O, Weber G H, Keränen S V E, Fowlkes C C, Hendriks C L L, Simirenko L, Shah N Y, Eisen M, Biggin M D, Hagen H, Sudar J D, Malik J, Knowles D and Hamann B 2006 Pointcloudxplore: Visual analysis of 3d gene expression data using physical views and parallel coordinates *Data Visualization 2006 (Proceedings of EuroVis 2006)* ed Santos B S, Ertl T and Joy K (Aire-la-Ville, Switzerland: Eurographics Association) pp 203–210

[17] Rübel O, Weber G, Keränen S V E, Fowlkes C C, Hendriks C L L, Simirenko L, Shah N Y, Eisen M B, Biggin M D, Hagen H, Sudar D, Malik J, Knowles D W and Hamann B 2006 *PointCloudXplore: A Visualization Tool for 3D Gene Expression Data* (*GI Lecture Notes in Informatics* vol S-4) (Bonn, Germany: Gesellschaft fuer Informatik (GI)) pp 107–117 lBNL-62336

[18] Rübel O, Weber G H, Huang M Y, Bethel E W, Biggin M D, Fowlkes C C, Hendriks C L, Keränen S V E, Eisen M B, Knowles D W, Malik J, Hagen H and Hamann B 2008 *IEEE Transactions on Computational Biology and Bioinformatics* LBNL-382E, to appear

[19] Jones C, Ma K L, Sanderson A and Myers L 2007 *Journal of Physics, Conference Series, Proceedings of SciDAC 2007* **78**

[20] Novotný M and Hauser H 2006 *IEEE Transactions on Visualization and Computer Graphics* **12** 893–900

[21] Stockinger K, Bethel E W, Campbell S, Dart E and Wu K 2006 Detecting Distributed Scans Using High-Performance Query-Driven Visualization *SC '06: Proceedings of the 2006 ACM/IEEE Conference on High Performance Computing, Networking, Storage and Analysis* (IEEE Computer Society Press)

[22] Ushizima D, Rübel O, Prabhat, Weber G, Bethel E W, Aragon C, Geddes C, Cormier-Michel E, Hamann B, Messmer P and Hagen H 2008 Automated Analysis for Detecting Beams in Laser Wakefield Simulations *2008 Seventh International Conference on Machine Learning and Applications, Proceedings of IEEE ICMLA'08* lBNL-960E

[23] Bremer P T, Bringa E M, Duchaineau M A, Gyulassy A G, Laney D, Mascarenhas A and Pascucci V 2007 *Journal of Physics: Conference Series (Proceedings of SciDAC 2007)* **78** doi:10.1088/1742-6596/78/1/012007

[24] Lorensen W E and Cline H E 1987 *Computer Graphics (Proceedings of ACM SIGGRAPH 87)* **21** 163–169 ISSN 0097-8930

[25] Nielson G M 2003 *IEEE Transactions on Visualization and Computer Graphics* **9** 341–351

[26] Milnor J W 1963 *Morse Theory* (Princeton, New Jersey: Princeton University Press) ISBN 0691080089

[27] Carr H, Snoeyink J and van de Panne M 2004 Simplifying flexible isosurfaces using local geometric measures *IEEE Visualization 2004* (IEEE) pp 497–504

[28] Gyulassy A, Natarajan V, Pascucci V, Bremer P T and Hamann B 2005 Topology-based simplification for feature extraction from 3D scalar fields *IEEE Visualization 2005* pp 535–542

[29] Reeb G 1946 *Comptes Rendus de l'Acadèmie des Sciences de Paris* **222** 847–849

[30] Boyell R L and Ruston H 1963 Hybrid techniques for real-time radar simulation *Proceedings of the 1963 Fall Joint Computer Conference* (IEEE) pp 445–458

[31] Edelsbrunner H, Harer J and Zomorodian A 2003 *Discrete & Computational Geometry* **30** 87–107

[32] Edelsbrunner H, Harer J, Natarajan V and Pascucci V 2003 Morse-smale complexes for piecewise linear 3-manifolds *Proceedings of the 19th ACM Symposium on Computational Geometry* pp 361–370

[33] Bremer P T, Edelsbrunner H, Hamann B and Pascucci V 2004 *IEEE Transactions Visualization and Computer Graphics* **10** 385–396

[34] Pascucci V, Cole-McLaughlin K and Scorzelli G 2005 Multi-resolution computation and presentation of contour trees Tech. Rep. UCRL-PROC-208680 Lawrence Livermore National Laboratory preliminary version appeared in the proceedings of the IASTED conference on Visualization, Imaging, and Image Processing (VIIP 2004), 2004, pp.452-290.

[35] Weber G H, Bremer P T and Pascucci V November/December 2007 *IEEE Transactions on Computer Graphics, Proceedings of Visualization 2007* **13** 1416–1423 ISSN 1077-2626 lBNL-63763

[36] Laney D, Bremer P T, Macarenhas A, Miller P and Pascucci V 2006 *IEEE Transcations on Visualization and Computer Graphics* **12** 1053–1060

[37] Gyulassy A, Natarajan V, Duchaineau M, Pascucci V, Bringa E M, Higginbotham A and Hamann B

November/December 2007 *IEEE Transactions on Computer Graphics, Proceedings of Visualization 2007* **13**

[38] Edelsbrunner H, Harer J, Natarajan V, and Pascucci V 2004 Local and global comparison of continuous functions *Proceedings of IEEE Visualization 2004* pp 275–280

[39] Day M S, Bell J B, Bremer P T, Pascucci V and Vince Beckner M L 2009 *Combustion and Flame* **156** 1035–1045 doi:10.1016/j.combustflame.2008.10.029

[40] Day M S and Bell J B 2000 *Combust. Theory Modelling* **4** 535–556

[41] Bell J B, Cheng R K, Day M S and Shepherd I G 2007 *Proc. Combust. Inst.* **31** 1309–1317

[42] Bremer P T, Weber G H, Pascucci V, Day M S and Bell J B 2009 *Transactions on Visualization and Computer Graphics* Accepted with minor revisions

[43] McLoughlin T, Laramee R S, Peikert R, Post F H and Chen M 2009 Over Two Decades of Integration-Based, Geometric Flow Visualization *Eurographics STAR - State of The Art Report* ed Pauly M and Greiner G pp 73–92

[44] Martin D F, Colella P and Graves D 2008 *Journal of Computational Physics* **227** 1863–1886

[45] Garth C, Tricoche X, Salzbrunn T and Scheuermann G 2004 Surface techniques for vortex visualization *Proc. Eurographics - IEEE TCVG Symp. on Visualization*

[46] Garth C, Krishnan H, Tricoche X, Tricoche T and Joy K I 2008 *IEEE TVCG* **14** 1404–1411 ISSN 1077-2626

[47] Green M, Rowley C and Haller G 2006 *J. Fluid Mech.* **to appear**

[48] Lekien F, Shadden S C and Marsden J E 2007 *Journal of Mathematical Physics* **48**

[49] Sadlo F and Peikert R 2008 Visualizing Lagrangian Coherent Structures and Comparison to Vector Field Topology *Topology-Based Methods in Visualization II* ed Hege H C, Polthier K and Scheuermann G (Springer) pp 15–30 ISBN 978-3-540-88605-1

[50] Garth C, Li G S, Tricoche X, Hansen C D and Hagen H 2007 Visualization of coherent structures in transient 2d flows *Topology-Based Methods in Visualization 2007* Mathematics + Visualization (Springer)

[51] Sovinec C, Glasser A, Gianakon T, Barnes D, Nebel R, Kruger S, Plimpton S, Tarditi A, Chu M and the NIMROD Team 2004 *J. Comp. Phys.* **195** 355

[52] Garth C, Gerhardt F, Tricoche X and Hagen H 2007 *IEEE Transactions on Visualization and Computer Graphics* **13** 1464–1471 ISSN 1077-2626

Enabling scientific teamwork

Mark Hereld*, Randy Hudson+, John Norris+, Michael E Papka, Thomas Uram*
*Mathematics and Computer Science, Argonne National Laboratory, Argonne, IL 60439, USA
+Flash Center, University of Chicago, Chicago, IL 60622, USA

E-mail: papka@anl.gov

Abstract. The Computer Supported Collaborative Work research community has identified that technology can be used to support distributed teams of researchers though common tools such as email, instant messaging, and conferencing environments alone are not enough. Building from a list of areas where it is believed technology can help support distributed teams we have divided our efforts into support of asynchronous and synchronous activities. This paper will describe two of our recent efforts to improve the productivity of distributed science teams. One effort focused on supporting the management and tracking of milestones and results, with the hope of helping manage information overload. The second effort is focused on providing an environment that supports real-time analysis of data. Both of these efforts are seen as add-ons to the existing collaborative infrastructure, developed to enhance the experience and remove the barrier of working at a distance.

1. Introduction

"Science is a collaborative enterprise, and the traditional environment for scientific collaboration is the laboratory. [1]" Scientists work together as part of a team to solve a common problem. Historically this has been working in the same physical space and for remote collaborations may have meant travelling to the remote laboratory as needed to support the collaboration. Modern information technology has been making the requirement of a physical space less of an issue. In the early 1990s William Wulf coined the term *collaboratory* in the National Research Council study "National Collaboratories: Applying Information Technologies for Scientific Research", where he lays out the vision of a research environment that was not bound by physical space, an environment that brings together researchers, resources and data to increase the scientific output and productivity of the teams [2]. Fifteen years later do we have infrastructure to enable scientific teamwork?

Stepping back and looking at what being part of a team brings to scientific research yields a variety of different answers; from variance in expertise to sounding board for the exchange of ideas. Dating back to the 1970s research has shown distributed teams perform below their full potential [3]. The work of Cummings and Kiesler looked at the outcome of the National Science Foundation's Knowledge and Distributed Intelligence research program and found that teams that were distributed across multiple universities (locations) performed worse in the generation of new ideas and knowledge [4]. They determined that technology such as email, instant messaging, or conferencing technology (both video and phone) did not provide any advantage to the distributed teams. Based on the data collected in the process of their research they where able to identify some requirements of technology that might help change this outcome. These requirements included: (1) tools to manage and track the trajectory of tasks over time, (2) tools to reduce information overload, (3) infrastructure to facilitate ongoing conversations, (4) infrastructure to encourage awareness with reasonable interruption for spontaneous talk, (5) tools that support simultaneous group decision making, and (6) tools and infrastructure capable of supporting presentations and meetings across a distance. Cummings and Kiesler go on to claim that "research on innovation in virtual organizations, and technology to support innovative teams, is critical to the future success of this type of work [5]."

The technology requirements laid out by Cummings and Kiesler, can be viewed from a work mode of either asynchronous and/or synchronous tasks. Beginning with asynchronous the requirements of 1, 2, 3, and 6 represent tasks that can be done without direct interaction with other team members. Synchronous activities 2, 4, 5, and 6 require or could require the direct interaction with other team members. The rest of the paper is divided into either asynchronous or synchronous tasks. This collection of efforts will address some of the requirements put forth in the previous paragraph and are examples of the use of technology to enable scientific teamwork. The rest of this paper will describe two of our recent efforts to improve the productivity of distributed science teams by addressing the technology requirements identified above. These efforts are described separately by whether they address asynchronous or synchronous teamwork.

2. Asynchronous tasks

In terms of asynchronous tasks we have been working with members of the University of Chicago's ASC / Alliances Center for Astrophysical Thermonuclear Flashes on the construction of tools to manage and track simulations and analysis over time while reducing the amount of information the individual researchers must keep track of in an attempt to support an ongoing dialogue about the science being studied. The results of this effort are presented in this section.

We have created a simulation-analysis management system that among other things allows a user to set up, start, stop and monitor simulations; analyze simulation output; catalog and retrieve old simulations; and prepare output for publication. This work has been guided by some fundamental beliefs. We believe the system needs to account for maximum automation; both in terms of its interactions with scientists but also in its ability to easily accommodate new tasks. It needs to be based on a single control interface that is accessible from as many platforms as possible without a reduction in capabilities. The system must be built using open standards when possible to ensure the ease at which it can interface with other infrastructure and be extended as needed. At all stages of the development the end users must be engaged to allow for proper input into the process and that all needs are met.

This section details the simulation-analysis management system. The system consists of components that capture and process simulation data and metadata *in situ* based primarily on log files generated during FLASH simulations, and a web application for interacting with the stored simulation data.

2.1. System components

The simulation-analysis management system consists of modular components to capture, store, verify, and process information during a simulation run. These components are assembled into workflows which we call *pipelines*. By examining the FLASH workflow, we have constructed a set of pipelines that capture the core FLASH processes. We also took care to preserve the modularity of the system, so that it could be easily employed for simulation-analysis management outside of FLASH.

Pipelines are used to collect data on, and create snapshots of, each recorded state of a simulation. They are also used to archive datasets created by a simulation and extract data for post-processing. Pipelines consist of the simulation code, a collector component, and an archiver component. The first process of the pipeline is the simulation itself. The *collector* and *archiver* processes follow next in the pipeline.

The collector is run on the system hosting the simulation. This program tracks the progress of the simulation via a central log file, and notes the current timestep that the simulation is in. All other data is collected by dynamically loaded tools, which the collector calls for each line of the central log file, as well as separate calls to mark the beginning and end of timesteps. The collector is started anytime after the simulation has been queued to run, even long after the simulation has ended. The program is capable of recovering from fatal conditions (e.g. being killed by an administrator). It can also verify the data integrity of a previously collected simulation. The user receives email notification of significant events, including simulation completion, simulation crash, and collector death.

The archiver is run on the system hosting the files to be archived, preferably by the creator of the files. This program tracks the progress of the simulation via the database, archiving files as they are added to the database. It can be started anytime after the collector has been started, and may, in the future, be started by the collector. Like the collector, the archiver is capable of recovering from fatal conditions. The user receives email notification of significant events such as archival completion and archiver death.

We introduce "soft" constraints, or suggestions, wherever appropriate to urge or remind users to use standard forms and categories wherever they can. For example, the collector prompts the user for a run description at run time to encourage the entering of a meaningful description versus a cut and paste of previous entries. The collector and archiver can use some sensible defaults for data they otherwise need from the user; for example, if given no destination path, the archiver will construct a unique path in a standard way.

In addition to the collector and the archiver, two additional supporting tools have been developed based on the needs of the users: the *associator* and the *verifier*. The associator is a simple command-line utility that allows the user to associate files, usually post-processing results, to a previously collected simulation and record the association in the database. Any files connected to a simulation in this manner also become available for archiving. The verifier represents the first of a generation of programs meant to maintain the accuracy of the information in the database. When run on a system which hosts files referenced in the database, the verifier confirms that each local file actually exists and updates the database as needed.

The database was designed to be as independent as possible from the peculiarities of a given simulation or domain science so that system could be reused. In the database are stored meta-data from several simulation files, user preferences, names and locations of data files (for, e.g., output from the VisIt visualization tool), a catalog of old simulations and links to files that users consider important (scripts and movies, e.g.). Our database is managed by MySQL. Django, a Python-based web framework, runs on the server, queries the database and returns results to the web browser. Django uses Matplotlib to create data plots and return them to the browser. With Django's administrative component, users annotate simulations and runs. Dojo is a JavaScript-based DHTML toolkit. We use Dojo for widgets in, and data access from, the client. We use Dojango, a Django project, to unite Django and Dojo.

To minimize the effort to make the FLASH pipeline usable by all computational scientists, the FLASH-particular code of the collector is isolated from its control module in the set of dynamically loaded tools. File types are defined in a database table rather than as a database enumerated type. Variables and their names are in separate tables of the database.

2.2. End user application

The end-user application is a web application that acts as a single place from which the computational scientist can manage most preparations for a simulation, the simulation itself, and post-processing and analysis. We want the user to be able to do at least the following via the web application:

- Set up, start, stop and monitor simulations
- Keep track of old simulations
- Edit annotations of simulations
- Launch scripts and view their output
- Prepare plots for publication
- Send plot data to other plotting tools

- Store simulation parameters and file locations
- View meta-data from the database
- Set preferences

This section describes the current state of the web application.

Simulations are composed of one or more individual runs, exactly one of which is the initial run. All other runs in the simulation are restarts from an earlier run. Since this naturally lends itself to a hierarchical representation, the application uses a tree list structure to present the simulations in the database to the user. Each restart is displayed as a child of its parent run (the run that generated the data used for the restart). The user can filter out simulation trees by date, owner, problem type, and several other criteria; navigate through the individual runs of each simulation; and select which runs to generate graphs from. The user can also choose which columns of the table are displayed and, if logged in, can edit tags, names and descriptions of runs and simulations.

When the user selects one or more runs and clicks the graph button, a curve for each simulation path containing a selected run is plotted on a graph page. Multiple data series can be displayed on the graph, with optional operations performed on the data prior to plotting, such as addition, subtraction, multiplication, division, modulo and two types of accumulation. The division operation, for example, can be used to convert mass units from grams to solar mass. An accumulator can integrate number of blocks over time steps. The display of graphs can be customized according to several criteria, such as data series selection, plot labels, the number of variables to display on the vertical axis, and whether run delimiters should appear. Often, simulations consist of multiple compute runs, with the results aggregated from all runs; run delimiters indicate these run boundaries on the graph, and can help the researchers identify problems that may have arisen between runs.

Graph data and display characteristics are encoded in the graph URL, so that the URL can be embedded in external sources (e.g., web pages) where they will be regenerated when accessed. Typically, graph URLs are constructed to capture a static view of the data, in which case the plots would not change over time. When monitoring ongoing simulations, however, FLASH researchers can view graphs generated from the most recent simulation data. These graphs can be integrated with project pages to provide current simulation state in the full context of a project.

This development has been guided by regular interactions with FLASH scientists using the system. As such, it is tailored to address many aspects of their everyday work, and reveals what they perceive as requirements for performing their science in a collaborative manner. Working in the system, users are aware of the activities of their cohorts on shared and non-shared projects. Outside the system, in their wiki-based project portal, they can view snapshots of data from past simulations, and dynamically updated graphs of running simulations.

3. Synchronous tasks

In terms of synchronous tasks we have been working to minimize the information overload that users often experience in terms of managing a number of applications as part of a single task, providing infrastructure to increase the sense of presence of remote collaborators, interactive visualization tools that support real-time group investigation of even the largest datasets, and infrastructure to support traditional activities that occur when a group is collocated at a distance. Many of these tasks build on the Access Grid infrastructure developed over the past ten years, which in turn build off a long history of collaboration infrastructure [6,7]. By leveraging the Access Grid, users have a single integrated point to connect to remote colleagues, organize data, and launch applications to interact with the data. Via the Access Grid, users can see if remote colleagues are active and their state, much like standard instant messaging technology in use today. Via the Access Grid a user can share any application on their desktop with a single remote user or any number of remote users. The implication here is that a desktop user running a parallel visualization of a dataset using ParaView can then share that output, the application itself with individuals around the world. The remote user has full control of the applications and the Access Grid provides the audio and video support to connect those users together.

Finally, since the initial deployment of the Access Grid, it has support for remote presentations and meetings at its very core. Now with over 50,000 downloads and deployments in more than seventy five different countries the Access Grid provides a platform to enable scientific teamwork.

In order to provide infrastructure for distributed collaborations to make real-time decisions we have developed VPCScreenIX. VPCScreenIX is an application for distributing screen content to multiple simultaneous remote users, allowing them to interact with the originating application as if it were running locally. While there have previously been many similar efforts, VPCScreenIX excels in its network efficiency, due to its use of streaming video as a delivery mechanism, and its ability to scale to many simultaneous users. These concerns are vital in the context of HPC-based visualization, to maximize the involvement of researchers who are often distant from the originating visualization applications and the huge data and compute resources they require.

3.1. Related efforts

Much effort has been dedicated to delivering remote access to computers and applications. Prominent among these efforts is VNC [8], for its performance, stability, openness, and wide availability. Remote Desktop is another notable effort, with an open specification and clients on many platforms, but server support typically limited to Windows. Many proprietary solutions exist, but have been omitted here.

VNC provides remote desktop access using a server and one or more clients. VNC uses a variety of encodings to optimize for bandwidth or quality. VNC uses the Remote Frame Buffer (RFB) protocol, an open specification that has been implemented in many applications on all major operating systems.

Synergy2 is open source software for sharing a mouse and keyboard between multiple computers that may be running different operating systems. Remote desktop support is outside of the scope of synergy2.

VPCScreen captures content from the screen and streams it as video to remote viewers. The video is encoded using a version of the H.261 codec modified to support arbitrary size video frames, called H.261as. The H.261 standard supports CIF (352x288) and QCIF (176x144) resolutions. H.261as overcomes this resolution limitation by applying the standard H.261 encoding to larger frames.

The ParaView Streaming Plug-in, developed at Argonne National Laboratory, enables researchers to share and discuss visualizations of their data by extending ParaView – an open source visualization application that runs on desktops as well as supercomputers – to stream its render area as video to remote receivers. The plug-in uses the H.261as codec described above. The user loads the plugin, specifies the stream destination address, begins streaming, and proceeds with normal use of the ParaView application. Remote sites start the viewing application – vic – with the same destination address to view the resulting video stream.

3.2. Application sharing

VPCScreenIX is a remote desktop application that uses streaming video to deliver content to remote viewers. Our goal was to develop an application-agnostic version of the ParaView Streaming Plug-in so that any application could be shared without modification. Streaming video was chosen as a delivery mechanism primarily because video codecs have been rigorously optimized to maximize framerate while minimizing bandwidth. Remote interaction support was adapted from the synergy2 project, which has largely solved the problem of cross-platform mouse/keyboard capture and synthesis.

The complete system (Figure 1) consists of three components: the IX library for capturing and synthesizing mouse and keyboard events; a video tool, vic, modified to support capture of mouse and keyboard events; and a modified version of VPCScreen that listens for events from remote viewers running vic.

Building on the mouse/keyboard sharing tool synergy2 described above, we developed an interaction library that solves a longstanding problem in collaborative applications: capturing mouse and keyboard events and sending them to remote systems, where they are synthesized into the system as if they occurred locally, while managing the complexity of this task across multiple platforms. This library includes an IXSender component, which can track windows on the display and capture mouse and keyboard events that occur inside them. The IXSender transmits these changes over TCP to the remote application, which has an integrated IXReceiver. The IXReceiver synthesizes these events into the local system. In the case of VPCScreenIX, events from the IXReceiver are input into the application whose window is being captured and transmitted. These events are processed in the context of the target application, effecting the intended actions.

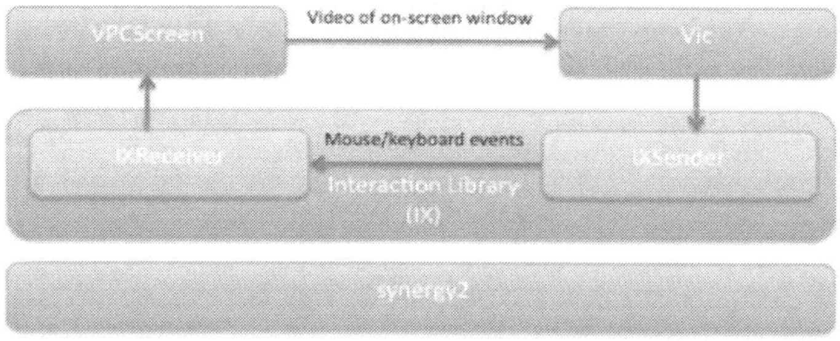

Figure 1. Architecture of the IX interaction library and its integration into VPCScreen and vic.

Users can select any visualization application (e.g., VisIT, VisTrails) from their desktop and stream its window as video. Furthermore, any application can be streamed, not just visualization applications, and the applications require no modification. The screenshots in Figure 2 show ParaView on MacOS being displayed on a Windows desktop.

We have integrated VPCScreenIX with the Access Grid to simplify starting applications. This integration enhances the existing remote collaboration by allowing participants to view and discuss the visualization, and does so with little user interaction. The application also benefits from the Access Grid's intelligent bridging infrastructure, a transparent substitute for multicast networking.

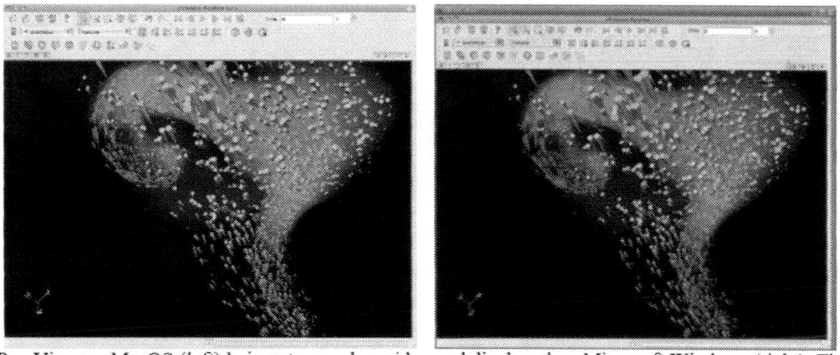

Figure 2. ParaView on MacOS (left) being streamed as video and displayed on Microsoft Windows (right). The Windows user can interact with the remote application natively.

VPCScreenIX is integrated as a plugin to the Access Grid Venue Client. The plugin adds buttons to the Venue Client toolbar for streaming a window, a region, or the entire screen. Clicking on one of these buttons launches VPCScreenIX in the selected mode, prompting the user to select a window or screen region as needed, and then begins to stream screen content to the established Access Grid video destination address.

The modified version of vic is included in the VPCScreenIXConsumerService, which replaces the standard video service for the Access Grid. It can view AG-native video streams and also supports interaction with remote sources when available. Using this service, users can see the video of their remote collaborators and video transmitted by VPCScreenIX, and interact with the VPCScreenIX streams with a single click.

3.3. Performance analysis

We studied the efficiency of VPCScreenIX versus VNC by devising a test to transmit a fixed size buffer with random noise variation in the output. The use of a random noise pattern is a fair approximation of the activity during use of a visualization application, with a significant portion of the screen content changing fairly rapidly. The machines used to conduct these tests were configured as follows:

- Sending machine: OSXvnc, MacOSX 10.5.6, 2.33 GHz Intel Core 2 Duo, 3GB RAM, ATI Radeon X1600
- Receiving machine: RealVNC, Windows XP, 2 x 3.4GHz Pentium, 2GB RAM, nVidia GeForce 7800 GTX
- VNC configuration: ZRLE (Zlib compressed Run Length Encoding), 32 bits per pixel

In our tests, we found that VPCScreenIX required about one fifth of the bandwidth required by OSXvnc. Processor utilization between the two was comparable, with VPCScreenIX requiring on average about 10% more CPU than OSXvnc. Measurements were made on the Mac using Apple's Instruments software.

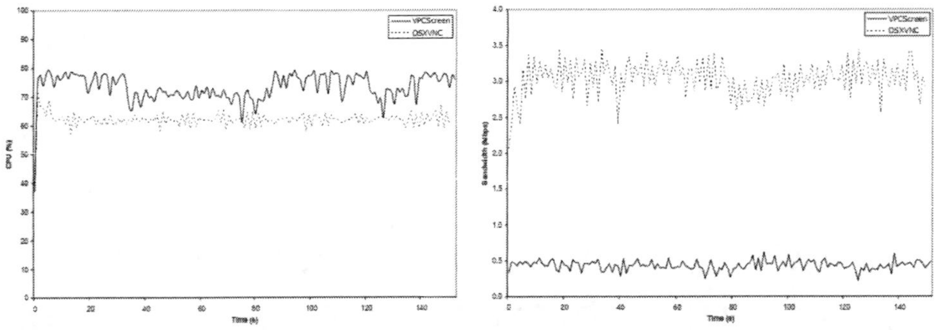

Figure 3. Comparison of CPU usage (left) and bandwidth usage (right) of VPCScreenIX and VNC. Measurements were taken using a transmitted image size of 1600x1200, while a 1450x820 video noise pattern was displayed.

3.4. User testing

Using the AccessGrid, we conducted a test of VPCScreenIX with climate scientists at Argonne National Laboratory (ANL) and Los Alamos National Laboratory (LANL), using ParaView to view an ocean modeling dataset. The scientists interacted with the dataset, using the Access Grid audio to discuss what was being shown and how to modify the visualization, while sharing a synchronized view of the application. The remote scientist was, in fact, unfamiliar with ParaView, but was successfully guided by the other through the steps necessary to develop the view of the data he desired.

This test was conducted using ParaView 3.4, running on Windows XP and transmitting over the esNET network. The receiving vic was run on Windows XP. The resolution of the video stream was 1619x864.

Following this test, the users concluded that this approach differed from their normal work, which typically involves PowerPoint to view static images of their data visualizations. They also said that it would be helpful for them to cooperatively develop their visualizations, and that it would contribute to a better shared understanding of the data and the problems at hand. Lastly, they commented that VPCScreenIX could be used with applications generally, including visualization tools used in their other collaborations.

4. Conclusions

Science is a group effort and it requires technology that supports the construction of the best scientific teams regardless of the team members' locations. We believe that as scientific efforts become larger and more distributed, technology should be used to break down the barriers that hinder working together at a distance. Guided by results from the computer supported collaborative work research community, we have performed the research, development and deployment of tools and infrastructure that address the areas of greatest need. We have described our efforts in terms of enabling asynchronous and synchronous teamwork. We have presented tools to help keep groups in sync by providing a framework for annotating results and to have the results automatically refreshed. We have built tools that allow for the real-time sharing of results to enable a decision-making environment that is independent of the location of team members. Leveraging the Access Grid environment, the integration and deployment of much of the described technology is simplified. Using the Access Grid also provides a familiar environment in which applications, data and colleagues are all integrated together. In our ongoing collaboration with science users, we have begun to overcome obstacles to effective science in distributed teams.

5. Acknowledgements

This work has been supported in part by the DOE, under contract DE-AC02-06CH11357, the NSF by grant OCI-0222509, NLM by contract N01-LM-3-3508 and the DOE-supported ASC / Alliance Center for Astrophysical Thermonuclear Flashes at the University of Chicago.

6. References

[1] T. A. Finholt, G. M. Olson, *From Laboratories to Collaboratories: A new Organizational Form for Scientific Collaboration*, **Psychological Science**, 8(1), pp. 28 – 26, 1997.

[2] V.G. Cerf, et al., **National Collaboratories: Applying Information Technologies for Scientific Research**, National Academy Press, 1993.

[3] I. D. Steiner, **Group Processes and Productivity**, Academic Press, 1972.

[4] J. N. Cummings and S. Kiesler, *Collaborative Research Across Disciplinary and Organizational Boundaries*, **Social Studies of Science**, 25(5), pp. 703-722, 2005.

[5] J. N. Cummings, S. Kiesler, *Who Collaborates Successfully? Prior Experience Reduces Collaboration Barriers in Distributed Interdisciplinary Research*, **Proceedings Of The ACM 2008 Conference On Computer Supported Cooperative Work**, pp. 437-446, 2008.

[6] T. L. Disz, R. Evard, M. W. Henderson, W. Nickless, R. Olson, M. E. Papka, and R. Stevens. *Designing The Future Of Collaborative Science: Argonne's Futures Laboratory*, **IEEE Parallel and Distributed Technology**, 3(2), pp.14–21, 1995.

[7] R. Stevens, *Access Grid: Enabling Group Oriented Collaboration On The Grid*, **The Grid: Blueprint for a New Computing Infrastructure**, pp. 191–199, Morgan Kaufmann, 1999.

[8] T. Richardson, Q. Stafford-Fraser, K.R. Wood, A. Hopper, *Virtual Network Computing*, **IEEE Internet Computing**, 2(1), pp. 33-38, 1998.

SciDAC 2009 IOP Publishing
Journal of Physics: Conference Series **180** (2009) 012086 doi:10.1088/1742-6596/180/1/012086

Tracking flame base movement and interaction with ignition kernels using topological methods

A Mascarenhas, R W Grout, C S Yoo and J H Chen

Combustion Research Department, Sandia National Lab, Livermore, CA 94550

E-mail: aamasca@sandia.gov

Abstract. We segment the stabilization region in a simulation of a lifted jet flame based on its topology induced by the Y_{OH} field. Our segmentation method yields regions that correspond to the flame base and to potential auto-ignition kernels. We apply a region overlap based tracking method to follow the flame-base and the kernels over time, to study the evolution of kernels, and to detect when the kernels merge with the flame. The combination of our segmentation and tracking methods allow us observe flame stabilization via merging between the flame base and kernels; we also obtain Y_{CH_2O} histories inside the kernels and detect a distinct decrease in radical concentration during transition to a developed flame.

1. Introduction

In this section, we will motivate our work, and briefly discuss prior work.

1.1. Motivation

Non-premixed turbulent jet flames are useful configurations to study phenomena relevant to practical energy conversion devices. Laboratory scale jet flames have been studied experimentally [1] and proven useful for developing combustion models. Computational studies of canonical configurations, albeit at much smaller scales, produce exceptionally detailed flow information. Presently, we investigate a 'direct numerical simulation' of an ethylene slot jet surrounded by a co-flow of hot oxidizer. The flame may be stabilized by a variety of mechanisms: here, the temperature is elevated, so that the mixture may auto-ignite. The stabilization is influenced by chemical reactions occurring between the jet exit and the flame base. If the reactants are cold, or the velocities are large, the flame base may be stabilized by a propagating edge flame [2] . Unlike the edge flame, an auto-igniting front does not propagate in the traditional sense, but instead moves relative to the advected fluid by 'sequential auto-ignition' [3]. This flame base has been observed to periodically exhibit displacement speeds indicative of an auto-ignition front [4].

To further understand the stabilization mode, we wish to connect upstream reaction with the behaviour of the flame base. For this chemistry, the location of the quasi-steady non-premixed flame is described by the Hydroxyl (Y_{OH}) mass fraction, and the auto-ignition process is reflected in the evolution of radical species, such as Formaldehyde (Y_{CH_2O}). Our interests are threefold. First, is the stabilization of the flame due to auto-ignition alone, or is there some interplay with a local propagation. Second, can we relate the effective displacement speed and position of the flame base to the nature of 'ignition kernels' formed upstream. Finally, can we relate the large scale flow structures to the propensity for formation of ignition kernels.

The first objective is relevant to describing the canonical configuration under consideration and framing the applicability of any conclusions drawn, the second and third together are steps in developing an effective combustion model to describe the stabilization mechanism. Accomplishing the second objective mandates a quantitative and unambiguous definition of the 'ignition kernels', and is the foremost motivation for applying topological methods to this problem. Further, this effort also begins to use a topological definition of the ignition kernels to explore the first objective.

1.2. Feature detection.

Topological based segmentation and feature detection methods have become increasingly popular. One of the most widely used topological structure is the *contour tree* [5, 6]. It encodes the topology of all level sets (also called isosurfaces) of a function. See [7, 8, 9] for applications. As described in Section 2, we base our segmentation method on [10]. This method employs the *merge tree*, a sub-structure of the contour tree, to represent the merging behaviour of level sets.

A closely related family of methods use the Morse-Smale complex to represent features. The Morse-Smale complex was first introduced for triangulations by Edelsbrunner et al. [11]; extensions to 3D can be found in [12, 13].

1.3. Feature tracking.

It is often of important to track features in time-dependent data sets. Samtaney et al. [14] apply methods from object tracking in image processing to feature tracking in scientific datasets. In follow-up work, Silver and Wang [15] use volume overlap of features to identify correspondences. As discussed in Section 2 we use the vertex overlap induced by a triangulation of space-time to connect features in time. Laney et al. [16] use similar ideas to track bubble structures in turbulent mixing. Reinders et al. [17] use motion prediction to improve matching accuracy. Ji et al. [18] connect isosurfaces using the 4D space-time isosurface that contains them.

2. Segmentation and tracking

In this section, we briefly describe our feature segmentation and tracking methods. Our feature segmentation is based on [10].

2.1. Segmentation

Given a smooth function $f : \mathbb{R}^n \to \mathbb{R}$ the *level set* of f at isovalue s is defined as all points in $\mathbb{R}^n$ with function value s. A connected component of a level set is called a *contour*. The *merge tree* of f records the merging behavior of the contours of f as the isovalue is swept top-to-bottom through the function range; see Figure 1. When the isovalue passes a maximum a new contour is born and a new leaf appears in the merge tree. When two contours meet at a saddle they merge into a single component represented in the merge tree as a joining of two branches.

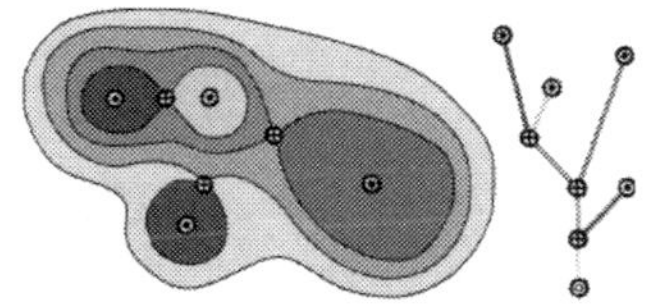

Figure 1: The merge tree (right) with branches colored to match their corresponding regions (left).

Each branch of the merge tree can be used to represent the subset of $\mathbb{R}^n$ defined by the union of all its contours, indicated by the colored regions in Figure 1. For a data point p, the value $S(p)$ is the id of its corresponding merge tree branch. We are interested in regions around high Y_{OH} values and we define these regions starting from the leaf branches of the merge tree.

We want two neighboring regions with similar Y_{OH} values to count as a single region. Following [10], we merge low persistence regions together and refine the merge tree arcs and the corresponding regions to create *segments*; see Figures 2(a) and 2(b). The *relevance* measure for an branch b with maximum u and saddle v, is defined as $rel(b) = (f(u) - f(v))/(f(u) - f_{min})$ where f_{min} is the global minimum of f. We are interested in regions around local maxima of Y_{OH}. Given a user defined threshold r, we can select regions around local maxima of Y_{OH} whose set of arcs B in the merge tree have relevance $rel(b) < r$ for all $b \in B$. Notice, that this final parameter choice represents a simple selection of a root branch b and its sub-tree all of whose branches have a lower relevance value. Denote this root b as the *leader* of the region, and say b *represents* its region. In Figure 3, the segment labeled S_t in the tree at time t is the leader of the region corresponding to the segments in the subtree rooted at S_t. Note that, unlike conventional tree diagrams, the layout of the merge tree has the leaves on top; the subtree rooted at S_t is obtained by walking up from S_t towards the maxima nodes.

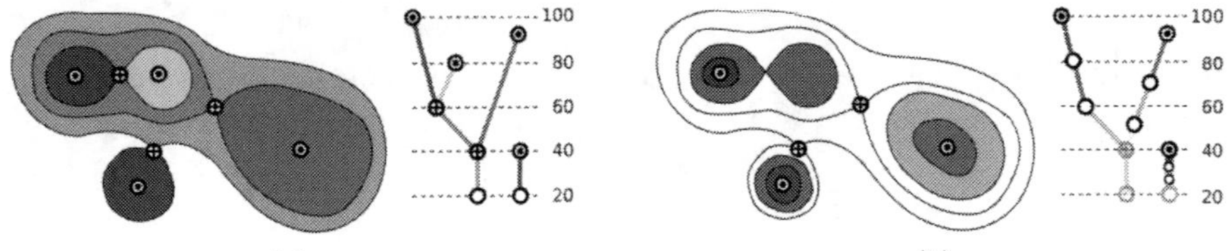

Figure 2: Merge tree adaptation: (a) The tree of Figure 2.1 cut at an absolute threshold of 20. The absolute function values are presented on the right side of the figure and we assume a global minimum at 0. Because we discard the lower part of the tree from just above the lowest saddle we obtain a forest of two independent merge trees; (b) The merge tree and the segments after simplification (yellow arc merged into red), refinement, and selection (colored arcs and regions).

Selecting flame components Because we are interested in selecting and tracking the flame and kernels, and because the flame is defined as the Y_{OH} level set of $s = 0.0005$, we have to modify our selection criterion to include these regions. This is easily accomplished: for every candidate leader arc b, we compare the local maximum in its subtree $f_{max}(b)$ with s. If $f_{max}(b) < s$, the region represented by b does not include a component of the flame, and we select b. If $f_{max}(b) > s$, we select the arc a in the subtree of b that contains the value s.

2.2. Segment tracking

In this section, we describe how we track segments over time. Our method is simple; we connect successive time-steps in a triangulation, employ a weighted overlap measure to detect segment correspondence, and encode this correspondence in a tracking graph. This graph can be used to extract tracking information for an arbitrary choice of the relevance parameter at each time-step. The arc weights can be used to select an edge based on a user selected overlap threshold.

The tracking graph. The data at each time-step is sampled on a regular grid. We decompose the unit 4-cube defined by diagonally opposite points p and q into 4-simplices along the diagonal. The algorithm for computing the tracking graph G_t between successive time-steps t and $t+1$, takes as input the segmented data for these time-steps and outputs a weighted graph encoding the correspondence between segments. We process every data point p at time t, and decompose the 4-cube containing p and its diagonally opposite point at time $t+1$. We then process each edge connecting p to data point q in $t+1$ and update the tracking graph: if arc $a = (S(p), S(q))$ belongs to G_t then increment its weight $w(a)$ by unity, else insert a with unit weight. The tracking graph G_t contains segment correspondence for any arbitrary choice of relevance for

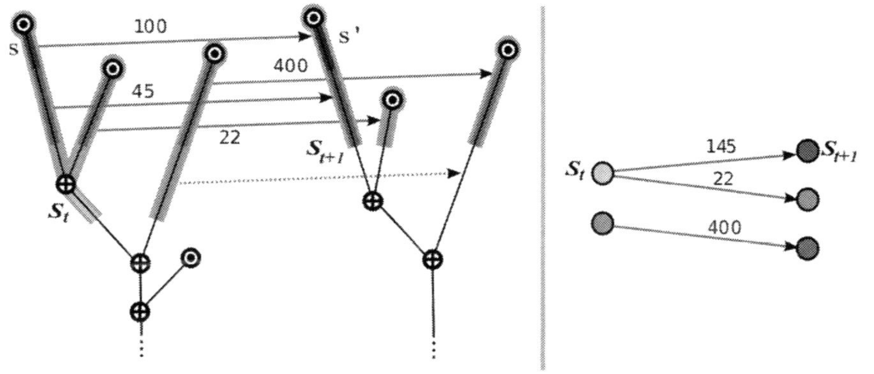

Figure 3: Left, portions of merge-trees at t and $t+1$. Selected segments are colored, tracking graph arcs between selected arcs are solid. The dotted arc does not connect to a selected segment at $t+1$. Right, tracking information extracted for segment leaders.

times t and $t+1$. Next we show how we extract the segment correspondence for a particular choice of relevance parameters.

Tracking with relevance selection. We describe the algorithm for computing the graph H_t which encodes the tracking information for the choice of relevance parameters R_t and R_{t+1}. Consider a region with leader segment S_t. Recall that the region is defined by all points whose segment id belongs to the subtree rooted at S_t. This collection of segments is connected to segments in $t+1$ some of which might be selected for relevance R_{t+1}. Denote the leader of such a segment by S_{t+1}. Initialize H_t to empty. For every segment s with leader S_t and arc $a = (s, s') \in G_t$ such that s' is selected, update arc $b = (S_t, S_{t+1})$ in H_t. The update inserts arc b if it does not exist in H_t or increases its weight by $w(a)$ if it does exist.

We illustrate this algorithm in Figure 3. The figure shows a merge-tree at times t and $t+1$ with arcs colored if the corresponding segments are selected. Segment S_t has two segments in its subtree that are connected to segments in the subtree of S_{t+1}; we accumulate their arc weights to obtain the arc between S_t and S_{t+1} with weight 145 shown at the right of the figure. Note that S_t splits into two segments at $t+1$ because a lower relevance value was chosen at $t+1$.

3. Results

In this section, we present the results of segmenting and tracking the flame base and kernels using the Y_{OH} field. The data consists of $2025 \times 1600 \times 400$ samples per time, and TTT time steps. For our experiments we focus on 30 timesteps of a $600 \times 200 \times 400$ block that includes part of one of the flame bases. Each timestep has a $2\mu s$ duration.

Segmentation and tracking. We segment the data for each timestep at a relevance of $R = 0.15$ and include all flame components by selecting segments that include the $Y_{OH} = 0.0005$ surface. The kernels are sufficiently large and slow moving that our overlap based tracking can successfully follow them over time. Figure 4(a) shows a zoomed in view of the tracking graph annotated with snapshots of the kernels and flame at various timesteps. As seen in the figure, components can merge and split before merging with the flame base and thereby stabilizing it. It appears that the upstream motion of the flame base is largely due to merging with kernels.

Tracking radical species. One of the goals of our analysis is to follow radical species concentrations in kernels as they develop so as to gain insight into the auto-ignition process.

The segmentation and tracking method described here enables this study. In our experiment, we follow the mass fraction of the Formaldehyde (Y_{CH_2O}) mass fraction at the local maximum in each kernel that is tracked. In Figure 4(b), we align the plots so that the times at which the kernels cross the flame threshold coincide. We observe that for almost all kernels the value of Y_{OH} mass fraction has a minimum at this time.

4. Conclusions

We have demonstrated that ignition kernels extracted using topological methods are instrumental in the stabilization of this flame. We observe that merging of the flame base with well developed kernels is responsible for the apparent upstream propagation of the flame base. We successfully track the kernels prior to merging with the flame base, allowing us to extract the evolution of radical species mass fractions. We have detected an interesting pattern of Y_{CH_2O} decrease near the time when the kernels transition to a developed flame. Further study of this phenomenon and other radical species behavior inside the tracked kernels will be in the future work.

References

[1] Barlow R S, Fiechtner G J, Carter C D and Chen J Y 2000 *Comb. Flame* **120** 549–569
[2] Ruetsch G R, Vervisch L and Liñán A 1995 *Phys. Fluids* **7** 1447–1454
[3] Echekki T and Chen J H 2003 *Comb. Flame* **134** 169–191
[4] Yoo C *et al.* 2008 *Ninth Turbulent Non-premixed Flames Workshop, Montreal, Canada*
[5] van Kreveld M J, van Oostrum R, Bajaj C L, Pascucci V and Schikore D 1997 Contour trees and small seed sets for isosurface traversal *Symp. on Comp. Geom.* pp 212–220
[6] Carr H, Snoeyink J and Axen U 2003 *Comput. Geom. Theory Appl.* **24** 75–94
[7] Fujishiro I, Takeshima Y, Azuma T and Takahashi S 200 *IEEE Comp. Graphics Appl.* **20**(5) 46–51
[8] Weber G *et al.* 2002 Exploring scalar fields using critical isovalues *Proc. IEEE Vis. '02* ed Gross M, Joy K I and Moorhead R J IEEE (Los Alamitos California: IEEE Comp. Society Press) pp 171–178
[9] Carr H, Snoeyink J and van de Panne M 2004 Simplifying flexible isosurfaces using local geometric measures *IEEE Vis. '04* (IEEE Comp. Society) pp 497–504
[10] Mascarenhas A, Grout R, Bremer P T, Pascucci V, Hawkes E and Chen J To be published *Topoinvis'09: Topological Methods in Data Analysis and Vis.*
[11] Edelsbrunner H, Harer J and Zomorodian A 2003 *Discrete Comput. Geom.* **30** 87–107
[12] Edelsbrunner H, Harer J, Natarajan V and Pascucci V 2003 Morse-Smale complexes for piecewise linear 3-manifolds *Proc. 19th Sympos. Comput. Geom.* pp 361–370
[13] Gyulassy A, Natarajan V, Pascucci V and Hamann B 2007 *IEEE Trans. on Comp. Graphics and Vis. (TVCG)* **13** 1440–1447
[14] Samtaney R, Silver D, Zabusky N and Cao J 1994 *IEEE Comp.* **27** 20–27
[15] Silver D and Wang X 1998 Tracking scalar features in unstructured datasets *Proc. IEEE Vis. '98* (IEEE Comp. Society Press) pp 79–86
[16] Laney D, Bremer P T, Mascarenhas A, Miller P and Pascucci V 2006 *IEEE Trans. Vis. and Comp. Graphics (TVCG) / Proc.of IEEE Vis.* **12** 1052–1060
[17] Reinders F, Post F H and Spoelder H J W 2001 *The Visual Comp.* **17** 55–71
[18] Ji G and Shen H W 2004 Efficient isosurface tracking using precomputed correspondence table *Proc. Symp. on Vis. (VisSym) '04* (Eurographics Association) pp 283–292

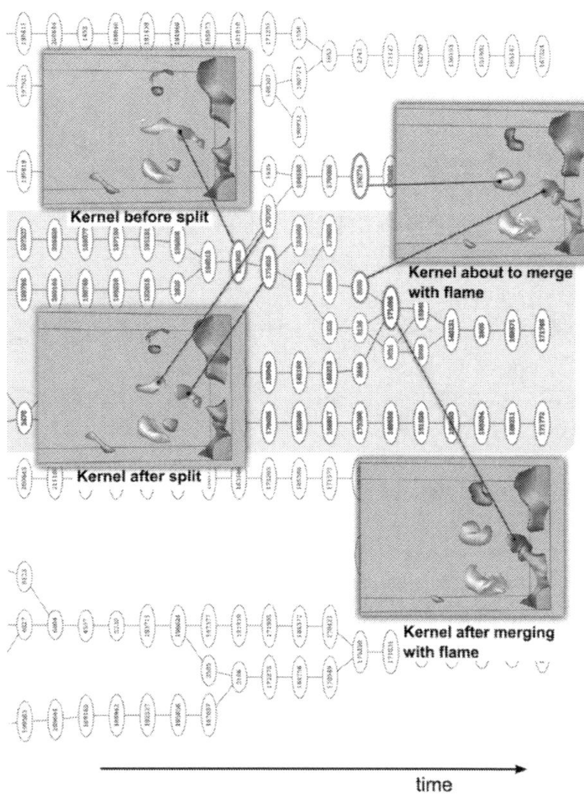

(a)

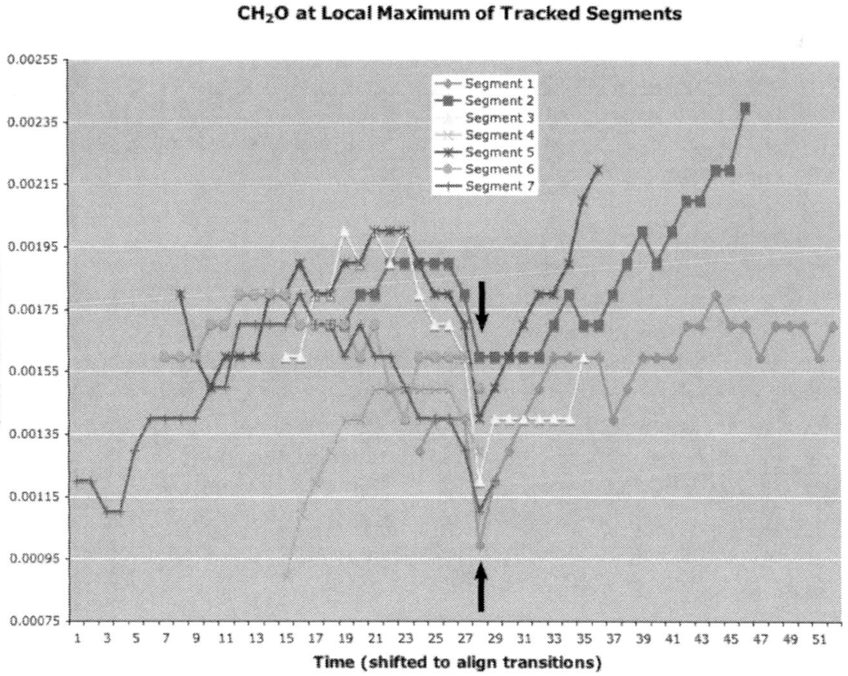

(b)

Figure 4: (a) Tracking graph showing kernels and flame base. Annotations show various topological changes experienced by the kernels and flame. (b) The mass fraction of Y_{CH_2O} inside tracked kernels. We align each individual kernels plot so that the time at which they transition to burning is on the same vertical line (show by arrows). We notice that the Y_{CH_2O} mass fraction drops at the transition time.

Journal of Physics: Conference Series **180** (2009) 012087 doi:10.1088/1742-6596/180/1/012087

Flow visualization in science and mathematics

Nelson Max[1], Carlos Correa, Chris Muelder, Shi Yan, Cheng-Kai Chen, and Kwan-Liu Ma
Department of Computer Science, University of California, Davis
1 Shields Ave., Davis California, 95616, USA

E-mail: max@cs.ucdavis.edu

Abstract. We present several methods for visualizing motion, vector fields, and flows, including polygonal surface advection, visibility driven transfer functions, feature extraction and tracking, and motion frequency analysis and enhancement. They are applied to chaotic attractors, turbulent vortices, supernovae, and seismic data.

1. Introduction

At the SciDAC Institute of Ultra-Scale Visualization, we are developing tools to visualize very large data sets, as well as basic visualization paradigms, algorithms, and methods. Here we report on several different visualization methods for 3D motion and flow fields. Visualizing 3D flow is difficult because representations tend to obscure each other, so we attempt to optimize visibility. We discuss several different representations, including advecting polygonal surfaces, contour surface volume rendering, and deforming textures. For interactive performance, we have implemented most of the techniques using hardware Graphics Processing Units (GPUs). Sections 2, 3, 4, and 5 below discuss, respectively, visualizing strange attractors, visibility driven transfer functions, feature tracking, and motion frequency analysis and enhancement.

2. Visualizing strange attractors

A strange attractor is an attractor (a limit set for trajectories) in a flow field whose trajectories show sensitive dependence on initial conditions. Since the initial discovery of the Lorenz attractor [1] in 1963, several other interesting examples from 3D steady flows have been studied, including the Rossler attractor [2], the Duffing attractor, and the Van der Pol attractor. More details on these attractors can be found in [3] and [4]. They look like smooth surfaces along two directions, but have a fractal behavior like a Cantor set in the other direction normal to these surfaces. It is possible to visualize these attractors by taking a random cloud of initial points, and then tracking them in the flow until they approach very close to the limit attractor. However, in looking at clouds of points, it is difficult to see the layered fractal structure. Therefore we have developed polygon-based visualizations, where we advect an initial triangular mesh towards the attractor. For more details of the methods described below, see Yan and Max [5].

[1] To whom any correspondence should be addressed.

2.1 Polygonal approximation

We start from an initial triangular mesh approximation, and let the flow advent the vertices towards the limit attractor, adaptively subdividing and retiling as necessary to track the change in surface shape. We use fourth-order Runge-Kutta integration to advect the vertices.

The Duffing and Van der Pol attractors represent 1D Newton-law motions with periodic forcing. To turn them into 3D flows, we wrap the relevant disk in the 2D position/velocity plane around a circle representing the forcing phase, to give a solid torus. A triangle mesh for the boundary of this torus then serves as the initial polygonal approximation. Figure 1 shows this torus approaching the Duffing attractor.

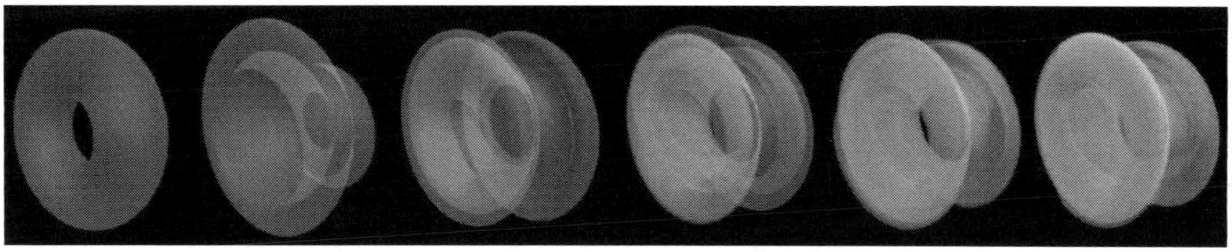

Figure 1. Six stages in the approach of an initial torus to the Duffing attractor.

For the Lorenz and Rossler attractors, we cannot use such a closed initial surface. One alternative would be to start with a disk somewhere near the attractor. But these two attractors loop around holes, and an initial large disk gets pulled and stretched by the flow near the holes because it has the wrong topology, while a smaller disk takes too long to spread out over the limit attractor. Therefore, we start with a random collection of points, move them for many integration steps until they approach the attractor, and then triangulate the resulting point cloud using the Cocone program of Dey *et al.* [6]. Figure 2*a* shows such a triangulation. Note that there is a gap where this program fails, as two sheets of the attractor approach too close to each other, and near this gap, there are incorrect triangles joining these two sheets. Figure 2*b* shows a modified surface where an annular sector near this gap has been removed, and then a collection of new pink triangles connecting one of the sheets across the gap has been added. Pink triangles have also been added to fill in some concave regions along the edges of the sheets. Figure 3 shows several stages in the advection of this surface, as is spreads over itself to create a fractal structure of multiple sheets.

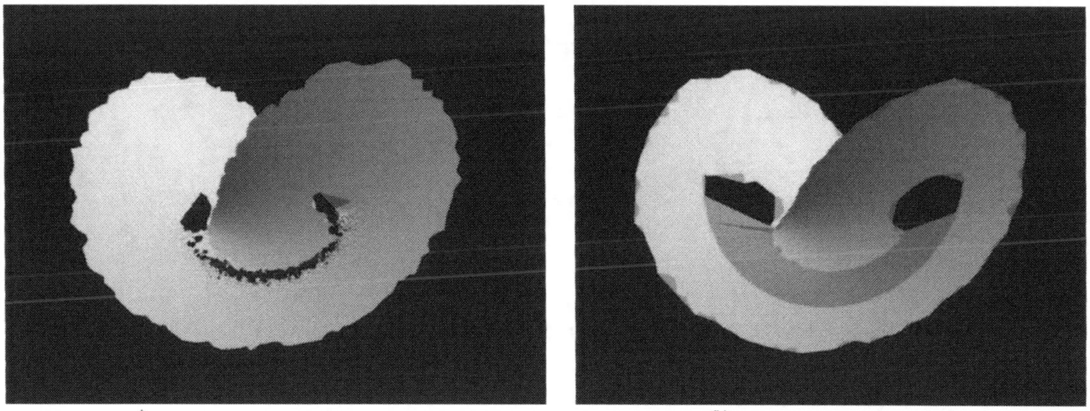

a) *b*)

Figure 2. *a*) A triangulation of points on the Lorenz attractor. *b*) The triangulation after an annular sector has been removed, and replaced by a sheet of pink triangles, and other pink triangles have been added to fill in concave regions.

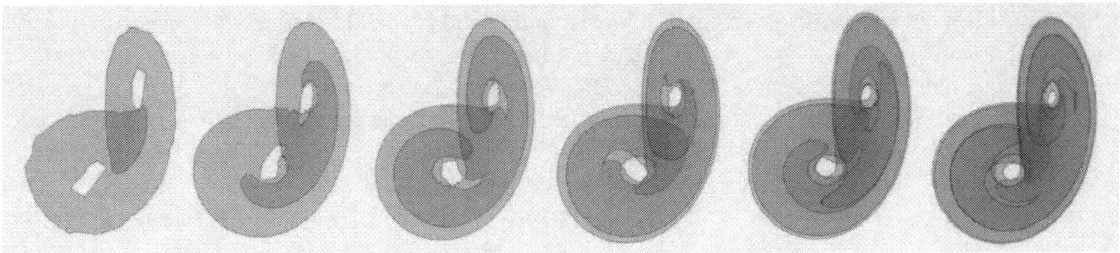

Figure 3. Six stages in the approach of an initial mesh to the Lorenz attractor.

2.2 Triangle subdivision

As the surface is advected by the flow, triangles will get stretched and bent, until they no longer approximate the desired surface. We deal with bent triangles by adaptively subdividing the edges of the mesh. We track the position of the midpoint of each edge in the original triangulation, and if this midpoint moves too far away from the edge, so that the angle between the segments joining it to the edge endpoints strays too far from 180° degrees, we subdivide the edge and the two triangles adjacent to it, and begin tracking the midpoints of any newly created edges. For steady flows like those defining these attractors, we can initialize these new midpoints at points inside triangles of the initial mesh.

2.3 Surface retiling

As the surface advection and subdivision proceeds, we soon get too many vertices to advect, and too many triangles for interactive display. Thus we periodically retile the surface, using a method similar to that of Turk [7]. We randomly insert new points on the surface, with probability density varying with the curvature of the surface approximation, by taking the probability of inserting a random point into a triangle proportional to a weighted sum of the triangle area and the average curvature estimates for its three vertices. To create a good spacing for these new points, we apply a repelling force between neighboring points on the same surface sheet, with a smaller force for points in high curvature triangles. For surfaces with borders, like those in figures 2 and 3, we add forces to move the new points away from the border, and then later add extra new border points along a Catmull-Rom spline interpolating the old border vertices. We do five iterations of the following two steps: *a*) move the non-border points in the direction of the sum of all the forces on them, and *b*) project them to the plane of their triangles. If they project outside their triangle, the neighboring triangles are recursively rotated up into this plane until the new point can be placed into one of them. All the triangles are retriangulated to include the new points as vertices. The new points are then moved to lie on a weighted least squares quadratic polynomial fit to the nearby old vertices on the same surface sheet, with the weights decreasing with distance from the new point. Then the old vertices are removed one by one, together with any triangles that use them, and the resulting holes are retriangulated. Finally, we iteratively consider all quadrilaterals formed by two triangles adjacent across an edge. If removing that edge and replacing it by the other diagonal of the quadrilateral would improve the triangulation, we perform this "edge flip".

Our strange attractors consist of many parallel closely spaced curved sheets, and if they are triangulated inconsistently, the sheets will appear to intersect. Therefore, every time we add a new random point P as described above, we look for intersections, with other nearby triangulated sheets, of the line through P normal to the surface. If any intersection points are found, they are paired to the point P with an attracting force that will tend to keep the two points aligned as they both move with the repulsion forces, and thus keep the triangulations aligned. Figure 4 shows a constant-forcing-phase cross section of the Duffing attractor, revealing the fractal structure of the closely spaced sheets.

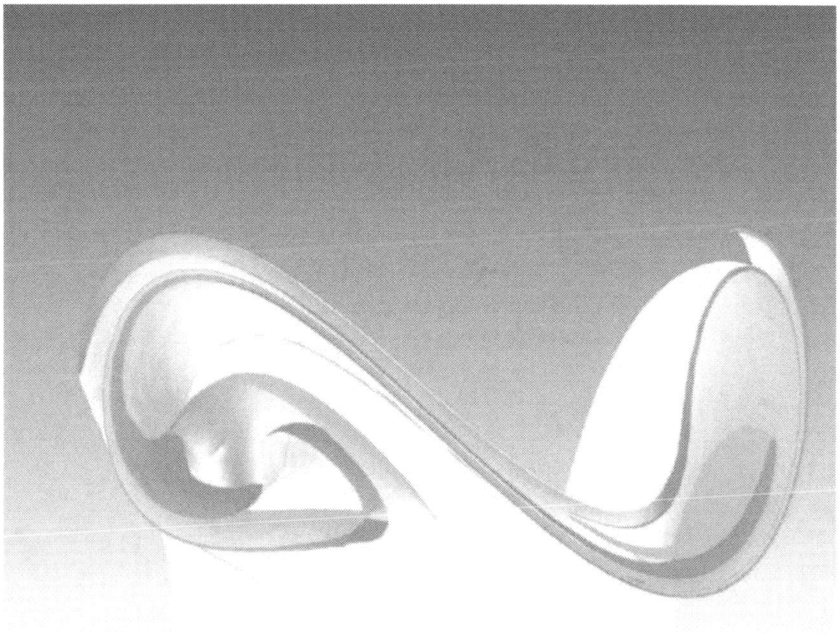

Figure 4. A cross section of the Duffing attractor.

3. Visibility Driven Transfer Functions

Scalar field volume visualization is also useful for flow visualization, to show scalar quantities like velocity magnitude or vorticity magnitude. Transfer functions are used to map the scalar quantity to color and opacity, which can then be integrated along the viewing rays, as explained in Max [8]. One popular paradigm for designing such transfer functions is to concentrate the opacity in specific narrow regions of the scalar value range, to give the appearance of concentric contour surfaces of different colors. Here, we use a sum of Gaussians opacity transfer function A of the form

$$A(x) = \sum_i \alpha_i \, g_{\mu_i,\sigma_i}(x)$$

where x is the scalar data value, $g_{\mu,\sigma}(x)$ is the Gaussian with mean μ and standard deviation σ, and the α_i are weighting factors. Figure 5a shows an example of such a visualization, of vorticity magnitude in a pseudo-spectral simulation of coherent vortex structures from [9]. Note that the inner yellow and red contour surfaces are visible only where the excessively opaque purple surface has been sliced away by the boundary of the data volume. In order to see these inner contours better, the purple surface must be made less opaque by changing its weight in the above equation. In figure 5b, the user has adjusted the opacity weights α_i and now the red and yellow contours are now more visible.

We have developed a system for automatically adjusting the α and σ parameters, to optimize the visibility of important ranges in the scalar function. The user first specifies the desired importance by another transfer function $O(x)$. The parameters are optimized to minimize a weighted average of the following two functions, summed over all the voxels p in the volume, where $x(p)$ is the scalar value at the voxel p. The first function E_S tries to match the transfer function to the user's importance function:

$$E_S(x(p)) = \left[A(x(p)) - O(x(p))\right]^2.$$

The second function E_V is used to maximize the visibility of the important voxels, and thus uses a minus sign in the function to be minimized:

$$E_V(x(p)) = -O(x(p))T(p)$$

where

$$T(p) = \exp[-\int A\left(x(r(t))\right) dt]$$

is the transparency along the ray $r(t)$ from the viewpoint to the voxel p, parametrized by arclength t. Thus, after summation over all the voxels p, the function to be minimized is

$$\sum_{p} [\beta_1 E_S\left(x(p)\right) + \beta_2 E_V\left(x(p)\right)]$$

where β_1 and β_2 are the weights for the matching and visibility terms, respectively.

It seems that it would be a difficult problem to evaluate this sum, since the visibility term involves an integral for every voxel. However, this term can be approximated by the same method used for producing the volume rendered image in GPU hardware. The volume is sliced in front to back order by multiple sampling planes parallel to the viewing screen, with spacing dt, 3D texturing hardware is used to access and interpolate the voxel data and the transfer functions, and the accumulated transparency integral is incrementally adjusted per viewing-ray sample by multiplication by $1. -A\big(x(r(t))\big)dt$, which is the first order Taylor approximation to $\exp[-A\left(x(r(t))\right) dt]$. The GPU can then multiply this accumulated transparency by $O\big(x(p)\big)$ and add this term to the output buffer for each ray through a pixel, in order to compute the visibility term to be optimized, in the same way that it multiplies the transparency by the color transfer function and accumulates the contribution to the ray color for creating a volume rendered image. We used the conjugate gradient method to optimize the opacity transfer function, and the GPU hardware to compute it during the optimization.

Figure 6 shows a visualization of the entropy in a supernova simulation with the highest entropies violet and blue and the lowest ones red. The optimized opacity transfer function reveals more of the important features. Note that this is a view dependent optimization, and may produce different results for different viewing angles. For more details see Correa and Ma [10].

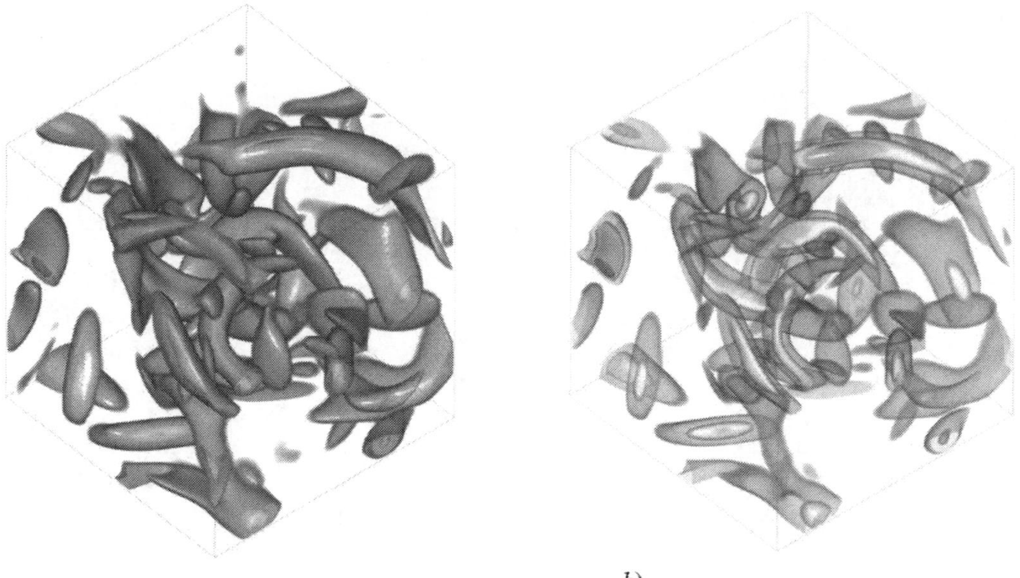

a) *b)*

Figure 5. Vorticity magnitude with: *a)* initial transfer function; *b)* user-adjusted transfer function.

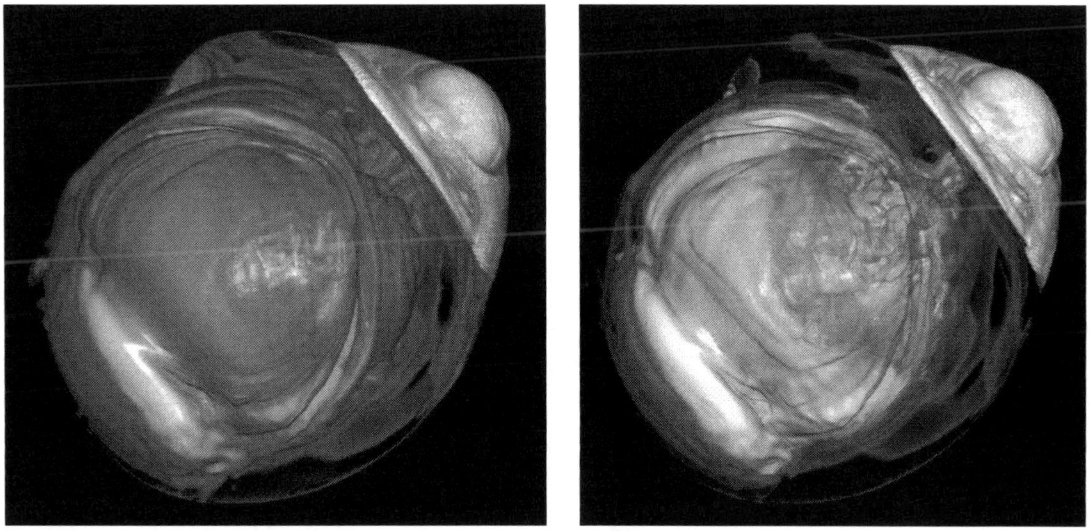

a) *b*)

Figure 6. Entropy in supernova simulation with: *a*) initial transfer functions; *b*) automatically optimized transfer functions.

4. Feature Tracking

One can visualize vortices as the regions where the vorticity magnitude, or some other scalar quantity computed from the velocity field, exceeds a threshold. To understand evolving vortices, we need to track them in time. Silver and Wang [11] developed a method that matches the regions in one time step to those in the next, using their volumes of overlap. This requires detecting the voxels that belong to each region, and to each pair of potentially overlapping regions. Although they did this efficiently using an octree data structure, their algorithm still needs to detect the voxels in the interior of the regions. We have developed a feature extraction and tracking algorithm that is based on adjusting the set of boundary voxels, those whose scalar values exceed the threshold, but which have adjacent neighbors whose values do not. Since it operates only near the boundaries of the tracked regions, it can be more efficient, especially if only one or a few selected regions are to be tracked, or if the motion is slow, so that only a few voxels need to be adjusted.

Our extraction and tracking method has two steps, prediction and adjustment. First, the motion of the region's center of gravity in previous frames is used to predict a translation of the previous frame's region boundary onto the current frame. Then region growing and shrinking adjusts the boundary to the new region or regions that it overlaps.

For the motion prediction, we use either a constant prediction, which does not translate the previous frame's region at all, a linear prediction which translates it by the motion of the center of gravity between the previous two frames, or a quadratic prediction, which fits a quadratic polynomial to the center of gravity of the previous three frames, and uses it to predict the translation. In order to compute the center of gravity incrementally, we maintain a count of the number of voxels in the region, and the sum of their coordinate vectors, and adjust the count and sum vector as voxels are added to or removed from the region. We also maintain voxel markings as inside, surface, or outside the region, and translate these markings by the predicted motion when moving to the next time step.

To incrementally adjust the vector sum, count, markings, and the set of boundary voxels for the next time step, we use a queue of voxels that are to be checked, which is initialized to the set of the predicted boundary voxels. While the queue is not empty, we remove a voxel V from it. If V satisfies the threshold condition for the new region, we add it to that region, and then for each neighboring voxel N that belongs to the region, if N is yet in the queue, we add it to the queue, and if any neighbors of V do not belong to the region, we add V to the boundary voxel set. If V does not belong to the region, we remove it from the region, and enqueue any neighbors that are currently marked as in the region. This adjustment automatically handles topological changes from region bifurcation, joining, and disappearance, but if all regions are to be tracked, methods which check the data values at all voxels are necessary to periodically check for the birth of new regions. The pseudocode below expresses this algorithm.

```
Initialize the queue with the predicted surface
while (the queue not empty)
   pop a voxel V off of the queue
   if (V's value is above the threshold value)
      mark V as in the region
      if (V is on the new surface)
            save V as part of the new surface
      for each neighbor N of V
            if (N is not marked as inside the region)
                  if (N is not in the queue)
                        enqueue N and increment the center of gravity data
   else // V's value is not above the threshold value
      mark V as not in the region and decrement the center of gravity data
      for each neighbor N of V
            if (N is marked as inside the region)
                  if (N is not in the queue) enqueue N
```

Figure 7 shows several vortices from the simulation from [9] shown in figure 5, identified by color and tracked over different time steps. For more details, see Muelder and Ma [12].

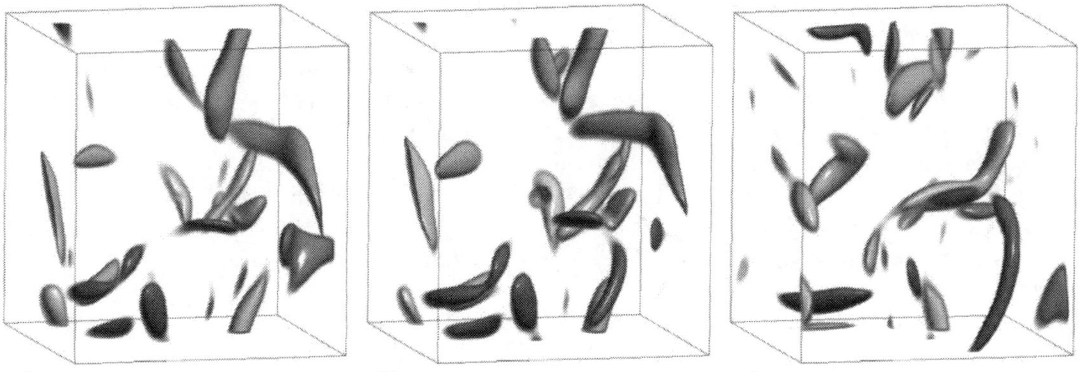

a) *b*) *c*)

Figure 7. Several time step frames from the vortex simulation of figure 5, with the tracked vortices distinguished by color. *a*) frame 0, *b*) frame 4, *c*) frame 16.

5. Motion Frequency Analysis and Enhancement

When studying time series of displacements from measurements or simulations of seismic waves and their effects on structures, it can be useful to analyse and visualize energy and displacement in separate temporal frequency bands. Here, we used fourth order Butterworth filters in the frequency domain to define high-pass and low-pass filtered data, and visualize the two components separately. The low-pass component represents the permanent displacement, and the high-pass component represents the transient shaking.

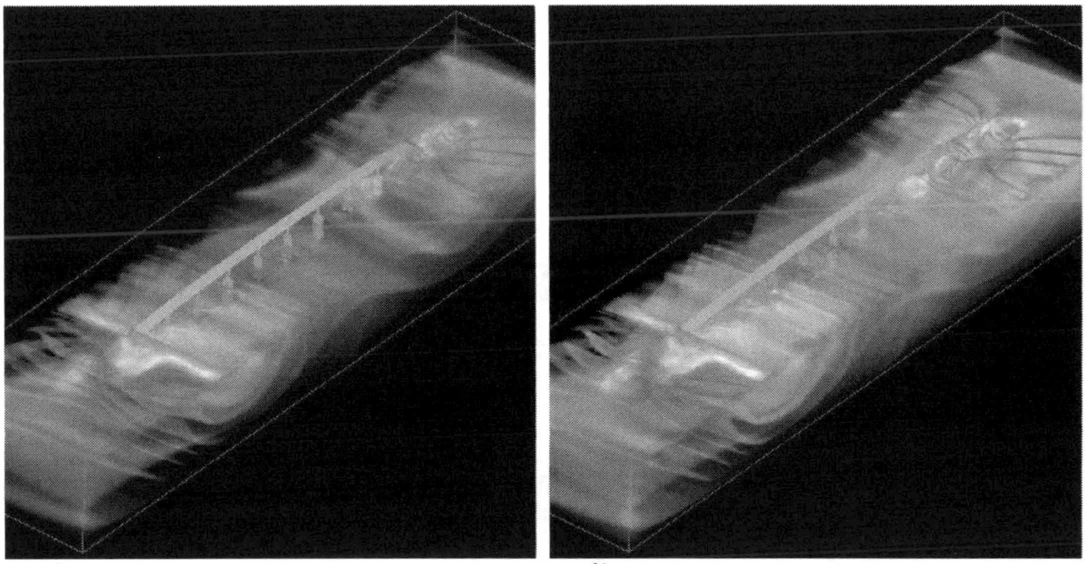

a) *b*)

Figure 8. Earthquake displacement: *a*) no enhancement, *b*) enhancement of high-pass component.

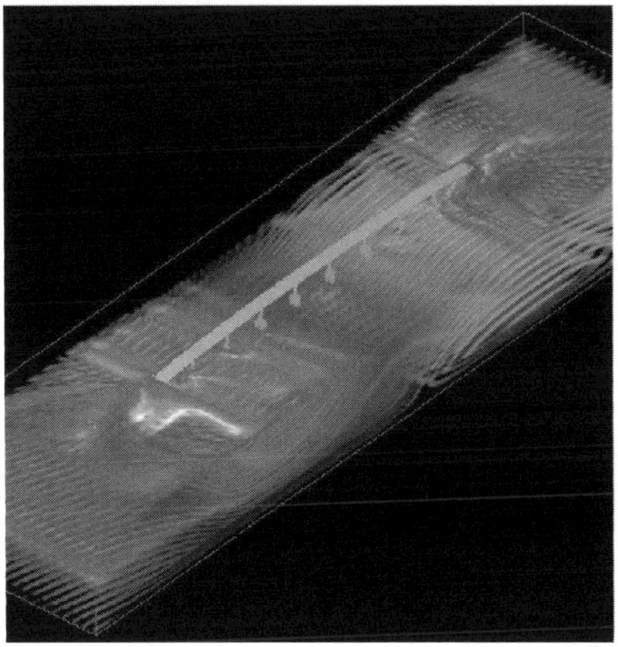

Figure 9. Earthquake data of figure 8, with the high-pass component visualized by a deforming tube texture.

We visualized data from a finite element simulation of the Humboldt Bay Middle Channel Bridge and the nearby river channel and banks, from Zhang *et al.* [13]. In figure 8*a*, the low-pass data is shown in orange, and the high-pass data is shown in blue. Each component was given its own opacity. In figure 8*b*, the high frequency component opacity was increased by a factor of 2.2, making it more visible. The data was given on a curvilinear Lagrangian grid of hexahedra, which, for visualization purposes, were divided into tetrahedra, so that the GPU ray tracing method of Weiler et al. [14] could be applied. For more details, see Chen et al. [15].

In figure 9, the high frequency component is indicated by a deforming texture of horizontal tubes. Instead of accessing actual 3D texture data, the tube opacity was evaluated analytically as a function of 3D position, modified by subtracting the time-varying displacement vector at that position, so as to enhance the motion of the tubes.

6. Conclusions and future work

We have discussed a number of different techniques, each developed to visualize different aspects of motion and flow, for various specific goals and applications. We are now applying our polygonal surface advection method to other simulation data, and have found that is has problems with turbulent flows. Edge lengths can become zero, and the advected surface can become too convoluted to compute and render interactively. We hope to develop methods of more aggressively simplifying the surface, and also to use point-based rendering, so that we do not need to maintain the connectedness topology for a polygonal mesh. The visibility driven transfer functions are view dependent, and could cause disorientation during interactive viewpoint changes, and we could develop viewpoint independent ones by optimizing over a representative collection of views.

Acknowledgements

This research was supported in part by the U.S. Department of Energy through the SciDAC program with Agreement No. DE-FC02-06ER25777 and DE-FG02-08ER54956, and in part by the Japan Society for the Promotion of Science.

References

[1] Lorenz E N, 1963 Deterministic nonperiodic flows *J. Atmospheric Sc.* **20** 130 - 41
[2] Rossler O E 1976 An equation for continuous chaos *Phys. Lett.* 1976 A **57** 379 - 80
[3] Abraham R H and Shaw C D 1963 *Dynamics: the geometry of behavior, Part two: chaotic behavior* Ariel Press
[4] Strogatz S H 1994 *Nonlinear dynamics and chaos* Westview Press
[5] Yan S and Max N 2009 Polygonal surface visualization of strange attractors, *Tech. report* CSE-2009-11. University of California, Davis
[6] Dey T K and Giesen J 2001 Detecting undersampling in surface reconstruction *Proc. ACM Symp. Computational Geometry* 257 - 63
[7] Turk G 1992 Re-tiling polygonal surfaces *Computer graphics* **26**(2) 55 - 64
[8] Max N 1995 Optical models for volume rendering *IEEE Trans. Visualization and Computer Graphics* **1**(2) 99 - 108
[9] Chen S-Y and Shan X 1992 High resolution turbulent simulation using the connection machine-2 *Computers in Phys.* **6**(6) 643 - 6
[10] Correa C and Ma K-L 2009 Visibility driven transfer functions *IEEE Pacific Visualization Symposium* Beijing 177 - 84
[11] Silver D and Wang X 1997 Tracking and visualizing turbulent 3D features *IEEE Trans. Visualization and Computer Graphics* **3**(2) 129 - 141
[12] Muelder C and MA K-L 2009 Interactive feature extraction and tracking by utilizing region coherency *IEEE Pacific Visualization Symposium* Beijing 17 - 24
[13] Zhang Y, Acero G, Conte J, Yang Z and Elgamal A 2004 Seismic reliability assessment of a bridge ground system *Proc. 13th World Conf. on Earthquake Engineering* Vancouver, B. C. Canada
[14] Weiler M, Kraus M, Merz M, and Ertl T 2003 Hardware-based ray casting for tetrahedral meshes *Proc. IEEE Visualization* 333 - 40
[15] Chen C-K, Correa C, and Ma K-L 2009 Frequency enhancements for visualizing 3D data *Tech. report* CSE-2009-4. University of California, Davis

Parallel visualization on leadership computing resources

T Peterka,[1] R B Ross,[1] H-W Shen,[2] K-L Ma,[3] W Kendall,[4] H Yu,[5]

[1]Mathematics and Computer Science Division, Argonne National Laboratory, Argonne, IL 60439, USA

[2]Department of Computer Science and Engineering, The Ohio State University, Columbus, OH 43210, USA

[3]Department of Computer Science, University of California at Davis, Davis, CA 95616, USA

[4]Department of Electrical Engineering and Computer Science, University of Tennessee at Knoxville, Knoxville, TN 37996, USA

[5]Sandia National Laboratories, California, Livermore, CA 94551, USA

E-mail: `tpeterka@mcs.anl.gov`

Abstract. Changes are needed in the way that visualization is performed, if we expect the analysis of scientific data to be effective at the petascale and beyond. By using similar techniques as those used to parallelize simulations, such as parallel I/O, load balancing, and effective use of interprocess communication, the supercomputers that compute these datasets can also serve as analysis and visualization engines for them. Our team is assessing the feasibility of performing parallel scientific visualization on some of the most powerful computational resources of the U.S. Department of Energy's National Laboratories in order to pave the way for analyzing the next generation of computational results. This paper highlights some of the conclusions of that research.

1. Introduction

Visual representations can provide compact summaries of large amounts of scientific data. As numerical simulations and experiments produce higher quantities of information, the challenge for visualization researchers is to continue to discover graphical methods that scale with the growth in data size and complexity. Since its recognition in 1989 as a core component in the scientific process [1], visualization has evolved with changing technologies, mainly influenced by consumer-grade graphics accelerators and relatively low-cost clustered architecture. Today, thanks to research in the field over the last twenty years, visualization has its own place in the science workflow pipeline shown in Figure 1. This is a data flow model—bytes move from one stage to the next. Data size in the hundreds of terabytes or petabytes, however, can overwhelm this model as the cost of managing data movement between stages overtakes the actual work done in the stages. The high cost of data movement is magnified by the relatively short duration of the visualization step. An alternative approach is to substitute some of the data flow with control flow—by moving some of the tasks to the data rather than the other way around.

This implies performing some or all of the analysis and visualization operations on the same supercomputers as the simulation. In the U.S. National Laboratories, the most powerful of these machines are called *leadership computing resources*. There are three potential benefits

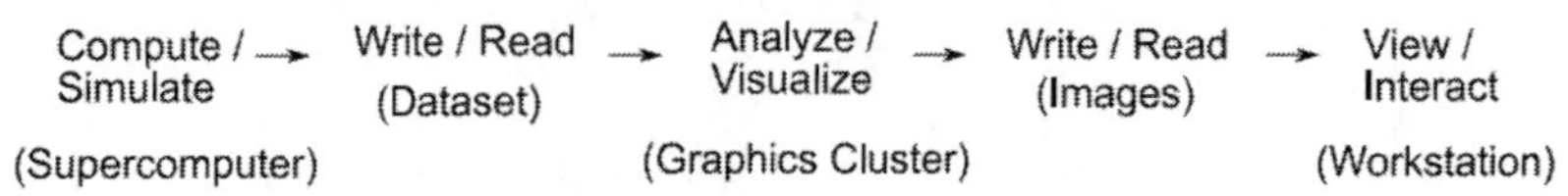

Figure 1. The stages in computational science: simulation, analysis, and interaction, in a customary data-flow pipeline. As the cost of data movement becomes prohibitive, performing visualization on machines traditionally reserved for computation can shorten overall time.

to this shift in performing visualization on leadership machines, compared to graphics clusters. First, end-to-end performance can improve because of reduced data movement. Second, this performance savings can afford higher quality visualization in the form of rendering improvement; for example, incorporating more complex lighting models or processing a larger portion of the total dataset in core. The third benefit is economic. As leadership machines continue to scale, deploying and operating proportional-size data analysis clusters becomes more costly. At some point, it will not be economically feasible for a leadership facility to maintain both types of architecture at these scales.

The U.S. Department of Energy has established Leadership Computing Facilities at Argonne and Oak Ridge National Laboratories. Each facility maintains a different supercomputing architecture: the Argonne Leadership Computing Facility (ALCF, [2]) has an IBM Blue Gene/P (BG/P) system while the Oak Ridge National Center for Computational Sciences (NCCS, [3]) maintains a Cray XT4 system. The BG/P has 160 K cores while the XT4 has approximately 31 K cores. The BG/P has access to a 5 petabyte PVFS parallel file system, while the XT4 is connected to a Lustre storage system.

As datasets grow in size and number of variables, the time that it takes to transfer each time-step (in the case of time-varying data) from storage to memory becomes significant. In fact, Peterka et al. [4, 5] showed that I/O can consume up to 90% of the total visualization time. One of the research directions of the SciDAC Institute for UltraScale Visualization [6] is to evaluate how parallel I/O techniques [7, 8] and libraries [9] can expedite the I/O stage in data-intensive visualization and analysis.

2. Selected Results of Case Studies

In one study, we benchmarked parallel volume rendering on the BG/P at the ALCF. The data for this test are simulation results produced by John Blondin of North Carolina State University and Anthony Mezzacappa of Oak Ridge National Laboratory. The VH-1 hydrodynamics code simulates the static accretion shock of a core-collapse supernova [10], and it generates a time-varying dataset. Each time step is a netCDF file with five variables.

The parallel volume rendering algorithm consists of three stages. In the first stage, I/O, the time-step file is read into memory by all processes in parallel. The second stage, rendering, has all processes concurrently performing software ray-casting through their respective data subdomains. At the end of the second stage, each process contains a volume-rendered image of its subdomain. The images are merged together in the third stage, called compositing. The relative time that each stage requires changes with the system scale; our experiments range from 64 processes to 32 K processes. The data size ranges from 1120^3 elements to 4480^3, and the image size ranges from 1024^2 to 4096^2 pixels.

A performance summary of three data sizes and system scales appears in Figure 2. The time axis represents the total end-to-end time needed to visualize a single time step, including I/O, rendering, and compositing. The 1120^3 dataset with 1024^2 image size scales to 16 K processes; beyond this point the subdomains are too small to justify using more cores. The upper two curves show scalability out to 32 K processes on larger data and image sizes. The uppermost curve represents a data size of 4480^3 with an image size of 4096^2; with 32 K processes, this is

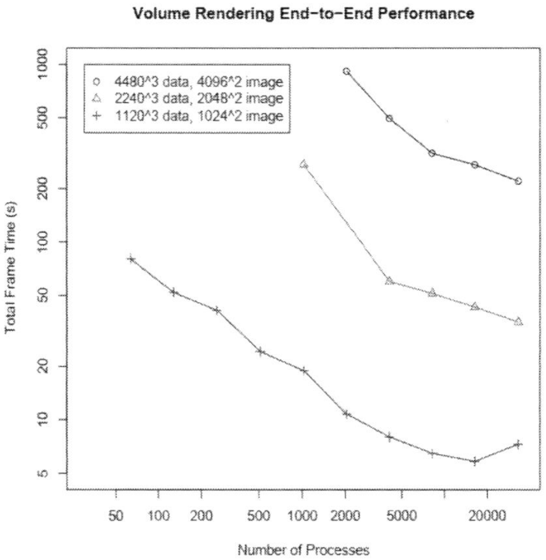

Figure 2. Scalability over a variety of data, image, and system sizes. Several performance points are available for each curve; one may select a smaller system scale, for example, based on resource availability, and still complete the visualization in a matter of seconds or minutes.

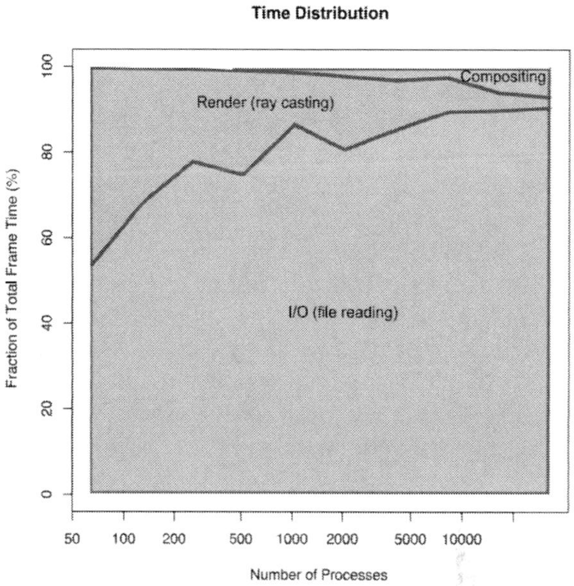

Figure 3. The relative percentage of time spent in I/O, rendering, and compositing as a function of system size. At large system scale, visualization is dominated by data movement: primarily by I/O and secondarily by compositing. The data size and image size for this test is 1120^3, and 1024^2, respectively, but the pattern shown is typical of all of our tests.

one of the largest in-core parallel visualization results published to date.

Figure 3 illustrates how the relative proportion of time spent in each stage changes as the system scale grows. At the left side of the plot, rendering occupies 50% of the total time, but as the system scale increases, the fraction of time spent in rendering is almost negligible. This is one reason why software rendering is practical on relatively low-power processors such as BG/P's. Most of the time is spent in I/O, and even compositing overtakes rendering at 8 K cores. This pattern is typical of all of our tests: large scale visualization is dominated by data movement.

Because I/O is such a critical part of the large-scale visualization process, our team devotes considerable effort to its study. Figure 4 shows the result of one such experiment. The diagrams represent a log of disk activity, when reading one variable out of the five variables contained in a netCDF dataset. The dark rectangles represent physical file blocks that are read in as a result of this collective I/O request. Starting with the original implementation (left), reorganizing the data layout and using a new 64-bit version of the netCDF format (right) resulted in a nearly three times shorter I/O time. Comparing the left and right diagrams, fewer blocks total blocks are accessed in the right diagram, and their physical placement on disk is contiguous. Our collaborators in the SciDAC SDM center provided an early release of the new netCDF format for this research.

One pattern that often occurs in scientific data organization is one time-step per file, resulting in many files for a single dataset. In a study on the Cray XT reading several hundred time-steps of NASA's Moderate Resolution Image Spectroradiometer (MODIS) dataset, our team is researching how to efficiently read all of these files in parallel into the memory of thousands of cores. This situation occurs, for example, when comparisons across time steps are required in the analysis. By using a greedy algorithm that assigns files or portions of files to processes in a round-robin fashion, and by placing files on all 144 Lustre *Object Storage Targets* (OSTs), we achieved the 28 GB/s aggregate read bandwidth shown in Figure 5. This is approximately 75% of the maximum peak bandwidth measured by the IOR benchmark.

Many application teams are experimenting with "I/O subsetting," or performing I/O operations from only a fraction of the nodes used for computation. We found that controlling the number of processes that perform the actual writing of the output image (*writers*) improves performance and memory usage. The right plot in Figure 6 shows that between 32 and 512 writers is appropriate. Figure 6 also shows that we are able to achieve better writing performance by controlling the subsetting through the application (right), rather than by supplying hints to MPI-IO (left). In both cases a minimum time occurs at 8 or 16 aggregators (the MPI-IO entities actually performing the writing on behalf of MPI processes), but by controlling the number of writers explicitly, this optimal setting is achievable in a wider range of configurations.

3. Summary

As leadership computing machines and the simulations that run on them extend to petascale and beyond, analysis of the results must scale accordingly. In order to meet this challenge, we are studying what circumstances favor performing more of the analysis and visualization operations directly on these flagship systems that have tens of thousands of cores, a high-throughput interconnect, and parallel storage infrastructure. As our research shows, these characteristics are critical at large scales because analysis and visualization are dominated by data movement. Through careful study of data movement patterns, especially to and from storage, entire datasets

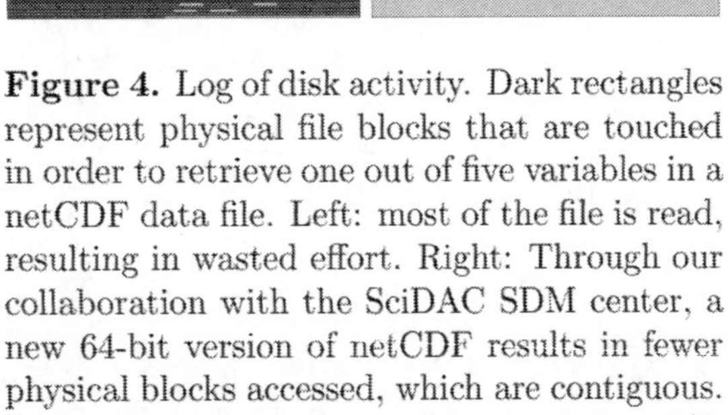

Figure 4. Log of disk activity. Dark rectangles represent physical file blocks that are touched in order to retrieve one out of five variables in a netCDF data file. Left: most of the file is read, resulting in wasted effort. Right: Through our collaboration with the SciDAC SDM center, a new 64-bit version of netCDF results in fewer physical blocks accessed, which are contiguous. This resulted in a three times speedup in I/O time.

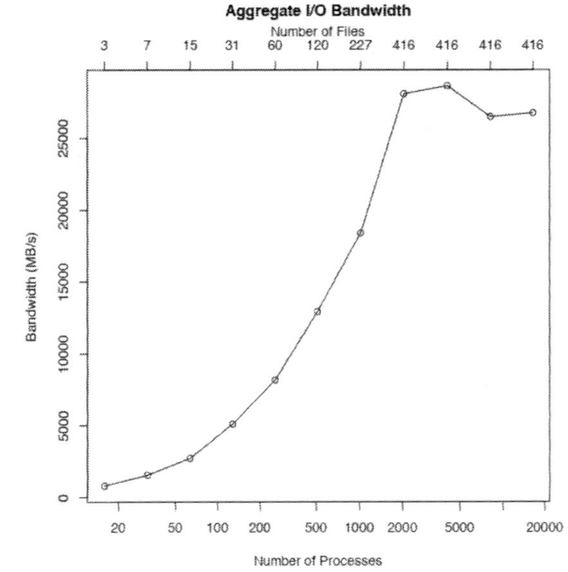

Figure 5. Aggregate bandwidth for reading up to 416 time steps of the MODIS full resolution dataset simultaneously into up to 16 K processes peaks at 28GB/s, or 75% of the IOR benchmark.

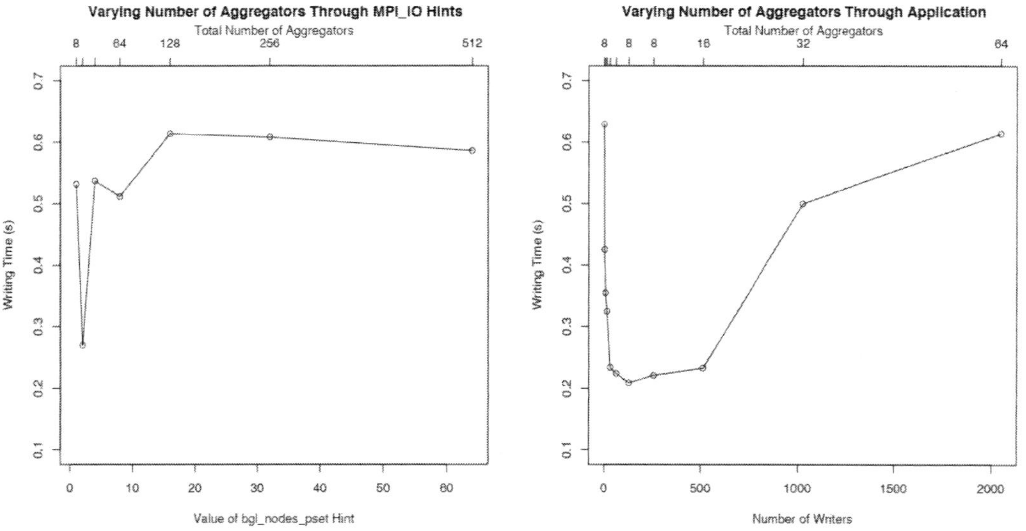

Figure 6. The effect of I/O subsetting, via MPI-IO hints (left) vs. using a custom communicator in the application code (right). As the right plot shows, manually assigning a number of processes to perform parallel image writing produces better a shorter time over a wider range of settings.

can be analyzed in-core and in full resolution. We are experimenting with other visualization techniques and more complex data sets that involve both scalar and vector time-varying data. We will continue to produce benchmarks that other researchers can use to perform parallel visualization on leadership computing resources.

Acknowledgments

We thank John Blondin and Tony Mezzacappa for their dataset and valuable feedback. We gratefully acknowledge the use of the resources of the Argonne Leadership Computing Facility at Argonne National Laboratory and the National Center for Computational Sciences at Oak Ridge National Laboratory. This work was supported by the Office of Advanced Scientific Computing Research, Office of Science, U.S. Department of Energy, under Contract DE-AC02-06CH11357. Work is also supported by DOE with agreement No. DE-FC02-06ER25777.

References

[1] DeFanti T A, Brown M D and McCormick B H 1989 *Computer* **22** 12–25 ISSN 0018-9162

[2] 2009 *Argonne Leadership Computing Facility* http://www.alcf.anl.gov/

[3] 2009 *National Center for Computational Sciences* http://www.nccs.gov/

[4] Peterka T, Yu H, Ross R and Ma K L 2008 Parallel volume rendering on the ibm blue gene/p *Proc. Eurographics Parallel Graphics and Visualization Symposium 2008* (Crete, Greece)

[5] Peterka T, Ross R, Yu H, Ma K L, Kenall W and Huang J 2008 Assessing improvements in the parallel volume rendering pipeline at large scale *Proc. SC 08 Ultrascale Visualization Workshop* (Austin TX)

[6] 2009 *SciDAC Institute for Ultra-Scale Visualization* http://ultravis.ucdavis.edu/

[7] Carns P, Ligon W B I, Ross R and Thakur R 2000 Pvfs: A parallel file system for linux clusters *Proc. 4th Annual Linux Showcase & Conference* (Atlanta, GA) p 28

[8] Thakur R, Gropp W and Lusk E 1999 Data sieving and collective i/o in romio *Proc. 7th Symposium on the Frontiers of Massively Parallel Computation* pp 182–189

[9] Li J, Liao W k, Choudhary A, Ross R, Thakur R, Gropp W, Latham R, Siegel A, Gallagher B and Zingale M 2003 Parallel netcdf: A high-performance scientific i/o interface *Proc. Supercomputing 2003* (Phoenix, AZ)

[10] Blondin J M, Mezzacappa A and DeMarino C 2003 *The Astrophysics Journal* **584** 971

Visualization of uncertainty and ensemble data: Exploration of climate modeling and weather forecast data with integrated ViSUS-CDAT systems

Kristin Potter[1], Andrew Wilson[2], Peer-Timo Bremer[3], Dean Williams[3], Charles Doutriaux[3], Valerio Pascucci[1], and Chris Johhson[1]

[1]Scientific Computing and Imaging Institue, University of Utah
[2]Sandia National Laboratories
[3]Lawrence Livermore National Laboratory

E-mail: kpotter@cs.utah.edu, atwilso@sandia.gov,
{bremer5,williams13,doutriaux1}@llnl.gov, {pascucci,crj}@sci.utah.edu

Abstract. Climate scientists and meteorologists are working towards a better understanding of atmospheric conditions and global climate change. To explore the relationships present in numerical predictions of the atmosphere, *ensemble datasets* are produced that combine time- and spatially-varying simulations generated using multiple numeric models, sampled input conditions, and perturbed parameters. These data sets mitigate as well as describe the uncertainty present in the data by providing insight into the effects of parameter perturbation, sensitivity to initial conditions, and inconsistencies in model outcomes. As such, massive amounts of data are produced, creating challenges both in data analysis and in visualization. This work presents an approach to understanding ensembles by using a collection of statistical descriptors to summarize the data, and displaying these descriptors using variety of visualization techniques which are familiar to domain experts. The resulting techniques are integrated into the ViSUS/Climate Data and Analysis Tools (CDAT) system designed to provide a directly accessible, complex visualization framework to atmospheric researchers.

1. Introduction

In order to better understand the long-term trends in climate change or the reliability of short-term weather predictions, it is essential to characterize the uncertainty of the data underlying scientific conclusions [9]. To this end, researchers are generating *ensemble data sets* which combine multiple runs of spatio-temporal simulations, using perturbed parameters and variations on initial conditions [6]. The resulting data sets provide a collection of estimations for each simulation variable, which in turn provides an understanding of the possible outcomes and the corresponding uncertainty. The main challenge to this work is addressing the increasing need to visualize uncertainty information [7, 12] in a manner that leads to improved understanding of the data.

The driving applications of this work reside in the fields of climate modeling and short term weather forecasting. Both areas combine processes underlying atmospheric behavior such as the global carbon cycle, atmospheric chemistry, vegetation, and ocean dynamics. The main

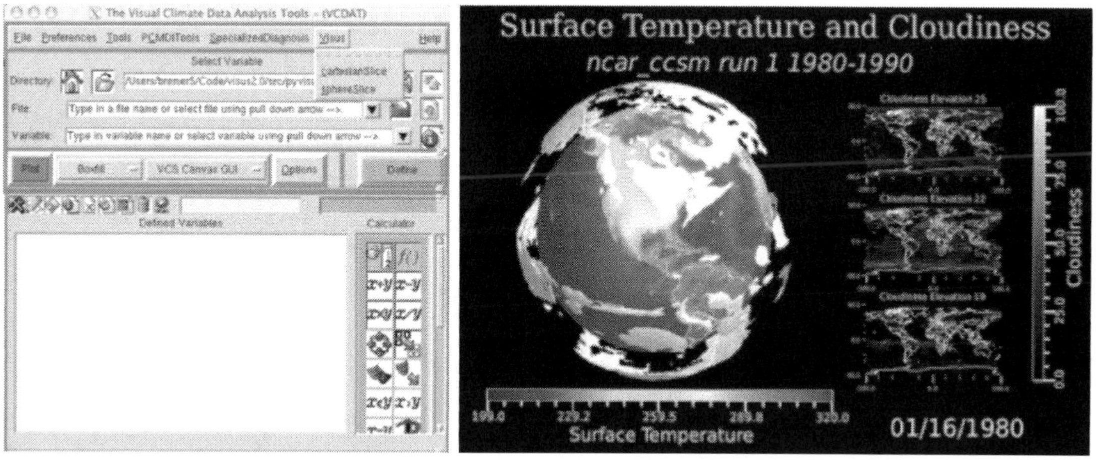

Figure 1. Screen shot of ViSUS 2.0 executed within CDAT. ViSUS/CDAT provides the user with a complex interface, multiple views, and a variety of visualization techniques for data exploration and analysis. (Left) The user interface of the VCDAT system. (Right) The data is visualized using 3D iso-contouring, slicing, and multiple 2D plots.

distinctions between the two applications stem from the spatial and temporal domains [5]. Climate models encompass the entire globe and are run over hundreds of years. The main goal is to understand global phenomena such as the impact of human activity on land and ocean temperatures or trends in natural disasters [1]. In contrast, short-range ensemble forecasting (SREF) data [2] focuses on North America, and predicts atmospheric variables such as temperature and humidity out to 87 hours (roughly three days). This data is used to predict near future weather patterns and is run on a daily basis, updating the initial conditions with observed data. While this work focuses on the challenges specific to the analysis and visualization of atmospheric data [11], the approaches presented here can be generalized to the many ensemble data sets becoming prevalent today.

ViSUS/CDAT, shown in Figure 1, is the framework used to develop the techniques for visualizing ensemble data and its corresponding uncertainty. This tool set is developed as a collaboration with the Earth System Grid Center for Enabling Technologies (ESG-CET), A Scalable and Extensible Earth System Model for Climate Change Science, and the climate modeling community in general. CDAT, or the Climate Data Analysis Tools, are a set of utilities specifically designed for the needs of climate researchers, providing advanced data analysis combined with the ability of reading specific data formats, supplying geospatial information and supporting a high level user interface. ViSUS is an integrated 3D visualization package, providing sophisticated visualization techniques seamlessly to the climate scientists. The incorporation of ViSUS with CDAT provides (among other things) a flexible system for the visualization of large ensemble datasets and is founded on the tight integration of 3D visualization tools, statistical data analysis techniques and compelling metaphors that facilitate the user intuition about the uncertainty of large, multidimensional, multivariate, time-varying datasets.

2. Ensembles and Uncertainty

An *ensemble data set* can be defined as a collection of multiple time-varying data sets (called *ensemble members*) that are generated by computational simulations of one or more state variables across space. The variation among the ensemble members arises from the use of different input conditions, simulation models, and simulation parameters. The generated data

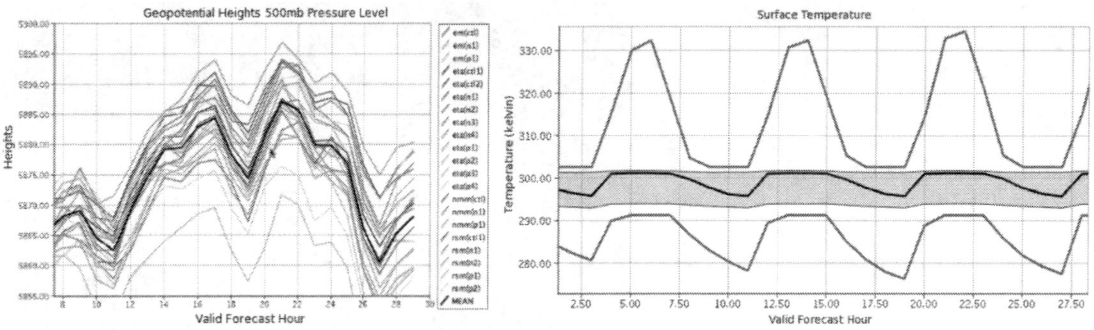

Figure 2. (Left) An example of the complexity of an ensemble data set. Here, the geopotential heights of the 500mb pressure level are shown at a single weather station across all valid forecast hours. Each model/perturbation group is shown using a different color, and the mean is shown in black. (Right) A quartile chart showing the range of surface temperatures across the valid forecast hours. The median of the data is shown in black, the inner quartile range shaded in grey, and the minimum and maximum ensemble members are in blue.

sets are multidimensional, since the spatial domain of the simulations can be 2 or 3D (or higher), multivariate, due to the possibly hundreds of state variables, and multivalued by collecting several values for each variable in space.

Ensemble data sets are chiefly useful as a tool to quantify and mitigate uncertainty and error in simulation results. Uncertainty can be described as the accuracy or confidence associated with the data. Errors can arise in the simulation through faulty estimations or measurements of the initial conditions, finite resolution and precision of the numerical model, sensitivity to input parameters, and from the nature of a numerical simulation as an approximate model of an incompletely understood real-world phenomenon. Ensembles alleviate the uncertainty present in a single, deterministic data set by providing a collection of outcomes, from which a best guess can be dervied with some level of confidence.

2.1. Challenges for Analysis and Visualization

The complex nature of the ensemble data creates challenges in data display due to the fact that it is multidimensional [3], multivariate [4, 10, 14], and multivalued [8]. For example, each of the four daily runs of the SREF ensemble contains 21 members comprising four models and eleven sets of input conditions. Each member contains 624 state variables at each of 24,000 grid points and includes 30 time steps. A single day's output contains 84 members, each of which is a complex data set that poses visualization challenges in its own right. When information from all members is displayed together, as in the plume chart in Figure 2, the result is visual chaos that conveys only a general notion of the behavior of the predicted variable. These challenges are exacerbated in more complex data sets such as the climate simulations that incorporate 24 different models instead of four.

3. Visualization of Ensembles

The most straightforward approach to reducing the complexity of ensemble data is to summarize through mean and standard deviation. Mean provides a general overview of the outcome of the ensemble by averaging across the members. Standard deviation reflects the variation of the ensemble and is representative of how well the mean characterizes the data. The combination of these two measures provides a strong indication of the behavior of the ensemble, and while they may not capture nuances of the underlying distribution, they are often appropriate since many observed quantities and natural phenomena are well modeled by a normal distribution [13].

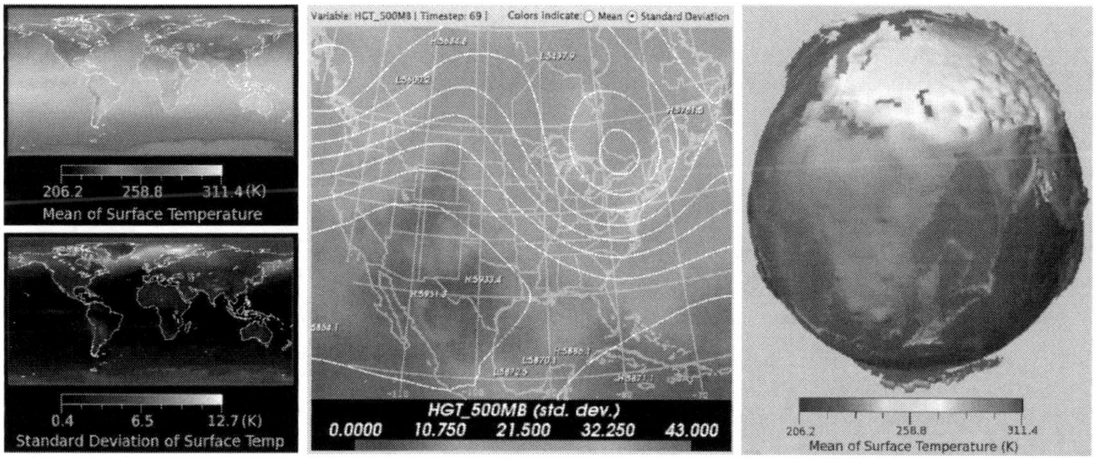

Figure 3. Visualizations of mean and standard deviation. (Left) Mean and standard deviation are visualized independently using color maps. (Center) Mean is presented through a color map, and standard deviation is shown as an overlaid contour. (Right) Standard deviation is mapped to a height field and mean is color mapped.

The ViSUS/CDAT framework provides a variety of techniques for visualizing mean and standard deviation, some of which are shown in Figure 3. Color maps, contours, and height fields are used independently or in combination to indicate the ensemble estimate as well as the confidence of that estimation. These visualizations provide a global understanding of the ensemble behavior, and identify regions where the results of the ensemble runs are questionable. In addition, ViSUS annotates the data with domain specific information, such as geographical and political maps, and latitude and longitude lines.

While mean and standard deviation are excellent tools for summarization, these measures discard large amounts of information available from ensembles data sets. Model biases, extrema, clusters, ranges, and other characteristics are hidden through this type of abstraction and, depending on the motivations of the driving application, may be of great importance. For this reason we provide additional visualization techniques to allow the user to further explore the ensemble.

Two dimensional charts present the data by removing the spatial component, either through summarization across space, or the selection of specific locations. This is helpful when specific locations in the data, or trends through time are of interest. The plume chart, Figure 2, left, shows the visualization of all ensemble members across time. While this plot reduces the overall data, it is still too visually cluttered to assist in data analysis beyond giving a notion of the general outcome. However, using a plot of this type to explore a specific model, or collection of model perturbations can help the user determine the sensitivity of a model to changes in parameters or if a model commonly predicts low or high values. The quartile chart, shown in Figure 2, right, simplifies the plume chart by expressing only the quartiles of the data. This type of chart conveys the range of the data and indicates where the central 50% of the data occurs. The median, or central value of the data is shown as a thick black line. This value is distinctive from the mean in that it is not affected by outlying data values and can indicate the most centrally positioned member.

Each of the visualization techniques presented here reduce the complexity of the data in some dimension. Mean/standard deviation displays show a single instance of time, and the charts compact the spatial domain. Depending on the needs of the researcher, the questions being asked of the data, and the scientific goals behind the ensemble, one paradigm may be more appropriate than another. The benefit of working within the VCAT framework is the ability to use more than one visualization method, thus exploiting the advantages of each. Within this system, the user can select a variable, choose the visualization technique, and interactively explore the data. Such a system provides a flexible environment for the investigation and visual analysis of ensemble data.

4. Conclusion

The rapid increase in computational capacity over the past decade has rendered ensemble data sets a viable tool for mitigating uncertainty. We believe that our work constitutes early progress toward the many new challenges posed by these large, complex and rich data sets. Future work will include the visualization of higher dimensional probability distribution functions, the use of non-normal distributions and higher order descriptive statistics, and the integration of statistical measures into more sophisticated visualization techniques.

5. Acknowledgements

This work was funded in part by the DOE SciDAC Visualization and Analytics Center for Enabling Technologies (www.vacet.org) and the NIH NCRR Center for Integrative Biomedical Computing (www.sci.utah.edu/cibc), NIH NCRR Grant No. 5P41RR012553-02.

Sandia is a multiprogram laboratory operated by Sandia Corporation, a Lockheed Martin Company, for the United States Department of Energy's National Nuclear Security Administration under contract DE-AC04-94AL85000.

References

[1] Climate of the 20th century experiment (20c3m). https://esg.llnl.gov:8443/index.jsp.
[2] Short-range ensemble forecasting. http://wwwt.emc.ncep.noaa.gov/mmb/SREF/SREF.html.
[3] A. Buja, D. Cook, and D. F. Swayne. Interactive high-dimensional data visualization. *Journal of Computational and Graphical Statistics*, 5(1):78–99, March 1996.
[4] R. Bürger and H. Hauser. Visualization of multi-variate scientific data. In *Eurographics 2007 STAR*, pages 117–134, 2007.
[5] G. Compo, J. Whitaker, and P. Sardeshmukh. Bridging the gap between climate and weather. http://www.scidacreview.org/0801/html/climate.html, 2008.
[6] T. Gneiting and A. Raftery. Atmospheric science: Weather forecasting with ensemble methods. *Science*, 310:248–249, October 2005.
[7] C. R. Johnson and A. R. Sanderson. A next step: Visualizing errors and uncertainty. *IEEE CG & A*, 23(5):6–10, 2003.
[8] A. Love, A. Pang, and D. Kao. Visualizing spatial multivalue data. *IEEE CG & A*, 25(3):69–79, May 2005.
[9] A. M. MacEachren, A. Robinson, S. Hopper, S. Gardner, R. Murray, M. Gahegan, and E. Hetzler. Visualizing geospatial information uncertainty: What we know and what we need to know. *Cartography and Geographic Information Science*, 32(3):139–160, July 2005.
[10] T. Mihalisin, J. Timlin, and J. Schwegler. Visualization and analysis of multi-variate data: a technique for all fields. In *IEEE Vis '91*, pages 171–178, 1991.
[11] T. Nocke, M. Fleshig, and U. Böhm. Visual exploration and evaluation of climate-related simulation data. In *IEEE 2007 Water Simulation Conference*, pages 703–711, 2007.
[12] A. Pang, C. Wittenbrink, and S. Lodha. Approaches to uncertainty visualization. *The Visual Computer*, 13(8):370–390, Nov 1997.
[13] J. Sivillo, J. Ahlquist, and Z. Toth. An ensemble forecasting primer. *Weather Forecasting*, 12:809–817, 1997.
[14] Z. Xie, S. Huang, M. Ward, and E. Rundensteiner. Exploratory visualization of multivariate data with variable quality. In *IEEE Symposium on Visual Analytics Science and Technology*, pages 183–190, 2006.

AUTHOR INDEX

AUTHOR INDEX

AUTHOR INDEX

AUTHOR INDEX

AUTHOR INDEX

Institute of Physics
Dirac House, Temple Back
Bristol BS1 6BE UK